This **visual, thought-provoking text**
helps students see the wide-reaching effects
of social problems as they examine their own
attitudes, beliefs, and behaviors

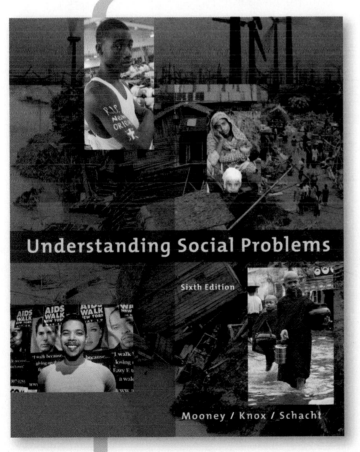

Whether the issue is crime, the high cost of health care, unemployment, prejudice and discrimination, pollution, or war and terrorism, everyone has been touched personally by social problems. This updated Sixth Edition of *Understanding Social Problems* uses a micro to macro approach to help students see how social problems touch us all, and it illustrates how their beliefs and actions (or inactions) perpetuate or alleviate social problems.

The authors lead students on a powerful journey that includes the stories of real people struggling with the challenges created by social problems. The book illustrates how social structure and culture contribute to society's problems and also explores alternative solutions and policies. New information on a host of interesting topics from today's headlines are included in the Sixth Edition, such as cyberbullying, the methamphetamine epidemic, the dangers of social networking, transnational crime, the debate over same-sex classes, foster care abuses, military health care, microcredit programs, immigrant guestworker programs, voluntary childlessness, and the acceleration of global warming and climate change.

Interesting and timely, the book presents a social problem topic in each chapter and frames it in a global as well as U.S. context. In every chapter, three major theoretical perspectives are applied to the social problem under discussion, and the consequences of the problem—as well as alternative solutions—are explored. Pedagogical features such as **The Human Side** and **Self and Society** enable students to grasp how social problems affect the lives of individuals and apply their understanding of social problems to their own lives.

Full of visual impact and the latest coverage and research, the Sixth Edition is a cutting-edge look at social problems

NEW! The Sixth Edition is now more vivid and engaging. Three new **Photo Essays** are included in this edition that discuss and visually depict social problems on a global scale.

- Chapter 2, **Modern Animal Food Production: Health and Safety Issues**
- Chapter 6, **Lack of Clean Water and Sanitation Among the Poor**
- Chapter 14, **Effects of Global Warming and Climate Change**

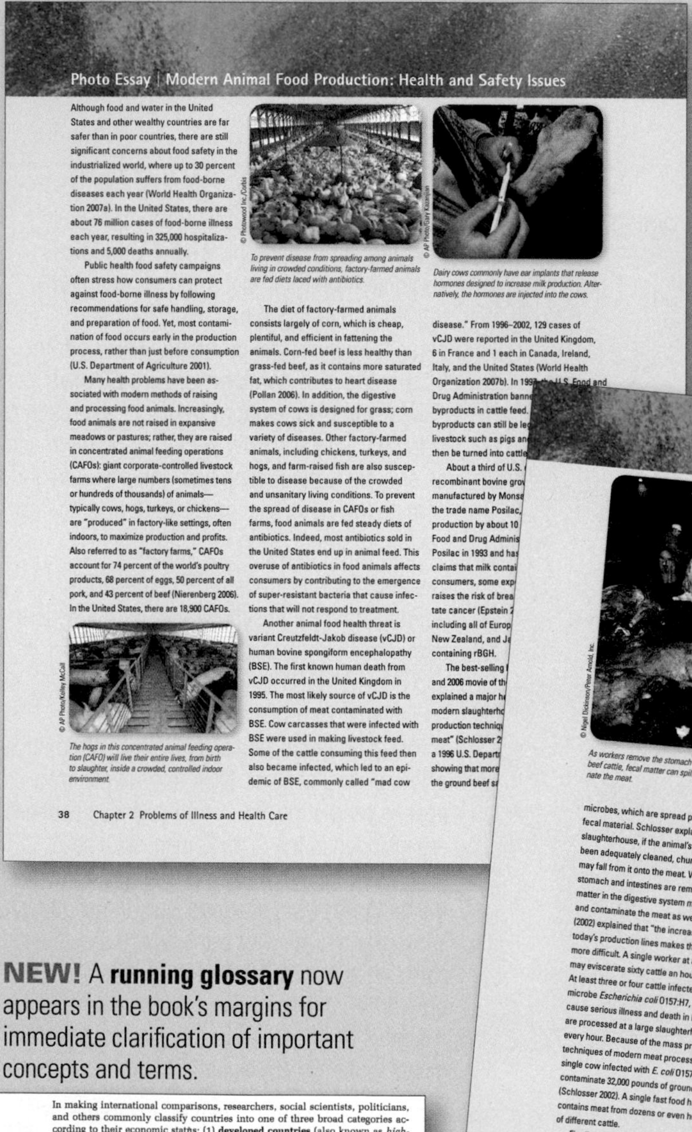

Photo Essay | Modern Animal Food Production: Health and Safety Issues

Although food and water in the United States and other wealthy countries are far safer than in poor countries, there are still significant concerns about food safety in the industrialized world, where up to 30 percent of the population suffers from food-borne diseases each year (World Health Organization 2007a). In the United States, there are about 76 million cases of food-borne illness each year, resulting in 325,000 hospitalizations and 5,000 deaths annually.

Public health food safety campaigns often stress how consumers can protect against food-borne illness by following recommendations for safe handling, storage, and preparation of food. Yet, most contamination of food occurs early in the production process, rather than just before consumption (U.S. Department of Agriculture 2001).

Many health problems have been associated with modern methods of raising and processing food animals. Increasingly, food animals are not raised in expansive meadows or pastures; rather, they are raised in concentrated animal feeding operations (CAFOs): giant corporate-controlled livestock farms where large numbers (sometimes tens or hundreds of thousands) of animals—typically cows, hogs, turkeys, or chickens—are "produced" in factory-like settings, often indoors, to maximize production and profits. Also referred to as "factory farms," CAFOs account for 74 percent of the world's poultry products, 68 percent of eggs, 50 percent of all pork, and 43 percent of beef (Nierenberg 2006). In the United States, there are 18,900 CAFOs.

© AP Photo/Kelley McCall

The hogs in this concentrated animal feeding operation (CAFO) will live their entire lives, from birth to slaughter, inside a crowded, controlled indoor environment.

© Photowood Inc./Corbis

To prevent disease from spreading among animals living in crowded conditions, factory-farmed animals are fed diets laced with antibiotics.

© AP Photo/Gary Kazanjian

Dairy cows commonly have ear implants that release hormones designed to increase milk production. Alternatively, the hormones are injected into the cows.

The diet of factory-farmed animals consists largely of corn, which is cheap, plentiful, and efficient in fattening the animals. Corn-fed beef is less healthy than grass-fed beef, as it contains more saturated fat, which contributes to heart disease (Pollan 2006). In addition, the digestive system of cows is designed for grass; corn makes cows sick and susceptible to a variety of diseases. Other factory-farmed animals, including chickens, turkeys, and hogs, and farm-raised fish are also susceptible to disease because of the crowded and unsanitary living conditions. To prevent the spread of disease in CAFOs or fish farms, food animals are fed steady diets of antibiotics. Indeed, most antibiotics sold in the United States end up in animal feed. This overuse of antibiotics in food animals affects consumers by contributing to the emergence of super-resistant bacteria that cause infections that will not respond to treatment.

Another animal food health threat is variant Creutzfeldt-Jakob disease (vCJD) or human bovine spongiform encephalopathy (BSE). The first known human death from vCJD occurred in the United Kingdom in 1995. The most likely source of vCJD is the consumption of meat contaminated with BSE. Cow carcasses that were infected with BSE were used in making livestock feed. Some of the cattle consuming this feed then also became infected, which led to an epidemic of BSE, commonly called "mad cow disease." From 1996–2002, 129 cases of vCJD were reported in the United Kingdom, 6 in France and 1 each in Canada, Ireland, Italy, and the United States (World Health Organization 2007b). In 199[illegible] the U.S. Food and Drug Administration banne[illegible] byproducts in cattle feed. [illegible] byproducts can still be le[illegible] livestock such as pigs an[illegible] then be turned into cattle[illegible]

About a third of U.S. [illegible] recombinant bovine gro[illegible] manufactured by Monsa[illegible] the trade name Posilac, [illegible] production by about 10 [illegible] Food and Drug Adminis[illegible] Posilac in 1993 and has[illegible] claims that milk contai[illegible] consumers, some exp[illegible] raises the risk of brea[illegible] tate cancer (Epstein 2[illegible] including all of Europ[illegible] New Zealand, and Ja[illegible] containing rBGH.

The best-selling [illegible] and 2006 movie of th[illegible] explained a major h[illegible] modern slaughterho[illegible] production techniqu[illegible] meat" (Schlosser 2[illegible] a 1996 U.S. Depart[illegible] showing that more[illegible] the ground beef s[illegible]

38 Chapter 2 Problems of Illness and Health Care

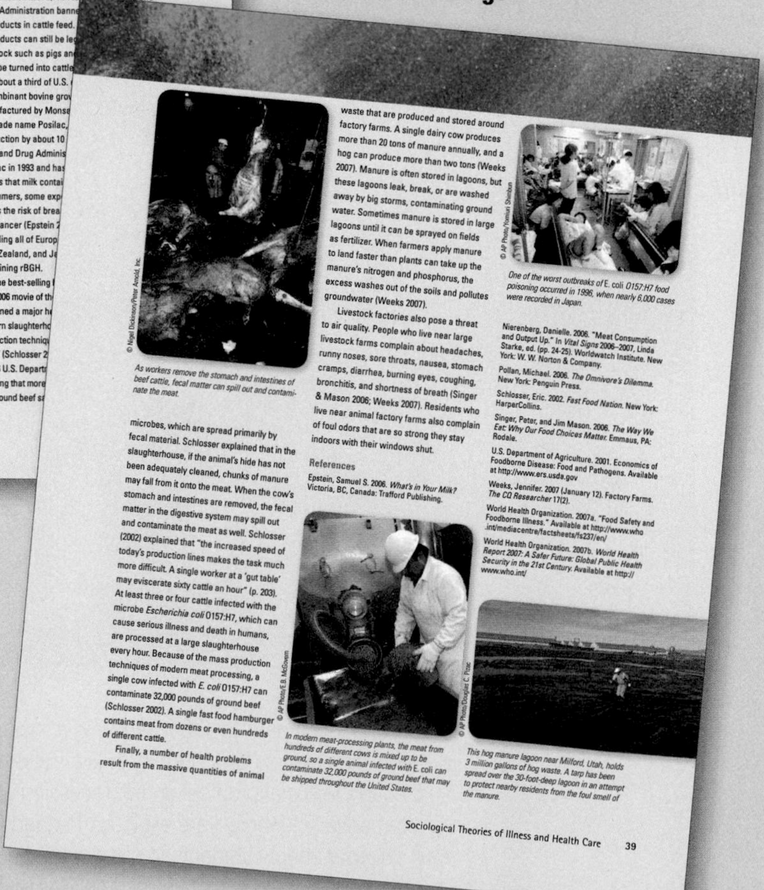

© Nigel Dickinson/Peter Arnold, Inc.

As workers remove the stomach and intestines of beef cattle, fecal matter can spill out and contaminate the meat.

microbes, which are spread primarily by fecal material. Schlosser explained that in the slaughterhouse, if the animal's hide has not been adequately cleaned, chunks of manure may fall from it onto the meat. When the cow's stomach and intestines are removed, the fecal matter in the digestive system may spill out and contaminate the meat as well. Schlosser (2002) explained that "the increased speed of today's production lines makes the task much more difficult. A single worker at a 'gut table' may eviscerate sixty cattle an hour" (p. 203). At least three or four cattle infected with the microbe *Escherichia coli* 0157:H7, which can cause serious illness and death in humans, are processed at a large slaughterhouse every hour. Because of the mass production techniques of modern meat processing, a single cow infected with *E. coli* 0157:H7 can contaminate 32,000 pounds of ground beef (Schlosser 2002). A single fast food hamburger contains meat from dozens or even hundreds of different cattle.

Finally, a number of health problems result from the massive quantities of animal waste that are produced and stored around factory farms. A single dairy cow produces more than 20 tons of manure annually, and a hog can produce more than two tons (Weeks 2007). Manure is often stored in lagoons, but these lagoons leak, break, or are washed away by big storms, contaminating ground water. Sometimes manure is stored in large lagoons until it can be sprayed on fields as fertilizer. When farmers apply manure to land faster than plants can take up the manure's nitrogen and phosphorus, the excess washes out of the soils and pollutes groundwater (Weeks 2007).

Livestock factories also pose a threat to air quality. People who live near large livestock farms complain about headaches, runny noses, sore throats, nausea, stomach cramps, diarrhea, burning eyes, coughing, bronchitis, and shortness of breath (Singer & Mason 2006; Weeks 2007). Residents who live near animal factory farms also complain of foul odors that are so strong they stay indoors with their windows shut.

© AP Photo/Yomiuri Shimbun

One of the worst outbreaks of E. coli *0157:H7 food poisoning occurred in 1996, when nearly 6,000 cases were recorded in Japan.*

References

Epstein, Samuel S. 2006. *What's in Your Milk?* Victoria, BC, Canada: Trafford Publishing.

Nierenberg, Danielle. 2006. "Meat Consumption and Output Up." In *Vital Signs* 2006–2007, Linda Starke, ed. (pp. 24-25). Worldwatch Institute. New York: W. W. Norton & Company.

Pollan, Michael. 2006. *The Omnivore's Dilemma.* New York: Penguin Press.

Schlosser, Eric. 2002. *Fast Food Nation.* New York: HarperCollins.

Singer, Peter, and Jim Mason. 2006. *The Way We Eat: Why Our Food Choices Matter.* Emmaus, PA: Rodale.

U.S. Department of Agriculture. 2001. Economics of Foodborne Disease: Food and Pathogens. Available at http://www.ers.usda.gov

Weeks, Jennifer. 2007 (January 12). Factory Farms. *The CQ Researcher* 17(2).

World Health Organization. 2007a. "Food Safety and Foodborne Illness." Available at http://www.who.int/mediacentre/factsheets/fs237/en/

World Health Organization. 2007b. *World Health Report 2007: A Safer Future: Global Public Health Security in the 21st Century.* Available at http://www.who.int/

© AP Photo/E.B. McGovern

In modern meat-processing plants, the meat from hundreds of different cows is mixed up to be ground, so a single animal infected with E. coli *can contaminate 32,000 pounds of ground beef that may be shipped throughout the United States.*

© AP Photo/Douglas C. Pizac

This hog manure lagoon near Milford, Utah, holds 3 million gallons of hog waste. A tarp has been spread over the 30-foot-deep lagoon in an attempt to protect nearby residents from the foul smell of the manure.

Sociological Theories of Illness and Health Care 39

NEW! A **running glossary** now appears in the book's margins for immediate clarification of important concepts and terms.

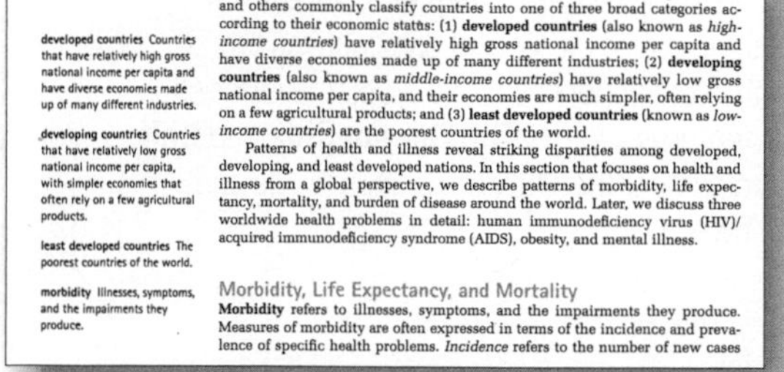

developed countries Countries that have relatively high gross national income per capita and have diverse economies made up of many different industries.

developing countries Countries that have relatively low gross national income per capita, with simpler economies that often rely on a few agricultural products.

least developed countries The poorest countries of the world.

morbidity Illnesses, symptoms, and the impairments they produce.

In making international comparisons, researchers, social scientists, politicians, and others commonly classify countries into one of three broad categories according to their economic status: (1) **developed countries** (also known as *high-income countries*) have relatively high gross national income per capita and have diverse economies made up of many different industries; (2) **developing countries** (also known as *middle-income countries*) have relatively low gross national income per capita, and their economies are much simpler, often relying on a few agricultural products; and (3) **least developed countries** (known as *low-income countries*) are the poorest countries of the world.

Patterns of health and illness reveal striking disparities among developed, developing, and least developed nations. In this section that focuses on health and illness from a global perspective, we describe patterns of morbidity, life expectancy, mortality, and burden of disease around the world. Later, we discuss three worldwide health problems in detail: human immunodeficiency virus (HIV)/acquired immunodeficiency syndrome (AIDS), obesity, and mental illness.

Morbidity, Life Expectancy, and Mortality

Morbidity refers to illnesses, symptoms, and the impairments they produce. Measures of morbidity are often expressed in terms of the incidence and prevalence of specific health problems. *Incidence* refers to the number of new cases

NEW! What Do You Think? sections appear repeatedly throughout each chapter, asking students to pause, think critically about an issue, and develop their own opinions about a particular topic presented in each chapter. These features explore questions such as:

- Should the federal government establish a national system of preschools for all 3- and 4-year-olds?

- Should pharmacists have the right to refuse to fill a prescription for birth control?

- Should children as young as 13 be sentenced to prison for life without the possibility of parole?

- Are gender roles a matter of nature or nurture?

- Should undocumented immigrants qualify for in-state tuition?

What Do You Think? A recent report by the British Equal Opportunities Commission found that, as in the United States, there is a high degree of occupational sex segregation in the United Kingdom. Men tend to be concentrated in areas such as construction, engineering, plumbing, and the like, whereas three-quarters of women are found in five occupational groups—cleaning, catering, caring, cashiering, and clerical—the five c's. These traditional areas of women's employment often command lower salaries than "men's work" despite requiring similar qualifications. The report concludes that "there is a clear financial incentive for women to choose training and work in sectors where men dominate the workforce, as pay tends to be higher in those areas" (EOC 2006, p. 1). Did a high school counselor or a college advisor ever try to direct you into traditional male or female areas of study? In what way, if any, are your own career goals influenced by society's definitions of male and female occupations?

What Do You Think? Physically, men and women are different from birth. Men, in general, are stronger, taller, and heavier and have more facial hair. Women develop breasts, have higher-pitched voices, menstruate, and bear children. But are these physical characteristics related to behavioral differences? Are women innately nurturers? Are men innately aggressive? Noting, for example, that most societies are (were) patriarchal, most would answer yes. Others, however, would be quick to note the role of socialization in traditional gender role assignment. Cross-cultural evidence is mixed. Some societies are characterized by traditional gender roles, in other societies androgyny dominates, and in still other societies traditional gender roles are reversed. What do you think? Are gender roles a matter of nature or nurture?

NEW! Test Yourself, a new feature at the end of each chapter, allows students to immediately assess their level of comprehension and understanding by answering multiple-choice and true and false questions from the chapter.

office or residence hall room that identifies them as individuals who are willing to provide a safe haven and support for LGBT people and those struggling with sexual orientation issues.

TEST YOURSELF

1. In 1996, which country became the first country in the world to include in its constitution a clause banning discrimination based on sexual orientation?
   a. South Africa
   b. Norway
   c. France
   d. The United States
2. In the United States, about one-fifth of gay male couple households and one-third of lesbian couple households have children present in the home.
   a. True
   b. False
3. Most reparative therapy programs allegedly achieve "conversion" to heterosexuality through
   a. uncovering repressed childhood sexual traumas and confronting the perpetrator
   b. hormonal treatments
   c. intensive psychotherapy and family therapy
   d. embracing evangelical Christianity and being "born again"
4. Why has Reverend Fred Phelps of the Westboro Baptist Church (Topeka, Kansas) picketed funerals of American servicemen and servicewomen who have died in the Iraq war?
   a. To protest the "Don't ask, don't tell" military policy
   b. To protest efforts to repeal the "Don't ask, don't tell" military policy
   c. To publicize his view that U.S. military casualties are God's way of punishing the United States for being nice to lesbians and gays
   d. To show support of Iraq's harsh penalties for homosexuality
5. The "contact hypothesis" explains why, in general, heterosexuals have more negative attitudes toward gay men and lesbian women if they have had prior contact with or know someone who is gay or lesbian.
   a. True
   b. False
6. The majority of Americans believe that homosexual individuals should have equal job opportunities.
   a. True
   b. False
7. A 2007 national poll found that more than half (55 percent) of Americans oppose same-sex marriage. However, the majority (56 percent) of _____ in the Pew Survey support gay marriage.
   a. Women
   b. Democrats
   c. Young adults ages 18–29
   d. Racial and ethnic minorities
8. African Americans and Muslims are more likely than lesbian, gay, and bisexual individuals to report being the victim of a hate crime.
   a. True
   b. False
9. More than half of Fortune 500 companies today offer same-sex domestic partner health benefits.
   a. True
   b. False
10. In which U.S. state can same-sex couples legally marry?
    a. None
    b. Hawaii
    c. California
    d. Massachusetts

Answers: 1.a, 2.a, 3.d, 4.c, 5.b, 6.a, 7.c, 8.b, 9.a, 10.d.

NEW! Throughout the book, **new research and information** ensures the text's currency. For example, Chapter 15, *Science and Technology*, highlights cyber-socializing, including new information on *MySpace* and *Facebook* and the dangers of social networking. Additionally, there are expanded sections on identity theft, state restrictions on abortion, e-commerce, and the stem cell debate.

The book also includes new information on:

- The plight of the uninsured and the effects of the high cost of health care on individuals and families

- The legal status of homosexuality and same-sex relationships globally and in the United States

- How marriage in America is changing

- U.S. and global wealth inequality

- The most recent crime statistics available from the National Crime Victimization Survey and the Uniform Crime Reports

- The Jena Six, and apologies and reparations as a means of achieving racial reconciliation

- New data on the acceleration of global warming and climate change

- And much more

Outstanding features encourage
critical thinking and self-examination of students' own attitudes and beliefs

Social Problems Research Up Close boxes present social science research and help students see the sociological enterprise from theory and data collection to findings and conclusions. These features expose students to a variety of studies and research methods and encourage critical thinking. Some topics examined in this feature include "How Marriage in America Is Changing," "Two-Faced Racism," "Self-Made Man," and "Campus Climate for GLBT People: A National Perspective."

Social Problems Research Up Close | Campus Climate for GLBT People: A National Perspective

Since the mid-1980s, numerous studies have documented the hostile climate that GLBT students, staff, faculty, and administrators experience on college campuses. A survey by the National Gay and Lesbian Task Force Policy Institute (Rankin 2003) is a significant contribution to the research literature in this area.

Sample and Methods

Because of the difficulty in identifying lesbian, gay, bisexual, and transgender individuals, Rankin (2003) used purposeful sampling of GLBT individuals and snowball sampling. Contacts were made with "out" GLBT individuals on campus, and they were asked to share the survey with other members of the GLBT community who were not so open about their sexual/gender identity.

Surveys, both paper-and-pencil and online versions, were administered to students, faculty, staff, and administrators at 14 colleges and universities from around the country. The surveys, which contained 35 questions and an additional space for respondents to provide commentary, were designed to collect data about (1) respondents' personal campus experiences as members of the GLBT community, (2) their perception of the climate for GLBT members of the academic community, and (3) their perceptions of institutional actions, including administrative policies and academic initiatives regarding GLBT issues and concerns on campus. A total of 1,869 usable surveys were returned, representing the following groups:
- 1,000 students (undergraduate and graduate), 150 faculty, and 467 staff/administrators
- 720 men, 848 women
- 326 people of color
- 572 gay people (mostly male), 458 lesbians, 334 bisexuals, and 68 transgendered individuals
- 825 "closeted" people

Findings and Discussion

Some of the findings of the Campus Climate Assessment are discussed in the following sections.

Lived Oppressive Experiences

Nineteen percent of the respondents reported that within the past year they had feared for their physical safety because of their sexual orientation or gender identity, and 51 percent had concealed their sexual or gender identity to avoid intimidation. GLBT people of color were more likely than white GLBT people to conceal their sexual or gender identity to avoid harassment. Many nonwhite respondents reported feeling out of place at predominantly white GLBT settings.

In the words of one student, "As a Chicana, I felt ostracized even more. Forget about feeling a sense of community when you're a member of two minority groups" (Rankin 2003, p. 25). The study also found that 27 percent of faculty, staff, and administrators and 40 percent of students indicated that they had concealed their sexual identity in the past year.

Respondents were also asked if they had experienced harassment in the past year. Harassment was defined as "conduct that has interfered unreasonably with their ability to work or learn on this campus or has created an offensive, hostile, intimidating working or learning environment." Undergraduate students were the most likely to have experienced harassment (36 percent), followed by faculty (27 percent), graduate students (23 percent), and staff (19 percent). Derogatory remarks were the most common form of harassment (89 percent). Other types of harassment included verbal harassment or threats, anti-GLBT graffiti, threats of physical violence, denial of services, and physical assault.

Perceptions of Anti-GLBT Oppression on Campus

About three-fourths of faculty, students, administrators, and staff rated the campus climate as homophobic. In contrast, most respondents rated the campus generally (not specific to GLBT people) as friendly, concerned, and respectful. Thus "even though respondents feel that the overall campus climate is hospitable, heterosexism and homophobia are still prevalent" (Rankin 2003, p. 31).

Institutional Actions

Participants were asked to respond to several questions about institutional actions regarding GLBT concerns on campus. Less than half (37 percent) agreed that "the college/university thoroughly addresses campus issues related to sexual orientation/gender identity," and only 22 percent agreed that "the curriculum adequately represents the contributions of GLBT persons." However, the majority of respondents agreed that their classrooms or their job sites were accepting of GLBT persons (63 percent) and that the college or university provided resources on GLBT issues and concerns (71 percent). These positive responses may be due in part to the inclusion of sexual orientation in the nondiscrimination policies of all but one of the participating colleges and universities. Several of the campuses also provided domestic partner benefits, had GLBT resource centers, and offered safe-space programs.

These findings indicate that intolerance and harassment of GLBT students, staff, faculty, and administrators continue to be prevalent on U.S. campuses. It is important to note that the colleges and universities that participated in this study are not representative of most institutions of higher learning in the United States and "may be among the most gay-friendly campuses in the country" (Rankin 2003, p. 3). Rankin explains that "all of the institutions who participated in this survey had a visible GLBT presence on campus, including, in most cases, a GLBT campus center. Most had sexual orientation nondiscrimination policies. As only 100 of the 5,500 U.S. colleges and universities have GLBT student centers, the 14 universities surveyed here are not representative of most institutions of higher education in the U.S." (p. 3). Thus Rankin suggests that the findings of this study "may significantly understate the problems facing GLBT students and staff at U.S. colleges and universities" (p. 3).

Self and Society | The Beliefs About Women Scale (BAWS)

The statements listed here describe different attitudes toward men and women. There are no right or wrong answers, only opinions. Indicate how much you agree or disagree with each statement, using the following scale: (A) strongly disagree, (B) slightly disagree, (C) neither agree nor disagree, (D) slightly agree, (E) strongly agree.

1. Women are more passive than men.
2. Women are less career-motivated than men.
3. Women don't generally like to be active in their sexual relationships.
4. Women are more concerned about their physical appearance than men are.
5. Women comply more often than men do.
6. Women care as much as men do about developing a job/career.
7. Most women don't like to express their sexuality.
8. Men are as conceited about their appearance as women are.
9. Men are as submissive as women are.
10. Women are as skillful in business-related activities as men are.
11. Most women want their partner to take the initiative in their sexual relationships.
12. Women spend more time attending to their physical appearance than men do.
13. Women tend to give up more easily than men do.
14. Women dislike being in leadership positions more than men do.
15. Women are as interested in sex as men are.
16. Women pay more attention to their looks than most men do.

17. Women are more easily influenced than men are.
18. Women don't like responsibility as much as men do.
19. Women's sexual desires are less intense than men's.
20. Women gain more status from their physical appearance than men do.

The Beliefs About Women Scale (BAWS) consists of 15 separate subscales; only four are used here. The items for these four subscales and coding instructions are as follows:

Subscales

1. Women are more passive than men (items 1, 5, 9, 13, 17).
2. Women are interested in careers less than men (items 2, 6, 10, 14, 18).
3. Women are less sexual than men (items 3, 7, 11, 15, 19).
4. Women are more appearance conscious than men (items 4, 8, 12, 16, 20).

Items 1–5, 7, 11–14, and 16–20 should be scored as follows: strongly agree, 2; slightly agree, 1; neither agree nor disagree, 0; slightly disagree, 1; and strongly disagree, 2. Items 6, 8, 9, 10, and 15 should be scored as follows: strongly agree, 2; slightly agree, 1; neither agree nor disagree, 0; slightly disagree, 1; and strongly disagree, 2. Scores range from 0 to 40; subscale scores range from 0 to 10. The higher your score, the more traditional your gender beliefs about men and women are.

Source: William E. Snell, Jr., Ph.D. (1997). College of Liberal Arts, Department of Psychology, Southeast Missouri State University. Reprinted with permission.

Self and Society boxes provide social surveys designed to help students assess their own attitudes, beliefs, knowledge, or behavior regarding some aspect of a social problem. In Chapter 5, *Family Problems*, the authors present an "Abusive Behavior Inventory" that invites students to assess the frequency of abusive behaviors in their own relationships, and in Chapter 6, *Poverty and Economic Inequality*, a "Food Security Scale" asks students to assess their own level of food security. Throughout the book, these features have been updated with new research and information.

The Human Side | Heart of the Uninsured

The following narrative is written by cardiologist Tim Garson (2007), who helped save the life of a child, only to lose her as a young adult because she had no health insurance to pay for her heart medication.

I think of her often—almost every day. We met when she was 5 years old. She had just come out of surgery after a long operation on her heart to change her color. Born a blue baby, she was now a normal pink.

It was my first night on call as a pediatric cardiology fellow. My job was to get her through the first night after surgery. She was fine for the first 3 hours. Suddenly, the dreaded "Dr. Garson, stat, ICU, bed 4." It was a cardiac arrest—just like on television's ER—nurses, doctors, chest compressions. Someone shouted "clear," and we did. The shock through the paddles caused her to jump—flat line. Then a beep came from her monitor and another and another. She was back. That night, her heart stopped three times, and three times science triumphed. By morning she was stable.

Her name was Ginny. The day she was discharged, we traded home phone numbers. I saw her monthly for a while and, as she improved, every 6 months in the clinic. We traded 10 sets of birthday and holiday cards. She did beautifully until her checkup at age 16, when we discovered a heart rhythm problem. I knew how devastating this problem could be, even leading to death. We tried a number of drugs and eventually found exactly what we needed: a once-a-day medication that was easy for Ginny to remember to take, although, unfortunately, it was very expensive.

Ginny did beautifully and was beaming when she introduced me to her friends at her high school graduation. She decided she wanted to work for a couple of years before going to college to help her parents with her medical bills (they both worked in a small grocery), but after she went through six job interviews, all with small businesses (no large employers in her town), no one would hire her. She told all of them that she had congenital heart disease. She told me that she was proud of how well she had done and wanted others to know. I couldn't tell her to lie about her heart disease, but she didn't need to volunteer the information. By the time she called to tell me, it was too late; there were no other jobs she could apply for in town, and she had no car to travel elsewhere.

If a person has a known high risk, such as Ginny, many employers will not employ him or her for fear of the effect on their insurance rate, despite the fact that this practice is illegal under the Americans with Disabilities Act. The more likely scenario is that the small business would not offer health insurance at all: In this country, fewer than half of small businesses with three to nine employees offer health insurance to their employees, compared with more than 90 percent of those with more than 200 employees. More than 70 percent of uninsured Americans are in families with at least one full-time worker.

One Sunday I received an early morning call. Sobbing . . . then "Tim, we've lost her!" Ginny's mom found her dead in bed. It didn't take long to figure out what had happened. Ginny's pill bottle was empty.

The last refill was 5 months before—just before her 19th birthday, when her Medicaid eligibility ended and her prescription drug coverage stopped. She had no job, and we guessed . . . that she was trying to save the family money.

I think of Ginny often. Almost every day. I never could understand how the "system" that had paid to fix her heart, and paid for her medicine, dropped her at 19. But that's the way it works. Medicaid and the State Children's Health Insurance Program (SCHIP) covers children of the poor, like Ginny, but between the ages of 19 and the Medicare age of 65, the so-called safety net has huge holes—and Ginny fell through.

The day of Ginny's funeral, I made the uninsured my personal issue and became actively involved in working toward health care reform. Efforts at reform are under way in states, and experimentation in both public programs and private insurance strategies (that would help small businesses afford insurance for people like Ginny) should be encouraged. The federal government can help by providing support such as that proposed in the yet-unpassed Health Partnership Act. In any of these public approaches, it is helpful to remember that in the United States, the uninsured are 46 million individuals, and each has a story to tell.

I think of Ginny often. Almost every day.

Adapted from Garson, Arthur "Tim". 2007. "Heart of the Uninsured." *Health Affairs* 26(1):227–231. Arthur "Tim" Garson Jr. is dean of the School of Medicine and vice president of the University of Virginia in Charlottesville, Virginia. The name of the patient in the story has been changed to protect privacy.

The book personalizes the social problem under discussion by including the experiences of real people who have been affected in **The Human Side** boxes, which appear in every chapter. The Human Side feature in Chapter 12, *Problems of Youth and Aging,* describes a son's plea for the continued incarceration of his mother's abuser, and in Chapter 13, *Population Growth and Urbanization,* two women explain why they have made the choice to be childfree.

As students are often faced with contradictory theories and study results, they sometimes walk away from a course in social problems without any real understanding of their causes and consequences. To address this problem, chapter sections titled **Understanding . . . [specific social problem]"** cap the body of each chapter just before the chapter summaries. These helpful sections synthesize the material presented in the chapter, summing up the present state of knowledge and theory on the chapter topic.

UNDERSTANDING ISSUES IN SEXUAL ORIENTATION

Recent years have witnessed a growing acceptance of lesbians, gay men, and bisexuals as well as increased legal protection and recognition of these marginalized populations. The advancements in gay rights are notable and include the Supreme Court's 2003 decriminalization of sodomy, the growing adoption of nondiscrimination policies covering sexual orientation, the increased recognition of domestic partnerships, the state civil union rights and responsibilities offered to same-sex couples in Vermont, Connecticut, and California, the Massachusetts ruling that same-sex marriage is allowable under the state constitution, and the election of the first openly gay bishop in the New Hampshire Episcopal diocese. But the winning of these battles in no way signifies that the war is over. Gay, lesbian, and bisexual individuals continue to be frequent victims of discrimination, harassment, and violence. In some countries homosexuality is formally condemned, with penalties ranging from fines to imprisonment and even death.

As both structural functionalists and conflict theorists note, nonheterosexuality challenges traditional definitions of family, child rearing, and gender roles. Every victory in achieving legal protection and social recognition for sexual orientation minorities fuels the backlash against them by groups who are determined to maintain traditional notions of family and gender. Often, this determination is rooted in and derives its strength from uncompromising religious ideology.

Outstanding teaching, presentation, and course management **tools**

PowerLecture CD-ROM

ISBN-10: 0-495-50813-6 • ISBN-13: 978-0-495-50813-7

This easy-to-use, one-stop digital library and presentation tool offers hundreds of images from the text, **JoinIn Student Response System, ExamView Computerized Testing**, and more.

See the insert at the front of this book for a complete description.

Instructor's Resource Manual

by Katheryn A. Dietrich, Texas A&M University and Blinn College

ISBN-10: 0-495-50812-8 • ISBN-13: 978-0-495-50812-0

The **Instructor's Resource Manual** provides learning objectives, key terms, lecture outlines, student projects, classroom activities, and Internet and **InfoTrac College Edition** exercises. Test items include multiple-choice and true/false questions with answers and page references, as well as short-answer and essay questions for each chapter. Each multiple-choice item has the question type (factual, applied, or conceptual) indicated. All questions are labeled as new, modified, or pickup so instructors know if the question is new to this edition of the test bank, modified but picked up from the previous edition of the test bank, or picked up straight from the previous edition of the test bank. Concise user guides for **InfoTrac College Edition** and the **Social Problems ABC Video Table of Contents** are included as appendices.

ABC® Videos

These ABC videos feature short, high-interest clips about current news events as well as historic raw footage going back 40 years. Perfect for discussion starters or to enrich your lectures. *See the insert at the front of this book for a complete listing of the videos.*

AIDS in Africa DVD

ISBN-10: 0-495-17183-2 • ISBN-13: 978-0-495-17183-6

Southern Africa has been overcome by a pandemic of unparalleled proportions. This documentary series focuses on the new democracy of Namibia and the many actions that are being taken to control HIV/AIDS. Included in this series are four documentary films created by the Periclean Scholars at Elon University:

1) *Young Struggles, Eternal Faith* focuses on caregivers in the faith community
2) *The Shining Lights of Opuwo* shows how young people share their messages of hope through song and dance
3) *A Measure of Our Humanity* describes HIV/AIDS as an issue related to gender, poverty, stigma, education, and justice
4) *You Wake Me Up* is a story of two HIV-positive women and their acts of courage helping other women learn to survive.

Our **exclusive online tools** make studying interactive and engergizing

CengageNOW™ for Social Problems

*Help your students save time, learn more, and succeed in the course with **CengageNOW for Social Problems**.*

*You can order access to **CengageNOW** automatically packaged with the text at no additional cost!*
On WebCT ISBN-10: 0-495-59818-6 • ISBN-13: 978-0-495-59818-3
On Blackboard ISBN-10: 0-495-59816-X • ISBN-13: 978-0-495-59816-9

Access to **CengageNOW** is web-based and packaged with every new copy of this text—all you need to do is order the access code, and your students are in! This powerful and interactive resource allows students to gauge their own unique study needs—then, it presents a personalized learning plan that helps them focus their study time on the concepts they most need to master. With the program's unique diagnostic quizzes written by Lois Sabol of Yakima Valley Community College and study plan, your students will quickly begin to optimize their study and get one step closer to success! See the inside front cover for details, or visit academic.cengage.com/sociology for details on packaging **CengageNOW** with Mooney's text.

InfoTrac® College Edition with InfoMarks™

Give your students anytime, anywhere access to reliable resources with **InfoTrac College Edition**, the online library! This fully searchable database offers 20 years' worth of full-text articles from almost 5,000 diverse sources, such as academic journals, newsletters, and up-to-the-minute periodicals. *Contact your Wadsworth Cengage Learning sales representative for details on packaging with the text.*

InSite for Writing and Research™

ISBN-10: 1-4130-2739-3
ISBN-13: 978-1-4130-2739-6

This all-in-one, online writing and research tool includes electronic peer review, an originality checker, an assignment library, help with common errors, and access to **InfoTrac College Edition**. **InSite** makes course management practically automatic! Visit academic.cengage.com/sociology for more information.

Companion Website

academic.cengage.com/sociology/mooney

The book's **Companion Website** includes chapter-specific resources for instructors and students. For instructors, the site offers a password-protected instructor's manual, Microsoft® PowerPoint® presentation slides, and more. For students, there is a multitude of text-specific study aids including the following: Tutorial practice quizzing that can be scored and emailed to the instructor, web links, **InfoTrac® College Edition** exercises, flash cards, MicroCase Online data exercises, crossword puzzles, Virtual Explorations, and much more!

*Order the text packaged with access to **CengageNOW**, **InfoTrac College Edition**, and **Insite for Writing and Research** by using ISBN-10: 0-495-49411-9 • ISBN-13: 978-0-495-49411-9*

Customize your course! Package one of these resources with the text for convenience and savings

Study Guide

ISBN-10: 0-495-50855-1 • ISBN-13: 978-0-495-50855-7

To order packaged with the text, use ISBN-10: 0-495-49407-0
ISBN-13: 978-0-495-49407-2

Written by Lori Fowler of Tarrant County College, the **Study Guide** includes learning objectives, brief and detailed chapter outlines, key terms, Internet activities, **InfoTrac College Edition** exercises, student projects and classroom activities, and practice tests consisting of multiple-choice and true/false questions with answers and page references, as well as short-answer questions and essay questions with page references to enhance and test your students' understanding of chapter concepts.

NEW!
Globalization: The Transformation of Social Worlds, Second Edition

D. Stanley Eitzen, Colorado State University, and Maxine Baca Zinn, Michigan State University
368 pages. 6 3/8 x 9 1/4. Paperbound.
ISBN-10: 0-495-50432-7 • ISBN-13: 978-0-495-50432-0

To order packaged with the text, use ISBN-10:
0-495-49410-0 • ISBN-13: 978-0-495-49410-2

See the insert at the front of this Instructor's Edition for a complete description.

NEW!
Social Problems: Readings with Four Questions, Third Edition

Joel M. Charon, Moorhead State University and Lee G. Vigilant, Minnesota State University at Moorhead
528 pages. 7 3/8 x 9 1/4. Paperbound.
ISBN-10: 0-495-50431-9 • ISBN-13: 978-0-495-50431-3

To order packaged with the text, use ISBN-10: 0-495-64452-8
ISBN-13: 978-0-495-64452-1

A unique and groundbreaking collection of 54 articles organized in 11 thematic sections, this new Third Edition combines a rigorous structural/conflict approach with a strong emphasis on the real-life experiences of those impacted by social problems. The articles include both classic and contemporary readings covering a wide range of issues in the United States and around the world. The introductory article, written by Joel M. Charon, introduces four questions that students are urged to apply throughout the reader: What is the problem? What makes the problem a "social problem"? What causes the problem? What can be done? These questions give students a consistent sociological framework to help them analyze the readings and think critically about social problems.

Extension: Wadsworth's Sociology Readings Collection

ISBN-10: 0-495-12802-3 • ISBN-13: 978-0-495-12802-1

Create your own customized reader for your introductory class drawing from dozens of classic and contemporary articles found on the exclusive Wadsworth TextChoice2 database. Create a customized reader just for your class containing as few as two or three seminal articles to more than a dozen edited selections. To build your own custom reader, visit our digital library: www.textchoice2.com.

Understanding Social Problems

Sixth Edition

Linda A. Mooney

David Knox

Caroline Schacht

East Carolina University

WADSWORTH
CENGAGE Learning™

Australia • Brazil • Japan • Korea • Mexico • Singapore • Spain • United Kingdom • United States

WADSWORTH
CENGAGE Learning

Understanding Social Problems,
Sixth Edition
Linda A. Mooney, David Knox, and
Caroline Schacht

Acquisitions Editor: Chris Caldeira

Development Editor: Jeremy Judson

Assistant Editor/Technology Project
Manager: Tali Beesley

Editorial Assistant: Erin Parkins

Marketing Manager: Michelle Williams

Marketing Assistant: Ileana Shevlin

Marketing Communications Manager:
Linda Yip

Project Manager, Editorial Production:
Cheri Palmer

Creative Director: Rob Hugel

Art Director: Caryl Gorska

Print Buyer: Judy Inouye

Permissions Editor: Roberta Broyer

Production Service: Graphic World Inc.

Text Designer: Jeanne Calabrese

Photo Researcher: Marcy Lunetta

Cover Designer: Dare Porter

Cover Images (clockwise from top to bot-
tom): © Peter Turnley/Corbis, © Saeed
Khan/Getty Images, © Adek Berry/Getty
Images, © Khin Maung Win/Getty
Images, © Mario Tama/Getty Images

Compositor: Graphic World Inc.

For product information and technology assistance, contact us at
Cengage Learning Customer & Sales Support, 1-800-354-9706

For permission to use material from this text or product,
submit all requests online at **cengage.com/permissions**

Further permissions questions can be emailed to
permissionrequest@cengage.com

Library of Congress Control Number: 2007940117

ISBN-13: 978-0-495-50428-3

ISBN-10: 0-495-50428-9

Wadsworth, Cengage Learning
10 Davis Drive
Belmont, CA 94002-3098
USA

Cengage Learning is a leading provider of customized learning solutions
with office locations around the globe, including Singapore, the United
Kingdom, Australia, Mexico, Brazil, and Japan. Locate your local office at:
international.cengage.com/region

Cengage Learning products are represented in Canada by
Nelson Education, Ltd.

For your course and learning solutions, visit **academic.cengage.com**

Purchase any of our products at your local college store or at our
preferred online store **www.ichapters.com**

Printed in the United States of America
1 2 3 4 5 6 7 12 11 10 09 08

*In memory of Morris Kolknicker, whose generosity
and compassion continue to inspire those who knew him.*

Brief Contents

Contents

PHOTO ESSAY: Effects
of Global Warming and
Climate Change **570**

Preface

Understanding Social Problems is intended for use in a college-level sociology course. We recognize that many students enrolled in undergraduate sociology classes are not sociology majors. Thus, we have designed our text with the aim of inspiring students—no matter what their academic major or future life path may be—to care about the social problems affecting people throughout the world. In addition to providing a sound theoretical and research basis for sociology majors, *Understanding Social Problems* also speaks to students who are headed for careers in business, psychology, health care, social work, criminal justice, and the nonprofit sector, as well as to those pursuing degrees in education, fine arts, and the humanities and to those who are "undecided." Social problems, after all, affect each and every one of us, directly or indirectly. And everyone, whether a leader in business or politics, a stay-at-home parent, or a student, can become more mindful of how his or her actions (or inactions) perpetuate or alleviate social problems. We hope that *Understanding Social Problems* not only informs but also inspires, planting seeds of social awareness that will grow no matter what academic, occupational, and life path students choose.

NEW TO THIS EDITION

The Sixth Edition of *Understanding Social Problems* includes new pedagogical features, including a running glossary and a Test Yourself feature at the end of every chapter. Each chapter also contains several sections titled *What Do You Think?* which are designed to engage students in critical thinking. This new edition also includes three new photo essays on modern animal food production (Chapter 2), lack of clean water and sanitation among the poor (Chapter 6), and effects of global warming (Chapter 14). Most of the chapter opening vignettes are new, and many of the chapter features (*The Human Side, Social Problems Research Up Close,* and *Self and Society*) have been changed. Finally, each chapter has new photos and new and updated figures and tables, as well as new and revised material, detailed as follows.

Chapter 1 (Thinking About Social Problems) includes an updated *Self and Society* feature with the newest statistics available on U.S. freshman attitudes toward select social problems. *The Human Side* also has been updated with new examples of college student activism, and the *Social Problems Research Up Close* feature now includes the latest statistics from the Centers for Disease Control and Prevention on high school students' reporting of sexual risk behaviors. This chapter also includes new *What Do You Think?* sections that ask the following questions: (1) Will definitions of social problems change over time as sources of information (e.g., the Internet) change?; (2) Do "solutions" to social problems, for example, busing to achieve racial integration, simply lead to other social problems?; (3) Should a sociologist's sources, as are a journalist's sources, be protected by law?; and (4) Should service learning, that is, volunteering in the community for academic credit, be a graduation requirement in college?

Chapter 2 (Problems of Illness and Health Care) includes a new *The Human Side* feature, "Heart of the Uninsured," and a new *Social Problems Research Up Close* feature about health risk behaviors among college students. There is increased coverage of the plight of the uninsured and the effects of the high cost of health care on individuals and families, a new section on workers' compensation, coverage of state-level health care reform, a new section on military health care, and a photo essay, "Modern Animal Food Production: Health and Safety Issues." This chapter includes new *What Do You Think?* sections: (1) Should obese children be considered victims of abuse and placed in protective custody?; (2) Should states be allowed to mandate the HPV vaccine for adolescent girls?; (3) Should colleges and universities have the right to dismiss students who seek treatment for suicidal thoughts and/or behavior?; and (4) Given that tobacco-related deaths grossly outnumber terrorist-related deaths, why hasn't the U.S. government waged a "war on tobacco" on a scale similar to its "war on terrorism?"

In **Chapter 3** (Alcohol and Other Drugs), statistics have been updated to include the most recent data available from the Monitoring the Future survey and the U.S. Department of Health and Human Services' National Survey on Drug Use and Health. A new table on commonly abused illegal drugs contains the pharmacological classification of the drug, commercial and street names, methods of administration, intoxication effects, and potential health hazards. A new *Self and Society* feature allows students to assess the negative consequences of their alcohol consumption, and a new *Social Problems Research Up Close* feature examines children's attitudes toward, beliefs about, and lifestyle associations with cigarette smoking. There are expanded sections on worldwide drug use, methamphetamine, the dangers of secondhand smoke, reinforcement theory, and sentencing differentials between crack and cocaine users. A debate between supply reduction versus demand reduction strategies of preventing illegal drug use has also been added. This chapter now includes *What Do You Think?* sections that address the following questions: (1) Should vouchers or other types of rewards be used as incentives to keep drug addicts "clean"?; (2) Is addiction a consequence of nature, nurture, or both?; (3) Should marijuana, like alcohol, be legal for those over the age of 21?; and (4) Given the many health consequences of smoking cigarettes and excessive alcohol consumption, why do these drugs remain legal in the United States?

Chapter 4 (Crime and Social Control) presents the most recent crime statistics available from the National Crime Victimization Survey and the Uniform Crime Reports. The new *Social Problems Research Up Close* feature examines fighting behavior among a sample of at-risk minority youth. A new *The Human Side* feature contains excerpts from a U.S. State Department report that documents the horrors of human trafficking through firsthand accounts of trafficking victims. There are expanded discussions of transnational crime, pornography and Internet solicitation of minors, prostitution, white-collar crime, computer crime, crime prevention programs, and corrections. There are also new sections on crime in mid-sized cities—a growing problem—and on international efforts to fight crime. New *What Do You Think?* sections address the following: (1) Should funding for AIDS prevention to other countries be contingent upon a loyalty oath against prostitution?; (2) Should states require special license plates for convicted sex offenders?; (3) What changes in crime rates might we expect given demographic trends over the next 50 years?; and (4) Should convicted felons be denied the right to vote?

Chapter 5 (Family Problems) includes updated information on changing structures and patterns in U.S. families and households, intimate partner violence, elder abuse, corporal punishment, the need for workplace and economic supports for families, causes and consequences of teenage childbearing, and the current status of sex education programs in the United States. This chapter contains new sections on "living apart together" relationships, "polyvictimization," and the use of computerized infant simulators in teenage pregnancy prevention programs. A new *Social Problems Research Up Close* feature looks at longitudinal research on how marriage in America is changing. New *What Do You Think?* sections in this chapter ask: (1) Why are adults with divorced parents more likely to cohabit before marriage than are adults with continuously married parents?; (2) Should smoking in a car when children are present be considered a form of child abuse as it inflicts harm on children in the form of secondhand smoke exposure?; (3) Should parents of minor children be required to complete a divorce education program before they can get a divorce?; (4) Should pharmacists have the legal right to refuse to fill a prescription for products such as birth control pills or Plan B emergency contraception?

Chapter 6 (Poverty and Economic Inequality) includes new data on U.S. and global income and wealth inequality, CEO compensation, and the characteristics of individuals receiving welfare. Topics new to this edition include the "couch-homeless," the housing crisis among the middle-class in suburbia, and microcredit and the Grameen Bank. A new photo essay looks at the lack of clean water and sanitation among the poor. New *What Do You Think?* sections address the following: (1) Why are the poor less likely to vote than those in higher income brackets?; (2) Why do students commonly think of a typical U.S. poor person as being a middle-age man, even though U.S. poverty statistics reveal that the higher poverty rates are among women, not men, and among youth, not middle-aged adults?; and (3) Would the federal response to the disaster left in the wake of Hurricane Katrina have been different if Katrina had devastated an area of the country where wealthier people resided?

In addition to updated information throughout **Chapter 7** (Work and Unemployment), this revised chapter also includes new discussions on FLA Watch (a Fair Labor Association watchdog group), weak U.S. labor laws, and benefits and disadvantages of labor unions for workers. A new *Social Problems Research Up Close* feature presents a Human Rights Watch report on Wal-Mart's anti-union strategies. New *What Do You Think?* sections ask the reader: (1) Are the *Oxford* and *Merriam-Webster* dictionary definitions of "McJob" accurate and fair?; (2) In what ways might high levels of production and consumption contribute to individual and social ills rather than to economic health and well-being?; (3) Do colleges and universities have an obligation to ensure that any apparel or items with their college or university logo be made in sweatshop-free conditions?; and (4) Are employers that provide at least partial pay to employees on maternity leave but no pay for paternity leave being unfair and discriminatory to men?

Chapter 8 (Problems in Education) includes results from a recent report from the Organization for Economic Cooperation and Development (OECD) that outlines education trends around the world. Also new to this chapter are sections on income-based integration of schools, the economic costs of dropouts, concerns over U.S. college students' level of preparedness, the debate over single-sex classrooms, efforts to bring Title IX in compliance with the No Child Left Behind Act, and the gender gap in higher education. This revised chapter has expanded sections on the inequality of educational attainment and includes an excerpt from

Jonathan Kozol's *The Shame of the Nation*. Additional information on bullying, cyberbullying, and freedom of speech issues also has been added. The *Social Problems Research Up Close* feature, which examines types and frequency of bullying behaviors among a sample of middle school and high school students, is also new. Finally, there are new *What Do You Think?* sections: (1) Should the federal government establish a national system of preschools for all 3- and 4-year-olds?; (2) Given the college gender gap, should colleges and universities establish special recruiting strategies for men?; (3) Should state law require students to remain in high school until the age of 21?; and (4) How can schools better balance the needs of special education students without further disadvantaging students from poor families?

Chapter 9, (Race, Ethnicity, and Immigration) includes expanded and updated discussions on the social construction of race, white privilege, anti-immigration vigilante groups, U.S. federal and state policies on illegal immigration, and the role of the Equal Employment Opportunity Commission in responding to employment discrimination. New topics include Asians as a "model minority," guestworker programs, immigrant "sanctuary cities," hate crime violence against American Indians, the Jena Six, and apologies and reparations as a means of achieving racial reconciliation. A new *Social Problems Research Up Close* feature presents research on "two-faced racism." *What Do You Think?* sections address the following: (1) What arguments can be made for discontinuing racial classification? What arguments can be made for continuing it?; (2) Should undocumented immigrants qualify for in-state college tuition?; (3) Are racial and ethnic jokes insulting and harmful to minority group members? Or are such jokes innocent and playful forms of fun?; (4) Why do many textbooks overlook or minimize the benefits women may receive from affirmative action?; and (5) Should the U.S. federal government offer an official apology and/or monetary or other reparations to African Americans for the slavery and legal segregation that occurred in the past?

New to **Chapter 10** (Gender Inequality) is an increased focus on global gender inequality. This revised chapter also includes increased coverage of gender issues related to boys and men, including an expanded section on the social psychological costs of traditional gender roles for boys. Sections on the gender pay gap, gender and student–teacher interactions, and gender discrimination have also been expanded. The chapter includes updated statistics as well as results from one of the largest studies of women and men in the media—the Global Media Monitoring Project. There is a new *Social Problems Research Up Close* feature ("Self-Made Man"), and a new *The Human Side* feature from Plan International's "Because I Am a Girl." This chapter also includes new *What Do You Think?* sections that address the following: (1) Assuming equal qualifications, is preferential treatment based on gender ever justified?; (2) Is there truly a "war against boys" or is the superior performance of girls in primary and secondary school a consequence of schools "catering" to girls?; and (3) Should transgendered individuals have the same rights and protections under the law as non-transgendered individuals?

In **Chapter 11** (Issues in Sexual Orientation), updated information is provided on the legal status of homosexuality and same-sex relationships globally and in the United States. This revised chapter includes updated survey data on same-sex marriage, attitudes toward homosexuality, and discrimination against gays and lesbians. New information also has been added to the sections on reparative therapy ("therapy" that purports to change homosexuals' sexual orientation), hate crime victimization, gay-straight alliances, and laws and policies that protect gays and lesbians from discrimination. New *What Do You Think?* sections address the following: (1) How is the experience of being a sexual orientation minority similar to the experience of being a racial or ethnic minority? How is it different?; (2) Why is it that in many countries male homosexuality is illegal, but female homosexuality is not?; (3) Is the higher acceptance of nonmonogamy among gay males due to their sexual orientation? Or their sex and gender?; and (4) Does social acceptance of homosexuality lead to the creation of laws that protect lesbians and gays? Or does the enactment of laws that protect lesbians and gays help to create more social acceptance of gays?

Chapter 12 (Problems of Youth and Aging) has the latest available statistics and data from the Administration on Aging and the State of the World's Children 2007 report. A UNICEF report on child well-being provides information on six dimensions of child well-being in economically advanced countries around the world. Additionally, the Foundation for Child Development's *Child and Youth Well-Being Index* documents the quality of life of children in the United States. There are expanded sections on the foster care system and co-housing for the elderly, and a new *The Human Side* that features a son's plea for the continued incarceration of his mother's abuser. There are new *What Do You Think?* sections on: (1) Should children as young as 13 be sentenced to prison for life without the possibility of parole?; (2) Does limiting housing availability by age violate the principles of fair housing as defined by the Fair Housing and Equal Opportunity Office?; and (3) Does the Elder Wisdom Circle provide a needed resource for children and young adults?

Chapter 13 (Population Growth and Urbanization) presents new statistics on population growth and fertility rates worldwide. New information has been added on urban poverty and the growth of slums, family planning services, and community development corporations. New to this chapter is a section on China's one-child policy. In this chapter's new *The Human Side* feature, two women explain why they have made the choice to be childfree. New *What Do You Think?* sections ask the following: (1) What could cities do to attract more families with children?; (2) If the United States offered more generous work-family benefits, such as paid parenting leave and government-supported child care, would the U.S. birth rate increase? Would such policies affect the number of children you would want to have?; (2) Should the United States provide funding to developing countries for medical facilities and equipment and training to health care personnel for the purpose of providing women with access to safe abortion?; and (4) If you lived in a community that had cycleways where motor vehicles were banned, would you be more likely to use a bicycle as a means of transportation?

Chapter 14 (Environmental Problems) includes new data on a wide range of topics, including global warming and climate change, the ozone hole, nuclear power, chemicals in the environment, disappearing species, environmental injustice, and ethanol. A new *Self and Society* feature focuses on attitudes toward global warming. A new *The Human Side* feature presents the story of a retired Marine who believes that his daughter's leukemia and death resulted from exposure to contaminated water at Camp Lejeune. This revised chapter also contains a photo essay on the effects of global warming and climate change. New *What Do You Think?* sections include those that address the following: (1) Do you think that organizers of Earth Day events should accept sponsorship from corporations with poor environmental records?; (2) Would you support a ban, or a tax, on plastic grocery bags in your community?; (3) Why are women more likely than men to worry about the environment, to be active in or sympathetic toward the environmental movement, and to give precedence to the environment over economic and energy concerns?; and (4) Is it better to purchase organic produce that has been trucked in from a distant state (no pesticides, but the transportation contributes to fossil fuel emissions) or locally grown produce from a farm that uses pesticides?

The revised **Chapter 15** (Science and Technology) highlights cyber-socializing, including information on MySpace.com and Facebook.com and the dangers of social networking. There are expanded sections on identity theft, state restrictions on abortion, e-commerce, and the stem cell debate. This revised chapter also features new discussions on the U.S. decline in science and technology supremacy, the implementation of the American Competitiveness Initiative, food and biotechnology, the 2007 ban on partial birth abortions, the 2007 Human Cloning Act, outsourcing, and Thomas Friedman's *The World Is Flat.* There is also new information on Web 2.0, biometrics, attention deficit traits, Internet censorship, and a typology of information and communication technology users. A new *Self and Society* feature allows students to compare their attitudes about abortion with a representative sample of U.S. adults. The new *What Do You Think?* feature highlights such issues as: (1) Would Americans be as accepting of high-tech gadgetry such as robots as the Japanese?; (2) Should women be required to view an ultrasound and/or listen to the fetal heartbeat before having an abortion?; and (3) Should Internet content be subject to the same federal regulations as newspapers and other print media?

Chapter 16 (Conflict, War, and Terrorism) highlights the costs of continued military presence in Iraq and Afghanistan, including information on military deaths, civilian casualties, and the financial cost of U.S. involvement. This chapter also includes a new section on the problems of small arms proliferation. The new *The Human Side* feature presents accounts from child soldiers about their forced conscription. A new *Self and Society* feature highlights the continued presence of nuclear weapons around the world and in U.S. society, despite the end of the Cold War. In addition, the section on nuclear nonproliferation has been updated to include recent developments in North Korea, Iran, and South Asia. New *What Do You Think?* sections address such controversial issues as: (1) Should universal military service after high school be required?; (2) Is the U.S. policy of preemptive attacks in cases of "imminent danger" justifiable?; (3) Is it appropriate for international forces to intervene on moral or humanitarian grounds in conflicts between sovereign nations and anti-government rebels?; and (4) Are there any situations in which it is appropriate to negotiate with terrorists?

FEATURES AND PEDAGOGICAL AIDS

We have integrated a number of features and pedagogical aids into the text to help students learn to think about social problems from a sociological perspective. It is our mission to help students not only apply sociological concepts to observed situations in their everyday lives and think critically about social problems and their implications, but also learn to assess how social problems relate to their lives on a personal level.

Exercises and Boxed Features

Self and Society. Each chapter includes a social survey designed to help students assess their own attitudes, beliefs, knowledge, or behaviors regarding some aspect of the social problem under discussion. In Chapter 5 (Family Problems), for example, the "Abusive Behavior Inventory" invites students to assess the frequency of various abusive behaviors in their own relationships.

The Human Side. In addition to the *Self and Society* boxed features, each chapter includes a boxed feature that further personalizes the social problems under discussion by describing personal experiences of individuals who have been affected by them. The *Human Side* feature in Chapter 9 (Race, Ethnicity, and Immigration), for example, describes the experience of an immigrant day laborer who was victimized by a violent hate crime.

Social Problems Research Up Close. In every chapter, these features present examples of social science research. These features demonstrate the sociological enterprise, from theory and data collection to findings and conclusions, thus exposing students to various studies and research methods.

Photo Essay. New to this edition, we now have three photo essays. Chapter 2 (Problems of Illness and Health Care) includes a photo essay titled "Modern Animal Food Production: Health and Safety Issues." In Chapter 6 (Poverty and Economic Inequality), a photo essay covers the topic "Lack of Clean Water and Sanitation Among the Poor." And in Chapter 14 (Environmental Problems) a photo essay depicts "Effects of Global Warming and Climate Change."

In-Text Learning Aids

Vignettes. Each chapter begins with a vignette designed to engage students and draw them into the chapter by illustrating the current relevance of the topic under discussion. Chapter 2 (Problems of Illness and Health Care), for instance, begins with a description of a family with no health insurance that must resort to receiving medical care from an annual free health clinic. Chapter 9 (Race, Ethnicity, and Immigration) begins with the story of talk radio host Don Imus, who was fired after calling the Rutgers University women's basketball team "nappy-headed hos" on a radio broadcast in April 2007.

Key Terms. Important terms and concepts are highlighted in the text where they first appear. To re-emphasize the importance of these words, they are listed at the end of every chapter and included in the glossary at the end of the text.

Running Glossary. New to the sixth edition is a running glossary that highlights the key terms in every chapter by putting the key terms and their definitions in the text margins.

What Do You Think? Sections. Each chapter contains several sections called *What Do You Think?* These sections invite students to use critical thinking skills to answer questions about issues related to the chapter content. For example, a *What Do You Think?* feature in Chapter 11 (Issues in Sexual Orientation) asks why in many countries male homosexuality is illegal, but female homosexuality is not. In Chapter 13 (Population Growth and Urbanization), a *What Do You*

Think? feature asks the reader if the U.S. birth rate would increase if the U.S. government instituted paid parenting leave and government-supported child care.

Glossary. All key terms are defined in the end-of-text glossary.

Understanding [specific social problem] Sections. All too often students, faced with contradictory theories and study results, walk away from social problems courses without any real understanding of their causes and consequences. To address this problem, chapter sections titled "Understanding . . . [specific social problem]" cap the body of each chapter just before the chapter summaries. Unlike the chapter summaries, these sections synthesize the material presented in the chapter, summing up the present state of knowledge and theory on the chapter topic.

SUPPLEMENTS

The sixth edition of *Understanding Social Problems* comes with a full complement of supplements designed with both faculty and students in mind.

Supplements for the Instructor

Instructor's Resource Manual with Test Bank. This supplement, written by Katheryn Dietrich of Texas A&M University and Blinn College, offers the instructor learning objectives, key terms, lecture outlines, student projects, classroom activities, and Internet and InfoTrac® College Edition exercises. Test items include multiple-choice and true-false questions with answers and page references, as well as short-answer and essay questions for each chapter. Each multiple-choice item has the question type (factual, applied, or conceptual) indicated. All questions are labeled as new, modified, or pickup, so instructors know if the question is new to this edition of the test bank, modified but picked up from the previous edition of the test bank, or picked up straight from the previous edition of the test bank. Concise user guides for InfoTrac College Edition and InfoMarks are included as appendices.

PowerLecture with JoinIn™ and ExamView®. This easy-to-use, one-stop digital library and presentation tool includes the following:

- Preassembled **Microsoft® PowerPoint® lecture slides** with graphics from the text, making it easy for you to assemble, edit, publish, and present custom lectures for your course.
- The PowerLectures CD-ROM, which includes video-based polling and quiz questions that can be used with the **JoinIn™ on TurningPoint®** personal response system.
- PowerLectures that also feature **ExamView® testing software,** which includes all the test items from the printed test bank in electronic format, enabling you to create customized tests of up to 250 items that can be delivered in print or online.

Videos. Adopters of *Understanding Social Problems* have several different video options available with the text.

ABC® Videos/DVD: Social Problems. This series of videos, comprised of footage from ABC broadcasts, is specially selected and arranged to accompany your Social Problems course. The segments may be used in conjunction with Wadsworth, Cengage Learning's Social Problems texts to help provide a real-world example to illustrate course concepts, or to instigate discussion. ABC Videos feature short, high-interest clips from current news events as well as historic raw footage going back 40 years. Clips are drawn from such programs as *World News Tonight, Good Morning America, This Week, PrimeTime Live, 20/20,* and *Nightline,* as well as numerous ABC News specials and material from the Associated Press Television News and British Movietone News collections. Contact your Cengage Learning representative for a complete listing of videos and policies.

AIDS in Africa DVD. Southern Africa has been overcome by a pandemic of unparalleled proportions. This documentary series focuses on Namibia, a new democracy, and the many actions that are being taken to control HIV/AIDS there. Included in this series are four documentary films created by the Project Pericles scholars at Elon University.

Wadsworth Sociology Video Library. This large selection of thought-provoking films is available to adopters based on adoption size.

Supplements for the Student

Study Guide. The study guide, written by Lori Fowler of Tarrant County College, includes learning objectives, brief and detailed chapter outlines, key terms, Internet activities, InfoTrac® College Edition exercises, student projects and classroom activities, practice tests consisting of multiple-choice and true-false questions with answers and page references, as well as short-answer questions and essay questions with page references to enhance and test student understanding of chapter concepts.

Online Resources

CengageNOW™ Personalized Study, a diagnostic tool (including a chapter-specific *Pre-Test, Individualized Study Plan,* and *Post-Test* written by Lois Sabol of Yakima Valley Community College) helps students master concepts and prepare for exams by creating a study plan based on the students' performance on the Pre-Test. Easily assign Personalized Study for the entire term, and, if you want, results will automatically post to your grade book. Order new student texts packaged with the access code to ensure that your students have four months of free access from the moment they purchase the text. Contact your local Wadsworth, Cengage Learning representative for ordering details. CengageNOW also features the most intuitive, easy-to-use online course management and study system on the market. It saves you time through its automatic grading and easy-to-use grade book and provides your students with an efficient way to study.

Extension: Wadsworth's Sociology Reader Collection. Create your own customized reader for your sociology class, drawing from dozens of classic and contemporary articles found on the exclusive Wadsworth, Cengage Learning TextChoice database. Using the TextChoice website (www.TextChoice.com), you can preview articles, select your content, and add your own original material. TextChoice will then produce your materials as a printed supplementary reader for your class.

Wadsworth's Sociology Home Page (academic.cengage.com/sociology). Here you will find a wealth of sociology resources such as Census 2000: A Student Guide for Sociology, Breaking News in Sociology, Guide to Researching Sociology on the Internet, Sociology in Action, and much more. Contained on the home page is the companion website for *Understanding Social Problems,* Sixth Edition.

Mooney/Knox/Schacht's *Understanding Social Problems* Companion Website (academic .cengage.com/sociology/mooney). This site provides access to useful learning resources for each chapter of the book. Instructors can also access password-protected instructor's manuals, PowerPoint lectures, and important sociology links. Click on the companion website to find useful learning resources for each chapter of the book. Some of these resources include

- Tutorial Practice Quizzes that can be scored and e-mailed to the instructor
- Web links
- Internet exercises
- Video exercises
- InfoTrac® College Edition exercises
- Flash cards of the text's glossary
- Crossword puzzles
- Essay questions
- Learning objectives
- Virtual explorations

And much more!

WebTutor™ for WebCT® or Blackboard®. Preloaded with content and available via access code when packaged with this text, WebTutor pairs all the content of this text's rich book companion website with all the sophisticated course management functionality of a WebCT or Blackboard product. You can assign materials (including online quizzes) and have the results flow automatically to your grade book.

InfoTrac® College Edition. Four months' access to this online database—featuring reliable, full-length articles from thousands of academic journals and periodicals—is available with this text at no additional charge! This fully searchable database now features stable, topically bookmarked InfoMarks® URLs to assist in research, plus InfoWrite critical thinking and writing tools. The database also offers 20 years' worth of full-text articles from almost 5,000 diverse sources, such as academic journals, newsletters, and up-to-the-minute periodicals, including *Time, Newsweek, Science, Forbes,* and *USA Today.* This incredible depth and breadth of material—available 24 hours a day from any computer with Internet access—makes conducting research so easy that your students will want to use it to enhance their work in every course!

Cengage Audio Study Products. Your students will enjoy the MP3-ready Audio Lecture Overviews for each chapter and a comprehensive audio glossary of key terms for quick study and review. Whether walking to class, doing laundry, or studying at their desk, students now have the freedom to choose when, where, and how they interact with their audio-based educational media. Contact your Wadsworth sales representative for more information.

ACKNOWLEDGMENTS

This text reflects the work of many people. We would like to thank the following for their contributions to the development of this text: Chris Caldeira, Acquisitions Editor; Jeremy Judson, Senior Development Editor; Tali Beesley, Assistant Editor; and Erin Parkins, Editorial Assistant. We would also like to acknowledge the support and assistance of Carol Jenkins (thanks, CJ) and Lee Maril. To each we send our heartfelt thanks. Special thanks also to George Glann, whose valuable contributions have assisted in achieving the book's high standard of quality from edition to edition.

Additionally, we are indebted to those who read the manuscript in its various drafts and provided valuable insights and suggestions, many of which have been incorporated into the final manuscript:

Joan Brehm
Illinois State University

Doug Degher
Northern Arizona University

Heather Griffiths
Fayetteville State University

Amy Holzgang
Cerritos College

Janét Hund
Long Beach City College

Kathrin Parks
University of New Mexico

Craig Robertson
University of North Alabama

Matthew Sanderson
University of Utah

Jacqueline Steingold
Wayne State University

William J. Tinney, Jr.
Black Hills State University

We are also grateful to the reviewers of the first, second, third, fourth, and fifth editions: David Allen, *University of New Orleans;* Patricia Atchison, *Colorado State University;* Wendy Beck, *Eastern Washington University;* Walter Carroll, *Bridgewater State College;* Deanna Chang, *Indiana University of Pennsylvania;* Roland Chilton, *University of Massachusetts;* Verghese Chirayath, *John Carroll University;* Margaret Chok, *Pellissippi State Technical Community College;* Kimberly Clark, *DeKalb College–Central Campus;* Anna M. Cognetto, *Dutchess Community College;* Robert R. Cordell, *West Virginia University at Parkersburg;* Barbara Costello, *Mississippi State University;* William Cross, *Illinois College;* Kim Davies, *Augusta State University;* Doug Degher, *Northern Arizona University;* Katherine Dietrich, *Blinn College;* Jane Ely, *State University of New York–Stony Brook;* William Feigelman, *Nassau Community College;* Joan Ferrante, *Northern Kentucky University;* Robert Gliner, *San Jose State University;* Roberta Goldberg, *Trinity College;* Roger Guy, *Texas Lutheran University;* Julia Hall, *Drexel University;* Millie Harmon, *Chemeketa Community College;* Madonna Harrington-Meyer, *University of Illinois;* Sylvia Jones, *Jefferson Community College;* Nancy Kleniewski, *University of Massachusetts, Lowell;* Daniel Klenow, *North Dakota State University;* Sandra Krell-Andre, *Southeastern Community College;* Pui-Yan Lam, *Eastern Washington University;* Mary Ann Lamanna, *University of Nebraska;* Phyllis Langton, *George Washington University;* Cooper Lansing, *Erie Community College;* Tunga Lergo, *Santa Fe Community College, Main Campus;* Dale Lund, *University of Utah;* Lionel Maldonado, *California State Univer-*

sity, *San Marcos;* Judith Mayo, *Arizona State University;* Peter Meiksins, *Cleveland State University;* JoAnn Miller, *Purdue University;* Clifford Mottaz, *University of Wisconsin–River Falls;* Lynda D. Nyce, *Bluffton College;* Frank J. Page, *University of Utah;* James Peacock, *University of North Carolina;* Barbara Perry, *Northern Arizona University;* Ed Ponczek, *William Rainey Harper College;* Donna Provenza, *California State University at Sacramento;* Cynthia Reynaud, *Louisiana State University;* Carl Marie Rider, *Longwood University;* Jeffrey W. Riemer, *Tennessee Technological University;* Cherylon Robinson, *University of Texas at San Antonio;* Rita Sakitt, *Suffolk County Community College;* Mareleyn Schneider, *Yeshiva University;* Paula Snyder, *Columbus State Community College;* Lawrence Stern, *Collin County Community College;* John Stratton, *University of Iowa;* D. Paul Sullins, *The Catholic University of America;* Joseph Trumino, *St. Vincent's College of St. John's University;* Robert Turley, *Crafton Hills College;* Alice Van Ommeren, *San Joaquin Delta College;* Joseph Vielbig, *Arizona Western University;* Harry L. Vogel, *Kansas State University;* Robert Weaver, *Youngstown State University;* Rose Weitz, *Arizona State University;* Bob Weyer, *County College of Morris;* Oscar Williams, *Diablo Valley College;* Mark Winton, *University of Central Florida;* Diane Zablotsky, *University of North Carolina.*

Finally, we are interested in ways to improve the text, and invite your feedback and suggestions for new ideas and material to be included in subsequent editions.

You can contact us at mooneyl@ecu.edu, knoxd@ecu.edu, or schachtc@ecu.edu.

About the Authors

Linda A. Mooney, Ph.D., is an associate professor of sociology at East Carolina University in Greenville, North Carolina. In addition to social problems, her specialties include law, criminology, and juvenile delinquency. She has published more than 30 professional articles in such journals as *Social Forces, Sociological Inquiry, Sex Roles, Sociological Quarterly,* and *Teaching Sociology.* She has won numerous teaching awards, including the University of North Carolina Board of Governor's Distinguished Professor for Teaching Award.

David Knox, Ph.D., is professor of sociology at East Carolina University. He has taught Social Problems, Introduction to Sociology, and Sociology of Marriage Problems. He is the author or co-author of 10 books and more than 60 professional articles. His research interests include various aspects of college student relationships, sexual values, and behavior.

Caroline Schacht, M.A., is a teaching instructor of sociology at East Carolina University. She has taught Introduction to Sociology, Deviant Behavior, Sociology of Food, Sociology of Education, Individuals in Society, and Courtship and Marriage. She has co-authored several textbooks in the areas of social problems, introductory sociology, courtship and marriage, and human sexuality.

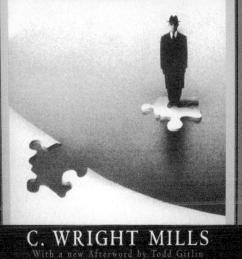

THE
SOCIOLOGICAL IMAGINATION
FORTIETH ANNIVERSARY EDITION

C. WRIGHT MILLS
With a new Afterword by Todd Gitlin

1

"Unless someone like
you cares a whole awful
lot, nothing is going to
get better. It's not."

Dr. Seuss, *The Lorax*

Thinking About Social Problems

**What Is a Social Problem? | Elements of Social Structure and Culture | The
Sociological Imagination | Theoretical Perspectives | Social Problems Research |
Goals of the Textbook | Understanding Social Problems | Chapter Review**

Only relatively recently have suicide bombers been considered a social problem to the U.S. public. More specifically, since the horror of September 11, 2001, terrorism in the United States has taken on new meaning. Here airport security guards inspect vehicles approaching the terminals.

In a 2006 Gallup poll a random sample of Americans were asked, "What is the most important problem facing this country today?" Leading problems included the war in Iraq, terrorism, the economy, unemployment, immigration, health care, education, the energy crisis, and poverty and homelessness. Moreover, survey results indicate that, overall, fewer than 30 percent of Americans were satisfied "with the way things are going in the country today." Compared with previous years, this number is quite low. Since the question was first asked in 1979, the average percentage of Americans saying that they were satisfied "with the way things are going in the country today" is 43 (Carroll 2006).

A global perspective on social problems is also troubling. In 1990 the United Nations Development Programme published its first annual *Human Development Report,* which measured the well-being of populations around the world according to a "human development index." This index measures three basic dimensions of human development: longevity, as measured by life expectancy at birth; knowledge (i.e., literacy, educational attainment); and a decent standard of living. The most recent report concluded that unless global action is taken immediately the "business as usual alternative will lead toward a world tarnished by mass poverty, divided by deep inequalities, and threatened by shared insecurities. In rich and poor countries alike future generations will pay a heavy price for failures of political leadership at this crossroads moment at the start of the twenty-first century" (Human Development Report 2006, p. 14).

Problems related to poverty and malnutrition, inadequate education, acquired immunodeficiency syndrome (AIDS) and other sexually transmitted diseases (STDs), inadequate health care, crime, conflict, oppression of minorities, environmental destruction, and other social issues are both national and international concerns. Such problems present both a threat and a challenge to our national and global society. The primary goal of this textbook is to facilitate increased awareness and understanding of problematic social conditions in U.S. society and throughout the world.

Although the topics covered in this book vary widely, all chapters share common objectives: to explain how social problems are created and maintained; to indicate how they affect individuals, social groups, and societies as a whole; and to examine programs and policies for change. We begin by looking at the nature of social problems.

WHAT IS A SOCIAL PROBLEM?

There is no universal, constant, or absolute definition of what constitutes a social problem. Rather, social problems are defined by a combination of objective and subjective criteria that vary across societies, among individuals and groups within a society, and across historical time periods.

Objective and Subjective Elements of Social Problems

Although social problems take many forms, they all share two important elements: an objective social condition and a subjective interpretation of that social condition. The **objective element of a social problem** refers to the existence of a social condition. We become aware of social conditions through our own life experience, through the media, and through education. We see the homeless,

objective element of a social problem Awareness of social conditions through one's own life experiences and through reports in the media.

hear gunfire in the streets, and see battered women in hospital emergency rooms. We read about employees losing their jobs as businesses downsize and factories close. In television news reports we see the anguished faces of parents whose children have been killed by violent youths.

The **subjective element of a social problem** refers to the belief that a particular social condition is harmful to society or to a segment of society and that it should and can be changed. We know that crime, drug addiction, poverty, racism, violence, and pollution exist. These social conditions are not considered social problems, however, unless at least a segment of society believes that these conditions diminish the quality of human life.

By combining these objective and subjective elements, we arrive at the following definition: A **social problem** is a social condition that a segment of society views as harmful to members of society and in need of remedy.

Variability in Definitions of Social Problems

Individuals and groups frequently disagree about what constitutes a social problem. For example, some Americans view the availability of abortion as a social problem, whereas others view restrictions on abortion as a social problem. Similarly, some Americans view homosexuality as a social problem, whereas others view prejudice and discrimination against homosexuals as a social problem. Such variations in what is considered a social problem are due to differences in values, beliefs, and life experiences.

Definitions of social problems vary not only within societies but also across societies and historical time periods. For example, before the 19th century it was a husband's legal right and marital obligation to discipline and control his wife through the use of physical force. Today, the use of physical force is regarded as a social problem rather than a marital right.

Tea drinking is another example of how what is considered a social problem can change over time. In 17th- and 18th-century England tea drinking was regarded as a "base Indian practice" that was "pernicious to health, obscuring industry, and impoverishing the nation" (Ukers 1935, cited by Troyer & Markle 1984). Today, the English are known for their tradition of drinking tea in the afternoon.

Because social problems can be highly complex, it is helpful to have a framework within which to view them. Sociology provides such a framework. Using a sociological perspective to examine social problems requires knowledge of the basic concepts and tools of sociology. In the remainder of this chapter we discuss some of these concepts and tools: social structure, culture, the "sociological imagination," major theoretical perspectives, and types of research methods.

What Do You Think? People increasingly are using information technologies to get their daily news with "traditional media outlets . . . struggling to hold their market share" (Saad 2007, p. 1). Although local television remains the number one source of news for most Americans, some research indicates that news on the Internet is beginning to replace television news as the primary source of information for many computer users (see Chapter 15). What role do the various media play in our awareness of social problems? Will definitions of social problems change as sources of information change and, if so, in what way?

subjective element of a social problem The belief that a particular social condition is harmful to society, or to a segment of society, and that it should and can be changed.

social problem A social condition that a segment of society views as harmful to members of society and in need of remedy.

AP/Wide World Photos

Whereas some individuals view homosexual behavior as a social problem, others view homophobia as a social problem. Here, participants carry a giant rainbow flag during a gay pride parade in Toronto, Canada. The 2006 Canadian census was revamped to include "same-sex married spouse" as a response option (Beeby 2005).

ELEMENTS OF SOCIAL STRUCTURE AND CULTURE

Although society surrounds us and permeates our lives, it is difficult to "see" society. By thinking of society in terms of a picture or image, however, we can visualize society and therefore better understand it. Imagine that society is a coin with two sides: On one side is the structure of society and on the other is the culture of society. Although each side is distinct, both are inseparable from the whole. By looking at the various elements of social structure and culture, we can better understand the root causes of social problems.

Elements of Social Structure

The **structure** of a society refers to the way society is organized. Society is organized into different parts: institutions, social groups, statuses, and roles.

structure The way society is organized including institutions, social groups, statuses, and roles.

institution An established and enduring pattern of social relationships.

social group Two or more people who have a common identity, interact, and form a social relationship.

primary groups Usually small numbers of individuals characterized by intimate and informal interaction.

secondary groups Involving small or large numbers of individuals, groups that are task oriented and are characterized by impersonal and formal interaction.

status A position that a person occupies within a social group.

Institutions. An **institution** is an established and enduring pattern of social relationships. The five traditional institutions are family, religion, politics, economics, and education, but some sociologists argue that other social institutions, such as science and technology, mass media, medicine, sports, and the military, also play important roles in modern society. Many social problems are generated by inadequacies in various institutions. For example, unemployment may be influenced by the educational institution's failure to prepare individuals for the job market and by alterations in the structure of the economic institution.

Social Groups. Institutions are made up of social groups. A **social group** is defined as two or more people who have a common identity, interact, and form a social relationship. For example, the family in which you were reared is a social group that is part of the family institution. The religious association to which you may belong is a social group that is part of the religious institution.

Social groups can be categorized as primary or secondary. **Primary groups,** which tend to involve small numbers of individuals, are characterized by intimate and informal interaction. Families and friends are examples of primary groups. **Secondary groups,** which may involve small or large numbers of individuals, are task oriented and are characterized by impersonal and formal interaction. Examples of secondary groups include employers and their employees and clerks and their customers.

Statuses. Just as institutions consist of social groups, social groups consist of statuses. A **status** is a position that a person occupies within a social group. The statuses we occupy largely define our social identity. The statuses in a family may consist of mother, father, stepmother, stepfather, wife, husband, child, and so on. Statuses can be either ascribed or achieved. An **ascribed status** is one that society assigns to an individual on the basis of factors over which the individual has no control. For example, we have no control over the sex, race, ethnic background, and socioeconomic status into which we are born. Similarly, we are assigned the status of child, teenager, adult, or senior citizen on the basis of our age—something we do not choose or control.

An **achieved status** is assigned on the basis of some characteristic or behavior over which the individual has some control. Whether you achieve the status of college graduate, spouse, parent, bank president, or prison inmate depends largely on your own efforts, behavior, and choices. One's ascribed statuses may affect the likelihood of achieving other statuses, however. For example, if you are born into a poor socioeconomic status, you may find it more difficult to achieve the status of college graduate because of the high cost of a college education.

Every individual has numerous statuses simultaneously. You may be a student, parent, tutor, volunteer fund-raiser, female, and Hispanic. A person's *master status* is the status that is considered the most significant in a person's social identity. Typically, a person's occupational status is regarded as his or her master status. If you are a full-time student, your master status is likely to be student.

Roles. Every status is associated with many **roles,** or the set of rights, obligations, and expectations associated with a status. Roles guide our behavior and allow us to predict the behavior of others. As a student, you are expected to attend class, listen and take notes, study for tests, and complete assignments. Because you know what the role of teacher involves, you can predict that your teacher will lecture, give exams, and assign grades based on your performance on tests.

A single status involves more than one role. For example, the status of prison inmate includes one role for interacting with prison guards and another role for interacting with other prison inmates. Similarly, the status of nurse involves different roles for interacting with physicians and with patients.

Elements of Culture

Whereas social structure refers to the organization of society, **culture** refers to the meanings and ways of life that characterize a society. The elements of culture include beliefs, values, norms, sanctions, and symbols.

Beliefs. **Beliefs** refer to definitions and explanations about what is assumed to be true. The beliefs of an individual or group influence whether that individual or group views a particular social condition as a social problem. Does secondhand smoke harm nonsmokers? Are nuclear power plants safe? Does violence in movies and on television lead to increased aggression in children? Our beliefs regarding these issues influence whether we view the issues as social problems. Beliefs influence not only how a social condition is interpreted but also the existence of the condition itself. For example, police officers' beliefs about their supervisors' priorities affected officers' problem-solving behavior and the time devoted to it (Engel & Worden 2003). The *Self and Society* feature in this chapter allows you to assess your own beliefs about various social issues and to compare your beliefs with a national sample of first-year college students.

Values. **Values** are social agreements about what is considered good and bad, right and wrong, desirable and undesirable. Frequently, social conditions are

> 66 When I fulfill my obligations as a brother, husband, or citizen, when I execute contracts, I perform duties that are defined externally to myself. . . . Even if I conform in my own sentiments and feel their reality subjectively, such reality is still objective, for I did not create them; I merely inherited them. 99
>
> Emile Durkheim
> Sociologist

ascribed status A status that society assigns to an individual on the basis of factors over which the individual has no control.

achieved status A status assigned on the basis of some characteristic or behavior over which the individual has some control.

role The set of rights, obligations, and expectations associated with a status.

culture The meanings and ways of life that characterize a society including beliefs, values, norms, sanctions, and symbols.

beliefs Definitions and explanations about what is assumed to be true.

values Social agreements about what is considered good and bad, right and wrong, desirable and undesirable.

Self and Society | Personal Beliefs About Various Social Problems

Indicate whether you agree or disagree with each of the following statements:

Statement		Agree	Disagree
1.	Federal military spending should be increased.	_____	_____
2.	The federal government is not doing enough to control environmental pollution.	_____	_____
3.	There is too much concern in the courts for the rights of criminals.	_____	_____
4.	Abortion should be legal.	_____	_____
5.	The death penalty should be abolished.	_____	_____
6.	Undocumented immigrants should be denied access to public education.	_____	_____
7.	Marijuana should be legalized.	_____	_____
8.	It is important to have laws prohibiting homosexual relationships.	_____	_____
9.	Colleges have the right to ban extreme speakers.	_____	_____
10.	The federal government should do more to control the sale of handguns.	_____	_____
11.	Racial discrimination is no longer a major problem in America.	_____	_____
12.	Realistically, an individual can do little to bring about changes in our society.	_____	_____
13.	Wealthy people should pay a larger share of taxes than they do now.	_____	_____
14.	Affirmative action in college admissions should be abolished.	_____	_____
15.	Same-sex couples should have the right to legal marital status.	_____	_____

PERCENTAGE OF FIRST-YEAR COLLEGE STUDENTS AGREEING WITH BELIEF STATEMENTS*

		PERCENTAGE AGREEING IN 2006		
STATEMENT NUMBER		**TOTAL**	**WOMEN**	**MEN**
1.	Military spending should be increased	32	30	36
2.	Federal government not doing enough to stop pollution	78	81	74
3.	Too much concern for criminals' rights	56	53	59
4.	Abortion should be legal	57	56	57
5.	Death penalty should be abolished	35	38	31
6.	Immigrants should be denied access to public schools	47	42	53
7.	Marijuana should be legalized	37	33	42
8.	Important to have laws prohibiting gay relationships	26	19	33
9.	Colleges should be able to ban speakers on campus	41	38	44
10.	Federal government should do more to control sale of handguns	74	81	66
11.	Racial discrimination no longer a problem	19	15	24
12.	Individuals can't influence social change	27	24	31
13.	Wealthy should pay higher taxes	58	58	58
14.	Affirmative action in college admissions should be abolished	47	42	53
15.	Same-sex couples should have legal right to marry	61	68	53

*Percentages are rounded.

Source: Pryor, John, Sylvia Hurtado, Victor Saenz, Jessica Korn, Jose Luis Santos, and William S. Korn. 2006. *The American Freshman: National Norms for Fall 2006*. Los Angeles, CA: Higher Education Research Institute.

viewed as social problems when the conditions are incompatible with or contradict closely held values. For example, poverty and homelessness violate the value of human welfare; crime contradicts the values of honesty, private property, and nonviolence; racism, sexism, and heterosexism violate the values of equality and fairness.

Values play an important role not only in the interpretation of a condition as a social problem but also in the development of the social condition itself. Sylvia Ann Hewlett (1992) explains how the American values of freedom and individualism are at the root of many of our social problems:

> There are two sides to the coin of freedom. On the one hand, there is enormous potential for prosperity and personal fulfillment; on the other are all the hazards of untrammeled opportunity and unfettered choice. Free markets can produce grinding poverty as well as spectacular wealth; unregulated industry can create dangerous levels of pollution as well as rapid rates of growth; and an unfettered drive for personal fulfillment can have disastrous effects on families and children. Rampant individualism does not bring with it sweet freedom; rather, it explodes in our faces and limits life's potential. (pp. 350–351)

Absent or weak values may contribute to some social problems. For example, many industries do not value protection of the environment and thus contribute to environmental pollution.

Norms and Sanctions. Norms are socially defined rules of behavior. Norms serve as guidelines for our behavior and for our expectations of the behavior of others.

There are three types of norms: folkways, laws, and mores. *Folkways* refer to the customs and manners of society. In many segments of our society it is customary to shake hands when being introduced to a new acquaintance, to say "excuse me" after sneezing, and to give presents to family and friends on their birthdays. Although no laws require us to do these things, we are expected to do them because they are part of the cultural traditions, or folkways, of the society in which we live.

Laws are norms that are formalized and backed by political authority. It is normative for a Muslim woman to wear a veil. However, in the United States failure to remove the veil for a driver's license photo is grounds for revoking the permit. Such is the case of a Florida woman who brought suit against the state, claiming that her religious rights were being violated because she was required to remove her veil for the driver's license photo (Candy 2003). She appealed the decision to Florida's District Court of Appeal and lost. The Court recognized, however, "the tension created as a result of choosing between following the dictates of one's religion and the mandates of secular law" (Associated Press 2006).

Some norms, called *mores,* have a moral basis. Violations of mores may produce shock, horror, and moral indignation. Both littering and child sexual abuse are violations of law, but child sexual abuse is also a violation of our mores because we view such behavior as immoral.

All norms are associated with **sanctions,** or social consequences for conforming to or violating norms. When we conform to a social norm, we may be rewarded by a positive sanction. These may range from an approving smile to a public ceremony in our honor. When we violate a social norm, we may be punished by a negative sanction, which may range from a disapproving look to the death penalty or life in prison. Most sanctions are spontaneous expressions of

norms Socially defined rules of behavior including folkways, mores, and laws.

sanctions Social consequences for conforming to or violating norms.

TABLE 1.1 Types and Examples of Sanctions

	POSITIVE	NEGATIVE
Informal	Being praised by one's neighbors for organizing a neighborhood recycling program.	Being criticized by one's neighbors for refusing to participate in the neighborhood recycling program.
Formal	Being granted a citizen's award for organizing a neighborhood recycling program.	Being fined by the city for failing to dispose of trash properly.

approval or disapproval by groups or individuals—these are referred to as informal sanctions. Sanctions that are carried out according to some recognized or formal procedure are referred to as formal sanctions. Types of sanctions, then, include positive informal sanctions, positive formal sanctions, negative informal sanctions, and negative formal sanctions (see Table 1.1).

Symbols. A **symbol** is something that represents something else. Without symbols, we could not communicate with each other or live as social beings.

The symbols of a culture include language, gestures, and objects whose meaning is commonly understood by the members of a society. In our society a red ribbon tied around a car antenna symbolizes Mothers Against Drunk Driving, a peace sign symbolizes the value of nonviolence, and a white hooded robe symbolizes the Ku Klux Klan. Sometimes people attach different meanings to the same symbol. The Confederate flag is a symbol of southern pride to some and a symbol of racial bigotry to others.

The elements of the social structure and culture just discussed play a central role in the creation, maintenance, and social response to various social problems. One of the goals of taking a course in social problems is to develop an awareness of how the elements of social structure and culture contribute to social problems. Sociologists refer to this awareness as the "sociological imagination."

THE SOCIOLOGICAL IMAGINATION

The **sociological imagination,** a term developed by C. Wright Mills (1959), refers to the ability to see the connections between our personal lives and the social world in which we live. When we use our sociological imagination, we are able to distinguish between "private troubles" and "public issues" and to see connections between the events and conditions of our lives and the social and historical context in which we live.

For example, that one person is unemployed constitutes a private trouble. That millions of people are unemployed in the United States constitutes a public issue. Once we understand that personal troubles such as human immunodeficiency virus (HIV) infection, criminal victimization, and poverty are shared by other segments of society, we can look for the elements of social structure and culture that contribute to these public issues and private troubles. If the various elements of social structure and culture contribute to private troubles and public issues, then society's social structure and culture must be changed if these concerns are to be resolved.

Rather than viewing the private trouble of being unemployed as a result of an individual's faulty character or lack of job skills, we may understand unem-

symbol Something that represents something else.

sociological imagination The ability to see the connections between our personal lives and the social world in which we live.

ployment as a public issue that results from the failure of the economic and political institutions of society to provide job opportunities to all citizens. Technological innovations emerging from the Industrial Revolution led to individual workers being replaced by machines. During the economic recession of the 1980s, employers fired employees so the firm could stay in business. Thus, in both these cases, social forces rather than individual skills largely determined whether a person was employed.

THEORETICAL PERSPECTIVES

Theories in sociology provide us with different perspectives with which to view our social world. A perspective is simply a way of looking at the world. A theory is a set of interrelated propositions or principles designed to answer a question or explain a particular phenomenon; it provides us with a perspective. Sociological theories help us to explain and predict the social world in which we live.

Sociology includes three major theoretical perspectives: the structural-functionalist perspective, the conflict perspective, and the symbolic interactionist perspective. Each perspective offers a variety of explanations about the causes of and possible solutions to social problems.

Structural-Functionalist Perspective

The structural-functionalist perspective is based largely on the works of Herbert Spencer, Emile Durkheim, Talcott Parsons, and Robert Merton. According to structural functionalism, society is a system of interconnected parts that work together in harmony to maintain a state of balance and social equilibrium for the whole. For example, each of the social institutions contributes important functions for society: Family provides a context for reproducing, nurturing, and socializing children; education offers a way to transmit a society's skills, knowledge, and culture to its youth; politics provides a means of governing members of society; economics provides for the production, distribution, and consumption of goods and services; and religion provides moral guidance and an outlet for worship of a higher power.

The structural-functionalist perspective emphasizes the interconnectedness of society by focusing on how each part influences and is influenced by other parts. For example, the increase in single-parent and dual-earner families has contributed to the number of children who are failing in school because parents have become less available to supervise their children's homework. As a result of changes in technology, colleges are offering more technical programs, and many adults are returning to school to learn new skills that are required in the workplace. The increasing number of women in the workforce has contributed to the formulation of policies against sexual harassment and job discrimination.

Structural functionalists use the terms *functional* and *dysfunctional* to describe the effects of social elements on society. Elements of society are functional if they contribute to social stability and dysfunctional if they disrupt social stability. Some aspects of society can be both functional and dysfunctional. For example, crime is dysfunctional in that it is associated with physical violence, loss of property, and fear. But according to Durkheim and other functionalists, crime is also functional for society because it leads to heightened awareness of shared moral bonds and increased social cohesion.

Sociologists have identified two types of functions: manifest and latent (Merton 1968). **Manifest functions** are consequences that are intended and commonly recognized. **Latent functions** are consequences that are unintended and often hidden. For example, the manifest function of education is to transmit knowledge and skills to society's youth. But public elementary schools also serve as babysitters for employed parents, and colleges offer a place for young adults to meet potential mates. The baby-sitting and mate-selection functions are not the intended or commonly recognized functions of education; hence they are latent functions.

What Do You Think? In viewing society as a set of interrelated parts, structural functionalists argue that proposed solutions to social problems may lead to other social problems. For example, urban renewal projects displace residents and break up community cohesion. Racial imbalance in schools led to forced integration, which in turn generated violence and increased hostility between the races. What are some other "solutions" that lead to social problems? Do all solutions come with a price to pay? Can you think of a solution to a social problem that has no negative consequences?

Structural-Functionalist Theories of Social Problems

Two dominant theories of social problems grew out of the structural-functionalist perspective: social pathology and social disorganization.

Social Pathology. According to the social pathology model, social problems result from some "sickness" in society. Just as the human body becomes ill when our systems, organs, and cells do not function normally, society becomes "ill" when its parts (i.e., elements of the structure and culture) no longer perform properly. For example, problems such as crime, violence, poverty, and juvenile delinquency are often attributed to the breakdown of the family institution; the decline of the religious institution; and inadequacies in our economic, educational, and political institutions.

Social "illness" also results when members of a society are not adequately socialized to adopt its norms and values. People who do not value honesty, for example, are prone to dishonesties of all sorts. Early theorists attributed the failure in socialization to "sick" people who could not be socialized. Later theorists recognized that failure in the socialization process stemmed from "sick" social conditions, not "sick" people. To prevent or solve social problems, members of society must receive proper socialization and moral education, which may be accomplished in the family, schools, churches, or workplace and/or through the media.

Social Disorganization. According to the social disorganization view of social problems, rapid social change disrupts the norms in a society. When norms become weak or are in conflict with each other, society is in a state of **anomie**, or normlessness. Hence people may steal, physically abuse their spouses or children, abuse drugs, commit rape, or engage in other deviant behavior because the

manifest functions Consequences that are intended and commonly recognized.

latent functions Consequences that are unintended and often hidden.

anomie A state of normlessness in which norms and values are weak or unclear.

norms regarding these behaviors are weak or conflicting. According to this view, the solution to social problems lies in slowing the pace of social change and strengthening social norms. For example, although the use of alcohol by teenagers is considered a violation of a social norm in our society, this norm is weak. The media portray young people drinking alcohol, teenagers teach each other to drink alcohol and buy fake identification cards (IDs) to purchase alcohol, and parents model drinking behavior by having a few drinks after work or at a social event. Solutions to teenage drinking may involve strengthening norms against it through public education, restricting media depictions of youth and alcohol, imposing stronger sanctions against the use of fake IDs to purchase alcohol, and educating parents to model moderate and responsible drinking behavior.

Conflict Perspective

The structural-functionalist perspective views society as composed of different parts working together. In contrast, the conflict perspective views society as composed of different groups and interests competing for power and resources. The conflict perspective explains various aspects of our social world by looking at which groups have power and benefit from a particular social arrangement. For example, feminist theory argues that we live in a patriarchal society—a hierarchical system of organization controlled by men. Although there are many varieties of feminist theory, most would hold that feminism "demands that existing economic, political, and social structures be changed" (Weir and Faulkner 2004, p. xii).

The origins of the conflict perspective can be traced to the classic works of Karl Marx. Marx suggested that all societies go through stages of economic development. As societies evolve from agricultural to industrial, concern over meeting survival needs is replaced by concern over making a profit, the hallmark of a capitalist system. Industrialization leads to the development of two classes of people: the bourgeoisie, or the owners of the means of production (e.g., factories, farms, businesses); and the proletariat, or the workers who earn wages.

The division of society into two broad classes of people—the "haves" and the "have-nots"—is beneficial to the owners of the means of production. The workers, who may earn only subsistence wages, are denied access to the many resources available to the wealthy owners. According to Marx, the bourgeoisie use their power to control the institutions of society to their advantage. For example, Marx suggested that religion serves as an "opiate of the masses" in that it soothes the distress and suffering associated with the working-class lifestyle and focuses the workers' attention on spirituality, God, and the afterlife rather than on worldly concerns such as living conditions. In essence, religion diverts the workers so that they concentrate on being rewarded in heaven for living a moral life rather than on questioning their exploitation.

❝ Underlying virtually all social problems are conditions caused in whole or in part by social injustice. **❞**

Pamela Ann Roby
Sociologist

Conflict Theories of Social Problems

There are two general types of conflict theories of social problems: Marxist and non-Marxist. Marxist theories focus on social conflict that results from economic inequalities; non-Marxist theories focus on social conflict that results from competing values and interests among social groups.

Marxist Conflict Theories. According to contemporary Marxist theorists, social problems result from class inequality inherent in a capitalistic system. A system of haves and have-nots may be beneficial to the haves but often translates into poverty for the have-nots. As we will explore later in this textbook, many social problems, including physical and mental illness, low educational achievement, and crime, are linked to poverty.

In addition to creating an impoverished class of people, capitalism also encourages "corporate violence." Corporate violence can be defined as actual harm and/or risk of harm inflicted on consumers, workers, and the general public as a result of decisions by corporate executives or managers. Corporate violence can also result from corporate negligence; the quest for profits at any cost; and willful violations of health, safety, and environmental laws (Reiman 2007). Our profit-motivated economy encourages individuals who are otherwise good, kind, and law abiding to knowingly participate in the manufacturing and marketing of defective brakes on American jets, fuel tanks on automobiles, and contraceptive devices (e.g., intrauterine devices [IUDs]). The profit motive has also caused individuals to sell defective medical devices, toxic pesticides, and contaminated foods to developing countries. As Eitzen and Baca Zinn note, the "goal of profit is so central to capitalistic enterprises that many corporate decisions are made without consideration for the consequences" (Eitzen & Baca Zinn 2000, p. 483).

Marxist conflict theories also focus on the problem of **alienation,** or powerlessness and meaninglessness in people's lives. In industrialized societies workers often have little power or control over their jobs, a condition that fosters in them a sense of powerlessness in their lives. The specialized nature of work requires workers to perform limited and repetitive tasks; as a result, the workers may come to feel that their lives are meaningless.

Alienation is bred not only in the workplace but also in the classroom. Students have little power over their education and often find that the curriculum is not meaningful to their lives. Like poverty, alienation is linked to other social problems, such as low educational achievement, violence, and suicide.

Marxist explanations of social problems imply that the solution lies in eliminating inequality among classes of people by creating a classless society. The nature of work must also change to avoid alienation. Finally, stronger controls must be applied to corporations to ensure that corporate decisions and practices are based on safety rather than on profit considerations.

Non-Marxist Conflict Theories. Non-Marxist conflict theorists, such as Ralf Dahrendorf, are concerned with conflict that arises when groups have opposing values and interests. For example, antiabortion activists value the life of unborn embryos and fetuses; pro-choice activists value the right of women to control their own bodies and reproductive decisions. These different value positions reflect different subjective interpretations of what constitutes a social problem. For antiabortionists the availability of abortion is the social problem; for pro-choice advocates the restrictions on abortion are the social problem. Sometimes the social problem is not the conflict itself but rather the way that conflict is expressed. Even most pro-life advocates agree that shooting doctors who perform abortions and blowing up abortion clinics constitute unnecessary violence and lack of respect for life. Value conflicts may occur between diverse categories of people, including nonwhites versus whites, heterosexuals versus homosexuals, young versus old, Democrats versus Republicans, and environmentalists versus industrialists.

alienation A sense of powerlessness and meaninglessness in people's lives.

Solving the problems that are generated by competing values may involve ensuring that conflicting groups understand each other's views, resolving differences through negotiation or mediation, or agreeing to disagree. Ideally, solutions should be win-win, with both conflicting groups satisfied with the solution. However, outcomes of value conflicts are often influenced by power; the group with the most power may use its position to influence the outcome of value conflicts. For example, when Congress could not get all states to voluntarily increase the legal drinking age to 21, it threatened to withdraw federal highway funds from those that would not comply.

Symbolic Interactionist Perspective

Both the structural-functionalist and the conflict perspectives are concerned with how broad aspects of society, such as institutions and large social groups, influence the social world. This level of sociological analysis is called *macro sociology*: It looks at the big picture of society and suggests how social problems are affected at the institutional level.

Micro sociology, another level of sociological analysis, is concerned with the social-psychological dynamics of individuals interacting in small groups. Symbolic interactionism reflects the micro-sociological perspective and was largely influenced by the work of early sociologists and philosophers such as Max Weber, George Simmel, Charles Horton Cooley, G. H. Mead, W. I. Thomas, Erving Goffman, and Howard Becker. Symbolic interactionism emphasizes that human behavior is influenced by definitions and meanings that are created and maintained through symbolic interaction with others.

Sociologist W. I. Thomas (1966) emphasized the importance of definitions and meanings in social behavior and its consequences. He suggested that humans respond to their definition of a situation rather than to the objective situation itself. Hence Thomas noted that situations that we define as real become real in their consequences.

Symbolic interactionism also suggests that our identity or sense of self is shaped by social interaction. We develop our self-concept by observing how others interact with us and label us. By observing how others view us, we see a reflection of ourselves that Cooley calls the "looking glass self."

Last, the symbolic interactionist perspective has important implications for how social scientists conduct research. German sociologist Max Weber argued that to understand individual and group behavior, social scientists must see the world through the eyes of that individual or group. Weber called this approach *verstehen,* which in German means "empathy." *Verstehen* implies that in conducting research, social scientists must try to understand others' view of reality and the subjective aspects of their experiences, including their symbols, values, attitudes, and beliefs.

> ❝ Each to each a looking glass, Reflects the other that doth pass. ❞
>
> **Charles Horton Cooley**
> Sociologist

Symbolic Interactionist Theories of Social Problems

A basic premise of symbolic interactionist theories of social problems is that a condition must be defined or recognized as a social problem for it to be a social problem. Based on this premise, Herbert Blumer (1971) suggested that social problems develop in stages. First, social problems pass through the stage of *societal recognition*—the process by which a social problem, for example, drunk driving, is "born." Second, *social legitimation* takes place when the social problem achieves recognition by the larger community, including the media, schools, and churches. As the visibility of traffic fatalities associated with alcohol increased, so did the legitimation of drunk driving as a social problem. The next stage in the development of a social problem involves *mobilization for action,* which occurs when individuals and groups, such as Mothers Against Drunk Driving, become concerned about how to respond to the social condition. This mobilization leads to the *development and implementation of an official plan* for dealing with the problem, involving, for example, highway checkpoints, lower legal blood-alcohol levels, and tougher drunk-driving regulations.

Blumer's stage-development view of social problems is helpful in tracing the development of social problems. For example, although sexual harassment and date rape occurred throughout the 20th century, these issues did not begin to receive recognition as social problems until the 1970s. Social legitimation of these problems was achieved when high schools, colleges, churches, employers, and the media recognized their existence. Organized social groups mobilized to develop and implement plans to deal with these problems. For example, groups successfully lobbied for the enactment of laws against sexual harassment and the enforcement of sanctions against violators of these laws. Groups also mobilized to provide educational seminars on date rape for high school and college students and to offer support services to victims of date rape.

Some disagree with the symbolic interactionist view that social problems exist only if they are recognized. According to this view, individuals who were victims of date rape in the 1960s may be considered victims of a problem, even though date rape was not recognized at that time as a social problem.

Labeling theory, a major symbolic interactionist theory of social problems, suggests that a social condition or group is viewed as problematic if it is labeled as such. According to labeling theory, resolving social problems sometimes involves changing the meanings and definitions that are attributed to people and situations. For example, so long as teenagers define drinking alcohol as "cool" and "fun," they will continue to abuse alcohol. So long as our society defines providing sex education and contraceptives to teenagers as inappropriate or immoral, the teenage pregnancy rate in the United States will continue to be higher than that in other industrialized nations.

Social constructionism is another symbolic interactionist theory of social problems. Similar to labeling theorists and symbolic interactionism in general, social constructionists argue that reality is socially constructed by individuals who interpret the social world around them. Society, therefore, is a social creation rather than an objective given. As such, social constructionists often question the origin and evolution of social problems. For example, most Americans define "drug abuse" as a social problem in the United States but rarely include alcohol or cigarettes in their discussion. A social constructionist would point to the historical roots of alcohol and tobacco use as a means of understanding their legal status. Central to this idea of the social construction of social problems are the media, universities, research institutes, and government agencies, which are often responsible for the public's initial "take" on the problem under discussion.

Table 1.2 summarizes and compares the major theoretical perspectives, their criticisms, and social policy recommendations as they relate to social problems.

TABLE 1.2 Comparison of Theoretical Perspectives

	STRUCTURAL FUNCTIONALISM	CONFLICT THEORY	SYMBOLIC INTERACTIONISM
Representative theorists	Emile Durkheim Talcott Parsons Robert Merton	Karl Marx Ralf Dahrendorf	George H. Mead Charles Cooley Erving Goffman
Society	Society is a set of interrelated parts; cultural consensus exists and leads to social order; natural state of society—balance and harmony.	Society is marked by power struggles over scarce resources; inequities result in conflict; social change is inevitable; natural state of society—imbalance.	Society is a network of interlocking roles; social order is constructed through interaction as individuals, through shared meaning, make sense out of their social world.
Individuals	Individuals are socialized by society's institutions; socialization is the process by which social control is exerted; people need society and its institutions.	People are inherently good but are corrupted by society and its economic structure; institutions are controlled by groups with power; "order" is part of the illusion.	Humans are interpretive and interactive; they are constantly changing as their "social beings" emerge and are molded by changing circumstances.
Cause of social problems?	Rapid social change; social disorganization that disrupts the harmony and balance; inadequate socialization and/or weak institutions.	Inequality; the dominance of groups of people over other groups of people; oppression and exploitation; competition between groups.	Different interpretations of roles; labeling of individuals, groups, or behaviors as deviant; definition of an objective condition as a social problem.
Social policy/ solutions	Repair weak institutions; assure proper socialization; cultivate a strong collective sense of right and wrong.	Minimize competition; create an equitable system for the distribution of resources.	Reduce impact of labeling and associated stigmatization; alter definitions of what is defined as a social problem.
Criticisms	Called "sunshine sociology"; supports the maintenance of the status quo; needs to ask "functional for whom?"; does not deal with issues of power and conflict; incorrectly assumes a consensus.	Utopian model; Marxist states have failed; denies existence of cooperation and equitable exchange; cannot explain cohesion and harmony.	Concentrates on micro issues only; fails to link micro issues to macro-level concerns; too psychological in its approach; assumes label amplified problem.

The study of social problems is based on research as well as on theory, however. Indeed, research and theory are intricately related. As Wilson (1983) stated:

> Most of us think of theorizing as quite divorced from the business of gathering facts. It seems to require an abstractness of thought remote from the practical activity of empirical research. But theory building is not a separate activity within sociology. Without theory, the empirical researcher would find it impossible to decide what to observe, how to observe it, or what to make of the observations. (p. 1)

SOCIAL PROBLEMS RESEARCH

Most students taking a course in social problems will not become researchers or conduct research on social problems. Nevertheless, we are all consumers of research that is reported in the media. Politicians, social activist groups, and organizations attempt to justify their decisions, actions, and positions by citing research results. As consumers of research, we need to understand that our personal experiences and casual observations are less reliable than generalizations based on systematic research. One strength of scientific research is that it is subjected to critical examination by other researchers (see this chapter's *Social Problems Research Up Close* feature). The more you understand how research is done, the better able you will be to critically examine and question research rather than to passively consume research findings. In the remainder of this section we discuss the stages of conducting a research study and the various methods of research used by sociologists.

Stages of Conducting a Research Study

Sociologists progress through various stages in conducting research on a social problem. In this section we describe the first four stages: formulating a research question, reviewing the literature, defining variables, and formulating a hypothesis.

Formulating a Research Question. A research study usually begins with a research question. Where do research questions originate? How does a particular researcher come to ask a particular research question? In some cases, researchers have a personal interest in a specific topic because of their own life experiences. For example, a researcher who has experienced spouse abuse may wish to do research on such questions as "What factors are associated with domestic violence?" and "How helpful are battered women's shelters in helping abused women break the cycle of abuse in their lives?" Other researchers may ask a particular research question because of their personal values—their concern for humanity and the desire to improve human life. Researchers who are concerned about the spread of HIV infection and AIDS may conduct research on questions such as "How does the use of alcohol influence condom use?" and "What educational strategies are effective for increasing safer sex behavior?" Researchers may also want to test a particular sociological theory, or some aspect of it, to establish its validity or conduct studies to evaluate the effect of a social policy or program. Research questions may also be formulated by the concerns of community groups and social activist organizations in collaboration with academic researchers. Government and industry also hire researchers to answer questions such as "How many children are victimized by episodes of violence at school?" and "What types of computer technologies can protect children against being exposed to pornography on the Internet?"

❝In science (as in everyday life) things must be believed in order to be seen as well as seen in order to be believed.**❞**

Walter L. Wallace
Social scientist

What Do You Think? In a free society there must be freedom of information. That is why the U.S. Constitution and, more specifically, the First Amendment protect journalists' sources. If journalists are compelled to reveal their sources, their sources may be unwilling to share information, and this would jeopardize the public's right to know. A journalist cannot reveal information given in confidence without permission from the source or a court order. Do you think sociologists should be granted the same protections as journalists? If a reporter at your school newspaper uncovered a scandal at your university, should he or she be protected by the First Amendment?

Reviewing the Literature. After a research question is formulated, the researcher reviews the published material on the topic to find out what is already known about it. Reviewing the literature also provides researchers with ideas about how to conduct their research and helps them formulate new research questions. A literature review serves as an evaluation tool, allowing a comparison of research findings and other sources of information, such as expert opinions, political claims, and journalistic reports.

Defining Variables. A **variable** is any measurable event, characteristic, or property that varies or is subject to change. Researchers must operationally define the variables they study. An *operational definition* specifies how a variable is to be measured. For example, an operational definition of the variable "religiosity" might be the number of times the respondent reports going to church or synagogue. Another operational definition of "religiosity" might be the respondent's answer to the question "How important is religion in your life?" (1, not important; 2, somewhat important; 3, very important).

Operational definitions are particularly important for defining variables that cannot be directly observed. For example, researchers cannot directly observe concepts such as "mental illness," "sexual harassment," "child neglect," "job satisfaction," and "drug abuse." Nor can researchers directly observe perceptions, values, and attitudes.

Formulating a Hypothesis. After defining the research variables, researchers may formulate a **hypothesis,** which is a prediction or educated guess about how one variable is related to another variable. The **dependent variable** is the variable that the researcher wants to explain; that is, it is the variable of interest. The **independent variable** is the variable that is expected to explain change in the dependent variable. In formulating a hypothesis, the researcher predicts how the independent variable affects the dependent variable. For example, Kmec (2003) investigated the impact of segregated work environments on minority wages, concluding that "minority concentration in different jobs, occupations, and establishments is a considerable social problem because it perpetuates racial wage inequality" (p. 55). In this example the independent variable is workplace segregation and the dependent variable is wages.

In studying social problems, researchers often assess the effects of several independent variables on one or more dependent variables. Jekielek (1998) examined the impact of parental conflict and marital disruption (two independent variables) on the emotional well-being of children (the dependent variable). Her research found that both parental conflict and marital disruption (separation or divorce) negatively affect children's emotional well-being. However, children in high-conflict, intact families exhibit lower levels of well-being than children who have experienced high levels of parental conflict, but whose parents divorce or separate.

Methods of Data Collection

After identifying a research topic, reviewing the literature, defining the variables, and developing hypotheses, researchers decide which method of data collection to use. Alternatives include experiments, surveys, field research, and secondary data.

Experiments. Experiments involve manipulating the independent variable to determine how it affects the dependent variable. Experiments require one or

variable Any measurable event, characteristic, or property that varies or is subject to change.

hypothesis A prediction or educated guess about how one variable is related to another variable.

dependent variable The variable that the researcher wants to explain; the variable of interest.

independent variable The variable that is expected to explain change in the dependent variable.

experiment A research method that involves manipulating the independent variable to determine how it affects the dependent variable.

Each chapter in this book contains a *Social Problems Research Up Close* box that describes a research report or journal article that examines some sociologically significant topic. Some examples of the more prestigious journals in sociology include the *American Sociological Review,* the *American Journal of Sociology,* and *Social Forces.* Journal articles are the primary means by which sociologists, as well as other scientists, exchange ideas and information. Most journal articles begin with *an introduction and review of the literature.* It is here that the investigator examines previous research on the topic, identifies specific research areas, and otherwise "sets the stage" for the reader. It is often in this section that research hypotheses, if applicable, are set forth. A researcher, for example, might hypothesize that the sexual behavior of adolescents has changed over the years as a consequence of increased fear of sexually transmitted diseases and that such changes vary on the basis of sex.

The next major section of a journal article is *sample and methods.* In this section the investigator describes the characteristics of the sample, if any, and the details of the type of research conducted. The type of data analysis used is also presented in this section (see Appendix). Using the sample research question, a sociologist might obtain data from the Youth Risk Behavior Surveillance Survey collected by the Centers for Disease Control and Prevention. This self-administered questionnaire is distributed biennially to more than 10,000 high school students across the United States.

The final section of a journal article includes the *findings and conclusions.* The findings of a study describe the results, that is, what the researcher found as a result of the investigation. Findings are then discussed within the context of the hypotheses and the conclusions that can be drawn. Often research results are presented in tabular form. Reading tables carefully is an important part of drawing accurate conclusions about the research hypotheses. In reading a table, you should follow the steps listed here (see table on next page):

1. *Read the title of the table and make sure that you understand what the table contains.* The title of the table indicates the unit of analysis (high school students), the dependent variable (sexual risk behaviors), the independent variables (sex and year), and what the numbers represent (percentages).

2. *Read the information contained at the bottom of the table, including the source and any other explanatory information.* For example, the information at the bottom of this table indicates that the data are from the Centers for Disease Control and Prevention, that "sexually active" was defined as having intercourse in the last three months, and that data on condom use were only from those students who were defined as being currently sexually active.

3. *Examine the row and column headings.* This table looks at the percentage of males and females, over four years, who reported ever having sexual intercourse, having four or more sex partners in a lifetime, being currently sexually active, and using condoms during the last sexual intercourse.

4. *Thoroughly examine the data in the table carefully, looking for patterns between variables.* As indicated in the table, in general, "risky" sexual behavior of males has gone down over the time period studied. Fewer males are ever or currently sexually active or have had more than four sex partners in a lifetime. Most importantly, condom use by sexually active males has increased from a low of 65.1 percent in 2001 to a high of 70.0 percent in 2005. Similarly, the percentage of females who are ever or currently sexually active has also decreased, and the percentage of females using a condom has increased. However, the percentage of females having four or more sex partners during their lifetime decreased between 1999 and 2003 but increased between 2003 and 2005.

5. *Use the information you have gathered in Step 4 to address the hypotheses.* Clearly, sexual practices, as hypothesized, have changed over time. For example, both males and females have a general increase in condom use during sexual intercourse. In addition, from a comparison of the data

more experimental groups that are exposed to the experimental treatment(s) and a control group that is not exposed. After the researcher randomly assigns participants to either an experimental group or a control group, she or he measures the dependent variable. After the experimental groups are exposed to the treatment, the researcher measures the dependent variable again. If participants have been randomly assigned to the different groups, the researcher may conclude that any difference in the dependent variable among the groups is due to the effect of the independent variable.

An example of a "social problems" experiment on poverty would be to provide welfare payments to one group of unemployed single mothers (experimen-

from 1999 to 2005, there is a tendency for changes in risky sexual behaviors to be greater for males than for females.

6. *Draw conclusions consistent with the information presented.* From the table can we conclude that sexual practices have changed over time? The answer is probably yes, although the limitations of the survey, the sample, and the measurement techniques used always should be considered. Can we conclude, however, that the observed changes are a consequence of the fear of sexually transmitted diseases? Although the data may imply it, having no measure of fear of sexually transmitted diseases over the time period studied, we would be premature to come to such a conclusion. More information, from a variety of sources, is needed. The use of multiple methods and approaches to study a social phenomenon is called *triangulation*.

Percentage of High School Students Reporting Sexual Risk Behaviors, by Sex and Survey Year

SURVEY YEAR	EVER HAD SEXUAL INTERCOURSE	FOUR OR MORE SEX PARTNERS DURING LIFETIME	CURRENTLY SEXUALLY ACTIVE*	CONDOM USED DURING LAST INTERCOURSE†
MALE				
1999	52.2	19.3	36.2	65.5
2001	48.5	17.2	33.4	65.1
2003	48.0	17.5	33.8	68.8
2005	47.9	16.5	33.3	70.0
FEMALE				
1999	47.7	13.1	36.3	50.7
2001	42.9	11.4	33.4	51.8
2003	45.3	11.2	34.6	57.4
2005	45.7	12.0	34.6	55.9

*Sexual intercourse during the three months preceding the survey.
†Among currently sexually active students.

Source: Centers for Disease Control and Prevention. 2006 (August 11). *Trends in HIV-Related Risk Behaviors Among High School Students—United States, 1991–2005.* Morbidity and Mortality Weekly Report 55, no. 31:851–854.

tal group) and no such payments to another group of unemployed single mothers (control group). The independent variable would be welfare payments; the dependent variable would be employment. The researcher's hypothesis would be that mothers in the experimental group would be less likely to have a job after 12 months than mothers in the control group.

The major strength of the experimental method is that it provides evidence for causal relationships, that is, how one variable affects another. A primary weakness is that experiments are often conducted on small samples, usually in artificial laboratory settings; thus the findings may not be generalized to other people in natural settings.

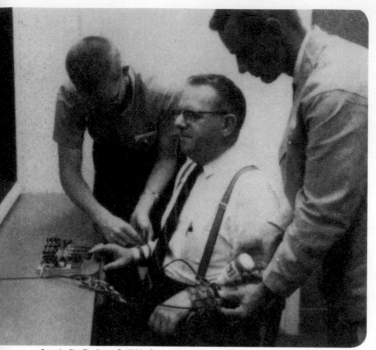

One of the most famous experiments in the social sciences, Stanley Milgram found that 65.0 percent of a sample of ordinary citizens were willing to use harmful electric shocks—up to 450 volts—on an elderly man with a heart condition simply because they were instructed by the experimenter to do so. It was later revealed that the man was not really receiving the shocks and that he had been part of the experimental manipulation. The experiment, although providing valuable information, raised many questions on the ethics of scientific research.

From the film *Obedience,* © 1968 by Stanley Milgram. Copyright renewed © 1993 by Alexandria Milgram and distributed by Penn State.

Surveys. Survey research involves eliciting information from respondents through questions. An important part of survey research is selecting a sample of those to be questioned. A **sample** is a portion of the population, selected to be representative so that the information from the sample can be generalized to a larger population. For example, instead of asking all abused spouses about their experience, the researcher could ask a representative sample of them and assume that those who were not questioned would give similar responses. After selecting a representative sample, survey researchers either interview people, ask them to complete written questionnaires, or elicit responses to research questions through computers.

❝My latest survey shows that people don't believe in surveys.**❞**

Laurence Peter
Humorist

survey research A research method that involves eliciting information from respondents through questions.

sample A portion of the population, selected to be representative so that the information from the sample can be generalized to a larger population.

What Do You Think? Imagine that you are doing research on the prevalence of cheating on examinations at your university or college. How would you get a random sample of the population? What variables do you think predict cheating; that is, what are some of the independent variables you would examine? How would you operationalize these variables? What are some of the problems associated with doing research on such a topic?

Interviews. In interview survey research, trained interviewers ask respondents a series of questions and make written notes about or tape-record the respondents' answers. Interviews may be conducted over the telephone or face to face. A recent Gallup Poll involved telephone interviews with a randomly selected national sample of more than 1,000 U.S. adults. One of the questions the interviewers asked was "How would you describe the current economic conditions in this country—would you say they are mostly good or mostly bad?" Responses were split almost equally: 53 percent thought the economy was "mostly good" and 45 percent "mostly bad," with the remainder having no opinion (Newport 2006).

One advantage of interview research is that researchers are able to clarify questions for the respondent and follow up on answers to particular questions. Researchers often conduct face-to-face interviews with groups of individuals who might otherwise be inaccessible. For example, some AIDS-related research attempts to assess the degree to which individuals engage in behavior that places them at high risk for transmitting or contracting HIV. Street youth and intravenous drug users, both high-risk groups for HIV infection, may not have a telephone or address because of their transient lifestyle. These groups may be accessible, however, if the researcher locates their hangouts and conducts face-to-face interviews. Research on drug addicts may also require a face-to-face interview survey design (Jacobs 2003).

The most serious disadvantages of interview research are cost and the lack of privacy and anonymity. Respondents may feel embarrassed or threatened when asked questions that relate to personal issues such as drug use, domestic violence, and sexual behavior. As a result, some respondents may choose not to participate in interview research on sensitive topics. Those who do participate may conceal or alter information or give socially desirable answers to the interviewer's questions (e.g., "No, I do not use drugs.").

Questionnaire. Instead of conducting personal or phone interviews, researchers may develop questionnaires that they either mail or give to a sample of respondents. Questionnaire research offers the advantages of being less expensive and less time consuming than face-to-face or telephone surveys. In addition, questionnaire research provides privacy and anonymity to the research participants. This reduces the likelihood that they will feel threatened or embarrassed when asked personal questions and increases the likelihood that they will provide answers that are not intentionally inaccurate or distorted. A recent study on the relationship between minority composition of the workplace and the likelihood of workplace drug testing is a case on point. Questionnaires were sent to union leaders of the Communication Workers of America (CWA), asking them about drug-testing policies at their local job sites. Analysis indicated that as the minority composition of the workplace goes up, the likelihood of *pre-employment testing* and *testing with cause* increases, whereas the likelihood of *random* drug testing decreases (Gee et al. 2006).

The major disadvantage of mail questionnaires is that it is difficult to obtain an adequate response rate. Many people do not want to take the time or make the effort to complete and mail a questionnaire. Others may be unable to read and understand the questionnaire.

"Talking" Computers. A new method of conducting survey research is asking respondents to provide answers to a computer that "talks." Romer et al. (1997) found that respondents rated computer interviews about sexual issues more favorably than face-to-face interviews and that the computer interviews were more reliable. Such increased reliability may be particularly valuable when one is conducting research on drug use, deviant sexual behavior, and sexual orientation because respondents reported that the privacy of computers was a major advantage.

"When I was younger I could remember anything—whether it happened or not."

Mark Twain
American humorist and writer

Field Research. Field research involves observing and studying social behavior in settings in which it occurs naturally. Two types of field research are participant observation and nonparticipant observation.

In participant observation research the researcher participates in the phenomenon being studied so as to obtain an insider's perspective on the people and/or behavior being observed. Palacios and Fenwick (2003), two criminologists, attended dozens of raves over a 15-month period to investigate the South Florida drug culture. In nonparticipant observation research the researcher observes the phenomenon being studied without actively participating in the group or the activity. For example, Dordick (1997) studied homelessness by observing and talking with homeless individuals in a variety of settings, but she did not live as a homeless person as part of her research.

Sometimes sociologists conduct in-depth detailed analyses or case studies of an individual, group, or event. For example, Fleming (2003) conducted a case study of young auto thieves in British Columbia. He found that unlike professional thieves, the teenagers' behavior was primarily motivated by thrill seeking—driving fast, the rush of a possible police pursuit, and the prospect of getting caught.

The main advantage of field research on social problems is that it provides detailed information about the values, rituals, norms, behaviors, symbols, beliefs, and emotions of those being studied. A potential problem with field research is that the researcher's observations may be biased (e.g., the researcher becomes too involved in the group to be objective). In addition, because field research is usually based on small samples, the findings may not be generalizable.

Secondary Data Research. Sometimes researchers analyze secondary data, which are data that have already been collected by other researchers or government agencies or that exist in forms such as historical documents; police reports; school records; and official records of marriages, births, and deaths. Caldas and Bankston (1999) used information from Louisiana's 1990 Graduation Exit Examination to assess the relationship between school achievement and television-viewing habits of more than 40,000 tenth graders. The researchers found that, in general, television viewing is inversely related to academic achievement for whites but has little or no effect on school achievement for African Americans. A major advantage of using secondary data in studying social problems is that the data are readily accessible, so researchers avoid the time and expense of collecting their own data. Secondary data are also often based on large representative samples. The disadvantage of secondary data is that the researcher is limited to the data already collected.

> 66 Feminists in all disciplines have demonstrated that objectivity has about as much substance as the emperor's new clothes. 99
>
> Connie Miller
> Feminist scholar

GOALS OF THE TEXTBOOK

This textbook approaches the study of social problems with several goals in mind.

field research Research that involves observing and studying social behavior in settings in which it occurs naturally.

1. *Providing an integrated theoretical background.* The book reflects an integrative theoretical approach to the study of social problems. More than one theoretical perspective can be used to explain a social problem because social problems usually have more than one cause. For example, youth crime is

linked to (1) an increased number of youths living in inner-city neighborhoods with little or no parental supervision (social disorganization), (2) young people having no legitimate means of acquiring material wealth (anomie theory), (3) youths being angry and frustrated at the inequality and racism in our society (conflict theory), and (4) teachers regarding youths as "no good" and treating them accordingly (labeling theory).

2. *Encouraging the development of a sociological imagination.* A major insight of the sociological perspective is that various structural and cultural elements of society have far-reaching effects on individual lives and social well-being. This insight, known as the *sociological imagination*, enables us to understand how social forces underlie personal misfortunes and failures and contribute to personal successes and achievements. Each chapter in this textbook emphasizes how structural and cultural factors contribute to social problems. This emphasis encourages you to develop your sociological imagination by recognizing how structural and cultural factors influence private troubles and public issues.

3. *Providing global coverage of social problems.* The modern world is often referred to as a "global village." The Internet, cell phones, and fax machines connect individuals around the world, economies are interconnected, environmental destruction in one region of the world affects other regions of the world, and diseases cross national boundaries. Understanding social problems requires an awareness of how global trends and policies affect social problems. Many social problems call for collective action involving countries around the world; efforts to end poverty, protect the environment, control population growth, and reduce the spread of HIV are some of the social problems that have been addressed at the global level. Each chapter in this book includes coverage of global aspects of social problems. We hope that attention to the global aspects of social problems will broaden students' awareness of pressing world issues.

4. *Providing an opportunity to assess personal attitudes, beliefs, and behaviors.* Each chapter in this textbook contains a section called *Self and Society,* which offers you an opportunity to assess your attitudes, beliefs, and behaviors regarding some aspect of the social problem discussed. Earlier in this chapter, the *Self and Society* feature allowed you to assess your beliefs about a number of social problems and to compare your beliefs with a national sample of first-year college students.

5. *Emphasizing the human side of social problems.* Each chapter contains a feature called *The Human Side,* which presents personal stories of how social problems have affected individual lives. By conveying the private pain and personal triumphs associated with social problems, we hope to elicit a level of understanding and compassion that may not be attained through the academic study of social problems alone. This chapter's *The Human Side* feature presents stories about how college students, disturbed by various social conditions, have participated in social activism.

6. *Highlighting social problems research.* In every chapter there are boxes called *Social Problems Research Up Close,* which present examples of social science research. These boxes demonstrate for students the sociological enterprise from theory and data collection to findings and conclusions. Examples of research topics covered include changes in marriages in America, racism among college students, and bullying and victimization among minority students.

Both structural functionalism and conflict theory address the nature of social change, although in different ways. Durkheim, a structural functionalist, argued that social change, if rapid, was disruptive to society and that the needs of society should take precedence over the desires of individuals; that is, social change should be slow and methodical regardless of popular opinion.

When it is slow and methodical, social change can be responsive to the needs of society and can contribute to the preferred state of society—that of equilibrium.

To conflict theorists, social change is a result of the struggle for power by different groups. Specifically, Marx argued that social change was a consequence of the struggle between different economic classes as each strove for supremacy. Marx envisioned social change as primarily a revolutionary process ultimately leading to a utopian society.

Social movements are one means by which social change is realized. A **social movement** is an organized group of individuals with a common purpose to either promote or resist social change through collective action. Some people believe that to promote social

change, one must be in a position of political power and/or have large financial resources. However, the most important prerequisite for becoming actively involved in improving levels of social well-being may be genuine concern and dedication to a social "cause." The following vignettes provide a sampler of college student activism—college students making a difference in the world.

- In 2006, 33 student activists "hit the road for a seven-week tour of 19 religious and military colleges that discriminate against gay and lesbian students" (Ferrara et al. 2006, p. 1). Although the bus was "tagged" with homophobic slogans and activists were arrested on six campuses, leader Jacob Reitan said the trip was "hugely positive." Activists met and talked with over 10,000 people including ten school presidents.

- Sean Sellers, a recent graduate of the University of Texas at Austin successfully led a "Boot the Bell" campaign in which "22 colleges and high schools either managed to remove a Taco Bell franchise from the campus or prevent one from being built" as part of a general protest against

sweatshops in the field (Berkowitz 2005, p. 1). Yum! Brands Inc., which owns Taco Bell as well as Kentucky Fried Chicken, Long John Silver's, and Pizza Hut, has agreed to increase the pay of tomato pickers and to improve working conditions of farm workers in general.

- At Middlebury College in Vermont students successfully convinced the administration that global warming is a real problem that needs to be addressed—immediately. Student activists convinced university officials that the college should invest $11 million in a biomass plant—a plant "fueled by wood chips, grass pellets, and a self-sustaining willow forest" (James 2007, p. 1). Further, one Middlebury group, after five days of protesting by 1,000 vocal activists, convinced Vermont Senator Bernie Sanders to reintroduce legislation in Congress that would reduce carbon emissions by 80 percent by the year 2050.

- Students at several colleges are petitioning their university administrations to buy "fair trade coffee"—coffee that is certified by monitors to have come from farmers who were paid a fair price for

7. *Encouraging students to take pro-social action.* Individuals who understand the factors that contribute to social problems may be better able to formulate interventions to remedy those problems. Recognizing the personal pain and public costs associated with social problems encourages some to initiate social intervention.

What Do You Think? Service learning is an evolving pedagogy that incorporates student volunteering with the dynamics of experiential learning and the rigors and structure of an academic curriculum. In its simplest form, service learning entails students volunteering in the community and receiving academic credit for their efforts. Universities and colleges are increasingly requiring service learning credits as a criterion for graduation. Do you think that all students should be required to engage in service learning? Why or why not?

social movement An organized group of individuals with a common purpose to either promote or resist social change through collective action.

their beans. Many of these students are members of Students for Fair Trade. As one student said, "This is easy activism." Students make their voices heard by buying coffee with a fair-trade certified label or by not buying coffee at all (Batsell 2002).

- Students at the University of California, Berkeley, came together to protest cutbacks in the campus's Ethnic Studies Department. The protest lasted more than a month with over 100 arrests and 6 hunger strikes. As a consequence of the students' activism, the university administration agreed to reopen a multicultural student center, hire eight tenure-track ethnic studies faculty over the next five years, and invest $100,000 in an Ethnic Research Center (Alvarado 2000).

- Students at Grinnell College in Iowa, in response to accusations of human rights violations of union workers in Coca-Cola bottling plants in Colombia (South America), formed an anti-Coke campaign. Using the official boycotting policy of the college, in November 2004 the student

initiative passed a boycott on all Coca-Cola products. Because Coca-Cola had an exclusive contract with Grinnell, Coke and Coke products continued to be sold on campus. However, wherever they are sold, there are now signs reading, "The Grinnell College student body has voted to boycott Coca-Cola products. This is a Coca-Cola product" (*Killer Coke* 2005).

- While a zoology major at the University of Colorado, Jeff Galus began the Animal Rights Student Group (website at http://www.colorado.edu/StudentGroups/animalrights/links.html). This organization focuses on informing the public about how animals are treated in research and which corporations use animals in testing their products.

- Students at more than 150 campuses are members of the anti-sweatshop movement, with many belonging to the Worker Rights Consortium (WRC). The WRC is a student-run watchdog organization that inspects factories worldwide, monitoring the monitors, as part of the anti-sweatshop movement. The WRC requires that member

schools agree to closely scrutinize manufacturers of collegiate apparel. The WRC also mandates "the protection of workers' health and safety, compliance with local labor laws, protection of women's rights, and prohibition of child labor, forced labor, and forced overtime" (Boston College 2003).

Students who are interested in becoming involved in student activism or who are already involved might explore the website for the Center for Campus Organizing (http://www.organizenow.net/cco/index.html)—a national organization that supports social justice activism and investigative journalism on campuses nationwide. The organization was founded on the premise that students and faculty have played critical roles in larger social movements for social justice in our society, including the Civil Rights movement, the anti–Vietnam War movement, the Anti-Apartheid movement, the women's rights movement, and the environmental movement. In 2006, almost half of all first-year college students participated in an organized demonstration (Pryor et al. 2006).

Individuals can make a difference in society by the choices they make. Individuals may choose to vote for one candidate over another, demand the right to reproductive choice or protest government policies that permit it, drive drunk or stop a friend from driving drunk, repeat a racist or sexist joke or chastise the person who tells it, and practice safe sex or risk the transmission of sexually transmitted diseases. Individuals can also make a difference by addressing social concerns in their occupational role as well as through volunteer work.

Although individual choices make an important impact, collective social action often has a more pervasive effect. For example, although individual parents discourage their teenage children from driving under the influence of alcohol, Mothers Against Drunk Driving contributed to the enactment of national legislation that has the potential to influence every U.S. citizen's decision about whether to use alcohol and drive. Schwalbe (1998) reminded us that we do not have to join a group or organize a protest to make changes in the world.

> "Activism pays the rent on being alive and being here on the planet. . . . If I weren't active politically, I would feel as if I were sitting back eating at the banquet without washing the dishes or preparing the food. It wouldn't feel right."
>
> Alice Walker
> Novelist

© Paul Fusco/Magnum Photos

One way to effect social change is through demonstrations. A U.S. survey of first-year college students revealed that 49.7 percent reported having participated in organized demonstrations in the last year (Pryor et al. 2006). Here, students march against the war in Iraq.

66In a certain sense, every single human soul has more meaning and value than the whole of history.99

Nicholas Berdyaev
Philosopher

We can change a small part of the social world single-handedly. If we treat others with more respect and compassion, if we refuse to participate in recreating inequalities, even in little ways, if we raise questions about official representation of reality, if we refuse to work in destructive industries, then we are making change. (p. 206)

UNDERSTANDING SOCIAL PROBLEMS

At the end of each chapter we offer a section titled "Understanding" in which we re-emphasize the social origin of the problem being discussed, the consequences, and the alternative social solutions. It is our hope that the reader will end each chapter with a "sociological imagination" view of the problem and with an idea of how, as a society, we might approach a solution.

Sociologists have been studying social problems since the Industrial Revolution. Industrialization brought about massive social changes: The influence of religion declined, and families became smaller and moved from traditional, rural communities to urban settings. These and other changes have been associated with increases in crime, pollution, divorce, and juvenile delinquency. As these social problems became more widespread, the need to understand their origins and possible solutions became more urgent. The field of sociology developed in response to this urgency. Social problems provided the initial impetus for the development of the field of sociology and continue to be a major focus of sociology.

There is no single agreed-on definition of what constitutes a social problem. Most sociologists agree, however, that all social problems share two important elements: an objective social condition and a subjective interpretation of that condition. Each of the three major theoretical perspectives in sociology—structural-functionalist, conflict, and symbolic interactionist—has its own notion of the causes, consequences, and solutions of social problems.

CHAPTER REVIEW

- **What is a social problem?**

Social problems are defined by a combination of objective and subjective criteria. The objective element of a social problem refers to the existence of a social condition; the subjective element of a social problem refers to the belief that a particular social condition is harmful to society or to a segment of society and that it should and can be changed. By combining these objective and subjective elements, we arrive at the following definition: A social problem is a social condition that a segment of society views as harmful to members of society and in need of remedy.

- **What is meant by the structure of society?**

The structure of a society refers to the way society is organized.

- **What are the components of the structure of society?**

  The components are institutions, social groups, statuses, and roles. Institutions are an established and enduring pattern of social relationships and include family, religion, politics, economics, and education. Social groups are defined as two or more people who have a common identity, interact, and form a social relationship. A status is a position that a person occupies within a social group and that can be achieved or ascribed. Every status is associated with many roles, or the set of rights, obligations, and expectations associated with a status.

- **What is meant by the culture of society?**

  Whereas social structure refers to the organization of society, culture refers to the meanings and ways of life that characterize a society.

- **What are the components of the culture of society?**

  The components are beliefs, values, norms, and symbols. Beliefs refer to definitions and explanations about what is assumed to be true. Values are social agreements about what is considered good and bad, right and wrong, desirable and undesirable. Norms are socially defined rules of behavior. Norms serve as guidelines for our behavior and for our expectations of the behavior of others. Finally, a symbol is something that represents something else.

- **What is the sociological imagination, and why is it important?**

  The sociological imagination, a term developed by C. Wright Mills (1959), refers to the ability to see the connections between our personal lives and the social world in which we live. It is important because when we use our sociological imagination, we are able to distinguish between "private troubles" and "public issues" and to see connections between the events and conditions of our lives and the social and historical context in which we live.

- **What are the differences between the three sociological perspectives?**

  According to structural functionalism, society is a system of interconnected parts that work together in harmony to maintain a state of balance and social equilibrium for the whole. The conflict perspective views society as composed of different groups and interests competing for power and resources. Symbolic interactionism reflects the micro-sociological perspective and emphasizes that human behavior is influenced by definitions and meanings that are created and maintained through symbolic interaction with others.

- **What are the first four stages of a research study?**

  The first four stages of a research study are formulating a research question, reviewing the literature, defining variables, and formulating a hypothesis.

- **How do the various research methods differ from one another?**

  Experiments involve manipulating the independent variable to determine how it affects the dependent variable. Survey research involves eliciting information from respondents through questions. Field research involves observing and studying social behavior in settings in which it occurs naturally. Secondary data are data that have already been collected by other researchers or government agencies or that exist in forms such as historical documents, police reports, school records, and official records of marriages, births, and deaths.

- **What is a social movement?**

  Social movements are one means by which social change is realized. A social movement is an organized group of individuals with a common purpose to either promote or resist social change through collective action.

TEST YOURSELF

1. Definitions of social problems are clear and unambiguous.
   a. True
   b. False
2. The social structure of society contains
   a. statuses and roles
   b. institutions and norms
   c. sanctions and social groups
   d. values and beliefs
3. The culture of society refers to its meaning and the ways of life of its members.
   a. True
   b. False
4. Alienation
   a. refers to a sense of normlessness
   b. is focused on by symbolic interactionist
   c. can be defined as the powerlessness and meaninglessness is people's lives
   d. is a manifest function of society
5. Blumer's stages of a social problems begins with
   a. mobilization for action
   b. societal recognition
   c. social legitimation
   d. development and implementation of a plan

6. The independent variable comes first in time; i.e., it precedes the dependent variable.
   a. True
   b. False
7. The third stage in defining a research study is
   a. formulating a hypothesis
   b. reviewing the literature
   c. defining the variables
   d. formulating a research question
8. A sample is a subgroup of the population—the group you actually give the questionnaire to.
   a. True
   b. False

9. Studying police behavior by riding along with patrol officers would be an example of
   a. participant observation
   b. nonparticipant observation
   c. field research
   d. both a and c
10. Goals of the book include
    a. providing global coverage of social problems
    b. highlighting social problems research
    c. encouraging students to take pro-social action
    d. all of the above

Answers: 1 b. 2 a. 3 a. 4 c. 5 b. 6 a. 7 c. 8 a. 9 d. 10 d.

KEY TERMS

achieved status
alienation
anomie
ascribed status
belief
culture
dependent variable
experiment
field research
hypothesis
independent variable

institution
latent function
manifest function
norm
objective element of a social problem
primary group
role
sample
sanction
secondary group
social group

social movement
social problem
sociological imagination
status
structure
subjective element of a social problem
survey research
symbol
value
variable

MEDIA RESOURCES

Understanding Social Problems, Sixth Edition Companion Website

academic.cengage.com/sociology/mooney
Visit your book companion website, where you will find flash cards, practice quizzes, Internet links, and more to help you study.

CENGAGENOW™

Just what you need to know NOW! Spend time on what you need to master rather than on information you already have learned. Take a pre-test for this chapter, and CengageNOW will generate a personalized study plan based on your results. The study plan will identify the topics you need to review and direct you to online resources to help you master those topics. You can then take a post-test to help you determine the concepts you have mastered and what you will need to work on. Try it out! Go to academic.cengage .com/login to sign in with an access code or to purchase access to this product.

2

Problems of Illness and Health Care

© Suzy Allman/Getty Images

At this annual three-day free medical clinic in Virginia, rural families, most with little or no health insurance, line up for hours to receive free health care. All services and medical supplies are donated.

At 2:00 a.m., a car drives into a field in rural southwestern Virginia. The field is large and dark. At the crack of dawn, the driver and two passengers get out of the car, walk a half a mile, and join a line of people that stretches a quarter of a mile across the field. Betty, a 29-year-old mother of six who works at a restaurant, is seriously overweight. Her 14-year-old daughter, Molly, has had such terrible tooth pain that she is unable to eat and has lost 15 pounds. Betty's boyfriend, Jake, who works at a dry cleaner, has had pain in his side off and on for nearly a year. Betty, Molly, and Jake wait in line for the gates to open for the Rural Area Medicine Clinic—a weekend event that occurs once a year for anyone who has no health insurance. During this 2-day event, about 1,500 volunteer doctors, nurses, dentists, and staff provide services to more than 6,000 uninsured women, men, and children. When Betty is examined, she finds out she has diabetes. Molly had such a severe dental infection that she had to have eight teeth pulled. And Jake learned that he has abdominal cancer that could have been detected much earlier with a physical examination (Garson 2007).

In the United States, lack of health insurance and the high cost of health care is, for millions of Americans, a pressing concern, and for some, literally a matter of life or death. In this chapter we address these and other problems of illness and health care in the United States and throughout the world. Taking a sociological look at health issues, we examine why some social groups experience more health problems than others and how social forces affect and are affected by health and illness. We begin by looking at patterns of health and illness around the world.

The Global Context: Patterns of Health and Illness Around the World

developed countries Countries that have relatively high gross national income per capita and have diverse economies made up of many different industries.

developing countries Countries that have relatively low gross national income per capita, with simpler economies that often rely on a few agricultural products.

least developed countries The poorest countries of the world.

In making international comparisons, researchers, social scientists, politicians, and others commonly classify countries into one of three broad categories according to their economic status: (1) **developed countries** (also known as *high-income countries*) have relatively high gross national income per capita and have diverse economies made up of many different industries; (2) **developing countries** (also known as *middle-income countries*) have relatively low gross national income per capita, and their economies are much simpler, often relying on a few agricultural products; and (3) **least developed countries** (known as *low-income countries*) are the poorest countries of the world.

Patterns of health and illness reveal striking disparities among developed, developing, and least developed nations. In this section that focuses on health and illness from a global perspective, we describe patterns of morbidity, life expectancy, mortality, and burden of disease around the world. Later, we discuss three worldwide health problems in detail: human immunodeficiency virus (HIV)/ acquired immunodeficiency syndrome (AIDS), obesity, and mental illness.

Morbidity, Life Expectancy, and Mortality

Morbidity refers to illnesses, symptoms, and the impairments they produce. Measures of morbidity are often expressed in terms of the incidence and prevalence of specific health problems. *Incidence* refers to the number of new cases of a specific health problem in a given population during a specified time period. *Prevalence* refers to the total number of cases of a specific health problem in a population that exists at a given time. For example, the *incidence* of HIV infection worldwide was 4.3 million in 2006; this means that there were 4.3 million people newly infected with HIV in 2006. In the same year the worldwide *prevalence* of HIV was nearly 40 million, meaning that nearly 40 million people worldwide were living with HIV infection in 2006 (Joint United Nations Programme on HIV/AIDS 2006).

Patterns of morbidity vary according to the level of development of a country. In less developed countries, where poverty and chronic malnutrition are widespread, infectious and parasitic diseases, such as HIV disease, tuberculosis, diarrheal diseases (caused by bacteria, viruses, or parasites), measles, and malaria are much more prevalent than in developed countries, where chronic diseases are the major health threat (Weitz 2006).

Wide disparities in **life expectancy**—the average number of years that individuals born in a given year can expect to live—exist between regions of the world (see Figure 2.1). Japan has the longest life expectancy (82 years), Swaziland has the lowest life expectancy (30 years), and 31 countries (primarily in Africa) have life expectancies of less than 50 years (UNICEF 2006).

As societies develop and increase the standard of living for their members, life expectancy increases and birthrates decrease (Weitz 2006). At the same time, the main causes of death and disability shift from infectious disease and high death rates among infants and women of childbearing age (owing to complications of pregnancy, unsafe abortion, or childbearing) to chronic, noninfectious illness and disease. This shift is referred to as the **epidemiological transition,** whereby low life expectancy and predominance of parasitic and infectious diseases shift to high life expectancy and predominance of chronic and degenerative diseases. As societies make the epidemiological transition, birthrates decline and life expectancy increases, so diseases that need time to develop, such as cancer, heart disease, Alzheimer's disease, arthritis, and osteoporosis, become more common, and childhood illnesses, typically caused by infectious and parasitic diseases, become less common, as do pregnancy-related deaths and health problems.

Today, the leading cause of **mortality,** or death, worldwide is cardiovascular disease (including heart disease and stroke), accounting for 30 percent of all deaths (World Health Organization 2006). In the United States the leading cause of death for both women and men is heart disease, followed by cancer and stroke (National Center for Health Statistics 2006). As shown in Table 2.1, U.S. mortality patterns vary by age. Later, we discuss how patterns of mortality are related to social factors, such as social class, sex, race or ethnicity, and education.

Mortality Rates Among Infants and Children. The **infant mortality rate,** the number of deaths of live-born infants under 1 year of age per 1,000 live births (in any given year), provides an important measure of the health of a population. In 2005 infant mortality rates ranged from an average of 97 in least developed nations to an average of 5 in industrialized nations. The U.S. infant mortality rate was 6, and 34 countries had infant mortality rates that were lower than that of the United States (UNICEF 2006). One of the major causes of infant death worldwide

morbidity Illnesses, symptoms, and the impairments they produce.

life expectancy The average number of years that individuals born in a given year can expect to live.

epidemiological transition A societal shift from low life expectancy and predominance of parasitic and infectious diseases to high life expectancy and predominance of chronic and degenerative diseases.

mortality Death.

infant mortality rate The number of deaths of live-born infants under 1 year of age per 1,000 live births (in any given year).

FIGURE 2.1
Life expectancy and
under-5 mortality rate by
region: 2005.
Source: UNICEF (2006).

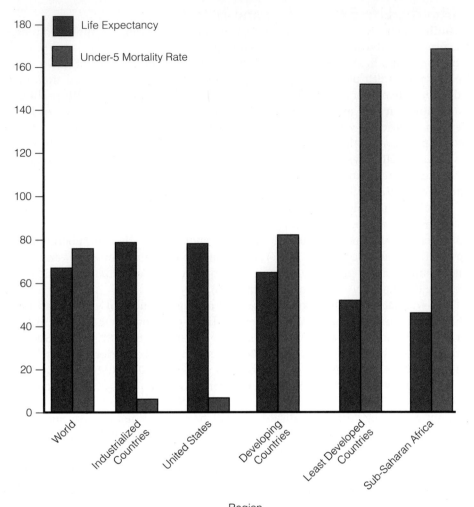

Region

TABLE 2.1 Top Three Causes of Death by Selected Age Groups:
United States, 2004

		LEADING CAUSES OF DEATH	
AGE (YEARS)	FIRST	SECOND	THIRD
1–4	Unintentional injuries	Congenital/chromosomal abnormalities	Cancer
5–14	Unintentional injuries	Cancer	Congenital/chromosomal abnormalities
15–24	Unintentional injuries	Homicide	Suicide
25–44	Unintentional injuries	Cancer	Heart disease
45 years and older	Heart disease	Cancer	Stroke

Source: National Center for Health Statistics (2006, Table 32).

is diarrhea, resulting from poor water quality and sanitation (Millennium Ecosystem Assessment 2005). Worldwide, only 58 percent of the population has access to adequate sanitation (Population Reference Bureau 2006). The major cause of U.S. infant deaths is disorders related to premature birth and low birth weight.

The **under-5 mortality rate,** another useful measure of child health, refers to the rate of deaths of children under age 5. Under-5 mortality rates range from an average of 153 in least developed nations to an average of 6 in industrialized countries. A major contributing factor to deaths of infants and children is undernutrition. In the developing world, one in four children under age 5 is underweight (UNICEF 2006). For these nutritionally deprived children, common childhood ailments such as diarrhea and respiratory infections can be fatal.

TABLE 2.2 Trained Childbirth Assistance and Lifetime Chance of Maternal Mortality by Region		
	PERCENTAGE OF BIRTHS ATTENDED BY SKILLED PERSONNEL	LIFETIME CHANCE OF MATERNAL MORTALITY
Developed countries	99	1 in 4,000
Developing countries	57	1 in 61
Sub-Saharan Africa	41	1 in 16
Source: UNICEF (2006).		

Maternal Mortality Rates. The **maternal mortality rate,** a measure of deaths that result from complications associated with pregnancy, childbirth, and unsafe abortion, also provides a sensitive indicator of the health status of a population. Maternal mortality is the leading cause of death and disability for women ages 15–49 in developing countries. The three most common causes of maternal death are hemorrhage (severe loss of blood), infection, and complications related to unsafe abortion.

Rates of maternal mortality show a greater disparity between rich and poor countries than any of the other societal health measures. Of the 529,000 annual maternal deaths worldwide, including 68,000 deaths from unsafe abortion, only 1 percent occur in high-income countries (World Health Organization 2005). Women's lifetime risk of dying from pregnancy or childbirth is highest in sub-Saharan Africa, where 1 in 16 women dies of pregnancy-related causes, compared with 1 in 4,000 women in developed countries (UNICEF 2006). High maternal mortality rates in less developed countries are related to poor-quality and inaccessible health care; most women give birth without the assistance of trained personnel (see Table 2.2). High maternal mortality rates are also linked to malnutrition and poor sanitation and to higher rates of pregnancy and childbearing at early ages. Women in many countries also lack access to family planning services and/or do not have the support of their male partners to use contraceptive methods such as condoms. Consequently, many women resort to abortion to limit their childbearing, even in countries where abortion is illegal and unsafe.

> ❝It is not uncommon for women in Africa, when about to give birth, to bid their older children farewell.❞
>
> United Nations Population Fund

Patterns of Burden of Disease

Another approach to measuring the health status of a population provides an indicator of the overall burden of disease on a population through a single unit of measurement that combines not only the number of deaths but also the impact of premature death and disability on a population (Murray & Lopez 1996). This comprehensive unit of measurement, called the *disability-adjusted life year* (DALY), reflects years of life lost to premature death and years lived with a disability. More simply, 1 DALY is equal to 1 lost year of healthy life.

under-5 mortality rate The rate of deaths of children under age 5.

maternal mortality rate A measure of deaths that result from complications associated with pregnancy, childbirth, and unsafe abortion.

Worldwide, tobacco is the leading cause of burden of disease (World Health Organization 2002). Hence tobacco has been called "the world's most lethal weapon of mass destruction" (SmokeFree Educational Services 2003, p. 1). The top 10 risk factors that contribute to the global burden of disease are underweight; unsafe sex; high blood pressure; tobacco; alcohol; unsafe water, sanitation, and hygiene; high cholesterol; indoor smoke from solid fuels; iron deficiency; and overweight (World Health Organization 2002).

What Do You Think? Data on deaths from international terrorism and tobacco-related deaths in 37 developed and Eastern European countries revealed that tobacco-related deaths outnumbered terrorist deaths by about a whopping 5,700 times (Thomson & Wilson 2005). The number of tobacco deaths was equivalent to the impact of a September 11, 2001, type terrorist attack every 14 hours! Given that tobacco-related deaths grossly outnumber terrorism-related deaths, why hasn't the U.S. government waged a "war on tobacco" on a scale similar to its "war on terrorism?"

SOCIOLOGICAL THEORIES OF ILLNESS AND HEALTH CARE

The sociological approach to the study of illness, health, and health care differs from medical, biological, and psychological approaches to these topics. Next, we discuss how the three major sociological theories—structural functionalism, conflict theory, and symbolic interactionism—contribute to our understanding of illness and health care.

Structural-Functionalist Perspective

The structural-functionalist perspective is concerned with how illness, health, and health care affect and are affected by other aspects of social life. For example, rather than look at individual reasons for suicide, the structural-functionalist approach looks for social patterns that may help to explain suicide rates. In one of the first scientific sociological research studies, Emile Durkheim (1951 [1897]) found that suicide rates were higher in countries characterized by lower levels of social integration and regulation—a finding that has been replicated in recent studies (Stockard & O'Brien 2002).

The structural-functionalist perspective draws attention to how changes in society affect health. For example, increased modernization and industrialization throughout the world has resulted in environmental pollution—a major health concern (see Chapter 14).

Just as social change affects health, health concerns may lead to social change. The emergence of HIV and AIDS in the U.S. gay male population was a force that helped unite and mobilize gay rights activists. Concern over the effects of exposure to tobacco smoke—the greatest cause of disease and death in the United States and other developed countries—has led to legislation banning smoking in public places (workplaces, clubs, restaurants, and/or bars) in about 20 states (Kumar 2007).

> **What Do You Think?** In 2005, the country of Bhutan became the first nation in the world to impose a national ban on the sale of tobacco and on smoking in public places (Robbins 2006). Do you think that such a ban would ever occur in the United States? Why or why not?

According to the structural-functionalist perspective, health care is a social institution that functions to maintain the well-being of societal members and, consequently, of the social system as a whole. Illness is dysfunctional in that it interferes with people performing needed social roles. To cope with nonfunctioning members and to control the negative effects of illness, society assigns a temporary and unique role to those who are ill—the sick role (Parsons 1951). This role carries with it an expectation that the person who is ill will seek and receive competent medical care, adhere to the prescribed regimen, and return as soon as possible to normal role obligations.

Finally, the structural-functionalist perspective draws attention to latent dysfunctions, or unintended and often unrecognized negative consequences of social patterns or behavior. For example, a latent dysfunction of widespread use of some drugs is the emergence of drug resistance, which occurs when drugs kill the weaker disease-causing germs while allowing variants resistant to the drugs to flourish. For generations the drug chloroquine was added to table salt to prevent malaria. But overuse led to drug-resistant strains of malaria, and now chloroquine is useless in preventing malaria (McGinn 2003). Another health-related example of a latent dysfunction is the unintended consequences of highly active antiretroviral therapy (HAART), which reduces the viral loads of HIV-positive patients and delays their progression to AIDS. One study found that HIV-positive young people engage in more unprotected sex, have more sexual partners, and are more likely to use illicit drugs than HIV-positive young people did before the availability of HAART (Lightfoot et al. 2005). The researchers explained that many HIV-infected individuals who are receiving antiretroviral therapy believe that unprotected sex is less risky because they have lower viral load levels. And, because this medication prolongs the lives of HIV-infected people, they might have more opportunities to transmit the virus to others. Engaging in more risky behavior and having more opportunities to infect others with HIV is an unintended negative consequence of antiretroviral therapy.

Conflict Perspective

The conflict perspective focuses on how wealth, status, power, and the profit motive influence illness and health care. Worldwide, populations living in poverty, with little power and status, experience more health problems and have less access to quality medical care (Feachum 2000).

The conflict perspective criticizes the pharmaceutical and health care industry for placing profits above people. In her book *Money-Driven Medicine,*

> "Health is what political leaders talk about if there's money at the end of the day."
>
> Peter Piot
> Executive director of the Joint United Nations Programme on HIV/AIDS (UNAIDS)

Maggie Mahar (2006) explains that power in our health care system has shifted from physicians, who are committed to putting their patients' interests ahead of their own financial interests, to corporations that are legally bound to put their shareholders' interests first. "Thus, many decisions about how to allocate health care dollars have become marketing decisions. Drugmakers, device makers, and insurers decide which products to develop based not on what patients need, but on what their marketers tell them will sell—and produce the highest profit" (Mahar 2006, p. xviii). For example, pharmaceutical companies' research and development budgets are spent not according to public health needs but rather according to calculations about maximizing profits. Because the masses of people in developing countries lack the resources to pay high prices for medication, pharmaceutical companies do not see the development of drugs for diseases of poor countries as a profitable investment. This explains why 90 percent of the $70 billion invested annually in health research and development by pharmaceutical companies and Western governments focuses on the health problems of the 10 percent of the global population living in developed industrialized countries (Thomas 2003).

Profits also compromise drug safety. Most pharmaceutical companies outsource their clinical drug trials (which assess drug effectiveness and safety) to Contract Research Organizations (CROs) in developing countries where trial subjects are plentiful, operating costs are low, and regulations are lax (Allen 2007). Because CROs can complete a clinical trial in less time and with less expense than a pharmaceutical company can, they offer millions of dollars in increased revenue per drug. The validity of the clinical trial results from CROs is questionable, however, because CROs can earn more money (in royalties and in future contracts) when the clinical trial results are favorable.

The profit motive also affects health via the food industry. See this chapter's Photo Essay for a look at health problems associated with modern food animal production—problems that stem largely from the concern of the food animal industry for economic efficiency and profit.

The conflict perspective points to ways in which powerful groups and wealthy corporations influence health-related policies and laws through financial contributions to politicians and political candidates and other means. After Merck & Co. received FDA approval for its drug Gardasil, a vaccine that protects against the two strains of human papillomavirus (HPV) that causes 70 percent of cervical cancers, Merck campaigned to make Gardasil mandatory for all 11- to 12-year-old girls. Mandatory Gardasil vaccines—at $360 per treatment—could generate billions of dollars in sales for Merck. To influence state legislators to pass bills requiring the Gardasil vaccine, Merck spent considerable sums of money on lobbying efforts. Merck also provided funding to Women In Government—a nonprofit organization consisting of female state legislators whose members helped introduce bills mandating the Gardasil vaccine for girls in about 20 states (Allen 2007).

What Do You Think? Bills have been introduced in a number of states that would require girls to receive the HPV vaccination by age 12 or sixth grade. Do you think that states have the right to require girls to get an HPV vaccination? If you had a 12-year-old daughter, would you want her to have the HPV vaccine?

Although the profit motive in the health care industry undermines quality and affordable health care, the profit motive can also contribute to positive changes in the U.S. health care system. In the global marketplace, large corporations struggle to compete with other companies in countries where the burden of providing health insurance does not fall on the employer. Concern for profit has led big business to join to the chorus of voices that are calling for U.S. health care reform.

Finally, conflict theorists also point to the ways in which health care and research are influenced by male domination and bias. When the male erectile dysfunction drug Viagra made its debut in 1998, women across the United States were outraged by the fact that some insurance policies covered Viagra (or were considering covering it), even though female contraceptives were not covered. The male-dominated medical research community has also been criticized for neglecting women's health issues and excluding women from major health research studies (Johnson & Fee 1997).

Symbolic Interactionist Perspective

Symbolic interactionists focus on (1) how meanings, definitions, and labels influence health, illness, and health care and (2) how such meanings are learned through interaction with others and through media messages and portrayals. According to the symbolic interactionist perspective of illness, "there are no illnesses or diseases in nature. There are only conditions that society, or groups within it, has come to define as illness or disease" (Goldstein 1999, p. 31). Psychiatrist Thomas Szasz (1970) argued that what we call "mental illness" is no more than a label conferred on those individuals who are "different," that is, those who do not conform to society's definitions of appropriate behavior.

Defining or labeling behaviors and conditions as medical problems is part of a trend known as **medicalization.** Initially, medicalization was viewed as occurring when a particular behavior or condition deemed immoral (e.g., alcoholism, masturbation, or homosexuality) was transformed from a legal problem into a medical problem that required medical treatment. The concept of medicalization has expanded to include (1) any new phenomena defined as medical problems in need of medical intervention, such as post-traumatic stress disorder, premenstrual syndrome, and attention-deficit/hyperactivity disorder and (2) "normal" biological events or conditions that have come to be defined as medical problems in need of medical intervention, including childbirth, menopause, and death.

Conflict theorists view medicalization as resulting from the medical profession's domination and pursuit of profits. A symbolic interactionist perspective suggests that medicalization results from the efforts of sufferers to "translate their individual experiences of distress into shared experiences of illness" (Barker 2002, p. 295). In her study of women with fibromyalgia (a pain disorder that has no identifiable biological cause), Barker (2002) suggested that the medicalization of symptoms and distress through a diagnosis of fibromyalgia gives sufferers a framework for understanding and validating their experience of distress.

medicalization Defining or labeling behaviors and conditions as medical problems.

Recent theorists have observed a shift from medicalization to **biomedicalization**—the view that medicine can not only control particular conditions but also *transform* bodies and lives. Examples of biomedical transformations include receiving an organ transplant, becoming pregnant through reproductive technology, and receiving artificial limbs (Clarke et al. 2003).

biomedicalization The view that medicine can not only control particular conditions but also transform bodies and lives.

Although food and water in the United States and other wealthy countries are far safer than in poor countries, there are still significant concerns about food safety in the industrialized world, where up to 30 percent of the population suffers from food-borne diseases each year (World Health Organization 2007a). In the United States, there are about 76 million cases of food-borne illness each year, resulting in 325,000 hospitalizations and 5,000 deaths annually.

Public health food safety campaigns often stress how consumers can protect against food-borne illness by following recommendations for safe handling, storage, and preparation of food. Yet, most contamination of food occurs early in the production process, rather than just before consumption (U.S. Department of Agriculture 2001).

Many health problems have been associated with modern methods of raising and processing food animals. Increasingly, food animals are not raised in expansive meadows or pastures; rather, they are raised in concentrated animal feeding operations (CAFOs): giant corporate-controlled livestock farms where large numbers (sometimes tens or hundreds of thousands) of animals— typically cows, hogs, turkeys, or chickens— are "produced" in factory-like settings, often indoors, to maximize production and profits. Also referred to as "factory farms," CAFOs account for 74 percent of the world's poultry products, 68 percent of eggs, 50 percent of all pork, and 43 percent of beef (Nierenberg 2006). In the United States, there are 18,900 CAFOs.

The hogs in this concentrated animal feeding operation (CAFO) will live their entire lives, from birth to slaughter, inside a crowded, controlled indoor environment.

To prevent disease from spreading among animals living in crowded conditions, factory-farmed animals are fed diets laced with antibiotics.

Dairy cows commonly have ear implants that release hormones designed to increase milk production. Alternatively, the hormones are injected into the cows.

The diet of factory-farmed animals consists largely of corn, which is cheap, plentiful, and efficient in fattening the animals. Corn-fed beef is less healthy than grass-fed beef, as it contains more saturated fat, which contributes to heart disease (Pollan 2006). In addition, the digestive system of cows is designed for grass; corn makes cows sick and susceptible to a variety of diseases. Other factory-farmed animals, including chickens, turkeys, and hogs, and farm-raised fish are also susceptible to disease because of the crowded and unsanitary living conditions. To prevent the spread of disease in CAFOs or fish farms, food animals are fed steady diets of antibiotics. Indeed, most antibiotics sold in the United States end up in animal feed. This overuse of antibiotics in food animals affects consumers by contributing to the emergence of super-resistant bacteria that cause infections that will not respond to treatment.

Another animal food health threat is variant Creutzfeldt-Jakob disease (vCJD) or human bovine spongiform encephalopathy (BSE). The first known human death from vCJD occurred in the United Kingdom in 1995. The most likely source of vCJD is the consumption of meat contaminated with BSE. Cow carcasses that were infected with BSE were used in making livestock feed. Some of the cattle consuming this feed then also became infected, which led to an epidemic of BSE, commonly called "mad cow

disease." From 1996–2002, 129 cases of vCJD were reported in the United Kingdom, 6 in France and 1 each in Canada, Ireland, Italy, and the United States (World Health Organization 2007b). In 1997, the U.S. Food and Drug Administration banned the use of cattle byproducts in cattle feed. However, cattle byproducts can still be legally fed to other livestock such as pigs and poultry, which can then be turned into cattle feed.

About a third of U.S. dairy cows are given recombinant bovine growth hormone (rBGH), manufactured by Monsanto and sold under the trade name Posilac, to increase milk production by about 10 percent. Although the Food and Drug Administration (FDA) approved Posilac in 1993 and has supported Monsanto's claims that milk containing rBGH is safe for consumers, some experts warn that rBGH raises the risk of breast, colon, and prostate cancer (Epstein 2006). Other countries, including all of Europe, Canada, Australia, New Zealand, and Japan, have banned milk containing rBGH.

The best-selling book, *Fast Food Nation*, and 2006 movie of the same title succinctly explained a major health problem related to modern slaughterhouse and meatpacking production techniques: "There is shit in the meat" (Schlosser 2002, p. 197). Schlosser cited a 1996 U.S. Department of Agriculture study showing that more than three-quarters of the ground beef sampled contained several

As workers remove the stomach and intestines of beef cattle, fecal matter can spill out and contaminate the meat.

microbes, which are spread primarily by fecal material. Schlosser explained that in the slaughterhouse, if the animal's hide has not been adequately cleaned, chunks of manure may fall from it onto the meat. When the cow's stomach and intestines are removed, the fecal matter in the digestive system may spill out and contaminate the meat as well. Schlosser (2002) explained that "the increased speed of today's production lines makes the task much more difficult. A single worker at a 'gut table' may eviscerate sixty cattle an hour" (p. 203). At least three or four cattle infected with the microbe *Escherichia coli* O157:H7, which can cause serious illness and death in humans, are processed at a large slaughterhouse every hour. Because of the mass production techniques of modern meat processing, a single cow infected with *E. coli* O157:H7 can contaminate 32,000 pounds of ground beef (Schlosser 2002). A single fast food hamburger contains meat from dozens or even hundreds of different cattle.

Finally, a number of health problems result from the massive quantities of animal waste that are produced and stored around factory farms. A single dairy cow produces more than 20 tons of manure annually, and a hog can produce more than two tons (Weeks 2007). Manure is often stored in lagoons, but these lagoons leak, break, or are washed away by big storms, contaminating ground water. Sometimes manure is stored in large lagoons until it can be sprayed on fields as fertilizer. When farmers apply manure to land faster than plants can take up the manure's nitrogen and phosphorus, the excess washes out of the soils and pollutes groundwater (Weeks 2007).

Livestock factories also pose a threat to air quality. People who live near large livestock farms complain about headaches, runny noses, sore throats, nausea, stomach cramps, diarrhea, burning eyes, coughing, bronchitis, and shortness of breath (Singer & Mason 2006; Weeks 2007). Residents who live near animal factory farms also complain of foul odors that are so strong they stay indoors with their windows shut.

References

Epstein, Samuel S. 2006. *What's in Your Milk?* Victoria, BC, Canada: Trafford Publishing.

In modern meat-processing plants, the meat from hundreds of different cows is mixed up to be ground, so a single animal infected with E. coli can contaminate 32,000 pounds of ground beef that may be shipped throughout the United States.

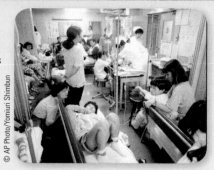

One of the worst outbreaks of E. coli O157:H7 food poisoning occurred in 1996, when nearly 6,000 cases were recorded in Japan.

Nierenberg, Danielle. 2006. "Meat Consumption and Output Up." In *Vital Signs* 2006–2007, Linda Starke, ed. (pp. 24-25). Worldwatch Institute. New York: W. W. Norton & Company.

Pollan, Michael. 2006. *The Omnivore's Dilemma.* New York: Penguin Press.

Schlosser, Eric. 2002. *Fast Food Nation.* New York: HarperCollins.

Singer, Peter, and Jim Mason. 2006. *The Way We Eat: Why Our Food Choices Matter.* Emmaus, PA: Rodale.

U.S. Department of Agriculture. 2001. Economics of Foodborne Disease: Food and Pathogens. Available at http://www.ers.usda.gov

Weeks, Jennifer. 2007 (January 12). Factory Farms. *The CQ Researcher* 17(2).

World Health Organization. 2007a. "Food Safety and Foodborne Illness." Available at http://www.who.int/mediacentre/factsheets/fs237/en/

World Health Organization. 2007b. *World Health Report 2007: A Safer Future: Global Public Health Security in the 21st Century.* Available at http://www.who.int/

This hog manure lagoon near Milford, Utah, holds 3 million gallons of hog waste. A tarp has been spread over the 30-foot-deep lagoon in an attempt to protect nearby residents from the foul smell of the manure.

The concepts of medicalization and biomedicalization suggest that conceptions of health and illness are socially constructed. It follows, then, that definitions of health and illness vary over time and from society to society. In some countries being fat is a sign of health and wellness; in others it is an indication of mental illness or a lack of self-control. Among some cultural groups, perceiving visions or voices of religious figures is considered a normal religious experience, whereas such "hallucinations" would be indicative of mental illness in other cultures. In 18th- and 19th-century America masturbation was considered an unhealthy act that caused a range of physical and mental health problems. Individuals caught masturbating were often locked up in asylums, treated with drugs (such as sedatives and poisons), or subjected to a range of interventions designed to prevent masturbation by stimulating the genitals in painful ways, preventing genital sensation, or deadening it. These physician-prescribed interventions included putting ice on the genitals; blistering and scalding the penis, vulva, inner thighs, or perineum; inserting electrodes into the rectum and urethra; cauterizing the clitoris by applying pure carbolic acid; circumcising the penis; and surgically removing the clitoris, ovaries, and testicles (Allen 2000). Today, most health professionals agree that masturbation is a normal, healthy aspect of sexual expression.

Symbolic interactionism draws attention to the effects that meanings and labels have on health behaviors and health-related policies. For example, as tobacco sales have declined in developed countries, transnational tobacco companies have looked for markets in developing countries, using advertising strategies that depict smoking as "an inexpensive way to buy into glamorous lifestyles of the upper or successful social class" (Egwu 2002, p. 44). In 2004 the Centers for Medicare & Medicaid Services decided to remove language in Medicare's coverage manual that states that obesity is not an illness (Stein & Connolly 2004). Labeling obesity as an illness means that treatment for obesity, ranging from joining weight-loss or fitness clubs to surgery and counseling, can be covered by Medicare.

Symbolic interactionists also focus on the stigmatization of individuals who are in poor health or who lack health insurance. A **stigma** refers to a discrediting label that affects an individual's self-concept and disqualifies that person from full social acceptance. (Originally, the word *stigma* referred to a mark burned into the skin of a criminal or slave.) The stigma associated with poor health often results in prejudice and discrimination against individuals with mental illnesses, drug addictions, physical deformities and impairments, missing or decayed teeth, obesity, HIV infection and AIDS, and other health conditions. Further, a study of uninsured U.S. adults found that "uninsured Americans . . . noted the stigma of lacking health insurance, citing medical providers who treat them like 'losers' because they are uninsured" (Sered & Fernandopulle 2005, p. 16).

The policy implications of stigmatization are clear: The stigma associated with health problems and/or lack of health insurance implies that the individual—rather than society—is responsible for his or her health. In U.S. culture, "sickness increasingly seems to be construed as a personal failure—a failure of ethical virtue, a failure to take care of oneself 'properly' by eating the 'right' foods or getting 'enough' exercise, a failure to get a Pap smear, a failure to control sexual promiscuity, genetic failure, a failure of will, or a failure of commitment—rather than society's failure to provide basic services to all of its citizens" (Sered & Fernandopulle 2005, p. 16). Next, we examine one of the most urgent health problems facing the world: HIV/AIDS.

stigma A discrediting label that affects an individual's self-concept and disqualifies that person from full social acceptance.

HIV/AIDS: A GLOBAL HEALTH CONCERN

One of the most urgent worldwide public health concerns is the spread of HIV, which causes AIDS. HIV/AIDS has killed more than 20 million people, and in 2006 nearly 40 million people worldwide were living with HIV infection. About one-quarter of people living with HIV do not know they are infected (Joint United Nations Programme on HIV/AIDS 2006; World Health Organization 2004).

HIV is transmitted through sexual intercourse, through sharing unclean intravenous needles, through perinatal transmission (from infected mother to fetus or newborn), through blood transfusions or blood products, and, rarely, through breast milk. Worldwide, the predominant mode of HIV transmission is through heterosexual contact (World Health Organization 2004).

Millions of children whose parents died of AIDS grow up in orphanages.

HIV/AIDS in Africa and Other Regions

HIV/AIDS is most prevalent in Africa, particularly sub-Saharan Africa. With slightly more than 10 percent of the world's population, Africa is home to 60 percent of individuals infected with HIV (Johnson 2007). About 1 in 12 African adults has HIV/AIDS, and as many as 9 of 10 HIV-infected people in sub-Saharan Africa do not know that they are infected (World Health Organization 2004). But HIV/AIDS also affects millions of people living in India and hundreds of thousands of people in China, the Mediterranean region, Western Europe, and Latin America. Eastern European countries and Central Asia are experiencing increasing rates of HIV infection, mainly from drug-injecting behavior and to a lesser extent from unsafe sex. The second highest HIV prevalence rate (after sub-Saharan Africa) is the Caribbean, where 2–3 percent of adults are infected with HIV (World Health Organization 2004).

The high rates of HIV in developing countries, particularly sub-Saharan Africa, are having alarming and devastating effects on societies. HIV/AIDS has reversed the gains in life expectancy made in sub-Saharan Africa, which peaked at 49 years in the late 1980s and fell to 46 years in 2005 (World Health Organization 2004). The HIV/AIDS epidemic creates an enormous burden for the limited health care resources of poor countries. Economic development is threatened by the HIV epidemic, which diverts national funds to health-related needs and reduces the size of a nation's workforce. AIDS deaths have left millions of orphans in the world; by 2010 25 million children are projected to be orphaned due to HIV/AIDS (World Health Organization 2004). Some scholars fear that AIDS-affected countries could become vulnerable to political instability as the growing number of orphans exacerbates poverty, and produces masses of poor young adults who are vulnerable to involvement in criminal activity and recruitment for insurgencies (Mastny & Cincotta 2005).

HIV/AIDS in the United States

According to the Centers for Disease Control and Prevention (2006, 2007a), more than 1 million people in the United States are living with HIV/AIDS; in about one-quarter of these people the infection is undiagnosed, and they are unaware of their infection. Among U.S. adults and adolescents, three-quarters (74 percent) of new HIV/AIDS diagnoses in 2005 were among men, with the remainder occurring among women. Among men with HIV/AIDS, the primary mode of transmission is through male-to-male sexual contact, followed by heterosexual contact and injection drug use. In a study of five U.S. cities, 25 percent of men who have sex with men were infected with HIV; nearly half of these infected men (48 percent) were unaware of their infection (Centers for Disease Control and Prevention 2005b). Among women with HIV/AIDS, the primary mode of transmission is through heterosexual contact, followed by injection drug use.

Nearly half (49 percent) of new HIV/AIDS diagnoses in 2005 were among African Americans, who make up about 13 percent of the U.S. population. Higher rates of HIV/AIDS among African Americans are partly due to the link between higher AIDS incidence and poverty. The nearly one in four African Americans who live in poverty experience limited access to high-quality health care and HIV infection prevention education. A recent study of HIV transmission among African American women in North Carolina found that women with HIV infection were more likely than noninfected women to be unemployed; to receive public assistance; to have had 20 or more lifetime sexual partners; or to have traded sex for drugs, money, or shelter (reported by Centers for Disease Control and Prevention 2005a).

Despite the widespread concern about HIV, many Americans—especially adolescents and young adults—engage in high-risk behavior. A national survey of college students found that only about half (52 percent) reported having used a condom the last time they had vaginal intercourse, and only 28 percent reported having used a condom the last time they had anal intercourse (American College Health Association 2007).

THE GROWING PROBLEM OF OBESITY

Obesity is increasingly being recognized as a major health problem throughout the industrialized world. A national public opinion poll found that 85 percent of adults in the United States believe that obesity is an epidemic (Trust for America's Health 2007). The United States has the highest prevalence of obesity among the developed nations (Children's Defense Fund 2006). Obesity, which can lead to heart disease, hypertension, diabetes, and other health problems, is the second biggest cause of preventable deaths in the United States (second only to tobacco use) (Stein & Connolly 2004). In 2005 an alarming report was published in the *New England Journal of Medicine* that suggested that over the next 50 years obesity will shorten the average U.S. life expectancy by at least 2–5 years, reversing the mostly steady increase in life expectancy that has occurred over the past two centuries (Olshansky et al. 2005).

The following statistics reflect the degree to which Americans are increasingly overweight or obese (as defined by body mass index) (National Center for Health Statistics 2006; Trust for America's Health 2007):

- The percentage of overweight (but not obese) adults has remained steady at 32–34 percent since the 1960s. But during that same period, the percentage of obese adults has jumped from 13 percent to 34 percent. This means that two-thirds of U.S. adults are either overweight or obese.
- The highest rate of obesity is seen among non-Hispanic black women (51 percent).
- The percentage of overweight adolescents ages 12–19 has more than tripled since the 1970s, from 5 percent to 17 percent.
- The percentage of overweight children ages 6–11 more than doubled since the 1970s, from 7 percent to 19 percent.

Although genetics and certain medical conditions contribute to many cases of overweight and obesity, two social and lifestyle factors that play a major role in the obesity epidemic are patterns of food consumption and physical activity level. National survey data show that less than one-third (30 percent) of U.S. adults (age 18 or older) engage in regular leisure-time physical activity (defined as moderate activity for 30 minutes or more at least 5 times a week or vigorous activity for 20 minutes at least 3 times a week) (National Center for Health Statistics 2006). One in five (22 percent) U.S. adults report that they do not engage in any physical activity (Trust for America's Health 2007). According to the National Center for Chronic Disease Prevention and Health Promotion (2004), more than one-third of youths in grades 9–12 do not engage in regular vigorous physical activity, and only one-third of high school students participate in daily physical education classes at school.

Americans are increasingly eating out at fast food and other restaurants where foods tend to contain more sugars and fats than foods consumed at home. In 1970 Americans spent one-third of their food dollars on food away from home; this amount grew to 47 percent in 2001 (Sturm 2005). A national study revealed that young adults ages 18–27 consumed fast food on average 2.5 times per week (Niemeier et al. 2006). Fast food consumption is strongly associated with weight gain and insulin resistance, suggesting that fast food increases the risk of obesity and type 2 diabetes (Niemeier et al. 2006; Pereira et al. 2005). Consumption of snack foods and sugary soft drinks has also increased. Among children ages 6–11 years, consumption of chips, crackers, popcorn, and/or pretzels tripled from the mid-1970s to the mid-1990s. Consumption of soft drinks doubled during the same period (Sturm 2005). As processed foods are increasingly marketed throughout the world, the consumption of foods high in fats and sweeteners is also increasing in developing nations. This changing pattern of food consumption, known as the *nutrition transition,* is contributing to a rapid rise in obesity and diet-related chronic diseases world-wide (Hawkes 2006).

© AP Photo/Scott Heppell

Childhood obesity is becoming more common throughout the developed world. At 8 years of age, Connor McCreaddie, shown here with his mother, weighed 218 pounds.

Cultural attitudes also play a role in obesity. In Black American culture, a "full physique" is considered an ideal body type. Black women with more-than-ample hips and plenty of "junk in the trunk" are considered attractive. Black women who are thin are criticized for being underweight ("Nobody but a dog wants a bone"). The cultural message to black Americans is "put some meat on those bones" (Bailey 2006).

Obesity is also related to socioeconomic status. In less developed countries, poverty is associated with undernutrition and starvation. In the United States, however, being poor is associated with an increased risk of being overweight or obese. High-calorie processed foods tend to be more affordable than fresh vegetables, fruits, and lean meats/fish. It is cheaper, for example, to buy three boxes of macaroni and cheese than one pound of skinless chicken breast and much cheaper to buy a large package of store-brand cookies than to buy five apples or pears. For individuals living at the lower end of the socioeconomic spectrum, overeating may be a way of compensating for the material comforts they cannot attain. In his research on obesity among black Americans, Dr. Eric Bailey (2006) makes the observation that if you are an African American trying to make it in society and you are not able to make it, food is one way to console yourself.

MENTAL ILLNESS: THE HIDDEN EPIDEMIC

mental health The successful performance of mental function, resulting in productive activities, fulfilling relationships with other people, and the ability to adapt to change and to cope with adversity.

mental illness All mental disorders, which are health conditions that are characterized by alterations in thinking, mood, and/or behavior associated with distress and/or impaired functioning and that meet specific criteria (such as level of intensity and duration) specified in *The Diagnostic and Statistical Manual of Mental Disorders.*

What it means to be mentally healthy varies across and within cultures. In the United States, **mental health** is defined as the successful performance of mental function, resulting in productive activities, fulfilling relationships with other people, and the ability to adapt to change and to cope with adversity (U.S. Department of Health and Human Services 2001). **Mental illness** refers collectively to all mental disorders, which are health conditions that are characterized by alterations in thinking, mood, and/or behavior associated with distress and/or impaired functioning and that meet specific criteria (such as level of intensity and duration) specified in the classification manual used to diagnose mental disorders, *The Diagnostic and Statistical Manual of Mental Disorders* (American

TABLE 2.3 Disorders Classified by the American Psychiatric Association

CLASSIFICATION	DESCRIPTION
Anxiety disorders	Disorders characterized by anxiety that is manifest in phobias, panic attacks, or obsessive-compulsive disorder
Dissociative disorders	Problems involving a splitting or dissociation of normal consciousness, such as amnesia and multiple personality
Disorders first evident in infancy, childhood, or adolescence	Disorders including mental retardation, attention-deficit/hyperactivity, and stuttering
Eating or sleeping disorders	Disorders including anorexia, bulimia, and insomnia
Impulse control disorders	Problems involving the inability to control undesirable impulses, such as kleptomania, pyromania, and pathological gambling
Mood disorders	Emotional disorders such as major depression and bipolar (manic-depressive) disorder
Organic mental disorders	Psychological or behavioral disorders associated with dysfunctions of the brain caused by aging, disease, or brain damage (such as Alzheimer's disease)
Personality disorders	Maladaptive personality traits that are generally resistant to treatment, such as paranoid and antisocial personality types
Schizophrenia and other psychotic disorders	Disorders with symptoms such as delusions or hallucinations
Somatoform disorders	Psychological problems that present themselves as symptoms of physical disease, such as hypochondria
Substance-related disorders	Disorders resulting from abuse of alcohol and/or drugs, such as barbiturates, cocaine, or amphetamines

Psychiatric Association 2000) (see Table 2.3). Mental illness is a "hidden epidemic" because the shame and embarrassment associated with mental problems discourage people from acknowledging and talking about them.

Extent and Impact of Mental Illness

Although transnational estimates of the prevalence of mental disorders vary, one study found a 40 percent lifetime prevalence of any mental disorder in the Netherlands and the United States, a 12 percent lifetime prevalence in Turkey, and a 20 percent lifetime prevalence in Mexico (WHO International Consortium in Psychiatric Epidemiology 2000). One-quarter (26 percent) of U.S. adults have a diagnosable mental disorder in any given year, and mental disorders are the leading cause of disability for individuals ages 15–44 (National Institute of Mental Health 2006). One of six Americans admit that poor mental health or emotional well-being kept them from doing their usual activities at least once during the last month. And 12 percent of Americans have visited a mental health professional, such as a psychologist, psychiatrist, or therapist, in the past 12 months (Newport 2004).

Untreated mental disorders can lead to poor educational achievement, lost productivity, unsuccessful relationships, significant distress, violence and abuse, incarceration, unemployment, and poverty. Half of students identified as

> **❝** It is no measure of health to be well adjusted to a profoundly sick society. **❞**
>
> Jiddu Krishnamurti

having emotional disturbances drop out of high school (Gruttadaro 2005). On any given day 150,000 people with severe mental illness are homeless, living on the streets or in public shelters (National Council on Disability 2002). As many as one in five adults in U.S. prisons and as many as 70 percent of youth incarcerated in juvenile justice facilities are mentally ill (Honberg 2005; Human Rights Watch 2003).

Mental disorders also contribute to mortality, with suicide being the fourth leading cause of death worldwide among 15- to 44-year-olds (Mercy et al. 2003). In the United States suicide is the third leading cause of death among people ages 15–24 (National Center for Health Statistics 2006). Most suicides in the United States (more than 90 percent) are committed by individuals with a mental disorder, most commonly a depressive or substance abuse disorder (National Institute of Mental Health 2006). In 2000 about half of the 1.7 million violent deaths that occurred in the world were the result of suicide, about one-third resulted from homicide, and one-fifth were from war injuries (Mercy et al. 2003). Suicides outnumber homicides two to one every year in the United States (Ezzell 2003).

Causes of Mental Disorders

Some mental illnesses are caused by genetic or neurological pathological conditions. However, social and environmental influences, such as poverty, history of abuse, or other severe emotional trauma, also affect individuals' vulnerability to mental illness and mental health problems. For example, iodine deficiency, common in poor countries, is believed to be the single most common preventable cause of mental retardation and brain damage (World Health Organization 2002). War within and between countries also contributes to mental illness. For example, experts predict that 16 percent of service members serving in Iraq and Afghanistan will develop post-traumatic stress disorder (Miller 2005) (see also Chapter 16 for a discussion of combat-related post-traumatic stress disorder). Depression, which can be caused by biochemical conditions, can also stem from cultural conditions. As Garfinkel and Goldbloom (2000) explained, "The radical shifts in society towards technology, changes in family and societal supports and networks and the commercialization of existence . . . may account for the current epidemic of depression and other psychiatric disorders" (p. 503). It may be safe to conclude that the causes of most mental disorders lie in some combination of genetic, biological, and environmental factors (U.S. Department of Health and Human Services 2001).

SOCIAL FACTORS AND LIFESTYLE BEHAVIORS ASSOCIATED WITH HEALTH AND ILLNESS

Public health education campaigns, articles in popular magazines, college-level health courses, and health professionals emphasize that to be healthy, we must adopt a healthy lifestyle. Despite these efforts, many individuals engage in high-risk health behaviors such as excessive alcohol consumption, cigarette smoking, unprotected sexual intercourse, and inadequate consumption of fruits and vegetables (see this chapter's *Social Problems Research Up Close* feature). However, health and illness are affected by more than personal lifestyle choices. In the following sections we examine how social factors such as globalization, social class and poverty, education, race, and gender affect health and illness.

Globalization

Globalization, broadly defined as the growing economic, political, and social interconnectedness among societies throughout the world, has eroded the boundaries that separate societies, creating a "global village." Globalization has had both positive and negative effects on health. On the positive side, globalized communications technology enhances the capacity to monitor and report on outbreaks of disease, disseminate guidelines for controlling and treating disease, and share scientific knowledge and research findings (Lee 2003). On the negative side, aspects of globalization such as increased travel and the expansion of trade and transnational corporations have been linked to a number of health problems.

Effects of Increased Travel on Health. Increased business travel and tourism have encouraged the spread of disease, such as the potentially fatal West Nile virus infection, which first appeared in the United States in 1999 and has spread to all 48 contiguous states (Centers for Disease Control and Prevention 2007b). Before 1999, the West Nile virus had never before been found in the United States. The most likely explanation of how the virus got to the United States is that it was introduced by an infected bird that was imported or an infected human returning from a country where the virus is common. Likewise, SARS (severe acute respiratory syndrome), was first diagnosed in South China in 2002 and within months spread to 29 countries, infecting thousands of individuals and killing more than 800.

Effects of Increased Trade and Transnational Corporations on Health. Increased international trade has expanded the range of goods available to consumers, but at a cost to global health. The increased transportation of goods by air, sea, and land contributes to pollution caused by the burning of fossil fuels. In addition, the expansion of international trade of harmful products such as tobacco, alcohol, and processed or "fast" foods is associated with a worldwide rise in cancer, heart disease, stroke, and diabetes (World Health Organization 2002).

Expanding trade has also facilitated the growth of transnational corporations that set up shop in developing countries to take advantage of lower labor costs and lax environmental and labor regulations (see also Chapter 7). Because of lax labor and human rights regulations, factory workers in transnational corporations—typically low-status uneducated women—are often exposed to harmful working conditions that increase the risk of illness, injury, and mental anguish (Hippert 2002). These workers are often exposed to toxic substances, lack safety equipment such as gloves and goggles, are denied bathroom breaks (which leads to bladder infections), and are assaulted at the workplace, because "physical brutality is frequently used as a mechanism of control on the production floor of the factory" (Hippert 2002, p. 863).

As a result of weak environmental laws in developing countries, transnational corporations are responsible for high levels of pollution and environmental degradation, which negatively affect the health of entire populations. Finally, the movement of factories out of the United States to other countries has resulted in significant loss of U.S. jobs in manufacturing and in the textile and

globalization The growing economic, political, and social interconnectedness among societies throughout the world.

The National College Health Assessment is a survey developed in 1998 by the American College Health Association to assess the health status of college students across the country. After briefly describing the sample and methods, we present selected findings of the 2006 National College Health Assessment.

Sample and Methods

Across North America, 123 postsecondary institutions self-selected to participate in the 2006 National College Health Assessment. Data from institutions that did not use random sampling techniques were not used, yielding a final sample of 94,806 students at 117 campuses (American College Health Association 2007). The overall response rate was 35 percent: 85.8 percent for schools using paper surveys and 23.2 percent for schools conducting web-based surveys. The survey contains about 300 questions that assess student health status and health problems, risk and protective behaviors, and health impediments to academic performance.

Selected Findings and Conclusions*

- *Most commonly reported health problems.* The most commonly reported health problems of college students are back pain and allergy problems (see Table 1).
- *Alcohol, tobacco, and other drug use.* The majority of college students (70 percent) reported having consumed alcohol in the

Table 1   Top 10 Self-Reported Health Problems Students Experienced in the Past School Year

HEALTH PROBLEM	RANK	PERCENTAGE
Back pain	1	46.6
Allergy problems	2	45.5
Sinus infection	3	28.8
Depression	4	17.8
Strep throat	5	13.2
Anxiety disorder	6	12.4
Asthma	7	11.2
Ear infection	8	9.3
Seasonal affective disorder (SAD)	9	8.1
Bronchitis	10	7.8

Adapted from: American College Health Association. 2007. "American College Health Association National College Health Assessment Spring 2006 Reference Group Data Report." *Journal of American College Health,* 55(4):195–226.

past 30 days. When students were asked how many alcoholic drinks they had the last time they partied or socialized, 26 percent reported having had 5–8 drinks and 12 percent reported consuming 9 or more alcoholic beverages. Excluding students who do not drink and students who do not drive from the analysis, one-third (34 percent) of college students reported that they drove after drinking any alcohol during the past 30 days. Among students who drink alcohol, nearly one in five (18 percent) said that in the past school year they physically injured themselves after drinking alcohol, and 14 percent reported having had

unprotected sex after consuming alcohol. The percentages of college students who used various substances in the past month are as follows: cigarettes, 83 percent; marijuana, 14 percent; amphetamines, 2 percent; and cocaine, 2 percent.

- *Sexual health and condom use.* About 5 percent of college students reported that they had a sexually transmitted infection, disease, or complication in the past year (genital warts/human papillomavirus [HPV], genital herpes, chlamydia, pelvic inflammatory disease, HIV, gonorrhea). Less than 1 percent (0.3 percent) reported having HIV infection, and 28 percent

apparel industry, which is significant because of the relationship between physical and mental illness and unemployment (Bartley et al. 2001).

Effects of International Free Trade Agreements on Health. The World Trade Organization (WTO) and regional trade agreements such as the North American Free Trade Agreement (NAFTA) establish rules aimed to increase international trade. These trade rules supersede member countries' laws and regulations, including those governing public health, if those laws or regulations create a barrier to trade. For example, the Methanex Corporation of Canada produces methanol, a component of methyl tertiary butyl ether (MTBE), a gas additive. When the state

Table 2   Reported Number of Times Students Experienced Mental Health Difficulties in the Past School Year

MENTAL HEALTH DIFFICULTY	0 TIMES PERCENTAGE	1–4 TIMES PERCENTAGE	5–8 TIMES PERCENTAGE	9 OR MORE TIMES PERCENTAGE
Felt things were hopeless	37.8	38.5	10.9	12.8
Felt overwhelmed by all you had to do	6.5	31.8	24.9	36.8
Felt exhausted (not from physical activity)	8.5	31.8	23.8	36.0
Felt very sad	20.6	45.4	15.7	18.3
Felt so depressed it was difficult to function	56.2	27.5	6.9	9.4
Seriously considered attempting suicide	90.7	7.3	0.9	1.1
Attempted suicide	98.7	1.0	0.1	0.1

Adapted from: American College Health Association. 2007. "American College Health Association National College Health Assessment Spring 2006 Reference Group Data Report." *Journal of American College Health,* 55(4):195–226.

reported ever having been tested for HIV. Although women are advised to get a gynecological exam annually, only 59 percent of college women reported having had a gynecological exam in the past year. Among sexually active students, only 52 percent said they used a condom the last time they had vaginal intercourse and only 28 percent used a condom during anal intercourse.

- *Nutrition, exercise, and weight.* Only 8 percent of college students reported that they ate 5 or more servings of fruits and vegetables daily; 44 percent reported that they exercised vigorously for at least 20 minutes or moderately for at least 30 minutes at least 3 days a week. Based on estimated body mass index (BMI) calculated from students' reported height and weight, 5 percent of college students are underweight, 64 percent are "healthy weight," 22 percent are "overweight," and 10 percent are obese.

- *Mental health.* Fifteen percent of college students report having been diagnosed with depression sometime in their lifetimes. Of these, 26 percent are currently receiving therapy for depression, and 37 percent are currently taking medication for depression. In the past year, 1 percent of students attempted suicide at least once; 9 percent seriously considered suicide at least once. Table 2 presents the number of times students reported experiencing various mental health difficulties in the past school year.

Discussion

The results of the National College Health Assessment reveal the extent and types of health problems and risk behaviors of college students. These data can be used to help colleges and universities design and implement health services that meet the needs of college students. The survey results also provide important baseline data for school administrations and health officials to use in measuring changes in health problems among U.S. college students.

*Percentages are rounded.

of California banned the use of MTBE because of its link to cancer, the Methanex Corporation initiated an approximately $1 billion lawsuit against the United States, claiming that California's ban of MTBE violates Chapter 11 of NAFTA. After a 5-year legal battle a closed tribunal sided with California and ruled against the Methanex Corporation in 2005, a major victory for environmentalists and Californians. Supporters of NAFTA say that the tribunal's decision demonstrates that U.S. trade agreements do not encroach on governments' right to enforce health and environmental regulations. However, other companies have succeeded in suing governments under NAFTA rules. In 2000 the U.S. Metaclad Company sued Mexico for $16 million because Mexico stopped the company from reopening a toxic waste dump that would contaminate people and the environment (Shaffer et al. 2005).

Another trade rule, the Agreement on Trade-Related Aspects of Intellectual Property Rights (TRIPS), mandates that all WTO member countries implement intellectual property rules that provide 20-year monopoly control over patented items, including medications. TRIPS limits the availability of generic drugs, thus contributing to higher drug costs. TRIPS also affects access to medications for life-threatening diseases in low-income countries. In 2005 India approved legislation that ended the decades-old practice of allowing drug companies to make low-cost generic drugs. This legislation, which resulted from the WTO's requirement that India enforce stricter patent rules for its pharmaceutical industry, is expected to curb the supply of affordable HIV/AIDS medications to impoverished nations (Mahapatra 2005).

Social Class and Poverty

In an address to the 2001 World Health Assembly, United Nations Secretary-General Kofi Annan stated that the biggest enemy of health in the developing world is poverty (United Nations Population Fund 2002). Poverty is associated with malnutrition, indoor air pollution, hazardous working conditions, lack of access to medical care, and unsafe water and sanitation (see also Chapter 6). Half of the urban population in Africa, Asia, Latin America, and the Caribbean have one or more diseases linked to inadequate water and sanitation (Millennium Ecosystem Assessment 2005).

In the United States low socioeconomic status is associated with higher incidence and prevalence of health problems, disease, and death. The percentage of Americans reporting fair or poor health and having disabling conditions is considerably higher among people living below the poverty line than for those with family income at least twice the poverty threshold (National Center for Health Statistics 2006). In the United States poverty is associated with higher rates of health-risk behaviors such as smoking, alcohol drinking, being overweight, and being physically inactive. The poor are also exposed to more environmental health hazards and, as we saw with Betty's family in the opening story of the chapter, the poor have unequal access to and use of medical care (Lantz et al. 1998). In addition, members of the lower class tend to experience high levels of stress and have few resources to cope with it (Cockerham 2005). Stress has been linked to a variety of physical and mental health problems, including high blood pressure, cancer, chronic fatigue, and substance abuse.

Just as poverty contributes to health problems, health problems contribute to poverty. Health problems can limit one's ability to pursue education or vocational training and to find or keep employment. The high cost of health care not only deepens the poverty of people who are already barely getting by but also can financially devastate middle-class families. Later in this chapter we look more closely at the high cost of health care and its consequences for individuals and families.

Poverty and Mental Health. Low socioeconomic status is also associated with increased risk of a broad range of psychiatric conditions. People living below the U.S. poverty line are roughly five times as likely as those with incomes twice the poverty line to have serious psychological distress (National Center for Health Statistics 2006). Two explanations for the link between social class and mental illness are the *causation* explanation and the *selection* explanation (Cockerham 2005). The selection explanation suggests that mentally ill individuals have difficulty achieving educational and occupational success and thus tend to drift to the lower class, whereas the mentally healthy are upwardly mobile. The *causation* explanation suggests that lower-class individuals experience greater adversity and stress as a result of their deprived and difficult living conditions, and this stress can reach the point at which the individual can no longer cope with daily living. Research that tested these two explanations found more support for the causation explanation (Hudson 2005).

Education

Although economic resources are important influences on health, "the strongest single predictor of good health appears to be education" (Cockerham 2004, p. 62). A *New York Times* report on the link between education and lifespan came to a similar conclusion: "The one social factor that researchers agree is consistently linked to longer lives in every country where it has been studied is education" (Kolata 2007, p. A1). Individuals with low levels of education are more likely to engage in health-risk behaviors such as smoking and heavy drinking. Women with less education are less likely to seek prenatal care and are more likely to smoke during pregnancy, which helps explain why low birth weight and infant mortality are more common among children of less educated mothers (Children's Defense Fund 2000).

In some cases lack of education means that individuals do not know about health risks or how to avoid them. A national survey in India found that only 18 percent of illiterate women had heard of AIDS, compared with 92 percent of women who had completed high school (Ninan 2003).

Gender

Gender affects the health of both women and men. As noted earlier, women in developing countries have high rates of mortality and morbidity as a result of the high rates of complications associated with pregnancy and childbirth. The low status of women in many less developed countries results in their being nutritionally deprived and having less access to medical care than men have (UNICEF 2006).

Sexual violence and gender inequality contribute to growing rates of HIV among girls and women. Although women are more susceptible to HIV infection for physical and biological reasons, increasing rates of HIV among women are also due to fact that many women, especially in African countries, do not have the social power to refuse sexual intercourse and/or to demand that their male partners use condoms (Lalasz 2004).

In the United States, violence against women is a public health concern: At least one in three women has been beaten, coerced into sex, or abused in some way—most often by someone the woman knows (Alan Guttmacher Institute 2004). "Although neither health care workers nor the general public typically thinks of battering as a health problem, woman battering is a major cause of injury, disability, and death among American women, as among women worldwide" (Weitz 2006, p. 62).

In the United States before the 20th century, the life expectancy of U.S. women was shorter than that of men because of the high rate of maternal mortality that resulted from complications of pregnancy and childbirth. In the United

States today life expectancy of women (80.4 years) is greater than that of men (75.2 years) (National Center for Health Statistics 2006). Lower life expectancy for U.S. men is due to a number of factors. Men tend to work in more dangerous jobs than women, such as agriculture and construction. Men are more likely than women to smoke cigarettes and to abuse alcohol and drugs but are less likely than women to visit a doctor and to adhere to medical regimens (Williams 2003). In addition, "beliefs about masculinity and manhood that are deeply rooted in culture . . . play a role in shaping the behavioral patterns of men in ways that have consequences for their health" (Williams 2003, p. 726). Men are socialized to be strong, independent, competitive, and aggressive and to avoid expressions of emotion or vulnerability that could be construed as weakness. These male gender expectations can lead men to take actions that harm themselves or to refrain from engaging in health-protective behaviors. For example, socialization to be aggressive and competitive leads to risky behaviors (such as dangerous sports, fast driving, and violence) that contribute to men's higher risk of injuries and accidents.

Gender and Mental Health. A review of research suggests that the prevalence of mental illness is higher among U.S. women than among U.S. men (Cockerham 2005). In 2005, U.S. women were more likely than men (3.9 percent versus 2.4 percent) to have experienced serious psychological distress during the past 30 days (National Center for Health Statistics 2006). Women and men differ in the types of mental illness they experience; rates of mood and anxiety disorders are higher among women and rates of personality and substance-related disorders are more common among men. Although women are more likely to attempt suicide, men are more likely to succeed at it because they use deadlier methods.

Biological factors may account for some of the gender differences in mental health. Hormonal changes during menstruation and menopause, for example, may predispose women to depression and anxiety, although evidence to support this explanation is "insufficient at present" (Cockerham 2005, p. 166). High testosterone and androgen levels in males may be linked to the greater prevalence of personality disorders in men, but again, research is not conclusive. Other explanations for gender differences in mental health focus on ways in which gender roles contribute to different types of mental disorders. For example, the unequal status of women and the strain of doing the majority of housework and child care may predispose women to experience greater psychological distress. Women may also be more likely to experience depression when their children leave home, because women are socialized to invest more in their parental role than men are.

Racial and Ethnic Minority Status

In the United States racial and ethnic minorities are more likely than non-Hispanic whites to rate their health as fair or poor (see Figure 2.2). Black U.S. residents, especially black men, have a lower life expectancy than white men (see Table 2.4). Black Americans are more likely than white Americans to die from stroke, heart disease, cancer, HIV infection, unintentional injuries, diabetes, cirrhosis, and homicide. Youth from racial/ethnic minority backgrounds and low socioeconomic status are more likely to be overweight and to engage in less healthy behaviors (Delva, O'Malley, & Johnston 2007); Black Americans have the

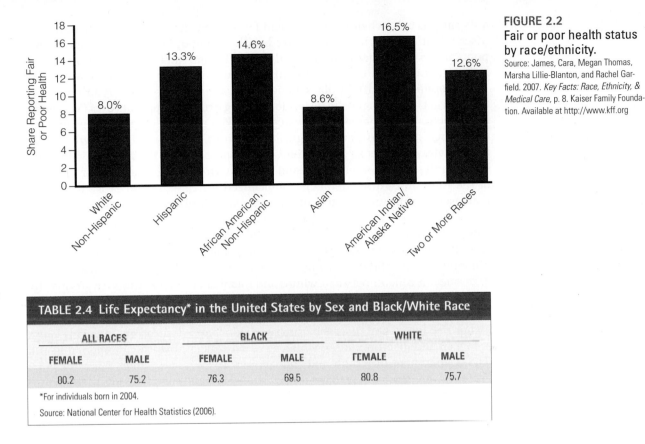

FIGURE 2.2
Fair or poor health status by race/ethnicity.
Source: James, Cara, Megan Thomas, Marsha Lillie-Blanton, and Rachel Garfield. 2007. *Key Facts: Race, Ethnicity, & Medical Care*, p. 8. Kaiser Family Foundation. Available at http://www.kff.org

TABLE 2.4 Life Expectancy* in the United States by Sex and Black/White Race

ALL RACES		BLACK		WHITE	
FEMALE	MALE	FEMALE	MALE	FEMALE	MALE
80.2	75.2	76.3	69.5	80.8	75.7

*For individuals born in 2004.

Source: National Center for Health Statistics (2006).

highest rate of obesity (James et al. 2007). U.S. blacks also have the highest rate of infant mortality of all racial and ethnic groups, largely because of higher rates of prematurity and low birth weight.

Compared with white Americans, Native Americans have higher death rates from motor vehicle injuries, diabetes, and cirrhosis of the liver (caused by alcoholism). Compared with non-Hispanic whites, Hispanics have more diabetes, high blood pressure, and lung cancer and have a higher risk of dying from violence, alcoholism, and drug use (Weitz 2006). Asian Americans typically have high levels of health, in large part due to the fact that they have the highest levels of income and education of any racial or ethnic U.S. minority group. Traditional Asian diets, which include lots of fish and vegetables, may also account for their higher levels of health.

Socioeconomic differences between racial and ethnic groups are largely responsible for racial and ethnic differences in health status (Cockerham 2004; Weitz 2006). One effect of the lower socioeconomic status of U.S. minorities is that they are less likely to have health insurance. Hispanic individuals are the most likely to be uninsured (32.7 percent), followed by American Indian/Alaska Natives (31.4 percent), Native Hawaiian/Other Pacific Islanders (21.7 percent), blacks (19.4 percent), Asians (16.1 percent), and non-Hispanic whites (10.7 percent) (DeNavas-Walt, Proctor, & Smith 2007). As we note elsewhere, compared with the insured, the uninsured are less likely to get timely and routine care and are more likely to be hospitalized for preventable conditions.

The poorer health of minorities is also related to the fact that minorities are more likely than whites to live in environments where they are exposed to haz-

> "Of all the forms of inequality, injustice in health care is the most shocking and inhumane."
>
> Martin Luther King Jr.
> Civil rights leader

ards such as toxic chemicals, dust, and fumes (see also Chapter 14). In addition, discrimination contributes to poorer health among oppressed racial and ethnic populations by restricting access to the quantity and quality of public education, housing, and health care. For example, a study of heart attack patients at 658 U.S. hospitals found that black patients were much less likely than white patients to get basic diagnostic tests, clot-busting drugs, or angioplasties (Vaccarino et al. 2005). In another study researchers looked at racial disparities in the rates of undergoing nine different surgical procedures. In the early 1990s whites had higher rates than blacks for all nine surgical procedures. By 2001 the difference between the rates among whites and blacks narrowed significantly for only one of the nine procedures, remained unchanged for three of the procedures, and increased significantly for five of the nine procedures (Jha et al. 2005).

Race, Ethnicity, and Mental Health. Medical sociologist William Cockerham (2005) reported that "almost all of the data and research that currently record differences in mental disorder between races show there is little or no significant difference in general between whites and members of racial minority groups" (p. 194). Differences that do exist are often associated more with social class than with race or ethnicity. However, some studies suggest that minorities have a higher risk for mental disorders, such as anxiety and depression, in part because of racism and discrimination, which adversely affect physical and mental health (U.S. Department of Health and Human Services 2001). In 2004, Hispanics (3.9 percent) were more likely to have experienced serious psychological distress during the past 30 days than blacks (3.3 percent) and whites (2.9 percent) (National Center for Health Statistics 2005). Minorities also have less access to mental health services, are less likely to receive needed mental health services, often receive lower quality mental health care, and are underrepresented in mental health research (U.S. Department of Health and Human Services 2001).

Family and Household Factors

Family and household factors are related to both physical and mental health. A study of adults in their 50s found that married people who live only with their spouse or with spouse and children had the best physical and mental health, whereas single women living with children had the lowest measures of health (Hughes & Waite 2002). Other research findings concur that married adults are healthier and have lower levels of depression and anxiety compared with adults who are single, divorced, cohabiting, or widowed (Mirowsky & Ross 2003; Schoenborn 2004). Two explanations for the association between marital status and health are the *selection* and the *causation* theories (also discussed earlier in explaining the link between poverty and health). In this context, the selection theory suggests that healthy individuals are more likely to marry and to stay married. The causation theory says that better health among married individuals results from the economic advantages of marriage and from the emotional support provided by most marriages—the sense of being cared about, loved, and valued (Mirowsky & Ross 2003).

For children, living in a two-parent household is associated with better health outcomes. A Swedish study found that children living with only one parent have a higher risk of death, mental illness, and injury than those in two-parent families, even when their socioeconomic disadvantage is taken into account (Hollander 2003).

PROBLEMS IN U.S. HEALTH CARE

The United States boasts of having the best physicians, hospitals, and advanced medical technology in the world, yet problems in U.S. health care remain a major concern on the national agenda. The World Health Organization's first-ever analysis of the world's health systems found that, although the United States spends a higher portion of its gross domestic product on health care than any other country, it ranks 37 out of 191 countries according to its performance (World Health Organization 2000). The report concluded that France provides the best overall health care among major countries, followed by Italy, Spain, Oman, Austria, and Japan. A more recent comparison of health care in six countries—Australia, Canada, Germany, New Zealand, the United Kingdom, and the United States—found that the U.S. ranks last on dimensions of access, patient safety, efficiency, and equity (Davis et al. 2007).

After presenting a brief overview of U.S. health care, we address some of the major health care problems in the United States—inadequate health insurance coverage, the high cost of medical care and insurance, the managed care crisis, and inadequate mental health care.

U.S. Health Care: An Overview

In the United States there is no one health care system; rather, health care is offered through various private and public means. In 2005, 27 percent of Americans were covered by government health insurance plans (Medicare, Medicaid, and military insurance) and 68 percent were covered by private insurance, most often employment-based (DeNavas-Walt et al. 2007) (see Figure 2.3).

In traditional health insurance plans the insured choose their health care provider, who is reimbursed by the insurance company on a fee-for-service basis. The insured individual typically must pay an out-of-pocket "deductible" (usually ranging from a few hundred to a thousand dollars or more per year per person for a "high deductible" plan) and then is often required to pay a percentage of medical expenses (e.g., 20 percent) until a maximum out-of-pocket expense amount is reached (after which insurance will cover 100 percent of medical costs up to a limit).

Health maintenance organizations (HMOs) are prepaid group plans in which a person pays a monthly premium for comprehensive health care services. HMOs attempt to minimize hospitalization costs by emphasizing preventive health care. *Preferred provider organizations (PPOs)* are health care organizations in which employers who purchase group health insurance agree to send their employees to certain health care providers or hospitals in return for cost discounts. In this arrangement health care providers obtain more patients but charge lower fees to buyers of group insurance.

Managed care refers to any medical insurance plan that controls costs through monitoring and controlling the decisions of health care providers. In many plans doctors must call a utilization review office to receive approval before they can hospitalize a patient, perform surgery, or order an expensive diagnostic test. Although the terms *HMO* and *managed care* are often used interchangeably, HMOs are only one form of managed care. Most Americans who have private insurance belong to some form of managed care plan. Recipients of Medicaid and Medicare may also belong to a managed care plan.

managed care Any medical insurance plan that controls costs through monitoring and controlling the decisions of health care providers.

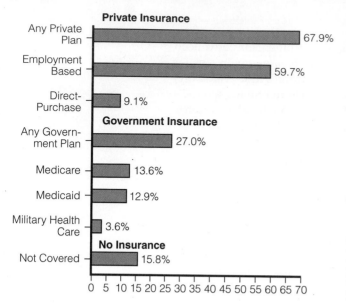

Private Insurance

- Any Private Plan — 67.9%
- Employment Based — 59.7%
- Direct-Purchase — 9.1%

Government Insurance

- Any Government Plan — 27.0%
- Medicare — 13.6%
- Medicaid — 12.9%
- Military Health Care — 3.6%

No Insurance

- Not Covered — 15.8%

0 5 10 15 20 25 30 35 40 45 50 55 60 65 70

FIGURE 2.3

Coverage by type of health insurance: 2006.

Source: DeNavas-Walt, Carmen, Bernadette D. Proctor, and Jessica Smith. 2007. *Income, Poverty, and Health Insurance Coverage in the United States: 2006*, p. 20. U.S. Census Bureau.

Medicare A federally funded program that provides health insurance benefits to the elderly, disabled, and those with advanced kidney disease.

Medicaid A public health insurance program, jointly funded by the federal and state governments, that provides health insurance coverage for the poor who meet eligibility requirements.

State Children's Health Insurance Program (SCHIP) A public health insurance program, jointly funded by the federal and state governments, that provides health insurance coverage for children whose families meet income eligibility standards.

Medicare. Medicare is funded by the federal government and reimburses the elderly and the disabled for their health care (see also Chapter 12). Individuals contribute payroll taxes to Medicare throughout their working lives and generally become eligible for Medicare when they reach 65, regardless of their income or health status. Medicare consists of four separate programs: Part A is hospital insurance for inpatient care, which is free, but enrollees may pay a deductible and a copayment, and coverage of home health nursing and hospice care is limited. Part B is a supplementary medical insurance program, which helps pay for physician, outpatient, and other services. Part B is voluntary and is not free; enrollees must pay a monthly premium as well as a copayment for services. Medicare does not cover long-term nursing home care, dental care, eyeglasses, and other types of services, which is why many individuals who receive Medicare also enroll in Part C, which allows beneficiaries to purchase private supplementary insurance that receives payments from Medicare. Part D is an outpatient drug benefit that is voluntary and requires enrollees to pay a monthly premium, meet an annual deductible, and pay coinsurance for their prescriptions. Critics of the Medicare prescription drug benefit argue that the drug coverage is inadequate and complicated and fails to lower the cost of prescription drugs. The Medicare prescription drug legislation provides billions of dollars in subsidies to HMOs and other managed care plans, paying them much more than it costs regular Medicare to provide the same services. These private plans can elect to cover a limited number of drugs and deny coverage for other drugs.

Medicaid and SCHIP. Medicaid, which provides health care coverage for the poor, is jointly funded by the federal and state governments (see also Chapter 6). Contrary to the belief that Medicaid covers all poor people, it does not. Eligibility rules and benefits vary from state to state, and in many states Medicaid provides health care only for the very poor who are well below the federal poverty level. In 1997 the **State Children's Health Insurance Program (SCHIP)** was created to expand health coverage to uninsured children, many of whom come from families with incomes too high to qualify for Medicaid but too low to afford private health insurance. Under this initiative states receive matching federal funds to provide medical insurance to uninsured children.

Although all poor children are eligible for Medicaid, many of their parents are not. States cannot receive matching federal funds to provide Medicaid to adults under age 65 without children, unless they are pregnant or disabled. As a result, more than 40 percent of low-income adults without children are uninsured. And not all children who are eligible for Medicaid are enrolled. In 2006 nearly one in five (19 percent) U.S. children living in poverty were not covered by health insurance (DeNavas-Walt et al. 2007). It is estimated that nearly three-quarters of uninsured children are eligible for Medicaid or SCHIP but are not enrolled. A recently enacted federal requirement that citizens supply documents to prove their citizenship and identity is likely to impede Medicaid enrollment (Kaiser Commission on Medicaid and the Uninsured 2007).

Workers' Compensation. **Workers' compensation** (also known as *workers' comp*) is an insurance program that provides medical and living expenses for people with work-related injuries or illnesses. Employers pay a certain amount into their state's workers' compensation insurance pool, and workers injured on the job can apply to that pool for medical expenses and for compensation for work days lost. In exchange for that benefit, workers cannot sue their employers for damages. However, not all employers acquire workers' compensation insurance, even in states where it is legally required. Further, many employees with work-related illness or injuries do not apply for workers' compensation benefits because (1) they fear getting fired for making a claim, (2) they are not aware that they are covered by workers' comp, and/or (3) the employer offers incentives (i.e., bonuses) to employees when no workers' comp claims are filed in a given period of time (Sered & Fernandopulle 2005). Even when employees file a workers' comp claim, the coverage the employee receives rarely covers the cost of the employee's injury or illness. "The typical scenario . . . is one in which the workers' comp insurance company delays accepting and paying the worker's claim. In the meantime, the injured employee racks up medical and other bills and then, in desperation, accepts a lump sum monetary settlement that does not come close to covering medical expenses or replacing lost wages" (Sered & Fernandopulle 2005, p. 94).

Military Health Care. Military health care includes Comprehensive Health and Medical Plan for Uniformed Services (CHAMPUS), Civilian Health and Medical Program of the Department of Veterans Affairs (CHAMPVA), and care provided by the Department of Defense and the Department of Veterans Affairs. It is commonly believed that one of the benefits of serving in the military is health care coverage. But 1.8 million, or one in eight (12.7 percent) nonelderly veterans lacked health coverage in 2004 (Lee 2007). And a series of *Washington Post* reports in 2007 brought attention to the abysmal conditions at Walter Reed Army Medical Center and military medical facilities and Veterans Administration (VA) hospitals around the country. Walter Reed's Building 18 was found to have mice, mold, and rot. Soldiers and veterans around the country reported that their home post medical treatment facility was characterized by "indifferent, untrained staff; lost paperwork; medical appointments that drop from computers; and long waits for consultations" (Hull & Priest 2007). Other reports of military medical facilities described peeling paint, asbestos, overflowing trash, fruit fly infestations, no nurses, and lack of blankets and linens.

Many veterans have complained that they have not received the benefits they deserve or that they have had long waits to get benefits. The VA has a backlog of 400,000 benefit claims. One veteran Marine from the Vietnam era said it took him 20 years to get the medical benefits he was entitled to (Hull & Priest 2007).

Mental health care for military personnel and military veterans is also inadequate. In a report by the American Psychological Association, more than 3 of 10 soldiers met the criteria for a "mental disorder," but far less than half of those in need (20–40 percent) sought help, either because of the stigma of having men-

workers' compensation Also known as workers' comp, an insurance program that provides medical and living expenses for people with work-related injuries or illnesses.

tal health problems or because help was not available. About 40 percent of active duty psychologist slots in the military are vacant. There are also not enough therapists to help families of those deployed and soldiers returning home from active duty. In addition, "Vietnam vets whose post-traumatic stress has been triggered by images of war in Iraq are flooding the system for help and are being turned away" (Hull & Priest 2007).

Inadequate Health Insurance Coverage

In 2006, most Americans—84.2 percent—had either private or government insurance, but that leaves 15.8 percent of the U.S. population—47 million Americans—lacking health insurance (DeNavas-Walt et al. 2007). This number is expected to grow to 56 million by 2013 (Gilmore & Kronick 2005). In a national poll, most Americans (95 percent) indicated that the rate of uninsurance in the United States is a serious problem (Toner & Elder 2007).

Disparities in Health Insurance Coverage. Whites are more likely than racial and ethnic minorities to have health insurance. As noted earlier, Hispanics have the largest percentage of uninsured with one-third (32.7 percent) of Hispanics lacking insurance, followed by American Indians/Alaska Natives (31.4 percent), Native Hawaiians and other Pacific Islanders (21.7 percent), blacks (19.4 percent), and Asians (16.1 percent) (DeNavas-Walt et al. 2007). In contrast, only 10.7 percent of non-Hispanic whites lack insurance (based on a 3-year average from 2004–2006).

Of all age groups, young adults ages 18–24 are the least likely to have health insurance. In 2006 nearly one in three young adults (29.3 percent) was uninsured (DeNavas-Walt et al. 2007).

What Do You Think? Many college students do not have health insurance, and each year hundreds of college students withdraw from school because of their inability to pay medical bills from accidents or unexpected illnesses. Some universities require students to have health insurance. Although mandatory health insurance may keep students from dropping out, it also adds to college bills and may prevent some individuals who cannot afford health insurance from enrolling in college. Do you think that universities should require students to have health insurance? Why or why not?

Health insurance status also varies by income and employment. The higher an individual's income, the more likely it is that the individual will have health insurance. And employed individuals are more likely than unemployed individuals to be insured. However, employment is no guarantee of health care coverage; in 2006, nearly 18 percent of full-time workers were uninsured (DeNavas et al. 2007). More than two-thirds of the uninsured live in a household with one full-time worker, and more than one-third of the uninsured in the United States have annual family incomes of more than $40,000 (Zeldin & Rukavina 2007). Not all businesses offer health benefits to their employees; in 2006, 61 percent of businesses offered health benefits to at least some of their employees—down

The 2007 release of *SiCKO,* Michael Moore's documentary on the U.S. health care industry, increased public awareness of the problems in the U.S. health care system.

from 69 percent in 2000 (Claxton et al. 2006). And even when employers offer health insurance, some employees are not eligible for health benefits because of waiting periods or part-time status. Some employees who are eligible may not enroll in employer-provided health insurance because they cannot afford to pay their share of the premiums.

Inadequate Insurance for the Poor. Many Americans believe that Medicaid and SCHIP—public health insurance programs for the poor—cover all low-income children, adults, and families. But Medicaid eligibility levels are set so low that many low-income adults are not eligible. Some states have waiting lists for Medicaid. Another problem is that because of the low reimbursement payments from Medicaid, many health care providers do not accept Medicaid patients. Although the number of uninsured children has fallen since SCHIP began in 1997, 30 percent of eligible children are not enrolled and remain uninsured (Urban Institute 2007).

Consequences of Inadequate Health Insurance. An estimated 18,000 deaths per year in the United States are attributable to lack of health insurance (Institute of Medicine 2004). Individuals who lack health insurance are less likely to receive preventive care, are more likely to be hospitalized for avoidable health problems, and are more likely have disease diagnosed in the late stages (Kaiser Commission on Medicaid and the Uninsured 2004). In a study of individuals who experienced an unintentional injury or a new chronic health problem, uninsured individuals reported receiving less medical care and poorer short-term changes in health than those with insurance (Hadley 2007).

In a national U.S. poll, 6 in 10 adults without insurance said that someone in their household went without medical care because of cost (Toner & Elder 2007). In the same poll, one-quarter of those with insurance said that someone in their household had gone without a medical test or treatment because insur-

FIGURE 2.4
Barriers to health care by
insurance status, 2003.*
Source: Kaiser Family Foundation (2005b).
*Experienced by respondent or member of
his or her family.

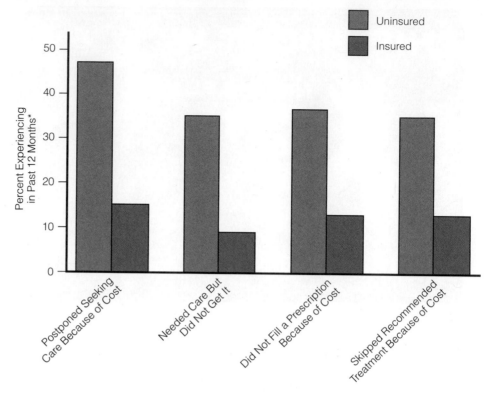

ance would not pay for it. As shown in Figure 2.4, individuals without insurance experience more barriers to health care than individuals with insurance.

Because most health care providers do not accept patients who do not have insurance, many uninsured individuals resort to using the local hospital emergency room. In a study of young adults with chronic health conditions, loss of health insurance resulted in decreased use of office-based physician services and a dramatic increase in visits to hospital emergency rooms (Scal & Town 2007). A federal law called the Emergency Medical Treatment and Active Labor Act requires hospitals to assess all patients who come to their emergency rooms to determine whether an emergency medical condition exists and, if it does, to stabilize the patient before transferring him or her to another facility. Uninsured hospital patients are almost always billed at a much higher cost than the prices negotiated by insurance companies.

Individuals who lack dental insurance commonly have untreated dental problems, which can lead to or exacerbate other health problems.

> Because they affect the ability to chew, untreated dental problems tend to exacerbate conditions such as diabetes or heart disease. . . . Missing and rotten teeth make it painful if not impossible to chew fruits, whole-grain foods, salads, or many of the fiber-rich foods recommended by doctors and nutrition experts. (Sered & Fernandopulle 2005, pp. 166–167)

In their book *Uninsured in America,* Sered and Fernandopulle (2005) described one interviewee who "covered her mouth with her hand during our entire interview because she was embarrassed about her rotting teeth" and another interviewee "used his pliers to yank out decayed and aching teeth"

(p. 166). The authors note that "almost every time we asked interviewees what their first priority would be if the president established universal health coverage tomorrow, the immediate answer was 'my teeth'" (p. 166).

The High Cost of Health Care

The United States spends more than twice as much per person for health care as other wealthy countries (Kaiser Family Foundation 2007a). Health care spending in the United States rose from $356 per person in 1970 to $6,697 in 2005 and is expected to rise to $12,320 by 2015. U.S. health care spending as a share of gross domestic product (GDP) grew from 7.2 percent in 1970 to 16.0 percent in 2005 and is expected to reach 20 percent of GDP by 2015 (Centers for Medicare & Medicaid Services 2007). Yet virtually every other wealthy nation has better health outcomes, as measured by life expectancy and infant mortality. Why does the United States spend so much on health care? And how do the high costs of health care affect individuals and families?

Several factors have contributed to escalating medical costs. These include increased longevity; excessive and inappropriate medical care; and the high costs of health care administration, drugs, doctors' fees, hospital services, medical technology, and health insurance.

Increased Longevity. Because of improved sanitation and medical advances, people are living longer today than in previous generations. People older than age 65 use medical services more than younger individuals and are also more likely to take prescription medicine on a daily basis (Kaiser Family Foundation 2005a). The average health care expense for the elderly U.S. population was $11,089 in 2002, compared with $3,352 per year for nonelderly adults (ages 19–64) (Stanton 2006).

People are not only living longer, they are also spending more of their lives with chronic diseases. A century ago, the average adult in Western nations spent only 1 percent of his or her life in illness, but today the average adult spends more than 10 percent of his or her life sick (Robbins 2006). Today people survive illnesses, conditions, and injuries that would have killed them a generation ago. Infants born prematurely who would not have survived a generation ago are kept alive today in hospital incubators. Individuals with HIV/AIDS are living longer today owing to the availability of new (and expensive) drugs. Individuals with kidney disease are receiving dialysis treatment and kidney transplants. And persons with heart disease are undergoing bypass surgery and other treatments that are extending their lives and their medical expenses.

Cost of Hospital Services, Doctors' Fees, and Medical Technology. High hospital costs and doctors' fees are factors in the rising costs of health care. The average visit to the emergency room costs a little more than $1,000 (Ehrenreich 2005). The use of expensive medical technology, unavailable just decades ago, also contributes to high medical bills. Consider the advancements in the treatment of preterm babies, for which very little could be done in 1950. By 1990, special ventilators, artificial pulmonary surfactant to help infant lungs develop, neonatal intensive care, and steroids for mother and/or baby became standard treatment for preterm babies in the United States (Kaiser Family Foundation

Many Americans travel to Mexico, Thailand, India, and other foreign countries to receive medical care at a lower price than what they would pay in the United States. This brochure advertises a service that not only includes affordable cosmetic surgery but also travel arrangements, resort and hotel reservations for post-hospital recuperation, and sightseeing tours.

2007b). Such technology greatly improved the life expectancy of preterm babies but also adds to health care costs.

Cost of Drugs. The high cost of drugs also contributes to health care costs. The United States pays 81 percent more for patented brand-name prescription drugs than Canada and six Western European nations (Sager & Socolar 2004). The high prices that Americans pay for prescription drugs partly explain why the pharmaceutical industry is among the most profitable industries in the United States.

Manufacturers argue that in other countries, where governments regulate prices, consumers pay too little for drugs. U.S. drug prices are high, claim drug-makers, because of the high cost of researching and developing new drugs. But most large drug companies pay substantially more for marketing, advertising, and administration than for research and development (Families USA 2007). A simple answer to why U.S. drug companies charge high prices is: because they can!

Cost of Health Insurance. From 2000–2006, health insurance premiums grew by 87 percent, far outpacing inflation (18 percent) and wage growth (20 percent). In 2006 the average annual premiums for employer-sponsored coverage were $4,242 for an individual and $11,480 for a family, with workers contributing, on average, $627 annually toward the cost of individual coverage and $2,973 toward family coverage (Claxton et al. 2006).

With the rising cost of medical insurance, companies are increasing the employees' share of the cost, decreasing the benefits, or not providing insurance at all. The cost of health care to businesses also affects the prices that consumers pay for goods and services. For example, in 2007, the cost of health benefits for employees at General Motors accounted for $1,783 of the sticker price of each new vehicle (Specter & Stoll 2007).

Cost of Health Care Administration. Health care administrative expenses are higher in the United States than in any other nation. According to the World Health Organization, 15 percent of the money paid to private health insurance companies for premiums goes to administrative expenses, compared with only 4 percent of the budgets of public insurance companies (i.e., Medicare and Medicaid) (reported by Krugman 2005). Insurance companies and for-profit HMOs spend between 20 percent and 30 percent of their budgets to cover the costs of stockholder dividends, lobbyists, huge executive salaries, marketing, and wasteful paperwork (Conyers 2003).

Consequences of the High Cost of Health Care for Individuals and Families. When 12-year-old Candice Jackson was hit by an uninsured driver while she was getting off a school bus in Windthorst, Texas, she spent 4 months in the hospital and incurred about $90,000 in uncovered medical bills. Her mother needed two knee replacements, adding another $20,000 to the family's medical bills. A few years later at age 16, Candice swerved off the road to avoid hitting a deer and sustained a head injury that required brain surgery. In anticipation of the medical bills from Candice's latest mishap, Candice's dad, Lanny Jackson, who works in the service center of a car dealership, feels forced to file for bankruptcy (Springen 2006).

Lanny Jackson is not alone. One study found that medical bills were a contributing factor in about half of all U.S. bankruptcies (Himmelstein et al. 2005). Ironically, many individuals who cannot pay their medical bills also cannot afford to file for bankruptcy, which can cost (in filing fees and attorney fees) more than $1,000. Some individuals and families cope with medical debt by taking out home equity loans, cashing out retirement accounts, and using credit cards.

Many individuals forgo needed medicine and/or medical care when they cannot afford to pay for it. For example, in a national sample of U.S. adults more than one-third said that in the past year they did not fill a prescription because of the cost, cut pills in half, or skipped doses to make a medication last (Kaiser Family Foundation 2005a). Forgoing medicine or medical care often exacerbates the medical condition, leading to even higher medical costs, or tragically, leading to death (see this chapter's *The Human Side* feature).

A more drastic measure involves breaking the law to receive free medical care in prison. This was the case for Larry Causey, age 57, who called the FBI and told them he was going to rob the post office in West Monroe, Louisiana. Then he went to the post office and handed a note to a teller demanding money. He left empty-handed and sat in his car until officers arrested him. Larry had no intention of committing robbery. Larry had cancer and could not afford cancer treatment, so he staged a robbery to get arrested and be put in jail, where he would receive medical treatment for his cancer ("Access to Free Health Care" 2001). Larry Causey's story is not as uncommon as we may think. "Sheriffs nationwide say they're also arresting people willing to trade their freedom for a free visit to the doctor" ("Access to Free Health Care" 2001).

Having insurance does not guarantee that one is protected against financial devastation resulting from illness or injury.

> These days, more and more families who think they are covered are discovering that the blanket is short. They may not be uninsured, but they are underinsured. Thanks to rising deductibles, hefty copayments, and caps on total reimbursements, some patients diagnosed with a chronic illness are finding that they cannot afford to fill all of the prescriptions that their doctor gives them. Others realize—too late—that when ICU care for a child with cancer costs $49,000 a day, a family can exceed a million-dollar cap on coverage in a matter of months. (Mahar 2006, p. xiv)

According to Families USA (2005), "the middle class, those with college degrees, decent jobs, health insurance—the group of people who feel secure and well-protected—are at high, and often highest, risk of being left penniless when serious illness hits" (p. 1). In a 2005 study that found that half of U.S. bankruptcies are attributable, at least in part, to medical bills, most of those bankrupted (75 percent) were insured when they first became sick or injured (Himmelstein et al. 2005). One cause of the high rate of medical bankruptcy among the insured is the high cost of copayments, deductibles, and exclusions. In addition, the link between coverage and employment means that insurance is often lost when it is needed the most—when workers lose their jobs because of medical problems.

The following narrative is written by cardiologist Tim Garson (2007), who helped save the life of a child, only to lose her as a young adult because she had no health insurance to pay for her heart medication.

I think of her often—almost every day. We met when she was 5 years old. She had just come out of surgery after a long operation on her heart to change her color. Born a blue baby, she was now a normal pink.

It was my first night on call as a pediatric cardiology fellow. My job was to get her through the first night after surgery. She was fine for the first 3 hours. Suddenly, the dreaded "Dr. Garson, stat, ICU, bed 4." It was a cardiac arrest—just like on television's ER—nurses, doctors, chest compressions. Someone shouted "clear," and we did. The shock through the paddles caused her to jump—flat line. Then a beep came from her monitor and another and another. She was back. That night, her heart stopped three times, and three times science triumphed. By morning she was stable.

Her name was Ginny. The day she was discharged, we traded home phone numbers. I saw her monthly for a while and, as she improved, every 6 months in the clinic. We traded 10 sets of birthday and holiday cards. She did beautifully until her checkup at age 16, when we discovered a heart rhythm problem. I knew how devastating this problem could be, even leading to death. We tried a number of drugs and eventually found exactly what we needed; a once-a-day medication that was easy for Ginny to remember to take, although, unfortunately, it was very expensive.

Ginny did beautifully and was beaming when she introduced me to her friends at her high school graduation. She decided she wanted to work for a couple of years before going to college to help her parents with her medical bills (they both worked in a small grocery), but after she went through six job interviews, all with small businesses (no large employers in her town), no one would hire her. She told all of them that she had congenital heart disease. She told me that she was proud of how well she had done and wanted others to know. I couldn't tell her to lie about her heart disease, but she didn't need to volunteer the information. By the time she called to tell me, it was too late; there were no other jobs she could apply for in town, and she had no car to travel elsewhere.

If a person has a known high risk, such as Ginny, many employers will not employ him or her for fear of the effect on their insurance rate, despite the fact that this practice is illegal under the Americans with Disabilities Act. The more likely scenario is that the small business would not offer health insurance at all: In this country, fewer than half of small businesses with three to nine employees offer health insurance to their employees, compared with more than 90 percent of those with more than 200 employees. More than 70 percent of uninsured Americans are in families with at least one full-time worker.

One Sunday I received an early morning call. Sobbing . . . then "Tim, we've lost her!" Ginny's mom found her dead in bed. It didn't take long to figure out what had happened. Ginny's pill bottle was empty.

The last refill was 5 months before—just before her 19th birthday, when her Medicaid eligibility ended and her prescription drug coverage stopped. She had no job, and we guessed . . . that she was trying to save the family money.

I think of Ginny often. Almost every day. I never could understand how the "system" that had paid to fix her heart, and paid for her medicine, dropped her at 19. But that's the way it works. Medicaid and the State Children's Health Insurance Program (SCHIP) covers children of the poor, like Ginny, but between the ages of 19 and the Medicare age of 65, the so-called safety net has huge holes—and Ginny fell through.

The day of Ginny's funeral, I made the uninsured my personal issue and became actively involved in working toward health care reform. Efforts at reform are under way in states, and experimentation in both public programs and private insurance strategies (that would help small businesses afford insurance for people like Ginny) should be encouraged. The federal government can help by providing support such as that proposed in the yet-unpassed Health Partnership Act. In any of these public approaches, it is helpful to remember that in the United States, the uninsured are 46 million individuals, and each has a story to tell.

I think of Ginny often. Almost every day.

Adapted from Garson, Arthur "Tim". 2007. "Heart of the Uninsured." *Health Affairs* 26(1):227–231. Arthur "Tim" Garson Jr. is dean of the School of Medicine and vice president of the University of Virginia in Charlottesville, Virginia. The name of the patient in the story has been changed to protect privacy.

Although the COBRA law allows people to continue their insurance coverage when they lose a job, the premiums for continued coverage may be unaffordable (often $10,000 a year or more) (PNHP 2005).

It is no surprise that when a 2006 Gallup poll asked Americans to name the "most important financial problem facing your family today" the most frequently cited answer, mentioned by 17 percent of Americans, was health care costs (Carroll 2006). And when a national sample of U.S. adults was asked to identify the most important problem in health or health care for the government to address, the most frequent response was the high cost of health care (Kaiser Family Foundation 2005a).

The Managed Care Crisis

In an attempt to control medical care costs, in the last few decades, the U.S. health care system has seen a dramatic rise in managed care. But Americans are concerned about the reduced quality of health care resulting from the emphasis on cost containment in managed care. Surveys found that more people have said that managed care plans do a "bad job" than a "good job" in serving customers (Kaiser Health Poll Report 2004). In a survey of physicians' views on the effects of managed care, the majority responded that managed care has negative effects on the quality of patient care because of limitations on diagnostic tests, length of hospital stay, and choice of specialists (Feldman et al. 1998). One former director of an HMO described the situation:

> I've seen from the inside how managed care works. I've been pressured to deny care, even when it was necessary. I have seen the bonus checks given to nurses and doctors for their denials. I have seen the medical policies that keep patients from getting care they need . . . and the inadequate appeal procedures. (Peeno 2000, p. 20)

Inadequate Mental Health Care

Since the 1960s, U.S. mental health policy has focused on reducing costly and often neglectful institutional care and on providing more humane services in the community. This movement, known as **deinstitutionalization,** had good intentions but has largely failed to live up to its promises. Only one in five U.S. children with mental illness is identified and receives treatment, and fewer than half of adults with a serious mental illness received treatment or counseling for a mental health problem during the past year (Gruttadaro 2005; Substance Abuse and Mental Health Services Administration 2003). Reasons for not seeking treatment include the stigma associated with mental illness (and thus with treatment for mental illness), fear and mistrust of treatment, cost of care (which is often not covered by health insurance), and lack of access to services.

Mental health services are often inaccessible, especially in rural areas. In most states services are available from "9 to 5"; the system is "closed" in the evenings and on weekends when many people with mental illness experience the greatest need. Across the nation people with severe mental illness end up in jails and prisons, homeless shelters, and hospital emergency rooms. Many children with untreated mental disorders drop out of school or end up in foster care or the juvenile justice system. As many as 70 percent of youths incarcerated in juvenile justice facilities have mental disorders (Honberg 2005). In a survey of 367 colleges and universities in the United States and Canada, most

deinstitutionalization The removal of individuals with psychiatric disorders from mental hospitals and large residential institutions to outpatient community mental health centers.

© Steven Lunetta Photography 2007

More than half of students who sought counseling at their college or university indicated on evaluation forms that counseling helped them to remain in school and 58 percent claim it helped to improve their academic performance (Gallagher 2007). Yet, only 58 percent of colleges/universities in the United States and Canada offer psychiatric services on campus.

(92 percent) counseling center directors believe that the number of college students with severe psychological problems has increased in recent years. Yet only 58 percent of colleges and universities offer psychiatric services on campus (Gallagher 2006).

Given the increasing growth of minority populations, another deficit in the mental health system is the inadequate number of mental health clinicians who speak the client's language and who are aware of cultural norms and values of minority populations (U.S. Department of Health and Human Services 2001).

The mental health system is also plagued by inadequate federal and state funding of public mental health centers, which results in rationing care to those most in need. Thus people must "hit bottom" before they can receive services.

What Do You Think? In 2004, Jordan Nott, a former student at George Washington University, was suspended and barred from campus after hospitalizing himself for depression and suicidal thoughts. The university charged Nott with violating its code of conduct by engaging in "endangering behavior" (Capriccioso 2006). In 2007, Virginia passed a bill to prevent public colleges and universities from dismissing students for attempting suicide or seeking mental health treatment for suicidal thoughts or behaviors (Sampson 2007). Do you think that suicidal students might avoid seeking mental health treatment if that puts them at risk for being dismissed from school?

STRATEGIES FOR ACTION: IMPROVING HEALTH AND HEALTH CARE

selective primary health care An approach to health care that focuses on using specific interventions to target specific health problems.

comprehensive primary health care An approach to health care that focuses on the broader social determinants of health, such as poverty and economic inequality, gender inequality, environment, and community development.

There are two broad approaches to improving the health of populations: selective primary health care and comprehensive primary health care (Sanders & Chopra 2003). **Selective primary health care** focuses on using specific interventions to target specific health problems, such as promoting condom use to prevent HIV infections and providing immunizations against childhood diseases to promote child survival. In contrast, **comprehensive primary health care** focuses on the broader social determinants of health, such as poverty and economic inequality, gender inequality, environment, and community development. Targeting specific health problems may be necessary, but not sufficient, for achieving long-term health gains. Sanders and Chopra (2003) emphasize that "only where health interventions are embedded within a comprehensive health care approach, including attention to social equity, health systems and human capacity development, can real and sustainable improvements in health status be seen" (p. 108).

As you read the following sections on improving maternal and infant health, preventing and alleviating HIV/AIDS, and fighting obesity, see whether you can identify which strategies represent selective primary health care approaches and which strategies are comprehensive. Also, be mindful that strategies to alleviate social problems discussed in subsequent chapters of this textbook are also important elements to a comprehensive primary health care approach.

Improving Maternal and Infant Health

As discussed earlier, pregnancy and childbirth are major causes of mortality and morbidity among women of reproductive age in the developing world. Access to family planning services, skilled birth attendants, affordable methods of contraception, and safe abortion services are important determinants of the well-being of mothers and their children (Save the Children 2002). Family planning reduces maternal mortality by reducing the number of unintended pregnancies and by enabling women to space births 2–3 years apart, which decreases infant mortality significantly (Murphy 2003). Since 1960 contraception use among married couples in developing countries has increased from 10–15 percent to 60 percent (United Nations Population Fund 2004), but there are still millions of women who do not have access to contraception and half of the world's pregnant women have no access to skilled care at childbirth. Although most reproductive health programs focus exclusively on women, some programs also reach out to men with services and education that enable them to share in the responsibility for reproductive health.

In many developing countries women's lack of power and status means that they have little control over health-related decisions (UNICEF 2006). Men make the decisions about whether or when their wives (or partners) will have sexual relations, use contraception, or use health services. Thus improving the status and power of women is an important strategy in improving their health. Promoting women's education increases the status and power of women to control their reproductive lives, exposes women to information about health issues, and also delays marriage and childbearing.

Improving maternal and infant health globally requires funding, but the cost is not prohibitive. The price of providing basic health services for mothers and infants in low-income countries is only $3 per person (Oxfam GB 2004). The question is, do the rich countries of the world have the political will to support efforts to protect the health and lives of women and infants in the developing world?

HIV/AIDS Prevention and Alleviation Strategies

As of this writing, there is no vaccine to prevent HIV infection. As researchers continue to work on developing such a vaccine, a number of other strategies are available to help prevent and treat HIV/AIDS.

HIV/AIDS Education and Access to Condoms. HIV/AIDS prevention efforts include educating populations about how HIV is transmitted and how to protect against HIV transmission and providing access to condoms as a means of preventing HIV transmission. Many people throughout the world remain uninformed or misinformed about HIV/AIDS. At least 30 percent of young people in a survey of 22 countries had never heard of AIDS, and in 17 countries surveyed, more than half of adolescents could not name a single method of protecting themselves against HIV infection (United Nations Population Fund 2002). A survey of U.S. adults found that significant numbers say that they do not know how HIV is transmitted or mistakenly believe that it is possible to transmit HIV through kissing (38 percent), sharing a drinking glass (18 percent), and touching a toilet seat (18 percent) (Kaiser Family Foundation 2004b). This same survey also found that 12 percent of U.S. adults did not know that there are drugs that can lengthen the lives of those with HIV infection and more than half (57 percent) did not know that a pregnant woman with HIV infection can take medication to reduce the risk of her baby being born with HIV infection.

HIV/AIDS education occurs in a variety of ways, including through media and public service announcements, faith-based groups, health care providers, and schools. Nearly 100 percent of U.S. parents of junior or senior high school students believe that HIV/AIDS is an appropriate topic for school sexuality programs (SIECUS 2004). With the HIV infection rate growing among the older than 50 population, HIV/AIDS education is also taking place in some senior centers (Goldberg 2005).

Some HIV/AIDS education is based on the ABC approach—a prevention strategy that involves three elements (Halperin et al. 2004):

A = Abstain. Young people who have not started sexual activity should be encouraged to abstain from or delay sexual activity to prevent HIV and other sexually transmissible infections as well as unwanted pregnancy.

B = Be faithful/reduce partners. After individuals become sexually active, returning to abstinence and remaining faithful to an uninfected partner are the most effective ways to avoid HIV infection.

C = Use condoms. People who have a sexual partner of unknown HIV status should be encouraged to practice correct and consistent use of condoms.

Providing education that advocates condom use and providing youth with access to condoms are controversial topics. Many conservatives believe that promoting use of condoms sends the "wrong message" that sex outside marriage is OK. Consequently, under the Bush administration federal support for "abstinence-only" education programs, which promote abstinence from sexual activity without teaching basic facts about contraception or providing access to contraception, expanded rapidly. But abstinence-only programs are criticized for failing to provide youth with potentially life-saving information. A report released by Representative Henry Waxman found that more than 80 percent of the abstinence-only curricula contain false, misleading, or distorted information about reproductive health (Waxman 2004). For example, the report indicated that

> many of the curricula misrepresent the effectiveness of condoms in preventing sexually transmitted diseases and pregnancy. One curriculum says that "the popular claim that 'condoms help prevent the spread of STDs,' is not supported by the data"; another states that "[i]n heterosexual sex, condoms fail to prevent HIV approximately 31 percent of the time." . . . These erroneous statements are presented as proven scientific facts. (Waxman 2004, p. i).

Another controversy involves the question of whether to provide condoms to prison inmates. Vermont and Mississippi allow condom distribution in prisons, as do Canada, most of Western Europe, and parts of Latin America. One deputy at the Los Angeles Sheriff's Department, which allows only homosexual inmates to receive condoms provided by a local nonprofit organization, said, "We're not promoting sex; we're promoting health" (Sanders 2005).

HIV Testing. Another strategy to curb the spread of HIV involves encouraging individuals to get tested for HIV infection so that they can modify their behavior (to avoid transmitting the virus to others) and so that they can receive early medical intervention, which can slow or prevent the onset of AIDS. An estimated one-fourth to one-third of HIV-infected Americans does not know that they are infected (Kaiser Family Foundation 2004a). About half (48 percent) of Americans report ever having been tested for HIV infection, including 20 percent who say they have been tested in the last year (Kaiser Family Foundation 2004c). Unfortunately, many individuals who have HIV infection continue to engage in risky behaviors, such as unprotected anal, genital, or oral sex and needle sharing (Diamond & Buskin 2000; Hollander 2005).

The Fight Against HIV/AIDS Stigma and Discrimination. The HIV/AIDS-related stigma stems from societal views that people with HIV/AIDS are immoral and shameful. The stigma associated with HIV/AIDS results in discrimination that can lead to loss of employment and housing, social ostracism and rejection, and lack of access to medical care. A survey of 1,000 physicians and nurses in Nigeria found that 1 in 10 admitted to refusing care for an HIV/AIDS patient or had denied HIV/AIDS patients admission to a hospital, and 20 percent believed that people living with HIV/AIDS have behaved immorally and deserved their fate (*HIV & AIDS Discrimination and Stigma* 2004). The stigma surrounding HIV/AIDS has also led to acts of violence against people perceived to be infected with HIV.

HIV/AIDS stigma and discrimination can deter people from getting tested for the disease, can make them less likely to acknowledge their risk of infection, and can discourage those who are HIV-positive from discussing their HIV status with their sexual and needle-sharing partners. One U.S. study found that HIV-infected teens rarely disclose their HIV status to even their close friends because they believe that having HIV is stigmatizing and that disclosure would cause their friends to fear, judge, and/or reject them (Suris et al. 2007). Combating the stigma and discrimination against people who are affected by HIV/AIDS is crucial to improving care, quality of life, and emotional health for people living with HIV and AIDS and to reducing the number of new HIV infections.

Former South African president Nelson Mandela's announcement in 2005 that his son, Makgatho Mandela, 54, had died of AIDS was a public attempt to fight the stigma associated with HIV/AIDS (Timberg 2005). The Miss HIV Stigma Free beauty pageant, first held in 2002 in Botswana, where more than one-third of adults are infected with HIV, combats the stigma by showing that HIV-infected individuals need not be ashamed and that with treatment they can lead productive lives (Goering 2005).

Fighting anti-gay prejudice and discrimination is also important in efforts to support the well-being of individuals diagnosed with HIV/AIDS. In Africa, where about one-half of the nations have laws that criminalize same-sex sexual behavior, fear of arrest drives gay Africans further underground, making them more difficult to reach for HIV interventions. "Fear of arrest prevents people from attending meetings or socializing in locations where their sexual identities become suspect. These are precisely the locations, however, where HIV prevention training, counseling, and materials (informational brochures, condoms . . . etc.) are available" (Johnson 2007, pp. 46–47).

Cynthia Leshomo is the 2005 winner of the Miss HIV Stigma Free beauty pageant for HIV-positive women. First held in 2002, the Miss HIV pageant is a way of showing that HIV-positive individuals need not be ashamed and that with treatment, they can look good and lead productive lives.

Per-Anders Pettersson/Getty Images

Needle Exchange Programs. Injection drug use accounts for most HIV cases in China, Russia, Iran, Afghanistan, Nepal, the Baltic states, and all of Central Asia as well as much of Southeast Asia and South America (Human Rights Watch 2005). To reduce transmission of HIV among injection drug users, their sex partners, and their children, some countries and U.S. communities have established **needle exchange programs** (also known as syringe exchange programs), which provide new, sterile syringes in exchange for used, contaminated syringes. Many needle exchange programs also provide other social and health services, such as referrals to drug counseling and treatment, HIV testing and screening for other sexually transmissible diseases, hepatitis vaccinations, and condoms. Needle exchange has been endorsed as an effective means of HIV prevention by the American Medical Association, the American Public Health Association, and the World Health Organization. Needle exchange programs also protect public health by providing safe disposal of potentially infectious syringes.

In Canada sterile injection equipment is available to drug users in pharmacies and through numerous needle exchange programs. In contrast, most U.S. states prohibit the sale or possession of sterile needles or syringes without a medical prescription.

In 2004, 184 needle exchange programs were operating in 36 states. Less than half of these programs receive public funding from local and/or state government (Centers for Disease Control and Prevention 2005c). The United States is the only country in the world to explicitly ban the use of federal funds for needle exchange (Human Rights Watch 2005).

Commenting on HIV/AIDS prevention, Altman (2003) noted, "The great irony is that we know how to prevent HIV transmission and it is neither technically difficult nor expensive. Most HIV transmission can be stopped by the widespread use of condoms and clean needles" (p. 41). Implementing these strategies, however, conflicts with religious and cultural beliefs and threatens the political power structure. Altman explained that "effective HIV prevention requires governments to acknowledge a whole set of behaviours—drug use, 'promiscuity,' homosexuality, commercial sex work—which they would often rather ignore, and a willingness to support, and indeed empower, groups practicing such behaviours" (p. 42).

Financial and Medical Aid to Developing Countries. Life-extending treatment for individuals infected with HIV is not affordable for many people in the developing world. Only 20 percent of the people who need HIV/AIDS drugs have access to them, and fewer than 10 percent of HIV-positive pregnant women are getting antiretroviral therapy that could not only extend their lives but also reduce the risk of transmitting HIV to their babies (Cowley 2006).

Developing countries—those hardest hit by HIV/AIDS—depend on aid from wealthier countries to help provide medications, HIV/AIDS education programs, and condoms. In 2002 the United Nations helped to create the Global Fund to Fight AIDS, Tuberculosis, and Malaria to help poor countries fight these diseases. But the biggest obstacle to fighting AIDS in Africa may not be lack of money but lack of health care personnel. At an Ethiopian hospital that serves the bulk of the country's patients receiving AIDS medication, two doctors and two nurses care for roughly 2,000 people. By contrast, the United States has 15 nurses for the same number of patients (Rosenberg 2005). The shortage of doctors and nurses in developing countries is partly due to the mass emigration of health professionals to wealthy countries. Reducing emigration by improving

needle exchange programs
Programs designed to reduce transmission of HIV by providing intravenous drug users with new, sterile syringes in exchange for used, contaminated syringes.

working conditions and wages for health workers in developing countries is an important piece of the fight against HIV/AIDS.

Fighting the Growing Problem of Obesity

In general, reducing and preventing obesity requires encouraging people to (1) eat a diet with sensible portions, with lots of high-fiber fruits and vegetables and with minimal sugar and fat, and (2) engage in regular physical activity. Some of the strategies to achieve these goals include the following:

- *Restrictions on advertisements.* The food industry spends an enormous amount of money advertising to children. In response to concerns about childhood obesity, Ireland has banned advertising of candy and fast food on television, Great Britain has banned junk-food advertising on children's television programming, and Sweden and Norway prohibit advertising that targets children. In the 1970s and 1980s, the U.S. Federal Trade Commission considered restrictions on junk-food advertising aimed at children, but those efforts were opposed by the food and advertising industries.

- *Public education.* A variety of public education strategies are being used to inform the public about the importance of exercise and diet and their effects on health. France requires that advertisements promoting processed, sweetened, or salted food and drinks on television, radio, billboards, and the Internet must include one of four health messages: "For your health, eat at least five fruits and vegetables a day," "For your health, undertake regular physical activity," "For your health, avoid eating too much fat, too much sugar, too much salt," or "For your health, avoid snacking between meals" (Combes 2007).

  Because many Americans do not know how many calories, fat grams, sodium, etc. are in the foods they eat, consumers are encouraged to read the nutritional labels on packaged foods. Proposed federal legislation called the Menu Education and Labeling (MEAL) Act, if passed, would require all chain restaurants to list nutritional information for all meals on the menu.

- *School nutrition and physical activity programs.* Observe students in U.S. school cafeterias across the country, and you will probably see many students consuming french fries, chips, sugary soft drinks, and other high-fat, high-sugar foods. Some schools are implementing nutrition programs that restrict the availability of "junk food" and provide nutrition education to students. In 2004 Arkansas became the first state to pass legislation banning vending machines (that sell soft drinks and snacks) from elementary schools (Mui 2005). In 2007, Alabama became the second state (following Mississippi) to ban soft drinks from elementary and middle schools (diet sodas are allowed in high schools) (Leech 2007). At least 16 states have passed legislation restricting the use of vending machines in schools and 24 states have recess/physical activity requirements (Cotton et al. 2007). One study of 5,200 fifth graders found that students from schools that participated in school-based healthy eating programs exhibited significantly lower rates of overweight and obesity, had healthier diets, and reported more physical activities than students from schools without nutrition programs (Veugelers & Fitzgerald 2005).

  Proposed federal legislation to combat obesity includes the Child Nutrition Promotion and School Lunch Protection Act of 2006, which would redefine what are considered "foods of minimal nutritional value" and restrict the sale of such foods in schools. The Childhood Obesity Reduction Act would create a Congressional Council to Combat Childhood Obesity. This council would provide grants to schools to develop and implement programs to increase exercise and improve nutritional choices.

- *Interventions to treat obesity.* Interventions to treat obesity include weight-loss or fitness clubs, nutrition and weight-loss counseling, weight-loss medications, and surgical procedures. In 2004, the U.S. government classified obesity as an illness so obesity treatments could be covered by Medicare. The proposed Medicaid Obesity Treatment Act would require Medicaid coverage of prescription drugs to treat obesity. Some private insurers also cover treatment for obesity; at least 7 states require it (Cotton et al. 2007).

U.S. Federal and State Health Care Reform

The United States is the only country in the industrialized world that does not have any mechanism for guaranteeing health care to its citizens. Other countries, such as Canada, Great Britain, France, Sweden, Germany, and Italy, have national health insurance systems, also referred to as *socialized medicine* and *universal health care.* **Socialized medicine,** or **universal health care,** refers to a state-supported system of health care delivery in which health care is purchased by the government and sold to the consumer at little or no additional cost. Despite differences in how socialized medicine works in various countries, in all systems of socialized medicine the government (1) directly controls the financing and organization of health services, (2) directly pays providers, (3) owns most of the medical facilities (Canada is an exception), (4) guarantees universal access to health care, and (5) allows private care for individuals who are willing to pay for their medical expenses (Cockerham 2004). Most countries with national health insurance allow or encourage private insurance as an upgrade to a higher class of service and a fuller range of services (Quadagno 2004). To the extent that health care is rationed in countries with national health insurance, rationing is done on the basis of medical need, not ability to pay.

The goals of health care-reform efforts in the United States generally fall into one of three categories: (1) creation of a universal health program; (2) expansion of existing government health insurance programs; and (3) implementing tax incentives and other strategies to make private insurance more affordable. Various health care reform efforts are being made at the federal and state levels.

Federal Health Care Reform. Since 1912, when Theodore Roosevelt first proposed a national health insurance plan, the idea of health care for all Americans has been advocated by the Truman, Nixon, Carter, and Clinton administrations. In a 2007 national poll, more than two-thirds (64 percent) of Americans said that the federal government should guarantee health insurance for all Americans (Toner & Elder 2007). In the same poll, nearly half (49 percent) said they would be willing to pay $500 or more per year in taxes so that all Americans could have health insurance.

A number of federal measures to reform health care have been proposed. The National Health Insurance Act, introduced by Representative John Conyers, would expand Medicare to every U.S. resident, creating a single-payer system in which a single tax-financed public insurance program replaces private insurance companies. Under this plan every U.S. resident would be issued a national

socialized medicine Also known as universal health care, a state-supported system of health care delivery in which health care is purchased by the government and sold to the consumer at little or no additional cost.

health insurance card, would receive all medically necessary services (including dentistry, eye care, mental health services, substance abuse treatment, prescription drugs, and long-term care), would have no copayments or deductibles, and would see the doctor of his or her choice. If this plan is adopted, it is estimated to save enough on administrative costs to provide coverage for all the uninsured and to substantially help the underinsured.

The insurance industry, not surprisingly, opposes the adoption of such a system because the private health insurance industry would be virtually eliminated. The health insurance industry's opposition to a single-payer universal health plan is matched only by the persistent efforts of those who advocate such a plan. At the federal level Representative Pete Stark (D-California) proposed an amendment to the U.S. Constitution to guarantee health care as a right for every American. The proposed amendment would say that "all persons shall enjoy the right to health care of equal high quality." Stark argued that "the health of every American is vital to their unalienable rights of 'life, liberty, and the pursuit of happiness' . . . To ensure these rights are fully enjoyed, we must be certain that every American can access quality health care—regardless of their income, race, education or job status" ("Stark Introduces Constitutional Amendment" 2005).

State-Level Health Care Reform. A number of states have taken action to increase health insurance coverage of their residents. Massachusetts passed landmark legislation requiring residents to be insured, similar to the requirement that all automobiles must be insured. The plan includes measures to expand Medicaid coverage for those who are eligible and to subsidize the cost of health insurance for low-income individuals. The remaining Massachusetts population must buy private insurance (using pretax dollars) through their employers or through a new state agency—the Commonwealth Care Health Insurance Connector. The uninsured face tax penalties; they lose their personal exemption and, by 2008, pay a penalty equal to half of what health insurance premiums would have cost. Employers who do not provide health insurance face annual penalties too—$295 per worker (Vock 2007). Other states, including Vermont, Maine, Minnesota, Illinois, Pennsylvania, Michigan, Hawaii, and California, have also taken steps to increase health insurance coverage of their residents. Some states have increased benefits to Medicaid clients who demonstrate healthy behavior, such as showing up for doctor's appointments, getting their children immunized, or following disease management programs. Arizona, Kansas, Montana, and West Virginia offer tax credits to small businesses that offer insurance to their employees, and Arkansas, New Mexico, and Oklahoma offer small businesses and the uninsured discounted coverage through the state.

Strategies to Improve Mental Health Care

Most mental disorders can be successfully treated with medications and/or psychotherapy or counseling. Yet, nearly half of all Americans who have a severe mental illness do not seek treatment (Substance Abuse and Mental Health Services Administration 2003).

Two areas for improving mental health care in the United States are eliminating the stigma associated with mental illness and improving health insurance coverage for treating mental disorders. Efforts in these areas will, it is hoped, promote the delivery of treatment to individuals suffering from mental illness.

> "Universal health care for every single American must not be a question of whether, it must be a question of how."
>
> Barack Obama
> Senator

Eliminating the Stigma of Mental Illness. The first White House conference on mental health called for a national campaign to eliminate the stigma associated with mental illness. Fearing the negative label of "mental illness" and the social rejection and stigmatization associated with mental illness, individuals are reluctant to seek psychological services (Komiya et al. 2000). In a study of 274 eighth graders, boys were more likely than girls to associate stigma with mental health service use (Chandra & Minkovitz 2006). For example, 40 percent of boys compared with 23 percent of girls agreed with the statement, "Seeing a counselor for emotional problems makes people think you are weird or different." In the same study, more boys than girls (38 percent versus 23 percent) reported that they were not willing to use mental health services. The most frequently cited reason was "Too embarrassed by what other kids would say." To assess your attitudes toward seeking professional psychological help, see this chapter's *Self and Society* feature.

Reducing the stigma associated with mental illness might be achieved through encouraging individuals to seek treatment and making treatment accessible and affordable. The surgeon general's report on mental health explained that

> effective treatment for mental disorders promises to be the most effective antidote to stigma. Effective interventions help people to understand that mental disorders are not character flaws but are legitimate illnesses that respond to specific treatments, just as other health conditions respond to medical interventions. (U.S. Department of Health and Human Services 1999, p. viii)

Real Men. Real Depression.

**It takes courage to ask for help.
These men did.**

This public education brochure on men and depression is available from the National Institute of Mental Health (www.nimh.nih.gov).

The National Alliance for the Mentally Ill (NAMI) has a Stigma-Busters campaign, whereby the public submits instances of media content that stigmatize individuals with mental illness to Stigma-Busters, which then investigates and conveys their concerns to media organizations and corporations, urging them to avoid stigmatizing portrayals of mental illness. For example, when McDonald's aired a radio commercial in 2004 stating, "I am hearing voices about the food at McDonald's. I am hearing voices telling me to go to McDonald's. I better listen to them. I am not crazy!" StigmaBusters wrote to McDonald's expressing disapproval of the ad (StigmaBusters Alert 2004).

Another tool used to reduce the stigma of mental illness is public education, such as the campaign called "Real Men, Real Depression," which includes print, television, and radio public service announcements. *Breaking the Silence,* a curriculum for elementary, middle, and high schools available through NAMI, uses true stories, activities, a board game, and posters to debunk myths about mental illness and sensitize students to the pain that words such as *psycho* and *schizo* and frightening or comic media images of mentally ill people can cause (Harrison 2002).

Eliminating Inequalities in Health Care Coverage for Mental Disorders. Another priority on the agenda to improve the nation's mental health care system involves eliminating the inequalities in health care coverage for mental disorders. The Mental Health Parity Act of 1996 was an important step in ending health care discrimination against individuals with mental illnesses by requiring equality between mental health care insurance coverage and other health care coverage—a concept known as **parity.** But the 1996 law only provided partial parity for the annual and lifetime limits between mental health coverage and medical/surgical coverage. The more recent Mental Health Parity Act of 2007 extends parity by including deductibles, copayments, out-of-pocket expenses, covered hospital days, and covered outpatient days. However, the law does not require insurers to provide mental health coverage, only parity. And the parity laws do not apply to employers with fewer than 50 employees. Some states have enacted mental health parity laws, which vary in their scope and application, but many of these laws do not address substance abuse, are limited to the more serious mental illnesses, or apply only to government employees.

UNDERSTANDING PROBLEMS OF ILLNESS AND HEALTH CARE

Although human health has probably improved more over the past half-century than over the previous three millennia, the gap in health between rich and poor remains wide and the very poor suffer appallingly (Feachum 2000). Poor countries need economic and material assistance to alleviate problems such as HIV/AIDS, high maternal and infant mortality rates, and malaria. The wealthy countries of the world do have resources to make a difference in the health of the

parity In health care, a concept requiring equality between mental health care insurance coverage and other health care coverage.

world. When a tsunami took the lives of thousands of people in 2004, media attention to the tragedy elicited an outpouring of aid throughout the world to help affected regions. Meanwhile, malaria, which has been referred to as "a silent tsunami," takes the lives of more than 150,000 African children each month, which is about the same as the death toll of the South Asian tsunami disaster (Sachs 2005). Yet malaria continues to receive comparatively scant public attention even though it is largely preventable with $5 mosquito bed nets, and treatable with medicines at roughly $1 per dose.

Although poverty may be the most powerful social factor affecting health, other social factors that affect health include globalization, increased longevity, family structure, gender, education, and race or ethnicity. But U.S. cultural values and beliefs view health and illness as being determined by individual behavior and lifestyle choices rather than by social, economic, and political forces. We agree that an individual's health is affected by the choices he or she makes—choices such as whether to smoke, exercise, eat a healthy diet, engage in risky sexual activity, use condoms, wear a seatbelt, and so on. However, the choices that individuals make are influenced by social, economic, and political forces that must be taken into account if the goal is to improve the health not only of individuals but also of entire populations.

By focusing on individual behaviors that affect health and illness, we often overlook social causes of health problems (Link & Phelan 2001). One of the most pressing social causes of health problems is environmental pollution and degradation. The Millennium Ecosystem Assessment (2005) synthesis report warned that pollution and depletion of the earth's natural resources are eroding ecosystems and could lead to an increase in existing diseases such as malaria and cholera as well as to an increased risk of new diseases emerging (see Chapter 14 for more on the health effects of environmental problems).

A sociological view of illness and health care also reminds us of the social *consequences* of health problems—consequences that potentially affect us all. In *Uninsured in America* (Sered & Fernandopulle 2005), the authors explain:

> If millions of American children do not have reliable, basic health care, all children who attend American schools are at risk through daily exposure to untreated disease. If millions of restaurant and food industry workers do not have health insurance, people preparing food and waiting tables are sharing their health problems with everyone they serve. . . . If tens of millions of Americans go without basic and preventive care, we all pay the bill when their health problems turn into complex medical emergencies necessitating expensive . . . treatment. And if we condone a health care system that contributes to the formation of an untouchable caste of the ill, infirm, and marginally employed, all of us, as a society, will lose the right to feel pride in the democratic values that we claim to cherish. (p. 20)

Finally, a sociological perspective looks for *social solutions* to problems of illness and health care, which include social policies and interventions at the local, national, and international level. At a 2005 meeting of the Commission on Human Rights in Geneva, 52 countries voted in favor of a resolution urging countries not only to commit to realizing the universal right to the highest attainable standards of physical and mental health but also to assist developing countries achieve the highest attainable standard of physical and mental health through financial and technical support and through training of personnel (UNOG 2005). The United States was the only country to vote against this resolution. National strategies to improve health range from U.S. efforts to pass

❝We cannot narrow the gap in health without addressing disparities in educational opportunity, employment, economic security, and housing.**❞**

Marion Wright Edelman
President of Children's
Defense Fund

federal universal health care legislation to national HIV/AIDS education programs. Local strategies to improve health range from banning trans fats in restaurants, schools, and/or supermarket bakeries and delis, as New York City; Philadelphia; Brookline, Massachusetts; and Montgomery County, Maryland, have done, to developing and implementing free health care clinics, such as the one described in the opening of this chapter.

Improvements in health care systems and health care delivery are also necessary to improve the health of populations around the world. Although certain changes in medical practices and policies may help to improve world health, "the health sector should be seen as an important, but not the sole, force in the movement toward global health" (Lerer et al. 1998, p. 18). Just as the social causes of problems of illness and health care are diverse, so must be the social solutions. Thus, a comprehensive approach to improving the health of a society requires addressing diverse issues such as poverty and economic inequality, gender inequality, population growth, environmental issues, education, housing, energy, water and sanitation, agriculture, and workplace safety.

Improving the health of the world also means seeking nonmilitary solutions to international conflicts. In addition to the deaths, injuries, and illnesses that result from combat, war diverts economic resources from health programs, leads to hunger and disease caused by the destruction of infrastructure, causes psychological trauma, and contributes to environmental pollution (Sidel & Levy 2002). Thus "the prevention of war . . . is surely one of the most critical steps mankind can make to protect public health" (White 2003, p. 228).

The World Health Organization (1946) defined **health** as "a state of complete physical, mental, and social well-being" (p. 3). Based on this definition, we conclude this chapter with the suggestion that the study of social problems is, essentially, the study of health problems, as each social problem is concerned with the physical, mental, and social well-being of humans and the social groups of which they are a part. As you read other chapters in this book, consider how the problems in each chapter affect the health of individuals, families, populations, and nations.

health According to the World Health Organization, "a state of complete physical, mental, and social well-being."

CHAPTER REVIEW

- **What are three measures that serve as indicators of the health of populations? Which health measure reveals the greatest disparity between developed and developing countries?**

  Measures of health that serve as indicators of the health of populations include morbidity (often expressed as incidence and prevalence rates), life expectancy, and mortality rates (including infant and under-5 childhood mortality rates and maternal mortality rates). Maternal mortality rates reveal the greatest disparity between developed and developing countries.

- **Which theoretical perspective criticizes the pharmaceutical and health care industry for placing profits above people?**

  The conflict perspective criticizes the pharmaceutical and health care industry for placing profits above people. For example, pharmaceutical companies' research and development budgets are spent not according to public health needs but rather according to calculations about maximizing profits. Because the masses of people in developing countries lack the resources to pay high prices for medication, pharmaceutical companies do not see the development of drugs for diseases of poor countries as a profitable investment.

- **Where is HIV/AIDS most prevalent in the world?**

HIV/AIDS is most prevalent in Africa, particularly sub-Saharan Africa. With slightly more than 10 percent of the world's population, Africa is home to 60 percent of individuals infected with HIV. About 1 in 12 African adults has HIV/AIDS and as many as 9 of 10 HIV-infected people in sub-Saharan Africa do not know that they are infected.

- **What is the second biggest cause of preventable deaths in the United States (second only to tobacco)?**

Obesity, which can lead to heart disease, hypertension, diabetes, and other health problems, is the second biggest cause of preventable deaths in the United States.

- **Why is mental illness referred to as a "hidden epidemic"?**

Mental illness is a "hidden epidemic" because the shame and embarrassment associated with mental problems discourage people from acknowledging and talking about mental illness. Because male gender expectations associate masculinity with emotional strength, men are particularly prone to deny or ignore mental problems.

- **What features of globalization have contributed to health problems?**

Features of globalization that have been linked to problems in health are increased transportation, the expansion of trade and transnational corporations, and free trade agreements.

- **What, according to former United Nations Secretary-General Kofi Annan, is the "biggest enemy of health in the developing world"?**

In an address to the 2001 World Health Assembly, United Nations Secretary-General Kofi Annan remarked, "The biggest enemy of health in the developing world is poverty." Approximately one-fifth of the world's population live on less than US$1 per day and nearly one-half live on less than US$2 per day. Poverty is associated with malnutrition, unsafe water and sanitation, indoor air pollution, hazardous working conditions, and lack of access to medical care.

- **According to a World Health Organization analysis of the world's health systems, which country provides the best overall health care?**

The World Health Organization found that France provides the best overall health care among major countries, followed by Italy, Spain, Oman, Austria, and Japan. The United States ranked 37 out of 191 countries, despite the fact that the United States spends a higher portion of its gross domestic product on health care than any other country.

- **What is the difference between selective primary health care and comprehensive primary health care?**

Selective primary health care focuses on using specific interventions to target specific health problems, such as promoting condom use to prevent HIV infections and providing immunizations against childhood diseases to promote child survival. In contrast, comprehensive primary health care focuses on the broader social determinants of health, such as poverty and economic inequality, gender inequality, environment, and community development.

- **What are the categories of health care-reform efforts in the United States?**

The goals of health care-reform efforts in the United States generally fall into one of three categories: (1) creation of a universal health program; (2) expansion of existing government health insurance programs; and (3) implementing tax incentives and other strategies to make private insurance more affordable.

- **How does the World Health Organization define health?**

Health, according to the World Health Organization, is "a state of complete physical, mental, and social well-being." Based on this definition, we suggest that the study of social problems is, essentially, the study of health problems, because each social problem is concerned with the physical, mental, and social well-being of humans and the social groups of which they are a part.

TEST YOURSELF

1. Worldwide, the leading cause of death is
   a. HIV/AIDS
   b. traffic accidents
   c. heart disease
   d. cancer
2. The United States has the lowest infant mortality rate in the world.
   a. True
   b. False

3. Data on deaths from international terrorism and tobacco-related deaths in 37 developed and Eastern European countries revealed that the number of tobacco deaths was equivalent to the impact of a September 11, 2001, type terrorist attack every
   a. 14 hours
   b. 14 days
   c. 14 weeks
   d. 14 months

4. Worldwide, the predominant mode of HIV transmission is through
   a. heterosexual contact
   b. needle-sharing
   c. same-sex sexual contact
   d. blood transfusions

5. In the United States, _____ of adults are either overweight or obese.
   a. 10 percent
   b. one-third
   c. two-thirds
   d. 20 percent

6. In the United States, there are many more suicides each year than homicides.
   a. True
   b. False

7. In a 2007 national poll, more than two-thirds (64 percent) of Americans said that _____ should guarantee health insurance for all Americans.
   a. employers
   b. states
   c. insurance companies
   d. the federal government

8. All U.S. children living below the poverty line are covered by publicly funded health insurance.
   a. True
   b. False

9. Of all age groups in the United States, young adults ages 18–24 are the least likely to have health insurance.
   a. True
   b. False

10. The United States is the only country in the industrialized world that does not have any mechanism for guaranteeing health care to its citizens.
    a. True
    b. False

Answers: 1 c. 2 b. 3 a. 4 a. 5 c. 6 a. 7 d. 8 b. 9 a. 10 a.

KEY TERMS

biomedicalization
comprehensive primary health care
deinstitutionalization
developed countries
developing countries
epidemiological transition
globalization
health
infant mortality rate
least developed countries

life expectancy
managed care
maternal mortality rate
Medicaid
medicalization
Medicare
mental health
mental illness
morbidity
mortality

needle exchange programs
parity
selective primary health care
socialized medicine
State Children's Health Insurance Program (SCHIP)
stigma
under-5 mortality rate
universal health care
workers' compensation

MEDIA RESOURCES

Understanding Social Problems, **Sixth Edition Companion Website**

academic.cengage.com/sociology/mooney
Visit your book companion website, where you will find flash cards, practice quizzes, Internet links, and more to help you study.

Just what you need to know NOW! Spend time on what you need to master rather than on information you already have learned. Take a pre-test for this chapter, and CengageNOW will generate a personalized study plan based on your results. The study plan will identify the topics you need to review and direct you to online resources to help you master those topics. You can then take a post-test to help you determine the concepts you have mastered and what you will need to work on. Try it out! Go to academic.cengage.com/login to sign in with an access code or to purchase access to this product.

© Kwame Zikomo/SuperStock

3

Substance abuse, the nation's number one preventable health problem, places an enormous burden on American society, harming health, family life, the economy, and public safety, and threatening many other aspects of life.

Robert Wood Johnson Foundation, Institute for Health Policy, Brandeis University

Alcohol and Other Drugs

The Global Context: Drug Use and Abuse | Sociological Theories of Drug Use and Abuse | Frequently Used Legal and Illegal Drugs | Societal Consequences of Drug Use and Abuse | Treatment Alternatives | Strategies for Action: America Responds | Understanding Alcohol and Other Drug Use | Chapter Review

Scott Krueger was athletic, intelligent, and handsome and what you'd call an all around "nice guy." A freshman at the Massachusetts Institute of Technology, he was a three-letter athlete and one of the top 10 students in his high school graduating class of more than 300. He was a "giver," not a "taker," tutoring other students in math after school while studying second-year calculus so he could pursue his own career in engineering. While at MIT he rushed a fraternity and celebrated his official acceptance into the brotherhood. The night he celebrated he was found in his room, unconscious, and after 3 days in an alcoholic coma he died. He was 18 years old (Moore 1997). In September 2000, MIT agreed to pay Scott's parents, Bob and Darlene Krueger, $4.75 million in a settlement over the death of their son and to establish a scholarship in his name. In 2002 an out-of-court settlement that paid the Kruegers $1.75 million was reached with Phi Gamma Delta and the fraternity officers. The defendants also agreed to produce an educational video ("Tell Me Something I Don't Know") on the dangers of drinking. This video is being shown on high school and college campuses across the United States (Crittenden 2002; Singley 2004). According to the National Institute on Alcohol Abuse and Alcoholism, more than 1,700 college students die every year from alcohol-related injuries (Park 2006).

Drug-induced death is just one of the many negative consequences that can result from alcohol and drug abuse. The abuse of alcohol and other drugs is a social problem when it interferes with the well-being of individuals and/or the societies in which they live—when it jeopardizes health, safety, work and academic success, family, and friends. But managing the drug problem is a difficult undertaking. In dealing with drugs, a society must balance individual rights and civil liberties against the personal and social harm that drugs promote—crack babies, suicide, drunk driving, industrial accidents, mental illness, unemployment, and teenage addiction. When to regulate, what to regulate, and who should regulate are complex social issues. Our discussion begins by looking at how drugs are used and regulated in other societies.

The Global Context: Drug Use and Abuse

Pharmacologically, a **drug** is any substance other than food that alters the structure or functioning of a living organism when it enters the bloodstream. Using this definition, everything from vitamins to aspirin is a drug. Sociologically, the term *drug* refers to any chemical substance that (1) has a direct effect on the user's physical, psychological, and/or intellectual functioning; (2) has the potential to be abused; and (3) has adverse consequences for the individual and/or society. Societies vary in how they define and respond to drug use. Thus drug use is influenced by the social context of the particular society in which it occurs.

Drug Use and Abuse Around the World
Globally, 5.0 percent of the world's population between the ages of 15 and 64—200 million people—reported using at least one illicit drug in the previous year (World Drug Report 2006). About half that number reported regular drug

drug Any substance other than food that alters the structure or functioning of a living organism when it enters the bloodstream.

AP/Wide World Photos

U.S. citizens visiting the Netherlands may be shocked or surprised to find people smoking marijuana and hashish openly in public. Pictured here is a tourist using a water pipe to smoke marijuana in a coffee shop.

use, that is, use at least once a month. According to the most recent report available:

> Cannabis remains by far the most widely used drug (some 162 million people), followed by amphetamine-type stimulants (some 35 million people), which include amphetamines (used by 25 million people) and Ecstasy (almost 10 million people). The number of opiate abusers is estimated at some 16 million people, of which 11 million are heroin abusers. Some 13 million are cocaine users. (World Drug Report 2006, p. 9)

The prevalence of drug use also varies dramatically by country. For example, the lifetime prevalence of illicit drug use varies from 46.0 percent of the adult population in the United States to 36.0 percent in England, 26.0 percent in Italy, 18.0 percent in Poland, and 9.0 percent in Sweden (ODCCP 2003). In Europe lifetime prevalence of illegal drug use *excluding* cannabis ranges from 12.0 percent in England to 2.0 percent in Finland. Moreover, 23.0 percent of European 15- to 16-year-olds have used marijuana at least once in their lifetime, compared with 35.0 percent of 15- to 16-year-olds in Canada, 43.0 percent in Australia, and 41.0 percent in the United States.

Global drug use also varies over time. Cannabis use and the use of amphetamine based-stimulants such as methamphetamine have increased dramatically over the past decade (World Drug Report 2006). Further, the use of opiate-based drugs (e.g., heroin), although remaining fairly stable over the past several years, has also increased in some parts of the world as a result of record levels of opium production, particularly in Afghanistan where narcotics trafficking accounts for half the economy (Moreau & Yousafzai 2006). Because of increased production levels, in 2006, 63 mayors of European cities were warned about the probable growth in the number of heroin-related deaths (UNODC 2006a).

Some of the differences in international drug use can be attributed to variations in drug policies. The Netherlands, for example, has had an official government policy of treating the use of drugs such as marijuana, hashish, and heroin as a health issue rather than a crime issue since the mid-1970s. In the first decade of the policy, drug use did not appear to increase. However, increases in marijuana use were reported in the early 1990s with the advent of "cannabis cafes." These coffee shops sell small amounts of marijuana for personal use and, presumably, prevent casual marijuana users from coming into contact with drug dealers (MacCoun & Reuter 2001; Drug Policy Alliance 2003). Some evidence suggests that marijuana use among Dutch youth is decreasing, rebutting those who would argue that liberal drug policies result in increased drug abuse (Sheldon 2000; Burke 2006).

Great Britain has also adopted a "medical model," particularly in regard to heroin and cocaine. As early as the 1960s, English doctors prescribed opiates and cocaine for their patients who were unlikely to quit using drugs on their own and for the treatment of withdrawal symptoms. By the 1970s, however, British laws had become more restrictive, making it difficult for either physicians or users to obtain drugs legally. Today, British government policy provides for limited distribution of drugs by licensed drug treatment specialists to addicts who might otherwise resort to crime to support their habits (Abadinsky 2004).

What Do You Think? A recent British report recommended that heroin addicts taking methadone, a synthetic drug used in the treatment of opiate addiction, be given "shopping vouchers and other rewards as incentives to stay clean" (Boseley 2007, p. 1). Would you recommend that vouchers, as part of a national strategy on drug control, be used in the United States? What other types of incentives would you recommend?

In stark contrast to such health-based policies, other countries execute drug users and/or dealers or subject them to corporal punishment, which may include whipping, stoning, beating, and torture. Such policies are found primarily in less developed nations, such as Pakistan and Malaysia, where religious and cultural prohibitions condemn the use of any type of drug, including alcohol and tobacco.

Drug Use and Abuse in the United States

According to government officials, there is a drug crisis in the United States—a crisis so serious that it warrants a multibillion-dollar-a-year "war on drugs." Americans' concern with drugs, however, has varied over the years. Ironically, in the 1970s, when drug use was at its highest, concern over drugs was relatively low. In 2005, when a sample of Americans were asked whether they worried about drug use "a great deal," "a fair amount," "only a little," or "not at all," 42.0 percent reported worrying a great deal and 23.0 percent reported worrying a fair amount (Lyons 2005).

As Table 3.1 indicates, use of illicit drugs in a person's lifetime is fairly common. Of people 12 years old and older, 46.1 percent reported using an illicit drug sometime in their lives. Marijuana and hashish had the highest occurrence in lifetime use (40.1 percent), with heroin having the lowest use (1.5 percent). Despite these relatively high numbers, particularly for marijuana and hashish, use of alcohol and tobacco is much more widespread than use of illicit drugs. Half of Americans age 12 and older reported being *current* alcohol drinkers, and an estimated 72 million Americans 12 and older reported *current* use of a tobacco product (U.S. Department of Health and Human Services 2006). In the United States cultural definitions of drug use are contradictory—condemning it on the one hand (e.g., heroin), yet encouraging and tolerating it on the other (e.g., alcohol). At various times in U.S. history many drugs that are illegal today were legal and readily available. In the 1800s and the early 1900s opium was routinely used in medicines as a pain reliever, and morphine was taken as

TABLE 3.1 Percentages Reporting Lifetime, Past Year, and Past Month Use of Illicit and Licit Drugs Among Individuals Age 12 or Older: 2005

DRUG	TIME PERIOD		
	LIFETIME	PAST YEAR	PAST MONTH
Any illicit drug[*]	46.1	14.4	8.1
Marijuana and hashish	40.1	10.4	6.0
Cocaine	13.8	2.3	1.0
Crack	3.3	0.6	0.3
Heroin	1.5	0.2	0.1
Hallucinogens	13.9	1.6	0.4
LSD	9.2	0.2	0.0
PCP	2.7	0.1	0.0
Ecstasy	4.7	0.8	0.2
Inhalants	9.4	0.9	0.3
Nonmedical use of any psychotherapeutic drug[†]	20.0	6.2	2.6
Pain relievers	13.4	4.9	1.9
Tranquilizers	8.7	2.2	0.7
Stimulants	7.8	1.1	0.4
Methamphetamine	4.3	0.5	0.2
Sedatives	3.7	0.3	0.1
Cigarettes	66.6	29.1	24.9
Alcohol	83.1[‡]	65.0[‡]	51.8

[*]"Any illicit drug" includes marijuana/hashish, cocaine (including crack), heroin, hallucinogens, inhalants, or any prescription type psychotherapeutic used nonmedically.

[†]"Nonmedical use" of any prescription-type pain reliever, tranquilizer, stimulant, or sedative; does not include over-the-counter drugs.

[‡]2003 percentages.

Source: U.S. Department of Health and Human Services (2004, 2006).

a treatment for dysentery and fatigue. Amphetamine-based inhalers were legally available until 1949, and cocaine was an active ingredient in Coca-Cola until 1906, when it was replaced with another drug—caffeine (Witters et al. 1992; Abadinsky 2004).

SOCIOLOGICAL THEORIES OF DRUG USE AND ABUSE

Most theories of drug use and abuse concentrate on what are called psychoactive drugs. These drugs alter the functioning of the brain, affecting the moods, emotions, and perceptions of the user. Such drugs include alcohol, cocaine, methamphetamine, heroin, and marijuana. **Drug abuse** occurs when acceptable social standards of drug use are violated, resulting in adverse physiological, psychological, and/or social consequences. For example, when an individual's drug use leads to hospitalization, arrest, or divorce, such use is usually considered abu-

drug abuse The violation of social standards of acceptable drug use, resulting in adverse physiological, psychological, and/or social consequences.

sive. Drug abuse, however, does not always entail drug addiction. Drug addiction, or **chemical dependency**, refers to a condition in which drug use is compulsive—users are unable to stop because of their dependency. The dependency may be psychological (the individual needs the drug to achieve a feeling of well-being) and/or physical (withdrawal symptoms occur when the individual stops taking the drug). For example, withdrawal from marijuana includes depression, anger, decreased appetite, and restlessness (Zickler 2003). In 2005, more than 22 million Americans, 9.1 percent of the population 12 or older, were defined as being dependent on or abusers of alcohol and/or other drugs. Of that number, 15.4 million (69.4 percent) were dependent on or abused alcohol only, 3.6 million (16.2 percent) were dependent on or abused illicit drugs but not alcohol, and 3.3 million (14.9 percent) were dependent on or abused both illicit drugs and alcohol (U.S. Department of Health and Human Services 2006). Individuals who are dependent on or abuse illicit drugs and/or alcohol are disproportionately male, American Indians or Alaska natives, between the ages of 18 and 25 (U.S. Department of Health and Human Services 2006).

© Partnership for a Drug-Free America®

This poster from the Office of National Drug Control Policy's National Youth Anti-Drug Media Campaign emphasizes the importance of a close relationship between parent and child in the fight against drug use by youths.

Various theories provide explanations for why some people use and abuse drugs. Drug use is not simply a matter of individual choice. Theories of drug use explain how structural and cultural forces as well as biological and psychological factors influence drug use and society's responses to it.

Structural-Functionalist Perspective

Structural functionalists argue that drug abuse is a response to the weakening of norms in society. As society becomes more complex and as rapid social change occurs, norms and values become unclear and ambiguous, resulting in **anomie**— a state of normlessness. Anomie may exist at the societal level, resulting in social strains and inconsistencies that lead to drug use. For example, research indicates that increased alcohol consumption in the 1830s and the 1960s was a response to rapid social change and the resulting stress (Rorabaugh 1979). Anomie produces inconsistencies in cultural norms regarding drug use. For example, although public health officials and health care professionals warn of the dangers of alcohol and tobacco use, advertisers glorify the use of alcohol and tobacco and the U.S. government subsidizes the alcohol and tobacco industries. Furthermore, cultural traditions, such as giving away cigars to celebrate the birth of a child and toasting a bride and groom with champagne, persist.

Anomie may also exist at the individual level, as when a person suffers feelings of estrangement, isolation, and turmoil over appropriate and inappropriate behavior. An adolescent whose parents are experiencing a divorce, who is separated from friends and family as a consequence of moving, or who lacks parental supervision and discipline may be more vulnerable to drug use because of such conditions. Thus, from a structural-functionalist perspective, drug use is a re-

chemical dependency A condition in which drug use is compulsive and users are unable to stop because of physical and/or psychological dependency.

anomie A state of normlessness in which norms and values are weak or unclear.

sponse to the absence of a perceived bond between the individual and society and to the weakening of a consensus regarding what is considered acceptable. Consistent with this perspective, a national poll of Americans age 18 years or older showed that peer pressure and lack of parental supervision were the two most common responses given for why teenagers take drugs (Pew Research Center 2002). Interestingly, in a survey of British youth, 20 percent responded that they had friends who "faked" drug use to "fit in" with their peers (BBC 2004).

Conflict Perspective

The conflict perspective emphasizes the importance of power differentials in influencing drug use behavior and societal values concerning drug use. From a conflict perspective drug use occurs as a response to the inequality perpetuated by a capitalist system. Societal members, alienated from work, friends, and family as well as from society and its institutions, turn to drugs as a means of escaping the oppression and frustration caused by the inequality they experience. Furthermore, conflict theorists emphasize that the most powerful members of society influence the definitions of which drugs are illegal and the penalties associated with illegal drug production, sales, and use.

For example, alcohol is legal because it is often consumed by those who have the power and influence to define its acceptability—white males (U.S. Department of Health and Human Services 2006). This group also disproportionately profits from the sale and distribution of alcohol and can afford powerful lobbying groups in Washington to guard the alcohol industry's interests. Because this group also commonly uses tobacco and caffeine, societal definitions of these substances are also relatively accepting. Conversely, minority group members disproportionately use crack cocaine. Now an issue before the U.S. Supreme Court, the criminal consequences associated with the use of this drug have been severe. As Taifa (2006) wrote:

> Among the most unjust inequities in our criminal justice system is the disparity between mandatory minimum sentences for those convicted of crack and powder cocaine offenses. Under federal law, possession of five grams of crack cocaine carries the same penalty as distribution of 500 grams of powder cocaine. Blacks comprise the vast majority of those convicted of crack cocaine offenses while the majority of those convicted of the latter are white. This disparity has led to inordinately harsh mandatory sentences disproportionately meted out to African-American defendants that are far more severe than sentences for comparable activity for white defendants. (p. 1)

The use of opium by Chinese immigrants in the 1800s provides a historical example. The Chinese, who had been brought to the United States to work on the railroads, regularly smoked opium as part of their cultural tradition. As unemployment among white workers increased, however, so did resentment of Chinese laborers. Attacking the use of opium became a convenient means of attacking the Chinese, and in 1877 Nevada became the first of many states to prohibit opium use. As Morgan (1978) observed:

> The first opium laws in California were not the result of a moral crusade against the drug itself. Instead, it represented a coercive action directed against a vice that was merely an appendage of the real menace—the Chinese—and not the Chinese per se, but the laboring "Chinamen" who threatened the economic security of the white working class. (p. 59)

" There are but three ways for the populace to escape its wretched lot. The first two are by route of the wine-shop or the church; the third is by that of the social revolution. **"**

Mikhail A. Bakunin
Anarchist and revolutionary

The criminalization of other drugs, including cocaine, heroin, and marijuana, follows similar patterns of social control of the powerless, political opponents, and/or minorities. In the 1940s marijuana was used primarily by minority group members, and users faced severe criminal penalties. But after white middle-class college students began to use marijuana in the 1970s, the government reduced the penalties associated with its use. Although the nature and pharmacological properties of the drug had not changed, the population of users was now connected to power and influence. Thus conflict theorists regard the regulation of certain drugs, as well as drug use itself, as a reflection of differences in the political, economic, and social power of various interest groups.

Symbolic Interactionist Perspective

Symbolic interactionism, which emphasizes the importance of definitions and labeling, concentrates on the social meanings associated with drug use. If the initial drug use experience is defined as pleasurable, it is likely to recur, and over time the individual may earn the label of "drug user." If this definition is internalized so that the individual assumes an identity of a drug user, the behavior will probably continue and may even escalate.

Drug use is also learned through symbolic interaction in small groups. In a study of binge drinking, researchers found that students who believed that their friends were binge drinking were more likely to binge-drink themselves (Weitzman et al. 2003). First-time users learn not only the motivations for drug use and its techniques but also what to experience. Becker (1966) explained how marijuana users learn to ingest the drug. A novice being coached by a regular user reported the experience:

> I was smoking like I did an ordinary cigarette. He said, "No, don't do it like that." He said, "Suck it, you know, draw in and hold it in your lungs . . . for a period of time." I said, "Is there any limit of time to hold it?" He said, "No, just till you feel that you want to let it out, let it out." So I did that three or four times. (p. 47)

Marijuana users not only learn the way to ingest the smoke but also to label the experience positively. When certain drugs, behaviors, and experiences are defined by peers as not only acceptable but also pleasurable, drug use is likely to continue.

> Because they (first-time users) think they're going to keep going up, up, up till they lose their minds or begin doing weird things or something. You have to like reassure them, explain to them that they're not really flipping or anything, that they're gonna be all right. You have to just talk them out of being afraid. (Becker 1966, p. 55)

Interactionists also emphasize that symbols can be manipulated and used for political and economic agendas. The popular DARE (Drug Abuse Resistance Education) program, with its anti-drug emphasis fostered by local schools and police, carries a powerful symbolic value that politicians want the public to identify with. "Thus, ameliorative programs which are imbued with these potent symbolic qualities (like DARE's links to schools and police) are virtually assured widespread public acceptance (regardless of actual effectiveness) which in turn advances the interests of political leaders who benefit from being associated with highly visible, popular symbolic programs" (Wysong et al. 1994, p. 461).

Biological and Psychological Theories

Drug use and addiction are probably the result of a complex interplay of social, psychological, and biological forces. Biological research has primarily concentrated on the role of genetics in predisposing an individual to drug use. Research indicates that severe, early-onset alcoholism may be genetically predisposed, with some men having 10 times the risk for addiction as those without a genetic predisposition. Interestingly, other problems such as depression, chronic anxiety, and attention deficit disorder are also linked to the likelihood of addiction. Nonetheless, researchers warn, "Nobody is predestined to be an alcoholic" (Firshein 2003).

Biological theories of drug use also hypothesize that some individuals are physiologically predisposed to experience more pleasure from drugs than others and, consequently, are more likely to be drug users. According to these theories, the central nervous system, which is composed primarily of the brain and spinal cord, processes drugs through neurotransmitters in a way that produces an unusually euphoric experience. Individuals not so physiologically inclined reported less pleasant experiences and are less likely to continue use (Jarvik 1990; National Institute on Alcohol Abuse and Alcoholism 2000).

Psychological explanations focus on the tendency of certain personality types to be more susceptible to drug use. Individuals who are particularly prone to anxiety may be more likely to use drugs as a way to relax, gain self-confidence, or ease tension. For example, research indicates that female adolescents who have been sexually abused or who have poor relationships with their parents are more likely to have severe drug problems (NIDA 2000).

Psychological theories of drug abuse also emphasize that drug use may be maintained by positive or negative reinforcement. Thus, for example, cocaine use may be maintained as a result of the rewarding "high" it produces—a positive reinforcement. Alternatively, heroin use, often associated with severe withdrawal symptoms, may continue as a result of a negative reinforcement, that is, the distress the user feels when faced with withdrawal (Abadinsky 2004). Reinforcement may come from a variety of sources including the media. In a recent study of the portrayal of alcohol use in 601 contemporary movies, researchers found that exposure to alcohol use in the movies was positively associated with early onset drinking (Sargant et al. 2006).

What Do You Think? Are alcoholism and other drug addictions a consequence of nature or nurture? If nurture, what environmental factors contribute to such problems, and what would you recommend in terms of prevention strategies? If nature, do you think that drug addiction is a consequence of biological factors alone? If you consume alcohol, what are some of your motivations for drinking?

FREQUENTLY USED LEGAL AND ILLEGAL DRUGS

More than 19.7 million people in the United States are illicit drug users, who represent 8.1 percent of the population age 12 and older. Users of illegal drugs, although varying by type of drug used, are more likely to be male, to be young, and to be a member of a minority group (U.S. Department of Health and Human Services 2006). Social definitions regarding which drugs are legal or illegal, however, have varied over time, circumstance, and societal forces. In the United States two of the most dangerous and widely abused drugs, alcohol and tobacco, are legal.

Alcohol

Americans' attitudes toward alcohol have had a long and varied history. Although alcohol was a common beverage in early America, by 1920 the federal government had prohibited its manufacture, sale, and distribution through the passage of the Eighteenth Amendment to the U.S. Constitution. Many have argued that Prohibition, like the opium regulations of the late 1800s, was in fact a "moral crusade" (Gusfield 1963) against immigrant groups who were more likely to use alcohol. The amendment had little popular support and was repealed in 1933. Today, the U.S. population is experiencing a resurgence of concern about alcohol. What has been called a "new temperance" has manifested itself in federally mandated 21-year-old drinking age laws, warning labels on alcohol bottles, increased concern over fetal alcohol syndrome and underage drinking, stricter enforcement of drinking and driving regulations (e.g., checkpoint traffic stops), and zero tolerance policies. Such practices may have had an effect on drinking norms. Between 2004 and 2005 the rate of past month alcohol use among 12- to 17-year-olds significantly decreased (U.S. Department of Health and Human Services 2006). Although some attack these policies as being too restrictive, in their absence (e.g., in Europe) youthful drinking is much higher (OJJDP 2005).

Nonetheless, alcohol remains the most widely used and abused drug in America. Over the past decade, although rates have declined for some age groups, the overall rate has increased both in terms of frequency and quantity of alcohol consumed (Jones 2006). Although most people who drink alcohol do so moderately and experience few negative effects (see this chapter's *Self and Society* feature), alcoholics are psychologically and physically addicted to alcohol and suffer various degrees of physical, economic, psychological, and personal harm.

The National Survey on Drug Use and Health, conducted by the U.S. Department of Health and Human Services, reported that 126 million Americans age 12 and older consumed alcohol at least once in the month preceding the survey; that is, they were current users (U.S. Department of Health and Human Services 2006). Of this number, 6.6 percent reported being heavy drinkers (defined as drinking five or more drinks per occasion on 5 or more days in the survey month) and 22.7 percent were **binge drinkers**—55 million people.

Even more troubling were the 10.8 million current users of alcohol who were 12–20 years old—underage drinkers—more than half of whom reported being heavy or binge drinkers (U.S. Department of Health and Human Services 2006). Although teen drinking has decreased in recent years, binge drinking in college continues to attract media attention and thus the public's attention. The alcohol consumed by binge drinkers represents 70.0 percent of all alcohol consumed by college students (Schemo 2002). Furthermore, the money spent annually by college students on alcohol, $5.5 billion, is more than what they spend on milk, soft drinks, coffee, tea, and books combined (MADD 2003). Many binge drinkers began drinking in high school, with almost one-third having their first drink before age 13. Research indicates that the younger the age of onset, the higher the probability is that an individual will develop a drinking disorder at some time in his or her life. For example, an individual's chance of becoming dependent on alcohol, as defined by the National Survey on Drug Use

binge drinking As defined by the U.S. Department of Health and Human Services, drinking five or more drinks on the same occasion on at least 1 day in the past 30 days prior to the National Survey on Drug Use and Health.

Self and Society | The Consequences of Alcohol Consumption

Indicate whether you have or have not experienced any of the following consequences of drinking in the year before completing this survey. When finished compare your responses to the percentages below.

of alcohol and other drugs on college campuses" (CORE 2006). The data below are for 2005 and represent the responses of a random sample of 33,379 undergraduate students at 53 U.S. colleges.

Consequence	Yes	No
1. Had a hangover.	____	____
2. Performed poorly on a test or other project.	____	____
3. Trouble with police or authorities.	____	____
4. Damaged property, pulled fire alarm, etc.	____	____
5. Got into an argument or fight.	____	____
6. Got nauseated or vomited.	____	____
7. Driven a car while under the influence.	____	____
8. Missed a class.	____	____
9. Been criticized by someone I know.	____	____
10. Thought I might have a problem.	____	____
11. Had a memory loss.	____	____
12. Done something I later regretted.	____	____
13. Been arrested for DWI/DUI.	____	____
14. Been taken advantage of sexually.	____	____
15. Taken advantage of another sexually.	____	____
16. Tried unsuccessfully to stop using.	____	____
17. Seriously thought about suicide.	____	____
18. Tried to commit suicide.	____	____
19. Been hurt or injured.	____	____

Consequence	Percent Reporting Consequence
1. Had a hangover.	62.8
2. Performed poorly on a test or other project.	21.8
3. Trouble with police or authorities.	13.9
4. Damaged property, pulled fire alarm, etc.	7.0
5. Got into an argument or fight.	31.0
6. Got nauseated or vomited.	53.8
7. Driven a car while under the influence.	26.3
8. Missed a class.	30.7
9. Been criticized by someone I know.	30.3
10. Thought I might have a problem.	10.4
11. Had a memory loss.	34.3
12. Done something I later regretted.	38.1
13. Been arrested for DWI/DUI.	1.4
14. Been taken advantage of sexually.	10.3
15. Taken advantage of another sexually.	3.0
16. Tried unsuccessfully to stop using.	4.8
17. Seriously thought about suicide.	4.0
18. Tried to commit suicide.	1.1
19. Been hurt or injured.	15.5

The above survey items are from the CORE Alcohol and Drug Survey. The survey "assesses the nature, scope, and consequences

Source: CORE Alcohol and Drug Survey. 2006. "Results." Available at http://www.siu.edu/~coreinst/

and Health, is 40.0 percent if the person's drinking began before the age of 13. The chances of being alcohol dependent also increase if an individual's parents (1) are alcoholics, (2) drink, (3) have a positive attitude about drinking, or (4) use discipline sporadically ("How Does Alcohol" 2003). Heavy teenage drinkers, as with their adult counterparts, are more likely to be white, non-Hispanic, and male (U.S. Department of Health and Human Services 2006).

Additional results from the National Survey on Drug Use and Health (U.S. Department of Health and Human Services 2006) include the following:

- The highest levels of both heavy and binge drinking are among 18- to 25-year-olds, peaking at age 21.
- Rates of alcohol use are higher among the employed than among the unemployed; however, patterns of heavy or binge drinking are highest among the unemployed.
- College graduates are less likely to be binge drinkers than high school graduates but are more likely to report alcohol use in the past month.
- Binge drinking is least likely to be reported by Asians and most likely to be reported by American Indians and Alaska Natives.

Still, I am not an alcoholic. As far as I can tell, I have no family history of alcoholism. I am not physically addicted to drinking, and I don't have the genetically based reaction to alcohol that addiction counselors call "a disease." In the nine years that I drank, I never hid bottles or drank alone, and I never spent a night in a holding cell awaiting DUI charges. Today, one glass of wine would not propel me into the type of bender where I'd wind up drinking whole bottles. While I have been to AA meetings, I don't go to them.

I am a girl who abused alcohol, meaning I drank for the explicit purpose of getting drunk, getting brave, or medicating my moods. In college, that abuse often took the form of binge drinking, which for women, means drinking four or more drinks in a row at least once during a span of two weeks. But frequently, before college and during it, more time would pass between rounds, and two or three drinks could get me wholly obliterated.

I wrote this book knowing that my alcohol abuse, though dangerous, was not unprecedented. Nor were the after effects I experienced as a result of it. Mine are ordinary experiences among girls and young women in both the U.S. and abroad, and I believe that very commonness makes them noteworthy.

In the past decade alone, girls have closed the gender gap in terms of drinking. I wrote this book because girls are drinking as much, and as early, as boys for the first time in history, because there has been a threefold increase in the number of women who get drunk at least ten times a month, and because a 2001 study showed 40 percent of college girls binge drink. When you factor in increased rates of depression, suicide, alcohol poisoning, and sexual assault, plus emerging research that suggests women who drink have greater chances of liver disease, reproductive disorders, and brain abnormalities, the consequences of alcohol abuse are far heavier for girls than boys.

I also wrote this book because I wanted to quash the misconceptions about girls and drinking: that girls who abuse alcohol are either masculine, sloppy, sexually available, or all of the above, that girls are drinking more and more often in an effort to compete with men, and that alcohol abuse is a lifestage behavior, a youthful excess that is not as damaging as other drugs. (pp. xv–xvi)

For many girls, alcohol abuse may be a stage that tapers off after the quarter-life mark. Many will be spared arrests, accidents, alcoholism, overdoses, and sexual assaults. A whole lot of them will have close calls, incidents they will recount with self mocking at dinner parties some fifteen years later. Some of them will have darker stories, memories or half memories or full-out blackouts, that they will store in the farthest corners of their mental histories and never disclose to their families or lovers. But I fear that women, even those women who escape the physical consequences of drinking, won't escape the emotional ones. I fear some sliver of panic, sadness, or self loathing will always stay with us. (p. xvii)

Nine years after I took my first drink, it occurs to me that I haven't grown up. I am missing so much of the equipment that adults should have, like the ability to sustain eye contact without flinching or letting my gaze roll slantwise to the floor. At this point in time, I should be able to hear my own unwavering voice rise in public without feeling my heart flutter like it's trying to take flight. I should be able to locate a point of conversation with the people I deeply long to know as my friends, like my memoirist neighbor or the woman in my reading group who carries the same tattered paperbacks that I do and wears the same foot-less tights. I should be able to stop self-censoring and smile when I feel like it. I should recognize happiness when I feel it expand in my gut. (pp. xvii–xviii)

In the end, I quit drinking because I didn't want to waste any more time picking up the pieces. I decided *smashed,* when it's used as a synonym for drunk, is a self-fulfilling prophecy. (p. ix)

Source: Zailckas, Koren. 2005. *Smashed: Story of a Drunken Girlhood.* New York: Viking. Reprinted with permission.

- Full-time college students between the ages of 18 and 22 years old are more likely to have used alcohol in the past month, binge drink, and drink more heavily than their peers not enrolled in college full-time.
- More males than females age 12–20 years reported current alcohol use, binge drinking, and heavy drinking.

Despite the fact that males are more likely to abuse alcohol than females, some evidence suggests that female drinking is on the rise (Armstrong and McCarroll 2004). This chapter's *The Human Side* feature describes *Smashed,* the story of a drunken girlhood (Zailckas 2005).

Many alcohol users reported using other controlled substances, but the more frequently a student binges, the higher the probability is of he or she reporting other drug use. Some evidence suggests that certain combinations of

According to the American Cancer Society (2006), about half of those who continue to smoke throughout their lives will die of a smoking-related disease. Further, the economic cost of cigarette smoking is astronomical—nearly $167 billion in health-related expenses annually. It is not surprising then that research on smoking and specifically smoking onset is plentiful. Most research on smoking onset, i.e., the age at which an individual begins to smoke, is focused on young adults or adolescents. Research on those younger than age 12 is rare but important—the younger a person is when he or she begins to smoke, the higher the probability of developing a nicotine addiction (American Cancer Society 2006). The present research examines "children's attitudes toward, belief about, and lifestyle associations with cigarette smoking" (Freeman, Brucks, & Wallendorf 2005, p. 1537).

Sample and Methods

Children were recruited from three racially, ethnically, and economically diverse elementary schools in the southwestern United States. Sample demographics were representative of the surrounding community with half of the respondents being female and 72.4 percent being non-Hispanic whites. Parental permission was obtained before children were included in the sample, and all children were in the second (n = 100) or fifth grade (n = 141); i.e., ages 7 and 8 or 11 and 12.

Using a child-friendly computer program, children were randomly assigned to view a set of six photographs and then were asked to pick the picture of the person who would like to smoke cigarettes the MOST and the person who would like to smoke cigarettes the LEAST. There were a total of 18 pictures that were divided into three subsets—six pictures each.

Within each set the images portrayed a range of physical attractiveness, sociability, and independence. For example, one of the sets included pictures of (1) a heavy-set woman standing in a field, (2) a thin, attractive female standing by a car, (3) a middle-aged man reading by a fireplace, (4) a young man playing lacrosse with some friends, (5) a young girl riding her bicycle, and (6) a young man sitting in a boat fishing by himself (p. 1539).

After each photo selection, trained interviewers asked the children to respond to probes concerning lifestyle associations. Queries included: "Tell me about him/her," "What does he/she like best about smoking cigarettes?" and "What does he/she not like about smoking cigarettes?" The interviewer followed up the initial responses until she or he "perceived that the participant had thoroughly articulated their perceptions of the person depicted in the selected picture and the person's motivations for liking (not liking) to smoke cigarettes" (p. 1540).

Finally, respondents were asked a series of structured survey items. Included were questions about the presence or absence of cigarette smoking in the home as well as a set of items designed to tap attitudes about the age-appropriateness of smoking, the negative consequences of smoking, and the instrumental benefits of smoking. Given the age of the respondents, possible agree-disagree responses included "yes," "maybe," "don't think so," and "no way."

drugs—for example, alcohol and cocaine—heighten the negative effects of either drug separately; that is, there is a negative drug interaction.

Tobacco

Although nicotine is an addictive psychoactive drug, and tobacco smoke has been classified by the Environmental Protection Agency as a Group A carcinogen, tobacco continues to be one of the most widely used drugs in the United States. According to a U.S. Department of Health and Human Services survey, 60.5 million Americans continue to smoke cigarettes—24.9 percent of the population age 12 and older. In 2005, 11.0 percent of the 12- to 17-year-old population reported smoking, a decline from 13.0 percent in 2002 (U.S. Department of Health and Human Services 2006). Interestingly, among 12- to 17-year-olds, three brands of cigarettes account for more than 50.0 percent of the tobacco market—Marlboro, Newport, and Camel. Use of all tobacco products, including smokeless tobacco (7.7 million users), cigars (13.6 million users), pipe tobacco (2.2 million users), and cigarettes, is higher for high school graduates than for college graduates, males, and Native Americans and Alaska Natives (U.S. Department of Health and Human Services 2006).

Findings and Conclusions

The results indicated that both second and fifth grader's selections of those who would like to smoke the MOST were not randomly distributed and were thus likely to be influenced by lifestyle associations. For example, the picture of a slender, professionally dressed woman leaning against a car elicited such responses as, "She thinks she's all that," "She's too cool not to smoke," and "She's just finished a stressful day at work, so she smokes to relax." There were no significant patterns of picture selections for those who would like to smoke the LEAST (only one of the six pictures resulted in a significant test statistic). Further, the presence of a smoker in the respondent's home was also unrelated to picture selection.

Responses to the open-ended probes were coded and analyzed. Not surprisingly, fifth graders were more articulate in identifying reasons for picture selection than second graders. Pictures of people who liked to smoke cigarettes the MOST were associated with certain motivations—they smoked to "feel better, alleviate stress, or overcome a bad mood; [to] feel cool or look cool; and . . . to have something to do" (p. 1542). Conversely, the picture of the individual who likes to smoke the LEAST (a young girl on a bicycle) was described as being too young to smoke, being already happy, and knowing that smoking has negative health consequences.

Finally, analysis of the survey items indicated that both second and fifth graders strongly agree with the negative consequences of smoking, i.e., that "smoking makes people look dumb," is "gross," and "makes people feel sick" and that they would be "upset if one of [their] friends started smoking" (p. 1542). Less agreement was found in terms of the instrumental benefits of smoking. Fifth graders, when compared with second graders, were significantly more likely to agree that smoking helps relax people and "cheers people up when they are in a bad mood" (p. 1542). The percentages of agreement by fifth graders for negative consequences and instrumental benefits were 38.3 percent and 29.8 percent, respectively.

In summarizing the results, the authors concluded that at an early age children have already developed lifestyle associations with smoking—people like to smoke out of boredom, to feel better, to relax, to be cool, and to relieve stress. Children are also very aware of the negative consequences of smoking, as well as some of the presumed instrumental benefits although the latter is truer for 10- and 11-year-olds than for 7- and 8-year-olds.

Children do not, however, have "well developed lifestyle associations with nonsmokers" (p. 1543). The lifestyle association attributed to the one picture consistently selected as someone who would like to smoke the LEAST (by fifth graders) was in part the fact that the woman was too young to smoke cigarettes. As the authors noted, in terms of smoking prevention, this is not a desirable outcome for it implies that there is an age at which smoking is appropriate and may "predispose young children to initiate tobacco use when they reach adolescence" (p. 1544). Thus, developing age-specific prevention programs may facilitate reducing tobacco use.

Source: Freeman, Dan, Merrie Brucks, and Melanie Wallendorf. 2005. "Young Children's Understandings of Cigarette Smoking." *Addiction* 100(6):1537–1545.

In 2005, 2.7 million youths between the ages of 12 and 17 reported past-month use of a tobacco product (U.S. Department of Health and Human Services 2006). Research evidence suggests that youth develop attitudes and beliefs about tobacco products at an early age (see this chapter's *Social Problems Research Up Close* feature). Advertising of tobacco products continues to have an influence on youth despite a 1998 legal settlement between the tobacco companies and the states that prohibited any tobacco company from taking "action, directly or indirectly, to target youth . . . in the advertising, promotion or marketing of tobacco products." Recent marketing of candy-flavored tobacco products (e.g., a pineapple-and-coconut-flavored cigarette called Kauai Kilada) is a case in point (Campaign for Tobacco-Free Kids 2005).

There is evidence that cigarette advertisers target minorities. "Studies have shown a higher concentration of tobacco advertising in magazines aimed at African Americans, such as *Jet* and *Ebony*, than in similar magazines aimed at a broader audience, such as *Time* and *People*. . . . From 1992 to 2000 smoking rates increased among African American twelfth graders from 8.7 percent to 14.2 percent" (American Heart Association 2003). Similarly, the tobacco industry is accused of developing advertising campaigns that target Hispanics and women. Advertising is but one venue criticized for the positive portrayal of tobacco use.

The Campaign for Tobacco-Free Kids calls the introduction of candy-flavored cigarettes and smokeless tobacco an "outrageous" tactic to lure youth into using tobacco products. Note the appeal to African-American youth and women in some of the packaging.

In a study by Goldstein, Soble, and Newman (1999), children's animated films (e.g., *Aladdin*) were found to portray smoking and alcohol use positively. Ironically, advertising by tobacco companies designed to *discourage* youthful smoking has been shown to be ineffective and, in some cases, may actually encourage underage smoking (Fox 2006).

Native Americans, who introduced it to the European settlers in the 1500s, first cultivated tobacco. The Europeans believed that tobacco had medicinal properties, and its use spread throughout Europe, ensuring the economic success of the colonies in the New World. Tobacco was initially used primarily through chewing and snuffing, but in time smoking became more popular, even though scientific evidence that linked tobacco smoking to lung cancer existed as early as 1859 (Feagin & Feagin 1994). However, it was not until 1989 that the U.S. Surgeon General concluded that tobacco products are addictive and that nicotine causes the dependency. Today, the hazards of tobacco use are well documented and have resulted in federal laws that require warning labels on cigarette packages and prohibit cigarette advertising on radio and television. Nonetheless, a recent report by the Harvard School of Public Health concludes that nicotine levels in cigarettes *increased* between 1997 and 2005. Says Dr. Howard Koh, one of the authors of the study, "It was systematic, it was pervasive, it involved all manufacturers, and it was by design" (Smith 2007, p. 1). According to the World Health Organization (2006), by the year 2020, 70.0 percent of tobacco-related deaths will take place in poor nations where many smokers are unaware of the health hazards associated with their behavior.

Marijuana

Marijuana is the most commonly used and most heavily trafficked illicit drug in the world (see Table 3.2 for a list of commonly abused illegal drugs, their commercial and street names, and their intoxication and health effects). Globally, there are 162 million marijuana users, representing 4.0 percent of the world's adult population. Regionally, marijuana is the most dominant illicit drug and its cultivation and consumption in North America are particularly high. For example, despite government eradication campaigns that destroyed 80.0 percent of Mexico's crop, the remaining 20.0 percent is sufficient to supply a large portion of the U.S. marijuana market (World Drug Report 2006).

Marijuana's active ingredient is THC (Δ^9-tetrahydrocannabinol), which in varying amounts can act as a sedative or a hallucinogen. When just the top of the marijuana plant is sold, it is called hashish. Hashish is much more potent than marijuana, which comes from the entire plant. Marijuana use dates back to 2737 B.C. in China, and marijuana has a long tradition of use in India, the Middle East, and Europe. In North America, hemp, as it was then called, was used to make rope and as a treatment for various ailments. Nevertheless, in 1937 Congress passed the Marijuana Tax Act, which restricted the use of marijuana; the law was passed as a result of a media campaign that portrayed marijuana users as "dope fiends" and, as conflict theorists note, was enacted at a time of growing sentiment against Mexican immigrants (Witters et al. 1992).

There are more than 14.6 million current marijuana users, representing 6.0 percent of the U.S. population age 12 and older (U.S. Department of Health and Human Services 2006). According to the Monitoring the Future survey, 44.8 percent of twelfth graders have used marijuana or hashish at least once in their lifetime, 33.6 percent used it in the last year, and 19.8 percent used it in the last month. This usage occurred despite the fact that 57.9 percent of the twelfth graders responded that smoking marijuana regularly is harmful. The study further showed that 8.1 percent of eighth graders and 16.8 percent of tenth graders reported use of an illicit drug in the past month, with marijuana being the most commonly reported. Although these numbers remain unacceptably high, they represent a 25.0 percent drop in current (i.e., use in the last month) teen marijuana use since 2001 (Monitoring the Future 2006).

Not surprisingly, teenage marijuana users who reported positive effects from use, such as "feeling happy," "getting really high," and "laughing a lot," were more likely to be addicted to marijuana. Almost 40.0 percent of young people between the ages of 16 and 21 who reported five or more positive responses were addicted to marijuana compared with 5.2 percent of those who had no positive responses (Eisner 2005).

Although the effects of alcohol and tobacco are, in large part, well known, the long-term psychological and physiological effects of marijuana are less understood. According to the Office of National Drug Control Policy (ONDCP 2005a), marijuana is associated with many negative health effects, including impaired memory, anxiety, panic attacks, and increased heart rate. Another important concern is that marijuana may be a **gateway drug,** the use of which causes progression to other drugs.

More likely, however, is the explanation that people who experiment with one drug are more likely to experiment with another. Indeed, most drug users are poly-drug users. Research has indicated that there is a strong "contemporaneous relationship between smoking cigarettes, drinking alcohol, smoking marijuana, and using cocaine" (Lee and Abdel-Ghany 2004, p. 454).

> 66 Everybody knew it's addictive. Everybody knew it causes cancer. We were all in it for the money. 99
>
> Victor Crawford
> Former tobacco lobbyist and smoker who developed lung cancer

What Do You Think? Many argue that the right of an adult to make an informed decision includes deciding to use illegal drugs. Similar to arguments about the legalization of prostitution and gambling, drug use, particularly marijuana use, is considered a victimless crime by many. Do you think marijuana, like alcohol, should be legal for those older than age 21? Older than age 18? Given its analgesic effect, should it be readily available to terminally ill patients in pain? Why or why not?

gateway drug A drug (e.g., marijuana) that is believed to lead to the use of other drugs (e.g., cocaine).

TABLE 3.2 Commonly Abused Illegal Drugs

SUBSTANCE	COMMERCIAL/STREET NAMES	HOW ADMINISTERED
Cannabis		
Hashish	Hash, hemp, boom, chronic, gangster	Swallowed, smoked
Marijuana	Grass, joint, mary jane, pot, reefer, weed, herb, blunt, skunk, dope	Swallowed, smoked
Depressants		
Barbiturates	Amytal, Nembutal, Seconal, Phenobarbital; barbs, reds, red birds, phennies, tooies, yellows, yellow jackets	Injected, swallowed
Benzodiazepines (other than flunitrazepam)	Ativan, Halcion, Librium, Valium, Xanax; candy, downers, sleeping pills, tranks	Swallowed, injected
Flunitrazepam	Rohypnol; forget-me pill, Mexican valium, R2, Roche, roofies, roofinol, rope, rophies	Swallowed, snorted
GHB	Gammahydroxybutyrate; G, Georgia home boy, grievous bodily harm, liquid ecstacy	Swallowed
Ketamine	Ketalar SV; cat, valiums, K, Special K, vitamin K	Injected, snorted, smoked
Hallucinogens		
LSD	Lysergic acid diethylamide; acid, blotter, boomers, cubes, microdot, yellow sunshines	Swallowed, absorbed through mouth tissue
Opioids		
Heroin	Diacetyl—morphine; brown sugar, dope, H, horse, junk, skag, skunk, smack, white horse	Injected, smoked, snorted
Opium	Laudanum, paregoric; big O, black stuff, block, gum, hop	Swallowed, smoked
Oxycodone HCL	OxyContin; Oxy, O.C., killer	Swallowed, snorted, injected
Stimulants		
Amphetamine	Biphetamine, Dexedrine; bennies, black beauties, crosses, hearts, LA turnaround, speed, truck drivers, uppers	Injected, swallowed, smoked, snorted
Cocaine	Cocaine hydrochloride; blow, bump, C, candy, Charlie, coke, crack, flake, rock, snow, toot	Injected, smoked, snorted
MDMA (methylenedioxy-methamphetamine)	Adam, clarity, Ecstasy, Eve, lover's speed, peace, STP, X, XTC	Swallowed
Methamphetamine	Desoxyn; chalk, crank, crystal, fire, glass, go fast, ice, meth, speed	Injected, swallowed, smoked, snorted
Other compounds		
Inhalants	Solvents (paint thinners, gasoline, glues), gases (butane, propane, aerosol propellants, nitrous oxide), nitrites (isoamyl, isobutyl, cyclohexyl); laughing gas, poppers, snappers, shippets	Inhaled through nose or mouth

Source: Adapted from the National Institute on Drug Abuse (2006b).

INTOXICATION EFFECTS	HEALTH EFFECTS
Euphoria, slowed thinking, slowed reaction time, confusion, impaired balance and coordination	Cough, respiratory infections, impaired memory and learning, increased heart rate, anxiety, panic attacks, addiction
Reduced anxiety, feeling of well-being, lowered inhibitions, slowed pulse and breathing, lowered blood pressure, poor concentration	Fatigue; confusion; impaired coordination, memory, and judgment; addiction; respiratory depression and arrest; death
Sedation, drowsiness	Depression, unusual excitement, fever, irritability, poor judgment, slurred speech, dizziness, life-threatening withdrawal
Sedation, drowsiness	Dizziness
	Visual and gastrointestinal disturbances, urinary retention, memory loss for the time under the drug's effects
	Drowsiness, nausea/vomiting, headache, loss of consciousness, loss of reflexes, seizures, coma, death
Pressure, impaired motor function	Memory loss, numbness, nausea/vomiting
Altered states of perception and feeling, nausea	Persisting perception disorder (flashbacks)
Increased body temperature, heart rate, and blood pressure; loss of appetite; sleeplessness, numbness, weakness, and tremors; persistent mental disorders	
Pain relief, euphoria, drowsiness	Nausea, constipation, confusion, sedation, respiratory depression and arrest, tolerance, addiction, unconsciousness, coma, death
Staggering gait	
Increased heart rate, blood pressure, and metabolism; feelings of exhilaration, energy, and increased mental alertness	Rapid or irregular heartbeat, reduced appetite, weight loss, heart failure, nervousness, insomnia
Rapid breathing	Tremor, loss of coordination, irritability, anxiousness, restlessness, delirium, panic, paranoia, impulsive behavior, aggressiveness, tolerance, addiction, psychosis
Increased temperature	Chest pain, respiratory failure, nausea, abdominal pain, strokes, seizures, headaches, malnutrition, panic attacks
Mild hallucinogenic effects, increased tactile sensitivity, empathic feelings	Impaired memory and learning, hyperthermia, cardiac toxicity, renal failure, liver toxicity
Aggression, violence, psychotic behavior	Memory loss, cardiac and neurological damage, impaired memory and learning, tolerance, addiction
Stimulation, loss of inhibition, nausea or vomiting, slurred speech, loss of motor coordination, wheezing	Unconsciousness, cramps, weight loss, muscle weakness, depression; memory impairment, damage to cardiovascular and nervous systems, sudden death

Cocaine

Cocaine is classified as a stimulant and, as such, produces feelings of excitation, alertness, and euphoria. Although prescription stimulants such as methamphetamine and dextroamphetamine are commonly abused, over the last 20 years societal concern over drug abuse has focused on cocaine. Its increased use, addictive qualities, physiological effects, and worldwide distribution have fueled such concerns. More than any other single substance, cocaine led to the present war on drugs.

Cocaine, which is made from the coca plant, has been used for thousands of years, but anticocaine sentiment in the United States did not emerge until the early 1900s, when it was primarily a response to the heavy use of cocaine among urban blacks, poor whites, and criminals (Witters et al. 1992; Thio 2007). Cocaine was outlawed in 1914 by the Harrison Narcotics Act, but its use and effects continued to be misunderstood. For example, a 1982 *Scientific American* article suggested that cocaine was no more habit forming than potato chips (Van Dyck & Byck 1982). As demand and then supply increased, prices fell from $100 a dose to $10 a dose, and "from 1978 to 1987 the U.S. experienced the largest cocaine epidemic in history" (Witters et al. 1992, p. 256).

Globally, cocaine use has decreased although production levels have remained fairly stable (World Drug Report 2006). Europe is the exception where cocaine use has increased dramatically in Italy, Spain, and England (UNODC 2006a). Cocaine, after marijuana, is the second most widely used illegal drug in North and South America (UNODC 2006b). According to the National Survey on Drug Use and Health, 2.4 million Americans 12 years and older are current cocaine users—a slight although not statistically significant increase from 2004 (U.S. Department of Health and Human Services 2006). Lifetime use of cocaine by high school seniors is 8.5 percent, by tenth graders is 4.8 percent, and by eighth graders is 3.7 percent. About half of all twelfth graders indicated that getting powdered cocaine is "fairly easy" or "very easy," with a near equal percentage responding that trying cocaine once or twice is a "great risk" (Monitoring the Future 2006).

Crack is a crystallized product made by boiling a mixture of baking soda, water, and cocaine. The result, also called rock, base, and gravel, is relatively inexpensive and was not popular until the mid-1980s. Crack is one of the most dangerous drugs to surface in recent years. Crack dealers often give drug users their first few "hits" free, knowing the drug's intense high and addictive qualities are likely to produce returning customers. An addiction to crack can take 6–10 weeks; an addiction to pure cocaine can take 3–4 years (Thio 2007).

According to the National Survey on Drug Use and Health, 3.3 percent of Americans 12 and older (7.9 million people) have tried crack cocaine once in their lifetime. Just over 682,000 are current users, compared with 467,000 in 2004. Results from the Monitoring the Future survey indicate that 3.9 percent of high school seniors reported using crack at least once in their lifetime, and half reported that using the drug just once or twice is a "great risk" (Monitoring the Future 2006).

Methamphetamine

Methamphetamine (meth, speed, crank) is a central nervous system stimulant that can be injected, snorted, smoked, or ingested orally and is highly addictive. It produces a short "rush" followed by periods of increased activity, decreased appetite, and a sense of well-being which can last between 20 minutes and 12

"Crack is a drug peddler's dream: it is cheap, easily concealed, and provides a short duration high that invariably leaves the user craving more.**"**

Tom Morganthau
Journalist

crack A crystallized illegal drug product produced by boiling a mixture of baking soda, water, and cocaine.

hours (ONDCP 2006a). As Thio (2007) noted, although the drug has only recently become popular, it is not new.

> During the Second World War, soldiers on both sides used it to reduce fatigue and enhance performance. Hitler was widely believed to be a meth addict. Later, in the 1960s, President John Kennedy also used the drug and soon it caught on among so-called "speed freaks." But, because it was extremely expensive as well as difficult to obtain, meth was never close to being as widely used as cocaine. (p. 276)

Today, methamphetamine is relatively inexpensive and easily obtained with more than 25 percent of high school seniors reporting that the drug is "fairly easy" or "very easy" to get (Monitoring the Future 2006). Ease of obtaining the drug is, in part, a result of the number of clandestine laboratories in the United States, the large amounts of methamphetamine smuggled into the United States from Mexico, and the ease of production. Because methamphetamine can easily be made from cold medications such as Sudafed, the U.S. Congress passed the Comprehensive Methamphetamine Control Act of 1996 that made obtaining the chemicals needed to make methamphetamine more difficult (ONDCP 2006a; Thio 2007). In 2006, the Combat Methamphetamine Epidemic Act, which further articulated standards for selling over-the-counter medications used in methamphetamine production, went into effect.

According to the National Survey on Drug Use and Health, 4.3 percent of the population 12 and older reported using methamphetamine at least once in their lifetime—10.4 million people—a slight decline from 2004. Current methamphetamine use among full-time college students (18- to 22-year-olds) significantly increased between 2004 (0.2 percent) and 2005 (0.5 percent), and 4.1 percent of all college students reported using methamphetamine at least once in their lifetime (U.S. Department of Health and Human Services 2006). Among students surveyed as part of the Monitoring the Future study, lifetime use by eighth, tenth, and twelfth graders was 2.7, 3.2, and 4.4 percent, respectively (Monitoring the Future 2006). Not surprisingly, when a sample of respondents were asked, "How concerned are you about the use or sale of 'crystal meth' in your local community? Very concerned, somewhat concerned, not too concerned, or not concerned at all?" 65 percent of Americans reported that they were very concerned (Carroll 2005). Furthermore, in a survey of state and local law enforcement agencies, 38.8 percent responded that "meth" is their greatest drug threat (U.S. Department of Justice 2007) (see Figure 3.1).

Other Drugs

Other drugs abused in the United States include "club drugs" (e.g., lysergic acid diethylamide and Ecstasy), heroin, prescription drugs (e.g., tranquilizers and amphetamines), and inhalants (e.g., glue).

Club Drugs. Club drugs is a general term for illicit, often synthetic drugs commonly used at nightclubs or all-night dances called raves. Club drugs include MDMA (3,4-methylenedioxy-*N*-methylamphetamine, Ecstasy), ketamine ("Special K"), LSD ("acid"), GHB (γ-hydroxybutyrate, "liquid ecstasy"), and Rohypnol (flunitrazepam, "roofies").

Ecstasy, the most common name for MDMA, is manufactured in and trafficked from Europe and is the most popular of the club drugs. It ranges in price from $20 to $30 a dose (Leshner 2003). There is some evidence that use of

club drugs A general term for illicit, often synthetic, drugs commonly used at nightclubs or all-night dances called "raves."

FIGURE 3.1
National Drug Threat Survey: Greatest drug threat as reported by state and local agencies, 2006.
Source: National Drug Threat Assessment (2007). Available at http://www.dea.gov/concern/18862/index.htm

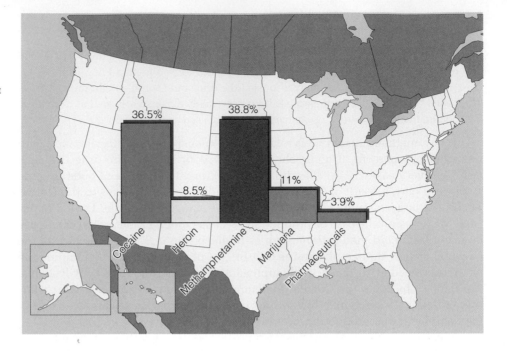

Ecstasy is growing throughout Europe (UNODC 2004) although, since 2002, use among youth in the United States has decreased by two-thirds (ONDCP 2006b). In 2005, 4.7 percent of the American population age 12 and older had used Ecstasy at least once in their lifetime (U.S. Department of Health and Human Services 2006). Ecstasy is associated with feelings of euphoria and inner peace, yet critics argue that as the "new cocaine," it can produce both long-term (e.g., permanent brain damage) and short-term (e.g., hyperthermia) negative effects (Cloud 2000; DEA 2000; ONDCP 2003a; Thio 2007). According to the Monitoring the Future survey (2006), many students are well aware of the dangers of Ecstasy. Approximately 59.3 percent of high school seniors reported that using Ecstasy once or twice is a "great risk."

Ketamine and LSD both produce visual effects when ingested. Use of ketamine, an animal tranquilizer, can also cause loss of long-term memory, respiratory problems, and cognitive difficulties. Ketamine use by eighth, tenth, and twelfth graders has decreased slowly but steadily over the last several years (Monitoring the Future 2006). LSD is a synthetic hallucinogen, although many other hallucinogens are produced naturally (e.g., peyote). In 2005, 22.4 million Americans, 9.2 percent of the population age 12 and older, reported lifetime use of LSD. Although twelfth graders have a higher past-month use of LSD (0.6 percent) than the general population age 12 and older (0.4 percent), LSD use among teenagers has decreased in recent years (U.S. Department of Health and Human Services 2006; Monitoring the Future 2006).

GHB and Rohypnol are often called **date-rape drugs** because of their use in rendering victims incapable of resisting sexual assaults. The Food and Drug Administration banned GHB, a central nervous system depressant, in 1990, although kits containing all the necessary ingredients to manufacture the drug continued to be available on the Internet. On February 18, 2000, President Clinton signed a bill that made GHB a controlled substance and thus illegal to

date–rape drugs Drugs that are used to render victims incapable of resisting sexual assaults.

manufacture, possess, or sell. Nonetheless, 1.1 percent of twelfth graders, 0.7 percent of tenth graders and 0.8 percent of eight graders reported past-year use of GHB (Monitoring the Future 2006).

Rohypnol, presently illegal in the United States, is lawfully sold by prescription in more than 50 countries for the short-term treatment of insomnia (Abadinsky 2004; ONDCP 2006c). It belongs to a class of drugs known as benzodiazepines, which also includes common prescription drugs such as Valium, Halcion, and Xanax. Rohypnol is a tasteless and odorless depressant. The effects of Rohypnol begin within 15–20 minutes; 1 milligram of the drug can incapacitate a victim for up to 12 hours (NIDA 2000; DEA 2000; ONDCP 2003b). Roofies are popular with high school and college students, costing only $5 a tablet. In 2006, 0.5 percent, 0.5 percent, and 1.1 percent of eighth, tenth, and twelfth graders, respectively, reported past year use of Rohypnol (Monitoring the Future 2006).

Heroin. Heroin (dope, H, smack, horse) is an analgesic—that is, a painkiller—and is the most commonly abused opiate drug. Most heroin comes from the poppy fields of Laos, Burma, Thailand, Afghanistan, Pakistan, Iran, Mexico, and Colombia (Thio 2007). Overall use of heroin is higher in Europe than in the United States (ODCCP 2003). Highly addictive, heroin can be injected, snorted, or smoked. If intravenous injection is used, the euphoric effects are felt within 7–8 seconds; if heroin is snorted or smoked, the effects are felt within 10–15 minutes (ONDCP 2006d).

According to the National Survey on Drug Use and Health, 3.5 million Americans age 12 and older—1.5 percent of the population—reported use of heroin sometime in their life (U.S. Department of Health and Human Services 2006). The number of Americans who are heroin dependent is, of course, much lower, ranging from 850,000 to 1 million addicts. In 2006 the rate of heroin use among eighth, tenth, and twelfth graders remained the same from the previous year, although the rate was substantially reduced from the record highs of the 1990s. When asked about the potential harm of heroin use, 59.1 percent of high school seniors said that trying heroin once or twice was a "great risk" (Monitoring the Future 2006). In addition to experiencing the negative repercussions that heroin use shares with all other drugs, heroin users are subjected to the risks of human immunodeficiency virus (HIV)/acquired immunodeficiency syndrome (AIDS) if they take the drug intravenously.

What Do You Think? Jeffrey Reiman wrote (2007, p. 37) that on the "basis of available scientific evidence, there is every reason to suspect that we do our bodies more damage, more irreversible damage, by smoking cigarettes and drinking liquor than by using heroin." That said, how would a social constructionist explain the legality of alcohol and tobacco products? Why do you think alcohol and tobacco products are legal, whereas heroin continues to be a target of the war on drugs?

Heroin may also be mixed with other drugs, often resulting in deadly combinations. For example, when heroin is mixed with fentanyl, a painkiller, the result is a lethal cocktail that has resulted in hundreds of deaths in urban areas throughout the midwest and northeast United States (Shenfeld 2006). In one day alone, 11 people in Chicago died from the drug. The mixture's street names say it all—"flat-liner," "executioner," "drop dead," "the exorcists," and "Al Capone." Heroin can be mixed with other drugs as well, including marijuana (an "A-bomb") and cocaine ("dragon rock") (ONDCP 2006d).

Psychotherapeutic Drugs. Estimates of lifetime use of psychotherapeutic drugs—that is, nonmedical use of any prescription pain reliever, stimulant, sedative, or tranquilizer—remained essentially the same between 2004 and 2005 (U.S. Department of Health and Human Services 2006). However, more than 6 million people, 2.6 percent of the population 12 and older, reported current use of a psychotherapeutic drug for nonmedical reasons. Of these users, 4.6 million use pain relievers, 1.8 million use tranquilizers, 1.1 million use stimulants, and 272,000 use sedatives. Those who reported prescription drug use for nonmedical reasons were then asked how they obtained the drugs. Nearly 60 percent responded that they received the drugs from friends and relatives for free (SAMHSA 2006).

Recently, there has been an increase in use of several types of psychotherapeutic drugs, particularly painkillers. One such drug is OxyContin (oxycodone HCl). The pharmacological characteristics of OxyContin make it a suitable heroin substitute, leading to one of its street names—"hillbilly heroin." As a prescription painkiller, OxyContin is often covered by health insurance plans, which contributes to its appeal. When insurance will no longer pay, it is not uncommon for people to switch to heroin, which is less expensive on the street than OxyContin.

According to the National Survey on Drug Use and Health, 334,000 Americans 12 and older reported using OxyContin in the last month—0.1 percent of the survey population (U.S. Department of Health and Human Services 2006). Annual use rates among eighth and tenth graders reached their highest levels in recent years—2.6 percent and 3.8 percent in 2006. Annual use for high school seniors increased between 2002 (4.0 percent) and 2005 (5.5 percent) before dropping to 4.3 percent in 2006 (Monitoring the Future 2006).

Several studies indicate that psychotherapeutic drug abuse among college students has been rising for several years (Whitten 2006). For example, it is estimated that 7.4 percent of college students have used the painkiller Vicodin (hydrocodone) without a prescription. Use of Vicodin among college students is higher among "men, whites, fraternity/sorority members, and at schools in the Northeast" (Whitten 2006, p. 1). Use of Vicodin among eighth, tenth, and twelfth graders is also high with 2006 annual use levels at 3.0 percent, 7.0 percent, and 9.7 percent, respectively (Monitoring the Future 2006).

Inhalants. Inhalants act on the central nervous system with users reporting a psychoactive, mind-altering effect (ONDCP 2006e). Common inhalants include adhesives (e.g., rubber cement), food products (e.g., vegetable cooking spray), aerosols (e.g., hair spray and air fresheners), anesthetics (ether), and cleaning agents (e.g., spot remover). In total, more than 1,000 household products are currently abused (ONDCP 2005b). More than 22.7 million people (9.4 percent) 12 and older have reported trying inhalants at least once in their lifetime (U.S. Department of Health and Human Services 2006). Youth, however, are particularly prone to inhalant use. According to Monitoring the Future, 16.1 percent of eighth graders, 13.3 percent of tenth graders, and 11.2 percent of twelfth graders reported lifetime use of an inhalant (Monitoring the Future 2006). Young people

66 The only livin' thing that counts is the fix. . . . Like I would steal off anybody—anybody, at all, my own mother gladly included. 99

Heroin addict

often use inhalants in the belief that they are harmless or that prolonged use is necessary for any harm to result. In fact, inhalants are dangerous because of their toxicity and can result in what is called sudden sniff death syndrome. Unfortunately, the percentage of youth who reported that inhalants are dangerous has decreased over the last 5 years.

SOCIETAL CONSEQUENCES OF DRUG USE AND ABUSE

Drugs are a social problem not only because of their adverse effects on individuals but also because of the negative consequences their use has for society as a whole. Everyone is a victim of drug abuse. Drugs contribute to problems within the family and to crime rates, and the economic costs of drug abuse are enormous. Drug abuse also has serious consequences for health at both the individual and the societal level.

Family Costs

The cost to families of drug abuse is incalculable. It is estimated that 8.3 million U.S. children—11 percent of all children—live with at least one parent in need of treatment for drug or alcohol dependency (Center for Substance Abuse Treatment 2006). Children raised in such homes have a higher probability of neglect and abuse, behavioral disorders, and absenteeism from school as well as lower self-concepts, learning problems, and increased risk of drug abuse (Easley & Epstein 1991; Associated Press 1999; "How Does Alcohol" 2003; Center for Substance Abuse Treatment 2006; SAMHSA 2007). Children of alcoholics are four times more likely to have alcohol or drug problems than children of nonalcoholics.

Nearly 5 million adults who abuse alcohol have children younger than age 18 living with them. Sixty percent of these parents are fathers; 38 percent are mothers. Parents who reported abusing alcohol in the past year are also more likely to report cigarette and illicit drug use than parents who did not report alcohol abuse in the previous year. They were also more likely to report "household turbulence," including yelling, serious arguments, and violence (NSDUH 2004). Such behaviors take a heavy emotional toll. In a survey of 902 U.S. adults with a family member (husband, wife, son, daughter, brother, sister, mother, or father) with a drug or alcohol addiction, 70 percent responded that their relatives' addiction had a negative impact on their emotional or mental health (Saad 2006).

Alcohol abuse is the single most common trait associated with wife abuse (Charon 2002). The more violent the interaction, the more likely it is that the husband has been drinking excessively. Furthermore, serious violence in the first year of marriage is twice as high among heavy drinkers than among social drinkers (Johnson 2003). In a study of 320 men who were married or living with someone, twice as many reported hitting their partner only after they had been drinking, compared with those who reported the same behavior while sober (Leonard and Blane 1992).

Crime Costs

The drug behavior of individuals arrested, incarcerated, and in drug-treatment programs provides evidence of a link between drugs and crime. For example, adults who have been arrested for serious violent or property offenses (see Chap-

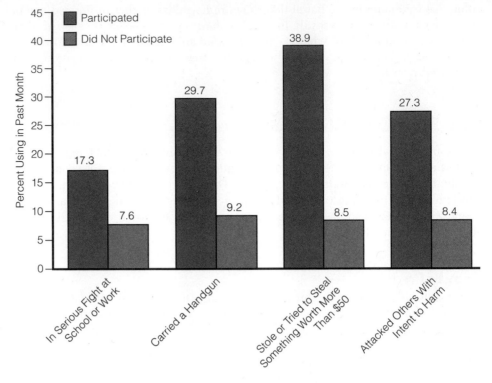

FIGURE 3.2
Past month illicit drug use among youths aged 12–17, by participation in fighting and delinquent behavior in the past year: 2005.
Source: U.S. Department of Health and Human Services (2006).

ter 4) are significantly more likely to have used an illicit drug in the past year (60.1 percent) compared with those not arrested (13.6 percent) (NSDUH 2006). Marijuana was the most commonly used illicit drug followed by cocaine, crack cocaine, hallucinogens, and methamphetamine. The National Crime Victimization Survey provides additional evidence of a crime-drug use link. Thirty percent of victims of violence reported that the offender was using drugs or alcohol (Bureau of Justice Statistics 2006).

The relationship between crime and drug use, however, is complex. Sociologists disagree as to whether drugs actually "cause" crime or whether, instead, criminal activity leads to drug involvement. Alternatively, as Siegel (2006) noted, criminal involvement and drug use can occur at the same time; that is, someone can take drugs and commit crimes out of the desire to engage in risk-taking behaviors. Furthermore, because both crime and drug use are associated with low socioeconomic status, poverty may actually be the more powerful explanatory variable. After extensive study of the assumed drug-crime link, Gentry (1995) concluded, "the assumption that drugs and crime are causally related weakens when more representative or affluent subjects are considered" (p. 491).

The National Survey on Drug Use and Health documents a relationship between delinquency and drug use. Youths were asked to report the number of times they had engaged in four categories of delinquency (see Figure 3.2). Youths who reported current drug involvement were more likely to have been involved in each of the behaviors measured. For example, 17.3 percent of those who reported being in a serious fight at school or at work used illicit drugs in the past month compared with 7.6 percent use by those who had not been in a serious fight at school or work (U.S. Department of Health and Human Services 2006).

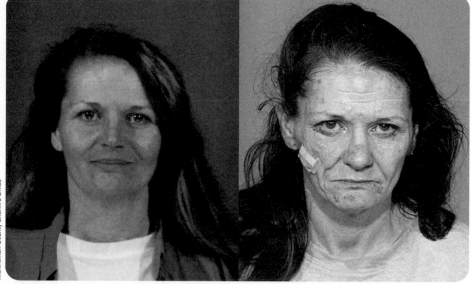

For those who use methamphetamine the physical transformation is remarkable. The time lapse between the before (left) and after (right) pictures of this methamphetamine user is only 3 years 5 months.

Multnomah County Sheriff's Office

In addition to the hypothesized crime-drug use link, some criminal offenses are defined by use of drugs: possession, cultivation, production, and sale of controlled substances; public intoxication; drunk and disorderly conduct; and driving while intoxicated. Driving while intoxicated is one of the most common drug-related crimes. According to the National Highway Traffic Safety Administration (2006) someone is killed in an alcohol-related motor vehicle accident every 31 minutes, and there is an alcohol-related motor vehicle accident resulting in non-fatal injury every 2 minutes. In fatal crashes, drivers are more likely to be male, to be between the ages of 21 and 24, and to be in a passenger car or a light truck.

Traffic fatalities in alcohol-related accidents totaled 16,919 in 2005—39 percent of all traffic fatalities. As shocking as this number is, alcohol-related traffic fatalities have actually decreased in recent years, 5 percent since 1995. Automobile accidents, however, remain the leading cause of death of young people between the ages of 16 and 20. The Federal Bureau of Investigation reported that over 1 million drivers are arrested each year for driving while under the influence—less than 1 percent of the 159 million self-reported incidents of alcohol-impaired driving by U.S. adults (Quinlan et al. 2005).

Based on data from the National Survey on Drug Use and Health, 13 percent of the population age 12 and older—31.7 million people—reported driving under the influence of alcohol at least once in the past year (U.S. Department of Health and Human Services 2006). Driving while under the influence of alcohol is related to age. For example, 8.3 percent of 16- and 17-year-olds reported driving while under the influence of alcohol, whereas 27.9 percent of 21- to 25-year-olds reported drinking and driving. After age 25, drunk driving rates decline with age.

Alcohol is not the only drug that impairs driving. Ten and one-half million people 12 years of age and older reported driving while under the influence of an illicit drug, representing 4.3 percent of the survey population. As with alcohol-impaired drivers, the rate is highest for 18- to 25-year-olds (U.S. Department of Health and Human Services 2006).

Economic Costs

The total economic cost of alcohol and drug use is billions of dollars. The cost of underage drinking alone is astronomical. According to a new study released by the Pacific Institute for Research and Evaluation (2007), $13.7 billion per year are associated with youth alcohol-related traffic accidents; the cost of youth violence resulting from alcohol use is $34.7 billion per year. The average total cost of alcohol-related problems *per underage drinker* is $4,680 a year.

Revenue is also lost to the underground economy because Americans spend an estimated $36 billion on cocaine, $11 billion on marijuana, $10 billion on heroin, and $5.4 billion on methamphetamine (ONDCP 2003c). Further, on-the-job drug use impairs performance and/or causes fatal accidents, which results in a loss of corporate revenues. This result has led to drug testing in many industries. For many employees such tests are routine, both as a condition of employment and as a requirement for employees to keep their jobs. Furthermore, the U.S. Supreme Court has held that students, particularly those in competitive activities including cheerleading and choir, may be subject to random drug testing. Although statistics on how many are not available, several school districts around the United States have also incorporated sobriety tests into the school routine (Healy 2005).

Other economic costs of drug abuse include the cost of homelessness, the cost of implementing and maintaining educational and rehabilitation programs, and the cost of health care. For example, the latest estimates of health care costs associated with smoking exceed $100 billion (Lindholm 2006). Also, the cost of fighting the war on drugs is likely to increase as organized crime groups develop new patterns of involvement (e.g., spread to other countries, use new technologies) in the illicit drug trade.

Health Costs

Some consumption of alcohol has been shown to be beneficial in that "moderate drinkers are generally healthier, live longer, and have lower death rates than abstainers" (Thio 2007, p. 304). However, the physical health consequences of abusing alcohol, tobacco, and other drugs are tremendous: shortened life expectancy; higher morbidity (e.g., cirrhosis of the liver, lung cancer); exposure to HIV infection (see Chapter 2), hepatitis, and other diseases through shared needles; a weakened immune system; birth defects such as fetal alcohol syndrome; drug addiction in children; and higher death rates.

Cigarette smoking is the leading preventable cause of disease and deaths in the United States. Of the more than 2.4 million U.S. deaths annually, over 440,000 are attributable to cigarette smoking alone. Smoking increases the risk of high blood pressure, blood clots, strokes, chronic obstructive pulmonary disease, atherosclerosis, and lung cancer, 90 percent of which is caused by smoking (American Cancer Society 2006; American Heart Association 2007). Worldwide, it is estimated that by the year 2020 more than 10 million tobacco-related deaths will occur annually.

Even nonsmokers are in danger of the health-related problems associated with tobacco use. According to a recent report by the U.S. Surgeon General, there are 126 million "involuntary smokers"; that is, people who are subject to secondhand smoke. One in five children is exposed to secondhand smoke at home where smoking bans, such as those in restaurants, workplaces, and the like, do not reach. According to the report, children who are subjected to secondhand smoke are more likely to suffer from sudden infant death syndrome (SIDS), lung and ear infections, and severe asthma (Neergaard 2006).

> **What Do You Think?** "Secondhand smoke is the most deadly form of child abuse in this country," said anti-smoking advocate John Banzhaf (Jamieson 2007, p. 2). Banzhaf is not alone—16 states are considering bans on smoking in vehicles in which a child is a passenger (DeFao 2007). In Bangor, Maine, a local ordinance was passed that bans smoking in cars whenever children under the age of 18 are present. Anti-smoking advocates applaud such legislation, yet others see this as another intrusion by the government into the lives of U.S. citizens. What do you think? Are so-called "nanny laws" a violation of Americans' civil rights?

Alcohol use is considered the third leading cause of preventable death in the United States, responsible for 75,766 deaths and as many as 2.3 million years of lost life (Jernigan 2005). Alcohol abuse "by minors is responsible for 3,200 deaths a year—four times more deaths than due to all illegal drug use combined" (Kohnle 2006, p. 1). Furthermore, maternal prenatal alcohol use is one of the leading preventable causes of birth defects and developmental disabilities in children. According to the Centers for Disease Control and Prevention, one of the most extreme effects of drinking while pregnant is **fetal alcohol syndrome,** a syndrome characterized by serious physical and mental handicaps, including low birth weight, facial deformities, mental retardation, and hearing and vision problems (Centers for Disease Control and Prevention 2004).

Heavy alcohol and drug use are also associated with negative consequences for an individual's mental health. Data on both male and female adults have shown that drug users are more likely to suffer from serious mental disorders, including anxiety disorders (e.g., phobias), depression, and antisocial personalities (ONDCP 2005c; U.S. Department of Health and Human Services 2006). Further, in a study by researchers at the Rand organization, it was found that "use of cigarettes and hard drugs at age 18 was associated with lower life satisfaction at age 29 . . ." (Bogart et al. 2006, p. 149). Although marijuana, the drug most commonly used by adolescents, is linked to short-term memory loss, learning disabilities, motivational deficits, and retarded emotional development, it was not linked to lower rates of life satisfaction in the Rand study.

The societal costs of drug-related health concerns are also extraordinary—an estimated $15.8 billion annually (ONDCP 2004). Health costs include medical services for drug users, the cost of disability insurance, medical costs of the effects of secondhand smoke, medical costs of the spread of AIDS, and medical costs of accident and crime victims as well as unhealthy infants and children. For example, cocaine use in pregnant women may lead to low-birth-weight babies, increased risk of spontaneous abortions, and abnormal placental functioning (Klutt 2000). In addition, women who smoke while using oral contraceptives have a greater risk of coronary heart disease and stroke than nonsmoking women who are taking oral contraceptives (American Heart Association 2007).

fetal alcohol syndrome A syndrome characterized by serious physical and mental handicaps as a result of maternal drinking during pregnancy.

FIGURE 3.3
Reasons for not receiving substance use treatment among persons 12 and older who needed and made an effort to get treatment but did not receive treatment and felt they needed treatment: 2004 and 2005 combined.
Source: U.S. Department of Health and Human Services (2006).

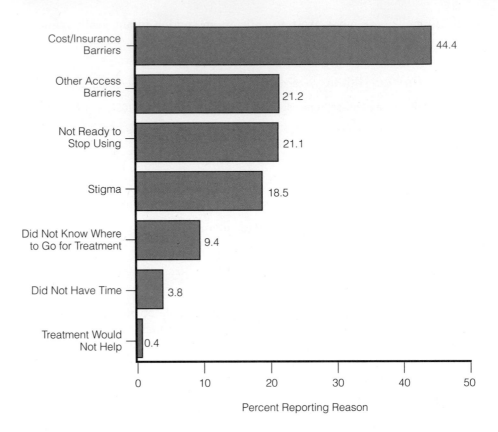

TREATMENT ALTERNATIVES

Prevention is always preferable to treatment. Prevention techniques fall into one of two categories (Hanson 2002). First is what may be called "risk and protective" strategies. Here, factors known to be associated with drug use (e.g., child abuse) are targeted and factors that help insulate a person from drug use (e.g., stable family) are encouraged. The second group of prevention techniques, rather than dealing with reducing the vulnerability of an individual, focuses on "the dynamics of the situations, beliefs, motives, reasoning and reactions that enter into the choice to abuse or not abuse drugs" (Hanson 2002, p. 3).

Treatment of drug users has become increasingly important, in part as a response to the greater need for treatment programs (U.S. Department of Health and Human Services 2006). In 2005, the number of people needing treatment for an alcohol or illicit drug problem reached 23.2 million, or 9.5 percent of the population age 12 and older; however, 20.9 million people also needed but did not receive treatment (see Figure 3.3) (U.S. Department of Health and Human Services 2006).

Individuals who are interested in overcoming chemical dependency have a number of treatment alternatives from which to choose. Some options include family therapy, counseling, private and state treatment facilities, community care programs, pharmacotherapy (i.e., use of treatment medications), behavior modification, drug maintenance programs, and employee assistance programs. Two commonly used techniques are inpatient or outpatient treatment and peer support groups.

Inpatient and Outpatient Treatment

Inpatient treatment refers "to the treatment of drug dependence in a hospital and includes medical supervision of detoxification" (McCaffrey 1998, p. 2). Most inpatient programs last between 30 and 90 days and target individuals whose withdrawal symptoms require close monitoring (e.g., alcoholics, cocaine addicts). Some drug-dependent patients, however, can be safely treated as outpatients. Outpatient treatment allows individuals to remain in their home and work environments and is often less expensive. In outpatient treatment the patient is under the care of a physician who evaluates the patient's progress regularly, prescribes needed medication, and watches for signs of a relapse.

The longer a patient stays in treatment, the greater is the likelihood of a successful recovery (NIDA 2006a). Other variables that predict success include support of family and friends, employer intervention, a positive relationship with therapeutic staff, and a program of recovery that addresses many of the needs of the patient. Although often assumed to be so, internal motivation may not be a prerequisite for change. Researchers supported by the National Institute on Drug Addiction and the Department of Veterans Affairs (VA) studied 2,095 men who were treated for alcohol or drug problems in 15 VA hospitals and then followed for a period of 5 years. Some of the men were voluntarily being treated while the treatment of others was court mandated. Investigators concluded that although the internal motivation of those who received court-ordered treatment was initially lower, 5 years later there were few differences between the two groups in terms of abstinence, recidivism, and employment (Kelly, Finney, & Moos 2005).

> **❝** Every $1 invested in addiction treatment programs yields a return of between $4 and $7 in reduced drug-related crime. **❞**
>
> *Principles of Drug Addiction Treatment,* National Institute of Drug Abuse

Peer Support Groups

Twelve-Step Programs. Both Alcoholics Anonymous (AA) and Narcotics Anonymous (NA) are voluntary associations whose only membership requirement is the desire to stop drinking or taking drugs. AA and NA are self-help groups in that nonprofessionals operate them, offer "sponsors" to each new member, and proceed along a continuum of 12 steps to recovery. Members are immediately immersed in a fellowship of caring individuals with whom they meet daily or weekly to affirm their commitment. Some have argued that AA and NA members trade their addiction to drugs for feelings of interpersonal connectedness by bonding with other group members. In a survey of recovering addicts, more than 50 percent reported using a self-help program such as AA in their recovery (Willing 2002). AA boasts over 100,000 groups where over 2 million members meet in 150 countries (Alcoholics Anonymous 2007).

Symbolic interactionists emphasize that AA and NA provide social contexts in which people develop new meanings. Others who offer positive labels, encouragement, and social support for sobriety surround abusers. Sponsors tell the new members that they can be successful in controlling alcohol and/or drugs "one day at a time" and provide regular interpersonal reinforcement for doing so. Some research indicates that mutual support programs work. For example, in a recent study assessing the effectiveness of such groups, Kelly, Stout, Zywiak, and Schneider (2006) concluded that involvement in such groups may be very successful—for both males and females—even when participation is limited.

Therapeutic Communities. In **therapeutic communities,** which house between 35 and 500 people for up to 15 months, participants abstain from drugs, develop marketable skills, and receive counseling. Synanon, which was established in

therapeutic communities
Organizations in which approximately 35–500 individuals reside for up to 15 months to abstain from drugs, develop marketable skills, and receive counseling

Don't get cuffed & stuffed.

Cops are cracking down.

DRUNK DRIVING
OVER THE LIMIT. UNDER ARREST.

Get-tough policies and increased domestic law enforcement is not just limited to illicit drug control. Recent campaigns by MADD and the National Highway Traffic Safety Administration (NHTSA) have focused on "cracking down" on drunk driving.

TABLE 3.3 National Priorities in the Fight Against Drugs	
PRIORITY	METHODS
Stopping drug use before it starts	Education and community outreach
Healing America's drug users	Get treatment resources where needed
Disrupting the markets	Attack the economic basis of the drug trade

Source: ONDCP (2007).

1958, was the first therapeutic community for alcoholics and was later expanded to include other drug users. More than 400 residential treatment centers are now in existence, including Daytop Village and Phoenix House, the largest therapeutic community in the country (Abadinsky 2004). Phoenix Houses serve more than 6,000 men, women, and teens a day in over 60 locations in 9 states (Phoenix House 2007). The longer a person stays at such a facility, the greater the chance he or she has of overcoming dependency. Symbolic interactionists argue that behavioral changes appear to be a consequence of revised self-definition and the positive expectations of others.

STRATEGIES FOR ACTION: AMERICA RESPONDS

Drug use is a complex social issue that is exacerbated by the structural and cultural forces of society that contribute to its existence. Although the structure of society perpetuates a system of inequality, creating in some the need to escape, the culture of society, through the media and normative contradictions, sends mixed messages about the acceptability of drug use. Thus trying to end drug use by developing programs, laws, or initiatives may be unrealistic. Nevertheless, numerous social policies have been implemented or proposed to help control drug use and its negative consequences with various levels of success.

Government Regulations

The largest social policy attempt to control drug use in the United States was Prohibition. Although this effort was a failure by most indicators, the government continues to develop programs and initiatives designed to combat drug use (see Table 3.3). In the 1980s the federal government declared a "war on drugs," which was based on the belief that controlling drug availability would limit drug use and, in turn, drug-related problems. In contrast to a **harm reduction** position, which focuses on minimizing the costs of drug use for both user and society (e.g., distributing clean syringes to decrease the risk of HIV infection), this "zero-tolerance" approach advocates get-tough law enforcement policies. For example, in New York, state law requires prison sentences for almost all drug offenders—first time or repeat—and limits judicial discretion in deciding what best serves the public's interest. Consequently, "Rockefeller Drug Laws," as they are called, have resulted in "excessively long and unnecessary prison sentences" for even the most minor drug offenders (Human Rights Watch 2007, p. 1). In response to public outcries and accusations of institutional racism (see Chapter 9), reform of Rockefeller Drug Laws and laws like them is under way in a number of states.

Yale law professor Steven Duke and coauthor Albert C. Gross, in their book *America's Longest War* (1994), argued that the war on drugs, much like Prohibition, has intensified other social problems: drug-related gang violence and turf wars, the creation of syndicate-controlled black markets, unemployment, the spread of AIDS, overcrowded prisons, corrupt law enforcement officials, and the diversion of police from other serious crimes. Concern about the war on drugs is not confined to the United States.

In 2005, the European Parliament officially acknowledged that the war on drugs was a failure: ". . . despite the policies carried out to date at international, European, and national levels, the production, consumption and sale of illicit substances . . . have reached extremely high levels in all the Member states, and faced with this failure it is essential that the EU [European Union] revise its general strategy on narcotic substances" (Drug Policy Alliance 2005, p. 1).

Further, a recent report assessing the war on drugs in Canada calls it an "abject failure," arguing that the conservative Canadian government has placed too much emphasis on law enforcement, which has siphoned off money that could have been used for a more "balanced approach" (Mickleburgh & Galloway 2007).

Consistent with conflict theory, still others have argued that the war on drugs unfairly targets minorities. Data analyzed by Human Rights Watch, an international human rights organization, indicates that, in general, black men are 11 times more likely to be incarcerated in state prisons for drug charges than are white men (Human Rights Watch 2007).

Despite concerns, the war on drugs continues at an astronomical societal cost—a projected $12.7 billion in 2007. The U.S. policy on fighting drugs is two-pronged. First is **demand reduction,** which entails reducing the demand for drugs through treatment and prevention (see Figure 3.4). For example, the Office of National Drug Control Policy has recently posted anti-drug videos on *YouTube,* a free web-based video site, as a means of educating youthful populations about the risks of drug use (ONDCP 2006g). The second strategy is **supply reduction.** A much more punitive strategy, supply reduction relies on international efforts, interdiction, and domestic law enforcement to reduce the supply of illegal drugs. Thirty-five percent of drug control spending in 2007 was focused on demand reduction and 65.0 percent on supply reduction. Since 2001, the amount of federal dollars spent on supply reduction has increased significantly (64.0 percent), whereas the amount of money directed toward demand reduction has increased only slightly (2.0 percent) (Katel 2006).

Many of the countries in which drug trafficking occurs are characterized by government corruption and crime, military coups, and political instability, making supply reduction difficult. Such is the case with Colombia, which supplies more than 90 percent of the cocaine in the United States. Despite U.S. efforts that include working closely with the Colombian government and channeling more than $100 million in support of counter-drug activities, coca cultivation in that country has actually increased in recent years. Such increases are, in part, a result of "narco-terrorist" groups who "battle for control of drug producing areas and use the profits from the drug trade to undermine Colombian democ-

harm reduction A recent public health position that advocates reducing the harmful consequences of drug use for the user as well as for society as a whole.

demand reduction One of two strategies in the U.S. war on drugs (the other is supply reduction), demand reduction focuses on reducing the demand for drugs through treatment, prevention, and research.

supply reduction One of two strategies in the U.S. war on drugs (the other is demand reduction), supply reduction concentrates on reducing the supply of drugs available on the streets through international efforts, interdiction, and domestic law enforcement.

FIGURE 3.4
Federal drug control
spending by function,
fiscal year 2007.
Source: National Drug Control Strategy
(2007).

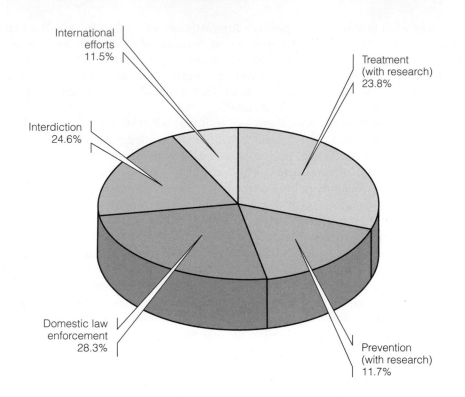

International
efforts
11.5%

Treatment
(with research)
23.8%

Interdiction
24.6%

Domestic law
enforcement
28.3%

Prevention
(with research)
11.7%

racy and the rule of law" (ONDCP 2006f). Interestingly, when American adults were asked in a national poll about financial assistance to foreign countries to fight drug trafficking, 42 percent responded that the United States was providing too much assistance (Pew Research Center 2002). Rather than foreign aid and military assistance, many argue that trade sanctions should be imposed in addition to crop eradication programs and interdiction efforts. Others, however, noting the relative failure of such programs in reducing the supply of illegal drugs entering the United States, argue that the war on drugs should be abandoned and that deregulation is preferable to the side effects of regulation.

Deregulation or Legalization: The Debate

Americans have mixed feelings about drugs and drug use. For example, only 34 percent believe that marijuana should be legal; however, 72 percent believe that recreational marijuana use should result in nothing more than a fine, and 47 percent have tried the drug (Time/CNN 2003). Given these results, it is not surprising that some advocate alternative measures rather than the punitive emphasis of the last several decades.

Deregulation is the reduction of government control over certain drugs. For example, although individuals must be 21 years old to purchase alcohol and 18 to purchase cigarettes, both substances are legal and can be purchased freely. Furthermore, in some states possession of marijuana in small amounts is now a misdemeanor rather than a felony, and in other states marijuana is lawfully used for medical purposes. In 1996 both Arizona and California passed acts known as "marijuana medical bills" that made the use and cultivation of marijuana, under a physician's orders, legal. Although prohibited by federal law, medical use of

deregulation The reduction of government control over, for example, certain drugs.

marijuana has also been approved in many other states including Alaska, Colorado, Maine, Nevada, Oregon, and Washington. Deregulation is popular in other countries as well. Canada has legalized the medical use of marijuana and has approved a cannabis spray that is administered through the mouth for treatment of pain in patients with multiple sclerosis (Grant 2005).

Proponents for the **legalization** of drugs affirm the right of adults to make an informed choice. They also argue that the tremendous revenues realized from drug taxes could be used to benefit all citizens, that purity and safety controls could be implemented, and that legalization would expand the number of distributors, thereby increasing competition and reducing prices. Drugs would thus be safer, drug-related crimes would be reduced, and production and distribution of previously controlled substances would be taken out of the hands of the underworld.

Those in favor of legalization also suggest that the greater availability of drugs would not increase demand, pointing to countries where some drugs have already been decriminalized. **Decriminalization,** or the removing of penalties for certain drugs, would promote a medical rather than criminal approach to drug use that would encourage users to seek treatment and adopt preventive practices. For example, making it a criminal offense to sell or possess hypodermic needles without a prescription encourages the use of nonsterile needles that spread infections such as HIV and hepatitis.

Opponents of legalization argue that it would be construed as government approval of drug use and, as a consequence, drug experimentation and abuse would increase. Furthermore, although the legalization of drugs would result in substantial revenues for the government, drug trafficking and black markets would still flourish because all drugs would not be decriminalized (e.g., crack). Legalization would also require an extensive and costly bureaucracy to regulate the manufacture, sale, and distribution of drugs. Finally, the position that drug use is an individual's right cannot guarantee that others will not be harmed. It is illogical to assume that a greater availability of drugs will translate into a safer society.

Collective Action

Social action groups, such as Mothers Against Drunk Driving (**MADD**), have successfully lobbied legislators to raise the drinking age to 21 and to provide harsher penalties for driving while impaired. MADD, with 3.5 million members and 600 chapters, has also put pressure on alcohol establishments to stop "two for one" offers and has pushed for laws that hold the bartender personally liable if he or she serves a person who is later involved in an alcohol-related accident. Even hosts in private homes can now be held liable if they allow underage guests to consume alcohol. "Social host" ordinances, as they are called, give police the authority to enter homes where it is suspected that underage drinking is occurring and fine adults, including parents, up to $2500 or more. Thirty-two states and hundreds of locales presently have social host laws (Ritter 2007).

MADD also has several national programs designed to increase public awareness of the problems associated with drinking and driving. For example, the Tie One On For Safety ribbon campaign encourages drivers to place a red ribbon on their vehicle as an indication of their commitment to driving safely.

legalization Making prohibited behaviors legal; for example, legalizing drug use or prostitution.

decriminalization The removal of criminal penalties for a behavior, as in the decriminalization of drug use.

MADD (Mothers Against Drunk Driving) A social action group committed to reducing drunk driving.

This program is in its 20th year of continuous operation. In addition, the popular "designate a driver" initiative, which focuses on the four deadliest holidays—Labor Day, the Christmas holiday season (November 24 through January 2), St. Patrick's Day, and Memorial Day—encourages motorists to designate nondrinking drivers at holiday times (MADD 2006).

Smokers, ex-smokers, and the families of victims of smoking are also taking collective action against tobacco companies. They charge that tobacco executives knew more than 50 years ago that tobacco was addictive and that they concealed this fact from both the public and the government. Furthermore, the groups charge that tobacco companies have manipulated nicotine levels in cigarettes with the intention of causing addiction. In a class action suit by more than 300,000 Florida smokers, a jury ordered the top five cigarette producers to pay $145 billion to the plaintiffs—the largest settlement to date.

Debate also concerns the truthfulness of tobacco companies' claims that some cigarettes are "light," "low tar," "ultra light," or "mild." Although a recent ruling prohibited the advertising and sale of such cigarettes given research indicating that they offer no health benefits over regular cigarettes, a judge recently blocked the decision until the appeal by the tobacco companies can be heard (Apuzzo 2006). However, in 2006, a federal judge "granted class action status to tens of millions of 'light cigarette' smokers for a potential $200 billion lawsuit against tobacco companies" (Hays 2006, p. 1).

What Do You Think? In 1998 there was a multibillion-dollar, multistate settlement against the tobacco companies. Settlement funds are paid yearly to the states and, in part, are to be used to help reduce smoking in the United States. How would you recommend the money be used; that is, what kind of smoking prevention programs would you advocate?

Finally, several initiatives have resulted in statewide referendums concerning the cost-effectiveness of government policies. For example, as a result of the Substance Abuse and Crime Prevention Act of 2000 (Proposition 36), California (as well as many other states) now requires that nonviolent first- and second-time minor drug offenders receive treatment, including job training, therapy, literacy education, and family counseling rather than jail time. The Act, passed by 61 percent of California voters, permanently changed state law. As of 2006, more than 150,000 people have benefited from Proposition 36, and California taxpayers have saved an estimated $1.3 billion. Despite overwhelming evidence of the success of the program, 2006 funding remains at 2000 levels—tantamount to a significant funding cut. Similarly, in 2006, voters in Arizona approved a measure that would undercut a "treatment instead of incarceration law" that had been passed in 1996. The new law permits judges to exclude offenders who have been arrested for possession of methamphetamine from treatment (Drug Policy Alliance 2006).

In addition, over the past decade voters and state governments have enacted 150 significant drug policy reforms. For example, Connecticut passed significant

❝ If some drunk gets out and kills your kid, you'd probably be a little crazy about it, too. **❞**

Kathy Prescott
Former MADD president

overdose prevention legislation, following in the footsteps of New Mexico. Texas and Kansas passed legislation providing for treatment instead of incarceration for first-time drug offenders, and Colorado reduced sentences for these offenders as well. Illinois passed legislation allowing for the sale of sterile syringes without a prescription and also mandated the gathering of race-based information during police traffic stops (Drug Policy Alliance 2007).

UNDERSTANDING ALCOHOL AND OTHER DRUG USE

In summary, substance abuse—that is, drugs and their use—is socially defined. As the structure of society changes, the acceptability of one drug or another changes as well. As conflict theorists assert, the status of a drug as legal or illegal is intricately linked to those who have the power to define acceptable and unacceptable drug use. There is also little doubt that rapid social change, anomie, alienation, and inequality further drug use and abuse. Symbolic interactionism also plays a significant role in the process: If people are labeled "drug users" and are expected to behave accordingly, then drug use is likely to continue. If people experience positive reinforcement of such behaviors and/or have a biological predisposition to use drugs, the probability of their drug involvement is even higher. Thus the theories of drug use complement rather than contradict one another.

Drug use must also be conceptualized within the social context in which it occurs. Many youths who are at high risk for drug use have been "failed by society"—they are living in poverty, victims of abuse, dependents of addicted and neglectful parents, and alienated from school. Despite the social origins of drug use, many treatment alternatives, emanating from a clinical model of drug use, assume that the origin of the problem lies within the individual rather than in the structure and culture of society. Although admittedly the problem may lie within the individual when treatment occurs, policies that address the social causes of drug abuse provide a better means of dealing with the drug problem in the United States.

As stated earlier, prevention is preferable to intervention, and given the social portrait of hard-drug users—young, male, and minority—prevention must entail dealing with the social conditions that foster drug use. Some data suggest that inner city adolescents are particularly vulnerable to drug involvement because of their lack of legitimate alternatives (Van Kammen & Loeber 1994):

> Illegal drug use may be a way to escape the strains of the severe urban conditions, and dealing illegal drugs may be one of the few, if not the only, way to provide for material needs. Intervention and treatment programs, therefore, should include efforts to find alternate ways to deal with the limiting circumstances of inner-city life, as well as create opportunities for youngsters to find more conventional ways of earning a living. (p. 22)

Social policies that deal with drug use have been predominantly punitive rather than preventive. Recently, however, there appears to be some movement toward educating the public and changing the culture of drugs. For example, a new media campaign by the Office of National Drug Control Policy features real teens "sharing their antidrug attitudes and commitments." The new program debuted on Fox's *American Idol* in an advertisement in which a 17-year-old girl read her original poetry about the dangers of drug use (ONDCP 2005d).

In this country and throughout the world, millions of people depend on legal drugs for the treatment of a variety of conditions, including pain, anxiety and nervousness, insomnia, overeating, and fatigue. Although drugs for these purposes are relatively harmless, the cultural message "better living through chemistry" contributes to alcohol and drug use and its consequences. But these and other drugs are embedded in a political and economic context that determines who defines what drugs and in what amounts are licit or illicit and what programs are developed in reference to them.

> **"**A child who reaches 21 without smoking, abusing alcohol or using drugs is virtually certain never to do so.**"**
>
> Joseph A. Califano Jr.
> President, Center on Addiction and Substance Abuse

CHAPTER REVIEW

- **What is a drug, and what is meant by drug abuse?**

  Sociologically, the term *drug* refers to any chemical substance that (1) has a direct effect on the user's physical, psychological, and/or intellectual functioning; (2) has the potential to be abused; and (3) has adverse consequences for the individual and/or society. Drug abuse occurs when acceptable social standards of drug use are violated, resulting in adverse physiological, psychological, and/or social consequences.

- **How do the three sociological theories of society explain drug use?**

  Structural functionalists argue that drug abuse is a response to the weakening of norms in society, leading to a condition known as anomie or normlessness. From a conflict perspective drug use occurs as a response to the inequality perpetuated by a capitalist system as societal members respond to alienation from their work, family, and friends. Symbolic interactionism concentrates on the social meanings associated with drug use. If the initial drug use experience is defined as pleasurable, it is likely to recur, and over time the individual may earn the label of "drug user."

- **What are the most frequently used legal and illegal drugs?**

  Alcohol is the most commonly used and abused legal drug in America. The use of tobacco products is also very high, with 25 percent of Americans reporting that they currently smoke cigarettes. Marijuana is the most commonly used illicit drug, with 162 million marijuana users, representing 4.0 percent of the world's population.

- **What are the consequences of drug use?**

  The consequences of drug use are fourfold. First is the cost to the family, often manifesting itself in higher rates of divorce, spouse abuse, child abuse, and child neglect. Second is the relationship between drugs and crime. Those arrested have disproportionately higher rates of drug use. Although drug users commit more crimes, sociologists disagree as to whether drugs actually "cause" crime or whether, instead, criminal activity leads to drug involvement. Third are the economic costs (e.g., loss of productivity), which are in the billions. Last are the health costs of abusing drugs, including shortened life expectancy; higher morbidity (e.g., cirrhosis of the liver and lung cancer); exposure to HIV infection, hepatitis, and other diseases through shared needles; a weakened immune system; birth defects such as fetal alcohol syndrome; drug addiction in children; and higher death rates.

- **What treatment alternatives are available for drug users?**

  Although there are many ways to treat drug abuse, two methods stand out. The inpatient-outpatient model entails medical supervision of detoxification and may or may not include hospitalization. Twelve-step programs such as Alcoholics Anonymous (AA) and Narcotics Anonymous (NA) are particularly popular, as are therapeutic communities. Therapeutic communities are residential facilities where drug users learn to redefine themselves and their behavior as a response to the expectations of others and self-definition.

- **What can be done about the drug problem?**

  First, there are government regulations limiting the use (e.g., the law establishing the 21-year-old drinking age) and distribution (e.g., prohibitions about importing drugs) of legal and illegal drugs. The government also imposes sanctions on those who violate drug regulations and provides treatment facilities for other offenders. Second, there are collective action groups—for example, Mothers Against Drunk Driving. Finally, there are local and statewide initiatives geared toward holding companies responsible for the consequences for their product—for example, class action suits against tobacco producers.

TEST YOURSELF

1. "Cannabis cafes" are commonplace throughout England.
   a. True
   b. False
2. The most used illicit drug in the world is
   a. heroin
   b. marijuana
   c. cocaine
   d. methamphetamine
3. What theory would argue that the continued legality of alcohol is a consequence of corporate greed?
   a. Structural functionalism
   b. Symbolic interactionism
   c. Reinforcement theory
   d. Conflict theory
4. Cigarettes smoking is
   a. the third leading cause of preventable death in the United States
   b. not addictive
   c. the most common use of tobacco products
   d. increasing

5. In the United States drinking is highest among young, nonwhite males.
   a. True
   b. False
6. According to the National Survey on Drug Use and Health, binge drinking is defined as five or more drinks per occasion on _____ or more days in a 1-month period.
   a. 1
   b. 2
   c. 5
   d. 10
7. The active ingredient in marijuana, THC, can act as a sedative or a hallucinogen.
   a. True
   b. False
8. In 2007, most federal drug control dollars were allocated to
   a. international efforts
   b. domestic law enforcement
   c. prevention and research
   d. treatment and research
9. Decriminalization refers to the removal of penalties for certain drugs.
   a. True
   b. False
10. The two-pronged drug control strategy of the U.S. government entails supply reduction and harm reduction.
    a. True
    b. False

Answers: 1 b. 2 b. 3 d. 4 c. 5 b. 6 a. 7 a. 8 b. 9 a. 10 b.

KEY TERMS

anomie
binge drinking
chemical dependency
club drugs
crack
date-rape drugs

decriminalization
demand reduction
deregulation
drug
drug abuse
fetal alcohol syndrome

gateway drug
harm reduction
legalization
MADD (Mothers Against Drunk Driving)
supply reduction
therapeutic communities

MEDIA RESOURCES

**_Understanding Social Problems,_ Sixth Edition
Companion Website**

academic.cengage.com/sociology/mooney
Visit your book companion website, where you will find flash cards, practice quizzes, Internet links, and more to help you study.

Just what you need to know NOW! Spend time on what you need to master rather than on information you already have learned. Take a pre-test for this chapter, and CengageNOW will generate a personalized study plan based on your results. The study plan will identify the topics you need to review and direct you to online resources to help you master those topics. You can then take a post-test to help you determine the concepts you have mastered and what you will need to work on. Try it out! Go to academic.cengage.com/login to sign in with an access code or to purchase access to this product.

"Unjust social arrange-
ments are themselves
a kind of extortion,
even violence."

John Rawls,
A Theory of Justice

Crime and Social Control

Jessica Lunsford was kidnapped, raped, and murdered by John Couey who, in 2007, was sentenced to death for his crimes.

Jessica Lunsford was a vivacious, happy third grader, until her body was found buried in a dark hole stuffed in two black trash bags with her favorite purple toy dolphin. Her hands had been tied behind her back with speaker wire. Her small fingers were poking out of the garbage bags that served as her coffin—signs of her efforts to get air. Kidnapped, raped, and buried alive by a convicted sex offender, 9-year-old Jessica Lunsford died in the way those who loved her most tried to protect her from—in the dark, abused, and afraid (Jessica Lunsford Foundation 2006; MSNBC 2007; CNN 2007a).

No one will ever know what it was like for Jessica to be abruptly awakened at 3:00 on that February 24 morning in 2005. The previous night she had done the things she normally did on Wednesday nights and then prepared for bed. She kissed her father good night and was tucked into bed by her grandmother. The door was left open a crack so the light from the family room could come in as she fell asleep (Kruse 2007).

High on cocaine, convicted sex offender John Evander Couey entered Jessie's dark bedroom in Homosassa, Florida, put his hand over her mouth, and told her she was going with him. Three weeks later, after Couey was captured in Georgia, Jessie's body was found not far from where she lived. In 2007, Couey was convicted of murder, rape, and kidnapping. Throughout the proceedings, Couey sat at the defense table and colored in a coloring book—something Jessica Lunsford will never again do.

Adam Walsh, Amber Hagerman, Megan Kanka, and Jessica Lunsford: tragically, these are the names of children whose abductions have led to changes in the criminal justice system. "The impacts of Jessica Lunsford's story have been huge. . . . Enormous good has come and thousands of children's lives are going to be saved because Jessica Lunsford lived," said Ernie Allen, president and chief executive officer of the National Center for Missing and Exploited Children (CNN 2007a, p. 1). On April 19, 2005, the state of Florida passed the *Jessica Lunsford Act*, a tough sex offender law that requires a "25-year minimum prison term for people convicted of certain sex crimes against children, and lifetime tracking by global positioning satellite once they're outside of prison" (Associated Press 2005a, p. 1). At least 18 other states have passed similar laws. Further, the U.S. Department of Justice in conjunction with state law enforcement agencies has begun tracking sex offenders who have failed to register their locations (MSNBC 2007).

The Department of Justice and the laws of the individual states are just one part of the massive bureaucracy called the criminal justice system, a system that often comes under public scrutiny, particularly in high-profile cases, such as the kidnapping, rape, and murder of Jessica Lunsford and the kidnapping of Shawn Hornbeck and Ben Ownby, who were found in 2007 in their abductor's Missouri apartment (CNN 2007b). In this chapter we examine the criminal justice system as well as theories, types, and demographic patterns of criminal behavior. The economic, social, and psychological costs of crime are also examined. The chapter concludes with a discussion of social control, including policies and prevention programs designed to reduce crime in the United States.

The Global Context: International Crime and Social Control

Several facts about crime are true throughout the world. First, crime is ubiquitous—there is no country where crime does not exist. Second, most countries have the same components in their criminal justice systems: police, courts, and prisons. Third, worldwide, adult males make up the largest category of crime suspects; and fourth, in all countries theft is the most common crime committed, whereas violent crime is a relatively rare event.

Even so, dramatic differences do exist in international crime rates, although comparisons are made difficult by variations in measurement and crime definitions (Siegel 2006). Because of these difficulties, Winslow and Zhang (2008, pp. 30–32) created a global crime database from data from the United Nations and Interpol—the international police agency. In their discussion of the "United States versus the World," the authors reveal some interesting and often counterintuitive findings. First, the United States does not have the highest crime rate in the world. The United States ranks 12th among 165 nations with Sweden, Denmark, Australia, and Great Britain, in rank order, each having a higher crime rate than the United States. And the differences are considerable. For example, although the U.S. crime rate is 4,160 reported crimes per 100,000 population, Sweden's crime rate, which is the highest in the world, is 9,604 reported crimes per 100,000 population.

Winslow and Zhang (2008) also examined crime rates by dividing them into types of crime—violent crime or property crime. Violent crime, as discussed later in the chapter, includes murder, rape, robbery, and aggravated assault. When one compares the United States to other countries, the United States once again is not in the top 10. With a violent crime rate of 504 reported crimes per 100,000 population, the United States ranks 19th in violent crime. Several developing countries (e.g., Namibia and Swaziland) as well as developed countries (e.g., Australia and Sweden) have higher violent crime rates than the United States. Property crimes show a similar pattern. Based on the global crime database created by Winslow and Zhang (2008), the United States ranks 13th in property crimes (car theft, burglary, and larceny) with Sweden, Denmark, Australia, and Great Britain topping the list.

Violent crime and property crimes represent just two types of crime that take place worldwide. Although we are concerned about these types of crimes and the possibility of victimization, Interpol has identified five global priority areas (Interpol 2007a): (1) drugs and criminal organizations (e.g., drug trafficking), (2) financial and high tech crimes (e.g., counterfeiting, fraud, and cyber-crime), (3) tracing of fugitives, (4) countering terrorism (discussed in Chapter 16), and (5) trafficking in human beings. Each of these priority areas contains a relatively new category of crimes—transnational crime. As defined by the U.S. Department of Justice, **transnational crime** is "organized criminal activity across one or more national borders" (U.S. Department of Justice 2003). The significance of transnational crime should not be minimized. As Shelley states (2007):

> Transnational crime will be a defining issue of the 21st century for policymakers—as defining as the Cold War was for the 20th century and colonialism was for the 19th. Terrorists and transnational crime groups will proliferate because these crime groups are major beneficiaries of globalization. They take advantage of increased travel, trade, rapid money movements, telecommunications and computer links, and are well positioned for growth. (p. 1)

For example, the Internet has led to an explosive growth in child pornography. In 2005 a United Nations expert on the subject told the 53-nation U.N. Commission on Human Rights that governments must act now to curb the proliferation of child pornography (Klapper 2005). Of late, several successes have been recorded. In 2007, the Austrian authorities reported uncovering an

transnational crime Criminal activity that occurs across one or more national borders.

The stories below come from a U.S. State Department report and are not representative of all the forms of human trafficking that take place around the world. They are, however, a glimpse into a world of suffering and servitude—a world that few of us understand and all of us fear. The material has not been edited and remains in its original state.

Central Africa

Mary, a 16-year-old demobilized child soldier forced to join an armed rebel group in Central Africa, remembers: "I feel so bad about the things that I did. It disturbs me so much that I inflicted death on other people. When I go home I must do some traditional rites because I have killed. I must perform these rites and cleanse myself. I still dream about the boy from my village whom I killed. I see him in my dreams, and he is talking to me, saying I killed him for nothing, and I am crying."

Cambodia

Neary grew up in rural Cambodia. Her parents died when she was a child, and, in an effort to give her a better life, her sister married her off when she was 17. Three months later she and her husband went to visit a fishing village. Her husband rented a room in what Neary thought was a guesthouse. But when she woke the next morning, her husband was gone. The owner of the house told her she had been sold by her husband for $300 and that she was actually in a brothel. For five years, Neary was raped by five to seven men every day. In addition to brutal physical abuse, Neary was infected with HIV and contracted AIDS. The brothel threw her out when she became sick, and she eventually found her way to a local shelter. She died of HIV/AIDS at the age of 23.

United Arab Emirates (U.A.E.)

Lusa is a 17 year-old orphan kidnapped in 2004 from her native Uzbekistan. Lusa's aunt engineered her abduction to Dubai using a cousin's passport, because the aunt wanted to take Lusa's apartment. In Dubai, Lusa was sold to a slavery and prostitution ring. When she was no longer useable in prostitution, the traffickers sent her to a psychiatric center. An Uzbek NGO (non-governmental organization) located her in Dubai. The NGO arranged to move her to a shelter, and they began working on her repatriation. Because she entered the U.A.E. illegally, on a false passport, the U.A.E. immigration service said she should serve a two-year prison sentence. Government officials and the enterprising NGO are negotiating Lusa's case.

Italy

Viola, a young Albanian, was 13 when she started dating 21-year-old Dilin, who proposed to marry her. They moved them to Italy where Dilin said he had cousins who could get him a job. Arriving in Italy, Viola's life changed forever. Dilin locked her in a hotel room and left her, never to be seen again. A group of men entered and began to beat Viola. Then, each raped her. The leader informed Viola that Dilin had sold her and that she had to obey him or she would be killed. For seven days Viola was beaten and repeatedly raped. Viola was sold a second time to someone who beat her head so badly she was unable to see for two days. She was told if she didn't work as a prostitute, her mother and sister in Albania would be raped and killed. Viola was forced to submit to prostitution until police raided the brothel in which she was held. She was deported to Albania.

---

international pornography ring in which suspects viewed online videos of children being sexually abused. The pornography ring was estimated to involve more than 2,300 suspects from 77 countries including the United States, Germany, France, South Africa, and Russia (Associated Press 2007a).

Human trafficking is another example of transnational crime. According to the U.S. State Department between 600,000 and 800,000 men, women, and children are trafficked across international borders annually, and millions more are trafficked within their own country (U.S. State Department 2006). Although the majority of persons are trafficked into commercial sexual exploitation, others are trafficked into forced labor and sexual servitude. This chapter's *The Human Side* feature provides an intimate portrait of trafficked persons and the horrific consequences of this crime.

crime An act, or the omission of an act, that is a violation of a federal, state, or local criminal law for which the state can apply sanctions.

SOURCES OF CRIME STATISTICS

The U.S. government spends millions of dollars annually to compile and analyze crime statistics. A **crime** is a violation of a federal, state, or local criminal law. For a violation to be a crime, however, the offender must have acted volun-

Lebanon

Silvia was a young, single, Sri Lankan mother seeking a better life for herself and her three-year-old son when she answered an advertisement for a housekeeping job in Lebanon. In the Beirut job agency, her passport was taken, and she was hired by a Lebanese woman who subsequently confined her and restricted her access to food and communications. Treated like a prisoner and beaten daily, Silvia was determined to escape. She jumped from a window to the street below, landing with such force that she is permanently paralyzed. She is now back in Sri Lanka. Today, she travels around the country telling her story so that others do not suffer a similar fate.

India

Shadir, a boy of 15 years, was offered a job that included good clothes and an education; he accepted. Instead of being given a job, Shadir was sold to a slave trader who took him to a remote village in India to produce hand-woven carpets. He was frequently beaten. He worked 12 to 14 hours a day and he was poorly fed. One day, Shadir was rescued by a NGO working to combat slavery. It took several days

for him to realize he was no longer enslaved. He returned to his village, was reunited with his mother, and resumed his schooling. Now Shadir warns fellow village children about the risks of becoming a child slave.

Turkey

Svetlana was a young Belarusian living in Minsk and looking for a job when she came upon some Turkish men who promised her a well-paying job in Istanbul. Once Svetlana crossed the border, her passport and money were taken and she was locked up. Svetlana and another foreign woman were sent to the apartment of two businessmen and forced into prostitution. Svetlana had other plans: In an attempt to escape, she jumped out of a window and fell six stories to the street below. According to Turkish court documents, customers did not take Svetlana to the hospital, they called the traffickers instead. These events led to her death. Svetlana's body lay unclaimed in the morgue for two weeks until Turkish authorities learned her identity and sent her body to Belarus. But Svetlana did not die in vain. Belarusian and Turkish authorities cooperated effectively to arrest and

charge those responsible for contributing to a death and for human trafficking.

Singapore

Karin, a young mother of two, was looking for a job in Sri Lanka when a man befriended her and convinced her that she could land a better job in Singapore as a waitress. He arranged and paid for her travel. A Sri Lankan woman met Karin upon arrival in Singapore, confiscated her passport, and took her to a hotel. The woman made it clear that Karin had to submit to prostitution to pay back the money it cost for her to be flown into Singapore. Karin was taken to an open space for sale in the sex market where she joined women from Indonesia, Thailand, India, and China to be inspected and purchased by men from Pakistan, India, China, Indonesia, and Africa. The men would take the women to nearby hotels and rape them. Karin was forced to have sex with an average of 15 men a day. She developed a serious illness and three months after her arrival was arrested by the Singaporean police during a raid on the brothel. She was deported to Sri Lanka.

Source: U.S. State Department (2006).

---

tarily and with intent and have no legally acceptable excuse (e.g., insanity) or justification (e.g., self-defense) for the behavior. The three major types of statistics used to measure crime are official statistics, victimization surveys, and self-report offender surveys.

Official Statistics

Local sheriffs' departments and police departments throughout the United States collect information on the number of reported crimes and arrests and voluntarily report them to the Federal Bureau of Investigation (FBI). The FBI then compiles these statistics annually and publishes them, in summary form, in the Uniform Crime Reports (UCR). The UCR lists **crime rates** or the number of crimes committed per 100,000 population, the actual number of crimes and the percentage of change over time, and clearance rates. **Clearance rates** measure the percentage of cases in which an arrest and official charge have been made and the case has been turned over to the courts.

These statistics have several shortcomings. For example, many incidents of crime go unreported. It is estimated that in 2005 only 38 percent of rapes and sexual assaults, 52 percent of robberies, 47 percent of assaults, and 56 percent

> 66 Ultimately, any crime statistic is only as useful as the reader's understanding of the processes that generated it. 99

Robert M. O'Brien
Sociologist

crime rate The number of crimes committed per 100,000 population.

clearance rate The percentage of crimes in which an arrest and official charge have been made and the case has been turned over to the courts.

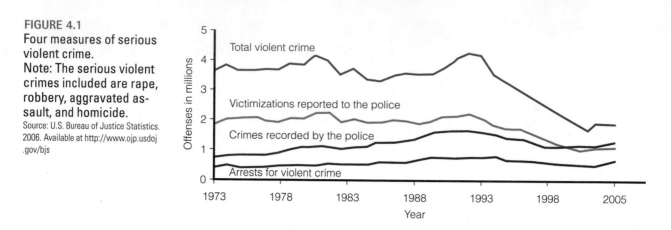

FIGURE 4.1
Four measures of serious violent crime.
Note: The serious violent crimes included are rape, robbery, aggravated assault, and homicide.
Source: U.S. Bureau of Justice Statistics. 2006. Available at http://www.ojp.usdoj.gov/bjs

of household burglaries were actually reported to the police (U.S. Bureau of the Census 2007). Even if a crime is reported, it may not be recorded by the police (see Figure 4.1). Alternatively, some rates may be exaggerated. Motivation for such distortions may come from the public (e.g., demanding that something be done), from political officials (e.g., election of a sheriff), and/or from organizational pressures (e.g., budget requests). For example, a police department may "crack down" on drug-related crimes in an election year. The result is an increase in the recorded number of these offenses for that year. Such an increase reflects a change in the behavior of law enforcement personnel, not a change in the number of drug violations. Thus official crime statistics may be a better indicator of what police are doing rather than of what criminals are doing.

Victimization Surveys

Victimization surveys ask people whether they have been victims of crime. The National Crime Victimization Survey, begun in 1973 and conducted annually by the U.S. Department of Justice, interviews nearly 150,000 people about their experiences as victims of crime. Interviewers collect a variety of information, including the victim's background (e.g., age, race and ethnicity, sex, marital status, education, and area of residence), relationship to the offender (stranger or nonstranger), and the extent to which the victim was harmed. For example, in 2005, the latest year for which victimization data are available, teens and young adults were the most likely to be the victims of violent crime and those older than age 65 were most likely to be the victims of property crime (Bureau of Justice Statistics 2006a). Further, victims of violent crime were disproportionately minorities with 49 percent of murder victims being white, and the remainder being African American, Asian, Pacific Islander, or Native American. Gender is related to crime victimization as well. Males are more likely to be victims of violent crime than females with the exception of sexual assault and rape. Although victimization surveys provide detailed information about crime victims, they provide less reliable data on offenders.

Self-Report Offender Surveys

Self-report surveys ask offenders about their criminal behavior. The sample may consist of a population with known police records, such as a prison population, or it may include respondents from the general population, such as college students.

Self and Society | Criminal Activities Survey

Read each of the following questions. If, since the age of 16, you have ever engaged in the behavior described, place a "1" in the space provided. If you have not engaged in the behavior, put a "0" in the space provided. After completing the survey, read the section on interpretation to see what your answers mean.

Questions

1. Have you ever been in possession of drug paraphernalia? ____
2. Have you ever lied about your age or about anything else when making application to rent an automobile? ____
3. Have you ever obtained a false ID in order to gain entry to a bar or event? ____
4. Have you ever tampered with a coin-operated vending machine or parking meter? ____
5. Have you ever shared, given, or shown pornographic material to someone under 18? ____
6. Have you ever begun and/or participated in a basketball, baseball, or football pool? ____
7. Have you ever used "filthy, obscene, annoying, or offensive" language while on the telephone? ____
8. Have you ever given or sold a beer to someone under the age of 21? ____
9. Have you ever been on someone else's property (land, house, boat, structure, etc.) without that person's permission? ____
10. Have you ever forwarded a chain letter with the intent to profit from it? ____
11. Have you ever improperly gained access to someone else's e-mail or other computer account? ____
12. Have you ever written a check for over $150 when you knew it was bad? ____

Interpretation

Each of the activities described in these questions represents criminal behavior that was subject to fines, imprisonment, or both under the laws of Florida in 2006. For each activity the following table lists the maximum prison sentence and/or fine for a first-time offender. To calculate your "prison time" and/or fines, sum the numbers corresponding to each activity you have engaged in.

OFFENSE	MAXIMUM PRISON SENTENCE	MAXIMUM FINE
1. Possession of drug paraphernalia	1 year	$1,000
2. Fraud	5 years	$5,000
3. Possession of false ID or driver's license	5 years	$5000
4. Fraud	2 months	$500
5. Protection of minors from obscenity	5 years	$5,000
6. Illegal gambling	2 months	$500
7. Harassing/obscene telecommunications	2 months	$500
8. Illegal distribution of alcohol	2 months	$500
9. Trespassing	1 year	$1,000
10. Illegal gambling	1 year	$1,000
11. Illegal misappropriation of cyber communication	5 years	$5,000
12. Worthless check	1 year	$1,000

Source: *Florida Criminal Code* (2006).

Self-report data compensate for many of the problems associated with official statistics but are still subject to exaggerations and concealment. The Criminal Activities Survey in this chapter's *Self and Society* feature asks you to indicate whether you have engaged in a variety of illegal activities.

Self-report surveys reveal that virtually every adult has engaged in some type of criminal activity. Why then is only a fraction of the population labeled criminal? Like a funnel, which is large at one end and small at the other, only a small proportion of the total population of law violators are ever convicted of a crime. For an individual to be officially labeled a criminal, his or her behavior (1) must become known to have occurred and (2) must come to the attention of the police who then file a report, conduct an investigation, and make an arrest; finally, (3) the arrestee must go through a preliminary hearing, an arraignment, and a trial and may or may not be convicted. At every stage of the process an offender may be "funneled" out. As Figure 4.1 indicates, the measures of crime used at various points in time lead to different results.

SOCIOLOGICAL THEORIES OF CRIME

Some explanations of crime focus on psychological aspects of the offender, such as psychopathic personalities, unhealthy relationships with parents, and mental illness. Other crime theories focus on the role of biological variables, such as central nervous system malfunctioning, stress hormones, vitamin or mineral deficiencies, chromosomal abnormalities, and a genetic predisposition toward aggression. Sociological theories of crime and violence emphasize the role of social factors in criminal behavior and societal responses to it.

Structural-Functionalist Perspective

According to Durkheim and other structural functionalists, crime is functional for society. One of the functions of crime and other deviant behavior is that it strengthens group cohesion: "The deviant individual violates rules of conduct that the rest of the community holds in high respect; and when these people come together to express their outrage over the offense . . . they develop a tighter bond of solidarity than existed earlier" (Erikson 1966, p. 4).

Crime can also lead to social change. For example, an episode of local violence may "achieve broad improvements in city services . . . be a catalyst for making public agencies more effective and responsive, for strengthening families and social institutions, and for creating public-private partnerships" (National Research Council 1994, pp. 9–10).

Although structural functionalism as a theoretical perspective deals directly with some aspects of crime, it is not a theory of crime per se. Three major theories of crime have developed from structural functionalism, however. The first, called strain theory, was developed by Robert Merton (1957) and uses Durkheim's concept of *anomie,* or normlessness. Merton argued that when legitimate means (e.g., a job) of acquiring culturally defined goals (e.g., money) are limited by the structure of society, the resulting strain may lead to crime.

Individuals, then, must adapt to the inconsistency between means and goals in a society that socializes everyone into wanting the same thing but provides opportunities for only some (see Table 4.1). *Conformity* occurs when individuals accept the culturally defined goals and the socially legitimate means of achieving them. Merton suggested that most individuals, even those who do not have easy access to the means and goals, remain conformists. *Innovation* occurs when an individual accepts the goals of society but rejects or lacks the socially legitimate means of achieving them. Innovation, the mode of adaptation most associated with criminal behavior, explains the high rate of crime committed by uneducated and poor individuals who do not have access to legitimate means of achieving the social goals of wealth and power.

Another adaptation is *ritualism,* in which the individual accepts a lifestyle of hard work but rejects the cultural goal of monetary rewards. The ritualist goes through the motions of getting an education and working hard, yet he or she is not committed to the goal of accumulating wealth or power. *Retreatism* involves rejecting both the cultural goal of success and the socially legitimate means of achieving it. The retreatist withdraws or retreats from society and may become an alcoholic, drug addict, or vagrant. Finally, *rebellion* occurs when an individual rejects both culturally defined goals and means and substitutes new goals and means. For example, rebels may use social or political activism to replace the goal of personal wealth with the goal of social justice and equality.

Whereas strain theory explains criminal behavior as a result of blocked opportunities, subcultural theories argue that certain groups or subcultures in society have values and attitudes that are conducive to crime and violence. Members of these groups and subcultures, as well as other individuals who interact with them, may adopt the crime-promoting attitudes and values of the group. For example, Kubrin and Weitzer (2003) found that retaliatory homicide is a response to subcultural norms of violence that exist in some neighborhoods.

But if blocked opportunities and subcultural values are responsible for crime, why don't all members of the affected groups become criminals? Control theory may answer that question. Hirschi (1969), consistent with Durkheim's emphasis on social solidarity, suggests that a strong social bond between individuals and the social order constrains some individuals from violating social norms. Hirschi identified four elements of the social bond: attachment to significant others, commitment to conventional goals, involvement in conventional activities, and belief in the moral standards of society. Several empirical tests of Hirschi's theory support the notion that the higher the attachment, commitment, involvement, and belief, the higher the social bond and the lower the probability of criminal behavior. For example, Laub, Nagan, and Sampson (1998) found that a good marriage contributes to the cessation of a criminal career. Warner and Rountree (1997) reported that local community ties, although varying by neighborhood and offense, decrease the probability that crimes will occur. Similarly, Van Wilsem, Wittebrood, and De Graaf (2006) conclude that criminal victimization is higher in neighborhoods characterized by social disorganization.

TABLE 4.1 Merton's Strain Theory

MODE OF ADAPTATION	CULTURALLY DEFINED GOALS	STRUCTURALLY DEFINED GOALS
1. Conformity	+	+
2. Innovation	+	−
3. Ritualism	−	+
4. Retreatism	−	−
5. Rebellion	±	±

+, acceptance of/access to; −, rejection of/lack of access to; ±, rejection of culturally defined goals and structurally defined means and replacement with new goals and means.

Source: Adapted with permission of The Free Press, a Division of Simon & Schuster Adult Publishing Group, from Robert K. Merton's *Social Theory and Social Structure* (1957). Copyright © 1957 by The Free Press; copyright renewed 1985 by Robert K. Merton. All rights reserved.

Conflict Perspective

Conflict theories of crime suggest that deviance is inevitable whenever two groups have differing degrees of power; in addition, the more inequality there is in a society, the greater the crime rate in that society. Social inequality leads individuals to commit crimes such as larceny and burglary as a means of economic survival. Other individuals, who are angry and frustrated by their low position in the socioeconomic hierarchy, express their rage and frustration through crimes such as drug use, assault, and homicide. In Argentina, for example, the soaring violent crime rate is hypothesized to be "a product of the enormous imbalance in income distribution . . . between the rich and the poor" (Pertossi 2000).

According to the conflict perspective, those in power define what is criminal and what is not, and these definitions reflect the interests of the ruling class. Laws against vagrancy, for example, penalize individuals who do not contribute to the capitalist system of work and consumerism. Furthermore, D'Alessio and Stolzenberg (2002, p. 178) found that "in cities with high unemployment, unemployed defendants have a substantially higher probability of pretrial detention" than employed defendants. Rather than viewing law as a mechanism that protects all members of society, conflict theorists focus on how laws are created by

> "There are two criminal justice systems in this country. There is a whole different system for poor people. It's the same courthouse, it's not separate, but it's not equal."
>
> **Paul Petterson**
> **Public defender**

To Marxists the cultural definition of women as property contributes to the high rates of female criminality and, specifically, involvement in prostitution, drug abuse, and petty theft. In 2005 there were 85,000 arrests for prostitution and commercial vice in the United States.

those in power to protect the ruling class. For example, wealthy corporations contribute money to campaigns to influence politicians to enact tax laws that serve corporate interests (Reiman 2007).

In addition, conflict theorists argue that law enforcement is applied differentially, penalizing those without power and benefiting those with power. For example, although the race of a victim should not matter, blacks are more likely to be arrested when involved in black-on-white crime than when involved in black-on-black crime (Eitle, D'Alessio, & Stolzenberg 2002). Moreover, female prostitutes are more likely to be arrested than are the men who seek their services. Unlike street criminals, corporate criminals are often punished by fines rather than by lengthy prison terms, and rape laws originated to serve the interests of husbands and fathers who wanted to protect their property—wives and unmarried daughters.

Societal beliefs also reflect power differentials. For example, "rape myths" are perpetuated by the male-dominated culture to foster the belief that women are to blame for their own victimization, thereby, in the minds of many, exonerating the offender. Such myths include the notion that when a woman says no she means yes, that "good girls" don't get raped, that appearance indicates willingness, and that women secretly want to be raped. Not surprisingly, in societies where women and men have greater equality, there is less rape.

What Do You Think? Using data from the General Social Survey, Barkan and Cohn (2005) investigated the relationship between racial prejudice and spending on criminal justice initiatives. The authors concluded that "the more that racially prejudiced whites perceive that African Americans are prone to violence, the more likely they are to want money spent to reduce crime" (p. 311). What is the independent variable? The dependent variable? In what way is the authors' conclusion consistent with the conflict perspective?

Symbolic Interactionist Perspective

Two important theories of crime emanate from the symbolic interactionist perspective. The first, labeling theory, focuses on two questions: How do crime and deviance come to be defined as such, and what are the effects of being labeled criminal or deviant? According to Howard Becker (1963):

> Social groups create deviance by making rules whose infractions constitute deviance, and by applying those rules to particular people and labeling them as outsiders. From

this point of view, deviance is not a quality of the act a person commits, but rather a consequence of the application by others of rules and sanctions to an "offender." The deviant is one to whom the label has successfully been applied; deviant behavior is behavior that people so label. (p. 238)

Labeling theorists make a distinction between **primary deviance,** which is deviant behavior committed before a person is caught and labeled an offender, and **secondary deviance,** which is deviance that results from being caught and labeled. After a person violates the law and is apprehended, that person is stigmatized as a criminal. This deviant label often dominates the social identity of the person to whom it is applied and becomes the person's "master status," that is, the primary basis on which the person is defined by others.

Being labeled as deviant often leads to further deviant behavior because (1) the person who is labeled as deviant is often denied opportunities for engaging in nondeviant behavior and (2) the labeled person internalizes the deviant label, adopts a deviant self-concept, and acts accordingly. For example, the teenager who is caught selling drugs at school may be expelled and thus denied opportunities to participate in nondeviant school activities (e.g., sports and clubs) and to associate with nondeviant peer groups. The labeled and stigmatized teenager may also adopt the self-concept of a "druggie" or "pusher" and continue to pursue drug-related activities and membership in the drug culture.

The assignment of meaning and definitions learned from others is also central to the second symbolic interactionist theory of crime, differential association. Edwin Sutherland (1939) proposed that through interaction with others, individuals learn the values and attitudes associated with crime as well as the techniques and motivations for criminal behavior. Individuals who are exposed to more definitions favorable to law violation (e.g., "crime pays") than to unfavorable ones (e.g., "do the crime, you'll do the time") are more likely to engage in criminal behavior. Thus children who see their parents benefit from crime or who live in high-crime neighborhoods where success is associated with illegal behavior are more likely to engage in criminal behavior.

Unfavorable definitions come from a variety of sources. Of particular concern of late is the role of video games in promoting criminal or violent behavior. One particular game, *Grand Theft Auto,* has players "head bashing, looting, drug-dealing, drive-by shooting, and running over innocent bystanders with a taxi" (Richtel 2003). In response to this and other violent video games, many states now require a video rating system that differentiates between cartoon violence, fantasy violence, intense violence, and sexual violence. In 2005, a multimillion-dollar suit was filed against the creators and marketers of the game in the wrongful deaths of three men, two of whom were police officers.

primary deviance Deviant behavior committed before a person is caught and labeled an offender.

secondary deviance Deviance that results from being caught and labeled as an offender.

index offenses Crimes identified by the FBI as the most serious, including personal or violent crimes (homicide, assault, rape, and robbery) and property crimes (larceny, motor vehicle theft, burglary, and arson).

TYPES OF CRIME

The FBI identifies eight index offenses as the most serious crimes in the United States. The **index offenses,** or street crimes as they are often called, can be against a person (called violent or personal crimes) or against property (see Table 4.2). Other types of crime include vice crime (such as drug use, gambling, and prostitution), organized crime, white-collar crime, computer crime, and juvenile delinquency. Hate crimes are discussed in Chapter 9.

TABLE 4.2 Index Crime Rates, Percentage Change, and Clearance Rates, 2005

	RATE PER 100,000, 2005	PERCENTAGE CHANGE IN RATE (2004–2005)	PERCENTAGE CLEARED, 2005
Violent crime			
Murder	5.6	+3.4	62.1
Forcible rape	32.2	−1.2	41.3
Robbery	150.8	+3.9	25.4
Aggravated assault	291.1	+1.8	55.2
Total	469.2	+2.3	45.5
Property crime			
Burglary	750.2	+0.5	12.7
Larceny/theft	2,342.6	−2.3	18.0
Motor vehicle theft	442.7	−0.2	13.0
Arson	26.9*	−2.7	17.9
Total†	3535.4	−1.5	16.3

*Arson rates per 100,000 are calculated independently because population coverage for arson is lower than for the other index offenses—1999 rate.

† Property crime totals do not include arson.

Source: FBI (2006).

Street Crime: Violent Offenses

The most recent data available from the FBI's Uniform Crime Reports indicate that the 2005 violent crime rate increased from the previous year by 3.4 percent. Remember, however, that crime statistics represent only those crimes *reported* to the police: 1.39 million violent crimes in 2005. Victim surveys indicate that a little over half of all violent crimes are actually reported to the police (U.S. Bureau of the Census 2007).

Violent crime includes homicide, assault, rape, and robbery. *Homicide* refers to the willful or non-negligent killing of one human being by another individual or group of individuals. Although homicide is the most serious of the violent crimes, it is also the least common, accounting for 1.2 percent of all violent crimes (FBI 2006). A typical homicide scenario includes a male killing a male with a handgun after a heated argument. The victim and offender are disproportionately young and of minority status. When a woman is murdered and the victim-offender relationship is known, she is most likely to have been killed by her husband or boyfriend (FBI 2006).

Mass murders have more than one victim in a killing event. In 2007, Cho Sueng-Hui, dubbed the "Virginia Tech killer," was responsible for one of the largest mass murders in U.S. history, killing more than 30 students and faculty on the college campus. Unlike mass murderers, serial killers kill consecutively over a long period of time. The most well-known serial killers, who were responsible for some of the most horrific episodes of homicide, are Ted Bundy, Kenneth Bianchi, and Jeffery Dahmer. More recently, Dennis Rader, the self-proclaimed "BTK" (bind, torture, kill) killer was captured. Accused of killing 10 people (2 men and 8 women) between 1974 and 1991, Rader was convicted of murder and received 10 consecutive life sentences with no chance of parole for 175 years (Coates 2005; Romano 2005).

Another form of violent crime, *aggravated assault*, involves attacking a person with the intent to cause serious bodily injury. Like homicide, aggravated assault occurs most often between members of the same race and, as with violent crime in general, is more likely to occur in warm weather months. In 2005 the assault rate was nearly 50 times the murder rate, with assaults making up an estimated 62.1 percent of all violent crimes (FBI 2006).

Rape is also classified as a violent crime and is also intraracial; that is, the victim and offender tend to be from the same racial group. The FBI definition of *rape* contains three elements: sexual penetration, force or the threat of force, and nonconsent of the victim. In 2005, 93,934 forcible rapes were reported in the United States, a slight decrease from the previous year (FBI 2006). Rapes are more likely to occur in warm months, in part because of the greater ease of victimization. People are outside more and later, doors are open, windows are unlocked, and so forth.

> **❝**Rape is the only crime in which the victim becomes the accused.**❞**
>
> Fred Adler
> Criminologist

Perhaps as much as 80 percent of all rapes are **acquaintance rapes**—rapes committed by someone the victim knows. Although acquaintance rapes are the most likely to occur, they are the least likely to be reported and the most difficult to prosecute. Unless the rape is what Williams (1984) calls a **classic rape**—that is, the rapist was a stranger who used a weapon and the attack resulted in serious bodily injury—women hesitate to report the crime out of fear of not being believed. The increased use of "rape drugs," such as Rohypnol, may lower reporting levels even further (see Chapter 3).

Robbery, unlike simple theft, also involves force or the threat of force or putting a victim in fear and is thus considered a violent crime. Officially, in 2005 more than 417,000 robberies took place in the United States. Robberies are most often committed by young adults with the use of a gun (FBI 2006). Robbers and thus robberies vary dramatically in type, from opportunistic robberies whose victims are easily accessible and that yield only a small amount of money to professional robberies of commercial establishments, such as banks, jewelry stores, and convenience stores. According to the FBI, in 2005, the average dollar value per robbery was $1,230 and per bank robbery was $4,169 (FBI 2006).

Street Crime: Property Offenses

Property crimes are those in which someone's property is damaged, destroyed, or stolen; they include larceny, motor vehicle theft, burglary, and arson. Property crimes have gone down since 1996, with a 13.9 percent decrease in the last decade. *Larceny*, or simple theft, accounts for more than two-thirds of all property arrests (FBI 2006). The average dollar value lost per larceny incident is $764. Examples of larcenies include purse-snatching, theft of a bicycle, pick pocketing, theft from a coin-operated machine, and shoplifting. In 2005 there were an estimated 6.8 million larcenies reported in the United States (FBI 2006), the most common index offense.

Larcenies involving automobiles and auto accessories are the largest category of thefts. However, because of the cost involved, *motor vehicle theft* is

acquaintance rape Rape committed by someone known to the victim.

classic rape Rape committed by a stranger, with the use of a weapon, resulting in serious bodily injury to the victim.

considered a separate index offense. Numbering more than 1.2 million in 2005, the motor vehicle theft rate has decreased 20.7 percent since 1996. Because of insurance requirements, vehicle theft is one of the most highly reported index crimes, and, consequently, estimates between the FBI's Uniform Crime Reports and the National Crime Victimization Survey are fairly compatible. Less than 14 percent of motor vehicle thefts are cleared.

Burglary, which is the second most common index offense after larceny, entails entering a structure, usually a house, with the intent to commit a crime while inside. Official statistics indicate that in 2005 more than 2.1 million burglaries occurred, a rate of 750 per 100,000 population. Most burglaries are residential rather than commercial and take place during the day when houses are unoccupied. The most common type of burglary is forcible entry, followed by unlawful entry.

Arson involves the malicious burning of the property of another. Estimating the frequency and nature of arson is difficult given the legal requirement of "maliciousness." Of the reported cases of arson, 43.6 percent involved structures (most of which were residential) and 29.0 percent involved movable property (e.g., boat or car), with the remainder being miscellaneous property (e.g., crops or timber). In 2005 the average dollar amount of damage as a result of arson was $14,910 (FBI 2006).

Vice Crime

Vice crimes, often thought of as crimes against morality, are illegal activities that have no complaining participant(s) and are often called **victimless crimes.** Examples of vice crimes include using illegal drugs, engaging in or soliciting prostitution, illegal gambling, and pornography.

Most Americans view drug use as socially disruptive (see Chapter 3). There is less consensus, however, nationally or internationally, that gambling and prostitution are problematic. For example, in the Netherlands prostitution is legal. Although a country of only 16 million, the Netherlands has an estimated 25,000–50,000 sex workers. Like other workers, sex workers in the Netherlands have access to the social security system and pay income tax ("Situation in the Netherlands" 2003). As many as 70 percent of prostitutes in the Netherlands are trafficked into the country (Morse 2006); many of whom "can hardly make ends meet in their own countries and are attracted by opportunities to make quick money in the West in order to support their families and save for a better future . . ." (Siegel 2005, p. 5).

Prostitution is illegal in the United States with the exception of several counties in Nevada. Despite its illegal status, it is a multimillion-dollar industry with 85,000 arrests for prostitution and commercial vice in 2005 (FBI 2006). Trafficking women for purposes of prostitution also occurs in the United States. In 2007, Juan Balderas pled guilty to charges of transportation for purposes of prostitution and importation of illegal aliens. Balderas' operation was just part of a larger smuggling ring in which women from Latin America were transported to the United States and forced to have sex with as many as 40 men a day. Brothels using women that Balderas and others smuggled into the United States were located in cities across the nation including Las Vegas, Atlanta, and New York City (Associated Press 2007b).

victimless crimes Illegal activities that have no complaining participant(s) and are often thought of as crimes against morality such as prostitution.

In the United States many states have legalized gambling, including casinos in Nevada, New Jersey, Connecticut, North Carolina, and other states, as well as state lotteries, bingo parlors, horse and dog racing, and jai alai. In addition, some have argued that there is little difference, other than societal definitions of acceptable and unacceptable behavior, between gambling and other risky ventures such as investing in the stock market. Conflict theorists are quick to note that the difference is who is making the wager.

Pornography, particularly Internet pornography, is a growing international problem. Regulation is made difficult by fears of government censorship and legal wrangling as to what constitutes "obscenity." For many, the concern with pornography is not its consumption per se but the possible effects of viewing or reading pornography—increased sexual aggression. Although the literature on this topic is mixed, Conklin concluded (2007, p. 221) that there is no "consistent evidence that nonviolent pornography causes sex crimes."

Organized crime refers to criminal activity conducted by members of a hierarchically arranged structure devoted primarily to making money through illegal means. Although often discussed under victimless crimes because of its association with prostitution, drugs, and gambling, organized crime groups often use coercive tactics. For example, organized crime groups may force legitimate businesses to pay "protection money" by threatening vandalism or violence.

The traditional notion of organized crime is the Mafia, a national band of interlocked Italian families. But members of many ethnic groups engage in organized crime in the United States.

> Chinese, Vietnamese, Korean, and Japanese gangs have been found on the East and West coasts, active in smuggling drugs and extorting money from businesses in their communities. Scores of other groups can be found in various cities: Israelis dominating insurance fraud in Los Angeles, Cubans running illegal gambling operations in Miami, Canadians engaging in gun smuggling and money laundering in Miami, Russians carrying out extortion and contract murders in New York. (Thio 2007, p. 374)

Organized crime also occurs in other countries. For example, with more than 90,000 members and associates in 3,000 crime groups, the Japanese Yakuza

organized crime Criminal activity conducted by members of a hierarchically arranged structure devoted primarily to making money through illegal means.

are one of the largest crime organizations in the world. The young men who join the Yakuza tend to be from the lower class and must undergo a training period of 5 years. During this apprenticeship members learn absolute loyalty to their superiors as well as the other norms and values of the group. The Yakuza are involved in drugs, illegal gambling, and prostitution, as well as several legitimate businesses. Interestingly, the Yakuza proudly display their name at their "corporate" headquarters, and recruits wear lapel pins identifying themselves as members (Thio 2004; Winslow & Zhang 2008).

Unlike traditional crime organizations that are hierarchically arranged, transnational crime organizations tend to be decentralized and less likely to operate through legitimate businesses. Transnational crime organizations, like transnational crime in general, directly or indirectly involve more than one country. Transnational crime organizations are a growing threat to the United States and to global security. As Wagley (2006) explained:

> The end of the Cold War—along with increasing globalization beginning in the 1990s—has helped criminal organizations expand their activities and gain global reach. Criminal networks are believed to have benefited from the weakening of certain government institutions, more open borders, and the resurgence of ethnic and regional conflicts across the former Soviet Union and many other regions. Transnational criminal organizations have also exploited expanding financial markets and rapid technological developments. (p. 1)

Transactional crime organizations are also less likely than the traditional crime "families" to develop around a family or ethnic structure. Transnational crime organizations are involved in many types of transnational crime including money laundering, narcotics, arms smuggling, and trafficking in persons. Further, terrorists are increasingly supporting themselves through transnational organized crime groups. For example, the 2003 bombing of a Madrid commuter train was financed through drug dealing (Wagley 2006).

White-Collar Crime

66 Criminal: A person with predatory instincts who has not sufficient capital to form a corporation. **99**

Howard Scott
Scientist and author

white-collar crime Includes both *occupational crime*, in which individuals commit crimes in the course of their employment, and *corporate crime*, in which corporations violate the law in the interest of maximizing profit.

White-collar crime includes both *occupational crime,* in which individuals commit crimes in the course of their employment, and *corporate crime,* in which corporations violate the law in the interest of maximizing profit. Occupational crime is motivated by individual gain. Employee theft of merchandise, or pilferage, is one of the most common types of occupational crime. Other examples include embezzlement, forgery and counterfeiting, and insurance fraud. Price fixing, antitrust violations, and security fraud are all examples of corporate crime, that is, crime that benefits the organization. In recent years several officers of major corporations, including Enron, WorldCom, Adelphia, and Imclone, have been charged with securities fraud, tax evasion, and insider trading. WorldCom engaged in what has been called the "largest accounting fraud in history," exaggerating its worth by $9 billion. Shareholders lost more than $3 billion, and 17,000 employees were in danger of losing their jobs (Ripley 2003). In part as a response to the widespread scandals of late, the U.S. Sentencing Commission has approved stiffer penalties for white-collar criminals, including a new sentencing formula (Lichtblau 2003). For example, founder and CEO of Adelphia Communications Corporation John

Ken Lay, CEO and founder of Enron, was convicted of 10 counts of fraud and conspiracy on May 25, 2006, in the first of several large corporate scandals (Pasha 2006). The collapse of Enron cost 4,000 employees their jobs, cost many individuals their life savings, and cost investors billions of dollars. The multimillionaire was facing 25–40 years in prison before his untimely death at age 64.

Rigas received a 15-year sentence and certain financial ruin as a result of his conviction for fraud and conspiracy (Sasseen 2006).

Nonetheless, many white-collar criminals go unpunished. First, many companies, not wishing the bad publicity surrounding a scandal, simply dismiss the parties involved rather than press charges. Second, many white-collar crimes, as traditional crimes, go undetected. In a recent survey of a representative sample of 1,600 U.S. households, the National White Collar Crime Center (NWCCC) found that nearly one in two households (46.5 percent) had been the victim of some type of white-collar crime in the last year. However, only 14.0 percent of the crimes were brought to the attention of a law enforcement agency (NWCCC 2006).

Third, federal prosecutions of white-collar criminals have decreased by 28.0 percent in the last 5 years. Few believe the decrease is a result of a lower prevalence of white-collar crime offenses. Two forces appear to be in operation. First, white-collar crimes are becoming increasingly complex, making prosecution a time-intensive endeavor. For example, in the Enron and World-Com cases federal prosecutors spent years going through millions of documents. With limited resources agencies such as the Securities and Trade Commission (SEC) are increasingly relying on companies to do their own independent investigations. Second, experts say that the decrease in prosecutions also represents a shift in priorities. For example, Marks (2006) argued that the federal government has shifted attention away from white-collar crimes as the need for homeland security has increased. Thus, as Figure 4.2 portrays, federal prosecutions of immigration violations have increased dramatically since the September 11, 2001, attacks, whereas white-collar crime prosecutions have steadily decreased (Marks 2006).

Corporate violence, a form of corporate crime, refers to the production of unsafe products and the failure of corporations to provide a safe working envi-

corporate violence The production of unsafe products and the failure of corporations to provide a safe working environment for their employees.

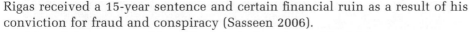

FIGURE 4.2
Number of federal criminal
prosecutions: 1993–2005.
Source: Marks (2006).

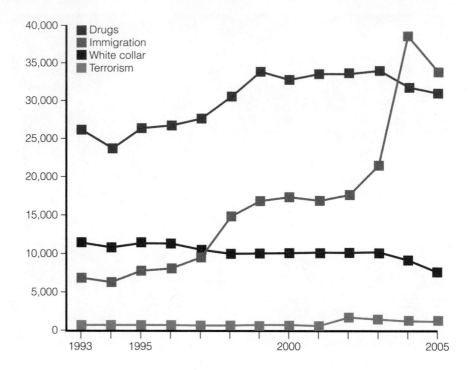

TABLE 4.3 Types of White–Collar Crime	
CRIMES AGAINST CONSUMERS	**CRIMES AGAINST EMPLOYEES**
Deceptive advertising	Health and safety violations
Antitrust violations	Wage and hour violations
Dangerous products	Discriminatory hiring practices
Manufacturer kickbacks	Illegal labor practices
Physician insurance fraud	Unlawful surveillance practices
CRIMES AGAINST THE PUBLIC	**CRIMES AGAINST EMPLOYERS**
Toxic waste disposal	Embezzlement
Pollution violations	Pilferage
Tax fraud	Misappropriation of government funds
Security violations	Counterfeit production of goods
Police brutality	Business credit fraud

ronment for their employees. Corporate violence is the result of negligence, the
pursuit of profit at any cost, and intentional violations of health, safety, and
environmental regulations. For example, after more than 1 year of recalls in 16
countries, Bridgestone/Firestone began a U.S. recall of more than 6.5 million
tires. The tires, many of which were standard equipment on the popular Ford
Explorer, had a 10-year history of tread separation. It was only after 88 U.S.
traffic deaths were linked to the defective tires, prompting a congressional investigation, that Ford and Bridgestone/Firestone acknowledged the overseas
recalls and the tires' questionable safety history (Pickler 2000). Subsequently,

Ford Motor Company was asked by a federal judge to turn over data on their 15-passenger van. Several deaths occurred as a result of the van rolling over, and Ford was accused of hiding evidence of the problem. According to the National Highway Traffic Safety Administration, more than 400 people have died in passenger van accidents since 1990 (Chicago Tribune 2003). Table 4.3 summarizes some of the major categories of white-collar crime.

A final problem in pursuing "corporate criminals" is the difficulty in assigning legal culpability to the offender. As Friedrichs (2004) noted:

> The absence of the direct intent to do harm, the difficulty in pinpointing the specific cause of the harm, the diffusion of responsibility for harm producing corporate decisions, and the economic and political clout of the corporations has combined to shield corporate employers from full fledged liability. (p. 72)

Some recent evidence, however, suggests that the white-collar crime culture of "boys will be boys" is being replaced, according to Marjorie Kelly, editor of *Business Ethics*, with a new and less tolerant "Puritanism" (Thomas 2005).

Computer Crime

Computer crime refers to any violation of the law in which a computer is the target or means of criminal activity. Sometimes called cyber-crime, computer crime is one of the fastest growing types of crime in the United States. It is also one of the most expensive, costing U.S. businesses an estimated $67.2 billion a year (Buechner 2006). Hacking, or unauthorized computer intrusion, is one type of computer crime. In just 1 month hackers successfully attacked the computer systems of Walt Disney World, Yahoo, eBay, and Amazon.com through "denial of service" invasions (Kong & Swartz 2000).

The increase in computer break-ins has also led to an increase in **identity theft,** the use of someone else's identification (e.g., social security number or birth date) to obtain credit or other economic rewards. In 2002 the number of identity thefts doubled from the previous year, making identity theft the most frequent complaint to the Federal Trade Commission (Lee 2003). Although mail theft is one of the most common modes of obtaining the needed information, new technologies have contributed to the increased rate of identity theft. For example, each week thousands of stolen credit card numbers are sold online in "membership only cyber-bazaars, operated largely by residents of the former Soviet Union who have become central players in credit card and identity theft" (Richtel 2002, p. 1). Recent accounts of unauthorized use of personal data from companies such as Lexis-Nexis, Bank of America, eBay, and ChoicePoint have led Congress to call for stricter treatment of "data brokers." In 2005 the personal information, including social security numbers, of nearly 100,000 University of Berkeley students and alumni was stolen (Associated Press 2005b). In 2006, a Department of Veterans Affairs official downloaded the personnel records of more than 26 million veterans to his laptop computer, which was then stolen, "exposing all the information necessary to swipe the identity of virtually every person released from military service since 1975" (Levy 2006).

Given the more than 10 million victims a year (Harrow 2005), it is not surprising that a survey of Americans indicates a rise in concern about identity theft and privacy issues in general (Cohen 2005). In response to such public opinion, in 2004 President Bush signed the Identity Theft Penalty Enhancement Act, which adds 2 years to prison sentences for those convicted of using

computer crime Any violation of the law in which a computer is the target or means of criminal activity.

identity theft The use of someone else's identification (e.g., social security number, birth date) to obtain credit or other economic rewards.

stolen credit cards or other personal information to commit a crime (McGuire 2004).

Identity theft is just one category of computer crime. Another category is Internet fraud. According to a 2006 report by the Internet Crime Complaint Center (ICCC 2006), the most common category of Internet fraud is Internet auction fraud, which represents 62.7 percent of all fraud complaints referred to law enforcement agencies. Internet auction fraud entails the misrepresentation of items for sale online and/or the nondelivery of products bought online. Other Internet fraud categories include pyramid schemes, investment fraud, credit card fraud, and counterfeit check schemes. In 2005 the ICCC received more than 220,000 complaints.

Another type of computer crime is Internet solicitation of minors. Forty-two states presently have laws that make online enticement of a child for sexual activity a crime. According to the National Center for Missing and Exploited Children (NCMEC), more than 30 million children younger than age 18 are on the "net" and one in seven "receives a sexual solicitation online which includes a request to engage in sexual activity, a request to engage in sexual talk, or a request to give out personal sexual information" (NCMEC 2007, p. 1). Websites such as MySpace and Facebook often attract such predators. In 2006 a 21-year-old man was arrested in Connecticut for the rape of a 14-year-old girl he had met on MySpace (Williams 2006).

What Do You Think? With increased concern over child sexual predators many states are legislating new and tougher laws. Ohio, for example, is considering a bill that would require "all habitual and child-oriented sex offenders" to display fluorescent green license plates (Associated Press 2007c, p. 1). Although some states require a special designation on driver's licenses, no state requires special license plates for sex offenders. What are some of the pros and cons of such a proposal?

Conklin (1998), Reid (2003), Schiesel (2005), and Siegel (2006) identify other types of computer crime:

- Two individuals were charged with the theft of 80,000 cellular phone numbers. Using a device purchased from a catalogue, the thieves picked up radio waves from passing cars, determined private cellular codes, reprogrammed computer chips with the stolen codes, and then, by inserting the new chips into their own cellular phones, charged calls to the original owners.
- Called "shaving," a programmer made $300 a week by programming a computer to round off each employee's paycheck down to the nearest dime and then to deposit the extra few pennies in the offender's account.
- A computer hacker broke into a telephone system and rigged the outcome of a radio station contest. Three hackers won a trip to Hawaii, a Porsche, and a cash prize.
- An oil company illegally tapped into another oil company's computer to get information that allowed the offending company to underbid the other company for leasing rights.

Females who join gangs often do so to win approval from their boyfriends who are gang members. Increasingly, though, females are forming independent "girl gangs." The most common type of female gang member remains, however, a female auxiliary to a male gang.

© A. Ramey/PhotoEdit

- After building a good credit record, a California man sold more than $800,000 worth of merchandise on eBay and never delivered the goods.
- While sitting in his car, a suspect was caught viewing child pornography he had just downloaded onto his laptop computer from a neighbor's wireless system.

Finally, as Conklin (2007, p. 144) explained, several aspects of computer technology lend themselves to criminal activities. Thieves who sell them on the black market often target computer chips, which are small and relatively valuable. Computer programs are pirated in violation of copyright infringement laws. Even cyber-extortion schemes are becoming increasingly common as hackers, demanding ransoms, take corporate homepages hostage. What is the result? Cyber-insurance, of course (Buechner 2006).

Juvenile Delinquency

In general, children younger than age 18 are handled by the juvenile courts, either as status offenders or as delinquent offenders. A *status offense* is a violation that can be committed only by a juvenile, such as running away from home, truancy, and underage drinking. A *delinquent offense* is an offense that would be a crime if committed by an adult, such as the eight index offenses. The most common status offenses handled in juvenile court are underage drinking, truancy, and running away. In 2005, 15.3 percent of all arrests (excluding traffic violations) were of offenders younger than age 18 (FBI 2006).

As is the case with adults, juveniles commit more property than violent offenses, but the number of violent juvenile offenses has increased in recent years. For example, between 2004 and 2005 juvenile arrests for murder climbed 19.9 percent and for robbery climbed 11.4 percent (FBI 2006). Americans are concerned about the high rate of juvenile violence, including violence in schools and gang-related violence. Gang-related crime is, in part, a function of two inter-

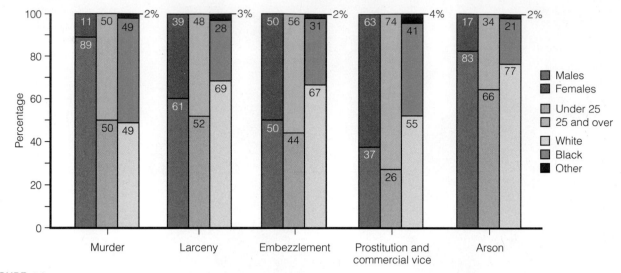

FIGURE 4.3
Percentage of arrests by sex, age, and race: 2005.
Source: FBI (2006).

related social forces: the increased availability of guns in the 1980s and the lucrative and expanding drug trade. It is estimated that the United States has 24,500 street gangs with 772,500 members, the highest proportion of which are racial and ethnic minorities (Shelden, Tracy, & Brown 2004). One of the most dangerous gangs in America is Mara Salvatrucha (MS-13), with members in more than 30 states (Campo-Flores 2005). Made up primarily of street-savvy Salvadorans and boasting a membership of more than 10,000, MS-13 has become one of the highest priorities of the FBI's criminal enterprise division.

DEMOGRAPHIC PATTERNS OF CRIME

Although virtually everyone violates a law at some time, individuals with certain demographic characteristics are disproportionately represented in the crime statistics. Victims, for example, are disproportionately young, lower-class, minority males from urban areas. Similarly, the probability of being an offender varies by gender, age, race, social class, and region (see Figure 4.3).

Gender and Crime

It is a universal truth that women everywhere are less likely to commit crime than men. In the United States, both official statistics and self-report data indicate that females commit fewer violent crimes than males. Why are females less likely to commit violent crimes? Called the *socialization hypothesis*, some would argue "girls are less violent than boys because they are controlled through subtle mechanisms, which include[s] learning that violence is incompatible with the meaning of their gender" (Heimer & DeCoster 1999, pp. 305–306).

In 2005 males accounted for 76.2 percent of all arrests, 82.1 percent of all arrests for violent crime, and 68.0 percent of all arrests for property crimes (FBI 2006). Not only are females less likely than males to commit serious crimes, but also the monetary value of female involvement in theft, property damage, and illegal drugs is typically less than that for similar offenses committed by males.

Nevertheless, a growing number of women have become involved in characteristically male criminal activities, such as gang-related crime and drug use.

The recent increase in crimes committed by females has led to the development of feminist criminology. Feminist criminology focuses on how the subordinate position of women in society affects their criminal behavior and victimization. For example, Chesney-Lind and Shelden (2004) reported that arrest rates for runaway juvenile females are higher than those for males not only because girls are more likely to run away as a consequence of sexual abuse in the home but also because police with paternalistic attitudes are more likely to arrest female runaways than male runaways. Feminist criminology thus adds insights into understanding crime and violence that are often neglected by traditional theories concentrating on gender inequality in society. Feminist criminology has also had an impact on public policy. Mandatory arrest for domestic violence offenders, the development of rape shield laws, public support for battered women's shelters, laws against sexual harassment, and the repeal of the spousal exception in rape cases are all, according to Winslow and Zhang (2008), outcomes of feminist criminology.

The subordinate position of women in the United States also affects their victimization rates. A report from the *Harvard School of Public Health* revealed that 70 percent of all female homicide victims in industrial countries are American (Harvard School of Public Health 2002). A female in the United States is five times more likely to be murdered than a female in Germany, eight times more likely to be murdered than a female in England, and three times more likely to be murdered than a female in Canada. Although males are almost four times more likely to be murdered than females, when a woman is murdered, it is most likely by an ex-boyfriend, husband, or other intimate partner (Bureau of Justice Statistics 2006b).

Age and Crime

In general, criminal activity is more prevalent among younger people than among older people. The highest arrest rates are for individuals younger than age 25. In 2005, 44.3 percent of all arrests in the United States were of people younger than age 25 (FBI 2006). Although those younger than age 25 made up over half of all arrests in the United States for crimes such as robbery, burglary, motor vehicle theft, and arson, those younger than age 25 were significantly less likely to be arrested for crimes such as fraud and forgery and counterfeiting. Those older than age 65 made up less than 1.0 percent of total arrests for the same year (FBI 2006).

Why is criminal activity more prevalent among individuals in their late teens and early 20s? One reason is that juveniles are insulated from many of the legal penalties for criminal behavior. Younger individuals are also more likely to be unemployed or employed in low-wage jobs. Thus, as strain theorists argue, they have less access to legitimate means for acquiring material goods.

Some research suggests, however, that high school students who have jobs become more, rather than less, involved in crime (Felson 2002). In earlier generations teenagers who worked did so to support themselves and/or their families. Today, teenagers who work typically spend their earnings on recreation and "extras," including car payments and gasoline. The increased mobility associated with having a vehicle also increases the opportunity for criminal behavior and reduces parental control.

Other hypothesized reasons for the age-crime relationship are also linked to specific theories of criminal behaviors. For example, conflict theorists would argue that teenagers and young adults have less power in society than their middle-aged and elderly counterparts. One manifestation of this lack of power is that the police, using a mental map of who is a "typical offender," are more likely to have teenagers and young adults in their suspect pool. With increased surveillance of teenagers and young adults comes increased detection of criminal involvement—a self-fulfilling prophecy.

Race, Social Class, and Crime

Race is a factor in who gets arrested. Minorities are disproportionately represented in official statistics. For example, although African Americans represent about 14 percent of the population, they account for more than 38 percent of all violent index offenses and 28.6 percent of all property index offenses (FBI 2006). In addition, an estimated 12 percent of all black males in their late 20s are in prison or jail compared with 1.7 percent of all white males in their late 20s (Bureau of Justice Statistics 2006c).

Nevertheless, it is inaccurate to conclude that race and crime are causally related. First, official statistics reflect the behaviors and policies of criminal justice actors. Thus the high rate of arrests, conviction, and incarceration of minorities may be a consequence of individual and institutional bias against minorities. For example, investigating race and ethnicity in sentencing outcomes, Steffensmeier and Demuth (2000) concluded that both race and ethnicity continue to be factors in the organizational decision-making process of the criminal justice system.

Furthermore, blacks are sent to prison for drug offenses at a rate 8.2 times higher than the rate for whites. If current trends continue, by 2020 two of every three black men between the ages of 18 and 34 will be in prison (Dickerson 2000; Fletcher 2000). These disturbing statistics have led to concerns about **racial profiling**—the practice of targeting suspects on the basis of race. Proponents of the practice argue that because race, like gender, is a significant predictor of who commits crime, the practice should be allowed. Opponents hold that racial profiling is little more than discrimination and should therefore be abolished.

Second, race and social class are closely related in that nonwhites are overrepresented in the lower classes. Because lower-class members lack legitimate means to acquire material goods, they may turn to instrumental, or economically motivated, crimes. In addition, although the "haves" typically earn social respect through their socioeconomic status, educational achievement, and occupational role, the "have-nots" more often live in communities where respect is

racial profiling The law enforcement practice of targeting suspects on the basis of race.

based on physical strength and violence, as subcultural theorists argue. For example, Kubrin (2005) examined the "street code" of inner city black neighborhoods by analyzing rap music lyrics. Her results indicate that song "lyrics instruct listeners that toughness and the willingness to use violence are central to establishing viable masculine identity, gaining respect, and building a reputation" (p. 375). This chapter's *Social Problems Research Up Close* feature examines violence and "being tough" in a sample of minority youth.

Thus the apparent relationship between race and crime may be, in part, a consequence of the relationship between these variables and social class. Philips (2002), in her investigation of white, black, and Latino homicide rates, concluded that

> it nonetheless remains clear from this study that a significant portion of the racial homicide differential could be reduced by improving socioeconomic conditions for minority populations. This conclusion provides some promising policy options. For example, improving levels of education, lowering levels of poverty, and reducing the extent of male unemployment among minority populations might well have an impact on levels of violence and reduce the striking racial homicide differential that currently exists in the United States. (p. 367)

A third hypothesis is that criminal justice system contact, which is higher for nonwhites, may actually act as the independent variable; that is, it may lead to a lower position in the stratification system. Kerley and colleagues (2004) found that "contact with the criminal justice system, especially when it occurs early in life, is a major life event that has a deleterious effect on individuals' subsequent income level" (p. 549).

Some research indicates, however, that even when social class backgrounds of blacks and whites are comparable, blacks have higher rates of criminality (D'Alessio & Stolzenberg 2003). In addition, to avoid the bias inherent in official statistics, researchers have compared race, class, and criminality by examining self-report data and victim studies. Barkan's (2006) findings indicated that although racial and class differences in criminal offenses exist, the differences are not as great as official data would indicate.

Region and Crime

In general, crime rates and, in particular, violent crime rates, are higher in metropolitan areas than in nonmetropolitan areas (Moore 2005). For example, in 2005, the violent crime rate in metropolitan statistical areas was 510 per 100,000 population; in cities in nonmetropolitan statistical areas it was 373.5 per 100,000 population (FBI 2006). Further, a recent survey by the Police Executive Research Forum found that murder rates have climbed by more than 10.0 percent in the nation's largest cities since 2004 (Jordan 2007).

Higher crime rates in urban areas result from several factors. First, social control is a function of small intimate groups that socialize their members to engage in law-abiding behavior, expressing approval for their doing so and disapproval for their noncompliance. In large urban areas people are less likely to know each other and thus are not influenced by the approval or disapproval of strangers. Demographic factors also explain why crime rates are higher in urban areas: large cities have large concentrations of poor, unemployed, and minority individuals.

Although large urban areas have higher crime rates, the recent increase in violent crime has disproportionately taken a toll on mid-sized cities—those with

A significant body of research suggests that fighting in adolescence is a fairly common event. For example, according to the Centers for Disease Control and Prevention (CDCP), in 2005, 35.9 percent of ninth through twelfth graders were involved in at least one physical altercation in the 12 months before they completed the CDCP survey. Further, 3.7 percent of the sample required treatment by a physician or nurse as a result of peer violence (Centers for Disease Control and Prevention 2006). Differential involvement—the tendency for one group to be more involved in delinquency than another—is also documented with black adolescents being more likely to instigate a fight, be the victim of a fight, and/or be involved in weapon-related violence compared with white, Asian, or Hispanic youth. Thus the present study is an important one for it examines the relationship between select explanatory variables (both risks and assets) and the likelihood of fighting among a sample of at-risk minority youth.

Sample and Methods

All respondents were in grades five through twelve and were enrolled in a central Alabama school system (Wright and Fitzpatrick 2006). A letter detailing the purpose of the study was sent home by the school district, and parents were asked to give permission for their child's inclusion in the study and referred to a sample questionnaire that was on file and available for review. The final sample consisted of 1,642 African American youth (51 percent female) with a median age of 14 years. Participation in the survey was voluntary and the response rate was 65 percent.

The dependent variable is fighting and was measured by asking respondents the frequency of their fighting in the last 30 days. Sociodemographic variables include sex, age, mother's and father's education, and mother's and father's occupational status. Risk factors, that is, factors associated with an increase

in the likelihood of fighting, include academic performance (poor grades), family intactness (zero or one parent in the home), parental violence (physical assault from parent or other adult guardian), and a composite measure of gang affiliation (being a member, being asked to be a member, or having friends who are members).

Asset variables are variables that are predicted to decrease incidents of fighting. The eight asset variables were divided into three categories. The first category is self-esteem measured by a respondent's sense of satisfaction, pride, worth, and respect. Parental involvement was measured by a respondent's (1) parents monitoring of where their child goes with their friends, (2) frequency of talking to parents about problems, and (3) frequency of eating dinner with the family. School involvement includes teacher attention, respondent's involvement in school activities and clubs, and self-reported happiness with school.

Results and Conclusions

Frequency of fighting was higher for elementary and middle school students than for high school students, with the highest fighting frequency occurring in middle schools. Across school type 15 percent or more of the students reported the highest response category of fighting—six or more times over the last 30 days.

Data analysis also indicates that fighting is negatively associated with family intactness and self-esteem, with two-parent homes and high self-esteem leading to lower probabilities of fighting. Alternatively, parental violence and gang affiliation is associated with increased probabilities of fighting. Talking to parents about problems and having parents who monitor activities with friends are significantly associated with decreased rates of fighting. Additional asset variables associated with decreased levels of fighting include being

happy with school, attention from teachers, and involvement with school clubs. As the authors note, interestingly, higher involvement in sports was associated with higher rates of fighting.

When multiple variables were analyzed at the same time, three of the four risk factors were significantly associated with higher rates of fighting—lower grades, exposure to violence in the family, and gang affiliation. Of the asset variables, a lack of parental monitoring and being unhappy at school were predictive of increased fighting behavior. Note that low self-esteem, when in the presence of other variables, is not associated with youthful fighting.

The authors concluded that the "risk and asset" model they present has practical implications in terms of organizing intervention techniques. Risk factors need to be "suppressed or eliminated" and asset factors need to be "encouraged or facilitated" (Wright and Fitzpatrick 2006, p. 260). For example, results from the present study show that "parental monitoring and being happy at school were associated with lower frequency of fighting, suggesting the importance of continued support for outreach to parents and further efforts to reduce or eliminate the community factors that promote proliferation of gangs" (Wright and Fitzpatrick 2006, p. 251).

Finally, Wright and Fitzpatrick (2006) noted the limitations of their research. First, respondents were all African American and thus the results from this study cannot be generalized to non-African-American samples. Second, the data are cross-sectional in nature, that is, looking at one point in time, and thus precluding an assessment of causality. Finally, and perhaps most importantly, nonresponding students (i.e., students who were absent, suspended, or expelled or did not secure parental permission to complete the survey) may have varied in some meaningful way from those who completed the survey.

Source: Wright and Fitzpatrick (2006).

populations between 400,000 and 1,000,000 such as Charlotte, North Carolina, St. Louis, Missouri, and Baltimore, Maryland (Kingsbury 2006). For example, Milwaukee, Wisconsin, had the largest jump in homicide rates in the United States—40.0 percent between 2004 and 2005. Kingsbury (2006) argued that the increase in homicide rates and violent crime in general is a consequence of three interrelated social forces. First, since September 11, 2001, homeland security efforts have drained needed federal funds previously used to help municipalities pay for police—so with less money there are fewer police. Second, those arrested in the 1980s and 1990s as part of the "war on drugs" are now beginning to be released on parole, putting more career criminals on city streets. Finally, many mid-sized cities have high unemployment rates, 7 percent in Milwaukee, for example, as manufacturing jobs are shipped overseas (see Chapter 7). On the basis of these three trends, Kingsbury (2006) concludes that violent crime rates are unlikely to decrease any time soon.

Crime rates also vary by region of the country (see Figure 4.4). In 2005 both violent and property crimes were highest in southern states followed by western, midwestern, and northeastern states. The murder rate is particularly high in the South, with 41.9 percent of all murders recorded in southern states (FBI 2006). The high rate of southern lethal violence has been linked to high rates of poverty and minority populations in the South, a southern "subculture of violence," higher rates of gun ownership, and a warmer climate that facilitates victimization by increasing the frequency of social interaction.

THE COSTS OF CRIME AND SOCIAL CONTROL

Crime often results in physical injury and loss of life. For example, in 2005 there were more than 16,600 victims of a homicide (FBI 2006). That number is dwarfed, however, by the deaths that take place as a consequence of white-collar crime. Criminologist Steven Barkin (2006), who collected data from a variety of sources, reported that annually there are

> (1) 56,425 workplace-related deaths from illness or injury; (2) 9,600 deaths from unsafe products; (3) 35,000 deaths from environmental pollution; and (4) 12,000 deaths from unnecessary surgery. Adding these figures together, 113,025 people a year die from corporate and professional crime and misconduct. (p. 388)

Moreover, the U.S. Public Health Service now defines violence as one of the top health concerns facing Americans. In addition to death and physical injury, crime also has economic, social, and psychological costs.

Economic Costs

Conklin (2007, p. 50) suggested that the financial costs of crime can be classified into at least six categories. First are *direct losses* from crime, such as the destruction of buildings through arson, of private property through vandalism, and of the environment by polluters. In 2005 the average dollar loss of destroyed or damaged property as a result of arson was $14,910 (FBI 2006). Second are costs associated with the *transferring of property*. Bank robbers, car thieves, and embezzlers have all taken property from its rightful owner at tremendous expense to the victim and society. For example, it is estimated that in 2005, $7.6 billion

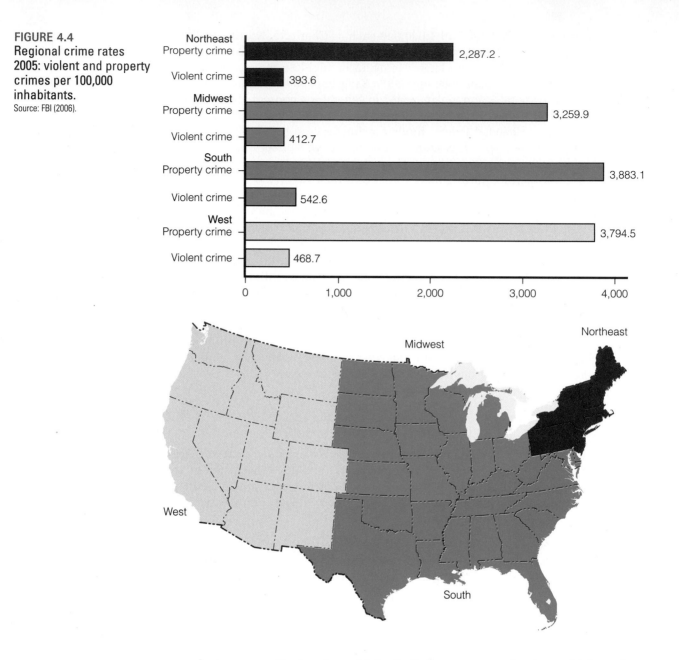

FIGURE 4.4
Regional crime rates 2005: violent and property crimes per 100,000 inhabitants.
Source: FBI (2006).

Northeast
Property crime — 2,287.2
Violent crime — 393.6

Midwest
Property crime — 3,259.9
Violent crime — 412.7

South
Property crime — 3,883.1
Violent crime — 542.6

West
Property crime — 3,794.5
Violent crime — 468.7

was lost as a result of motor vehicle theft; the average value per vehicle at the time of the theft was $6,173 (FBI 2006).

A third major cost of crime is that associated with *criminal violence*, for example, the medical cost of treating crime victims (over $5 billion annually) or the loss of productivity of injured workers (Surgeon General 2002). Fourth are the costs associated with the production and sale of illegal goods and services, that is, *illegal expenditures*. The expenditure of money on drugs, gambling, and prostitution diverts funds away from the legitimate economy and enterprises and lowers property values in high-crime neighborhoods. Fifth is the cost of *prevention and protection*—the billions of dollars spent on locks and safes, surveillance cameras, guard dogs, and the like. It is estimated that Americans spend $65 billion annually on self-protection items (Surgeon General 2002).

Finally, there is the cost of *social control*—the criminal justice system, law enforcement, litigative and judicial activities, corrections, and victims' assistance. The cost of the criminal justice system is estimated to be $90 billion annually and growing (Surgeon General 2002). Reasons for such growth include increases in the rates of arrest and conviction, changes in the sentencing structure, public attitudes toward criminals, the growing number of young males, and the war on drugs. Regardless of the cause, however, the staggering cost of public institutions has led to the "privatization" of prisons, whereby the private sector increasingly supplies needed prison services.

What is the total economic cost of crime? One estimate suggested that the total cost of crime in the United States is more than $1.7 trillion a year (Anderson 1999). Although costs from "street crimes" are staggering, the costs from "crimes in the suites," such as tax evasion, fraud, false advertising, and antitrust violations, are greater than the cost of the FBI index crimes combined (Reiman 2007). For example, Barkan (2006), using an FBI estimate, reported that the total cost of property crime and robbery is $17.1 billion annually. This is less than the $44 billion price tag for employee theft alone.

> **"** The possibility of being a victim of a crime is ever present on my mind. . . . Thinking about it is as natural as . . . breathing. **"**
>
> 40-year-old woman in New York City

Social and Psychological Costs

Crime entails social and psychological costs as well as economic costs. One such cost—fear—is dependent upon individual perceptions of crime as a problem. When a random sample of Americans were asked whether there was more or less crime today than in the previous year, 68 percent reported that there was more crime, 16 percent responded there was less crime, and 8 percent responded that the amount of crime was about the same (Saad 2006). Interestingly, perceptions of crime locally were more positive than perceptions of crime nationally. Just over half (51 percent) of Americans believe that there is more crime in their "area" than there was a year ago, 30 percent believe there is less crime than a year ago, the remainder believing that the crime rates in their "area" have remained about the same.

Not only do Americans worry about crime at the aggregate level, but they also worry about crime at the individual level. When a random sample of Americans were asked the extent to which they worry about crime, half reported that they frequently/occasionally worry about having their house burglarized, 47 percent frequently/occasionally worry about having their car stolen or broken into, and 40 percent frequently/occasionally worry about a school-aged child being physically harmed at school (Carroll 2006).

Fear of crime may be fueled by media presentations that may not accurately reflect the crime picture. For example, in a content analysis of local television portrayals of crime victims and offenders, whites were overrepresented as victims and blacks were overrepresented as offenders compared with official statistics (Dixon & Linz 2000).

And how do most Americans deal with their fear of street crime? According to a Gallup Poll, the most common method of dealing with fear of victimization is to "avoid going to certain places/neighborhoods you might otherwise want to go to" (Carlson 2005, p. 2). Other methods of dealing with fear of crime include, in order, keeping a dog for protection, having a burglar alarm installed in your home, buying a gun for protection of yourself and your home, carrying mace or pepper spray, carrying a gun for defense, and carrying a knife for defense.

White-collar crimes also take a social and psychological toll at both the individual and the societal level. Rosoff, Pontell, and Tillman (2002) state that white-collar crime can produce "feelings of cynicism among the public, remove an essential element of trust from everyday interaction, de-legitimatize political institutions, and weaken respect for the law" (p. 346). In addition, the authors argue that white-collar crime "encourages and facilitates" other types of crime; that is, "there is a connection, both direct and indirect, between 'crime in the suites' and 'crime in the streets'" (p. 346).

STRATEGIES FOR ACTION: CRIME AND SOCIAL CONTROL

Clearly, one way to combat crime is to attack the social problems that contribute to its existence. Moreover, when a random sample of Americans were asked which of two views came closer to their own in dealing with the crime problem, increasing law enforcement or resolving social problems, the majority of respondents, 65.0 percent, selected resolving social problems (Gallup Poll 2007a).

In addition to policies that address social problems, numerous social programs have been initiated to alleviate the crime problem. These policies and programs include local initiatives, criminal justice policies, legislative action, and international efforts in the fight against crime.

Local Initiatives

Youth Programs. Early intervention programs acknowledge that it is better to prevent crime than to "cure" it once it has occurred. *Fight Crime: Invest in Kids* is a nonpartisan, nonprofit anticrime organization made up of more than 3,000 law enforcement leaders and violence survivors (Fight Crime 2007). *Fight Crime: Invest in Kids*

> takes a hard nosed look at crime prevention strategies, informs the public and policymakers about those findings, and urges investment in programs proven effective by research. . . . [The] organization focuses on high quality early education programs, prevention of child abuse and neglect, after-school programs for children and teens, and interventions to get troubled kids back on track. (p. 1)

One such program is the Perry Preschool Project. After a group of children were randomly assigned to either a control group or an experimental group, the experimental-group members received academically oriented interventions for 1–2 years, frequent home visits, and weekly parent-teacher conferences. When the control and experimental groups were compared, the experimental group had better grades, higher rates of high school graduation, lower rates of unemployment, and fewer arrests (Murray, Guerra, & Williams 1997).

In recognition of the link between juvenile delinquency and adult criminality, many anticrime programs are directed to at-risk youths. These prevention strategies, including youth programs such as Boys and Girls Clubs, are designed to keep young people "off the streets," provide a safe and supportive environment, and offer activities that promote skill development and self-esteem. According to Gest and Friedman (1994), housing projects with such clubs report 13 percent fewer juvenile crimes and a 25 percent decrease in the use of crack. More recently, the *Helping Families Initiative* in Mobile County, Alabama, targets children who are "at risk" and intervenes with the assistance of the District Attorney's office, social workers, police officers, teachers, and parents. The program focuses on ". . . the roughly 60 percent of children who have been suspended for serious violations such as fighting and bringing drugs to school, but haven't been arrested or adjudicated . . ." (Maxwell 2006, p. 28). Intervention entails everything from family counseling to transportation services. Although assessment of the program is difficult, school officials note that suspension rates have decreased since the program began.

Finally, many youth programs are designed to engage juveniles in noncriminal activities and integrate them into the community. In *Weed and Seed*, a program under the Department of Justice, "law enforcement agencies and prosecutors cooperate in 'weeding out' violent criminals and drug abusers . . . and 'seed' much-needed human services including prevention, intervention, treatment and neighborhood restoration programs" (Weed and Seed 2007, p. 1). As part of the program, "safe havens" are established in, for example, schools, where multiagency services are provided for youth.

Community Programs. Neighborhood watch programs involve local residents in crime prevention strategies. For example, MAD DADS (Men Against Destruction—Defending Against Drugs and Social-Disorder) patrol the streets in high-crime areas of the city on weekend nights, providing positive adult role models and fun community activities for troubled children. Members also report crime and drug sales to police, paint over gang graffiti, organize gun buyback programs, and counsel incarcerated fathers. At present, there are 65,000 MAD DADS in 60 chapters located in 16 states (MAD DADS 2007). In 2006, 10,000 communities in 50 states—more than 34 million people—participated in "National Night Out," a crime prevention event in which citizens, businesses, neighborhood organizations, and local officials joined together in outdoor activities to heighten awareness of neighborhood problems, promote anticrime messages, and strengthen community ties (National Night Out 2007).

Mediation and victim–offender dispute–resolution programs are also increasing, with more than 3,000 such programs in the United States and Canada. The growth of these programs is a reflection of their success rate: two-thirds of cases referred result in face-to-face meetings, 95 percent of these cases result in a written restitution agreement, and 90 percent of the written restitution agreements are completed within 1 year. It must be noted, however, that nationally only 20 percent to 30 percent of all court-ordered restitution is actually paid (Victim-Offender Reconciliation Program 2006).

Finally, the Internet has provided several ways to fight crime locally. Community-sponsored Internet sites routinely post names, descriptions, and photographs of wanted criminals. Other community efforts have made the transition from traditional media (e.g., television) to cyber-crime fighting. One such case is McGruff and the "Take a bite out of crime" campaign that now hosts on its home page, among other things, a campaign to "take a bite out of cyber-crime." As part of this initiative, McGruff, the crime-fighting dog, has a host of new priorities including the fight against cyber-bullying, identity theft, spam, worms, spyware, viruses, and online predators (NCPC 2007).

Children and adults march down a busy street during a peace march against violence Saturday, June 9, 2001, in South Central Los Angeles. Nearly two dozen organizations took part in the march.

Criminal Justice Policy

The criminal justice system is based on the principle of **deterrence,** that is, the use of harm or the threat of harm to prevent unwanted behaviors. The criminal justice system assumes that people rationally choose to commit crime, weighing the rewards and consequences of their actions. Thus the recent emphasis on "get tough" measures holds that maximizing punishment will increase deterrence and cause crime rates to decrease. Research indicates, however, that the effectiveness of deterrence is a function of not only the severity of the punishment but also the certainty and swiftness of the punishment. Furthermore, get-tough policies create other criminal justice problems, including overcrowded prisons and, consequently, the need for plea-bargaining and early-release programs. California prisons are so crowded that officials are asking inmates to voluntarily transfer to out-of-state prisons (Warren 2007).

Law Enforcement Agencies. In 2005 the United States had 673,146 full-time law enforcement officers and 295,924 full-time civilian employees, yielding an estimated 3.5 law enforcement personnel per 1,000 inhabitants (FBI 2006). However, only 69.5 percent of all law enforcement personnel are sworn officers—2.4 per 1,000 inhabitants—the remainder being support staff (e.g., correctional officers, clerks, meter attendants). With fewer police on the streets (Kingsbury 2006), many people are concerned about the ability of the police to effectively fight crime. In a recent Gallup Poll (2007a) when respondents were asked, "How much confidence do you have in the ability of the police to protect you from violent crime?" over one-third of the respondents replied "not very much" or "none at all." In addition, accusations of racial profiling, police brutality, and discriminatory arrest practices have shaken public confidence in the police.

In response to such trends, the Crime Control Act of 1994 established the Office of Community Oriented Policing Services (COPS). Called the "most important development in policing in the past quarter century" (Skogan & Roth 2004, p. xvii), community-oriented policing involves collaborative efforts among the police, the citizens of a community, and local leaders. As part of community policing efforts, officers speak to citizen groups, consult with social agencies, and enlist the aid of corporate and political leaders in the fight against neighborhood crime (COPS 2007).

Officers using community policing techniques often employ "practical approaches" to crime intervention. Such solutions may include what Felson (2002) called "situational crime prevention." Felson argued that much crime could be prevented simply by minimizing the opportunity for its occurrence. For example, cars could be outfitted with unbreakable glass, flush-sill lock buttons, an audible reminder to remove keys, and a high-security lock for steering columns (Felson 2002). These techniques and community-oriented policing in general have been fairly successful. An assessment indicated that visible community policing is positively associated with quality-of-life measures at both the individual level (e.g., perception of personal safety) and the neighborhood level (e.g., perception of neighborhood gang activity) (Reisig & Parks 2004).

Rehabilitation Versus Incapacitation. An important debate concerns the primary purpose of the criminal justice system: Is it to rehabilitate offenders or to incapacitate them through incarceration? Both rehabilitation and incapacitation are concerned with *recidivism rates*, or the extent to which criminals commit another crime. Advocates of **rehabilitation** believe that recidivism can be reduced

deterrence The use of harm or the threat of harm to prevent unwanted behaviors.

rehabilitation A criminal justice philosophy that argues that recidivism can be reduced by changing the criminal through such programs as substance abuse counseling, job training, education, etc.

by changing the criminal, whereas proponents of **incapacitation** think that recidivism can best be reduced by placing the offender in prison so that he or she is unable to commit further crimes against the general public.

Societal fear of crime has led to a public emphasis on incapacitation and a demand for tougher mandatory sentences, a reduction in the use of probation and parole, support of a "three strikes and you're out" policy, and truth-in-sentencing laws. However, these tough measures have recently come under attack for two reasons. First, research indicates that incarceration may not deter crime. For example, a study of California parolees found that 70 percent are back behind bars within 2 years (Petrillo 2006). Second is the accusation that get-tough measures, such as California's "three strikes and you're out" policy, are patently unfair (Irwin 2005). Leandro Andrade stole $153 worth of children's videos from K-Mart and was sentenced to 50 years to life in prison. Is that equitable? As unjust as it sounds, those in favor of the policy would be quick to note that the man had prior convictions for burglary and shoplifting. Andrade appealed his sentence to the U.S. Supreme Court, which held that California's three-strikes law did not violate the constitutional ban on grossly disproportionate sentences ("Major Rulings," 2003). In 2004 Proposition 66, which would have, in effect, repealed California's three-strikes policy, was defeated by a slim margin. A similar reform initiative is likely to come before California voters in 2008 (Maclachlan 2006).

Although incapacitation is clearly enhanced by longer prison sentences, deterrence, as discussed previously, and rehabilitation may not be (Irwin 2005). Rehabilitation assumes that criminal behavior is caused by sociological, psychological, and/or biological forces rather than being solely a product of free will. If such forces can be identified, the necessary change can be instituted. Rehabilitation programs include education and job training, individual and group therapy, substance abuse counseling, and behavior modification. As Siegel (2006) noted, from this perspective "even the most hardened criminal may be helped by effective institutional treatment plans and services" (p. 506). Such is the case with California's Department of Correction's *Preventing Parolee Crime Program* (PPCP). PPCP provides "literacy training, employment services, housing assistance, and substance abuse treatment to tens of thousands of parolees" (Zhang, Roberts, & Callanan 2006, p. 551). Although reductions in reincarceration rates and the number of parolee absconders were modest, the potential savings to California taxpayers is substantial.

Not all rehabilitation programs are state operated. *Second Chance* is a community-based rehabilitation program for parolees in Encanto, California. Funded by private donations, this nonprofit organization helps ex-offenders live crime-free lives through job and life skill training. An evaluation of the program found that 70 percent of the graduates stayed out of jail. In California, the cost of incarcerating an inmate is about $34,000 a year; the cost of rehabilitating an inmate in *Second Chance* is $4,000 (Petrillo 2006).

Corrections. While the debate between rehabilitation and incarceration continues, the number of inmates continues to grow. At year's end 2005, 2.2 million

AP/Wide World Photos

According to the U.S. Bureau of Justice Statistics, there were 2,193,798 prisoners held in federal or state prisons or local jails on December 31, 2005. Typically, individuals convicted of felonies are confined in prisons, whereas people who have committed misdemeanors and have sentences of 1 year or less are confined in jails.

incapacitation A criminal justice philosophy that argues that recidivism can be reduced by placing the offender in prison so that he or she is unable to commit further crimes against the general public.

FIGURE 4.5
Prison population rates per 100,000 and rank in world (out of 215 countries) as of January 2007.

Source: International Centre for Prison Studies. 2007. Available at http://www.kcl.ac.uk/depsta/rel/icps/world-prison-pop-seventh.pdf

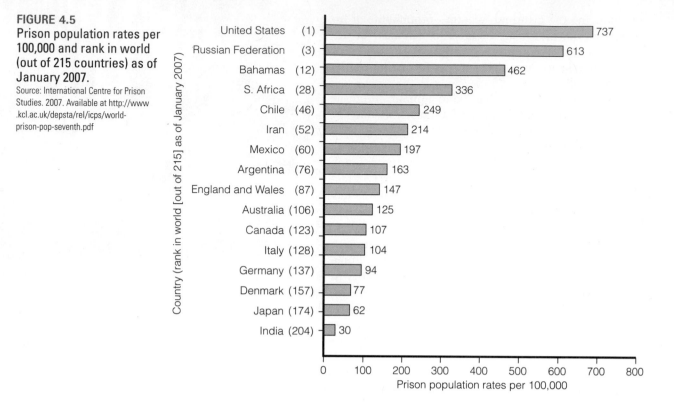

inmates were housed in the nation's federal, state, and local institutions (Bureau of Justice Statistics 2006d). Jail officials were also supervising 72,000 women and men who remained in the community in work release programs, on probation, or in other alternative programs (Bureau of Justice Statistics 2006d).

Minorities continue to be disproportionately incarcerated in the United States, with blacks and Hispanics comprising 60.0 percent of all state and federal prisoners. Black men are 6.6 times more likely than white men to be imprisoned; Hispanic men are 2.5 times more likely to be imprisoned than white men. Much of the racial disparity in incarceration rates is a consequence of the war on drugs, which disproportionately targets minorities (Fellner 2006). Men far outnumber women in prison—14 men for every 1 woman—although over the next 5 years the number of female inmates is projected to grow at a faster rate than the number of male inmates (Ohlemacher 2007).

More than 9.25 million men and women are held in penal institutions around the world (*News Highlights* 2007). The United States has the highest incarceration rate in the world (see Figure 4.5) and it comes with a $60 billion price tag, up from $9 billion just two decades ago ("Death Behind Bars" 2005; *News Highlights* 2007). U.S. incarceration rates are projected to increase 13.0 percent by the year 2011—three times faster than the overall population growth in the United States. The cost to states is estimated to be an additional $27 million (Ohlemacher 2007).

Increases in the prison population have usually been met with building of more prisons, but concerns with budget deficits have caused many states to "close prisons, lay off guards, and consider shortening sentences" (Butterfield 2002). **Probation** entails the conditional release of an offender who, for a specific time period and subject to certain conditions, remains under court supervision

probation The conditional release of an offender who, for a specific time period and subject to certain conditions, remains under court supervision in the community.

in the community (Siegel 2006). In 2005, 4.2 million adult women and men were on probation at the federal, state, or local level, about half of whom had been convicted of a felony and the other half convicted of a misdemeanor. Twenty-eight percent were under court supervision for a drug offense (Bureau of Justice Statistics 2006e).

What Do You Think? States have the right to establish qualifications for voter eligibility. Some states, such as Florida, Kentucky, and Virginia, have lifetime bans on voting by convicted felons. Others, such as New York, Connecticut, and New Jersey, do not allow inmates or those on parole to vote. Although over the last decade 16 states have "loosened" voting restrictions, a record number of Americans—5.3 million—are denied the right to vote because of criminal records (Eckholm 2006). Do you think that convicted felons should be denied the right to vote? What about inmates and parolees? Do you think that present restrictions constitute institutional discrimination, given the disproportionate number of minorities who are behind bars?

Capital Punishment. With **capital punishment** the state (the federal government or a state) takes the life of a person as punishment for a crime. Thirteen of the 38 states in the United States that allow capital punishment have recently put executions on hold. The majority of these states have halted executions because of concerns over lethal injection. One concern is the question of whether or not death by lethal injection violates the Eighth Amendment's prohibition against cruel and unusual punishment. For example, in 2007, a district court judge held that Tennessee's lethal injection procedures "present a substantial risk of unnecessary pain" that could "result in a terrifying, excruciating death" (Schelzig 2007, p. 1).

The second issue concerns the role of physicians in state executions. According to Vu (2007, p. 1), the "American Medical Association is adamant that it is a violation of medical ethics for doctors to participate in, or even be present at, executions." Ironically, as Deborah Denno, a professor at Fordham University School of Law, noted, the very people who can help assure that the death penalty does not violate the Eighth Amendment want nothing to do with injecting the deadly drugs (Vu 2007).

In 2006, 53 executions took place in 14 states, with more than 3,374 inmates being on death row (Peterson 2007). In the same year, 2,148 people were executed in 22 countries despite the global trend toward abolition of the death penalty (Amnesty International 2007). For example, more than 128 countries have eliminated state executions in law or practice, including Australia, Canada, Italy, Denmark, and Ireland. The United States was the last country in the world to reject the execution of offenders who committed their crimes before the age of 18 (Wallis 2005). In 2006, four countries accounted for 94 percent of all executions worldwide—China, Iran, Saudi Arabia, and the United States (Amnesty International 2007).

Proponents of capital punishment argue that executions of convicted murderers are necessary to convey public disapproval and intolerance for such

> **"**I feel morally and intellectually obligated to concede that the death penalty experiment has failed.**"**
>
> Justice Harry Blackmun
> U.S. Supreme Court Justice

capital punishment The state (the federal government or a state) takes the life of a person as punishment for a crime.

heinous crimes. Those against capital punishment believe that no one, including the state, has the right to take another person's life and that putting convicted murderers behind bars for life is a "social death" that conveys the necessary societal disapproval.

Proponents of capital punishment also argue that it deters individuals from committing murder. Critics of capital punishment hold, however, that because most homicides are situational and are not planned, offenders do not consider the consequences of their actions before they commit the offense. Critics also point out that the United States has a higher murder rate than many Western European nations that do not practice capital punishment and that death sentences are racially discriminatory. For example, a study of capital punishment in the United States between 1973 and 2002 found that "... minority death row inmates convicted of killing whites face higher execution probabilities than other capital offenders" (Jacobs et al. 2007, p. 610).

Capital punishment advocates suggest that executing a convicted murderer relieves the taxpayer of the costs involved in housing, feeding, guarding, and providing medical care for inmates. Opponents of capital punishment argue that the principles that decide life and death issues should not be determined by financial considerations. In addition, taking care of convicted murderers for life may actually be less costly than sentencing them to death, because of the lengthy and costly appeals process for capital punishment cases (Death Penalty Information Center 2007).

Nevertheless, those in favor of capital punishment argue that it protects society by preventing convicted individuals from committing another crime, including the murder of another inmate or prison official. One study of the deterrent effect of capital punishment concluded that each execution is associated with at least eight fewer homicides (Rubin 2002). Opponents contend, however, that capital punishment may result in innocent people being sentenced to death. Since 1976, 123 death row inmates have been exonerated (Quindlen 2006). In addition, a report by the Justice Project titled *A Broken System* showed reversible errors in two of every three death penalty cases reviewed over the study period (Herbert 2002). Stating that the system was riddled with error, in 2002 the governor of Illinois commuted the sentence of 167 inmates from the death penalty to life in prison or less (Wilgoren 2003; Napolitano 2004). Acknowledging that the system is "far from fixed," the new governor of Illinois continued the moratorium (Peterson 2007). Furthermore, some hypothesize that the recent reduction in U.S. executions is a result of exonerations based on new DNA evidence. Others maintain, however, that the reduction is merely a reflection of the decade decline in homicide rates (Lane 2004; Peterson 2007).

Legislative Action

Gun Control. Fueled by the Columbine school shooting and other recent images of children as both gun victims and offenders, in May 2000 tens of thousands of women descended on Washington, D.C., demanding that something be done about gun violence. Although the impact of the Million Mom March is still unknown, most Americans, and particularly women, support some restriction on handguns. When a national sample of U.S. adults were asked, "Would you like to see gun laws in this country made more strict, less strict, or remain as they are?" 49 percent responded more strict, 14 percent less strict, and 35 percent remain as they are (Gallup Poll 2007b).

Those against gun control argue that not only do citizens have a constitutional right to own guns but also that more guns may actually lead to less crime as would-be offenders retreat in self-defense when confronted (Lott 2003). Advocates of gun control, however, insist that the 250 million privately owned firearms in the United States, one-third of them handguns (Conklin 2007), significantly contribute to the violent crime rate in the United States and distinguish the country from other industrialized nations.

After a 7-year battle with the National Rifle Association (NRA), gun control advocates achieved a small victory in 1993 when Congress passed the Brady Bill. The law initially required a 5-day waiting period on handgun purchases so that sellers can screen buyers for criminal records or mental instability. The law was amended in 1998 to include an instant check of buyers and their suitability for gun ownership. Under federal law, those prohibited from purchasing guns include

> people convicted of or under indictment for felony charges, fugitives, the mentally ill, those with dishonorable military discharges, those who have renounced U.S. citizenship, illegal aliens, illegal drug users, and those convicted of domestic violence misdemeanors or who are under domestic violence restraining orders. . . . (Siegel 2006, p. 49)

Today, the law requires background checks of not just handgun users but also those who purchase rifles and shotguns.

In addition to federal regulations, cities and states can create other restrictions. Such is the case in Washington, D.C., where for the last 30 years gun ownership has virtually been banned (Williams 2007). Only residents with permits such as police, security guards, and the like, can possess handguns and rifles. In what has been called the most important gun control case in 70 years, in 2007 a federal appeals court held that the District of Columbia's prohibition against gun ownership was in violation of the Second Amendment, which reads "A well-regulated militia, being necessary to the security of a free state, the right of people to keep and bear arms, shall not be infringed." The decision, along with the killings at Virginia Tech, has revived the debate over the meaning of the Second Amendment and the intent of the drafters of the Constitution.

Other Legislation. Major legislative initiatives have been passed in recent years, including the 1994 Crime Control Act, which created community policing, "three strikes and you're out," and truth-in-sentencing laws. Significant crime-related legislation presently before Congress includes the following bills (Orator 2007):

- *Crack-Cocaine Equitable Sentence Act.* This act is intended to eliminate the disparities in punishment for possession of crack cocaine versus possession of other forms of cocaine.
- *Second Chance for Ex-Offenders Act.* If passed, this legislation would permit expungement of criminal records for certain nonviolent offenses.
- *Violent Crime Reduction Act.* This act amends the Violent Crime Control and Law Enforcement Act of 1994 in an effort to further reduce gang activity and violent crime.
- *Keeping the Internet Devoid of Sexual Predators Act (KIDS Act).* This bill, among other things, would require that convicted sex offenders register online identifiers.
- *Federal Death Penalty Abolition Act.* This bill, if passed, would abolish the death penalty under federal law.

International Efforts in the Fight Against Crime

Europol is the European law enforcement organization that handles criminal intelligence. Unlike the FBI, Europol officers do not have the power to arrest; they predominantly provide support services for law enforcement agencies of European Union member countries. The support services provided by Europol include "fast information exchange, sophisticated intelligence analysis, co-ordination, expertise and training" (Europol 2006, p. 1). Europol, in conjunction with law enforcement agencies in member states, fights against transnational crimes such as illicit drug trafficking, child pornography, human trafficking, money laundering, and counterfeiting of the euro.

Interpol, the International Criminal Police Organization, was established in 1923 and is the world's largest international police organization with 186 member countries (Interpol 2007b). Similar to Europol, Interpol provides support services for law enforcement agencies of member nations. It has three core functions (Interpol 2007c). First, Interpol operates a worldwide police communication network that operates 24 hours a day, 7 days a week. This network "provides police around the world with a common platform through which they can share crucial information about criminals and criminality" (p. 1). Second, Interpol's extensive databases (e.g., DNA profiles, names, and photographs) ensure that police get the information they need to investigate existing crime and prevent new crime from occurring. Finally, Interpol provides emergency support services and operational activities to law enforcement personnel in the field. To that end, a *Command and Coordination Center* fields inquiries from member nations and is the first line of defense for countries in crisis.

Under the United Nations Office of Drugs and Crime (UNODC) is the Crime Programme, the U.N. central office for international "crime prevention, criminal justice, and criminal law reform" (UNODC 2007, p. 1). The UNODC Crime Programme "defines and promotes internationally recognized principles in such areas as independence of the judiciary, protection of victims, alternatives to imprisonment, treatment of prisoners, police use of force, mutual legal assistance and extradition" (p. 1). The UNODC Crime Programme focuses on crime in developing and transitional nations and, like Europol and Interpol, is particularly concerned with transnational organized crime groups, human trafficking, and political corruption. Moreover, the UNODC Crime Programme advises countries in the development of international law conventions such as the *United Nations Convention Against Transnational Organized Crime.*

Finally, the International Centre for the Prevention of Crime (ICPC) is a consortium of policy makers, academicians, governmental officials, and practitioners from all over the world (ICPC 2007). Located in Montreal, Canada, members of ICPC

> exchange experiences, consider emerging knowledge, and improve policies and programmes in crime prevention and community safety. The ICPC staff monitor developments, provide direct assistance to members, and contribute to public knowledge and understanding in the field. (ICPC 2007, p. 1)

In fulfilling such tasks, the ICPC seeks to (1) raise awareness of and access to crime prevention knowledge; (2) enhance community safety; (3) facilitate the sharing of crime prevention information between countries, cities, and justice systems; and (4) respond to calls for technical assistance.

UNDERSTANDING CRIME AND SOCIAL CONTROL

What can we conclude from the information presented in this chapter? Research on crime and violence supports the contentions of both structural functionalists and conflict theorists. Inequality in society, along with the emphasis on material well-being and corporate profit, produces societal strains and individual frustrations. Poverty, unemployment, urban decay, and substandard schools—the symptoms of social inequality—in turn lead to the development of criminal subcultures and conditions favorable to law violation. Furthermore, criminal behavior is encouraged by the continued weakening of social bonds among members of society and between individuals and society as a whole, the labeling of some acts and actors as "deviant," and the differential treatment of minority groups by the criminal justice system.

There has been a general decline in crime over the last decade (although there have been recent increases), making it tempting to conclude that get-tough criminal justice policies are responsible for the reductions. Other valid explanations exist and are likely to have contributed to the falling rates: changing demographics, community policing, stricter gun control, and a reduction in the use of crack cocaine. Nonetheless, "nail 'em and jail 'em" policies have been embraced by citizens and politicians alike. Get-tough measures include building more prisons and imposing lengthier mandatory prison sentences on criminal offenders. Advocates of harsher prison sentences argue that "getting tough on crime" makes society safer by keeping criminals off the streets and by deterring potential criminals from committing crime.

> **"**To do justice, to break the cycle of violence, to make Americans safer, prisons need to offer inmates a chance to heal like a human, not merely heel like a dog.**"**
>
> Richard Stratton
> Former inmate

Prison sentences, however, may not always be effective in preventing crime. In fact, they may promote it by creating an environment in which prisoners learn criminal behavior, values, and attitudes from each other. With more than 650,000 inmates from state and federal facilities being released into the general population each year—1,780 a day—punitive policies may be a shortsighted temporary fix (NGA 2007).

Rather than getting tough on crime after the fact, some advocate getting serious about prevention. Re-emphasizing the values of honesty and, most important, taking responsibility for one's actions is a basic line of prevention with which most agree. The movement toward **restorative justice,** a philosophy primarily concerned with repairing the victim–offender–community relation, is in direct response to the concerns of an adversarial criminal justice system that encourages offenders to deny, justify, or otherwise avoid taking responsibility for their actions.

Restorative justice holds that the justice system, rather than relying on "punishment, stigma and disgrace" (Siegel 2006, p. 275), should "repair the harm" (Sherman 2003, p. 10). Key components of restorative justice include restitution to the victim, remedying the harm to the community, and mediation. Restorative justice programs have been instituted in several states, including Vermont and New York. In 2002 the United Nations Economic and Social Council endorsed the "Basic Principles on the Use of Restorative Justice Programmes in Criminal Matters" around the world (United Nations 2003).

restorative justice A philosophy primarily concerned with reconciling conflict between the victim, the offender, and the community.

CHAPTER REVIEW

- **Are there any similarities between crime in the United States and crime in other countries?**

  All societies have crime and have a process by which they deal with crime and criminals; that is, they have police, courts, and correctional facilities. Worldwide, most offenders are young males and the most common offense is theft; the least common offense is murder.

- **How can we measure crime?**

  There are three primary sources of crime statistics. First are official statistics, for example, the FBI's Uniform Crime Reports, which are published annually. Second are victimization surveys designed to get at the "dark figure" of crime, crime that is missed by official statistics. Finally, there are self-report studies that have all the problems of any survey research. Investigators must be cautious about whom they survey and how they ask the questions.

- **What sociological theory of criminal behavior blames the schism between the culture and structure of society for crime?**

  Strain theory was developed by Robert Merton (1957) and uses Durkheim's concept of *anomie,* or normlessness. Merton argued that when legitimate means (e.g., a job) of acquiring culturally defined goals (e.g., money) are limited by the structure of society, the resulting strain may lead to crime. Individuals, then, must adapt to the inconsistency between means and goals in a society that socializes everyone into wanting the same thing but provides opportunities for only some.

- **What are index offenses?**

  Index offenses, as defined by the FBI, include two categories of crime: violent crime and property crime. Violent crimes include murder, robbery, assault, and rape; property crimes include larceny, car theft, burglary, and arson. Property crimes, although less serious than violent crimes, are the most numerous.

- **What is meant by white-collar crime?**

  White-collar crime includes two categories: occupational crime, that is, crime committed in the course of one's occupation; and corporate crime, in which corporations violate the law in the interest of maximizing profits. In occupational crime the motivation is individual gain.

- **How do social class and race affect the likelihood of criminal behavior?**

  Official statistics indicate that minorities are disproportionately represented in the offender population. Nevertheless, it is inaccurate to conclude that race and crime are causally related. First, official statistics reflect the behaviors and policies of criminal justice actors. Thus the high rate of arrests, conviction, and incarceration of minorities may be a consequence of individual and institutional bias against minorities. Second, race and social class are closely related in that nonwhites are overrepresented in the lower classes. Because lower-class members lack legitimate means to acquire material goods, they may turn to instrumental, or economically motivated, crimes. Thus the apparent relationship between race and crime may, in part, be a consequence of the relationship between these variables and social class.

- **What are some of the economic costs of crime?**

  First are direct losses from crime, such as the destruction of buildings through arson or of the environment by polluters. Second are costs associated with the transferring of property (e.g., embezzlement). A third major cost of crime is that associated with criminal violence, for example, the medical cost of treating crime victims. Fourth are the costs associated with the production and sale of illegal goods and services. Fifth is the cost of prevention and protection. Finally, there is the cost of the criminal justice system, law enforcement, litigation and judicial activities, corrections, and victims' assistance.

- **What is the present legal status of capital punishment in this country?**

  Although at present capital punishment is legal in 38 states, several states are questioning its constitutionality and thus have halted executions. Of those states that have a moratorium, the most common legal issue is whether or not death by lethal injection violates the Eighth Amendment's prohibition against cruel and unusual punishment. There is also federal legislation pending that would abolish the death penalty under federal law.

TEST YOURSELF

1. The United States has the highest violent crime rate in the world.
   a. True
   b. False

2. The Uniform Crime Reports is a compilation of data from
   a. U.S. Census Bureau
   b. law enforcement agencies
   c. victimization surveys
   d. the Department of Justice
3. According to _____ crime results from the absence of legitimate opportunities as limited by the social structure of society.
   a. Hirschi
   b. Marx
   c. Merton
   d. Becker
4. Which of the following is not an index offense?
   a. Drug possession
   b. Homicide
   c. Rape
   d. Burglary
5. The economic costs of white-collar crime outweigh the costs of traditional street crime.
   a. True
   b. False
6. Women everywhere commit less crime than men.
   a. True
   b. False

7. Probation entails
   a. early release from prison
   b. a suspended sentence
   c. court supervision in the community in lieu of incarceration
   d. incapacitation of the offender
8. Europol is an advisory and support law enforcement agency for European Union members.
   a. True
   b. False
9. Public opinion generally indicates that respondents
   a. are in favor of the death penalty
   b. prefer solving social problems that cause crime rather than increasing the number of police on the streets
   c. have faith in the police
   d. all of the above
10. The United States has the highest incarceration rate in the world.
    a. True
    b. False

Answers: 1 b. 2 b. 3 c. 4 a. 5 a. 6 a. 7 c. 8 a. 9 d. 10 a.

KEY TERMS

acquaintance rape
capital punishment
classic rape
clearance rate
computer crime
corporate violence
crime
crime rate

deterrence
identity theft
incapacitation
index offenses
organized crime
primary deviance
probation

racial profiling
rehabilitation
restorative justice
secondary deviance
transnational crime
victimless crimes
white-collar crime

MEDIA RESOURCES

Understanding Social Problems, Sixth Edition
Companion Website

academic.cengage.com/sociology/mooney
Visit your book companion website, where you will find flash cards, practice quizzes, Internet links, and more to help you study.

CENGAGENOW™

Just what you need to know NOW! Spend time on what you need to master rather than on information you already have learned. Take a pre-test for this chapter, and CengageNOW will generate a personalized study plan based on your results. The study plan will identify the topics you need to review and direct you to online resources to help you master those topics. You can then take a post-test to help you determine the concepts you have mastered and what you will need to work on. Try it out! Go to academic.cengage.com/login to sign in with an access code or to purchase access to this product.

5

Family Problems

> "We must recognize that there are healthy as well as unhealthy ways to be single or to be divorced, just as there are healthy and unhealthy ways to be married."
>
> Stephanie Coontz,
> Family historian

The Global Context: Families of the World | Changing Patterns in U.S. Families and Households | Sociological Theories of Family Problems | Violence and Abuse in Intimate and Family Relationships | Strategies for Action: Preventing and Responding to Violence and Abuse in Intimate and Family Relationships | Problems Associated with Divorce | Strategies for Action: Strengthening Marriage and Alleviating Problems of Divorce | Teenage Childbearing | Strategies for Action: Interventions in Teenage Childbearing | Understanding Family Problems | Chapter Review

USA Today asked film director Luc Jacquet what the subjects of his box-office hit documentary film *March of the Penguins* have most in common with humans. Jacquet answered, "They form couples and are faithful. It's the only way they can raise the chick under extreme conditions" (quoted in Saad 2006). Like penguins, humans often go to great lengths to find a mate, and most choose to enter the institution of marriage, which typically includes the expectation of faithfulness. However, pair-bonded human relationships in modern societies today are formed primarily to meet needs for love and companionship, rather than to rear children. And unlike penguins, humans rear their young in a variety of family contexts, including polygamous families, single-parent families, same-sex families, divorced families, living together families, and stepfamilies.

In this chapter, we examine the diversity of couples and families across the globe and changing patterns in U.S. families and households. We then turn our attention to family problems, including violence and abuse in families, divorce and its aftermath, and teenage childbearing, focusing on these issues in the United States. We emphasize at the outset that many of the problems families face, such as health problems, poverty, job-related problems, drug and alcohol abuse, and discrimination, are dealt with in other chapters in this text.

The Global Context: Families of the World

The U.S. Census defines *family* as a group of two or more persons related by blood, marriage, or adoption. Sociology offers a broader definition of family: a **family** is a kinship system of all relatives living together or recognized as a social unit. This broader definition recognizes foster families, unmarried same-sex and opposite-sex couples and families, and any relationships that function and feel like a family. As we describe in the following section, family forms and patterns vary worldwide.

Monogamy and Polygamy

In many countries, including the United States, the only legal form of marriage is **monogamy**—a marriage between two partners. A common variation of monogamy is **serial monogamy**—a succession of marriages in which a person has more than one spouse over a lifetime but is legally married to only one person at a time.

Some societies practice **polygamy**—a form of marriage in which one person may have two or more spouses. The most common form of polygamy, known as **polygyny**, involves one husband having more than one wife. Although polygyny is diminishing worldwide, it is still practiced in some Asian and African societies, where monogamy is nevertheless still the predominant marriage form (Adams 2004). The second and less common form of polygamy is **polyandry**—the concurrent marriage of one woman to two or more men.

In the United States polygamy is illegal and is often referred to as **bigamy**—the criminal offense of marrying one person while still being legally married to another. Polygamous marriages do exist in the United States among some fundamentalist Mormon splinter groups, a phenomenon depicted in *Big Love*—an HBO television series featuring a polygamous family. Polygamy in the United States

family A kinship system of all relatives living together or recognized as a social unit, including adopted persons.

monogamy Marriage between two partners; the only legal form of marriage in the United States.

serial monogamy A succession of marriages in which a person has more than one spouse over a lifetime but is legally married to only one person at a time.

polygamy A form of marriage in which one person may have two or more spouses.

polygyny A form of marriage in which one husband has more than one wife.

polyandry The concurrent marriage of one woman to two or more men.

bigamy The criminal offense of marrying one person while still legally married to another.

© HBO/The Kobal Collection

also occurs among some immigrants who come from countries where polygamy is accepted, such as Mali and Ghana and other West African countries. It is estimated, for example, that thousands of New Yorkers are involved in polygamous marriages (Bernstein 2007). Immigrants who practice polygamy generally keep their lifestyle a secret because under U.S. immigration law, polygamy is grounds for deportation.

Role of Women in the Family

The roles of women in families also vary across societies. In some societies wives are commonly expected to be subservient to their husbands (e.g., rural Pakistan and India). In Northern European countries, specifically Norway and Sweden, women and men tend to have **egalitarian relationships,** in which partners share decision making and assign family roles based on choice rather than on traditional beliefs about gender (Lindsey 2005).

In most of the world domestic chores are primarily the responsibility of women. In a study of husbands and wives in 13 nations, in all but 1 nation (Russia), respondents reported that women performed most of the household labor (Davis & Greenstein 2004). The percentage of wives reporting that they "always" or "usually" did the housework ranged from 98 percent in Japan to 38 percent in Russia; the percentage of husbands reporting that wives always or usually did the housework ranged from 93 percent in Japan to 31 percent in Russia. In the United States 60 percent of husbands and 67 percent of wives reported that the wives always or usually did the housework. When husbands perform domestic tasks, they are often viewed as "helping" their wives and not as performing a male role.

Social Norms Related to Childbearing

Social norms concerning childbearing also vary widely throughout the world. In every society, women learn that their role includes having children. But compared with less developed societies where social expectations for women to

egalitarian relationship
A relationship in which partners share decision making and assign family roles based on choice rather than on traditional beliefs about gender.

have children are strong, in developed countries with high levels of gender equality, social norms concerning childbearing are more flexible; women may view having children as optional—as a personal choice.

Norms about out-of-wedlock childbirth also vary across the globe. Although more than one-third of all U.S. births are to unmarried women, compared with some countries, the United States has a low proportion of nonmarital births. Two-thirds of births in Iceland and at least half of births in Norway and Sweden are out of wedlock. Other countries with higher proportions of nonmarital births than the United States include Denmark, France, the United Kingdom, and Finland. In other countries, such as Germany, Italy, Greece, and Japan, less than 15 percent of births occur out of wedlock (Ventura & Bachrach 2000). In India, it is almost unheard of for a Hindu woman to have a child outside marriage; unwed childbearing would bring great shame to the woman and her family (Laungani 2005).

Same-Sex Relationships

Norms and policies concerning same-sex intimate relationships also vary around the world. In some countries homosexuality is punishable by imprisonment or even death (see Chapter 11).

In 2001 the Netherlands became the first country in the world to offer legal marriage to same-sex couples, and in 2003 Belgium became the second. In 2005 Canada and Spain legalized same-sex marriage, followed by South Africa in 2006. In 2004 Massachusetts became the first U.S. state to offer same-sex marriage. In the United States the 1996 federal Defense of Marriage Act defines marriage as the union between one man and one woman and denies federal recognition of same-sex marriages. As of Summer 2007, 45 states have laws or state constitutional amendments that ban marriage between same-sex couples. Five states have broad family recognition laws (civil union or domestic partnership laws) that extend to same-sex couples nearly all the rights and responsibilities extended to married couples under state law: Vermont, California, New Jersey, New Hampshire, and Oregon. Three states (Hawaii, Maine, and Washington) and the District of Columbia offer more limited rights and protections to same-sex couples (see also Chapter 11).

Sixteen countries recognize same-sex couples for the purposes of immigration, a benefit not yet granted by the United States. Several countries—Denmark, Finland, France, Germany, Hungary, Iceland, Israel, Norway, Portugal, Sweden, Switzerland, and the United Kingdom (except Scotland)—grant same-sex couples legal rights and protections that are more limited than marriage. For example, in France, registered same-sex (and opposite-sex) couples can enter a type of marital arrangement called *Pacte civil de solidarite* (civil solidarity pact), or PAC, which grants them the right to file joint tax returns, extend their Social Security coverage to each other, and receive the same health, employment, and welfare benefits as legal spouses. It also commits the couple to assume joint responsibility for household debts. Most couples (60 percent) opting for PACs in France are heterosexual couples (Demian 2004).

It is clear from the previous discussion that families are shaped by the social and cultural context in which they exist. As we discuss the family issues addressed in this chapter—violence and abuse, divorce, and teenage childbearing—we refer to social and cultural forces that shape these events and the attitudes surrounding them. Next, we look at changing patterns and structures of U.S. families and households.

FIGURE 5.1
Median age at first mar-
riage, by sex, selected
years: United States.
Source: U.S. Census Bureau (2006b).

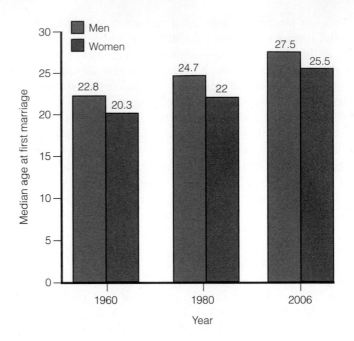

CHANGING PATTERNS IN U.S. FAMILIES AND HOUSEHOLDS

Over the last century dramatic changes have occurred in the patterns and structures of U.S. families and households. A **family household,** as defined by the U.S. Census Bureau, consists of two or more individuals related by birth, marriage, or adoption who reside together. **Nonfamily households** can consist of one person who lives alone, two or more people as roommates, or cohabiting heterosexual or homosexual couples involved in a committed relationship. The Census Bureau considers households composed of heterosexual cohabiting couples and same-sex couples as "nonfamily" households, even though such couples function economically and emotionally as a family. Some of the significant changes in U.S. families and households that have occurred over the past several decades include the following:

- *Increased singlehood and older age at first marriage.* U.S. adults are more likely to live alone today than in the past. The proportion of households consisting of one person living alone rose from 17 percent in 1970 to 26 percent in 2005 (U.S. Census Bureau 2006a). In part, this is due to the fact that U.S. women and men are staying single longer. From 1960 to 2006, the median age at first marriage increased from about 20 to 26 for women and from about 23 to 28 for men (U.S. Census Bureau 2006b) (see Figure 5.1). Today, 13.1 percent of women and 18.5 percent of men ages 40–44 have never been married—the highest figures in this nation's history (U.S. Census Bureau 2006b). In effect, individuals are delaying marriage.
- *Delayed childbearing.* Women are also delaying having children. Between 1981 and 2003, the birthrate for women ages 40–44 more than doubled (Hamilton, Martin, & Sutton 2004). First-birth rates for women ages 30 to 34 years, 35 to 39 years, and 40 to 44 years increased considerably from 2002 to 2003, by 7 percent, 12 percent, and 11 percent, respectively (Hamilton et al. 2004).

family household As defined by the U.S. Census Bureau, two or more individuals related by birth, marriage, or adoption who reside together.

nonfamily household A household that consists of one person who lives alone, two or more people as roommates, or cohabiting heterosexual or homosexual couples involved in a committed relationship.

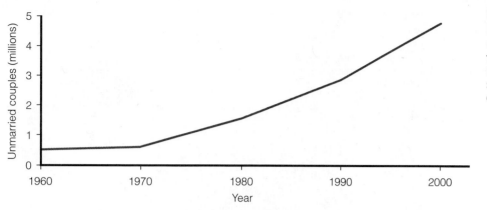

FIGURE 5.2
Number of cohabiting, unmarried couples of the opposite sex by year, United States.
Source: U.S. Census Bureau (2000) and earlier reports in the same series.

• *Increased heterosexual and same-sex cohabitation.* Many U.S. adults who are technically "single" are in long-term committed cohabiting relationships with a partner. Nationally, 9 percent of coupled households are unmarried partner households (Simmons & O'Connell 2003). From 1960 to 2000 the number of cohabiting unmarried couples skyrocketed (see Figure 5.2). The percentage of people who cohabited with their spouses before marriage more than doubled between 1980 and 2000, rising from 16 percent to 41 percent (Amato et al. 2007).

What Do You Think? Adults with divorced parents are more likely to cohabit before marriage than are adults with continuously married parents (Amato et al. 2007). Why do you think this is so?

Couples live together to assess their relationship, to reduce or share expenses, or to avoid losing pensions from previously deceased partners. Most Americans view cohabitation as a normal, acceptable stage in the courtship process (Thornton & Young-DeMarco 2001). For some, cohabitation is an alternative to marriage. Most of the 5.5 million cohabiting couples in 2000 were heterosexual couples, but about 1 in 9 had partners of the same sex (Simmons & O'Connell 2003).

Increased cohabitation among adults means that children are increasingly living in families that may function as two-parent families but do not have the social or legal recognition that married-couple families have. About 4 in 10 (43 percent) opposite-sex unmarried partner households, one-fifth (22.3 percent) of gay male couples, and one-third (34.3 percent) of lesbian couples have children present in the home (Simmons & O'Connell 2003). When children are denied a legal relationship to both parents because of the parents' unmarried status and/or sexual orientation, they may be denied Social Security survivor benefits, health care insurance, or the ability to have either parent authorize medical treatment in an emergency, among other protections.

Some states, cities, counties, and employers allow unmarried partners (same-sex and/or heterosexual partners) to apply for a **domestic partnership** designation, which grants them some legal entitlements, such as health in-

domestic partnership A status granted to unmarried couples, including gay and lesbian couples, by some states, counties, cities, and workplaces that conveys various rights and responsibilities.

Actress Helena Bonham Carter and director Tim Burton, who have been in a relationship since 2001 and have a son together, are a "living apart together" couple. They live in adjoining houses in London.

© AP Photo/Laurent Emmanuel

surance benefits and inheritance rights that have traditionally been reserved for married couples. Eight states and the District of Columbia have approved second-parent adoption for lesbian and gay parents either by statute or state appellate court rulings, which means that it is granted in all counties statewide. Lesbian and gay parents also have been granted second-parent adoptions in some counties in 18 other states.

- *More interracial/interethnic unions.* Most dating, living together, and married couples in the United States involve two individuals of the same race and ethnicity. However, interracial/interethnic couples are growing in number. Between 1980 and 2000 the proportion of interracial or interethnic marriages more than doubled, from 4 percent in 1980 to 9 percent in 2000 (Amato et al. 2007). Likewise, attitudes toward interracial relationships have changed in recent decades. In a 2007 national survey, more than 8 in 10 U.S. adults (83 percent) agree that "it's all right for blacks and whites to date," up from 8 percent in 1987 (Pew Research Center 2007). Among younger people—those born since 1977—94 percent say it is all right for blacks and whites to date.

- *A new family form: Living apart together.* Family scholars have identified an emerging family form in which couples—married or unmarried—live apart in separate residences. This new family form, known as **living apart together (LAT) relationships**, has been identified as a new social phenomenon in several western European countries as well as in the United States (Lara 2004; Levin 2004). Couples may choose this family form for a number of reasons, including the desire to maintain a measure of independence and to avoid problems that may arise from living together.

- *Increased births to unmarried women.* The percentage of births to unmarried women increased from 18.4 percent of total births in 1980 to 30.1 percent in 1991 to 36.8 percent in 2005 (Ventura 1995; Hamilton, Martin, & Ventura 2006). Childbearing by unmarried women had reached record levels in 2005 (see Figure 5.3). Among black women in the United States, more than two-thirds of births are to unmarried women (see Figure 5.4).

Having a baby outside marriage has become more socially acceptable and does not carry the stigma it once did. In the 1950s and 1960s more than half

living apart together (LAT) relationships An emerging family form in which couples—married or unmarried—live apart in separate residences.

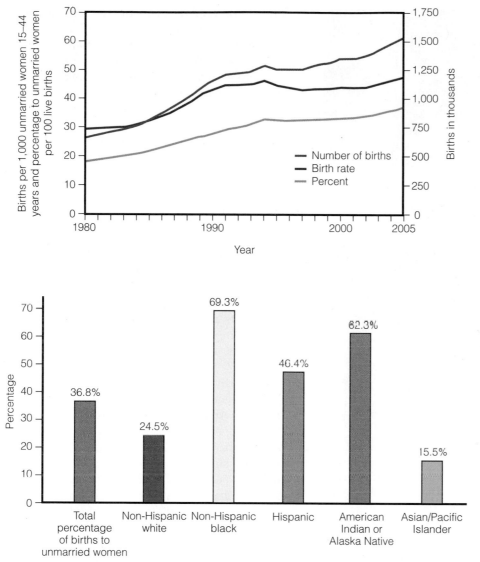

FIGURE 5.3
Number of births, birth rate, and percentage of births to unmarried women: United States, 1980–2005.
Source: Hamilton et al. (2006).

FIGURE 5.4
Percentage of all births to unmarried women by race and Hispanic origin: United States, 2005.
Source: Hamilton et al. (2006).

of U.S. women who gave birth to a baby conceived out of wedlock married before the birth of the baby. In the 1990s less than one in four (23 percent) of such women married before the birth of the baby (Ventura & Bachrach 2000). In a national Gallup survey of U.S. adults, only 37 percent said it was "very important" for couples to marry if they had a child together (Saad 2006).

Family scholar Stephanie Coontz (1997) emphasized the fact that we must be careful not to overdramatize the increase because "much illegitimacy was covered up in the past and reporting methods have become much more sophisticated" (p. 29). In addition, although more than one-third of all U.S. births are to unmarried women, not all unwed mothers are single parents. Half of new unwed mothers are cohabiting with the fathers when their children are born (Sigle-Rushton & McLanahan 2002).

- *Increased single-parent families.* The rise in both divorce and nonmarital births has resulted in an increase in single-parent families. From 1970 to 2003

Getty Images

Not all nonmarital births imply the absence of a father. Actress Goldie Hawn and actor Kurt Russell have been in a committed cohabiting relationship for more than 20 years. Although their child Wyatt was born "out of wedlock," he has been raised in a stable, loving family with his mother and his father. Kurt Russell also helped Goldie raise her two children (Kate and Oliver) from a previous marriage.

divorce rate The number of divorces per 1,000 population.

the proportion of single-mother families grew from 12 percent to 26 percent and single-father families grew from 1 percent to 6 percent (Fields 2004). As noted earlier, many children in what are considered to be "single-parent families" have two parental adults in the home who are living in an unmarried cohabiting relationship: 16 percent of children living with single fathers and 9 percent of children living with single mothers also live with their parents' partners (Forum on Child and Family Statistics 2000).

- *Increased divorce and blended families.* The **divorce rate**—the number of divorces per 1,000 population—doubled from 1950 to its peak around 1980, increasing from a rate of 2.6 to 5.2. In nearly every year since the early 1980s, the divorce rate has decreased, and in 2005 it was 3.6 (Munson & Sutton 2006; U.S. Census Bureau 2007). Despite the decline in divorce rates since the 1980s, more than one-quarter (28 percent) of U.S. adults have been divorced; among 50 to 64 year olds, 45 percent have been divorced (Saad 2006).

  Most divorced individuals remarry and create blended families, traditionally referred to as stepfamilies. An estimated one-quarter of all children born in the United States will live with a stepparent before they reach adulthood (Mason 2003). Federal, state, and private sector policies have not kept pace with the concerns of modern stepfamilies. In most states stepparents have no obligation to support their stepchildren during the marriage, nor do they have any right of custody or control. In the event of divorce stepparents usually have no rights to custody or even visitation and no obligation to pay child support. Stepchildren have no right of inheritance in the event of the stepparent's death (unless the stepparent has specified such inheritance in a will) (Mason 2003).

- *Increased employment of mothers.* Employment of married women with children under age 18 rose from 24 percent in 1950 to 40 percent in 1970 to 66 percent in 2005 (Gilbert 2003; Bureau of Labor Statistics 2006). According to the Bureau of Labor Statistics (2006), among single-parent families in 2005, the mother was employed in 71 percent of those families maintained by women. In nearly two-thirds (61 percent) of U.S. married-couple families with children younger than under age 18, both parents were employed. Less than one-third (31 percent) of these families involved an employed father and an unemployed mother, which means that the idea of the traditional two-parent family in which the husband is a breadwinner and the wife is a homemaker does not reflect the reality of American families. Yet work, school, and medical care in the United States tend to be organized around the expectation that every household has a full-time mother at home who is available to transport children to medical appointments, pick up children from school on early dismissal days, and stay home when a child is sick (Coontz 1992).

The Marital Decline and Marital Resiliency Perspectives on the American Family

Do the recent transformations in American families signify a collapse of marriage and family in the United States? Does the trend toward diversification of family forms mean that marriage and family are disintegrating, falling apart, or even disappearing? Or, as in other aspects of social life, has family simply undergone transformations in response to changes in socioeconomic conditions, gender roles, and cultural values? The answers to these questions depend on whether we adopt the marital decline perspective or the marital resilience perspective.

According to the **marital decline perspective,** (1) personal happiness has become more important than marital commitment and family obligations and (2) the decline in lifelong marriage and the increase in single-parent families have contributed to a variety of social problems, such as poverty, delinquency, substance abuse, violence, and the erosion of neighborhoods and communities (Amato 2004). According to the **marital resiliency perspective,** "poverty, unemployment, poorly funded schools, discrimination, and the lack of basic services (such as health insurance and child care) represent more serious threats to the well-being of children and adults than does the decline in married two-parent families" (Amato 2004, p. 960). According to this perspective, many marriages in the past were troubled, but because divorce was not socially acceptable, these problematic marriages remained intact. Rather than the view of divorce as a sign of the decline of marriage, divorce provides a second chance for happiness for adults and an escape from dysfunctional and aversive home environments for many children.

Family scholar Stephanie Coontz (1997) observed that "marriage is certainly a transformed institution, and it plays a smaller role than ever before in organizing social and personal life" (p. 31). Although women and men are marrying later than they did in the past and although their marriages may not last as long as they once did, most people eventually do get married. About 95 percent of women and men in their early 60s have been married at least once (Hacker 2003). Among young adults (ages 18–29), three-quarters (76 percent) are either currently married, were previously married, or plan to marry some day; another 13 percent are living with a partner (Saad 2006). And despite high rates of cohabitation, living together tends to be a precursor to marriage rather than a permanent substitute (Goldstein & Kenney 2001).

A national Gallup survey shows that Americans rank their family as the most important aspect of life (Moore 2003). And in a national survey of first-year U.S. college students, about three-quarters of both women and men indicated that "raising a family" was an "essential" or "very important" objective (American Council on Education and University of California 2006). Most U.S. adults continue to say that they have "old-fashioned values about family and marriage," although the percentage endorsing this statement has declined from 87 percent in 1987 to 76 percent in 2007 (Pew Research Center 2007). Most U.S. adults, for example, value monogamy within marriage; only 4 percent say that a married man and woman having an affair is morally acceptable (Saad 2006).

> 66 We recognize today that children can be effectively raised in many different family systems and that it is the emotional climate of the family, rather than its kinship structure, that primarily determines a child's emotional well-being and healthy development. 99

David Elkind
Child development specialist

marital decline perspective
A pessimistic view of the current state of marriage that includes the beliefs that (1) personal happiness has become more important than marital commitment and family obligations and (2) the decline in lifelong marriage and the increase in single-parent families have contributed to a variety of social problems.

marital resiliency perspective
A view of the current state of marriage that includes the beliefs that (1) poverty, unemployment, poorly funded schools, discrimination, and the lack of basic services (such as health insurance and child care) represent more serious threats to the well-being of children and adults than does the decline in married two-parent families and (2) divorce provides a second chance for happiness for adults and an escape from dysfunctional and aversive home environments for many children.

Although the high rate of marital dissolution seems to suggest a weakening of marriage, divorce may also be viewed as resulting from placing a high value on marriage, such that a less than satisfactory marriage is unacceptable. In effect, people who divorce may be viewed not as incapable of commitment but as those who would not settle for a bad marriage. Indeed, the expectations that young women and men have of marriage have changed. Whereas once the main purpose of marriage was to have and raise children, today women and men want marriage to provide adult intimacy and companionship (Coontz 2000).

The high rate of out-of-wedlock childbirth and single parenting is also not necessarily indicative of a decline in the value of marriage. In interviews with a sample of low-income single women with children, most women said they would like to be married but just have not found "Mr. Right" (Edin 2000).

In interview after interview, mothers stressed the seriousness of the marriage commitment and their belief that "it should last forever." Thus, it is not that mothers held marriage in low esteem, but rather the fact that they held it in such high esteem that convinced them to forego marriage, at least until their prospective marriage partner could prove himself worthy economically or they could find another partner who could (Edin 2000, pp. 120–121).

Low-income single mothers in Edin's (2000) study were reluctant to marry the father of their children because these men had low economic status, traditional notions of male domination in household and parental decisions, and patterns of untrustworthy and even violent behavior. Given the low level of trust these mothers have of men and given their view that husbands want more control than the women are willing to give them, women realize that a marriage that is also economically strained is likely to be conflictual and short-lived. "Interestingly, mothers say they reject entering into economically risky marital unions out of respect for the institution of marriage, rather than because of a rejection of the marriage norm" (Edin 2000, p. 130).

Is the well-being of a family measured by the degree to which that family conforms to the idealized married, two-parent, stay-at-home mom model of the 1950s? Or is family well-being measured by function rather than form? As suggested by family scholars Mason, Skolnick, and Sugarman (2003), "the important question to ask about American families . . . is not how much they conform to a particular image of the family, but rather how well do they function—what kind of love, care, and nurturance do they provide?" (p. 2).

Finally, it is important to have a perspective that takes into account the historical realities of families. Family historian Stephanie Coontz (2004) explained:

> I have spent much of my career as a historian explaining to people that many things that seem new in family life are actually quite traditional. Two-provider families, for example, were the norm through most of history. Stepfamilies were more numerous in much of history than they are today. There have been several times and places when cohabitation, out-of-wedlock births, or nonmarital sex were more widespread than they are today. Divorce was higher in Malaysia during the 1940s and 1950s than it is today in the United States. Even same-sex marriage, though comparatively rare, has been accepted in some cultures under certain conditions. (p. 974)

In sum, it is clear that the institution of marriage and family has undergone significant changes in the last few generations. What is not as clear is whether these changes are for the better or for the worse. As this chapter's *Social Problems Research Up Close* feature discusses, changes in marriage and family may be viewed as neither all good nor all bad, but are perhaps a more complex mix.

Coontz (2005a) noted, "Marriage has become more joyful, more loving, and more satisfying for many couples than ever before in history. At the same time it has become optional and more brittle" (p. 306).

SOCIOLOGICAL THEORIES OF FAMILY PROBLEMS

Three major sociological theories—structural functionalism, conflict theory, and symbolic interactionism—help to explain different aspects of the family institution and the problems in families today.

Structural-Functionalist Perspective

The structural-functionalist perspective views the family as a social institution that performs important functions for society, including producing new members, regulating sexual activity and procreation, socializing the young, and providing physical and emotional care for family members. According to the structural-functionalist perspective, traditional gender roles contribute to family functioning: Women perform the "expressive" role of managing household tasks and providing emotional care and nurturing to family members, and men perform the "instrumental" role of earning income and making major family decisions.

According to the structural-functionalist perspective, the high rate of divorce and the rising number of single-parent households constitute a "breakdown" of the family institution that has resulted from rapid social change and social disorganization. The structural-functionalist perspective views the breakdown of the family as a primary social problem that leads to secondary social problems such as crime, poverty, and substance abuse.

Structural-functionalist explanations of family problems examine how changes in other social institutions contribute to family problems. For example, a structural-functionalist view of divorce examines how changes in the economy (such as more dual-earner marriages) and in the legal system (such as the adoption of "no-fault" divorce) contribute to high rates of divorce. Changes in the economic institution, specifically falling wages among unskilled and semi-skilled men, also contribute to both intimate partner abuse and the rise in female-headed single-parent households (Edin 2000).

The structural-functionalist perspective is also concerned with latent dysfunctions—unintended and unrecognized negative consequences. For example, one of the latent dysfunctions of marriage-promotion programs is that they may encourage battered women to stay in their abusive relationships.

> By stigmatizing single parents, stigmatizing divorce, or encouraging women to believe that they are harming their children if they leave their partners, [marriage-promotion] programs make it more difficult for some women to leave violent relationships or encourage them, intentionally or not, to remain with abusive partners. (Family Violence Prevention Fund 2005, p. 5)

Conflict and Feminist Perspectives

Conflict theory focuses on how capitalism, social class, and power influence marriages and families. Feminist theory is concerned with how gender inequalities influence and are influenced by marriages and families. Feminists are critical of the traditional male domination of families—a system known as **patriarchy**—that is reflected in the tradition of wives taking their husband's

patriarchy A male-dominated family system that is reflected in the tradition of wives taking their husband's last name and children taking their father's name.

In *Alone Together: How Marriage in America Is Changing,* Paul Amato and colleagues (2007) examined survey data from 1980 and 2000 to see how marital quality changed over that 20-year period. One of the key questions the authors attempted to answer is: Has marriage declined in its ability to provide satisfying and enduring relationships for adults? Or, has marriage become an even stronger and more satisfying arrangement for couples today compared to in the past?

Sample and Methods

Two sets of data were used for this study. To collect the 1980 data, telephone interviewers used a random-digit dialing procedure to locate a national sample of 2,034 married individuals 55 years and younger who were living with their spouses. The overall response rate was 65 percent. Data from 2000 also came from telephone interviews with 2,100 married individuals from a national random sample of households, with an overall response rate of 63 percent. The interview questions were designed to measure five dimensions of marital quality: (1) marital happiness; (2) marital interaction; (3) marital conflict; (4) marital problems; and (5) divorce proneness. To assess whether marital quality changed between 1980 and 2000, the researchers compared the responses from 1980 with those from 2000. They also looked at gender differences to see whether marital quality for wives and husbands differed.

Selected Findings and Discussion

Analysis of the 1980 and 2000 data suggests that marriages have become stronger and more satisfying in some respects and weaker and less satisfying in other respects. Overall, two dimensions of marital quality improved (conflict and problems), one dimension deteriorated (interaction), and two dimensions were unchanged (happiness and divorce proneness).

1. *Marital happiness:* The mean level of marital happiness reported by spouses in 1980 was equivalent to the level of marital happiness reported in 2000. In both years, the majority of spouses (74 percent in 1980; 68 percent in 2000) rated their marriage as better than average.

2. *Marital interaction:* All five measures of marital interaction showed a decline from 1980 to 2000. Compared with spouses in 1980, spouses in 2000 were less likely to report that their husband/wife almost always accompanied them while visiting friends, shopping, eating their main meal of the day, going out for leisure, and working around the home.

3. *Marital conflict:* Of the five items that measure marital conflict, three declined significantly between 1980 and 2000: reports of disagreeing with one's spouse "often" or "very often," reports of violence ever occurring in the marriage, and reports of marital violence within the previous 3 years. No change was observed in the percentage of spouses who reported arguments over the division of household labor or in the number of serious quarrels in the previous 2 months.

4. *Marital problems:* Compared with spouses in 1980, spouses in 2000 were less likely to report marital problems in the following four areas: getting angry easily, having feelings that are hurt easily, experiencing jealousy, and being domineering. There was little change in the percentage of spouses who reported that getting in trouble with police, drinking/drugs, or extramarital sex were problems in their marriage.

5. *Divorce proneness:* Although the average level of divorce proneness was stable between 1980 and 2000, responses to many of the individual items changed. To explain the difference in the pattern of divorce proneness across the two samples, the researchers divided the spouses into three groups: those who reported a low degree of divorce proneness, those who reported a moderate degree of divorce proneness, and those who reported a high degree of

last name and children taking their father's name. Patriarchy is also reflected in the view that wives and children are the property of husbands and fathers.

The overlap between conflict and feminist perspectives is evident in views on how industrialism and capitalism have contributed to gender inequality. With the onset of factory production during industrialization, workers—mainly men—left the home to earn incomes and women stayed home to do unpaid child care and domestic work. This arrangement resulted in families founded on what Engels calls "domestic slavery of the wife" (quoted by Carrington 2002, p. 32). Modern society, according to Engels, rests on gender-based slavery, with women doing household labor for which they receive neither income nor status while men leave the home to earn an income. Times have certainly changed since Engels made his observations, with most wives today leaving the home to earn incomes. However, wives employed full-time still do the bulk of unpaid domes-

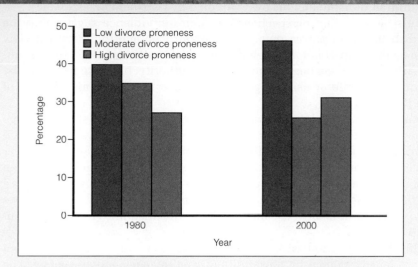

Percentage of respondents reporting low, medium, and high levels of divorce proneness, 1980 and 2000.
Source: Amato et al. (2007).

divorce proneness. In 1980, the percentages of spouses in the low, medium, and high divorce proneness groups did not vary as much as in 2000. As shown in the figure, the percentage of spouses who reported either a low or a high degree of divorce proneness increased from 1980 to 2000, whereas the percentage reporting moderate divorce proneness declined.

6. *Gender differences:* Responses to items measuring marital interaction, marital problems, and divorce proneness were similar for husbands and wives. Two items measuring marital happiness revealed a gender difference: Between 1980 and 2000 wives (but not husbands) reported greater happiness with the amount of understanding received from their spouses. Wives also reported more happiness with their husbands' work around the house, whereas husbands reported less happiness with their wives' work around the house. Regarding the measure of marital conflict, both husbands and wives reported declines in violence, but this change was greater for husbands than for wives.

In sum, this research suggests that between 1980 and 2000, "marriages became more peaceful, with fewer disagreements, less aggression, and fewer interpersonal sources of tension between spouses" (Amato et al. 2007, p. 68). Although overall levels of marital satisfaction did not seem to change from 1980 to 2000, the results for divorce proneness were mixed, with an increase in the proportion of both stable and unstable marriages. The finding that most concerned the researchers was that between 1980 and 2000, the lives of husbands and wives became more separate, as spouses shared fewer activities. Citing previous research that has found that spouses who spend less time together are less happy in their marriage and more likely to divorce, Amato et al. suggested that "it is possible that the gradual decline in marital interaction between 1980 and 2000 will erode future marital happiness and increase subsequent levels of marital instability" (p. 69).

Source: Amato, Paul R., Alan Booth, David R. Johnson, and Stacy J. Rogers. 2007. *Alone Together: How Marriage in America Is Changing.* Cambridge, MA: Harvard University Press. Used by permission.

tic labor, and women are more likely than men to compromise their occupational achievement to take on child care and other domestic responsibilities. The continuing unequal distribution of wealth that favors men contributes to inequities in power and fosters economic dependence of wives on husbands. When wives do earn more money than their husbands (which is the case in 30 percent of marriages), the divorce rate is higher—the women can afford to leave abusive or inequitable relationships (Jalovaara 2003).

Economic factors have also influenced norms concerning monogamy. In societies in which women and men are expected to be monogamous within marriage, there is a double standard that grants men considerably more tolerance for being nonmonogamous. Engels explained that monogamy arose from the concentration of wealth in the hands of a single individual—a man—and from the need to bequeath this wealth to children of his own, which requires that his wife be monogamous. The "sole exclusive aims of monogamous marriage were to make the man supreme in the family and to propagate, as the future heirs to his wealth, children indisputably his own" (quoted by Carrington 2002, p. 32).

Feminist and conflict perspectives on domestic violence suggest that the unequal distribution of power among women and men and the historical view of women as the property of men contribute to wife battering. When wives violate or challenge the male head-of-household's authority, the male may react by "disciplining" his wife or using anger and violence to reassert his position of power in the family.

Although modern gender relations within families and within society at large are more egalitarian than in the past, male domination persists, even if less obvious. Lloyd and Emery (2000) noted that "one of the primary ways that power disguises itself in courtship and marriage is through the 'myth of equality between the sexes.' . . . The widespread discourse on 'marriage between equals' serves as a cover for the presence of male domination in intimate relationships . . . and allows couples to create an illusion of equality that masks the inequities in their relationships" (pp. 25–26).

Conflict theorists emphasize that social programs and policies that affect families are largely shaped by powerful and wealthy segments of society. The interests of corporations and businesses are often in conflict with the needs of families. Corporations and businesses strenuously fought the passage of the 1993 Family and Medical Leave Act, which gives people employed full-time for at least 12 months in companies with at least 50 employees up to 12 weeks of unpaid time off for parenting leave, illness or death of a family member, and elder care. Government, which is largely influenced by corporate interests through lobbying and political financial contributions, enacts policies and laws that serve the interests of for-profit corporations rather than families.

Symbolic Interactionist Perspective

Symbolic interactionism emphasizes that human behavior is influenced by meanings and definitions that emerge from small-group interaction. Divorce, for example, was once highly stigmatized and informally sanctioned through the criticism and rejection of divorced friends and relatives. As societal definitions of divorce became less negative, however, the divorce rate increased. The social meanings surrounding single parenthood, cohabitation, and delayed childbearing and marriage have changed in similar ways. As the definitions of each of these family variations became less negative, the behaviors became more common.

The symbolic interactionist perspective is concerned with how labels affect meaning and behavior. For example, when a noncustodial divorced parent (usually a father) is awarded "visitation" rights, he may view himself as a visitor in his children's lives. The meaning attached to the visitor status can be an obstacle to the father's involvement because the label "visitor" minimizes the importance of the noncustodial parent's role and results in conflict and emotional turmoil for fathers (Pasley & Minton 2001). Fathers' rights advocates suggest replacing the term *visitation* with terms such as *parenting plan* or *time-sharing arrangement,* because these terms do not minimize either parent's role.

Symbolic interactionists also point to the effects of interaction on one's self concept, especially the self-concept of children. In a process called the "looking glass self," individuals form a self-concept based on how others interact with them. Family members, such as parents, grandparents, siblings, and spouses, are significant others who have a powerful effect on our self-concepts. For example, negative self-concepts may result from verbal abuse in the family, whereas positive self-concepts may develop in families in which interactions are supportive and loving. The importance of social interaction in children's developing self-concept suggests a compelling reason for society to accept rather than stigmatize nontraditional family forms. Imagine the effect on children who are called "illegitimate" or who are teased for having two moms or dads.

The symbolic interactionist perspective is useful in understanding the dynamics of domestic violence and abuse. For example, some abusers and their victims learn to define intimate partner violence as an expression of love (Lloyd 2000). Emotional abuse often involves using negative labels (e.g., "stupid," "whore," or "bad") to define a partner or family member. Such labels negatively affect the self-concept of abuse victims, often convincing them that they deserve the abuse. In the next section we discuss violence and abuse in intimate and family relationships, noting the scope, causes, and consequences of this troubling social problem.

VIOLENCE AND ABUSE IN INTIMATE AND FAMILY RELATIONSHIPS

Although intimate and family relationships provide many individuals with a sense of intimacy and well-being, for others these relationships involve physical violence, verbal and emotional abuse, sexual abuse, and/or neglect. Indeed, in U.S. society people are more likely to be physically assaulted, abused and neglected, sexually assaulted and molested, or killed in their own homes rather than anywhere else and by other family members rather than by anyone else (Gelles 2000). Before reading further, you may want to take the Abusive Behavior Inventory in this chapter's *Self and Society* feature.

Intimate Partner Violence and Abuse

Abuse in relationships can take many forms, including emotional and psychological abuse, physical violence, and sexual abuse. **Intimate partner violence (IPV)** refers to actual or threatened violent crimes committed against individuals by their current or former spouses, cohabiting partners, boyfriends, or girlfriends.

Prevalence and Patterns of Intimate Partner Violence. Globally, one woman in every three has been subjected to violence in an intimate relationship (United Nations Development Programme 2000). In the United States, most acts of intimate partner violence (85 percent) are committed against women; however, 22 percent of men have experienced physical, sexual, or psychological IPV during their lifetimes (compared with 29 percent of women) (National Center for Injury Control and Prevention 2006). When women assault their male partners, these assaults tend to be acts of retaliation or self-defense (Johnson 2001). Most research on intimate partner abuse has been conducted on heterosexuals, but more than one-third of gay women and gay men in one study reported physical violence in their relationships in the past year (McKenry et al. 2004).

❝ O. J. is going to kill me and get away with it. **❞**

Nicole Brown Simpson
Murdered wife of O. J. Simpson

intimate partner violence (IPV) Actual or threatened violent crimes committed against individuals by their current or former spouses, cohabiting partners, boyfriends, or girlfriends.

Self and Society | Abusive Behavior Inventory

Circle the number that best represents your closest estimate of how often each of the behaviors happened in your relationship with your partner or former partner during the previous 6 months (1, never; 2, rarely; 3, occasionally; 4, frequently; 5, very frequently).

1. Called you a name and/or criticized you.   1 2 3 4 5
2. Tried to keep you from doing something you wanted to do (e.g., going out with friends, going to meetings).   1 2 3 4 5
3. Gave you angry stares or looks.   1 2 3 4 5
4. Prevented you from having money for your own use.   1 2 3 4 5
5. Ended a discussion with you and made the decision himself/herself.   1 2 3 4 5
6. Threatened to hit or throw something at you.   1 2 3 4 5
7. Pushed, grabbed, or shoved you.   1 2 3 4 5
8. Put down your family and friends.   1 2 3 4 5
9. Accused you of paying too much attention to someone or something else.   1 2 3 4 5
10. Put you on an allowance.   1 2 3 4 5
11. Used your children to threaten you (e.g., told you that you would lose custody, said he/she would leave town with the children).   1 2 3 4 5
12. Became very upset with you because dinner, housework, or laundry was not done when he/she wanted it done or done the way he/she thought it should be.   1 2 3 4 5
13. Said things to scare you (e.g., told you something "bad" would happen, threatened to commit suicide).   1 2 3 4 5

14. Slapped, hit, or punched you.   1 2 3 4 5
15. Made you do something humiliating or degrading (e.g., begging for forgiveness, having to ask his/her permission to use the car or to do something).   1 2 3 4 5
16. Checked up on you (e.g., listened to your phone calls, checked the mileage on your car, called you repeatedly at work).   1 2 3 4 5
17. Drove recklessly when you were in the car.   1 2 3 4 5
18. Pressured you to have sex in a way you didn't like or want.   1 2 3 4 5
19. Refused to do housework or child care.   1 2 3 4 5
20. Threatened you with a knife, gun, or other weapon.   1 2 3 4 5
21. Spanked you.   1 2 3 4 5
22. Told you that you were a bad parent.   1 2 3 4 5
23. Stopped you or tried to stop you from going to work or school.   1 2 3 4 5
24. Threw, hit, kicked, or smashed something.   1 2 3 4 5
25. Kicked you.   1 2 3 4 5
26. Physically forced you to have sex.   1 2 3 4 5
27. Threw you around.   1 2 3 4 5
28. Physically attacked the sexual parts of your body.   1 2 3 4 5
29. Choked or strangled you.   1 2 3 4 5
30. Used a knife, gun, or other weapon against you.   1 2 3 4 5

Scoring:

Add the numbers you circled and divide the total by 30 to find your score. The higher your score, the more abusive your relationship.

The inventory was given to 100 men and 78 women equally divided into groups of abusers/abused and nonabusers/nonabused. The men were members of a chemical dependency treatment program in a veterans' hospital and the women were partners of these men. Abusing or abused men earned an average score of 1.8; abusing or abused women earned an average score of 2.3. Nonabusing or nonabused men and women earned scores of 1.3 and 1.6, respectively.

Source: Shepard, Melanie F., and James A. Campbell, 1992. "The Abusive Behavior Inventory: A Measure of Psychological and Physical Abuse." *Journal of Interpersonal Violence,* September 1992, 7(3): 291–305. Inventory is on pages 303–304. Used by permission of Sage Publications, 2455 Teller Road, Newbury Park, CA 91320.

Data suggest that intimate partner violence has decreased considerably in recent years. In research that compared marriages in 1980 with marriages in 2000, Amato et al. (2007) found that reports of violence ever occurring in the marriage (slapping, hitting, kicking, or throwing things) decreased from 21 percent in 1980 to 12 percent in 2000. Other data reveal that the rate of nonfatal

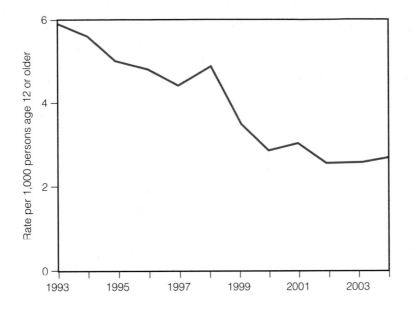

FIGURE 5.5
Nonfatal intimate part-
ner victimization rate,
1993–2004.
Source: Catalano & Shannon (2006).

intimate partner violence against women declined by nearly half from 1993 to 2001 (Bureau of Justice Statistics 2007), although that downward trend leveled off in more recent years (see Figure 5.5). The number of men murdered by intimate partners dropped by 71 percent between 1976 and 2004, and the number of women murdered by intimate partners reached the lowest level recorded in 2004 (Fox & Zawitz 2006).

Factors associated with higher rates of intimate partner victimization against women include being young (ages 16–24), being black, and earning lower incomes. The rate of nonfatal intimate partner violence is highest among separated couples (see Figure 5.6). Characteristics of abusive partners include alcohol or substance abuse, a history of trauma, limited support systems, emotional dependency and jealousy, male unemployment (which creates feelings of inadequacy), and having a traditional role-relationship ideology (Umberson et al. 2003; Loy et al. 2005).

Four patterns of partner violence have been identified: common couple violence, intimate terrorism, violent resistance, and mutual violent control (Johnson & Ferraro 2003).

1. *Common couple violence* refers to occasional acts of violence arising from arguments that get "out of hand." Common couple violence usually does not escalate into serious or life-threatening violence.
2. *Intimate terrorism* is violence that is motivated by a wish to control one's partner and involves the systematic use of not only violence but also economic subordination, threats, isolation, verbal and emotional abuse, and other control tactics. Intimate terrorism is almost entirely perpetrated by men and is more likely to escalate over time and to involve serious injury.
3. *Violent resistance* refers to acts of violence that are committed in self-defense. Violent resistance is almost exclusively perpetrated by women against a male partner.
4. *Mutual violent control* is a rare pattern of abuse "that could be viewed as two intimate terrorists battling for control" (Johnson & Ferraro 2003, p. 169).

FIGURE 5.6
Nonfatal intimate part-
ner victimization rate for
females by marital status,
1993–2004.
Source: Catalano & Shannon (2006).

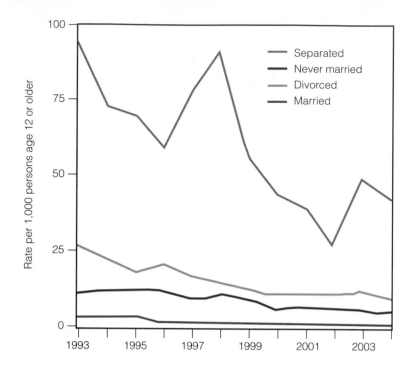

Intimate partner abuse also takes the form of sexual aggression, which refers to sexual interaction that occurs against one's will through use of physical force, threat of force, pressure, use of alcohol or drugs, or use of position of authority. In 2005, 7 in 10 female rape or sexual assault victims stated that the offender was an intimate partner, other relative, or friend or acquaintance (Bureau of Justice Statistics 2007). A national study found that about 3 percent of college women experienced a completed or attempted rape during a college year (Fisher, Cullen, & Turner 2000). Nearly 90 percent of the sexually assaulted college women knew the person who assaulted them. Based on data from the National Violence Against Women survey, half of the women raped by an intimate partner and two-thirds of the women physically assaulted by an intimate partner had been victimized multiple times (Rand 2003).

Effects of Intimate Partner Violence and Abuse. Battering results in physical, psychological, economic, and marital consequences. Each year, intimate partner violence results in 2 million injuries and 1,300 deaths nationwide (National Center for Injury Prevention and Control 2006). In recent years, about one-third of female murder victims and 3 percent of male murder victims in the United States were killed by intimates (Bureau of Justice Statistics 2007). On average, between 1993 and 2004 half of female victims and more than one-third of male victims of nonfatal IPV suffered injuries from their victimization (Catalano 2006). Many battered women are abused during pregnancy, resulting in a high rate of miscarriage and birth defects. Psychological consequences for victims of intimate partner violence can include depression, anxiety, suicidal thoughts and attempts, lowered self-esteem, inability to trust men, fear of intimacy, and substance abuse (National Center for Injury Prevention and Control 2006).

Battering also interferes with women's employment. Some abusers prohibit their partners from working. Other abusers "deliberately undermine women's employment by depriving them of transportation, harassing them at work, turning off alarm clocks, beating them before job interviews, and disappearing when they promise to provide child care" (Johnson & Ferraro 2003, p. 508). Battering also undermines employment by causing repeated absences, impairing women's ability to concentrate, and lowering their self-esteem and aspirations.

Women who have experienced physical or sexual abuse as adults or children are also less likely to be married or in a stable, long-term relationship. In a study of more than 2,000 women living in low-income neighborhoods, 42 percent of women who did not report abuse were married, compared with only 22 percent of the women who reported past abuse (Cherlin et al. 2005). Abuse, whether physical or emotional, is also a factor in many divorces, which often results in a loss of economic resources. Women who flee an abusive home and who have no economic resources may find themselves homeless.

Many children who witness domestic violence get involved by yelling, calling for help, or intervening to try to stop the abuse (Edleson et al. 2003). Children who witness domestic violence are at risk for emotional, behavioral, and academic problems as well as future violence in their own adult relationships (Parker et al. 2000; Kitzmann et al. 2003). Children may also commit violent acts against a parent's abusing partner.

Why Do Some Adults Stay in Abusive Relationships? Adult victims of abuse are commonly blamed for tolerating abusive relationships and for not leaving the relationship as soon as the abuse begins. But from the point of view of the victims, there are compelling reasons to stay. These reasons include love, emotional dependency, commitment to the relationship, hope that things will get better, the view that violence is legitimate because they "deserve" it, guilt, fear, economic dependency, and feeling stuck.

Few and Rosen (2005) interviewed 28 women (7 black and 21 white) who were victims of chronic abuse from a male dating partner and found that 24 of them had played a caretaker role in their families of origin that made them more vulnerable to being seduced by abusive boyfriends and to becoming trapped by their commitment to rescue them. One woman in the study reported that she tended to be attracted to needy men and tried to make them feel good about themselves. She explained, "I always was a rescuer in my family. I felt that I was rescuing him [her boyfriend] and taking care of him. He never knew what it was like to have a good, positive home environment, so I was working hard to create that for him" (Few & Rosen 2005, p. 272). Women in this study also reported that witnessing abuse in their parents' adult relationships taught them the notion that oppression and abuse of power were normal within intimate relationships.

Most of the women in Few and Rosen's (2005) study reported feeling pressure to be in a serious relationship or to have a husband and feared that if they ended the abusive relationship, they might not find someone else and would end up alone. In another study of why abused women stay in abusive relationships, the most important reason reported by women in these relationships was fear of loneliness (Hendy et al. 2003).

Six of the seven black women in Few and Rosen's (2005) study talked about the scarcity of eligible black men as the reason some black women settle for abusive men. Another reason black women remained in their abusive relationships was to prove that black relationships were not inherently dysfunctional. One black woman explained, "There's so much negativity out there about Black relationships. Our relationships are always portrayed as being so adversarial. . . .

We have so few positive images of Black healthy relationships. . . . So you got to settle . . . as much as you can" (Few & Rosen 2005, p. 273). In addition, abused black women who hesitated to seek help from police cited a distrust of police and talked explicitly about the historical mistreatment of black men by police in the South. Christina, a black woman, who reluctantly called the police as a last resort, explained,

> He was going to lose everything if I made this report. One, because he was Black. Two, because he was a Black man in a very White town. I was afraid he would be mistreated by these hick officers . . . that he would lose his job. . . . I decided to call the police and found when I did, none of my girlfriends would support me. (Few & Rosen 2005, p. 273)

Some of Christina's black girlfriends told her that she was a "traitor" for calling the police. None of her girlfriends would testify that her boyfriend "terrorized" her at parties.

Some victims of intimate partner abuse stay because they fear retribution from their abusive partner if they leave. Research has found that leaving a marriage characterized by high levels of verbal or physical aggression is associated with detrimental effects on well-being because conflicts can continue after the divorce, especially when children or other issues keep the ex-spouses tied to each other (Kalmijn & Monden 2006). And many victims delay leaving a violent home because they fear the abuser will hurt or neglect a family pet (Fogle 2003).

Victims also stay because abuse in relationships is usually not ongoing and constant but rather occurs in cycles. The **cycle of abuse** involves a violent or abusive episode followed by a makeup period when the abuser expresses sorrow and asks for forgiveness and "one more chance." The honeymoon period may last for days, weeks, or even months before the next outburst of violence occurs.

Child Abuse

Child abuse refers to the physical or mental injury, sexual abuse, negligent treatment, or maltreatment of a child under the age of 18 by a person who is responsible for the child's welfare. The most common form of child maltreatment is **neglect**—the caregiver's failure to provide adequate attention and supervision, food and nutrition, hygiene, medical care, and a safe and clean living environment (see Figure 5.7).

cycle of abuse A pattern of abuse in which a violent or abusive episode is followed by a makeup period when the abuser expresses sorrow and asks for forgiveness and "one more chance," before another instance of abuse occurs.

child abuse The physical or mental injury, sexual abuse, negligent treatment, or maltreatment of a child younger than age 18 by a person who is responsible for the child's welfare.

neglect A form of abuse involving the failure to provide adequate attention, supervision, nutrition, hygiene, health care, and a safe and clean living environment for a minor child or a dependent elderly individual.

What Do You Think? Because second-hand smoke in vehicles is hazardous to children, Arkansas passed legislation in 2006 that banned smoking in vehicles containing children strapped in car seats. Louisiana followed, becoming the second state to ban smoking in vehicles carrying a child in a car seat. Maine went even further by banning smoking in vehicles carrying anyone under 18. Other states are considering similar legislation. Do you think that smoking in a car when children are present should be considered a form of child abuse as it inflicts harm on children in the form of second-hand smoke exposure? Should smoking with children in the car be banned in the United States?

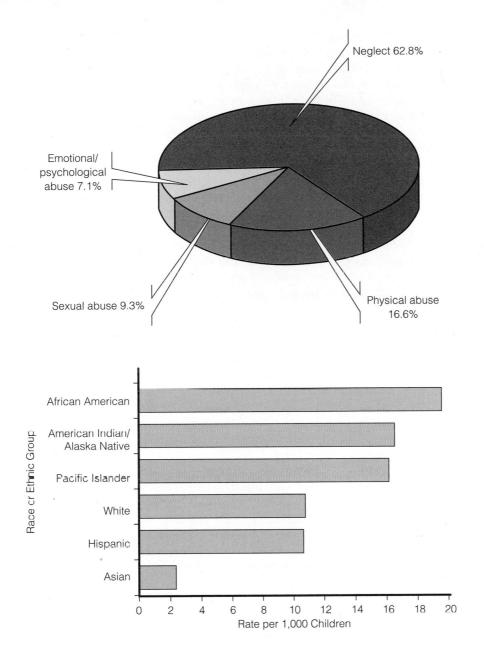

FIGURE 5.7
Types of child maltreatment, 2005.
Source: U.S. Department of Health and Human Services, Administration on Children, Youth, and Families (2007).

Neglect 62.8%

Emotional/
psychological
abuse 7.1%

Sexual abuse 9.3%

Physical abuse
16.6%

FIGURE 5.8
Rates of child abuse and neglect by race and ethnicity: United States, 2005.
Source: U.S. Department of Health and Human Services, Administration on Children, Youth, and Families (2007).

African American

American Indian/
Alaska Native

Pacific Islander

White

Hispanic

Asian

Race or Ethnic Group

0   2   4   6   8   10   12   14   16   18   20
Rate per 1,000 Children

Children at the highest risk for victimization include those in the age group birth to 3 years, children with disabilities, and some minority children (U.S. Department of Health and Human Services, Administration on Children, Youth, and Families 2007). Although half of child abuse and neglect victims in 2005 were white, *rates* of victimization are higher among some minority children (see Figure 5.8).

Perpetrators of child abuse are most often the parents of the victim. Parents who are at the greatest risk of child maltreatment are those who are socially isolated, poor or unemployed, or young and single, have a mental illness, lack an understanding of children's needs and child development, and have a history of domestic abuse (National Center for Injury Prevention and Control 2007).

Victimization and maltreatment of children occurs outside as well as within the family. Many children who experience child abuse are "polyvictims" in that they experience multiple forms of trauma and maltreatment. Finkelhor, Ormrod, & Turner (2007) studied a national sample of children ages 2–17 to assess polyvictimization and its effects. In this study the authors looked at various types of victimization, including sexual victimization, physical assault, property victimization (e.g., theft and burglary), witnessing/indirect victimization (e.g., witnessing domestic violence or murder), and peer/sibling victimization (e.g., bullying). Children in this study experienced an average of three types of victimization and more than one in five children (22 percent) experienced at least four types of victimization in a 1-year period. The most common victimizations were peer and sibling assaults, witnessing nonweapon assaults, emotional bullying, and theft.

Effects of Child Abuse. The effects of child abuse and neglect vary according to the type(s), frequency, and intensity of the abuse or neglect. Physical injuries sustained by child abuse cause pain, disfigurement, scarring, physical disability, and death. In 2005 an estimated 1,460 U.S. children died of abuse or neglect (U.S. Department of Health and Human Services, Administration on Children, Youth, and Families 2007). Most of these children (77 percent) were younger than age 4. Two-thirds of murders of children younger than age 5 are committed by a parent or other family member (Bureau of Justice Statistics 2007). Head injury is the leading cause of death in abused children (Rubin et al. 2003). **Shaken baby syndrome,** whereby the caretaker, most often the father, shakes a baby to the point of causing the child to experience brain or retinal hemorrhage, most often occurs in response to a baby, typically younger than 6 months, who will not stop crying (Ricci et al. 2003; Smith 2003). Battered or shaken babies often suffer permanent disabilities.

In addition to the risk of immediate injury and death, abuse during childhood is associated with depression, low academic achievement, smoking, alcohol and drug abuse, eating disorders, obesity, teen pregnancy, sexually transmitted diseases, sexual promiscuity, low self-esteem, aggressive behavior, juvenile delinquency, adult criminality, suicide, and experiencing abuse victimization as an adult (Administration for Children and Families 2003; National Center for Injury Prevention and Control 2007). Among females early forced sex is associated with decreased self-esteem, increased levels of depression, running away from home, alcohol and drug use, and multiple sexual partners (Jasinski, Williams, & Siegel 2000; Whiffen, Thomson, & Aube 2000). Compared to nonabused peers, sexually abused girls are also more likely to experience teenage pregnancy, to have higher numbers of sexual partners in adulthood, and to acquire sexually transmitted infections and experience forced sex (Browning & Laumann 1997; Stock et al. 1997). Women who were sexually abused as children also report a higher frequency of post-traumatic stress disorder (Spiegel 2000) and suicide ideation (Thakkar et al. 2000). A review of the research suggests that sexual abuse of boys produces many of the same reactions that sexually abused girls experience, including depression, sexual dysfunction, anger, self-blame, suicidal feelings, guilt, and flashbacks (Daniel 2005). Married adults who were physically and sexually abused as children report lower marital satisfaction, higher stress, and lower family cohesion than married adults with no abuse history (Nelson & Wampler 2000).

Effects of child sexual abuse are likely to be severe when the sexual abuse is forceful, is prolonged, and involves intercourse and when the abuse is per-

shaken baby syndrome
A form of child abuse whereby the caretaker shakes a baby to the point of causing the child to experience brain or retinal hemorrhage.

petrated by a father or stepfather (Beitchman et al. 1992). Not only has the child been violated physically, but she or he also has lost an important social support. One woman who had been sexually abused by her father described feeling that she had lost her father; he was no longer a person to love and protect her (Spiegel 2000).

It is important to note that studies concerned with single forms of child victimization may overestimate the negative effects of these single forms because they do not take into account the effects of polyvictimization. In one study of child victims, over 92 percent of the rape victims and 76 percent of the dating violence victims were polyvictims—they had experienced at least four types of victimization in the last year (Finkelhor et al. 2007). Polyvictims in this study were more likely to exhibit anxiety and depression than victims of single forms of victimization, and when the effects of polyvictimization were taken into account, the effects of individual types of victimization were eliminated or greatly reduced. For example, sexual victimization by itself was associated with high levels of anxiety, depression, and anger. But when polyvictimization was taken into account, the association between sexual assault and negative symptoms dropped significantly (Finkelhor et al. 2007).

Although the most common form of elder abuse is neglect, elders are also victims of physical violence.

Elder Abuse, Parent Abuse, and Sibling Abuse

Domestic violence and abuse may involve adults abusing their elderly parents or grandparents, children abusing their parents, and siblings abusing each other.

Elder Abuse. A survey of abuse of U.S. adults age 60 and older found that annually there are 8.3 reports of elder abuse for every 1,000 older Americans (Teaster et al. 2006). **Elder abuse** includes physical abuse, sexual abuse, psychological abuse, financial abuse (such as improper use of the elder person's financial resources), and neglect. The most common form of elder abuse is neglect—failure to provide basic health and hygiene needs, such as clean clothes, doctor visits, medication, and adequate nutrition. Neglect also involves unreasonable confinement, isolation of elderly family members, lack of supervision, and abandonment.

Older women are far more likely than older men to suffer from abuse/neglect. Two of three cases of elder abuse reported to state adult protective services involved women (Teaster et al. 2006). This same survey found that most cases of elder abuse (89 percent) occur in a domestic setting. Most perpetrators were adult children (33 percent), followed by other family members (22 percent) and spouses/intimate partners (11 percent). Risk factors associated with elder abuse include social isolation, a history of a poor-quality relationship between the abused and abuser, a pattern of family violence, and a history of mental health or substance abuse problems in the abuser (Newton 2005). As the proportion of elderly people in the U.S. population increases (see Chapter 12), it is not surprising that reports of elder abuse increased significantly (20 percent) between 2000 and 2004 (Teaster et al. 2006).

elder abuse The physical or psychological abuse, financial exploitation, or medical abuse or neglect of the elderly.

Parent Abuse. Some parents are victimized by their children's violence, ranging from hitting, kicking, and biting to pushing a parent down the stairs and using a weapon to inflict serious injury to or even kill a parent. More violence is directed against mothers than against fathers, and sons tend to be more violent toward parents than are daughters (Ulman 2003). In most cases of children being violent toward their parents, the parents had been violent toward the children.

Sibling Abuse. The most prevalent form of abuse in families is sibling abuse. Ninety-eight percent of the females and 89 percent of the males in one study reported having been emotionally abused by a sibling, and 88 percent of the females and 71 percent of the males reported having been physically abused by a sibling (Simonelli et al. 2002). Sexual abuse also occurs in sibling relationships.

Factors Contributing to Intimate Partner and Family Violence and Abuse

Research suggests that cultural, community, and individual and family factors contribute to domestic violence and abuse.

Cultural Factors. Violence in the family stems from our society's acceptance of violence as a legitimate means of enforcing compliance and solving conflicts at personal, national, and international levels (Viano 1992). Violence and abuse in the family may be linked to cultural factors, such as violence in the media (see Chapter 4), gender inequality and gender socialization, and acceptance of corporal punishment.

Gender Inequality and Gender Socialization. In patriarchal societies "the beating of women and children has been taken for granted, with both men and women accepting it as normal" (Adams 2004, p. 1079). In sub-Saharan Africa, for example, widespread spousal abuse is linked to cultural views of women. One Nigerian woman who was beaten unconscious by her husband explained, "He doesn't believe I have any rights of my own. . . . If I say no, he beats me" (quoted by LaFraniere 2005, p. A1). According to Nigeria's minister for women's affairs, "It is like it is a normal thing for women to be treated by their husbands as punching bags. . . . The Nigerian man thinks that a woman is his inferior. . . . Even when they marry out of love, they still think the woman is below them and they do whatever they want" (quoted by LaFraniere 2005, p. A1).

In the United States before the late 19th century a married woman was considered the property of her husband. A husband had a legal right and marital obligation to discipline and control his wife through the use of physical force. This traditional view of women as property may contribute to men's doing with their "property" as they wish. A recent study of men in battering intervention programs found that about half of the men viewed battering as acceptable in certain situations (Jackson et al. 2003).

The view of women and children as property also explains marital rape and father-daughter incest. Historically, the penalties for rape were based on property rights laws designed to protect a man's property—his wife or daughter—from rape by other men; a husband or father "taking" his own property was not considered rape (Russell 1990). In the past a married woman who was raped by her husband could not have her husband arrested because marital rape was not considered a crime. Today, marital rape is considered a crime in all 50 states.

Traditional male gender roles have taught men to be aggressive and to be dominant in male-female relationships. Male abusers are likely to hold traditional attitudes toward women and male-female roles (Lloyd & Emery 2000). Traditional male gender socialization also discourages men from verbally expressing their feelings, which increases the potential for violence and abusive behavior (Umberson et al. 2003). Anderson (1997) found that men who earn less money than their partners are more likely to be violent toward them: "Disenfranchised men then must rely on other social practices to construct a masculine image. Because it is so clearly associated with masculinity in American culture, violence is a social practice that enables men to express a masculine identity" (p. 667). Traditional female gender roles have also taught women to be submissive to their male partner's control.

Acceptance of Corporal Punishment. **Corporal punishment** is the intentional infliction of pain for a perceived misbehavior (Block 2003). Many mental health professionals and child development specialists argue that it is ineffective and damaging to children. Children who experience corporal punishment display more antisocial behavior, are more violent, and have an increased incidence of depression as adults (Straus 2000). Yet many parents accept the cultural tradition of spanking as an appropriate form of child discipline.

More than 90 percent of parents of toddlers reported using corporal punishment (Straus 2000). Parents of black children are more likely to use corporal punishment than parents of white or Latino children (Grogan-Kaylor & Otis 2007). Although not everyone agrees that all instances of corporal punishment constitute abuse, undoubtedly, some episodes of parental "discipline" are abusive.

Community Factors. Community factors that contribute to violence and abuse in the family include social isolation and inaccessible or unaffordable community services, such as health care, day care, elder care, and respite care facilities.

Living in social isolation from extended family and community members increases a family's risk for abuse. Isolated families are removed from material benefits, care-giving assistance, and emotional support from extended family and community members.

Failure to provide medical care to children and elderly family members (a form of neglect) is sometimes a result of the lack of accessible or affordable health care services in the community. Failure to provide supervision for children and adults may result from inaccessible day care and elder care services. Without elder care and respite care facilities, socially isolated families may not have any help with the stresses of caring for elderly family members and children with special needs.

Individual and Family Factors. Individual and family factors associated with intimate partner and family violence and abuse include a history of family violence, drug and alcohol abuse, and poverty. Men who witnessed their fathers abusing their mothers and women who witnessed their mothers abusing their fathers are more likely to become abusive partners themselves (Heyman & Slep 2002; Babcock, Miller, & Siard 2003). Individuals who were abused as children are more likely to report being abused in an adult domestic relationship (Heyman & Slep 2002). Mothers who have been sexually abused as children are more likely to physically abuse their own children (DiLillo, Tremblay,

corporal punishment The intentional infliction of pain for a perceived misbehavior.

& Peterson 2000). Although a history of abuse is associated with an increased likelihood of being abusive as an adult, most adults who were abused as children do not continue the pattern of abuse in their own relationships (Gelles 2000).

Alcohol use is reported as a factor in 50–75 percent of incidents of physical and sexual aggression in intimate relationships (Lloyd & Emery 2000). Alcohol and other drugs increase aggression in some individuals and enable the offender to avoid responsibility by blaming his or her violent behavior on drugs/alcohol.

Although abuse in adult relationships occurs among all socioeconomic groups, it is more prevalent among the poor. Studies show that at least 50–60 percent of women receiving welfare have experienced physical abuse by an intimate partner, compared with 22 percent of the general population (Family Violence Prevention Fund 2005). However, Kaufman and Zigler (1992) noted that "although most poor people do not maltreat their children, and poverty, per se, does not cause abuse and neglect, the correlates of poverty, including stress, drug abuse, and inadequate resources for food and medical care, increase the likelihood of maltreatment" (p. 284).

STRATEGIES FOR ACTION: PREVENTING AND RESPONDING TO VIOLENCE AND ABUSE IN INTIMATE AND FAMILY RELATIONSHIPS

Next, we look at strategies for preventing and responding to violence and abuse. These include **primary prevention** strategies, which target the general population; **secondary prevention** strategies, which target groups at high risk for family violence and abuse; and **tertiary prevention** strategies, which target families who have experienced abuse (Gelles 1993; Harrington & Dubowitz 1993).

Primary Prevention Strategies

Specific abuse-prevention strategies include public education and media campaigns, which may help to reduce domestic violence by conveying the criminal nature of domestic assault and offering ways to prevent abuse. Other abuse-prevention efforts focus on parent education to teach parents realistic expectations about child behavior and methods of child discipline that do not involve corporal punishment. For example, Mental Health America (2003) distributes a fact sheet on alternatives to spanking (see Table 5.1). In 1979 Sweden became the first country in the world to ban corporal punishment in all settings, including the home. Other countries have followed Sweden's lead, and in 2007, the Netherlands became the 18th country to ban corporal punishment (Global Initiative to End All Corporal Punishment of Children 2007). In the United States it is legal in all 50 states for a parent to spank, hit, belt, paddle, whip, or otherwise inflict punitive pain on a child, so long as the corporal punishment does not meet the individual state's definition of child abuse. Corporal punishment is also permitted in public schools in 24 states and in private schools in every state except Iowa and New Jersey.

Another abuse-prevention strategy involves reducing violence-provoking stress by reducing poverty and unemployment and providing adequate housing, child-care programs and facilities, nutrition, medical care, and educational opportunities. However, rather than strengthening the supports for poor families with children, welfare reform legislation enacted in 1996 limits cash assistance

primary prevention Family violence prevention strategies that target the general population.

secondary prevention Family violence prevention strategies that target groups thought to be at high risk for family violence.

tertiary prevention Family violence prevention strategies that target families who have experienced family violence.

TABLE 5.1 Effective Discipline Techniques for Parents: Alternatives to Spanking

Punishment is a "penalty" for misbehavior, but discipline is a method of teaching a child right from wrong. Alternatives to physical discipline include the following:

1. Be a positive role model.

Children learn behaviors by observing their parents' actions, so parents must model the ways in which they want their children to behave. If a parent yells or hits, the child is likely to do the same.

2. Set rules and consequences.

Make rules that are fair, realistic, and appropriate to a child's level of development. Explain the rules to children along with the consequences of not following them. If children are old enough, they can be included in establishing the rules and consequences for breaking them.

3. Encourage and reward good behavior.

When children are behaving appropriately, give them verbal praise and occasionally reward them with tangible objects, privileges, or increased responsibility.

4. Create charts.

Charts to monitor and reward behavior can help children learn appropriate behavior. Charts should be simple and should focus on one behavior at a time, for a certain length of time.

5. Give time-outs.

A "time-out" involves removing a child from a situation following a negative behavior. This can help the child calm down, end the inappropriate behavior, and reenter the situation in a positive way. Explain what the inappropriate behavior is, why the time-out is needed, when the time-out will begin, and how long it will last. Set an appropriate length of time for the time-out based on age and level of development, usually just a few minutes.

Source: Based on Mental Health America. 2003. "Effective Discipline Techniques for Parents: Alternatives to Spanking." Strengthening Families Fact Sheet. http://www.nmha.org. Reprinted with permission.

---

to poor single parents to 2 consecutive years with a 5-year lifetime limit (some exceptions are granted) and forces women into the labor force. Many women going from welfare to work will experience greater hardships as a result of a loss of food stamp benefits, increases in federal housing rent, loss of Medicaid benefits, cost of transportation to work, and child-care costs (Edin & Lein 1997). Some women forced to go to work and unable to afford child care will leave their children unattended, increasing child neglect.

Secondary Prevention Strategies

Some families have a higher risk of experiencing violence and abuse. Such high-risk families include low-income families; single-parent families; teen-parent families; and families headed by adults with a history of mental illness, substance abuse, or prior child abuse victimization. Secondary prevention strategies, designed to prevent abuse from occurring in high-risk families, include parent education programs, parent support groups, individual counseling, substance abuse treatment, and home visiting programs.

Tertiary Prevention Strategies

Between 1993 and 2004, about 21 percent of female victims and 10 percent of male victims of nonfatal IPV contacted a private or government agency for assistance (Catalano 2006). The National Domestic Violence Hotline (1-800-799-SAFE) is a 24-hour, toll-free service that provides crisis assistance and local domestic violence shelter and safe house referrals for callers across the country. Shelters provide abused women and their children with housing, food, and counseling services. Safe houses are private homes of individuals who volunteer to provide temporary housing to abused persons who decide to leave their violent homes. Some communities have abuse shelters for victims of elder abuse. Some programs offer a safe shelter for pets of domestic violence victims. Because one in four victims reports a delay in leaving dangerous domestic situations because of concerns over the safety of a pet, some domestic abuse agencies have paired with veterinary schools, humane societies, and community organizations to help victims and their pets escape violent homes (Fogle 2003).

Children who are abused in the family may be removed from their homes and placed in foster care or in the care of another family member, such as a grandparent. However, every state has various types of family preservation programs to prevent family breakup when desirable and possible without jeopardizing the welfare of children in the home. **Family preservation programs** are in-home interventions for families who are at risk of having a child removed from the home because of abuse or neglect.

Alternatively, a court may order an abusing spouse or parent to leave the home. Abused spouses or cohabiting partners may obtain a restraining order that prohibits the perpetrator from going near the abused partner. About half of the states and Washington, D.C., now have mandatory arrest policies that require police to arrest abusers, even if the victim does not want to press charges. However, many victims of intimate partner violence do not report the violence to law enforcement authorities. Reasons that victims do not report intimate partner violence to the police include (1) believing that such violence is a private or personal matter, (2) fearing retaliation, (3) viewing the violence as a "minor" crime, (4) protecting the offender, and (5) believing that the police will not help or will be ineffective (Catalano 2006).

Treatment for abusers—which may be voluntary or mandated by the court—typically involves group and/or individual counseling, substance abuse counseling, and/or training in communication, conflict resolution, and anger management. Treatment for men who sexually abuse children typically involves cognitive behavior therapy (changing the thoughts that lead to sex abuse) and medication to reduce the sex drive (Stone 2004).

Men who stop abusing their partners learn to take responsibility for their abusive behavior, develop empathy for their partner's victimization, reduce their dependency on their partners, and improve their communication skills (Scott & Wolfe 2000). However, evaluations of batterer intervention programs found no significant differences between treatment and control groups on re-offense rates or men's attitudes toward domestic violence (Jackson et al. 2003).

PROBLEMS ASSOCIATED WITH DIVORCE

family preservation programs In-home interventions for families who are at risk of having a child removed from the home because of abuse or neglect.

The United States has the highest divorce rate among Western nations. Despite the decline in divorce rates in recent years, 40 percent of first marriages end in divorce and 60 percent of those marriages involve children (Kimmel 2004). Divorce is considered problematic not only because of the negative effects it has on children but also because of the difficulties it causes for adults. However, in

"Our pre-nup guarantees that we'll be together forever. In the event of divorce, you and I get custody of each other!"

some societies legal and social barriers to divorce are considered problematic because such barriers limit the options of spouses in unhappy and abusive marriages. Ireland did not allow divorce under any condition until 1995, and Chile did not allow divorce until 2004.

Even when divorce is a legal option, social barriers often prevent spouses from divorcing. Hindu women, for example, experience great difficulty leaving a marriage, even when the husband is abusive. Loss of status, possible loss of custody of her children, homelessness, poverty, and being labeled a "loose" woman are enough to keep women locked in the "confines of a tyrannous family as silent sufferers" (Laungani 2005, p. 88).

> 66 It is now widely accepted that men and women have the right to expect a happy marriage, and that if a marriage does not work out, no one has to stay trapped. 99
>
> Sylvia Ann Hewlett
> Family advocate

Social Causes of Divorce

When we think of why a particular couple gets divorced, we typically think of a number of individual and relationship factors that might have contributed to the marital breakup: incompatibility in values or goals; poor communication; lack of conflict resolution skills; sexual incompatibility; extramarital relationships; substance abuse; emotional or physical abuse or neglect; boredom; jealousy; and difficulty coping with change or stress related to parenting, employment, finances, in-laws, and illness. But understanding the high rate of divorce in U.S. society requires awareness of how the following social and cultural factors contribute to marital breakup:

1. *Changing function of marriage.* Before the Industrial Revolution the institution of marriage was a unit of economic production and consumption that

TABLE 5.2 Factors That Decrease Women's Risk of Separation or Divorce During the First 10 Years of Marriage

FACTOR	PERCENT DECREASE IN RISK OF DIVORCE OR SEPARATION
Annual income over $50,000 (vs. under $25,000)	30
Having a baby 7 months or more after marriage (vs. before marriage)	24
Marrying over 25 years of age (vs. under 18)	24
Having an intact family of origin (vs. having divorced parents)	14
Religious affiliation (vs. none)	14
Some college (vs. high school dropout)	13

Source: Whitehead and Popenoe (2005).

was largely organized around producing, socializing, and educating children. But the institution of marriage has changed over the last few generations:

Marriage changed from a formal institution that meets the needs of the larger society to a companionate relationship that meets the needs of the couple and their children and then to a private pact that meets the psychological needs of individual spouses. (Amato et al. 2007, p. 70)

When spouses do not feel that their psychological needs—for emotional support, intimacy, affection, love, personal growth—are being met in the marriage, they may consider divorce with the hope of finding a new partner to fulfill these affectional needs.

2. *Increased economic autonomy of women.* As noted earlier, before 1940 most wives were not employed outside the home and depended on their husband's income. Today, about two-thirds of married women are in the labor force (Bureau of Labor Statistics 2006). A wife who is unhappy in her marriage is more likely to leave the marriage if she has the economic means to support herself (Jalovaara 2003). An unhappy husband may also be more likely to leave a marriage if his wife is self-sufficient and can contribute to the support of the children.

3. *Increased work demands and economic stress.* Another factor influencing divorce is increased work demands and the stresses of balancing work and family roles. Workers are putting in longer hours, often working overtime or taking second jobs. And as discussed in Chapters 6 and 7, many families struggle to earn enough money to pay for rising housing, health care, and child-care costs. Financial stress can cause marital problems. Whitehead and Popenoe (2005) reported that couples with an annual income under $25,000 are 30 percent more likely to divorce than couples with incomes over $50,000 (see Table 5.2).

4. *Dissatisfaction with marital division of labor.* Many employed parents, particularly mothers, come home to work a **second shift**—the work involved in caring for children and household chores (Hochschild 1989). Wives are more likely than husbands to perceive the marital division of labor—household chores and child care—as unfair (Nock 1995). This perception of unfairness can lead to marital tension and resentment, as reflected in the following excerpt:

My husband's a great help watching our baby. But as far as doing housework or even taking the baby when I'm at home, no. He figures he works five days a week; he's not going to come home and clean. But he doesn't stop to think that I work seven days a week. Why should I have to come home and do the housework without help from anybody else? My husband and I have been through this over and over again. Even if he would just pick up from the kitchen table and stack the dishes for me, that would make a big difference. He does nothing. On his weekends off, I have to provide a sitter for the baby so he can go fishing. When I have a day off, I have the baby all day long

second shift The household work and child care that employed parents (usually women) do when they return home from their jobs.

without a break. He'll help out if I'm not here, but the minute I am, all the work at home is mine. (quoted by Hochschild 1997, pp. 37–38)

Women are increasingly looking for egalitarianism in relationships. Women want to be equal partners in their marriages, not just in earning income but in sharing the work of household chores, child rearing, and marital communication and in making decisions for the family. They are also looking for partners who are considerate and dependable (O'Reilly et al. 2005). Frustrated by men's lack of participation in marital work, women who desire relationship egalitarianism may see divorce as the lesser of two evils (Hackstaff 2003).

5. *Liberalized divorce laws.* Before 1970 the law required a couple who wanted a divorce to prove that one of the spouses was at fault and had committed an act defined by the state as grounds for divorce—adultery, cruelty, or desertion. In 1969 California became the first state to initiate **no-fault divorce,** which permitted a divorce based on the claim that there were "irreconcilable differences" in the marriage. No-fault divorce law has contributed to the U.S. divorce rate by making divorce easier to obtain. Although the U.S. divorce rate started climbing before California instituted the first no-fault divorce law, the widespread adoption of such laws has probably contributed to its continued escalation. Today, all 50 states recognize some form of no-fault divorce.

6. *Increased individualism.* U.S. society is characterized by **individualism**—the tendency to focus on one's individual self-interests and personal happiness rather than on the interests of one's family and community. "Marital commitment lasts only as long as people are happy and feel that their own needs are being met" (Amato 2004, p. 960). Belief in the right to be happy, even if it means getting divorced, is reflected in social attitudes toward divorce: two-thirds (67 percent) of U.S. adults report that divorce is morally acceptable (Saad 2006). **Familism**—the view that the family unit is more important than individual interests—is still prevalent among Asian Americans and Mexican Americans, which helps to explain why the divorce rate is lower among these groups than among whites and African Americans (Mindel, Habenstein, & Wright 1998).

7. *Increased life expectancy.* Finally, more marriages today end in divorce, in part, because people live longer than they did in previous generations. Because people live longer today than in previous generations, "till death do us part" involves a longer commitment than it once did. Indeed, one can argue that "marriage once was as unstable as it is today, but it was cut short by death not divorce" (Emery 1999, p. 7).

Consequences of Divorce

Divorce is considered a social problem because of the distress and difficulties associated with it. When parents have bitter and unresolved conflict and/or if one parent is abusing the child or the other parent, divorce may offer a solution to family problems. But divorce often has negative effects for ex-spouses and their children and also contributes to problems that affect society as a whole.

Physical and Mental Health Consequences. In a review of research on the consequences of divorce for adults, Amato (2003) cited numerous studies showing that divorced individuals have more health problems and a higher risk of mortality than married individuals; and divorced individuals also experience lower levels of psychological well-being, including more unhappiness, depression,

no-fault divorce A divorce that is granted based on the claim that there are irreconcilable differences within a marriage (as opposed to one spouse being legally at fault for the marital breakup).

individualism The tendency to focus on one's individual self-interests and personal happiness rather than on the interests of one's family and community.

familism The view that the family unit is more important than individual interests.

anxiety, and poorer self-concepts. Both divorced and never-married individuals are, on average, more distressed than married people because unmarried people are more likely than married people to have low social attachment, low emotional support, and increased economic hardship (Walker 2001). Some research suggests that divorce leads to higher levels of depressive symptoms for women, but not for men (Kalmijn & Monden 2006), especially when there are young children in the family (Williams & Dunne-Bryant 2006). This finding is probably due to the increased financial and parenting strains experienced by divorced mothers who have custody of young children.

However, Amato (2003) cited studies in which divorced individuals report higher levels of autonomy and personal growth than married individuals do. For example, many divorced mothers report improvements in career opportunities, social lives, and happiness after divorce; some divorced women report more self-confidence, and some men report more interpersonal skills and a greater willingness to self-disclose. Some studies showed that for people in a poor-quality marriage, divorce has a less negative or even a positive effect on well-being (Amato 2003, Kalmijn & Monden 2006). However, leaving a bad marriage does not always result in increased well-being. For example, leaving a marriage characterized by high levels of physical or verbal aggression "may not increase well-being because the divorce is a trigger for even more problems after the divorce" (Kalmijn & Monden 2006, p. 1210). In sum, some men and women experience a decline in well-being after divorce; others experience an improvement.

Economic Consequences. Compared with married individuals, divorced individuals have a lower standard of living, have less wealth, and experience greater economic hardship, although this difference is considerably greater for women than for men (Amato 2003). In general, the economic costs of divorce are greater for women and children because women tend to earn less than men (see Chapter 10) and because mothers devote substantially more time to household and childcare tasks than fathers do. The time women invest in this unpaid labor restricts their educational and job opportunities as well as their income. Men are less likely than women to be economically disadvantaged after divorce because they continue to profit from earlier investments in education and career.

After divorce, both parents are responsible for providing economic resources to their children. However, some nonresident parents fail to provide child support. In some cases failure to pay child support is not due to fathers being "deadbeats" but rather to the fact that many fathers are "dead broke." About one-third of nonresident fathers in 1999 lived in households with incomes below the poverty line, or their personal income was below the poverty level for a single person (Sorensen & Oliver 2002). Not surprisingly, poor fathers are less likely to pay child support. In 1999, 70 percent of poor fathers and 28 percent of nonpoor fathers failed to pay child support (Sorensen & Oliver 2002). More than one-quarter of poor fathers who paid child support in 1999 spent half or more of their personal income on child support.

Effects on Children and Young Adults. Parental divorce is a stressful event for children and is often accompanied by a variety of stressors, such as continuing conflict between parents, a decline in the standard of living, moving and perhaps changing schools, separation from the noncustodial parent (usually the father), and parental remarriage. These stressors place children of divorce at higher risk for a variety of emotional and behavioral problems. Reviews of research on the consequences of divorce for children have found that children with divorced parents score lower on measures of academic success, psychological adjustment, self-concept, social competence, and long-term health; they also have higher levels of aggressive behavior and depression (Amato 2003; Wallerstein 2003).

Many of the negative effects of divorce on children are related to the economic hardship associated with divorce. Economic hardship is associated with less effective and less supportive parenting, inconsistent and harsh discipline, and emotional distress in children (Demo, Fine, & Ganong 2000).

One study found that divorce in one generation has adverse effects not only on that generation's children but also on future grandchildren who are not yet born (Amato & Cheadle 2005). Divorce in the first generation was associated with lower education, more marital discord, more divorce, and greater parent-child tensions in the second generation, which contributed to lower education, more marital discord, and weaker ties with parents in the third generation.

Despite the adverse effects of divorce on children, current research findings suggested that "most children from divorced families are resilient, that is, they do not suffer from serious psychological problems" (Emery, Sbarra, & Grover 2005, p. 24). Other researchers who studied the effects of divorce on children concluded that "most offspring with divorced parents develop into well-adjusted adults," despite the pain they feel associated with the divorce (Amato & Cheadle 2005, p. 191).

Divorce can also have positive consequences for children and young adults. In highly conflictual marriages divorce may actually improve the emotional well-being of children relative to staying in a conflicted home environment (Jekielek 1998). In interviews with 173 grown children whose parents divorced years earlier, Ahrons (2004) found that most of the young adults reported positive outcomes for their parents as well as for themselves. More than half of the young adults in this study reported that their relationships with their fathers actually improved after the divorce. In this chapter's *The Human Side* feature, one of the interviewees in Ahrons' study describes her experience of her parents' divorce.

Most divorced individuals remarry, which necessitates (for children) adaptation to new parents and step-siblings. Research confirms that of children growing up in stepfamilies, "80 percent . . . are doing well. Children in stable stepfamilies look very much like those raised in stable first families" (Pasley 2000, p. 6).

Effects on Father–Child Relationships. Children who live with their mothers may have a damaged relationship with their nonresidential father, especially if he becomes disengaged from their lives. Although some research has found that young adults whose parents divorced are less likely to report having a close relationship with their father compared with children whose parents are together (DeCuzzi, Knox, & Zusman 2004), in a study of 173 adult children of divorce, more than half felt that their relationships with their fathers improved after the divorce (Ahrons 2004). Children may benefit from having more quality time with their fathers after parental divorce. Some fathers report that they became more active in the role of father after divorce. One father commented:

> Since my divorce I have been able to take my children on camping trips alone. We have spent weeks in the wilderness talking, cooperating, sharing in ways that would never have been possible on a "family" trip. I am sad for the divorce but the bonding with my children has been an unforeseen advantage. (Author's files)

❝ Divorce is definitely not a single event but a long-lasting process of radically changing family relationships that begins in the failing marriage, continues through the often chaotic period of the marital rupture and its immediate aftermath, and extends even further, often over many years of disequilibrium. **❞**

Judith S. Wallerstein
Divorce researcher and scholar

Dr. Constance Ahrons conducted interviews with 173 grown children whose parents divorced years earlier. These interviews became the basis for her book We're Still Family: What Grown Children Have to Say about Their Parents' Divorce *(2004). This* Human Side *feature contains an excerpt from this book in which one of the interviewees talks about her parents' divorce.*

Sure, I would have liked to have had that perfect family that's on the cover of every magazine at Christmas. None of my friends had this perfect family but it's the one that every kid imagines the most popular kid at school has. I was only seven when my parents separated and I don't really remember much about what it was like when we all lived together, but I remember feeling sad and confused when they told me.

My parents were really young when they got married . . . it's hard for me to even imagine them together. I think the divorce was a good decision, a necessary one, and I think we're all better off because of it today. I'm pretty lucky because my mom and dad told me that no matter what happened between them they both still loved me. . . . I always knew I was very valued. In some ways I think the divorce made both my parents really emphasize how much they cared about me. Some friends of mine with married parents didn't know where they stood in terms of their parents' affection or felt neglected or had pretty bad living situations.

I'm not saying it was always easy. I remember times when my parents disagreed about some decision that involved me and I felt caught in the middle. Sometimes I felt angry about the scheduling and going back and forth. I remember feeling really jealous when my mom told me her boyfriend Dan was moving in. I was surprised when my dad told me he was getting remarried and I really resented my stepmom and her kids. Now I'm really close with my stepmom and I think she makes a much better mate for my dad than my mom did. I'm also close with my "stepdad," even though he and my mom never married and he's now married to someone else. It's confusing to explain all the relationships, and I used to be embarrassed about it, but now I feel lucky to have four parents. They were all there at my college graduation and I think it's widened my view of what I think a family is . . . it's helped me to communicate better and more freely with people who are important to me.

Source: Ahrons, Constance, 2004. *We're Still Family: What Grown Children Have to Say about Their Parents' Divorce* (pp. 23–34). New York: HarperCollins.

The mother's attitude toward the father's continued contact with the child can have a dramatic effect on the father-child relationship. "If the mother approves of the close contact between her child and his/her father, the child will benefit both from the continued affection of the father and from the parental harmony. If the mother disapproves of the father's influence, the child, feeling torn by conflicting loyalties, may fail to benefit" (Wallerstein 2003, pp. 76–77).

Some divorced mothers not only fail to encourage their children's relationships with their fathers but also actively attempt to alienate the children from their father. (We note that some divorced fathers do likewise.) Thus some children of divorce suffer from **parental alienation syndrome (PAS)**, defined as an emotional and psychological disturbance in which children engage in exaggerated and unjustified denigration and criticism of a parent. "Children of PAS show negative parental reactions and perceptions which can be grossly exaggerated. . . . Put simply, they profess rejection and hatred of a previously loved parent, most often in the context of divorce and child custody conflicts" (Family Court Reform Council of America 2000). Parental alienation syndrome has been described as a form of "psychological kidnapping," whereby one parent manipulates children's psyches to make them hate and reject the other parent. Children who suffer from PAS are victims of a form of child abuse in which one parent essentially brainwashes the child to hate the other parent. A parent may alienate his or her child from the other parent by engaging in the following behaviors (Schacht 2000):

parental alienation syndrome (PAS) An emotional and psychological disturbance in which children engage in exaggerated and unjustified denigration and criticism of a parent.

- Minimizing the importance of contact and relationship with the other parent.
- Being rude to the other parent; refusing to speak to or tolerate the presence of the other parent, even at events important to the child; refusing to allow the other parent near the home for drop-off or pick-up visitations.
- Failing to express concern for missed visits with the other parent.
- Failing to display any positive interest in the child's activities or experiences during visits.
- Expressing disapproval or dislike of the child's spending time with the other parent and refusing to discuss anything about the other parent ("I don't want to hear about . . .") or selective willingness to discuss only negative matters.
- Making innuendos and accusations against the other parent, including statements that are false.
- Demanding that the child keep secrets from the other parent.
- Destruction of gifts or memorabilia from the other parent.
- Promoting loyalty conflicts (e.g., offering an opportunity for a desired activity that conflicts with scheduled visitation).

Long-term effects of PAS on children are extremely serious and can include long-term depression, inability to function, guilt, hostility, alcoholism and other drug abuse, and other symptoms of internal distress (Family Court Reform Council of America 2000). Indeed, the effects on the rejected parent are equally devastating.

Some noncustodial divorced fathers discontinue contact with their children as a coping strategy for managing emotional pain (Pasley & Minton 2001). Many divorced fathers are overwhelmed with feelings of failure, guilt, anger, and sadness over the separation from their children (Knox 1998). Hewlett and West (1998) explained that "visiting their children only serves to remind these men of their painful loss, and they respond to this feeling by withdrawing completely" (p. 69). Divorced fathers commonly experience the legal system as favoring the mother in child-related matters. One divorced father commented:

> I believe that the system [judges, attorneys, etc.] have [sic] little or no consideration for the father. At some point the system creates an environment where the father loses any natural desire to see his children because it becomes so difficult, both financially and emotionally. At that point, he convinces himself that the best thing to do is wait until they are older. (quoted by Pasley & Minton 2001, p. 242)

As we have seen, the effects of divorce on adults and children are mixed and variable. In a review of research on the consequences of divorce for children and adults, Amato (2003) concluded that "divorce benefits some individuals, leads others to experience temporary decrements in well-being that improve over time, and forces others on a downward cycle from which they might never fully recover" (p. 206).

STRATEGIES FOR ACTION: STRENGTHENING MARRIAGE AND ALLEVIATING PROBLEMS OF DIVORCE

Two general strategies for responding to the problems of divorce are those that prevent divorce by strengthening marriages and those that strengthen post-divorce families.

"A successful marriage requires falling in love many times, always with the same person."

Mignon McLaughlin

Strategies to Strengthen Marriage and Prevent Divorce

A growing "marriage movement" involves efforts by some religious leaders, policy makers, therapists, and educators to strengthen marriage and prevent divorce through a number of strategies. These efforts include premarital and marriage education, covenant marriage and divorce law reform, and provision of workplace and economic supports. Amato et al. (2007) explained,

> Policies to strengthen marital quality and stability are based on consistent evidence that happy and stable marriages promote the health, psychological well-being, and financial security of adults . . . as well as children. . . . Moreover, recent research suggests that a large proportion of marriages that end in divorce are not deeply troubled, and that many of these marriages might be salvaged if spouses sought assistance for relationship problems . . . and stayed the course through difficult times. (pp. 245–246)

Marriage Education. Marriage education, also known as family life education, includes various types of workshops and classes that (1) teach relationship skills, communication, and problem solving; (2) convey the idea that sustaining healthy marriages requires effort; and (3) convey the importance of having realistic expectations of marriage, commitment, and a willingness to make personal sacrifices (Hawkins et al. 2004). An alternative or supplement to face-to-face family life education is web-based education, such as the Forever Families website, a faith-based family education website (Steimle & Duncan 2004).

The Federal Healthy Marriage Initiative provides federal funds to support research and programs to encourage healthy marriages and promote involved and responsible fatherhood. Funds may be used for a variety of activities, including marriage and premarital education, public advertising campaigns that promote healthy marriage, high school programs on the value of marriage, marriage mentoring programs, and parenting skills programs. Some states have passed or considered legislation that requires marriage education in high schools or that provides incentives (such as marriage license fee reductions) to couples who complete a marriage education program.

Marriage education is also promoted by faith-based groups. The Community Marriage Policy strategy asks religious officials (who perform 75 percent of all U.S. weddings) to follow the Common Marriage Policy of the American Roman Catholic Church, which includes the following five components (Browning 2003): (1) a 6-month minimum marriage preparation period; (2) the use of a premarital questionnaire to identify problems or potential problems the couple may have; (3) the practice of mentoring engaged and newlywed couples; (4) the use of marriage education for engaged and married couples for the purpose of exploring the relationship, identifying problems, and learning effective communication and conflict resolution techniques; and (5) engagement ceremonies held before the entire congregation (Browning 2003).

Covenant Marriage and Divorce Law Reform. With the passing of the 1996 Covenant Marriage Act, Louisiana became the first state to offer two types of marriage contracts: (1) the standard marriage contract that allows a no-fault divorce (after a 6-month separation) or (2) a **covenant marriage,** which permits divorce only under condition of fault (e.g., abuse, adultery, or felony conviction) or after a 2-year separation. Couples who choose a covenant marriage must also get premarital counseling. Variations of the covenant marriage have also been adopted in Arizona and Arkansas. Only 3 percent of couples in states

covenant marriage A type of marriage (offered in a few states) that requires premarital counseling and that permits divorce only under condition of fault or after a marital separation of more than 2 years.

with covenant marriage laws have chosen the covenant marriage option (Coontz 2005b).

Advocates of the covenant marriage believe that such marriages will strengthen marriages and decrease divorce. However, critics argue that covenant marriage may increase family problems by making it more emotionally and financially difficult to terminate a problematic marriage and by prolonging the exposure of children to parental conflict (Applewhite 2003).

Several states are considering divorce reform legislation that is intended to decrease the number of divorces by making divorce harder to obtain by extending the waiting period required before a divorce is granted or requiring proof of fault (e.g., adultery or abuse). Opponents argue that divorce law reform measures would increase acrimony between divorcing spouses (which harms the children as well as the adults involved), increase the legal costs of getting a divorce (which leaves less money to support any children), and delay court decisions on child support and custody and distribution of assets.

Workplace and Economic Supports. The most important pro-marriage and divorce-prevention measures may be those that maximize employment and earnings. Given that research finds a link between financial hardship and marital quality, policies to strengthen marriage should include a focus on the economic well-being of poor and near-poor couples and families (Amato et al. 2007). "Policy makers should recognize that any initiative that improves the financial security and well being of married couples is a pro-marriage policy" (Amato et al. 2007, p. 256). Supports such as job training, employment assistance, flexible workplace policies that decrease work-family conflict, affordable child care, and economic support, such as the earned income tax credit are discussed in Chapters 6 and 7. In addition, policy makers should take a hard look at policies that penalize poor couples for marrying. Poor couples who marry are often penalized by losing Medicaid benefits, food stamps, and other forms of assistance.

Strategies to Strengthen Families During and After Divorce

When one or both marriage partners decide to divorce, what can the couple do to achieve a "friendly divorce" and minimize the negative consequences of divorce for their children? According to Ahrons (2004), it is the post-divorce conflict between parents and not the divorce itself that is traumatic for children. A review of the literature on the effects of parental conflict on children suggests that children who are exposed to high levels of parental conflict are at risk for anxiety, depression, and disruptive behavior; they are more likely to be abusive toward romantic partners in adolescence and adulthood and are likely to have higher rates of divorce and maladjustment in adulthood (Grych 2005). Two strategies that help to reduce parental conflict and promote cooperative parenting after divorce are divorce mediation and divorce education programs.

divorce mediation A process in which divorcing couples meet with a neutral third party (mediator) who assists the individuals in resolving issues such as property division, child custody, child support, and spousal support in a way that minimizes conflict and encourages cooperation.

Divorce Mediation. In **divorce mediation** divorcing couples meet with a neutral third party, a mediator, who helps them resolve issues of property division, child support, child custody, and spousal support (i.e., alimony) in a way that minimizes conflict and encourages cooperation. In a longitudinal study researchers compared two groups of divorcing parents who were petitioning for a court custody hearing: parents who were randomly assigned to try mediation and those who were randomly assigned to continue the adversarial court process (Emery et al. 2005). If mediation did not work, the parents in the mediation group could still go to court to resolve their case. The study found that the parents who participated in mediation were much more likely than the parents who did not to settle their custody dispute outside court. The researchers also found that mediation can speed settlement, save money, and increase compliance. Most important, mediation led to improved relationships between nonresidential parents and children as well as between divorced parents 12 years after the dispute settlement. An increasing number of jurisdictions and states have mandatory child custody mediation programs, whereby parents in a custody or visitation dispute must attempt to resolve their dispute through mediation before a court will hear the case.

Divorce Education Programs. Another trend aimed to strengthen post-divorce families is the establishment of divorce education programs that emphasize the importance of cooperative co-parenting for the well-being of children. Parents are taught about children's reactions to divorce, learn nonconflictual co-parenting skills, and learn how to avoid negative behavior toward their ex-spouse. In some programs children are taught that they are not the cause of the divorce, learn how to deal with grief reactions to divorce, and learn techniques for talking to parents about their concerns. Grych (2005) reports that court-connected programs for divorcing couples are available in nearly half the counties in the United States.

What Do You Think? Many counties and some states (e.g., Arizona and Hawaii) require divorcing spouses to attend a divorce education program, whereas it is optional in other jurisdictions. Do you think that parents of minor children should be required to complete a divorce education program before they can get a divorce? Why or why not?

TEENAGE CHILDBEARING

In the 1950s, when most teen mothers were married and were expected to be stay-at-home wives, teenage childbearing was not a public concern. Teenage births today are considered problematic because most teenage births occur outside wedlock and because early parenthood interferes with the acquisition of education and job-related skills and is associated with negative outcomes for teen parents and their children (Mauldon 2003).

In recent years the U.S. teenage birth rate (per 1,000) dropped steadily from 60 in 1990 to 40 in 2005 (see Figure 5.9). This decline has been attributed to teens delaying first intercourse, increased use of contraception, and education about HIV and pregnancy prevention (As-Sanie, Gantt, & Rosenthal 2004).

Nevertheless, 1 million U.S. females between the ages of 15 and 19 become pregnant annually. About half these teens carry their babies to term, 35 percent have an abortion, and the remainder miscarry. Most teens today who carry their babies to term (95 percent) keep the baby rather than place it for adoption (Jorgensen 2000). Teenage birth rates are highest for Hispanic teens and lowest for Asian/Pacific Islander teens (see Figure 5.10).

> " For many disadvantaged teenagers, childbearing reflects—rather than causes—the limitations of their lives. "
>
> Ellen W. Freeman and Karl Rickels
> Researchers

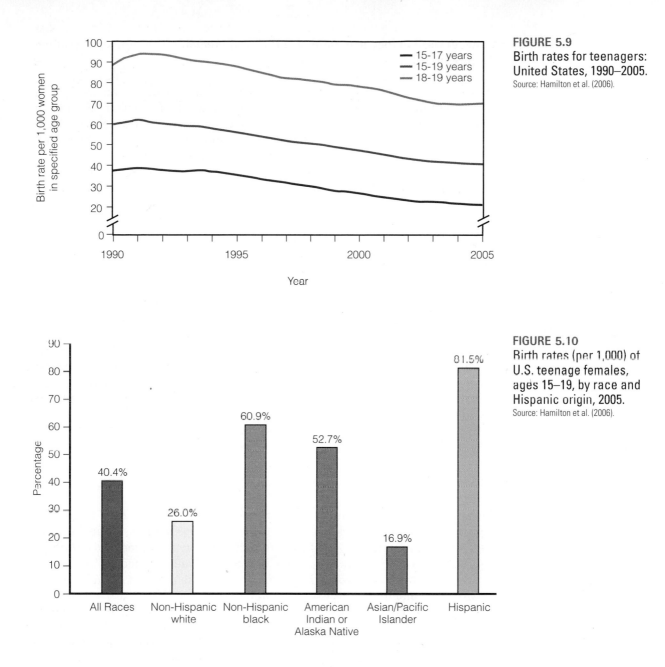

FIGURE 5.9
Birth rates for teenagers:
United States, 1990–2005.
Source: Hamilton et al. (2006).

FIGURE 5.10
Birth rates (per 1,000) of
U.S. teenage females,
ages 15–19, by race and
Hispanic origin, 2005.
Source: Hamilton et al. (2006).

The consequences of teenage childbearing—low educational achievement and poverty—are also contributing factors to teenage childbearing. In addition, teenage childbearing is associated with a higher risk of poor health outcomes for children.

Low Educational Achievement

Low academic achievement is both a contributing factor and a potential outcome of teenage parenthood. Teenage females who do poorly in school may have little hope of success and achievement in pursuing educational and occupational goals, and they may think that their only remaining option for a meaningful role in life is to become a parent.

Becoming a teen parent also tends to curtail future academic achievement. In one study, eighth-grade students who later became teenage parents had significantly lower test scores, were more likely to have had behavior problems in school, had lower educational aspirations for the future, and were more likely to have been held back at least one grade in school than their peers who did not become teenage parents (Mollborn 2007). When individuals who became parents in their teens were compared with those who did not, those who became parents in their teens had, at age 26, an average of 2 years less school (11.9 versus 13.9 years of school) (Mollborn 2007). In other words, at age 26, teen parents had, on average, a high school degree, whereas nonparents had attained 2 years of post-secondary education.

The children of teen parents are also at risk for low academic achievement. A study based on a large nationally representative sample of U.S. children found that children born to mothers ages 17 and younger began kindergarten with lower levels of school readiness—including lower math and reading scores, language and communication skills, social skills, and physical and emotional well-being—compared with children born to older mothers (Terry-Humen et al. 2005). When the researchers controlled for the mother's marital status and socioeconomic status, these effects were diminished but still important. However, other researchers who studied the effects of teen parenting on children's well-being found that "teen parenting had little or no effect on children's performance on standardized academic tests; correlations between early fertility and children's test scores reflect individual and family background factors rather than the causal influence of early childbearing itself" (Levine, Emery, & Pollack 2007, p. 116).

Poverty

Teens from low socioeconomic backgrounds are more likely than those from higher socioeconomic backgrounds to become teenage parents. Although many teen mothers experience poverty before becoming pregnant, teenage childbearing tends to exacerbate the problems that these disadvantaged young women already face. Most teenagers who become parents are not married and have no means of economic support or have limited earning capacity. Teen mothers often rely on the support of their own parents or rely on public assistance. Having a child at a young age makes it difficult for the teen parent to attain academic credentials, compete in the job market, and break out of what for many is a cycle of poverty that includes their children. One analysis found that the decline in teen birth rates over the 1990s accounts for 26 percent of the overall decline in the number of young children (younger than age 6) in poverty between 1995 and 2002 (National Campaign to Prevent Teen Pregnancy 2005).

Poor Health Outcomes

Compared with older pregnant women, pregnant teenagers are less likely to receive timely prenatal care and to gain adequate weight. Teens are also more likely to smoke and use alcohol and drugs during pregnancy (Jorgensen 2000; Ventura, Curtin, & Matthews 2000). As a consequence of these and other factors, infants born to teenagers are at higher risk of low birth weight, of premature birth, and of dying in the first year of life.

STRATEGIES FOR ACTION: INTERVENTIONS IN TEENAGE CHILDBEARING

Interventions in teenage childbearing include efforts to prevent teenage pregnancies through sex education and access to contraceptive services and to provide various types of support to teenage parents and their children.

Sexuality Education and Access to Contraceptive Services

Despite the decline in U.S. teen pregnancy rates over the past few decades, the United States has one of the highest teen pregnancy rates in the developed world—almost twice as high as those of England, Wales, and Canada, and eight times as high as those of the Netherlands and Japan (Guttmacher Institute 2006). In Northern and Western Europe low teen pregnancy and birth rates are attributed to the widespread availability and use of effective contraception among sexually active teens (Singh & Darroch 2000).

In the United States, there has been more emphasis in recent years on providing abstinence-based sex education than on providing teens with access to contraceptive services. More than $80 million in federal and state dollars are spent on abstinence education annually (Trenholm et al. 2007). To receive federal funding for abstinence sex education, states may not use the funds to promote condom or contraceptive use, and teachers are required to teach ideas such as bearing children outside of wedlock is harmful to society and "likely to have harmful psychological and physical effects" (Huffstutter 2007).

Supporters of abstinence-only programs believe that promoting condoms sends the "wrong message" that sex outside of marriage is OK. Critics of abstinence-only sex education argue that abstinence-only programs do not protect the 46 percent of all 15- to 19-year-olds in the United States who have had sex at least once against pregnancy or sexually transmissible disease (Guttmacher Institute 2006). A national study of U.S. youth in grades 9–12 found that among the sexually active students, only 63 percent reported that either they or their partner had used a condom during their last sexual intercourse (Centers for Disease Control and Prevention 2006). Nearly one-fifth (17 percent) of sexually active females ages 15–19 and 9 percent of males in the same age

The East Los Angeles Men's Health Center, a project of Bienvenidos Children's Center, Inc. http://www.bienvenidos.org

This brochure, distributed by the National Latino Fatherhood and Family Initiative, is a teen pregnancy prevention effort targeting young Latino men.

group said that they used no method of contraception the last time they had sex (Kaiser Family Foundation 2005).

Several states, including Wisconsin, Ohio, Connecticut, Rhode Island, Montana, and New Jersey, have chosen to forgo federal sex education funding because of federal dictates that the money be used exclusively for teaching chastity. Critics of abstinence-only programs often cite a report that found that more than 80 percent of the abstinence-only curricula contain false, misleading, or distorted information about reproductive health (Waxman 2004). For example, the report found that "many of the curricula misrepresent the effectiveness of condoms in preventing sexually transmitted diseases and pregnancy" (Waxman 2004, p. i). Critics also point to a scientific evaluation of abstinence education programs commissioned by the Congress that found that youth who received abstinence education were no more likely than control group youth to have abstained from sex and, among those who reported having had sex, they had similar numbers of sexual partners and had initiated sex at the same mean age (Trenholm et al. 2007). Contrary to concerns raised by some critics of abstinence education, youth who received abstinence education were no more likely to have engaged in unprotected sex than control group youth.

According to the Sex Information and Education Council of the United States, most parents want schools to provide **comprehensive sexuality education** that includes topics such as contraception, sexually transmitted diseases, HIV/AIDS, and disease-prevention methods, as well as the benefits of abstinence (SIECUS 2004). Yet comprehensive sex education is not universally taught. As of Spring 2007, 19 states and the District of Columbia mandate that public schools teach sex education (Guttmacher Institute 2007a). Many states, including some that do not mandate sex education, have requirements on how abstinence and contraception are treated when taught:

- 22 states require that abstinence be stressed when taught as part of sex education; 10 states require simply that it be covered during instruction.
- 14 states and the District of Columbia require that sex education programs cover contraception; no state requires that it be stressed.

Most sexually active teens in the United States cannot obtain contraceptive services in their schools and find significant barriers to obtaining contraception elsewhere. Although 21 states and the District of Columbia allow minors to consent to contraceptive services without a parent's involvement, 25 states allow minors access to such services only under certain conditions (e.g., if the minor is married, has had a previous pregnancy, or is referred by a physician or clergy) (Guttmacher Institute 2007b). Four states have no specific policy on minors' access to contraceptive services.

Proposed legislation that would require parental notification for minors obtaining prescription contraception from federally funded family planning clinics further threatens minors' access to contraception. In a survey of teenage females, one in five teens said that they would use no contraception or rely on withdrawal if parental notification was required (Jones et al. 2005).

Teens younger than age 18 are also restricted from over-the-counter access to Plan B, the only Food and Drug Administration (FDA)–approved product for emergency contraception. This type of contraception is used to prevent pregnancy after unprotected intercourse occurs and must be taken within 72 hours of unprotected intercourse. In 2006, the FDA approved Plan B as an over-the-

comprehensive sexuality education Sex education that includes topics such as contraception, sexually transmitted diseases, HIV/AIDS, and disease-prevention methods as well as the benefits of abstinence.

counter medication for those 18 and older; however, it remains a prescription drug for minors (Guttmacher Institute 2007c).

> **What Do You Think?** Another barrier to access to contraception for all U.S. women is the refusal of some pharmacists to fill prescriptions for birth control. Four states—Arkansas, Georgia, Mississippi, and South Dakota—allow pharmacists to refuse to dispense contraceptives (Guttmacher Institute 2007c). Do you think pharmacists should have the legal right to refuse to fill a prescription for products such as birth control pills or Plan B emergency contraception?

Computerized Infant Simulators

Some teen pregnancy prevention programs use computerized infant simulators to give adolescents a realistic view of parenting. Computerized infant simulators are realistic, life-sized computerized "dolls" that are programmed to cry at random intervals (typically between 8 and 12 times in 24 hours) with crying periods lasting typically between 10 and 15 minutes. The "baby" stops crying only when the caregiver "attends" to the doll by inserting a key into a slot in the infant simulator's back until it stops crying. The infant simulator records data, including the amount of time the caregiver took to attend to the infant (insert the key) and any instances of "rough handling," such as dropping, hitting, or shaking the doll. Participants who are found to neglect or handle the doll roughly may receive a private counseling session and may be required to take a parenting class.

An evaluation of a computerized infant simulator program with adolescents found that the program was effective in changing perceptions of the time and

© AP Photo/Independent Record, Mark Goldstein

Computerized infant simulators, such as the one pictured here, are used in parenting education as well as teenage pregnancy-prevention programs.

effort involved in caring for an infant and in recognizing the significant effect having a baby has on all aspects of one's life (de Anda 2006). Nearly two-thirds of the adolescent participants reported that the program helped change their minds about using birth control.

Resources and Assistance to Teenage Parents

Various types of resources and assistance to teenage parents can help them and their children overcome the disadvantages associated with this family form. For example, although teenage parents tend to fall behind in educational attainment, this need not be the case. "If they are provided with enough material resources, contemporary teenage parents may be able to go quite far in school, despite their initial socioeconomic and educational disadvantage" (Mollborn 2007, p. 102). Teenage parents benefit from assistance with child care, financial support, and housing.

Other programs that can assist teenage mothers and their children include prenatal programs to help ensure the health of the mother and baby and public welfare, such as WIC (the Special Supplemental Food Program for Women, Infants, and Children) and TANF (Temporary Assistance to Needy Families).

Increase Men's Involvement with Children

Strategies to increase and support fathers' involvement with their children are relevant to both children of teen mothers and children of divorce. At the federal and state levels fatherhood initiative programs encourage fathers' involvement with children through a variety of means (U.S. Department of Health and Human Services 2000). These include promoting responsible fatherhood by improving work opportunities for low-income fathers, increasing child support collections, providing parent education training for men, supporting access and visitation by noncustodial parents, and involving boys and young men in teenage pregnancy prevention and early parenting programs. Because teenage parents are less likely than older parents to use positive and effective child-rearing techniques, parent education programs for teen mothers and fathers are an important component of improving the lives of young parents and their children.

UNDERSTANDING FAMILY PROBLEMS

Family problems can best be understood within the context of the society and culture in which they occur. Although domestic violence, divorce, and teenage pregnancy and parenthood may appear to result from individual decisions, these decisions are influenced by myriad social and cultural forces.

The impact of family problems, including divorce, abuse, and teenage childbearing, is felt not only by family members but also by society at large. Family members experience life difficulties such as poverty, school failure, low self-esteem, and mental and physical health problems. Each of these difficulties contributes to a cycle of family problems in the next generation. The impact on society includes public expenditures to assist single-parent families and victims of domestic violence and neglect, increased rates of juvenile delinquency, and lower worker productivity.

For some the solution to family problems implies encouraging marriage and discouraging other family forms, such as single parenting, cohabitation, and same-sex unions. But many family scholars argue that the fundamental issue is making sure that children are well cared for, regardless of their parents' marital status or sexual orientation. Some even suggest that marriage is part of the problem, not part of the solution. Martha Fineman of Cornell Law School said, "This obsession with marriage prevents us from looking at our social problems and addressing them. . . . Marriage is nothing more than a piece of paper, and yet we rely on marriage to do a lot of work in this society: It becomes our family policy, our police in regard to welfare and children, the cure for poverty" (quoted by Lewin 2000, p. 2).

Strengthening marriage is a worthy goal because strong marriages offer many benefits to individuals and their children. However, "strengthening marriage does not have to mean a return to the patriarchal family of an earlier era. . . . Indeed, greater marital stability will only come about when men are willing to share power, as well as housework and child care, equally with women" (Amato 1999, p. 184). And strengthening marriage does not mean that other family forms should not also be supported. In their book *Joined at the Heart,* Al and Tipper Gore (2002) suggested that the first and most important step to helping families is to change our way of thinking about families so that our view of family encompasses those who are connected emotionally and committed to one another as family—those who are "joined at the heart" (p. 327). The reality is that the postmodern family comes in many forms, each with its strengths, needs, and challenges. Given the diversity of families today, social historian Stephanie Coontz (2004) suggested that "the appropriate question. . . is not what single family form or marriage arrangement we would prefer in the abstract, but how we can help people in a wide array of different committed relationships minimize their shortcomings and maximize their solidarities" (p. 979). She further argued that

> If we withdrew our social acceptance of alternatives to marriage, marriage itself might suffer. . . . The same personal freedoms that allow people to expect more from their married lives also allow them to get more out of staying single and give them more choice than ever before in history about whether or not to remain together. (Coontz 2005a, p. 310)

Efforts to prevent teenage childbearing are aimed to protect both teen parents and their children from negative outcomes such as poverty and poor health. However, it is important to keep in mind that negative outcomes of teen childbearing are largely the result of preexisting social disadvantages of the teens who become parents. If we mistakenly assume that negative outcomes are caused by teen parenting by itself, we may neglect other teens with disadvantaged backgrounds who do not become teen parents, but who nevertheless suffer and whose children later suffer similar negative outcomes (Levine et al. 2007). The underlying causes of teen childbearing—poverty, economic inequality, and problems in education—need to be addressed for the well-being of *all* youth, not just teen parents. Finally, an important aspect of preventing teen childbearing is to prevent teen pregnancy. Research, as well as the experience of many European countries, suggests that providing teens with contraceptives is effective in reducing teen pregnancy. As long as U.S. sex education policy encourages abstinence-only sex education, without providing teens with access to contraceptive services, U.S. teenage pregnancy rates will continue to be higher than need be.

The three family problems emphasized in this chapter—domestic violence and abuse, problems of divorce, and teenage parenthood—have something in common; economic hardship and poverty can be a contributing factor and a consequence of each of these problems. In the next chapter, we turn our attention to poverty and economic inequality—problems that are at the heart of many other social ills.

CHAPTER REVIEW

- **What are some examples of diversity in families around the world?**

Some societies recognize monogamy as the only legal form of marriage, whereas other societies permit polygamy. Societies also vary in their policies regarding same-sex couples and their norms regarding the roles of women, men, and children in the family.

- **What are some of the major changes in U.S. households and families that have occurred in the past several decades?**

Some of the major changes in U.S. households and families that have occurred in recent decades include increased singlehood and older age at first marriage, delayed childbearing, increased heterosexual and same-sex cohabitation, more interracial/interethnic unions, the emergence of living apart together (LAT) relationships, increased births to unmarried women, increased single-parent families, fewer children living in married-couple families, increased divorce and blended families, and increased employment of married mothers. According to the marital decline perspective, the recent transformations in American families signify a collapse of marriage and family in the United States. According to the marital resiliency perspective, poverty, unemployment, poorly funded schools, discrimination, and the lack of basic services (such as health insurance and child care) are more harmful to the well-being of children and adults than is the decline in married two-parent families.

- **Feminist theories of family are most similar to which of the three main sociological theories: structural functionalism, conflict theory, or symbolic interactionism?**

Feminist theories of family are most aligned with conflict theory. Both feminist and conflict theories are concerned with how gender inequality influences and results from family patterns.

- **What are the four patterns of partner violence identified by Johnson and Ferraro (2003)?**

The four patterns of partner violence are (1) common couple violence (occasional acts of violence arising from arguments that get "out of hand"); (2) intimate terrorism (violence that is motivated by a wish to control one's partner); (3) violent resistance (acts of violence that are committed in self-defense); and (4) mutual violent control (both partners battling for control).

- **What are the differences between primary prevention, secondary prevention, and tertiary prevention strategies with regard to preventing and responding to domestic violence and abuse?**

Primary prevention strategies target the general population, secondary prevention strategies target groups at high risk for family violence and abuse, and tertiary prevention strategies target families who have experienced abuse.

- **Why do many abused adults stay in abusive relationships?**

Adult victims of abuse are commonly blamed for choosing to stay in their abusive relationships. From the point of view of the victim, reasons to stay in the relationship include love, emotional dependency, commitment to the relationship, hope that things will get better, the view that violence is legitimate because they "deserve" it, guilt, fear, economic dependency, feeling stuck, and fear of loneliness. Some victims stay because they fear the abuser will abuse or neglect a pet.

- **What are some of the effects of divorce on children?**

Reviews of recent research on the consequences of divorce for children find that children with divorced parents score lower on measures of academic success, psychological adjustment, self-concept, social competence, and long-term health and that they have higher levels of aggressive behavior and depression. Such effects are related to the economic hardship associated with divorce, the reduced parental supervision resulting from divorce, and parental conflict during and after divorce. In highly conflictual marriages divorce may actually improve the emotional well-being of children relative to staying in a conflicted home environment.

- **What is divorce mediation?**

  In divorce mediation divorcing couples meet with a neutral third party, a mediator, who helps them resolve issues of property division, child custody, child support, and spousal support in a way that minimizes conflict and encourages cooperation. In some states, counties, and jurisdictions, divorcing couples who are disputing child custody issues are required to participate in divorce mediation before their case can be heard in court.

- **Why is teenage childbearing considered a social problem?**

  Teenage childbearing is considered a social problem because of the adverse consequences for teenage mothers and their children, including (1) increased risk of poverty for single mothers and their children, (2) risk of poor health outcomes for babies born to teenage women, and (3) risk of dropping out of school for teenage mothers and for low academic achievement of their children. Poverty and low educational achievement are also factors that contribute to teenage childbearing.

- **How does the European approach to teenage sexuality compare with the U.S. approach?**

  The European approach to teenage sexual activity involves providing widespread confidential and accessible contraceptive services to adolescents. Although sex education is provided in schools throughout the United States, most programs emphasize abstinence and do not provide students with access to contraception. Research suggests that comprehensive sexuality education that includes topics such as abstinence, sexually transmitted diseases, HIV/AIDS, contraception, and disease-prevention methods is more effective for preventing pregnancy, as well as disease.

TEST YOURSELF

1. The United States has the highest proportion of nonmarital births of any country in the world.
   a. True
   b. False
2. Feminists are critical of which of the following?
   a. Egalitarianism
   b. Symbolic interactionism
   c. Marital resiliency perspective
   d. Patriarchy
3. Two perspectives on the state of marriage in the United States are the marital decline perspective and the marital _____ perspective.
   a. health
   b. resiliency
   c. incline
   d. stability
4. In the United States, people are more likely to be physically assaulted, sexually assaulted and molested, or killed by _____ than by anyone else.
   a. a family member
   b. an employee
   c. a stranger
   d. a friend
5. Most adults who were abused as children continue the pattern of abuse in their own relationships.
   a. True
   b. False
6. Which of the following is the most prevalent form of abuse in families?
   a. Sexual abuse by a father
   b. Sexual abuse by an uncle or cousin
   c. Verbal abuse by a mother
   d. Sibling abuse
7. Most divorced individuals remarry.
   a. True
   b. False
8. Couples who choose a covenant marriage license in Louisiana
   a. can only get divorced if they have no children
   b. must pay a "marriage tax" of $2,000, which they get back, with interest, after they are married for 10 years
   c. must be separated for 2 years before they can get a divorce
   d. must pay a "divorce tax" of $2,000 if they file for divorce (in addition to court and legal fees)

9. Most teens today who give birth place their babies for adoption.
   a. True
   b. False

10. Research suggests that adolescents who are exposed to abstinence-only sex education are less likely to become sexually active in their teens than are adolescents who do not receive abstinence-only sex education.
    a. True
    b. False

Answers: 1 b. 2 d. 3 b. 4 a. 5 b. 6 d. 7 a. 8 c. 9 b. 10 b.

KEY TERMS

bigamy
child abuse
comprehensive sexuality education
corporal punishment
covenant marriage
cycle of abuse
divorce mediation
divorce rate
domestic partnership
egalitarian relationship
elder abuse
familism

family
family household
family preservation programs
individualism
intimate partner violence (IPV)
living apart together (LAT)
   relationships
marital decline perspective
marital resiliency perspective
monogamy
neglect
no-fault divorce

nonfamily household
parental alienation syndrome (PAS)
patriarchy
polyandry
polygamy
polygyny
primary prevention
second shift
secondary prevention
serial monogamy
shaken baby syndrome
tertiary prevention

MEDIA RESOURCES

Understanding Social Problems, Sixth Edition Companion Website

academic.cengage.com/sociology/mooney
Visit your book companion website, where you will find flash cards, practice quizzes, Internet links, and more to help you study.

CENGAGENOW™

Just what you need to know NOW! Spend time on what you need to master rather than on information you already have learned. Take a pre-test for this chapter, and CengageNOW will generate a personalized study plan based on your results. The study plan will identify the topics you need to review and direct you to online resources to help you master those topics. You can then take a post-test to help you determine the concepts you have mastered and what you will need to work on. Try it out! Go to academic.cengage.com/login to sign in with an access code or to purchase access to this product.

© Irving Olson, 2007

6

"We are the first genera-
tion that can look ex-
treme poverty in the eye,
and say this and mean
it—we have the cash,
we have the drugs, we
have the science. Do we
have the will to make
poverty history?"

Bono, U2 (rock music group)

Poverty and Economic Inequality

The Global Context: Poverty and Economic Inequality Around the World | Sociologi-
cal Theories of Poverty and Economic Inequality | Economic Inequality and Poverty
in the United States | Consequences of Poverty and Economic Inequality | Strategies
for Action: Antipoverty Programs, Policies, and Proposals | Understanding Poverty
and Economic Inequality | Chapter Review

Washington, DC, the capital of one of the wealthiest nations in the world, has one of the highest rates of poverty in the United States.

Not far from the Capitol Building, 60-year-old John Treece pondered his life in deep poverty as he left a local food pantry with two bags of free groceries. Plagued by arthritis and back problems from years of manual labor, Treece has been unable to find a full-time job for 15 years. He's tried to get Social Security disability benefits, but the Social Security Administration disputes his injuries and work history. Treece earns a little more than $5,000 a year doing odd jobs. His clothes are tattered and he lives in a $450-a-month room in a boarding house in a high-crime neighborhood. Treece does not go hungry, thanks to food stamps, the food pantry, and help from relatives. But items that require cash, such as toothpaste, soap, and toilet paper are harder to come by. "Sometimes it makes you want to do the wrong thing, you know," said Treece, referring to crime. "But I ain't a kid no more. I can't do no time. At this point, I ain't got a lotta years left." Despite his poor circumstances, Treece is positive and grateful for what he has. "I don't ask for nothing. . . . I just thank the Lord for this day and ask that tomorrow be just as blessed" (quoted in Pugh 2007, n.p.).

In this chapter we examine the extent of poverty globally and in the United States, focusing on the consequences of poverty for individuals, families, and societies. We present theories of poverty and economic inequality and consider strategies for rectifying economic inequality and poverty.

The Global Context: Poverty and Economic Inequality Around the World

Who are the poor? Are rates of world poverty increasing, decreasing, or remaining stable? The answers depend on how we define and measure poverty.

Defining and Measuring Poverty

Poverty has traditionally been defined as the lack of resources necessary for material well-being—most importantly food and water, but also housing, land, and health care. This lack of resources that leads to hunger and physical deprivation is known as **absolute poverty.** In contrast, **relative poverty** refers to a deficiency in material and economic resources compared with some other population. Although many lower-income Americans, for example, have resources and a level of material well-being that millions of people living in absolute poverty can only dream of, they are relatively poor compared with the American middle and upper classes.

absolute poverty The lack of resources that leads to hunger and physical deprivation.

relative poverty A deficiency in material and economic resources compared with some other population.

Various measures of poverty are used by governments, researchers, and organizations. Next, we describe international and U.S. measures of poverty.

International Measures of Poverty. The World Bank sets a "poverty threshold" of $1 per day to compare poverty in most of the developing world, $2 per day in Latin America, $4 per day in Eastern Europe and the Commonwealth of Inde-

TABLE 6.1 Measures of Human Poverty in Developing and Industrialized Countries

	LONGEVITY	KNOWLEDGE	DECENT STANDARD OF LIVING
For developing countries	Probability at birth of not surviving to age 40	Adult illiteracy	A composite measure based on 1. Percentage of people without access to safe water 2. Percentage of people without access to health services 3. Percentage of children younger than 5 who are underweight
For industrialized countries	Probability at birth of not surviving to age 60	Adult functional illiteracy rate	Percentage of people living below the income poverty line, which is set at 50% of median disposable income

Source: Adapted from UNDP (2000).

pendent States (CIS), and $14.40 per day in industrial countries (which corresponds to the income poverty line in the United States). Another poverty measure is based on whether individuals are experiencing hunger, which is defined as consuming less than 1,960 calories a day.

In industrial countries national poverty lines are sometimes based on the median household income of a country's population. According to this relative poverty measure, members of a household are considered poor if their household income is less than 50 percent of the median household income in that country.

Recent poverty research has concluded that poverty is multidimensional and includes dimensions such as food insecurity, poor housing, unemployment, psychological distress, powerlessness, hopelessness, vulnerability, and lack of access to health care, education, and transportation (Narayan 2000). To capture the multidimensional nature of poverty, the United Nations Development Programme developed a composite measure of poverty: the **human poverty index (HPI)** (UNDP 1997). Rather than measure poverty by income, three measures of deprivation are combined to yield the HPI: (1) deprivation of a long, healthy life, (2) deprivation of knowledge, and (3) deprivation in decent living standards. As shown in Table 6.1, the HPI for developing countries (HPI-1) is measured differently from the HPI for industrialized countries (HPI-2). Among the 18 industrialized countries for which the HPI-2 was calculated, Sweden has the lowest level of human poverty (6.5 percent), followed by Norway (7.0 percent) and the Netherlands (8.2 percent) (UNDP 2006). The industrialized countries with the highest rates of human poverty are Italy (29.9 percent), Ireland (17 percent), and the United States (16 percent).

U.S. Measures of Poverty. In 1964 the Social Security Administration devised a poverty index based on data that indicated that families spent about one-third of their income on food. The official poverty level was set by multiplying food costs by three. Since then, the poverty level has been updated annually for inflation but has otherwise remained unchanged. Poverty thresholds differ by the number of adults and children in a family and by the age of the family head of household, but is the same across the continental United States (see Table 6.2). Anyone living in a household with pretax income below the official poverty line is considered

TABLE 6.2 Poverty Thresholds: 2006 (Householder Younger Than 65 Years Old)

HOUSEHOLD MAKEUP	POVERTY THRESHOLD
One adult	$10,488
Two adults	$13,500
One adult, one child	$13,896
Two adults, one child	$16,227
Two adults, two children	$20,444

Source: U.S. Census Bureau (2007).

> "Human poverty is more than income poverty—it is the denial of choices and opportunities for living a tolerable life."
>
> UNDP
> *Human Development Report 1997*

human poverty index (HPI)
A measure of poverty based on measures of deprivation of a long, healthy life; deprivation of knowledge; and deprivation in decent living standards.

"poor." Individuals living in households with incomes that are above the poverty line, but not very much above it, are classified as "near poor," and those living in households with income below 50 percent of the poverty line live in "deep poverty," also referred to as "severe poverty". A common working definition of "low-income" households is households with incomes that are between 100 percent and 200 percent of the federal poverty line or up to twice the poverty level.

The U.S. poverty line has been criticized as underestimating the extent of material hardship in the United States. The poverty line is based on the assumption that low-income families spend one-third of their household income on food. That was true in the 1950s, but because other living costs (e.g., housing, medical care, child care, and transportation) have risen more rapidly than food costs, low-income families today spend far less than one-third of their income on food. In addition, the current poverty measure is a national standard that does not reflect the significant variation in the cost of living from state to state and between urban and rural areas.

Researchers at the National Center for Children in Poverty have determined that, across the country, families on average need a minimum income that is about twice the poverty line (roughly $40,000 for a family of four) (Cauthen & Fass 2007). In cities, the figure is higher ($50,000); in rural areas, the figure is lower ($30,000). When a 2007 Gallup poll asked the American public to estimate the minimum amount of yearly income a family of four would need "to get along in your local community," the average answer was $52,000 (rounded to the nearest thousand) (Jones 2007). This figure varied by region: $61,000 in the East, $46,000 in the Midwest, and $49,000 in the South. There was also variation between urban areas ($48,000), suburban areas ($58,000), and rural areas ($42,000).

Another shortcoming of the official poverty line measurement is that it is based on pretax income so tax burdens that affect the amount of disposable income available to meet basic needs are disregarded. On the other hand, it underestimates income for some families because it does not count the federal Earned Income Tax Credit many families receive. Family assets, such as savings and property, are also excluded in official poverty calculations, and noncash government benefits that assist low-income families—food stamps, Medicaid, and housing and child care assistance—are not taken into account.

The Extent of Global Poverty and Economic Inequality

Globally, 2.5 billion people—more than one-fourth of the world's population—live on less than $2 a day and about 1 billion people—1 in 6 people on this planet—live on less than $1 a day (World Bank 2007). In sub-Saharan Africa nearly half the population lives on less than $1 a day (UNDP 2006). Every day, nearly 1 in 5 (18 percent) of the world's population goes hungry. In South Asia 1 in 4 goes hungry, and in sub-Saharan Africa as many as 1 in 3 goes hungry (UNDP 2003).

Global economic inequality has reached unprecedented levels. The most comprehensive study on the world distribution of household wealth offers the following facts on wealth inequality worldwide (Davies et al. 2006):

- The richest 1 percent of adults in the world own 40 percent of global household wealth; the richest 2 percent of adults own more than half of global wealth; and the richest 10 percent of adults own 85 percent of total global wealth.

- The poorest half of the world adult population owns barely 1 percent of global wealth.
- Households with per adult assets of $2,200 are in the top half of the world wealth distribution; assets of $61,000 per adult places a household in the top 10 percent, and assets of more than $500,000 per adult places a household in the richest 1 percent worldwide.
- Although North America has only 6 percent of the world adult population, it accounts for one-third (34 percent) of all household wealth worldwide. More than one-third (37 percent) of the richest 1 percent of individuals in the world reside in the United States.
- The degree of wealth inequality in the world is as if 1 person in a group of 10 takes 99 percent of the total pie and the other 9 people in the group share the remaining 1 percent.

In 2000 the average income of the richest 20 countries was 37 times that of the poorest 20 countries—a gap that doubled in the past 40 years (World Bank 2001). From 1960 to 2002 income per person in the world's poorest countries rose only slightly, from $212 to $267, whereas income in the richest 20 nations tripled from $11,417 to $32,339 (Schifferes 2004). Although inequality between nations accounts for most of the inequality in global distribution of income, within-nation income differences are growing (Goesling 2001).

Among the developed countries of the world, the United States has the greatest degree of income inequality and the highest rate of poverty. Among the 19 member countries of the Organisation for Economic Development (OECD), the United States had the second highest per capita income in 2004 (Norway had the highest), but it also had the highest poverty rate and child poverty rate. But income inequality is much greater in poor countries than it is in the United States. The income share of the top 10 percent of Americans is 30 percent of national income—the highest share among developed nations. But in several undeveloped countries in Asia, Latin America, and Africa, the top 10 percent of income earners receive more than 40 percent of national income (Sanderson & Alderson 2005).

SOCIOLOGICAL THEORIES OF POVERTY AND ECONOMIC INEQUALITY

The three main theoretical perspectives in sociology—structural functionalism, conflict theory, and symbolic interactionism—offer insights into the nature, causes, and consequences of poverty and economic inequality.

Structural-Functionalist Perspective

According to the structural-functionalist perspective, poverty results from institutional breakdown: economic institutions that fail to provide sufficient jobs and pay, educational institutions that fail to provide adequate education in low-income school districts, family institutions that do not provide two parents, and government institutions that do not provide sufficient public support. These institutional breakdowns create a "culture of poverty" whereby, over time, the poor develop norms, values, beliefs, and self-concepts that contribute to their own plight. According to Lewis, the **culture of poverty** is characterized by female-centered households, an emphasis on gratification in the present rather than in the future, and a relative lack of participation in society's major institutions (Lewis 1966). "The people in the culture of poverty have a strong feeling of marginality, of helplessness, of dependency, of not belonging. . . . Along with this feeling of powerlessness is a widespread feeling of inferiority, of personal unworthiness" (Lewis 1998, p. 7). Early sexual activity, unmarried parenthood, joblessness, reliance on public assistance, illegitimate income-producing activities (e.g., selling drugs), and substance abuse are common among the **underclass**—people living in persistent poverty. The culture of poverty view emphasizes that the behaviors, values, and attitudes exhibited by the chronically poor are transmitted from one generation to the next, perpetuating the cycle of poverty. Behaviors, values, and attitudes of the underclass emerge from the constraints and blocked opportunities that have resulted largely from the failure of the economic institution to provide employment, as jobs moved out of inner-city areas to the suburbs (Wilson 1996; Jargowsky 1997; Van Kempen 1997).

> Where jobs are scarce . . . and where there is a disruptive or degraded school life purporting to prepare youngsters for eventual participation in the workforce, many people eventually lose their feeling of connectedness to work in the formal economy; they no longer expect work to be a regular, and regulating, force in their lives. . . . These circumstances also increase the likelihood that the residents will rely on illegitimate sources of income, thereby further weakening their attachment to the legitimate labor market. (Wilson 1996, pp. 52–53)

From the late 1960s through the 1980s, poverty became more and more concentrated in inner-city neighborhoods, and conditions in those neighborhoods steadily deteriorated. But during the economic boom of the late 1990s, when unemployment was low, poverty became less concentrated (Kingsley & Pettit 2003). This decrease in the concentration of poverty during strong economic times supports the view that economic conditions underlie the culture of poverty.

From a structural-functionalist perspective economic inequality within a society can be beneficial for society. A system of unequal pay, argued Davis and Moore (1945), motivates people to achieve higher levels of training and education and to take on jobs that are more important and difficult by offering higher rewards for higher achievements. If physicians were not offered high salaries, for example, who would want to endure the arduous years of medical training and long, stressful hours at the hospital? However, this argument is criticized on the grounds that many important occupational roles, such as child-care workers, are poorly paid (the average salary of a child-care worker is less than $19,000 per year) (Bureau of Labor Statistics 2007), whereas many individuals in nonessential roles (e.g., professional sports stars and entertainers) earn astronomical sums of money. The structural-functionalist argument that CEO pay is high because of the risks and responsibilities of the job falls apart when one considers that the average CEO pay is 56 times the pay of a U.S. Army general with 20 years of experience (Anderson et al. 2004). If pay is based on risk and responsibility, does it make sense that the annual pay of the first 919 U.S. soldiers who were killed in Iraq was about equal to the combined pay of just five average U.S. CEOs?

culture of poverty The set of norms, values, and beliefs and self-concepts that contribute to the persistence of poverty among the underclass.

underclass A persistently poor and socially disadvantaged group.

Forbes magazine reported that Apple CEO Steven Jobs earned $646.6 million in total compensation in 2006. In contrast, a U.S. Army general makes an annual salary of between $168,000 and $204,000.

Conflict Perspective

Karl Marx (1818–1883) proposed that economic inequality results from the domination of the *bourgeoisie* (owners of the means of production) over the *proletariat* (workers). The bourgeoisie accumulate wealth as they profit from the labor of the proletariat, who earn wages far below the earnings of the bourgeoisie. Modern conflict theorists recognize that the power to influence economic outcomes comes not only from ownership of the means of production but also from management position, interlocking board memberships, control of media, and financial contributions to politicians.

For example, wealthy corporations use financial political contributions to influence politicians to enact policies that benefit the wealthy. Laws and policies that favor the rich, such as tax breaks that benefit the wealthy, are sometimes referred to as **wealthfare.** Laws and policies that benefit corporations, such as low-interest government loans to failing businesses and special subsidies and tax breaks to corporations, are known as **corporate welfare.**

Between 2001 and 2003, 252 of America's largest and most profitable corporations avoided paying state income taxes on nearly two-thirds of their U.S. profits—at a cost to state governments of $42 billion (McIntyre 2005a). In that 3-year period, 71 of the 252 companies paid no state income tax at all in at least 1 year, despite having made $86 billion in pretax profits in those no-tax years. Some companies (e.g., AT&T, Boeing, Eli Lilly, Toys "R" Us, and Merrill Lynch) paid no state tax over the entire 3-year period. In addition, 275 large U.S. corporations paid, on average, only 17.3 percent of their U.S. profits in federal income taxes in 2002—less than half the 35 percent rate that the tax code requires. From 2001 to 2003, 82 of these 275 corporations enjoyed at least 1 year in which they paid no federal income tax, despite pretax profits in those no-tax years of $102 billion (McIntyre 2005b). Corporate income taxes in the United States have fallen so much in the last few decades that they are nearly the lowest among the world's developed countries. U.S. corporate taxes were 4 percent of gross do-

wealthfare Laws and policies that benefit the rich.

corporate welfare Laws and policies that benefit corporations.

mestic product (GDP) in 1965. This figure fell to 1.5 percent of GDP in 2003, whereas in other wealthy industrialized countries that belong to the OECD, corporate taxes were 3 percent of GDP (Citizens for Tax Justice 2005).

Federal tax cuts enacted by the G. W. Bush administration also benefit the wealthy. In 2007, households in the top 1 percent of the income scale saved more than $52,000 as a result of Bush's tax cuts, whereas middle-income families saved $678, and the lowest fifth of households saved only $71 (Citizens for Tax Justice 2007).

Conflict theorists also note that throughout the world "free-market" economic reform policies have been hailed as a solution to poverty. Yet, although such economic reform has benefited many wealthy corporations and investors, it has also resulted in increasing levels of global poverty. As companies relocate to countries with abundant supplies of cheap labor, wages decline. Lower wages lead to decreased consumer spending, which leads to more industries closing plants, going bankrupt, and/or laying off workers (downsizing). These result in higher unemployment rates and a surplus of workers, enabling employers to lower wages even more.

What Do You Think? The poor in the United States have low rates of voting and thus have minimal influence on elected government officials and the policies they advocate. Why do you think the poor are less likely to vote than those in higher income brackets? What strategies might be effective in increasing voter participation among the poor?

Symbolic Interactionist Perspective

Symbolic interactionism focuses on how meanings, labels, and definitions affect and are affected by social life. This view calls attention to ways in which wealth and poverty are defined and the consequences of being labeled "poor." Individuals who are viewed as poor—especially those receiving public assistance (i.e., welfare)—are often stigmatized as lazy, irresponsible, and lacking in abilities, motivation, and moral values. Wealthy individuals, on the other hand, tend to be viewed as capable, motivated, hardworking, and deserving of their wealth.

The symbolic interactionist perspective also focuses on the meanings of being poor. A qualitative study of more than 40,000 poor women and men in 50 countries around the world explored the meanings of poverty from the perspective of those who live in poverty (Narayan 2000). One of the study's findings is that the experience of poverty involves psychological dimensions such as powerlessness, voicelessness, dependency, shame, and humiliation.

Meanings and definitions of wealth and poverty vary across societies and across time. Although many Americans think of poverty in terms of income level, for millions of people poverty is not primarily a function of income but of their alienation from sustainable patterns of consumption and production. For indigenous women living in the least developed areas of the world poverty and wealth are determined primarily by access to and control of their natural resources (such as land and water) and traditional knowledge, which are the sources of their livelihoods (Susskind 2005).

By global standards the Dinka, the largest ethnic group in the sub-Saharan African country of Sudan, are among the poorest of the poor, being among the least modernized peoples of the world. In Dinka culture wealth is measured in large part by how many cattle a person owns. To the Dinka cattle have a social, moral, and spiritual value as well as an economic value. In Dinka culture a man pays an average "bridewealth" of 50 cows to the family of his bride. Thus men use cattle to obtain a wife to beget children, especially sons, to ensure continuity of their ancestral lineage, and, according to Dinka religious beliefs, to strengthen their linkage with God.

Although modernized populations might label the Dinka as poor, the Dinka view themselves as wealthy. As one Dinka elder explained, "It is for cattle that we are admired, we, the Dinka. . . . All over the world, people look to us because of cattle . . . because of our great wealth; and our wealth is cattle" (Deng 1998, p. 107).

Definitions of poverty also vary within societies. For example, in Ghana men associate poverty with a lack of material assets, whereas for women poverty is defined as food insecurity (Narayan 2000).

ECONOMIC INEQUALITY AND POVERTY IN THE UNITED STATES

The United States is a nation of tremendous economic variation, ranging from the very rich to the very poor. Signs of this disparity are visible everywhere, from opulent mansions perched high above the ocean in California to shantytowns in the rural South where people live with no running water or electricity.

Economic Inequality in the United States

Wide disparities exist in the incomes of Americans. In 2005, the top 1 percent of U.S. households with the highest incomes received 21.8 percent of all income, representing their largest share of national income since 1928 (Johnston 2007). The top 10 percent of Americans collected nearly half (48.5 percent) of all reported income in 2005 (Johnston 2007). Although total reported income in the United States increased almost 9 percent in 2005 (the most recent year for which data are available as of this writing), average incomes for the bottom 90 percent dropped $172 (0.6 percent) from the year before, whereas the incomes of the top 1 percent rose to an average of more than $1.1 million each, an increase of more than $139,000 (or about 14 percent) (Johnston 2007). The top 300,000 Americans in 2005 collectively received as much income as the bottom 150 million Americans. Per person, the top group received 440 times as much income as the average person in the bottom half earned, nearly doubling the gap from 1980 (Johnston 2007).

Income inequality in the United States is reflected in comparisons of the average income of CEOs with that of production workers. In 2005 the average total compensation for CEOs of 350 leading U.S. corporations was $11.6 million. The ratio of CEO pay to average worker pay was 411-to-1 in 2005. Although this is smaller than the peak of 525-to-1 in 2000, it is nearly 10 times as large as the 1980 ratio of 42-to-1 (Anderson et al. 2006). Between 1990 and 2005, average CEO pay rose almost 300 percent (after adjusting for inflation), whereas the aver-

"Money can't buy happiness, but if I had a big house, fancy car
and a giant plasma TV, I wouldn't mind being unhappy!"

age worker's pay rose less than 5 percent, and the minimum wage worker's pay *dropped* by more than 9 percent. If the federal minimum wage had grown at the same rate as CEO pay since 1990, it would have been $22.61 in 2005 instead of $5.15. Likewise, if the average annual salary of production workers had increased at the same rate as CEO pay since 1990, the average worker in 2005 would have made $108,138 rather than $28,314 (Anderson et al. 2006).

The distribution of wealth is much more unequal than the distribution of wages or income. **Wealth** refers to the total assets of an individual or household minus liabilities (mortgages, loans, and debts). Wealth includes the value of a home, investment real estate, cars, unincorporated business, life insurance (cash value), stocks, bonds, mutual funds, trusts, checking and savings accounts, individual retirement accounts (IRAs), and valuable collectibles.

Nearly three-quarters (73 percent) of Americans agree with the statement, "Today it's really true that the rich get richer while the poor get poorer" (Pew Research Center 2007). And they are right; the wealth of Americans has become increasingly concentrated over time. In the early 1960s, the average level of wealth held by the wealthiest one-fifth of U.S. households was 15 times that of the overall median; by 2004 it was more than 23 times (Mishel, Bernstein, & Allegretto 2007). The wealthiest 5 percent of U.S. households held more than one-third (34.3 percent) of all net worth and 42.2 percent of all net financial assets. At the bottom end of the distribution, the 90 percent of households with the lowest incomes received 57.5 percent of all income, but held just 28.7 percent and 19.1 percent of all net worth and net financial assets, respectively (Mishel et al. 2007).

Patterns of Poverty in the United States

Although poverty is not as widespread or severe in the United States as it is in many other parts of the world, poverty is nevertheless a significant social problem in the United States. In 2005, 37 million Americans—12.6 percent of the U.S. population—lived below the poverty line (DeNavas-Walt, Proctor, & Lee 2006). In

wealth The total assets of an individual or household minus liabilities.

Photo courtesy of the author

Children are more likely than adults to live in poverty.

the same year, nearly 16 million Americans—43 percent of the nation's 37 million poor people—were living in deep or severe poverty. This is the highest rate of deep poverty in the United States since 1975 (Pugh 2007). According to research by Mark Rank at the University of Wisconsin, more than half (58 percent) of Americans between the ages of 20 and 75 will spend at least 1 year in poverty, and one in three Americans will experience a full year of extreme poverty at some point in his or her adult life (Pugh 2007).

Poverty rates vary considerably among the states, from 5.6 percent in New Hampshire (in 2005) to 20.1 percent in Mississippi (U.S. Census Bureau 2006a). Poverty rates vary according to age, education, sex, family structure, race and ethnicity, and labor force participation.

Age and Poverty. Children are more likely than adults to live in poverty (see Table 6.3). More than one-third (34.9 percent) of the U.S. poor population are children (DeNavas-Walt et al. 2006). Compared with other industrialized countries, the United States has the highest child poverty rate (and highest overall poverty rate) (Mishel et al. 2007).

Since the late 1950s the poverty rate among the elderly has experienced a downward trend, largely as a result of more Social Security benefits and the growth of private pensions (see also Chapter 12). In 1959 the poverty rate among the U.S. elderly population was 35.2 percent; this rate fell to 25.3 percent in 1969, 15.2 percent in 1979, and 11.4 percent in 1989 and reached

> **❝**It would appear that for most Americans the question is no longer if, but rather when, they will experience poverty. In short, poverty has become a routine and unfortunate part of the American life course.**❞**
>
> Mark Rank
> University of Wisconsin

TABLE 6.3 U.S. Poverty Rates by Age, 2005

AGE (YEARS)	POVERTY RATE
Younger than 18	17.6
18–64	11.1
65 and older	10.1
All ages	12.6

Source: DeNavas-Walt et al. (2006).

Economic Inequality and Poverty in the United States   **219** a record low of 9.7 percent in 1999. The elderly poverty rate in 2005 was 10.1 percent (DeNavas-Walt et al. 2006).

What Do You Think? In our sociology classes we introduce the topic of U.S. poverty by asking students to think of an image of a person who represents poverty in America and to draw that imaginary person. Students are asked to give the person a name (to indicate their sex) and to write down the age of the person. In every semester and in every class most students draw a picture of a middle-aged man. Yet U.S. poverty statistics reveal that the higher poverty rates are among women, not men, and among youth, not middle-aged adults. Why do you think the most common image of a U.S. poor person is a middle-aged man?

Sex and Poverty. Women are more likely than men to live below the poverty line—a phenomenon referred to as the **feminization of poverty.** The 2005 poverty rates for U.S. women and men were 14.1 percent and 11.1 percent, respectively (U.S. Census Bureau 2006b). As discussed in Chapter 10, women are less likely than men to pursue advanced educational degrees and tend to have low-paying jobs, such as service and clerical jobs. However, even with the same level of education and the same occupational role, women still earn significantly less than men. Women who are minorities and/or who are single mothers are at increased risk of being poor.

Education and Poverty. Education is one of the best insurance policies for protecting an individual against living in poverty. In general, the higher a person's level of educational attainment, the less likely that person is to be poor (see also Chapter 8). Adults without a high school diploma are the most vulnerable to being poor, followed by those with a high school diploma but no college degree (see Figure 6.1).

Family Structure and Poverty. Poverty is much more prevalent among female-headed single-parent households than among other types of family structures (see Figure 6.2). The relationship between family structure and poverty helps to explain why women and children have higher poverty rates than men (see also Chapter 5). In 2004, children living in female-householder families with no husband present experienced a much higher poverty rate (42 percent) than did children in married-couple families (9 percent) (Federal Interagency Forum on Child and Family Statistics 2006).

In other industrialized countries poverty rates of female-headed families are lower than those in the United States. Unlike the United States, other developed countries offer a variety of supports for single mothers, such as income supplements, tax breaks, universal child care, national health care, and higher wages for female-dominated occupations.

feminization of poverty The disproportionate distribution of poverty among women.

Race or Ethnicity and Poverty. As displayed in Figure 6.3, poverty rates are higher among blacks, Hispanics, and Native American/Alaska Natives than among non-

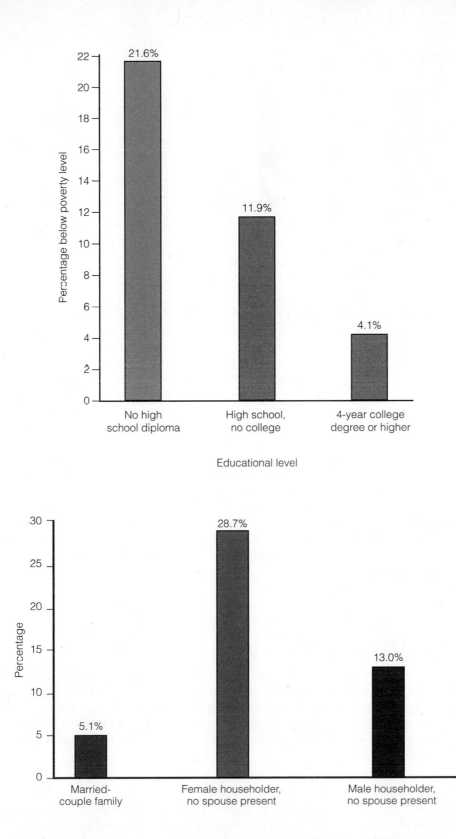

FIGURE 6.2
U.S. poverty rates by family structure: 2005.
Source: DeNavas-Walt et al. (2006).

FIGURE 6.3
U.S. poverty rates by race
and Hispanic origin: 2005.
*Three-year average (2003–2005).
Source: DeNavas-Walt et al. (2006).

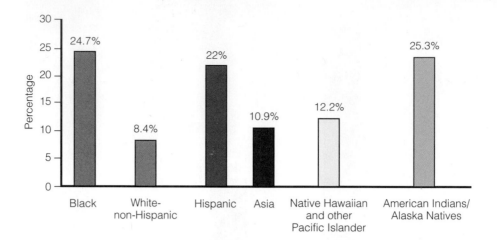

Hispanic whites. As discussed in Chapter 9, past and present discrimination has contributed to the persistence of poverty among minorities. Other contributing factors include the loss of manufacturing jobs from the inner city, the movement of whites and middle-class blacks out of the inner city, and the resulting concentration of poverty in predominantly minority inner-city neighborhoods (Massey 1991; Wilson 1987, 1996). Finally, blacks and Hispanics are more likely to live in female-headed households with no spouse present—a family structure that is associated with high rates of poverty.

Labor Force Participation and Poverty. A common image of the poor is that they are jobless and unable or unwilling to work. Although the poor in the United States are primarily children and adults who are not in the labor force, many U.S. poor are classified as working poor. The **working poor** are individuals who spend at least 27 weeks per year in the labor force (working or looking for work), but whose income falls below the official poverty level. In 2005, 36.8 percent of all U.S. poor (ages 16 and older) worked; 11.4 percent worked year-round full-time (U.S. Census Bureau 2006b).

CONSEQUENCES OF POVERTY AND ECONOMIC INEQUALITY

Poverty is associated with health problems and hunger, increased vulnerability from natural disasters, problems in education, problems in families and parenting, and housing problems. These various problems are interrelated and contribute to the perpetuation of poverty across generations, feeding a cycle of intergenerational poverty. In addition, poverty and economic inequality breed social conflict and war.

working poor Individuals who spend at least 27 weeks per year in the labor force (working or looking for work) but whose income falls below the official poverty level.

Health Problems, Hunger, and Poverty

Poverty is a major factor in many of the world's health problems. In developing countries absolute poverty is associated with hunger and malnutrition, high rates of maternal and infant deaths, indoor air pollution from heating and cook-

Texas A&M, photograph by Damian Medina

Students at universities across the country participate in Hunger Banquets—an event created by Oxfam, an international nongovernmental organization dedicated to eliminating hunger and poverty. The intent of the banquet is to increase people's awareness of the economic and nutritional inequalities through an experiential exercise. Based on worldwide statistics, 55 percent of the attendees (representing the poorest people of the world) are randomly chosen to receive rice and water, 30 percent (representing the world's middle-income population) receive beans and rice, and the wealthiest 15 percent receive a full-course meal.

ing fumes, and unsafe water and sanitation (see this chapter's Photo Essay) (World Health Organization 2002). More than 850 million people go hungry each day, equivalent to the combined populations of North America, Japan, and Europe (Halweil 2005). Every day 30,000 children die as a result of poverty (Oxfam 2006).

In the United States low socioeconomic status is associated with higher incidence and prevalence of health problems, disease, and death (Malatu & Schooler 2002). Despite the increasing rates of obesity among Americans, many U.S. poor do not have enough to eat. A Department of Agriculture report found that 11 percent (12.6 million households) of U.S. households were "food insecure" at some time during 2005 (which means that the household had difficulty providing enough food for all their members due to a lack of resources) and 3.9 percent of households had very low food security (some household members had reduced food intake and disrupted normal eating patterns due to lack of resources) (Nord, Andrews, & Carlson 2006). Assess your own degree of food security in this chapter's *Self and Society* feature.

Poor children and adults also receive inadequate and inferior health care, which exacerbates their health problems. Finally, poverty is linked to higher levels of mental health problems, including stress, depression, and anxiety (Leventhal & Brooks-Gunn 2003).

Economic inequality also affects psychological and physical health. Streeten (1998) cited research suggesting that "perceptions of inequality translate into psychological feelings of lack of security, lower self-esteem, envy, and unhappiness, which, either directly or through their effects on life-styles, cause illness" (p. 5). Poor and middle-income adults who live in states with the greatest gap between the rich and the poor are much more likely to rate their own health as poor or fair than people who live in states where income is more equitably distributed (Kennedy et al. 1998).

> **"**The social class you belong to really matters—it determines your health, how long you live, where you live, your exposure to crime, your success in school, and the likely success of your children.**"**
>
> Lawrence Mishel
> Economic Policy Institute

In this Photo Essay, we focus on one of the most devastating aspects of being among the poorest of poor: lacking access to clean water and sanitation. More than 1 billion people lack access to clean water, and 2.6 billion lack access to adequate sanitation (United Nations Development Programme 2006). This is a major health problem; water-related diseases cause 3 million deaths per year (Oxfam 2006). Each year, 1.5 million children under age 5 die from diarrhea resulting from unsafe water and lack of sanitation (UNICEF 2006).

In many poor areas of the world, people urinate and defecate in the street, along the roadside, in buckets, or in plastic bags that are tied up and thrown in ditches or along the road. Used in this way, plastic bags are known as "flying toilets." The "flying toilet" is the primary means of human waste disposal in Kibera, Nairobi (United Nations Development Programme 2006).

> The conditions here are terrible. . . . There is sewage everywhere. Some people have pit latrines, but they are shallow and overflow when it rains. Most people use buckets and plastic bags for toilets and the children use the streets and yards. Our children suffer all the time from diarrhea and other diseases because it is so filthy. (resident of Kibera, quoted in United Nations Development Programme 2006, p. 38)

Three-quarters of the world's rural population obtain water from a communal source. In a survey of households in 23 countries, about half of households spend more than 30 minutes per trip collecting water; more than one-fifth spend more than 1 hour to obtain water (UNICEF 2006). To collect enough water for drinking, food preparation, personal hygiene, house cleaning, and laundry, a household of five needs at least 32 gallons of water per day (a little more than 6 gallons per person) (Satterthwaite & McGranahan 2007). This is equivalent to carrying six heavy suitcases of water every day. Average water consumption per person in the United States is about 150 gallons.

Lack of easy access to water and sanitation places more of a burden on women and girls, who are expected to fetch water as part of their gender role. In Burkina Faso, women

The Musca sorbens *fly, which breeds in human feces, causes 2 million new cases of blindness-causing trachoma each year in the developing world. The flies burrow into human eyes, which causes decades of repeat infections. Victims describe the infections as feeling as if they have thorns in their eyes.*

commonly walk 1–2.5 hours daily to get water for their families to drink and use for cooking and cleaning; they carry more than 40 pounds of water on their head (Vidyasagar 2007). Women are also expected to take care of children who are sick, often as a result of unclean water and poor sanitation. These time demands on girls and women interfere with income-producing activities and education. A 10-year-old girl in El Alto, Bolivia, was standing in line waiting to get water from a standpipe when she said: "Of course I wish I were in school. I want to learn to read and to write—and I want to be there with my friends. But how can I? My mother needs me to get water, and the standpipe here is only open from 10 to 12. You have to get in line early because so many people come here" (United Nations Development Programme 2006, p. 47).

Cultural norms in many countries require that women not be seen defecating, which

Imagine having to share this one toilet with more than 1,000 other people.

In many poor areas of the world, it is common for residents to defecate in plastic bags that they dump in ditches or throw on the roadside.

Whereas Americans and other wealthy populations take running water for granted, poor populations must leave their homes to get water, often from a water kiosk or standpipe where they may have to wait in a long line.

Due to the high fees and unequal service associated with private water companies, plans to privatize water are often met with public protest.

Sulabh International Social Service Organization is an NGO that has constructed and maintains more than 7,500 pay and use toilet facilities known as Sulabh Complexes throughout India. For a nominal fee (which is waived for children, the disabled, the elderly, and anyone who cannot afford the fee), people can use public toilets, bath, and laundry facilities.

forces them to limit their food and water intake so they can relieve themselves in the dark of night in fields or roadsides. One woman in Bangladesh explained, "Men can answer the call of nature anytime they want . . . but women have to wait until darkness" (United Nations Development Programme 2006, p. 48). Delaying bodily functions can cause liver infection and acute constipation, and going out in darkness to eliminate places women at risk for physical attack.

Municipalities are typically required to provide or arrange for the provision of water supply and sanitation services, but in many countries, public services are inadequate because of lack of investment, corruption, and political interference. Since the 1990s, the World Bank and the International Monetary Fund (IMF) have pushed for

privatization of public services, requiring poor countries to privatize in return for aid, loans, and debt relief. Privatization means that for-profit companies (largely multinational companies) own and/or manage water and sanitation services. Privatization usually results in price hikes to consumers and does little to improve access to water and sanitation for the poor.

Other efforts to improve water and sanitation access for poor populations include reforming public utilities and implementing various community management schemes. In some underserved areas, small-scale entrepreneurs transport and sell water and operate standpipes or water kiosks. The cost of water is high for poor families, but in some areas, there are no other options for obtaining water. Nonprofit nongovernmental

organizations (NGOs) are also involved in global efforts to improve water access and sanitation throughout the developing world.

Improving access to clean water and sanitation for poor populations is one of the most important priorities in the fight against poverty. For every $1 invested in water and sanitation, $3–$4 are saved in health spending or through increased productivity (Oxfam 2006). But more importantly, access to clean water and sanitation is a basic human right that should be enjoyed by every woman, man, and child.

References

Oxfam. 2006 (November). *Our Generation's Choice.* Oxfam Briefing Paper. Available at http://www.oxfam.org

Satterthwaite, David, and Gordon McGranahan. 2007. "Providing Clean Water and Sanitation." In *2007 State of the World: Our Urban Future*, L. Starke, ed. (pp. 26–45). New York: W. W. Norton & Company.

Sulabh International. 2006. *Profile, Aims, and Objectives.* Available at http://www.sulabhinternational.org

UNICEF. 2006 (September). *Progress for Children: A Report Card on Water and Sanitation.* Number 5. Available at http://www.unicef.org

United Nations Development Programme. 2006. *Human Development Report 2006: Beyond Scarcity: Power, Poverty, and the Global Water Crisis.* New York: Palgrave Macmillan.

Vidyasagar, D. 2007. "Global Minute: Water and Health—Walking for Water and Water Wars." *Journal of Perinatology* 27:56–58.

In many poor countries, women and girls walk long distances to get water for their families; they may carry more than 40 pounds of water on their head.

Self and Society | Food Security Scale

The U.S. Department of Agriculture conducts national surveys to assess the degree to which U.S. households experience food security, food insecurity, and food insecurity with hunger. To assess your own level of food security, respond to the following items and use the scoring key to interpret your results.

1. In the last 12 months, the food that (I/we) bought just didn't last, and (I/we) didn't have money to get more.
   - (a) **Often true**
   - (b) **Sometimes true**
   - (c) Never true

2. In the last 12 months, (I/we) couldn't afford to eat balanced meals.
   - (a) **Often true**
   - (b) **Sometimes true**
   - (c) Never true

3. In the last 12 months, did you ever cut the size of your meals or skip meals because there wasn't enough money for food?
   - (a) **Yes**
   - (b) No (skip Question 4)

4. If you answered yes to Question 3, how often did this happen in the last 12 months?
   - (a) **Almost every month**
   - (b) **Some months but not every month**
   - (c) Only 1 or 2 months

5. In the last 12 months, did you ever eat less than you felt you should because there wasn't enough money to buy food?
   - (a) **Yes**
   - (b) No

6. In the last 12 months, were you ever hungry but didn't eat because you couldn't afford enough food?
   - (a) **Yes**
   - (b) No

Scoring and Interpretation: The answer responses in boldface type indicate affirmative responses. Count the number of affirmative responses you gave to the items, and use the following scoring key to interpret your results.

Number of Affirmative Responses and Interpretation

0 or 1 item: *Food secure* (In the last year you have had access to enough food for an active, healthy life.)

2, 3, or 4 items: *Food insecure* (In the last year you have had limited or uncertain availability of food and have been worried or unsure you would get enough to eat.)

5 or 6 items: *Food insecure with hunger evident* (In the last year you have experienced more than isolated occasions of involuntary hunger as a result of not being able to afford enough food.)

If you scored as food insecure (with or without hunger), you might consider exploring whether you are eligible for public food assistance (e.g., food stamps) or whether there is a local food assistance program (e.g., food pantry or soup kitchen) that you could use.

Source: Based on the short form of the 12-month Food Security Scale found in Bickel et al. (2000).

Natural Disasters and Poverty

Although natural disasters such as hurricanes, tsunamis, floods, and earthquakes strike indiscriminately—rich and poor alike—the poor are more vulnerable to devastation from such disasters. After a tsunami devastated a large part of South and Southeast Asia in December 2004—killing thousands and tearing apart families, villages, and communities—Oxfam director Barbara Stocking noted that "it is not mere chance that most of those who died or have been left homeless and destitute were already among the world's poorest. Poor families are always much more severely affected by natural disasters. . . . They live in flimsier homes" (Stocking 2005, n.p.).

The poor have few to no resources to help them avoid or cope with natural disasters. Many poor people in the path of Hurricane Katrina lacked a means of transportation to evacuate. And when the poor lose their homes and their livelihoods in a natural disaster, they do not have the resources to rebuild. For example, when poor fishing communities lost their boats and nets—their very means of survival—in the Asian tsunami, they had no bank accounts or insurance policies to replace their losses.

Reuters/Ami Dave/Landov

Natural disasters, such as the December 2004 tsunami, are more devastating to the poor, who live in flimsier housing, have little to no infrastructure, and lack resources to cope with and recover from devastation.

Poverty also affects natural disaster relief efforts. Because the poorest of the tsunami victims lived in areas with weak or nonexistent infrastructure, hundreds of thousands of tsunami victims were cut off from aid because they did not live near a road or airport. In addition, just as upper-class passengers on the *Titanic* were given priority access to lifeboats, some reports claimed that the tsunami-affected areas that catered to well-off tourists received more assistance than the thousands of poor people who lived in the villages (Roberts 2005). Similarly, in the wake of Hurricane Katrina, some of the local poor residents waited at the back of an evacuation line while 700 guests and employees of the Hyatt Hotel were bused out first (Dowd 2005). Other victims of Hurricane Katrina were stranded for days without food, water, and medical supplies. Five days after the hurricane hit, 20,000 hungry, dehydrated, desperate people stranded in the rain-soaked and sweltering hot Louisiana Superdome in New Orleans, where overflowed toilets forced people to relieve themselves in hallways and stairwells, waited to be evacuated. For days, in some cases a week or more, some hurricane victims waited to be rescued from rooftops or from neck-high floodwater in their attics. For many, rescue came too late. Scores of media interviews and editorials suggested that the government's slow response was at least partly due to the fact that Katrina's victims were predominantly poor. A church pastor lamented, "I think a lot of it has to do with race and class. The people affected were largely poor people. Poor, black people" (quoted by Gonzalez 2005).

What Do You Think? Do you think that the federal response to the disaster left in the wake of Hurricane Katrina would have been different if Katrina had devastated an area of the country where wealthier people resided? Do you think, for example, that residents of Hollywood, California, or Long Island, New York, would have been stranded for days on their rooftops with signs saying, "HELP ME"?

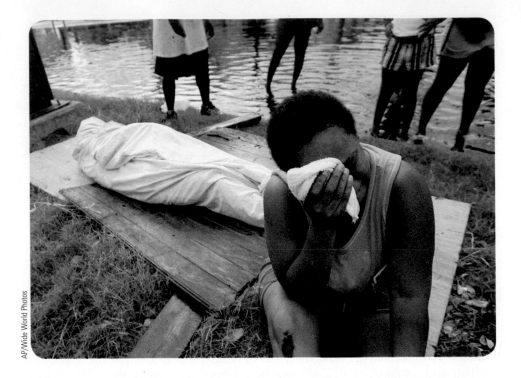

Many of the more than 1,300 people who died in the wake of Hurricane Katrina were poor.

AP/Wide World Photos

Educational Problems and Poverty

In recent years access to education has risen in developing countries. However, in many countries most children from the poorest households have no schooling. A study of 35 countries in West and Central Africa and in South Asia found that in 10 countries half or more of 15- to 19-year-olds from poor households never completed first grade (United Nations Population Fund 2002). "Children aged 6–14 from the wealthiest 20 percent of households are substantially more likely to be enrolled in school than children from the poorest 40 percent of households in almost all [developing] countries" (p. 47).

In the United States children living in poverty are more likely to suffer academically than are children who are not poor. "Overall, poor children receive lower grades, receive lower scores on standardized tests, are less likely to finish high school, and are less likely to attend or graduate from college than are nonpoor youth" (Seccombe 2001, p. 323). Health problems associated with childhood poverty, including poorer vision, lead poisoning, asthma, and inadequate nutrition, contribute to poor academic performance (Rothstein 2004). The poor often attend schools that are characterized by lower quality facilities, overcrowded classrooms, and a higher teacher turnover rate (see also Chapter 8). Children living in poor inner-city ghettos "have to contend with public schools plagued by unimaginative curricula, overcrowded classrooms, inadequate plant and facilities, and only a small proportion of teachers who have confidence in their students and expect them to learn" (Wilson 1996, p. xv). Because poor parents have less schooling on average than do nonpoor parents, they may be less able to encourage and help their children succeed in school. However, research suggests that family income is a stronger predictor of ability and achievement outcomes than are measures of parental schooling or family structure

(Duncan & Brooks-Gunn 1997). Poor parents have fewer resources to provide their children with books, computers, travel, and other goods and experiences that promote cognitive development and educational achievement (Sobolewski & Amato 2005). And with the skyrocketing costs of tuition and other fees, many poor parents cannot afford to send their children to college. The cost for low-income families of sending a child to a 4-year public college or university rose from 13 percent of family income in 1980 to 25 percent in 2000 (Washburn 2004).

Poverty also presents obstacles to educational advancement among poor adults. Women and men who want to further their education to escape poverty may have to work while attending school or may be unable to attend school because of unaffordable child care, transportation, and/or tuition, fees, and books.

Family Stress and Parenting Problems Associated with Poverty

Poverty is both a cause and a consequence of family problems. For example, domestic violence causes some women to flee from their homes and live in poverty without the economic support of their husbands. The stresses associated with low income contribute to substance abuse, domestic violence, child abuse and neglect, divorce, and questionable parenting practices. For example, economic stress is associated with greater marital discord (Sobolewski & Amato 2005), and couples with incomes less than $25,000 are 30 percent more likely to divorce than couples with incomes greater than $50,000 (Whitehead & Popenoe 2004). Child neglect is more likely to be found with poor parents who are unable to afford child care or medical expenses and leave children at home without adult supervision or fail to provide needed medical care. Poor parents are more likely than other parents to use harsh physical disciplinary techniques, and they are less likely to be nurturing and supportive of their children (Mayer 1997; Seccombe 2001).

Another family problem associated with poverty is teenage pregnancy. Poor adolescent girls are more likely to have babies as teenagers or to become young single mothers. Early childbearing is associated with numerous problems, such as increased risk of premature or low-birth-weight babies, dropping out of school, and lower future earning potential as a result of lack of academic achievement. Luker (1996) noted that "the high rate of early childbearing is a measure of how bleak life is for young people who are living in poor communities and who have no obvious arenas for success" (p. 189). For poor teenage women who have been excluded from the American dream and disillusioned with education, "childbearing . . . is one of the few ways . . . such women feel they can make a change in their lives" (p. 182).

> Having a baby is a lottery ticket for many teenagers: it brings with it at least the dream of something better, and if the dream fails, not much is lost. . . . In a few cases it leads to marriage or a stable relationship; in many others it motivates a woman to push herself for her baby's sake; and in still other cases it enhances the woman's self-esteem, since it enables her to do something productive, something nurturing and socially responsible. . . . To the extent that babies can be ill or impaired, mothers can be unhelpful or unavailable, and boyfriends can be unreliable or punitive, childbearing can be just another risk gone wrong in a life that is filled with failures and losses. (Luker 1996, p. 182)

TABLE 6.4 Characteristics of the Homeless in U.S. Cities

- 51 percent are single men
- 30 percent are families with children
- 16 percent are mentally ill
- 13 percent are employed
- 9 percent are veterans
- 17 percent are single women
- 2 percent are unaccompanied minors
- 26 percent are substance abusers

Source: U.S. Conference of Mayors (2006).

Housing Problems and Homelessness

The following description of housing in a low-income inner-city neighborhood is not atypical of U.S. housing conditions for the poor:

> From the outside, Jamal's building looks like an ordinary house that has seen better days. . . . But once you walk through the front door, all resemblance to a real home disappears. . . . The building has been broken up into separate living quarters, a rooming house with whole families squeezed into spaces that would not even qualify as bedrooms in most homes. . . . Six families take turns cooking their meals in the only kitchen. . . . The plumbing breaks down without warning. . . . Windows . . . are cracked and broken, pieced together by duct tape that barely blocks the steady, freezing draft blowing through on a winter evening. Jamal is of the opinion that for the princely sum of $300 per month, he ought to be able to get more heat. (Newman 1999, pp. 3–9)

Many poor families and individuals live in housing units that lack central heating and air conditioning, sewer or septic systems, and electric outlets in one or more rooms; many have no telephones. Housing units of the poor are also more likely to have holes in the floor, leaky roofs, and open cracks in the walls or ceiling. In addition, poor individuals are more likely than the nonpoor to live in high-crime neighborhoods.

Even substandard housing would be a blessing to the hundreds of thousands of men, women, and children in the United States who are homeless on a given day. Over the course of a lifetime, an estimated 9–15 percent of the U.S. population becomes homeless (Hoback & Anderson 2007). Lack of affordable housing is a major cause of homelessness. Other causes of homelessness include mental illness and the lack of mental health services, substance abuse and the lack of substance abuse services, low-paying jobs, unemployment, domestic violence, poverty, and prison release (see Table 6.4).

Homeless individuals live on the street or outdoors; others live in homeless shelters (see this chapter's *The Human Side* feature). Individuals who do not have a home of their own and who stay at the home of family or friends are known as **couch-homeless,** also referred to as "precariously housed" or "couch-surfers." On any given night, an estimated 1.65 percent of the U.S. population is couch-homeless (Hoback & Anderson 2007).

In recent years there has been a surge in unprovoked violent attacks against homeless individuals. Over a 6-year time period, 614 violent acts against homeless individuals in the United States have been documented, resulting in 189 deaths (National Coalition for the Homeless 2007). In most cases the attacks are by teenage and young adult males. In Toms River, New Jersey, five high school students were charged with beating a 50-year-old homeless man nearly to death with pipes and baseball bats as he slept in the woods. In Spokane, Washington, a 22-year-old man was charged with first-degree murder in the case of a one-legged 50-year-old homeless man who was set on fire in his wheelchair on a downtown street; he died of his burns. And in Nashville, Tennessee, two men, ages 21 and 22, were charged with shoving a 32-year-old homeless woman who was sleeping on a boat ramp into the river and leaving her to drown (Lewan 2007).

> "Homeless people are the newest minority group in America that is 'OK' to hate and hurt. It's as though, somehow, they're viewed as less deserving, less human than the rest of us."
>
> Michael Stoops
> Executive director of the National Coalition for the Homeless

couch–homeless Individuals who do not have a home of their own and who stay at the home of family or friends.

More than 70,000 people in the United States alone are homeless each night.

What Do You Think? At least 46 states have hate crime laws that provide tougher penalties when violent crimes are based on prejudice against someone because of that person's race, religion, or ethnic background. As of this writing (August 2007), six states are considering adding homeless status to their hate crime laws. Do you think that violent acts toward homeless individuals should be categorized as hate crimes and subject to harsher penalties? Why or why not?

Intergenerational Poverty

As we have seen, problems associated with poverty, such as health and educational problems, create a cycle of poverty from one generation to the next. Poverty that is transmitted from one generation to the next is called **intergenerational poverty.** In a study of intergenerational poverty using a national longitudinal survey of families, Corcoran and Adams (1997) found considerable mobility out of childhood poverty: three-quarters of white poor children and more than half of black poor children escaped poverty in early adulthood. However, both white and black children in poor families were still much more likely to be poor in early adulthood than were children raised in nonpoor families.

Intergenerational poverty creates a persistently poor and socially disadvantaged population, referred to as the underclass. Although the underclass is stereotyped as being composed of minorities living in inner-city or ghetto communities, the underclass is a heterogeneous population that includes poor whites living in urban and nonurban communities (Alex-Assensoh 1995).

William Julius Wilson attributes intergenerational poverty and the underclass to a variety of social factors, including the decline in well-paid jobs and

intergenerational poverty
Poverty that is transmitted from one generation to the next.

While a sociology graduate student at Columbia University, Gwendolyn Dordick undertook a study of homeless people living in New York City. She spent 15 months with four groups of homeless people: inhabitants of a large bus terminal, residents of a shantytown, occupants of a large public shelter, and clients of a small, church-run, private shelter. Dordick published the results of her study in her book Something Left to Lose: Personal Relations and Survival Among New York's Homeless *(1997). In this* Human Side *feature we present some of Dordick's observations, as well as excerpts from her conversations with homeless individuals she encountered at the Station and the Shanty.*

The Station

The Station, located on Manhattan's West Side, encompasses bus terminals and depots, ticketing windows, shops, and fast food restaurants. Scattered among the commuters and visitors are homeless women and men and young adolescent boys and girls who have either run away or were kicked out by their guardians. One homeless man described living on the edge:

> Living on the streets makes you do a lot of things that you wouldn't normally do. . . .

Comin' into this environment I've done a lot of things I said I wouldn't. . . . There was some people that came along in a van and just threw sandwiches on the street and I picked them up and ate them. . . . The guilt almost killed me . . . but my stomach said, "Hey, listen, you better eat this food." (Dordick 1997, pp. 5–6)

Homeless individuals often rely on one another, offering each other companionship, friendship, and protection. Although they have little to share, they often share what they have with their fellow homeless friends.

> These people, when I got down here, these people reached out to me because they knew, they already knew what it was like. They're not afraid to help their fellow man. As soon as I got down here I met Ron and the fellows and they didn't push me away. I mean I didn't know where to go, I didn't know where to eat, I didn't know where to sleep. They just invited me right in. And ever since then at least I've been healthy, and I've been clean since I've met them. (Dordick 1997, p. 13)

Homeless individuals are often treated harshly by police. One homeless man la-

mented: "You may have an invalid laying down here. He's got problems and the [Station] cops will come up and kick him. Like he's an animal with no rights" (Dordick 1997, p. 11).

Pregnancy is common among homeless women who do not have access to or cannot afford contraception. Pregnancy can have disastrous consequences for these women, who fear having a baby and having to care for a baby. According to one man in the Station:

> It's one thing being homeless, but pregnant and homeless? Some women have their babies right out here; others get rid of them. . . . Some of them abort theirself by sticking hangers up their vaginas. I've seen that myself. This young girl didn't want a baby and she stuck a hanger up her vagina. She had to go to the hospital. (Dordick 1997, p. 25)

The Shanty

A barricaded makeshift community of 20 or so residents sits on a formerly vacant lot visible from the nearby streets and a bridge that crosses the East River. The Shanty consists of 15 makeshift dwellings, or "huts" as they are called by the residents, which are made of a variety of discarded materials such as pieces

---

their movement out of urban areas, the resultant decline in the availability of marriageable males able to support a family, declining marriage rates and an increase in out-of-wedlock births, the migration of the middle class to the suburbs, and the effect of deteriorating neighborhoods on children and youth (Wilson 1987, 1996).

War and Social Conflict

Poverty and economic inequality are often root causes of conflict and war within and between nations. Poorer countries are more likely than wealthier countries to be involved in civil war, and countries that experience civil war tend to become and/or remain poor. Armed conflict and civil war are generally more likely to occur in countries with extreme and growing inequalities between ethnic groups (United Nations 2005). In the developing world most of the people recruited for armed conflict are unemployed. "They don't have education oppor-

of wood and boards, cardboard, mattresses, fabric, and plastic tarps.

The materials are fastened with nails, twine, or fabric. One of the residents has tapped into a source of electricity by running a wire from a lamppost into several huts, providing electricity for light and heat. Dordick noted that, as in the case of the residents of the Station, welfare plays a minimal role in the lives of residents of the Shanty. "So difficult is negotiating the system that most forgo their entitlements" (Dordick 1997, p. 58). One resident explained:

> I don't get welfare. I just can't . . . do it. I hate those people in there. They make you . . . sit and ask you questions that don't make any sense. . . . You're homeless but you have to have an address. What kind of shit is that? Give me a break. They want you to get so . . . upset that you do get up and walk out. (Dordick 1997, p. 58)

Although drug use is common among residents of the Shanty, using drugs in public is a violation of norms of "etiquette." One resident explained that using drugs in the presence of a nonuser is disrespectful: "For me to just take my works out and shoot, I would feel uncomfortable in front of you. It's not right . . . very disrespectful. God forbid, I could

be an influence. I could cause you to do it" (Dordick 1997, pp. 71–72).

Although survival among the homeless requires hustling, buying, trading, and selling, not everything is for sale. Some belongings have sentimental value that outweighs their economic value. One resident of the Shanty treasures a small gold key:

> There's a golden rule about gifts. You treasure them. You don't give them away, you don't sell them. I have right here a little key, a skeleton key. A little kid handed it to me four years ago. And every time somebody see that and say, "what is that?" I say it's a key to the world. I wouldn't give it to anyone if it was given to me. I treasure it. (Dordick 1997, p. 70)

Friendships and love relationships among the homeless suffer from the stresses of drug addictions and impoverished conditions. Nevertheless, Dordick explained that the homeless "survive through their personal relationships" (Dordick 1997, p. 193). Relationships are critical to securing the material resources needed to survive and to creating—to the extent that it is possible—a safe and secure environment. One resident of the Shanty, Richie, conveys the importance of a love relationship:

Regardless of what people might think and say, most of us that might have a woman it's all we want. We don't look for anyone else. We really don't. . . . I'm happy just to take care of my woman. . . . I happen to love my girl. . . . Really, I know it sounds corny, but that's the truth. . . . As a matter of fact, we're gonna be married soon.

A Final Note

Virtually all the homeless individuals who Dordick encountered expressed the desire to escape homelessness and be self-reliant—and they want to be understood. In the words of one homeless man at the Station:

> You never see me sleep in the street. I worked 32 years of my life. Went to prison in '85. . . . I was brought up with a certain degree of independence. . . . And now all I need is two dollars to go and sit in a movie all night long. My pride is too good to beg. I don't want you to help me, Miss. I want you to understand me. (Dordick 1997, p. 5)

Source: Excerpted and reprinted by permission from Dordick, Gwendolyn. 1997. *Something Left to Lose: Personal Relations and Survival among New York's Homeless.* Philadelphia: Temple University Press.

tunities and they don't really see what the future holds for them other than war and misery" (World Population News Service 2003, p. 4). Tanzania president Benjamin Mkapa said that "countries with impoverished, disadvantaged and desperate populations are breeding grounds for present and future terrorists" (quoted by Schifferes 2004, n.p.). A United Nations report suggested that "the most effective conflict prevention strategies . . . are those aimed at achieving reductions in poverty and inequality, full and decent employment for all, and complete social integration" (United Nations 2005, p. 94).

Not only does poverty breed conflict and war, but war also contributes to poverty. War devastates infrastructures, homes, businesses, and transportation systems. In the wake of war, populations often experience hunger and homelessness.

In the United States the widening gap between the rich and the poor may lead to class warfare (hooks 2000). Briggs (1998) asked how long the United States can maintain social order "when increasing numbers of persons are left out of the banquet while a few are allowed to gorge?" (p. 474). Although Karl Marx predicted that the have-nots would revolt against the haves, Briggs did not foresee a revival of Marxism. "The means of surveillance and the methods of suppression by the governments of industrialized states are far too great to offer any prospect of success for such endeavors" (p. 476). Instead, Briggs predicted that American capitalism and its resulting economic inequalities would lead to social anarchy—a state of political disorder and weakening of political authority.

STRATEGIES FOR ACTION: ANTIPOVERTY PROGRAMS, POLICIES, AND PROPOSALS

In this section we provide an overview of government programs designed to alleviate poverty, including various types of welfare and public assistance, the earned income tax credit, and policies and proposals that involve increasing wages. We also look at faith-based initiatives and international responses to poverty.

Government Public Assistance and Welfare Programs in the United States

Many public assistance programs stipulate that households are not eligible for benefits unless their income and/or assets fall below a specified guideline. Programs that have eligibility requirements based on income are called **means-tested programs.** Government public assistance programs designed to help the poor include Supplemental Security Income, Temporary Assistance to Needy Families, food programs, housing assistance, medical care, educational assistance, child care, child support enforcement, and the earned income tax credit (EITC).

Supplemental Security Income. Supplemental Security Income (federal SSI), administered by the Social Security Administration, provides a minimum income to poor people who are age 65 or older, blind, or disabled. Under the 1996 welfare reforms the definition of disability has been sharply restricted, and the eligibility standards have been tightened.

Temporary Assistance to Needy Families. Before 1996 a cash assistance program called Aid to Families with Dependent Children (AFDC) provided single parents (primarily women) and their children with a minimum monthly income. In 1996 Congress passed the Personal Responsibility and Work Opportunity Reconciliation Act (PRWORA), commonly referred to as "welfare reform," which ended AFDC and replaced it with a program called **Temporary Assistance to Needy Families (TANF).** In 2005 TANF families received an average monthly amount (of cash and cash equivalent assistance) of $370 (Office of Family Assistance 2007). Within 2 years of receiving benefits adult TANF recipients must be either employed or involved in work-related activities, such as on-the-job training, job search, and vocational education. A federal lifetime limit of 5 years is set for families receiving benefits, and able-bodied recipients ages 18–50 without dependents have a 2-year lifetime limit (19 states have set shorter limits). Some exceptions to these rules are made for individuals with disabilities, victims of domestic violence, residents in high unemployment areas, and those caring for young children. To qualify for TANF benefits, unwed mothers younger than age 18 are re-

means-tested programs Assistance programs that have eligibility requirements based on income.

Temporary Assistance to Needy Families (TANF) A federal cash welfare program that involves work requirements and a 5-year lifetime limit.

quired to live in an adult-supervised environment (e.g., with their parents) and to receive education and job training. Legal immigrants who entered the United States before August 22, 1996, can receive TANF, but those who entered after this date can receive services only after they have been in the country for 5 years.

Food Assistance. The largest food assistance program in the United States is food stamps, followed by school meals and the Special Supplemental Food Program for Women, Infants, and Children (WIC). The food stamp program issues monthly benefits through coupons or electronic benefits transfer (EBT), using a plastic card similar to a credit card and a personal identification number (PIN). In 2005 the typical household receiving food stamps had a gross income of $648 per month, did not receive cash welfare benefits, and received a monthly food stamp benefit of $209 (31 percent received the maximum benefit of $499.00 for a family of four) (Barrett 2006).

USDA Food Stamp Program

A SMALL REASON TO FIND OUT IF YOU QUALIFY FOR FOOD STAMPS.

Call **1-800-221-5689**

United States Department of Agriculture
Food and Nutrition Service
USDA is an equal opportunity provider and employer.

USDA

This poster is a public service message designed to encourage economically distressed families to apply for food stamps.

To supplement food stamps, school meals, and WIC, many communities have food pantries (which distribute food to poor households), "soup kitchens" (which provide cooked meals on-site), and food assistance programs for the elderly population (such as Meals-on-Wheels). Despite the various forms of food assistance, a significant share of poor U.S. children (18 percent of young children and 12 percent of school-age children) receives no food assistance (Zedlewski & Rader 2005).

Housing Assistance. Housing costs are a major burden for the poor. Half of the working poor spend at least 50 percent of their income on housing (Grunwald 2006). Federal housing assistance programs include public housing, Section 8 housing, and other private project-based housing.

The **public housing** program, initiated in 1937, provides federally subsidized housing that is owned and operated by local public housing authorities (PHAs). To save costs and avoid public opposition, high-rise public housing units were built in inner-city projects. These have been plagued by poor construction, managerial neglect, inadequate maintenance, and rampant vandalism.

> Distressed public housing subjects families and children to dangerous and damaging living environments that raise the risks of ill health, school failure, teen parenting, delinquency, and crime—all of which generate long-term costs that taxpayers ultimately bear. . . . These severely distressed developments are not just old, outmoded, or run down. Rather, many have become virtually uninhabitable for all but the most vulnerable and desperate families. (Turner et al. 2005, pp. 1–2)

public housing Federally subsidized housing that is owned and operated by local public housing authorities (PHAs).

The HOPE VI Urban Demonstration Program was established in 1992 to transform the nation's most distressed public housing projects by rebuilding the physi-

In recent years housing prices in Fairfax County, Virginia, grew 12 times as fast as household incomes. In 2006, the county's median family would have to spend more than half of its income to afford the county's median home. In response to this housing crisis, Fairfax began offering housing subsidies to families earning $90,000 a year.

cal structure of public housing developments, expanding the opportunities of its residents, and building a sense of community among residents. This federal housing initiative has demolished 80,000 units of the worst public housing and built mixed-income housing developments in their place (Grunwald 2006).

Rather than building new housing units for low-income families, **Section 8 housing** consists of existing housing. With Section 8 housing, federal rent subsidies are provided either to tenants (in the form of certificates and vouchers) or to private landlords. Other private project-based housing includes privately owned housing units that do not receive rent subsidies but receive other federal subsidies, such as interest rate reductions. Unlike public housing that confines low-income families to high-poverty neighborhoods, the aim with Section 8 housing and other private project-based housing is to disperse low-income families throughout the community. However, because of opposition by residents in middle-class neighborhoods, most Section 8 housing units remain in low-income areas.

The level of housing assistance available is sorely inadequate to meet the housing needs of low-income Americans. For every low-income family that receives federal housing assistance, there are three eligible families without it (Grunwald 2006).

Lack of affordable housing is not just a problem for the poor living in urban areas. "The problem has climbed the income ladder and moved to the suburbs, where service workers cram their families into overcrowded apartments, college graduates have to crash with their parents, and firefighters, police officers and teachers can't afford to live in the communities they serve" (Grunwald 2006). A major barrier to building affordable housing is zoning regulations that set minimum lot size requirements, density restrictions, and other controls. Referred to as "snob zoning," such zoning regulations serve the interests of upper middle-class suburbanites who want to maintain their property values and keep out the "riffraff"—the lower income segment of society who would presumably hurt the character of the community. Thus, one answer to the housing problem is to change zoning regulations that exclude affordable housing. Fairfax County, Virginia, is one of more than 100 communities that have adopted "inclusionary zoning," which requires developers to reserve a percentage of units for affordable housing (Grunwald 2006).

What Do You Think? The number of spaces in homeless shelters is grossly inadequate to accommodate the numbers of homeless individuals, which means that hundreds of thousands of homeless people have no place to be, except in public. Many cities have passed laws that prohibit homeless people from begging as well as sleeping and even sitting in public. Do you think that such laws unfairly punish the homeless? Or are these laws necessary to protect the public?

Section 8 housing A housing assistance program in which federal rent subsidies are provided either to tenants (in the form of certificates and vouchers) or to private landlords.

Medicaid. The largest U.S. public medical care assistance program is Medicaid, which provides medical services and hospital care for the poor through reimbursements to physicians and hospitals. However, many low-income individuals and families do not qualify for Medicaid and either cannot afford health insurance or cannot pay the deductible and copayments under their insurance plan. Out-of-pocket medical expenses for poor adult Medicaid beneficiaries have grown twice as fast as their incomes in recent years, causing low-income beneficiaries to cut back on essential medical care (Center on Budget and Policy Priorities 2005).

In the earlier AFDC welfare program all recipients were automatically entitled to Medicaid. Under the TANF program states decide who is eligible for Medicaid; eligibility for cash assistance does not automatically convey eligibility for Medicaid. A provision of the 1996 welfare reform legislation guarantees welfare recipients at least 1 year of transitional Medicaid when leaving welfare for work (see also Chapter 2).

Educational Assistance. Educational assistance includes Head Start and Early Head Start programs and college assistance programs (see also Chapter 8). Head Start and Early Head Start programs provide educational services for disadvantaged infants, toddlers, and preschool-age children and their parents. Evaluations of Head Start and Early Head Start programs indicate that they improve children's cognitive, language, and social-emotional development and strengthen parenting skills (Administration for Children and Families 2002). According to the Children's Defense Fund (2003), every $1 invested in high-quality early childhood care and education saves as much as $7 by increasing the likelihood that children will be literate, go to college, and be employed and by decreasing the likelihood that they will drop out of schools, be dependent on welfare, or be arrested for criminal activity.

To alleviate economic barriers for low-income individuals wanting to attend college, the federal government offers grants, loans, and work opportunities. The Pell grant program aids students from low-income families. The guaranteed student loan program enables college students and their families to obtain low-interest loans with deferred interest payments. The federal college work-study program provides jobs for students with "demonstrated need."

Child-Care Assistance. In the United States lack of affordable, good child care is a major obstacle to employment for single parents and a tremendous burden on dual-income families and employed single parents. The cost of child care for a 4-year-old child ranges from about $3,000 to nearly $10,000 per year. Child-care fees for an infant are even higher, ranging from nearly $4,000 to more than $13,000 a year (National Association of Child Care Resource and Referral Agencies 2006).

Some public- and private-sector programs and policies provide limited assistance with child care. The Dependent Care Assistance Plan provisions of the 1981 Economic Recovery Tax Act permits individuals to exclude the value of employer-provided child-care services from their gross income. However, few employers provide on-site child care or subsidies for child care. At the same time Congress increased the amount of the child-care tax credit and modified the federal tax code to allow taxpayers to shelter pretax dollars for child care in "flexible spending plans." The Family Support Act of 1988 offered additional funding for child-care services for the poor (in conjunction with mandatory work requirements). The Child Care and Development Block Grant, which became law in 1990, targeted child-care funds to low-income groups, and the PRWORA appropriated funds for child care. But child-care assistance is inadequate; in 2005, 17 states had waiting lists for child-care assistance, and many families earn more than the eligibility limit, but not enough to afford child-care expenses (National Association of Child Care Resource and Referral Agencies 2006).

Child Support Enforcement. To encourage child support from absent parents, the PRWORA requires states to set up child support enforcement programs, and single parents who receive TANF are required to cooperate with child support enforcement efforts. The welfare reform law established a Federal Case Registry and National Directory of New Hires to track delinquent parents across state lines, increased the use of wage withholding to collect child support, and allowed states to seize assets and to revoke driving licenses, professional licenses, and recreational licenses of parents who fall behind in their child support.

Earned Income Tax Credit. The federal **earned income tax credit (EITC),** created in 1975, is a refundable tax credit based on a working family's income and number of children. The EITC is designed to offset Social Security and Medicare payroll taxes on working poor families and to strengthen work incentives. In 2006 an eligible family of four with two children could receive a credit of up to $4,536. Almost one of every six families who file federal income tax returns claims the federal EITC, which lifts more children out of poverty than any other program (Llobrera & Zahradnik 2004).

In 2007, 20 states and the District of Columbia offered a state EITC that supplements the federal credit and works as a rebate for state taxes paid by low-income working people. In addition, local governments in Montgomery County, Maryland, and Denver, Colorado, San Francisco, California, and New York City, New York, offer their own version of EITCs.

Welfare in the United States: Myths and Realities

In a 2007 national survey of U.S. adults, nearly 7 in 10 (69 percent) agreed that the government has a responsibility "to take care of people who can't take care of themselves"—up from 57 percent in 1994 (Pew Research Center 2007). Nevertheless, negative attitudes toward welfare assistance and welfare recipients are not uncommon (Epstein 2004). For example, more than two-thirds (69 percent) of U.S. adults believe that "poor people have become too dependent on government assistance programs" (Pew Research Center 2007). What are some of the common myths about welfare that perpetuate negative images of welfare and welfare recipients?

Myth 1. People who receive welfare are lazy, have no work ethic, and prefer to have a "free ride" on welfare rather than work.

Reality. Most recipients of TANF cash benefits and food stamps are children and therefore are not expected to work. Unemployed adult welfare recipients experience a number of barriers that prevent them from working, including poor health, job scarcity, lack of transportation, lack of education, and/or the desire to stay home and care for their children (which often stems from the inability to pay for child care or the lack of trust in child-care providers) (Zedlewski 2003).

earned income tax credit (EITC) A refundable tax credit based on a working family's income and number of children.

Welfare recipients who stay home to care for children *are* doing very important work: parenting. "Raising children is work. It requires time, skills, and commitment. While we as a society don't place a monetary value on it, it is work that is invaluable—and indeed, essential to the survival of our society" (Albelda & Tilly 1997, p. 111).

It is also important to note that many adults receiving public assistance are participating in work activities, including job training/education, job searches, and employment. Over half (56.5 percent) of adults receiving TANF in 2005 participated in work activities; 1 in 5 (21.3 percent) was employed, earning an average monthly income of $668 (Office of Family Assistance 2007). More than one-fourth (29 percent) of food stamp recipients in 2005 had earnings, typically $648 per month (Barrett 2006).

Finally, most adult welfare recipients would rather be able to support themselves and their families than rely on public assistance. The image of a welfare "freeloader" lounging around enjoying life is far from the reality of the day-to-day struggles and challenges of supporting a household on a monthly TANF check of $370, which was the average monthly cash and cash-equivalent assistance to TANF families in 2005 (Office of Family Assistance 2007).

TABLE 6.5 Percentage of Individuals Living Below the Poverty Level in Households That Receive Means-Tested Assistance: 2005

TYPE OF ASSISTANCE	PERCENTAGE
Any type of assistance	67.8
Cash assistance	20.9
Food stamps	37.9
Medicaid	54.7
Public or subsidized housing	17.3

Source: U.S. Census Bureau (2006a).

Myth 2. Most welfare mothers have large families with many children.

Reality. In 2005 the average number of individuals in TANF families was 2.4, including an average of only 1.8 children (Office of Family Assistance 2007). The average size of households receiving food stamps in 2005 was 2.3 (Barrett 2006).

Myth 3. Welfare benefits are granted to many people who are not really poor or eligible to receive them.

Reality. Although some people obtain welfare benefits through fraudulent means, it is much more common for people who are eligible to receive welfare not to receive benefits. Only about half of families poor enough to qualify for TANF receive monthly cash assistance, and only 6 of 10 of individuals eligible for the food stamp program receive benefits (Fremstad 2004; U.S. Department of Agriculture 2006). In 2005, 67.8 percent of individuals below the poverty line lived in households that received government means-tested assistance; nearly one-third received no benefits (see Table 6.5).

A main reason for not receiving benefits is lack of information; people do not know they are eligible. Many people who are eligible for public assistance do not apply for it because they do not want to be stigmatized as lazy people who just want a "free ride" at the taxpayers' expense—their sense of personal pride prevents them from receiving public assistance. Others want to avoid the administrative and transportation hassles involved in obtaining assistance (Zedlewski et al. 2003). Finally, some individuals who are eligible for public assistance do not receive it because it is not available. In cities across the United States, thousands of eligible low-income households are on waiting lists for public housing assistance because there are not enough public housing units available (U.S. Conference of Mayors 2006).

Minimum Wage Increase and "Living Wage" Laws

In 2007, the federal minimum wage increased from $5.15 to $5.85 an hour—the first increase in a decade. The minimum wage will increase 70 cents each summer until 2009, when all minimum wage jobs will pay at least $7.25 an hour. Some states have established a minimum wage that is higher than the federal minimum wage. As of August 2007, 28 states and the District of Columbia have mandated a minimum wage that is higher than the federal $5.85.

Many cities and counties throughout the United States have **living wage laws** that require state or municipal contractors, recipients of public subsidies or tax breaks, or, in some cases, all businesses to pay employees wages that are significantly above the federal minimum, enabling families to live above the poverty line. Research findings show that businesses that pay their employees a living wage have lower worker turnover and absenteeism, reduced training costs, higher morale and higher productivity, and a stronger consumer market (Kraut, Klinger, & Collins 2000).

As more individuals receiving welfare (TANF) reach their time limits and are forced to enter the job market, it is increasingly important to provide jobs that pay a living wage. As shown in this chapter's *Social Problems Research Up Close* feature, single mothers who work in low-wage jobs often have more hardships than those who are dependent on welfare.

Faith-Based Services for the Poor

Under President G. W. Bush's **Faith-Based and Community Initiative,** faith-based (i.e., religious) organizations compete for federal funding for programs that serve the needy, such as homeless services and food aid programs. Critics point out that the faith-based initiative is not a serious effort to help the poor but rather a political tool to engender support among Bush's conservative base. "Bush made it clear from the beginning that no new money would be allocated for social programs aimed at the needy. Instead, the existing pie would simply be carved up in different ways, with religious groups getting bigger slices" (Boston 2005, n.p.). Indeed, although Bush has pushed for increased funding of religious-based groups, he has also proposed deep cuts for many traditional antipoverty programs, such as public housing subsidies and food stamps.

Critics are also concerned about the degree to which the faith-based initiative violates the separation of church and state and affects the rights of clients seeking services. "How are the lives of the jobless improved when they are told they won't get work until they first get right with God?" (Boston 2005, n.p.). Although religious groups are prohibited from using government funding to promote religion, this policy is difficult to police. Furthermore, Bush seems to endorse proselytizing in faith-based social services by comments such as those made in a January 2004 speech in New Orleans when he said, "We want to fund programs that save Americans one soul at a time." Faith-based groups that impose religion on social service clients risk losing their funding, however. In 2005 a federal judge blocked the Bush administration from providing future grants to a faith-based mentoring program that promoted religious worship and instruction in its program (Associated Press 2005). It is of some reassurance, then, that religious organizations that express their religiosity more publicly and place a higher value on evangelizing are less likely to apply for and receive government funding than religious organizations that have less religious policies and practices (Ebaugh, Chafetz, & Pipes 2005).

living wage laws Laws that require state or municipal contractors, recipients of public subsidies or tax breaks, or, in some cases, all businesses to pay employees wages that are significantly above the federal minimum, enabling families to live above the poverty line.

Faith-Based and Community Initiative A program in which faith-based (i.e., religious) organizations compete for federal funding for programs that serve the needy, such as homeless services and food aid programs.

As welfare recipients reach the time limit established by welfare legislation of 1996 for receiving welfare benefits, they are forced into the workforce. But as individuals leave welfare for work, they often find themselves in low-paying jobs, with no or few benefits. How do individuals in low-income jobs compare with those dependent on welfare in terms of their well-being? And how do both low-wage earners and welfare recipients survive on income that does not meet their basic needs? Researchers Kathryn Edin and Laura Lein (1997) conducted research to answer these questions.

Sample and Methods

The sample consisted of 379 African-American, white, and Mexican-American single mothers from four cities (Chicago, San Antonio, Boston, and Charleston, South Carolina).

The mothers either received welfare cash assistance (Aid to Families with Dependent Children, or AFDC) ($N = 214$) or were nonrecipients who held low-wage jobs earning $5 to $7 an hour between 1988 and 1992 ($N = 165$). Edin and Lein (1997) used a "snowball sampling" technique in which each mother who was interviewed was asked to refer researchers to one or two friends who might also participate in interviews. Nearly 90 percent of the mothers contacted agreed to be interviewed.

Edin and Lein collected data by conducting multiple semistructured in-depth interviews with women in the sample. Interview topics included the mothers' income and job experience, types and amount of welfare benefits they received, spending behavior, housing situation, use of medical care and child care, and hardships the women and their children experienced because of lack of financial resources.

Interviewing the mothers more than once was an important research strategy in gathering accurate information. Mothers who were unclear about their expenditures in the first interview could keep careful track of what they spent between interviews and give a more precise accounting of their spending in a later interview. Also, some mothers who insisted that they received no child support later revealed that the child's father "helped out" every week by providing cash. "Most mothers termed absent fathers' cash contributions as 'child support' only if it was collected by the state" (Edin & Lein 1997, p. 13).

Findings and Conclusions

Low-wage-earning single mothers had a higher monthly reported income than welfare-reliant mothers. However, the expenses of wage-earning mothers were also higher. This is because employed mothers usually have to pay for child care, transportation to work, and additional clothing to wear to work. If newly employed mothers have a federal housing subsidy, some of their new income is spent on the increase in the rent they must pay. And employed mothers are usually not eligible for Medicaid, which means that they have more out-of-pocket medical expenses and often go uninsured.

The monthly expenses of both groups of women exceeded their reported monthly income, forcing women to use various strategies to make ends meet. Cash welfare and food stamps covered only three-fifths of welfare-reliant mothers' expenses. The main job of low-wage-earning mothers covered only 63 percent of their expenses. Edin and Lein (1997) found that women relied on three basic strategies to make ends meet: work in the formal, informal, or underground economy; cash assistance from absent fathers, boyfriends, relatives, and friends; and cash assistance or help from agencies, community groups, or charities in paying overdue bills. Welfare recipients had to keep their income-generating activities hidden from their welfare caseworkers and other government officials. Otherwise, their welfare checks would be reduced by nearly the same amount as their earnings. Many of the wage-earning mothers also concealed income generated "on the side" to maintain eligibility for food stamps, housing subsidies, or other benefits that would have been reduced or eliminated if they had reported this additional income.

Most of the single mothers in the study described experiencing serious material hardship during the previous 12 months. Material hardships included not having enough food and clothes, not receiving needed medical care, not having health insurance, having the utilities or phone cut off, not having a phone, and being evicted and/or homeless. An important finding was that wage-reliant mothers experienced more hardship than welfare-reliant mothers. In addition to the increased financial pressures of child-care costs, transportation, health care, and work clothing, employed mothers worried about not providing adequate supervision of their children and struggled with balancing work and parenting responsibilities, especially when their children were sick. Nevertheless, almost all the mothers said they would rather work than rely on welfare. They believed that work provided important psychological benefits and increased self-esteem, avoided the stigma of welfare, and enabled them to be good role models for their children.

Source: Based on Edin and Lein (1997).

TABLE 6.6 The Millennium Development Goals
1. Eradicate extreme poverty and hunger.
2. Achieve universal primary education.
3. Promote gender equality and empower women.
4. Reduce child mortality.
5. Improve maternal health.
6. Combat HIV/AIDS, malaria, and other diseases.
7. Ensure environmental sustainability.
8. Develop a global partnership for development.

International Responses to Poverty

In 2000 leaders from 191 United Nations member countries pledged to achieve eight **Millennium Development Goals**—an international agenda for reducing poverty and improving lives. One of the Millennium Development Goals (MDGs) is to reduce by half the proportion of people who live on $1 a day and who suffer from hunger. As can be seen in Table 6.6, several other MDGs involve alleviating problems related to poverty, such as disease, child and maternal mortality, and lack of access to education. The target date for achieving most of the MDGs is 2015, with 1990 as the benchmark. According to the World Bank (2007), all regions except for sub-Saharan Africa are "on track" to reach the MDG of reducing extreme poverty by half (from its 1990 level) by 2015. Extreme poverty—the proportion of people living on less than $1 a day—fell from 29 percent in 1990 to 18 percent in 2004 and is expected to fall to 12 percent by 2015. Approaches for achieving poverty reduction throughout the world include promoting economic growth, investing in "human capital," providing financial aid and debt cancellation, and providing microcredit programs that provide loans to poor people.

Promoting Economic Growth. In low-income countries, increases in gross domestic product are associated with reductions in poverty. In 2006 growth in the gross domestic product of low-income countries averaged 5.9 percent—up from an average of 4.0 percent in 2001–2005 (World Bank 2007). An expanding economy creates new employment opportunities and increased goods and services. As employment prospects improve, individuals are able to buy more goods and services. The increased demand for goods and services, in turn, stimulates economic growth. As emphasized in Chapters 13 and 14, economic development requires controlling population growth and protecting the environment and natural resources, which are often destroyed and depleted in the process of economic growth.

However, economic growth does not always reduce poverty; in some cases it increases it. Policies that involve cutting government spending, privatizing basic services, liberalizing trade, and producing goods primarily for export may increase economic growth at the national level, but the wealth ends up in the hands of the political and corporate elite at the expense of the poor. Growth does not help poverty reduction when public spending is diverted away from meeting the needs of the poor and instead is used to pay international debt, finance military operations, and support corporations that do not pay workers fair wages and that are hostile to unionization. The World Bank lends $30 billion a year to developing nations to pay primarily for roads, bridges, and industrialized agriculture that mostly benefit corporations. "Relatively little attention or money has been given to developing basic social services, building schools and clinics, and building decent public sanitation and clean water systems in some of the world's poorest countries" (Mann 2000, p. 2). Thus "economic growth, though essential for poverty reduction, is not enough. Growth must be pro-poor, expanding the opportunities and life choices of poor people" (UNDP 1997, pp. 72–73).

Millennium Development Goals Eight goals that comprise an international agenda for reducing poverty and improving lives.

Because three-fourths of poor people in most developing countries depend on agriculture for their livelihoods, economic growth to reduce poverty must include raising the productivity of small-scale agriculture operations. Not only does improving the productivity of small-scale agriculture operations create employment, but it also reduces food prices.

Investing in Human Capital. Promoting economic development in a society requires having a productive workforce. Yet in many poor countries large segments of the population are illiterate and without job skills and/or are malnourished and in poor health. Thus a key feature of poverty reduction strategies involves investing in human capital. The term **human capital** refers to the skills, knowledge, and capabilities of the individual. Investments in human capital involve programs and policies that provide adequate nutrition, sanitation, housing, health care (including reproductive health care and family planning), and educational and job training.

Poor health is both a consequence and a cause of poverty; improving the health status of a population is a significant step toward breaking the cycle of poverty. A cross-country comparison of children living in households that survive on $1 a day found that children living in countries with higher levels of per capita public spending on health had significantly lower levels of mortality and malnutrition (Wagstaff 2003).

Investments in education are also critical for poverty reduction. Increasing the educational levels of a population better prepares individuals for paid employment and for participation in political affairs that affect poverty and other economic and political issues. Improving the educational level and overall status of women in developing countries is also associated with lower birthrates, which in turn fosters economic development.

Providing Financial Aid and Debt Cancellation. To pay for investments in human capital, poor countries depend on financial aid from wealthier countries. To meet the MDG of halving poverty by 2015, the United Nations recommends that wealthier countries allocate just 0.7 percent of national income to aid. But only 5 of 22 major donors (none of them from the most powerful Group of Seven [G7] nations) are meeting that target. (The G7 countries are the United States, Japan, Germany, France, Great Britain, Italy, and Canada.) Of 22 OECD countries the United States is the *least* generous donor in terms of aid as a proportion of its national income (Oxfam 2005).

Another way to help poor countries invest in human capital and reduce poverty is to provide debt relief. In poor countries with large debts money needed for health and education is instead spent on debt repayment. Poor countries are often caught in a vicious cycle, paying out more in debt repayments than they receive in aid. Canceling the debts of 32 of the poorest countries would cost citizens of the richest countries of the world just $2.10 a year for each citizen for 10 years (Oxfam 2005).

Microcredit Programs. The old saying "It takes money to make money" explains why many poor people are stuck in poverty: they have no access to financial resources and services. **Microcredit programs** refer to the provision of loans to people who are generally excluded from traditional credit services because of their low socioeconomic status. Microcredit programs give poor people the

> "Overcoming poverty is not a gesture of charity. It is an act of justice."
>
> Nelson Mandela

human capital The skills, knowledge, and capabilities of the individual.

microcredit programs The provision of loans to people who are generally excluded from traditional credit services because of their low socioeconomic status.

Muhammad Yunus won the 2006 Nobel Peace Prize for his pioneering work in starting the Grameen Bank, which has more than 2,000 branches in more than 65,000 villages. It has provided loans to more than 6 million poor people, 96% of whom are women.

© AP Photo/Fabian Bimmer

financial resources they need to become self-sufficient and to contribute to their local economies.

The Grameen Bank in Bangladesh, started in 1976, has become a model for the more than 3,000 microcredit programs which, by the end of 2004, served more than 92 million poor clients (Roseland & Soots 2007). To get a loan from the Grameen Bank, borrowers must form small groups of five people "to provide mutual, morally binding group guarantees in lieu of the collateral required by conventional banks" (Roseland & Soots 2007, p. 160). Initially, only two of the five group members are allowed to apply for a loan. When the initial loans are repaid, the other group members may apply for loans.

UNDERSTANDING POVERTY AND ECONOMIC INEQUALITY

After George W. Bush was re-elected in the 2004 presidential election, numerous commentators remarked that Bush's victory was largely due to his focus on "values" and the appeal that this focus had for the American public, especially those with strong religious ties. In his campaign Bush emphasized values such as freedom and democracy, of marriage being defined as one man and one woman, and of the sanctity of life (implying disapproval of abortion and stem-cell research). However, Economic Policy Institute president Lawrence Mishel reminded us that "economic inequality and how it is addressed is as much a 'values' issue as any of those that are more frequently discussed" (2005, p. 1).

❝If all of the afflictions of the world were assembled on one side of the scale and poverty on the other, poverty would outweigh them all.**❞**

Exodus Rabbah, Mishpatim 31:14

As we have seen in this chapter, economic prosperity has not been evenly distributed; the rich have become richer while the poor have become poorer. Meanwhile, the United States has implemented welfare reform measures that essentially weaken the safety net for the impoverished segment of the population—largely children. The welfare reform legislation of 1996 has achieved its goal of reducing welfare rolls across the country. From 1996 to 2005 the number of families receiving TANF dropped by 57 percent (U.S. Department of Health and Human Services 2006). Advocates of welfare reform argue that transitions from welfare to work benefit children by creating positive role models in their working mothers, promoting maternal self-esteem, and fostering career advancement and higher family earnings. Critics of welfare reform argue that reforms increase stress on parents, force young children into inadequate child care, reduce parents' abilities to monitor the behavior of their adolescents, and deepen the poverty of many families. Of those who leave TANF, only 60 percent are able to find employment, and of those only a fraction are able to earn a living wage (Parisi, Grice, & Taquino 2003). Although the long-term effects of welfare reform are not yet known, one study of the impact of welfare reform on children concluded that reforms will help some children and hurt others (Duncan & Chase-Lansdale 2001). As we discuss in the next chapter ("Work and Unemployment"), many of the jobs available to those leaving welfare for work are low paying, have little security, and offer few or no benefits. Without decent wages and without adequate assistance in child care, housing, health care, and transportation, many families who leave welfare for work will find that their situation becomes worse, not better.

A common belief among U.S. adults is that the rich are deserving and the poor are failures. Blaming poverty on the individual rather than on structural and cultural factors implies not only that poor individuals are responsible for their plight but also that they are responsible for improving their condition. If we hold individuals accountable for their poverty, we fail to make society accountable for making investments in human capital that are necessary to alleviate poverty. Such human capital investments include providing health care, adequate food and housing, education, child care, and job training. Economist Lewis Hill (1998) believed that "the fundamental cause of perpetual poverty is the failure of the American people to invest adequately in the human capital represented by impoverished children" (p. 299). By blaming the poor for their plight, we also fail to recognize that there are not enough jobs for those who want to work and that many jobs do not pay wages that enable families to escape poverty. And last, blaming the poor for their condition diverts attention away from the recognition that the wealthy—individuals and corporations—receive far more benefits in the form of wealthfare or corporate welfare, without the stigma of welfare.

Ending or reducing poverty begins with the recognition that doing so is a worthy ideal and an attainable goal. Imagine a world where everyone had comfortable shelter, plentiful food, clean water and sanitation, adequate medical care, and education. If this imaginary world were achieved and if absolute poverty were effectively eliminated, what would be the effects on social problems such as crime, drug abuse, family problems (e.g., domestic violence, child abuse, and divorce), health problems, prejudice and racism, and international conflict? In the current global climate of conflict and terrorism, we might consider that "reducing poverty and the hopelessness that comes with human deprivation is perhaps the most effective way of promoting long-term peace and security" (World Bank 2005).

According to one source, the cost of eradicating poverty worldwide would be only about 1 percent of global income—and no more than 2–3 percent of national income in all but the poorest countries (UNDP 1997). Certainly the costs of allowing poverty to continue are much greater than that.

CHAPTER REVIEW

- **What is the difference between absolute poverty and relative poverty?**

Absolute poverty refers to a lack of basic necessities for life, such as food, clean water, shelter, and medical care. In contrast, relative poverty refers to a deficiency in material and economic resources compared with some other population.

- **How is poverty measured?**

The World Bank sets a "poverty threshold" of $1 per day to compare poverty in most of the developing world, $2 per day in Latin America, $4 per day in Eastern Europe and the Commonwealth of Independent States (CIS), and $14.40 per day in industrial countries (which corresponds to the income poverty line in the United States). Another poverty measure is based on whether individuals are experiencing hunger, which is defined as consuming less than 1,960 calories a day. According to measures of relative poverty, members of a household are considered poor if their household income is less than 50 percent of the median household income in that country. Each year, the U.S. federal government establishes "poverty thresholds" that differ by the number of adults and children in a family and by the age of the family head of household. Anyone living in a household with pretax income below the official poverty line is considered "poor." To capture the multidimensional nature of poverty, the United Nations Development Programme developed the human poverty index that combines three measures of deprivation: (1) deprivation of a long, healthy life, (2) deprivation of knowledge, and (3) deprivation in decent living standards.

- **Which sociological perspective criticizes wealthy corporations for using financial political contributions to influence politicians to enact policies that benefit corporations and the wealthy?**

The conflict perspective is critical of wealthy corporations that use financial political contributions to influence laws and policies that favor corporations and the rich. Such laws and policies, sometimes referred to as wealthfare or corporate welfare, include low-interest government loans to failing businesses and special subsidies and tax breaks to corporations.

- **In the United States what age group has the highest rate of poverty?**

U.S. children are more likely than adults to live in poverty. More than one-third of the U.S. poor population are children. Child poverty rates are much higher in the United States than in Canada or in any other Western European industrialized country.

- **What are some of the consequences of poverty and economic inequality for individuals, families, and societies?**

Poverty is associated with health problems and hunger, increased vulnerability from natural disasters, problems in education, problems in families and parenting, and housing problems. These various problems are interrelated and contribute to the perpetuation of poverty across generations, feeding a cycle of intergenerational poverty. In addition, poverty and economic inequality breed social conflict and war.

- **What are some of the U.S. government public assistance programs designed to help the poor?**

Government public assistance programs designed to help the poor include Supplemental Security Income, Temporary Assistance to Needy Families (TANF), food programs (such as school meal programs and food stamps), housing assistance, Medicaid, educational assistance (such as Pell grants), child care, child support enforcement, and the earned income tax credit (EITC).

- **What are three common myths about welfare and welfare recipients?**

Common myths about welfare and welfare recipients are (1) that welfare recipients are lazy, have no work ethic, and prefer to have a "free ride" on welfare rather than work; (2) that most welfare mothers have large families with many children; and (3) that welfare benefits are granted to many people who are not really poor or eligible to receive them.

- **What are four general approaches for achieving poverty reduction throughout the world?**

  Approaches for achieving poverty reduction throughout the world include promoting economic growth, investing in "human capital," providing financial aid and debt cancellation to nations, and providing microcredit programs that provide loans to poor people.

TEST YOURSELF

1. According to your text, which of the following countries has the lowest level of poverty as measured by the human poverty index?
   a. Sweden
   b. Italy
   c. United States
   d. France
2. According to the official U.S. poverty threshold guidelines, a single adult earning $12,000 a year is considered "poor."
   a. True
   b. False
3. The degree of wealth inequality in the world is as if one person in a group of 10 takes _____ of the total pie and the other 9 people in the group share the remaining pie.
   a. one-quarter
   b. one-half
   c. 70 percent
   d. 99 percent
4. According to the structural-functionalist perspective, CEO pay is high because
   a. of the risks and responsibilities of the job
   b. most CEOs are men

   c. there are few of them compared to the number of workers
   d. they have financial influence over politicians
5. Corporate income taxes in the United States have risen so much in the past few decades that they are nearly the highest among the world's developed countries.
   a. True
   b. False
6. Wealth in the United States is more equally distributed today than it was in the 1960s.
   a. True
   b. False
7. What age group in the United States has the highest rate of poverty?
   a. Younger than 18
   b. 18–29
   c. 30–55
   d. Older than 55
8. According to your text, the wealthy are hardest hit by natural disasters because they have more to lose than do the poor.
   a. True
   b. False
9. For every low-income family that receives federal housing assistance, there are three eligible families without it.
   a. True
   b. False
10. Which federal program lifts more children out of poverty than any other program?
    a. Public and Section 8 housing
    b. TANF
    c. EITC
    d. Food stamps

Answers: 1 a. 2 b. 3 d. 4 a. 5 b. 6 b. 7 a. 8 b. 9 a. 10 c.

KEY TERMS

absolute poverty
corporate welfare
couch-homeless
culture of poverty
earned income tax credit (EITC)
Faith-Based and Community
    Initiative
feminization of poverty
human capital

human poverty index (HPI)
intergenerational poverty
living wage laws
means-tested programs
microcredit programs
Millennium Development Goals
public housing
relative poverty

Section 8 housing
Temporary Assistance to Needy
    Families (TANF)
underclass
wealth
wealthfare
working poor

MEDIA RESOURCES

Understanding Social Problems, **Sixth Edition
Companion Website**

academic.cengage.com/sociology/mooney
Visit your book companion website, where you will find flash cards, practice quizzes, Internet links, and more to help you study.

 Just what you need to know NOW! Spend time on what you need to master rather than on information you already have learned. Take a pre-test for this chapter, and CengageNOW will generate a personalized study plan based on your results. The study plan will identify the topics you need to review and direct you to online resources to help you master those topics. You can then take a post-test to help you determine the concepts you have mastered and what you will need to work on. Try it out! Go to academic.cengage .com/login to sign in with an access code or to purchase access to this product.

"When a man tells you that he got rich through hard work, ask him whose."

Don Marquis, Journalist

Work and Unemployment

The Global Context: The Economy in the 21st Century | Sociological Theories of Work and the Economy | Problems of Work and Unemployment | Strategies for Action: Responses to Workers' Concerns | Understanding Work and Unemployment | Chapter Review

Franklin County, Ohio

Timothy Bowers, 62, was sentenced in Columbus, Ohio, to 3 years in prison on bank robbery charges. Mr. Bowers had demanded money from a bank teller and then gave the money to a guard as he waited for the police to arrive and arrest him. In court, he told the judge that he committed the crime because he couldn't find a job and wanted to be in prison until he can collect Social Security ("Jobless Robber Gets What He Wanted—Prison" 2006).

Unemployment can lead some people to desperation. Timothy Bowers committed a bank robbery because he couldn't find a job and wanted to be in prison until he was old enough to collect Social Security.

Unemployment is just one of many concerns regarding the well-being of workers throughout the world. In this chapter we examine problems of unemployment and other problems involving work such as slavery and forced labor, sweatshop labor, health and safety hazards in the workplace, job dissatisfaction and alienation, work-family concerns, and declining labor strength and representation. We begin by looking at the global economy.

The Global Context: The Economy in the 21st Century

In 2007 the European Union (EU), the largest single trading bloc in the world, which represents one-fourth of the world's wealth, accepted two new member nations (Romania and Bulgaria). Residents of EU countries can buy and sell goods and services in any of the 27 member countries without tariff barriers, and most of the EU countries share a common currency, the euro. The EU reflects the increasing globalization of economic institutions. The **economic institution** refers to the structure and means by which a society produces, distributes, and consumes goods and services.

In recent decades innovations in communication and information technology have spawned the emergence of a **global economy**—an interconnected network of economic activity that transcends national borders and spans the world. The globalization of economic activity means that increasingly our jobs, the products and services we buy, and our nation's political policies and agendas influence and are influenced by economic activities occurring around the world. After summarizing the two main economic systems in the world—capitalism and socialism—we describe how industrialization and post-industrialization have changed the nature of work, and we look at the emergence of free trade agreements and transnational corporations.

economic institution The structure and means by which a society produces, distributes, and consumes goods and services.

global economy An interconnected network of economic activity that transcends national borders.

Socialism and Capitalism

Socialism is an economic system characterized by state ownership of the means of production and distribution of goods and services. In a socialist economy the government controls income-producing property. Theoretically, goods and services are equitably distributed according to the needs of the citizens. Socialist economic systems emphasize collective well-being rather than individualistic pursuit of profit.

Under **capitalism** private individuals or groups invest capital (money, technology, machines) to produce goods and services to sell for a profit in a competitive market. Whereas socialism emphasizes social equality, capitalism emphasizes individual freedom. Capitalism is characterized by economic motivation through profit, the determination of prices and wages primarily through supply and demand, and the absence of government intervention in the economy. More people are working in a capitalist economy today than ever before in history (Went 2000). Critics of capitalism argue that it creates too many social evils, including alienated workers, poor working conditions, near-poverty wages, unemployment, a polluted and depleted environment, and world conflict over resources.

In reality, there are no pure socialist or capitalistic economies. Rather, most countries have mixed economies, incorporating elements of both capitalism and socialism. Most developed countries, for example, have both private-owned and state-owned enterprises, as well as a social welfare system. The U.S. economy is dominated by capitalism, but there are elements of socialism in our welfare system and in government subsidies to industry.

Industrialization, Post–Industrialization, and the Changing Nature of Work

The nature of work has been shaped by the Industrial Revolution, the period between the mid-18th century and the early 19th century when the factory system was introduced in England. **Industrialization** dramatically altered the nature of work: Machines replaced hand tools; and steam, gasoline, and electric power replaced human or animal power. Industrialization also led to the development of the assembly line and an increased division of labor as goods began to be mass produced. The development of factories contributed to the emergence of large cities where the earlier informal social interactions dominated by primary relationships were replaced by formal interactions centered on secondary groups. Instead of the family-centered economy characteristic of an agricultural society, people began to work outside the home for wages.

Post-industrialization refers to the shift from an industrial economy dominated by manufacturing jobs to an economy dominated by service-oriented, information-intensive occupations. Post-industrialization is characterized by a highly educated workforce, automated and computerized production methods, increased government involvement in economic issues, and a higher standard of living.

Primary, Secondary, and Tertiary Work Sectors. The three fundamental work sectors (primary, secondary, and tertiary) reflect the major economic transforma-

socialism An economic system characterized by state ownership of the means of production and distribution of goods and services.

capitalism An economic system characterized by private ownership of the means of production and distribution of goods and services for profit in a competitive market.

industrialization The replacement of hand tools, human labor, and animal labor with machines run by steam, gasoline, and electric power.

post-industrialization The shift from an industrial economy dominated by manufacturing jobs to an economy dominated by service-oriented, information-intensive occupations.

tions in society—the Industrial Revolution and the Post-Industrial Revolution. The *primary work sector* involves the production of raw materials and food goods. In less developed regions, including East Asia, South Asia, and sub-Saharan Africa, agricultural jobs make up half or more of all jobs, whereas in developed countries, agriculture represents only 3.2 percent of total employment (International Labour Organization 2007).

The *secondary work sector* involves the industrial production of manufactured goods from raw materials (e.g., paper from wood). Industry represents about one-fifth (21.3 percent) of total global employment (International Labour Organization 2007).

The *tertiary work sector* includes professional, managerial, technical support, and service jobs. The transition to a post-industrialized society is marked by a decrease in manufacturing jobs and an increase in service and information technology jobs in the tertiary work sector. In 2006, service jobs represented 40 percent of total global employment and, for the first time in history, overtook the share of agriculture (38.7 percent of global employment) (International Labour Organization 2007). Between 1994 and 2000 employment in manufacturing averaged above 17 million, but this number has steadily declined to about 14.2 million in 2006 (Bureau of Labor Statistics 2007a).

McDonaldization of the Workplace. Sociologist George Ritzer (1995) coined the term **McDonaldization** to refer to the process by which the principles of the fast food industry are being applied to more and more sectors of society, particularly the workplace. McDonaldization involves four principles:

1. *Efficiency.* Tasks are completed in the most efficient way possible by following prescribed steps in a process overseen by managers.
2. *Calculability.* Quantitative aspects of products and services (such as portion size, cost, and the time it takes to serve the product) are emphasized over quality.
3. *Predictability.* Products and services are uniform and standardized. A Big Mac in Albany is the same as a Big Mac in Tucson. Workers behave in predictable ways. For example, servers at McDonald's learn to follow a script when interacting with customers.
4. *Control through technology.* Automation and mechanization are used in the workplace to replace human labor.

The principles of McDonaldization are not new. Henry Ford's assembly line was designed to produce automobiles efficiently and predictably by using technology to replace human labor. But the degree to which these rational principles characterize the workplace today is unprecedented.

What are the effects of McDonaldization on workers? In a McDonaldized workplace employees are not permitted to use their full capabilities, be creative, or engage in genuine human interaction. Workers are not paid to think, just to follow a predetermined set of procedures. Because human interactions are unpredictable and inefficient (they waste time), "we're left with either no interaction at all, such as at ATMs, or a 'false fraternization.' Rule number 17 for Burger King workers is to smile at all times" (Ritzer, quoted by Jensen 2002, p. 41). Workers also may feel that they are merely extensions of the machines they operate. The alienation that workers feel—the powerlessness and meaninglessness that characterizes a "McJob"—may lead to dissatisfaction with one's job and,

McDonaldization The process by which principles of the fast food industry (efficiency, calculability, predictability, and control through technology) are being applied to more and more sectors of society, particularly the workplace.

more generally, with one's life. Worker dissatisfaction and alienation are discussed later in this chapter.

What Do You Think? The slang term "McJob" has been added to recent editions of several dictionaries. For example, the *Oxford English Dictionary* defines "McJob" as "an unstimulating, low-paid job with few prospects, especially one created by the expansion of the service sector." *Merriam-Webster* defines "McJob" as "a low-paying job that requires little skill and provides little opportunity for advancement." Not surprisingly, McDonald's is opposed to such characterizations of their employment conditions and has fought back with an advertising campaign in which posters at more than 1,200 restaurants play up "the positive aspects of working for McDonald's," and include the phrase, "Not bad for a McJob." In 2007, McDonald's offered its employees "the opportunity" to sign a petition urging dictionaries to define "McJob" in a more positive way. Do you think that the *Oxford* and *Merriam-Webster* dictionary definitions of "McJob" are accurate and fair?

The Globalization of Trade and Free Trade Agreements

Just as industrialization and post-industrialization changed the nature of economic life, so has the globalization of trade—the expansion of trade of raw materials, manufactured goods, and agricultural products across national and hemispheric borders. The first set of global trade rules were adopted through the General Agreement on Tariffs and Trade (GATT) in 1947. GATT members met periodically to revise trade agreements in negotiations called "rounds." In 1995 the World Trade Organization (WTO) replaced GATT as the organization overseeing the multilateral trading system.

In the 1980s and early 1990s U.S. officials began negotiating regional free trade agreements that would open doors to U.S. goods in neighboring countries and reduce the massive U.S. trade deficit, which had grown from $25.3 billion in 1980 to $122 billion in 1985 (Schaeffer 2003). **Free trade agreements** are pacts between two countries or among a group of countries that make it easier to trade goods across national boundaries. Free trade agreements reduce or eliminate foreign restrictions on exports, reduce or eliminate tariffs (or taxes) on imported goods, and prevent U.S. technology from being copied and used by competitors through protection of "intellectual property rights." Treaties such as the Canada–U.S. Free Trade Agreement, the North American Free Trade Agreement (NAFTA), the Free Trade Area of the Americas (FTAA), and the Central American Free Trade Agreement (CAFTA) are designed to accomplish these trade goals.

U.S. officials have also used Section 301 of the Trade Acts of 1984 and 1988 to force trade negotiations with individual countries. If U.S. trade officials determine that other countries have denied U.S. corporations "reasonable" access to domestic markets, sold their goods in the United States at below-market prices, or failed to protect the patents and copyrights of U.S. companies, Section 301

free trade agreements Pacts between two countries or among a group of countries that make it easier to trade goods across national boundaries by reducing or eliminating restrictions on exports and tariffs (or taxes) on imported goods and protecting intellectual property rights.

allows the United States to impose retaliatory sanctions and tariffs on goods from these countries (Schaeffer 2003).

Through GATT and the WTO, free trade agreements, and Section 301, U.S. trade officials have expanded trading opportunities, benefiting large export manufacturing and service industries in the global north, specifically aircraft, auto, computer, pharmaceutical, and entertainment industries in Western Europe, the United States, and Japan. But trade globalization also hurt the U.S. steel and textile-apparel industries and the workers employed in them, small businesses that cannot compete with large retail chain stores, supermarkets, franchises, and small farmers (Schaeffer 2003). Since NAFTA was signed in 1993, the growth of U.S. exports supported 1 million U.S. jobs, but the growth of imports from Mexico and Canada displaced production that would have supported 2 million jobs. Thus NAFTA has resulted in the loss of 1 million U.S. jobs, two-thirds of which are in the manufacturing industries (Scott & Ratner 2005).

Foreign workers have also been hurt by trade agreements. Since NAFTA took effect, real wages of Mexican manufacturing workers have fallen, and more than 1 million agricultural jobs in Mexico have been lost (Scott & Ratner 2005).

As discussed in Chapters 2 and 14, free trade agreements also undermine the ability of national, state, and local governments to implement environmental and food or product safety policies.

Transnational Corporations

Although free trade agreements have increased business competition around the world, resulting in lower prices for consumers for some goods, they have also opened markets to monopolies (and higher prices) because they have facilitated the development of large-scale transnational corporations. **Transnational corporations,** also known as *multinational corporations,* are corporations that have their home base in one country and branches, or affiliates, in other countries. The number of transnational corporations more than doubled from about 35,000 in 1990 to about 75,000 in 2005 (Roach 2007). Among the world's largest economies, 29 are companies, rather than countries (Roach 2007).

Transnational corporations provide jobs for U.S. managers, secure profits for U.S. investors, and help the United States compete in the global economy. Transnational corporations benefit from increased access to raw materials, cheap foreign labor, and the avoidance of government regulations.

> By moving production plants abroad, business managers may be able to work foreign employees for long hours under dangerous conditions at low pay, pollute the environment with impunity, and pretty much have their way with local communities. Then the business may be able to ship its goods back to its home country at lower costs and bigger profits. (Caston 1998, pp. 274–275)

transnational corporations
Also known as multinational corporations, corporations that have their home base in one country and branches, or affiliates, in other countries.

Transnational companies can also avoid or reduce tax liabilities by moving their headquarters to a "tax haven." When Halliburton, a U.S.-based multinational corporation with operations in over 120 countries and the Pentagon's largest private contractor operating in Iraq, announced in 2007 that it was moving its corporate headquarters from Texas to Dubai, critics accused the company of avoiding U.S. taxes. Indeed, Dubai's tax-free zones have lured about one-quarter of Fortune 500 companies to establish corporate headquarters there (Buncombe 2007).

The savings that big business reaps from cheap labor abroad are not passed on to consumers. "Corporations do not outsource to far-off regions so that U.S.

consumers can save money. They outsource in order to increase their margin of profit" (Parenti 2007). For example, shoes made by Indonesian children working 12-hour days for 13 cents an hour, cost only $2.60 but still sold for $100 or more in the United States. In 2006, the share of U.S. national income going to wages and salaries was at its *lowest level on record,* with data going back to 1929. The share of national income reflected in corporate profits, in contrast, was at its *highest level on record* (Aron-Dine & Shapiro 2007).

Transnational corporations contribute to the trade deficit in that more goods are produced and exported from outside the United States than from within. Transnational corporations also contribute to the budget deficit, because the United States does not get tax income from U.S. corporations abroad, yet transnational corporations pressure the government to protect their foreign interests; as a result, military spending increases. Third, transnational corporations contribute to U.S. unemployment by letting workers in other countries perform labor that could be performed by U.S. employees. Finally, transnational corporations are implicated in an array of other social problems, such as poverty resulting from fewer jobs, urban decline resulting from factories moving away, and racial and ethnic tensions resulting from competition for jobs.

Halliburton—the Pentagon's largest private contractor operating in Iraq—is a multinational corporation with operations in more than 120 countries. In 2007, Halliburton announced it was moving its corporate headquarters from Texas to Dubai—a tax-free zone that has lured about one-quarter of Fortune 500 companies to set up corporate headquarters there.

SOCIOLOGICAL THEORIES OF WORK AND THE ECONOMY

Numerous theories in economics, political science, and history address the nature of work and the economy. In sociology structural functionalism, conflict theory, and symbolic interactionism serve as theoretical lenses through which we may better understand work and economic issues and activities.

Structural-Functionalist Perspective

According to the structural-functionalist perspective, the economic institution is one of the most important of all social institutions. It provides the basic necessities common to all human societies, including food, clothing, and shelter. By providing for the basic survival needs of members of society, the economic institution contributes to social stability. After the basic survival needs of a society are met, surplus materials and wealth may be allocated to other social uses, such as maintaining military protection from enemies, supporting political and religious leaders, providing formal education, supporting an expanding population, and providing entertainment and recreational activities. Societal development is dependent on an economic surplus in a society (Lenski & Lenski 1987).

Although the economic institution is functional for society, elements of it may be dysfunctional. For example, before industrialization agrarian societies had a low degree of division of labor, in which few work roles were available to members of society. Limited work roles meant that society's members shared similar roles and thus developed similar norms and values (Durkheim 1966 [1893]).

> **"** Free trade does not, as evidenced in CAFTA, mean fair trade. **"**
>
> Stephen F. Lynch
> U.S. Congressman

In contrast, industrial societies are characterized by many work roles, or a high degree of division of labor, and cohesion is based not on the similarity of people and their roles but on their interdependence. People in industrial societies need the skills and services that others provide. The lack of common norms and values in industrialized societies may result in *anomie*—a state of normlessness—which is linked to a variety of social problems, including crime, drug addiction, and violence (see Chapters 3 and 4).

The structural-functionalist perspective is also concerned with how changes in one aspect of society affect other aspects. How, for example, does the level of economic development of a country affect the subjective life satisfaction and core values of its population? Research indicates that as one moves from subsistence-level economies in developing countries to advanced industrialized societies, there is a large increase in the percentage of the population who consider themselves happy or satisfied with their lives as a whole. Above a certain economic point, however, the curve levels off. In other words, "moving from a starvation level to a reasonably comfortable existence makes a big difference," but once this level is reached, further economic development does not increase subjective well-being (Inglehart 2000, p. 219). Economic development level also affects core values, because economic insecurity breeds xenophobia and deference to authority, whereas a sense of basic security fosters values such as self-expression (rather than deference to authority) and not only a tolerance of cultural diversity but also a sense that cultural differences are stimulating and exotic (Inglehart 2000).

What Do You Think? The economic health of a country is commonly measured by how much the country is producing (the total value of goods and services) and how much money consumers are spending on the purchase of goods and services. In what ways might high levels of production and consumption contribute to individual and social ills rather than to health and well-being?

Conflict Perspective

According to Karl Marx, the ruling class controls the economic system for its own benefit and exploits and oppresses the working masses. Whereas structural functionalism views the economic institution as benefiting society as a whole, conflict theory holds that capitalism benefits an elite class that controls not only the economy but other aspects of society as well—the media, politics and law, education, and religion.

The conflict perspective is critical of ways that the government caters to the interests of big business at the expense of workers, consumers, and the public interest. This system of government that serves the interests of corporations—known as **corporatocracy**—involves ties between government and business. For example, George W. Bush, the first president with an MBA (master's degree in business administration), is a former Texas oilman, and Dick Cheney was the CEO of Halliburton, the world's largest oil field services company, until he joined the Bush ticket in 2000. The majority of Bush's cabinet and advisers have ties to a number of corporations, including General Motors, Bank of

corporatocracy A system of government that serves the interests of corporations and that involves ties between government and business.

America, Chevron, Sears, Goldman Sachs, and Boeing (Center for Responsive Politics 2005a).

Corporate interests also find their way into politics through large political contributions and "soft money," which is money that flows through a loophole to provide political parties, candidates, and contributors a means to evade federal limits on political contributions. Critics of this system of campaign financing argue that corporations and interest groups purchase political influence through financial contributions. Because of the high cost of political campaigns, many candidates rely on funds from special interest groups, who then expect (and often get) special treatment in the form of lower business taxes, environmental loopholes, subsidies, or lower standards of consumer and worker protection. For example, the Bankruptcy Abuse Prevention and Consumer Protection Act, which was passed by the House of Representatives in 2005, makes it more difficult for individuals to escape from debt by filing for bankruptcy protection. Finance and credit companies, which benefit from the new bankruptcy law, contributed more than $8 million (64 percent to Republicans) in the 2004 election cycle, and spent millions more on federal lobbying (Center for Responsive Politics 2005b).

A survey of business leaders' views on political fund-raising found that the main reasons U.S. corporations make political contributions are fear of retribution and to buy access to lawmakers ("Big Business for Reform" 2000). Although 75 percent of the surveyed business leaders said that political donations give them an advantage in shaping legislation, nearly three-quarters (74 percent) said that business leaders are pressured to make large political donations. Half of the executives said that their colleagues "fear adverse consequences for themselves or their industry if they turn down requests" for contributions.

Penalties for violating health and safety laws in the workplace provide an example of legal policy that favors corporate interests. Suppose that a corporation is guilty of a serious violation of health and safety laws, in which "serious violation" is defined as one that poses a substantial probability of death or serious physical harm to workers. What penalty do you think that corporations should pay for such a violation? According to a report by the AFL-CIO (2005), serious violations of workplace health and safety laws carry an average penalty of only $873. As discussed in Chapter 4, penalties for corporate crimes tend to be much less severe than those applied to individuals who violate the law. For example, under federal law, causing the death of a worker by willfully violating safety rules—a misdemeanor with a 6-month maximum prison term—is a less serious crime than harassing a wild burro on federal lands, which is punishable by 1 year in prison (Barstow & Bergman 2003).

Some social critics believe that the influence of corporations in U.S. government is so strong that "the corporate leaders run the system more than the president does. The Republicans tend to be more closely tied to the interests of certain corporations, but the leaders of the corporatocracy will find some way to render ineffective any president who fails to advocate for what they want or who tries to stand in their way" (Perkins, quoted by MacEnulty 2005, p. 10). The pervasive influence of corporate power in government exists not only in the United States but also throughout the world. On a global scale the policies of the International Monetary Fund (IMF) and the World Bank pressure developing countries to open their economies to foreign corporations, promoting export production at the expense of local consumption, encouraging the exploitation of labor as a means of attracting foreign investment, and hastening the degradation of natural resources as countries sell their forests and minerals to earn money to pay back loans.

In his book *Confessions of an Economic Hit Man,* John Perkins (2004) described his prior job as an "economic hit man"—a highly paid professional who would convince leaders of poor countries to accept huge loans (primarily from the World Bank) that were much bigger than the country could possibly repay. The loans would be used to help develop the country by paying for needed infrastructure, such as roads, electrical plants, airports, shipping ports, and industrial plants. One of the conditions of the loan was that the borrowing country had to give 90 percent of the loan back to U.S. companies (such as Halliburton or Bechtel) to build the infrastructure. The result: The wealthiest families in the country benefit from additional infrastructure and the poor masses are stuck with a debt they cannot repay. The United States uses the debt as leverage to ask for "favors," such as land for a military base or access to natural resources such as oil. According to Perkins, large corporations want "control over the entire world and its resources, along with a military that enforces that control" (quoted by MacEnulty 2005, p. 10).

Symbolic Interactionist Perspective

According to symbolic interactionism, the work role is a central part of a person's identity. When making a new social acquaintance, one of the first questions we usually ask is "What do you do?" The answer largely defines for us who that person is. For example, identifying a person as a truck driver provides a different social meaning than identifying someone as a physician. In addition, the title of one's work status—maintenance supervisor or university professor—also gives meaning and self-worth to the individual. An individual's job is one of his or her most important statuses; for many it represents a "master status," that is, the most significant status in a person's social identity.

Symbolic interactionism emphasizes the fact that attitudes and behavior are influenced by interaction with others. The applications of symbolic interactionism in the workplace are numerous: Employers and managers are concerned with using interpersonal interaction techniques that elicit the attitudes and behaviors they want from their employees; union organizers are concerned with using interpersonal interaction techniques that persuade workers to unionize; and personnel in job-training programs are concerned with using interpersonal interaction techniques that are effective in motivating participants.

From a symbolic interactionist perspective we might look at the meanings of work that members of a society learn. In the United States work and the rewards of work are viewed as part of the American Dream. Does work play a major part in your view of the American Dream? Assess your views by reading this chapter's *Self and Society* feature.

PROBLEMS OF WORK AND UNEMPLOYMENT

In this section we examine unemployment and other problems associated with work. Problems of workplace discrimination based on gender, race and ethnicity, and sexual orientation are addressed in Chapters 9, 10, and 11. Poverty, minimum wage, and living wage issues are discussed in Chapter 6. Here, we

> **"**No race can prosper 'til it learns there is as much dignity in tilling a field as in writing a poem.**"**
>
> Booker T. Washington
> Address to the Atlanta Exposition, September 18, 1895

Self and Society | How Do You Define the American Dream?

From the following list, choose two items that you personally believe best represent or define the American Dream. After you make your choices, compare your responses to those of a national sample of U.S. adults.

Owning a home
Having a family
Obtaining a quality education for your children
Being financially secure
Living in freedom
Having a secure retirement
Enjoying good health
Living in a good community
Having a good job

Comparison: In a national survey of U.S. adults, respondents were asked to select two items from the given list that best represent their definition of the American Dream (National League of Cities 2004). As shown in the following table, the most common choice was "living in freedom" (33 percent), followed by "being financially secure" (26 percent).

ITEM	PERCENTAGE OF U.S. ADULTS WHO CHOSE ITEM
Living in freedom	33
Being financially secure	26
Obtaining a quality education for your children	17
Having a family	17
Enjoying good health	16
Having a good job	9
Owning a home	9
Living in a good community	6
Having a secure retirement	6

discuss problems concerning forced labor, sweatshop labor, health and safety hazards in the workplace, job dissatisfaction and alienation, work-family concerns, unemployment and underemployment, and labor unions and the struggle for workers' rights.

Forced Labor and Slavery

Worldwide at least 12.3 million people are victims of forced labor (International Labour Organization 2005a). **Forced labor,** also known as *slavery,* refers to any work that is performed under the threat of punishment and is undertaken involuntarily. Slavery expert Kevin Bales (1999) explained that the resurgence of slavery around the world is linked to three main factors: (1) rapid growth in population, especially in the developing world; (2) social and economic changes that have displaced many rural dwellers to urban centers and their outskirts, where people are powerless and jobless and are vulnerable to exploitation and slavery; and (3) government corruption that allows slavery to go unpunished, even though it is illegal in every country.

Forced labor exists all over the world but is most prevalent in India, Pakistan, Bangladesh, and Nepal. Most forced laborers work in agriculture, mining, prostitution, and factories. Forced laborers produce goods we use every day, including sugar from the Dominican Republic, chocolate from the Ivory Coast, paper clips from China, carpets from Nepal, and cigarettes from India. Between 40 percent and 50 percent of all victims of forced labor are children (International Labour Organization 2005a). Child labor is discussed in Chapter 12.

forced labor Also known as slavery, any work that is performed under the threat of punishment and is undertaken involuntarily.

© Mark Peterson/Corbis

Forced prison labor is a type of forced labor that is controlled by the state. Forced prison labor is particularly widespread in China.

Forms of Slavery and Forced Labor. Modern slavery is different from the old form most people know, which is **chattel slavery.** In chattel slavery slaves were considered property that could be bought and sold. In the past the high cost of purchasing a slave (about $40,000 in today's money) gave the master incentive to provide a minimum standard of care to ensure that the slave would be healthy enough to work and generate profit for the long term. Today, slaves are cheap, costing an average of $100 (Cernasky 2002). Because they are so cheap and abundant, slaves are no longer a major investment worth maintaining. If slaves become ill or injured, too old to work, or troublesome to the slaveholder, they are dumped or killed and replaced with another slave (Cernasky 2002).

Although chattel slavery still exists in some areas, most forced laborers today are not "owned" but are rather controlled by violence or the threat of violence. The most common form of forced labor today is called *bonded labor*. Bonded laborers are usually illiterate, landless, rural, poor individuals who take out a loan simply to survive or to pay for a wedding, funeral, medicines, fertilizer, or other necessities. Debtors must work for the creditor to pay back the loan, but often they are unable to repay it. Creditors can keep debtors in bondage indefinitely in two main ways. First, they can charge the debtors illegal fines (for workplace "violations" or for poorly performed work) or charge laborers for food, tools, and transportation to the work site while keeping wages too low for the debt to ever be repaid. Alternatively, creditors can claim that all the labor performed by the debtor is collateral for the debt and cannot be used to reduce it (Miers 2003).

Another common form of forced labor involves luring individuals with the promise of a good job and instead holding them captive and forcing them to work. Migrant workers are particularly vulnerable because if they try to escape and report their abuse, they risk deportation. Organized crime rings are sometimes involved in the international trafficking of human beings, which often flows from developing nations to the West. A form of forced labor most common

chattel slavery An old form of slavery in which slaves are considered property that can be bought and sold.

in South Asia is sex slavery, in which girls are forced into prostitution by their own husbands, fathers, and brothers to earn money to pay family debts. Other girls are lured by offers of good jobs and then are forced to work in brothels under the threat of violence.

Another type of forced labor is conducted by the state or military. Forced military service, which has been reported in parts of Africa, and forced prison labor (common in China) are examples of state and military forced labor.

Slavery and Forced Labor in the United States. Kevin Bales estimates that there are 100,000–150,000 slaves in the United States today, mostly because of human trafficking for domestic work, migrant farm labor, or work in the sex industry (Cockburn 2003). Migrant workers are tricked into working for little or no pay as a means of repaying debts from their transport across the U.S. border, similar to debt bondage in South Asia. Traffickers posing as employment agents lure women into the United States with the promise of good jobs and education but then place them in "jobs" where they are forced to do domestic or sex work.

Sweatshop Labor

A U.S. Department of Labor investigation of the Daewoosa Samoa garment factory in American Samoa—a factory that produces men's sportswear for JCPenney—found that garment factory workers lived and worked under conditions of poor sanitation, malnutrition, electrical hazards, fire hazards, machinery hazards, illegally low wages, sexual harassment and invasion of privacy, workplace violence and corporal punishment, and overcrowded barracks in which two workers were forced to share each bed (National Labor Committee 2001). Female workers reported that the company owner routinely entered their barracks to watch them shower and dress. Workers reported incidents in which security guards slapped and kicked workers. The food provided to the workers at the Daewoosa Samoa garment factory consisted of a watery broth of rice and cabbage.

The workers at the Daewoosa Samoa garment factory are among the millions of people worldwide who work in **sweatshops**—work environments that are characterized by less-than-minimum-wage pay, excessively long hours of work (often without overtime pay), unsafe or inhumane working conditions, abusive treatment of workers by employers, and/or the lack of worker organizations aimed at negotiating better working conditions. Sweatshop labor conditions occur in a wide variety of industries, including garment production, manufacturing, mining, and agriculture.

At a U.S. Senate panel hearing, members of a Senate subcommittee heard testimony from those who witnessed sweatshop conditions at overseas factories that produce goods for U.S. companies (Tate 2007). A former textile worker in Bangladesh described conditions at a company called Harvest Rich, where clothing is sewn for the U.S. firms including Wal-Mart and JCPenney. She testified that hundreds of children, some as young as 11 years old, were illegally working at Harvest Rich, sometimes for up to 20 hours a day. "Before clothing shipments had to leave for the United States, there are often mandatory 19 to 20-hour shifts from 8:00 a.m. to 3:00 or 4:00 a.m. . . . The workers would sleep on the factory floor for a few hours before getting up for their next shift in the morning. If they did anything wrong, they were beaten every day." Workers had 2 days off a month and were paid $3.20 a week.

" Everything that is really great and inspiring is created by the individual who can labor in freedom. **"**

Albert Einstein

sweatshops Work environments that are characterized by less-than-minimum-wage pay, excessively long hours of work (often without overtime pay), unsafe or inhumane working conditions, abusive treatment of workers by employers, and/or the lack of worker organizations aimed to negotiate better working conditions.

Sweatshop labor commonly occurs in the garment industry.

A worker in Colombia's flower industry, who was employed at a plant owned by the U.S. company Dole, described workers being exposed to hazardous pesticides, the firing of sick workers, forced pregnancy testing for women, and strong-arm union busting tactics by companies. Other testimony revealed conditions at a plant in Jordan:

> At the Western garment factory, which made fleece jackets for Walmart [sic], there were 14 or 15-year-old kids working 18 or 20 hour shifts. . . . They worked from 8:00 in the morning until midnight or until 4:00 a.m. And they did this seven days a week. They did not get paid for first four months of 2006, they did not receive one cent in wages. They were working as slave labor. When they passed out they were struck by rulers to wake them up. There were four girls who were raped by management. (Tate 2007)

Many products in the U.S. consumer market are made under sweatshop conditions. An investigative report on working conditions in five Chinese factories that produce products for Disney, Wal-Mart, Kmart, Mattel, and McDonald's revealed sweatshop conditions that violate Chinese labor laws (Students and Scholars Against Corporate Misbehavior 2005). Workers are forced to work grueling 12- to 15-hour days, earning just 33–41 cents an hour. In some factories women are denied their legal maternity rights. Workers are housed in overcrowded dorm rooms and fed horrible food at the factory canteen. Workers are charged for the housing and food provided at the factory (even if they live and eat elsewhere), which often costs them one-fifth to one-third of their monthly wages. Some factories have no fans and become oppressively hot. Workers often faint from exhaustion and the unbearably stifling heat. Some workers are exposed to strong-smelling gases from working with glue, with no protective masks or ventilation system. Crushed fingers and other injuries are common in some factory departments. Workers have no health insurance, no pension, and no right to freedom of association or to organize.

Sweatshop Labor in the United States. Sweatshop conditions in overseas industries have been widely publicized. However, many Americans do not realize the extent to which sweatshops exist in the United States. The Department of Labor estimates that more than half of the country's 22,000 sewing shops violate minimum wage and overtime laws and that 75 percent violate safety and health laws ("The Garment Industry" 2001). Most garment workers in the United States are immigrant women who typically work 60–80 hours a week, often earning less than minimum wage with no overtime, and many face verbal and physical abuse.

Immigrant farm workers, who process 85 percent of the fruits and vegetables grown in the United States, also work under sweatshop conditions. Many live in substandard and crowded housing provided by their employer and lack access to safe drinking water as well as bathing and sanitary toilet facilities. Farm workers commonly suffer from heat exhaustion, back and muscle strains, injuries resulting from the use of sharp and heavy farm equipment, and illness resulting from pesticide exposure (Austin 2002). Working 12-hour days under hazardous conditions, farm workers have the lowest annual family incomes of any U.S. wage and salary workers, and more than 60 percent of them live in poverty (Thompson 2002). Problems associated with immigrant labor are discussed further in Chapter 9.

Health and Safety Hazards in the U.S. Workplace

Many workplaces are safer today than in generations past. Nevertheless, fatal and disabling occupational injuries and illnesses still occur in troubling numbers. The incidence of illnesses resulting from hazardous working conditions is probably much higher than the reported statistics show, because long-term latent illnesses caused by, for example, exposure to carcinogens often are difficult to relate to the workplace and are not adequately recognized and reported.

Workplace Fatalities. The International Labour Organization estimates that 1.1 million workers worldwide die on the job or from occupational disease each year ("Editorial" 2000). In 2005, 5,702 U.S. workers—93 percent of whom were men—died of fatal work-related injuries (Bureau of Labor Statistics 2006). The most common type of job-related fatality involves transportation accidents (see Figure 7.1). Industries with the highest rates (per 100,000 workers) of fatal injuries include agriculture/forestry/fishing and hunting, mining, transportation, and construction.

Occupational Illnesses and Nonfatal Injuries. The Bureau of Labor Statistics (2004) reported 4.4 million nonfatal occupational injuries and illnesses in private industry in 2003—a rate of 5 cases per 100 full-time workers. Sprains and strains are the most common nonfatal occupational injury or illness involving days away from work. Laborers and material movers and truck drivers suffer the most injuries and illnesses involving days away from work, followed by nursing aides, orderlies, and attendants, who commonly experience back strains from lifting and moving patients (Bureau of Labor Statistics 2005a).

The most common types of workplace illness are disorders associated with repeated motion or trauma, such as carpal tunnel syndrome (a wrist disorder that can cause numbness, tingling, and severe pain), tendinitis (inflammation of the tendons), and noise-induced hearing loss. Such disorders—referred to by a

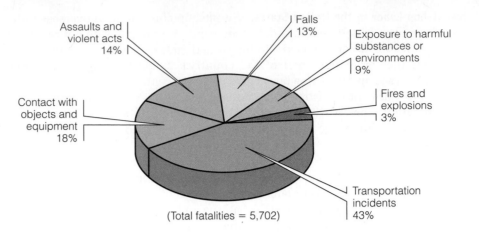

FIGURE 7.1
Causes of workplace fatalities, 2005.
Source: Bureau of Labor Statistics (2006).

Assaults and violent acts 14%

Contact with objects and equipment 18%

Falls 13%

Exposure to harmful substances or environments 9%

Fires and explosions 3%

Transportation incidents 43%

(Total fatalities = 5,702)

number of terms, including **cumulative trauma disorders** and **repetitive motion disorders**—are muscle, tendon, vascular, and nerve injuries that result from repeated or sustained actions or exertions of different body parts. Jobs that are associated with high rates of upper-body cumulative trauma disorders include computer programming, manufacturing, meatpacking, poultry processing, and clerical and office work. Repetitive motion disorders are classified as illnesses, not injuries, because they are not sudden, instantaneous traumatic events. Carpal tunnel syndrome results in more days absent from work than fractures or amputations.

Job Stress. Another work-related health problem is job stress. In a national sample of U.S. employees more than one-fourth (26 percent) felt "overworked" and 27 percent felt "overwhelmed" by how much work they had to do often or very often in the past month (Galinsky et al. 2005). A national Gallup poll found that nearly one-third of respondents (31 percent) said they are somewhat or completely dissatisfied with the amount of on-the-job stress in their jobs (Gallup Organization 2006). Sources of workplace stress include stresses embedded in tasks and work roles, interpersonal relationships in the workplace, difficulty balancing work and family, the physical work environment, workplace harassment, and the stress of adapting to changes in workplace technology (Barling, Kelloway, & Frone 2005).

Prolonged job stress, also known as **job burnout,** can cause or contribute to physical and mental health problems, such as high blood pressure, ulcers, headaches, anxiety, and depression. Taking time off to heal and "recharge one's batteries" is not an option for many workers: One-half of the U.S. workforce has no paid sick leave and one-fourth has no paid vacation (Watkins 2002). And pressures in the workplace often mean that workers bring their jobs with them on vacation. One in five employees works on his or her laptop computer while on vacation, 40 percent check office e-mail, and 50 percent check voicemail (Fram 2007).

Job Dissatisfaction and Alienation

A 2005 Gallup poll found that 5 percent of non-Hispanic whites, 10 percent of Hispanics, and 16 percent of blacks are somewhat or very dissatisfied with their job or the work they do (Gallup Organization 2005). One source of dissatisfac-

cumulative trauma disorders Also known as repetitive motion disorders, muscle, tendon, vascular, and nerve injuries that result from repeated or sustained actions or exertions of different body parts.

job burnout Prolonged job stress that can cause or contribute to high blood pressure, ulcers, headaches, anxiety, depression, and other health problems.

This man, vacationing on the Greek island of Santorini, is among the one in five U.S. workers who works while on vacation.

tion is declining wages. In 2005 nearly one-fourth (24.5 percent) of U.S. workers earned poverty-level hourly wages (Mishel, Bernstein, & Sylvia Allegretto 2006). Workers in low-wage, low-status jobs with few or no benefits and little job security are vulnerable to feelings of dissatisfaction not only with their jobs but also with their limited housing and lifestyle options. In this chapter's *The Human Side* feature some of the dissatisfactions associated with low-wage work are expressed.

One form of job dissatisfaction is a feeling of **alienation.** Work in industrialized societies is characterized by a high degree of division of labor and specialization of work roles. As a result, workers' tasks are repetitive and monotonous and often involve little or no creativity. Limited to specific tasks by their work roles, workers are unable to express and utilize their full potential—intellectual, emotional, and physical. According to Marx, when workers are merely cogs in a machine, they become estranged from their work, the product they create, other human beings, and themselves. Marx called this estrangement "alienation." As we discussed earlier, the McDonaldization of the workplace also contributes to alienation.

Alienation usually has four components: powerlessness, meaninglessness, normlessness, and self-estrangement. Powerlessness results from working in an environment in which one has little or no control over the decisions that affect one's work. Meaninglessness results when workers do not find fulfillment in their work. Workers may experience normlessness if workplace norms are unclear or conflicting. For example, many companies that have family leave policies informally discourage workers from using them, or workplaces that officially promote nondiscrimination in reality practice discrimination. Alienation also involves a feeling of self-estrangement, which stems from the workers' inability to realize their full human potential in their work roles and lack of connections to others.

> 66 Clearly the most unfortunate people are those who must do the same thing over and over again, every minute, or perhaps twenty to the minute. They deserve the shortest hours and the highest pay. 99
>
> John Kenneth Galbraith
> American economist

alienation The condition that results when workers perform repetitive, monotonous work tasks, and they become estranged from their work, the product they create, other people, and themselves.

Barbara Ehrenreich's best-selling book, *Nickel and Dimed,* describes the day-to-day struggles of low-wage work and surviving on a low-wage income.

Barbara Ehrenreich, a successful journalist with a Ph.D., wondered how anyone could survive on low-income wages: $6 to $7 an hour. To find out, she lived for one year on wages she earned doing low-wage work, moving from Florida to Maine to Minnesota, taking jobs as a waitress, hotel maid, house cleaner, nursing home aide, and Wal-Mart sales person, and living in the cheapest lodging she could find that offered an acceptable level of safety. Ehrenreich chronicled her experiences in Nickel and Dimed: On (Not) Getting By in America—*a book that became a New York Times bestseller. In this* Human Side *feature we present excerpts from an interview of Barbara Ehrenreich by Jamie Passaro in which Ehrenreich talks about her experiences as a low-wage earner and her viewpoints concerning low-wage work in America (Passaro 2003).*

Passaro: What surprised you most during your months of low-wage work?

Ehrenreich: It was a surprise to me how challenging these jobs were. I was expecting that I would be doing dull, repetitive work, that I would be bored out of my mind.

Instead I was struggling all the time, physically and mentally, to master these jobs. At Wal-Mart I had to memorize the locations of hundreds of clothing items so I could put everything back in its exact place. In the nursing home I had about fifteen minutes to learn the names and dietary requirements of thirty patients. It took all the concentration I had. So I no longer use the word *unskilled* to describe any job.

Passaro: How do you think your experience would have been different if you were a man?

Ehrenreich: A lot of low-wage jobs are really for either sex now because, as heavy industry declines, the "masculine" jobs of the past are not there anymore. There are men working at Wal-Mart and in restaurants and in nursing homes. The only difference for me is that a man probably would not have been as fearful as I was about living in a creepy residential motel with no privacy or security. . . .

Passaro: When you went back to your middle-class life after working low-wage jobs, how were you different?

Ehrenreich: I was more impatient with affluent people who don't see these problems or who aren't particularly interested and brush them off. . . .

Passaro: Why do you think class inequality is such a taboo subject in the mainstream media?

Ehrenreich: It undercuts the American myth that anybody can become rich, that it's just a matter of personal ability and determination. . . . We like to tell ourselves that everybody is equal. To admit that large numbers of people are systematically held back is hard, because it means upward mobility is not an option for everybody. But that's the way it is. There are just too many

Work–Family Concerns

Spouses, parents, and adult children caring for elderly parents struggle to balance their work and family responsibilities. In nearly two-thirds (62 percent) of married couples with children younger than age 18 and in more than half (56 percent) of married couples with children younger than age 6, both parents are employed. In addition, 72 percent of women in female-headed single-parent households and 84 percent of men in male-headed single-parent households are employed (Bureau of Labor Statistics 2007b).

A major concern of employed parents is arranging and paying for child care. About 3.3 million children younger than age 13 (15 percent of 6- to 12-year-olds) are left without adult supervision for some period of time each week (Vandivere et al. 2003). In some two-parent households spouses or partners work different shifts so that one adult can be home with the children. However, working different shifts strains marriage relationships, because the partners rarely have time off together.

things pressing poor people down, keeping them where they are. . . . The poor have become "invisiblized" in our society. They're given very little mention in the news and entertainment media. You just don't hear about them. The media system is fed by corporate advertising, and advertisers want "good demographics"—that is, they want to reach mostly the upper middle-class. . . .

Passaro: Many editors claim the middle class isn't interested in reading about poverty or the working poor. So why did *Nickel and Dimed* grab the attention of the media and the middle-class people who are presumably reading it?

Ehrenreich: One reason is that *Nickel and Dimed* is very personal and subjective, not preachy. It's not about the poor in general. It's just about me trying to survive. So people who are completely unfamiliar with the world of low-wage work can see it through the eyes of someone who is somewhat like them. . . .

Passaro: Do you think that we should boycott chain stores and restaurants that don't pay a living wage?

Ehrenreich: And then where are you going to shop or eat? At an upscale restaurant where the busboys and the dishwashers still earn little above minimum wage and the coffee beans have been picked by children in Central America? A lot of people come up to me and say, "I'll never go to Wal-Mart again." Well, terrific. So you go to a nice little boutique, which also pays its retail clerks seven dollars an hour and maybe gets its very expensive clothes from sweatshops, too. You could pay two hundred dollars for a dress that some poor seamstress made five dollars for sewing. These problems are so widespread, it's hard for me to see how boycotting a single business would help much. That said, if a boycott were called on some particular business, and there were a focused campaign surrounding it, I would respect it. . . .

Passaro: I feel guilty wherever I shop.

Ehrenreich: There's no avoiding that guilt. What are you going to do? Weave your own cloth like Gandhi tried to do? What we can do to help hardworking people trapped in poverty is fight for increasing social benefits, universal health insurance, and a universal child-care subsidy.

We can demand that cities build affordable housing. . . . Another possibility would be to tell the courts to get serious about enforcing the law against firing people for union activity. That's the law, but it's not enforced. You could also join the living-wage movement, which is using whatever leverage it has to convince individual cities to raise wages. . . .

Passaro: Education is often seen as the best way to move people out of poverty, yet menial jobs are always going to exist. Does the nature of these jobs need to change?

Ehrenreich: I get a little annoyed when someone says, "What's wrong with these people? Why don't they get an education?" Well, great, but then who's going to take care of your elderly grandmother in the nursing home? Who's going to wait on you when you go to a restaurant or a discount store? These are important jobs, jobs that need to be done, jobs that take intelligence and concentration and sometimes a great deal of compassion. Why don't we just pay people decently for doing them?

Source: Passaro, Jamie 2003 (January). "Fingers to the Bone: Barbara Ehrenreich on the Plight of the Working Poor," *The Sun*, pp. 4–10. Used by permission.

Another work-family concern involves caring for elderly family members. More than one-third (35 percent) of workers say that they have provided care for a relative or in-law age 65 or older in the past year (Bond et al. 2002).

Two-thirds of working parents feel they do not have enough time to spend with their families (Galinsky, Bond, & Hill 2004). U.S. workers clock more hours than workers in any other Organization for Economic Development (OECD) country except New Zealand. Full-time workers in the United States averaged 46.2 weeks of work per year (in 2004), or 10.2 weeks more than workers in Sweden, who worked the fewest weeks of all the OECD countries (Mishel et al. 2006). More time at work means less time to care for and be with one's family.

One reason that the average U.S. worker puts in more hours on the job than the average European worker is that the typical European worker enjoys significantly more paid vacations each year (6–8 weeks) than the typical U.S. worker, who has an average of 16.6 paid vacation days (Galinsky et al. 2005). In Europe, the minimum vacation is, by law, 4–5 weeks per year, whereas the United States

> **"**Don't sacrifice your life to work and ideals. The most important things in life are human relations. I found that out too late.**"**
>
> Katharin de Susannah Prichard
> Australian author

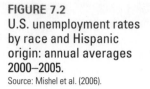

FIGURE 7.2
U.S. unemployment rates
by race and Hispanic
origin: annual averages
2000–2005.
Source: Mishel et al. (2006).

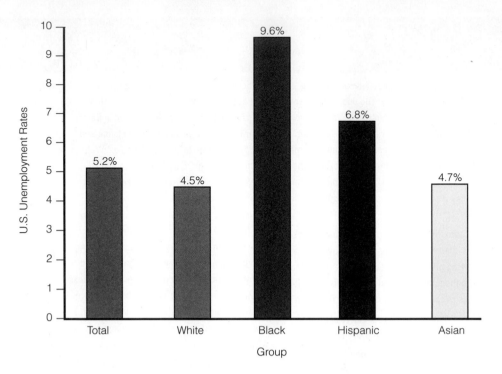

has no mandated vacation time. The typical European workweek also tends to be shorter (such as the 35-hour week in France). And there are possible cultural differences: U.S. workers tend to be more willing to work longer hours to achieve higher earnings, whereas European workers tend to be more willing to sacrifice potential earnings to have more leisure time (OECD 2004).

Unemployment and Underemployment

The International Labour Organization (2007) reported that in 2006 an estimated 195 million people worldwide—6.3 percent of the labor force—were unemployed. The Middle East and North Africa remains the region with the highest unemployment rate in the world (12.2 percent in 2006). Nearly half of the jobless worldwide are young people ages 15–24 (United Nations 2005).

Measures of **unemployment** in the United States consider an individual to be unemployed if he or she is currently without employment, is actively seeking employment, and is available for employment. In 2000 the U.S. unemployment rate dipped to a 31-year low of 4 percent but rose following the events of September 11, 2001, and in 2005 the unemployment rate was 5.1 percent (Mishel et al. 2006). Rates of unemployment are higher among racial and ethnic minorities (see Figure 7.2) and among those with lower levels of education (see Chapter 8).

The causes of unemployment are varied and complex. One cause of U.S. unemployment is **job exportation,** the relocation of jobs to other countries where products can be produced more cheaply. **Automation,** or the replacement of human labor with machinery and equipment, also contributes to unemployment.

Another cause of unemployment is increased global competition. In 2005 General Motors, the world's largest automaker, announced that it would cut 25,000 jobs, or about 23 percent of its workforce, largely because of increased competition in the automobile market. The practice of laying off large numbers

unemployment To be currently without employment, actively seeking employment, and available for employment, according to U.S. measures of unemployment.

job exportation The relocation of jobs to other countries where products can be produced more cheaply.

automation The replacement of human labor with machinery and equipment.

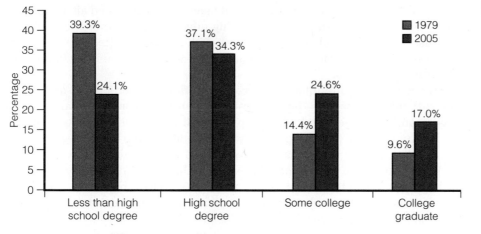

FIGURE 7.3
Shares of long-term
unemployment by
education.
Source: Mishel et al. (2006).

of employees—called **downsizing**—is a common occurrence in the corporate world.

Long-Term Unemployment. The **long-term unemployment rate** refers to the share of the unemployed who have been out of work for 27 weeks or more. In 2005, 1 in 5 (19.6 percent) of the unemployed individuals in the United States had been out of work for 6 months or more, up from 8.6 percent in 1979 (Mishel et al. 2006). Although less educated workers are most vulnerable to unemployment, from 1979 to 2005 the share of long-term unemployment actually decreased for workers with less education (see Figure 7.3). In 1979, those with less than a high school degree made up 39.3 percent of the long-term unemployed, but the share dropped to 24.1 percent in 2005. Meanwhile, the share for workers with a college degree rose from 9.2 percent to 17.0 percent (Mishel et al. 2006). This means that higher levels of education are no longer providing insulation against long-term joblessness.

Underemployment. Unemployment figures do not include "discouraged" workers, who have given up on finding a job and are no longer looking for employment. **Underemployment** is a broader term that includes unemployed workers as well as (1) those working part-time but who wish to work full-time ("involuntary" part-timers), (2) those who want to work but have been discouraged from searching by their lack of success ("discouraged" workers), and (3) others who are neither working nor seeking work but who indicate that they want and are available to work and have looked for employment in the last 12 months. The underemployment rate, which tends to be higher than the unemployment rate, was 8.9 percent in 2005 (Mishel et al. 2006).

Effects of Unemployment on Individuals, Families, and Societies. Unemployment causes serious difficulties for individuals and families. About one-third of people who are laid off drop out of the workforce; they are neither employed or actively looking for work. About one-third of people find jobs that, 2 years after their layoffs, pay 15–20 percent less than the jobs that they had before. And about one-third find jobs that, 2 years later, pay the same or more than they had earned before (Uchitelle 2006). Workers who lose their jobs and receive unemployment insurance receive less than 40 percent of their prior wages (Stettner &

downsizing The practice of laying off large numbers of employees.

long-term unemployment rate The share of the unemployed who have been out of work for 27 weeks or more.

underemployment Unemployed workers as well as (1) those working part-time but who wish to work full-time, (2) those who want to work but have been discouraged from searching by their lack of success, and (3) others who are neither working nor seeking work but who want and are available to work and have looked for employment in the last year.

Allegretto 2005). Typically, unemployment benefits are exhausted after 6 months. Long-term unemployment can have lasting effects, such as increased debt, diminished retirement and savings accounts (which are depleted to meet living expenses), and/or relocation from secure housing and communities to unfamiliar places to find a job.

But even when individuals who are laid off find another job in 1 or 2 weeks, they still suffer damage to their self-esteem from having been told they are no longer wanted or needed at their workplace. And being fired affects worker trust and loyalty in future jobs. Employees who are not fired during a mass layoff are also affected, as they worry that "their job could be next" (Uchitelle 2006).

For workers who receive health insurance through their employer, losing a job can mean losing health insurance coverage (see also Chapter 2). Unemployment is a risk factor for substance abuse, poor health, and homelessness, which affects families as well as individuals. When an adult is unemployed, other family members are often compelled to work more hours to keep the family afloat. In addition, unemployment, along with bleak job prospects, propels some individuals to turn to illegitimate, criminal sources of income, such as theft, drug dealing, and prostitution. In families, unemployment is also a risk factor for child and spousal abuse and marital instability.

Plant closings and large-scale layoffs affect communities by lowering property values and depressing community living standards. High numbers of unemployed adults create a drain on societies that provide support to those without jobs. The high numbers of young adults without jobs create a risk for crime, violence, and unrest (United Nations 2005). As discussed in Chapters 6 and 13, unemployed young adults are targeted for recruitment into terrorist groups.

Labor Unions and the Struggle for Workers' Rights

Having a job is no guarantee of having favorable working conditions and receiving decent pay and benefits. **Labor unions,** which originally developed to protect workers and represent them at negotiations between management and labor, are declining in membership.

Benefits and Disadvantages of Labor Unions to Workers. Labor unions have played an important role in fighting for fair wages and benefits, healthy and safe work environments, and other forms of worker advocacy. In 2006 the median weekly earnings of full-time wage and salary workers who were union members was $833, compared with a median of $642 for nonunion workers (Bureau of Labor Statistics 2007c). Unionized workers also received insurance and pension benefits worth more than those of nonunion employees, and union workers also got more paid time off than nonunion workers.

Labor unions are also influential in achieving better working conditions. For example, the United Food and Commercial Workers (UFCW), the country's largest union representing poultry-processing workers, was instrumental in the formation of an Occupational Safety and Health Administration (OSHA) rule that established a federal workplace "potty" policy governing when employees can use the bathroom while on the job. According to UFCW international president Doug H. Dority, "For years workers in food processing industries have had to suffer the indignity of being denied the right to go to the bathroom when needed, just to maintain ever increasing assembly-line speeds" ("New OSHA

labor unions Worker advocacy organizations that originally developed to protect workers and represent them at negotiations between management and labor.

Policy Relieves Employees" 1998, p. 8). Dority claimed that "poultry processors often have no other choice than to relieve themselves where they stand on the assembly line because their floor boss will not let them leave their workstation" (p. 8). Now, OSHA mandates that employers must make toilet facilities available so that employees can use them when they need to.

The increasing number of women in labor unions, which nearly doubled from 20 percent to 39 percent between 1960 and 1998, has helped to strengthen labor unions' advocacy for women ("Labor's 'Female Friendly' Agenda" 1998). For example, a number of unions have been successful in bargaining for expanded family leave benefits, subsidized child care, elder care, and pay equity.

One of the disadvantages of unions is that members must pay dues and other fees, and these dues have been rising in recent years. Average annual dues for the 15 largest unions in 2004 ranged from $210 to $830, not including additional fees known as "special assessments." High union dues are problematic in light of the high salaries many union leaders make. In 2004, Donald Doser, president of the International Union of Operating Engineers, was the highest paid union leader, earning more than $800,000 (Brenner 2007).

Another disadvantage for unionized workers is the loss of individuality. Unionized workers are members of an overall bargaining unit in which the majority rules. Decisions made by the majority may conflict with individual employee's specific employment needs.

Declining Union Density.
The strength and membership of unions in the United States have declined over the last several decades. **Union density**—the percentage of workers who belong to unions—grew in the 1930s and peaked in the 1940s and 1950s, when 35 percent of U.S. workers were unionized. In 2006 the percentage of American workers belonging to unions had fallen to 12 percent, its lowest point in decades (Bureau of Labor Statistics 2007b).

One reason for the decline in union representation is the loss of manufacturing jobs, which tend to have higher rates of unionization than other industries. Job growth has occurred in high technology and financial services, where unions have little presence. In addition, globalization has led to layoffs and plant closings at many unionized work sites as a result of companies moving to other countries to find cheaper labor. A major reason why union representation has declined is that corporations take active measures to keep workers from unionizing, and weak U.S. labor laws fail to support and protect unionization.

Corporate Antiunion Activities.
In the 1960s and 1970s U.S. corporations mounted an offensive attack on labor unions, "aiming to tame them or maim them" (Gordon 1996, p. 207). Corporations hired management consultants to help them develop and implement antiunion campaigns. They threatened unions with decertification, fired union leaders and organizers, and threatened to relocate their plants unless the unions and their members "behaved."

> One management consultant firm . . . was unusually blunt in broadcasting its methods. A late-1970s blurb promoting its manual promised: "We will show you how to screw your employees (before they screw you)—how to keep them smiling on low pay—how to maneuver them into low-pay jobs they are afraid to walk away from." (Gordon 1996, p. 208)

At least 23,000 workers each year are fired or discriminated against at their workplace because of involvement in union-related activity (Bonior 2006).

" The Labor Movement: the folks who brought you the weekend. "

Bumper sticker

union density The percentage of workers who belong to unions.

A study of 62 union campaigns in the Chicago metropolitan area showed that, among employers faced with organizing campaigns (Mehta & Theodore 2005):

- 30 percent of employers fire pro-union workers.
- 49 percent of employers threaten to close a worksite when workers try to form a union, but only 2 percent actually do.
- 51 percent of employers coerce workers into opposing unions with bribery or favoritism.
- 82 percent of employers hire high-priced union-busting consultants to fight union organizing drives.
- 91 percent of employers force employees to attend one-on-one antiunion meetings with their supervisors.

This chapter's *Social Problems Research Up Close* features a Human Rights Watch research report that details Wal-Mart's antiunion strategies.

Weak U.S. Labor Laws. The 1935 National Labor Relations Act (NLRA) is the primary federal labor law in the United States. The NLRA guarantees the right to unionize, bargain collectively, and to strike to private sector employees. However, in addition to excluding public sector workers, the law excludes agricultural and domestic workers, supervisors, railroad and airline employees, and independent contractors. As a result, millions of workers do not have the right under U.S. law to negotiate their wages, hours, or employment terms.

In addition, changes in U.S. labor law over the years have eroded workers' right to freedom of association. Originally, labor law required employers to grant a demand for union recognition if a majority of workers signed a card indicating they wanted a union. But since 1947, employers can reject workers' demand for unionization and force a National Labor Relations Board (NLRB) election, which requires about one-third of workers to petition for the Board to hold the election. "The company then uses the time leading up to the election to focus its campaign against union formation, while disallowing opportunities for opposing views" (Human Rights Watch 2007, p. 18).

Although it is illegal to fire workers for engaging in union activities, there are few consequences for employers that do so.

> If you get fired for trying to organize, for example, you can apply to the NLRB. If the NLRB finds that you were illegally fired, the employer has to give you back-pay for the time you were fired—minus any money that you may have earned at another job. As you can imagine, most people who are fired for trying to organize will, in fact, get another job somewhere, so there's no compensation for them at all. Then all the employer has to do is post on a bulletin board at the work site that they won't do it again. So there are effectively no sanctions, and it is in the employers' interest to fire people. They really don't suffer many consequences for doing so, and firing a leading union supporter sends a very powerful message to the rest of the employees. The message is: if you too try to lead an organizing campaign, you are going to lose your job; and if you vote for a union, you could lose your job. (Bonior 2006)

In addition, there is a backlog of thousands of cases of unfair labor practices by employers, and workers often wait about 2 years from the filing of a charge until the NLRB resolves a case, discouraging many workers from filing charges (International Confederation of Free Trade Unions 2006).

Labor Union Struggles Around the World. International norms established by the United Nations and the International Labour Organization declare the rights of workers to organize, negotiate with management, and strike (Human Rights Watch 2000). In European countries labor unions are generally strong. However, in many less-developed countries and in countries undergoing economic transition, workers and labor unions struggle to have a voice in matters of wages and working conditions.

The International Confederation of Free Trade Unions (ICFTU) publishes an Annual Survey of Violations of Trade Union Rights that describes severe abuses of workers' rights in countries around the world. This annual survey is frightening documentation of the lack of workers' rights around the world and of the abuses of governments and employers in the continued suppression of workers' rights. The survey results in the last few years showed that between 100 and several hundred trade unionists are killed each year. Several thousands more are imprisoned, beaten in demonstrations, tortured by security forces or others, and often sentenced to long prison terms. And each year hundreds of thousands of workers lose their jobs merely for attempting to organize a trade union.

Colombia remains the most dangerous place in the world to be a trade unionist, with 70 people killed in 2005 for their trade union activity (International Confederation of Free Trade Unions 2006). In some cases the families of the unionist murder victims were also killed. Most reported cases of trade unionist assassinations are not properly investigated, and the murderers are not caught or punished.

Trade union rights are violated by both employers and governments. Governments in numerous countries hamper trade union activity or strike action and fail to enforce existing national and international laws that protect workers' rights. Some countries, such as Oman, Saudi Arabia, and Burma, do not recognize the right to form trade unions. Other countries, such as China, Egypt, and Syria, impose a trade union monopoly. And some governments, such as those in Belarus and Moldova (countries in Eastern Europe), try to coerce workers into joining the government-supported union. Eager to secure financial benefits from participating in the global market, governments see trade unions as an obstacle to their economic development.

STRATEGIES FOR ACTION: RESPONSES TO WORKERS' CONCERNS

Government, private business, human rights organizations, labor organizations, college student activists, and consumers play important roles in responding to the concerns of workers. Next we look at responses to slavery and sweatshop labor, health and safety concerns, work-family policies and programs, workforce development programs, efforts to strengthen labor, and challenges to corporate power and economic globalization.

Efforts to End Slavery

More than 50 years ago the United Nations stated in Article 4 of its Universal Declaration of Human Rights that "no one shall be held in slavery or servitude; slavery and the slave trade shall be prohibited in all their forms." Yet slavery persists

Wal-Mart is the largest company in the world, with more than $350 billion in revenue and more than $11 billion in profits in the fiscal year ending January 2007. With nearly 4,000 stores nationwide, Wal-Mart employs more workers in the United States than any other private employer. None of the more than 1.3 million U.S. Wal-Mart employees is in a labor union (although Wal-Mart has recognized unions at some of its international operations, including those in Argentina, Brazil, Mexico, England, Japan, and China). Research conducted by Human Rights Watch (2007) reveals the many legal and illegal antiunion tactics used by Wal-Mart.

Wal-Mart's strategy to prevent union formation is complex and multifaceted. The company does not engage in massive anti-union firings nor announce to workers that their store will close if a union is formed. Instead, the company uses myriad more subtle tactics that, bit by bit, chip away at—and sometimes devastate—workers' right to organize. (p. 3)

In this feature, we describe the methods and selected findings of the Human Rights Watch (2007) research report on Wal-Mart's anti-union strategies.

Methods and Sample

The organization Human Rights Watch interviewed 41 current and former Wal-Mart workers and managers between 2004 and early 2007 who were employed at Wal-Mart during or after union organizing campaigns. These workers and managers were selected because they had first-hand experience with the company's responses to attempts to orga-nize workers. Some supported the union, some were opposed, and others were ambivalent. Human Rights Watch representatives also met with labor lawyers and union organiz-ers, analyzed labor law violation legal cases against Wal-Mart, and reviewed numerous publications addressing working conditions at Wal-Mart.

Human Rights Watch contacted Wal-Mart three times in writing to obtain the company's views and to request meetings with company officials to hear their perspectives. Wal-Mart repeatedly refused to meet with Human Rights Watch and provided very limited responses in writing and over the phone.

Selected Findings

Human Rights Watch found that Wal-Mart campaigned aggressively against union formation, through a number of strategies that violate workers' right to choose freely whether to organize.

The Manager's Toolbox

Wal-Mart's "Manager's Toolbox" contains instructions to managers on preventing union formation. For example, the Toolbox instructs managers that to keep unions out of the company's U.S. stores, they must monitor mo-rale of their employees using a list of morale indicators. The Toolbox warns managers:

If your responses to the morale survey indicate the facility may have low morale, then you could be vulnerable to a union-organizing attempt. Now is the time to fix them! Address your Associates' issues! Don't wait for a union to volunteer to [f]ix the morale problems for you. (Human Rights Watch 2007, p. 77)

The Toolbox also includes a section titled, "The Facts on Unions," that presents information that managers should provide to their workers about unions. This information is designed to emphasize limitations and possible drawbacks of union formation. For example, managers are instructed the following:

Do tell associates if a union is voted in, everything (their wages, benefits and working conditions) would go on the bargaining table. It is much like the game show LET'S MAKE A DEAL! They could get more, they could get the same, or they could get less. Regardless, they will be responsible for dues, fees, fines, and assessments. . . . (Human Rights Watch 2007, p. 78)

The Toolbox also describes Wal-Mart's Open Door Policy as the "greatest barrier" to worker organizing. The Open Door Policy states that workers may discuss concerns with managers and, if they are not satis-fied with the response, take their concerns to any level of management without fear of retaliation. Although this policy sounds like a worker-friendly business practice, Wal-Mart cites the policy to workers as a key reason why workers do not need what Wal-Mart calls "third-party representation," asking workers why they would want to pay a union to speak for them when they have the Open Door Policy through which they can speak for themselves for free.

Censorship of Information

Human Rights Watch found that Wal-Mart has limited workers' access to informa-tion about the benefits of union formation.

throughout the world. The international community has drafted treaties on slavery, but many countries have yet to ratify and implement the different treaties.

One strategy to fight slavery is punishment. Slave traffickers often avoid punishment because, as a former official of the U.S. Agency for International Development explained, "government officials in dozens of countries assist,

Wal-Mart has banned union representatives from handbilling—passing out informational leaflets—outside its stores, while allowing representatives of nonunion organizations to remain. The company has confiscated union literature from employees and from break room tables and prohibited employees from distributing union flyers, while permitting nonunion information. Wal-Mart managers have prohibited employees from discussing unions or even talking with co-workers about wages and working conditions.

Threats and Surveillance

Wal-Mart has illegally threatened workers with serious consequences if they form a union, including loss of benefits, such as raises. The company has also used several illegal methods to gather information about union activity while simultaneously pressuring workers to stop organizing. "The company has coercively interrogated workers about their and their co-workers' union sympathies through direct and often hostile questioning and sent managers to eavesdrop on discussions among employees . . ." (p. 9). Former workers and managers from a Wal-Mart store in Kingman, Arizona, disclosed that Wal-Mart has also monitored union activity by focusing security cameras on areas where union organizing is most active.

One former Wal-Mart loss prevention worker explained that during the organizing drive at his store, "we were to monitor any activity that we thought might be organized . . . and place cameras in certain areas. . . ." (p. 10). He added that, in particular, they were supposed to focus the cameras on union leader Brad Jones and "report back what we saw. We needed to find a reason to fire Brad" (p. 10).

Wal-Mart employees in Aiken, South Carolina, told Human Rights Watch that they believe the 2001 union campaign at their store failed because of the climate of fear created by the company's response. One worker explained,

> I think the union failed because a lot of them were scared to come forward, scared for their jobs. That's exactly the reason I didn't sign up. . . . I never went to any union meetings. I was scared to. . . . Some of the girls, other associates, would say that if Wal-Mart would get wind of [my involvement with] the union, I'd be fired. (p. 11)

Wal-Mart's Union Hotline

When managers detect organizing activity at their stores, they are required to call Wal-Mart's Union Hotline to report union activity to headquarters. Almost immediately, Wal-Mart sends one or more members of its Labor Relations Team to the store to implement an aggressive antiunion campaign in conjunction with store management. Labor Relations Team members and store managers hold small- and large-group meetings that workers are strongly urged to attend. At the meetings, management describes the negative consequences of organizing and often shows videos to reinforce this message. Managers and the videos tell workers that union dues and other union fees are expensive and that workers may lose wages and benefits as a result of unionization. They also warn workers that Wal-Mart can permanently replace workers who go on strike. 'The videos dramatize the anti-union message by showing an example of a picket line that turns violent, characterizing unions as antiquated organizations, and portraying union organizers as harassing and bothersome people" (p. 7). A Wal-Mart worker from Greeley, Colorado, told Human Rights Watch:

> Anyone who saw the video would know it was anti-union, but they called it an educational video, which it wasn't. It made me pretty upset. . . . They depicted the organizing committee as people who'll be on your butt forever 'till you sign the card, and that's not how we are. We're not going to make you do something you don't want to do. Home office kept saying this is an educational class. It should not just give one side of the union. They should give pro and con, not just con. It's not fair to the associates. (p. 8)

Conclusion

The Human Rights Watch research report on Wal-Mart's labor practices documents the various strategies Wal-Mart uses to keep its U.S. workers from unionizing. Although Wal-Mart is not alone among U.S. companies in its efforts to combat union formation and in violating weak U.S. labor laws and taking advantage of ineffective labor law enforcement, "Wal-Mart stands out for the sheer magnitude and aggressiveness of its anti-union apparatus and actions" (Human Rights Watch 2007, p. 4).

Source: Human Rights Watch. 2007 (May).

overlook, or actively collude with traffickers" (quoted by Cockburn 2003, p. 16). In many countries the justice system is more likely to jail or expel sex slaves than to punish traffickers ("Sex Trade Enslaves Millions of Women, Youth" 2003). In 25 countries, however, slave trafficking is actively prosecuted and treated as a serious crime.

In the United States the Victims of Trafficking and Violence Protection Act, passed by Congress in 2000, protects slaves against deportation if they testify against their former owners. Convicted slave traffickers in the United States are subject to prison sentences, as shown in the following examples (Cockburn 2003):

- Louisa Satia and Kevin Waton Nanji each received 9 years for luring a 14-year-old girl from Cameroon with promises of schooling and then isolating her in their Maryland home, raping her, and forcing her to work as their domestic servant for 3 years.
- Sardar and Nadira Gasanov were sentenced to 5 years each for recruiting women from Uzbekistan with promises of jobs, taking their passports, and forcing them to work in strip clubs and bars in Texas.
- Juan, Ramiro, and Jose Ramos each received 10–12 years for transporting Mexicans to Florida and forcing them to work as fruit pickers.

U.S. corporations are also being held accountable for enterprises that involve forced labor and other human rights and labor violations. In 2003 the Unocal oil company became the first corporation in history to stand trial in the United States for human rights violations abroad (George 2003). Unocal was accused of involvement in a pipeline project that used Myanmar (formerly Burma) military personnel to provide "security" for a natural gas pipeline project in the remote Yadana region near the Thai border. According to the Ninth U.S. Circuit Court of Appeals, the soldiers' true role was to force villagers in the pipeline region to work without pay—a modern form of slavery. The military also forced villagers living along the pipeline route to relocate without compensation, raped and assaulted villagers, and imprisoned and/or executed those who opposed them. In 2004 Unocal announced that it had reached a settlement with the parties who alleged that Unocal was complicit in the human rights violations committed by the Myanmar military. In 2005, Unocal was purchased by ChevronTexaco.

Responses to Sweatshop Labor

The Fair Labor Association (FLA), established in 1996, is a coalition of companies, universities, and nongovernmental organizations (NGOs) that works to promote adherence to international labor standards and improve working conditions worldwide. In 2007, 20 leading brand-name apparel and footwear companies voluntarily participated in FLA's monitoring system, which inspects their overseas factories and requires them to meet minimum labor standards, such as not requiring workers to work more than 60 hours a week. In addition, 194 colleges and universities require their collegiate licensees (companies that manufacture logo-carrying goods for colleges and universities) to participate in FLA's monitoring system.

In its first few years of operation the FLA was criticized for allowing firms to select and directly pay their own monitors and to have a say in which factories were audited. In 2002 the FLA responded to these criticisms by taking much more control over external monitoring, with the FLA staff selecting factories for evaluation, choosing the monitoring organization, and requiring that inspections be unannounced (O'Rourke 2003). The FLA continues to be criticized, however, for having low standards in allowing below-poverty wages and excessive overtime and for requiring that only a small percentage of a manufacturer's

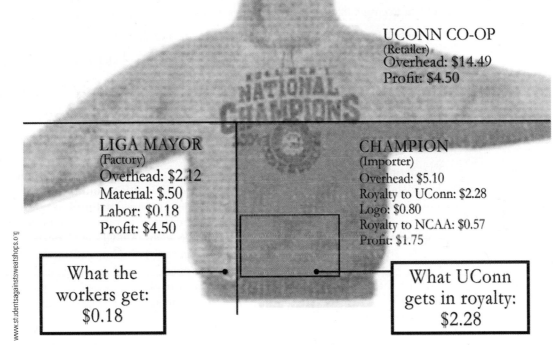

STORE PRICE: $37.99

UCONN CO-OP
(Retailer)
Overhead: $14.49
Profit: $4.50

LIGA MAYOR
(Factory)
Overhead: $2.12
Material: $.50
Labor: $0.18
Profit: $4.50

CHAMPION
(Importer)
Overhead: $5.10
Royalty to UConn: $2.28
Logo: $0.80
Royalty to NCAA: $0.57
Profit: $1.75

What the workers get: $0.18

What UConn gets in royalty: $2.28

www.studentsagainstsweatshops.org

This UConn men's championship sweat shirt, for sale at the UConn Co-op, was sewn by workers at a factory in Mexico who earn 18 cents per garment—less than one-tenth of what UConn makes in royalties. Profit, overhead, and other expenses along the supply chain push the retail sales price up to $37.99.

supplier factories be inspected each year. Critics also suggest that companies use their participation in the FLA as a marketing tool. Once "certified" by the FLA, companies can sew a label into their products saying that the products were made under fair working conditions (Benjamin 1998). In 2006, United Students Against Sweatshops created "FLA Watch" to "expose the truth about the Fair Labor Association . . . and . . . the FLA's ongoing failure to defend the rights of workers" ("FLA Watch" 2007a). The home page of the FLA Watch website explains, "The FLA purports to be an 'independent' monitor of working conditions in the apparel industry. But the organization is funded and controlled by the very corporations that have been repeatedly found to be sweatshop violators." FLA Watch further accuses the Fair Labor Association of being "nothing more than a public relations mouthpiece for the apparel industry. Created, funded, and controlled by Nike, Adidas, and other leading sweatshop abusers, the FLA is a classic case of the 'fox guarding the hen house'" (FLA Watch 2007b).

Student Activism. United Students Against Sweatshops is a student activism group that is affiliated with The Worker Rights Consortium (WRC)—a nonprofit organization working to ensure that factories that produce clothing and other goods bearing school logos respect basic rights of workers, such as the right to unionize and to receive living wages and overtime pay. More than 140 colleges and universities are affiliated with the WRC and participate in a Designated Suppliers Program. This program requires brands that are licensed to make university apparel to be produced in factories in which employees are paid a living wage and are represented by a union or other form of employee representation.

> **What Do You Think?** Do you think that colleges and universities have an obligation to ensure that any apparel or items with their college or university logo be made in sweatshop-free conditions? Does your college or university participate in either the Fair Labor Association or the Designated Suppliers Program? Do you think most students care if the college/university logo clothing or products they buy are made under sweatshop conditions?

Legislation. Perhaps the most effective strategy against sweatshop work conditions is legislation. In the United States, at least 160 U.S. cities, counties, and districts, plus 6 states have passed "sweatfree" procurement laws that prohibit public entities (such as schools, police, and fire departments) from purchasing uniforms and apparel made under sweatshop conditions ("Sweat-Free Update" 2007).

In 2006, the Decent Working Conditions and Fair Competition Act was introduced in the U.S. Congress—the first proposed federal anti-sweatshop legislation in the United States. The bill was re-introduced in 2007. If passed, this legislation will prohibit the import, export, or sale of sweatshop goods in the United States.

> Up to this point, it has been the companies that have demanded and won all sorts of enforceable laws—intellectual property and copyright laws backed up by sanctions—to defend their corporate trademarks, labels, and products. Yet, the corporations have long said that extending similar laws to protect the human rights of the 16-year-old girl in Bangladesh who sews the garment would be "an impediment to free trade." Under this distorted sense of values, the label is protected, but not the human being, the worker who makes the product. (National Labor Committee 2007)

International Anti-Sweatshop Efforts. The international community is also involved in efforts to improve working conditions and end sweatshop labor. The Clean Clothes Campaign is an international campaign in 9 European countries, focused on improving working conditions in the global garment and sportswear industries. The goals of the Clean Clothes Campaign are:

- Put pressure on companies to take responsibility for ensuring decent working conditions.
- Support workers, trade unions, and NGOs in producer countries.
- Raise awareness among consumers by providing information about working conditions in the global garment and sportswear industry, so that citizens can use their power as consumers.
- Explore legal possibilities for improving working conditions, and lobby for legislation to promote good working conditions and for laws that would compel governments and companies to become ethical consumers.

Pressure from opponents of sweatshop labor and consumer boycotts of products made by sweatshop labor have resulted in some improvements in factories that make goods for companies such as Nike and Gap, which have cut back on child labor, use less dangerous chemicals, and require fewer employees to work 80-hour weeks. At many factories supervisors have stopped hitting employees, have improved ventilation, and have stopped requiring workers to obtain permission to use the toilet. But improvements are not widespread, and oppressive forms of labor continue throughout the world. According to the National Labor Committee, two areas where "progress seems to grind to a halt" are efforts to form unions and efforts to achieve wage increases (Greenhouse 2000).

Responses to Worker Health and Safety Concerns

Over the past few decades health and safety conditions in the U.S. workplace have improved as a result of media attention, demands by unions for change, more white-collar jobs, and regulations by OSHA. Through OSHA the government develops, monitors, and enforces health and safety regulations in the workplace. Since OSHA was created three decades ago, workplace fatalities have dropped by 75 percent ("Editorial" 2000). But much work remains to be done to improve worker safety and health. Inadequate funding leaves OSHA unable to do its job effectively, with only 2,138 federal and state inspectors responsible for monitoring and enforcing job safety laws at 8 million workplaces (AFL-CIO 2005). At current staffing levels it would take federal OSHA employees 108 years to inspect each workplace under its jurisdiction just once.

Because "the task of monitoring and enforcement simply cannot be effectively carried out by a government administrative agency," Kenworthy (1995) suggested that the United States follow the example of many other industrialized countries: Turn over the bulk of responsibility for health and safety monitoring to the workforce (p. 114). Worker health and safety committees are a standard feature of companies in many other industrialized countries and are mandatory in most of Europe. These committees are authorized to inspect workplaces and cite employers for violations of health and safety regulations.

Business and industry often fight efforts to improve safety and health conditions in the workplace. For example, in 1999, after a 10-year struggle between labor and business, OSHA issued ergonomic standards requiring employers to implement ergonomic programs in jobs where musculoskeletal disorders occur. **Ergonomics** refers to the designing or redesigning of the workplace to prevent and reduce cumulative trauma disorders. According to OSHA, the proposed ergonomic standards would prevent 4.6 million workers over the next 10 years from experiencing painful, potentially debilitating work-related musculoskeletal disorders (U.S. Department of Labor 2001). However, business and industry representatives pressured Congress and President George W. Bush to repeal the ergonomic standard soon after Bush took office in 2000. Now, there are no regulations requiring employers to assess ergonomic hazards in the workplace (such as excessive repetition or poor workstation design) or to take steps to reduce these hazards.

Research suggests that working beyond 50 hours a week jeopardizes health and safety in the workplace. As a result, most countries limit work hours to less than 48 hours, and about half of countries have a 40-hour limit. However, more than one in five (22 percent) of the global workforce works more than 48 hours per week (Lee, McCann, & Messenger 2007). Thus, efforts to improve health and safety of workers worldwide include establishing and enforcing laws that limit work hours.

ergonomics The designing or redesigning of the workplace to prevent and reduce cumulative trauma disorders.

In developing countries governments fear that strict enforcement of workplace regulations will discourage foreign investment ("Editorial" 2000). Investment in workplace safety in developing countries, whether by domestic firms or foreign multinationals, is far below that in the rich countries. Unless global standards of worker safety are implemented and enforced in *all* countries, millions of workers throughout the world will continue to suffer under hazardous work conditions. Low unionization rates and workers' fears of losing their jobs—or their lives—if they demand health and safety protections leave most workers powerless to improve their working conditions.

Behavior-Based Safety Programs. A controversial health and safety strategy used by business management is behavior-based safety programs. Instead of examining how work processes and conditions compromise health and safety on the job, **behavior-based safety programs** direct attention to workers themselves as the problem. Behavior-based safety programs claim that 80–96 percent of job injuries and illnesses are caused by workers' own carelessness and unsafe acts (Frederick & Lessin 2000). These programs focus on teaching employees and managers to identify, "discipline," and change unsafe worker behaviors that cause accidents and encourage a work culture that recognizes and rewards safe behaviors.

Critics contend that behavior-based safety programs divert attention away from the employer's failure to provide safe working conditions. They also say that the real goal of behavior-based safety programs is to discourage workers from reporting illness and injuries. Workers whose employers have implemented behavior-based safety programs describe an atmosphere of fear in the workplace, such that workers are reluctant to report injuries and illnesses for fear of being labeled an "unsafe worker." At one factory that had implemented a behavior-based safety program, when a union representative asked workers during shift meetings to raise their hands if they were afraid to report injuries, about half of 150 workers raised their hands (Frederick & Lessin 2000). Worried that some workers feared even raising their hand in response to the question, the union representative asked a subsequent group to write yes on a piece of paper if they were afraid to report injuries. Seventy percent indicated they were afraid to report injuries. Asked why they would not report injuries, workers said, "We know that we will face an inquisition," "We would be humiliated," and "We might be blamed for the injury."

A rule issued by OSHA protects workers by prohibiting discrimination against an employee for reporting a work-related fatality, injury, or illness ("Workers at Risk" 2003). This rule also prohibits discrimination against an employee for filing a safety and health complaint or asking for health and safety records.

behavior-based safety programs A strategy used by business management that attributes health and safety problems in the workplace to workers' behavior, rather than to work processes and conditions.

Work–Family Policies and Programs

The increase in women in the workforce over the past several decades (see Table 7.1) has been accompanied by an increase in government and company policies designed to help women and men balance their work and family roles. Such policies are referred to by a number of terms, including "work-family," "work-life," and "family-friendly" policies.

Federal and State Family and Medical Leave Initiatives. In 1993 President Clinton signed into law the **Family and Medical Leave Act (FMLA),** which requires all companies with 50 or more employees to provide eligible workers (who work at least 25 hours a week and have been working for at least 1 year) with up to 12 weeks of job-protected, *unpaid* leave so that they can care for a seriously ill child, spouse, or parent; stay home to care for their newborn, newly adopted, or newly placed child; or take time off when they are seriously ill. Yet 40 percent of the workforce is not covered by the FMLA (Watkins 2002). Lower-wage earners are the least likely to have family and medical leave benefits and typically have few if any resources to fall back on in times of family illness or crisis.

TABLE 7.1 Percentage of U.S. Women Age 16 and Older in the Labor Force: 1970–2005

YEAR	PERCENTAGE OF WOMEN IN LABOR FORCE
1970	43.3
1980	51.5
1990	57.5
2005	59.3

Source: Bureau of Labor Statistics (2005b).

In 2002 California became the first state in the country to adopt a comprehensive family leave policy that provides workers up to 6 weeks of time off with about 55 percent of their regular pay while caring for a newborn or newly adopted child or when a family member is seriously ill (Watkins 2002). A number of other states have adopted state family leave policies.

In 2004 the Healthy Families Act was introduced into Congress. This act would require all employers with at least 15 employees to provide 7 days of paid sick leave annually for full-time employees who work at least 30 hours a week (or 1,500 hours a year). One study found that, if passed, the Healthy Families Act would save employers money, largely because of reduced employee turnover (Lovell 2005).

The proposed federal Family Leave Insurance bill, introduced in Congress in 2004 and 2005, would provide partial wage replacement ($250 a week) and job protection for eligible workers caring for (1) a newborn, newly adopted, or newly placed foster child or (2) a seriously ill child, spouse, parent, or parent-in-law. If passed, this policy would be funded by a payroll deduction of 2 cents per hour worked.

Worldwide, 164 countries have laws that guarantee workers paid maternity leave. The only countries that do not have such laws are the United States, Papua New Guinea, and Swaziland (Heymann 2006).

Employer–Provided Work–Family Policies. Aside from government-mandated work-family policies, some corporations and employers have "family-friendly" work policies and programs, including unpaid or paid family and medical leave, child-care assistance, assistance with elderly parent care, and flexible work options. Not surprisingly, organizations with higher proportions of full-time female employees are more likely to offer flexible scheduling, unpaid parental leave, and dependent care assistance (Davis & Kalleberg 2006).

A survey of more than 1,000 employers (with 50 or more employees) found that only 7 percent provide child care at or near the worksite, only 3 percent of companies pay for their employees' child-care costs, but nearly half (45 percent) have Dependent Care Assistance Plans that help employees pay for child care with pretax dollars (Bond et al. 2006). This survey also found that although elder care is not required by the federal FMLA, 79 percent of employers say they provide paid or unpaid time off without jeopardizing their jobs for employees to care for an elderly family member.

66 Our leaders talk as though they value families, but act as though families were a last priority. 99

Sylvia Hewlett and Cornel West
Family advocates

Family and Medical Leave Act (FMLA) A federal law that requires companies with 50 or more employees to provide eligible workers (who work at least 25 hours a week and have been working for at least 1 year) with up to 12 weeks of job-protected, unpaid leave so that they can care for an ill child, spouse, or parent; stay home to care for their newborn, newly adopted, or newly placed child; or take time off when they are seriously ill.

Offering employees more flexibility in their work hours helps parents balance their work and family demands. Flexible work arrangements, which benefit child-free workers as well as employed parents, include flextime, a compressed workweek, and teleworking. **Flextime** allows the employee to begin and end the workday at different times so long as 40 hours of work per week are maintained. Although 8 in 10 working women say control over their work hours is an important benefit, most (64 percent) do not have control over their work hours (AFL-CIO 2004). A **compressed workweek** allows employees to condense their work into fewer days (e.g., four 10-hour days each week).

Telework refers to flexible and alternative work arrangements that involve the use of information technology. Telework allows employees to work part- or full-time at home or at a satellite office. About 8 percent of U.S. workers have an employer that allows them to telecommute at least 1 day per month and roughly 20 percent of the workforce (including the self-employed) engages in telework (WorldatWork 2007). Studies have shown that employee satisfaction among teleworkers is higher than that for their nonteleworking counterparts, largely because the flexibility that telework allows enables workers to balance work and family demands (Lovelace 2000).

Barriers to Employees' Use of Work–Family Benefits. One study of more than 1,000 employers with 50 or more employees found that 30 percent of employers that are covered by the FMLA report offering *fewer* than the mandated 12 weeks of family leave (Bond et al. 2005). Perhaps the employers were unaware of their responsibilities under the federal FMLA, or they were deliberately violating the law. Even when work-family policies and programs are available, many employees do not take advantage of them. One barrier to using work-family benefits is lack of awareness: Many employees do not know what benefits are available to them. In a survey of employees covered by the FMLA, only 38 percent correctly reported that the FMLA applied to them and about half said they did not know whether it did (Cantor et al. 2001). In a study of employees in seven organizations, more than two-thirds were unaware of or mistaken about at least one work-life policy or practice. These employees believed that a policy was in operation when in fact it was not or that a policy that did formally exist did not (Still & Strang 2003).

Other barriers that discourage employee use of work-family benefits are related to economic and job security. Many workers who are covered by the FMLA do not use it because they cannot afford to take leave without pay. One survey found that 88 percent of workers who needed time off but did not take it said they would have taken leave if they could have received pay during their absence; nearly one-third of workers who needed leave but did not take it cited worry about losing their job as a reason for not taking leave (Cantor et al. 2001). Professor Phyllis Moen (2003) explained, "Many are afraid to take advantage of existing [work-family] options because doing so might signal a less-than-full

flextime A work arrangement that allows employees to begin and end the workday at different times so long as 40 hours per week are maintained.

compressed workweek A work arrangement that allows employees to condense their work into fewer days (e.g., four 10-hour days each week).

telework A work arrangement involving the use of information technology that allows employees to work part- or full-time at home or at a satellite office.

commitment to their jobs, exacting costs in future promotions, salary increases, and even job security" (p. 335).

Corporate initiatives in work-family benefits are a step in the right direction. However, companies that innovate in the area of work life by adopting many family-friendly programs do not show particularly high levels of benefit use among their employees (Still & Strang 2003). Although several magazines and professional organizations give awards and recognition to companies that offer family-friendly benefits, Moen (2003) suggested that "a better gauge of the worker friendliness of corporations might be the proportion of their workforces actually using them" (p. 335). In addition, "corporate leaders, along with those handing out rewards, need to recognize that job security is also a key component of friendliness or worklife quality" (Moen 2003, p. 335).

Workforce Development and Job-Creation Programs

The International Labour Organization (2003) estimated that at least 1 billion new jobs are needed in the coming decade to meet the United Nations goal of halving extreme poverty by 2015. Developing a workforce and creating jobs involves far-reaching efforts, including those designed to improve health and health care, alleviate poverty and malnutrition, develop infrastructures, and provide universal education.

> **Nothing will work unless you do.**
>
> Maya Angelou

In the United States workforce development programs include the following approaches (Martinson & Holcomb 2007):

Service-focused employment preparation. Focuses on strategies to improve the employability of hard-to-employ individuals through identifying special needs and providing a combination of targeted interventions (i.e., substance abuse treatment, domestic violence services, prison release reintegration assistance, mental health services, homelessness services, and others) in combination with employment services.

Subsidized employment. Involves subsidizing wages paid to employees with public funds. Employers are typically from the public or nonprofit sectors but can include for-profit businesses as well. In addition to receiving a paycheck, participants also receive a range of other supports and assistance.

Efforts to prepare high school students for work include the establishment of technical and vocational high schools and high school programs and school-to-work programs. School-to-work programs involve partnerships between business, labor, government, education, and community organizations that help prepare high school students for jobs (Leonard 1996). Although school-to-work programs vary, in general, they allow high school students to explore different careers, and they provide job skill training and work-based learning experiences, often with pay (Bassi & Ludwig 2000).

Educational attainment is often touted as the path to employment and economic security. Although education and skill development are important for individuals to realize their potential and to become better informed and more productive citizens, education alone is not the answer to unemployment and economic insecurity. More than two-thirds of jobs in the United States require little education (Martinson & Holcomb 2007); thus it is important to ensure that all jobs pay a minimum "living wage" (see also Chapter 6).

Economist Jared Bernstein (2006) suggested that the cultural focus on improving education "blames the victim" for his or her joblessness or economic hardships. "It removes the responsibility from policy makers and . . . from the inequities embedded in our market outcomes as well as in our education system itself, and places it squarely on the . . . individual" (Bernstein 2006, p. 56). In addition, Bernstein noted that the economy fails to produce enough jobs for those at all levels of education, including college graduates. The answer to joblessness is not more educated individuals; it is more jobs. One way to create more jobs is direct public-sector job creation, such as jobs to improve the nation's infrastructure and jobs that are geared toward energy independence (Bernstein 2006).

But as long as our economy allows people who work full-time to earn poverty-level wages, having a job is not necessarily the answer to economic self-sufficiency. As David Shipler (2005) explained in his book *The Working Poor:*

> A job alone is not enough. Medical insurance alone is not enough. Good housing alone is not enough. Reliable transportation, careful family budgeting, effective parenting, effective schooling are not enough when each is achieved in isolation from the rest. There is no single variable that can be altered to help working people move away from the edge of poverty. Only where the full array of factors is attacked can America fulfill its promise. (p. 11)

Efforts to Strengthen Labor

More than two-thirds of U.S. adults say that "labor unions are necessary to protect the working person" (Pew Research Center 2007). Although efforts to strengthen labor are viewed as problematic to corporations, employers, and some governments, such efforts have the potential to remedy many of the problems facing workers.

After numerous unsuccessful attempts to unionize workers at Wal-Mart, labor unions found a creative way to advocate for workers' rights at the chain—they helped form a new and unusual type of workers' association aimed at improving working conditions at Wal-Mart Stores Inc. The new group, called the Wal-Mart Workers Association, is not a labor union, but an organization of current and past Wal-Mart employees that advocates for decent wages and benefits and fair and equal treatment of all workers and provides assistance to past and present Wal-Mart workers who have been victims of discrimination or unfair treatment.

In an effort to strengthen their power, some labor unions have merged with one another. Labor union mergers result in higher membership numbers, thereby increasing the unions' financial resources, which are needed to recruit new members and to withstand long strikes. Because workers must fight for labor protections within a globalized economic system, their unions must cross national boundaries to build international cooperation and solidarity. Otherwise, employers can play working and poor people in different countries against each other. An example of international union cooperation occurred when leaders from 21 unions in 11 countries on 5 continents resolved to form a global union network at International Paper Company (IP), the largest paper company in the world. One union leader remarked, "IP crosses national borders in search of the highest profits, and the unions . . . have resolved to match that corporate globalization with a globalization of workers' solidarity" ("Unions Forge Global Network" 2002).

Strengthening labor unions requires combating the threats and violence against workers who attempt to organize or who join unions. One way to do this is to pressure governments to apprehend and punish the perpetrators of such violence. Although about 3,500 trade unionists were murdered between 1990 and

College student groups across the country have participated in boycotts against Coca-Cola in protest of the violence against union leaders at Colombian Coca-Cola plants.

Kirschner/United Students Against Sweatshops

2005, only 600 cases were investigated, resulting in just 6 convictions (Moloney 2005). Another tactic is to stop doing business with countries where government-sponsored violations of free trade union rights occur.

In the United States the NLRB and the courts play an important role in upholding workers' rights to unionize and sanctioning employers who violate these rights. The NLRB has the authority to issue job reinstatement and "back-pay" orders or other remedial orders to workers wrongfully fired or demoted for participating in union-related activities. The NLRB and the courts have held that employer threats to close the plant if the union succeeds in organizing can be unlawful under certain circumstances (Bronfenbrenner 2000). For example, *Guardian Industries Corp. v. NLRB* held that it was unlawful for a supervisor to say to an employee, "If we got a union in there, we'd be in the unemployment line." However, under the employer free-speech provisions of the Taft-Hartley Act, the courts have permitted the employer to predict a plant closing in situations in which it is based on an objective assessment of the economic consequences of unionization.

Proposed legislation called the Employee Free Choice Act would allow workers to sign a card stating that they want to be represented by a union. If a majority of the employees in any workplace sign such a card, the company would then have to recognize the union and bargain over terms and conditions of employment. In March 2007, the U.S. House of Representatives passed the Employee Free Choice Act and, as of this writing, the bill is pending in the U.S. Senate, but President George W. Bush has promised to veto the legislation if it passes the Senate.

Challenges to Corporate Power and Globalization
Challenges to corporate power and globalization include campaign finance reform and the antiglobalization movement.

In 2007, hundreds of South Koreans protested a free trade agreement with the United States, arguing that an increase in U.S. imports would hurt Korean businesses and threaten their livelihoods.

© AP Photo/Yonhap, Seo Myung-kon

Campaign Finance Reform. In the United States advocates for campaign finance reform have challenged the power that corporations have in influencing laws and policies. Campaign finance reform efforts were rewarded when Congress passed the McCain-Feingold bill known as the Bipartisan Campaign Reform Act of 2002. This law helps to remove the corrupting influence of soft money from federal elections so that corporations, labor unions, and wealthy donors will no longer be able to buy political influence and access. The law also prohibits corporations and unions from funding broadcast ads, run shortly before elections, that are designed to influence the election or defeat of candidates. Such ads may be funded by individual contributions, but broadcast stations must keep a public record of political ads and who paid for them.

The Antiglobalization Movement. Challenges to corporate globalization have also taken root in the United States and throughout the world. Antiglobalization activists have targeted the World Trade Organization, the International Monetary Fund (IMF), and the World Bank as forces that advance corporate-led globalization at the expense of social goals such as justice, community, national sovereignty, cultural diversity, ecological sustainability, and workers' rights. Some of the more notable actions of the antiglobalization movement include the 1999 protest in Seattle against the policies of the World Trade Organization, the protest against the 2000 meeting of the International Monetary Fund and the World Bank in Washington, D.C., and the 2003 demonstration in Miami to protest the Free Trade Area of the Americas. In 2005, organized opposition to the Central American Free Trade Agreement occurred in countries such as Guatemala, Costa Rica, and Honduras. And in 2007, hundreds of South Koreans protested against a trade agreement between the United States and South Korea. Media attention to these events contributes to the growing worldwide awareness of the forces of corporate globalization and its social, environmental, and economic effects.

UNDERSTANDING WORK AND UNEMPLOYMENT

On December 10, 1948, the General Assembly of the United Nations adopted and proclaimed the Universal Declaration of Human Rights. Among the articles of that declaration are the following:

> Article 23. Everyone has the right to work, to free choice of employment, to just and favourable conditions of work and to protection against unemployment.
>
> Everyone, without any discrimination, has the right to equal pay for equal work.
>
> Everyone who works has the right to just and favourable remuneration ensuring for himself and his family an existence worthy of human dignity, and supplemented, if necessary, by other means of social protection.
>
> Everyone has the right to form and to join trade unions for the protection of his interests.
>
> Article 24. Everyone has the right to rest and leisure, including reasonable limitation of working hours and periodic holidays with pay.

More than half a century later, workers around the world are still fighting for these basic rights as proclaimed in the Universal Declaration of Human Rights.

To understand the social problems associated with work and unemployment, we must first recognize that corporatocracy—the ties between government and corporations—serves the interests of corporations over the needs of workers. We must also be aware of the roles that technological developments and post-industrialization play on what we produce, how we produce it, where we produce it, and who does the producing. With regard to what we produce, the United States is moving away from producing manufactured goods to producing services. With regard to production methods, the labor-intensive blue-collar assembly line is declining in importance, and information-intensive white-collar occupations are increasing. Although some people argue that the growth of multinational corporations brings economic growth, jobs, lower prices, and quality products to consumers throughout the world, others view global corporations as exploiting workers, harming the environment, dominating public policy, and degrading cultural values. "One thing is for certain—global corporations are an inescapable presence in the modern world and will be so for the foreseeable future" (Roach 2007, p. 2).

Decisions made by U.S. corporations about what and where to invest influence the quantity and quality of jobs available in the United States. As conflict theorists argue, such investment decisions are motivated by profit, which is part of a capitalist system. Profit is also a driving factor in deciding how and when technological devices will be used to replace workers and increase productivity. If goods and services are produced too efficiently, however, workers are laid off and high unemployment results. When people have no money to buy products, sales slump, recession ensues, and social welfare programs are needed to support the unemployed. When the government increases spending to pay for its social programs, it expands the deficit and increases the national debt. Deficit spending and a large national debt make it difficult to recover from the recession, and the cycle continues.

What can be done to break the cycle? Those adhering to the classic view of capitalism argue for limited government intervention on the premise that business will regulate itself by means of an "invisible hand" or "market forces." For example, if corporations produce a desired product at a low price, people will buy it, which means workers will be hired to produce the product, and so on.

Ironically, those who support limited government intervention also sometimes advocate government intervention to bail out failed banks and to lend money to troubled businesses. Such government help benefits the powerful segments of our society. Yet when economic policies hurt less powerful groups, such as minorities, there has been a collective hesitance to support or provide social welfare programs. It is also ironic that such bail-out programs, which contradict the ideals of capitalism, are needed because of capitalism. For example, the profit motive leads to multinationalization, which leads to unemployment, which leads to the need for government programs. The answers are as complex as the problems.

CHAPTER REVIEW

- **The United States is described as a "post-industrialized" society. What does that mean?**

  Post-industrialization refers to the shift from an industrial economy dominated by manufacturing jobs to an economy dominated by service-oriented, information-intensive occupations. The U.S. post-industrialized economy is characterized by a highly educated workforce, automated and computerized production methods, increased government involvement in economic issues, and a higher standard of living.

- **What are transnational corporations?**

  Transnational corporations are corporations that have their home base in one country and branches, or affiliates, in other countries. Transnational corporations dominate the world economy today. In less than 20 years the number of transnational corporations has increased from 7 to more than 45,000, and the top 100 economies around the world are transnational corporations rather than nations.

- **What are the four principles of McDonaldization?**

  The four principles of McDonaldization are (1) efficiency, (2) predictability, (3) calculability, and (4) control through technology.

- **According to data from the World Value Survey, how does the level of economic development of a country affect the subjective life satisfaction of its population?**

  Data from the World Value Survey indicate that as one moves from subsistence-level economies in developing countries to advanced industrialized societies, there is a large increase in the percentage of the population who consider themselves happy or satisfied with their lives as a whole. But once a society moves from a starvation level to a reasonably comfortable existence, the increase in life satisfaction levels off.

- **Does slavery still exist today? If so, where?**

  Forced labor, commonly known as slavery, exists today all over the world, including the United States, but it is most prevalent in India, Pakistan, Bangladesh, and Nepal. Most forced laborers work in agriculture, mining, prostitution, and factories.

- **What is the most common cause of job-related fatality and nonfatal job-related illness or injury?**

  The most common type of job-related fatality involves transportation accidents. Sprains and strains are the most common nonfatal occupational injury or illness involving days away from work.

- **What are some of the challenges that workers face in balancing work and family?**

  Workers often struggle to find the time and energy to care for elderly parents and children and to have time with their families.

- **What are some of the causes of unemployment?**

  Causes of unemployment include job exportation (the relocation of jobs to other countries), automation (the replacement of human labor with machinery and equipment), increased global competition, and downsizing (the practice of laying off large numbers of employees).

- **How does unionization benefit employees?**

  Compared with nonunion workers, union workers have higher average wages, receive more insurance and pension benefits, and get more paid time off.

- **What is ergonomics?**

  Ergonomics refers to the designing or redesigning of the workplace to prevent and reduce cumulative trauma disorders. After OSHA instituted ergonomic standards in the workplace to help prevent painful, potentially debilitating work-related musculoskeletal disorders, business and industry representatives pressured Congress and President George W. Bush to repeal the ergonomic standard soon after Bush took office in 2000.

- **What is the federal Family and Medical Leave Act?**

  In 1993 President Clinton signed into law the Family and Medical Leave Act (FMLA), which requires all companies with 50 or more employees to provide eligible workers (who work at least 25 hours a week and have been working for at least 1 year) with up to 12 weeks of job-protected, unpaid leave so that they can care for a seriously ill child, spouse, or parent; stay home to care for their newborn, newly adopted, or newly placed child; or take time off when they are seriously ill.

TEST YOURSELF

1. Half or more of all jobs in less developed regions are _____ jobs.
   a. agricultural
   b. information technology
   c. health sector
   d. education or government
2. According to your text, corporations outsource labor to other countries so that U.S. consumers can save money on cheaper products.
   a. True
   b. False
3. According to a 2005 report by the AFL-CIO (2005), serious violations of workplace health and safety laws carry an average penalty of
   a. $873
   b. $8,773
   c. $78,337
   d. $788,333
4. The most common type of job-related fatality involves which of the following?
   a. Mining accidents
   b. Construction accidents
   c. Transportation accidents
   d. Homicide
5. Among most U.S. married couples with children younger than age 6, only one parent is employed.
   a. True
   b. False
6. How much paid vacation time (per year) do European workers typically have?
   a. None
   b. 1 week
   c. 2–3 weeks
   d. 6–8 weeks
7. Which is the most dangerous country in the world to be involved in trade union activities?
   a. China
   b. Ireland
   c. Egypt
   d. Colombia
8. More than one in five of the global workforce works more than 48 hours per week.
   a. True
   b. False
9. More than two-thirds of U.S. jobs require high levels of education.
   a. True
   b. False
10. More than two-thirds of U.S. adults say that "labor unions are necessary to protect the working person."
    a. True
    b. False

Answers: 1 a. 2 b. 3 a. 4 c. 5 b. 6 d. 7 d. 8 a. 9 b. 10 a.

KEY TERMS

alienation
automation
behavior-based safety programs
capitalism
chattel slavery
compressed workweek
corporatocracy
cumulative trauma disorders
downsizing
economic institution
ergonomics

Family and Medical Leave Act
(FMLA)
flextime
forced labor
free trade agreements
global economy
industrialization
job burnout
job exportation
labor unions
long-term unemployment rate

McDonaldization
post-industrialization
repetitive motion disorders
socialism
sweatshops
telework
transnational corporations
underemployment
unemployment
union density

MEDIA RESOURCES

Understanding Social Problems, Sixth Edition
Companion Website

academic.cengage.com/sociology/mooney
Visit your book companion website, where you will
find flash cards, practice quizzes, Internet links, and
more to help you study.

Just what you need to know NOW!
Spend time on what you need to
master rather than on information
you already have learned. Take a
pre-test for this chapter, and CengageNOW will generate a personalized study plan based on your results. The
study plan will identify the topics you need to review
and direct you to online resources to help you master
those topics. You can then take a post-test to help you
determine the concepts you have mastered and what you
will need to work on. Try it out! Go to academic.cengage
.com/login to sign in with an access code or to purchase
access to this product.

© AP Photo/CTK, Petra Masova

"The education of a nation, instead of being confined to a few schools and universities for the instruction of the few, must become the national care and expense for the formation of the many."

John Adams, Second president of the United States

Problems in Education

The Global Context: Cross-Cultural Variations in Education | Sociological Theories of Education | Who Succeeds? The Inequality of Educational Attainment | Problems in the American Educational System | Strategies for Action: Trends and Innovations in American Education | Understanding Problems in Education | Chapter Review

A freshman at one of the cutting edge high schools in Baltimore, Maryland, Mariya "has enough talent to fill stadiums" (Toppo 2006, p. 1). At 15, she is already considered a gifted poet. She is the type of student every teacher wants in her or his classroom—bright and inquisitive, a "superstar" with a promising future.

Yet, Mariya is at home in bed—she didn't go to school today, or yesterday, or the day before. She has missed 35 of 62 days this semester and has already failed the ninth grade. Perhaps it's the morning sickness. Or perhaps, like her mother before her, she has just given up. At 15 and pregnant, Mariya's mother dropped out of school. This time it will be different, she insists. Mariya *will* graduate from high school—baby and all.

But day after day, with her grandmother and little sister by her side, Mariya remains at home. She studies her growing belly wishing she could finish the year and go on to 10th grade. But, realizing how much time she has already missed, she just sighs. "If I fail, I fail." She's tired. "I just don't feel like going to school."

Dropping out is just one of the many issues that must be addressed in today's schools. Students continue to graduate from high school unable to read, work simple math problems, or write grammatically correct sentences. Graduates discover that they are ill prepared for corporations that demand literate, articulate, informed employees. Teachers leave the profession because of uncontrollable discipline problems, inadequate pay, and overcrowded classrooms. Students and teachers alike are "dumbing down"—lowering their standards, expectations, and role performances to fit increasingly undemanding and unresponsive systems of learning.

And yet it is education that is often claimed as a panacea—the cure-all for poverty and prejudice, drugs and violence, war and hatred, and the like. Can one institution, riddled with problems, be a solution to other social problems? In this chapter we focus on this question and on what is being called an educational crisis. We begin with a look at education around the world.

The Global Context: Cross-Cultural Variations in Education

Looking only at the American educational system might lead one to conclude that most societies have developed some method of formal instruction for their members. After all, the United States has more than 100,000 schools, 4.4 million primary and secondary school teachers and college faculty, 5 million administrators and support staff, and 72.1 million students (NCES 2006a). In reality, many societies have no formal mechanism for educating the masses. There are 781 million illiterate adults around the world, and 100 million children have little or no access to schools (UNESCO 2005, 2006). Figure 8.1 portrays youth illiteracy rates, by sex, around the world. Note that, in general, girls and youth in South and West Asia and sub-Saharan Africa are the most likely to be illiterate.

Education at a Glance, a publication of the Organization for Economic Cooperation and Development (OECD), reports education statistics on 30 countries (OECD 2006). Some interesting findings are revealed. First, in general, educa-

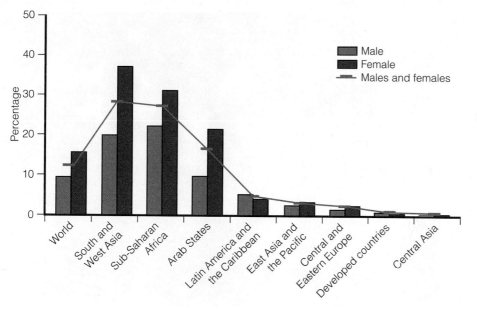

FIGURE 8.1
World youth illiteracy rates
by region and sex, 2006.
Source: UNESCO (2006).

tional levels are rising. For example, many countries saw an increase in the proportion of young people attending colleges and universities. However, large numbers of people do not graduate from tertiary institutions. In Mexico, New Zealand, and the United States a little more than half of those enrolled in colleges and universities receive degrees, whereas 80 percent or more do so in Korea, Japan, and Ireland.

Second, achievement levels in mathematics vary significantly by country. In the United States, as well as in Greece, Mexico, and Portugal, at least 25 percent of students fail to perform up to established proficiency standards. In Finland fewer than 7 percent perform below this threshold. Students from disadvantaged backgrounds are, on average, 3.5 times more likely to underperform compared with their advantaged counterparts.

Third, in 23 of the 30 OECD countries studied, the number of children of compulsory education age will decrease over the next decade. In Korea, for example, children aged 5–14 will decline by as much as 29 percent between 2005 and 2015, reducing demands on primary and secondary schools in that country. Fourth, education spending, in general, is rising. However, cumulative spending on a child's education varies dramatically by country. On the top tier are countries that spend at least $100,000 per child on primary and secondary education and include the United States, Denmark, Italy, Norway, Iceland, Luxembourg, Austria, and Switzerland. On the bottom tier are Mexico, Poland, Turkey, and the Slovak Republic, with each cumulatively spending less than $40,000 per child.

Fifth, educational expectancy—"the number of years of study over a lifetime based on present patterns of participation"—has increased (OECD 2006, p. 6). In the 30 OECD countries the average educational expectancy is 17 years, and it is more than 20 years in the United Kingdom, Sweden, and Australia. Increases are attributable to two trends: an increase in preschool enrollment at ages 3 and 4, and an increase in tertiary school enrollment. Interestingly, there has been a rapid rise in the number of students enrolling in universities and colleges out-

Oprah Winfrey's Leadership Academy for Girls in South Africa is a boarding school for girls from some of the poorest areas of South Africa. Over 150 students out of 3,500 applicants in grades 7–12 were selected based on leadership and academic skills following an intensive testing and interview process.

© AP Photo/Denis Farrell

side their home country. Fifty-two percent of these students enroll in colleges and universities in four countries—France, Germany, the United Kingdom, and the United States.

Sixth, educational attainment, as defined by OECD, is the average number of years that today's 25- to 64-year-old has spent in formal education. It ranges from 8.5 years in Portugal to 14 years in Norway with an overall average of 11.9 years. OECD countries are, however, primarily developed countries. Other resources indicate that the range of educational attainment worldwide is actually much greater. For example, the average years of schooling for those older than age 15 in China is 6.4 years, in India it is 5.1 years, in Iraq it is 4 years, and in the Sudan it is 2.1 years (NationMaster 2006).

Differences also exist in the everyday operations of a school. Some countries empower professionals to organize and operate their school systems. Japan, for example, hires professionals to develop and implement a national curriculum and to administer nationwide financing for its schools. In contrast, school systems in the United States are often run at the local level by school boards and PTAs composed of laypeople. In effect, local communities raise their own funds and develop their own policies for operating the school system in their area; the result is a lack of uniformity from district to district. Thus parents who move from one state to another often find that the quality of education available to their children differs radically. In countries with professionally operated and institutionally coordinated schools, such as Japan, teaching has traditionally been a prestigious and respected profession. Consequently, Japanese students are attentive and obedient to their teachers. In the United States the phrase "Those who can, do; those who can't, teach" reflects the lack of esteem for the teaching profession. The insolence and defiance among students in American classrooms are evidence of the disrespect that many American students have for their teachers.

SOCIOLOGICAL THEORIES OF EDUCATION

The three major sociological perspectives—structural functionalism, conflict theory, and symbolic interactionism—are important in explaining different aspects of American education.

Structural-Functionalist Perspective

According to structural functionalism, the educational institution serves important tasks for society, including instruction, socialization, the sorting of individuals into various statuses, and the provision of custodial care (Sadovnik 2004). Many social problems, such as unemployment, crime and delinquency, and poverty, can be linked to the failure of the educational institution to fulfill these basic functions (see Chapters 4, 6, and 7). Structural functionalists also examine the reciprocal influences of the educational institution and other social institutions, including the family, political institution, and economic institution.

Instruction. A major function of education is to teach students the knowledge and skills that are necessary for future occupational roles, self-development, and social functioning. Although some parents teach their children basic knowledge and skills at home, most parents rely on schools to teach their children to read, spell, write, tell time, count money, and use computers. As discussed later, many U.S. students display a low level of academic achievement. The failure of schools to instruct students in basic knowledge and skills both causes and results from many other social problems.

Socialization. The socialization function of education involves teaching students to respect authority—behavior that is essential for social organization (Merton 1968). Students learn to respond to authority by asking permission to leave the classroom, sitting quietly at their desks, and raising their hands before asking a question. Students who do not learn to respect and obey teachers may later disrespect and disobey employers, police officers, and judges.

The educational institution also socializes youth into the dominant culture. Schools attempt to instill and maintain the norms, values, traditions, and symbols of the culture in a variety of ways, such as celebrating holidays (e.g., Martin Luther King Jr. Day, Thanksgiving); requiring students to speak and write in standard English; displaying the American flag; and discouraging violence, drug use, and cheating.

As the number and size of racial and ethnic minority groups have increased, American schools are faced with a dilemma: Should public schools promote only one common culture, or should they emphasize the cultural diversity reflected in the U.S. population? Some evidence suggests that most Americans believe that schools should do both—they should promote one common culture and emphasize diverse cultural traditions (Elam, Rose, & Gallup 1994). **Multicultural education**—that is, education that includes all racial and ethnic groups in the school curriculum—promotes awareness and appreciation for cultural diversity (also see Chapter 9). In 2004, two-thirds of a sample of U.S. adults responded that it was "very important" that colleges and universities teach students to participate in a diverse society (Orfield & Lee 2006).

> **“** The beautiful thing about learning is that nobody can take it away from you. **”**
>
> B. B. King
> Musician

multicultural education
Education that includes all racial and ethnic groups in the school curriculum, thereby promoting awareness and appreciation for cultural diversity.

FIGURE 8.2
Unemployment rate of
individuals age 25 years
or older by highest level of
education, 2007.
Source: U.S. Department of Labor, Bureau
of Labor Statistics, Office of Employment
and Unemployment Statistics (2007).

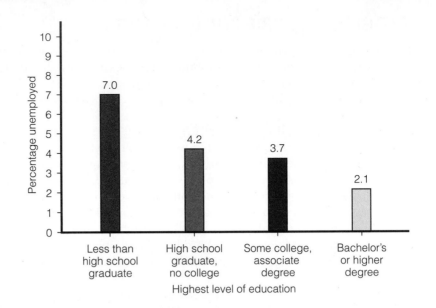

Sorting Individuals into Statuses. Schools sort individuals into statuses by providing credentials for individuals who achieve various levels of education at various schools within the system. These credentials sort people into different statuses—for example, "high school graduate," "Harvard alumna," and "English major." In addition, schools sort individuals into professional statuses by awarding degrees in fields such as medicine, engineering, and law. The significance of such statuses lies in their association with occupational prestige and income— the higher one's education, the higher one's income. Furthermore, unemployment rates are tied to educational status as seen in Figure 8.2.

Custodial Care. The educational system also serves the function of providing custodial care (Merton 1968), which is particularly valuable to single-parent and dual-earner families and the likely reason for the increase in enrollments of 3- and 4-year-olds. In 1964, 10 percent of 3- and 4-year-olds were enrolled in preschool; 54 percent were enrolled in 2004 (U.S. Census Bureau 2007). Ironically, Yale researchers found that state-supported preschools are expelling youngsters at three times the rate of public elementary and secondary schools, primarily for aggressive behavior (Dobbs 2005).

The school system provides supervision and care for children and adolescents until they are 18 years old—12 years of school, totaling almost 13,000 hours per pupil! Yet some school districts are increasing class hours. The Knowledge Is Power program in Houston, Texas, requires students to attend school several weeks in the summer, on alternate Saturdays, and from 7:25 a.m. to 5:00 p.m. during the regular school year. Some of the motivations behind the "more time" movement are working parents, the hope that increased supervision will reduce delinquency rates, higher educational standards that require longer hours of study, and a significant and positive association between hours of instruction and academic achievement (Wilgoren 2001; Mathews 2007).

Conflict Perspective

Conflict theorists emphasize that the educational institution solidifies the class positions of groups and allows the elite to control the masses. Although the official goal of education in society is to provide a universal mechanism for achievement, in reality educational opportunities and the quality of education are not equally distributed.

Conflict theorists point out that the socialization function of education is really indoctrination into a capitalist ideology (Sadovnik 2004). In essence, students are socialized to value the interests of the state and to function to sustain it. Such indoctrination begins in kindergarten. Rosabeth Moss Kanter (1972) coined the term "the organization child" to refer to the child in nursery school who is most comfortable with supervision, guidance, and adult control. Teachers cultivate the organization child by providing daily routines and rewarding those who conform. In essence, teachers train future bureaucrats to be obedient to authority.

In addition, to conflict theorists education serves as a mechanism for **cultural imperialism**, or the indoctrination into the dominant culture of a society. When cultural imperialism exists, the norms, values, traditions, and languages of minorities are systematically ignored. A Mexican-American student recalls his feelings about being required to speak English (Rodriguez 1990):

© AP Photo/Michael Conroy

To cover financial costs associated with the increasing needs of educational programs, school systems often find it necessary to contract with major corporations such as Nike.

> When I became a student, I was literally "remade"; neither I nor my teachers considered anything I had known before as relevant. I had to forget most of what my culture had provided, because to remember it was a disadvantage. The past and its cultural values became detachable, like a piece of clothing grown heavy on a warm day and finally put away. (p. 203)

Conflict theorists are also quick to note that learning is increasingly a commercial enterprise as necessary financial support for equipment, laboratories, and technological upgrades are funded by corporations anxious to bombard students with advertising and other procapitalist messages. For example, a 2005 Government Accountability Office (GAO) report found that "nearly 75 percent of U.S. high schools, 65 percent of middle schools, and 30 percent of elementary schools had exclusive contracts [with soda companies] for the 2003–2004 school year" (Pinson 2006).

Finally, the conflict perspective focuses on what Kozol (1991) called the "savage inequalities" in education that perpetuate racial disparities. Kozol documented gross inequities in the quality of education in poorer districts, largely composed of minorities, compared with districts that serve predominantly white middle-class and upper-middle-class families. Kozol revealed that schools in poor districts tend to receive less funding and to have inadequate facilities, books, materials, equipment, and personnel. For example, disadvantaged children are more likely to attend schools with fewer certified or experienced teachers and to have fewer advanced placement courses than schools attended by more advantaged children (Viadero 2006).

cultural imperialism The indoctrination into the dominant culture of a society.

Symbolic Interactionist Perspective

Whereas structural functionalism and conflict theory focus on macro level issues, such as institutional influences and power relations, symbolic interactionism examines education from a micro level perspective. This perspective is concerned with individual and small-group issues, such as teacher-student interactions and the self-fulfilling prophecy.

Teacher–Student Interactions. Symbolic interactionists have examined the ways that students and teachers view and relate to each other. For example, children from economically advantaged homes may be more likely to bring to the classroom social and verbal skills that elicit approval from teachers. From the teachers' point of view middle-class children are easy and fun to teach: They grasp the material quickly, do their homework, and are more likely to "value" the educational process. Children from economically disadvantaged homes often bring fewer social and verbal skills to those same middle-class teachers, who may, inadvertently, hold up social mirrors of disapproval. Teacher disapproval contributes to lower self-esteem among disadvantaged youth.

Self-Fulfilling Prophecy. The **self-fulfilling prophecy** occurs when people act in a manner consistent with the expectations of others. For example, a teacher who defines a student as a slow learner may be less likely to call on that student or to encourage the student to pursue difficult subjects. He or she may also be more likely to assign the student to lower ability groups or curriculum tracks (Riehl 2004). As a consequence of the teacher's behavior, the student is more likely to perform at a lower level.

A classic study by Rosenthal and Jacobson (1968) provided empirical evidence of the self-fulfilling prophecy in the public school system. Five elementary school students in a San Francisco school were selected at random and identified for their teachers as "spurters." Such a label implied that they had superior intelligence and academic ability. In reality, they were no different from the other students in their classes. At the end of the school year, however, these five students scored higher on their intelligence quotient (IQ) tests and made higher grades than their classmates who were not labeled as spurters. In addition, the teachers rated the spurters as more curious, interesting, and happy and more likely to succeed than the nonspurters. Because the teachers expected the spurters to do well, they treated the students in a way that encouraged better school performance.

WHO SUCCEEDS? THE INEQUALITY OF EDUCATIONAL ATTAINMENT

self-fulfilling prophecy A concept referring to the tendency for people to act in a manner consistent with the expectations of others.

Figure 8.3 shows the extent of the variation in highest level of education attained by individuals 25 years of age and older in the United States. As noted earlier, conflict theory focuses on such variations in discussions of education inequalities. Educational inequality is based on social class and family background, race and ethnicity, and gender. Each of these factors influences who succeeds in school.

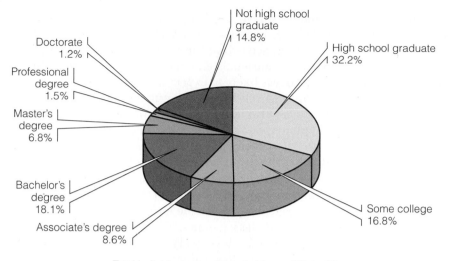

Doctorate 1.2%
Professional degree 1.5%
Master's degree 6.8%
Bachelor's degree 18.1%
Associate's degree 8.6%
Not high school graduate 14.8%
High school graduate 32.2%
Some college 16.8%

Total individuals age 25 and older — 189.4 million

FIGURE 8.3
Highest level of education attained by individuals age 25 years old and older, 2005.
Percentages do not sum to 100 percent because of rounding.
Source: NCES (2006a).

Social Class and Family Background

One of the best predictors of educational success and attainment is socioeconomic status. Children whose families are in middle and upper socioeconomic brackets are more likely to perform better in school and to complete more years of education than children from lower socioeconomic class families. For example, in 2005, 54 percent of fourth graders receiving federally subsidized meals scored below "basic" proficiency in reading for their grade level. The comparable number for students not poor enough to receive federally subsidized meals was 24 percent (Poliakoff 2006). Muller and Schiller (2000) reported that students from higher socioeconomic backgrounds are more likely to enroll in advanced mathematics course credits and to graduate from high school—two indicators of future educational and occupational success. In addition, compared with low-income students, high-income students are six times more likely to graduate with a bachelor's degree in 5 years (Toppo 2004). Moreover, of top-tier colleges and universities, less than 3 percent of enrolled students come from the lowest socioeconomic group (Viadero 2006).

What Do You Think? Research suggests that poor children who attend high-quality preschools are "less likely to drop out of school, repeat grades, or need special education [and] as adults, are less likely to commit crimes, more likely to be employed, and likely to have higher earnings" (Olson 2007a, p. 30). However, access to preschools varies by family income—the wealthier the family, the higher the probability a child will attend preschool. Further, although 38 states have state-financed preschools, funding levels vary from a high of $9,000 per child in New Jersey to a low of $721 per child in Maryland (Rolnick & Grunewald 2007). Given the above, should the federal government establish a nationalized program of preschools for all 3- and 4-year-olds? If so, should all preschool teachers be required to have a bachelor's degree?

Head Start Begun in 1965 to help preschool children from the most disadvantaged homes, Head Start provides an integrated program of health care, parental involvement, education, and social services for qualifying children.

Families with low incomes have fewer resources to commit to educational purposes. Low-income families have less money to buy books or computers or to pay for tutors and lessons in activities such as dance and music and are less likely to take their children to museums and zoos (Rothstein 2004). Parents in low-income brackets are also less likely to expect their children to go to college, and their behavior may lead to a self-fulfilling prophecy.

Disadvantaged parents are also less involved in their children's education. For example, 90 percent of nonpoor children are read to frequently by a family member, compared with 78 percent of poor children (NCES 2006b). Low-income parents are also less likely to engage their children in conversation and, when they do, they are more likely to "give orders to their children, mirroring what they face daily in their own worlds of work" (Nelson 2006, p. 1). Although working-class parents may value the education of their children, in contrast to middle- and upper-class parents, they are intimidated by their child's schools and teachers and are less likely to attend teacher conferences or be members of the PTA (Lareau 1989; Kahlenberg 2006). Furthermore, new teachers in low-income schools are more likely to report having difficulty engaging parents in the educational process, compared with new teachers in higher-income schools (MetLife 2005). Because low-income parents are often low academic achievers themselves, their children are exposed to parents who have limited language and academic skills. Children learn the limited language skills of their parents, which restricts their ability to do well academically. Low-income parents may be unable to help their children with their math, science, and English homework because they often do not have the academic skills to do the assignments.

Children from poor families also have more health problems and nutritional deficiencies. In 1965 Project **Head Start** began to help preschool children from the most disadvantaged homes. Head Start provides an integrated program of health care, parental involvement, education, and social services. Today, 906,000 3- to 5-year-olds are enrolled in Head Start (Head Start 2007). Graduates of Head Start "score better on intelligence and achievement tests, their health status is better, and they have the socio-emotional traits to help them adjust to school" (Zigler, Styfco, & Gilman 2004, p. 341).

Despite the apparent success of Head Start, in 2003 President Bush mandated that all 4-year-olds in the Head Start program participate in a nationwide testing assessment called the National Reporting System (NRS) (CDF 2003a; NABE 2005). The evaluation was designed to measure whether and how English- and Spanish-speaking children were learning and compared facilities between local Head Start programs. Some feared that the assessment was a precursor to dismantling the program. The Children's Defense Fund, a child advocacy group, in response to such fears, aired television advertisements that stated, "Call Congress . . . Tell Congress Head Start's not broken, so don't break it" (CDF 2003b).

However, a 2005 report released by the GAO found that the Bush administration's testing initiative, the NRS, failed to meet "professional standards." Specifically, the GAO found that the Head Start Bureau failed to follow the rigorous measures necessary to ensure the validity and reliability of the assessment tool, concluding that the NRS could not be counted on to "provide reliable information on children's progress . . . especially the Spanish speaking children" (NABE 2005).

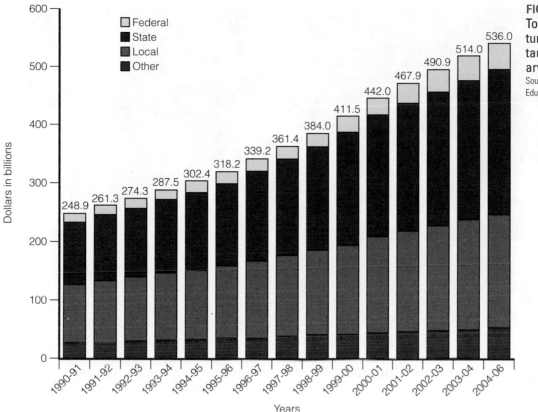

FIGURE 8.4
Total U.S. expenditures for elementary and secondary education.
Source: U.S. Department of Education (2007).

Assessments of Early Head Start, a program for infants and toddlers from low-income families, have already been conducted. Mathematica Policy Research, Inc. (MPR) conducted a 7-year national evaluation of Early Head Start programs culminating in a 2004 report to Congress. In summarizing the report, MPR concludes that "participating children perform significantly better in cognitive, language, and social-emotional development than their peers who do not participate. The program also had important impacts on many aspects of parenting and the home environment, and supported parents' progress toward economic self-sufficiency" (MPR 2007, p. 1).

Lack of adequate funding for Head Start has long been a problem, as is equality of educational funding in general. Children who live under lower socioeconomic conditions receive fewer public educational resources. Schools that serve low socioeconomic districts are largely overcrowded and understaffed, are less likely to have teachers with advanced degrees and more likely to have higher rates of teacher turnover, and are lacking adequate building space and learning materials (CDF 2004; Viadero 2006; Price 2007).

The U.S. tradition of decentralized funding means that local schools depend on local taxes, usually property taxes. Although the amount varies by state and municipality, in 2004–2005, about 40 percent of school funding came from local sources (see Figure 8.4) (U.S. Department of Education 2007). The amount of money available in each district varies by the socioeconomic status of the district. For example, in New York, schools teaching the poorest students receive

As shown above, the per pupil expenditure varies dramatically by socioeconomic status of the school district. In 2005–2006, the average per pupil expenditure was $8,701.

$2,319 per student less than schools teaching the wealthiest students (Wiener & Pristoop 2006). This system of depending on local communities for financing has several consequences:

- Low socioeconomic status school districts are poorer because less valuable housing means lower property values; in the inner city houses are older and more dilapidated; and less desirable neighborhoods are hurt by "white flight," with the result that the tax base for local schools is lower in deprived areas.
- Low socioeconomic status school districts are less likely to have businesses or retail outlets where revenues are generated; such businesses have closed or moved away.
- Because of their proximity to the downtown area, low socioeconomic status school districts are more likely to include hospitals, museums, and art galleries, all of which are tax-free facilities. These properties do not generate revenues.
- Low socioeconomic status neighborhoods are often in need of the greatest share of city services; fire and police protection, sanitation, and public housing consume the bulk of the available revenues. Precious little is left over for education in these districts.
- In low socioeconomic status school districts a disproportionate amount of the money has to be spent on maintaining the school facilities, which are old and in need of repair, so less is available for the children themselves.

Although the state provides additional funding to supplement local taxes, this funding is not always enough to lift schools in poorer districts to a level that even approximates the funding available to schools in wealthier districts. For example, *Leandro v. State* (1997), a landmark case that eventually made its way to North Carolina's Supreme Court, addressed the argument that the quality of a child's education should not be dependent on the wealth of the family and the community into which the child is born or resides. The Court ruled in two different decisions that the state constitution required that all schools in North Carolina must provide adequate resources to fully educate disadvantaged (poor, special education, and limited English proficiency) students and that the state had not met that requirement (NCCAI 2005). In 2006, testing data gathered by

North Carolina's Department of Public Instruction indicated that 19 high schools had not met statewide testing requirements for achievement in reading and math. Researchers noted, among other things, that most of the 19 high schools were attended primarily by minority students from high poverty areas with a vast majority of the students qualifying for free and reduced lunch programs. Similar to findings in other states, these students were typically taught by novice teachers, many of whom were not fully licensed in the core subject areas they were teaching (Action for Children 2007).

Race and Ethnicity

Up from 22 percent in 1972, in 2004, 43 percent of public school students were racial or ethnic minorities (NCES 2006b). In comparison to whites, Hispanics and blacks are less likely to succeed in school at almost every level. As early as the start of kindergarten, Hispanics and blacks have lower mean achievement scores than other racial or ethnic groups in reading and mathematics (NCES 2006b). By fourth grade, 41 percent of whites compared with 15 percent of Hispanics and 13 percent of blacks are reading at grade level; by eighth grade, 35 percent of whites, 12 percent of Hispanics, and 7 percent of blacks are performing at grade level in mathematics (CDF 2006). It is important to note, however, that socioeconomic status interacts with race and ethnicity (Lareau & Horvat 2004). Because race and ethnicity are so closely tied to socioeconomic status, it appears that race or ethnicity alone can determine school success. Although race and ethnicity also have independent effects on educational achievement or lack thereof (Bankston & Caldas 1997; Jencks & Phillips 1998; NCES 2006b), their relationship is largely a result of the association between race and ethnicity and socioeconomic status.

As Table 8.1 indicates, educational attainment has increased over time and varies by race and ethnicity. In general, the high school graduation gap between racial and ethnic groups is narrowing; the college graduation gap, however, is getting wider—whites and Asians on one side and Hispanics and African Americans on the other.

One reason that some minority students have academic difficulty is that they did not learn English as their native language (see also Chapter 9). For example, between 1979 and 2004, the number of 5- to 17-year-old children who spoke a language other than English in their homes increased from 3.8 to 9.9 million, that is, from 9 percent of all children in that age group to 19 percent of all children in that age group. The number of children who spoke English with some difficulty also increased during the same time period—from 3 percent (1.3 million) to 5 percent (2.8 million) of all 5- to 17-year-olds. According to *The Condition of Education,* the most recent data available indicated that of those who do not speak English in the home or have difficulty speaking English the most common language spoken is Spanish (NCES 2006b). Sixty-seven percent of Hispanic school-aged children compared with 5 percent of black and white school-aged children report speaking a language other than English in the home (U.S. Census Bureau 2007). Because of high birthrates, increased immigration, and low levels of private school enrollment, Hispanics accounted for 64 percent of the students added to public school enrollment between 1993 and 2003, the decade with the most concentrated change (Fry 2006). Blacks accounted for 23 percent of the increase and Asians 11 percent, whereas white enrollment declined by 1 percent.

TABLE 8.1 Educational Attainment by Race, Ethnicity, and Sex, 1970 and 2005

	1970		2005	
	Males	**Females**	**Males**	**Females**
Four years of high school or more				
White	54.0	55.0	85.2	86.2
Black	30.1	32.5	81.1	81.2
Hispanic	37.9	34.2	58.0	58.9
Asian	NA	NA	90.4	85.1
Total	51.9	52.8	84.9	85.4
Four years of college or more				
White	14.4	8.4	29.4	26.7
Black	4.2	4.6	16.1	18.8
Hispanic	7.8	4.3	11.8	12.1
Asian	NA	NA	54.0	46.7
Total	13.5	8.1	28.9	26.5

NA = Not available.

Source: U.S. Census Bureau (2007).

To help American children who do not speak English as their native language, some educators advocate **bilingual education**—teaching children in both English and their non-English native language. To facilitate bilingual education, many school systems across the nation have implemented a program called **total immersion.** In this program elementary students in particular receive literacy and communication instruction totally in Spanish, for example, thus enabling students to communicate with one another in both languages in and out of the classroom. Advocates claim that bilingual education results in better academic performance of minority students, enriches all students by exposing them to different languages and cultures, and enhances the self-esteem of minority students. Critics argue that bilingual education limits minority students and places them at a disadvantage when they compete outside the classroom, reduces the English skills of minorities, costs money, and leads to hostility with other minorities who are also competing for scarce resources.

Another factor that hurts minority students academically is that many tests used to assess academic achievement and ability are biased against minorities. Questions on standardized tests often require students to have knowledge that is specific to the white middle-class majority culture, and students for whom English is not their native language are seriously disadvantaged.

In addition to being hindered by speaking a different language and being from a different cultural background, minority students in white school systems are also disadvantaged by overt racism and discrimination. Much of the educational inequality experienced by poor children results because a high percentage of them are also nonwhite or Hispanic. Discrimination against minority students takes the form of unequal funding, as discussed earlier, as well as racial profiling and school segregation.

bilingual education In the United States, teaching children in both English and their non-English native language.

total immersion An educational program, in which students, particularly elementary age students, receive literacy and communication instruction entirely in a foreign language, usually Spanish.

© Michael Newman/PhotoEdit

The debate over bilingual education is likely to grow. By 2040 less than half of all school-age children will be non-Hispanic whites.

Studies indicate that minority students, and specifically black students, may be the victims of what is being called "learning while black" (Morse 2002). The allegation is

> not unlike police who stop people on the basis of race; teachers and school officials discipline black students more often—and more harshly—than whites. The result: black students are more likely to slip behind in their studies and abandon school all together—if they're not kicked out first. (Morse 2002, p. 50)

One study of 11,000 middle school students found that black students were more than twice as likely as whites to be sent to the principal's office or to be suspended. Even in preschool African Americans are twice as likely to be expelled as their white counterparts (Dobbs 2005). Although the debate is likely to continue, it must be noted that differences in discipline patterns are not necessarily a consequence of racism. They may reflect, for example, differences in behavior.

In 1954 the U.S. Supreme Court ruled in *Brown v. Board of Education* that segregated education was unconstitutional because it was inherently unequal. Despite this ruling, many schools today are racially segregated. In 1966 a landmark study titled *Equality of Educational Opportunity* (Coleman et al. 1966) revealed the extent of segregation in U.S. schools. In this study of 570,000 students and 60,000 teachers in 4,000 schools, the researchers found that almost 80 percent of all schools attended by whites contained 10 percent or fewer blacks and that whites outperformed minorities (excluding Asian Americans) on academic tests. Coleman and colleagues emphasized that the only way to achieve quality education for all racial groups was to desegregate the schools. This recommendation, known as the **integration hypothesis,** advocated busing to achieve racial balance.

Despite the Coleman report, court-ordered busing, and an emphasis on the equality of education, public schools remain largely segregated. As shown in Table 8.2, most black and Hispanic U.S. students attend schools that are predominantly minority in enrollment. An ongoing study by the Civil Rights Proj-

integration hypothesis A theory that the only way to achieve quality education for all racial and ethnic groups is to desegregate the schools.

TABLE 8.2 Racial Composition of Schools Attended by the Average Student of Each Race, 2003–2004

RACE IN EACH SCHOOL	RACIAL COMPOSITION (PERCENT) OF SCHOOLS ATTENDED BY AVERAGE:				
	WHITE STUDENT	BLACK STUDENT	LATIN STUDENT	ASIAN STUDENT	NATIVE AMERICAN STUDENT
White	78	30	28	45	44
Black	9	53	12	12	7
Latino	9	13	55	20	11
Asian	3	3	5	22	3
Native American	1	1	1	1	35
Total	100	100	100	100	100

Source: Orfield and Lee (2006).

ect at Harvard University has found that since the early 1980s there has been a trend toward re-segregation, particularly for Hispanic students (Orfield & Lee 2006). Similarly, Kozol (2005) argued that there has been a "restoration of apartheid schooling in America."

> Many Americans I meet who live far from our major cities and have no first-hand knowledge of realities in urban public schools seem to have a rather vague and general impression that the great extremes of racial isolation they recall as matters of grave national significance some 35 or 40 years ago have gradually, but steadily, diminished in more recent years. The truth, unhappily, is that the trend for well over a decade now has been precisely the reverse. Schools that were already deeply segregated 25 or 30 years ago . . . are no less segregated now, while thousands of other schools that had been integrated either voluntarily or by force of law have since been rapidly resegregating both in northern districts and in broad expanses of the south (p. 18).

Research has documented the harmful effects of this continued practice. After examining the reading and mathematics achievement levels of a nationally representative sample of high school students, Roscigno (1998) concluded that "school racial composition matters . . . in the direction one would expect, even with class composition and other familial and educational attributes accounted for. Attending a black segregated school continues to have a negative influence on achievement" (p. 1051). Conversely, attending a desegregated school, while having no negative achievement effect on white students, tends to have positive effects on black and Hispanic students in terms of learning and graduation rates (Orfield & Lee 2006).

Recently, two important cases concerning school desegregation were decided by the U.S. Supreme Court. Plaintiffs in both cases questioned the constitutionality of race-based integration, arguing that students should not be denied their choice of school to maintain racial balance (Price 2007). The Court, siding with the plaintiffs, held that public school systems "can not seek to achieve or maintain integration through measures that take explicit account of a student's race" (Greenhouse 2007, p. 1). The court's decisions reflect a general trend toward using socioeconomic or income-based integration rather than race-based integration variables. Kahlenberg (2006) advocated this approach for several reasons.

First, "socioeconomic integration more directly and effectively achieves the first aim of racial integration: raising the achievement of students" (Kahlenberg 2006, p. 10). Second, socioeconomic integration, because of the relationship between race and income, achieves racial integration and racial integration, and, in turn, fosters racial tolerance and social cohesion. Lastly, unlike race-based integration that is subject to "strict scrutiny" by the government, school assignments based on socioeconomic status are perfectly legal. In 2006, 40 school districts affecting more than 2.5 million students were using race-neutral policies to determine school assignments.

Gender

Worldwide women receive less education than men. An estimated 780 million adults in the world are illiterate, and two-thirds of them are women (UNESCO 2006; EFA Global Monitoring Report 2007). Further, according to a United Nations report, 115 million children worldwide are not in school and the majority of them are girls (Foulkes 2005). Although progress in reducing the education gender gap has been made, gender parity in primary and secondary schools has not been achieved. For example, of the 181 countries with 2004 data available, one-third have not achieved gender parity in primary schools, and two-thirds of the 177 countries with data available on secondary schools have not achieved gender parity (EFA Global Monitoring Report 2007). The United Nations' millennium goal of universal primary education for both boys and girls by 2015, although in sight, will be difficult to achieve without expanded funds and opportunities.

Historically, U.S. schools have discriminated against women. Before the 1830s U.S. colleges accepted only male students. In 1833 Oberlin College in Ohio became the first college to admit women. Even so, in 1833 female students at Oberlin were required to wash male students' clothes, clean their rooms, and serve their meals and were forbidden to speak at public assemblies (Fletcher 1943, Flexner 1972).

In the 1960s the women's movement sought to end sexism in education. Title IX of the Education Amendments of 1972 states that no person shall be discriminated against on the basis of sex in any educational program receiving federal funds. These guidelines were designed to end sexism in the hiring and promoting of teachers and administrators. Title IX also sought to end sex discrimination in granting admission to college and awarding financial aid. Finally, the guidelines called for an increase in opportunities for female athletes by making more funds available to their programs.

In 2006, the U.S. Department of Education, in order to bring Title IX into compliance with provisions of the No Child Left Behind Act, gave "public school districts broad new latitude to expand the number of single-sex classes and schools . . ." (Schemo 2006, p. 1). Single-sex classes had traditionally been confined to physical education, extracurricular activities, and sex education classes. When single-sex schools were available, districts were compelled to offer a single-sex school for the other sex student. Now the law allows single-sex schools to be offered to students of one sex as long as students of the other sex have an option to attend a "comparable coeducational" program in another school. Theoretically, the new law allows schools to cater to the needs of both girls and boys by providing students an equal opportunity for success in all subjects. However, critics argue that if racial segregation is unacceptable—sex segregation should be as well.

Although gender inequality in education continues to be a problem worldwide, the push toward equality has had some effect. For example, in 1970 nearly twice as many men as women had 4 years of college or more—8.1 percent compared with 13.5 percent. By 2005, 26.5 percent of women and 28.9 percent of men had 4 years of college or more (see Table 8.1).

What Do You Think? In 1870, nearly six times as many men graduated from college as women (Fagan 2007). However, in the fall of 2004, 57 percent of college enrollees were women—more than ever before. Further, the U.S. Department of Education estimates that by the school year 2013–2014, female graduates will outnumber male graduates by more than 300,000. The greatest disparity in higher education is between minority men and minority women. What do you think is causing the U.S. gender gap in higher education and what are its ramifications? Should colleges and universities establish special recruiting strategies for men and particularly minority men?

Traditional gender roles account for many of the differences in educational achievement and attainment between women and men. As noted in Chapter 10, schools, teachers, and educational materials reinforce traditional gender roles in several ways. Some evidence suggests, for example, that teachers provide less attention and encouragement to girls than to boys and that textbooks tend to stereotype females and males in traditional roles (Evans & Davies 2000; Spade 2004; EFA Global Monitoring Report 2007).

Studies of academic performance indicate that females tend to lag behind males in math and science (Mead 2006). One explanation is that women experience workplace discrimination in these areas, and this restricts their occupational and salary opportunities. The perception of restricted opportunities, in turn, negatively affects academic motivation and performance among girls and women (Baker & Jones 1993).

Most of the research on gender inequality in the schools focuses on how female students are disadvantaged in the educational system. But what about male students? For example, schools fail to provide boys with adequate numbers of male teachers to serve as positive role models. To remedy this, some school systems actively recruit male teachers, especially in the elementary grades where female teachers are in the majority.

The problems that boys bring to school may indeed require schools to devote more resources and attention to them. More than 70 percent of students with learning disabilities such as dyslexia are male, as are 80 percent of students identified as having serious emotional problems or autism. Boys are also more likely than girls to have speech impairments, to be labeled as mentally retarded, to exhibit discipline problems, to drop out of school or be expelled, and to feel alienated from the learning process (see this chapter's *Self and Society* feature and assess your own school alienation rate) (Bushweller 1995; Goldberg 1999; Sommers 2000; Dobbs 2005; Mead 2006). As discussed in Chapter 10, the argument that girls have been educationally shortchanged has recently come under

Self and Society | The Student Alienation Scale

Indicate your agreement with each statement by selecting one of the responses provided:

1. It is hard to know what is right and wrong because the world is changing so fast.
___ Strongly agree ___ Agree ___ Disagree ___ Strongly disagree

2. I am pretty sure my life will work out the way I want it to.
___ Strongly agree ___ Agree ___ Disagree ___ Strongly disagree

3. I like the rules of my school because I know what to expect.
___ Strongly agree ___ Agree ___ Disagree ___ Strongly disagree

4. School is important in building social relationships.
___ Strongly agree ___ Agree ___ Disagree ___ Strongly disagree

5. School will get me a good job.
___ Strongly agree ___ Agree ___ Disagree ___ Strongly disagree

6. It is all right to break the law as long as you do not get caught.
___ Strongly agree ___ Agree ___ Disagree ___ Strongly disagree

7. I go to ball games and other sports activities at school.
___ Always ___ Most of the time ___ Some of the time ___ Never

8. School is teaching me what I want to learn.
___ Strongly agree ___ Agree ___ Disagree ___ Strongly disagree

9. I go to school parties, dances, and other school activities.
___ Always ___ Most of the time ___ Some of the time ___ Never

10. A student has the right to cheat if it will keep him or her from failing.
___ Strongly agree ___ Agree ___ Disagree ___ Strongly disagree

11. I feel like I do not have anyone to reach out to.
___ Always ___ Most of the time ___ Some of the time ___ Never

12. I feel that I am wasting my time in school.
___ Always ___ Most of the time ___ Some of the time ___ Never

13. I do not know anyone that I can confide in.
___ Strongly agree ___ Agree ___ Disagree ___ Strongly disagree

14. It is important to act and dress for the occasion.
___ Always ___ Most of the time ___ Some of the time ___ Never

15. It is no use to vote because one vote does not count very much.
___ Strongly agree ___ Agree ___ Disagree ___ Strongly disagree

16. When I am unhappy, there are people I can turn to for support.
___ Always ___ Most of the time ___ Some of the time ___ Never

17. School is helping me get ready for what I want to do after college.
___ Strongly agree ___ Agree ___ Disagree ___ Strongly disagree

18. When I am troubled, I keep things to myself.
___ Always ___ Most of the time ___ Some of the time ___ Never

19. I am not interested in adjusting to American society.
___ Strongly agree ___ Agree ___ Disagree ___ Strongly disagree

20. I feel close to my family.
___ Always ___ Most of the time ___ Some of the time ___ Never

21. Everything is relative and there just aren't any rules to live by.
___ Strongly agree ___ Agree ___ Disagree ___ Strongly disagree

22. The problems of life are sometimes too big for me.
___ Always ___ Most of the time ___ Some of the time ___ Never

23. I have lots of friends.
___ Always ___ Most of the time ___ Some of the time ___ Never

24. I belong to different social groups.
___ Strongly agree ___ Agree ___ Disagree ___ Strongly disagree

Interpretation

This scale measures four aspects of alienation: powerlessness, or the sense that high goals (e.g., straight A's) are unattainable; meaninglessness, or lack of connectedness between the present (e.g., school) and the future (e.g., job); normlessness, or the feeling that socially disapproved behavior (e.g., cheating) is necessary to achieve goals (e.g., high grades); and social estrangement, or lack of connectedness to others (e.g., being a "loner"). For items 1, 6, 10, 11, 12, 13, 15, 18, 19, 21, and 22, the response indicating the greatest degree of alienation is "strongly agree" or "always." For all other items the response indicating the greatest degree of alienation is "strongly disagree" or "never."

Source: Mau, Rosalind Y., 1992. "The Validity and Devolution of a Concept: Student Alienation." *Adolescence,* 27(107): 739–740. Used by permission of Libra Publishers, Inc., 3089 Clairemont Drive, Suite 383, San Diego, California 92117.

attack; some academicians charge that it is boys, not girls, who have been left behind (Sommers 2000). Others argue that, although it is clear that some boys are in desperate need of help,

> The hysteria about boys is partly a matter of perspective. While most of society has finally embraced the idea of equality for women, the idea that women might actually surpass men in some areas (even as they remain behind in others) seems hard for many people to swallow. Thus, boys are routinely characterized as "falling behind" even as they improve in absolute terms. (Mead 2006, p. 3)

PROBLEMS IN THE AMERICAN EDUCATIONAL SYSTEM

When a random sample of Americans were asked, "Overall, how satisfied are you with the quality of education students receive in kindergarten through grade 12 in the U.S. today?" over half responded "somewhat dissatisfied" or "completely dissatisfied" (Gallup Poll 2007). This is particularly troublesome, given that total expenditures for elementary and secondary schools and other related programs have increased dramatically to more than $500 billion in 2005 (NCES 2007a). Consistent with the public's concerns is the recent emphasis on educational reform and the problems the Bush administration hopes to address—low academic achievement, high dropout rates, questionable teacher training, and school violence. These and other problems contribute to the widespread concern over the quality of education in the United States.

“The more you know, the more you know you don't know.”

Unknown

Low Levels of Academic Achievement

A recent report by the National Assessment of Adult Literacy (NAAL) describes the literacy abilities of a representative sample of U.S. adults 16 years of age or older (NAAL 2007). Three types of literacy were measured—*prose* (e.g., reading a newspaper article), *document* (e.g., reading a transportation schedule), and *quantitative* (e.g., balancing a checkbook)—and evaluated. Literacy levels included below basic, basic, intermediate, or proficient. Data for 2003, the most recent data available, indicated there have been few changes since 1992—the last year in which the assessment was conducted. According to the report, approximately 30 million people in the United States are estimated to have below basic literacy skills including seven million non-English speakers.

Among children illiteracy tends to be highest among students who attend the poorest schools, although apathy and ignorance can be found among students from more affluent schools as well. For example, an ABC News special titled *Burning Questions: America's Kids—Why They Flunk* began with the following interview with students from middle-class high schools:

Interviewer: Do you know who's running for president?
First Student: Who, run? Ooh. I don't watch the news.
Interviewer: Do you know when the Vietnam War was?
Second Student: Don't even ask me that. I don't know.
Interviewer: Which side won the Civil War?
Third Student: I have no idea.
Interviewer: Do you know when the American Civil War was?
Fourth Student: 1970.

However, results from the National Assessment of Educational Progress (NAEP), a nationwide testing effort, show some improvement over time. Although varying by socioeconomic status and race and ethnicity, in general, mathematics scores of fourth and eighth graders have increased between 1990 and 2005. Writing scores for fourth and eighth graders have also improved, although changes for twelfth graders were not statistically significant (NCES 2006b). It is important to note, however, that even statistically significant increases, although a step in the right direction, may mask poor performances. For example, as noted, mathematics scores for eighth graders have significantly increased between 1990 and 2005. However, the percentage of eighth graders who are "proficient" in mathematics (meaning a "solid academic performance") is less than one-third. Similarly, the percentage of twelfth graders who are proficient in science was only 18 percent in 2005—a decrease from 21 percent in 1996, and the percentage of eighth graders who are proficient in reading is only 31 percent (NCES 2006b).

Further, U.S. students are still outperformed by many of their foreign counterparts—something particularly troubling in a knowledge-based global economy. For example, among the 30 OECD countries, only four countries have lower high school graduation rates than the United States—New Zealand, Mexico, Turkey, and Spain. Further, based upon the Program for International Assessment (PISA), U.S. 15-year-olds perform below the OECD country average in all subjects except reading. For example, 25 percent of 15-year-old U.S. students failed "to demonstrate the most basic mathematical skills" (Schleicher 2006, p. 79). The result? U.S. 15-year-olds placed 27th out of 30 countries on mathematical achievement.

International comparisons in higher education are also revealing. Although the United States ranks second among the 30 OECD countries in terms of the percentage of 35- to 64-year-olds with a college degree, it ranks seventh in terms of the percentage of 25- to 34-year-olds with a college degree (Callan 2007). Thus, whereas other countries, most notably Canada, Japan, and Korea, have increased their college graduation rates in recent years, the United States graduation rate of young adults remains stagnant. Further, although ranking high in the percentage of high school graduates *participating* in 2- or 4-year degree programs, the U.S. ranks 16th out of 30 countries in college *completions*. Not surprisingly, in a recent survey of 4,700 educators in 127 schools in 12 urban school districts, nearly one-quarter of public school teachers responded that "success in college would elude most students in their schools" (Toppo 2007, p. 1). Moreover, based on the 2005 results of the ACT, a college admissions test, 70 percent of would-be college students are unprepared for college in reading, writing, and mathematics (Spence 2007). The failure of schools to properly prepare young people for college has taken an economic toll. The United States spends $1.4 billion a year in remedial courses at community colleges for high school graduates who have not learned the basic skills necessary to succeed in college (Olson 2007b).

School Dropouts

Nearly 90 percent of U.S. adults believe that the dropout rate in this country is a very serious problem (Thornburgh 2006). There are two kinds of dropout rates. The *status dropout rate* is the percentage of an age group that is not in school and has not earned a high school degree or its equivalent. In 2004, the status dropout rate for 16- to 24-year-olds was 10 percent, down from 15 percent in

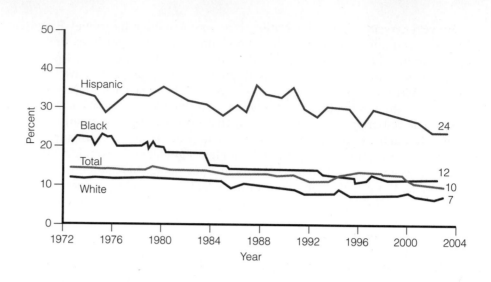

FIGURE 8.5
Status dropout rates of
16- to 24-year-olds, by
race/ethnicity, 1972–2004.
Source: NCES (2006b).

1972. The status dropout rate varies considerably by race and ethnicity with, for example, whites having the lowest status dropout rate and Hispanics the highest (see Figure 8.5) (NCES 2006b).

The *event dropout rate* is the percentage of students who drop out of high school each year. Although there is some debate as to what the event dropout rate is, most researchers set it at between 30 percent and 50 percent (Thomas & Date 2006; Thornburgh 2006). Low-income students are more than twice as likely to drop out as their middle-class counterparts and are six times as likely to drop out as their peers from high-income families. Event dropout rates also vary by race and ethnicity, with whites and Asian/Pacific Islander youth having lower dropout rates than African Americans or Hispanics (NCES 2006b). Minorities and low-income students are also more likely to drop out of 4-year colleges than their white middle- and upper-class counterparts (Leonhardt 2005).

A report by the Educational Testing Service, titled *One-Third of a Nation,* found that three factors in combination explain more than half of the variation in dropout rates. First are socioeconomic indicators—parent's income, parent's occupation, and the like. Second is the number of parents in the home. Having only one parent in the home increases the likelihood of a child dropping out of school. Finally, the number of times a student has changed schools is positively related to dropping out, that is, the higher the number of different schools attended, the higher the likelihood of dropping out (Barton 2005).

A study funded by the Bill and Melinda Gates Foundation helped to explain why students drop out and what can be done to keep them in school. Using focus groups and a survey of more than 500 dropouts, the report provided the much needed student perspective (Bridgeland, DiIulio, & Morison 2006). Respondents identified five major reasons for dropping out of high school: (1) "classes were not interesting," (2) "missed too many days and can't catch up," (3) "spent time with people who were not interested in school," (4) "had too much freedom and not enough rules," and (5) "was failing in school" (p. 3). The former students were also asked what could be done to improve a student's chances of remaining in school. Eighty-one percent responded "opportunities for real world-learning (internships, service-learning, etc.) to make classrooms more relevant" and "better teachers who keep classes interesting" (p. 13). Other top responses included smaller classes and more individual attention, better communication between parents and schools, and increased supervision at home and at school to ensure students attend classes. Seventy-four percent said they regretted dropping out and would do things differently if given the opportunity; 81 percent said that they knew the importance of graduating from high school in terms of success later in life (p. 3).

What Do You Think? A recent study assessing the economic costs associated with dropouts concluded that if the number of high school dropouts were cut in half, U.S. taxpayers would realize a savings (e.g., lower government spending on welfare) of $45 billion a year (Levin et al. 2006, p. 1). Yet, only one state, New Mexico, legally requires high school graduation, and only 17 states require attendance until the age of 18 (Keller 2006). Recently, the National Education Association, the nation's largest teacher's union, proposed that states make it illegal for a student to leave high school without a degree before the age of 21. What do you think? Should states require students to attend high school until the age of 21? How would you enforce such a law?

The economic and social consequences of dropping out of school are significant. Dropouts are more likely than those who complete high school to be unemployed (see Figure 8.2) and to earn less when they are employed (NCES 2006b). Individuals who do not complete high school are also more likely to engage in criminal activity, to have poorer health and lower rates of political participation, and to require more government services such as welfare and health care assistance (NCES 2006b; Thomas & Date 2006). Although federal funding for "second chance" opportunities has decreased over the years, General Educational Development (GED) certification is still one of the most commonly sought alternative credentials. Nationally, more than 40 percent of students who drop out of school complete their high school degree at some point in time (Entwisle, Alexander, & Olson 2004).

A relatively new initiative for dropouts is called early or middle school college. Typically, students who have dropped out or are at risk of dropping out are admitted to community colleges or, in some cases, 4-year degree programs. They receive a secondary school education, earn a high school degree, and often accrue college credits. Students are offered tutoring services, private teacher conferences, a low student-teacher ratio, individualized attention, and child care if needed. Such programs have been credited with lowering dropout rates of high-risk students (e.g., students who have failing grades, emotional or family problems, or high absenteeism) (Manzo 2005).

Crime, Violence, and School Discipline

In 2007, a lone gunman, a Virginia Tech student, entered a dormitory and classroom building and killed 32 students and faculty before taking his own life. It was the single deadliest one-man shooting rampage in American history (Apuzzo

Following the nation's largest massacre on a campus in which 32 students and faculty members were killed, students and faculty at Virginia Tech gathered for graduation. Degrees were awarded posthumously on May 12, 2007, to 26 students. "Short was their stay on this mortal stage. Great was their impact," said President Charles Steger of the slain students during the commencement ceremonies.

© AP Photo/Carolyn Kaster

bullying Inherent in a relationship between individuals, groups, or individuals and groups, bullying entails an imbalance of power that exists over a long period of time in which the more powerful intimidate or belittle others.

2007). Despite the horror of this and other school shootings, the chance of a student being killed at school is quite rare. For example, according to a report produced by the National Center for Educational Statistics and the Bureau of Justice Statistics, in the 2004–2005 school year there were 56 million students enrolled in primary and secondary schools. During the same time period there were 21 school-related homicides and 7 suicides—one homicide or suicide per 2 million students between the ages of 5 and 18 (NCES 2007b).

In 2004, there were 1.4 million nonfatal crimes committed against 12- to 18-year-olds, with the most common, 62 percent, being theft. In the same year, there were 583,000 violent crimes—most of them minor offenses. However, 18 percent of the violent crimes reported were serious in nature (e.g., rape, aggravated assault, and robbery). Gender differences were also noted. Male students were more likely to report being threatened or injured on school property, more likely to carry a weapon to school, and more likely to fight on school premises compared with their female counterparts. Although the percentage of schools reporting one or more violent incidents increased between the 1999–2000 and 2003–2004 school years, in absolute numbers violent school crime has decreased (NCES 2007b).

Certain school characteristics are associated with an increase in the probability of a student being the victim of a crime. For example, 21 percent of all 12- to 18-year-olds surveyed in the *School Crime Supplement* to the National Crime Victimization Survey report attending schools in neighborhoods with gang activity. These students are more likely to be victimized than students in neighborhoods without gang activity (NCES 2005). In addition, students who report knowing another student who brought a gun to school or who actually saw a student with a gun have higher rates of victimization. Finally, the prevalence of alcohol and other drugs in a school is indirectly related to school violence. Schools in neighborhoods with higher gang activity have students who are more likely to report alcohol and drug availability (Addington et al. 2004).

Discipline problems are also disruptive of the learning environment. When asked, public school principals reported frequent (i.e., occurring at least once a week) school disciplinary problems: verbal abuse of teachers, disorder in classrooms, disrespect for teachers, gang activities, and bullying (NCES 2007b). Of late, bullying has become one focus of the efforts to stop school violence. **Bullying** is characterized by an "imbalance of power that exists over a long period of time between two individuals, two groups, or a group and an individual in which the more powerful intimidate or belittle others" (Hurst 2005, p. 1) (see this chapter's *Social Problems Research Up Close* feature). Although many countries have had antibullying policies in place for decades, U.S. policies and research on the topic are comparatively new. One research report concluded that television viewing is associated with bullying behavior. At age 4 each hour of television viewed daily is associated with a 9 percent increased risk of bullying by age 7. Moreover, increased rates of bullying were also associated with decreased cognitive stimulation. For example, elementary-school bullies were less likely to be read to by a parent than their nonbullying peers (Zimmerman et al. 2005).

With technology, a new type of bullying has emerged—cyberbullying. **Cyberbullying** refers to the use of electronic devices (e.g., websites, e-mail, instant messaging, or text messaging) to send or post negative or hurtful messages or images about an individual or a group (Kharfen 2006). Capable of reaching wider audiences and thus doing more harm, some states and school districts have begun creating policies to deal with cyberbullying. The problems in creating such policies are twofold (Chaker 2007). First, because many of the activities surrounding cyberbullying occur at home, schools must be careful not to overstep their boundaries. However, administrators and teachers argue that if the behavior impacts the learning environment then school officials have an obligation to intervene.

The second problem concerns First Amendment rights to free speech and whether or not schools can suppress student speech off campus (Chaker 2007). A recent U.S. Supreme Court decision may have implications for cyberbullying regulations. Joseph Frederick, who was on a public sidewalk across from school grounds, held up a banner with the words "Bong Hits 4 Jesus" on it. The principal ordered him to take the sign down but he refused, resulting in a 10-day suspension for violating the school's policy on promoting illegal drugs. Although a lower court agreed that Frederick's First Amendment rights were violated, the school board appealed and the U.S. Supreme Court held that a school *can* suppress student speech, even if it occurs off school grounds, when it (1) undermines the school's anti-drug policy and (2) takes place at a school-sanctioned event (Carvin 2007). Although cyberbullying clearly meets the first criterion of a violation taking place, in this case a violation of a school's antibullying regulations, arguing that online activities from home are "school sanctioned" may prove to be more difficult.

Cyberbullying, like bullying in general, is pervasive and damaging. Research indicates that (Safe School Initiative 2004; Arbelo & McDermott 2005; Andersen 2006; Kharfen 2006; NCES 2007b):

- Every day an estimated 160,000 children do not go to school because of fear of being bullied.
- In 2005, 28 percent of students aged 12–18 reported having been bullied at school.
- More than 13 million children aged 6–17 are victims of cyberbullying and more than 2 million of these victims do not tell anyone about the harassment.
- Victims suffer from long-term emotional damage including post-traumatic stress disorder.
- Many school shooters felt that they were bullied, persecuted, or injured by others.
- Children who are bullies by the age of 8 are three times more likely to be convicted of a crime by the age of 30.

Students are not the only people victimized in schools. Annually, on the basis of a 5-year average, teachers were the victims of approximately 90,000 violent crimes, including rape, sexual assault, aggravated and simple assault, and robbery. Central city rather than rural or suburban teachers are the most likely to be threatened or attacked. In addition, teachers at secondary schools are more likely to be the victim of a violent crime than teachers at elementary schools. Further, public school teachers are more likely to be threatened or physically attacked than private school teachers (NCES 2005, 2007b).

cyberbullying The use of electronic devices (e.g., websites, e-mail, instant messaging, text messaging) to send or post negative or hurtful messages or images about an individual or a group.

Research on bullying generally falls into two categories. First, researchers study the *consequences* of bullying for the victim and the perpetrator. For example, bullying victims suffer from depression and low self-esteem, whereas bullies have enhanced rates of violence, crime, and alcohol and drug abuse (Arbelo & McDermott 2005; Andersen 2006). Second, researchers study the variables associated with *being* a bully or a victim. The hope is that if we can identify the characteristics of a bully or a victim early on we can develop successful interventions that will reduce the prevalence of the behavior. To that end, Peskin, Tortolero, and Markham (2006) investigate variables associated with being a bully, a victim, or both, in a sample of middle and high school students.

Sample and Methods

Students from eight predominantly black and Hispanic secondary schools located in a large urban school district in Texas were selected for study. Classes were sampled by grade, resulting in a sample size of 1,413 respondents and a response rate of 52 percent. Nearly 60 percent of the sample was females. Middle school students (sixth, seventh, and eighth graders) comprised 56 percent of the sample, ninth graders comprised 11 percent of the sample,

and 10th through 12th graders comprised 32 percent of the sample. Sixty-four percent of the sample described themselves as Hispanic, with the remainder self-identifying as African American. Parental permission for inclusion in the study was required.

Students were asked about their participation in bullying and their rates of victimization using response options from never to 7 or more times in the last 30 days. Response options were then collapsed into two categories: 0 to 2 times and 3 or more times. Eight types of bullying behaviors were measured, i.e., students were asked their frequency of (1) upsetting other students for the fun of it, (2) group teasing, (3) harassing, (4) teasing, (5) rumor spreading, (6) starting arguments, (7) getting others to fight, and (8) excluding others. There were four student victimization variables: (1) called names, (2) picked on by others, (3) made fun of, and (4) got hit and pushed.

In addition to measuring the frequency of bullying and victimization events, students were classified into one of four categories.

> ... a student was classified as a bully if he/she participated in at least two of the "bullying" behaviors at least three

times in the last 30 days. Victims were classified as those students who reported that at least one of the "victim" behaviors happened to them at least three times in the last 30 days. Four mutually exclusive categories were constructed: (1) bullies; (2) victims; (3) those who reported both bullying and being a victim (bully-victim); and (4) students reporting neither behaviors. (p. 471)

Findings and Conclusions

Seven percent of the sample was defined as bullies, 12 percent as victims, and 5 percent as bully-victims (p. 472). Grade level was found to be significantly related to bully/victim status. The prevalence of bullying is highest in the ninth grade (11.5 percent) and lowest in the sixth grade (4.9 percent) and 10th grade (6.2 percent). Thus, the prevalence of bullying increases from the sixth to ninth grade and decreases thereafter—a curvilinear relationship. Alternatively, the highest rate of victimization is in the sixth grade with 20.8 percent of sixth graders meeting the criteria for being a victim. The lowest rate of being a victim was among 11th and 12th graders (7.5 percent). Bully-victims also varied by grade level with the highest levels occurring in the 11th and

In response to crime, violence, and other disciplinary problems, schools throughout the country have police officers patrolling the halls, require students to pass through metal detectors before entering school, and conduct random locker searches. Video cameras set up in classrooms, cafeterias, halls, and buses purportedly deter some school violence. Other means of dealing with disciplinary problems include suspension, expulsion, peer mediation, and transferring students to specialized schools. A relatively new way to deal with school "troublemakers," the alternative school, houses students who have committed a variety of offenses while allowing them to continue with their education. In the 2000–2001 school year 39 percent of all U.S. school districts had an alternative school—66 percent of urban districts, 41 percent of suburban districts, and 35 percent of rural districts (U.S. Census Bureau 2007).

12th grades (7.9 percent) and lowest levels in the sixth grade (3.8 percent) and ninth grade (3.8 percent).

There were no significant relationships between the dependent variable and gender. Males and females were equally likely to report being bullies, victims, and bully-victims. However, race/ethnicity was significantly associated with the dependent variable. Blacks, compared with Hispanics, were more likely to be bullies, victims, and bully-victims. For example, although only 3.7 percent of Hispanics reported being bully-victims, 8.6 percent of blacks reported being bully-victims.

When specific bullying behaviors are examined, "upsetting students for the fun of it" was the most common kind of bullying activity followed by teasing, group teasing, and starting arguments. The lowest prevalence of bullying behaviors was getting others to fight. This analysis revealed gender differences. Males were significantly more likely to participate in harassing and teasing behaviors. With the exception of rumor spreading, African Americans were more likely to be involved in each of the remaining bullying behaviors. The only statistically significant difference related to grade is among eleventh and twelfth graders, who are significantly

more likely to group tease than students in other grades.

Victims report that the most common form of bullying is name calling followed by being made fun of. Males were significantly more likely to be hit and pushed than females, and blacks are significantly more likely to be made fun of, called names, or be hit and pushed than their Hispanic counterparts. Further, eleventh and twelfth graders were statistically more likely to report being picked on or made fun of than students in other class levels.

In summary, the estimates of bullying and victimization in the present investigation are comparable to prevalence estimates in other U.S. studies. That said, the results of the present research suggest the need for early intervention and the direction it should take.

> While steps to decrease physical types of bullying may be targeted largely at males, steps to reduce verbal and relational types should be targeted at all students. . . . Interventions should be developed in middle school as the prevalence of these behaviors seems to peak as students begin high school. . . . In our study, teasing and name calling were

most prevalent; thus, targeted actions for the reduction of these behaviors may be a focus for intervention activities. (p. 479)

The authors note several limitations of their study. High-risk students may have been excluded from the sample either because of failure to return the consent form or absence from class on the day the survey was administered. Additionally, the response rate, although similar to those in comparable studies, was fairly low. Further, the classification scheme, i.e., the criteria by which a student was categorized as a bully, victim, or bully-victim, was unique to this study, making comparisons with other classification schemes difficult.

Finally, the authors note that future research should concentrate on the role of race/ethnicity in the prevalence of bullying and victimization. Rather than simply noting racial/ethnic differences, however, research must begin to articulate the way in which racial/ethnic differences impact "the content" of bullying behavior. It is only then that we can fully appreciate the role of social factors "in the development of bullying and victimization problems" (p. 480).

Source: Peskin, Tortolero, and Markham (2006).

Inadequate School Facilities and Programs

A report titled *America's Crumbling Infrastructure Eroding the Quality of Life* documents the troubling conditions that exist in U.S. schools. According to the report, the American Society of Civil Engineers' report card on America's infrastructure resulted in an overall grade of D for U.S. schools (ASCE 2005). Further, the American Federation of Teachers (AFT) characterizes many of today's schools as having

> . . . rodent infestation, mice droppings, fallen ceiling tiles, poor lighting, mold that has caused mushrooms to grow, crumbling exterior walls, asbestos, severely overcrowded classrooms and hallways, freezing rooms in the winter and extreme heat in the summer, old carpeting, clogged bathroom toilets and no stall doors, inadequate circuit breakers causing frequent outages, and poor ventilation. (AFT 2006, p. 4)

More and more school buildings and facilities are in need of repair. Mold, defective ventilation systems, faulty plumbing, and the like are not uncommon. Still, quality education is expected to continue in the classrooms despite such deplorable conditions.

© Mark Richards/PhotoEdit, Inc.

There is a considerable body of research that documents the harmful effects of such surroundings. For example, in a survey of school principals nearly half indicated that harmful environmental factors negatively impact the quality of instruction (Chaney & Lewis 2007). Air quality, noise, overcrowding, and the like affect a child's ability to learn, a teacher's ability to teach, and a staff member's ability to be effective. Billions of dollars are needed to provide the best possible conditions for public schools and, for the most part, this burden falls on the states and their individual localities. For example, between 1995 and 2004, public school districts spent $600 billion on school construction building more than 12,000 new schools and completing more than 130,000 renovation and improvement projects (BEST 2006). However, research indicates that infrastructure spending is not uniform. Data indicate that over the decade studied, "the lowest investment ($4,140 per student) was made in the poorest communities, while the highest investment ($11,500 per student) was made in the high income communities" (BEST 2006, p. 5). The report concludes that although school spending has increased dramatically over the decade studied, poor schools—the schools with the greatest needs—received the lowest facilities investment.

In addition to concerns over the school environment, lunch programs have also come under attack. For example, there are serious concerns about the sale of soft drinks in school cafeterias in light of increasing rates of child obesity (see Chapter 2). In 2006, Bill Clinton's Alliance for a Healthier Generation in conjunction with the American Heart Association brokered an agreement between the top three beverage companies—Pepsi, Coca-Cola, and Cadbury Schweppes. The agreement, which was voluntary, states that by 2009, high-calorie sodas will be removed from school cafeterias and replaced by 100 percent fruit juices, water, and low and nonfat milk (Burros & Warner 2006). The problem is that many school districts have signed contracts with bottlers that they simply cannot afford to get out of. Such is the case in Racine, Wisconsin, where school officials chose not to remove the sugary drinks from a local high school when they discovered they would have to pay Pepsi $200,000 in penalties for violating their contract (Shin 2007).

Clinton's next focus will be on cafeteria lunches and vending machine snacks. As a nonprofit organization, Alliance for a Healthier Generation is forbidden to lobby the government, so the former president is planning to "negotiate directly with catering companies, purchasers, and school nutritionists" (Kluger 2006, p. 25). Meanwhile, the Child Nutrition and WIC Reauthorization Act of 2004 remains in effect. The Act requires that public, charter, and private schools receiving federal funds establish nutrition goals and guidelines for all food served and/or available on campuses. In a further attempt to address healthier eating and promote physical activity, the development of school wellness policies is also required as part of the legislation (Samuels 2006).

Special education programs pose yet another problem for school systems. Before 1997, when the Individuals with Disabilities Education Act (IDEA) was adopted, more than 1 million students with disabilities were excluded from public schools, and 3.5 million did not receive appropriate services (NCD 2000). With the implementation of IDEA, which was reaffirmed in 2004, as of 2005 more than 6.7 million children and young people with disabilities, ages 3 through 21, qualified for educational interventions under IDEA (IDEAData 2006). Pursuant to IDEA mandates, public school special education programs are now required to provide guaranteed access to education for disabled children. These specially designed instructional programs are structured to meet the unique needs of each child at no cost to the parent. As part of the program, by law, if the school district cannot meet the unique needs of the child, it becomes the school system's fiscal responsibility to pay for the child's education in an appropriate private school placement. Recently, however, two questions have emerged. First, who decides if the program offered in the public school is appropriate for the student's needs? Second, should a student have to attend the public school and participate in the program before moving to a private placement? These questions are at the root of a major case presently before the Supreme Court (Samuels 2007).

What Do You Think? All students, whether in regular classrooms or special education classes, are required by law to have a free and appropriate education. Special education students have the right to be placed in an alternative private setting paid for by the school system when the school system cannot provide for their unique needs. However, this can often mean that the special education students are receiving a higher quality, more individualized education than those mainstream students who are in overcrowded classrooms in poor school districts with very little funding to address their basic needs (Berger 2007). How do you balance the unique needs of the special education student with the basic needs of the poor mainstream student? How would you allocate funds?

Compounding the problems associated with placement of students in special education is the well-understood fact that many teachers are opting to leave the special education classrooms for traditional classrooms or to pursue different fields entirely. With a lack of certified teachers in this area, a critical shortage has occurred nationwide, thus affecting the quality of education taking place in special education classrooms.

Recruitment and Retention of Quality Teachers

School districts with inadequate funding and facilities, low salaries, lack of community support, and minimal professional development have difficulty attracting and retaining qualified school personnel. This problem has placed the nation's schools at odds with a growing problem in the face of higher accountability standards. With a national average turnover rate of 15.7 percent in 2004, each year school systems open the new academic year without enough teachers to cover all their classes (WEP 2005). Further, research indicates that in an average school year, approximately 1,000 teachers resign from their positions every day with more than 30 percent of new teachers leaving the profession within the first 5 years (Alliance for Excellent Education 2005; Honawar 2007). Experts predict that the United States will need more than 2 million new teachers over the next decade (NEA 2006a).

In addition to the growing concerns related to recruitment and retention, poorer districts are also faced with the demand for talented teachers who can meet the needs of children from diverse backgrounds and of varying abilities. The number of minority teachers who can serve as role models, have similar life experiences, and have similar language and cultural backgrounds is far too few for the number of minority students. Recruiting and retaining quality teachers in poverty-level schools is critical to the success of its students. For example, Hanushek, Rivkin, and Kain (2005) reported that if a child from a poor family has a good teacher for 5 consecutive years it would close the achievement gap between that child and a child from a higher income family.

Nonetheless, several studies have shown that poorer communities do not attract high-quality teachers. Not only are these low-income, high-minority districts likely to employ beginning teachers with less than 3 years' experience, they are also 77 percent more likely to assign these teachers to areas outside their specialty. Further, minority students are twice as likely to be taught by substitute teachers than white students (Barton 2004). In addition, these high-poverty schools experience turnover rates that are nearly one-third higher than the national rate, leading to a shortage of qualified teachers and contributing significantly to the high rate of attrition nationally (CDF 2004).

Even those who remain in the teaching profession are not necessarily competent or effective. For example, there is evidence that those who choose teaching as a career, on average, have lower college entrance exam scores than the average college student (NCES 2003a). In addition, a national survey of principals revealed that 90 percent believe that U.S. colleges and universities are doing such a poor job of training teachers that many of the graduates are not properly prepared to go into the classroom (Winter 2005). Moreover, a U.S. Department of Education survey found that less than 36 percent of current teachers report feeling "very well prepared" to initiate curriculum and performance standards, and less than 20 percent feel ready to meet the needs of the diverse student population (*A Quality Teacher* 2002).

In an effort to place "highly qualified" teachers in the classroom, many states have implemented mandatory competency testing as part of state licensure re-

> **"**There is growing consensus among researchers and educators that the single most important factor in determining student performance is the quality of his or her teachers.**"**
>
> **Alliance for Excellent Education**

Across the nation, school system representatives attend teacher job fairs at colleges and other venues to recruit highly qualified teachers for their local school districts.

quirements. Most states require prospective teachers to pass the Praxis Series, which tests general knowledge as well as specialty area competency (ETS 2007). The need for teachers who are officially classified as "highly qualified" is tied to national mandates that place an emphasis on the importance of having licensed teachers in the classroom. To be considered "highly qualified," all teachers of core academic subjects "must attain a bachelor's degree or better, must obtain full state certification, and must demonstrate knowledge in the subjects taught (U.S. Department of Education 2006). Some research suggests that teacher licensure has a direct impact on student achievement, with students learning more from certified teachers than from noncertified teachers (Viadero 2005).

The National Board for Professional Teaching Standards (NBPTS) has also created and implemented a plan designed to "retain, reward, and advance" those teachers who are accomplished in their respective fields. According to the NBPTS, studies show that, in general, students of national board–certified teachers perform better on standardized tests and have shown greater testing gains than students of teachers who are not national board certified (NBPTS 2007). Teachers who have a bachelor's degree and have been in the classroom for 3 or more years are eligible to seek this higher level of certification, a level that certifies them nationwide to teach in their specialty area.

In an effort to meet the demands of placing teachers in classrooms while facing teacher shortages, states are now allowing skilled professionals who have an interest in teaching but did not receive a teaching degree to enter the teaching profession. Called *lateral entry* by some states, the program allows the person to obtain a lateral entry teaching license while actually teaching in the classroom. With this approach the teacher is provided with orientation training by the school system as well as assistance in obtaining the license needed to remain in the classroom.

In addition, more than half of the states have adopted **alternative certification programs,** whereby college graduates with degrees in fields other than edu-

alternative certification program A program whereby college graduates with degrees in fields other than education can become certified if they have "life experience" in industry, the military, or other relevant jobs.

cation can become certified if they have "life experience" in industry, the military, or other relevant jobs. Teach for America, a program originally conceived by a Princeton University student in an honors thesis, is an alternative teacher education program with the aim of recruiting liberal arts graduates into teaching positions in economically deprived and socially disadvantaged schools. After completing an 8-week training program, recruits are placed as full-time teachers in rural and inner-city schools. Critics argue that these programs may place unprepared personnel in schools.

STRATEGIES FOR ACTION: TRENDS AND INNOVATIONS IN AMERICAN EDUCATION

Americans consistently rank improving education as one of their top priorities. Recent attempts to improve schools include raising graduation requirements, barring students from participating in extracurricular activities if they are failing academic subjects, lengthening the school year, and prohibiting dropouts from obtaining driver's licenses. In addition, there is a nationwide movement to eliminate **social promotion,** the passing of students from grade to grade even if they are failing. However, educational reformers are calling for changes that go beyond get-tough policies that maintain the status quo.

National Educational Policy

President Bush's education plan, as established by the No Child Left Behind (NCLB) Act of 2001, was signed into law in January 2002. The federally funded plan is organized around four principles. The first principle is *accountability.* Each year every state is required to test the math and reading abilities of third-through eighth-grade students. Parents, administrators, and others have access to the data, and states issue a report card rating the performance of each school, teacher, and student.

The second principle is *flexibility.* The new law permits federal funds to be transferred between programs and local schools have more input into how federal funds are used. *Expanding options for parents* is the third principle. After a school is identified as failing, parents are able to transfer their child to a better performing or charter school in the same school district. Also, supplemental funds that can be used for a tutor, summer school programs, or any other school service are provided to children in failing schools.

The last principle is using *teaching methods* that are known to work. Money is provided to "reading first"—a presidential plan to help children read. Teacher quality should improve because local schools are better able to recruit and retain excellent teachers. Last, local schools are able to use federal funds "for hiring new teachers, increasing teacher pay, improving teacher training and development and other uses" (*Fact Sheet* 2003). According to the plan, by 2006, only "highly qualified" teachers would be in the classroom (see this chapter's *The Human Side* feature).

However, at the end of 2006, not all states had met this requirement. Further, soon after the implementation of NCLB, school districts quickly realized the problems they faced as a result of the plan. For example, to make adequate yearly progress (AYP), the factor by which a school is measured, all student groups (e.g., blacks, whites, Native Americans, limited English proficiency, dis-

social promotion The passing of students from grade to grade even if they are failing.

abled, or others) must attain a set level of achievement in both reading and math. According to the Act, if one student group does not reach the set levels, the entire school receives a failing grade and is in danger of being severely sanctioned (DPI 2007). Because of the tremendous disparities between student groups, NCLB regulations were changed to allow for alternative testing of disabled and limited English proficiency students; deadlines for making AYP in reading and math were also extended to 2014 for all states (Olson & Hoff 2006).

In addition to concerns about testing accountability, critics of the law also argue that it unfairly burdens the states, which must absorb the financial cost of its provisions. Thus, for example, a Connecticut Department of Education report concluded that the state would have to spend $41.6 million by 2008 to carry out all the testing requirements mandated by NCLB (Archer 2005a). As a result, in August 2005 the Connecticut Attorney General filed a lawsuit against the U.S. Department of Education, arguing that the Act itself contains prohibitions against unfunded federal mandates (Official Press Release 2005). Although overruled at the lower court level, Connecticut continues to pursue its lawsuit against the U.S. Department of Education (Attorney General 2007). Other legal maneuvers have also taken place. For example, the National Education Association, the largest organization of teachers in the nation, has joined schools in several districts to bring the first federal lawsuit against the U.S. Department of Education for failing to provide funding for NCLB initiatives. Similar to Connecticut's suit, *Pontiac v. Spellings* is still being pursued in the federal courts (NEA 2007b).

Proponents of the law hold that such legal maneuvers mask the intent of the legislation—to improve education in the United States. For example, in response to the Connecticut lawsuit, one U.S. Department of Education official noted that Connecticut's students "are suffering from one of the largest achievement gaps in the nation" (Archer 2005b, p. 2). Although Connecticut students are at the top of the performance ladder nationwide, low socioeconomic students and minority students score significantly lower than their high socioeconomic counterparts.

Results of empirical tests of NCLB are mixed. For example, a report by the Northwestern Evaluation Association, a nonprofit independent research group, compared reading and math scores of thousands of students in multiple states and school districts. Two dependent variables were measured: *achievement level* (i.e., how is this year's fourth-grade class doing compared to last year's fourth-grade class?) and *achievement growth* (i.e., how is this child doing in the fourth grade compared to how she or he was doing in the third grade?) (Cronin et al. 2005). Results indicate that, although achievement levels in both math and reading have increased under NCLB, achievement growth, particularly for minorities, has actually decreased.

A study by Hall and Kennedy (2006) also found conflicting results. On the one hand, their analysis indicated that, overall, achievement gains were found in elementary schools where reading scores showed gains in 27 of 31 states and math scores improved in 29 of 32 states. However, the picture in middle schools and high schools is less clear with a combination of gains and losses in both reading and math. Although minority and low-income students showed achievement gains, whites and higher income students, in general, showed more dramatic gains, thus increasing the achievement gap between the two groups.

Although the primary goal of a teacher is to educate, a teacher's impact on a student can be life changing. For example, in a survey of secondary school students, 95 percent reported that they had a teacher who has made a positive difference in their lives. Furthermore, the students indicate that these teachers helped them to do better in school, introduced them to new ideas, and helped them to pursue their individual interests (MetLife 2005). The following narrative poignantly describes a classroom teacher's efforts to make a difference in an environment driven by accountability standards.

It has been about twelve years now. So many things have changed, but I will never forget the first day of administering a new state-mandated science experiment as part of a week-long assessment. I am not a scientist, but as an elementary school teacher at the time, science was one of the many topics I had to cover in my classroom. So, here we were, about to embark on this great new scientific experiment, recording for the history books of the state's Department of Education just how effectively my students, and other students across the state, could log this amazing phenomenon they were about to witness.

According to the detailed instructions, some raisins were supposed to "dance" on the top of the surface of a container as two unknown ingredients were mixed together and poured over the raisins sitting in the bottom of a clear cup. Now, the instructions indicated that for ten minutes, I was to record the time on the chalkboard in 15 second intervals. The purpose was to allow the students to have the most accurate information possible for recording in their group journal.

So, with "I Heard It Through the Grapevine" playing through my head, for 10 minutes I dutifully recorded the time in 15 second intervals. During that 10 minutes, the student interest went from rapt attention to doodling on the test papers to taking out books to read. During that excruciating 10 minutes, absolutely nothing happened. No dancing, not even a flutter. I later learned it was not just in my room that this newfangled test failed. No one in my building had any success either. Further, no one in the entire county had dancing raisins. As a matter of fact, the raisins refused to dance at every school in the entire state! And it was only Monday morning at 9:00, with two more tasks that day plus three each day for the next four days.

Up until about two years ago, I was sworn to secrecy and wouldn't have been allowed to share the above secure testing information with anyone. However, in a field that operates on a perpetual pendulum, one must accept that bureaucracy will forever be dabbling with ways that some believe to be on the cutting edge of meeting the educational needs of children. Hence, the testing format that started a little over ten years ago has now become obsolete, and a newer state-mandated testing format under No Child Left Behind has been put into motion.

Despite the gradual wariness and loss of confidence that many veteran teachers have about state assessments, we must keep our focus elsewhere. As a seasoned, and oft-times battle-weary veteran teacher, I have come to realize that an educator must maintain an unwavering focus on why he or she has entered this field. What a teacher does has to be done for the sake of the student and no one else. Those of us who have the heart and passion to teach have to remember the motto, "To teach is to touch lives forever." A teacher must realize that he or she will have a tremendous impact on the life of a child, and vice versa.

You will be sobered when you attend the funeral of a student who has died in an auto accident and you find yourself attempting to comfort the parents who are younger than you are, trying to remind them of the joy that their child brought to your classroom. You will be angered and, perhaps, stunned when you read the police blotter section of the local newspaper, and see the names of your former

Despite the Bush administration's claim that all 50 states have successfully implemented NCLB and that achievement gaps are narrowing (White House Fact Sheet 2007), critics continue to argue that the Act's effectiveness is questionable. Although scheduled for reauthorization in 2007, experts predict that the process will entail a long and protracted debate that will continue until well after the 2008 elections (Olson & Hoff 2006).

Character Education

Research indicates that cheating is a fairly common event among students in the United States. For example, Donald McCabe, an expert on the subject, estimates that 50 percent of U.S. undergraduate students admit to "serious cheating" on a paper (e.g., plagiarism), and 22 percent admit to cheating on an exam or test

students listed under the arrests for DUI, drug possession, and assault and battery. Good kids; challenging kids; it matters not. They were your kids, in your classrooms, and here they are.

For many students, we *become* the parent. Imagine my shock in my second year of teaching when a parent of a problem child I was on the phone with hung up on me right after saying, *"Mr. Jenkins, when my child is home with me she is my problem. When she's in that school with you, she's yours."* Since when did parents stop being parents during school hours?

What we do, we must do for the children. But it has become more and more difficult in recent years, especially in light of the pressures of No Child Left Behind. Never before have I seen such a cloud hanging over doom-and-gloom staff meetings as we compare test scores and study where we must be by this time next year. After all, funding and school recognition are at stake—all because of test results. In some circles the NCLB (No Child Left Behind) program has been coined the NTLTT (No Teachers Left To Teach) program, as we witness increased numbers of teachers opting to take early retirement instead of trying to keep up with the rigors and requirements of becoming and remaining "highly qualified" in our respective fields.

I get more and more frustrated when I hear of test results and "Annual Yearly Progress" reports as we blindly graduate unprepared students. After all, if we graduate those who can't make the grade, once they're gone, our scores are sure to increase. It troubles me to hear the constant threats of "teacher accountability" based on the success of their assigned students; accountability that is measured by test performance. It is especially disturbing when I have over 180 students with as many as 40 in a class. It makes me angry when I feel pressured to teach within a prescribed curriculum based on desired testing outcomes. Whatever happened to teaching students to be successful in life; to helping them develop life skills instead of testing skills?

I teach American History. About 95 percent of the people we talk about in my classroom are dead. My job is to help 13-year-olds understand the significance of the events that played out in the lives of these dead people—the Founding Fathers of our nation—to help them see how those who lived immediately before and after our Founding Fathers served to mold and make this country great—the idol and envy of so many others. I want to teach with my God-given gifts and talents and try to show these youngsters why I believe it is important to know something about their amazing past. They say that history repeats itself and

I have seen proof of it over and over again. I have seen state testing come and go by the wayside and am now facing the pressures and uncertainty of another attempt to find the most appropriate way to teach children. While my school, county, state, and even country attempt to find the answer to how to educate most effectively, I just want to be left alone to teach and reach the children in my own way. Forget the administrative pressure, state tests, and government programs. Raisins may dance, or they may fall flat as they did twelve years ago, but, in the long run, it doesn't really matter. Kids will always be kids despite the drastic changes we may see in them. They need role models in their lives—especially when your 68-minute class may be more time spent with that kid in one day than they spend with their parent(s) in a whole week.

Yes, the pendulum is changing directions again, but the kids will not be. If I can hang in there another eight years or so, I just might see it swing back again the other way. In the meantime, I only have 180 days to make an impact on some needy kids' lives before I must send them away and prepare for the next batch. Time waits for no one; especially kids. Reach them while you can, as they won't always be there for you.

Source: Jenkins (2006).

(Takach 2006). Further, a recent survey of high school students by the Josephson Institute of Ethics found that about 33 percent of high school students admit to copying all or part of an Internet document for a classroom assignment at least once (Abdul-Alim 2006).

Particularly worrisome is the fairly common belief that "copying questions and answers from a test" is not cheating. As one student explains:

What's important is getting ahead. The better grades you have, the better school you get into, the better you're going to do in life. And if you learn to cut corners to do that, you're going to be saving yourself time and energy. In the real world, that's what's going to be going on. The better you do, that's what shows. It's not how moral you were in getting there. (Slobogin 2002, p. 1)

To many educators and the general public, statements like this signify the need for **character education.**

Eighty percent of Americans rate moral values in the United States as only fair or poor and 8 in 10 believe that moral values are getting worse rather than better (Gallup 2007). Despite these results, many school curricula neglect this side of education—the moral and interpersonal aspects of developing as an individual and as a member of society. President Bush's educational reform policy includes support for character education, and research indicates that it has an impact on student outcomes. In a review of 15 years of research on the subject, Battistich (2005) concluded that "high quality character education . . . is not only effective at promoting the development of good character, but it is a promising approach to the prevention of a wide range of contemporary problems . . . includ[ing] aggressive and anti-social behaviors, drug use, precocious sexual activity, criminal activities, academic under-achievement, and school failure" (p. 1).

Service learning programs, one type of character education, are increasingly popular at universities and colleges nationwide. Service learning programs are community-based initiatives in which students volunteer in the community and receive academic credit for doing so. Studies on student outcomes have linked service learning to enhanced civic responsibility and moral reasoning, a reduction of risky behaviors, and higher levels of self-esteem (Jacobs 1999; Independent Sector 2002; Ramierz-Valles & Brown 2003). How does the public feel about character education? In a national poll 76 percent of respondents favored requiring schools to teach about values and morality (NPR 2003). Character education also occurs to some extent in schools that have peer mediation and conflict resolution programs. Such programs teach the value of nonviolence, collaboration, and helping others as well as skills in interpersonal communication and conflict resolution.

Education and Computer Technology

Computers in the classroom allow students to access large amounts of information. The proliferation of computers both in school and at home may mean that teachers will become facilitators and coaches rather than sole providers of information. Not only do computers enable students to access enormous amounts of information, including that from the World Wide Web, but they also allow students to progress at their own pace. However, computer technology is not equally accessible to all students. For example, although most students have access to computers at school, access to computers at home varies dramatically by parent's education, parent's income, and race and ethnicity. In that respect schools "bridge the digital divide" between poor minority students and their more advantaged counterparts (DeBell & Chapman 2006).

The Enhancing Education through Technology program is part of the NCLB Act. The goals of the program are threefold: (1) to improve student achievement through the use of technology resources, (2) to ensure that teachers integrate technology into the curriculum in such a way as to improve student achievement, and (3) to help students become technically literate by the eighth grade (NCES 2003b). Thus, two of the three goals of the program are to increase academic achievement through the use of technological resources.

Recently, a study mandated by Congress assessed the use of educational technology in improving student achievement. Reading and math scores on standardized tests of students using educational technology products selected

character education Education that emphasizes the moral and interpersonal aspects of an individual.

Students work on their laptops at the Philadelphia School of the Future. Sponsored as part of a public/private partnership between the City of Philadelphia and Microsoft, this model for future schools opened on September 6, 2006.

for their presumed effectiveness were compared with reading and math scores on standardized tests of students not using the educational software. The performances of 9,424 students in 132 school districts were evaluated. The results indicate that there were no statistically significant achievement differences between students in classes where the software was used and students in classrooms where the software was not used (NCES 2007c).

Similarly, a review of studies on distance education classes indicated that there are few significant differences in student outcomes between distance learning and traditional classroom instruction (Cavanaugh et al. 2004). **Distance learning** separates the student and the teacher by time and/or place. They are, however, linked by some communication technology (e.g., videoconferencing, e-mail, real-time chat room, or closed-circuit television). In the fall of 2005, there were more than 3.2 million college students taking at least one online course (Allen & Seaman 2006). Further, in 2006, about 700,000 precollegiate students, mostly in high school, were enrolled in at least one online or blended course, i.e., a course that combines computerized lessons with traditional classroom instruction (Cavanagh 2007).

Online education often serves a segment of the population that would not otherwise be able to attend school—older, married, full-time employees, and persons from remote areas. Thus, for example, the number of public school online course offerings is higher in rural areas than in towns, cities, or urban areas (Bausell & Klemick 2007). Further, some research suggests that online learning benefits those who have historically been disadvantaged in the classroom. For example, a study by DeNeui and Dodge (2006) indicates that females in a blended course were more likely to use a learning management system (e.g., Blackboard)

distance learning Learning in which, by time or place, the learner is separated from the teacher.

than males and to significantly outperform them as measured by final grades in an Introduction to Psychology class.

School Choice

Traditionally, children have gone to school in the district where they live. School vouchers, charter schools, home schooling, and private schools provide parents with alternative school choices for their children. **School vouchers** are tax credits that are transferred to the public or private school that parents select for their child. At present, there is national voucher legislation before Congress. The America's Opportunity Scholarships for Kids Act, according to the National Education Association, would provide federal funding for states, school districts, and nonprofit organizations to offer scholarships to disadvantaged students in underperforming schools to attend private schools (NEA 2006b).

Proponents of the voucher system argue that it reduces segregation and increases the quality of schools because the schools must compete for students to survive. Opponents argue that vouchers increase segregation because white parents use the vouchers to send their children to private schools with few minorities. Research by Saporito (2003) supports this contention:

> Findings show that white families avoid schools with higher percentages of nonwhite students. The tendency of white families to avoid schools with higher percentages of nonwhites cannot be accounted for by other school characteristics such as test scores, safety, or poverty rates. I also find that wealthier families avoid schools with higher poverty rates. The choices of whites and wealthier students lead to increased racial and economic segregation in the neighborhood schools that these children leave. (p. 181)

Vouchers, opponents argue, are also unfair to economically disadvantaged students who are not able to attend private schools because of the high tuition. In response to such criticisms, proponents note that the targeted value of a voucher—about $3,000—would be more than sufficient. For example, U.S. Department of Education statistics indicate that the average cost of tuition for private schools, both primary and secondary, is $3,116 (Boaz & Barrett 2004).

Opponents also argue that the use of vouchers for religious schools violates the constitutional guarantee of separation of church and state. However, the U.S. Supreme Court, in reviewing the voucher program in Cleveland, Ohio, held that the use of tax dollars for enrollment in religious schools is not unconstitutional. In the 2006–2007 school year, approximately 6,000 students were enrolled in the Cleveland voucher program. According to the National School Board Association, with vouchers worth as much as $3,450, the voucher program in Cleveland "diverts up to $19 million a year from a Cleveland public school fund aimed at educating disadvantaged students" (NSBA 2007, p. 1).

Vouchers can also be used for charter schools. **Charter schools** originate in contracts, or charters, which articulate a plan of instruction that must be approved by local or state authorities. Although charter schools can be funded by foundations, universities, private benefactors, and entrepreneurs, many are supported by tax dollars. It is estimated that more than 3,900 charter schools are operating in 41 states and that they are serving 1.1 million students (CER 2006). Charter schools, like school vouchers, were designed to expand schooling options and to increase the quality of education through competition. Evidence suggests that "charter schools produce student learning gains comparable to those of conventional schools, despite resource limitations" (Rand 2003, p. 1).

school vouchers Tax credits that are transferred to the public or private school that parents select for their child.

charter schools Schools that originate in contracts, or charters, which articulate a plan of instruction that must be approved by local or state authorities.

Some parents are choosing not to send their children to school at all but to teach them at home. In 2003 more than 1 million students in the United States were home-schooled, a significant increase from 850,000 in 1999 (NCES 2006c). For some parents, **home-schooling** is part of a fundamentalist movement to protect children from perceived non-Christian values in the public schools. Other parents are concerned about the quality of their children's education and their safety. How does being schooled at home instead of attending public school affect children? Some evidence suggests that home-schooled children perform as well as or better than their institutionally schooled counterparts (Webb 1989; Winters 2001).

Another choice parents can make is to send their children to a private school. About 11.7 percent of all students are enrolled in private elementary or private high schools (CAPE 2007). The primary reason parents send their children to private schools is for religious instruction. The second most common reason is the belief that private schools are superior to public schools in terms of academic achievement. Evidence may support this belief. Students in private schools generally perform higher on achievement tests than their public school counterparts (NCES 2003c). It may be, however, that students from private schools do better on standardized tests not because of private school attendance but because, in general, they are more likely to come from wealthier and more educated families, two variables that are directly associated with school achievement (Mendez 2005). Parents also choose private schools for their children to have greater control over school policy, to avoid busing, or to obtain a specific course of instruction, such as dance or music.

What Do You Think? Given the less-than-stellar performance of many U.S. schools, some have argued that what is needed is privatization. **Privatization** entails states hiring corporations to operate local institutions. For example, Philadelphia schools are managed by seven different private providers including Edison Schools, the largest of the for-profit corporations. Edison is hoping to help failing schools and failing students while making a return for investors. Proponents argue that market-driven education will result in a better "product"; critics, however, are ideologically opposed to companies making a profit from public school children. Do you think public schools would be more efficiently run by private corporations? Why or why not?

UNDERSTANDING PROBLEMS IN EDUCATION

What can we conclude about the educational crisis in the United States? Any criticism of education must take into account the fact that just over a century ago the United States had no systematic public education system at all. Many American children did not receive even a primary school education. Instead, they worked in factories and on farms to help support their families. Whatever education they received came from the family or the religious institution. In the mid-1800s educational reformer Horace Mann advocated at least 5 years of

home-schooling The education of children at home instead of in a public or private school.

privatization A practice in which states hire businesses to provide services or operate local schools.

mandatory education for all U.S. children. Mann believed that mass education would function as the "balanced wheel of social machinery" to equalize social differences among members of an immigrant nation. His efforts resulted in the first compulsory education law in 1852, which required 12 weeks of attendance by school-age children each year. By World War I every state mandated primary school education, and by World War II secondary education was compulsory as well.

Public schools are supposed to provide all U.S. children with the academic and social foundations necessary to participate in society in a productive and meaningful way. But as conflict theorists note, for many children the educational institution perpetuates an endless downward cycle of failure, alienation, and hopelessness. Legally required protections can help. For example, in 2004, *Williams v. California,* a class action suit against the state of California on behalf of more than 1 million low-income students, was settled. Lawyers for the plaintiffs had argued that the students were permanently disadvantaged as a result of the schools they attended. The settlement, which is funded by a $1 billion appropriation, held that (1) all students must have needed textbooks and materials, (2) every classroom must be clean, safe, and in good condition, and (3) every teacher must meet standards set down by the state of California and federal law (UCLA 2005).

Breaking the downward cycle, however, is likely to become increasingly difficult as student enrollments swell and diversify, a consequence of the "baby-boom echo" and changing immigration patterns. Between 1989 and 2009 elementary school enrollment will increase 12 percent, high school enrollment will increase 19 percent, and full-time college student enrollment will increase 11 percent (U.S. Department of Education 2000).

The public, however, is supportive of government spending on education. In a national survey on spending priorities, education was the number one priority with 74 percent of respondents saying that the government spends too little on education (NORC 2007). Legislation such as the College Student Relief Act of 2007 (this bill would reduce the interest rate on subsidized undergraduate loans from 6.8 percent to 3.4 percent over a 5-year period) is essential (NEA 2007b). But even with financial support, as structural functionalists argue, education alone cannot bear the burden of improving our schools.

Jobs must be provided for those who successfully complete their education. If not, students will have few incentives to exert any effort. Rosenbaum (2002) explained:

> What is missing from current practices is a mechanism for creating and conveying signals that tell students the value of their present actions in achieving desirable career goals. Other countries produce such signals with linkage mechanisms. . . . The German system provides a clear mechanism that makes the relationship between school performance and career option totally obvious [through an apprenticeship program]. . . . Students know that apprenticeships lead to respected occupations, and that school grades affect selection into apprenticeships. (p. 488)

Finally, as a society we must attend to and be cognizant of the importance of early childhood development. As Poliakoff (2006) noted,

> Children's physical, emotional, and cognitive development are profoundly shaped by the circumstances of their pre-school years. Before some children are even born, birth

weight, lead poisoning, and nutrition have taken a toll on their capacity for academic achievement. Other factors—excessive television watching, little exposure to conversation or books, parents who are absent or distracted, inadequate nutrition—further compromise their early development. (p. 10)

We must provide support to families so that children grow up in healthy, safe, and nurturing environments. Children are the future of our nation and of the world. Whatever resources we provide to improve the lives and education of children are sure to be wise investments in our collective future.

CHAPTER REVIEW

- **Do all countries educate their citizens?**

  No. Many societies have no formal mechanism for educating the masses. As a result, worldwide, there are 781 million illiterate adults around the world and 100 million children have little or no access to schools. The problem of illiteracy is greater in developing countries than in developed nations, and illiteracy rates are higher for women than for men (see Figure 8.1).

- **According to the structural-functionalist perspective, what are the functions of education?**

  There are four major functions. The first is instruction—that is, teaching students knowledge and skills. The second is socialization that, for example, teaches students to respect authority. The third is sorting individuals into statuses by providing them with credentials. The fourth function is custodial care—a babysitting agency of sorts.

- **What is a self-fulfilling prophecy?**

  A self-fulfilling prophecy occurs when people act in a manner consistent with the expectations of others.

- **What variables predict school success?**

  Three variables tend to predict school success. Socioeconomic status predicts school success: the higher the socioeconomic status, the higher the likelihood of school success. Race predicts school success, with nonwhites and Hispanics having more academic difficulty than whites and non-Hispanics. Gender also predicts success, although it varies by grade level.

- **What are some of the conclusions of the study summarized in the *Social Problems Research Up Close* feature?**

  The results of the study indicate that there are more victims of bullying than bullies, and that bullying behavior is highest in the ninth grade and lowest in the sixth grade and tenth grade. The highest rate of victimization is in the sixth grade. Males and females were equally likely to report being bullies, victims, and bully-victims. Blacks, compared to Hispanics, were more likely to be bullies, victims, and bully-victims. "Upsetting students for the fun of it" was the most common kind of bullying activity followed by teasing, group teasing, and starting arguments.

- **What are some of the problems associated with the American school system?**

  One of the main problems is the lack of student achievement in our schools—particularly when U.S. data are compared with data from other industrialized countries. Minority dropout rates are high, and school violence, crime, and discipline problems continue to be a threat. School facilities are in need of repair and renovations, and personnel, including teachers, have been found to be deficient.

- **What are some of the solutions being considered to address problems in the schools?**

  The No Child Left Behind Act of 2001 is the primary means by which educational problems in the United States are being addressed. This Act focuses on accountability through testing, flexibility of funding, expanding options for parents, and, perhaps most important, improving the quality of teaching in the schools. Reauthorization of the Act is presently under consideration.

- **What are the arguments for and against school choice?**

  Proponents of school choice programs argue that they reduce segregation and that schools which have to compete with one another will be of a higher quality. Opponents argue that school choice programs increase segregation and treat disadvantaged students unfairly. Low-income students cannot afford to go to private schools, even with vouchers. Furthermore, those opposed to school choice are quick to note that using government vouchers to help pay for religious schools is unconstitutional.

TEST YOURSELF

1. All societies have some formal mechanism to educate their citizenry.
   a. True
   b. False
2. According to structural functionalists, which of the following is not a major function of education?
   a. Teach students knowledge and skills
   b. Socialize students into the dominant culture
   c. Indoctrinate students into the capitalist ideology
   d. Provide custodial care for children
3. The number of children speaking a language other than English in the home has decreased in recent years.
   a. True
   b. False
4. In a recent study, students identified five major reasons for dropping out of high school. Which of the following is a reason given for dropping out of school?
   a. Classes were not interesting
   b. Had to drop out to work
   c. Pregnancy
   d. Hated school

5. School policies governing cyberbullying are difficult to make because
   a. Much of cyberbullying activities take place at home
   b. Cyberbullying speech may be protected under the First Amendment
   c. Enforcement of the policies would be impossible
   d. Both a and b are correct
6. Public school special education programs are required to provide guaranteed access to education for disabled children.
   a. True
   b. False
7. Over the next decade, the need for teachers is likely to
   a. Decrease because of the baby-boom echo
   b. Increase because of immigration patterns
   c. Remain stable
   d. Decrease because of the aging population
8. Which of the following is not one of the four principles of the No Child Left Behind Act?
   a. Accountability
   b. Longer school days
   c. Parental involvement
   d. Highly qualified teachers
9. Computer software in the classroom has been shown to significantly increase academic achievement.
   a. True
   b. False
10. School vouchers are tax credits that are transferred to the public or private school that parents select for their child.
    a. True
    b. False

Answers: 1 b. 2 c. 3 b. 4 a. 5 d. 6 a. 7 b. 8 b. 9 b. 10 a.

KEY TERMS

alternative certification program
bilingual education
bullying
character education
charter schools
cultural imperialism

cyberbullying
distance learning
Head Start
home-schooling
integration hypothesis
multicultural education

privatization
school vouchers
self-fulfilling prophecy
social promotion
total immersion

MEDIA RESOURCES

Understanding Social Problems, Sixth Edition
Companion Website

academic.cengage.com/sociology/mooney
Visit your book companion website, where you will find flash cards, practice quizzes, Internet links, and more to help you study.

 Just what you need to know NOW! Spend time on what you need to master rather than on information you already have learned. Take a pre-test for this chapter, and CengageNOW will generate a personalized study plan based on your results. The study plan will identify the topics you need to review and direct you to online resources to help you master those topics. You can then take a post-test to help you determine the concepts you have mastered and what you will need to work on. Try it out! Go to academic.cengage .com/login to sign in with an access code or to purchase access to this product.

9

"The 21st century will be the century in which we redefine ourselves as the first country in world history which is literally made up of every part of the world."

Kenneth Prewitt,
Census Bureau director

Race, Ethnicity, and Immigration

Talk radio host Don Imus ignited a storm of controversy after calling the Rutgers University women's basketball team "nappy-headed hos" on a radio broadcast in April 2007. Although Imus was known for making racial slurs throughout his radio talk-show host career, this time was different: the incident was widely viewed on the Internet site YouTube.com, creating a public outcry. Two days after the controversial remark, Imus apologized. "My characterization [of the team] was thoughtless and stupid. . . . And we're sorry." Three days later, he added: "Here's what I have learned: that you can't make fun of everybody, because some people don't deserve it. And because the climate on this program has been what it has been for 30 years doesn't mean that it has to be that broadcast for the next five years or whatever, because that has to change . . . and I understand that" (quoted in Clemmitt 2007, p. 499). Imus was initially penalized for his offensive remarks by being suspended for 2 weeks, but when advertisers began pulling their sponsorship from Imus' show, both networks that carried it—CBS Radio and MSNBC TV—fired him. The outcome was hailed by some as a long-overdue response to the racist cultural climate perpetuated by comedians, rappers, and some talk-show hosts. However, others criticized Imus' firing as an attack against freedom of speech and argued that Imus was unfairly singled out among many media figures who use racially insensitive language.

In 2007, talk radio host Don Imus was fired after he called the players on Rutgers women's basketball team "nappy-headed hos."

The insulting remarks made by Don Imus are just one small example of how U.S. minority groups are represented and treated by the white majority. A **minority group** is a category of people who have unequal access to positions of power, prestige, and wealth in a society and who tend to be targets of prejudice and discrimination. Minority status is not based on numerical representation in society but rather on social status. For example, before Nelson Mandela was elected president of South Africa, South African blacks suffered the disadvantages of a minority, even though they were a numerical majority of the population.

In this chapter we focus on prejudice and discrimination, their consequences for racial and ethnic minorities, and the strategies designed to reduce these problems. We also examine issues related to U.S. immigration, because immigrants often bear the double burden of being minorities *and* foreigners who are not welcomed by many native-born Americans. We begin by examining racial and ethnic diversity worldwide and in the United States, emphasizing first that the concept of race is based on social rather than biological definitions.

The Global Context: Diversity Worldwide

A first-grade teacher asked the class, "What is the color of apples?" Most of the children answered red. A few said green. One boy raised his hand and said "white." The teacher tried to explain that apples could be red, green, or sometimes golden, but never white. The boy insisted his answer was right and finally said, referring to the apple, "Look inside" (Goldstein 1999). Like apples, human beings may be similar on the "inside," but they are often classified into categories according to external appearance. After examining the social construction

minority group A category of people who have unequal access to positions of power, prestige, and wealth in a society and who tend to be targets of prejudice and discrimination.

of racial categories, we review patterns of interaction among racial and ethnic groups and examine racial and ethnic diversity in the United States.

The Social Construction of Race

The concept of **race** refers to a category of people who are believed to share distinct physical characteristics that are deemed socially significant. Cultural definitions of race have taught us to view race as a scientific categorization of people based on biological differences between groups of individuals. However, "races are not biologically real but are cultural and social inventions created in specific cultural, historical, and political contexts. . . . Races are not scientifically valid because there are no objective, reliable, meaningful criteria scientists can use to construct or identify racial groupings" (Mukhopadhyay, Henze, & Moses 2007, pp. 1, 5).

Historically, in the United States, racial classification has been primarily based on skin color and secondarily on hair texture as well as the size and shape of the eyes, lips, and nose. But distinctions among human populations are graded, not abrupt. Skin color is not black or white but rather ranges from dark to light with many gradations of shades. Noses are not either broad or narrow but come in a range of shapes. Physical traits such as these, as well as hair color and other characteristics, come in an infinite number of combinations. For example, a person with dark skin can have any blood type and can have a broad nose (a common combination in West Africa), a narrow nose (a common combination in East Africa), or even blond hair (a combination found in Australia and New Guinea) (Cohen 1998).

Further, skin color, hair texture, and facial features are only a few of the many traits that vary among human beings. What if we classified people into racial categories based on eye color instead of skin color? Or hair color? Or blood type? (There are no racial blood types; blood types cut across races.) Is there any scientific reason for selecting certain traits over others in determining racial categories? The answer is "No."

Some physical variations among people are the result of living for thousands of years in different geographic regions. According to anthropologists, modern humans evolved in Africa about 100,000 years ago and developed dark skin from the natural skin pigment, melanin, to protect against the sun's ultraviolet radiation, levels of which are high near the equator. When humans began migrating to regions farther from the equator lighter skin developed because of the reduced exposure to ultraviolet radiation. But more importantly lighter skin was necessary to allow enough UV rays to penetrate the skin to produce vitamin D, which is important for human health (Mukhopadhyay et al. 2007).

The science of genetics also challenges the notion of race. Geneticists have discovered that the genes of any two unrelated persons, chosen at random from around the globe, are 99.9 percent alike (Ossorio & Duster 2005). Furthermore, "most human genetic variation—approximately 85 percent—can be found between any two individuals from the same group (racial, ethnic, religious, etc.). Thus, the vast majority of variation is within-group variation" (Ossorio & Duster 2005, p. 117). Classifying people into different races fails to recognize that over the course of human history, migration and intermarriage have resulted in the blending of genetically transmitted traits. Thus there are no "pure" races; people in virtually all societies have genetically mixed backgrounds (Keita & Kittles 1997). And contrary to what some people believe, all humans belong to the same species.

❝I hope that people will finally come to realize that there is only one 'race'—the human race—and that we are all members of it. **❞**

Margaret Atwood
Writer

race A category of people who are believed to share distinct physical characteristics that are deemed socially significant.

The American Anthropological Association has passed a resolution stating that "differentiating species into biologically defined 'races' has proven meaningless and unscientific" (Etzioni 1997, p. 39).

> To summarize, races are unstable, unreliable, arbitrary, culturally created divisions of humanity. This is why scientists . . . have concluded that race, as scientifically valid biological divisions of the human species, is fiction not fact. (Mukhopadhyay et al. 2007, p. 14)

As clear evidence that race is a social rather than a biological concept, different societies construct different systems of racial classification, and these systems change over time. For example, "at one time in the not too distant past in the United States, Italians, Greeks, Jews, the Irish, and other 'white' ethnic groups were not considered to be white. Over time . . . the category of 'white' was reshaped to include them" (Rothenberg 2002, p. 3).

The significance of race is not biological but social and political, because race is used to separate "us" from "them" and becomes a basis for unequal treatment of one group by another. Despite the increasing acceptance that "there is no biological justification for the concept of 'race'" (Brace 2005, p. 4), its social significance continues to be evident throughout the world.

What Do You Think? Do you think the time will ever come when a racial classification system will no longer be used? Why or why not? What arguments can be made for discontinuing racial classification? What arguments can be made for continuing it?

Patterns of Racial and Ethnic Group Interaction

When two or more racial or ethnic groups come into contact, one of several patterns of interaction occurs; these include genocide, expulsion, colonialism, segregation, acculturation, pluralism, and assimilation. These patterns of interaction occur when the groups exist in the same society or when different groups from different societies come into contact. Although not all patterns of interaction between racial and ethnic groups are destructive, author and Mayan shaman Martin Prechtel reminded us, "Every human on this earth, whether from Africa, Asia, Europe, or the Americas, has ancestors whose stories, rituals, ingenuity, language, and life ways were taken away, enslaved, banned, exploited, twisted, or destroyed" (quoted by Jensen 2001, p. 13).

Genocide refers to the deliberate, systematic annihilation of an entire nation or people. The European invasion of the Americas, beginning in the 16th century, resulted in the decimation of most of the original inhabitants of North and South America. Some native groups were intentionally killed, whereas others fell victim to diseases brought by the Europeans. In the 20th century Hitler led the Nazi extermination of 12 million people, including 6 million Jews, in what is known as the Holocaust. More recently, in the early 1990s ethnic Serbs attempted to eliminate Muslims from parts of Bosnia—a process they called "ethnic cleansing." In 1994 genocide took place in Rwanda when Hutus slaughtered hundreds of thousands of Tutsis (called "cockroaches" by the Hutus)—an event depicted in the 2004 film *Hotel Rwanda*. And as this book goes to press, genocide is occurring in the Darfur region of Sudan, where the Sudanese govern-

genocide The deliberate, systematic annihilation of an entire nation or people.

In 1994 Hutus in Rwanda committed genocide against the Tutsis, resulting in 800,000 deaths.

ment, using Arab *Janjaweed* militias, its air force, and organized starvation, is systematically killing the black Sudanese population (see also Chapter 16).

Expulsion occurs when a dominant group forces a subordinate group to leave the country or to live only in designated areas of the country. The 1830 Indian Removal Act called for the relocation of eastern tribes to land west of the Mississippi River. The movement, lasting more than a decade, has been called the Trail of Tears because tribes were forced to leave their ancestral lands and endure harsh conditions of inadequate supplies and epidemics that caused illness and death. After Japan's attack on Pearl Harbor in 1941, President Franklin Roosevelt authorized the removal of any people considered threats to national security. All people on the West Coast with at least one-eighth Japanese ancestry were transferred to evacuation camps surrounded by barbed wire, where 120,000 Japanese Americans experienced economic and psychological devastation. In 1979 Vietnam expelled nearly 1 million Chinese from the country as a result of long-standing hostilities between China and Vietnam.

Colonialism occurs when a racial or ethnic group from one society takes over and dominates the racial or ethnic group(s) of another society. The European invasion of North America, the British occupation of India, and the Dutch presence in South Africa before the end of apartheid are examples of outsiders taking over a country and controlling the native population. As a territory of the United States, Puerto Rico is essentially a colony whose residents are U.S. citizens, but they cannot vote in presidential elections unless they move to the mainland.

Segregation refers to the physical separation of two groups in residence, workplace, and social functions. Segregation can be *de jure* (Latin meaning "by law") or *de facto* ("in fact"). Between 1890 and 1910 a series of U.S. laws, which came to be known as **Jim Crow laws,** were enacted to separate blacks from whites by prohibiting blacks from using "white" buses, hotels, restaurants, and drinking fountains. In 1896 the U.S. Supreme Court (in *Plessy v. Ferguson*) supported de jure segregation of blacks and whites by declaring that "separate but equal" facilities were constitutional. Blacks were forced to live in separate neighborhoods and attend separate schools. Beginning in the 1950s various rulings overturned these Jim Crow laws, making it illegal to enforce racial segregation. Although de jure segregation is illegal in the United States, de facto segregation still exists in the tendency for racial and ethnic groups to live and go to school in segregated neighborhoods.

Acculturation refers to adopting the culture of a group different from the one in which a person was originally raised. Acculturation may involve learning the dominant language, adopting new values and behaviors, and changing the spelling of the family name. In some instances acculturation may be forced, as in the California decision to discontinue bilingual education and force students to learn English in school.

expulsion Occurs when a dominant group forces a subordinate group to leave the country or to live only in designated areas of the country.

colonialism Occurs when a racial or ethnic group from one society takes over and dominates the racial or ethnic group(s) of another society.

segregation The physical separation of two groups in residence, workplace, and social functions.

Jim Crow laws Between 1890 and 1910, a series of U.S. laws enacted to separate blacks from whites by prohibiting blacks from using "white" buses, hotels, restaurants, and drinking fountains.

acculturation The process of adopting the culture of a group different from the one in which a person was originally raised.

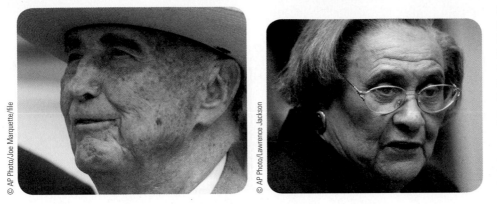

Strom Thurmond (left) served as governor of South Carolina and as a U.S. Senator and was known as being a segregationist who opposed the Civil Rights Act of 1957. Shortly after his death in 2003, Essie Mae Washington-Williams (right) publicly revealed that she was Thurmond's illegitimate daughter, born to a 16-year-old African American maid when Thurmond was 22. Thurmond met Essie Mae when she was 16. He helped her pay for college and continued to provide financial support to her in her adult years.

Pluralism refers to a state in which racial and ethnic groups maintain their distinctness but respect each other and have equal access to social resources. In Switzerland, for example, four ethnic groups—French, Italians, Swiss Germans, and Romansch—maintain their distinct cultural heritage and group identity in an atmosphere of mutual respect and social equality. In the United States the political and educational recognition of multiculturalism reflects efforts to promote pluralism.

Assimilation is the process by which formerly distinct and separate groups merge and become integrated as one. Assimilation is sometimes referred to as the "melting pot," whereby different groups come together and contribute equally to a new, common culture. Although the United States has been referred to as a melting pot, in reality, many minorities have been excluded or limited in their cultural contributions to the predominant white Anglo-Saxon Protestant tradition.

Assimilation can be of two types: secondary and primary. *Secondary assimilation* occurs when different groups become integrated in public areas and in social institutions, such as neighborhoods, schools, the workplace, and in government. *Primary assimilation* occurs when members of different racial or ethnic groups are integrated in personal, intimate associations, as with friends, family, and spouses. Nineteen states had **antimiscegenation laws** banning interracial marriage until 1967, when the Supreme Court (in *Loving v. Virginia*) declared these laws unconstitutional. (The term *miscegenation* comes from the Latin *miscere* "to mix" and *genus* "kind.")

Between 1980 and 2000 the proportion of interracial or interethnic marriages more than doubled, from 4 percent in 1980 to 9 percent in 2000 (Amato et al. 2007). Since 1960 the number of black–white married couples has increased 5-fold, the number of Asian–white married couples has increased more than 10-fold, and the number of Hispanics married to non-Hispanics has tripled. This rise in nontraditional unions is related to the increased independence that adult children have from their parents and communities of origin as well as to increased societal acceptance of mixed marriages (Rosenfeld & Kim 2005). The percentage of U.S. adults who agree that "it's all right for blacks and whites to date" increased from 48 percent in 1987 to 76 percent in 2002 and 83 percent in 2007 (Pew Research Center 2007) (see Figure 9.1). Among younger people—those born since 1977—94 percent say it is all right for blacks and whites to date.

pluralism A state in which racial and ethnic groups maintain their distinctness but respect each other and have equal access to social resources.

assimilation The process by which minority groups gradually adopt the cultural patterns of the dominant majority group.

antimiscegenation laws Laws banning interracial marriage until 1967, when the Supreme Court (in *Loving v. Virginia*) declared these laws unconstitutional.

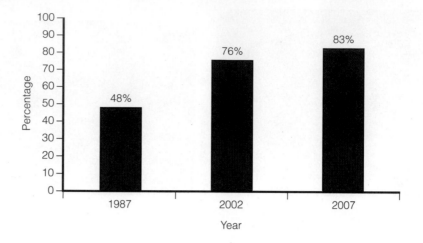

FIGURE 9.1
Percentage of U.S. adults who agree that "it's all right for blacks and whites to date."
Source: Pew Research Center (2007).

RACIAL AND ETHNIC GROUP DIVERSITY AND RELATIONS IN THE UNITED STATES

The racial and ethnic characteristics of the U.S. population are becoming increasingly diversified. In four states—California, New Mexico, Hawaii, and Texas—minority populations outnumber non-Hispanic whites. In these states Hispanics are the largest minority group, except for Hawaii, where the largest minority group is Asian Americans ("National Briefing" 2005).

Racial Diversity in the United States

The first census in 1790 divided the U.S. population into four groups: free white males, free white females, slaves, and other persons (including free blacks and Indians). To increase the size of the slave population, the *one-drop rule* appeared, which specified that even one drop of "Negroid" blood defined a person as black and therefore eligible for slavery. In 1960 the census recognized only two categories: white and nonwhite. In 1970 the census categories consisted of white, black, and "other" (Hodgkinson 1995). In 1990 the U.S. Bureau of the Census recognized four racial classifications: (1) white, (2) black, (3) American Indian, Aleut, or Eskimo, and (4) Asian or Pacific Islander. The 1990 census also included the category of "other." Beginning with the 2000 census the Office of Management and Budget required federal agencies to use a minimum of five race categories: (1) white, (2) black or African American, (3) American Indian or Alaska Native, (4) Asian, and (5) Native Hawaiian or other Pacific Islander. In addition, respondents to federal surveys and the census now have the option of officially identifying themselves as being more than one race rather than checking only one racial category (see Figure 9.2). Figure 9.3 presents 2005 census data on the racial composition of the United States.

Mixed-Race Identity. About 4.5 percent of U.S. male–female married couples and 10 percent of unmarried couples are interracial (Fields 2004). The new census option for identifying as "mixed race" avoids putting children of mixed-race parents in the difficult position of choosing the race of one parent over the other when filling out race data on school and other forms. It also avoids impairment

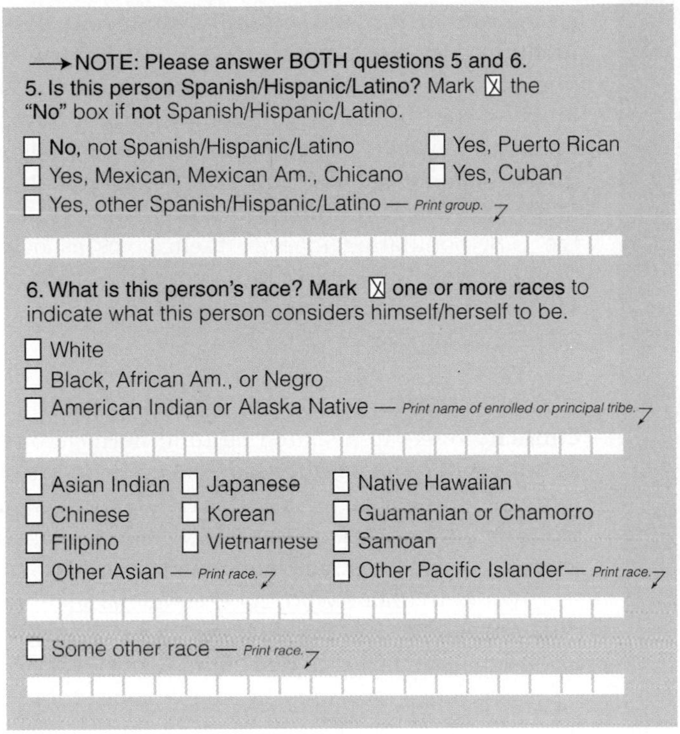

NOTE: Please answer BOTH questions 5 and 6.

5. Is this person Spanish/Hispanic/Latino? Mark ☒ the "No" box if not Spanish/Hispanic/Latino.

☐ No, not Spanish/Hispanic/Latino   ☐ Yes, Puerto Rican
☐ Yes, Mexican, Mexican Am., Chicano ☐ Yes, Cuban
☐ Yes, other Spanish/Hispanic/Latino — *Print group.* ⤸

6. What is this person's race? Mark ☒ one or more races to indicate what this person considers himself/herself to be.

☐ White
☐ Black, African Am., or Negro
☐ American Indian or Alaska Native — *Print name of enrolled or principal tribe.* ⤸

☐ Asian Indian ☐ Japanese   ☐ Native Hawaiian
☐ Chinese   ☐ Korean   ☐ Guamanian or Chamorro
☐ Filipino   ☐ Vietnamese ☐ Samoan
☐ Other Asian — *Print race.* ⤸   ☐ Other Pacific Islander— *Print race.* ⤸

☐ Some other race — *Print race.* ⤸

FIGURE 9.2
Reproduction of questions on race and Hispanic origin from the 2000 census.
Source: U.S. Census Bureau, Census 2000 questionnaire.

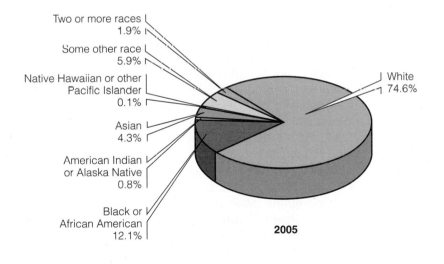

Two or more races 1.9%
Some other race 5.9%
Native Hawaiian or other Pacific Islander 0.1%
Asian 4.3%
American Indian or Alaska Native 0.8%
Black or African American 12.1%
White 74.6%

2005

FIGURE 9.3
Race composition of the United States: 2005.
Source: U.S. Census Bureau (2005).

of children's self-esteem and social functioning that comes from choosing the racial category of "other." Such a category implies that the society does not recognize and respect mixed-race individuals, and thus "children growing up within mixed families may feel ashamed of their 'irregular' racial makeup and may experience rejection and alienation in the wider social community" (Zack 1998, p. 23).

Some critics of the new mixed-race option are concerned that the wide-scale recognition of mixed-race identity will decrease the numbers within minority

Although Barack Obama's background is biracial—he is the son of a black Kenyan immigrant father and a white Kansas native mother—he identifies as a black African American.

66 I'm often asked, Where am I from? Or what am I? And when I answer U.S. (or American), I'm told, 'No, really, tell me.' 99

Conan Hom
Asian American resident
of Lexington, Massachusetts

ethnicity A shared cultural heritage or nationality.

groups and disrupt the solidarity and loyalty based on racial identification. For example, what will happen to organizations and movements devoted to equal rights for blacks if much of the "black" population acquires a new mixed-race identity? However, the 2000 census data suggest that the mixed-race option will not have the large national impact that critics fear. As shown in Figure 9.3, only 1.9 percent of the U.S. population identify themselves as being of more than one race, although about 5 percent of U.S. black people say that they are multiracial (Schmitt 2001b).

Ethnic Diversity in the United States

Ethnicity refers to a shared cultural heritage or nationality. Ethnic groups can be distinguished on the basis of language, forms of family structures and roles of family members, religious beliefs and practices, dietary customs, forms of artistic expression such as music and dance, and national origin.

Two individuals with the same racial identity may have different ethnicities. For example, a black American and a black Jamaican have different cultural, or ethnic, backgrounds. Conversely, two individuals with the same ethnic background may identify with different races. Although most Hispanics and Latinos are white, in the 2000 census 2 percent of Latinos identified themselves as black (Navarro 2003).

The current Census Bureau classification system does not allow people of mixed Hispanic or Latino ethnicity to identify themselves as such. Individuals with one Hispanic and one non-Hispanic parent still must say that they are either Hispanic or not Hispanic. In addition, Hispanics must select one country of origin, even if their parents are from different countries.

U.S. citizens come from a variety of ethnic backgrounds. The largest ethnic population in the United States is of Hispanic origin. (The terms "Hispanic" and "Latino" are used interchangeably here.) More than one in eight (13.3 percent) people in the United States are Hispanic or Latino and two-thirds (66.9 percent) of all U.S. Hispanics or Latinos are of Mexican origin (Ramirez & de la Cruz 2003). Other Hispanic Americans have origins in Central and South America (14.3 percent), Puerto Rico (8.6 percent), and Cuba (3.7 percent).

The use of racial and ethnic labels is often misleading and imprecise. The ethnic classification of "Hispanic/Latino," for example, lumps together disparate groups such as Puerto Ricans, Mexicans, Cubans, Venezuelans, Colombians, and others from Latin American countries. The racial term *American Indian* includes more than 300 separate tribal groups that differ enormously in language, tradition, and social structure. The racial label *Asian* includes individuals from China, Japan, Korea, India, the Philippines, or one of the countries of Southeast Asia. And what about people who are Asian but who live in the United States? The term "Asian American" is used to describe people with Asian racial features who are born in the United States, as well as those who immigrate to the United States. Columbia University professor Derald Wing Sue, an Asian American who was born in the United States, said, "When I get out of a cab after having a conversation with a white cab driver, they'll say something like, 'Boy, you speak excellent English. . . . From their perspective, that's meant as a compliment, but another hidden meaning is being communicated, and that is that I am a perpetual foreigner in my own land" (quoted in Tanneeru 2007).

Race and Ethnic Group Relations in the United States

Despite significant improvements over the last two centuries, race and ethnic group relations continue to be problematic. The racial divide in the United States sharpened in 2005 in the wake of Hurricane Katrina, which left victims—who were predominantly black and poor—waiting for days to be rescued from their flooded attics or rooftops or to be evacuated from overcrowded "shelters" where there was no food, water, medical supplies, or working toilets. Following this crisis, a national survey found that 7 in 10 blacks (71 percent) said that the Katrina disaster shows that racial inequality remains a major problem in the United States, whereas a majority of whites (56 percent) said that this was not a particularly important lesson of Katrina. The same survey found that most blacks (77 percent), compared with only 17 percent of whites, believe that the government's response to the disaster would have been faster if most of Katrina's victims had been white (Pew Research Center 2005).

Following Hurricane Katrina in 2005, a national survey found that the majority of blacks believe that the government's response to the crisis would have been faster if most of Katrina's victims had been white.

Racial tensions concerning the treatment of African Americans spiked again in 2007 after a series of racially charged events occurred in Jena, Louisiana, which began with the hanging of nooses on the "white tree" (where only white students gathered) on the Jena High School campus the day after black students sat under the tree. Three white students who were responsible for hanging the nooses received a punishment of only 3 days of in-school suspension for their "prank." But to blacks, this prank was a serious act of racial hate that recalled the history of lynching of blacks in the United States. The black/white racial tensions stemming from this prank and the failure of the school board and superintendent to take it seriously led to fights between black and white students. Six black students, known as the *Jena Six,* were arrested and charged with crimes related to their alleged involvement in the assault of a white student. Five of the six students were charged with attempted murder. Critics accused District Attorney Reed Walters, who is white, of prosecuting blacks more harshly than whites. After massive protests in Jena and nationwide, charges were reduced to aggravated second-degree battery in four of the cases.

In response to a question asking whether relations between blacks and whites will always be a problem for the United States or whether a solution will eventually be worked out, 42 percent of U.S. adults said that race relations will always be a problem (Gallup Organization 2007a). A Gallup poll survey asked a national sample of U.S. adults to rate relations between various groups in the United States. Results of this survey are presented in Table 9.1. Relations between various racial and ethnic groups are influenced by prejudice and dis-

❝The problem is . . . we don't believe we are as much alike as we are. Whites and blacks, Catholics and Protestants, men and women. If we saw each other as more alike, we might be very eager to join in one big human family in this world, and to care about that family the way we care about our own.**❞**

Morrie Schwartz
Sociologist (from *Tuesdays with Morrie*)

An estimated 20,000–25,000 protesters marched in Jena, Louisiana, in September 2007 to protest the unfair and harsh charges against six black students dubbed the "Jena Six."

TABLE 9.1 Perceptions of Race and Ethnic Relations in the United States, 2006

A national sample of U.S. adults was asked to rate relations between various groups in the United States. The results are depicted in this table.

	VERY OR SOMEWHAT GOOD	VERY OR SOMEWHAT BAD
Whites and blacks	64%	34%
Whites and Hispanics	60%	37%
Blacks and Hispanics	48%	37%
Whites and Asians	83%	12%

Percentages do not add to 100 because of "no opinion" responses.

Source: Gallup Organization (2007a).

crimination (discussed later in this chapter). Race and ethnic relations are also complicated by issues concerning immigration—the topic we turn to next.

IMMIGRANTS IN THE UNITED STATES

The growing racial and ethnic diversity of the United States is largely the result of immigration as well as the higher average birthrates among many minority groups. Immigration generally results from a combination of "push" and "pull" factors. Adverse social, economic, and/or political conditions in a given country "push" some individuals to leave that country; whereas favorable social, economic, and/or political conditions in other countries "pull" some individuals to those countries. Before reading further, you may want to assess your attitudes toward U.S. immigrants and immigration in this chapter's *Self and Society* feature.

U.S. Immigration: A Historical Perspective

For the first 100 years of U.S. history all immigrants were allowed to enter and become permanent residents. The continuing influx of immigrants, especially those coming from nonwhite, non-European countries, created fear and resentment among native-born Americans, who competed with immigrants for jobs and who held racist views toward some racial and ethnic immigrant populations. Increasing pressures from U.S. citizens to restrict or halt entirely the immigration of various national groups led to legislation that did just that. America's open door policy on immigration ended in 1882 with the Chinese Exclusion Act, which suspended for 10 years the entrance of the Chinese to the United States and declared Chinese ineligible for U.S. citizenship. The Immigration Act of 1917 required all immigrants to pass a literacy test before entering the United States. And in 1921 the Johnson Act introduced a limit on the number of immigrants who could enter the country in a single year, with stricter limitations for certain countries (including those in Africa and the Near East). The 1924 Immigration Act further limited the number of immigrants allowed into the United States and completely excluded the Japanese. Other federal immigration laws include the 1943 Repeal of the Chinese Exclusion Act, the 1948 Displaced Persons Act (which permitted refugees from Europe), and the 1952 Immigration and Naturalization Act (which permitted a quota of Japanese immigrants).

In the 1960s most immigrants were from Europe, but today most immigrants are from Latin America (predominantly Mexico) or Asia (see Figure 9.5). In 2005 more than 1 in 10 U.S. residents (12.4 percent) were born in a foreign country (U.S. Census Bureau 2005) (see Table 9.2), and 1 in 5 U.S. children, age 18 and younger, is the child of an immigrant (Fix & Capps 2002).

Being a U.S. Immigrant: Challenges and Achievements

Many immigrants struggle to adjust to life in the United States. Despite the prejudice, discrimination, and lack of social support they experience, many foreign-born U.S. residents work hard to succeed educationally and occupation-

On May 1, 2006, thousands of immigrant workers did not show up for work to show the nation how valuable immigrant labor is for the U.S. economy. Across the country, many businesses closed for the day; others experienced high rates of employee absenteeism.

Self and Society | Attitudes Toward U.S. Immigrants and Immigration

For each of the following questions, choose the answer that best reflects your attitudes. After answering all the questions, compare your answers with the results of a national Gallup poll sample of U.S. adults.

1. On the whole, do you think immigration is a good thing or a bad thing for this country?
   (a) Good thing
   (b) Bad thing

2. Do you think that legal immigrants (a) mostly help the economy by providing low-cost labor or (b) mostly hurt the economy by driving wages down for many Americans?
   (a) Mostly help
   (b) Mostly hurt

3. Do you think that illegal immigrants (a) mostly help the economy by providing low-cost labor or (b) mostly hurt the economy by driving wages down for many Americans?
   (a) Mostly help
   (b) Mostly hurt

4. Which comes closer to your point of view: (a) Illegal immigrants in the long run become productive citizens and pay their fair share of taxes, or (b) illegal immigrants cost the taxpayers too much by using government services like public education and medical services?
   (a) Pay fair share
   (b) Cost too much

5. Which comes closer to your point of view: (a) illegal immigrants mostly take jobs that American workers want, or (b) illegal immigrants mostly take low-paying jobs Americans don't want?
   (a) Take jobs that American workers want
   (b) Take jobs that American workers do not want

Comparison Data

A Gallup poll found the following attitudes toward U.S. immigrants and immigration (see Figure 9.4).

Source: Gallup Organization (2007b)

TABLE 9.2 Foreign-Born Population and Percentage of Total Population for the United States, 1890–2005

YEAR	NUMBER (IN MILLIONS)	PERCENTAGE OF TOTAL
2005	35.6	12.4
1990	19.8	7.9
1970	9.6	4.7
1950	10.3	6.9
1930	14.2	11.6
1890	9.2	14.8

Source: Lollock (2001) and U.S. Census Bureau (2005).

---

ally. Although immigrants are less likely than U.S. natives to graduate from high school, the percentage of foreign-born residents with a bachelor's degree or more education (27.3 percent) is nearly equivalent to that of the native-born population (27.2 percent) (Larsen 2004).

Immigrant employment is relatively concentrated in a small number of sectors, including building and grounds maintenance, food preparation, and construction (Holzer 2005). In a national survey about half (51 percent) of nonimmigrants said that recent immigrants take jobs away from Americans who want them; 15 percent said that they did not get a job because it was given to an immigrant instead (NPR, Kaiser Family Foundation, & Kennedy School of Govern-

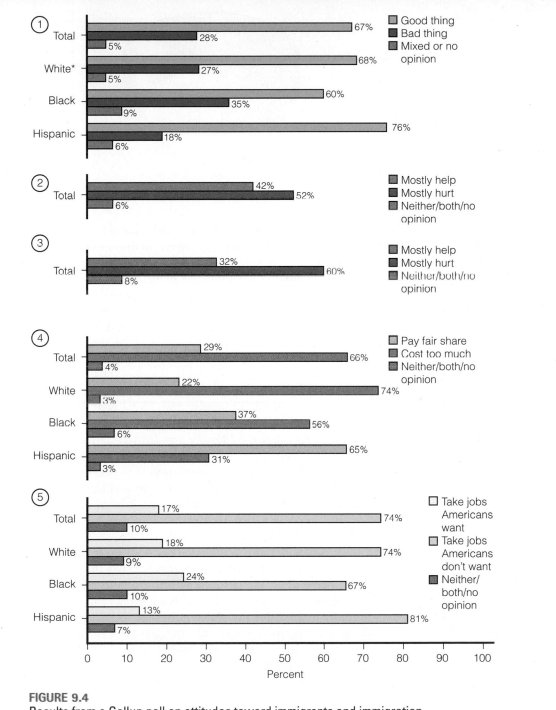

FIGURE 9.4
Results from a Gallup poll on attitudes toward immigrants and immigration.
Source: Gallup organization (2007b).
*"White" refers to non-Hispanic whites.

FIGURE 9.5
U.S. foreign-born residents
by region of birth, 2005.
Source: U.S. Census Bureau (2005).

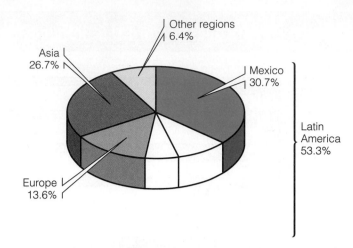

ment 2005). Although immigrant labor displaces some U.S. workers, over the long term immigrants have modest negative effects on the employment of less educated U.S. workers and generate significant benefits for the U.S. economy. Immigrants provide labor in sectors where shortages might occur otherwise, and immigrant labor helps reduce the prices of some products and services, such as housing and food (Holzer 2005). In North Carolina, the state with the highest growth of immigrants in the 1990s, one county commissioner noted that without immigration "our economy would shut down" (Foust, Grow, & Pascual 2002, p. 82).

Compared with the native-born U.S. population, immigrants are more likely to be unemployed and to live in poverty (Larsen 2004). Almost 43 percent of immigrants work at jobs that pay less than $7.50 an hour, compared with 28 percent of all workers (National Immigration Law Center 2007). Yet low-income immigrants are less likely than low-income natives to use benefits such as Medicaid, Temporary Assistance to Needy Families (TANF), and food stamps. In part, this is because many noncitizen immigrants are ineligible for federal public benefit programs (although some states provide assistance to immigrants who are not eligible for federally funded services). In addition, many immigrants do not know that they—and/or their children—may be eligible to receive welfare benefits, or they are afraid that seeking benefits will have a negative effect on their legal status and citizenship (Capps et al. 2005).

Further, many immigrants pay Social Security taxes, even though the workers will not collect Social Security payments because they are not citizens. The states bear the cost of social services, education, and medical services for the immigrant population. However, research suggests that the economic benefits that immigrants provide for the states outweigh the costs associated with supporting them. For example, a study of immigrants in North Carolina found that over the prior 10 years, Latino immigrants had cost the state $61 million in a variety of benefits—but were responsible for more than $9 billion in state economic growth (Beirich 2007).

Guestworker Programs

The United States has two guestworker programs that allow employers to import unskilled labor for temporary or seasonal work lasting less than 1 year: the H-2A program for agricultural work and the H-2B program for non-agricultural work.

To obtain approval from the Department of Labor to bring in guestworkers, employers must certify that (1) there are not sufficient U.S. workers who are able, willing, qualified, and available to perform the work and (2) the wages and working conditions of U.S. workers similarly employed will not be "adversely affected" by the importation of guestworkers. H-2 visas generally do not permit guestworkers to bring their families to the United States. In 2005 the United States issued about 89,000 H-2B visas and about 32,000 H-2A visas (SPLC 2007a). Most workers came from Mexico, followed by Jamaica and Guatemala.

A Southern Poverty Law Center report revealed that the guestworker program constitutes a "modern-day system of indentured servitude" (SPLC 2007a, p. 2). To work under this program, guestworkers often incur debts ranging from $500 to more than $10,000 to pay for visas, travel costs, and recruiter fees. When they arrive at their job, their employer often takes their identity documents (passports and Social Security cards) to ensure that workers do not leave before their contract is fulfilled. Employers often use threats to control their workers. "In one case where workers refused to work until they received their pay after not having been paid in several weeks, the employer responded by threatening to call immigration and declare that the workers had 'abandoned' their work and were thus 'illegal' workers" (SPLC 2007a, p. 16). Guestworkers are often paid substantially less than the minimum wage and are rarely paid overtime pay, despite working well over 40 hours a week. A Guatemalan guestworker in the forestry industry said, "Our pay would come out to approximately $25 for a 12-hour workday" (SPLC 2007a, p. 19). Although guestworkers perform some of the most difficult and dangerous jobs in America, many who are injured on the job are unable to obtain medical treatment and workers' compensation benefits. Although employers hiring H-2A workers are required to provide them with free housing, the quality of housing is often "seriously substandard, even dangerous" and located in isolated rural areas where workers are dependent on their employers for transportation to work, grocery stores, and banks (for which employers often charge exorbitant fees). In some cases, guestworkers are literally locked up in their living quarters. Unable to obtain legal assistance, guestworkers who are abused and denied their legal rights must endure the abuse or try to escape in a foreign land without passports, money, or tickets home.

❝This guestworker program's the closest thing I've ever seen to slavery.**❞**

Charles Rangel
House Ways and
Means Committee chair

Illegal Immigration

Many immigrants come to the United States illegally, without going through legal channels such as the H-2 visa program. There are an estimated 9.3 million undocumented immigrants (i.e., "illegal immigrants") in the United States, representing about one-fourth of the total foreign-born population (Passel, Capps, & Fix 2004). More than half of undocumented immigrants (57 percent) are Mexican (see Table 9.3).

Illegal immigration is an ongoing concern in the United States. Many U.S. citizens are concerned that (1) illegal immigrants burden taxpayers who pay for services such as schools and health care for illegal immigrants, (2) too many people are coming into this country, (3) illegal immigration increases the likelihood of terrorism, and (4) "the wrong kinds of people are coming into the country" (NPR et al. 2005).

Border Crossing. The U.S. Border Patrol is charged with deterring people from illegally crossing the border into the United States and apprehending those who

TABLE 9.3 Undocumented U.S. Immigrants by Country of Origin

COUNTRY OF ORIGIN	PERCENTAGE OF UNDOCUMENTED IMMIGRANTS
Mexico	57
Other Latin American countries	23
Asia	10
Europe and Canada	5
Other parts of the world	5

Source: Passel et al. (2004).

do. In addition, some groups of U.S. citizens have taken action to try to prevent illegal border crossing. For example, in 2003 a Texas rancher invited a vigilante group called Ranch Rescue to his property to stop undocumented Latinos from crossing his land. Fatima Leiva and Edwin Mancia, Salvadorans, were among a group of immigrants traveling on foot when members of Ranch Rescue detained them. During the detention, Mancia was struck on the back of the head and attacked by a Rottweiler dog owned by a Ranch Rescue member. Although Ranch Rescue purports to be protecting the United States from illegal immigration, the underlying motive for their anti-immigrant actions is hate. The president and spokesperson for Ranch Rescue has described Mexicans as "dog turds" who are "ignorant, uneducated and desperate for a life in a decent nation because the one [they] live in is nothing but a pile of dog [excrement] made up of millions of little dog turds" (quoted in SPLC 2005a, p. 3). Between 2005 and 2007, 250 new organizations that target immigrants—called "nativist extremist" organizations—have sprung up across the country, but largely in states that border Mexico. Members of these organizations, some of them armed, engage in vigilante round-ups of undocumented Latino immigrants ("The Year in Hate" 2007).

Despite efforts to seal the U.S.–Mexican border, illegal border crossings occur every day. Some people cross (or attempt to cross) the U.S.–Mexican border with the help of a *coyote*—a hired guide who typically charges $2,000–$3,000 to lead people across the border. Crossing the border illegally involves a number of risks, including death from drowning (e.g., while trying to cross the Rio Grande) or dehydration.

Undocumented Immigrants in the Workforce. An estimated 5 percent of U.S. workers are undocumented immigrants (Passel et al. 2004). Virtually all undocumented men are in the labor force. Their labor force participation exceeds that of men who are legal immigrants or who are U.S. citizens because undocumented men are less likely to be disabled, retired, or in school. Undocumented women are less likely to be in the labor force (62 percent) than undocumented men because they are more likely to be stay-at-home mothers (Passel et al. 2004). "The vast majority of illegal immigrants leave their home countries to work hard, save their money, then return to their homeland. . . . These individuals do not travel their difficult and dangerous journeys searching for a welfare handout; they immigrate to work" (Maril 2004, pp. 11–12).

Undocumented workers often do work that U.S. workers are unwilling to do. Workers routinely work 60 or more hours per week and earn less than the minimum wage of $5.15 per hour. They are not paid overtime and have no benefits. Illegal workers who cleaned Wal-Mart stores generally worked 7 nights a week, 364 days a year and were often locked in stores (Greenhouse 2005). Immigrant workers who are legally admitted to the United States to work under the temporary foreign worker visa program also endure abusive labor conditions. "Because of language barriers and their vulnerable status under immigration laws, these workers may be the most exploited in the nation" (Bauer, quoted in SPLC 2005b) (see also Chapter 7).

Policies Regarding Illegal Immigration. In 1986 Congress approved the Immigration Reform and Control Act, which made hiring illegal immigrants an illegal act punishable by fines and even prison sentences. In 2005 Wal-Mart agreed to pay a record $11 million to settle charges that it used hundreds of illegal immigrants to clean its stores (Greenhouse 2005). Nevertheless, undocumented workers continue to enter and work in the United States.

Recent concerns about illegal immigration have led to numerous policy proposals and stepped up efforts to apprehend "illegal aliens." In 2006, U.S. Customs and Border Protection, an agency within the Department of Homeland Security, apprehended 1.3 million illegal aliens (U.S. Customs and Border Security 2007). In the same year, 6,000 National Guard Troops were deployed to help the Border Patrol. In the Secure Fence Act of 2006, President G. W. Bush authorized the construction of a 700-mile fence along the 2,100-mile U.S. border with Mexico. Public opposition to the building of a border fence has come from landowners who do not want to sell their land to the government; environmentalists who are concerned about the damage the fence project will do to endangered plants and animals in the Lower Rio Grande Valley National Wildlife Refuge; business owners who are concerned about the border fence's impact on the economy; and others who say that a fence will not effectively stop illegal immigration, will cost taxpayers more than $2 billion, and will damage diplomatic relations with Mexico.

In 2007, an immigration reform bill stalled in Congress. The bill proposed that illegal immigrants could obtain a "Z visa" after paying a $5,000 fine. The bill also proposed a "point system" that would emphasize immigrants' education, language, and job skills over family ties in awarding green cards required to be legally employed. Some critics of the bill said the proposed legislation would destroy families, others argued that it threatened U.S. workers, and still others objected to providing what they viewed as "amnesty" to those who had illegally entered the United States. As shown in Table 9.4, most U.S. adults favor some means of allowing illegal immigrants to become U.S. citizens.

Frustrated by lack of federal measures to control illegal immigration, at least 18 states and more than 25 cities and counties have passed various laws and ordinances related to illegal immigration (Bazar 2007). In 2006, Hazelton, Pennsylvania, became the first city to take business licenses away from employers who hire undocumented workers and to fine landlords $1,000 a day for renting to illegal immigrants. In 2007, Arizona passed one of the toughest illegal immigration bills in the country. Under the new Arizona law, employers that knowingly hire illegal immigrants would have their business license suspended or revoked. Opponents say state and local immigration laws are unconstitutional because they override federal law, they do not provide an adequate appeals pro-

TABLE 9.4 Preferred Response to Illegal Immigration in the United States	
Require them to leave the United States and not allow them to return.	14%
Require them to leave, but allow them to return temporarily to work.	6%
Require them to leave, but allow them to return and become citizens if meet certain requirements.	42%
Allow them to remain in the United States and become citizens if meet certain requirements.	36%
Source: Saad (2007).	

cess for employers and workers, and such laws could lead to discrimination against anyone who looks or sounds foreign.

In contrast to states that have taken a hard-line approach to illegal immigration, more than 30 major cities, including San Francisco, New York City, and Newark, have declared themselves "sanctuary cities" for undocumented immigrants. These cities have "sanctuary policies" that prohibit police from asking about immigrant status without criminal cause and instruct city employees not to notify the federal government of the presence of undocumented immigrants living or working in their communities.

Some churches have also played a role in providing a sanctuary to undocumented immigrants. In August 2006, Elvira Arellano, a Mexican citizen who lived illegally in the United States, took refuge in the Adalberto United Methodist Church in an attempt to avoid deportation by U.S. Immigration and Customs Enforcement. Arellano, whose young son is a U.S. citizen, said she should not have to choose between leaving her son in the states and taking him to Mexico. Arellano was arrested in August 2007 when she left the church and traveled to California and was deported to Mexico. Elvira Arellano's deportation has ignited a debate concerning immigration policies involving illegal immigrants whose children are U.S. citizens. Arellano is the president of La Familia Latina Unida (United Latino Family), a group that lobbies for families that could be split as a result of deportation.

What Do You Think? Although the Supreme Court ruled in 1982 that states must educate illegal immigrants through the twelfth grade, states are divided in their laws regarding supporting illegal immigrants through in-state college tuition. Since 2001 10 states have passed laws permitting certain undocumented students to receive in-state tuition (National Immigration Law Center 2006). Immigrants' rights advocates argue that it is in the best interest of the United States to help illegal immigrant youth achieve a college education. But opponents worry that granting in-state tuition to undocumented immigrants would (1) be a financial burden to taxpayers, (2) make it harder for legal citizens to get accepted into state universities, and (3) open the door to granting other benefits now reserved for state residents. Do you think that undocumented immigrants should qualify for in-state college tuition? Why or why not?

Becoming a U.S. Citizen

Foreign-born residents of the United States may or may not apply for and be granted U.S. citizenship. In 2000, of all foreign-born U.S. residents, 35 percent were **naturalized citizens** (immigrants who applied and met the requirements for U.S. citizenship); 65 percent were not U.S. citizens (Lollock 2001).

To become a U.S. citizen, immigrants must have been lawfully admitted for permanent residence; must have resided continuously as a lawful permanent U.S. resident for at least 5 years; must be able to read, write, speak, and understand basic English (certain exemptions apply); and must show that they have

naturalized citizens Immigrants who apply for and meet the requirements for U.S. citizenship.

After taking refuge in a Chicago church for 1 year to avoid deportation to Mexico, Elvira Arellano was arrested in California and deported to Mexico, leaving behind her 8-year-old son, a U.S. citizen.

"good moral character" (U.S. Citizenship and Immigration Services 2003). Applicants who have been convicted of murder or an aggravated felony are permanently denied U.S. citizenship. In addition, applicants are denied citizenship if in the last 5 years they have engaged in any one of a variety of offenses, including prostitution, illegal gambling, controlled substance law violation (except for a single offense of possession of 30 grams or less of marijuana), habitual drunkenness, willful failure or refusal to support dependents, and criminal behavior involving "moral turpitude." To become a U.S. citizen, one must take the oath of allegiance and swear to support the Constitution and obey U.S. laws, renounce any foreign allegiance, and bear arms for the U.S. military or perform services for the U.S. government when required. Finally, applicants for U.S. citizenship must pass an examination on U.S. government and history administered by the U.S. Citizenship and Immigration Services.

SOCIOLOGICAL THEORIES OF RACE AND ETHNIC RELATIONS

Some theories of race and ethnic relations suggest that individuals with certain personality types are more likely to be prejudiced or to direct hostility toward minority group members. Sociologists, however, concentrate on the impact of the structure and culture of society on race and ethnic relations. Three major sociological theories lend insight into the continued subordination of minorities.

Structural-Functionalist Perspective

Structural functionalists emphasize that each component of society affects the stability of the whole. From this perspective, racial and ethnic inequality is dysfunctional for society (Williams & Morris 1993; Schaefer 1998). A society that practices discrimination fails to develop and utilize the resources of minority members. Prejudice and discrimination aggravate social problems, such as crime and violence, war, unemployment and poverty, health problems, family problems, urban decay, and drug use—problems that cause human suffering as well as impose financial burdens on individuals and society.

Picca and Feagin (2007) explained how "the system of racial oppression in the United States affects not only Americans of color but white Americans and society as a whole" (p. 271):

> Whites lose when they have to pay huge taxes to keep people of color in prisons because they are not willing to remedy patterns of unjust enrichment and . . . to pay to expand education, jobs, or drug-treatment programs that would be less costly. They lose by driving long commutes so they do not have to live next to people of color in cities. . . . They lose when white politicians use racist ideas and arguments to keep from passing legislation that would improve the social welfare of all Americans. Most of all, whites lose . . . by not having in practice the democracy that they often celebrate to the world in their personal and public rhetoric. (p. 271)

The structural-functionalist analysis of manifest and latent functions also sheds light on issues of race and ethnic relations. For example, the manifest function of the civil rights legislation in the 1960s was to improve conditions for racial minorities. However, civil rights legislation produced an unexpected negative consequence, or latent dysfunction. Because civil rights legislation supposedly ended racial discrimination, whites were more likely to blame blacks for their social disadvantages and thus perpetuate negative stereotypes such as "blacks lack motivation" and "blacks have less ability" (Schuman & Krysan 1999).

Conflict Perspective

The conflict perspective examines how competition over wealth, power, and prestige contributes to racial and ethnic group tensions. Consistent with this perspective, the "racial threat" hypothesis views white racism as a response to perceived or actual threats to whites' economic well-being or cultural dominance by minorities.

For example, between 1840 and 1870 large numbers of Chinese immigrants came to the United States to work in mining (the California Gold Rush of 1848), railroads (the transcontinental railroad, completed in 1860), and construction. As Chinese workers displaced whites, anti-Chinese sentiment rose, resulting in increased prejudice and discrimination and the eventual passage of the Chinese Exclusion Act of 1882, which restricted Chinese immigration until 1924. More recently, white support for Proposition 209—a 1996 resolution passed in California that ended state affirmative action programs—was higher in areas with larger Latino, African American, or Asian American populations, even after controlling for other factors (Tolbert & Grummel 2003). In other words, opposition to affirmative action programs that help minorities was higher in areas with greater racial and ethnic diversity, suggesting that whites living in diverse areas felt more threatened by the minorities.

In another study researchers interviewed individuals in white racist Internet chat rooms to examine the extent to which people would advocate interracial violence in response to alleged economic and cultural threats (Glaser, Dixit, &

Green 2002). The researchers posed three scenarios that might be perceived as threatening: interracial marriage, minority in-migration (i.e., blacks moving into one's neighborhood), and job competition (i.e., competing with a black person for a job). Respondents' reactions to interracial marriage were the most volatile, followed by in-migration. The researchers concluded that violent ideation among white racists stems from perceived threats to white cultural dominance and separateness rather than from perceived economic threats.

Furthermore, conflict theorists suggest that capitalists profit by maintaining a surplus labor force, that is, by having more workers than are needed. A surplus labor force ensures that wages will remain low because someone is always available to take a disgruntled worker's place. Minorities who are disproportionately unemployed serve the interests of the business owners by providing surplus labor, keeping wages low, and, consequently, enabling them to maximize profits. Some of immigration's biggest supporters are business leaders who want to keep wages low. Sociology professor Stephen Klineberg noted that "the accelerating immigration of Hispanics has meant massively downward pressure on their wages, negotiating power, and ability to confront their employers over violations of their rights" (quoted by Greenhouse 2003, n.p.).

Conflict theorists also argue that the wealthy and powerful elite fosters negative attitudes toward minorities to maintain racial and ethnic tensions among workers. So long as workers are divided along racial and ethnic lines, they are less likely to join forces to advance their own interests at the expense of the capitalists. In addition, the "haves" perpetuate racial and ethnic tensions among the "have-nots" to deflect attention away from their own greed and exploitation of workers.

Symbolic Interactionist Perspective

The symbolic interactionist perspective focuses on the social construction of race and ethnicity—how we learn conceptions and meanings of racial and ethnic distinctions through interaction with others—and how meanings, labels, and definitions affect racial and ethnic groups. We have already explained that contemporary race scholars agree that there is no scientific, biological basis for racial categorizations. However, people have learned to think of racial categories as real, and, as the *Thomas Theorem* suggests, if things are defined as real, they are real in their consequences. Ossorio and Duster (2005) explain:

> People often interact with each other on the basis of their beliefs that race reflects physical, intellectual, moral, or spiritual superiority or inferiority. . . . By acting on their beliefs about race, people create a society in which individuals of one group have greater access to the goods of society—such as high-status jobs, good schooling, good housing, and good medical care—than do individuals of another group. (p. 119)

The labeling perspective directs us to consider the role that negative stereotypes play in race and ethnicity. **Stereotypes** are exaggerations or generalizations about the characteristics and behavior of a particular group. When Americans in a 1990 National Opinion Research Center poll were asked to evaluate various racial and ethnic groups, blacks were rated least favorably (Shipler 1998). Most of the respondents labeled blacks as less intelligent than whites (53 percent), lazier than whites (62 percent), and more likely than whites to prefer being on

stereotypes Exaggerations or generalizations about the characteristics and behavior of a particular group.

welfare to being self-supporting (78 percent). Negative stereotyping of minorities leads to a self-fulfilling prophecy. As Schaefer (1998) explained:

> Self-fulfilling prophecies can be devastating for minority groups. Such groups often find that they are allowed to hold only low-paying jobs with little prestige or opportunity for advancement. The rationale of the dominant society is that these minority individuals lack the ability to perform in more important and lucrative positions. Training to become scientists, executives, or physicians is denied to many subordinate group individuals, who are then locked into society's inferior jobs. As a result, the false definition becomes real. The subordinate group has become inferior because it was defined at the start as inferior and was therefore prevented from achieving the levels attained by the majority. (p. 17)

Even stereotypes that appear to be positive can have negative effects. The view of Asian Americans as a "model minority" involves the stereotypes of Asian Americans as excelling in academics and occupational success. These stereotypes mask the struggles and discrimination experienced by many Asian Americans and also put enormous pressure on Asian-American youth to live up to the social expectation of being a high academic achiever (Tanneeru 2007).

The symbolic interactionist perspective is concerned with how individuals learn negative stereotypes and prejudicial attitudes through language. Different connotations of the colors white and black, for example, may contribute to negative attitudes toward people of color. The white knight is good, and the black knight is evil; angel food cake is white, and devil's food cake is black. Other negative terms associated with black include black sheep, black plague, black magic, black mass, blackballed, and blacklisted. The continued use of derogatory terms such as *Jap, Gook, Spic, Frog, Kraut, Coon, Chink, Wop, Towelhead, Kike,* and *Mick* also confirms the power of language in perpetuating negative attitudes toward minority group members. Similarly, advocates for immigrant rights suggest that the term *illegal aliens* is derogatory; they prefer the term *undocumented worker* as a more neutral term.

> **What Do You Think?** Are racial and ethnic jokes insulting and harmful to minority group members? Or are such jokes innocent and playful forms of fun? Have you ever felt offended by hearing a racial or ethnic joke? How did you respond to the joke-teller?

In the next section we explore the concepts of prejudice and racism in more depth and discuss ways in which socialization and the media perpetuate negative stereotypes and prejudicial attitudes toward racial and ethnic groups.

PREJUDICE AND RACISM

Prejudice refers to negative attitudes and feelings toward or about an entire category of people. Prejudice can be directed toward individuals of a particular religion, sexual orientation, political affiliation, age, social class, sex, race, or ethnicity. **Racism** is the belief that race accounts for differences in human char-

66 Blacks know that racially charged language that seems innocuous to whites can be extremely serious. If it's in a police officer's head, it can get you killed. **99**

Joe Feagin
Sociologist

66 I'm not against Blacks. I'm against all nonwhites. **99**

Ku Klux Klan member

prejudice Negative attitudes and feelings toward or about an entire category of people.

racism The belief that race accounts for differences in human character and ability and that a particular race is superior to others.

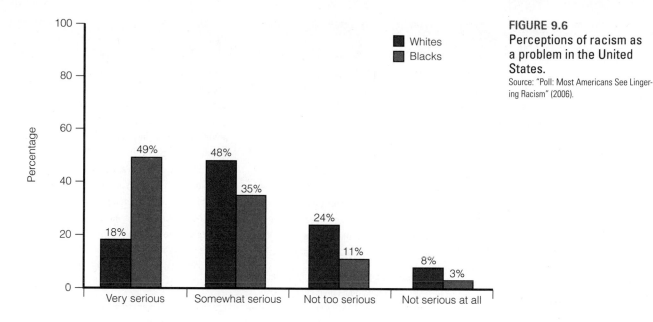

FIGURE 9.6
Perceptions of racism as a problem in the United States.
Source: "Poll: Most Americans See Lingering Racism" (2006).

acter and ability and that a particular race is superior to others. In a national sample of U.S. adults, almost half (49 percent) of blacks and 18 percent of whites said racism in the United States is a "very serious" problem (see Figure 9.6) ("Poll: Most Americans See Lingering Racism" 2006). This chapter's *Social Problems Research Up Close* feature suggests that racism in the United States is more common than many people realize.

Aversive and Modern Racism

Compared with traditional, "old-fashioned" prejudice, which is blatant, direct, and conscious, contemporary forms of prejudice are often subtle, indirect, and unconscious. Two variants of these more subtle forms of prejudice are aversive racism and modern racism.

Aversive Racism. **Aversive racism** represents a subtle, often unintentional, form of prejudice exhibited by many well-intentioned white Americans who possess strong egalitarian values and who view themselves as nonprejudiced. The negative feelings that aversive racists have toward blacks and other minority groups are not feelings of hostility or hate but rather feelings of discomfort, uneasiness, disgust, and sometimes fear (Gaertner & Dovidio 2000). Aversive racists may not be fully aware that they harbor these negative racial feelings; indeed, they disapprove of individuals who are prejudiced and would feel falsely accused if they were labeled prejudiced. "Aversive racists find Blacks 'aversive,' while at the same time find any suggestion that they might be prejudiced 'aversive' as well" (Gaertner & Dovidio 2000, p. 14).

    Another aspect of aversive racism is the presence of pro-white attitudes, as opposed to anti-black attitudes. In several studies respondents did not indicate that blacks were worse than whites, only that whites were better than blacks (Gaertner & Dovidio 2000). For example, blacks were not rated as being lazier than whites, but whites were rated as being more ambitious than blacks.

aversive racism A subtle form of prejudice that involves feelings of discomfort, uneasiness, disgust, fear, and pro-white attitudes.

Before the 1960s, racism in the United States was blatant and pervasive, and it was not uncommon to hear whites publicly and unapologetically express racist views against blacks and other nonwhites. In more recent years, survey research indicates a substantial decline in the percentage of whites who express racist views and a corresponding increase in the percentage of whites who profess to be tolerant of racial diversity and "colorblind."

In this feature, we present a brief synopsis of research in which sociologists Leslie Picca and Joe Feagin examined accounts of racism in the everyday life of college students. This research, published in a book titled *Two-Faced Racism,* revealed that "whites tend to have 'two faces' when it comes to their racial views, commentaries, and actions" (Picca & Feagin 2007, p. 19).

Sample and Methods

In 2002–2003, Picca and Feagin, with the help of numerous college and university faculty members, recruited 934 college students to voluntarily participate in a research project that required students to keep a diary or journal of "everyday events and conversations that deal with racial issues, images, and understandings" (p. 31). Students were instructed to be detailed in their accounts, to conceal all identities (including their own) and disguise all names and identifiers of persons they write about, and to include their reactions and perceptions to the events they write about. Students did not write their own name on the journals, but did indicate their gender, racial group, and age. Instructors who recruited students for this research decided whether to use the journal writing as a class assignment or as extra credit. The majority of students (63 percent) were from colleges and universities in the South, 19 percent were from the Midwest, 14 percent were from the West, and 4 percent were from the Northeast. Most of the participating students were 18–25 years of age, although there were many in their late 20s and a few were older. Rates of participation reported by the 23 instructors varied from

0 to 80 percent (15–30 percent average). The present study is based on journals written by 626 white students (68 percent women and 32 percent men).

Selected Findings

About three-quarters of the roughly 9,000 journal entries submitted by white college students described events that were clearly racist. In relatively few of these accounts, students described their own racist actions or thoughts, as in the following diary entry:

Kristi: When I went to pick up the laundry, I saw a young black man sitting in the driver's side of a mini-van with the engine running. My first thought was that he was waiting for a friend to rob the store and he was the getaway driver. Even worse, I had to look into the store to see what was going on and what (or who) he was waiting for. . . . I am so embarrassed and saddened by my thinking and I suppose I could even omit this from my journal but it is too important to try to pretend that I don't have thoughts like this that pop into my head ostensibly from nowhere. (p. 1)

More frequently, students described the racist comments or actions of friends, family members, acquaintances, and strangers. In a small number of journal accounts, students described antiracist actions by whites.

Students wrote about racist accounts targeted at African Americans, Latinos, Asian Americans, Native Americans, Jewish Americans, and Middle Easterners. Accounts of anti-Asian American racism were more prevalent on the west coast, and anti-Native American racism was more common in the Rocky Mountain states, but in all geographic regions, the most frequently targeted minority group was black Americans. The researchers noted, "A very large proportion of the harshest and most negative images, comments, and actions in the journals targeted black people using hostile racial epithets of various kinds" (p. 247).

In analyzing the racist accounts described in the journals, the researchers drew on the

work of sociologist Erving Goffman, who used theatrical metaphors to describe and understand human behavior. Goffman suggested that when people are "backstage" in private situations they act differently than they do when they are in "frontstage" public situations. Picca and Feagin's research reveals that the expression of racist attitudes has gone backstage to private settings where whites are among other whites, especially family and friends because "many (but by no means all) whites realize it is no longer socially acceptable to be blatantly racist in frontstage areas" (p. xi). Consider the following diary entries:

Lisa: Today I went to go for some coffee with some of these girls I know. As we were waiting [in line] a discussion arose about one girl who had been randomly roomed with an African American for the fall semester. The other two girls gasped when they heard this exclaiming that they didn't know what they would do if they had been so unfortunate. One girl said, "I would be out of there so fast. Black people just don't live the same way we do." She then continued to explain how when she got the name of her roommate, she was so worried that it sounded like a black name. She said she was so relieved to come to school and realized she was white. Meanwhile, every single person on line was listening to this conversation. I was so worried that some of them would get offended. (p. 172)

Hannah: Three of my friends (a white girl and two white boys) and I went back to my house to drink a little more before we ended the night. My one friend, Dylan started telling jokes. . . . Dylan said: "What's the most confusing day of the year in Harlem?" "Father's Day . . . Who's your Daddy?" Dylan also referred to black people as "Porch monkeys." Everyone laughed a little, but it was obvious that we all felt a little less comfortable when he was telling jokes like that. My friend Dylan is not a racist person. He has more black friends than I do, that's why I was surprised he so freely said something like that. Dylan would never have said something like that around anyone who was a minority. . . . It is this sort of "joking" that helps to keep racism alive today. People know the places they have to be politically correct and most people will be. However, until this sort of "behind-the-scenes" racism comes to an end, people will always harbor those stereotypical views that are so prevalent in our country. This kind of joking really does bother me, but I don't know what to do about it. I know that I should probably stand up and say I feel uncomfortable when my friends tell jokes like that, but I know my friends would just get annoyed with me and say that they obviously don't mean anything by it. (pp. 17–18)

Both Lisa's and Hannah's journal entries illustrate the social norm that defines expressions of racism as acceptable in private "backstage" settings, and inappropriate in public "frontstage" settings. Hannah's journal entry also exemplifies a theme of "white innocence" whereby white students offer excuses for and minimize the significance of racist events. Despite Dylan's racist joke-telling, Hannah asserts that he is not a racist person. In their journal accounts of racist behavior, white college students often defended whites who make racist comments as "nice guys" or "nice people." As illustrated in the following journal entry, whites often view racist comments not as "real racism," but as harmless remarks that are not to be taken seriously.

Sam: On Sunday night I had a discussion with Frank, a white college male, about racism in our building. I asked him if he felt like there was any in the hall and he told me that he hadn't observed any "real" racism and he replied that while he had heard racist jokes, he didn't see any "Klansmen or burning crosses" so he didn't take it to be a serious problem. I asked him why he didn't consider racist jokes to be as serious a problem as racial violence. He said

that as long as nobody was directly being hurt, either by words or by more physical means, then it shouldn't be considered real racism. (p. 103)

Sometimes whites think they are in a safe "backstage" setting when making racial remarks, but then discover otherwise, as exemplified by the following journal entry:

Dana: On Thursday . . . my friend Megan (a white female) went on her first date with Steve. As their conversation began they discussed typical first date topics like family, friends, home, etc. Somehow Megan began talking about how squirrels in the Bronx are black and so people call them "squiggers." When Steve did not laugh Megan wondered why he did not think it was funny. As the conversation progressed and Steve began to talk about his family, he revealed to Megan that his dad is black and mom is white. Right then, Megan realized why Steve did not find her joke to be so funny. Megan felt horrible. Without realizing it, Megan hurt someone that appeared to be just as white as she was. . . . (p. 195)

Although white women also display racist behavior, the journal entries in this research suggest that "white men disproportionately make racist jokes and similar racially barbed comments" (p. 133).

Carissa: The white men got on the subject of so-called nicknames for black people. Some mentioned were porch monkeys, jigaboos, tree swingers, etc. The one thing I took notice of was that not one girl made a comment. Most of them just seemed to stare off and pretend to not hear anything. Is this because women are more sensitive? Or are they just afraid to express their true feelings? (p. 133)

Some students made comments in their journals or on the cover sheet that point to a general lack of awareness of racism. One student, for example, wrote, "This assignment

made me realize how many racial remarks are said every single day and I usually never catch any of them or pay close attention" (p. 39). Another student described her discussion with another classmate:

Gabrielle: As we talked we began discussing the contents of our journals and we both came to the conclusion that neither of us ever realized how many racial comments we hear every day. Talking with my friend gave me a new perspective on just how much I am surrounded by racial issues and comments, and it made me realize how often I simply ignore or don't even notice certain comments that should bother me. (p. 275)

Many students were offended by the racist remarks they recorded in their journals, but did not express their disapproval and, so, participated in racism as a "bystander." Some students indicated that after participating in the journal writing assignment, they are more aware of the need to intervene in social situations where racism is being expressed.

Kyle: As my last entry in this journal, I would like to express what I have gained out of this assignment. I watched my friends and companions with open eyes. I was seeing things that I didn't realize were actually there. By having a reason to pick out the racial comments and actions I was made aware of what is really out there. Although I noticed that I wasn't partaking in any of the racist actions or comments, I did notice that I wasn't stopping them either. I am now in a position to where I can take a stand and try to intervene in many of the situations. (p. 275)

Conclusion

Picca and Feagin's study of college students' journals reveals that racism is still a pervasive feature of U.S. society, especially in "back-stage" social contexts. The researchers also conclude: "Our data probably underestimate the severity and extent of the racist performances that whites engage in behind closed doors. Numerous students indicated that they suppressed, or thought about suppressing, racial events that they were involved in or observed" (pp. 33–34). For example, one student wrote:

As I am getting ready to turn in this assignment I looked back over some of the entries . . . and thought that I should change a lot of them for fear of whoever

is reading them might be offended even though I was very reserved in some of the accounts of what happened. (p. 34)

Based on the research presented in *Two-Faced Racism* and other current qualitative research on racism, Picca and Feagin suggest that "the majority of whites still participate in openly racist performances in the backstage arena . . . and do not define such performances

as problematic and deserving of action aimed at eradication" (p. 22). The researchers further note that "most of our college student diarists did not, or could not, see the connection between the everyday racial performances they recorded and the unjust discrimination and widespread suffering endured by people of color in society generally" (p. 28).

Source: Picca and Feagin (2007).

Gaertner and Dovidio (2000) explain that "aversive racists would not characterize blacks more negatively than whites because that response could readily be interpreted by others or oneself to reflect racial prejudice" (p. 27). Compared with anti-black attitudes, pro-white attitudes reflect a more subtle prejudice that, although less overtly negative, is still racial bias.

Modern Racism. Like aversive racism, **modern racism** involves the rejection of traditional racist beliefs, but a modern racist displaces negative racial feelings onto more abstract social and political issues. The modern racist believes that serious discrimination in the United States no longer exists, that any continuing racial inequality is the fault of minority group members, and that demands for affirmative action for minorities are unfair and unjustified. "Modern racism tends to 'blame the victim' and places the responsibility for change and improvements on the minority groups, not on the larger society" (Healey 1997, p. 55). Like aversive racists, modern racists tend to be unaware of their negative racial feelings and do not view themselves as prejudiced.

> ❝If we were to wake up some morning and find that everyone was the same race, creed and color, we would find some other causes for prejudice by noon.❞
>
> George Aiken

What Do You Think? In a poll conducted for CNN, respondents were asked, "Do you consider yourself to be in any way racially biased or not?" ("Poll: Most Americans See Lingering Racism" 2006). What percentage of whites and what percentage of blacks do you think said that they considered themselves to be racially biased? The answer is 13 percent of whites and 12 percent of blacks in the CNN poll considered themselves to be racially biased. What do you think of these data?

modern racism A subtle form of racism that involves the belief that serious discrimination in America no longer exists, that any continuing racial inequality is the fault of minority group members, and that the demands for affirmative action for minorities are unfair and unjustified.

Learning to Be Prejudiced: The Role of Socialization and the Media

Psychological theories of prejudice focus on forces within the individual that give rise to prejudice. For example, the frustration-aggression theory of prejudice (also known as the scapegoating theory) suggests that prejudice is a form of hostility that results from frustration. According to this theory, minority groups serve as convenient targets of displaced aggression. The authoritarian-personality theory of prejudice suggests that prejudice arises in people with a certain personality type. According to this theory, people with an authoritarian personality—who are highly conformist, intolerant, cynical, and preoccupied with power—are prone to being prejudiced.

Rather than focus on the individual, sociologists focus on social forces that contribute to prejudice. Earlier we explained how intergroup conflict over wealth, power, and prestige gives rise to negative feelings and attitudes that serve to protect and enhance dominant group interests. In the following discussion we explain how prejudice is learned through socialization and the media.

Learning Prejudice through Socialization. Although most researchers agree that the majority of children learn conceptions of racial and ethnic distinctions by the time they are about 6 years old, Van Ausdale and Feagin (2001) suggest that children as young as 3 years old have acquired prejudicial attitudes:

> Well before they can speak clearly, children are exposed to racial and ethnic ideas through their immersion in and observation of the large social world. Since racism exists at all levels of society and is interwoven in all aspects of American social life, it is virtually impossible for alert young children either to miss or ignore it. . . . Children are inundated with it from the moment they enter society. (pp. 189–190)

In the socialization process individuals adopt the values, beliefs, and perceptions of their family, peers, culture, and social groups. Prejudice is taught and learned through socialization, although it need not be taught directly and intentionally. Parents who teach their children to not be prejudiced yet live in an all-white neighborhood, attend an all-white church, and have only white friends may be indirectly teaching negative racial attitudes to their children. Socialization can also be direct, as in the case of parents who use racial slurs in the presence of their children or who forbid their children from playing with children from a certain racial or ethnic background. Children can also learn prejudicial attitudes from their peers. The telling of racial and ethnic jokes among friends, for example, perpetuates stereotypes that foster negative racial and ethnic attitudes.

Prejudice and the Media. The media contribute to prejudice by portraying minorities in negative and stereotypical ways—or by not portraying them at all. In the 2003–2004 television season 73 percent of characters on prime-time television were white, and 16 percent were African American (Glaubke & Heintz-Krowles 2004). The percentage of Latino characters was 6.5 percent—a notable increase from 3 percent in 1999. Only 3 percent were Asian, and less than 1 percent were Arab, Middle Eastern, Indian, or Pakistani. There were no Native American characters. An analysis of character portrayals in the 2003–2004 prime-time television season revealed that Latinos were more likely than any other group to be cast in low-status occupations, such as domestic worker, and as criminals. Nearly half (46 percent) of Arabs and Middle Easterners were cast as criminals.

Another media form that is used to promote hatred toward minorities and to recruit young people to the white power movement is "white power music," which contains anti-Semitic, racist, and homophobic lyrics. A CD that was distrib-

Byron Calvert operates Panzerfaust Records, one of the nation's largest "white power" music labels.

© AP Photo/The Minnesota Public Radio, Jeff Horwich

uted to thousands of middle and high school students by the neo-Nazi record label Panzerfaust Records contains the following lyrics from the group the Bully Boys: "Whiskey bottles/baseball bats/pickup trucks/and rebel flags/we're going on the town tonight/hit and run/let's have some fun/we've got jigaboos on the run." And in a song called "Wrecking Ball," the band H8Machine advises kids to "destroy all your enemies," promising that "the best things come to those who hate" ("Neo-Nazi Label Woos Teens" 2004). The Southern Poverty Law Center has identified 123 domestic and 227 international white power bands that promote hate and intolerance through their music ("White Power Bands" 2002).

The Internet also spreads messages of hate toward minority groups through the websites of various white supremacist and hate group organizations. After Hurricane Katrina in 2005, white supremacists posted hundreds of messages on the Internet expressing hopes that blacks in New Orleans would be wiped out. One message suggested that "they pile up all the niggers and use them as human sand bags against the rising storm surge" (Potok 2005, n.p.).

The Southern Poverty Law Center, a nonprofit organization that combats hate, intolerance, and discrimination, identified 566 hate websites in 2006 ("Hate Websites" 2007). And in 2007, about 12,000 white supremacist propaganda videos, hate music concert videos, and Holocaust denial videos were posted on video-sharing websites, such as YouTube (Mock 2007). Hate websites use sophisticated graphics, music, and entertaining games to lure children, teenagers, and adults (Nemes 2002). Once individuals are hooked on racist ideology, they use Internet discussion groups or chat rooms to connect with other like-minded individuals with whom they can share their racist views and have those views reinforced.

Although a number of countries, including Germany and France, have passed laws prohibiting hate speech on the Internet, in the United States free speech is protected under the First Amendment. Thus, many German neo-Nazi white supremacist groups use U.S.-based Internet servers to avoid prosecution under German law (Kaplan & Kim 2000).

DISCRIMINATION AGAINST RACIAL AND ETHNIC MINORITIES

discrimination Actions or practices that result in differential treatment of categories of individuals.

Whereas prejudice refers to attitudes, **discrimination** refers to actions or practices that result in differential treatment of categories of individuals. National survey research finds that most U.S. adults agree that racial discrimination persists in the United States. Yet, one-third say discrimination against blacks is rare

(Pew Research Center 2007). Although prejudicial attitudes often accompany discriminatory behavior or practices, one can be evident without the other.

Individual Versus Institutional Discrimination

Individual discrimination occurs when individuals treat other individuals unfairly or unequally because of their group membership. Individual discrimination can be overt or adaptive. In **overt discrimination** the individual discriminates because of his or her own prejudicial attitudes. For example, a white landlord may refuse to rent to a Mexican American family because of his or her own prejudice against Mexican Americans. Or a Taiwanese American college student who shares a dorm room with an African American student may request a roommate reassignment from the student housing office because he or she is prejudiced against blacks.

Suppose that a Cuban American family wants to rent an apartment in a predominantly non-Hispanic neighborhood. If the landlord is prejudiced against Cubans and does not allow the family to rent the apartment, that landlord has engaged in overt discrimination. But what if the landlord is not prejudiced against Cubans but still refuses to rent to a Cuban family? Perhaps that landlord is engaging in **adaptive discrimination,** or discrimination that is based on the prejudice of others. In this example the landlord may fear that if he or she rents to a Cuban American family, other renters who are prejudiced against Cubans may move out of the building or neighborhood and leave the landlord with unrented apartments. Overt and adaptive individual discrimination can coexist. For example, a landlord may not rent an apartment to a Cuban family because of his or her own prejudices and the fear that other tenants may move out.

Institutional discrimination occurs when normal operations and procedures of social institutions result in unequal treatment of and opportunities for minorities. Institutional discrimination is covert and insidious and maintains the subordinate position of minorities in society. For example, the practice of businesses moving out of inner-city areas results in reduced employment opportunities for America's highly urbanized minority groups. In the retail industry traditional warehouses designed to house inventory are being replaced by large distribution centers that require more space, thus encouraging location away from central cities to lower priced land in less populated areas. A study by the U.S. Equal Employment Opportunity Commission (EEOC) found that as areas become less populated, the percentage of minority workers declines (EEOC 2004).

Institutional discrimination also occurs in education. When schools use standard intelligence tests to decide which children will be placed in college preparatory tracks, they are limiting the educational advancement of minorities whose intelligence is not fairly measured by culturally biased tests developed from white middle-class experiences. And the funding of public schools through local tax dollars results in less funding for schools in poor and largely minority school districts.

Institutional discrimination is also found in the criminal justice system, which more heavily penalizes crimes that are more likely to be committed by minorities. For example, the penalties for crack cocaine, more often used by minorities, have traditionally been higher than those for other forms of cocaine use, even though the same prohibited chemical substance is involved. As conflict theorists emphasize, majority group members make rules that favor their own group.

individual discrimination The unfair or unequal treatment of individuals because of their group membership.

overt discrimination Discrimination that occurs because of an individual's own prejudicial attitudes.

adaptive discrimination Discrimination that is based on the prejudice of others.

institutional discrimination Discrimination in which the normal operations and procedures of social institutions result in unequal treatment of minorities.

Next, we look at discrimination in employment, housing, education, and politics. Finally, we expose the extent and brutality of physical and verbal violence against minorities.

Employment Discrimination

When a national sample of U.S. adults was asked, "Do you feel that racial minorities in this country have equal job opportunities as whites?" more than half (53 percent) said "No" (Gallup Organization 2007a). Despite laws against it, discrimination against minorities occurs today in all phases of the employment process, from recruitment to interview, job offer, salary, promotion, and firing decisions.

A sociologist at Northwestern University studied employers' treatment of job applicants in Milwaukee, Wisconsin, by dividing job applicant "testers" into four groups: blacks with a criminal record, blacks without a criminal record, whites with a criminal record, and whites without a criminal record (Pager 2003). Applicant testers, none of whom actually had a criminal record, were trained to behave similarly in the application process and were sent with comparable résumés to the same set of employers. The study found that white applicants with no criminal record were the most likely to be called back for an interview (34 percent) and that black applicants with a criminal record were the least likely to be called back (5 percent). But surprisingly, white applicants *with* a criminal record (17 percent) were more likely to be called back for an interview than were black applicants *without* a criminal record (14 percent)! The researcher concluded that "the powerful effects of race thus continue to direct employment decisions in ways that contribute to persisting racial inequality" (Pager 2003, p. 960).

Discrimination in hiring may be unintended. For example, many businesses rely on their existing employees to refer new recruits when a position opens up. Word-of-mouth recruitment is inexpensive and efficient; some companies offer bonuses to employees who bring in new recruits. But this traditional recruitment practice tends to exclude minority workers, because they often do not have a network of friends and family members in higher positions of employment who can recruit them (Schiller 2004).

Employment discrimination contributes to the higher rates of unemployment and lower incomes of blacks and Hispanics compared with those of whites (see Chapters 6 and 7). Lower levels of educational attainment among minority groups account for some, but not all, of the disadvantages they experience in employment and income. As shown in Figure 9.7, average lifetime earnings of whites are higher than those for blacks and Hispanics at the same level of educational attainment (Day & Newburger 2002). Although the lifetime earnings of Asians and Pacific Islanders with advanced degrees are equivalent to those of whites, at every other level of educational attainment, whites earn more than Asians and Pacific Islanders.

Workplace discrimination also includes unfair treatment. In one workplace with a large Hispanic workforce, Hispanic workers were selected each week to clean the lunchroom without being paid for that work (Greenhouse 2003). One Hispanic worker said that this treatment was a matter of dignity and that it made the workers feel humiliated.

Since the terrorist attacks of September 11, 2001, workplace discrimination against individuals who are (or who are perceived to be) Muslim, Arab, Middle

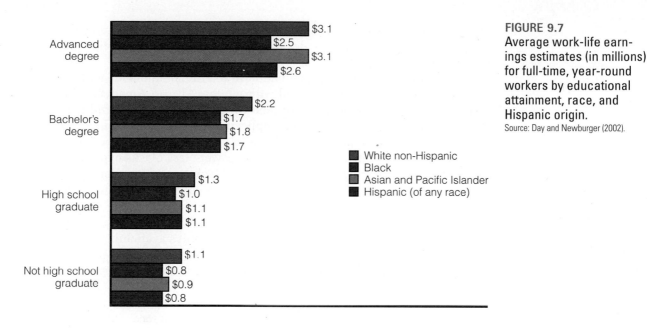

FIGURE 9.7
Average work-life earnings estimates (in millions) for full-time, year-round workers by educational attainment, race, and Hispanic origin.
Source: Day and Newburger (2002).

Eastern, South Asian, or Sikh has increased. The most common complaints of post-9/11 backlash reported to the EEOC are alleged harassment and firing (EEOC 2003).

Housing Discrimination and Segregation

Before the 1968 Federal Fair Housing Act and the 1974 Equal Credit Opportunity Act, discrimination against minorities in housing and mortgage lending was as rampant as it was blatant. Banks and mortgage companies commonly engaged in "redlining"—the practice of denying mortgage loans in minority neighborhoods on the premise that the financial risk was too great, and the ethical standards of the National Association of Real Estate Boards prohibited its members from introducing minorities into white neighborhoods. Instead, realtors practiced "geographic steering," whereby they discouraged minorities from moving into certain areas by showing them homes only in minority neighborhoods.

Although housing discrimination is illegal today, it is not uncommon. To assess discrimination in housing, researchers use a method called "paired testing." In a paired test two individuals—one minority and the other nonminority—are trained to pose as home seekers, and they interact with real estate agents, landlords, rental agents, and mortgage lenders to see how they are treated. The testers are assigned comparable or identical income, assets, and debt as well as comparable or identical housing preferences, family circumstances, education, and job characteristics. A paired testing study of housing discrimination in 23 metropolitan areas found that in the rental market whites were more likely to receive information about available housing units and had more opportunities to inspect available units than did blacks and Hispanics (Turner et al. 2002). The incidence of discrimination was greater for Hispanic renters than for black renters. The same study found that in the home sales market, white home buyers were more likely to be able to inspect available homes and to be shown homes in more predominantly non-Hispanic white neighborhoods than were comparable black and Hispanic buyers. Whites were also more likely to receive information and assistance with financing.

In a study of housing discrimination in the Philadelphia area, Massey and Lundy (2001) found that compared with whites, African Americans were less likely to have a rental agent return their calls, less likely to be told that a unit was available, more likely to pay application fees, and more likely to have credit mentioned as a potential problem in qualifying for a lease. These racial effects were exacerbated by sex and class. Lower-class blacks experienced less access to rental housing than middle-class blacks, and black females experienced less access than black males. Lower-class black females were the most disadvantaged group. They experienced the lowest probability of contacting and speaking to a rental agent and, even if they did make contact, they faced the lowest probability of being told of a housing unit's availability. Lower-class black females also faced the highest chance of paying an application fee. On average, lower-class black females were assessed $32 more per application than white middle-class males.

In other research on mortgage lending discrimination, minorities were less likely to receive information about loan products, they received less time and information from loan officers, they were often quoted higher interest rates, and they had higher loan denial rates than whites, other things being equal (Turner & Skidmore 1999).

Despite continued housing discrimination, homeownership rates among minorities and low-income groups increased substantially in the 1990s, reaching record rates in many central cities. However, minority and low-income homeowner rates still lag behind the overall homeownership rate. Also, many of the gains in minority and low-income homeownership rates are due to increases in *subprime lending*—higher-fee, higher-interest-rate loans offered to borrowers who have poor (or nonexistent) credit records (Williams, Nexiba, & McConnell 2005).

Although homeownership rates among racial minorities have increased, residential segregation of racial and ethnic groups persists. Analysis of the 2000 census data revealed that although blacks and whites lived in neighborhoods that were slightly more integrated than they were in 1990, people still lived in largely segregated neighborhoods. In 2000 an average white person living in a metropolitan area (which includes city dwellers and suburban residents) lived in a neighborhood that is about 80 percent white and 7 percent black (Schmitt 2001a). In contrast, the average black person lived in a neighborhood that is 33 percent white and 51 percent black. A 2005 Gallup poll also found that Americans tend to live in neighborhoods largely populated by people of similar racial or ethnic backgrounds: 86 percent of non-Hispanic whites, 66 percent of blacks, and 61 percent of Hispanics reported living in areas where there are many people from their own backgrounds (Carroll 2005).

For years sociologists have known that U.S. minorities, who are disproportionately represented among the poor, tend to be segregated in concentrated areas of low-income housing, often in inner-city areas of concentrated poverty (Massey & Denton 1993). After Hurricane Katrina in 2005 citizens across the United States and the world were shocked by the degree of poor, segregated communities such as the Lower Ninth Ward of New Orleans, where 98 percent of residents were black and 36 percent lived below the poverty line.

❝The winds of Katrina blew away the flimsy veil that long has shielded most Americans from the ugly reality of our nation's continuing problems with race, class, and poverty.❞

Mark Potok
Southern Poverty Law Center

Educational Discrimination and Segregation

Both institutional discrimination and individual discrimination in education negatively affect racial and ethnic minorities and help to explain why minorities (with the exception of Asian Americans) tend to achieve lower levels of academic attainment and success (see also Chapter 8). Institutional discrimination is evidenced by inequalities in school funding—a practice that disproportionately hurts minority students (Kozol 1991). Nearly half of school funding comes from local taxes. In 2002 the federal government supplied 8 percent of educational expenditures and state government contributed 48 percent; local governments provided the remainder (Schiller 2004). Because minorities are more likely than whites to live in economically disadvantaged areas, they are more likely to go to schools that receive inadequate funding. Inner-city schools, which serve primarily minority students, receive less funding per student than do schools in more affluent, primarily white areas.

For example, inequalities in school funding resulted in New York City receiving $2,000 less per pupil than Buffalo, Rochester, Syracuse, and Yonkers. New York State Supreme Court Judge Leland DeGrasse found that New York State's system of school funding violated federal civil rights laws because it disproportionately hurt minority students (more than 70 percent of the state's Asian, black, and Hispanic students live in New York City) (Goodnough 2001).

Another institutional education policy that is advantageous to whites is the policy that gives preference to college applicants whose parents or grandparents are alumni. The overwhelming majority of alumni at the highest ranked universities and colleges are white. Thus white college applicants are the primary beneficiaries of these so-called legacy admissions policies. About 10–15 percent of students in most Ivy League colleges and universities are children of alumni. Harvard University accepts about 11 percent of its overall applicant pool, but for legacy applicants the admission rate is 40 percent (Schmidt 2004). As a result of pressure from state lawmakers and minority rights activists, in 2004 Texas A&M University became the first public college to abandon its legacy admittance policy.

Minorities also experience individual discrimination in the schools as a result of continuing prejudice among teachers. One college student completing a teaching practicum reported that some of the teachers in her school often spoke about children of color as "wild kids who slam doors in your face" (Lawrence 1997, p. 111). In a survey conducted by the Southern Poverty Law Center, 1,100 educators were asked whether they had heard racist comments from their colleagues in the past year. More than one-fourth of survey respondents answered yes ("Hear and Now" 2000). It is likely that teachers who are prejudiced against minorities discriminate against them, giving them less teaching attention and less encouragement.

Racial and ethnic minorities are also treated unfairly in educational materials, such as textbooks, which often distort the history and heritages of people of color (King 2000). For example, Zinn (1993) observed, "To emphasize the heroism of Columbus and his successors as navigators and discoverers, and to deemphasize their genocide, is not a technical necessity but an ideological choice. It serves, unwittingly—to justify what was done" (p. 355).

Finally, racial and ethnic minorities are largely isolated from whites in an increasingly segregated school system. A study by the Civil Rights Project at Harvard University found that U.S. schools in the 2000–2001 school year were more segregated than they were in 1970 (Orfield 2001). The upward trend in school segregation is due to large increases in minority student enrollment, continuing white flight from urban areas, the persistence of housing segregation, and the termination of court-ordered desegregation plans. Court-mandated busing became a means to achieve equality of education and school integration in the early 1970s after the Supreme Court (in *Swann v. Charlotte-Mecklenberg*) endorsed busing to desegregate schools. But in the 1990s lower courts lifted desegregation orders in dozens of school districts (Winter 2003a). And in 2007, the United States Supreme Court issued a landmark ruling that race cannot be a factor in the assignment of children to public schools. In a bitterly divided 5 to 4 vote, the Court invalidated voluntary school desegregation plans after hearing a case involving schools in Seattle, Washington, and Louisville, Kentucky, that used race when assigning some students to schools in an effort to end racial segregation. The decision jeopardizes similar plans in hundreds of districts nationwide, and it further restricts how public school systems may achieve racial diversity.

Political Discrimination

Historically, African Americans have been discouraged from political involvement by segregated primaries, poll taxes, literacy tests, and threats of violence. However, tremendous strides have been made since the passage of the 1965 Voting Rights Act, which prohibited literacy tests and provided for poll observers. Blacks have won mayoral elections in Atlanta, Cleveland, Washington, DC, and New York City, and the governorship in Virginia. However, racial minorities and Hispanics continue to be underrepresented in political positions and voting participation.

Discrimination against racial and ethnic minorities in the U.S. political process persists. After "voting irregularities" in the 2000 national elections, the National Association for the Advancement of Colored People (NAACP) and several civil rights groups filed a lawsuit in Florida to eliminate unfair voting practices (NAACP 2001). In the 2000 election thousands of black voters complained that they were wrongfully turned away from the polls or had trouble casting their ballots. Complaints were not limited to Florida; black voters in about a dozen other states reported similar unfair treatment (NAACP 2000). Numerous incidents of voter intimidation and suppression in predominantly African American communities were also reported in the 2004 presidential election (Fitrakis & Wasserman 2004).

Hate Crimes

66 Race hate isn't human nature; race hate is the abandonment of human nature. 99

Orson Welles

hate crime An unlawful act of violence motivated by prejudice or bias.

In June 1998 James Byrd, Jr., a 49-year-old father of three, was walking home from a niece's bridal shower in the small town of Jasper, Texas. According to police reports, three white men riding in a gray pickup truck saw Byrd, a black man, walking down the road and offered him a ride. The men reportedly drove down a dirt lane and, after beating Byrd, chained him to the back of the pickup truck and dragged him for 2 miles down a winding, narrow road. The next day, police found Byrd's mangled and dismembered body. The three men who were arrested had ties to white supremacist groups. James Byrd had been brutally murdered simply because he was black.

The murder of James Byrd exemplifies a **hate crime**—an unlawful act of violence motivated by prejudice or bias. Examples of hate crimes, also known as

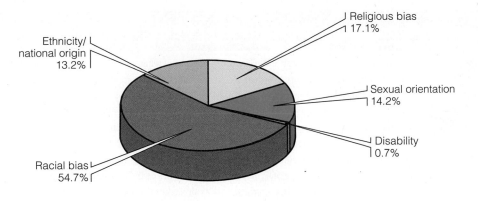

Religious bias
17.1%

Ethnicity/
national origin
13.2%

Sexual orientation
14.2%

Disability
0.7%

Racial bias
54.7%

Total number of single-bias hate crimes: 7,160

"bias-motivated crimes" and "ethnoviolence," include intimidation (e.g., threats), destruction of or damage to property, physical assault, and murder.

From the first year that FBI hate crime data were published in 1992, most hate crimes have been based on racial bias (see Figure 9.8). Keep in mind that FBI hate crime data undercount the actual number of hate crimes because (1) not all U.S. jurisdictions report hate crimes to the FBI (reporting is voluntary), (2) it is difficult to prove that crimes are motivated by hate or prejudice, (3) law enforcement agencies shy away from classifying crimes as hate crimes because it makes their community "look bad," and (4) victims are often reluctant to report hate crimes to the authorities.

In the United States, the rate of violence against American Indians is higher than that among any other minority group (Buchanan 2007). Hate crime violence against American Indians is especially prevalent in "border towns" adjacent to Indian reservations.

After the terrorist attacks of September 11, 2001, hate crimes against individuals perceived to be Muslim and/or Middle Eastern increased significantly. Perpetrators of these hate crimes may have felt that they were acting in defense of a perceived threat. A poll by the Council on American-Islamic Relations showed that more than half of America's 7 million Muslims (57 percent) said they experienced bias or discrimination after September 11 (Morrison 2002). On the positive side the same poll found that most American Muslims (79 percent) also experienced special kindness or support from friends or colleagues of other faiths. Acts of kindness include verbal assurances, support, and even offers to help guard local mosques and Islamic schools.

More recently, violence against U.S. immigrants has become a growing national problem. This chapter's *The Human Side* feature describes the hate crime victimization of a Hispanic immigrant in Georgia—a state with one of the fastest growing immigrant populations.

Motivations for Hate Crimes. Levin and McDevitt (1995) found that the motivations for hate crimes were of three distinct types: thrill, defensive, and mission. Thrill hate crimes are committed by offenders who are looking for excitement and attack victims for the "fun of it." Defensive hate crimes involve offenders who view their attacks as necessary to protect their community, workplace, or college campus from "outsiders" or to protect their racial and cultural purity

AP/Wide World Photos

Domingo Lopez Vargas, an immigrant day laborer, was brutally beaten in a hate crime in Canton, Georgia.

Domingo Lopez Vargas left his dirt-poor Guatemalan farm village to come to the United States, where he hoped to earn decent money for his wife and nine children. After picking oranges in Florida, he moved to Georgia, where the booming construction business lured immigrant workers. Unlike many of his compadres, Lopez had legal status, which helped him find steady work hanging doors and windows. When work dried up, Lopez joined the more than 100,000 jornaleros—day laborers—who wait for landscaping and construction jobs on street corners and in front of convenience stores all across Georgia. Usually there are plenty of pickup trucks that swing by, offering $8 to $12 an hour for digging, planting, painting, or hammering. But this day, nada. By late afternoon, Lopez had tired of waiting in the cold, so he walked up the street to pick up a few things at a grocery store.

"I got milk, shampoo and toothpaste," Lopez recalled. "When I was leaving the store, this truck stopped right in front of me and said, 'Do you want to work?' . . . I said, yes, how much? They said nine dollars an hour. I didn't ask what kind of job. I just wanted to work, so I said yes."

Until that afternoon, Lopez said, "Americans had always been very nice to me"— which might explain why he wasn't concerned that the four guys in the pickup truck looked awfully young to be contractors. Or why he didn't think twice about being picked up so close to sunset. "I took the offer because I know sometimes people don't stop working until 9 at night," he says.

The four young men, all high school students, drove Lopez to a remote spot strewn with trash. "They told me to pick up some plastic bags that were on the ground. I thought that was my job, to clean up the trash. But when I bent over to pick it up, I felt somebody hit me from behind with a piece of wood, on my back." It was just the start of a 30-minute pummeling that left Lopez bruised and bloody from his thighs to his neck. "I thought I was dying," he said. "I tried to stand up but I couldn't." Finally, after he handed over all the cash in his wallet, $260, along with his Virgin Mary pendant, the teenagers sped away.

As a result of injuries incurred by the beating, Lopez could not work for 4 months, and he was left with $4,500 in medical bills. Sometimes he still puzzles over his attackers' motives. "They were young," he speculates, "and maybe they didn't have enough education. Or maybe their families . . . taught them to kill people, and that is what they have learned."

Having sworn off day labor, Lopez works night shifts now, cutting up chickens at the nearby Tyson plant, wincing through the pain that shoots up his right arm when he lowers the boom on a bird. But it's only temporary, he says. "I called my wife and told her what happened. She told me to move back to Guatemala. I wanted to, but I didn't have enough money to go back and the police officers told me not to move out of the country because they will still need me to work on the case. After the case is finished, I want to go back to my family."

Source: Adapted from Moser (2004).

from being "contaminated" by interracial marriage and childbearing. A study of white racists in Internet chat rooms found that the topic of interracial marriage was more likely than other topics (such as blacks moving into one's neighborhood and competing for one's job) to elicit advocacy of violence (Glaser et al. 2002). For example, one respondent said, "better kill her. kill him and her. pull a oj. . . . im not kidding. I would do it if it was my sister. i would gladly go to prison then live a free life knowing some mud babies were calling me uncle whitey" (p. 184).

Mission hate crimes are perpetrated by white supremacist group members or other offenders who have dedicated their lives to bigotry. The Ku Klux Klan, the first major racist white supremacist group in the United States, began in

Tennessee shortly after the Civil War. Klansmen have threatened, beaten, mutilated, and lynched blacks as well as whites who dared to oppose them. Other racist groups known to engage in hate crimes are the Identity Church Movement, the neo-Nazis, and the skinheads. The Southern Poverty Law Center identified 844 hate groups in the United States in 2006—a 40 percent increase since 2000, when there were 602 hate groups in the United States (SPLC 2007b).

Hate on Campus. Karl Nichols, a white residence hall director at the University of Mississippi, learned about racism in college, but not in the classroom and not from a textbook. Two chunks of asphalt were hurled through Nichols's dormitory room window along with a note warning, "You're going to get it, you God-forsaken nigger-lover" ("Hate on Campus" 2000, p. 10). The next night, someone attempted to set Nichols's door on fire. According to the university's investigation of these incidents, Nichols "may have violated racist taboos . . . by openly displaying his affinity for African American individuals and black culture, by dating black women, by playing black music . . . and by promoting diversity" in his dormitory (p. 10).

Karl Nichols is not alone. At Brown University in Rhode Island, a black student was beaten by three white students who told her she was a "quota" who did not belong at a university. At the State University of New York in Binghamton, an Asian American student was left with a fractured skull after a racially motivated assault by three students. Two students at the University of Kentucky—one white, one black—were crossing the street just off campus when they were attacked by 10 white men. The attackers yelled racist slurs at the black student and choked him until he could not speak or move. The assailants called the white student a "nigger lover" as they broke his hand and nose. "I definitely thought I was going to lose my life," the black student said later. The white student was shocked by the incident, commenting, "I didn't know that much hate existed" ("Hate on Campus" 2000, p. 7).

According to the FBI, more than 1 in 10 hate crimes (13.5 percent in 2005) occur at schools or colleges (FBI 2006). Far more common than hate crimes are "bias incidents," which are events that are not crimes but still can have the same negative and divisive effects. Howard J. Ehrlich, director of the Prejudice Institute in Baltimore, estimates that each year one-fourth of racial and ethnic minority college students and up to 5 percent of white college students are targets of bias-motivated name-calling, e-mails, telephone calls, verbal aggression, and other forms of psychological intimidation (Willoughby 2003).

Hate Group Members in the Military. In December 1995, two members of the 82nd Airborne at Fort Bragg, North Carolina, who belonged to a white supremacist skinhead gang shot and killed a black couple in a random, racially motivated double murder that shocked the nation. These hate crime murders led to congressional hearings and a major investigation of extremism in the military. The killers were sentenced to life in prison, and 19 other members of the 82nd Airborne were dishonorably discharged for neo-Nazi gang activities.

The Fort Bragg murders were not the first instance of military personnel involvement in hate groups. In the late 1980s and early 1990s a number of cases came to light in which extremists in the military were caught diverting stolen firearms and explosives to neo-Nazi and white supremacist organizations, conducting guerrilla training for paramilitary racist militias, and murdering non-white civilians. In 1986, after the discovery that active-duty soldiers were providing

❝Hate groups send their guys into the U.S. military because the U.S. military has the best weapons and training.❞

T. J. Leyden
Former racist skinhead and Marine who recruited inside the Marine Corps for the skinhead group Hammerskins

guerrilla training and stolen military weapons to a paramilitary Ku Klux Klan group led by a former Green Beret, then Secretary of Defense Casper Weinberger issued the military's first directive concerning hate group activity: "Military personnel must reject participation in white supremacy, neo-Nazi and other such groups which espouse or attempt to create overt discrimination. Active participation, including public demonstrations, recruiting and training members, and organizing or leading such organizations is utterly incompatible with military service" (quoted in Holthouse 2006). But this directive did not put an end to hate group activity among some military personnel, so in 1996, the Pentagon finally launched an investigation and crackdown. One general at Fort Lewis, Washington, ordered all 19,000 soldiers to be strip-searched for hate group tattoos.

According to investigations by the Southern Poverty Law Center, the military's tough stance against hate group affiliations among military personnel has relaxed since the recent war in Iraq and Afghanistan and the pressure to maintain enlistment numbers. Department of Defense investigator Scott Barfield said,

> Recruiters are knowingly allowing neo-Nazis and white supremacists to join the armed forces, and commanders don't remove them from the military even after we positively identify them as extremists or gang members. . . . Last year, for the first time, they didn't make their recruiting goals. They don't want to start making a big deal again about neo-Nazis in the military, because then parents who are already worried about their kids signing up and dying in Iraq are going to be even more reluctant about their kids enlisting if they feel they'll be exposed to gangs and white supremacists. (quoted in Holthouse 2006)

In 1 year, Barfield, who is based at Fort Lewis, identified and submitted evidence on 320 extremists there in the past year, but only two were discharged. The Army's Criminal Investigation Division conducts an annual assessment of extremist and gang activity among army personnel. "Every year, they come back with 'minimal activity,' which is inaccurate," said Barfield. "It's not epidemic, but there's plenty of evidence we're talking numbers well into the thousands, just in the Army" (quoted in Holthouse 2006).

STRATEGIES FOR ACTION: RESPONDING TO PREJUDICE, RACISM, AND DISCRIMINATION

Because racial and ethnic tensions exist worldwide, strategies for combating prejudice, racism, and discrimination globally require international cooperation and commitment. The World Conference Against Racism, Racial Discrimination, Xenophobia, and Related Intolerance, held in Durban, South Africa, in 2001 exemplifies international efforts to reduce racial and ethnic tensions and inequalities and to increase harmony among the various racial and ethnic populations of the world. Unfortunately, the U.S. delegation to this conference withdrew because of the expectation that hateful language would be used against Israel (because of the Israeli-Palestinian conflict).

In the following sections, we discuss the Equal Employment Opportunity Commission's role in responding to employment discrimination and examine the issue of affirmative action in the United States. We also discuss educational strategies to promote diversity and multicultural awareness and appreciation in schools. Finally, we look at apologies and reparations as a means of achieving racial reconciliation.

The Equal Employment Opportunity Commission

The **Equal Employment Opportunity Commission (EEOC),** a U.S. federal agency charged with ending employment discrimination in the United States, is responsible for enforcing laws against discrimination, including Title VII of the 1964 Civil Rights Act that prohibits employment discrimination on the basis of race, color, religion, sex, or national origin. The EEOC investigates, mediates, and may file lawsuits against private employers on behalf of alleged victims of discrimination. For example, after receiving and investigating complaints from across the country that Walgreen's discriminated against African American retail management and pharmacy employees in promotion, compensation, and assignment, the EEOC filed a discrimination suit against Walgreen's that was settled for $20 million in monetary relief for an estimated 10,000 class members (EEOC 2007b).

The most frequently filed claims with the EEOC are allegations of race discrimination, racial harassment, or retaliation from opposition to racial discrimination. In Fiscal Year 2006, the EEOC received 27,238 charges alleging race-based discrimination, accounting for 36 percent of the agency's private sector caseload (EEOC 2007a). Because of budget shortfalls and staffing declines, the EEOC has a backlog of cases in the tens of thousands (Lee 2006).

In 2007, the EEOC launched a national initiative to combat racial discrimination in the workplace. The goals of this initiative, called E-RACE (Eradicating Racism And Colorism from Employment), are to (1) identify factors that contribute to race and color discrimination, (2) explore strategies to improve the administrative processing and litigation of race and color discrimination cases, and (3) increase public awareness of race and color discrimination in employment.

Affirmative Action

Affirmative action refers to a broad range of policies and practices in the workplace and educational institutions to promote equal opportunity as well as diversity. Affirmative action is an attempt to compensate for the effects of past discrimination and prevent current discrimination against women and racial and ethnic minorities. Vietnam veterans and people with disabilities may also qualify under affirmative action policies. Although the largest category of affirmative action beneficiaries is women, the majority of students in two sociology classes did not know that women were covered by affirmative action (Beeman, Chowdhry, & Todd 2000).

> **What Do You Think?** Although women are the largest category designated to benefit from affirmative action, a survey of 35 introductory sociology texts published in the 1990s found that nearly 90 percent of the texts did not mention affirmative action in their sections on gender inequality, and only 20 percent of texts included women in their definitions of affirmative action (Beeman et al. 2000). Why do you think many textbooks overlook or minimize the benefits women may receive from affirmative action?

Equal Employment Opportunity Commission (EEOC) A U.S. federal agency charged with ending employment discrimination in the United States that is responsible for enforcing laws against discrimination, including Title VII of the 1964 Civil Rights Act that prohibits employment discrimination on the basis of race, color, religion, sex, or national origin.

affirmative action A broad range of policies and practices in the workplace and educational institutions to promote equal opportunity as well as diversity.

Federal Affirmative Action. Affirmative action policies developed in the 1960s from federal legislation that required any employer (universities as well as businesses) who received contracts from the federal government to make "good faith efforts" to increase the pool of qualified minorities and women (U.S. Department of Labor 2002). Such efforts can be made by expanding recruitment and training programs. Hiring decisions are to be made on a nondiscriminatory basis.

Affirmative Action in Higher Education. The Supreme Court's 1974 ruling in *University of California Board of Regents v. Bakke* marked the beginning of the decline of affirmative action. Alan Bakke, a white male, had applied to the University of California at Davis medical school and was rejected, even though his grade point average and score on the medical school admissions test were higher than those of several minority applicants who had been admitted. The medical school had established fixed racial quotas, guaranteeing admission to 16 minority applicants regardless of their qualifications. Bakke claimed that such quotas discriminated against him as a white male and that the University of California had violated his Fourteenth Amendment right to equal protection under the law. The Supreme Court ruled in Bakke's favor by a 5 to 4 vote (showing how split the court was), concluding that the University of California unwittingly engaged in "reverse discrimination," which was unconstitutional. Affirmative action programs, the court ruled, could not use fixed quotas in admission, hiring, or promotion policies. However, the court affirmed the right for universities and employers to consider race as a factor in admission, hiring, and promotion to achieve diversity.

Since the *Bakke* case, numerous legal battles have challenged affirmative action (Olson 2003). In *Hopwood v. Texas* Cheryl Hopwood and three other white individuals who had been denied admission to the University of Texas Law School claimed that their Fourteenth Amendment rights to equal protection under the law had been violated by the university's affirmative action admission policies. The Fifth Circuit Court of Appeals decided in 1997 that the University of Texas could no longer use race as a factor in awarding financial aid, admitting students, and hiring and promoting faculty. Higher courts refused to review the decision. But in 2003 the U.S. Supreme Court, after hearing the appeals from two white applicants who applied but were not accepted to the University of Michigan, affirmed the right of colleges to consider race in admissions, but the Court rejected Michigan's use of a point system to do so. According to the University of Michigan undergraduate admissions procedure, minority status provided 20 points on a 150-point scale for admission. The Court found that the problem with the point system was that for some applicants it turned race into the decisive factor instead of just one of many factors (Winter 2003b). In response, the University of Michigan abandoned its point system and created an undergraduate admissions policy similar to its law school admissions policy, which may serve as a model for how other universities can achieve a diverse student body while following court guidelines. In the new "holistic review" approach the university considers the unique circumstances of each student, prioritizing academics and treating all other factors, including race, equally. Applicants are now required to write more essays, including one on cultural diversity. Other universities, including the University of Wisconsin, University of Washington, and University of California, have also adopted "holistic admissions" procedures to increase their minority student populations.

Some universities, such as the University of Texas at Austin, use the "10 percent plan" as a way to maintain minority enrollment. Graduates in the top 10 percent of their high school classes are admitted automatically to the public college or university of their choice; standardized test scores and other factors are not considered.

In 2006, Michigan voters decided to ban affirmative action in university admissions, becoming the fifth state to do so (following California, Florida, Texas, and Washington). The continued weakening of affirmative action in higher education will necessitate creative admissions policies, such as holistic admissions and the 10 percent plan to ensure access of minorities to higher education.

Attitudes Toward Affirmative Action. Although affirmative action remains a divisive issue among Americans, national survey data reveal that support for affirmative action has increased in recent years (Pew Research Center 2007) (see Table 9.5). Women and blacks are more likely to support affirmative action than are men and whites. Among first-year college students, less than half (42 percent) agreed that "affirmative action in college admissions should be abolished" (Pryor et al. 2006).

Public opinion poll results are influenced by how survey questions are worded and framed. Survey questions that ask whether respondents favor "affirmative action programs for women and minorities" elicit more favorable responses than questions that ask about affirmative action for minorities only (Paul 2003). In addition, terms such as "affirmative action," "equal," and "opportunity" in survey questions yield more support for affirmative action policies, whereas terms such as "special preferences," "preferential treatment," and "quotas" tend to lessen support.

Supporters of affirmative action suggest that such policies have many social benefits. In a review of more than 200 scientific studies of affirmative action, Holzer and Neumark (2000) concluded that these policies produce benefits for women, minorities, and the overall economy. Holzer and Neumark (2000) found that employers who adopt affirmative action increase the relative number of women and minorities in the workplace by an average of 10–15 percent. Since the early 1960s affirmative action in education has contributed to an increase in the percentage of blacks attending college by a factor of 3 and the percentage of blacks enrolled in medical school by a factor of 4. Black doctors choose more often to practice medicine in inner cities and rural areas serving poor or minority patients than their white medical school classmates do (Holzer & Neumark 2000). Increasing the numbers of minorities in educational and professional positions also provides positive role models for other, especially younger, minorities "who can identify with them and form realistic goals to occupy the same roles themselves" (Zack 1998, p. 51).

Opponents of affirmative action suggest that such programs constitute reverse discrimination, which hurts whites. However, 2000 data showed that there are 1.3 million unemployed blacks and 112 million employed whites; thus if every unemployed black worker in the United States were to displace a white worker, only 1 percent of white workers would be af-

> "In order to get beyond racism, we must first take account of race. There is no other way. And in order to treat some persons equally, we must treat them differently."
>
> Harry A. Blackmun
> U.S. Supreme Court Justice

TABLE 9.5 Percentage of U.S. Adults Who Favor Affirmative Action*: 1995 and 2007

	1995	2007
Black	94	93
White	53	65
White women	59	71
White men	46	59

*Percentage who say they favor "affirmative action programs to help blacks, women, and other minorities get better jobs and education."

Source: Pew Research Center (2007)

fected. Because affirmative action pertains only to job-qualified applicants, the actual percentage of affected whites would be a fraction of 1 percent (Plous 2003). The main causes of unemployment among the white population are corporate downsizing, computerization and automation, and factory relocations outside the United States, not affirmative action.

Some critics of affirmative action argue that it undermines the self-esteem of women and minorities. Although affirmative action may have this effect in rare cases, in many cases affirmative action can raise the self-esteem of women and minorities by providing them with opportunities for educational advancement and employment (Plous 2003). Another criticism of affirmative action is that it fails to help the most impoverished of minorities—those whose deep and persistent poverty impairs their ability to compete not only with whites but also with other more advantaged minorities (Wilson 1987).

Opposition to affirmative action threatens the future of such policies and programs and the future educational and occupational opportunities of minorities. After California passed Proposition 209—an initiative to abolish affirmative action—black student enrollment in the UCLA law school declined from 10.3 percent in 1996 to 1.4 percent in 2000 (Greenberg 2003). Black and Latino enrollment has also significantly declined in the University of California, Berkeley undergraduate program and law school; the University of California, San Diego School of Medicine; and the University of Texas Law School.

It is ironic that President George W. Bush, who does not support affirmative action, was himself a beneficiary of a long-standing policy that gives preferential treatment to college applicants. When President Bush applied to highly selective Yale University in 1964, he had a C average in high school, but he was nevertheless admitted to Yale because he was a legacy applicant—that is, the son and grandson of distinguished alumni.

Educational Strategies

In a national Gallup poll, respondents were asked, "What do you think is the most important thing that could improve the situation of blacks in the United States today?" The most frequent answer among both blacks and whites was "improved education system" (Gallup Organization 2007a). One way to improve the educational system is to reduce or eliminate disparities in school funding. As noted earlier, schools in poor districts—which predominantly serve minority students— have traditionally received less funding per pupil than do schools in middle- and upper-class districts (which predominantly serve white students). In recent years more than two dozen states have been forced by the courts to come up with a new system of financing schools to increase inadequate funding of schools in poor districts (Goodnough 2001). Other educational strategies focus on reducing prejudice, racism, and discrimination and fostering awareness and appreciation of racial/ethnic diversity. These strategies include multicultural education, "whiteness studies," and efforts to increase diversity among student populations.

Multicultural Education in Schools and Communities. In schools across the nation **multicultural education**, which encompasses a broad range of programs and strategies, works to dispel myths, stereotypes, and ignorance about minorities; to promote tolerance and appreciation of diversity; and to include minority groups in the school curriculum (see also Chapter 8). With multicultural education the school curriculum reflects the diversity of U.S. society and fosters an

"Teach tolerance. Because open minds open doors for all our children."

Radio public service announcement

multicultural education Educational programs and strategies designed to dispel myths, stereotypes, and ignorance about minorities; promote tolerance and appreciation of diversity; and include minority groups in the school curriculum.

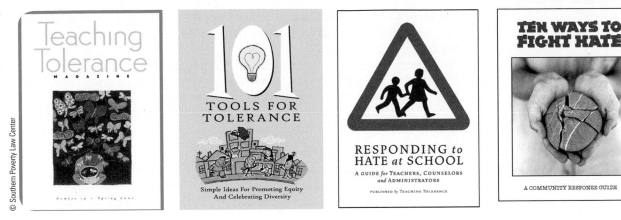

awareness and appreciation of the contributions of different racial and ethnic groups to U.S. culture. The Southern Poverty Law Center's program Teaching Tolerance publishes and distributes materials and videos designed to promote better human relations among diverse groups. These materials are sent to schools, colleges, religious organizations, and a variety of community groups across the nation.

Many colleges and universities have made efforts to promote awareness and appreciation of diversity by offering courses and degree programs in racial and ethnic studies and by sponsoring multicultural events and student organizations. A national survey by the Association of American Colleges and Universities found that 54 percent of colleges and universities required students to take at least one course that emphasizes diversity and another 8 percent were in the process of developing such a requirement (Humphreys 2000). Evidence points to a number of positive outcomes for both minority and majority students who take college diversity courses, including increased racial understanding and cultural awareness, increased social interaction with students who have backgrounds different from their own, improved cognitive development, increased support for efforts to achieve educational equity, and higher satisfaction with their college experience (Humphreys 1999).

Whiteness Studies. Traditionally, studies of and courses on race have focused on the social disadvantages experienced by racial minorities, while ignoring or minimizing the other side of the racial inequality equation—the social advantages conferred upon whites. Courses in Whiteness Studies, which are being offered in many colleges and universities, focus on increasing awareness of white privilege—an awareness that is limited among white students. "White privilege is so fundamental as to be largely invisible, expected, and normalized" (Picca & Feagin 2007, p. 243). In an often-cited paper called "White Privilege: Unpacking the Invisible Knapsack," Peggy McIntosh (1990) likened white privilege to an "invisible weightless knapsack" that whites carry with them without awareness of the many benefits inside the knapsack. Just a few of the many benefits McIntosh associates with being white include being able to

- Go shopping without being followed or harassed.
- Be assured that my skin color will not convey that I am financially unreliable when I use checks or credit cards.

More than 600,000 teachers receive *Teaching Tolerance* magazine—a free resource for tolerance education in the classroom (Dees 2000). Other free materials available from the Southern Poverty Law Center (http://www.spi-center.org) include *101 Tools for Tolerance, Responding to Hate at School,* and *Ten Ways to Fight Hate.*

- Swear, or dress in second-hand clothes, or not answer letters without having people attribute these choices to the bad morals, poverty, or the illiteracy of my race.
- Avoid ever being asked to speak for all the people of my racial group.
- Be sure that if a traffic cop pulls me over or if the IRS audits my tax return, it is not because I have been singled out because of my race.
- Take a job with an affirmative action employer without having coworkers on the job suspect that I got it because of race.

Whiteness studies "serve to rectify something wrong with the way we study race in America: By traditionally focusing on minority groups, the implicit message that scholarship projects is that nonwhites are 'deviant,' that's why they are studied" (Conley, 2002, n.p.). As explained by Professor Gregory Jay (2005):

> Whiteness Studies attempts to trace the economic and political history behind the invention of "whiteness," to challenge the privileges given to so-called "whites," and to analyze the cultural practices (in art, music, literature, and popular media) that create and perpetuate the fiction of "whiteness." . . . "Whiteness Studies" is an attempt to think critically about how white skin preference has operated systematically, structurally, and sometimes unconsciously as a dominant force in American—and indeed in global—society and culture. (n.p.)

Diversification of College Student Populations. Recruiting and admitting racial and ethnic minorities in institutions of higher education can foster positive relationships among diverse groups and enrich the educational experience of all students—minority and nonminority alike (American Council on Education & American Association of University Professors 2000). Psychologist Gordon Allport's (1954) "contact hypothesis" suggested that contact between groups is necessary for the reduction of prejudice between group members. For example, native-born Americans with higher levels of contact with immigrants tend to have more positive views of immigrants and immigration than those with less contact (NPR et al. 2005). Ensuring a diverse student population can provide students with opportunities for contact with different groups and can thereby reduce prejudice. One study found that students with the most exposure to diverse populations during college had the most cross-racial interactions 5 years after leaving college (Gurin 1999).

Retrospective Justice Initiatives: Apologies and Reparations

In 2003, Brown University President Ruth J. Simmons appointed a Steering Committee on Slavery and Justice to investigate and issue a public report on the university's historical relationship to slavery and the trans-Atlantic slave trade. The Steering Committee's research into the history of Brown University concluded that "there is no question that many of the assets that underwrote the University's creation and growth derived, directly and indirectly, from slavery and the slave trade" (Brown University Steering Committee on Slavery and Justice 2007, p. 13). The committee's final report included various recommendations for ways that Brown University should acknowledge and make amends for its past ties to the slave trade.

Various governments around the world have issued official apologies for racial and ethnic oppression. After World War II, West Germany signed a reparations agreement with Israel in which West Germany agreed to pay Israel for the enslavement and persecution of Jews during the Holocaust and to compensate for Jewish property that was stolen by the Nazis.

In the United States, various forms of reparations were offered to Indian tribes to compensate for land that had been taken by force or deception. However, the U.S. government has never issued an official apology to Native Americans for the atrocities committed against them. President Gerald Ford and Congress apologized to Japanese Americans for their internment during World War II, and reparations of $20,000 were granted to each surviving internee who was a U.S. citizen or legal resident alien at time of internment. In 1993, President Bill Clinton apologized to native Hawaiians for overthrowing the government of their nation. In 1994, the state of Florida offered monetary compensation to the survivors and descendents of the 1923 "Rosewood massacre," in which a white mob attacked and murdered black residents in Rosewood, Florida, and set fire to the town. And in 1997, the U.S. government offered monetary reparations to surviving victims of the Tuskegee Syphilis Study, in which blacks suffering from syphilis were denied medical treatment.

What Do You Think? As of this writing, no U.S. president or Congress has issued an apology for the government's role in slavery and legal segregation of African Americans. Some scholars argue that achieving black/white racial reconciliation in the United States requires that the U.S. government not only issue an official apology but also provide substantial monetary and other reparations to African Americans (Brooks 2004). Do you think that the U.S. federal government should offer an official apology to African Americans for the slavery and legal segregation that occurred in the past? Do you think African Americans should receive monetary or other reparations? Why or why not?

The growing movement to redress past large-scale violations of human rights is based on the moral principles of taking responsibility for and attempting to rectify past wrongdoings. Supporters of the reparative justice movement believe that the granting of apologies and reparations to groups that have been mistreated promotes dialogue and healing, increases awareness of present inequalities, and stimulates political action to remedy current injustices. Some who are opposed to this movement argue that "the quest for historical redress, and for monetary reparations in particular, is just one more symptom of the 'culture of complaint,' of the elevation of victimhood and group grievance over self-reliance and common nationality," whereas others claim that "preoccupation with past injustice is a distraction from the challenge of present injustice" (Brown University Steering Committee on Slavery and Justice 2007, p. 39). Advocates of restitution programs face opposition from those who question whether restitution programs are morally justified. Many Americans view the enslavement of African Americans as a closed chapter in U.S. history that has nothing to do with the current status of black Americans. Consider the following letter sent to the Brown University Steering Committee on Slavery and Justice (2007):

> You disgust me, as you disgust many other Americans. Slavery was wrong, but at that time it was a legal enterprise. It ended, case closed. You cite slavery's effects as being the reason that black people are so far behind, but that just illustrates your ignorance. Black people, here and now, are behind because some can't keep their hands off drugs, or guns, or can't move forward, can't get off welfare, can't do the simple things to improve their life. . . . They don't deserve money, they deserve a boot in the backside over and over until they can find their own way. . . . (p. 9)

In addition, proponents of proposals for restitution face difficult questions about what form reparations should take, who the beneficiaries of the reparations should be, who should bear the costs of payment, and whether a reparative payment should be a single-shot event or a continuing program (Posner & Vermeule 2003).

The Brown University Steering Committee on Slavery and Justice (2007) examined "retrospective justice initiatives" from around the world and concluded that the most successful generally combined three elements: (1) formal acknowledgement of an offense; (2) a commitment to truth telling, to ensure that the relevant facts are uncovered, discussed, and properly memorialized; and (3) the making of some form of amends in the present to give material substance to expressions of regret and responsibility. In the Committee's view, "reparative justice is not an invitation to 'wallow in the past' but a way for societies to come to terms with painful histories and move forward" (Brown University Steering Committee on Slavery and Justice 2007, p. 39).

UNDERSTANDING RACE, ETHNICITY, AND IMMIGRATION

❝I have a dream that my four little children will one day live in a nation where they will not be judged by the color of their skin, but by the content of their character.❞

Martin Luther King, Jr.
Civil rights leader

After considering the material presented in this chapter, what understanding about race and ethnic relations are we left with? First, we have seen that racial and ethnic categories are socially constructed; they are largely arbitrary, imprecise, and misleading. Although some scholars suggest that we abandon racial and ethnic labels, others advocate adding new categories—multiethnic and multiracial—to reflect the identities of a growing segment of the U.S. and world population.

Conflict theorists and structural functionalists agree that prejudice, discrimination, and racism have benefited certain groups in society. But racial and ethnic disharmony has created tensions that disrupt social equilibrium. Symbolic interactionists note that negative labeling of minority group members, which is learned through interaction with others, contributes to the subordinate position of minorities.

Prejudice, racism, and discrimination are debilitating forces in the lives of minorities and immigrants. Despite these negative forces, many minority group members succeed in living productive, meaningful, and prosperous lives. But many others cannot overcome the social disadvantages associated with their minority status and become victims of a cycle of poverty (see Chapter 6). Minorities are disproportionately poor, receive inferior education and health care, and, with continued discrimination in the workplace, have difficulty improving their standard of living.

Achieving racial and ethnic equality requires alterations in the structure of society that increase opportunities for minorities—in education, employment and income, and political participation. In addition, policy makers concerned with racial and ethnic equality must find ways to reduce the racial and ethnic wealth gap and to foster wealth accumulation among minorities (Conley 1999). Social class is a central issue in race and ethnic relations. Professor and activist bell hooks (2000) (who spells her name in all lowercase) warned that focusing on issues of race and gender can deflect attention away from the larger issue of class division that increasingly separates the haves from the have-nots. Addressing class inequality must, suggests hooks, be part of any meaningful strategy to reduce inequalities suffered by minority groups. Civil rights activist Lani Guinier, in an interview with Paula Zahn on the *CBS Evening News* on July 18, 1998, suggested that "the real challenge is to . . . use race as a window on issues of class, issues of gender, and issues of fundamental fairness, not just to talk about race as if it's a question of individual bigotry or individual prejudice. The issue is more than about making friends—it's about making change." But, as Shipler (1998) noted, making change requires members of society to recognize that change is necessary, that there is a problem that needs rectifying:

> One has to perceive the problem to embrace the solutions. If you think racism isn't harmful unless it wears sheets or burns crosses or bars blacks from motels and restaurants, you will support only the crudest anti-discrimination laws and not the more refined methods of affirmative action and diversity training. (p. 2)

Public awareness of inequality—racial and economic—became heightened after the tragic events of Hurricane Katrina and the government's slow response to the predominantly poor and black victims of the disaster. U.S. Representative Jesse Jackson Jr. described what he witnessed in the wake of Katrina: "I saw 5,000 African Americans on the I-10 causeway, desperate, perishing, dehydrated, babies dying. It looked like Africans in the hull of a slave ship. It was so ugly and so obvious. . . . We have great tolerance for black suffering and black marginalization" (quoted by Jackson 2005, n.p.). If Katrina results in decreased tolerance for racial and economic inequality and the suffering it brings, perhaps the more than 1,000 lives lost in the disaster will not have been in vain.

CHAPTER REVIEW

- **What is a minority group?**

  A minority group is a category of people who have unequal access to positions of power, prestige, and wealth in a society and who tend to be targets of prejudice and discrimination. Minority status is not based on numerical representation in society but rather on social status.

- **What is meant by the idea that race is socially constructed?**

  The concept of race refers to a category of people who are believed to share distinct physical characteristics that are deemed socially significant. Races are cultural and social inventions; they are not scientifically valid because there are no objective, reliable, meaningful criteria scientists can use to identify racial groupings. Different societies construct different systems of racial classification, and these systems change over time. The significance of race is not biological but social and political, because race is used to separate "us" from "them" and becomes a basis for unequal treatment of one group by another.

- **What are the various patterns of interaction that may occur when two or more racial or ethnic groups come into contact?**

  When two or more racial or ethnic groups come into contact, one of several patterns of interaction occurs, including genocide, expulsion, colonialism, segregation, acculturation, pluralism, and assimilation.

- **Beginning with the 2000 census, what are the five race categories used to identify the race composition of the United States?**

  Beginning with the 2000 census, the five race categories are (1) white, (2) black or African American, (3) American Indian or Alaska Native, (4) Asian, and (5) Native Hawaiian or other Pacific Islander. In addition, respondents to federal surveys and the census now have the option of officially identifying themselves as being of more than one race, rather than checking only one racial category.

- **What is an ethnic group?**

  An ethnic group is a population that has a shared cultural heritage or nationality. Ethnic groups can be distinguished on the basis of language, forms of family structures and roles of family members, religious beliefs and practices, dietary customs, forms of artistic expression such as music and dance, and national origin. The largest ethnic population in the United States is Hispanics or Latinos.

- **What percentage of the U.S. population (in 2005) was born outside the United States?**

  In 2005 more than 1 in 10 U.S. residents (12.4 percent) were born in a foreign country.

- **What were the manifest function and latent dysfunction of the civil rights movement?**

  The manifest function of the civil rights legislation in the 1960s was to improve conditions for racial minorities. However, civil rights legislation produced an unexpected consequence, or latent dysfunction. Because civil rights legislation supposedly ended racial discrimination, whites were more likely to blame blacks for their social disadvantages and thus perpetuate negative stereotypes such as "blacks lack motivation" and "blacks have less ability."

- **How does contemporary prejudice differ from more traditional, "old-fashioned" prejudice?**

  Traditional, old-fashioned prejudice is easy to recognize, because it is blatant, direct, and conscious. More contemporary forms of prejudice are often subtle, indirect, and unconscious. In addition, racist expressions have gone "backstage" to private social settings.

- **Is it possible for an individual to discriminate without being prejudiced?**

  Yes. In overt discrimination an individual discriminates because of his or her own prejudicial attitudes. But sometimes individuals who are not prejudiced discriminate because of someone else's prejudice. For example, a store clerk may watch black customers more closely because the store manager is prejudiced against blacks and has instructed the employee to follow black customers in the store closely. Discrimination based on someone else's prejudice is called adaptive discrimination.

- **Are U.S. schools segregated?**

  Racial and ethnic minorities are largely isolated from whites in an increasingly segregated school system. A study by the Civil Rights Project at Harvard University found that U.S. schools in 2000–2001 were more segregated than they were in 1970. The upward trend in school segregation is due to large increases in minority student enrollment, continuing white flight from urban areas, the persistence of housing segregation, and the termination of court-ordered desegregation plans.

- **According to FBI data, the majority of hate crimes are motivated by what kind of bias?**

  Since the FBI began publishing hate crime data in 1992, the majority of hate crimes have been based on racial bias. However, the rate of violence against American Indians is higher than that among any other minority group. Hate crime violence against American Indians is especially prevalent in "border towns" adjacent to Indian reservations.

- **What is the role of the Equal Employment Opportunity Commission (EEOC) in combating employment discrimination?**

  The Equal Employment Opportunity Commission (EEOC) is responsible for enforcing laws against discrimination, including Title VII of the 1964 Civil Rights Act that prohibits employment discrimination on the basis of race, color, religion, sex, or national origin. The EEOC investigates, mediates, and may file lawsuits against private employers on behalf of alleged victims of discrimination. The most frequently filed claims with the EEOC are allegations of race discrimination, racial harassment, or retaliation from opposition to racial discrimination.

- **What group constitutes the largest beneficiary of affirmative action policies?**

  Affirmative action policies are designed to benefit racial and ethnic minorities, women, and, in some cases, Vietnam veterans and people with disabilities. The largest category of affirmative action beneficiaries is women.

- **What are Whiteness Studies?**

  Courses in Whiteness Studies, which are being offered in many colleges and universities, focus on increasing awareness of white privilege—an awareness that is limited among white students.

- **According to the Brown University Steering Committee on Slavery and Justice, successful retrospective justice initiatives contain what three elements?**

  The Brown University Steering Committee on Slavery and Justice examined retrospective justice initiatives from around the world and concluded that the most successful generally combined three elements: (1) formal acknowledgement of an offense; (2) a commitment to truth telling, to ensure that the relevant facts are uncovered, discussed, and properly memorialized; and (3) the making of some form of amends in the present to give material substance to expressions of regret and responsibility.

TEST YOURSELF

1. All humans belong to the same species.
   a. True
   b. False
2. Which of the following occurs when a racial or ethnic group from one society takes over and dominates the racial or ethnic group(s) of another society?
   a. Secondary assimilation
   b. Colonialism
   c. Genocide
   d. Pluralism
3. Minority populations outnumber non-Hispanic whites in each of the following states except
   a. California
   b. Texas and New Mexico
   c. Hawaii
   d. Florida
4. A Southern Poverty Law Center report reveals that the guestworker program constitutes a "modern-day system of _____."
   a. amnesty
   b. colonialism
   c. indentured servitude
   d. paid apprenticeship
5. Which minority group in the United States is considered a "model minority"?
   a. women
   b. black women
   c. Scandinavian immigrants
   d. Asian Americans
6. Which of the following has/have moved "backstage"?
   a. racist behavior
   b. undocumented immigrants
   c. multicultural education
   d. affirmative action
7. Under U.S. law, hate groups are not allowed to express racial or ethnic hatred on Internet websites.
   a. True
   b. False
8. Which of the following is responsible for enforcing laws against discrimination?
   a. Local police departments
   b. Equal Employment Opportunity Commission
   c. U.S. Department of Labor
   d. The Supreme Court
9. In a national Gallup poll, respondents were asked, "What do you think is the most important thing that could improve the situation of blacks in the United States today?" The most frequent answer among both blacks and whites was "improved education system."
   a. True
   b. False
10. The U.S. government has never issued an official apology to Native Americans for the atrocities committed against them.
    a. True
    b. False

Answers: 1 a. 2 b. 3 d. 4 c. 5 d. 6 a. 7 b. 8 b. 9 a. 10 a.

KEY TERMS

acculturation
adaptive discrimination
affirmative action
antimiscegenation laws
assimilation
aversive racism
colonialism
discrimination
ethnicity
Equal Employment Opportunity
    Commission (EEOC)

expulsion
genocide
hate crime
individual discrimination
institutional discrimination
Jim Crow laws
minority group
modern racism
multicultural education

naturalized citizens
overt discrimination
pluralism
prejudice
race
racism
segregation
stereotypes

MEDIA RESOURCES

***Understanding Social Problems,* Sixth Edition**
Companion Website

academic.cengage.com/sociology/mooney
Visit your book companion website, where you will
find flash cards, practice quizzes, Internet links, and
more to help you study.

Just what you need to know NOW!
Spend time on what you need to
master rather than on information
you already have learned. Take a
pre-test for this chapter, and CengageNOW will generate a personalized study plan based on your results. The
study plan will identify the topics you need to review
and direct you to online resources to help you master
those topics. You can then take a post-test to help you
determine the concepts you have mastered and what you
will need to work on. Try it out! Go to academic.cengage
.com/login to sign in with an access code or to purchase
access to this product.

10

Gender Inequality

Captain of the high school track team. Champion sprinter. Fluent in German. Played the clarinet. High school class valedictorian. Wing leader, ROTC. Accepted at West Point. Graduated second in class. Third generation Army. Commissioned officer. Platoon leader. Decorated soldier.

Male or female? In the 1950s the above characteristics would have described a male. Today, it is more difficult to tell, and in this case it is a female. But she is no ordinary female!

Emily Tatum Perez was born in Germany, the daughter of African-American and Hispanic parents (Partlow & Parker 2006; Portraits of Sacrifice 2006; Thornburgh 2006). At 15 her family moved to Fort Washington, MD, where she attended Oxon Hills High School in an area known for gangs, drugs, and violence. Still, she excelled—leading her track team, graduating at the top of her class, and inspiring others. When a family member was diagnosed with human immunodeficiency virus (HIV)-acquired immunodeficiency syndrome (AIDS), Emily started a successful AIDS ministry at her local church. The American Red Cross later honored her for her volunteering efforts.

Since childhood, Emily had wanted to join the Army. In September 2001, she entered West Point just 2 weeks before the tragic events of 9/11. Despite dramatic changes to curriculum and courses of study now geared to combat in Iraq and Afghanistan, Emily continued to rise to the top of her class. She led the West Point track team into the record books with her talent as a sprinter and triple jump competitor. Emily held the second highest rank in her senior class. And, when Emily was named Brigade Command Sergeant Major, she became the highest-ranking minority woman in the history of West Point.

Emily graduated with "The Class of 9/11" with a bachelor of science in sociology. She left West Point and was commissioned as a 2nd Lieutenant in the U.S. Army and in December of 2005 was sent to Iraq as a Medical Service Corps Officer. On September 26, 2006, Emily Jazmin Tatum Perez, 23, an extraordinary soldier, an extraordinary woman, returned to her beloved West Point where she was buried, the first female graduate to die in the war in Iraq.

Clearly, Emily Perez could do it all. But not long ago she would not have had the opportunities she had to excel. West Point did not admit women to the academy until 1976, one indication of the gender inequality that existed then—and now as women continue to be barred from direct ground combat. The term *gender inequality,* however, raises the question: unequal in what way? Depending on the issue, both women and men are victims of inequality. When income, career advancement, household work, and sexual harassment are the focus, women are most often disadvantaged. But when life expectancy, mental and physical illness, and access to one's children after divorce are considered, it is often men who are disadvantaged. In this chapter we seek to understand inequalities for *both* genders.

In this chapter we look at **sexism**—the belief that innate psychological, behavioral, and/or intellectual differences exist between women and men and that these differences connote the superiority of one group and the inferiority of the other. As with race and ethnicity, such attitudes often result in prejudice and discrimination at both the individual and institutional levels. Individual discrimination is reflected by the physician who will not hire a male nurse because he or she believes that women are more nurturing and empathetic and are there-

sexism The belief that innate psychological, behavioral, and/or intellectual differences exist between women and men and that these differences connote the superiority of one group and the inferiority of the other.

fore better nurses. Institutional discrimination—that is, discrimination built into the fabric of society—is exemplified by the difficulty many women experience in finding employment; they may have no work history and few job skills as a consequence of living in traditionally defined marriages.

Discerning the basis for discrimination is often difficult because the different types of minority statuses may intersect. For example, elderly African-American and Hispanic women are more likely to receive lower wages and to work in fewer prestigious jobs than younger white women. They may also experience discrimination if they are "out" as homosexuals. Such **double or triple (multiple) jeopardy** occurs when a person is a member of two or more minority groups. In this chapter, however, we emphasize the impact of gender inequality.

Gender refers to the social definitions and expectations associated with being female or male and should be distinguished from **sex,** which refers to one's biological identity. In most Western cultures we take for granted that there are two categories of gender. However, in many other societies three and four genders have been recognized. For example, many Polynesian cultures recognize a third gender called the mahū—individuals who take on the work roles of members of the opposite sex (Nanda 2000). Other societies recognize hermaphrodites (individuals born with ambiguous genitalia) as a third gender. In the United States, the majority of these babies have surgery to "correct" their genitalia, thus keeping them within the traditional two-gender system.

What Do You Think? Physically, men and women are different from birth. Men, in general, are stronger, taller, and heavier and have more facial hair. Women develop breasts, have higher-pitched voices, menstruate, and bear children. But are these physical characteristics related to behavioral differences? Are women innately nurturers? Are men innately aggressive? Noting, for example, that most societies are (were) patriarchal, most would answer yes. Others, however, would be quick to note the role of socialization in traditional gender role assignment. Cross-cultural evidence is mixed. Some societies are characterized by traditional gender roles, in other societies androgyny dominates, and in still other societies traditional gender roles are reversed. What do you think? Are gender roles a matter of nature or nurture?

The Global Context: The Status of Women and Men

double or triple (multiple) jeopardy The disadvantages associated with being a member of two or more minority groups.

gender The social definitions and expectations associated with being female or male.

sex A person's biological classification as male or female.

There is no country in the world in which women and men have equal status. Although much progress has been made in closing the gender gap in areas such as education, health care, employment, and government, gender inequality is still prevalent throughout the world.

The World Economic Forum assessed the gender gap in 115 countries by measuring the extent to which women have achieved equality with men in four areas: economic participation and opportunity, educational attainment,

TABLE 10.1 Gender Gap Rankings: Top and Bottom 10 Countries and the United States*

COUNTRY	OVERALL RANKING	OVERALL SCORE*	ECONOMIC PARTICIPATION AND OPPORTUNITY RANKING	EDUCATIONAL ATTAINMENT RANKING	HEALTH AND SURVIVAL RANKING	POLITICAL EMPOWERMENT RANKING
Top 10 countries						
Sweden	1	0.8133	9	22	70	1
Norway	2	0.7994	11	14	61	2
Finland	3	0.7958	8	17	1	3
Iceland	4	0.7813	17	49	92	1
Germany	5	0.7524	32	31	36	5
Philippines	6	0.7516	4	1	1	16
New Zealand	7	0.7509	14	16	69	11
Denmark	8	0.7462	19	1	76	13
United Kingdom	9	0.7365	37	1	63	12
Ireland	10	0.7335	47	1	81	9
United States	**23**	**0.7042**	**3**	**66**	**1**	**66**
Bottom 10 countries						
Mauritania	106	0.5833	93	103	1	106
Morocco	107	0.5826	102	99	90	92
Iran	108	0.5802	113	79	52	109
Egypt	109	0.5785	108	90	66	111
Benin	110	0.5778	55	113	86	76
Nepal	111	0.5477	100	109	111	102
Pakistan	112	0.5433	112	110	112	37
Chad	113	0.5246	65	115	56	91
Saudi Arabia	114	0.5241	115	93	54	115
Yemen	115	0.4594	114	114	48	113

*All overall scores are reported on a scale of 0 to 1, with 1 representing maximum gender equality.

Source: Hausmann, Tyson, & Zahidi (2007).

health and survival, and political empowerment (Hausmann, Tyson, & Zahidi 2007). Table 10.1 presents the overall scores and rankings of (1) the 10 countries with the smallest gender gap (i.e., the least gender inequality), (2) the 10 countries with the largest gender gap (i.e., the most gender inequality), and (3) the United States, which did not rank in the top 10 or the bottom 10 but ranked number 23 of the 115 countries studied. Ties were possible. For example, Finland and the Philippines are tied for first place on the composite measure of health and survival. Further, note that the overall score approximates the proportion of the gender gap a country has *closed*—in the United States, 0.7042 or 70.42 percent.

© David Turnley/Corbis

In traditional Muslim societies, women are forbidden to show their faces or other parts of their bodies when in public. As pictured, Muslim women wear a veil to cover their faces and a *chador*, a floor-length loose-fitting garment, to cover themselves from head to toe. Although some women adhere to this norm out of fear of repercussions, many others believe veiling was first imposed on Muhammad's wives out of respect for women and the desire to protect them from unwanted advances. There are more than half a million Muslims living in the United States.

Gender inequality varies across cultures, not only in its extent or degree but also in its forms. For example, in the United States gender inequality in family roles commonly takes the form of an unequal division of household labor and child care, with women bearing the heavier responsibility for these tasks. In other countries forms of gender inequality in the family include the expectation that wives ask their husbands for permission to use birth control (see Chapter 13), the practice of aborting female fetuses in cultures that value male children over female children, and unequal penalties for spouses who commit adultery, with wives receiving harsher punishment. For example, in Afghanistan in 2005 a 29-year-old woman was stoned to death for committing adultery, whereas the man accused of committing adultery with her was allegedly whipped 100 times and freed (Amnesty International 2005).

A global perspective on gender inequality must take into account the different ways in which such inequality is viewed. For example, many non-Muslims view the practice of Muslim women wearing a headscarf in public as a symbol of female subordination and oppression. To Muslims who embrace this practice (and not all Muslims do), wearing a headscarf reflects the high status of women and represents the view that women should be respected and not treated as sexual objects.

Similarly, cultures differ in how they view the practice of female genital cutting (FGC), also known as female genital mutilation or female circumcision. There are several forms of FGC, ranging from a symbolic nicking of the clitoris to removal of the clitoris and labia and partial closure of the vaginal opening by stitching the two sides of the vulva together, leaving only a small opening for the passage of urine and menstrual blood. After marriage the sealed opening is reopened to permit intercourse and childbearing. After childbirth the woman's vulva is often stitched back together.

Most FGC procedures are done by nonmedical personnel using unsterilized blades or string. Health risks associated with FGC include pain, hemorrhage, infection, shock, scarring, and infertility. Worldwide, 140 million girls and young women have experienced FGC and an additional 2 million are at risk each year (Plan International 2007). FGC is a common practice in many countries in the northern half of sub-Saharan Africa as well as in Egypt and Yemen. The prevalence of FGC among women ranges from 5 percent in Niger to 99 percent in Guinea (UNICEF 2005).

People from countries in which FGC is not the norm generally view this practice as a barbaric form of violence against women. For example, an Ethiopian immigrant in the United States was convicted of aggravated battery and cruelty to children for using a pair of scissors to remove his daughter's clitoris. He was sentenced to 10 years in prison (Haines 2006). In countries in which it commonly occurs, FGC is viewed as an important and useful practice. In some countries it is considered a rite of passage that enhances a woman's status. In other countries it is aesthetically pleasing. For others FGC is a moral imperative based on religious beliefs (Yoder, Abderrahim, & Zhuzhuni 2004).

Inequality in the United States

Although attitudes toward gender equality are becoming increasingly liberal, the United States has a long history of gender inequality (you can assess your own beliefs about gender equality in this chapter's *Self and Society* feature). Women have had to fight for equality: the right to vote, equal pay for comparable work, quality education, entrance into male-dominated occupations, and legal equality. As shown in Table 10.1, the World Economic Forum (Hausmann et al. 2007)—based on its assessment of women's economic participation and opportunities, political empowerment, educational attainment, and health and survival—ranks the United States only 23rd in the world in terms of gender equality. Most U.S. citizens agree that American society does not treat women and men equally: Women have lower incomes, hold fewer prestigious jobs, earn fewer graduate degrees, and are more likely than men to live in poverty.

Men are also victims of gender inequality. In 1963 sociologist Erving Goffman wrote that in the United States there is only

> one complete unblushing male . . . a young, married, white, urban, northern heterosexual, Protestant father of college education, fully employed, of good complexion, weight and height, and a recent record in sports. . . . Any male who fails to qualify in one of these ways is likely to view himself . . . as unworthy, incomplete, and inferior. (p. 128)

Although standards of masculinity have relaxed, Williams (2000) argued that masculinity is still based on "success"—at work, on the athletic field, on the streets, and at home—which must be constantly maintained and proven, therefore placing enormous pressure on boys and men.

When U.S. college students were asked to list the best and worst things about being the opposite sex, the same qualities, although in opposite categories, emerged (Cohen 2001). For example, what males list as the best thing about being female (e.g., free to be emotional), females list as the worst thing about being male (e.g., not free to be emotional). Similarly, what females list as the best thing about being male (e.g., higher pay), males listed as the worst thing about being female (e.g., lower pay). As Cohen noted (2001), although "some differences are

The statements listed here describe different attitudes toward men and women. There are no right or wrong answers, only opinions. Indicate how much you agree or disagree with each statement, using the following scale: (A) strongly disagree, (B) slightly disagree, (C) neither agree nor disagree, (D) slightly agree, (E) strongly agree.

1. Women are more passive than men.

2. Women are less career-motivated than men.

3. Women don't generally like to be active in their sexual relationships.

4. Women are more concerned about their physical appearance than men are.

5. Women comply more often than men do.

6. Women care as much as men do about developing a job/career.

7. Most women don't like to express their sexuality.

8. Men are as conceited about their appearance as women are.

9. Men are as submissive as women are.

10. Women are as skillful in business-related activities as men are.

11. Most women want their partner to take the initiative in their sexual relationships.

12. Women spend more time attending to their physical appearance than men do.

13. Women tend to give up more easily than men do.

14. Women dislike being in leadership positions more than men do.

15. Women are as interested in sex as men are.

16. Women pay more attention to their looks than most men do.

17. Women are more easily influenced than men are.

18. Women don't like responsibility as much as men do.

19. Women's sexual desires are less intense than men's.

20. Women gain more status from their physical appearance than men do.

The Beliefs About Women Scale (BAWS) consists of 15 separate subscales; only four are used here. The items for these four subscales and coding instructions are as follows:

Subscales

1. Women are more passive than men (items 1, 5, 9, 13, 17).

2. Women are interested in careers less than men (items 2, 6, 10, 14, 18).

3. Women are less sexual than men (items 3, 7, 11, 15, 19).

4. Women are more appearance conscious than men items 4, 8, 12, 16, 20).

Items 1–5, 7, 11–14, and 16–20 should be scored as follows: strongly agree, 2; slightly agree, 1; neither agree nor disagree, 0; slightly disagree, 1; and strongly disagree, 2. Items 6, 8, 9, 10, and 15 should be scored as follows: strongly agree, 2; slightly agree, 1; neither agree nor disagree, 0; slightly disagree, 1; and strongly disagree, 2. Scores range from 0 to 40; subscale scores range from 0 to 10. The higher your score, the more traditional your gender beliefs about men and women are.

Source: William E. Snell, Jr., Ph.D. (1997). College of Liberal Arts, Department of Psychology, Southeast Missouri State University. Reprinted with permission.

exaggerated or oversimplified . . . we identif[ied] a host of ways in which we 'win' or 'lose' simply because we are male or female" (p. 3).

SOCIOLOGICAL THEORIES OF GENDER INEQUALITY

Both structural functionalism and conflict theory concentrate on how the structure of society and, specifically, its institutions contribute to gender inequality. However, these two theoretical perspectives offer opposing views of the development and maintenance of gender inequality. Symbolic interactionism, on the other hand, focuses on the culture of society and how gender roles are learned through the socialization process.

Structural-Functionalist Perspective

Structural functionalists argue that preindustrial society required a division of labor based on gender. Women, out of biological necessity, remained in the home performing functions such as bearing, nursing, and caring for children. Men, who were physically stronger and could be away from home for long periods of time, were responsible for providing food, clothing, and shelter for their families. This division of labor was functional for society and over time became defined as both normal and natural.

Industrialization rendered the traditional division of labor less functional, although remnants of the supporting belief system still persist. Today, because of day care facilities, lower fertility rates, and the less physically demanding and dangerous nature of jobs, the traditional division of labor is no longer as functional. Thus modern conceptions of the family have, to some extent, replaced traditional ones—families have evolved from extended to nuclear, authority is more egalitarian, more women work outside the home, and greater role variation exists in the division of labor. Structural functionalists argue, therefore, that as the needs of society change, the associated institutional arrangements also change.

Conflict Perspective

Many conflict theorists hold that male dominance and female subordination are shaped by the relationships men and women have to the production process. During the hunting and gathering stage of development, males and females were economic equals, each controlling his or her own labor and producing needed subsistence. As society evolved to agricultural and industrial modes of production, private property developed and men gained control of the modes of production, whereas women remained in the home to bear and care for children. Inheritance laws that ensured that ownership would remain in their hands furthered male domination. Laws that regarded women as property ensured that women would remain confined to the home.

As industrialization continued and the production of goods and services moved away from the home, the male–female gaps continued to grow—women had less education, lower incomes, and fewer occupational skills and were rarely owners. World War II necessitated the entry of a large number of women into the labor force, but in contrast with previous periods, many of them did not return to the home at the end of the war. They had established their own place in the workforce and, facilitated by the changing nature of work and technological advances, now competed directly with men for jobs and wages.

Conflict theorists also argue that continued domination by males requires a belief system that supports gender inequality. Two such beliefs are (1) that women are inferior outside the home (e.g., they are less intelligent, less reliable, and less rational) and (2) that women are more valuable in the home (e.g., they have maternal instincts and are naturally nurturing). Thus, unlike structural functionalists, conflict theorists hold that the subordinate position of women in society is a consequence of social inducement rather than biological differences that led to the traditional division of labor.

Symbolic Interactionist Perspective

Although some scientists argue that gender differences are innate, symbolic interactionists emphasize that through the socialization process both females and males are taught the meanings associated with being feminine and masculine.

> **"**People call me a feminist whenever I express sentiments that differentiate me from a doormat or a prostitute.**"**
>
> Rebecca West
> Author and feminist

Women as well as girls are often portrayed provocatively as a means of selling a product or service. This billboard is a good example of the cultural emphasis placed on women's physical appearance.

© Richard Levine/Alamy

Gender assignment begins at birth as a child is classified as either female or male. However, the learning of gender roles is a lifelong process whereby individuals acquire society's definitions of appropriate and inappropriate gender behavior.

Gender roles are taught by the family, in the school, in peer groups, and by media presentations of girls and boys and women and men (see the discussion on the social construction of gender roles later in this chapter). Most important, however, gender roles are learned through symbolic interaction as the messages that others send us reaffirm or challenge our gender performances. As Lorber (1998) noted:

> Gender is so pervasive that in our society we assume it is bred into our genes. Most people find it hard to believe that gender is constantly created and recreated out of human interaction, out of social life, and is the texture and order of social life. Yet gender, like culture, is a human production that depends on everyone constantly "doing gender." (p. 213)

Feminist theory, although also consistent with a conflict perspective, incorporates many aspects of symbolic interactionism. Feminists argue that conceptions of gender are socially constructed as societal expectations dictate what it means to be female or what it means to be male. Thus, for example, women are generally socialized into **expressive roles** (i.e., nurturing and emotionally supportive roles) and males are more often socialized into **instrumental roles** (i.e., task-oriented roles). These roles are then acted out in countless daily interactions as boss and secretary, doctor and nurse, football player and cheerleader "do gender." Feminists also hold that gender "is a central organizing factor in the social world and so must be included as a fundamental category of analysis in sociological research" (Renzetti & Curran 2003, p. 8). Feminists, noting that the impact of the structure and culture of society is not the same for different groups of women and men, encourage research on gender that takes into consideration the differential effects of age, race and ethnicity, and sexual orientation.

expressive roles Roles into which women are traditionally socialized, i.e., nurturing and emotionally supportive roles.

instrumental roles Roles into which men are traditionally socialized, i.e., task-oriented roles.

GENDER STRATIFICATION: STRUCTURAL SEXISM

As structural-functionalists and conflict theorists agree, the social structure underlies and perpetuates much of the sexism in society. **Structural sexism,** also known as institutional sexism, refers to the ways the organization of society and specifically its institutions subordinate individuals and groups based on their sex classification. Structural sexism has resulted in significant differences in the education and income levels, occupational and political involvement, and civil rights of women and men.

Education and Structural Sexism

Literacy rates worldwide indicate that women are less likely than men to be able to read and write, with millions of women being denied access to even the most basic education (ILO 2007a). For example, on average, women in South Asia have only half as many years of education as their male counterparts. In addition, worldwide, for every 100 boys who do not attend elementary school, there are 115 girls who do not attend elementary school (UNICEF 2007).

In 2005 few differences existed between men and women in their completion rates of high school and college degrees (U.S. Bureau of the Census 2007). In fact, in recent years most U.S. colleges and universities have had a higher percentage of women than men enrolling directly from high school (see Chapter 8). This trend is causing some concern that many young American men may not have the education to compete in today's global economy. Although differences are narrowing, men are still more likely to go on to complete a graduate or professional degree than are women. As Figure 10.1 indicates, dramatic differences also appear in the types of advanced degrees that men and women earn. For example, women receive 68 percent of all psychology doctorates, but they earn only 18 percent of all doctorates in engineering (National Science Foundation 2007). Although the general trend is for men to have higher levels of education than women, in recent years the educational levels of African-American women have increased at a faster rate than those of African-American men. In 2004, 6 of 10 African-American undergraduate students younger than age 25 were women (ACE 2006). Further, African-American women earn 50 percent more PhDs, JDs, MDs, and DDSs than African-American men (Urban Institute 2003).

One explanation for why women earn fewer advanced degrees than men is that women are socialized to choose marriage and motherhood over long-term career preparation. From an early age women are exposed to images and models of femininity that stress the importance of domestic family life. In a study of college women who were asked about their future plans, Hoffnung (2004) found that the majority wanted a career, marriage, and children. Many expected to go on for an advanced degree. However, 7 years after the study those who had become mothers—often unexpectedly—were less likely to have pursued their education than those who did not have children.

Structural limitations also discourage women from advancing in higher education. For example, women seeking academic careers may find that securing a tenure-track position is more difficult for them than it is for men. In addition, McBrier (2003), in examining sex inequality in law schools, found that women progress through the professorial ranks at a slower pace than men. She attributes the differences to a variety of variables, including family and geographic constraints, social capital differences, degree prestige, and prior work

structural sexism The ways in which the organization of society, and specifically its institutions, subordinate individuals and groups based on their sex classification.

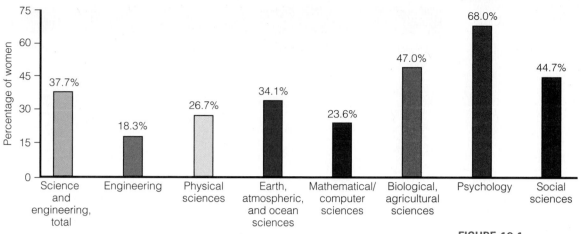

FIGURE 10.1
Science and engineering doctorates awarded to women, 2005.
Source: National Science Foundation (2007).

history. Similarly, a study of sociologists found that among those who earned their PhDs in 1997 both males and females were equally as likely to acquire a tenure-track position. However, those with children were less likely than sociologists without children to have done so (American Sociological Association 2004).

Long, Allison, and McGinnis (1993) examined the promotions of 556 men and 450 women with PhDs in biochemistry. They found that women were less likely to be promoted to associate or full professor, were held to a higher standard than men, and were particularly disadvantaged in more prestigious departments. In addition, a committee at the Massachusetts Institute of Technology found that women scientists were allocated less laboratory space, were excluded from search committees, and were encouraged to teach more than their male counterparts at the same rank—all factors that adversely affect promotion (Lawler 1999). Finally, even in public schools, women make up 75 percent of all classroom teachers but only about 50 percent of principals and assistant principals (U.S. Bureau of the Census 2005).

Work and Structural Sexism

According to the International Labor Office (ILO), in 2006, women made up 40 percent of the world's *total* labor force (ILO 2007a). In the same year, 57 percent of all *women* participated in the labor force. Women's labor force participation rates vary dramatically by region and country within region. For example, 71.1 percent of all women in North America participate in the labor force compared to 32 percent of all women in the Middle East and North Africa (ILO 2007b). Further, 70.0 percent of all U.S. women participate in the labor force compared with 44 percent of Mexican women (U.S. Census Bureau 2007).

Worldwide, women tend to work in jobs that have little prestige and low or no pay, where no product is produced, and where they are the facilitators for others. Women are also more likely to hold positions of little or no authority within the work environment and to have more frequent and longer periods of unemployment (Athreya 2003; WHO 2006; ILO 2007a, 2007b). Although improvements are being made—35 percent of all U.S. women are now in managerial or professional occupations—women of color are even less likely to hold

TABLE 10.2 Highly Sex-Segregated Occupations, 2005

FEMALE-DOMINATED OCCUPATIONS	PERCENTAGE OF FEMALE WORKERS
Child care workers	95
Dental hygienists	97
Dietitians	95
Elementary and middle school teachers	82
Librarians	85
Paralegals and legal assistants	86
Prekindergarten and kindergarten teachers	98
Receptionists	92
Registered nurses	92
Secretaries	97
Speech therapists	92
Teacher assistants	90
Travel agents	77

MALE-DOMINATED OCCUPATIONS	PERCENTAGE OF MALE WORKERS
Airplane pilots and navigators	95
Architects	76
Automobile mechanics	98
Civil engineers	87
Clergy	85
Construction workers	97
Dentists	78
Firefighters	97
Grounds maintenance workers	92
Lawyers	70
Mechanical engineers	94
Physicians	68
Police officers	86

Source: U.S. Bureau of the Census (2007).

positions of power (EEOC 2003; IWPR 2006a). In an investigation of female and male African-American and white firefighters, black women were the most subordinated group, as black males and white females relied on their superordinate gender and race statuses, respectively (Yoder & Aniakudo 1997).

No matter what the job, if a woman does it, it is likely to be valued less than if a man does it (Barko 2003). For example, in the early 1800s 90 percent of all clerks were men and being a clerk was a prestigious profession. As the job became more routine, in part because of the advent of the typewriter, the pay and prestige of the job declined and the number of female clerks increased. Today, 91 percent of clerks are female (U.S. Census Bureau 2007), and the position is one of relatively low pay and prestige.

The concentration of women in certain occupations and men in other occupations is referred to as **occupational sex segregation** (see Table 10.2). For example, women are overrepresented in semiskilled and unskilled occupations, and men are disproportionately concentrated in professional, administrative, and managerial positions. In some occupations sex segregation has decreased in recent years. For example, between 1983 and 2005 the percentage of female physicians doubled from 16 percent to 32 percent, female dentists increased from 7 percent to 22 percent, and female clergy increased from 6 percent to 15 percent (U.S. Bureau of the Census 2007). Although the pace is slow, increasingly men are applying for jobs traditionally held by women, leading to such terms as *mannies* (male nannies) and *murses* (male nurses) (Cullen 2003). Some evidence suggests that men seeking traditionally female jobs have an edge in hiring and promotions—called the **glass elevator effect** (Williams 2007). However in some fields the trend has slowed or even reversed; for example, public school teaching is even more dominated by women today than it was two decades ago (U.S. Bureau of the Census 2007). Women are still heavily represented in low-prestige, low-wage, **pink-collar jobs** that offer few benefits.

What Do You Think? A recent report by the British Equal Opportunities Commission found that, as in the United States, there is a high degree of occupational sex segregation in the United Kingdom. Men tend to be concentrated in areas such as construction, engineering, plumbing, and the like, whereas three-quarters of women are found in five occupational groups—cleaning, catering, caring, cashiering, and clerical—the five c's. These traditional areas of women's employment often command lower salaries than "men's work" despite requiring similar qualifications. The report concludes that "there is a clear financial incentive for women to choose training and work in sectors where men dominate the workforce, as pay tends to be higher in those areas" (EOC 2006, p. 1). Did a high school counselor or a college advisor ever try to direct you into traditional male or female areas of study? In what way, if any, are your own career goals influenced by society's definitions of male and female occupations?

occupational sex segregation The concentration of women in certain occupations and men in other occupations.

glass elevator effect The tendency for men seeking or working in traditionally female occupations to benefit from their minority status.

pink-collar jobs Jobs that offer few benefits, often have low prestige, and are disproportionately held by women.

Sex segregation in occupations continues for several reasons (Martin 1992; Williams 1995; Renzetti & Curran 2003). First, cultural beliefs about what is an "appropriate" job for a man or a woman still exist. Cejka and Eagly (1999) reported that the more college students believe that an occupation is male or female dominated, the more they attribute success in that occupation to masculine or feminine characteristics. In addition, males and females continue to be socialized to learn different skills and acquire different aspirations. A government report documents that work at age 12 is sex segregated; girls babysit and boys do lawn work (U.S. Bureau of Labor Statistics 2000). Furthermore, a study of 14- to 16-year-olds found that, although both males and females held gender-equitable ideas about men's and women's roles in society, they enrolled in gender-typical subjects in school and aspired to gender-typical occupations (Tinklin et al. 2005). Finally, Skuratowicz and Hunter (2004), in studying personnel practices in a U.S. bank, found that employees were encouraged to apply for newly created customer resource manager (CRM) positions with pamphlets picturing women. In contrast, customers were informed of new personal banker (PB) positions with video displays picturing men in business suits. Managers assumed that men would have little interest in the CRM position, which had no direct authority over other employees but included increased customer contact, and would be more attracted to the PB positions, which involved selling financial products and working on commission. Not surprisingly, after the employment process was completed, the CRMs were predominately female and the PBs were mostly male.

Opportunity structures differ as well. For example, women and men, upon career entry, are often channeled by employers into gender-specific jobs that carry different wages and promotion opportunities. However, even women in higher-paying jobs may be victimized by a **glass ceiling**—an often invisible barrier that prevents women and other minorities from moving into top corporate positions. When 1,200 executives in eight countries were asked about the existence of a glass ceiling, 70 percent of women and 57 percent of men responded that they believed that a glass ceiling existed and that it was preventing women from excelling in business (Clark 2006). Women and minorities have different social networks than do white men, which contributes to this barrier. White men in high-paying jobs are more likely to have interpersonal connections with individuals in positions of authority (Padavic & Reskin 2002). Women also may be excluded by male employers and by those employees who fear that the prestige of their profession will be compromised by the entrance of women or by those who simply believe that "the ideal worker is normatively masculine" (Martin 1992, p. 220). In addition, women often find that their opportunities for career advancement are adversely affected after returning from family leave. For example, female lawyers returning from maternity leave found their career mobility stalled after being reassigned to less prestigious cases (Williams 2000). Not surprisingly, one study found that in the United States 49 percent of women in high-level managerial positions had no children; 19 percent of men similarly situated had no children (Hewlett 2002).

Finally, because family responsibilities primarily remain with women, working mothers may feel pressure to choose professions that permit flexible hours and career paths, sometimes known as mommy tracks (Moen & Yu 2000). Thus, for example, women dominate the field of elementary education, which permits them to be home when their children are not in school. Nursing, also dominated by women, often offers flexible hours. Although the type of career pursued may be the woman's choice, it is a **structured choice**—a choice among limited options as a result of the structure of society. Interestingly, Epstein (2007) noted that when men wish to limit their work hours to assist with family responsibilities they are severely disciplined and may face discrimination.

glass ceiling An invisible barrier that prevents women and other minorities from moving into top corporate positions.

structured choice Choices that are limited by the structure of society.

Income and Structural Sexism

Women are twice as likely as men to earn at or below minimum wage (U.S. Bureau of Labor Statistics 2007). Furthermore, in 2005 full-time working women had median weekly earnings of $612, compared with full-time weekly earnings

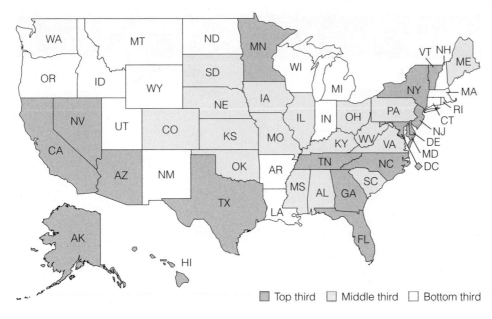

FIGURE 10.2
Ratio of women's to men's earnings across the United States, 2005.
Note: Ratio of median annual earnings between women and men, aged 16 and older, who work full-time, year-round.
Source: Institute for Women's Policy Research (2006b).

☐ Top third ☐ Middle third ☐ Bottom third

of $771 for men (U.S. Department of Labor 2006). Although the U.S. gender gap in pay has decreased over the years, from women making 60 percent of what a man made in 1980 to nearly 80 percent of what a man makes today, the movement toward pay equality has recently slowed down. Moreover, at the current rate of progress the gender gap in pay will not be closed for another 50 years (IWPR 2006a; 2007). As Figure 10.2 illustrates, the gender gap in pay also varies by state. For example, in the District of Columbia the full-time working woman earns, on average, 86 percent of what the full-time working man earns, whereas in Wyoming the average full-time working woman earns 61 percent of what the average full-time working man earns (IWPR 2006b).

In general, the higher one's education is, the higher one's income. Yet, even when men and women have identical levels of educational achievement and both work full-time, women, on average, earn significantly less than men (see Table 10.3).

> . . . just one year after college graduation, women earn only 80 percent of what their male counterparts earn. Ten years after graduation, women fall further behind, earning only 69 percent of what men earn. Even when controlling for hours, occupation, parenthood, and other factors known to affect earnings, the research indicates that one-quarter of the pay gap remains unexplained and is likely due to sex discrimination. (AAUW 2007, p. 1)

Racial differences also exist. Although women in general earn 77 percent as much as men, African-American and Hispanic-American women earn just 71 percent and 58 percent, respectively, of men's salaries (NOW 2007a). Even among celebrities a significant income gap exists. According to *Forbes* magazine's Celebrity 100 issue, the top female athlete (Maria Sharapova) earned 21 percent of what the top male athlete (Tiger Woods) earned (Goldman & Blakeley 2006).

In an investigation of the gender pay gap, Kilbourne et al. (1994) analyzed data from the National Longitudinal Survey that included more than 5,000

TABLE 10.3 Effects of Education and Sex on Median Weekly Earnings of Full-Time Workers, 2005

	MEDIAN WEEKLY EARNINGS	
EDUCATIONAL ATTAINMENT	MEN	WOMEN
Total, 25 years old and older	771	612
Less than 1 year of high school	413	321
Four years of high school (no diploma)	492	382
High school graduate	652	493
Some college, no degree	753	570
Associate degree	791	614
Bachelor's degree	1,071	813
Master's degree	1,333	983
Professional degree	1,558	1,131
Doctorate degree	1,536	1,214

Source: U.S. Department of Labor (2006).

66We know that we can do what men can do, but we still don't know that men can do what women can do. That's absolutely crucial. We can't go on doing two jobs.99

Gertrude Stein
Author and feminist

devaluation hypothesis The hypothesis that women are paid less because the work they perform is socially defined as less valuable than the work performed by men.

emotion work Work that involves caring for, negotiating, and empathizing with people.

human capital hypothesis The hypothesis that female–male pay differences are a function of differences in women's and men's levels of education, skills, training, and work experience.

comparable worth The belief that individuals in occupations, even in different occupations, should be paid equally if the job requires "comparable" levels of education, training, and responsibility.

women and 5,000 men. They concluded that occupational pay is gendered and that "occupations lose pay if they have a higher percentage of female workers or require nurturant skills" (p. 708). Cohen and Huffman (2003) also found that in occupations with high female representation, men performing the same job as women earned higher salaries. Budig (2003) reported that regardless of the female–male distribution in the job, men are paid more than women.

Two hypotheses are frequently cited in the literature for why the income gender gap continues to exist. One hypothesis is called the **devaluation hypothesis.** It argues that women are paid less because the work they perform is socially defined as less valuable than the work performed by men. Guy and Newman (2004) argued that these jobs are undervalued in part because they include a significant amount of **emotion work**—that is, work that involves caring, negotiating, and empathizing with people, which is rarely specified in job descriptions or performance evaluations. The other hypothesis, the **human capital hypothesis,** argues that female–male pay differences are a function of differences in women's and men's levels of education, skills, training, and work experience. Although these explanations are difficult to test, most researchers agree that there is more support for the devaluation hypothesis than for the human capital hypothesis (Padavic & Reskin 2002; England & Li 2006). One way researchers test these hypotheses is to compare the earnings of individuals in occupations with comparable worth but different gender compositions. **Comparable worth** refers to the belief that individuals in occupations, even in different occupations, should be paid equally if the job requires "comparable" levels of education, training, and responsibility. For example, in a comparable worth lawsuit, nurses successfully sued the City of Denver for paying them less than other employees (e.g., tree trimmers, sign painters) who had less education (Nelson & Bridges 1999).

TABLE 10.4 Percentage of Women Elected by Level and Type of Government Position, 2005

LEVEL OF GOVERNMENT/ POSITION	NUMBER OF SEATS	NUMBER OF WOMEN	PERCENTAGE HELD BY WOMEN
U.S. president	1	0	0.0
U.S. vice president	1	0	0.0
U.S. Congress	535	86	16.1
House	435	70	16.1
Senate	100	16	16.0
Governors	50	9	16.0
State legislators	7,382	1,734	23.5

Source: CAWP (2007).

Politics and Structural Sexism

Women received the right to vote in the United States in 1920 with the passage of the Nineteenth Amendment. Even though this amendment went into effect more than 85 years ago, women still play a rather minor role in the political arena. In general, the more important the political office is, the lower the probability that a woman will hold it. Although women constitute 52 percent of the population, the United States has never had a woman president or vice president, and, until 1993, when a second woman was appointed, the United States had only one female Supreme Court justice. The highest-ranking women ever to serve in U.S. government have been former Secretary of State Madeleine Albright, current Speaker of the House Nancy Pelosi, and current Secretary of State Condoleezza Rice. Appointed in 2005, Rice is also the first African-American woman to serve in such a capacity. In 2005 women represented only 16 percent of all governors and held only 16 percent of all U.S. congressional seats (see Table 10.4). Worldwide, the percentage of national and local legislative seats held by women is 16 percent (IDEA 2006).

In response to the under-representation of women in the political arena, some countries have instituted electoral quotas. Today, there are 74 countries "where quotas have been implemented in the constitution, regulations and laws, or where political parties have implemented their own internal quotas" (IDEA 2006, p. 1). In India a 1993 amendment held one-third of all seats in local contests for women. Eight thousand women were elected. A 1996 law in Brazil required that a minimum of 20 percent of each party's candidates be women. Countries with similar policies include Finland, Germany, Mexico, South Africa, and Spain (Sheehan 2000). In addition, in 2005 the Transitional Administrative Law in Iraq identified a target goal of 25 percent female representation in the Iraqi parliament. Exceeding that goal, 31 percent of subsequently elected members of parliament were women.

The relative absence of women in politics, as in higher education and in high-paying, high-prestige jobs in general, is a consequence of structural limitations. Running for office requires large sums of money, the political backing of powerful individuals and interest groups, and a willingness of the voting public to elect women. Thus minority women have even greater structural barriers to election and, not surprisingly, represent an even smaller percentage of elected officials. Women's lack of representation among political leaders begins early. In the last 5 years less than one-third of student government offices at midwestern universities were held by women (Miller 2004). Nonetheless, when a nationwide random sample of 1,229 adults was asked, "Would you vote for a woman as president if she were qualified?" 92 percent responded "yes." Thus Hillary Clinton is the leading candidate for the Democratic Party in the upcoming 2008 presidential election. Interestingly, however, when respondents were asked, "Is America ready for a woman president?" only 55 percent responded "yes" (CBS News 2006).

Would an increase in female participation make a difference in politics? Voting patterns among women suggest that it might (CAWP 2006). Women tend to be more likely to vote Democratic than men and to express more concern over health care, child care, education, poverty, and homelessness (Renzetti & Curran 2003; UNICEF 2007). Men focus more on the federal deficit, taxes, energy, defense, and foreign policy. Women are less supportive of the death penalty than men are and are more supportive of gun control laws. Furthermore, women politicians report that they feel a responsibility to represent both their elected constituency and women in general. As Congresswoman Marge Roukema noted (Hawkesworth et al. 2001):

> I didn't really want to be stereotyped as the woman legislator. . . . I wanted to deal with things like banking and finance. But I learned very quickly that if the women like me in Congress were not going to attend to some of these family concerns, whether it was for jobs or children, pension equity, or whatever, then they weren't going to be attended to. So I quickly shed those biases that I had and said, "Well nobody else is going to do it; I'm going to do it." (p. 6)

Interestingly, some research suggests that gender differences in partisanship and public policy attitudes begin as early as adolescence (Fridkin & Kenney 2007).

Civil Rights, the Law, and Structural Sexism

In many countries, victims of gender discrimination cannot bring their cases to court. This is not true in the United States. The 1963 Equal Pay Act and Title VII of the 1964 Civil Rights Act make it illegal for employers to discriminate in wages or employment on the basis of sex. Nevertheless, such discrimination still occurs as evidenced by the thousands of grievances filed each year with the Equal Employment Opportunity Commission (EEOC)—23,247 grievances in 2006 (EEOC 2007a).

In one of the largest employment discrimination suits ever filed, Wal-Mart Stores, Inc. has been named in a class-action suit in which 1.5 million former and current female employees allege that management discriminated against them by (1) paying male employees more and (2) denying promotions to women. When faced with lawsuits of this nature, employers use various techniques to justify their employment practices. Wal-Mart, for example, stated that the stores were independently owned, thus supporting their contention that there was no company-wide policy or practice of discrimination in place (Kravets 2007). Another technique used by employers to justify differences in pay is the use of different job titles for the same type of work. Repeatedly the courts have ruled, however, that jobs that are "substantially equal," regardless of title, must result in equal pay. In the Wal-Mart case, if the courts ultimately rule in favor of the plaintiffs, Wal-Mart Stores, Inc. faces the possibility of having to pay billions of dollars in damages (Kravets 2007).

Discrimination, although illegal, takes place at both the institutional and the individual levels (see Chapter 9). Institutional discrimination includes screening devices designed for men, hiring preferences for veterans, the practice of promoting from within an organization based on seniority, and male-dominated recruiting networks (Reskin & McBrier 2000). For example, the Augusta National Golf Club, home of the Masters Golf Tournament and a virtual "who's who of the corporate world," has refused to change its policy forbidding women from joining, despite years of political pressure from women's groups and negative publicity (National Council of Women's Organizations 2005; Hack 2006). One of the most blatant forms of individual discrimination is sexual harassment, discussed later in this chapter.

Discrimination takes place in other forms as well. Women are discriminated against on the basis of family responsibility—sometimes called "maternal profiling" (NOW 2007b). In the United States women often have difficulty obtaining home mortgages or rental property because they have lower incomes, shorter work histories, and less collateral. Until fairly recently, husbands who raped their wives were exempt from prosecution. Even today, some states require a legal separation agreement and/or separate residences for a raped wife to receive full protection under the law. Women in the military have traditionally been restricted in the duties they can perform, and, finally, since the U.S. Supreme Court's 1973 *Roe v. Wade* decision, which made abortion legal, the right of a woman to obtain an abortion has steadily been limited and narrowed by subsequent legislative acts and judicial decisions (Connolly 2005). The debate has continued with several recent court decisions (see Chapter 15).

What Do You Think? The arguments are as old as the "battle of the sexes": (1) women who are working, particularly married women, are simply making money for the "extras," whereas men who are working are working to support their families; (2) women who are working will get pregnant and quit so there is no need to invest in their professional development; and (3) a woman's primary concern is her family not her career—hiring or promoting women is unfair to her career-minded male counterpart. Each of these arguments justifies treating men and women differently. Assuming equal qualifications, is preferential treatment in the workplace ever justified? If so, under what conditions?

THE SOCIAL CONSTRUCTION OF GENDER ROLES: CULTURAL SEXISM

cultural sexism The ways in which the culture of society perpetuates the subordination of an individual or group based on the sex classification of that individual or group.

As social constructionists note, structural sexism is supported by a system of cultural sexism that perpetuates beliefs about the differences between women and men. **Cultural sexism** refers to the ways the culture of society—its norms, values, beliefs, and symbols—perpetuate the subordination of an individual or group because of the sex classification of that individual or group.

For example, the *belief* that females are less valuable than males has serious consequences. China has 20 percent fewer girls than boys in the birth to 4-year-old age group. China's "missing girl" phenomenon, a growing problem, is a consequence of selective abortions of female fetuses and premature death of female infants as a result of withholding nourishment and health care (Banister 2003). Similarly, it is estimated that a half a million girls are aborted each year in India. A recent report showed that educated Indian women are more likely to abort unwanted female fetuses than, as often assumed, uneducated, poor Indian women. The wealthier, educated women are the ones who have access to the ultrasound technology necessary to determine the sex of their baby (Rao 2006). Cultural sexism takes place in a variety of settings, including the family, the school, and the media as well as in everyday interactions.

Family Relations and Cultural Sexism

From birth, males and females are treated differently. The toys that male and female children receive convey different messages about appropriate gender behavior. For example, in a study of leisure time activities of 5- to 13-year-old boys and girls, girls listed dolls as their favorite toys and boys listed blocks, Legos, and the like as their favorite toys (Cherney & London 2006). Recently, retail giant Toys 'R' Us, after much criticism, removed store directories labeled "Boy's World" and "Girl's World." Similarly, toy manufacturer Mattel came under fire after producing a pink, flowered Barbie computer for girls and a blue Hot Wheels computer for boys. The social significance of the gender-specific computers and the public criticism came after it was revealed that the accompanying software packages were different—the boys' package had more educational titles (Bannon 2000).

Household Division of Labor. Girls and boys work within the home in approximately equal amounts until the age of 18, when the female-to-male ratio begins to change (Robinson & Bianchi 1997; United Nations Population Fund 2003). In a study of household labor in 10 Western countries, Bittman and Wajcman (2000) reported that "women continue to be responsible for the majority of hours of unpaid labor," ranging from a low of 70 percent in Sweden to a high of 88 percent in Italy (p. 173). A recent study in Mexico found that women in the paid labor force performed household chores for an additional 33 hours of work each week; men contributed 6 hours a week to domestic chores (UNICEF 2007).

In the United States married women report performing between two-thirds and three-quarters of the household work (United Nations 2000). The fact that women, even when working full-time, contribute significantly more hours to home care than men is known as the "second shift" (Hochschild 1989). In a University of Michigan study of 50 pregnant couples, researcher Marlena Studer asked spouses to identify how much child care they each expected to perform after the baby was born. Compared with their husbands' projections, working and stay-at-home mothers anticipated that they would perform almost three times as many hours of child care per week than their husbands (UM News Service 2005). To the extent that women work the second shift more than men, they convey traditional conceptions of gender to their children that are often carried

66 Women weren't born Democrat, we weren't born Republican, and we weren't born yesterday. 99

Patricia Ireland
President, National Organization
for Women, 1991–2001

According to Hochschild, women are expected to work "second shifts" by having gainful outside employment as well as performing household chores and child care once they arrive home after a day's work. Research indicates that women contribute significantly more hours to home and child care and have less access to leisure time than their male counterparts.

Corbis Premium =F/Alamy

into adulthood. For example, Cunningham (2001) found that boys are particularly more likely to perform housework as teenagers when they are exposed to nontraditional divisions of household labor. Similarly, Evertsson (2006) reported that Swedish girls and boys in two-parent homes are more likely to engage in atypical housework if the parent of the same sex also engages in nontraditional household chores.

Three explanations for the continued traditional division of labor emerge from the literature. The first explanation is the *time-availability approach.* Consistent with the structural-functionalist perspective, this position claims that role performance is a function of who has the time to accomplish certain tasks. Because women are more likely to be at home, they are more likely to perform domestic chores.

A second explanation is the *relative resources approach.* This explanation, consistent with a conflict perspective, suggests that the spouse with the least power is relegated the most unrewarding tasks. Because men have more education, higher incomes, and more prestigious occupations, they are less responsible for domestic labor. Thus, for example, because women earn less money than men do, on average, they turn down overtime and other work opportunities to take care of children and household responsibilities, which subsequently reduces their earnings potential even further (Williams 2000).

Gender role ideology, the final explanation, is consistent with a symbolic interactionist perspective. It argues that the division of labor is a consequence of traditional socialization and the accompanying attitudes and beliefs. Women and men have been socialized to perform various roles and to expect their partners to perform other complementary roles. Women typically take care of the house and men the yard. These patterns begin with household chores assigned to girls and boys and are learned through the media, schools, books, and toys. A test of the three positions found that although all three had some support, gender role ideology was the weakest of the three in predicting work allocation (Bianchi et al. 2000). It should be noted that the three positions are not mutually exclusive. For example, Mannino and Deutsch (2007) in their longitudinal study of 81 married women found support for the relative resources and gender role ideology perspectives.

The School Experience and Cultural Sexism

Sexism is also evident in the schools. It can be found in the books students read, the curricula and tests they are exposed to, their extracurricular activities, and the different ways teachers interact with students.

Textbooks. The bulk of research on gender images in textbooks and other instructional materials documents the way males and females are portrayed stereotypically. For example, Purcell and Stewart (1990) analyzed 1,883 storybooks used in schools and found that they tended to depict males as clever, brave, adventurous, and income-producing and females as passive and as victims. Females were more likely to be in need of rescue and were also depicted in fewer occupational roles than males. Witt (1996), in a study of third-grade textbooks from six publishers, reported that little girls were more likely to be portrayed as having both traditionally masculine *and* feminine traits, whereas little boys were more likely to be pictured as having masculine characteristics only. These results are consistent with research that suggests that boys are much less free to explore gender differences than girls and with Purcell and Stewart's conclusion that boys are often depicted as having "to deny their feelings to show their manhood" (1990, p. 184). Finally, in a study of 200 popular children's picture books, researchers found that fathers were largely absent and, when they did appear, they were withdrawn and inept as parents (Anderson & Hamilton 2005). Although some evidence suggests that the frequency of male and female textbook characters is increasingly equal, portrayals of girls and boys largely remain stereotypical (Evans & Davies 2000; Black 2006).

Curricula and Testing. The differing expectations and/or encouragement that females and males receive contribute to their varying abilities, as measured by standardized tests, in disciplines such as reading, math, and science. In kindergarten, boys and girls' scores in reading and math are essentially the same. By third grade, boys significantly outperform girls in math and science, and girls outperform boys in reading. The gender gaps in these disciplines continue through high school (Dee 2006). Are such differences a matter of aptitude? The possibility was raised by the former president of Harvard University in discussing why women do not excel in science and mathematic fields at the same rate as men. Women's groups and academics around the country criticized him for not recognizing structural and cultural factors, including his own administration, during which tenure offers to women decreased by two-thirds (Winters 2005).

Interestingly, a poll revealed that a majority of Americans believe that women and men are equally good at math and science. Even those who believe that men are better at these subjects are split as to whether differences are the result of innate or societal factors (Jones 2005). There is, however, little doubt that social factors play a role. In an experiment at the University of Waterloo, male and female college students, all of whom said they were good in math, were shown either gender-stereotyped or gender-neutral advertisements. When, subsequently, female students who had seen the female-stereotyped advertisements took a math test, they performed not only lower than women who had seen the gender-neutral advertisements but also lower than their male counterparts (Begley 2000). Research also indicates that standardized tests themselves are biased. The format of the tests—almost exclusively timed and multiple-choice—favors males, according to some theorists (Smolken 2000; Davis 2007). Finally, women faculty at the Massachusetts Institute of Technology (MIT) and Harvard report that their female students were discouraged from entering science fields by the hostile environment in which they saw their female faculty working (Lawler 1999).

Extracurricular Activities. Encouragement to participate in sports and extracurricular activities is gender biased, despite Title IX of the 1972 Educational Amendments Act, which prohibits officials from "tracking" students by sex (Orecklin 2003). Although women's and girls' participation is now at an all-time high with, for example, more than 166,000 women participating in NCAA sporting events, differences remain in the sports that males and females play (U.S. Bureau of the Census 2007). Men are more likely to participate in competitive sports that emphasize traditional male characteristics such as winning, aggression, physical strength, and dominance. Women are more likely to participate in sports that emphasize individual achievement (e.g., figure skating) or cooperation (e.g., synchronized swimming). This choice may in part be a result of peer pressure. Alley and Hicks (2005) found that adolescents, when asked to read descriptions of male and female peers, rated these individuals as less masculine or less feminine when they participated in cross-gender sports (e.g., ballet for boys and karate among girls). Boys and girls differ in other extracurricular activities as well. For example, girls are more likely to participate in cultural activities than are boys (Dumais 2002).

> "In the school room, more than any other place, does the difference of sex, if there is any, need to be forgotten."
>
> Susan B. Anthony
> Feminist

FIGURE 10.3
As this picture from the U.S. Bureau of Labor Statistics' Kid's Careers website illustrates, males are less likely than females to be encouraged to choose gender-atypical areas of study.
Source: U.S. Bureau of Labor Statistics (2005).

Teacher–Student Interactions. Sexism is also reflected in the way that teachers treat their students. Millions of young girls are subjected to sexual harassment by male teachers who then fail them when they refuse the teachers' sexual advances (Quist-Areton 2003). After interviewing 800 adolescents, parents, and teachers in three school districts in Kenya, Mensch and Lloyd (1997) reported that teachers were more likely to describe girls as lazy and unintelligent. "And when the girls do badly," the researchers remarked, "it undoubtedly reinforces teachers' prejudices, becoming a vicious cycle." Similarly, in the United States Sadker and Sadker (1990) observed that elementary and secondary school teachers pay more attention to boys than to girls. Teachers talk to boys more, ask them more questions, listen to them more, counsel them more, give them more extended directions, and criticize and reward them more frequently.

However, Sommers (2000), in a book titled *The War against Boys*, argued that it is "boys, not girls, on the weak side of the educational gender gap" (p. 14). Noting that boys are at a higher risk for learning disabilities and that they lag behind in reading and writing scores, Sommers argued that the belief that females are educationally shortchanged is simply untrue. Kimmel (2006), on the other hand, argued that the "war against boys" is a myth and that gender differences in achievement are, at least in part, a consequence of gender role socialization, not reverse discrimination. For example, boys tend to overestimate their abilities and stay in classes in which they are less capable of succeeding that in turn skews their achievement levels downward.

What Do You Think? Whatever you call it—"the war against boys," "the boy crisis," "the trouble with boys"—it is, as Mead (2006) noted, a compelling story for it "touches on Americans' deepest insecurities, ambivalences, and fears about gender roles and the 'battle of the sexes'" (p. 3). Some argue that what seems like bad news is actually good news—boys are not doing worse, girls are simply doing better (Mead 2006). Others hold that it is a zero-sum game where there are winners and losers. From elementary school to high school girls outperform boys, and there must be something or someone to blame—feminism, mothers, female teachers, the school environment, something. What do you think? Why are girls excelling? Do schools cater to girls? Do you think that boys believe that doing well in school is too "feminine"?

Gender-based definitions are not the only things that affect student outcomes. The effect of teachers' gender on school performance was tested by Dee (2006) using the National Education Longitudinal Survey. Analyzing data from a nationally representative sample of eighth grade boys and girls and their teachers, he found that (1) when the teacher is a female, boys are more likely to be seen as disruptive, (2) when the teacher is a female, girls are less likely to be seen as either disruptive or inattentive, (3) when the teacher is male, female students are less likely to perceive a class as useful, to ask questions in class, or to look forward to the class, and (4) when the teacher is female, regardless of the subject, boys are less likely to report looking forward to the class. In conclusion, the author noted that, "simply put, girls have better educational outcomes when taught by women, and boys are better off when taught by men" (p. 71) (see also Chapter 8).

Media, Language, and Cultural Sexism

Another concern voiced by social scientists is the extent to which the media portray females and males in a limited and stereotypical fashion and the impact of such portrayals. For example, in the acclaimed documentary *Killing Us Softly 3: Advertising's Image of Women,* Jean Kilbourne (2000) demonstrates how advertisements in the United States routinely turn women into objects, sexualize young girls, portray women as solely responsible for child care (and men as incompetent caregivers), and promote violence against women. Perhaps most surprisingly, the patterns Kilbourne detected have not changed since her first documentary in 1979.

Similarly, a study of 6 months of Sunday morning television talk shows in the United States found that women were almost nonexistent as panelists (Common Dreams 2005). Even *Sesame Street* until the introduction of Abby Cadabby never had a female star the likes of Elmo or Big Bird (Ms. Piggy was a muppet, but not a *Sesame Street* character) (Dominus 2006). Finally, *Essence* magazine launched a campaign in 2005 against sexism in rap music. This campaign was in response to inflammatory images such as rapper Nelly swiping a credit card down a woman's buttocks and the rapper Ludacris shown about to bite a woman's leg on a compact disc cover.

Men are also victimized by media images. A recent study of 1,000 adults found that two-thirds of the respondents thought that women in television advertisements were pictured as "intelligent, assertive and caring," whereas men

Abby Cadabby, the newest edition to the *Sesame Street* lineup, is a 3-year-old "fairy-in-training" and is the first new female muppet in 13 years. According to her homepage, Abby will bring greater diversity to *Sesame Street* and will expand opportunities to explore people from diverse backgrounds.

© AP Photo/Mike Derer

were portrayed as "pathetic and silly" (Abernathy 2003). In a study of beer and liquor ads, Messner and Montez de Oca (2005) concluded that, although beer ads of the 1950s and 1960s focused on men in their work roles and only depicted women as a reflection of men's home lives, present-day ads portray young men as "bumblers" and "losers" and women as "hotties" and "bitches." In addition, a study of 1,200 children between the ages of 10 and 17 (Dittrich 2002) found that, when asked to name their most admired TV characters, boys' and girls' top five slots were filled by men. The children associated "worrying about appearance and weight, crying, whining, and weakness" with female TV characters; playing sports, wanting to be kissed or to have sex, and being a leader were associated with being a male character.

One of the largest studies of women and men in the media is the Global Media Monitoring Project (Gallagher 2006). Nearly 13,000 news stories on television, on the radio, and in newspapers were monitored by 40,000 news personnel in 76 countries. The results indicate that

- Women are dramatically under-represented as news subjects, and, when they are the subject of news, they are most often "stars" or ordinary women rather than figures of authority or experts.
- Although the number of female reporters has increased in recent years, female reporters are still a minority in electronic (37 percent) and print media (29 percent).
- The older a woman is, the less likely she is to be on television—for example, by the age of 50 only 7 percent of newscasters are women.
- Male journalists are more likely to work on the "hard" news (e.g., politics) whereas women are more often assigned "soft stories" (e.g., social issues).
- Just 10 percent of all news stories focus specifically on women; news on gender equality/inequality is almost nonexistent.

As with media images, both the words we use and the way we use them can reflect gender inequality. The term *nurse* carries the meaning of "a woman who . . ." and the term *engineer* suggests "a man who . . . ." Terms such as *broad, old maid,* and *spinster* have no male counterparts. Sexually active teenage females are described by terms carrying negative connotations, whereas terms for equally sexually active male teenagers are considered complimentary among today's youth. Language is so gender stereotyped that the placement of male or female before titles is sometimes necessary, as in the case of "female police officer" or "male prostitute." Furthermore, as symbolic interactionists note, the embedded meanings of words carry expectations of behavior.

Virginia Sapiro (1994) has shown how female–male differences in communication style reflect the structure of power and authority relations between men and women. For example, women are more likely to use disclaimers ("I could be wrong but . . .") and self-qualifying tags ("That was a good movie, wasn't it?"), reflecting less certainty about their opinions. Communication differences between women and men also reflect different socialization experiences. Women are more often passive and polite in conversation; men are less polite, interrupt more often, and talk more (Renzetti & Curran 2003). Finally, in a study of computer-mediated communication of college students' postings for an introduction to psychology class, Guiller and Durndell (2006) reported that women were more likely than men "to make attenuated contributions and express agreement, whereas males were more likely than females to make authoritative contributions and express disagreement" (p. 368).

Roman Catholic Women priests

Roman Catholic women are seen with the sacraments of communion as part of their ordination ceremony as "women priests" in direct conflict with the doctrines of the Roman Catholic Church. Faced with excommunication for ecclesiastical disobedience, women have participated in five ordination ceremonies around the world since 2002. In contrast, other denominations embrace women as religious leaders as evidenced by the recent ordination of Katharine Jefferts Schori as the first female chief pastor and presiding bishop of the Episcopal Church of the United States.

Religion and Cultural Sexism

Most Americans claim membership in a church, synagogue, or mosque. Research indicates that women attend religious services more often, rate religion as more important to their lives, and are more likely to believe in an afterlife than are men (Davis, Smith, & Marsden 2002; Smith 2006; Adams 2007). In general, religious teachings have tended to promote traditional conceptions of gender. For example, in 1998 the Southern Baptist Convention *officially* took the stance that a wife should "submit graciously" to her husband, assume her "God-given responsibility to respect her husband, and [to] serve as his helper." Former President Jimmy Carter, a lifelong active Southern Baptist, left the church in protest over this statement (White 2000). Although Gallagher (2004) found that most evangelical Christians believe that marriage should be considered an equal partnership, she also found that these same Evangelicals continue to believe that the male is the head of the household.

Orthodox Jewish women are not counted as part of the *minyan,* or quorum, required at prayer services, are not allowed to read from the Torah, and are required to sit separately from men at religious services. In addition, Catholic Church doctrine forbids the use of artificial forms of contraception, and Muslim women are required to be veiled in public at all times (Renzetti & Curran 2003). Women cannot serve as ordained religious leaders in the Catholic Church, in Orthodox Jewish synagogues, or in Islamic temples. Despite this "stained glass ceiling" (Adams 2007), women are increasingly active in leadership positions in churches and temples across the nation. For example, according to the Hartford Institute for Religion Research, in 1977 there were 1,311 women in head positions in eight "faith groups." By 2000, that number was 16,994 (Rossiter 2007). Nonetheless, today in the United States 85 percent of clergy are male (U.S. Census Bureau 2007), and even in denominations that allow for the ordination of women, these female clergy often do not hold the same status as their male counterparts and are often limited in their duties (Renzetti & Curran 2003).

However, religious teachings are not all traditional in their beliefs about women and men. Quaker women have been referred to as the "mothers of feminism" because of their active role in the early feminist movement. Reform Judaism has allowed ordination of women as rabbis for more than 25 years and gays and lesbians for more than 15 years. In addition, the Women-Church movement—a coalition of feminist faith-sharing groups composed primarily of Catholic women—offers feminist interpretations of Christian teachings. Within many other religious denominations, individual congregations choose to interpret their religious teachings from an inclusive perspective by replacing masculine pronouns in hymns, the Bible, and other religious readings (Anderson 1997; Renzetti & Curran 2003).

SOCIAL PROBLEMS AND TRADITIONAL GENDER ROLE SOCIALIZATION

Cultural sexism, transmitted through the family, school, media, and language, perpetuates traditional gender role socialization. Gender roles, however, are slowly changing. As one commentator observed (Fitzpatrick 2000), "The hard lines that once helped to define masculine [and feminine] identity are blurring. Women serve in the military, play pro basketball, run corporations and govern. Men diet, undergo cosmetic surgery, bare their souls in support groups, and cook" (p. 1).

Despite this **gender tourism** (Fitzpatrick 2000), most research indicates that traditional gender roles remain dominant, particularly for males who, in general, have less freedom to explore the gender continuum. Social problems that result from traditional gender socialization include the feminization of poverty, social-psychological and health costs, and conflict in relationships.

What Do You Think? A transgender individual is a person whose sense of gender identity—masculine or feminine—is inconsistent with his or her biological (sometimes called chromosomal) sex (male or female). Transgender is not a sexual orientation and transgender individuals may have any sexual orientation—heterosexual, homosexual, or bisexual. Transsexuals are transgender individuals "who have changed or are in the process of changing [their] . . . physical sex to conform to [their] . . . internal sense of gender identity" (HRC 2005, p. 7). Do you think that transgender individuals, including transsexuals, should have the same legal protection that non-transgender individuals have?

gender tourism The recent tendency for definitions of masculinity and femininity to become less clear, resulting in individual exploration of the gender continuum.

The Feminization of Poverty

Seventy percent of the 1.5 billion people who live on $1 a day or less are women (Plan International 2007). Further, women are more likely to be unemployed than men, with more than 80 million women worldwide actively seeking employment in 2006—a 23 percent increase from a decade ago (ILO 2007a). Contributing to the problem is the increase in female-headed households. For example, one-third of the households in developing nations are headed by women (World Global Issues 2003). Many of these households are headed by young women with dependent children and older women who have outlived their spouses—two of the groups most likely to be poor. Worldwide, women "work twice as many hours as men, . . . receive only one-tenth of the world's income, and own less than a hundredth of the world's property" (Palmberg 2005, p. 1).

A report card of U.S. efforts to reduce poverty among women was released by U.S. Women Connect, a nonprofit activist group. Although the United States received a grade of B for placing women in decision-making positions, it received an F for efforts to reduce female poverty. Citing federal statistics, the group reported that although the overall poverty rate in the United States has decreased, female poverty has increased over the last 5 years (Winfield 2000). In 2004 the poverty rate of U.S. females was 13.9 percent, compared with 11.5 percent for males (U.S. Bureau of the Census 2007). As noted earlier, both individual and institutional discrimination contribute to the economic plight of women.

© AP/David Longstreath

Increasingly, the poor of the world are women. Called the "feminization of poverty," the tendency for women to be disproportionately poor is evidenced most notably in developing nations where suitable housing, clean water, food, health care, and sanitary living conditions are scarce.

Traditional gender role socialization also contributes to poverty among women. Women are often socialized to put family ahead of their education and careers. Women are expected to take primary responsibility for child care, which contributes to the alarming rate of single-parent poor families in the United States. Hispanic and black female-headed households are the poorest of all families headed by a single woman (U.S. Bureau of the Census 2007). In addition, a study of the relationship between marital status, gender, and poverty in the United States, Australia, Canada, and France indicated that in all four countries never-married women are more likely to live in poverty than ever-married women (Nichols-Casebolt & Krysik 1997).

Social–Psychological and Other Health Costs

Many of the costs of traditional gender socialization are social-psychological in nature. Reid and Comas-Diaz (1990) noted that the cultural subordination of women results in women having low self-esteem and being dissatisfied with their roles as spouses, homemakers/workers, mothers, and friends. In a study of self-esteem among more than 1,160 students in grades 6 through 10, girls were significantly more likely to have "steadily decreasing self-esteem," whereas boys were more likely to fall into the "moderate and rising" self-esteem group (Zimmerman et al. 1997).

The disparity between boys and girls, men and women, seems to be growing. A study comparing college women in 1966 to those in 1996 found that women's attitudes about their bodies and their overall self-esteem were markedly more negative in 1996 than in 1966, whereas men's levels were the same at both points in time (Sondhaus et al. 2001). It is not surprising, then, that women are twice as likely as men to be clinically depressed at some point in their lives (WHO 2006).

Adolescent girls are also more likely to be dissatisfied with their looks including physical attractiveness, appearance, and body weight than adolescent boys (Bowker, Gadbois, & Cornock 2003). An estimated 10 million females are fighting eating disorders such as anorexia and bulimia, and 88 percent of patients undergoing cosmetic surgery are women (NEDA 2006; Plan International 2007). Interestingly, satisfaction with body image is predictive of self-esteem. According to a report by Perrin and colleagues, both adolescent boys (with the exception of African-American boys) and adolescent girls report that the happier they are with their bodies, the happier they are with themselves (Reuters 2007).

Not all researchers have found that women have a more negative self-concept than men. Summarizing their research on the self-concepts of women and men in the United States, Williams and Best (1990) found "no evidence of an appreciable difference" (p. 153). They also found no consistency in the self-concepts of women and men in 14 countries: "In some of the countries the men's perceived self was noticeably more favorable than the women's, whereas in others the reverse was found" (p. 152). Furthermore, Burger and Solano (1994) reported that women are becoming more assertive and desirous of controlling their own lives rather than merely responding to the wishes of others or the limitations of the social structure.

Men are also affected by traditional gender socialization (Gupta 2003). Men experience enormous cultural pressure to be successful in their work and to earn a high income. Sanchez and Crocker (2005) found that among college-age women *and* men, the more participants were invested in traditional ideals of gender, the lower their self-concept and psychological well-being were. Furthermore, among working-class men and boys, "being tough" compensates for the lack of financial success (Williams 2000). In addition, Hochschild (2001) identified a group of self-proclaimed "overtime hounds" who worked long hours for both the money and to prove they were "man enough" to do it. Traditional male socialization also discourages males from expressing emotion and asking for help—part of what William Pollack (2000a) calls the **boy code.**

> *(13-year-old)* Boys are supposed to shut up and take it, keep it all in. It's harder for them to release or vent without feeling girly. And that can drive them to shoot themselves. (p. 1)

> *(16-year-old)* I don't cry in front of people. I'll do it sometimes when I'm in my room or talking to one of my good friends on the phone. But not in front of the guys. I feel really embarrassed when I cry in front of guys. They look at you as a wimp. (p. 7)

> *(17-year-old)* I always have and probably always will hold things in. I know that this is not the best thing to do, and I know that I can talk about it whenever I would like to, but I don't. (p. 10)

> *(16-year-old)* I think the other boys, like me, hold emotions in. I think they also have difficulty talking about things that bother them. Boys are expected to be tough. They are told to shake it off when they are hurt. They are thought of as being weak when they cry. . . . (p. 11)

Much of the boy code continues into adulthood as described in journalist Norah Vincent's book *Self-Made Man* (2006) (see this chapter's *Social Problems Research Up Close* feature).

❝ As I talked to boys across America, I'm struck by how trapped they feel. Our culture puts boys in a gender straightjacket. **❞**

William Pollack
Psychologist

boy code A set of societal expectations that discourage males from expressing emotion, weakness, or vulnerability, or asking for help.

On average, men in the United States die about 5 years earlier than women, although gender differences in life expectancy have been shrinking (U.S. Bureau of the Census 2007). Traditional male gender socialization is linked to higher rates of cirrhosis of the liver, most cancers, homicides, drug- and alcohol-induced deaths, suicide, and firearm and motor vehicle accidents (U.S. Bureau of the Census 2007). At every stage of life "American males have poorer health and a higher risk of mortality than females" (Gupta 2003, p. 84). Men engage in risky behaviors more often than women do—from smoking, drinking, and abusing drugs to not wearing a seat belt and working in dangerous environments (Gupta 2003). Being married improves men's health more so than it does women's (Williams & Umberson 2004), in large part because wives encourage their husbands to take better care of themselves.

Although men have higher rates of HIV/AIDS worldwide, the disease disproportionately affects women in many areas of the world, and in some age groups (see Chapter 2). For example, in sub-Saharan Africa 58 percent of those infected are women. Further, three-quarters of infected 15- to 24-year-olds are young women and girls (Plan International 2007). Women's inequality contributes to the spread of the disease (Heyzer 2003; WHO 2007). First, in many of these societies "women lack the power in relationships to refuse sex or negotiate protected sex" (Heyzer 2003, p. 1). Second, women are often the victims of rape and sexual assault, with little social or legal recourse. Third, gender norms often dictate that men have more sexual partners than women, putting women at greater risk. Finally, some women turn to prostitution as a means of supporting themselves and their children. With little education and no training or work history, they have few options. Ironically, it is women who care for those who are ill, often dropping out of school or quitting jobs to do so, furthering their subordinate role (McGregor 2003).

Are gender differences in morbidity and mortality a consequence of socialization differentials or physiological differences? Although both nature and nurture are likely to be involved, social rather than biological factors may be dominant. As part of the "masculine mystique," men tend to engage in self-destructive behaviors—heavy drinking and smoking, poor diets, lack of exercise, stress-related activities, higher drug use, and a refusal to ask for help. Men are more likely to have the diagnosis of antisocial personality disorder—a disorder characterized by a pervasive disregard for the law and the rights of others (WHO 2006). Men are also more likely to work in hazardous occupations than women are. Women's higher rates of depression are also likely to be rooted in traditional gender roles. The heavy burden of child care and household responsibilities, the gender pay gap and occupation gap, and fewer socially acceptable reactions to stress (e.g., it is more acceptable for males than females to drink alcohol) contribute to gender differences in depression.

Conflict in Relationships

Worldwide, gender inequality influences relationships. Research indicates that the distribution of resources affects the decision-making process. For example, a study by the International Food Policy Research Institute found that "assets brought to marriage have an impact on bargaining power within the marriage"—the more assets, the more power (United Nations Population Fund 2003). In five of the developing countries studied—Bangladesh, Ethiopia, Ghana, the Philippines, and South Africa—men brought more land and other assets to the marriage

Journalist Norah Vincent wanted to know what it is like to be a man. So, for 18 months, she observed and participated in the world of men, not as Norah Vincent, but as "Ned"—the male persona she created for the purpose of her research. Norah Vincent describes her journey into manhood in the book *Self-Made Man* (2006).

Sample and Methods

Norah Vincent used a method of research known as participant observation—a method in which the researcher participates in the phenomenon being studied to obtain an insider's perspective of the people and/or behavior being observed. The main advantage of participant observation research is that it provides detailed qualitative information about the behavior, norms, values, rituals, beliefs, and emotions of those being studied. However, participant observation research is based on small samples, so the findings may not be generalizable. The researcher's observations may also be biased by that particular researcher's values and perspectives. Norah Vincent admits that her book is not a scientific or objective study; rather, it is "one person's observations about her own experience" (p. 17). She also admits that her "observations are full of my own prejudices and preconceptions" and she describes her book as "a travelogue as much as anything else . . . a six-city tour of an entire continent, a woman's-eye view of one guy's approximated life, not an authoritative guide to the whole vast and variegated terrain of manhood in America" (p. 17).

To participate in and observe the world of men, Norah sought the help of a make-up artist to transform herself into Ned. She learned how to apply an artificial beard stubble, wore masculine-looking glasses, had her hair cut into a man's hairstyle, wore men's clothing, and stuffed her pants with a prosthetic penis known as a "packable softie." Norah bulked up her upper body muscle through 6 months of weight lifting and increased protein consumption, and she flattened her breasts by wearing a flat-front sports bra two sizes too small. Norah, who already had a deep voice, hired a voice coach to help her speak like a man.

Having achieved a male identity, Ned spent 18 months exploring what it means to be a man by (1) joining a men's bowling team, (2) going to men's strip clubs, (3) dating women, (4) living at a monastery, (5) working at a sales job, and (6) attending a men's movement group.

Findings and Conclusion

Examples of Ned's experiences and observations include the following:

1. Bowling with the Guys. One of Ned's first outings was to the local bowling alley, where he played in a Monday night men's league. It is here that he was introduced to male interaction cues.

> Our evenings together always started out slowly with a few grunted hellos that among women would have been interpreted as rude. This made my female antennae twitch a little. Were they pissed off at me about something? But among these guys no interpretation was necessary. Everything was out and aboveboard, never more, never less, than what was on anyone's mind. If they were pissed at you, you'd know it. These gruff greetings were indicative of nothing so much as fatigue and appropriate male distance. . . . They were coming from long, wearing workdays, usually filled with hard physical labor and the slow, soul-deadening deprecation that comes of being told what to do all day. . . . They didn't have the energy for pretense. (pp. 29–30)

2. Strip Clubs. Ned went to strip clubs to get a glimpse of "a substratum of the male sexual psyche that most women either don't know about, don't want to know about, or both" (p. 65). After weeks of observing, and interacting with various strippers and other patrons, Ned offers the following reflections:

> I'd been inside a part of the male world that most women and even a lot of men never see, and I'd seen it as just another one of the boys. In those places male sexuality felt like something you weren't supposed to feel but did, like something heavy you were carrying around and had nowhere to

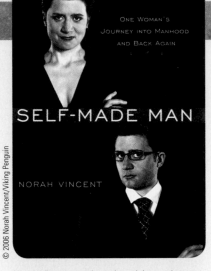

Norah Vincent conducted participant observation research by posing as "Ned" and experiencing various aspects of how men interact and experience the world.

unload except in the lap of some damaged stranger, and then only for five minutes. . . . It wasn't nearly so simple as men objectifying women. . . . Nobody won . . . nobody was more or less victimized than anyone else. The girls got money. The men got an approximation of sex and flirtation. But in the end everyone was equally debased by the experience. (pp. 90–91)

Ned recounted an interaction he had with Phil, a male acquaintance who went with Ned to a strip club. Phil said, "I go to some of these bars . . . and I say to myself, these girls were somebody's daughter. Somebody put them to bed, somebody kissed them and hugged them and gave them love and now they're in this pit." Ned responded, "Or, maybe someone didn't." "Yeah," Phil nodded, "I've thought about that too" (p. 91).

3. Dating. As a lesbian in real life, Norah Vincent had dated women. But dating women as Ned provided unique insights into male/female relationships. Ned found his dates at bars and on the Internet.

Ned found that most of the women he dated "were carrying the baggage of previous hurts at the hands of men, which in many cases had prejudiced them unfairly against the male sex" (p. 100). Ned described that as a "man" dating women,

> . . . I often felt attacked, judged, on the defensive. Whereas with the men I met and befriended as Ned there was a presumption of innocence—that is, you're a good guy until you prove otherwise—with women there was quite often a presumption of guilt: you're a cad like every other guy until you prove otherwise. (p. 101)

Ned experienced the difficulty that heterosexual men have in living up to the ideals of women, observing that "women wanted a take-control man, at the same time, they wanted a man who was vulnerable to them, a man who would show his colors and open his doors, someone expressive, intuitive, attuned" (p. 111). Ned noted, ". . . a man is expected to be modern . . . to support feminism . . . to see and treat women as equals in every respect . . . but," Ned continued, "he is on the other hand often still expected to be traditional at the same time, to treat a lady like a lady, to lead the way and pick up the check." (p. 112)

4. A Catholic Monastery. In the weeks Ned spent at a Catholic monastery, he observed men in a context in which women and sexual expression were, at least in principle, excluded from the male experience. Even in this context, however, male gender role expectations were evident, as in the following excerpt:

> . . . I made the mistake once of referring to one of the other monks as cute—the kind of thing that women say all the time about sweet elderly gentleman like the one to whom I was referring . . . But as soon as the offending remark was out of my mouth, Father Jerome pounced on it, sneering. "He's not cute. You don't call other men cute." (p. 144)

On another occasion, Ned was at the dinner table with the monks and he told Father Richard that he looked very good for his age.

> As soon as the remark came out of my mouth, everyone at the table stopped eating mid-forkful and looked at me as if I had three heads. Father Richard . . . said a very suspicious, squint-eyed "thank you," and looked away, clearly embarrassed. But the implication from other quarters was clear: "What the hell's wrong with you, kid? Don't you know that properly socialized males don't behave that way with each other?" (pp. 144–145)

5. Work. Ned interviewed for several jobs and landed a job as a salesman. One of his coworkers advised him, "You're a man. . . . You gotta pitch like a man." Ned learned that in sales work, women and men had different strategies.

> Girls pitched differently. They flirted. They cajoled and smiled and eased their way into the sales underhandedly, which was exactly how I'd started out trying to do it. I'd tried initially to ask for the sale the way I asked for food in a restaurant as a woman, or the way I asked for help at a gas station—pleadingly. But coming from a man this was off-color. It didn't work. It bred contempt in both men and women. . . . People see weakness in a woman and they want to help. They see weakness in a man and they want to stamp it out. (p. 213)

6. Men's Movement Group. Ned attended a men's movement group in which 25–30 guys met once a month. He noted that hugging was a central part of the group:

> Most men don't tend to share much physical affection with their male friends, so here the guys made a point of hugging each other long and hard at every possible opportunity as a way of offsetting what the world had long deprived them of, and what they in turn had been socialized to disallow themselves. (p. 232)

Based on his experiences in the men's group, Ned arrived at the following conclusion:

> . . . these guys were trying to find the love that their fathers had been unable to give them, or possibly the love that the entire culture had conspired to keep men from giving each other. . . . They had a profound need for other men's love. . . . They needed a man's affection and respect, a man's approval, and a man's shared perspective on their feelings. Having a mother's or a woman's love just wasn't and could never be the same. It couldn't fill the hole. (p. 241)

Concluding Thoughts

In the concluding chapter of *Self-Made Man,* Norah explains that the experience of being a guy was not what she had expected.

> I had thought that by being a guy I would get to do all the things I didn't get to do as a woman, things I'd always envied about boyhood when I was a child: the perceived freedoms of being unafraid in the world. . . . But when it actually came to the business of being Ned I rarely felt free at all. (pp. 275–276)

As Ned, Norah found that "somebody is always evaluating your manhood. Whether it's other men, other women, even children. And everybody is always on the lookout for your weakness or your inadequacy" (p. 276). Norah described the male role as a "straightjacket . . . that is no less constrictive than its feminine counterpart. You're not allowed to be a complete human being. Instead you get to be a coached jumble of stoic poses. You get to be what's expected of you" (p. 276).

In conclusion, Norah Vincent's study challenges the notion that white men have all the advantages. "White manhood in America isn't the standard anymore by which women and all other minorities are being measured and found wanting. . . . It's just another set of marching orders, another stereotype to inhabit." Although the Woman's Movement has largely liberated U.S. women from the stereotypical traditional female role, Norah Vincent notes that "men haven't had their movement yet. . . . And they're due for it, as are the women who live with, fight with, take care of and love them" (p. 287).

Source: Norah Vincent. 2006. *Self-Made Man.* New York: Viking Penguin. Used by permission.

and thus had more power in the decision-making process. It should be noted, however, that more than half of women's work is unpaid and thus is not viewed as household income. When the value of household work is included in income totals, women contribute between 40 percent and 60 percent of household income (United Nations Population Fund 2003). Similarly, an investigation of household decision making in 30 developing countries revealed that women's influence in the decision-making process is greater if she controls some of the income and assets. The study concluded that "in only ten of the countries surveyed did 50 percent or more of women participate in all household decisions, including those taken in regard to their own healthcare, major household purchases, daily household spending, and visits with family or relatives outside of the household" (UNICEF 2007, p. 16).

Gender inequality also has an effect on relationships in the United States. For example, negotiating work and home life can be a source of relationship problems. Whereas men in traditional versus dual-income relationships are more likely to report being satisfied with household task arrangements, women in dual-income families are the most likely to be dissatisfied with household task arrangements (Baker, Krigen, & Riley 1996). Furthermore, performing an inequitable proportion of the housework and child care results in higher levels of anger among married women than among men (Ross & Van Willigen 1996).

Although the contribution of men to household tasks has increased over the last several decades (Hook 2006), research indicates that men have more leisure time than women. For example, Mattingly and Bianchi (2003) examined the quantity and quality of free time experienced by women and men in the United States. They concluded that men experience 30 minutes more free time per day than women. A difference of 30 minutes might seem small, but consider that it translates into men having 164 hours more free time per year than women, which is equivalent to 4 weeks of paid vacation. Additionally, men were more likely to have pure free time rather than contaminated free time, i.e., free time contaminated by nonleisure activities (e.g., watching television while cooking or taking care of children).

We must also consider, of course, the practical difficulties of raising a family, having a career, and maintaining a happy relationship with a significant other. In one survey more than 80 percent of both men and women responded that changing gender roles makes it more difficult to have a successful marriage (Morin & Rosenfeld 2000). Successfully balancing work, marriage, and children may require a number of strategies, including (1) a mutually satisfying distribution of household labor, (2) rejection of stereotypical roles such as "super mom" and "breadwinner dad," (3) seeking outside help from others (e.g., child care providers or domestic workers), and (4) a strong commitment to the family unit. Finally, violence in relationships is gender specific (see Chapters 4 and 5). Although men are more likely to be victims of violent crime, women are more likely to be victims of rape and domestic violence. Violence against women reflects male socialization that emphasizes aggression and dominance over women. Male violence is a consequence of gender socialization and a definition of masculinity that holds that "as long as nobody is seriously hurt, no lethal weapons are employed, and especially within the framework of sports and games—football, soccer, boxing, wrestling—aggression and violence are widely accepted and even encouraged in boys" (Pollack 2000b, p. 40).

STRATEGIES FOR ACTION: TOWARD GENDER EQUALITY

In recent decades there has been an increasing awareness of the need to increase gender equality throughout the world. Strategies to achieve this end have focused on empowering women in social, educational, economic, and political spheres and improving women's access to education, nutrition, and health care and basic human rights. But as we will see in the following section on grassroots movements, there is also a men's movement that is concerned with gender inequities and issues facing men.

Grassroots Movements

Efforts to achieve gender equality in the United States have been largely fueled by the feminist movement. Despite a conservative backlash, feminists, and to a lesser extent men's activist groups, have made some gains in reducing structural and cultural sexism in the workplace and in the political arena.

Feminism and the Women's Movement. Feminism is the belief that women and men should have equal rights and responsibilities. The U.S. feminist movement began in Seneca Falls, New York, in 1848 when a group of women wrote and adopted a women's rights manifesto modeled after the Declaration of Independence. Although many of the early feminists were primarily concerned with suffrage, feminism has its "political origins . . . in the abolitionist movement of the 1830s," when women learned to question the assumption of "natural superiority" (Anderson 1997, p. 305). Early feminists were also involved in the temperance movement, which advocated restricting the sale and consumption of alcohol, although their greatest success was the passing of the Nineteenth Amendment in 1920, which recognized women's right to vote.

The rebirth of feminism almost 50 years later was facilitated by a number of interacting forces: an increase in the number of women in the labor force, the publication of Betty Friedan's book *The Feminine Mystique,* an escalating divorce rate, the socially and politically liberal climate of the 1960s, student activism, and the establishment of the Commission on the Status of Women by John F. Kennedy. The National Organization for Women (NOW) was established in 1966 and remains the largest feminist organization in the United States, with more than 500,000 members in 550 chapters across the country. One of NOW's hardest fought battles is the struggle to win ratification of the equal rights amendment (ERA), which states that "equality of rights under the law shall not be denied or abridged by the United States, or by any state, on account of sex." The proposed amendment to the Constitution passed both the House of Representatives and the Senate in 1972 but failed to be ratified by the required 38 states by the 1979 deadline, later extended to 1982. Several lawmakers, however, have introduced legislation that would remove time limits on the ratification process. Should this legislation pass, only 3 more states would be needed to ratify the amendment, which, to date, has been ratified in 35 states. The National Council of Women's Organizations, representing more than 200 organizations with a collective membership of more than 12 million members, has initiated a "3-state strategy" which has "sparked the current vigorous and rapidly growing campaign to achieve the ERA" (ERA 2007, p. 1).

> **"**Feminism is a socialist, anti-family, political movement that encourages women to leave their husbands, kill their children, practice witchcraft, destroy capitalism and become lesbians.**"**
>
> Pat Robertson
> Author and evangelist

feminism The belief that men and women should have equal rights and responsibilities.

Supporters of the ERA argue that its opponents used scare tactics—saying that the ERA would lead to unisex bathrooms, mothers losing custody of their children, and mandatory military service for women—to create a conservative backlash. Susan Faludi, in *Backlash: The Undeclared War against American Women* (1991), contended that contemporary arguments against feminism are the same as those levied against the movement 100 years ago and that the negative consequences predicted by opponents of feminism (e.g., women unfulfilled and children suffering) have no empirical support.

Today, a new wave of feminism is being led by young women and men who grew up with the benefits won by their mothers, but who are shocked by the stoning to death of a woman accused of adultery in Afghanistan, the continuing practice of female genital cutting in at least 25 countries, the fact that millions of women throughout the world lack access to modern contraception, and the fact that, even in one of the "freest" countries in the world, U.S. women face increasing restrictions on abortion, can be denied prescription contraception by a pharmacist who morally objects, earn less than men earn for doing the same job, and experience alarming rates of date rape and intimate partner abuse. Young feminists also grapple with many of the issues their mothers faced, in particular, how to balance work and family life (Orenstein 2001). These young feminists are more inclusive than their predecessors, welcoming all who champion the cause of global equality. Not surprisingly, the new feminists are likely to attract a more diverse group of supporters than their predecessors because future feminist efforts focus on gender equality rather than "gender sameness" (Parker 2000). Some observers, however, note that the new diversity of the women's movement may contribute to "a tension within feminism between the felt urgency to present concerns and grievances as a single unified group of women, and the need to give voice to the variations in concerns and grievances that exist among feminists on the basis of race and ethnicity, social class, sexual orientation, age, physical ability/disability, and a host of other factors" (Renzetti & Curran 2003, p. 25).

The Men's Movement. As a consequence of the women's rights movement, men began to re-evaluate their own gender status. As with any grassroots movement, the men's movement has a variety of factions. One of the early branches of the men's movement is known as the mythopoetic men's movement, which began after the publication of Robert Bly's (1990) *Iron John*—a fairy tale about men's wounded masculinity that was on the *New York Times* bestseller list for more than 60 weeks (Zakrzewski 2005). Participants in the mythopoetic men's movement meet in men-only workshops and retreats to explore their internal masculine nature, male identity, and emotional experiences through the use of stories, drumming, dance, music, and discussion. These groups are concerned with improving father-son and male friendship relationships, with overcoming homophobia that interferes with such relationships, and with exploring male forms of intimacy and emotion. A participant observation study of a mythopoetic men's group found that participants, in general, are white middle-class men who feel that they have little emotional support; question relationships with their fathers and sons; and are overburdened by responsibilities, unsatisfactory careers, and the demands of an overly competitive society (Schwalbe 1996).

The men's movement also includes men's organizations that advocate gender equality and work to make men more accountable for sexism, violence, and homophobia. For example, the National Organization for Men Against Sexism

The men's movement grew out of a reaction to the women's movement and a need to explore men's roles. In general, most men's movements support the goals of seeking improvement in marital relationships, supporting feminism and gender equality, improving interaction with other males, and engaging in more quality interaction with their children.

© Phil Schermeister/Corbis

(NOMAS) was founded in 1975 to promote a perspective for enhancing men's lives that is pro-feminist, gay-affirmative, and anti-racist and that is committed to justice on a broad range of social issues including class, age, religion, and physical abilities (NOMAS 2007). Pro-feminist men's groups besides NOMAS include the national network of Men's Resource Centers.

Other men's groups oppose feminism and view the feminist agenda as an organized form of male bashing. The agenda of antifeminist men's groups is to maintain and promote traditional gender ideology and roles. For example, the Promise Keepers, part of a Christian men's movement, and Louis Farrakhan's Nation of Islam, have often been criticized as patriarchal and antifeminist (Renzetti & Curran 2003).

Some men's groups focus on issues concerning children and fathers' rights. Groups such as the American Coalition for Fathers and Children, Dads Against Discrimination, the National Coalition of Free Men (NCFM), and Fathers4Justice are attempting to change the social and legal bias against men in divorce and child custody decisions, which tend to favor women. Fathers4Justice, a fathers' rights group in Great Britain and Canada, has received international media attention by engaging in publicity stunts and acts of civil disobedience. In one such incident member Jason Hatch dressed as Batman, climbed over the fence around Buckingham Palace, and perched on a palace balcony.

Other concerns on the agenda of some men's rights groups include the domestic violence committed against men by women, false allegations of child sexual abuse, wrongful paternity suits, and the oppressive nature of restrictive masculine gender norms. Just as women have fought against being oppressed by expectations to conform to traditional gender stereotypes, men are beginning to want the same freedom from traditional gender expectations. For example, men who enter nontraditional work roles, such as nurse and primary school teacher, are often stigmatized for participating in "feminine" work. A study of men in nontraditional work roles found that these men commonly experience embarrassment, discomfort,

Strategies for Action: Toward Gender Equality   421

shame, and disapproval from friends and peers (Simpson 2005). Just as the women's movement has fought for women's right to work in male-dominated professions, some men's rights activists want men to be able to work as a nurse, stay home with the kids, order quiche in a restaurant, and drink a frozen piña colada in a bar without being ridiculed for violating male gender norms.

Public Policy

A number of statutes have been passed to help reduce gender inequality. They include the Equal Pay Act of 1963, Title VII of the Civil Rights Act of 1964, Title IX of the Educational Amendments Act of 1972, the Pregnancy Discrimination Act of 1978, the Family and Medical Leave Act of 1993, the Victims of Trafficking and Violence Protection Act of 2000, and the Violence Against Women Reauthorization Act of 2005. Further, there are several relevant bills pending including the Local Law Enforcement Hate Crimes Prevention Act of 2007 which, for the first time, will protect individuals against hate-motivated violence based on gender and gender identity (NOW 2007c). Recently, public policy has focused on two issues—sexual harassment and affirmative action.

Sexual Harassment. Sexual harassment is a form of sex discrimination that violates Title VII of the 1964 Civil Rights Act. The U.S. EEOC (2007b) defines **sexual harassment** in the workplace as "unwelcome sexual advances, requests for sexual favors, and other verbal or physical conduct of a sexual nature when this conduct affects an individual's employment, unreasonably interferes with an individual's work performance, or creates an intimidating, hostile, or offensive work environment" (p. 1). Sexual harassment can be of two types: (1) *quid pro quo,* in which an employer requires sexual favors in exchange for a promotion, salary increase, or any other employee benefit, and (2) the existence of a hostile environment that unreasonably interferes with job performance, as in the case of sexually explicit comments or insults being made to an employee. Common examples of sexual harassment include unwanted touching, the invasion of personal space, making sexual comments about a person's body or attire, and sexual joke telling (Uggen & Blackstone 2004). A newer form of sexual harassment is Internet harassment, in which sexually explicit e-mails are sent to coworkers or images of an individual are posted on a website (Barak 2005).

The victim as well as the harasser may be a woman or a man, although adult women are the most frequent targets of sexual harassment (Uggen & Blackstone 2004). Not surprisingly, women who work in male-dominated occupations and blue-collar jobs are more likely to experience sexual harassment (Jackson & Newman 2004).

In 1998 the U.S. Supreme Court extended protection to victims of same-sex harassment. The victim of sexual harassment does not have to be the person harassed but could be anyone affected by the offensive conduct. In 2005 the California Supreme Court ruled (in *Miller v. Department of Corrections*) that a supervisor who has a consensual affair with a subordinate can be sued for sexual harassment by an employee who is not involved in the affair if the employee can demonstrate that widespread sexual favoritism was severe or pervasive enough to alter his or her working conditions and create a hostile working environment (Ison 2005).

Sexual harassment occurs in a variety of settings, including the workplace, public schools, military academies, and college campuses. For example, an in-

sexual harassment In reference to workplace harassment, when an employer requires sexual favors in exchange for a promotion, salary increase, or any other employee benefit and/or the existence of a hostile environment that unreasonably interferes with job performance.

dependent commission found that university officials knew of accusations of sexual harassment by members of the University of Colorado football program since 1997 and did little about them until victims went public in 2004 (National Women's Law Center 2004).

What Do You Think? Although the EEOC definition of sexual harassment typically refers to workplace harassment, it can occur in a variety of settings. A recently released report from the American Association of University Women titled *Drawing the Line: Sexual Harassment on Campus* (AAUW, 2006) surveyed more than 2,000 U.S. undergraduate college students. The results indicate that (1) two-thirds of undergraduates report having been the victim of sexual harassment, (2) females were more likely to say they were upset, ashamed, embarrassed, or humiliated as a result of the harassment than males, (3) a majority of the harassers admitted to thinking that sexual harassment is funny, and (4) although few of the students reported the harassment, nearly half wish there was an office or person to contact about the unwanted advances. Have you ever been the victim of sexual harassment? Did you report it and, if so, to whom? What would you do to reduce the incidents of sexual harassment on your campus?

The EEOC suggests that prevention is the best tool to eliminate sexual harassment. Schools and workplaces are encouraged to (1) adopt the policy that sexual harassment will not be tolerated, (2) provide sexual harassment training to administrators and employees, and (3) establish complaint or grievance procedures and take immediate action when an employee or student complains of sexual harassment. Any person who experiences sexual harassment on the job can file a charge of discrimination at any EEOC office. In fiscal year 2004 the EEOC received 12,025 charges of sexual harassment; 15 percent were filed by males (EEOC 2007b).

Affirmative Action. As discussed in Chapter 9, **affirmative action** refers to a broad range of policies and practices to promote equal opportunity as well as diversity in the workplace and on campuses. Affirmative action policies, developed in the 1960s from federal legislation, require that any employer (universities as well as businesses) that receives contracts from the federal government must make "good faith efforts" to increase the number of female and other minority applicants. Such efforts can be made through expanding recruitment and training programs and by making hiring decisions on a nondiscriminatory basis. In an investigation of the effectiveness of various efforts to increase managerial diversity in the workplace Kalev, Kelly, and Dobbin (2006) concluded that

> Although inequality in attainment at work may be rooted in managerial bias and the social isolation of women and minorities, the best hope for remedying it may lie in practices that assign organizational responsibility for change. . . . Structures that

affirmative action A broad range of policies and practices to promote equal opportunity as well as diversity in the workplace and on campuses.

embed accountability, authority, and expertise (affirmative action plans, diversity committees and taskforces, diversity managers and departments) are the most effective means of increasing the proportions of white women, black women, and black men in private sector management. (p. 611)

People commonly associate affirmative action with racial and ethnic minorities; some people are not aware that affirmative action benefits women. Indeed, the largest category of affirmative action beneficiaries is women (Beeman, Chowdhry, & Todd 2000).

As discussed in Chapter 9, numerous legal battles have occurred to challenge affirmative action. In 1996 a California ballot initiative abolished racial and sexual preferences in government programs, which included state colleges and universities (Chavez 2000). Voters in Washington State passed a similar initiative in 1998, and in 1999 the governor of Florida signed an executive order ending that state's affirmative action program. However, in 2003 the U.S. Supreme Court held that universities have a "compelling interest" in a diverse student population and therefore may take minority status into consideration when making admissions decisions ("Major Rulings of the 2002–2003 Term," 2003). In 2006, Michigan voters approved Proposal 2 which "bans the use of affirmative action programs that give preferential treatment" on the basis of race, color, gender, ethnicity, or national origin in government contracts, employment, or higher education (Lewis 2007, p. 1). Several lawsuits contesting implementation of the legislation have been filed. Those in favor of equal opportunity fear that anti-affirmative action activists will place similar bans on other state ballots for the 2008 elections (Erickson 2007).

International Efforts

International efforts to address problems of gender inequality date back to the 1979 Convention on the Elimination of All Forms of Discrimination Against Women (CEDAW), often referred to as the International Women's Bill of Rights, adopted by the United Nations in 1979. More than 185 countries ratified the Women's Bill of Rights, including every country in Europe and South and Central America. The United States is the only industrialized country that has not ratified the document (Amnesty International 2007).

Another significant international effort occurred in 1995 when representatives from 189 countries adopted the Beijing Declaration and Platform for Action at the Fourth World Conference on Women sponsored by the United Nations. The platform reflects an international commitment to the goals of equality, development, and peace for women everywhere. The platform identifies strategies to address critical areas of concern related to women and girls, including poverty, education, health, violence, armed conflict, and human rights. Every 5 years the United Nations has sponsored a conference to monitor progress in addressing these concerns. At its 2005 conference delegates voted to encourage member nations to pay particular attention to problems of AIDS among women and girls, the status of women and girls in the West Bank and the Gaza Strip and Afghanistan, the trafficking of women and girls for prostitution, and the economic advancement of women (United Nations 2005). This chapter's *The Human Side* feature describes some of the trials and tribulations of women and girls all over the world.

In addition to international efforts, individual countries have instituted programs or policies designed to combat sexism and gender inequality. For ex-

66 Freedom, especially freedom for women, is more than the absence of oppression. It's the right to speak and vote and worship freely. Human rights require the rights of women. 99

Laura Bush
First Lady of the United States

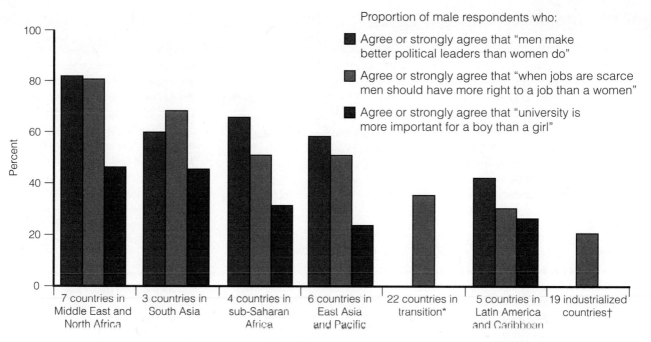

FIGURE 10.4
Men's Discriminatory Attitudes Toward Women by Region of the World.
Note: UNICEF calculations based on data from World Values Survey, 1999–2004.
*Countries such as Croatia, Albania, Latvia, etc. (incomplete data).
†Countries such as the United States, England, France, etc. (incomplete data).
Source: UNICEF 2007.

ample, Japan implemented the Basic Law for a Gender-Equal Society, a "blueprint for gender equality in the home and workplace" (Yumiko 2000, p. 41). However, a United Nations report ranked Japan behind all other industrialized countries in women's empowerment. For example, Japanese women earn just 44 percent of what men earn and are more than 90 percent of the 8 million part-time workers (Faiola 2007).

A South African Bill of Rights prohibits discrimination on the basis of, among other things, gender, pregnancy, and marital status (International Women's Rights Project 2000), and China has established the Program for the Development of Chinese Women, which focuses on empowering women in the areas of education, human rights, health, child care, employment, and political power (Women's International Network 2000). Finally, hailed as the most important legislation in gender equality in more than 30 years, the United Kingdom has recently passed the Gender Equality Duty Act which states that all officials in England, Scotland, and Wales "must demonstrate that they are promoting equality for women and men and that they are eliminating sexual discrimination and harassment" (EOC 2007).

UNDERSTANDING GENDER INEQUALITY

Gender roles and the social inequality they create are ingrained in our social and cultural ideologies and institutions and are therefore difficult to alter. For example, in almost all societies women are primarily responsible for child care and men are primarily responsible for military service and national defense (World Bank 2003). Nevertheless, as we have seen in this chapter, growing attention to gender issues in social life has spurred some change. Women who have traditionally been expected to give domestic life first priority are now finding it

Drawing from a number of sources, Plan International's report, *Because I Am a Girl* (2007), dramatically describes the plight of young women and girls in the world today. Although some of the quotes have been shortened, all of the information accurately reflects the original narratives. Where known, the speaker's name, age, and country of origin have been listed. Compare the experiences described below to that of young women and girls in the United States.

I never ever understand why boys and girls are not equal to each other. In rural areas elders think that girls are born to give birth and to marry and for cleaning the house. Girls who live in rural areas . . . are not sent to schools. Their parents are not aware of the changing world yet. (p. 5)

Girl, 15, Turkey

Khadja was my older sister. She died two years ago. She wasn't even 20 years old. . . . She was only 14 years when she married but all the girls in our community marry very young. . . . As the pregnancy advanced, my sister's husband wanted her to rest but my aunt refused, saying that Khadja was not the only woman who ever got pregnant. One day her waters broke when she was splitting wood. She carried on as if nothing had happened because she didn't understand what this meant. A couple of days later, Khadja had horrible pains. We did not take her to the hospital, which was far from the village. She died two days later, without anyone trying anything to save her. I think that the baby died inside her. My mother said this must have been meant to be, but deep down she has never accepted it and she still suffers. (p. 29)

Amadou, Mali

My parents used to think that I was their property. They used to abuse me, using

words which I cannot repeat without making me cry. (p. 33)

Girl, 13, Bangladesh

I am the one who does all the house work. . . . I do the cooking and take care of the household items. [My brother] just eats and goes outside to play. (p. 33)

Girl, 10, Ethiopia

We are five children, two older boys and three younger girls. But we the younger sisters are the housemaids every Saturday and Sunday while our two older brothers have no work during weekends because they thought that as boys, they have no responsibility to do household chores. (p. 34)

Barbie, 15, Philippines

In some of our communities, when a girl starts showing signs of maturity she becomes the focus in that community and the next thing to happen is to initiate her in the secret society without even asking her consent and finally giving her hand in marriage to whoever her family pleases. They don't consider the age. All they know is that their child is well matured and should be married. Sometimes the men they get married to are much older than their parents in fact, but as tradition demands they just have to obey. Often and again girl children are often forced to marry to chiefs especially if they are beautiful and reside in a village. (p. 44)

Konima, 18, Sierra Leone

To stop this inhuman attitude towards girls, there should be stringent laws against the practice of child marriages, and both the governments and the civil societies should initiate awareness raising campaigns at every community on gender equity and the evil consequences of child marriages. (p. 44)

Savitha, 14, India

Every girl child in some corner cries in silence because of marriage in early ages. You may not be among those who cry for freedom, those who wish to study further but are forced to marry an old ugly man for money. But I have seen and faced forced marriage, and I understand how horrible it is. I escaped a forced marriage, because I always went against my parents and society. Sometimes you just need to stand on your own. (p. 49)

Girl, 20, Sri Lanka

School is a good thing. If you go to school, you will become a female teacher, a minister. However many parents say that it is not good to send girls to school. . . . I have many things to do when I come back home even if I am tired. I sweep the floor, I go to buy things for my mother, and I play with my brother. I do not have much time to do my homework. (p. 51)

Ballovi Eliane, 10, Benin

Unfortunately, in my beautiful country Cameroon, potable water is not found everywhere. . . . In the villages, people trek for kilometers to fetch drinking water from wells or running streams. Worse still is the fact that this unworthy task is assigned to only young girls and women who are victims of gender discriminations. The young boy is privileged to have good education, while the girls go to fetch water from streams. One often sees them with big basins of water on their heads in the early mornings, afternoons and evenings to fetch water while the boys play football forgetting that they need water to take a bath. "After all" they say "why worry when God has blessed us with one or more sisters to relieve us of this task." Without water she will not be able to perform other household duties, such as laundry, cooking, and washing of dishes. In Bertoua, in my neighbourhood, after school or early in the morning, you will meet on your way a group of girls queuing up before the only village well waiting to fetch water. (p. 69)

Ida, 16, Bertoua

I married at age 12, before I even had my first period. I am from a lower caste family and I never attended school. We cannot afford nutritious food or a decent house to live in. I have three children—two daughters and a son. My last childbirth was especially difficult—I cannot describe for you how much I suffered during that time. I still feel weak and I look like an old woman. I have enormous awful days in my life. (p. 75)

Ganga, 19, Nepal

Girls are the ones that are involved in cleaning the surroundings and disposing rubbish. Most times girls contract deadly diseases while disposing waste. They contract diseases like cholera which kills in a short time. (p. 83)

Judith, 14, Zambia

Until recently, I had to work from 8 am until 10 pm each day. We get only two days off a month. I walk to work and back because I cannot afford to take a bus or bicycle rickshaw. The factory is three kilometers away and it takes 30 minutes to walk. I normally get home at 10.30 pm. . . . If we want to use the bathroom, we have to get permission from the supervisor and he monitors the time. If someone makes a mistake, the supervisor docks four or five hours of overtime wage, or lists her as absent, taking the whole day's wage. In my factory there is no childcare, no medical facilities. The women don't receive maternity benefits. We have to work overtime, but we are always cheated on our overtime pay . . . it is very crowded,

very hot and badly ventilated. The water we have to drink is dirty. The workers often suffer from diarrhea, jaundice, kidney problems, anemia, and eye pain. Our seats have no backs and since we have to work long hours, we suffer from backaches and shoulder pain. I cannot support myself with the wage I am getting. . . . Because we have to work very long hours, seven days a week, we have no family life, no personal life, no social life. (p. 88)

Nasrin Akther, Bangladesh

If I did something the employer didn't like, she would grab my hair and hit my head on the wall. She would say things like, "I don't pay you to sit and watch TV! You don't wash the dishes well. I pay your mother good money and you don't do anything [to deserve it]." . . . Once I forgot clothes in the washer, and they started to smell, so she grabbed my head and tried to stick it in the washing machine. (p. 95)

Saida B, 15, Casablanca, Morocco

Some relatives come from the cities to the village and take these girls to towns promising to help them learn a trade but end up introducing them into prostitution, forced labour and other illegal activities. This usually happens to girls who come from a poor family background. . . . Female relatives who are accomplices promised their parents that they will sponsor them in school or a trade, only to end up introducing the child into prostitution. Since the girls might not have the transport or means to return to the village they have no choice than to give in. (p. 101)

Violet, 15, and Martha, 16, Cameroon

I am 16 years old. One day, in March 2004, I was collecting firewood for my family when three armed men on camels came and surrounded me. They held me down

and tied my hands and raped me one after the other. When I arrived home, I told my family what happened. They threw me out of the home and I had to build my own hut away from them. I was engaged to a man and I was so much looking forward to getting married. After I got raped, he did not want to marry me and broke off the engagement because he said I was now disgraced and spoilt. It is the worst thing for me. . . . When I was eight months' pregnant from the rape, the police came to my hut and forced me with their guns to go to the police station. They asked me questions so I told them I had been raped. They told me that as I was not married I would deliver the baby illegally. They beat me with a whip on the chest and back and put me in jail. There were other women in jail who had the same story. During the day, we had to walk to the well four times a day to get the policemen water, clean and cook for them. At night I was in one small cell with 23 other women. I had no other food than what I could find during my work during the day. And the only water was what I drank at the well. I stayed 10 days in the jail and now I have to pay the fine, 20,000 Sudanese dinars ($65) (they asked me.) My child is now two months old. (p. 109)

Girl, 16, Darfur

Violence against children, especially girls, has crossed all limits. . . . People feel that a girl is meant to be used—either as a doormat, a maid, a birth-giving machine or as a source of physical pleasure. Something CONCRETE seriously needs to be done to change the current scenario because now a girl does not feel safe even in her own house, let alone the streets. (p. 115)

Girl, 16, India

Source: Plan International (2007).

Definitions of appropriate gender roles change over time. Fifty years ago, women playing professional basketball would have been unimaginable. Men's roles are also changing, although more slowly, as seen in this depiction of a men's locker room.

acceptable to be more ambitious in seeking a career outside the home. Men's roles are also changing. Men who have traditionally been expected to be aggressive and task oriented are now expected to be more caring and nurturing. Increasingly, both women and men are embracing **androgyny**—the blending of both masculine and feminine characteristics. The concept of androgyny implies that both masculine and feminine characteristics and roles are equally valued. However, "achieving gender equality . . . is a grindingly slow process, since it challenges one of the most deeply entrenched of all human attitudes" (Lopez-Claros & Zahidi 2005, p. 1). Although varying by region of the world, discriminatory attitudes continue to persist as illustrated in Figure 10.4 (UNICEF 2007). Further, in a survey of U.S. college freshmen, one in five (27 percent) agreed with the statement, "The activities of a married woman are best confined to the home and family" (Sax et al. 2004).

Although traditionally it has been women who have fought for gender equality, men also have much to gain. They too are victims of discrimination and gender stereotyping. Eliminating gender stereotypes and redefining gender in terms of equality does not mean simply liberating women but liberating men and our society as well. "What we have been talking about is allowing people to be more fully human and creating a society that will reflect that humanity. Surely that is a goal worth striving for" (Basow 1992, p. 359). Regard-

androgyny Having both feminine and masculine characteristics.

less of whether traditional gender roles emerged out of biological necessity, as the structural functionalists argue, or out of economic oppression, as the conflict theorists hold, or both, it is clear that, today, gender inequality carries a high price: poverty, loss of human capital, feelings of worthlessness, violence, physical and mental illness, and death. Perhaps we have reached a time when we need to ask ourselves, are the costs of gender inequality too high to continue to pay?

CHAPTER REVIEW

- **Does gender inequality exist worldwide?**

  There is no country in the world in which men and women are treated equally. Although women suffer in terms of income, education, and occupational prestige, men are more likely to suffer in terms of mental and physical health, mortality, and the quality of their relationships.

- **How do the three major sociological theories view gender inequality?**

  Structural functionalists argue that the traditional division of labor was functional for preindustrial society and over time has become defined as both normal and natural. Today, however, modern conceptions of the family have replaced traditional ones to some extent. Conflict theorists hold that male dominance and female subordination evolved in relation to the means of production—from hunting and gathering societies in which females and males were economic equals to industrial societies in which females were subordinate to males. Symbolic interactionists emphasize that through the socialization process both females and males are taught the meanings associated with being feminine and masculine.

- **What is meant by the terms *structural sexism* and *cultural sexism*?**

  Structural sexism refers to the ways in which the organization of society, and specifically its institutions, subordinate individuals and groups based on their sex classification. Structural sexism has resulted in significant differences between education and income levels, occupational and political involvement, and civil rights of women and men. Structural sexism is supported by a system of cultural sexism that perpetuates beliefs about the differences between women and men. Cultural sexism refers to the ways the culture of society—its norms, values, beliefs, and symbols—perpetuates the subordination of an individual or group because of the sex classification of that individual or group.

- **What is the difference between the glass ceiling and the glass elevator?**

  The glass ceiling is an often invisible barrier that prevents women and other minorities from moving into top corporate positions. The glass elevator, on the other hand, refers to the tendency for men seeking traditionally female jobs to have an edge in hiring and promotion practices.

- **What are some of the problems caused by traditional gender roles?**

  First is the feminization of poverty. Women are socialized to put family ahead of education and careers, a belief that is reflected in their less prestigious occupations and lower incomes. Second are social-psychological and health costs. Women tend to have lower self-esteem and higher rates of depression and eating disorders than men. Men, on average, die about 5 years earlier than women and have higher rates of cirrhosis of the liver, most cancers, homicide, drug- and alcohol-induced deaths, suicide, and firearm and motor vehicle accidents. Last, gender inequality influences the nature of relationships. For example, although men are more likely to be victims of violent crime, women are more likely to be victims of domestic violence.

- **What strategies can be used to end gender inequality?**

  Grassroots movements, such as feminism and the women's rights movement and the men's rights movement, have made significant inroads in the fight against gender inequality. Their accomplishments, in part, have been the result of successful lobbying for passage of laws concerning sex discrimination, sexual harassment, and affirmative action. Besides these national efforts, international efforts continue as well. One of the most important is the Convention to Eliminate All Forms of Discrimination Against Women (CEDAW), also known as the International Women's Bill of Rights, which was adopted by the United Nations in 1979.

TEST YOURSELF

1. The United States is the most gender equal nation in the world.
   a. True
   b. False
2. Symbolic interactionist argue that
   a. male domination is a consequence of men's relationship to the production process
   b. gender inequality is functional for society
   c. gender roles are learned in the family, in the school, in peer groups, and in the media
   d. women are more valuable in the home than in the workplace
3. Recent trends indicate that more males enter college from high school than females.
   a. True
   b. False
4. The devaluation hypothesis argues that female–male pay differences are a function of women's and men's different levels of education, skills, training, and work experience.
   a. True
   b. False
5. The glass ceiling is an often invisible barrier that prevents women and other minorities from moving into top corporate positions.
   a. True
   b. False
6. Which of the following has the highest percentage of U.S. female politicians?
   a. U.S. House of Representatives
   b. U.S. Senate
   c. Governor offices
   d. State legislatures
7. The Global Media Monitoring project found that
   a. The number of female reporters has decreased in recent years.
   b. Women are dramatically underrepresented as news subjects.
   c. Older women were more likely to appear on television than younger women.
   d. Male journalists are more likely to work on stories dealing with social issues than women journalists.
8. The feminization of poverty refers to
   a. the tendency for women to be caretakers of the poor
   b. feminists' criticism of public policy on poverty
   c. the disproportionate number of women who are poor
   d. gender role socialization of poor women
9. *Quid pro quo* sexual harassment refers to the existence of a hostile working environment.
   a. True
   b. False
10. Feminists would argue which of the following?
    a. The ERA should be part of the Constitution
    b. The passage of the Nineteenth Amendment was a mistake
    c. Occupational sex segregation benefits everyone
    d. NOMAS is a radical, sexist organization

Answers: 1 b. 2 c. 3 b. 4 b. 5 a. 6 d. 7 b. 8 c. 9 b. 10 a.

KEY TERMS

affirmative action
androgyny
boy code
comparable worth
cultural sexism
devaluation hypothesis
double or triple (multiple) jeopardy
emotion work

expressive roles
feminism
gender
gender tourism
glass ceiling
glass elevator effect
human capital hypothesis
instrumental roles

occupational sex segregation
pink-collar jobs
sex
sexism
sexual harassment
structural sexism
structured choice

MEDIA RESOURCES

Understanding Social Problems, **Sixth Edition**
Companion Website

academic.cengage.com/sociology/mooney
Visit your book companion website, where you will find flash cards, practice quizzes, Internet links, and more to help you study.

 Just what you need to know NOW! Spend time on what you need to master rather than on information you already have learned. Take a pre-test for this chapter, and CengageNOW will generate a personalized study plan based on your results. The study plan will identify the topics you need to review and direct you to online resources to help you master those topics. You can then take a post-test to help you determine the concepts you have mastered and what you will need to work on. Try it out! Go to academic.cengage .com/login to sign in with an access code or to purchase access to this product.

11

Issues in Sexual Orientation

The Global Context: A World View of Laws Pertaining to Homosexuality | Homosexuality and Bisexuality in the United States: A Demographic Overview | The Origins of Sexual Orientation Diversity | Sociological Theories of Sexual Orientation Issues | Heterosexism, Homophobia, and Biphobia | Discrimination Against Sexual Orientation Minorities | Strategies for Action: Reducing Antigay Prejudice and Discrimination | Understanding Issues in Sexual Orientation | Chapter Review

"Homophobia alienates mothers and fathers from sons and daughters, friend from friend, neighbor from neighbor, Americans from one another. So long as it is legitimated by society, religion, and politics, homophobia will spawn hatred, contempt, and violence, and it will remain our last acceptable prejudice."

Byrne Fone

As a junior and senior at Homewood-Flossmoor High School in the suburbs of Chicago, Myka Held played a key role in leading a campaign to promote tolerance of gay and lesbian students. The campaign involved selling gay-friendly t-shirts to students and teachers and having as many people as possible wear the t-shirts to school on a designated day. The t-shirts say, "gay? fine by me." The shirts were purchased through a nonprofit organization (http://www.finebyme.org) that was started by students at Duke University. "I think it's really important for gay people out there to know that there are straight people who support them," Ms. Held said (quoted in Puccinelli 2005). "I have always supported equal rights for every person and have been disgusted by discrimination and prejudice. As a young Jewish woman, I believe it is my duty to stand up and support minority groups. . . . In my mind, fighting for gay rights is a proxy for fighting for every person's rights" (Held 2005). More than 100 students and 3 brave teachers wore the t-shirts in the first year of the campaign, and in the second year more than 200 students showed their support by buying and wearing the t-shirts. "Since shirt day, I have become more interested in politics and seeing how just one or two people can make a difference," said Held. "What I did was on a small scale, it's true, but the difference . . . I made in opening the minds of some students is important" (Held 2005).

Fine By Me, photo by Michael Yaksich

At the Homewood-Flossmoor high school in the suburbs of Chicago, Myka Held was involved in a campaign to promote gay tolerance by selling t-shirts that say, "gay? fine by me" and having students and teachers wear them on the same day at school. The "gay? fine by me" t-shirt campaign was started by students at Duke University.

As the opening vignette illustrates, fighting prejudice and discrimination against sexual orientation minorities is an issue not just for lesbians, gays, and bisexuals but for all those who value fairness and respect for human beings in all their diversity. In this chapter we examine prejudice and discrimination toward sexual orientation minorities—nonheterosexual individuals: homosexual (or gay) women (also known as lesbians), homosexual (or gay) men, and bisexual individuals.

It is beyond the scope of this chapter to explore how sexual diversity and its cultural meanings vary throughout the world. Rather, in this chapter we focus on Western conceptions of diversity in sexual orientation. The term **sexual orientation** refers to the classification of individuals as heterosexual, bisexual, or homosexual, based on their emotional and sexual attractions, relationships, self-identity, and lifestyle. **Heterosexuality** refers to the predominance of emotional and sexual attraction to individuals of the other sex. **Homosexuality** refers to the predominance of emotional and sexual attraction to individuals of the same sex, and **bisexuality** is emotional and sexual attraction to members of both sexes. Lesbians, gays, and bisexuals, sometimes referred to collectively as the **lesbigay population,** are considered part of a larger population referred to as **transgendered individuals.** Transgendered individuals include "a range of people whose gender identities do not conform to traditional notions of masculinity and femininity" (Cahill, Mitra, & Tobias 2002, p. 10). Transgendered individuals include not only homosexuals and bisexuals but also cross-dressers, transvestites, and

sexual orientation The classification of individuals as heterosexual, bisexual, or homosexual, based on their emotional and sexual attractions, relationships, self-identity, and lifestyle.

heterosexuality The predominance of emotional and sexual attraction to individuals of the other sex.

homosexuality The predominance of emotional and sexual attraction to individuals of the same sex.

bisexuality Emotional and sexual attraction to members of both sexes.

transsexuals. Cross-dressers are individuals, usually heterosexual men, who occasionally dress in the clothing of the other sex; transvestites are homosexual men who occasionally dress as women; and transsexuals are individuals who have undergone hormone treatment and sex-reassignment surgery to change their gender identity (McCammon, Knox, & Schacht 2004). Much of the current literature on the treatment and political and social agendas of the lesbigay population includes other members of the transgendered community; hence the term **LGBT** or **GLBT** is often used to refer collectively to lesbians, gays, bisexuals, and transgendered individuals.

After summarizing the legal status of lesbians and gay men around the world, we discuss the prevalence of homosexuality, heterosexuality, and bisexuality in the United States, review explanations for sexual orientation diversity, and apply sociological theories to better understand societal reactions to sexual diversity. Then, after detailing the ways in which nonheterosexuals are victimized by prejudice and discrimination, we end the chapter with a discussion of strategies to reduce antigay prejudice and discrimination.

What Do You Think? How is the experience of being a sexual orientation minority similar to the experience of being a racial or ethnic minority? How is it different?

The Global Context: A World View of Laws Pertaining to Homosexuality

Homosexual behavior has existed throughout human history and in most, perhaps all, human societies (Kirkpatrick 2000). A global perspective on laws and social attitudes regarding homosexuality reveals that countries vary tremendously in their treatment of same-sex sexual behavior—from intolerance and criminalization to acceptance and legal protection. At least 85 member states of the United Nations criminalize consensual same-sex behavior among adults (Ottosson 2007). In the majority of these countries, both male and female homosexuality are illegal; in 38 of these countries only male homosexuality is illegal. Legal penalties vary for violating laws that prohibit homosexual sexual acts. In many countries, homosexuality is punishable by prison sentences and/or corporal punishment, such as whipping or lashing. In some countries people found guilty of engaging in same-sex sexual behavior may receive the death penalty (see Table 11.1).

What Do You Think? Why do you think that in many countries male homosexuality is illegal, but female homosexuality is not?

In general, countries throughout the world are moving toward increased legal protection of sexual orientation minorities, as discrimination on the basis of

lesbigay population A collective term sometimes used to refer to lesbians, gays, and bisexuals.

transgendered individuals A range of people whose gender identities do not conform to traditional notions of masculinity and femininity, including homosexuals, bisexuals, cross-dressers, transvestites, and transsexuals.

LGBT or GLBT Terms often used to refer collectively to lesbians, gays, bisexuals, and transgendered individuals.

sexual orientation has become part of a broad international human rights agenda (Sanders 2007). At least 22 countries have national laws that ban various forms of discrimination against gays, lesbians, and bisexuals (International Gay and Lesbian Human Rights Commission 1999). In 1996 South Africa became the first country in the world to include in its constitution a clause banning discrimination based on sexual orientation. Fiji, Ecuador, and Portugal also have constitutions that ban discrimination based on sexual orientation (Sanders 2007). In 2007, the United Nations Humans Rights Council launched the *Yogyakarta Principles on the Application of International Law in Relation to Issues of Sexual Orientation and Gender Identity*. Adopted at a meeting of international legal experts in the Indonesian city of Yogyakarta, these groundbreaking principles propose international legal standards under which governments should end violence, abuse, and discrimination against lesbian, gay, bisexual, and transgendered people and ensure full equality.

In recent years legal recognition of same-sex relationships has become more widespread. In 2001 the Netherlands became the first country in the world to offer full legal marriage to same-sex couples. Same-sex married couples and opposite-sex married couples in the Netherlands are treated identically, with two exceptions. Unlike opposite-sex marriages, same-sex couples married in the Netherlands are unlikely to have their marriages recognized as fully legal abroad. Regarding children, parental rights will not automatically be granted to the nonbiological spouse in gay couples. To become a fully legal parent, the spouse of the biological parent must adopt the child. In 2003 Belgium legalized same-sex marriages. In 2005 Spain, Canada, and South Africa became the third, fourth, and fifth countries, respectively, to legalize same-sex marriage.

Other countries recognize same-sex **registered partnerships** or "civil unions," which are federally recognized relationships that convey most but not all the rights of marriage (some countries also offer registered partnerships or civil unions to opposite-sex couples). Federally recognized registered partnerships or civil unions for same-sex couples are available in Australia, Belgium, Brazil, Canada, Denmark, Finland, France, Germany, Great Britain, Greenland, Hungary, Iceland, Israel, Italy, New Zealand, Norway, Portugal, Spain, Sweden, and in Mexico City and the northeastern state of Coahuila, Mexico. As we discuss later in this chapter, a number of U.S. states give legal recognition to same-sex couples.

Human rights treaties and transnational social movement organizations have increasingly asserted the rights of people to engage in same-sex relations. International organizations such as Amnesty International, which resolved in

TABLE 11.1 Countries in Which Homosexual Acts Are Subject to the Death Penalty
• Afghanistan
• Iran
• Mauritania
• Pakistan
• Saudi Arabia
• Sudan
• United Arab Emirates
• Yemen
• Some parts of Nigeria, Somalia, and the Chechen Republic in Russia

Source: Ottosson (2006).

© AP/Wide World Photos

Gert Kasteel, left, and Dolf Pasker were among the world's first same-sex couples to marry legally under Dutch law after the Netherlands became the first country to allow same-sex marriages.

registered partnerships Federally recognized relationships that convey most but not all the rights of marriage.

On May 17, 2007, lesbian, gay, bisexual, and transgendered groups in more than 50 countries commemorated the International Day Against Homophobia, an initiative launched in 2005 that commemorates the day in 1990 when the World Health Organization removed homosexuality from its roster of disorders.

© Vincent Yu/AP Photos

1991 to defend those imprisoned for homosexuality, the International Lesbian and Gay Association (founded in 1978), and the International Gay and Lesbian Human Rights Commission (founded in 1990) continue to fight antigay prejudice and discrimination. The United Nations Human Rights Commission has proposed the Resolution on Sexual Orientation and Human Rights. This landmark resolution recognizes the existence of sexual orientation–based discrimination around the world, affirms that such discrimination is a violation of human rights, and calls on all governments to promote and protect the human rights of all people, regardless of their sexual orientation (International Gay and Lesbian Human Rights Commission 2004). Despite the worldwide movement toward increased acceptance and protection of homosexual individuals, the status and rights of lesbians and gays in the United States continues to be one of the most divisive issues in U.S. society.

HOMOSEXUALITY AND BISEXUALITY IN THE UNITED STATES: A DEMOGRAPHIC OVERVIEW

Before looking at demographic data concerning homosexuality and bisexuality in the United States, it is important to understand the ways in which identifying or classifying individuals as homosexual, gay, lesbian, and bisexual is problematic.

Sexual Orientation: Problems Associated with Identification and Classification

The classification of individuals into sexual orientation categories (e.g., gay, straight, bisexual, lesbian, homosexual, or heterosexual) is problematic for a number of reasons (Savin-Williams 2006). First, because of the social stigma associated with nonheterosexual identities, many individuals conceal or falsely portray their sexual orientation identities to protect themselves against prejudice and discrimination. But more important, distinctions among sexual orientation categories are simply not as clear-cut as many people would believe. Consider the early research on sexual behavior by Kinsey and his colleagues (1948, 1953), who found that although 37 percent of men and 13 percent of women had had at least one same-sex sexual experience since adolescence, few of the individuals reported exclusive homosexual behavior. These data led Kinsey to conclude that heterosexuality and homosexuality represent two ends of a sexual orientation continuum and that most individuals are neither entirely homosexual nor entirely heterosexual but fall somewhere along this continuum. In other words, most people are, to some degree, bisexual.

Sexual orientation classification is complicated by the fact that many people who are sexually attracted to or have had sexual relations with individuals of the same sex do not view themselves as homosexual or bisexual. For example, "research conducted across different cultures and historical periods (including present-day Western culture) has found that many individuals develop passionate infatuations with same-gender partners in the absence of same gender sexual desires . . . whereas others experience same-gender sexual desires that never manifest themselves in romantic passion or attachment" (Diamond 2003, p. 173).

Consider the findings of a national survey of U.S. adults that investigated (1) sexual attraction to individuals of the same sex, (2) sexual behavior with people of the same sex, and (3) homosexual self-identification (Michael et al. 1994). This survey found that 4 percent of women and 6 percent of men said that they are sexually attracted to individuals of the same sex, and 4 percent of women and 5 percent of men reported that they had had sexual relations with a same-sex partner after age 18. Yet less than 3 percent of the men and less than 2 percent of women in this study identified themselves as homosexual or bisexual (Michael et al. 1994). A more recent survey of more than 4,000 men in New York City found that, among sexually active men who also reported a sexual identity, 4 percent reported a gay identity but 12 percent reported same-sex sexual behavior in the past year; almost 10 percent who self-identified as straight had at least one sexual encounter with another man during the previous year (Pathela et al. 2006). Approximately 70 percent of straight-identified men who have sex with men reported being married; 10 percent of married men reported same-sex behavior during the preceding year.

A final difficulty in labeling a person's sexual orientation is that an individual's sexual attractions, behavior, and identity may change across time. For example, in a longitudinal study of 156 lesbian, gay, and bisexual youth, 57 percent consistently identified as gay/lesbian and 15 percent consistently identified as bisexual over a 1-year period, but 18 percent transitioned from bisexual to lesbian or gay (Rosario et al., 2006).

The Prevalence of Nonheterosexual Adults in the United States

Despite the difficulties inherent in categorizing individuals into sexual orientation categories, recent data reveal the prevalence of individuals who identify as lesbian, gay, or bisexual and of cohabiting same-sex couples living in the United

> "Although it is typically presumed that heterosexual individuals only fall in love with other-gender partners and gay/lesbian individuals only fall in love with same-gender partners, this is not always so."
>
> Lisa M. Diamond
> Psychologist

Residents of America's first gay and lesbian retirement community, Palms of Manasota, in Palmetto, Florida.

Palms of Manasota

TABLE 11.2 Sexual Identity of U.S. College Students*	
Heterosexual	94.1%
Gay/Lesbian	2.0%
Bisexual	2.4%
Transgender	0.1%
Unsure	1.4%
*Based on a sample of 94,806 students at 117 campuses.	
Source: American College Health Association (2007).	

States. Estimates of the number of gay, lesbian, and bisexual adults living in the United States range from 8.8 million (from the American Community Survey) to 15.3 million (from the market research firm Harris Interactive) (Human Rights Campaign 2007a).

As noted earlier, in the national survey by Michael et al. (1994), less than 3 percent of the men and less than 2 percent of women in this study identified themselves as homosexual or bisexual. Smith and Gates (2001) estimated that between 4 percent and 5 percent of the total U.S. adult population is gay. A 2004 national poll found that about 5 percent of U.S. high school students identify as lesbian or gay (Curtis 2004). And a national study of U.S. college students found that 4.4 percent identified either as gay, lesbian, or bisexual (see Table 11.2). An estimated 1–3 million Americans older than age 65 are gay, lesbian, bisexual, or transgendered—a number that is expected to double over the next quarter century (Cahill 2007).

Prevalence of Same-Sex Unmarried Couple Households in the United States

The 2000 census found that about 1 in 9 (or 594,000) unmarried-partner households in the United States involve partners of the same sex (Simmons & O'Connell 2003). Nationally, 51 percent of same-sex couples had male partners. About one-fifth (22.3 percent) of gay male couple households and one-third (34.3 percent) of lesbian couple households have children present in the home (Simmons & O'Connell 2003).

Data from the 2000 census also revealed that 99.3 percent of U.S. counties reported same-sex cohabiting partners, compared with 52 percent of counties in 1990 (Bradford et al. 2002). Same-sex couples are more likely to live in metropolitan areas than in rural areas. However, the largest proportional increases in the number of same-sex couples self-reporting in 2000 compared with 1990 came in rural, sparsely populated states. States with the highest percentage of same-sex unmarried partners of all coupled households include California, Massachusetts, Vermont, and New York (Bradford et al. 2002; Simmons & O'Connell 2003).

Why are data on the numbers of U.S. GLBT adults and couples relevant? Primarily these data are important because census numbers on the prevalence of GLBT adults and couples can influence laws and policies that affect gay individuals and their families. In anticipation of the 2000 census, the National Gay and Lesbian Task Force Policy Institute and the Institute for Gay and Lesbian Strategic Studies conducted a public education campaign urging people to "out" themselves on the 2000 census. The slogan was, "The more we are counted, the more we count" (Bradford et al. 2002, p. 3). "The fact that the Census documents the actual presence of same-sex couples in nearly every state legislative and U.S. congressional district means anti-gay legislators can no longer assert that they have no gay and lesbian constituents" (Bradford et al. 2002, p. 8).

THE ORIGINS OF SEXUAL ORIENTATION DIVERSITY

One of the prevailing questions regarding sexual orientation centers on its origin or "cause." Questions about the causes of sexual orientation are typically concerned with the origins of homosexuality and bisexuality. Because heterosexuality is considered normative and "natural," causes of heterosexuality are rarely considered.

In much of the biomedical and psychological research on sexual orientation an attempt is made to identify one or more causes of sexual orientation diversity. The driving question behind this research is, Is sexual orientation inborn or is it learned or acquired from environmental influences? Although a number of factors have been correlated with sexual orientation, including genetic factors, gender role behavior in childhood, and fraternal birth order, there is no single theory that can explain diversity in sexual orientation (McCammon et al. 2004).

Beliefs About What "Causes" Homosexuality

Aside from what "causes" homosexuality, sociologists are interested in what people *believe* about the "causes" of homosexuality. Most gays believe that homosexuality is an inherited, inborn trait. In a national study of homosexual men

FIGURE 11.1
Should Homosexuality Be Considered an Acceptable Alternative Lifestyle? based on views about origin of homosexuality.
Source: Saad (2007).

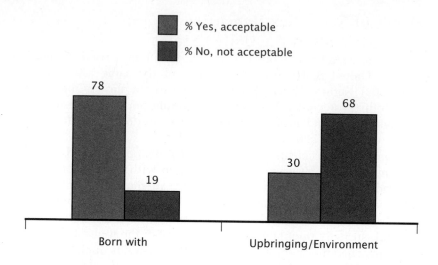

90 percent believe that they were born with their homosexual orientation; only 4 percent believe that environmental factors are the sole cause (Lever 1994). The percentage of Americans who believe that homosexuality is something a person is born with increased from 19 percent in 1989 to 42 percent in 2007 (Saad 2007).

Individuals who believe that homosexuality is biologically based tend to be more accepting of homosexuality (see Figure 11.1). In contrast, "those who believe homosexuals choose their sexual orientation are far less tolerant of gays and lesbians and more likely to conclude homosexuality should be illegal than those who think sexual orientation is not a matter of personal choice" (Rosin & Morin 1999, p. 8).

Can Homosexuals Change Their Sexual Orientation?

Individuals who believe that homosexuals choose their sexual orientation tend to think that homosexuals can and should change their sexual orientation. Various forms of **reparative therapy** or **conversion therapy** are dedicated to changing homosexuals' sexual orientation. Some religious organizations sponsor "ex-gay ministries," which claim to "cure" homosexuals and transform them into heterosexuals through prayer and other forms of "therapy." In recent years, the ex-gay movement has adopted a new approach, targeting youths viewed as exhibiting signs of "prehomosexuality" with both "preventive" measures and conversion techniques. Most reparative therapy programs allegedly achieve "conversion" to heterosexuality through embracing of evangelical Christianity and being "born again" (Cianciotto & Cahill 2007).

Consider the following examples:

- One church counselor told a gay woman that God could heal her and make her "whole again." The counselor laid hands on her to rid her of the demonic "spirit of homosexuality," gave her Bible verses to memorize, and worked on her "femininity" by teaching her how to apply makeup, encouraging her to grow her hair long, and instructing her to replace her old jeans, sweatpants, and gym shorts with skirts and dresses (Human Rights Campaign 2000).

- A 21-year-old gay man struggling with his sexual orientation reported that his strict religious family, upon learning of his orientation, threatened to

reparative therapy or conversion therapy Various therapies that are aimed at changing homosexuals' sexual orientation.

disown him and have him excommunicated unless he changed. He went into an ex-gay program voluntarily because he could not bear to lose his family, but other young men, he said, were forced into the program after being kidnapped. The program "counselors" strapped him to a chair with electrodes and sensors and showed him pictures of nude men, shocking him when he became aroused; they continued this treatment until he did not respond. He finally fled the program after being sexually abused by a male orderly. The experience was traumatizing and left him with feelings of self-hatred and fear and thoughts of suicide.

- Sixteen-year-old Zach was enrolled by his parents in a Christian camp-like program to change him into a heterosexual. The program, called Refuge, discourages homosexual behavior by imposing the following rules on its participants: no secular music, no more than 15 minutes per day behind a closed bathroom door, no contact with any practicing homosexual, no masturbation, and (no joke) no Calvin Klein underwear (Buhl 2005).

Critics of reparative therapy and ex-gay ministries take a different approach: "It is not gay men and lesbians who need to change . . . but negative attitudes and discrimination against gay people that need to be abolished" (Besen 2000, p. 7). The American Psychiatric Association, the American Psychological Association, the American Academy of Pediatrics, the American Counseling Association, the National Association of School Psychologists, the National Association of Social Workers, and the American Medical Association agree that homosexuality is not a mental disorder and needs no cure—that sexual orientation *cannot* be changed and that efforts to change sexual orientation do not work and may, in fact, be harmful (Human Rights Campaign 2000; Potok 2005). According to Lambda Legal (the nation's largest organization working for the civil rights of the LGBT population), ex-gay programs and conversion therapy practitioners can be held liable for the harm they cause to clients who can sue for malpractice, consumer fraud, false advertising, or child abuse and neglect laws for minors forced to attend an ex-gay program (Cianciotto & Cahill 2007).

Close scrutiny of reports of "successful" reparative therapy reveal that (1) many claims come from organizations with an ideological perspective on sexual orientation rather than from unbiased researchers, (2) the treatments and their outcomes are poorly documented, and (3) the length of time that clients are followed after treatment is too short for definitive claims to be made about treatment success (Human Rights Campaign 2000). Indeed, at least 13 ministries of Exodus International—the largest ex-gay ministry network that claims to include over 170 ex-gay programs in 17 countries—have closed because their directors reverted to homosexuality (Fone 2000). Michael Bussee, who in 1976 helped start Exodus International, said, "After dealing with hundreds of people, I have not met one who went from gay to straight. Even if you manage to alter someone's sexual behavior, you cannot change their true sexual orientation" (quoted by Holthouse 2005, p. 14). Bussee worked to help "convert" gay people for 3 years, until he and another male Exodus employee fell in love and left the organization.

Wayne Besen (2007) explained that "ex-gays" are individuals who are in denial about their sexual orientation. Such denial is reflected in the explanation of ex-gay ministries given by John Smid, leader of the ex-gay ministry Love in Action, to a reporter for *The Memphis Flyer*. Pointing to a gold-colored wall, Smid said: "I'm looking at that wall and suddenly I say it's blue. Someone else comes along and says, 'No, it's gold,' But I want to believe that wall is blue. Then

"Ex-gay ministries are one of the most elaborate and well-funded hoaxes ever perpetrated on the American people."

Wayne Besen
Author, activist

God comes along and He says, 'You're right, John, it's blue,' That's the help I need. God can help me make that wall blue" (cited in Besen 2007, p. 21).

SOCIOLOGICAL THEORIES OF SEXUAL ORIENTATION ISSUES

Sociological theories do not explain the origin or "cause" of sexual orientation diversity; rather, they help to explain societal reactions to homosexuality and bisexuality and ways in which sexual identities are socially constructed.

Structural-Functionalist Perspective

Structural functionalists, consistent with their emphasis on institutions and the functions they fulfill, emphasize the importance of monogamous heterosexual relationships for the reproduction, nurturance, and socialization of children. From a structural-functionalist perspective homosexual relations, as well as heterosexual nonmarital relations, are defined as "deviant" because they do not fulfill the main function of the family institution—producing and rearing children. Clearly, however, this argument is less salient in a society in which (1) other institutions, most notably schools, have supplemented the traditional functions of the family, (2) reducing (rather than increasing) population is a societal goal, and (3) same-sex couples can and do raise children.

Some structural functionalists argue that antagonisms between heterosexuals and homosexuals disrupt the natural state, or equilibrium, of society. Durkheim, however, recognized that deviation from society's norms can also be functional. As Durkheim observed, deviation "may be useful as a prelude to reforms which daily become more necessary" (Durkheim 1993, p. 66). Specifi-

Many same-sex couples rear children in a stable, nurturing environment, which negates the argument that homosexual relations do not fulfill the family institution's main function of producing and rearing children.

© AP Photo/Douglas C. Pizac

cally, the gay rights movement has motivated many people to reexamine their treatment of sexual orientation minorities and has produced a sense of cohesion and solidarity among members of the gay population (although bisexuals have often been excluded from gay and lesbian communities and organizations). Gay activism has been instrumental in advocating more research on human immunodeficiency virus (HIV) and acquired immunodeficiency syndrome (AIDS), more and better health services for HIV and AIDS patients, protection of the rights of HIV-infected individuals, and HIV/AIDS public education. Such HIV/AIDS prevention strategies and health services benefit the society as a whole.

The structural-functionalist perspective is concerned with how changes in one part of society affect other aspects. With this focus on the interconnectedness of society we note that urbanization has contributed to the formation of strong social networks of gays and bisexuals. Cities "acted as magnets, drawing in gay migrants who felt isolated and threatened in smaller towns and rural areas" (Button, Rienzo, & Weld 1997, p. 15). Given the formation of gay communities in large cities, it is not surprising that the gay rights movement first emerged in large urban centers.

Other research has demonstrated that the worldwide increase in liberalized national policies on same-sex relations and the lesbian and gay rights social movement has been influenced by three cultural changes: the rise of individualism, increasing gender equality, and the emergence of a global society in which nations are influenced by international pressures (Frank & McEneaney 1999).

Individualism "appears to loosen the tie between sex and procreation, allowing more personal modes of sexual expression" (p. 930).

> Whereas once sex was approved strictly for the purpose of family reproduction, sex increasingly serves to pleasure individualized men and women in society. This shift has involved the casting off of many traditional regulations on sexual behavior, including prohibitions of male-male and female-female sex. (p. 936)

Gender equality involves the breakdown of sharply differentiated sex roles, thereby supporting the varied expressions of male and female sexuality. Globalization permits the international community to influence individual nations. For example, when Zimbabwe president Robert Mugabe pursued antihomosexual policies in 1995, 70 members of the U.S. Congress signed a letter asking him to halt his antihomosexual campaign. Many international organizations and human rights associations joined the protest. The pressure of international opinion led Zimbabwe's Supreme Court to rule in favor of lesbian and gay groups' right to organize.

The structural-functionalist perspective is also concerned with latent functions, or unintended consequences. A latent function of the gay rights movement is increased opposition to gay rights. For example, after the Supreme Court decision to strike down a Texas sodomy law in 2003, public support (as measured by Gallup opinion polls) for gay rights dipped, suggesting that the Supreme Court ruling produced a backlash of public opposition to gay rights (Saad 2007).

Conflict Perspective

Conflict theorists, particularly those who do not emphasize a purely economic perspective, note that the antagonisms between heterosexuals and nonheterosexuals represent a basic division in society between those with power and those without power. When one group has control of society's institutions and resources, as in the case of heterosexuals, they have the authority to dominate other groups. The recent battle over gay rights is just one example of the political struggle between those with power and those without it.

A classic example of the power struggle between gays and straights took place in 1973 when the American Psychiatric Association (APA) met to revise its classification scheme of mental disorders. Homosexual activists had been appealing to the APA for years to remove homosexuality from its list of mental illnesses but with little success. The view of homosexuals as mentally ill contributed to their low social prestige in the eyes of the heterosexual majority. In 1973 the APA's board of directors voted to remove homosexuality from its official list of mental disorders. The board's move generated a great deal of resistance from conservative APA members and was put to a referendum, which resulted in affirmation of the board's decision (Bayer 1987).

More recently, gays and lesbians have been waging a political battle to win civil rights protections in the form of laws prohibiting discrimination on the basis of sexual orientation (discussed later in this chapter). Conflict theory helps to explain why many business owners and corporate leaders oppose civil rights protection for gays and lesbians. Employers fear that such protection would result in costly lawsuits if they refused to hire homosexuals, regardless of the reason for their decision. Business owners also fear that granting civil rights protections to homosexual employees would undermine the economic health of a community by discouraging the development of new businesses and even driving out some established firms (Button et al. 1997).

However, many companies are recognizing that implementing antidiscrimination policies that include sexual orientation is good for the "bottom line." The majority (88 percent) of Fortune 500 companies have included sexual orientation in their nondiscrimination policies, and employers are increasingly offering benefits to domestic partners of LGBT employees (Human Rights Campaign 2007a). Gay-friendly work policies help employers maintain a competitive edge in recruiting and maintaining a talented and productive workforce. Companies are also competing for the gay and lesbian dollar. GLBT consumers had an estimated total buying power of $660 billion in 2006. And a 2007 survey found that 88 percent of GLBT adults and 70 percent of heterosexual adults were likely "to consider a brand that is known to provide equal workplace benefits for all of their employees, including gay and lesbian employees" (cited in Human Rights Campaign 2007a, p. 9).

In summary, conflict theory frames the gay rights movement and the opposition to it as a struggle over power, prestige, and economic resources. Recent trends toward increased social acceptance of homosexuality may, in part, reflect the corporate world's competition for the gay and lesbian consumer dollar.

Symbolic Interactionist Perspective

Symbolic interactionism focuses on the meanings of heterosexuality, homosexuality, and bisexuality; how these meanings are socially constructed; and how they influence the social status and self-concepts of nonheterosexual individuals. The meanings we associate with same-sex relations are learned from society—from family, peers, religion, and the media. The negative meanings associated with homosexuality are reflected in the current slang use of the phrase "That's so gay" or "You're so gay," which is meant to convey that something or someone is stupid or worthless (Kosciw & Diaz 2005).

Historical and cross-cultural research on homosexuality reveals the socially constructed nature of homosexuality and its meaning. Although many Americans assume that same-sex romantic relationships have always been taboo in our society, during the 19th century "romantic friendships" between women were encouraged and were regarded as preparation for a successful marriage. The nature of these friendships bordered on lesbianism. President Grover Cleveland's sister Rose wrote to her friend Evangeline Whipple in 1890: "It makes me heavy with emotion . . . all my whole being leans out to you. . . . I dare not think of your arms" (Goode & Wagner 1993, p. 49).

The symbolic interactionist perspective also points to the effects of labeling on individuals. Once individuals become identified or labeled as lesbian, gay, or bisexual, that label tends to become their **master status.** In other words, the dominant heterosexual community tends to view "gay," "lesbian," and "bisexual" as the most socially significant statuses of individuals who are identified as such. Esterberg (1997) noted that "unlike heterosexuals, who are defined by their family structures, communities, occupations, or other aspects of their lives, lesbians, gay men, and bisexuals are often defined primarily by what they do in bed. Many lesbians, gay men, and bisexuals, however, view their identity as social and political as well as sexual" (p. 377).

Negative social meanings associated with homosexuality also affect the self-concepts of nonheterosexual individuals. **Internalized homophobia**—a sense of personal failure and self-hatred among lesbians and gay men resulting from social rejection and stigmatization—has been linked to increased risk for depression, substance abuse and addiction, anxiety, and suicidal thoughts (Gilman et al. 2001; Bobbe 2002). Unlike other marginalized minorities who have the support of their families and churches, lesbian and gay individuals are often rejected by their religion and sometimes by their families as well. When Reverend Mel White, a closeted gay Christian man who was nearly driven to suicide after two decades of struggling to save his marriage and his soul with "reparative therapies" finally came out to his mother, her response was, "I'd rather see you at the bottom of that swimming pool, drowned, than to hear this" (White 2005, p. 28).

Labels also affect heterosexuals' attitudes. When Gallup tested alternative terms for referring to gay Americans, it found that using the label "gays and lesbians" results in somewhat more favorable, pro-gay responses than does using the term "homosexual" (Saad 2005). The negative meanings associated with homosexuality are explored in the following section.

HETEROSEXISM, HOMOPHOBIA, AND BIPHOBIA

The United States, along with many other countries throughout the world, is predominantly heterosexist. **Heterosexism** refers to "the institutional and societal reinforcement of heterosexuality as the privileged and powerful norm" (SIECUS 2000). Heterosexism is based on the belief that heterosexuality is superior to homosexuality; it results in prejudice and discrimination against homosexuals and bisexuals. Prejudice refers to negative attitudes, whereas discrimination refers to behavior that denies individuals or groups equality of treatment. Before reading further, you may wish to complete this chapter's *Self and Society* feature, which assesses your behaviors toward individuals you perceive to be homosexual.

master status The status that is considered the most significant in a person's social identity.

internalized homophobia A sense of personal failure and self-hatred among some lesbians and gay men due to social rejection and stigmatization.

heterosexism The institutional and societal reinforcement of heterosexuality as the privileged and powerful norm, based on the belief that heterosexuality is superior to homosexuality.

This questionnaire is designed to examine which of the following statements most closely describes your behavior during past encounters with people you thought were homosexuals. Rate each of the following self-statements as honestly as possible by choosing the frequency that best describes your behavior: (1) never, (2) rarely, (3) occasionally, (4) frequently, or (5) always.

1. I have spread negative talk about someone because I suspected that he or she was gay. _____
2. I have participated in playing jokes on someone because I suspected that he or she was gay. _____
3. I have changed roommates and/or rooms because I suspected my roommate to be gay. _____
4. I have warned people who I thought were gay and who were a little too friendly with me to keep away from me. _____
5. I have attended antigay protests. _____
6. I have been rude to someone because I thought that he or she was gay. _____
7. I have changed seat locations because I suspected the person sitting next to me to be gay. _____
8. I have had to force myself to stop from hitting someone because he or she was gay and very near me. _____
9. When someone I thought to be gay has walked toward me as if to start a conversation, I have deliberately changed directions and walked away to avoid him or her. _____
10. I have stared at a gay person in such a manner as to convey to him or her my disapproval of his or her being too close to me. _____
11. I have been with a group in which one (or more) person(s) yelled insulting comments to a gay person or group of gay people. _____
12. I have changed my normal behavior in a restroom because a person I believed to be gay was in there at the same time. _____
13. When a gay person has "checked" me out, I have verbally threatened him or her. _____
14. I have participated in damaging someone's property because he or she was gay. _____

15. I have physically hit or pushed someone I thought was gay because he or she brushed his or her body against me when passing by. _____
16. Within the past few months, I have told a joke that made fun of gay people. _____
17. I have gotten into a physical fight with a gay person because I thought he or she had been making moves on me. _____
18. I have refused to work on school and/or work projects with a partner I thought was gay. _____
19. I have written graffiti about gay people or homosexuality. _____
20. When a gay person has been near me, I have moved away to put more distance between us. _____

Scoring: The revised Self-Report of Behavior Scale is scored by totaling the number of points endorsed on all items (never, 1; rarely, 2; occasionally, 3; frequently, 4; always, 5), yielding a range from 20 to 100 total points. The higher the score, the more negative are the attitudes toward homosexuals.

Comparison Data

The Self-Report of Behavior Scale was originally developed by Sunita Patel (1989) in her psychology master's thesis research at East Carolina University. College men (from a university campus and from a military base) were the original participants (Patel et al. 1995). The scale was revised by Shartra Sylivant (1992), who used it with a coed high school student population, and by Tristan Roderick (1994), who involved college students to assess the scale's psychometric properties. The scale was found to have high internal consistency. Two factors were identified: (1) a passive avoidance toward homosexuals and (2) active or aggressive reactions.

In the study of Roderick et al. (1998) the mean score for 182 college women was 24.76. The mean score for 84 men was significantly higher, 31.60. In Sylivant's (1992) high school sample, the mean scores were 33.74 for young women and 44.40 for young men.

The revised Self-Report of Behavior Scale is reprinted by the permission of the students and faculty who participated in its development: S. Patel, S. L. McCammon, T. E. Long, L. J. Allred, K. Wuensch, T. Roderick, and S. Sylivant.

Homophobia

homophobia Negative attitudes and emotions toward homosexuality and those who engage in same-sex sexual behavior.

The term **homophobia** is commonly used to refer to negative attitudes and emotions toward homosexuality and those who engage in it. Homophobia is not necessarily a clinical phobia (i.e., one involving a compelling desire to avoid the feared object despite recognizing that the fear is unreasonable). Other terms that refer to negative attitudes and emotions toward homosexuality include *homonegativity* and *antigay bias*.

In general, individuals who are more likely to have negative attitudes toward homosexuality are older, are Republican, attend religious services regularly, are less educated, live in the South or Midwest, and reside in small rural towns (Loftus, 2001; Curtis 2003; Page 2003; Saad 2007). Public opinion surveys also indicate that men are more likely than women to have negative attitudes toward gays (Saad 2007). But many studies on attitudes toward homosexuality do not distinguish between attitudes toward gay men and attitudes toward lesbians. Research that has assessed attitudes toward male versus female homosexuality has found that heterosexual women and men hold similar attitudes toward lesbians, but that men are more negative toward gay men (Louderback & Whitley 1997; Price & Dalecki 1998).

Cultural Origins of Homophobia. Why do many Americans disapprove of homosexuality? Antigay bias has its roots in various aspects of U.S. culture.

Religion. Most Americans who view homosexuality as unacceptable say that they object on religious grounds (Rosin & Morin 1999). Indeed, conservative Christian ideology has been identified as the best predictor of homophobia (Plugge-Foust & Strickland 2000). Many religious leaders teach that homosexuality is sinful and prohibited by God. The Roman Catholic Church rejects all homosexual expression and resists any attempt to validate or sanction the homosexual orientation. Some fundamentalist churches have endorsed the death penalty for homosexual people and teach the view that AIDS is God's punishment for engaging in homosexual sex. An organization of Christian fundamentalists claimed that the destruction brought on by Hurricane Katrina in 2005 was God's judgment against New Orleans for holding the annual gay Southern Decadence party (Curtis 2005).

The Westboro Baptist Church (Topeka, Kansas), headed by the antigay Reverend Fred Phelps, maintains a website called godhatesfags.com. Members of this church have held antigay demonstrations near the funerals of people who

Fred Phelps, pastor of Westboro Baptist Church, and his followers are picketing the funeral of a soldier killed in Iraq. Phelps says that God has punished America with improvised explosive devices for being nice to gays and lesbians.

have died from AIDS, carrying signs reading "Gays Deserve to Die." Phelps and his followers have also picketed funerals of American servicemen and servicewomen who have died in the Iraq war, saying that U.S. military casualties are God's way of punishing the United States for being nice to lesbians and gays (Intelligence Briefs 2006).

Theologians and religious scholars have different viewpoints on the Bible's position on homosexuality. Many scholars believe that key passages actually are denouncing orgies and prostitution—or in the case of the town of Sodom, inhospitality—and not homosexuality. Two Old Testament passages that do condemn homosexual acts are found amid a long list of religious prohibitions, including eating pork and wearing mixed fabrics—rules that have been abandoned by most contemporary Christians (Potok 2005).

Some religious groups, such as the Quakers and the United Church of Christ (UCC), are accepting of homosexuality, and other groups have made reforms toward increased acceptance of lesbians and gays. In the early 1970s the UCC became the first major Christian church to ordain an openly gay minister, and in 2005 the UCC became the largest Christian denomination to endorse same-sex marriages. Some Episcopal priests perform "ceremonies of union" between same-sex couples; some Reform Jewish groups sponsor gay synagogues (Fone 2000). In June 2003 the Reverend V. Gene Robinson was elected bishop of the Episcopal diocese of New Hampshire, becoming the first openly gay bishop in the church's history. Although the official position of the United Methodist Church is one that condemns homosexuality, some Methodist ministers advocate acceptance of and equal rights for lesbians and gay men (*The United Methodist Church and Homosexuality,* 1999). Acceptance of gays and lesbians is even found among Southern Baptists, one of the more conservative Christian religions. The Association of Welcoming and Affirming Baptists, organized in 1993, is a network of nearly 50 congregations and other church groups that advocate for the inclusion of LGBT individuals within the Baptist community of faith.

Marital and Procreative Bias. Many societies have traditionally condoned sex only when it occurs in a marital context that provides for the possibility of producing and rearing children. Although a handful of several states have granted legal recognition to same-sex couples, only one—Massachusetts—allows same-sex couples to be legally married. Furthermore, even though assisted reproductive technologies make it possible for gay individuals and couples to have children, many people believe that these advances should be used only by heterosexual married couples.

Concern About HIV and AIDS. Although most cases of HIV and AIDS worldwide are attributed to heterosexual transmission, the rates of HIV and AIDS in the United States are much higher among gay and bisexual men than among other groups. Lesbians, incidentally, have a very low risk for sexually transmitted HIV—a lower risk than heterosexual women. Nevertheless, many people associate HIV and AIDS with homosexuality and bisexuality. In 1987, 43 percent of U.S. adults agreed that "AIDS might be God's punishment for immoral sexual behavior." In 2007, just 23 percent agreed with this statement (Pew Research Center 2007). This association between AIDS and homosexuality has fueled antigay sentiments.

Rigid Gender Roles. Antigay sentiments also stem from rigid gender roles. When Cooper Thompson (1995) was asked to give a guest presentation on male roles

© AP/Wide World Photos

When the Reverend V. Gene Robinson was elected bishop of the Episcopal diocese in New Hampshire in 2003, he became the first openly gay bishop in the church's history.

at a suburban high school, male students told him that the most humiliating put-down was being called a fag. The boys in this school gave Thompson the impression that they were expected to conform to rigid, narrow standards of masculinity to avoid being labeled in this way.

From a conflict perspective heterosexual men's subordination and devaluation of gay men reinforces gender inequality. "By devaluing gay men . . . heterosexual men devalue the feminine and anything associated with it" (Price & Dalecki 1998, pp. 155–156). Negative views toward lesbians also reinforce the patriarchal system of male dominance. Social disapproval of lesbians is a form of punishment for women who relinquish traditional female sexual and economic dependence on men. Not surprisingly, research findings suggest that individuals with traditional gender role attitudes tend to hold more negative views toward homosexuality (Louderback & Whitley 1997).

Myths and Negative Stereotypes. Prejudice toward homosexuals can also stem from some of the myths and negative stereotypes regarding homosexuality. One negative myth about homosexuals is that they are sexually promiscuous and lack "family values," such as monogamy and commitment to relationships. Although some homosexuals do engage in casual sex, as do some heterosexuals, many homosexual couples develop and maintain long-term committed relationships. Between 64 percent and 80 percent of lesbians report that they are in a committed relationship at any given time, and between 46 percent and 60 percent of gay men report being in a committed relationship (Cahill et al. 2002). According to research, about one-third of gay male couples are monogamous, but most nonmonogamous gay couples are open with each other about their outside sexual activities (Advocate 2002, LaSala 2004, Shernoff 2006). Furthermore, although a majority of male couples are not sexually exclusive, they are emotionally monogamous—a concept one researcher coined as "monogamy of the heart" (LaSala 2005).

> **What Do You Think?** According to research on the topic, nonmonogamy generally is more accepted in the gay male subculture than in the heterosexual society or in the lesbian subculture. Why do you think this is so? Do you think the higher acceptance of nonmonogamy among gay males is explained by their sexual orientation? Or their sex and gender?

Another myth that is not supported by data is that homosexuals, as a group, are child molesters. In fact, 95 percent of all reported incidents of child sexual abuse are committed by heterosexual men (SIECUS 2000). Most often the abuser is a father, stepfather, or heterosexual relative of the family. Research cited by Cahill and Jones (2003) indicated that a child's risk of being molested by his or her relative's heterosexual partner is more than 100 times greater than the risk of being molested by someone who is homosexual or bisexual. Furthermore, "when a man abuses a young girl, the problem is not heterosexuality. . . . Similarly, when a priest sexually abuses a boy or underage teen, the problem is not homosexuality. The problem is child abuse" (Cahill & Jones 2003, p. 1).

Some of the negative myths and stereotypes about lesbians and gays can be traced to the antigay propaganda of Paul Cameron in the 1970s and 1980s. A former psychology instructor at University of Nebraska, Cameron began publishing pseudoscientific pamphlets "proving" that gay people commit more serial murders, molest more children, and intentionally spread diseases. Cameron's antigay pamphlets, distributed largely in fundamentalist churches, also claimed that gay people "were obsessed with consuming human excrement, allowing them to spread deadly diseases simply by shaking hands with unsuspecting strangers or using public restrooms" (Moser 2005, p. 14).

Biphobia

Just as the term *homophobia* is used to refer to negative attitudes toward gay men and lesbians, **biphobia** refers to "the parallel set of negative beliefs about and stigmatization of bisexuality and those identified as bisexual" (Paul 1996, p. 449). Although individuals identified as both homosexuals and bisexuals are often rejected by heterosexuals, women and men identified as bisexuals also face rejection from many homosexual individuals. Thus bisexuals experience "double discrimination."

Biphobia includes negative stereotyping of bisexuals, the exclusion of bisexuals from social and political organizations of lesbians and gay men, and fear and distrust of, as well as anger and hostility toward, people who identify themselves as bisexual (Firestein 1996). Individuals with negative attitudes toward bisexual individuals often believe that bisexuals are actually homosexuals who are in denial about their true sexual orientation or are trying to maintain heterosexual privilege (Israel & Mohr 2004). Bisexual individuals are sometimes viewed as heterosexuals who are looking for exotic sexual experiences. One negative stereotype that encourages biphobia is the belief that bisexuals are, by definition, nonmonogamous. However, many bisexual women and men prefer and have long-term committed monogamous relationships.

Lesbians seem to exhibit greater levels of biphobia than do gay men. This is because many lesbian women associate their identity with a political stance against sexism and patriarchy. Some lesbians view heterosexual and bisexual women who "sleep with the enemy" as traitors to the feminist movement.

Increased Public Acceptance of Homosexuality and Support of Gay Rights

As shown in Table 11.3, over the past few decades attitudes toward homosexuality in the United States have become more accepting, and support for protecting civil rights of gays and lesbians has increased (Loftus 2001; Saad 2007). U.S. adults most likely to say they view homosexuality as an acceptable lifestyle are younger (18–34), are Democrat or Independent, are women, and do not attend worship services (or attend infrequently) (see Table 11.4) (Saad 2007).

Part of the explanation for these changing attitudes is the increasing levels of education in the U.S. population, because individuals with more education tend to be more liberal in their attitudes toward homosexuality. Increased contact between homosexuals and heterosexuals and positive depictions of gay and lesbian individuals in the media also contribute to increased acceptance of homosexuality.

biphobia Negative attitudes and emotions toward bisexuality and people who identify as bisexual.

TABLE 11.3 Changes in Attitudes Toward Homosexuality and Gay Rights

	1982	1996	2007
Percentage of U.S. adults who agree that homosexuality should be considered an acceptable lifestyle	34	44	57
homosexuals should have equal rights in terms of job opportunities	59	84	89
marriages between same-sex couples should be recognized by the law as valid, with the same rights as traditional marriages		27	46

Source: Saad (2007).

TABLE 11.4 Homosexuality as an Acceptable Alternative Lifestyle

	YES (%)	NO (%)
Men	53	44
Women	61	35
18–34 years	75	23
35–54 years	58	39
55+ years	45	51
Republican	36	58
Independent	60	36
Democrat	72	27
Worship services		
Attend weekly	33	64
Attend nearly weekly/monthly	57	40
Attend less often/never	74	22

Source: Saad (2007).

Increased Contact Between Homosexuals and Heterosexuals. Greater acceptance of homosexuality may also be due to increased personal contact between heterosexuals and openly gay individuals as more gay and lesbian Americans are coming out to their family, friends, and coworkers. Psychologist Gordon Allport's (1954) "contact hypothesis" asserts that contact between groups is necessary for the reduction of prejudice. In general, heterosexuals have more favorable attitudes toward gay men and lesbian women if they have had prior contact with or know someone who is gay or lesbian (Mohipp & Morry 2004).

Contact with openly gay individuals reduces negative stereotypes and ignorance and increases support for gay and lesbian equality (Wilcox & Wolpert 2000). A national poll of U.S. adults found that 4 in 10 Americans have close friends or relatives who are gay (Neidorf & Morin 2007). In another poll, more than half of the respondents (56 percent) said they have a friend or acquaintance who is gay or lesbian, nearly one-third (32 percent) said they work with someone who is gay or lesbian, and nearly one-fourth (23 percent) said that someone in their family is gay or lesbian (Newport 2002). A national poll of U.S. high

© Kevin Winter/Getty Images for The Billies

After former professional basketball player John Amaechi told the world he was gay, he said that 95 percent of the correspondence he received was "overwhelmingly supportive and positive." He described the remaining 5 percent as "unbelievably, viscerally, frighteningly negative."

school students found that 16 percent of students have a gay or lesbian person in their family, and 72 percent know someone who is gay or lesbian (Curtis 2004). One woman describes how her brother's coming out of the closet changed her views on homosexuality (Yvonne 2004):

I was raised in a devout born-again Christian family. I was raised as a child to believe that homosexuality was evil and a perversion. When I was growing up, I used to wonder if any of the kids at my school could be gay. I couldn't imagine it could be so. As it happens, there was a gay individual even closer than I imagined. My brother Tommy came out of the closet in the early 1990s. After he came out, I had to confront my own denial about the fact that, in my heart of hearts, I had always known that Tommy was gay. Over the years, his partner Rod has come to be a loved and cherished member of our family, and we have all had to confront the prejudices and stereotypes we have held onto for so long about sexual orientation. It seems to me that coming out of the closet is the greatest weapon that gays and lesbians have. If my own brother had never come out, my family would never have been forced to confront the deep-seated prejudices we were raised with. . . . I am still a Christian, but my husband (who also has a gay brother) and I attend a church that truly puts the teachings of Christ into practice—teachings about love, tolerance and inclusivity. As a Christian who grew up with a gay family member, I know that the propaganda put forth by the religious right on this issue is founded in fear, hatred and prejudice. None of these are values taught by Jesus!

Gays and Lesbians in the Media. Another explanation for the increasing acceptance of gays and lesbians is the positive depiction of gays and lesbians in the popular media. In 1998 Ellen DeGeneres came out on her sitcom *Ellen,* and by 2008 many television viewers had seen gay and lesbian characters in television shows such as *Buffy the Vampire Slayer, Queer Eye for the Straight Guy, Six Feet Under*, and *The L Word.* "Honest, non-stereotyped and diverse portrayals of gays and lesbians in prime time can offer youth a realistic representation of the gay community . . . [and] can offer positive role models for gay and lesbian youth" (Miller et al. 2002, p. 21).

One study found that college students reported lower levels of antigay prejudice after watching television shows with prominent homosexual characters (e.g., *Six Feet Under* and *Queer Eye for the Straight Guy*) (Schiappa, Gregg, & Hewes 2005). The researchers propose that just as Allport's contact hypothesis suggests that intergroup contact may reduce prejudice, the "parasocial contact hypothesis" suggests that contact with individuals through the media has similar effects.

DISCRIMINATION AGAINST SEXUAL ORIENTATION MINORITIES

sodomy Oral and anal sexual acts.

In June 2003 a Supreme Court decision in *Lawrence v. Texas* invalidated state laws that criminalize **sodomy**—oral and anal sexual acts. This historic decision overruled a 1986 Supreme Court case (*Bowers v. Hardwick*), which up-

In the past decade, television viewers have had exposure to more gay and lesbian characters in television shows. Showtime's *The L Word* features lesbian characters.

held a Georgia sodomy law as constitutional. The 2003 ruling, which found that sodomy laws were discriminatory and unconstitutional, removes the stigma and criminal branding that sodomy laws have long placed on GLBT individuals. Before this historic ruling was made, sodomy was illegal in 13 states (Alabama, Florida, Idaho, Kansas, Louisiana, Mississippi, Missouri, North Carolina, Oklahoma, South Carolina, Texas, Utah, and Virginia) and four states (Kansas, Missouri, Oklahoma, and Texas) targeted only same-sex acts. Penalties for engaging in sodomy ranged from a $200 fine to 20 years' imprisonment. In states that criminalized both same- and opposite-sex sodomy, sodomy laws were usually not used against heterosexuals but were used primarily against gay men and lesbians (ACLU 1999).

Like other minority groups in U.S. society, homosexuals and bisexuals experience various forms of discrimination. Next, we look at sexual orientation discrimination in the workplace and in housing, in the military, in marriage and parenting, in violent expressions of hate, and in treatment by police.

Discrimination in the Workplace and in Housing

The percentage of Americans saying that homosexuals should have equal job opportunities grew from 56 percent in 1977 to 74 percent in 1992 to 88 percent in 2003 (Saad 2005). The percentage of U.S. adults who agree that "school boards ought to have the right to fire teachers who are known homosexuals" declined from 51 percent in 1987 to 28 percent in 2007 (Pew Research Center 2007).

Yet as of 2007, it was still legal in 30 states to fire, decline to hire or promote, or otherwise discriminate against an employee because of his or her sexual orientation (Human Rights Campaign 2007a).

In a survey of gay, lesbian, and bisexual residents of Topeka, Kansas, 16 percent of respondents reported that they had been denied employment because of their sexual orientation, 11 percent reported being denied a promotion, 15 percent reported that they had been fired, and 35 percent reported that they had received harassing letters, e-mail, or faxes at work (Colvin 2004). Two respondents in the study reported the following experiences (p. 3):

> My job found out that I was a lesbian and my "friend" that comes in every night was my girlfriend. She was told not to come in anymore or I would be fired. And later because she came in I was fired.

> As soon as my . . . boss suspected I was gay, he harassed me until I took a job with another state agency. Prior to that I had three outstanding employee evaluations, but he couldn't find anything I did right.

What Do You Think? The Faith-Based Initiative allows religious organizations to receive federal funding to pay for the delivery of a wide range of social services. In 2005 the U.S. House of Representatives passed an amendment to the School Readiness Act that allows taxpayer-funded faith-based organizations that provide preschool Head-Start programs to fire Head-Start teachers for being lesbian or gay. Conservative groups argue that such discrimination is necessary for religious freedom and to maintain the religious character of a faith-based organization. Although such discrimination is allowed with private funds, the legality of allowing faith-based groups to discriminate in employment funded by taxpayer monies is highly controversial. Do you think that religious organizations that receive federal funds to provide social services under the Faith-Based Initiative should be allowed to discriminate in hiring?

Like other minority group members, LGBT individuals are sometimes denied the option to rent or buy residential property because of prejudice against them. Although some states ban sexual orientation discrimination in housing, in most states there are no laws protecting LGBT individuals from housing discrimination. In addition, same-sex families are not allowed to qualify as a "family" when applying for public housing, which decreases their chances of being eligible for public housing assistance.

Discrimination in the Military

A Pew Research Center (2006) survey found that the majority of Americans (60 percent) favor gays serving openly in the military. Nevertheless, sexual orientation discrimination occurs in the military. In 1993 President Clinton instituted a "Don't ask, don't tell" policy in which recruiting officers are not allowed to ask about sexual orientation and homosexuals are encouraged not to volunteer such information. More than 11,000 service members have been discharged under "Don't ask, don't tell" since it took effect, and countless others have decided not to join the military due to this discriminatory law (Servicemembers Legal Defense Network 2007). The "Don't ask, don't tell" policy also provided an additional means for servicemen to harass lesbian service members by threatening to "out" those who refused their advances or threatening to report them, thus ending their careers (Human Rights Watch 2001).

A Government Accounting Office report found that the "Don't ask, don't tell" policy has cost nearly $200 million for the replacement and training of personnel who were discharged as a result of the policy (Human Rights Campaign 2005a). The report also found that nearly 800 specialists with critical military skills have been discharged, including 54 linguists who specialize in Arabic. As of this writing, gay rights advocates are pushing for the passage of the Military Readiness Enhancement Act, which would repeal the "Don't ask, don't tell" policy.

Most European countries, including France, Germany, Spain, Switzerland, and Denmark, have lifted their bans on gays in the military. Great Britain has gone a step further by actively encouraging gays to enlist (Lyall 2005).

Discrimination in Marriage

Before the 2003 Massachusetts Supreme Court ruling in *Goodridge v. Department of Public Health,* no state had declared that same-sex couples have a constitutional right to be legally married. In response to growing efforts to secure legal recognition of same-sex couples, opponents of same-sex marriage have prompted antigay marriage legislation. In 1996 Congress passed and President Clinton signed the **Defense of Marriage Act,** which states that marriage is a "legal union between one man and one woman" and which denies federal recognition of same-sex marriage. In effect, this law allows states to either recognize or not recognize same-sex marriages performed in other states. As of this writing, 26 states had banned gay marriage through a constitutional amendment and 19 states had a state law banning same-sex marriage (Human Rights Campaign 2007b). Many of these states ban not only same-sex marriage, but other forms of partner recognition, such as domestic partnerships and civil unions (National Gay and Lesbian Task Force 2007). Some states ban domestic partnerships for opposite-sex couples as well. These broader measures, known as "Super DOMAs," potentially endanger employer-provided domestic partner benefits, joint and second-parent adoptions, health care decision-making proxies, or any policy or document that recognizes the existence of same-sex partnership, as well as unmarried opposite-sex partnerships (Cahill & Slater 2004).

> **"**If the love of two people, committing themselves to each other exclusively for the rest of their lives, is not worthy of respect, then what is?**"**
>
> Andrew Sullivan
> Editor and gay rights advocate

A group of conservative Republicans introduced the Federal Marriage Amendment, which would amend the U.S. Constitution to define marriage as being a union between a man and a woman. The Federal Marriage Amendment did not pass in 2004 in the Senate or the House, but supporters vowed to continue the fight. The Federal Marriage Amendment was reintroduced in the 109th Congress in 2005 as the Marriage Protection Amendment. If passed, the constitutional amendment would deny marriage and probably also civil union and domestic partnership rights to same-sex couples (LAWbriefs 2005a). This chapter's *The Human Side* feature presents one woman's plea to protect same-sex families and to oppose a federal marriage constitutional amendment.

Defense of Marriage Act Federal legislation that states that marriage is a legal union between one man and one woman and denies federal recognition of same-sex marriage.

Arguments in Favor of Same-Sex Marriage. Advocates of same-sex marriage argue that banning same-sex marriages or refusing to recognize same-sex marriages granted in other states is a violation of civil rights that denies same-sex couples the many legal and financial benefits that are granted to heterosexual married couples. For example, married couples have the right to inherit from a spouse

In April 2005 Dr. Kathleen Moltz, an assistant professor at Wayne State University School of Medicine, testified against the passage of the Federal Marriage Amendment before a United States Senate Judiciary Committee. The following paragraphs are excerpts from her testimony.

Members of the Subcommittee: . . . I am here as the mother of two beautiful children whose welfare I am trying desperately to protect, as the partner of the wonderful woman with whom I share my life, and as a pediatrician who has taken the oath "first, do no harm." In 1990 . . . my partner Dahlia Schwartz and I became a couple. After several years together, we were married in a traditional Jewish wedding ceremony in 1996. . . . This was years before any state recognized marriage rights for same-sex couples. Dahlia and I were together for several more years before we decided to have children. We are now proud parents to our daughter Aliana, and our son Itamar. Dahlia carried Itamar through a difficult pregnancy, a precipitous labor and an emergency C-section. At several times during the delivery, I was asked to leave the room—something that a different-sex spouse would not have to go through. I cannot put into the words the agoniz-

ing emotions of not being able to be present when my partner and child were in medical distress.

Immediately after he was born, Itamar experienced temperature regulation problems, rapid breathing and hypoglycemia. . . . The pediatrician on staff refused to discuss Itamar's condition with me because I was not his "real mother." What is a "real mother" if not the person who would lay down her life for her child? I am a real mother to my children, and so is Dahlia. Fortunately, since then I have adopted Itamar through second-parent adoption, and Dahlia has adopted Aliana. Both children now have the benefit of a legal relationship with two parents.

On May 21, 2004, Dahlia and I were legally married in Massachusetts, where we had lived for eight years. . . . In June 2004 we moved to Michigan, where I took a job as a pediatric endocrinologist at Wayne State University. . . . I moved my family to Michigan so that I could take a job that would allow Dahlia to stay home with the kids. The domestic partner health benefits that Wayne State provided made this possible, particularly because Dahlia has a continuing medical condition that makes health insurance a necessity. I would not have taken the job without the domestic

partnership benefits. The move also allowed me to be near my parents—our children's grandparents—who live eight houses down from us. . . .

Not long after we moved to Michigan, the state became embroiled in a campaign to pass Proposal 2, an amendment to the state constitution that would ban marriage rights for same-sex couples. When our daughter asked what it was all about, we told her that there were people who believed that we couldn't really be a family. We told her that we thought this was silly because, obviously, we are a family—we share love, children and a commitment to raising healthy, happy kids. When the results of the election came in, Aliana asked about the outcome. We told her that the amendment had passed. With tears in her eyes, she asked, "Does this mean our family has to split up?" Like children do, our 4-year-old went straight to the heart of the issue. The voters had sent us a message that day: you are not a family.

We were dismayed and stunned by the results of the Michigan election and spent days wondering which of our neighbors and colleagues thought that our family should not have equal rights. . . . When anti-gay groups from outside our state tried to use the

who dies without a will, to avoid inheritance taxes between spouses, to make crucial medical decisions for a partner and to take family leave to care for a partner in the event of the partner's critical injury or illness, to receive Social Security survivor benefits, and to include a partner in his or her health insurance coverage. Other rights bestowed on married (or once married) partners include assumption of spouse's pension, bereavement leave, burial determination, domestic violence protection, reduced rate memberships, divorce protections (such as equitable division of assets and visitation of partner's children), automatic housing lease transfer, and immunity from testifying against a spouse (Sullivan 2003). Finally, unlike 17 other countries that recognize same-sex couples for immigration purposes, the United States does not recognize same-sex couples in granting immigration status because such couples are not considered "spouses."

Another argument for same-sex marriage is that it would promote relationship stability among gay and lesbian couples. "To the extent that marriage pro-amendment to take away the health benefits insurance I obtain through my work, I could not sit idly by. Throughout the campaign, supporters of the amendment insisted that the amendment had nothing to do with the health benefits that families like mine receive. In fact, their brochure even claimed that it was "only about marriage." But as soon as the amendment passed, it became a weapon to take away the health insurance upon which many families—including my own—rely.

. . . I am concerned that the Federal Marriage Amendment, which is very similar to Michigan's amendment, will be used to deny equal benefits nationwide. . . . Ironically, the same people who promoted this amendment favor policies that permit or encourage one parent to stay at home with children. But under the false pretense of protecting marriage, this law might force us either to move away from grandparents and family or to deprive our kids of precious time with their parents. No one has been able to explain to me how even one marriage is protected by this unfair, discriminatory law.

I have also heard that the Michigan amendment—and the [proposed] federal amendment . . . is necessary to "protect" marriage as a sacred institution. . . . As an American with great respect for our Constitution, I don't understand why federal law should play a role in defining for the various religions which marriages are "sacred." . . . [N]o religious denomination can ever be forced to perform marriages that don't meet its standards. For instance, rabbis cannot be compelled to perform interfaith marriages even though the laws of every state allow them.

. . . Finally, I have heard that marriage must be "protected" from families like mine for the good of children. As a pediatrician, I know that this is completely unsupported by any scientific fact. Every piece of creditable medical evidence I can find, every study, indicates that children with lesbian and gay parents do just as well as their peers. Every major medical, psychiatric and psychological association that has issued an opinion on the subject endorses increasing, not removing, legal protection of gay and lesbian families. Their endorsement is based on a commitment to protecting the health and welfare of children and their families. In short . . . it is of the utmost importance to extend, not to remove, legal protection to the children and to both parents in gay and lesbian families.

. . . I don't know what harm . . . a constitutional amendment might cause. I fear that families like mine, with young children, will lose health benefits; will be denied common decencies like hospital visitation when tragedy strikes; will lack the ability to provide support for one another in old-age. I fear that my loving, innocent children will face hatred and insults implicitly sanctioned by a law that brands their family as unequal. I know that these sweet children have already been shunned and excluded by people claiming to represent values of decency and compassion.

I also know what such an amendment will not do. It will not help couples who are struggling to stay married. It will not assist any impoverished families struggling to make ends meet or to obtain health care for sick children. It will not keep children with their parents when their parents see divorce as their only option. It will not help any single American citizen to live life with more decency, compassion or morality. In the coming months and debates, I urge you to consider both the medical evidence and the experiences of families like mine with an open heart and an open mind. . . . And I pray, and my family prays, that in dealing with our precious Constitution, you will follow the dictates of the oath that binds my profession: first, do no harm.

Source: Testimony from Dr. Kathleen Moltz, April 2005, U.S. Senate Judiciary Committee.

vides status, institutional support, and legitimacy, gay and lesbian couples, if allowed to marry, would likely experience greater relationship stability" (Amato 2004, p. 963). This outcome, argues Amato, would be beneficial to the children of same-sex parents. Without legal recognition of same-sex families, children living in gay- and lesbian-headed households are denied a range of securities that protect children of heterosexual married couples. These include the right to get health insurance coverage and Social Security survivor benefits from a non-biological parent. In some cases children in same-sex households lack the automatic right to continue living with their nonbiological parent should their biological mother or father die (Tobias & Cahill 2003). It is ironic that the same pro-marriage groups that stress that children are better off in married couple families disregard the benefits of same-sex marriage to children.

There are religious-based arguments in support of same-sex marriage. In a sermon titled "The Christian Case for Gay Marriage," Jack McKinney (2004) interprets Luke 4: "Jesus is saying that one of the most fundamental religious tasks is to stand with those who have been excluded and marginalized. . . . [Jesus] is determined to stand with them, to name them beloved of God, and to dedicate his life to seeing them empowered." McKinney goes on to ask, "Since when has it been immoral for two people to commit themselves to a relationship of mutual love and caring? No, the true immorality around gay marriage rests with the heterosexual majority that denies gays and lesbians more than 1,000 federal rights that come with marriage." As noted earlier, in 2005 the United Church of Christ became the largest Christian denomination to endorse same-sex marriages.

Finally, a cross-cultural and historical view of marriage and family suggests that marriage is a social construction that comes in many forms. In response to President Bush's call for a constitutional amendment banning gay marriage as a threat to civilization, the American Anthropological Association released the following statement:

> The results of more than a century of anthropological research on households, kinship relationships, and families, across cultures and through time, provide no support whatsoever for the view that either civilization or viable social orders depend upon marriage as an exclusively heterosexual institution. Rather, anthropological research supports the conclusion that a vast array of family types, including families built upon same-sex partnerships, can contribute to stable and humane societies. (n.d.)

Arguments Against Same-Sex Marriage. Whereas advocates of same-sex marriage argue that so long as same-sex couples cannot be legally married, they will not be regarded as legitimate families by the larger society, opponents do not want to legitimize homosexuality as a socially acceptable lifestyle. Opponents of same-sex marriage who view homosexuality as unnatural, sick, and/or immoral do not want their children to view homosexuality as an accepted "normal" lifestyle.

Opponents of same-sex marriage commonly argue that such marriages would subvert the stability and integrity of the heterosexual family. However, Sullivan (1997) suggested that homosexuals are already part of heterosexual families:

> [Homosexuals] are sons and daughters, brothers and sisters, even mothers and fathers, of heterosexuals. The distinction between "families" and "homosexuals" is, to begin with, empirically false; and the stability of existing families is closely linked to how homosexuals are treated within them. (p. 147)

Many opponents of same-sex marriage base their opposition on their religious views. In a Pew Research Center national poll, the majority of Catholics and Protestants oppose legalizing same-sex marriage, whereas the majority of secular respondents favor it (Green 2004). However, churches have the right to say no to marriage for gays in their congregations, but marriage is still a *civil* option that does not require religious sanctioning.

Finally, opponents of gay marriage point to public opinion polls that suggest that the majority of Americans are against same-sex marriage. A 2007 national poll found that more than half (55 percent) of Americans oppose same-sex marriage—down from 65 percent in 1996 (Pew Research Center 2007). However, the majority (56 percent) of young adults ages 18–29 in the Pew survey support gay marriage and a national survey of college freshmen found that more than two-thirds (68 percent) agree that "same-sex couples should have the right to legal marital status" (Pryor et al. 2006).

"...THEN PEOPLE WILL SAY, 'WHY CAN'T DOGS GET MARRIED?'... AND THEN, 'WHY CAN'T CATS AND DOGS MARRY?...'"

Discrimination in Child Custody and Visitation

Several respected national organizations—including the Child Welfare League of America, the American Psychological Association, the American Psychiatric Association, and the National Association of Social Workers—have taken the position that a parent's sexual orientation is irrelevant in determining child custody (Landis 1999). In a review of research on family relationships of lesbians and gay men, Patterson (2001) concluded that "the greater majority of children with lesbian or gay parents grow up to identify themselves as heterosexual" and that "concerns about possible difficulties in personal development among children of lesbian and gay parents have not been sustained by the results of research" (p. 279). Patterson (2001) noted that the "home environments provided by lesbian and gay parents are just as likely as those provided by heterosexual parents to enable psychosocial growth among family members" (p. 283). Nevertheless, some court judges are biased against lesbian and gay parents in custody and visitation disputes. For example, in 1999 the Mississippi Supreme Court denied custody of a teenage boy to his gay father and instead awarded custody to his heterosexual mother who remarried into a home "wracked with domestic violence and excessive drinking" (*Custody and Visitation* 2000, p. 1).

Discrimination in Adoption and Foster Care

National survey data show that the percentage of U.S. adults who favor allowing gays to adopt children increased from 38 percent in 1999 to 46 percent in 2006 (Pew Research Center 2006). There are two types of adoption: joint adoption in which a couple jointly adopts a child and second-parent adoption (also known as stepparent adoption) in which one partner adopts the child of a partner. Most adoptions by gay people have been by individual gay men or lesbians who adopt the biological children of their partners. Such second-parent or stepparent adoptions ensure that the children can enjoy the benefits of having two legal parents, especially if one parent dies or becomes incapacitated.

Laws that govern adoption vary widely among the states, and in many cases, adoption decisions are made on a case-by-case basis. Same-sex couples are prohibited from adopting in Florida, Mississippi, and Utah. Joint adoption is allowed statewide in 11 states (and the District of Columbia) and in 2 states, same-sex couples have successfully petitioned for joint adoption (Human Rights Campaign 2007c). Second-parent adoption is permitted in 9 states, and in 15 states, same-sex couples have successfully petitioned for second-parent adoption in some jurisdictions (Human Rights Campaign 2007d).

Nebraska and Utah have policies prohibiting gays from being foster parents; policies banning gay foster parenting were recently removed by the Department of Social Services in Missouri and overturned by the Arkansas state Supreme Court (Gates et al. 2007).

What Do You Think? Discrimination against gays also occurs in the area of assisted reproduction. In 2001, Guadalupe Benitez—a lesbian woman—sued two doctors who refused to artificially inseminate her for religious reasons. Benitez claims that the doctors violated California's antidiscrimination laws that protect gays and lesbians. Although some states allow doctors to refuse to provide certain services such as abortion, contraception, and sterilization, doctors have not been granted the right to refuse services only to certain segments of the adult population. As of this writing, the case is still in the courts (Parker 2007). Do you think that doctors who provide a service should be able to use religious freedom as grounds to pick and choose who they will provide that service to?

Hate Crimes Against Sexual Orientation Minorities

On October 6, 1998, Matthew Shepard, a 21-year-old student at the University of Wyoming, was abducted and brutally beaten. He was found tied to a wooden ranch fence by two motorcyclists who had initially thought he was a scarecrow. His skull had been smashed, and his head and face had been slashed. The only apparent reason for the attack: Matthew Shepard was gay. On October 12 Shepard died of his injuries. Media coverage of his brutal attack and subsequent death focused nationwide attention on hate crimes against sexual orientation minorities.

Other incidents of brutal hate crimes against gays include the murder of Billy Jack Gaither in Alabama in February 1999. Two men who claimed to be angry over a sexual advance made by Gaither plotted his murder, beat Gaither to death with an ax handle, and then burned his body on a pyre of old tires. In July

❝The prejudice against gays and lesbians today is meaner, nastier—a vindictiveness that's rooted in hatred, not ignorance.**❞**

Betty DeGeneres
Comedienne Ellen DeGeneres's mother

This man continues to march in Jerusalem's Gay Pride event after he was stabbed by an antigay protester.

1999 at Fort Campbell, Kentucky, Private First Class Barry Winchell was beaten to death with a baseball bat by another soldier because Winchell was perceived to be homosexual.

In eighteenth-century America, when laws against homosexuality often included the death penalty, violence against gays and lesbians was widespread and included beatings, burnings, various kinds of torture, and execution (Button et al. 1997). Although such treatment of sexual orientation minorities is no longer legally condoned, gays, lesbians, and bisexuals continue to be victimized by hate crimes. Anti-LGBT hate crimes are crimes against individuals or their property that are based on bias against the victim because of his or her perceived sexual orientation or gender identity. Such crimes include verbal threats and intimidation, vandalism, sexual assault and rape, physical assault, and murder.

Lesbian, gay, and bisexual individuals are more likely than African Americans and Muslims to report being the victim of a hate crime. A report released by the Williams Institute at UCLA showed the following rates of hate crime victimization (Stotzer 2007):

- 8 in 100,000 African Americans report being the victim of hate crime
- 12 in 100,000 Muslims report being the victim of hate crime
- 13 in 100,000 gay men, lesbians, and bisexuals report being the victim of hate crime

According to the FBI Uniform Crime Reports, there were 1,213 victims of sexual orientation hate crimes in 2005, representing 7.3 percent of all hate crime victims reported to the FBI (FBI 2006). But as discussed in Chapter 9, FBI hate crime statistics underestimate the incidence of hate crimes. The National Coalition of Anti-Violence Programs (2007) reported that there were 1,834 victims of antigay hate crimes in 2005, which dropped to 1,672 in 2006. And in 2006 there were 11 anti-LGBT murders. And because it is not uncommon for heterosexual men and women to be mistaken for gay men and lesbians, 1 in 10 victims (10 percent) of anti-LGBT violence in 2006 identified as being heterosexual.

Antigay Hate and Harassment in Schools and on Campuses. A national survey of students ages 13 through 18 and secondary school teachers found that actual or perceived sexual orientation is one of the most common reasons that students are harassed by their peers, second only to physical appearance (Harris Interactive and GLSEN 2005). Other findings from this survey, published in a report titled *From Teasing to Torment: School Climate in America,* revealed the extent of antigay behavior in U.S. schools:

- 52 percent of teens frequently hear students make homophobic remarks.
- 69 percent of teens frequently hear students say "that's so gay" or "you're so gay"; expressions in which "gay" is meant to mean something bad or devalued.
- 90 percent of LGBT teens (vs. 62 percent of non-LGBT teens) have been verbally or physically harassed or assaulted during the past year because of their perceived or actual appearance, gender, sexual orientation, gender expression, race/ethnicity, disability, or religion.

Another national survey of 1,732 LGBT students ages 13–20 found that (Kosciw & Diaz 2005):

- Nearly two-thirds (64 percent) reported feeling unsafe at school because of their sexual orientation.
- Nearly two-thirds (64 percent) reported that they had been verbally harassed at least some of the time in school in the past year because of their sexual orientation.
- Over one-third (38 percent) of students had experienced physical harassment and nearly one-fifth (18 percent) had been physically assaulted at school because of their sexual orientation.

Given the harsh treatment of LGBT youth in school settings, it is not surprising that 40 percent of gay youth report that their schoolwork is negatively affected by conflicts over their sexual orientation (GLSEN 2000). More than one-fourth of gay youth drop out of school—usually to escape the harassment, violence, and alienation they endure there (Chase 2000). Lesbian, gay, and bisexual youth who report high levels of victimization at school also have higher levels of substance use, suicidal thoughts, and sexual risk behaviors than heterosexual peers who report high levels of at-school victimization (Bontempo & D'Augelli 2002). A survey of youths' risk behavior conducted by the Massachusetts Department of Education in 1999 found that 30 percent of gay teens attempted suicide in the previous year, compared with 7 percent of their straight peers (reported in Platt 2001).

Antigay hate is also common among college students (see this chapter's *Social Problems Research Up Close* feature). In a survey of 484 young adults at six community colleges in California, 10 percent reported physically assaulting or threatening people whom they believed to be homosexual and 24 percent reported calling homosexuals insulting names (Franklin 2000). The researcher concluded from the findings of the study that, overall, many young adults believe that antigay harassment and violence is socially acceptable.

Since the mid-1980s, numerous studies have documented the hostile climate that GLBT students, staff, faculty, and administrators experience on college campuses. A survey by the National Gay and Lesbian Task Force Policy Institute (Rankin 2003) is a significant contribution to the research literature in this area.

Sample and Methods

Because of the difficulty in identifying lesbian, gay, bisexual, and transgender individuals, Rankin (2003) used purposeful sampling of GLBT individuals and snowball sampling. Contacts were made with "out" GLBT individuals on campus, and they were asked to share the survey with other members of the GLBT community who were not so open about their sexual/gender identity.

Surveys, both paper-and-pencil and online versions, were administered to students, faculty, staff, and administrators at 14 colleges and universities from around the country. The surveys, which contained 35 questions and an additional space for respondents to provide commentary, were designed to collect data about (1) respondents' personal campus experiences as members of the GLBT community, (2) their perception of the climate for GLBT members of the academic community, and (3) their perceptions of institutional actions, including administrative policies and academic initiatives regarding GLBT issues and concerns on campus. A total of 1,669 usable surveys were returned, representing the following groups:

- 1,000 students (undergraduate and graduate), 150 faculty, and 467 staff/administrators
- 720 men, 848 women
- 326 people of color
- 572 gay people (mostly male), 458 lesbians, 334 bisexuals, and 68 transgendered individuals
- 825 "closeted" people

Findings and Discussion

Some of the findings of the Campus Climate Assessment are discussed in the following sections.

Lived Oppressive Experiences

Nineteen percent of the respondents reported that within the past year they had feared for their physical safety because of their sexual orientation or gender identity, and 51 percent had concealed their sexual or gender identity to avoid intimidation. GLBT people of color were more likely than white GLBT people to conceal their sexual or gender identity to avoid harassment. Many nonwhite respondents reported feeling out of place at predominantly white GLBT settings.

In the words of one student, "As a Chicana, I felt ostracized even more. Forget about feeling a sense of community when you're a member of two minority groups" (Rankin 2003, p. 25). The study also found that 27 percent of faculty, staff, and administrators and 40 percent of students indicated that they had concealed their sexual identity to avoid discrimination in the past year.

Respondents were also asked if they had experienced harassment in the past year. Harassment was defined as "conduct that has interfered unreasonably with your ability to work or learn on this campus or has created an offensive, hostile, intimidating working or learning environment." Undergraduate students were the most likely to have experienced harassment (36 percent), followed by faculty (27 percent), graduate students (23 percent), and staff (19 percent). Derogatory remarks were the most common form of harassment (89 percent). Other types of harassment included verbal harassment or threats, anti-GLBT graffiti, threats of physical violence, denial of services, and physical assault.

Perceptions of Anti-GLBT Oppression on Campus

About three-fourths of faculty, students, administrators, and staff rated the campus climate as homophobic. In contrast, most respondents rated the campus generally (not specific to GLBT people) as friendly, concerned, and respectful. Thus "even though respondents feel that the overall campus climate

is hospitable, heterosexism and homophobia are still prevalent" (Rankin 2003, p. 31).

Institutional Actions

Participants were asked to respond to several questions about institutional actions regarding GLBT concerns on campus. Less than half (37 percent) agreed that "the college/university thoroughly addresses campus issues related to sexual orientation/gender identity," and only 22 percent agreed that "the curriculum adequately represents the contributions of GLBT persons." However, the majority of respondents agreed that their classrooms or their job sites were accepting of GLBT persons (63 percent) and that the college or university provided resources on GLBT issues and concerns (71 percent). These positive responses may be due in part to the inclusion of sexual orientation in the nondiscrimination policies of all but one of the participating colleges and universities. Several of the campuses also provided domestic partner benefits, had GLBT resource centers, and offered safe-space programs.

These findings indicate that intolerance and harassment of GLBT students, staff, faculty, and administrators continue to be prevalent on U.S. campuses. It is important to note that the colleges and universities that participated in this study are not representative of most institutions of higher learning in the United States and "may be among the most gay-friendly campuses in the country" (Rankin 2003, p. 3). Rankin explains that "all of the institutions who participated in this survey had a visible GLBT presence on campus, including, in most cases, a GLBT campus center. Most had sexual orientation nondiscrimination policies. As only 100 of the 5,500 U.S. colleges and universities have GLBT student centers, the 14 universities surveyed here are not representative of most institutions of higher education in the U.S." (p. 3). Thus Rankin suggests that the findings of this study "may significantly understate the problems facing GLBT students and staff at U.S. colleges and universities" (p. 3).

Police Mistreatment of Sexual Orientation Minorities

Data collected by the National Coalition of Anti-Violence Programs (2007) suggested that two-thirds of antigay hate crimes are *not* reported to the police. One reason that LGBT victims of hate crimes do not report these crimes to the police is the perception that the police will not be helpful and may even inflict further abuses. Among victims of antigay hate crimes who did report the crime to the police, victims described the police as "courteous" 49 percent of the time, "indifferent" 34 percent of the time, verbally abusive 11 percent of the time, and physically abusive 6 percent of the time (National Coalition of Anti-Violence Programs 2007). A report by Amnesty International (2005) found that police mistreatment and abuse of LGBT people are widespread in the United States. Types of police mistreatment revealed in the report include targeted and discriminatory enforcement of laws against LGBT people, verbal abuse, inappropriate pat-down and strip searches, failure to protect LGBT people in holding cells, inappropriate response or failure to respond to hate crimes or domestic violence calls, sexual harassment and abuse, and physical abuse. In one case a 31-year-old gay African-American man who was arrested by Chicago police after an altercation with his landlord alleged that officers handcuffed him and raped him with a billy club that had been dipped in cleaning fluid. The victim received an out-of-court settlement of $20,000.

Some police departments have taken initiatives to improve their practices regarding the treatment of LGBT individuals. For example, Washington, DC, created the Gay and Lesbian Liaison Unit, and many police departments provide training on hate crimes against LGBT individuals. However, 28 percent of police departments in the Amnesty International (2005) study do not provide any training in LGBT issues.

Effects of Antigay Bias and Discrimination on Heterosexuals

Lesbians, gays, bisexuals, and transgendered individuals are not the only victims of anti-LGBT prejudice and discrimination. In the following list, we explain how heterosexuals are also victims of anti-LGBT bias and discrimination.

1. *Restriction of male gender expression.* Because of the antigay climate, heterosexuals, especially males, are hindered in their own self-expression and intimacy in same-sex relationships. "The threat of victimization (i.e., antigay violence) probably also causes many heterosexuals to conform to gender roles and to restrict their expressions of (nonsexual) physical affection for members of their own sex" (Garnets, Herek, & Levy 1990, p. 380). Homophobic epithets frighten youth who do not conform to gender role expectations, leading some youth to avoid activities that they might otherwise enjoy and benefit from (e.g., arts for boys and athletics for girls) (GLSEN 2000).

2. *Dysfunctional sexual behavior.* Some cases of rape and sexual assault are related to homophobia and compulsory heterosexuality. For example, college men who participate in gang rape, also known as "pulling train," entice each other into the act "by implying that those who do not participate are unmanly or homosexual" (Sanday 1995, p. 399). Homonegativity also encourages early sexual activity among adolescent men. Adolescent male virgins are often teased by their male peers, who say things like "You mean you don't do it with

> " Injustice anywhere is a threat to justice everywhere. We are caught in an inescapable network of mutuality, tied in a single garment of destiny. Whatever affects one directly affects all indirectly. "
>
> Martin Luther King Jr.
> Civil rights leader

girls yet? What are you, a fag or something?" Not wanting to be labeled and stigmatized as a "fag," some adolescent boys "prove" their heterosexuality by having sex with girls.

3. *Loss of rights for individuals in unmarried relationships.* Social disapproval of same-sex relationships in the United States has contributed to the passage of state constitutional amendments that prohibit same-sex marriage, which can also result in denying rights and protections to opposite-sex unmarried couples. For example, in 2005 Judge Stuart Friedman of Cuyahoga County (Ohio) agreed that a man who was charged with assaulting his girlfriend could not be charged with a domestic violence felony because the Ohio state constitutional amendment granted no such protections to unmarried couples (Human Rights Campaign 2005b). As we mentioned earlier, some antigay marriage measures also threaten the provision of domestic partnership benefits to unmarried heterosexual couples.

4. *Heterosexual victims of hate crimes and harassment.* As we discussed earlier in this chapter, hate crimes are crimes of perception, meaning that victims of antigay hate crimes may not be homosexual; they may just be perceived as being homosexual. Recall that 1 in 10 victims of antigay hate crimes in 2006 was heterosexual (National Coalition of Anti-Violence Programs 2007). And in a survey of students across the nation, 13 percent of non-LGBT students reported that they have been called names, teased, bullied, or hurt at school because people thought they were gay (Harris Interactive and GLSEN 2005).

5. *Fear and grief.* Many heterosexual family members and friends of homosexuals live with the fear that their lesbian or gay friend or family member could be victimized by antigay prejudice and discrimination. They may also be afraid of being mistreated themselves simply for having a gay or lesbian friend or family member. Youth with gay and lesbian family members are often taunted by their peers. And finally, imagine the grief experienced by the family and friends of victims of antigay hate, such as the mother of Matthew Shepard.

6. *School shootings.* Antigay harassment has also been a factor in many of the school shootings in recent years. In March 2001, 15-year-old Charles Andrew Williams fired more than 30 rounds in a San Diego suburban high school, killing 2 and injuring 13 others. A woman who knew Williams reported that the students had teased Williams and called him gay (Dozetos 2001). According to the Gay, Lesbian, and Straight Education Network (GLSEN), Williams's story is not unusual. Referring to a study of harassment of U.S. students that was commissioned by the American Association of University Women, a GLSEN report concluded, "For boys, no other type of harassment provoked as strong a reaction on average; boys in this study would be less upset about physical abuse than they would be if someone called them gay" (Dozetos 2001).

STRATEGIES FOR ACTION: REDUCING ANTIGAY PREJUDICE AND DISCRIMINATION

Many of the efforts to change policies and attitudes regarding sexual orientation minorities have been spearheaded by organizations such as the Human Rights Campaign (HRC), the National Gay and Lesbian Task Force (NGLTF), the Gay and Lesbian Alliance Against Defamation (GLAAD), the International Lesbian

66 Resistance begins with people confronting pain, whether it's theirs or somebody else's, and wanting to do something to change it. 99

bell hooks
Author

and Gay Association (ILGA), the Lambda Legal Defense and Education Fund, the National Center for Lesbian Rights, and GLSEN. But the effort to reduce antigay prejudice and discrimination is not just a "gay agenda"; many heterosexuals and mainstream organizations (such as the National Education Association) have worked on this agenda as well.

Many of the advancements in gay rights have been the result of political action and legislation. Barney Frank (1997), an openly gay U.S. representative, emphasized the importance of political participation in influencing social outcomes. He noted that demonstrative and cultural expressions of gay activism, such as "gay pride" celebrations, marches, demonstrations, or other cultural activities promoting gay rights, are important in organizing gay activists. However, he warned:

> Too many people have seen the cultural activity as a substitute for democratic political participation. In too many cases over the past decades we have left the political arena to our most dedicated opponents [of gay rights], whose letter writing, phone calling, and lobbying have easily triumphed over our marching, demonstrating, and dancing.

> The most important lesson . . . for people who want to make America a fairer place is that politics—conventional, boring, but essential politics—will ultimately have a major impact on the extent to which we can rid our lives of prejudice. (p. xi)

Next, we look at efforts to reduce employment discrimination against sexual orientation minorities, provide recognition and support to lesbian and gay families, and include sexual orientation in hate crime legislation. We also provide an overview of educational policies and programs designed to reduce intolerance and support nonheterosexual students in schools and on campuses.

Reducing Employment Discrimination Against Sexual Orientation Minorities

Most Americans are in favor of equal rights for homosexuals in the workplace (Saad 2007). Yet, as of this writing, federal law protects individuals from discrimination only on the basis of race, religion, national origin, sex, age, and disability. A bill called the **Employment Nondiscrimination Act (ENDA),** introduced in Congress in 1994, would make it illegal to discriminate in the workplace on the basis of sexual orientation. ENDA would not apply to religious organizations, businesses with fewer than 15 employees, or the military. ENDA failed to pass the Senate when it came up for a vote in 1996. In 2007, ENDA was amended to prohibit employment discrimination on the basis of gender expression and gender identity—in effect, expanding employment protections to all transgendered individuals, including cross-dressers and transsexuals. But to get enough votes to pass the bill, these expanded protections were removed before ENDA came up for a vote in the House, which passed the bill in November 2007. Although many advocates for equal rights were disappointed that the final ENDA bill did not include protections for all transgendered individuals, the passing of the bill in the House was an important step forward and may pave the way for future progress toward protecting all transgendered individuals from employment discrimination. At the time of this writing, the ENDA bill is pending in the Senate.

Efforts are also being made to end discrimination in the military. The Military Readiness Enhancement Act, reintroduced in Congress in 2007, would re-

Employment Nondiscrimination Act (ENDA) A bill that would make it illegal to discriminate based on sexual orientation; as of August 2007 this bill had failed to pass the Senate.

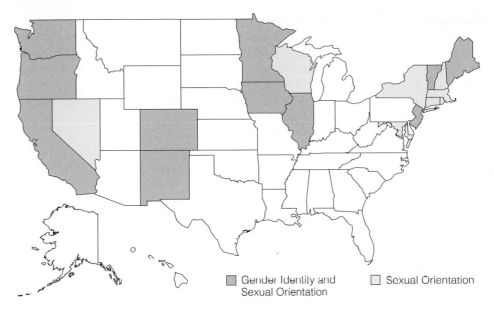

FIGURE 11.3
States that prohibit workplace discrimination based
on sexual orientation and
gender identity.
Source: Human Rights Campaign (2007a).

Gender Identity and Sexual Orientation **Sexual Orientation**

States that prohibit workplace discrimination based on sexual orientation and gender identity (12 states and the District of Columbia):

California (prohibited discrimination based on gender identity in 2003, sexual orientation 1992); Colorado (2007); Illinois (2005); Iowa (2007); Maine (2005); Minnesota (1993); New Jersey (gender identity 2006, sexual orientation 1992); New Mexico (2003); Oregon (2008); Rhode Island (gender identity 2001, sexual orientation 1995); Vermont (gender identity 2007, sexual orientation 1992); Washington (2006); and the District of Columbia (gender identity 2006, sexual orientation 1973).

States that prohibit workplace discrimination based on sexual orientation (8 states):

Connecticut (policy instituted in 1991), Hawaii (1991), Maryland (2001), Massachusetts (1989), Nevada (1999), New Hampshire (1997), New York (2002) and Wisconsin (1982).

peal "Don't ask, don't tell" and replace it with a statute banning discrimination on the basis of sexual orientation.

With the absence of federal legislation prohibiting antigay discrimination, some state and local governments and some private employers have taken measures to prohibit employment discrimination based on sexual orientation. The scope of these measures varies, from prohibiting discrimination only in public employment to comprehensive protection against discrimination in public and private employment, education, housing, public accommodations, credit, and union practices. In 2007, for the first time in the history of the gay rights movement, more than half (52 percent) of the U.S. population lived in jurisdictions that ban discrimination on the basis of sexual orientation (Foreman 2007).

Local and State Bans on Antigay Employment Discrimination. In 1974 Minneapolis became the first municipality to ban antigay job discrimination. By 2007, 171 cities and counties prohibited sexual orientation discrimination in the private workplace (Human Rights Campaign 2007a). In 1982 Wisconsin became the first state to prohibit antigay employment discrimination. By 2007, 20 states (and the District of Columbia) had laws banning sexual orientation discrimination in the workplace (both public and private sectors) (see Figure 11.3).

Nondiscrimination Policies in the Workplace. In 1975 AT&T became the first employer to add sexual orientation to its nondiscrimination policy. By 2007, 88 percent of Fortune 500 companies included sexual orientation in their equal employment opportunity or nondiscrimination policies. And 90 percent of colleges and universities on *U.S. News and World Report* annual list of the top 125 colleges and universities prohibit discrimination based on sexual orientation (Human Rights Campaign 2007a).

Effects of Nondiscrimination Laws and Policies. A survey of 126 U.S. cities and counties that had implemented laws or policies prohibiting discrimination on the basis of sexual orientation found that such laws and policies sometimes had negative effects and "resulted in divisions in the community, greater controversy or tension, or the increased mobilization of those opposed to gay rights" (Button et al. 1997, p. 120). However, the more commonly cited effects were positive and included the following:

- Reduced discrimination against lesbians and gays in public employment and other institutions covered by the law.
- Increased feelings of security, comfort, and acceptance by lesbians and gay men, resulting in lesbians and gay men being more likely to "come out of the closet"— to reveal their sexual orientation to others (or to at least stop hiding it).
- An increased sense of legitimacy among gays and lesbians that helped them overcome "internalized homophobia."

What Do You Think? Do you think that social acceptance of homosexuality leads to the creation of laws that protect lesbians and gays? Or does the enactment of laws that protect lesbians and gays help to create more social acceptance of gays?

Providing Legal Recognition and Support to Lesbian and Gay Couples

In 2003 the Massachusetts Supreme Court ruled (in *Goodridge v. Department of Public Health*) that denying same-sex couples the protections, benefits, and obligations of civil marriage violated the basic premises of individual liberty and equality in the Massachusetts Constitution (Cahill & Slater 2004). As of this writing, no other state has affirmed the right of same-sex couples to be legally married.

What Do You Think? In Massachusetts, of the more than 5,000 same-sex couples who got married in the first year after a court order legalizing such marriages went into effect, 63 percent were lesbian couples and 36 percent were gay male couples (Johnston 2005). Why do you think there was a significantly higher percentage of lesbian marriages than gay male marriages?

GLASBERGEN

**"Same-sex marriage is nothing new.
We've been having the same sex for 25 years."**

Some states have passed legislation to allow same-sex couples legal status that entitles them to many of the same rights and responsibilities as married heterosexual couples (see Table 11.5). For example, in Vermont, Connecticut, New Jersey, and (effective in 2008) New Hampshire, same-sex couples can apply for a civil union license. A **civil union** is a legal status parallel to civil marriage under state law. Civil union status entitles same-sex couples to almost all of the rights and responsibilities available under state law to married couples. Unlike marriage for heterosexual couples, the rights of partners in same-sex civil unions are not recognized by federal law, so they do not have the more than 1,000 federal protections that go along with civil marriage, nor is their legal status recognized in other states.

Some states, counties, cities, and workplaces allow unmarried couples, including gay couples, to register as **domestic partners** (or "reciprocal beneficiaries" in Hawaii). The rights and responsibilities granted to domestic partners vary from place to place but may include coverage under a partner's health and pension plan, rights of inheritance and community property, tax benefits, access to married student housing, child custody and child and spousal support obligations, and mutual responsibility for debts.

In 1991 the Lotus Development Corporation became the first major American firm to extend domestic partner recognition to gay and lesbian employees. More than half (53 percent) of Fortune 500 companies today offer same-sex domestic partner health benefits (Human Rights Campaign 2007a). However, even when companies offer domestic partner benefits to same-sex partners of employees, these benefits are usually taxed as income by the federal government, whereas spousal benefits are not.

Another attempt to provide equal benefits to same-sex couples is the Uniting American Families Act—a bill introduced in Congress in 2005 that would

civil union A legal status that entitles same-sex couples who apply for and receive a civil union certificate to nearly all of the benefits available to married couples.

domestic partners A status granted to unmarried couples, including gay and lesbian couples, by some states, counties, cities, and workplaces that conveys various rights and responsibilities.

TABLE 11.5 States That Recognize Same-Sex Marriage

STATE	SAME-SEX RELATIONSHIP RECOGNITION
California	Effective January 1, 2005, the state's domestic partner registry was expanded to confer almost all of the state-level spousal rights and responsibilities on registered couples.
Connecticut	Effective October 1, 2005, same-sex couples in Connecticut may enter into civil unions with almost all of the benefits of marriage under state law.
District of Columbia	Effective July 8, 2002, registered domestic partners are entitled to the same rights as legal family members to visit their domestic partners in the hospital and make decisions concerning the treatment of a domestic partner's remains after the partner's death. In 2004 and 2006, the law was enhanced to allow registered domestic partners to add each other to a real estate deed without paying recording taxes and to have spousal rights in the areas of spousal immunity, inheritance, spousal support, and public assistance.
Hawaii	Effective July 8, 1997, the state offers "reciprocal beneficiary" status to same-sex couples, with limited benefits relative to married couples in Hawaii. Afforded benefits include certain rights and obligations associated with survivorship, inheritance, property ownership, and insurance.
Maine	Effective July 30, 2004, a state domestic partner registry provides registered couples with inheritance rights, next-of-kin status, victim's compensation, and priority in guardian and conservator rights.
Massachusetts	Effective May 17, 2004, the state offers civil marriage licenses to same-sex couples.
New Hampshire	As of January 1, 2008, the state will offer civil unions, providing same-sex couples the same rights, responsibilities, and obligations under state laws that opposite sex married couples receive.
New Jersey	Effective February 19, 2007, same sex couples in New Jersey may enter into civil unions, providing same-sex couples the same rights, responsibilities, and obligations under state laws as opposite-sex married couples.
Oregon	As of January 1, 2008, the state will offer domestic partnerships to same-sex couples, providing the same rights, responsibilities and obligations under state laws that opposite-sex married couples receive.
Vermont	Effective July 1, 2000, same-sex couples in Vermont may enter into civil unions that entitle the couples to the more than 300 state-level rights and responsibilities extended to opposite-sex spouses.
Washington	Effective July 22, 2007, same-sex couples in Washington may register as domestic partners and receive certain limited rights, including inheritance rights, hospital visitation, and health care decision-making rights.

Source: Human Rights Campaign (2007a).

enable foreign same-sex partners of U.S. citizens or permanent residents to seek immigration status. As noted earlier, 17 other countries throughout the world recognize same-sex couples for immigration purposes.

Protecting Gay and Lesbian Parental Rights

One in three lesbian women has given birth, and one in six gay men has fathered or adopted a child (Gates et al. 2007). As noted earlier, 2000 census data revealed that one-third (33 percent) of female same-sex householders and 22 percent of male same-sex householders were living with their children (younger than 18 years of age) (Simmons & O'Connell 2003). A number of policies and rulings reflect the increasing provision of parental rights and responsibilities to gay and lesbian parents and co-parents.

- In an unprecedented ruling the California Board of Equalization granted head-of-household tax status to a nonbiological lesbian coparent ("NCLR

> "It makes no moral or legal sense to deny children the right to be raised by married parents on the basis of their parents' sexual orientation."
>
> Paul Amato
> Family scholar

Wins Equal Tax Benefits" 2000). This is the first ruling by any state tax board that provides equitable tax status for gay and lesbian families.

- A Massachusetts court ruled that two women may be listed as "mother" on a birth certificate when one of the women donated the egg for the child and the other carried the child (LAWbriefs 2000).

- A Pennsylvania court ordered a nonbiological, nonadoptive parent to pay child support for the five children she jointly parented with her former partner (LAWbriefs 2003).

- In 2005 the Supreme Court of Pennsylvania overturned a lower court decision that found it to be in the best interest of a child not to have any contact with her nonbiological lesbian mother because the child's biological mother had alienated the child from her. The court noted that a heterosexual parent who alienated the child would not be rewarded for such behavior and that "this scenario is equally applicable" (LAWbriefs 2005a).

- The West Virginia Supreme Court awarded a woman custody of the child she and her deceased lesbian partner raised together. The deceased partner's parents had fought for custody of the child, and, after being denied by a trial court, an appellate court reversed the decision and awarded custody to the grandparents. The Supreme Court overruled the appellate court decision, recognizing the parental rights of the woman who parented the child for his entire life (LAWbriefs 2005b).

- In Tennessee Christy Berry had primary custody of her young son after her divorce from the child's father. The father sought a change in custody, claiming that Ms. Berry's sexual orientation would harm their child. A trial court granted the father's request for custody, but an appeals court reversed the decision, noting that the boy was doing well with his mother and that there was no evidence her sexual orientation had any adverse effect on him (LAWbriefs 2005b).

Antigay Hate Crimes Legislation

Hate crime laws call for tougher sentencing when prosecutors can prove that the crime committed was a hate crime. As of May 2007, 32 states and the District of Columbia had hate crime laws that include sexual orientation, 13 states had hate crime laws that did not include sexual orientation, 1 state had hate crime laws that addressed crimes motivated by bias or prejudice but did not list categories, and 4 states had no hate crime laws (National Gay and Lesbian Task Force 2007).

The 1969 federal hate crimes law covers hate crimes based on race, religion, and national origin. For years gay rights advocates lobbied Congress to add federal hate crime law protections based on sexual orientation, gender, gender identity, and disability. The Local Law Enforcement Hate Crimes Prevention Act of 2007 (also called the Matthew Shepard Act) would expand the federal hate-crime law to include crimes motivated by bias against a victim's actual or perceived sexual orientation, gender, gender identity, or disability. The bill would also provide funds to help local and state investigations and prosecutions of hate crimes. In May 2007, the Local Law Enforcement Hate Crimes Prevention Act passed the House and, as of this writing, is being considered in the Senate. President Bush has suggested he may veto the legislation if it reaches his desk, despite the fact that the majority of U.S. adults favor the bill (Newport 2007).

Educational Strategies: Policies and Programs in the Public Schools

If schools are to promote the health and well-being of all students, they must address the needs and promote acceptance of gay, lesbian, and bisexual youth. Three strategies for attaining these goals are to (1) include sexual orientation diversity in sex education programs, (2) implement policies against antigay harassment in school, and (3) establish gay-straight alliances in middle and high schools.

Sex Education. In a national survey of U.S. adults, only 19 percent said that schools should not teach about homosexuality at all (NPR/Kaiser/Kennedy School Poll 2004). Although most Americans want teachers to address the topic of homosexuality, a slight majority wants them to do so in a neutral way. More than half (52 percent) of U.S. adults said that teachers should teach "only what homosexuality is, without discussing whether it is wrong or acceptable," compared with 18 percent who said schools should teach that homosexuality is wrong. Only 8 percent of adults in this survey said schools should teach that homosexuality is acceptable.

As of 2006, only one state (New Jersey) and the District of Columbia mandated that homosexuality be taught in public school sex education classes; one state (Louisiana) banned teachers from discussing homosexuality in the classroom (SIECUS 2006). Only one state (California) required sex education teachers to convey a positive message about homosexuality in that they must respect all relationships. And 8 states required sex education teachers to teach negative messages about homosexuality and/or same-sex sexual activity. For example, some states require teachers to refer to homosexuality and/or same-sex sexual activity as a public health risk or must state that homosexuality is not acceptable to the general public.

Protecting Students from Harassment. One strategy for promoting tolerance for diversity among students involves establishing and enforcing rules that prohibit antigay harassment in the schools. Only 11 states and the District of Columbia have laws, regulations, or policies that prohibit harassment and/or discrimination based on sexual orientation in schools, and 14 states have a law that prohibits bullying in school (Human Rights Campaign 2005b). In the absence of state laws prohibiting antigay harassment in schools, some schools have their own policies against such behavior. At the federal level, Congresswoman Linda Sanchez from California introduced The Safe Schools Improvement Act (H.R. 3132) in 2007—a federal bill requiring school districts to adopt policies prohibiting bullying and harassment based on race, gender, religion, sexual orientation, and gender identity/expression, among others.

One resource for creating a "harassment-free" climate is GLSEN—the leading national organization fighting against harassment and discrimination in K–12 schools. GLSEN conducts training for school staffs around the country and has developed the faculty training program of the Massachusetts Department of Education: the Safe Schools for Gay and Lesbian Students program, the first statewide effort with the aim of ending homophobia in schools (GLSEN 2000). The National Education Association has also implemented national training programs to educate teachers in every state about the role they can and must play to stop antigay harassment in their schools (Chase 2000).

Schools that do not protect students against harassment can face legal challenges. A number of court rulings have held school districts responsible for failing to protect LGBT students from discrimination, violence, and harassment and have ordered school districts to pay between $40,000 and $1 million in damages (Cianciotto & Cahill 2003).

Gay-Straight Alliances. **Gay-straight alliances (GSAs)** are school-sponsored clubs for gay teens and their straight peers. GSAs create safe, supportive environments for gay youth and their allies. Most clubs also organize school events to increase acceptance of sexual minorities and reduce antigay harassment. For example, GSAs organize a Day of Silence during which students do not speak for an entire day in recognition of the daily harassment endured by LBGT students. In an effort to combat antigay remarks commonly made in school, members of one GSA in Hamilton, Wisconsin, designed classroom posters that read, "I just heard you say 'That's so gay.' What I think you meant was. . .," followed by a list of adjectives such as ridiculous, silly, absurd, and foolish (Kilman 2007). After the posters were displayed, students reported that "that's so gay" remarks decreased. The first GSA was established in 1988, and today there are more than 3,500 GSAs in U.S. schools (GLSEN 2007).

Campus Policies and Programs Dealing with Sexual Orientation

D'Emilio (1990) suggested that colleges and universities have the ability and the responsibility to promote gay rights and social acceptance of homosexual people:

> For reasons that I cannot quite fathom, I still expect the academy to embrace higher standards of civility, decency, and justice than the society around it. Having been granted the extraordinary privilege of thinking critically as a way of life, we should be astute enough to recognize when a group of people is being systematically mistreated. We have the intelligence to devise solutions to problems that appear in our community. I expect us also to have the courage to lead rather than follow. (p. 18)

Student groups have been active in the gay liberation movement since the 1960s. Because of the activism of students and the faculty and administrators who support them, nearly 400 U.S. colleges and universities have nondiscrimination policies that include sexual orientation (Singh & Wathington 2003). Other measures to support the LGBT college student population include gay and lesbian studies programs, social centers, and support groups, as well as campus events and activities that celebrate diversity. Many campuses have Safe Zone or Ally programs designed to visibly identify students, staff, and faculty who support the LGBT population. The Safe Zone and Ally programs may require a training session that provides a foundation of knowledge needed to be an effective ally to LGBT students and those questioning their sexuality. Participants in Safe Zone or Ally programs display some type of sign or placard outside their office or residence hall room that identifies them as individuals who are willing to provide a safe haven, a listening ear, and support for LGBT people and those struggling with sexual orientation issues (Safe Zone Resources 2003).

gay-straight alliances (GSAs) School-sponsored clubs for gay teens and their straight peers.

Logos of Safe Zone programs at four universities and colleges. From left to right: Pennsylvania State University, University of North Carolina at Chapel Hill, Purdue University, and Cornell College.

(from left to right) Courtesy of Pennsylvania State University, Courtesy of University of North Carolina at Chapel Hill, Courtesy of Purdue University, Courtesy of Cornell College

UNDERSTANDING ISSUES IN SEXUAL ORIENTATION

Recent years have witnessed a growing acceptance of lesbians, gay men, and bisexuals as well as increased legal protection and recognition of these marginalized populations. The advancements in gay rights are notable and include the Supreme Court's 2003 decriminalization of sodomy, the growing adoption of nondiscrimination policies covering sexual orientation, the increased recognition of domestic partnerships, the state civil unions offered to same-sex couples in Vermont, Connecticut, New Jersey, and New Hampshire, the Massachusetts ruling that same-sex marriage is allowable under the state constitution, and the election of the first openly gay bishop in the New Hampshire Episcopal diocese. But the winning of these battles in no way signifies that the war is over. Gay, lesbian, and bisexual individuals continue to be frequent victims of discrimination, harassment, and violence. In some countries homosexuality is formally condemned, with penalties ranging from fines to imprisonment and even death.

As both structural functionalists and conflict theorists note, nonheterosexuality challenges traditional definitions of family, child rearing, and gender roles. Every victory in achieving legal protection and social recognition for sexual orientation minorities fuels the backlash against them by groups who are determined to maintain traditional notions of family and gender. Often, this determination is rooted in and derives its strength from uncompromising religious ideology.

As symbolic interactionists note, meanings associated with homosexuality are learned. Powerful individuals and groups opposed to gay rights focus their efforts on maintaining the negative meanings of homosexuality to keep the gay, lesbian, and bisexual population marginalized. Since taking office, President George W. Bush has issued more than 250 proclamations, including those honoring National School Lunch Week and Wright Brothers Day, but he has refused to recognize Gay Pride Month with an official proclamation (Tobias & Cahill 2003). Bush did issue a presidential proclamation for Marriage Protection Week as part of an effort to define marriage as a union between a woman and a man, thus denying marriage rights to same-sex couples. Another direct attack on gay rights is the proposal of the Marriage Protection Amendment, which would amend the Constitution to define marriage as being between one woman and one man. New York Senator Hillary Rodham Clinton noted that the push for the amendment represents the first time an attempt has been made to amend the Constitution to specifically deny rights to any group of individuals. She also suggested that the current campaign against gay rights is a political strategy to drive wedges between Americans and to divert the public's attention away from social problems. "They'd rather talk about taking away rights and undermining the ability of Americans to live their own lives, to have their own families," said Clinton, "than to talk about the miserable economy, to talk about their miserable foreign policy, to talk about their rollback of environmental laws and workers' rights, education and health care!" (quoted in Johnston 2003).

Hillary Rodham Clinton suggests that politicians who support the Federal Marriage Amendment are driving a wedge between Americans and diverting the public's attention away from social problems. "They'd rather talk about taking away rights and undermining the ability of Americans to live their own lives, to have their own families," said Clinton, "than to talk about the miserable economy, to talk about their miserable foreign policy, to talk about their rollback of environmental laws and workers' rights, education and health care!"

Political efforts to undermine gay rights and recognition must contend with the fact that "gay is everywhere": 2000 census data reveal that 99.3 percent of U.S. counties reported same-sex cohabiting partners, compared with 52 percent of counties in 1990. With the gay population being more "out" than ever before, more heterosexual individuals are reporting having family, work-related, or friendship or acquaintance ties to lesbian and gay individuals. This increased contact, as well as gay-straight alliances in schools and more frequent portrayals of lesbians and gays in the media, have inspired many heterosexual individuals to support the gay rights movement.

But, as one scholar noted, "The new confidence and social visibility of homosexuals in American life have by no means conquered homophobia. Indeed it stands as the last acceptable prejudice" (Fone 2000, p. 411). True, the American public is becoming increasingly supportive of gay rights. But as Yang (1999) pointed out, as the antigay minority diminishes in size, "it often becomes more dedicated and impassioned" (p. ii).

> Our task in the coming years is to get the heterosexual Americans who support our cause to feel as passionately outraged by the injustices we face and to be as strongly motivated to act in support of our rights as our adversaries are in their opposition to our rights. (p. iii)

CHAPTER REVIEW

- **Are there any countries in which homosexuality is illegal?**

  Yes, at least 85 member states of the United Nations criminalize consensual same-sex behavior among adults. In 38 of these countries only male homosexuality is illegal; female homosexuality is not. Legal penalties for homosexuality vary from prison sentences and/or corporal punishment (such as whipping or lashing) to the death penalty.

- **Are there any countries in the world that have national laws that ban discrimination against gays, lesbians, and bisexuals?**

  Yes, more than 20 countries have national laws that ban discrimination based on sexual orientation.

- **Is there any country where same-sex couples can be legally married?**

  Yes. In 2001 the Netherlands became the first country in the world to offer legal marriage to same-sex couples. Belgium, Spain, Canada, and South Africa as well as the state of Massachusetts, have also legalized same-sex marriage.

- **In what ways is the classification of individuals into sexual orientation categories problematic?**

  Classifying individuals into sexual orientation categories is problematic for a number of reasons: (1) Some people conceal or falsely portray their sexual orientation identity; (2) attractions, love, behavior, and self-identity do not always match; and (3) sexual orientation can change over time.

- **What is the relationship between beliefs about what "causes" homosexuality and attitudes toward homosexuality?**

  Individuals who believe that homosexuality is biologically based or inborn tend to be more accepting of homosexuality. In contrast, individuals who believe that homosexuals choose their sexual orientation are less tolerant of gays and lesbians.

- **What is the position of the American Psychiatric Association, the American Psychological Association, the American Academy of Pediatrics, and the American Medical Association on conversion or reparative therapy for gays and lesbians?**

  These organizations agree that sexual orientation cannot be changed and that efforts to change sexual orientation (conversion or reparative therapy) do not work and may, in fact, be harmful.

- **What three cultural changes have influenced the worldwide increase in liberalized national policies on same-sex relations and the gay rights movement?**

  The worldwide increase in liberalized national policies on same-sex relations and the gay rights social movement has been influenced by (1) the rise of individualism, (2) increasing gender equality, and (3) the emergence of a global society in which nations are influenced by international pressures.

- **What is internalized homophobia? What are the effects of internalized homophobia on lesbian and gay individuals?**

  Internalized homophobia is a sense of personal failure and self-hatred among lesbian and gay individuals resulting from social rejection and stigmatization. Internalized homophobia has been linked to increased risk for depression, substance abuse and addiction, anxiety, and suicidal thoughts.

- **What factor has been identified as the best predictor of homophobia?**

  Conservative Christian ideology has been identified as the best predictor of homophobia.

- **Bisexuals experience "double discrimination." What does that mean?**

  Bisexual individuals face rejection not only from heterosexuals but also from many homosexual individuals. Hence, bisexuals experience "double discrimination."

- **In the United States, what factors help to explain the increased acceptance of homosexuality and support for gay rights in the last decade?**

  Factors that help explain the increased acceptance of homosexuality and support of gay rights include the increasing levels of education among U.S. adults, the positive depiction of gays and lesbians in the popular media, and increased contact between heterosexuals and openly gay and lesbian individuals.

- **In June 2003, what was the Supreme Court's decision in *Lawrence v. Texas*?**

  In *Lawrence v. Texas*, the Supreme Court ruled that sodomy laws (laws that criminalized oral and anal sexual acts) were discriminatory and unconstitutional, removing the stigma and criminal branding that sodomy laws have long placed on GLBT individuals. Before this historic ruling, sodomy was illegal in 13 states; in 4 states, sodomy laws targeted only same-sex acts. In states that criminalized both same- and opposite-sex sodomy, sodomy laws were usually not used against heterosexuals but were used primarily against gay men and lesbians.

- **Is employment discrimination based on sexual orientation illegal in all 50 states?**

  As of 2007, only 20 states ban employment discrimination based on sexual orientation. This means it is legal in 30 states to discriminate against someone on the basis of sexual orientation. As of this writing, a federal bill called the Employment Nondiscrimination Act (ENDA), which would ban employment discrimination against individuals on the basis of sexual orientation, is pending before Congress.

- **What is the "Don't ask, don't tell" policy?**

  In 1993 President Bill Clinton instituted a "Don't ask, don't tell" policy in which recruiting officers are not allowed to ask about sexual orientation and homosexuals are encouraged not to volunteer such information. More than 11,000 service members have been discharged under "Don't ask, don't tell" since it took effect, and countless others have decided not to join the military because of this discriminatory law.

- **What is the Marriage Protection Amendment?**

  The Marriage Protection Amendment, previously known as the Federal Marriage Amendment, is a proposal to amend the U.S. Constitution to define marriage as being a union between a man and a woman.

- **Same-sex couples want the same legal and financial benefits that are granted to heterosexual married couples. What are some of these benefits?**

  Married couples have the right to inherit from a spouse who dies without a will, to avoid inheritance taxes between spouses, to make crucial medical decisions for a partner and to take family leave to care for a partner in the event of the partner's critical injury or illness, to receive Social Security survivor benefits, and to include a partner in his or her health insurance coverage. Other rights bestowed on married (or once married) partners include assumption of a spouse's pension, bereavement leave, burial determination, domestic violence protection, reduced rate memberships, divorce protections (such as equitable division of assets and visitation of partner's children), automatic housing lease transfer, immunity from testifying against a spouse, and spousal immigration eligibility.

- **How common is it for anti-LGBT hate crimes to involve heterosexual victims?**

  Because it is not uncommon for heterosexual men and women to be mistaken for gay men and lesbians, 1 in 10 victims (10 percent) of anti-LGBT violence in 2006 identified as being heterosexual.

- **According to a national survey of students ages 13 through 18 and secondary school teachers, what are the two most common reasons that students are harassed by their peers?**

  The most common reason students are harassed by their peers is physical appearance; the second most common reason for being harassed is actual or perceived sexual orientation.

- **A report by Amnesty International (2005) found that police mistreatment and abuse of LGBT people are widespread in the United States. What types of police mistreatment were revealed in this report?**

  Police mistreatment of LGBT individuals reported by Amnesty International included targeted and discriminatory enforcement of laws against LGBT people, verbal abuse, inappropriate pat-down and strip searches, failure to protect LGBT people in holding cells, inappropriate response or failure to respond to hate crimes or domestic violence calls, sexual harassment and abuse, and physical abuse.

- **What are "civil unions" and in what states may same-sex couples enter into civil unions?**

  A civil union is a legal status parallel to civil marriage under state law that entitles same-sex couples to almost all the rights and responsibilities available under state law to married couples. Same-sex couples may enter into civil unions in Vermont, Connecticut, New Jersey, and, effective in 2008, New Hampshire.

- **What are three strategies that promote the health and well-being of all students and address the needs and promote acceptance of GLBT youth?**

  Three strategies discussed in your text are to (1) include sexual orientation diversity in sex education programs, (2) implement policies against antigay harassment in school, and (3) establish gay-straight alliances (GSAs) in middle and high schools.

- **What are Safe Zone or Ally programs?**

  Safe Zone or Ally programs are designed to visibly identify students, staff, and faculty who support the LGBT population. Participants in Safe Zone or Ally programs display a sign or placard outside their office or residence hall room that identifies them as individuals who are willing to provide a safe haven and support for LGBT people and those struggling with sexual orientation issues.

TEST YOURSELF

1. In 1996, which country became the first country in the world to include in its constitution a clause banning discrimination based on sexual orientation?
   a. South Africa
   b. Norway
   c. France
   d. The United States

2. In the United States, about one-fifth of gay male couple households and one-third of lesbian couple households have children present in the home.
   a. True
   b. False

3. Most reparative therapy programs allegedly achieve "conversion" to heterosexuality through
   a. uncovering repressed childhood sexual trauma and confronting the perpetrator
   b. hormonal treatments
   c. intensive psychotherapy and family therapy
   d. embracing evangelical Christianity and being "born again"

4. Why has Reverend Fred Phelps of the Westboro Baptist Church (Topeka, Kansas) picketed funerals of American servicemen and servicewomen who have died in the Iraq war?
   a. To protest the "Don't ask, don't tell" military policy
   b. To protest efforts to repeal the "Don't ask, don't tell" military policy
   c. To publicize his view that U.S. military casualties are God's way of punishing the United States for being nice to lesbians and gays
   d. To show support of Iraq's harsh penalties for homosexuality

5. The "contact hypothesis" explains why, in general, heterosexuals have more negative attitudes toward gay men and lesbian women if they have had prior contact with or know someone who is gay or lesbian.
   a. True
   b. False

6. The majority of Americans believe that homosexual individuals should have equal job opportunities.
   a. True
   b. False

7. A 2007 national poll found that more than half (55 percent) of Americans oppose same-sex marriage. However, the majority (56 percent) of _____ in the Pew Survey support gay marriage.
   a. Women
   b. Democrats
   c. Young adults ages 18–29
   d. Racial and ethnic minorities

8. African Americans and Muslims are more likely than lesbian, gay, and bisexual individuals to report being the victim of a hate crime.
   a. True
   b. False

9. More than half of Fortune 500 companies today offer same-sex domestic partner health benefits.
   a. True
   b. False

10. In which U.S. state can same-sex couples legally marry?
    a. None
    b. Hawaii
    c. California
    d. Massachusetts

Answers: 1 a. 2 a. 3 d. 4 c. 5 b. 6 a. 7 c. 8 b. 9 a. 10 d.

KEY TERMS

biphobia
bisexuality
civil union
conversion therapy
Defense of Marriage Act
domestic partners
Employment Nondiscrimination Act (ENDA)

gay-straight alliances (GSAs)
GLBT
heterosexism
heterosexuality
homophobia
homosexuality
internalized homophobia
lesbigay population

LGBT
master status
registered partnerships
reparative therapy
sexual orientation
sodomy
transgendered individuals

MEDIA RESOURCES

***Understanding Social Problems,* Sixth Edition**
Companion Website

academic.cengage.com/sociology/mooney
Visit your book companion website, where you will find flash cards, practice quizzes, Internet links, and more to help you study.

Just what you need to know NOW! Spend time on what you need to master rather than on information you already have learned. Take a pre-test for this chapter, and CengageNOW will generate a personalized study plan based on your results. The study plan will identify the topics you need to review and direct you to online resources to help you master those topics. You can then take a post-test to help you determine the concepts you have mastered and what you will need to work on. Try it out! Go to academic.cengage .com/login to sign in with an access code or to purchase access to this product.

12

Problems of Youth and Aging

"The quality of a nation is reflected in the way it recognizes that its strength lies in its ability to integrate the wisdom of elders with the spirit and vitality of its children and youth."

Margaret Mead, anthropologist

© jozef sedmak/iStockphoto.com

"When I was 17, I . . . found that my mother's legal rights to me had been terminated seven years earlier . . . without her knowledge or mine. Suddenly I realized that I had no legal mother, that I belonged to nobody."

Jackie—placed in foster care at age 7 after her grandmother died and her mother was unable to care for her (Pew 2004, p. 1).

"I was only six when I went into foster care. I remember vividly just sitting outside the courthouse . . . my birth mother crying. And then suddenly, I was living somewhere else, in some house I didn't know. No one told me anything. For five years, no one told me anything. . . . I would feel like I was just being passed around and not really knowing what was going on. . . . I didn't even know what rights I had . . . if I had any."

Luis—separated from his younger brother and sister and placed in foster care after he was observed by his kindergarten teacher snorting a line of sugar through a straw (Pew 2004, pp. 2, 9).

A recent campaign initiated by the Pew Charitable Trust called "Kids Are Waiting" emphasizes the number of children in foster care who are waiting for a suitable placement or reunification with their parents or guardians. In 2005, there were over a half a million children in the U.S. foster care system.

Both Jackie and Luis are victims of the foster care system—a broken system if you will—a system intended to temporarily care for children until reunification with parents or adoption can take place. But much like the care of the elderly, the foster care system is fraught with abuses and institutional violations. That is not the only similarity shared by the young and the old. Both groups are often the victims of stereotyping, age discrimination, and poverty. Both groups suffer disproportionately from depression and are at higher risks for suicide and physical and mental abuse. Both groups are also major population segments of American society. In this chapter we examine the problems and potential solutions associated with youth and aging. We begin by looking at age in a cross-cultural context.

The Global Context: Youth and Aging Around the World

According to a recent report, "[S]ince the beginning of recorded human history, young children have outnumbered older people. Very soon this will change" (U.S. State Department 2007, p. 6). By 2020 there will be equal proportions of the global population that are younger than 5 years of age and older than 65 years of age; thereafter, there will be a higher proportion of the elderly than of the very young. For example, it is estimated that by 2050, 16 percent of the world's population will be 65 years and older, whereas only 6 percent of the world's population will be younger than 5 years of age.

The populations of different countries, of course, age at different rates. Between 2006 and 2030 the percentage of the elderly in more developed countries is expected to grow by 51 percent; in less developed countries the proportion of the elderly is expected to grow by 140 percent (U.S. State Department 2007).

Differences in the treatment of the young and old are also associated with whether the country is more developed or less developed. For example, although proportionately more elderly live in more developed countries than in less developed ones, these societies have fewer statuses for the elderly to occupy. Their positions as caretakers, homeowners, employees, and producers are often usurped by those aged 18–64. Paradoxically, the more primitive the society, at least historically, the more likely that society is to practice *senilicide*—the killing of the elderly. In some societies the elderly are considered a burden and are left to die or, in some cases, actually killed. For example, in Malawi (a country in Southeast Africa) the elderly are killed for their body parts, which bring a high price on the black market ("Elderly Become 'Muti' Targets" 2003).

Not all societies treat the elderly as a burden. Scandinavian countries provide government support for in-home care workers for the elderly who can no longer perform tasks such as cooking and cleaning. Eastern cultures such as the Japanese revere the elderly, in part, because of their presumed proximity to honored ancestors. Japanese elders sit at the head of the table, enter a room and bathe first, and are considered the heads of families. Japan has the fastest growing proportion of the elderly of any country in the world growing from 20 percent of the population in 2006 to about 30 percent of the population in 2030 (U.S. State Department 2007).

Societies also differ in the way they treat children. In less developed societies children work as adults, marry at a young age, and pass from childhood directly to adulthood with no recognized period of adolescence. In contrast, in industrialized nations children are often expected to attend school for 12–16 years and, during this time, to remain financially and emotionally dependent on their families.

Because of this extended period of dependence, the United States treats "minors" and adults differently. There is a separate justice system for juveniles, and there are age limits for driving, drinking alcohol, joining the military, entering into a contract, marrying, dropping out of school, and voting. These limitations would not be tolerated if placed on individuals on the basis of sex or race. Hence **ageism**—the belief that age is associated with certain psychological, behavioral, and/or intellectual traits, at least with reference to children—is significantly more tolerated than sexism or racism in the United States.

Despite this differential treatment, people in the United States are fascinated with youth and being young. This was not always the case. The elderly were once highly valued in the United States, particularly older men who headed families and businesses. Younger men even powdered their hair, wore wigs, and dressed in a way that made them look older. It should be remembered, however, that in 1900, the average life expectancy in the United States was 47 and more than half the population was younger than the age of 16. Being old was rare and respected; to some it was a sign that God looked upon the individual favorably.

One theory argues that the shift from valuing the old to valuing the young took place during the transition from an agriculturally based society to an industrial one. Land, which was often owned by elders, became less important, as did their knowledge and skills about land-based economies. With industrialization, technological skills, training, and education became more important than land-ownership. Called **modernization theory,** this position argues that as a society becomes more technologically advanced, the position of the elderly declines (Hillier & Barrow 2007).

ageism The belief that age is associated with certain psychological, behavioral, and/or intellectual traits.

modernization theory A theory claiming that as society becomes more technologically advanced, the position of the elderly declines.

YOUTH AND AGING

Age is largely socially defined. Cultural definitions of "old" and "young" vary from society to society, from time to time, and from person to person. For example, in ancient Greece or Rome, where the average life expectancy was 20 years, one was old at 18; similarly, one was old at age 30 in medieval Europe and at age 40 in the United States in 1850.

Age is also a variable that has a dramatic impact on one's life (Matras 1990):

- Age determines one's life experiences because the date of birth determines the historical period in which a person lives. Twenty years ago cell phones and palm-sized computers could not have been imagined.
- Different ages are associated with different developmental stages (physiological, psychological, and social) and abilities. Ben Franklin observed, "At 20 years of age the will reigns; at 30 the wit; at 40 judgment."
- Age defines roles and expectations of behavior. The expression "act your age" implies that some behaviors are not considered appropriate for people of certain ages.
- Age influences the social groups to which one belongs. Whether one is part of a sixth-grade class, a labor union, or a senior's bridge club depends on one's age.
- Age defines one's legal status. Sixteen-year-olds can get a driver's license, 18-year-olds can vote and get married without their parents' permission, and 65-year-olds are eligible for full Social Security benefits.

Childhood, Adulthood, and Elderhood

Every society assigns different social roles to different age groups. **Age grading** is the assignment of social roles and opportunities, given an individual's chronological age (Hillier & Barrow 2007). Although the number of age grades varies by society, most societies make at least three distinctions: childhood, adulthood, and elderhood. Table 12.1 gives the percentages of the U.S. population by year for each of these three age groups.

Childhood. The period of childhood in our society is from birth through age 17 and is often subdivided into infancy, childhood, and adolescence. Infancy has always been recognized as a stage of life, but the social category of childhood

---

TABLE 12.1 Percentage of U.S. Population in Three Age Groups: 1950, 2000, 2006, and 2050

YEAR	ALL AGES	PERCENTAGE YOUNGER THAN 18 YEARS	PERCENTAGE 18–64 YEARS	PERCENTAGE 65 YEARS AND OLDER
1950	100.0	31.3	60.6	8.2
2000	100.0	25.7	61.9	12.4
2006	100.0	25.0	63.0	12.0
2050	100.0	23.7	56.0	20.3

Data are for the resident population. Data for 1950 exclude Alaska and Hawaii.
Source: U.S. Bureau of the Census (1980 [includes data for 1950], 2000, 2007b).

age grading The assignment of social roles and opportunities given an individual's chronological age.

developed only after industrialization, urbanization, and modernization took place. Before industrialization infant mortality was high because of the lack of adequate health care and proper nutrition. Once infants could be expected to survive infancy, the concept of childhood emerged, and society began to develop norms in reference to children. In the United States child labor laws prohibit children from being used as cheap labor, educational mandates require that children attend school until the age of 16, and federal child pornography laws impose severe penalties for the sexual exploitation of children.

Adulthood. The period from age 18 through age 64 is generally subdivided into young adulthood, adulthood, and middle age. Each of these statuses involves dramatic role changes related to entering the workforce, getting married, and having children. The concept of middle age is relatively recent and developed as life expectancy extended. Some people in this phase are known as members of the **sandwich generation** because they are often emotionally and economically responsible for both their young children and their aging parents.

Elderhood. At age 65 one is likely to be considered elderly, a category that is often subdivided into the young-old, the old, and the old-old. Membership in one of these categories does not necessarily depend on chronological age. The healthy, active, independent elderly—increasing in number—are often considered to be the young-old, whereas the old-old are less healthy, less active, and more dependent.

SOCIOLOGICAL THEORIES OF AGE INEQUALITY

Three sociological theories help to explain age inequality and the continued existence of ageism in the United States. These theories—structural functionalism, conflict theory, and symbolic interactionism—are discussed in the following sections.

Structural-Functionalist Perspective

Structural functionalism emphasizes the interdependence of society—how one part of a social system interacts with other parts to benefit the whole. From a structural-functionalist perspective the elderly must gradually relinquish their roles to younger members of society. This transition is viewed as natural and necessary to maintain the integrity of the social system. The elderly gradually withdraw as they prepare for death, and society withdraws from the elderly by segregating them in housing such as retirement villages and nursing homes. In the interim the young have learned through the educational institution how to function in the roles surrendered by the elderly. In essence, a balance in society is achieved whereby the various age groups perform their respective functions: The young go to school, adults fill occupational roles, and the elderly, with obsolete skills and knowledge, disengage. As this process continues, each new group moves up and replaces another, benefiting society and all its members.

This theory is known as **disengagement theory** (Hillier & Barrow 2007). Some researchers no longer accept this position as valid, however, given the

sandwich generation The generation that has the responsibility of simultaneously caring for their children and their aging parents.

disengagement theory A theory that the elderly disengage from productive social roles in order to relinquish these roles to younger members of society.

increased number of elderly individuals who remain active throughout life (Riley 1987). In contrast to disengagement theory, **activity theory** emphasizes that the elderly disengage in part because they are structurally segregated and isolated, not because they have a natural tendency to do so (Quadagno 2005). For those elderly individuals who remain active, role loss may be minimal. In studying 1,720 respondents who reported using a senior center in the previous year, Miner, Logan, and Spitze (1993) found that those who attended the center were less disengaged and more socially active than those who did not.

Conflict Perspective

The conflict perspective focuses on age grading as another form of inequality because both the young and the old occupy subordinate statuses. Some conflict theorists emphasize that individuals at both ends of the age continuum are superfluous to a capitalist economy. Children are untrained, inexperienced, and neither actively producing nor consuming in an economy that requires both. Similarly, the elderly, although once working, are no longer productive and often lack required skills and levels of education. Both the young and the old are considered part of what is called the *dependent population;* that is, they are an economic drain on society. Hence children are required to go to school in preparation for entry into a capitalist economy, and the elderly are forced to retire.

Other conflict theorists focus on how different age strata represent different interest groups that compete with one another for scarce resources. Debates about funding for public schools, child health programs, Social Security, and Medicare largely represent conflicting interests of the young versus the old.

What Do You Think? In 2006, Unilever, under its Dove soap brand, conducted a global survey about women, beauty, and age (Dove 2007). A representative sample of more than 1,400 women between the ages of 50 and 64 in nine countries (United States, Canada, Mexico, Brazil, the United Kingdom, Italy, Germany, France, and Japan) were interviewed. The results indicated that more than 90 percent of women believe that (1) the media and advertising do not portray images of women older than 50 realistically and (2) ". . . society is less accepting of appearance considerations for women older than 50 than their younger counterparts, with showcasing one's body the least acceptable" (p. 1). In response to such results, Unilever has launched an advertising initiative called the "Campaign for Real Beauty." The initiative showcases women older than 50 in a variety of poses using Dove beauty products. Says marketing director Kathy O'Brien, "Dove seeks to create an attitudinal change in the anti-aging category—from negative and fear-driven to affirmative and hope-driven" (p. 1). What do you think of the Dove "pro-aging" campaign? Would a similar campaign for men older than 50 be effective? What do you think was Unilever's motivation for beginning the Dove campaign?

activity theory A theory that the elderly disengage, in part, because they are structurally segregated and isolated with few opportunities to engage in active roles.

Symbolic Interactionist Perspective

The symbolic interactionist perspective emphasizes the importance of examining the social meaning and definitions associated with age. Teenagers are often portrayed as lazy, aimless, and awkward. The elderly are also defined in a number of stereotypical ways, contributing to a host of myths surrounding the inevitability of physical and mental decline. Table 12.2 identifies some of these myths.

Media portrayals of elderly people contribute to their negative image. Young people are typically portrayed in active, vital roles and are often overrepresented in commercials and television programming (Bauder 2006). In contrast, elderly individuals are portrayed as difficult, complaining, and burdensome and are often underrepresented in commercials. A study of elderly characters in popular 1940s through 1980s films concluded, "older individuals of both genders were portrayed as less friendly, having less romantic activity, and enjoying fewer positive outcomes than younger characters at a movie's conclusion" (Brazzini et al. 1997, p. 541). Media images are powerful; a study found that

> "Since it is the Other within us who is old, it is natural that the revelation of our age should come to us from outside, from others."
>
> Simone de Beauvoir
> Feminist author

TABLE 12.2 Myths and Facts About the Elderly

CATEGORY	MYTH	FACT
Health	The elderly are always sick; most are in nursing homes.	More than 85 percent of the elderly are healthy enough to engage in their normal activities. Less than 5 percent are confined to a nursing home.
Mental status	The elderly are senile.	Although some of the elderly learn more slowly and forget more quickly, most remain oriented and mentally intact. Only 20–25 percent develop Alzheimer's disease or some other incurable form of brain disease. Senility is not inevitable as one ages.
Isolation	The elderly are alone and isolated from family members.	Most older people have regular contact with friends and family; many see at least one of their children once a week.
Employment	The elderly are inefficient employees.	Although only 19.8 percent of men and 11.5 percent of women, age 65 and older, are still employed, those who continue to work are efficient workers. Compared with younger workers, the elderly have lower job turnover, fewer accidents, and less absenteeism. Older workers also report higher satisfaction in their work.
Politics	The elderly are not politically active.	In 2004 individuals age 65 and older were more likely to be registered to vote and/or to vote than any other age group.
Sexuality	Sexual satisfaction disappears with age.	Many elderly people report active and satisfying sex lives. For example, of couples age 75 years old and older, more than 25 percent reported having sexual intercourse once a week.
Adaptability	The elderly cannot adapt to new working conditions.	A high proportion of the elderly are flexible in accepting change in their occupations and earnings. Adaptability depends on the individual: Many young people are set in their ways, and many older people adapt to change readily.

Sources: Palmore (1984), Binstock (1986), Seeman & Adler (1998), Toner (1999), Federal Elections Commission (2005), Quadagno (2005), AOA (2007), Hillier and Barrow (2007), and U.S. Bureau of the Census (2007b).

Despite age grading, many singers such as Stevie Nicks, Mick Jagger, and Cher continue to perform and are commercially successful. Tina Turner, almost 60 in this picture, defies age stereotypes.

children as young as 5 years old have already developed negative stereotypes of the elderly (EurekAlert 2003) (see this chapter's *Social Problems Research Up Close* feature).

The elderly are also portrayed as childlike in terms of clothes, facial expressions, temperament, and activities—a phenomenon known as **infantilizing elders**. For example, young and old are often paired together. A promotional advertisement for the movie *Just You and Me, Kid,* with Brooke Shields and George Burns, described it as "the story of two juvenile delinquents." In *Grumpy Old Men* the characters played by Jack Lemmon and Walter Matthau get "fussy" when they get tired, and the media focus on images of Santa visiting nursing homes and local elementary school children teaching residents arts and crafts. Elderly individuals are often depicted in role reversal, cared for by their adult children, as in the television situation comedies *Golden Girls, Frasier,* and *King of Queens.* Finally, a new search engine designed for aging baby boomers is called—what else—Cranky.com (Liedtke 2007).

Negative stereotypes and media images of the elderly engender **gerontophobia,** a shared fear or dread of elderly persons, which may create a self-fulfilling prophecy. For example, in a 20-year study of 600 respondents, researchers found that elderly people with positive perceptions of aging (e.g., seeing the elderly as wise) lived, on average, 7½ years longer than those who had a negative image of aging (e.g., considering most elderly people to be senile) (Ramirez 2002).

PROBLEMS OF YOUTH

The number of people younger than the age of 18—73.7 million in 2006—is the largest in history and will grow to 80 million by the year 2020 (U.S. Bureau of the Census 2007a). Although some evidence indicates that conditions for children

infantilizing elders The portrayal of the elderly in the media as childlike in terms of clothes, facial expressions, temperament, and activities.

gerontophobia Fear or dread of the elderly.

© Reuters/NewMedia Inc./Corbis

The extent to which minorities, including the elderly, are stereotyped in the media is often the subject of social science research. However, the following study by Donlan and colleagues (2005) goes a step further than simply documenting the existence of such stereotypes. In their study the researchers ask important questions regarding the impact of lifetime television exposure and the effects of maintaining a TV character impression diary on, among other things, views of the elderly and projected future media consumption.

Sample and Methods

Seventy respondents between the ages of 60 and 92 were recruited from a cross-section of New England communities. Their average age was 73, and, reflecting the general trend for women to live longer than men, 64 percent of the sample were women. Seventy percent of the participants were retired, and, of those who were employed, two-thirds worked part-time.

Participants were randomly assigned to one of two conditions. Members of the control group were asked to keep a "media diary" of television viewing for a week's time. For 7 days respondents recorded the name of each show they watched and the channel, time, and type of program (e.g., drama or comedy). The intervention group, that is, the experimental group, in addition to recording the same information as the control group, also responded to several questions about the portrayal of the elderly. Specifically, each day they were asked to select "one older character that made the strongest impression" on them and answer questions about that character's portrayed level of health, role in the program (e.g., major or minor part), and estimated age. After completing the viewing diaries, all respondents were asked to answer questions concerning their views on aging and estimated frequency of future television viewing.

There are two independent variables in this study. The first is what the researchers call lifetime television exposure, as measured by asking respondents the age at which they began watching television and the number of hours, in general, they watched television per week. The second independent variable is the experimental manipulation, that is, having respondents in the intervention group keep an "impression" diary of selected older television characters.

Attitude toward old age is the first dependent variable. Respondents were asked the extent to which negative aging words (e.g., wrinkled, grumpy, or senile) matched or did not match their views of the elderly. This was rated on a 7-point scale from "does not match at all" to "completely matches." Donlan and colleagues (2005) hypothesized that "elders with higher exposure to television will harbor more negative images of aging than those with lower exposure to television" (p. 309).

The second dependent variable was measured by asking respondents about the portrayal of the elderly on television. For example, respondents were asked, "When older persons appear on television as characters, how are they most often portrayed?" Responses ranged on a 5-point scale from "very negatively" to "very positively." Donlan and coworkers hypothesized that "an intervention based on maintaining a diary of viewing impressions will increase elders' awareness that

---

may be improving, numerous problems remain (see Table 12.3). Recently, for example, UNICEF conducted a comprehensive assessment of children's lives in economically advanced nations. Forty indicators measured six dimensions of child well-being: (1) material well-being, (2) health and safety, (3) educational well-being, (4) family and peer relationships, (5) behaviors and risks, and (6) subjective well-being. Of the 21 nations for which complete data were available, the United States scored second to last on the overall measure of child well-being. Only the United Kingdom scored lower. Further, the United States was in the bottom third of the rankings for five (educational well-being was the exception) of the six dimensions (UNICEF 2007a). The top one-third countries included the Netherlands, Sweden, Denmark, Finland, Spain, Switzerland, and Norway.

Child Labor

child labor Involves children performing work that is hazardous, that interferes with a child's education, or that harms a child's health or physical, mental, social, or moral development.

Child labor involves a child performing work that is hazardous, that interferes with a child's education, or that harms a child's health or physical, mental, social, or moral development (ILO 2006). Even though virtually every country in the world has laws that limit or prohibit the extent to which children can be

television programming presents the old in a stereotypical and infrequent manner" (p. 309).

Findings and Conclusions

Supporting the first hypothesis, Donlan and colleagues (2005) found that lifetime television exposure was significantly and directly associated with negative age stereotypes; that is, the higher the lifetime television exposure, the greater the negative attitudes toward the elderly. In fact, lifetime television exposure was a better predictor of negative age stereotypes than the respondent's age, education, or health. As Donlan and coworkers conclude, "The impersonal programming of television seemed to have a greater impact on images-of-aging formation than did more personal factors, such as depression and self-rated health" (p. 314).

The second hypothesis was also supported. Participants in the intervention group, compared with the control group participants, were more likely to be aware of the infrequency of older characters on television. They were also more likely to report their intention to decrease television viewing in the future. However, there was no significant difference between partici-

pants in the intervention group and participants in the control group with regard to their views on aging. This is not surprising, however. As Donlan and colleagues note, it would be unrealistic to expect significant changes in the way respondents view the elderly as a result of 1 week of heightened awareness.

Furthermore, the analysis revealed that participants in the intervention group were more likely to acknowledge the negative portrayal of the elderly on television after the intervention than before the intervention. This finding held only for comedies, not dramas. In explaining this result, the investigators noted that comedy programming has long stereotyped the elderly. In fact, one intervention group participant stated that, in reference to a popular situation comedy character, "The portrayal of this character is negative—but it's a comedy" (Donlan et al. 2005, p. 314).

In addition to allowing hypothesis testing, the media diaries provide a qualitative glimpse into perceptions of the elderly by older television viewers. For example, the omission of the elderly was noted in a variety of program types.

A 71-year-old male viewer of television news noted, "When people are interviewed about different matters, older people were left out." Similarly, a 60-year-old game show viewer indicated that, "They should try and get older Americans as contestants." [And] according to a 68-year-old homemaker who watched more than 45 hours of television a week, "I feel like we've been ignored. I feel like we're non-existent." (Donlan et al. 2005, pp. 314–315)

Finally, the researchers acknowledged that the results may not be applicable to other countries. For example, several of the participants noted that BBC (British Broadcasting Corporation) programming portrays the elderly in a more positive light. As one participant commented in reference to a British comedy shown on public television in the United States, "I wish America could get a show going like this that has some intelligence to it—not some idiotic nonsense. In America older parents are always portrayed as the outlaw in-laws."

---

employed, child labor persists throughout the world. A report by the International Labor Office estimated that there are 218 million school-age children who are child laborers, although the number has decreased in recent years (ILO 2006). The largest reductions have been in Latin America and the Caribbean where the number of child laborers has decreased by two-thirds over the last 4 years. The least progress has been made in sub-Saharan African where, according to the report, child labor remains at alarmingly high rates.

Child laborers work in factories, workshops, construction sites, mines, quarries, and fields, on deep-sea fishing boats, at home, on the street, and on the battlefield where, globally, 250,000 child soldiers endure the rigors of armed conflict (SOWC 2007). Child laborers make bricks, shoes, soccer balls, fireworks and matches, furniture, toys, rugs, and clothing. They work in the manufacturing of brass, leather goods, and glass. They tend livestock and pick crops. In Egypt more than 1 million children, ages 7–12, work each year in cotton-pest management. They endure routine beatings by their foremen as well as exposure to heat and pesticides (Human Rights Watch 2001). Children typically earn the equivalent of $1 per day and work from 7:00 a.m. to 6:00 p.m. daily, with one midday break, 7 days a week.

" Child labor is a scar on the world's conscience. **"**

UNICEF report

TABLE 12.3 Each Day in America

1 mother dies in childbirth.

4 children are killed by abuse or neglect.

5 children or teens commit suicide.

8 children or teens are killed by firearms.

33 children or teens die from accidents.

77 babies die before their first birthdays.

192 children are arrested for violent crimes.

383 children are arrested for drug abuse.

906 babies are born at low birth weight.

1,153 babies are born to teen mothers.

1,672 public school students receive corporal punishment.*

1,839 babies are born without health insurance.

2,261 high school students drop out.*

2,383 children are confirmed as abused or neglected.

2,411 babies are born into poverty.

2,494 babies are born to mothers who are not high school graduates.

4,017 babies are born to unmarried mothers.

4,302 children are arrested.

17,132 public school students are suspended.*

*Based on calculations per school day (180 days of 7 hours each).
Source: CDF (2007b).

Illegal and oppressive employment of children also occurs in the United States in restaurants, grocery stores, meat-packing plants, sweatshops in urban garment districts, and in agriculture. Between 300,000 and 800,000 U.S. child workers labor on commercial farms alone, frequently under dangerous and grueling conditions (Human Rights Watch 2003). Child farm workers in the United States often work 12-hour days, sometimes beginning at 3:00 or 4:00 a.m. Many are exposed to dangerous pesticides that cause cancer and brain damage, with short-term symptoms including rashes, headaches, dizziness, nausea, and vomiting. They often work in 100° F temperatures without adequate access to drinking water and are sometimes forced to work without access to toilets or hand-washing facilities (Human Rights Watch 2003). Agriculture is also one of the most dangerous occupations, and children (as well as adults) sustain high rates of injury from work with knives, other sharp tools, and heavy equipment. Despite federal prohibitions, even youth employed in service and retail sectors are exposed to harmful conditions and dangerous equipment (e.g., paper balers, box crushers, and dough mixers) (Runyan et al. 2007). It is estimated that 230,000 workers younger than age 18 are injured on the job every year in the United States, and 65 die from their workplace injuries (NCL 2007).

Child Prostitution and Trafficking. One of the worst forms of child labor is child prostitution and child trafficking (see Chapter 4). Worldwide, it is estimated that there are 1 million child prostitutes; in the United States the estimate is 300,000 (Dorman 2001; United Nations 2003a).

In some countries, particularly where the human immunodeficiency virus (HIV)/acquired immunodeficiency syndrome (AIDS) pandemic is rampant, orphaned children work as prostitutes to support themselves (SOWC 2007). In poor countries the sexual services of children are often sold by their families in an attempt to get money. Some children are kidnapped or lured by traffickers with promises of employment, only to end up in a brothel. For example, promised a job as a waitress, this 13-year-old girl was brought to the United States and forced to work as a "sex slave" (Allen 2002):

> She was a virgin, and she didn't know what they were talking about, but she knew it was bad, so she refused. She was then brutally gang-raped to induct her into the business. For the next six months, she was forced to service 10–15 men a day. Twice she was impregnated, twice forced to have an abortion, and twice she was back in the brothel the next day. The traffickers circulated her through trailer brothels and private parties, where she was passed around. She was pistol-whipped and raped if she resisted. The girl was rescued when two girls ran to neighbors who called police. She had multiple sexually transmitted diseases, scar tissue from the forced abortions, and

© Tomas Bravo/Reuters/Corbis

Street children are often subject to police brutality and violence. Many are held for long periods of time without due process. It is estimated that there are 200 million street children worldwide. In this picture, two plainclothes policemen take a youth into custody.

was addicted to drugs and alcohol. She had posttraumatic stress syndrome, including severe depression and suicidal thoughts. She was physically, mentally, emotionally and spiritually broken. (p. 1)

The U.S. Immigration and Naturalization Service identified 250 brothels in 26 American cities in which forced prostitutes, including children, are taken. Americans also engage in child-sex tourism abroad, particularly in Southeast Asia. Of 240 identified cases in which legal action was taken against foreigners for sexually abusing children in this region, researchers found that about one-fourth of the violators were from the United States (Dorman 2001).

Orphaned and Street Children

Worldwide, there are millions of orphans—143 million in the developing world alone (SOWC 2007). Many are kept in institutions, some are living on the streets, and still others, with no means of supporting themselves, turn to sex work, crime, or drugs. In many institutions children are simply warehoused with little food, clean water, or medical care. Even in institutions that do provide such necessities, many of the children are emotionally neglected or otherwise subject to inhumane treatment (Human Rights Watch 2005). Recently, the number of orphans has increased dramatically for at least three reasons.

First is the devastation of the HIV/AIDS pandemic. Globally, the number of children orphaned by the disease, predominantly in developing nations, is 15 million, and this number is expected to grow to 25 million by 2010 ("AIDS Creating Global 'Orphans Crisis,'" 2003; Avert 2007). More than 12 million of these children live in sub-Saharan Africa where an estimated 9 percent of all children have lost one or both parents to HIV/AIDS (Avert 2007). Second, children are increasingly orphaned from armed conflicts. For example, the civil war in Rwanda is responsible for hundreds of thousands of orphans, many of whom are now street children. Yet a third reason is natural disasters, such as the tsunami in December 2004, which left hundreds of thousands of children orphaned and homeless.

Globally, according to UNICEF, more than 200 million children live on the streets—the equivalent of two-thirds the population of the United States. Many of these children have never known a home, a bathroom, a kitchen, or a safe place to sleep (Lyst 2007). Some are "addicted to inhalants, such as cobbler's glue, which offers them an escape from reality in exchange for a host of physical and psychological problems" (Casa Alianza 2003, p. 1). Still others, resorting to prostitution and vulnerable to rape, have HIV/AIDS. For example, of Mexico's 2 million street children at least 7 percent are HIV-positive. In many cities street children are driven outside city limits and deserted, arrested, imprisoned, and even murdered—all in the name of "social cleansing." Unfortunately, the number of street children is predicted to increase as the global population continues to grow and urbanization spreads (SOWC 2007).

Children's Rights

Historically, children have had little control over their lives. They have been "double dependent" on both their parents and the state. Indeed, colonists in America regarded children as property. Beginning in the 1950s, however, the view that children should have greater autonomy became popular and was codified in several legal decisions and international treaties. In 1959 the United Nations General Assembly approved the Declaration on the Rights of the Child, which held that health care, housing, and education, as well as freedom from abuse, neglect, and exploitation, are fundamental rights of children.

A second measure, the Convention on the Rights of the Child (CRC), further articulated the rights of children and was adopted by the United Nations in 1989. The CRC is recognized for the establishment of international norms in the treatment of children—protection from violence, abuse, abduction, hazardous employment, and exploitation; adequate nutrition; free primary education; adequate health care; and equal treatment regardless of gender, race, and cultural background (UNICEF 2007b, p. 1). Countries ratifying the document have made significant improvements in the lives of children. Based on the principles of the CRC, children in Rwanda are no longer in adult detention facilities but have been moved to newly established juvenile centers. In Germany and Belgium state officials have been given greater authority to prosecute those who sexually exploit children, and in El Salvador, Bolivia, and Peru new juvenile justice codes have been enacted.

Only two countries, the United States and Somalia, have failed to ratify the Convention (UNICEF 2007b). The United States may not have signed the pact because of Article 11, which requires that children be assured "the highest attainable standard of health." Some people are concerned that accepting such a position would result in cases being brought to court that the United States is simply unwilling, at present, to hear. The most recent provisions of the Convention deal with the involvement of children in armed conflicts, the sale of children, child prostitution, and child pornography (United Nations 2003b).

Children are both discriminated against and granted special protections under the law. Although legal mandates require that children go to school until

66 Today we live in a world where almost everyone agrees that anyone below 18 years old is a child and has the right to special care and protection. 99

UNICEF
Voices of Youth

age 16, other laws provide a separate justice system whereby children have limited legal responsibility based on their age status. For example, in 2005 the U.S. Supreme Court prohibited execution of convicted criminals who committed their crimes before the age of 18. In a 5–4 decision Justice Kennedy, writing for the majority, argued "the age of 18 is the point where society draws the line for many purposes between childhood and adulthood. It is, we conclude, the age at which the line for death eligibility ought to rest" (Totenberg 2005, p. 1).

Finally, in what is being called one of the most important court decisions on children's rights in recent years, a federal court has held that abused and neglected children have the constitutional right to legal representation. The decision, resulting from a class action suit against Georgia's child welfare agency, will amend Georgia law, which previously held that abused and neglected children had a right to an attorney only when the state had entered a plea to terminate parental rights (Dewan 2005).

What Do You Think? In 2007, California's Senate Public Safety Committee passed the *Juvenile Life Without Parole Reform Act* which would eliminate life without parole sentencing for children younger than 18 and replace it with a sentence of 25 years to life. The new sentencing guidelines would give youthful offenders the possibility of parole after 25 years assuming compelling evidence of rehabilitation. The United States is one of the few countries in the world that sentences children to life without the possibility of parole. Currently there are more than 2,000 children serving life sentences without the possibility of parole in U.S. prisons. In all other countries combined, there are a total of 12 youth offenders serving a sentence of life in prison without parole (Human Rights Watch 2007). What do you think? Should children as young as 13 be sentenced to life in prison without the possibility of parole? At what age would you draw the line in terms of who should or should not be sentenced to life in prison without parole?

Foster Care and Adoption

According to the Administration for Children and Families, as of September 2005 there were more than 513,000 children in the U.S. foster care system (ACF 2006). Most children are placed in foster care because of parental abuse or neglect (CWLA 2005). Although black and Hispanic children are disproportionately represented in the foster care system, we cannot conclude that child abuse and neglect are more likely in black and Hispanic households (see Chapter 5). For example, because of the tendency for minority families to have lower incomes than white families, relatives may be less able to provide a home for the abused or neglected foster child.

The average number of months in foster care is 28.6. Because older children are less likely to be adopted, the average age of a child in foster care is 10, with children younger than 1 representing less than 6 percent of the total number of

In 2006, Michael and Sharon Gravelle were convicted of felony child endangerment, misdemeanor child endangerment, and misdemeanor child abuse. The prosecutor contended that the Gravelles locked some of their 11 adopted special needs children in alarm-rigged cages at night to control their behavior, disciplined another with a sawed-off broomstick, and forced yet another child to live in the bathroom and sleep in the bathtub for 81 nights because of bed-wetting. The children have since been removed from the home and are now in foster care.

foster care children (ACF 2006). Many of these children have special physical or emotional needs or are part of a sibling group, which makes adoption difficult. Many others "age-out" of the system; in 2005, 25,000 children were forced to leave foster care because of their age. As one former foster child states, "I turned 18 a month before I graduated from high school. The day after graduation I was kicked out of my foster home, where I had been living for two years. I was 18, a high school graduate on my way to college in the fall, and I was homeless" (Pew 2007, p. 1). Children who age out of the system are more likely to be incarcerated within 2 years of leaving foster care, more than 20 percent will become homeless at some time in their lives, only half will graduate from high school, and less than 3 percent will receive college degrees.

The foster care system has grown over the years, with federal reimbursement to states increasing from $1.2 billion in 1989 to $6.2 billion in 2004 (PBS 2005). Despite increased funding, stories of abused, neglected, and missing children in the foster care system remain. For example, in 2004 the nonprofit advocacy group Children's Rights filed a class action suit against the state of Mississippi on behalf of 3,500 foster children. According to the suit, the state has not met federal or its own state standards in nearly a decade, leading to abuse and neglect, and the deaths of several children. With settlement of the suit in 2007, the agreement requires that the state of Mississippi develop a plan to reform the foster care system; if the plan is inadequate, a district court will decide what remedy should be granted (Children's Rights 2007).

Poverty and Economic Discrimination

There are 2.2 billion children in the world, and nearly half of them live in poverty (SOWC 2007). Worldwide it is estimated that more than one-third of poor children live in developing countries. The highest poverty rates are in sub-Saharan Africa, where 65 percent of the children are poor, and in South Asia, where 59 percent of the children are poor. Most of these children live in rural areas; one in five has no access to safe water, and one in seven has no access to health services (World Youth Report 2005; SOWC 2007).

Of the major industrialized nations of the world, the United States leads in child poverty, with a rate of 17.0 percent, or one of every six children living in poverty in 2004 (CDF 2006). Further, the number of American children in *extreme* poverty—that is, those living in families with incomes one-half of the poverty level or less—grew twice as fast as child poverty in general. In 2004, 33.0 percent of black children, 29.0 percent of Hispanic children, and 10.0 percent of non-Hispanic white children were considered poor (Forum on Child and Family Statistics 2006).

The effects of childhood poverty are far-reaching (see Chapter 6). Studies have shown that child poverty is related to school failure (CDF 2006), negative involvement with parents (Harris & Marmer 1996), stunted growth, reduced cognitive abilities, limited emotional development (Brooks-Gunn & Duncan 1997; CDF 2004), and a higher likelihood of dropping out (Duncan et al. 1998) or being held back from progressing to the next grade level in school (SOWC 2007). Poverty is also the best single predictor of child abuse and neglect (CDF 2006).

Three decades ago the elderly were the poorest age group in the United States, but today it is children ("Snapshots of the Elderly," 2000). Although Social Security, Medicare, housing subsidies, and Supplemental Security Income (SSI) keep millions of elderly persons out of poverty, the United States has no universal government policy to protect children. Further, federal spending on children is shrinking. The Urban Institute reported that in 1960, 20 percent of domestic federal spending was on children; in 2006 15 percent was on children; and by 2017, an estimated 13 percent will be spent on children (Carasso, Steuerle, & Reynolds 2007). In addition, recent welfare reform has led to cutbacks in the few programs that do benefit children. For example, immediately after passage of the 1996 Welfare Reform Act, the number of children receiving food stamps decreased by nearly one-third. Since 2000, the number of children receiving food stamps has increased although it has yet to achieve pre-1996 levels (Child Trends 2006).

Children are also the victims of discrimination in terms of employment, age restrictions, wages, training programs, and health benefits. Traditionally, children worked on farms and in factories but were displaced during the Industrial Revolution. In 1938 Congress passed the Fair Labor Standards Act, which required factory workers to be at least 16 years old. Although the law was designed to protect children, it was also discriminatory in that it prohibited minors from having free access to jobs and economic independence. Today, 14- and 15-year-olds are restricted in the number of hours per day and hours per week they can work; those younger than 14 are, in general, prohibited from working.

Children, Violence, and the Media

Gang violence, child abuse, and crime are all-too-common childhood experiences. For example, the results of a survey of ninth through twelfth graders conducted by the Centers for Disease Control and Prevention revealed that as many as 6 percent of students stay home from school on 1 or more days because they feel unsafe at school and/or on their way to or from school (YRBSS 2006).

According to the Children's Defense Fund, U.S. children are 12 times more likely to die from gunfire, 16 times more likely to be murdered by a gun, and

NDOT Photograph

The Children's Memorial Flag is part of a larger initiative to reduce violence against children. The paper doll-like figures are solid in color with the exception of a white chalk outline of a missing child "symbolizing the thousands of children lost to violence" (CWLA 2007). The flag was created by a 16-year-old Alameda County, California, student and is flown in all 50 states on National Children's Memorial Flag Day.

9 times more likely to die from a firearm accident than children in 25 other industrialized nations *combined* (CDF 2003a). In 2004, the gun death toll in the United States for children and teens was 2,825—more than the number of military personnel killed in Iraq and Afghanistan since the wars began through December 2006 (CDF 2007a). Child abuse and neglect remain at epidemic levels with an estimated 1 million children victimized annually; children younger than age 18 account for 15.3 percent of all arrests, excluding traffic violations (ACF 2007; FBI 2006).

School violence and particularly the tragic deaths of 12 students and 1 teacher at Columbine High School have focused attention on the relationship between youth violence, guns, and the media. In fact, when a random sample of Americans was asked the causes of school violence, 68 percent responded that "the portrayal of violence and use of guns in today's entertainment and music" was an extremely or very important factor (Gallup 2004). The typical child watches 28 hours of television a week. By the time the child is 18, he or she will have seen 16,000 murders and 200,000 other acts of violence on television. Ironically, the Center for Media and Public Affairs reported that commercial television for children is 50–60 times more violent than prime-time TV with, for example, cartoons averaging 80 violent acts an hour (CDF 2003b).

"New media"—computers, the Internet, video games, and the like—have also come under attack. For example, with children playing video games, on average, 50 minutes a day (Kaiser Family Foundation 2005), there is some fear that violent video games may lead to aggressive behavior and indeed research supports such fears. For example, a study by Anderson and Dill (2000) showed that "exposure to violent video games . . . increases violent behavior both in the short term (e.g., laboratory aggression) and in the long term (e.g., delinquency)." In 2005 the brother of a murder victim filed a multimillion-dollar wrongful death suit against the makers and marketers of *Grand Theft Auto,* arguing that the alleged offender was programmed for violence as a result of playing the video game. Several states are now considering banning the sale of violent video games to those younger than age 17 (Bradley 2005). Additionally, Germany, the third largest market for computer gaming in the world, is considering legislation that would make it a criminal offense for any developer, retailer, or player to make, sell, or use any video game which features "cruel violence" (Benoit 2007).

Children's Health

In addition to the problems already discussed, children have a variety of health problems, both mental and physical. One in five children, for example, has a diagnosable mental disorder, such as depression or schizophrenia (see Chapter 2). Many children, particularly girls, have eating disorders, such as anorexia nervosa or bulimia. It is estimated that 1 percent of young girls will develop anorexia and of that number, 10 percent will die of the disease. Suicide is the third most common cause of death of 10- to 14-year-olds, 15- to 19-year-olds, and 20- to 24-year-olds (NIMH 2006). Suicide rates among the young have tripled since 1960. More youth die from suicide than from heart disease, cancer, HIV/AIDS, stroke, birth defects, and chronic lung disease combined (NMHA 2003).

Worldwide, the most common causes of children's deaths are complications at birth, pneumonia, diarrhea, and malaria (Save the Children 2007). In the least developed countries, 33 percent of children younger than age 5—42 million children—are moderately or severely under weight (SOWC 2007). Yet, tragically, in the United States, child obesity is a serious and growing problem. According to the American Heart Association, "childhood obesity is one of the most critical public health problems today and threatens to reverse the last half century's gains in reducing cardiovascular (CVD) disease and death" (AHA 2005). The prevalence of overweight children has increased from 5 percent in 1980 to 16 percent in 2005, with rates as high as 20 percent for African-American and Hispanic children. In addition to an increased risk of a variety of health problems, overweight children are less likely to have friends, have more difficulty networking, and, if teased about their weight, are more likely to commit suicide.

Kids in Crisis

Childhood is a stage of life that is socially constructed by structural and cultural forces of the past and present. The old roles for children as laborers and farm helpers are disappearing, yet no new roles have emerged. While being bombarded by the media, children must face the challenges of dealing with an uncertain future, peer culture, divorce, poverty, and crime. Parents and public alike fear that children are becoming increasingly involved with sex, drugs, alcohol, and violence. Some even argue that childhood as a stage of life is disappearing.

Despite the many problems associated with childhood, the condition of children in the United States is better than the public's perception of it. For example, 12 percent of children lack health insurance, but 93 percent of a national sample of adults believed that 20–30 percent of children lacked health insurance. In addition, almost half the adults sampled believed that 30 percent of children live in poverty when in fact 17 percent live in poverty. The significance of these discrepancies lies in the public's influence on social policy. If the public has an inaccurate picture of the state of children in the United States, policy makers may be influenced incorrectly (Peterson 2003).

Although some gains are being made, conditions remain intolerable for millions of children in the United States and around the world: homelessness, sexual exploitation, low birth weight, hunger, childhood depression, absent or inadequate health care, and dangerous living conditions (e.g., foster care and child abuse). Annually, the Foundation for Child Development issues a comprehensive report on how children are doing in the United States. Called the Child and Youth Well-Being Index (CWI), the latest report concluded that:

> After eight years of steady improvement from 1994 through 2002, the quality of life of American's children and youth appears to be at a standstill. . . . The overall stall in children's quality of life—underscored by the dramatic decline in children's health as well as persistent ethnic and racial disparities in the areas of education and poverty—sends a strong signal that America should be doing more to improve children's lives. (Foundation for Child Development 2007, p. 3)

Such an investment, particularly in preventive initiatives, is not only humane but also cost-effective, because millions of dollars that would be spent on solving child-related problems after they occur could be saved.

DEMOGRAPHICS: THE GRAYING OF AMERICA

In recent years the ratio of children to old people has changed dramatically (see Table 12.1 and Figure 12.1). In 2005, the most recent year for which data are available, there were 36.8 million persons 65 years of age or older representing 12.4 percent of the population or about one in eight Americans, and their numbers are growing. Between 1995 and 2005 the percentage of persons 65 and older increased by 9.4 percent. During the same time period, the number of people between 45 and 64, those who will turn 65 in the next two decades, increased by 40 percent (AOA 2007).

These statistics reflect three significant demographic changes. First, 76 million baby boomers, born between 1946 and 1964, are getting older. They represent the widest part of the three pyramids in Figure 12.1. Second, life expectancy has increased as a result of a general trend toward modernization, including better medical care, sanitation, and nutrition. For example, a child born in 2004 can expect to live 79 years—30 years more than a child born in 1900 (AOA 2007). Finally, decreasing birth rates mean fewer children and a higher percentage of the elderly. For example, in Japan low birth rates and rising life expectancies have contributed to the highest proportion of elderly persons in the world (Statistical Handbook of Japan 2006).

What Do You Think? Clearly the world's population is getting older and, as such, there is a tendency to concentrate on the social problems associated with demographic aging. There is, however, much to be gained from this rapidly expanding population (AARP 2007). Older people have a wealth of experience—a wisdom of sorts. That is the basis of the Elder Wisdom Circle (EWC), an online group of 600 "cyber-grandparents" between the ages of 60 and 105 who collectively have more than 45,000 years of experience (EWC 2007). Each month they receive an estimated 3,500 letters, more than one-third of which are from teenagers and young adults (Stich 2006). Three-quarters of the letters deal with relationship issues. The cyber-grandparents have even written a book—*The Elder Wisdom Circle Guide for a Meaningful Life*. Would you ever write to or take advice from an EWC advisor? Do you talk to your own grandparents about problems you are having? What do you think are some of the benefits or problems associated with the EWC?

Age and Race and Ethnicity

In 2005, 18.5 percent of the elderly population were racial or ethnic minorities. Minority populations are projected to represent 23.6 percent of the U.S. elderly population by 2020, up from 16.4 percent in 2000. Although the elderly white population is projected to increase 74 percent between 2004 and 2030, the elderly minority population is expected to increase 183 percent in the same time period (AOA 2004). The highest growth is among Hispanics (254 percent), followed by

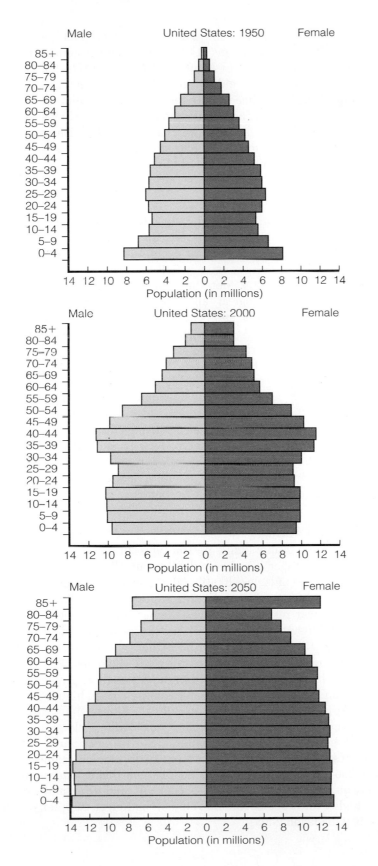

FIGURE 12.1
U.S. population pyramids.
Source: U.S. Bureau of the Census (2002).

Advertising Archives, Inc.

Women, as they get older, are often portrayed as sexually unattractive, something men are rarely subjected to. Advertising contributes to this image by idealizing youth in advertisements that advocate the use of wrinkle creams, hair dyes, and getting one's "girlish figure" back.

Asians and Pacific Islanders (208 percent), African Americans (147 percent), and American Indians, Eskimos, and Aleuts (143 percent) (AOA 2007). The growth of the elderly minority population is a consequence of higher fertility rates and immigration patterns, particularly among Hispanics. Given their higher rates of diabetes, arthritis, and cardiovascular disease, the increased numbers of older minorities present a unique challenge to health care providers.

Age and Sex

In the United States elderly women outnumber elderly men. The sex ratio for individuals age 65 and older is 139 women for every 100 men; for those 85 and older the sex ratio is 218 women for every 100 men (AOA 2007). Men die at an earlier age than women for both biological and sociological reasons: heart disease, stress, and occupational risk (see Chapters 2 and 10).

At age 65 men can expect to live an additional 16.8 years, whereas women can expect to live an additional 19.8 years (AOA 2007). The fact that women live longer results in a sizable number of elderly women who are poor. Not only do women, in general, make less money than men, but also older women may have spent their savings on their husband's illness and, as homemakers, they may receive fewer Social Security benefits. In addition, retirement benefits and other major sources of income may be lost with a husband's death. Seventy percent of all elderly poor are women, half of whom were not poor before the death of their husbands.

Age and Social Class

Social class influences how long a person lives. In general, higher social class is associated with longer life, fewer debilitating illnesses, greater number of social contacts and friends, less likelihood of one defining oneself as "old," and greater likelihood of success in adapting to retirement. Higher social class is also related to fewer residential moves, higher life satisfaction, more leisure time, and healthier self-ratings. Of individuals age 65 and older, 26 percent of those with annual incomes of $35,000 or higher reported that their health was "excellent," whereas only 10 percent of those with incomes of less than $10,000 rated their health as such. Functional limitations such as problems with walking, dressing, and bathing were also less common among higher income groups (Seeman & Adler 1998). In a study of elderly Chinese living in Hong Kong, Zhang and Woo (2003) found that income was significantly and positively associated with better general health, better cognitive functioning, lower depression, more restful sleep, and fewer visits to the doctor. Higher educational levels are also correlated with better health—the higher one's education, the more likely an individual is to exercise, eat right, and drink and smoke less. In short, the higher one's socioeconomic status, the longer, happier, and healthier one's life is (Quadagno 2005).

500   Chapter 12 Problems of Youth and Aging

PROBLEMS OF THE ELDERLY

The increase in the number of the elderly worldwide presents a number of institutional problems. The **dependency ratio**—the number of societal members who are younger than 18 or 65 and older compared with the number of people who are between 18 and 64—is increasing. In 2000 there were 63 "dependents" for every 100 people between the ages of 18 and 64. By 2050 the estimated ratio will be 78–100 (Quadagno 2005). This dramatic increase and the general movement toward global aging may lead to a shortage of workers and military personnel, foundering pension plans, and declining consumer markets. It may also lead to increased taxes as governments struggle to finance elder-care programs and services, heightening intergenerational tensions as societal members compete for scarce resources. In addition to these macro-level concerns, elderly individuals face a number of challenges of their own. This chapter's *Self and Society* feature tests your knowledge of the elderly and some of these concerns.

Work and Retirement

What one does (occupation), for how long (work history), and for how much (wages) are important determinants of retirement income. Indeed, employment is important because it provides the foundation for economic resources later in life. Yet for the elderly who want to work, entering and remaining in the labor force may be difficult because of negative stereotypes, lower levels of education, reduced geographic mobility, fewer employable skills, and discrimination. Nonetheless, as retiree benefits decline and recession increases, older Americans are working longer than ever before. Since the mid-1990s, older Americans have become the fastest-growing group in the labor force, with those older than age 55 expected to make up 19.1 percent of all workers by 2012 (Porter & Walsh 2005).

One way that older employees remain in the workforce is through phased retirement. **Phased retirement** allows workers to ease into retirement by reducing hours worked per day, days worked per week, or months worked per year. For example, many universities allow faculty to work part-time for half salary while collecting partial pension benefits. Phased retirement is not only beneficial for retirees, who may not want to continue to work full-time, but also good for employers, who benefit from the skills and knowledge of older employees. Interestingly, a survey of 1,400 U.S. employees found that younger and older workers have very different perceptions about whether or not employers value older workers. For example, 54 percent of older workers believe that their employers value employees older than 50, whereas only 25 percent of younger workers said that their company values employees older than 50 (BLR 2006).

In 1967 Congress passed the Age Discrimination in Employment Act (ADEA), which was designed to ensure continued employment for people between the ages of 40 and 65. In 1986 the upper limit was removed, making mandatory retirement illegal in most occupations. Currently, under ADEA "it is unlawful to discriminate against a person because of his/her age with respect to any term, condition, or privilege of employment, including hiring, firing, promotion, layoff, compensation, benefit, job assignments, and training (EEOC 2007a, p. 1). Nevertheless, thousands of cases of age discrimination occur annually. In 2006, there were 16,548 age discrimination claims filed with the U.S.

dependency ratio The number of societal members who are under 18 or are 65 and older compared to the number of people who are between 18 and 64.

phased retirement Allows workers to ease into retirement by reducing hours worked per day, days worked per week, or months worked per year.

Equal Opportunity Commission (EECO 2007b). Not surprisingly, when a sample of U.S. opinion leaders (e.g., political leaders, academicians, and nongovernmental organization officials) were asked the extent to which discrimination against older persons is a concern 71 percent responded that it is a "big" or "moderate" problem (AARP 2007).

Although employers cannot advertise a position by age, they can state that the position is an "entry level" one or that "2–3 years' experience" is required. Despite strong evidence to the contrary, a study conducted by the National Council on the Aging revealed, "50 percent of employers surveyed believed that older workers cannot perform as well as younger workers" (Reio & Sanders-Reto 1999, p. 12). In addition, if displaced, older workers remain unemployed longer than younger workers; they often have to accept lower salaries than they earned earlier and may be more likely to give up looking for work (Goldberg 2000).

Retirement is a relatively recent phenomenon. Before passage of the Social Security laws, individuals continued to work into old age. Today, Social Security payments are usually limited to those older than 65 years of age (under some circumstances, a person can begin collecting benefits at age 62), moving to age 67 by the year 2022. However, in a survey, when a sample of retired respondents was asked at what age they left the workforce, 68 percent responded that they had retired before the age of 65 (Carroll 2005).

Retirement is difficult in the United States because "work" is often equated with "worth." A job structures one's life and provides an identity; retirement

❝Age-based retirement arbitrarily severs productive persons from their livelihood, squanders their talent, scars their health, strains an already overburdened social security system, and drives many elderly people into poverty and despair. Ageism is as odious as racism and sexism.❞

Norman Vincent Peele
Author and motivational speaker

often culturally signifies the end of one's productivity. Retirement may also involve a dramatic decrease in personal income. Six in 10 older Americans are worried about "not having enough money for retirement" (Jones 2005). The desire to remain financially independent, a lack of confidence in the Social Security system, increased educational levels and technological skills, and the desire to continue working have led to a recent reduction in early retirements. In addition, as noted, the number of people who stop working once they retire is also decreasing as retirees open their own businesses, work part-time, continue their educations, volunteer in the community, and begin second careers (Porter & Walsh 2005).

Poverty

In 2005, the median income for individuals 65 years old and older was $15,696, and 26.7 percent of the elderly reported incomes of $10,000 or less (AOA 2007). Poverty among the elderly varies dramatically by sex, race, ethnicity, marital status, and age: women, minorities, those who are single or widowed, and the old-old are most likely to be poor (AOA 2007; Hillier & Barrow 2007). Eight percent of elderly whites are poor, compared with 23.2 percent of elderly blacks and 19.9 percent of elderly Hispanics (AOA 2007).

Elderly women, particularly widows, and specifically black and Hispanic widows, are more likely to be poor than elderly men (Angel, Jimenez, & Angel 2007). Nearly half of elderly Hispanic women and 4 of 10 elderly black women are living in poverty. The comparable percentage for white women is half that of minorities. Minority women are more likely to have been working in low-paying jobs with no retirement plan, to have little or no savings, and to have fewer resources to fall back on. For example, older black households have an estimated median net worth of $42,600 compared with older white households, which have an estimated median household net worth of $213,000 (Federal Interagency Forum on Aging-Related Statistics 2007). For many minority women Social Security is the sole source of support. The highest poverty rate, more than 46 percent, is among elderly Hispanic women who live alone or with non-relatives (AOA 2007).

Social Security, actually titled "Old Age, Survivors, Disability, and Health Insurance," is a major source of income for many elderly persons. When Social Security was established in 1935, it was never intended to be a person's sole economic support in old age; rather, it was meant to supplement other savings and assets. Figure 12.2, however, indicates that those in the lowest fifth of the income quintiles, i.e., the poorest elderly, are the most likely to have Social Security as their primary source of income. As you move from the lowest fifth to the highest fifth income quintile, Social Security becomes less important as a source of income and earnings, pensions, and assets become more prominent. For the highest fifth, i.e., the richest elderly, Social Security is only 19 percent of their total income (Federal Interagency Forum on Aging-Related Statistics 2007). Today, 96 percent of Americans age 65 and older who work in paid employment or who are self-employed receive Social Security payments. The average monthly check for a retiree in 2006 was $1,002 (Social Security 2007).

Although Social Security keeps many elderly persons from being poor, the Social Security system has been criticized for being based on the number of years of paid work and preretirement earnings. Hence women and minorities, who often earn less during their employment years, receive less in retirement

FIGURE 12.2
Aggregate income for the population age 65 and older, by source and income quintile, 2004.
Source: Federal Interagency Forum on Aging-Related Statistics (2007).

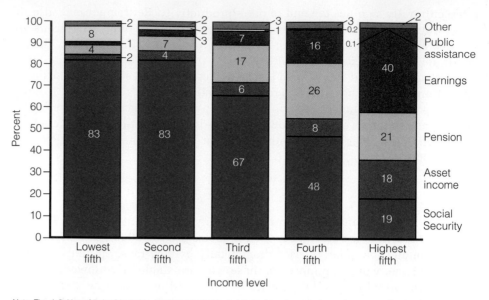

Note: The definition of "other" includes, but is not limited to, public assistance, unemployment compensation, worker's compensation, alimony, child support, and personal contributions. Quintile limits are $10,339 for the lowest quintile, $16,363 for the second quintile, $25,587 for the third quintile, $44,129 for the fourth quintile, and open-ended for the highest quintile.

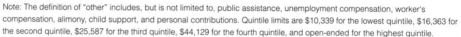

Reference population: These data refer to the civilian noninstitutionalized population.

benefits. Another concern is whether funding for the Social Security system will be adequate to provide benefits for the increased numbers of elderly in the next decades. In the next 25 years the number of Americans of retirement age will *increase* by 100 percent, whereas the ratio of workers who contribute to the Social Security system will *decrease* by an estimated 60 percent (Sloan 2005). Fear of the demise of Social Security has led the Bush administration to recommend that privatization be implemented over a 7-year period from 2010 to 2017 at a total cost of $712 billion (Sloan 2006). Such a system would allow workers to put the money they now have deducted from their payroll for Social Security (FICA) into a personally owned and invested retirement account. Despite some claims that the system is collapsing, surveys indicate that most Americans are opposed to such a plan (Gallup 2005; Morin & VandeHei 2005).

Health Issues

The biology of aging is called **senescence.** Senescence follows a universal pattern but does not have universal consequences. "Massive research evidence demonstrates that the aging process is neither fixed nor immutable. Biologists are now showing that many symptoms that were formerly attributed to aging—for example, certain disturbances in cardiac function or in glucose metabolism in the brain—are instead produced by disease" (Riley & Riley 1992, p. 221). Biological functioning is also intricately related to social variables. Altering lifestyles, activities, and social contacts affects mortality and morbidity. For example, a study of 302 volunteers showed that regular routine activities (e.g., household chores, running errands, and gardening) were related to longevity. Those who lived a more sedentary lifestyle, on average, died about 6 years sooner than their active counterparts (Taylor 2006).

senescence The biology of aging.

Biological changes are consequences of either **primary aging,** caused by physiological variables such as cellular and/or molecular variation (e.g., gray hair), or secondary aging. **Secondary aging** entails changes attributable to poor diet, lack of exercise, increased stress, and the like. Secondary aging exacerbates and accelerates primary aging.

Alzheimer's disease, an example of primary aging, received national attention when former president Ronald Reagan announced that he had the disease. Named for German neurologist Alois Alzheimer, the debilitating disease has affected both the mental and the physical condition of 5 million Americans and will affect a projected 16 million by 2050. According to the Alzheimer's Association, the costs of the disease are nearly $150 billion per year and growing. Costs include nursing home and medical expenses as well as costs associated with loss of job productivity for family members who serve as caregivers (Saul 2007). The major symptoms of Alzheimer's are a loss of "cognitive functioning (memory, learning, reaction time, and word usage), disorientation, declines in self-care, and inappropriate social behavior such as violent outbursts" (Hillier & Barrow 2007, p. 288). In a web log titled "Mondays with Mother: An Alzheimer's Story," Anne Robertson, a published author, poignantly describes her feelings as she visits her mother's nursing home (2007).

> I wonder if I will always be able to find her in her eyes . . . those windows of the soul. Now when I visit, I can still find her there . . . when I really look . . . when we pray and I look deep and say, "I love you." It is like she is trapped in this body that goes around and puts tissues in her shoes or looks for clouds in dresser drawers. But she is in there still, wondering how many years she will live life in this foggy, other world before she is allowed to come out into the sun.

Citing the tendency for Alzheimer's to run in families, several studies have linked the disease to variant genes (Wade 2007).

In 2005 more than one-third (38.3 percent) of the noninstitutionalized elderly rated their health as excellent or very good; 22.8 percent of blacks and 28.4 percent of Hispanics rated their health as excellent or very good, compared with 40.9 percent of older whites (AOA 2007). Although elderly Americans are in better health than in previous generations, health often declines with age, with most older Americans having at least one chronic condition. The most common chronic conditions reported include hypertension, arthritis, and heart disease (see Figure 12.3).

Older people account for 36 percent of all hospital stays, averaging 5.6 days per year compared with 4.8 days for people of all ages (AOA 2007). Health is a major quality-of-life issue for the elderly, especially because they face higher medical bills with reduced incomes. Older Americans spend two times as much on health care as their younger counterparts, and more than half of the amount is spent on insurance. The poor elderly, often women and/or minorities, spend an even higher proportion of their resources on health care.

Medicare (see Chapter 2) was established in 1966 to provide medical coverage to those older than age 65 and currently insures more than 42.4 million people—95 percent of the 65 and older population (AOA 2007; Kaiser Family Foundation 2007). Although it is widely assumed that the government pays the medical bills of the elderly, the elderly are responsible for as much as 25 percent of their total health costs. Medicare, for example, does not pay for most immunizations, dental care and dentures, eye examinations and glasses, and hearing

> **"** If exercise could be put in a pill, it would be the number one anti-aging medicine. **"**
>
> Robert Butler
> Gerontologist

primary aging Biological changes associated with aging due to physiological variables such as cellular and molecular variation.

secondary aging Biological changes associated with aging that can be attributed to poor diet, lack of exercise, increased stress, and the like.

FIGURE 12.3
Percentage of people age 65 and older who reported having selected chronic conditions, by sex, 2003–2004.
Source: Federal Interagency Forum on Aging-Related Statistics (2007).

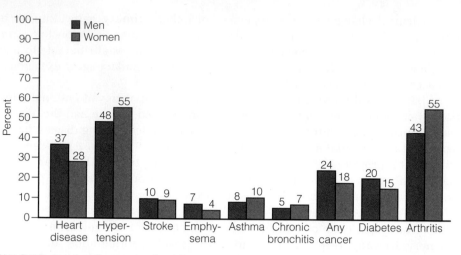

Note: Data are based on a 2-year average from 2003–2004. The question used to estimate the percentage of people who report having arthritis is "Have you EVER been told by a doctor or other health professional that you have some form of arthritis, rheumatoid arthritis, gout, lupus, or fibromyalgia?" This differs from the questions that were asked to estimate the percentage of people who report having "arthritic symptoms" in *Older Americans 2004.*

Reference population: These data refer to the civilian noninstitutionalized population.

aids and hearing examinations—the very services needed by elderly persons (Hillier & Barrow 2007). In 2004 people older than age 65 spent $4,193 on out-of-pocket health care expenditures, compared with $2,664 for the general population (AOA 2007). The difference between Medicare benefits and the actual cost of medical care is called the **medigap.** Almost three-fourths of Medicare recipients have some type of medigap insurance coverage. In addition, a prescription drug benefit was added to Medicare as part of a 2003 legislative package designed to improve and modernize the program (Quadagno 2005). The drug benefit is called the Medicare Part D Prescription Plan and, in 2007, covered "75 percent of the first $2,400 a person spends on drugs (after a $265 deductible), but after that is spent, coverage pauses until the yearly expenses reach $5,451 at which point Medicare kicks back in" (Safire 2007, p. 1).

Because health is associated with income, the poorest old are often the most ill: they receive less preventive medical care, have less knowledge about health care issues, and have limited access to health care delivery systems. Further, as health care costs have increased so too have the out-of-pocket expenses associated with medical services (PRB 2007). **Medicaid** is a federal- and state-funded program for those who cannot afford to pay for medical care. However, eligibility requirements often disqualify many of the elderly poor, often minorities and women.

Living Arrangements

The elderly live in a variety of contexts, depending on their health and financial status. Most elderly individuals do not want to be institutionalized, preferring to remain in their own homes or in other private households with friends and relatives (see Figure 12.4). Of the noninstitutionalized elderly population, 54.8 percent live at home with their spouse, although that number decreases with age and varies significantly by sex (AOA 2007). The homes of the elderly, however, are usually older, located in inner-city neighborhoods, in need of repair, and

medigap The difference between Medicare benefits and the actual cost of medical care.

Medicaid A public assistance program designed to provide health care for the poor.

often too large to be cared for easily. For example, in 2005 the median year of construction of homes occupied by older residents was 1966, and nearly half were in need of repair (AOA 2007).

Although many of the elderly poor live in government housing or apartments with subsidized monthly payments, the wealthier elderly often live in retirement communities. These are often planned communities, located in states with warmer climates, and expensive. These communities offer various amenities and activities, have special security, and are restricted by age. One criticism of these communities is that they segregate the elderly from the young and discriminate against younger people by prohibiting them from living in certain areas.

Those who cannot afford retirement communities or who may not be eligible for subsidized housing often live with relatives in their own home or in the homes of others. It is estimated that more than 22 million people provide informal care for aging family members and that by 2030, an additional 5.7–6.6 million will be needed (Jackson 2003; Butler 2007). Over 60 percent are women, many of whom are daughters who care not only for their elderly parents but also for their own children.

Other living arrangements include shared housing, modified independent living arrangements, and nursing homes. With shared housing people of different ages live together in the same house, apartment, or condominium; they have separate bedrooms but share a common kitchen and dining area. They often share chores and financial responsibilities. In modified independent living arrangements elderly persons live in their own house, apartment, or condominium within a planned community where special services, such as meals, transportation, and home repairs, are provided. Skilled or semiskilled health care professionals are available on the premises, and the residences have call buttons so help can be summoned in case of an emergency.

What Do You Think? One type of modified independent living is called co-housing (Abrahms 2006). The concept of co-housing began in Denmark in the 1970s and was introduced to the United States in 2005 when the first senior co-housing development opened its doors in Davis, California. Co-housing entails semicommunal living in which individuals have separate living units but share a "common house" which, at a minimum, includes a kitchen, a dining area, and a meeting room. Senior co-housing developments, as many housing units for the elderly in general, usually restrict residency by age, i.e., only those older than a certain age limit can live there. Do you think that it is fair to limit housing availability by age? Does it violate the principles of fair housing as defined by the Fair Housing and Equal Opportunity Office, which states that its mission is to create "equal housing opportunities for all persons living in America by administering laws that prohibit discrimination in housing on the basis of race, color, religion, sex, national origin, age, disability, and familial status?" (FHEO 2006, p. 1).

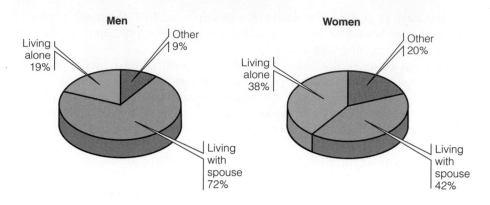

FIGURE 12.4
Living arrangements of persons 65 and older, 2005.
Source: AOA (2007).

Men

Living alone 19%

Other 9%

Living with spouse 72%

Women

Living alone 38%

Other 20%

Living with spouse 42%

Nursing homes are residential facilities that provide full-time nursing care for residents. Nursing homes may be private or public. Private facilities are expensive and are operated for profit by an individual or a corporation. Public facilities are nonprofit and are operated by a government agency, religious organization, or the like. The average annual cost for all nursing homes is $74,806 and for a semiprivate room is $65,385 (AAHSA 2007). The probability of being in such an extended care facility is associated with race, age, and sex: whites, the old-old, and women are more likely to be in residence. Elderly persons with chronic health problems are also more likely to be admitted to nursing homes. Nursing homes vary dramatically in cost, services provided, and quality of care. One study found that of the 17,000 nursing homes in the United States, 90 percent were understaffed (Pear 2002). Understaffing is a problem that will continue through the decades as the older than 85 population increases to a projected 21 million in 2050 (Clemmitt 2006).

The fastest growing facilities for the elderly are assisted-living quarters—39,500 facilities in 2007—at an annual cost of $32,572 (AAHSA 2007). Assisted-living facilities, although not as "full service" as nursing homes, offer private living units with the assurance of an around-the-clock staff. For example, the typical assisted living resident is an 85-year-old woman who, although mobile, needs assistance with several activities of daily living such as bathing and dressing (NCAL 2006). Currently, assisted living facilities are licensed by state governments (NCAL 2007). Moreover, under the federal Older Americans Act every state is required to have an ombudsman who acts as an advocate for residents of nursing homes and assisted living facilities (ORC 2007).

nursing homes Public or private residential facilities that provide full-time nursing care for residents.

elder abuse The physical or psychological abuse, financial exploitation, medical abuse, and/or neglect of the elderly.

Victimization and Abuse

Elder abuse refers to physical or psychological abuse, financial exploitation, medical abuse, and/or neglect of the elderly (Hooyman & Kiyak 1999; Green 2003; Hillier & Barrow 2007). In one California case an adult protective services (APS) worker found a

70-year-old diabetic woman hungry and shoeless. A pan of rotten chicken fat sat on the stove. The smell of spent crack cocaine permeated the air. A man and a disheveled woman emerged from a filthy bedroom and quickly left. Alone with the APS worker, the frightened woman wept and said, "Thank God. I knew he would send somebody to help me. I've prayed and prayed for so long." She had been repeatedly hit and kept in her room, she said, and described being walked around the neighborhood and portrayed as the caregiver's sick grandmother in a ploy to beg for money. (Flaherty 2004, p. 1)

In the United States alone the number of older Americans who are abused in some way is estimated to be more than 2.1 million (APA 2007). Even in Japan, where the elderly are traditionally revered, incidents of elder abuse are on the rise. Although elder abuse can take place in private homes by family members, the elderly, like children, are particularly vulnerable to abuse when they are institutionalized. In 1990 the Nursing Home Reform Act was passed, establishing various rights for nursing care residents: the right to be free of mental and physical abuse, the right not to be restrained unless necessary as a safety precaution, the right to choose one's physician, and the right to receive mail and telephone communication (Harris 1990, p. 362). Despite this "bill of rights," a recent Congressional report found that federal health officials issued "minimal penalties" when nursing home violations occurred. Even in the face of the most serious violations (e.g., abuse or neglect leading to death) many nursing homes remained open. For example, in Michigan, a nursing home remained open after it had been repeatedly cited for "poor quality care, poor nutrition services, medication errors, and employing people who had been convicted of abusing patients" (Pear 2007a, p. 1).

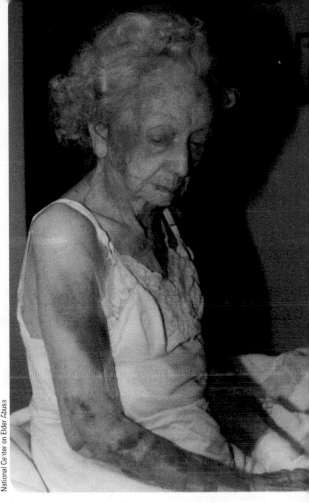

If you suspect elder abuse or are concerned about the well-being of an older person, call your state abuse hotline immediately.

Whether the abuse occurs in the home or in an institution, the victim is most likely to be female, widowed, and white, with a limited income, and in her mid- to late 70s. The abuser tends to be an adult child or spouse of the victim who misuses alcohol. Some research suggests that the perpetrator of the abuse is more often an adult child who is financially dependent on the elderly victim (Mack & Jones 2003). Whether the abuser is an adult child or a spouse may simply depend on the person with whom the elder victim lives. This chapter's *The Human Side* feature describes an ever-increasing form of abuse—financial exploitation by a caretaker.

Many of the problems of the elderly are compounded by their lack of interaction with others, loneliness, and inactivity. This is particularly true for the old-old. The elderly are also segregated in nursing homes and retirement communities, separated from family and friends, and isolated from the flow of work and school. As with most problems of the elderly, the problems of isolation, loneliness, and inactivity are not randomly distributed. They are higher among the elderly poor, women, and minorities. A cycle is perpetuated—being poor and old results in being isolated and engaging in fewer activities. Such withdrawal affects health, which makes the individual less able to establish relationships or participate in activities.

To celebrate her 90th birthday, Dr. Elisabeth Shanks—a spry, independent, retired psychiatrist—went parasailing (Breton 2007). Most people half her age wouldn't contemplate such a move. Yet, such was the vivacious spirit of Dr. Shanks—a spirit that would soon be stolen from her. In 2006, Patricia Murtaugh, a 58-year-old "friend" of Dr. Shanks, was convicted of second-degree grand larceny after deceiving Dr. Shanks and stealing thousands of dollars from her ("Queens: Woman Sentenced" 2006). When, in March 2007, Murtaugh became eligible for parole, Jonathan Shanks, the only child of Dr. Shanks, wrote the following letter to the parole board.

19 March 2007

Re: Parole Hearing for Patricia Murtaugh—Inmate # 06G0955

To Whom It May Concern:

I am the son and only child of Elisabeth Shanks, one of Patty Murtaugh's victims. I apologize for sending this to you on the eve of Murtaugh's parole hearing, but I learned of it just moments ago.

Briefly, the actions of Murtaugh have caused my mother irreparable harm. They have caused her emotional and physical suffering from which she will never recover. I am not alleging any physical contact by Murtaugh—but Murtaugh's actions have caused my mother's physical condition to rapidly deteriorate to a state where she is almost bedridden.

Before Murtaugh's theft of my mother's money, my mother was an extremely independent woman. Well into her 90s, she lived alone in her apartment in the Bronx. She did all her own grocery shopping at near-by stores and prepared her own meals. Because her hearing was deteriorating and she did not want to lose the ability to communicate, she was taking a sign-language course in mid-town Manhattan. She traveled alone on the bus and on the subway to these classes. She would have been remarkably spry for a woman a decade younger. For a woman her age, her ability to live and travel without assistance was amazing.

What Murtaugh did changed that. My mother's condition has deteriorated dramatically since she was victimized. She now requires an attendant be with her 24 hours a day. She cannot move from her bed, even to the bathroom without assistance. Her will to live has also deteriorated. She constantly asks why Murtaugh did this to her.

From my conversations with the Assistant District Attorney who prosecuted her case, Greg Pavlides, I have learned that Murtaugh insists she had my mother's permission to open a joint bank account and that she is in a state of denial about the harm she has caused.

I assure you, Murtaugh was never given permission to have access to my mother's money. Murtaugh phoned me in Canada where I live to try to get my approval for Power of Attorney. I told her that there was no need for that. I asked only that she make sure that my mother was writing monthly checks for her rent and utility bills. In phone calls with my mother, she told me she had no intention of giving Murtaugh access to her bank accounts.

Murtaugh took advantage of a vulnerable elderly woman—solely for her own financial gain. She also exploited a family connection to gain my mother's trust, some thing [sic] I find almost as reprehensible as the theft itself (Murtaugh's mother was a friend of my mother's sister—although Murtaugh herself never had any prior contact with either my mother or my mother's sister).

What Do You Think? An article published in the *American Sociological Review* indicated that, on average, Americans could name only two people with whom they would discuss important personal matters (McPherson, Smith-Lovin, Brashears 2006). Twenty-five percent reported having no one to confide in, a significant increase from two decades ago when only 10 percent of respondents had no close relationships. Social isolation is linked to a variety of health issues including higher morbidity and lower life expectancy (Harder 2006). Although the elderly do benefit from staying socially connected, so do the young. For example, a study of college students revealed that having a small circle of friends or feeling lonely weakens the immune system, making it more difficult to fight off infections. Do you have a large group of acquaintances but few friends? How many people can you turn to in a crisis? What do you think is responsible for the relationship between health and social connections?

Murtaugh stole my mother's future and destroyed any chance of my mother enjoying the few remaining years of life she has left. My mother spent her entire life helping others. She worked as a physician long past the retirement age of most of her colleagues. She also planned and saved for her future. Now, her future is bleak.

As I have mentioned, her physical and emotional health has suffered. Not only does she now require round-the-clock care, it appears she will be unable to spend her remaining years in the apartment she has loved living in for so many years.

Murtaugh stole almost every penny my mother saved during her career. What little Murtaugh missed is rapidly being spent on home-care. When it is gone, my mother will no longer be able to afford to live in her apartment and will have to go to a state-sponsored nursing home. Unfortunately, I am not familiar with what institutions are available in the United States, but from my knowledge of state-sponsored nursing homes in Canada, they are a far cry from the conditions my mother had hoped to spend her remaining years. It is not an understatement to say

Murtaugh ruined my mother's life—and stole her future. Murtaugh took what my mother had worked so hard to save, destroyed any chance of living her final years in her own home, and made her into just a shadow, both physically and emotionally, of what she had been, just months before.

In addition to just learning of the parole hearing, I have also just learned that my mother's 94-year-old sister has flown from England where she lives so that she can testify about the terrible impact Murtaugh's actions have had on my mother. Unfortunately, a death in the family means she will not be able to attend the hearing.

So, it is up to me to try to convince you, to urge you, to require Murtaugh to serve her full sentence. From what I have been told, Murtaugh continues to refuse to accept the seriousness of her actions. She has made no attempt to make any restitution other than what was ordered by the court. Our calculations are that she stole close to $200,000 from my mother. The District Attorney was unable to trace all the money that disappeared from my mother's account and accepted a guilty plea to a lesser amount. But even this amount

is considerable and the court-ordered $2,500 my mother received is negligible compared to what was taken.

Given the amount stolen, the lack of any attempt at restitution, the lack of remorse, the abuse of family trust (the use of her mother's relationship with my mother's sister), and the devastating effect this has had on my mother, I ask that Murtaugh's request for parole be denied.

Thank you for assistance in this matter.

Jonathan Shanks (2007)

Postscript: On March 20, 2007, the New York State Division of Parole wrote, "Merit release and parole are denied. . . . This panel concludes that if you are released at this time there exists a reasonable probability that you will not live and remain at liberty without further violations of the law. All factors considered, including the multiple victims involved and the particularly vulnerable nature of the victims, your release would not be in the best interests of the community at this time" (NYS Division of Parole 2007).

Quality of Life

Although some elderly persons do suffer from declining mental and physical functioning, many others do not. Being old does not mean being depressed, poor, and sick. Nearly half of Americans 80 years and older are in good physical and mental health, and most live on their own. Even those in their 90s, although often suffering from declining physical health, still enjoy good mental health and live relatively independently (Saad 2006). Interestingly, in a study of caretakers' perceptions of the quality of life of nursing home residents, Mittal and colleagues (2007) reported that residents perceive their quality of life more positively than caretakers do!

It is estimated that between 3 percent and 5 percent of the elderly population suffer from depression (AP 2006). Of those who are depressed two social factors tend to be in operation. One is society's negative attitude toward the elderly. Words and phrases such as "old," "useless," and "a has-been" reflect cultural connotations of the elderly that influence feelings of self-worth. The roles of the elderly also lose their clarity. How is a retiree supposed to feel or act? What does a retiree do? As a result, the elderly become dependent on external validation that may be weak or absent.

66 For age is opportunity, no less than youth itself;

although in another dress.

And as the evening twilight fades away,

The sky is filled with stars, invisible by day. 99

Henry Wadsworth Longfellow
Poet

The second factor contributing to depression among the elderly is the process of growing old. This process carries with it a barrage of stressful life events, all converging in a relatively short time period. These include health concerns, retirement, economic instability, loss of significant other(s), physical isolation, job displacement, and increased awareness of the inevitability of death as a result of physiological decline. All of these events converge on elderly persons to increase the incidence of depression and anxiety and perhaps affect the decision not to prolong life.

The official position of the American Medical Association is that physicians must respect a patient's decision to forgo life-sustaining treatment but should not participate in physician-assisted suicide. When a random sample of Americans was asked whether physician-assisted suicide was morally acceptable or morally wrong, respondents were sharply divided with responses equally split between the two alternatives (Saad 2007). Oregon is the only state that permits physician-assisted suicide through its Death with Dignity Act. The Death with Dignity Act "allows terminally-ill Oregonians to end their lives through the voluntary self-administration of lethal medications, expressly prescribed by a physician for that purpose" (Death with Dignity Act 2006). Since the legislation was passed in 1997, nearly 300 patients have died under the terms of the law. In 2006, the mean age of the 46 patients who died was 74 years (Annual Report 2007).

STRATEGIES FOR ACTION: GROWING UP AND GROWING OLD

Activism by or on behalf of children or the elderly has been increasing in recent years, and, as the numbers of children and the elderly grow, such activism is likely to escalate and to be increasingly successful. For example, global attention on the elderly led to 1999 being declared the International Year of Older Persons, and "the first nearly universally ratified human rights treaty in history" deals with children's rights (United Nations 1998, p. 1). Such activism takes several forms, including collective action through established organizations and the exercise of political and economic power.

Collective Action

Countless organizations are working on behalf of children, including the Children's Defense Fund, UNICEF, Children's Partnership, Children Now, Save the Children, and the Children's Action Network. Many successes take place at the local level, where parents, teachers, corporate officials, politicians, and citizens join together in the interest of children. In Kentucky AmeriCorp volunteers raised the reading competency of underachieving youths 116 percent in just 6 months; in Dayton, Ohio, behavioral problems of students at an elementary school were significantly reduced once "character education" was introduced (see Chapter 8); and in the Los Angeles Crenshaw High School student entrepreneurs established "Food from the Hood," a garden project that is expected to earn $50,000 in profit.

Some programs combine the interests of both the young and the old. The Adopt-a-Grandparent Program arranges for children to visit nursing home residents, and the Foster Grandparent Program pairs children with special needs with low-income elderly. In addition, the Children's Defense Fund, the Child

Protests are not limited to the young. The elderly have already been active in protesting changes to Social Security and other government policies that they believe are detrimental. As the number of elderly grows, social activism is likely to increase.

Welfare League of America, the AARP, and the National Council on the Aging joined together in 1986 to create Generations United, a "national membership organization focused solely on improving the lives of children, youth, and older people through intergenerational strategies, programs, and public policies" (Generations United 2007a, p. 1). Representing more than 70 million Americans, it is one of the few national organizations that acts as an advocate on behalf of the young and the old (Generations United 2007b).

More than 1,000 organizations are directed to realizing political power, economic security, and better living conditions for the elderly. One of the earliest and most radical groups is the Gray Panthers, founded in 1970 by Margaret Kuhn. The Gray Panthers were responsible for revealing the unscrupulous practices of the hearing aid industry, persuading the National Association of Broadcasters to add "age" to "sex" and "race" in the Television Code of Ethics statement on media images, and eliminating the mandatory retirement age. In view of these successes, it is interesting that the Gray Panthers, with only 40,000 members, is a relatively small organization compared with the AARP.

The AARP, founded in 1958, has more than 37 million members (10 times that of the National Rifle Association) age 50 and older and has an annual budget of more than $800 million and nearly 2,000 employees (Birnbaum 2005; AARP 2006). Services of the AARP include discounted mail-order drugs, investment opportunities, travel information, volunteer opportunities, and medigap coverage. In 2008, AARP will begin offering health maintenance organization (HMO) memberships to Medicare recipients, as well as a managed care plan (known as a preferred provider organization [PPO]) to members younger than age 64 (Pear 2007b). The AARP is the largest volunteer organization in the United States with the exception of the Roman Catholic Church. Not surprisingly, it is one of the most powerful lobbying groups in Washington. Some observers note that the AARP is such a powerful lobbying group that "it holds the key to how or whether Social Security will be restructured" (Birnbaum 2005, p. 1).

Political Power

Children are unable to hold office, to vote, or to lobby political leaders. Child advocates, however, acting on behalf of children, have wielded considerable political influence in areas such as child care, education, health care reform, and crime prevention. In addition, most Americans support funding of such programs. A Children's Defense Fund study showed that the majority of those surveyed favored federally funded after-school programs and subsidized child care even if it meant raising taxes (CDF 2000).

As conflict theorists emphasize, the elderly compete with the young for limited resources. They have more political power than the young and more political power in some states than in others. In Florida there is concern that the elderly may eventually wield too much political power and act as a voting bloc, demanding excessive services at the cost of other needy groups. For example, if the elderly were concentrated in a particular district, they could block tax increases for local schools. To the extent that future political issues are age based and that the elderly are able to band together, their political power may increase as their numbers grow over time. By 2030 almost half of all adults in developed countries and two-thirds of all voters will be near or at retirement age. The growing political power of the elderly is already becoming evident; the Netherlands has a political party called the Pension Party.

Economic Power

Although children have little economic power, the economic power of the elderly has grown considerably in recent years, leading one economist to refer to the elderly as a "revolutionary class" (Thurow 1996). The 2005 median income for men age 65 and older was $21,784 and for women age 65 and older it was $12,495. Households headed by individuals age 65 and older had a median income of $37,765 (AOA 2004). Although these incomes are significantly lower than the incomes for men and women between 45 and 54 years of age, income is only one source of economic power. The facts that many elderly persons own their homes, have substantial savings and investments, enjoy high levels of disposable income, and are growing in number contribute to their increased importance as consumers.

Finally, in many parts of the country the elderly have become a major economic power as part of what is called the **mailbox economy.** The mailbox economy refers to the tendency for a substantial portion of local economies to be dependent on pension and Social Security checks received in the mail by older residents (Atchley 2000). States with the highest proportion of elderly individuals—as a percentage of each state's population—include Florida (16.8 percent), West Virginia (15.3 percent), Pennsylvania (15.2 percent), North Dakota (14.7 percent), Iowa (14.7 percent), Maine (14.6 percent), South Dakota (14.2 percent), and Rhode Island (13.9 percent) (AOA 2007) (see Figure 12.5).

mailbox economy The tendency for a substantial portion of local economies to be dependent on pension and Social Security checks received in the mail by older residents.

Fighting Discrimination Through Law

The number of age-based complaints to the U.S. Equal Employment Opportunity Commission (EEOC) has risen as the population has aged. As discussed earlier in the chapter, the ADEA was passed in 1967 and protects workers 40 years of age and older (Hillier & Barrow 2007). In 1996 the U.S. Supreme Court heard *O'Connor v. Consolidated Coin Corp.,* which held that workers do not

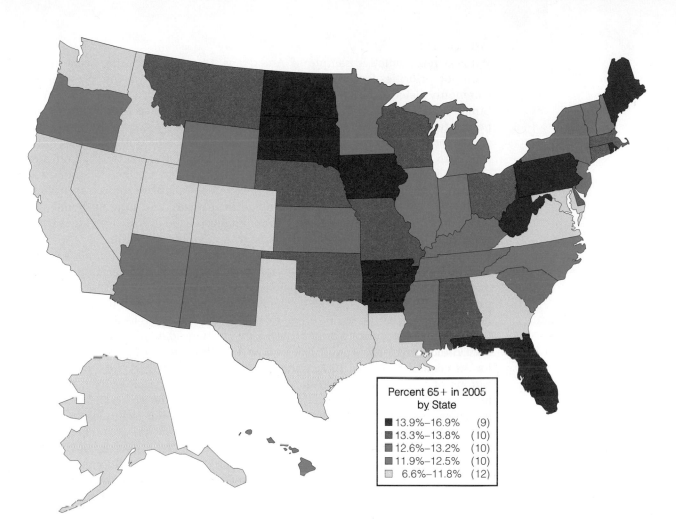

FIGURE 12.5

Persons 65 and older as a percentage of the total population of each state.
Source: AOA (2007).

have to prove that they were replaced by someone younger than 40 to have a finding of age discrimination. One of the most significant cases was heard in 2003—*EEOC and Arnett et al. v. CalPERS* (California Public Employees' Retirement System). In this case the EEOC recovered $250 million for California public safety officers who had been discriminated against on the basis of age. Age discrimination suits have the lowest success rate of any of the eight protected classes, including sex, race, and religious discrimination (AARP 2003). However, in 2005 the U.S. Supreme Court made winning age bias suits a little easier. The Court ruled that, for an age discrimination lawsuit to be successful, workers who bring the suits against their employers do not have to prove that the discrimination was intentional (Greenhouse 2005). Finally, in 2007 the U.S. Supreme Court agreed to hear an age discrimination suit against FedEx in which the plaintiffs argued that performance review measures violated the federal prohibition against age discrimination (Anderson 2007).

Government Policy

When a representative sample of Americans 18 years and older were asked about top priorities for the federal government, health care was the second most commonly listed priority preceded only by the war in Iraq (NBC News/Wall Street Journal 2007). As mentioned earlier, Medicaid provides health care for the poor; **Medicare** is a federally funded public insurance program originally designed for people over the age of 65 (see Chapter 2). In 1988 Congress passed the Medicare Catastrophic Coverage Act (MCCA), which was the most significant change in Medicare since its establishment in 1966. The new benefits included unlimited hospitalization, an upper limit on the amount of money recipients would pay for physicians' services, home health care and nursing home services, and unlimited hospice care. These changes were particularly significant because many of the illnesses of the elderly are chronic in nature.

The reforms were financed by increasing monthly medical premiums by $4 a month and by imposing an annual fee based on a person's federal income tax bracket. The maximum premium paid was $800 per person and $1,600 per couple (Harris 1990; Torres-Gil 1990). The AARP initially supported the reforms but later withdrew its support, as did many other organizations. The additional monies paid were simply not worth the new benefits, they contended. The AARP also argued that the elderly should not have to bear the burden of reforms necessary for the general public. In 1989, under pressure from the AARP and other organizations of the elderly, Congress repealed the MCCA. Pressured by public demands and fear of bankruptcy, Congress is currently in the process of reforming health care policy (see Chapter 2), including Medicare. For example, the Medicare Prescription Drug, Improvement, and Modernization Act of 2003 extended prescription drug coverage to the elderly and the disabled beginning in 2006.

Other government initiatives for the elderly include the Older Americans Act Amendments of 2000 (AOA 2001). The Older Americans Act Amendments of 2000 created the National Family Caregiver Support Program. This program helps "hundreds of thousands of family caregivers of older loved ones who are ill or who have disabilities" (AOA 2003). Among other things the National Family Caregiver Support Program provides respite care, supplemental services, and individual counseling and training for caregivers (AOA 2003). In 2006, the Older Americans Act was again reauthorized. Among other provisions, the age of eligibility was lowered from 60 to 55 for grandparents and other relatives raising children under the National Family Caregiver Support Program (Generations United 2007b).

Although Congress is presently considering several bills related to the elderly, two are particularly noteworthy. The first is the Elder Justice Act of 2007, which, if passed, will oversee services and prevention related to elder abuse, including monitoring long-term care facilities. The Act, originally introduced in 2005, would also establish an Office of Elder Justice under the Department of Health and Human Services (Elder Justice Act 2007). The second act is the Senior Nutrition Act of 2007. By altering the eligibility criteria, this Act will assist the neediest of the elderly in getting food through the commodity supplemental food program. As proposed, to participate in the program a person must be at least 60 years of age, be eligible for food stamps, and have a household income "that is less than or equal to 185 percent of the most recent Federal Poverty Income Guidelines published by the Department of Health and Human Services" (Senior Nutrition Act of 2007).

Medicare A federally funded public insurance program that provides health insurance benefits to the elderly, disabled workers and their dependents, and persons with advanced kidney disease.

In 1997 Congress passed the State Children's Health Insurance Program (SCHIP), which "is the largest single expansion of health insurance coverage for children in more than 30 years" (U.S. Department of Health and Human Services 1999, p. 1). SCHIP was designed for families who cannot afford health insurance but whose incomes are too high to qualify for Medicaid (see Chapter 2). In conjunction with Medicaid, SCHIP provides health insurance coverage to more than 30 million children. Nonetheless, "chronic budget shortfalls, often confusing enrollment processes, and dramatic variations in eligibility and coverage from state to state" leave 9 million children in America uninsured and millions more underinsured (CDF 2007b, p. 1). Although SCHIP is presently being considered for reauthorization, several other health care initiatives for children are also being reviewed. For example, the All Healthy Children Act would ensure comprehensive health benefits for all children under the age of 18 regardless of residence and would improve children's access to health care by paying a higher proportion of private insurance payment rates (CDF 2007c).

Finally, several important child-centered legislative initiatives have been passed recently including the Adam Walsh Child Protection and Safety Act of 2006. The Act expands the National Sex Offender Registry, strengthens penalties for crimes against children, and makes it harder for sex predators to reach children on the Internet (White House 2006). It also establishes a new National Child Abuse Registry and requires that a background check be conducted on all adoptive and foster parents before they are able to take custody of a child.

> **"**No man or woman stands as tall as one who stoops to help a child.**"**
>
> Teddy Roosevelt
> Former U.S. president

UNDERSTANDING YOUTH AND AGING

What can we conclude about youth and aging in American society? Age is an ascribed status and, as such, is culturally defined by role expectations and implied personality traits. Society regards both the young and the old as dependent and in need of the care and protection of others. Society also defines the young and the old as physically, emotionally, and intellectually inferior. As a consequence of these and other attributions, both age groups are sociologically a minority with limited opportunity to obtain some or all of society's resources.

Although both the young and the old are treated as minority groups, different meanings are assigned to each group. In general, in the United States the young are more highly valued than the old. Structural functionalists argue that this priority on youth reflects the fact that the young are preparing to take over important statuses, whereas the elderly are relinquishing them. Conflict theorists emphasize that in a capitalist society, both the young and the old are less valued than more productive members of society. Conflict theorists also point out the importance of propagation, that is, the reproduction of workers, which may account for the greater value placed on the young than on the old. Finally, symbolic interactionists describe the way images of the young and the old intersect and are socially constructed.

The collective concerns for the elderly and the significance of defining ageism as a social problem have resulted in improved economic conditions for the elderly. Currently, the elderly population is one of society's more powerful minorities. Research indicates, however, that despite their increased economic status, the elderly are still subject to discrimination in areas such as housing, employment, and medical care and are victimized by systematic patterns of stereotyping, abuse, and prejudice.

In contrast, the position of children in the United States remains tragic, with one in six children living in poverty. Wherever there are poor families, there are poor children, many of whom are educated in inner-city schools, live in dangerous environments, and lack basic nutrition and medical care. In addition, age-based restrictions limit the entry of these children into certain roles (e.g., employee) and demand others (e.g., student). Although most of society's members would agree that children require special protections, concerns regarding quality-of-life issues and rights of self-determination are only recently being debated. Age-based decisions are potentially harmful. If budget allocations were based on indigence rather than age, more resources would be available for those truly in need. Furthermore, age-based decisions could eventually lead to intergenerational conflict. Government assistance is a zero-sum relationship—the more resources one group gets, the fewer resources another group receives. For example, between 2007 and 2017, it is projected that "spending on children under current law is scheduled to shrink relative to other programs that have more rapid, built-in growth and thereby command ever-increasing shares of projected government revenues" (Carasso et al. 2007, p. 2).

Social policies that allocate resources on the basis of need rather than age would shift the attention of policy makers to remedying social problems rather than serving the needs of special interest groups. Age should not be used to cause negative effects on an individual's life any more than race, ethnicity, gender, or sexual orientation is. Although elimination of all age barriers or requirements is unrealistic, a movement toward assessing the needs of individuals and their abilities would be more consistent with the American ideal of equal opportunity for all.

CHAPTER REVIEW

- **What problems do the young and old have in common?**

  Among others, both the young and the old are victims of stereotypes, physical abuse, age discrimination, and poverty.

- **What age distinctions are commonly made in most societies?**

  Most societies make a distinction between childhood, adulthood, and elderhood. Childhood, often subdivided into infancy, childhood, and adolescence, is usually thought of as extending from birth to 17 years old; adulthood extends from 18 to 64 years old, and elderhood begins at 65 years of age.

- **What is disengagement theory?**

  According to disengagement theory, to achieve a balanced society, the elderly must relinquish their roles to younger members. Thus the young go to school, adults fill occupational roles, and the elderly, with obsolete skills and knowledge, disengage.

- **What are some of the problems of youth?**

  Problems of the young include (1) child labor, (2) orphaned children, (3) street children, (4) limited civil rights, (5) poverty, (6) foster care and adoption, (7) economic discrimination, (8) violence by and against children (e.g., gang violence, child abuse, and crime), and (9) health concerns.

- **What are some of the conclusions of the UNICEF study on child well-being?**

  The study, which was conducted in economically advanced nations, measured 40 indicators of six dimensions of child well-being. Of the 21 nations for which complete data were available, the United States scored second to last on the overall measure of child well-being. Only the United Kingdom scored lower. Further, the United States was in the bottom third of the rankings for five (educational well-being was the exception) of the six dimensions.

- **What three independent variables affect the consequences of aging?**

  Race, sex, and social class affect the consequences of aging. Racial minorities, women, and the lower classes are more likely to suffer adversely from the aging process.

- **What are some of the problems of the elderly?**

  For the elderly who want to work, entering and remaining in the labor force may be difficult because of negative stereotypes, lower levels of education, reduced geographic mobility, fewer employable skills, and discrimination. Retirement may also be difficult because of role transition and lowered income. Poverty is a problem as well, with 26.7 percent of the elderly reporting incomes of less than $10,000. Some elderly also have chronic health problems, including depression. Finally, elder abuse, although also present in private homes, is particularly problematic in institutionalized settings.

- **In terms of strategies for action, what are some of the similarities between the young and the old?**

  Both the young and the old have organizations working on their behalf—for example, the Children's Defense Fund and UNICEF and the AARP and the Gray Panthers. All groups have advocates working for their political and economic interests, although the elderly, according to conflict theorists, already wield considerable political and economic power. Finally, both the young and the old have benefited from government policies, such as the Older Americans Act, Medicare, and the State Children's Health Insurance Program (SCHIP).

- **What is the Adam Walsh Child Protection and Safety Act of 2006?**

  The Act expands the National Sex Offender Registry, strengthens penalties for crimes against children, and makes it harder for sex predators to reach children on the Internet. It also establishes a new National Child Abuse Registry and requires that a background check be conducted on all adoptive and foster parents before they are able to take custody of a child.

TEST YOURSELF

1. Disengagement theory argues that
   a. the value of the elderly decreases as technology becomes more advanced
   b. the elderly withdraw from socially productive roles and relinquish them to younger members of society
   c. because of structural limitations the elderly have few opportunities for active roles
   d. the meaning associated with the elderly is learned in the socialization process
2. What two countries have failed to ratify the Convention on the Rights of the Child?
   a. England and Egypt
   b. Nigeria and Kenya
   c. the United States and Somalia
   d. France and Germany
3. In 2050, there will be more old people than very young people.
   a. True
   b. False
4. Which of the following is not a reason for the aging population in the United States?
   a. The aging baby boomers
   b. Increased life expectancy
   c. Decreased birth rates
   d. The increased fertility rate
5. The results of the Child and Youth Well-Being Index (CWI) indicate that
   a. the well-being of children in the United States has increased
   b. the well-being of children in the United States has decreased
   c. the well-being of children in the United States has, after years of increasing, stalled
   d. the CWI does not measure the well-being of a child in the United States
6. The poorest of the poor elderly are white widowed women who live alone.
   a. True
   b. False
7. Medicaid is a public assistance program designed to provide health care to the elderly.
   a. True
   b. False
8. Alzheimer's disease
   a. is an example of primary aging
   b. will decrease over the next several decades
   c. is an example of secondary aging
   d. does not affect an individual's physical condition

9. With the exception of the Catholic Church, the AARP is the largest volunteer organization in the United States.
   a. True
   b. False

10. Spending on children given current programs and policies is expected to shrink relative to other expenditures.
    a. True
    b. False

Answers: 1 b. 2 c. 3 a. 4 d. 5 c. 6 b. 7 b. 8 a. 9 a. 10 a.

KEY TERMS

activity theory
age grading
ageism
child labor
dependency ratio
disengagement theory
elder abuse

gerontophobia
infantilizing elders
mailbox economy
Medicaid
Medicare
medigap
modernization theory

nursing homes
phased retirement
primary aging
sandwich generation
secondary aging
senescence

MEDIA RESOURCES

Understanding Social Problems, **Sixth Edition**
Companion Website

academic.cengage.com/sociology/mooney
Visit your book companion website, where you will find flash cards, practice quizzes, Internet links, and more to help you study.

Just what you need to know NOW! Spend time on what you need to master rather than on information you already have learned. Take a pre-test for this chapter, and CengageNOW will generate a personalized study plan based on your results. The study plan will identify the topics you need to review and direct you to online resources to help you master those topics. You can then take a post-test to help you determine the concepts you have mastered and what you will need to work on. Try it out! Go to academic.cengage.com/login to sign in with an access code or to purchase access to this product.

© AP Photo/Bikas Das

13

"Population may be the key to all the issues that will shape the future: economic growth; environmental security; and the health and well-being of countries, communities, and families."

Nafis Sadik, Executive director, UN Population Fund

Population Growth and Urbanization

The Global Context: A World View of Population Growth and Urbanization | **Sociological Theories of Population Growth and Urbanization** | **Social Problems Related to Population Growth and Urbanization** | **Strategies for Action: Responding to Problems of Population Growth, Population Decline, and Urbanization** | **Understanding Problems of Population Growth and Urbanization** | **Chapter Review**

In August 2007 a bridge on Interstate 35W in Minnesota collapsed into the Mississippi River, killing at least a dozen people and injuring many more. Two years before this tragic event, the Interstate 35W bridge was deemed "structurally deficient." Indeed, federal government reports indicated that in 2006 nearly a quarter of the nation's 600,000 major bridges (longer than 20 feet) were "structurally deficient" or "functionally obsolete" (McLaughlin 2007). Some observers attributed the tragedy that occurred on Interstate 35W to the government's failure to allocate adequate funding for the repair and maintenance of our nation's bridges. But another contributing factor is our growing population, and the increasing demands such growth places on our nation's infrastructure, especially in urban areas where population growth is highest. Bridges that were built decades ago are now carrying more passengers, and more weight, than ever. In only 10 years, from 1995 to 2005, the weight load on urban highways increased by half (Lohn 2007).

The collapse of Interstate 35W could be blamed on the government's failure to provide adequate funding for the repair and maintenance of our nation's bridges. But another contributing factor is the increasing demands our growing population places on our nation's infrastructure, especially in urban areas where population growth is highest.

In this chapter, we focus on problems associated with population growth. Population growth affects virtually every aspect of social life and "is the single most important set of events ever to occur in human history" (Weeks 2005, p. 4). Because the growth of the world's population is occurring largely in urban areas, we also explore the problems associated with the ever-increasing urbanization of our planet.

The Global Context: A World View of Population Growth and Urbanization

Although thousands of years passed before the world's population reached 1 billion, in less than 300 years the population exploded from 1 billion to 6 billion (see Figure 13.1). World population was 1.6 billion when we entered the 20th century and 6.1 billion when we entered the 21st century.

After presenting a brief overview of the history of world population growth, current population trends, and future population projections, we review the development of urbanization and describe the current state of urbanization throughout the world.

World Population: History, Current Trends, and Future Projections

Humans have existed on this planet for at least 200,000 years. For 99 percent of human history population growth was restricted by disease and limited food supplies. Around 8000 B.C. the development of agriculture and the domestication of animals led to increased food supplies and population growth, but even

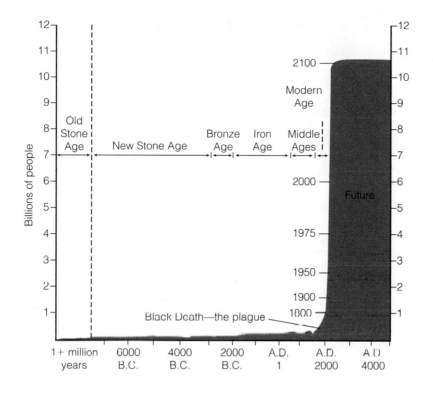

FIGURE 13.1
World population growth through history.
Source: McFalls (2007).

then harsh living conditions and disease still put limits on the rate of growth. This pattern continued until the mid-18th century, when the Industrial Revolution improved the standard of living for much of the world's population. The improvements included better food, cleaner drinking water, improved housing and sanitation, and advances in medical technology, such as antibiotics and vaccinations against infectious diseases—all of which contributed to rapid increases in population.

Population Doubling Time. Population **doubling time** is the time required for a population to double from a given base year if the current rate of growth continues. It took several thousand years for the world's population to double from 4 million to 8 million, a few thousand years to double from 8 million to 16 million, about 1,000 years to double from 16 million to 32 million, and less than 1,000 years to double to 64 million. The next doubling of the world's population occurred between the European Renaissance and the Industrial Revolution in a time span of about 400 years. But after 1750 the doubling of world population took little more than 100 years, and the next doubling took less time. The most recent doubling (from 3 billion in 1960 to 6 billion in 1999) took only about 40 years (Weeks 2005).

Although world population will continue to grow in the coming decades, it may never double in size again. In fact, because fertility rates have dropped around the world, "we're finally at a point where it's possible that a child born today will live to see the stabilization of world population" (Nelson 2004, p. 17).

Current Population Trends and Future Projections. World population is projected to grow from 6.7 billion in 2007 to 9.2 billion in 2050 (United Nations 2007).

doubling time The time required for a population to double in size from a given base year if the current rate of growth continues.

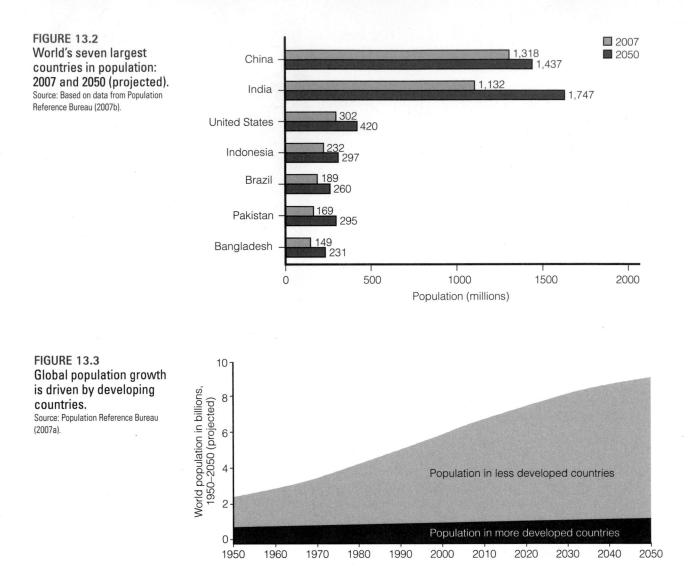

FIGURE 13.2
World's seven largest countries in population: 2007 and 2050 (projected).
Source: Based on data from Population Reference Bureau (2007b).

FIGURE 13.3
Global population growth is driven by developing countries.
Source: Population Reference Bureau (2007a).

Although China is the most populated country in the world today, India will become the most populated country by 2050 (see Figure 13.2).

Ninety-nine percent of world population growth is in developing countries, mostly in Africa and Asia (see Figure 13.3). As the size of a country's population grows, so does its **population density,** or the number of people per unit of land area. Overall, population density is higher in less-developed countries (see Table 13.1), which means that people in less-developed countries live in more crowded conditions. The population density of India, for example, is 344 people per square kilometer, compared with 31 people per square mile in the United States. To get an idea of how crowded the population is in India, consider that India has one-third the land area of the United States but nearly four times the population. Imagine if the U.S. population quadrupled. Then imagine that this quadrupled population all lived in the eastern third of the United States. That will give you

population density The number of people per unit of land area.

an idea of how crowded living conditions are in India. And more than 20 countries have a higher population density than that of India (Population Reference Bureau 2007b).

Higher population growth in developing countries is largely due to higher **total fertility rates**—the average lifetime number of births per woman in a population. As shown in Table 13.2, the least developed regions of the world have the highest rates of fertility.

Population size is affected not only by fertility rates but also by immigration. Much of the population growth occurring in the United States is due to high immigration rates. In other countries populations are changing as a result of human immunodeficiency virus (HIV)/ acquired immunodeficiency syndrome (AIDS). For example, despite their high fertility rates, the four African countries most affected by HIV/AIDS—Botswana, Lesotho, Namibia, and Swaziland—are expected to experience a decline in population over the coming decades as deaths (resulting from AIDS) outnumber births.

Will there be an end to the rapid population growth that has occurred in recent decades? Will the population of the world stabilize? Although some predict that population will stabilize around the middle of the 21st century, no one knows for sure. There has been a significant reduction in fertility rates around the world, from a global average of 5 children per woman in the 1950s to 2.55 children in 2007 (United Nations 2007). To reach population stabilization, fertility rates throughout the world would need to achieve what is called "replacement level," whereby births would replace, but not outnumber, deaths. **Replacement-level fertility** is 2.1 births per woman, that is, slightly more than 2 because not all female children will live long enough to reach their reproductive years. More than 50 countries have already achieved below-replacement fertility rates, and by 2050 the average fertility rate worldwide is projected to be below replacement level. However, even if every country in the world achieved replacement-level fertility rates, populations would continue to grow for several decades because of **population momentum**—continued population growth as a result of past high fertility rates that have resulted in a large number of young women who are currently entering their childbearing years. So despite the below-replacement fertility rates in more-developed regions, population in these regions is expected to continue to grow until about 2030 and then to begin to decline. The U.S. population, however, will continue to increase through 2050 because of immigration.

In sum, there are two population trends occurring simultaneously that, on the surface, appear to be contradictory: (1) The total number of people on this planet is rising and is expected to continue to increase over the coming decades; and (2) about 40 percent of the world's population lives in countries in which couples have so few children that the countries' populations are likely to decline over the coming years (Population Reference Bureau 2004). These countries include China, Japan, and most European countries. As we discuss later in this chapter, each of these trends presents a set of problems and challenges.

TABLE 13.1 Population Density

AREA	POPULATION DENSITY (PEOPLE PER SQUARE KILOMETER)
World	49
More-developed countries	27
Less-developed countries	65

Source: Population Reference Bureau (2007b).

TABLE 13.2 Fertility Rates by Region

More-developed	1.6
Less-developed (excluding China)	3.3
Africa	5.0
North America	2.0
Latin America/Caribbean	2.5
Asia	2.4
Europe	1.5
Oceania	2.1

Source: Population Reference Bureau (2007b).

66 Are we experiencing a population explosion or birth dearth? The answer may be both. **99**

Population Reference Bureau

total fertility rate The average lifetime number of births per woman in a population.

replacement-level fertility 2.1 births per woman (slightly more than 2 because not all female children will live long enough to reach their reproductive years).

population momentum Continued population growth as a result of past high fertility rates that have resulted in a large number of young women who are currently entering their childbearing years.

The region of the world with the highest fertility rate is Africa, where women have an average of five children in their lifetime.

An Overview of Urbanization Worldwide and in the United States

As early as 5000 B.C., cities of 7,000–20,000 people existed along the Nile, Tigris-Euphrates, and Indus River valleys. But not until the Industrial Revolution in the 19th century did **urbanization,** the transformation of a society from a rural to an urban one, spread rapidly.

As population has increased, so has the proportion of people living in urban areas. An **urban area** is a spatial concentration of people whose lives are centered around nonagricultural activities. Although countries differ in their definitions of "urban," most countries designate places with 2,000 people or more as being urbanized. According to the U.S. census definition, an "urban population" consists of individuals living in cities or towns of 2,500 or more inhabitants. An "urbanized area" refers to one or more places and the adjacent densely populated surrounding territory that together have a minimum population of 50,000.

The number of cities, as well as the share of population that lives in cities, has grown at an incredible pace. Worldwide, the number of cities with more than 1 million inhabitants grew from 86 in 1950 to 400 in 2006, and by 2015 there will be at least 550 (Davis 2006). The number of **megacities**—urban areas with 10 million residents or more—is also increasing. In 1950 there was only one city with more than 10 million inhabitants: New York City. By 2015, 22 cities are projected to have more than 10 million people; 17 of these will be in less-developed countries (United Nations 2006).

Worldwide, the share of the global population living in urban areas has increased from 29 percent in 1950 to about 50 percent in 2008 and is expected to reach nearly 60 percent by 2030 (United Nations 2006). The year 2008 marks a historic milestone: For the first time in history more than half of the world's population will be living in urban areas (United Nations Population Fund 2007).

The first wave of urbanization took place in North America and Europe over two centuries, from 1750 to 1950 when the percentage of population that lived in urban areas increased from 10 to 52 percent. In the second wave of urbanization, in less-developed regions, the urban population will increase from 18 percent in 1950 to about 56 percent in 2030. Although three-quarters of the population in more-developed regions already lives in urban areas (see Figure 13.4), most population growth expected between 2000 and 2030 will be in urban areas of less-developed regions.

About 60 percent of the increasing urban population is the result of births that occur in urban areas. The remaining 40 percent is due to a combination of migration from rural to urban areas and the reclassification of areas from rural to urban (United Nations Population Fund 2007). Rural dwellers migrate to urban areas to flee war or natural disasters or to find employment. As foreign corporate-controlled commercial agriculture displaces traditional subsistence farming in poor rural areas, peasant farmers flock to the city looking for employment. Some rural dwellers migrate to urban areas in search of a better job—one

urbanization The transformation of a society from a rural to an urban one.

urban area A spatial concentration of people whose lives are centered around nonagricultural activities.

megacities Urban areas with 10 million residents or more.

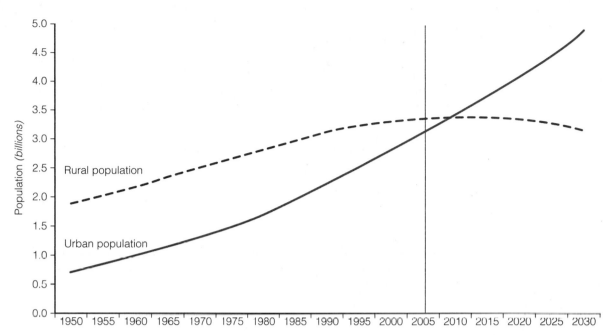

FIGURE 13.4
Urban and rural population
of the world, 1950–2030.
Source: United Nations (2006).

that has higher wages and better working conditions. Governments have also stimulated urban growth by spending more to improve urban infrastructures and services while neglecting the needs of rural areas (Clark 1998).

History of Urbanization in the United States. Urbanization of the United States began as early as the 1700s, when most major industries were located in the most populated areas, including New York City, Philadelphia, and Boston. Unskilled laborers, seeking manufacturing jobs, moved into urban areas as industrialization accelerated in the 19th century. The "pull" of the city was not the only reason for urbanization, however. Technological advances were making it possible for fewer farmers to work the same amount of land. Thus "push" factors were also involved—making a living as a farmer became more and more difficult as technology, even then, replaced workers.

Urban populations continued to multiply as a large influx of European immigrants in the late 1800s and early 1900s settled in U.S. cities. This influx was followed by a major migration of southern rural blacks to northern urban areas. People were lured to the cities by the promise of employment and better wages and urban amenities such as museums, libraries, and entertainment. Immigrants are also attracted to cities with large ethnic communities that provide a familiar cultural environment in which to live and work.

Suburbanization. In the late 19th century railroad and trolley lines enabled people to live outside the city and still commute into the city to work. As more and more people moved to the **suburbs**—urban areas surrounding central cities—the United States underwent **suburbanization.** As city residents left the city to live in the suburbs, cities lost population and experienced **deconcentration,** or the redistribution of the population from cities to suburbs and surrounding areas.

suburbs Urban areas surrounding central cities.

suburbanization The development of urban areas outside central cities, and the movement of populations into these areas.

deconcentration The redistribution of the population from cities to suburbs and surrounding areas.

Many factors have contributed to suburbanization and deconcentration. After World War II many U.S. city dwellers moved to the suburbs out of concern for the declining quality of city life and the desire to own a home on a spacious lot. Suburbanization was also spurred by racial and ethnic prejudice, as the white majority moved away from cities that, because of immigration, were becoming increasingly diverse. Mass movement into suburbia was encouraged by the federal interstate highway system (financed by the government under the guise of ensuring national defense), the affordability of the automobile, and the dismantling of metropolitan mass transit systems (Lindstrom & Bartling 2003). In the 1950s Veterans Administration and Federal Housing Administration loans made housing more affordable, enabling many city dwellers to move to the suburbs. Suburb dwellers who worked in the central city could commute to work or work in a satellite branch of a company in suburbia that was connected to the main downtown office. As increasing numbers of people moved to the suburbs, so did businesses and jobs. Without a strong economic base, city services and the quality of city public schools declined, which further increased the exodus from the city.

What Do You Think? In cities across the country families with children are leaving the city to move to the suburbs where they can afford a bigger house with more space. Consequently, many cities are experiencing a decline in their population of children (Egan 2005). How might significant reductions in the youth population affect cities? What could cities do to attract more families with children?

U.S. Metropolitan Growth and Urban Sprawl. Simply defined, a **metropolitan area,** also known as a **metropolis,** is a densely populated core area together with adjacent communities. Metropolises have grown rapidly in the United States. As new areas reach the minimum required city or urbanized area population and as adjacent towns, cities, and counties satisfy the requirements for inclusion in metropolitan areas, both the number and size of metropolitan areas have grown. Most Americans live in one of the more than 360 metropolitan areas in the nation. One U.S. state—New Jersey—is entirely occupied by metropolitan areas as designated by the U.S. census.

The growth of metropolitan areas is often referred to as **urban sprawl**—the ever increasing outward growth of urban areas. Urban sprawl results in the loss of green open spaces, the displacement and endangerment of wildlife, traffic congestion and noise, and pollution liabilities—problems that we discuss later in this chapter.

Those who enjoy the conveniences and amenities of urban life but who find large metropolitan areas undesirable may choose to live in a **micropolitan area**—a small city (between 10,000 and 50,000 people) located beyond congested metropolitan areas. These areas are large enough to attract jobs, restaurants, community organizations, and other benefits yet are small enough to elude traffic jams, high crime rates, and high costs of housing and other living expenses associated with larger cities.

metropolitan area Also known as a metropolis, a densely populated core area together with adjacent communities.

metropolis From the Greek meaning "mother city." See also *metropolitan area.*

urban sprawl The ever increasing outward growth of urban areas that results in the loss of green open spaces, the displacement and endangerment of wildlife, traffic congestion and noise, and pollution.

micropolitan area A small city (between 10,000 and 50,000 people) located beyond congested metropolitan areas.

SOCIOLOGICAL THEORIES OF POPULATION GROWTH AND URBANIZATION

The three main sociological perspectives—structural functionalism, conflict theory, and symbolic interactionism—can be applied to the study of population and urbanization.

Structural–Functionalist Perspective

Structural functionalism focuses on how changes in one aspect of the social system affect other aspects of society. For example, the **demographic transition theory** of population describes how industrialization has affected population growth. According to this theory, in traditional agricultural societies high fertility rates are necessary to offset high mortality and to ensure continued survival of the population. As a society becomes industrialized and urbanized, improved sanitation, health, and education lead to a decline in mortality. The increased survival rate of infants and children along with the declining economic value of children leads to a decline in fertility rates. About one-third of the world's countries have completed the demographic transition—the progression from a population with short lives and large families to one in which people live longer and have smaller families (Cincotta, Engelman, & Anastasion 2003). Many low-fertility countries have entered what is known as a "second demographic transition," in which fertility falls below the two-child replacement level. This second demographic transition has been linked to greater educational and job opportunities for women, increased availability of effective contraception, and the rise of individualism and materialism (Population Reference Bureau 2004).

Urbanization plays a significant role in the demographic transition. Because health care delivery is more cost-effective in cities than in rural areas, governments prioritize urban health clinics. With greater access to health care urban dwellers are the first to experience declines in infant mortality and fertility. A study of contraceptive use in Kenya revealed that women who lived in urban areas were more likely to have used contraception (Kimuna & Adamchak 2001). Recent declines in fertility throughout Africa are mostly an urban phenomenon (Cincotta et al. 2003).

Structural functionalists view the development of urban areas as functional for societal development. Although cities initially served as centers of production and distribution, today they are centers of finance, administration, education, health care, and information.

Urbanization is also dysfunctional, because it leads to increased rates of anomie, or normlessness, as the bonds between individuals and social groups become weak (see also Chapters 1 and 4). Whereas in rural areas social cohesion is based on shared values and beliefs, in urban areas social cohesion is based on interdependence created by the specialization and social diversity of the urban population. Anomie is linked to higher rates of deviant behavior, including crime, drug addiction, and alcoholism. Overcrowding, poverty, the rapid spread of infectious disease, and environmental destruction are also considered dysfunctions associated with urbanization.

demographic transition theory A theory that attributes population growth patterns to changes in birth rates and death rates associated with the process of industrialization.

Conflict Perspective

The conflict perspective focuses on how wealth and power, or the lack thereof, affect population problems. In 1798 Thomas Malthus predicted that the population would grow faster than the food supply and that masses of people were destined to be poor and hungry. According to Malthusian theory, food shortages would lead to war, disease, and starvation, which would eventually slow population growth. However, conflict theorists argue that food shortages result primarily from inequitable distribution of power and resources (Livernash & Rodenburg 1998).

Conflict theorists also note that population growth results from pervasive poverty and the subordinate position of women in many less-developed countries. Poor countries have high infant and child mortality rates. Hence women in many poor countries feel compelled to have many children to increase the chances that some will survive into adulthood. Their subordinate position prevents many women from limiting their fertility. In many developing countries a woman must get her husband's consent before she can receive any contraceptive services (UNICEF 2006). Thus, according to conflict theorists, population problems result from continued economic and gender inequality.

Power and wealth also affect the development and operations of urban areas. The capitalistic pursuit of wealth contributed to the development of cities, because capitalism requires that the production and distribution of goods and services be centrally located, thus, at least initially, leading to urbanization. Today, global capitalism and corporate multinationalism, in search of new markets, cheap labor, and raw materials, have largely spurred urbanization of the developing world. Capitalism also contributes to migration from rural areas into cities because peasant farmers who have traditionally produced goods for local consumption are being displaced by commercial agriculture that is geared to producing fruits, flowers, and vegetables for export to the developed world. Displaced from their traditional occupations, peasant farmers are flocking to cities to find employment (Clark 1998).

The conflict perspective also focuses on how individuals and groups with wealth and power influence decisions that affect urban populations. For example, according to citizens' groups working to stop urban sprawl in Central and Eastern Europe, city officials may be bribed to approve a new shopping mall or other development project (Sheehan 2001). In addition, deteriorating conditions in U.S. inner cities are often ignored because the residents of inner cities lack the wealth, power, and status to solicit government spending on needed infrastructure and services, such as sidewalk and road repairs, street cleaning, and beautification projects.

Symbolic Interactionist Perspective

The symbolic interactionist perspective focuses on how meanings, labels, and definitions learned through interaction affect population problems. For example, many societies are characterized by **pronatalism**—a cultural value that promotes having children. Throughout history many religions have worshiped fertility and recognized it as being necessary for the continuation of the human race. In many countries religions prohibit or discourage birth control, contraceptives, and abortion. Women in pronatalistic societies learn through interaction with others that deliberate control of fertility is socially unacceptable. Women who use contraception in communities in which family planning is not socially accepted face ostracism by their community, disdain from relatives and friends, and even divorce and abandonment by their husbands (Women's Studies Project 2003). However, once some women learn new definitions of fertility control,

❝ Cities have been ignored because they are blacker, browner, poorer, and more female than the rest of the nation. **❞**

Julianne Malveaux
Writer and scholar

pronatalism A cultural value that promotes having children.

they become role models and influence the attitudes and behaviors of others in their personal networks (Bongaarts & Watkins 1996).

The symbolic interaction perspective can also be applied to understanding how urban life affects interaction patterns and social relationships. The classical and modern views represent different observations of how urban living affects social relationships.

Classical Theoretical View. Cities have the reputation of being cold and impersonal. George Simmel observed that urban living involved an overemphasis on punctuality and individuality and a detached attitude toward interpersonal relationships (Wolff 1978). It is not difficult to find evidence to support Simmel's observations: New Yorkers pushing each other to get onto the subway during rush hour; motorists cursing each other in Los Angeles traffic jams; and Chicago residents ignoring the homeless man asleep on the sidewalk.

Louis Wirth (1938), a second-generation student of Simmel, argued that urban life is disruptive for both families and friendships. He believed that because of the heterogeneity, density, and size of urban populations, interactions become segmented and transitory, resulting in weakened social bonds. Wirth held that as social solidarity weakens, people exhibit loneliness, depression, stress, and antisocial behavior.

Modern Theoretical View. In contrast to Wirth's pessimistic view of urban areas, Herbert Gans (1984) argued that cities do not interfere with the development and maintenance of functional and positive interpersonal relationships. Among other communities Gans studied an Italian urban neighborhood in Boston and found such neighborhoods to be community oriented and marked by close interpersonal ties. Rather than finding the social disorganization described by Wirth, Gans observed that kinship and ethnicity helped bind people together. Intimate small groups with strong social bonds characterized these enclaves. Thus Gans saw the city as a patchwork quilt of different neighborhoods or urban villages, each of which helped individuals deal with the pressures of urban living.

A Theoretical Synthesis. Fisher (1982) interviewed more than 1,000 respondents in various urban areas and found evidence for both the classical and the modern theoretical views of urbanism. From the classical perspective he found that heterogeneity (the diversity among urban residents) does make community integration and consensus difficult—community cohesion is less tight. Ties that do exist are less often kin related than in nonurban areas and are more often based on work relationships, memberships in voluntary and professional organizations, and proximity to neighbors.

Fisher also found, however, that the diversity of urban populations facilitates the development of subcultures that have a sense of community ties. For example, large urban areas include diverse groups such as gays, ethnic and racial minorities, and artists. These individuals find each other and develop their own unique subcultures.

Other research has also found support for a synthesis of the classical and modern views of urban life. Tittle (1989) found that the larger the size of the community, the weaker a respondent's social bonds and the higher his or her anonymity, tolerance, alienation, and reported incidence of deviant behavior were. Tittle also found that racial and ethnic groups created their own sense of community in urban neighborhoods.

SOCIAL PROBLEMS RELATED TO POPULATION GROWTH AND URBANIZATION

Next, we examine some of the social problems related to population growth and urbanization, including environmental problems, poverty and unemployment, global insecurity, poor maternal and infant health, transportation and traffic problems, and the effects of sprawl on wildlife and human health. First, we consider the problems faced by countries where fertility rates are low and population is declining.

Problems Associated with Below-Replacement Fertility

In more than one-third of the world's countries—including China, Japan, and all European countries—fertility rates have fallen well below the 2.1 children replacement level (Butz 2005). Because these low fertility rates will eventually lead to population decline, some reports have sounded an alarm about the possibility of a "birth dearth." Low fertility rates lead not only to a decline in population size but also to an increasing proportion of the population comprising elderly members. A birth dearth eventually results in fewer workers to support the pension, social security, and health care systems for the elderly. Below-replacement fertility rates also raise concern about the ability of a country to maintain a productive economy, because there may not be enough future workers to replace current workers as they age and retire.

Environmental Problems and Resource Scarcity

As we discuss in Chapter 14, population growth places increased demands on natural resources, such as forests, water, cropland, and oil and results in increased waste and pollution. Over the past 50 years Earth's ecosystems have been degraded more rapidly and extensively than in any other comparable period of time in human history (Millennium Ecosystem Assessment 2005).

The countries that suffer most from shortages of water, farmland, and food are developing countries with the highest population growth rates. For example, about one-third of the developing world's population (1.7 billion people) live in countries with severe water stress (Weiland 2005). However, countries with the largest populations do not necessarily have the largest impact on the environment. This is because the impact that each person makes on the environment— each person's **environmental footprint**—is determined by the patterns of production and consumption in that person's culture. The environmental footprint of an average person in a high-income country is about six times bigger than that of someone in a low-income country and many more times bigger than that in the least developed nations (United Nations Population Fund 2004). For example, even though the U.S. population is only one-fourth the size of India's, its environmental footprint is about three times bigger—the United States releases about three times as much carbon into the atmosphere each year as does India. Hence, although population growth is a contributing factor in environmental

environmental footprint The impact that each person makes on the environment.

problems, patterns of production and consumption are at least as important in influencing the effects of population on the environment.

Urban Poverty and Unemployment

Poverty and unemployment are problems that plague countries with high population growth as well as urban areas in both rich and poor countries. Less-developed, poor countries with high birth rates do not have enough jobs for a rapidly growing population, and land for subsistence farming becomes increasingly scarce as populations grow. In some ways poverty leads to high fertility, because poor women are less likely to have access to contraception and are more likely to have large families in the hope that some children will survive to adulthood and support them in old age. But high fertility also exacerbates poverty, because families have more children to support and national budgets for education and health care are stretched thin.

Until recently, all measures of poverty, whether based on income, consumption, or expenditure, showed that poverty was deeper and more widespread in rural areas than in cities. In general, urban areas have offered better access to health care, education and information, basic infrastructure, and jobs. However, poverty today is increasing more rapidly in urban areas than in rural areas (United Nations Population Fund 2007).

In the United States the highest rates of poverty are in the central cities, in part because many central cities face unemployment rates that are much higher than the national average. In the United States and other industrialized countries urban unemployment and poverty are partly the results of deindustrialization, or the loss and/or relocation of manufacturing industries. Since the 1970s many urban factories have closed or relocated, forcing blue-collar workers into unemployment. When prospects of finding decent employment are low, the resulting feelings of frustration and worthlessness can lead to drug use, crime, and violence (Rodriquez 2000).

> **66** Many of today's problems in the inner-city ghetto neighborhoods—crime, family dissolution, welfare, low levels of social organization, and so on—are fundamentally a consequence of the disappearance of work. **99**
>
> **William Julius Wilson**
> Sociologist

Urban Housing and Sanitation Problems

In both developed and less-developed countries the urban poor struggle to find affordable decent housing. Many cities are experiencing a housing crisis, because the number of low-income renters has increased while the number of low-cost rental units has dropped. Housing that is available and affordable is often substandard, characterized by outdated plumbing and wiring, overcrowding, rat infestations, toxic lead paint, and fire hazards (see also Chapter 6).

Low-income housing tends to be concentrated in inner-city areas of extreme poverty, but jobs are increasingly moving to the suburbs, and central-city residents often lack transportation to get to these jobs. The more affluent suburbs restrict development of affordable housing to keep out "undesirables" and maintain their high property values. Suburban zoning regulations that require large lot sizes, minimum room sizes, and single-family dwellings serve as barriers to low-income development in suburban areas (Orfield 1997).

Concentrated areas of poverty and poor housing in urban areas are called **slums.** In the United States slums that are occupied primarily by African Americans are known as **ghettos,** and those occupied primarily by Latinos are called **barrios.** Nearly one in three city dwellers, a billion people, a sixth of the world's population, live in slums characterized by overcrowding; little employment;

slums Concentrated areas of poverty and poor housing in urban areas.

ghettos In the United States, slums that are occupied primarily by African Americans.

barrios In the United States, slums that are occupied primarily by Latinos.

Nearly one in three city dwellers—almost 1 billion people—live in slums characterized by overcrowding, little employment, and poor water, sanitation, and health care services.

© David Turnley/Corbis

and poor water, sanitation, and health care services (United Nations Population Fund 2007).

In sub-Saharan Africa, urbanization has become virtually synonymous with slum growth; nearly three-quarters (72 percent) of the urban population in sub-Saharan Africa live in slum conditions (Davis 2006). Worldwide, the growth of slums is outpacing the growth of urbanization.

> Thus, the cities of the future, rather than being made out of glass and steel . . . are instead largely constructed out of crude brick, straw, recycled plastic, cement blocks, and scrap wood. Instead of cities of light soaring toward heaven, much of the twenty-first century urban world squats in squalor, surrounded by pollution, excrement, and decay. (Davis, 2006, p. 19)

Global Insecurity

A Population Institute report titled *Breeding Insecurity: Global Security Implications of Rapid Population Growth,* warns that rapid population growth is a contributing factor to global insecurity, including civil unrest, war, and terrorism (Weiland 2005). The report points out that many developing countries are characterized by a "youth bulge"—a high proportion of 15- to 29-year-olds relative to the adult population. Youth bulges result from high fertility rates and declining infant mortality rates, a common pattern in developing countries today. Youth bulges are growing rapidly throughout Africa and the Middle East. For example, in Iraq young adults make up nearly half (47 percent) of the adult population. The combination of a youth bulge with other characteristics of rapidly growing populations, such as resource scarcity, high unemployment rates, poverty, and rapid urbanization, sets the stage for political unrest. "Large groups of unemployed young people, combined with overcrowded cities and lack of access to farmland and water create a population that is angry and frustrated with the status quo and thus is more likely to resort to violence to bring about change" (Weiland 2005, p. 3).

Poor Maternal, Infant, and Child Health

As noted in Chapter 2, maternal deaths (deaths related to pregnancy and childbirth) are the leading cause of mortality for reproductive-age women in the developing world. Having several children at short intervals increases the chances of premature birth, infectious disease, and death for the mother or the baby. Childbearing at young ages (teens) has been associated with anemia and hemorrhage, obstructed and prolonged labor, infection, and higher rates of infant mortality (Zabin & Kiragu 1998). In developing countries one in four children is born unwanted, increasing the risk of neglect and abuse. In addition, the more children a woman has, the fewer the parental resources (parental income and time and maternal nutrition) and social resources (health care and education) available to each child. The adverse health effects of high fertility on women and children are, in themselves, compelling reasons for providing women with family planning services. "Reproductive health and choice are often the key to a woman's ability to stay alive, to protect the health of her children and to provide for herself and her family" (Catley-Carlson & Outlaw 1998, p. 241).

Transportation and Traffic Problems

Urban areas are often plagued with transportation and traffic problems. A study of U.S. urban areas found that in 2005 drivers experienced 38 hours of delays—up from 14 hours in 1982 (Schrank & Lomax 2007). This study revealed that a total of 2.9 billion gallons of fuel were wasted in 2005 because of traffic congestion.

Many public roads in urban areas are afflicted with what some call *autosclerosis*—"clogged vehicular arteries that slow rush hour traffic to a crawl or a stop, even when there are no accidents or construction crews ahead" ("Bridge to the 21st Century," 1997, p. 1). According to Jan Lundberg, director of the Alliance for a Paving Moratorium, "the average vehicle speed for cross town traffic in New York City is less than six miles per hour—slower than it was in the days of horse-drawn buggies" (quoted in Jensen 2001, p. 6). And "traffic jams in Atlanta have been so entangled that babies have been born in traffic standstills, and some desperate drivers have had to leave their cars to relieve themselves behind roadside bushes" (Shevis 1999, p. 2).

Cars and light trucks are the largest single source of air pollution (Union of Concerned Scientists 2003). Air pollution and traffic congestion have been major forces that drive residents and businesses away from densely populated urban areas (Warren 1998). Health problems associated with congested traffic include stress, respiratory problems, and death. Breathing the air in some cities in the world is equivalent to smoking two packs of cigarettes per day (Brown 2006).

© Peter Johnson/Corbis

Worldwide, pregnancy is the leading cause of death for young women ages 15–19. Most (95 percent) maternal deaths occur in Africa and Asia. This woman in sub-Saharan Africa has a 1 in 16 risk of dying in pregnancy or childbirth, compared to a 1 in 2,800 chance for a woman in a developed country.

> "Everybody says that living in the inner city is dangerous, but the truth is that, if you take car crashes into account, the suburbs are statistically far more dangerous places to live."
>
> Jan Lundberg
> Director of the Alliance for
> a Paving Moratorium

In 2005, drivers on U.S. urban roadways wasted 2.9 billion gallons of fuel due to traffic congestion.

© EGDigital/iStockphoto

And more than 20 million people are severely injured or killed on the world's roads each year (World Health Organization 2003). Indeed, far more people are killed and injured in automobile accidents than by violent crime. Therefore it has been argued that, despite higher crime rates in the inner city, the suburbs are the more dangerous place to live, because suburbanites "drive three times as much, and twice as fast, as urban dwellers" (Durning 1996, p. 24).

In addition to the day-to-day concerns of traffic congestion and air pollution from vehicles, cities are faced with protecting their public transportation systems from terrorist attacks, such as the 2004 bombing of a commuter train in Madrid and the 2005 bombings of London's public transit system. Finally, Hurricanes Katrina and Rita in 2005 reminded city officials that they must have a transportation plan in the event of an evacuation.

Effects of Sprawl on Wildlife and Human Health

The spread of urban and suburban areas increasingly is replacing natural habitats with pavement, buildings, and human communities. The loss of open green space, trees, and plant life affects animals whose homes are turned into parking lots, shopping centers, office buildings, and housing developments. According to the U.S. Fish and Wildlife Service, habitat loss resulting from urban and suburban sprawl is the number one reason that wildlife species are becoming increasingly endangered (Shevis 1999) (see also Chapter 14).

Evidence of wildlife displacement resulting from sprawl is found across the nation. Coyotes, normally found only in the West and in Appalachia, are now being sighted in every state (Shevis 1999). Other displaced species include bear, Canada geese, and deer.

Sprawl also poses health hazards to humans. As more wooded areas are cleared for development, deer are displaced. "With no place to go, they bound into suburban backyards in search of food and water and across highways, frequently injuring themselves and causing harm to drivers. . . . Deer cause an es-

Health experts agree that most Americans are overweight and do not get enough exercise. As discussed in Chapter 2, obesity has reached epidemic proportions and is the second leading cause of preventable deaths in the United States. Although eating too much of the wrong kinds of foods is a major cause of obesity, lack of physical activity also plays a major role.

Here, we describe the first national study to investigate whether our physical activity levels, weight, and health are related to the type of place in which we live (Ewing et al. 2003). This study assesses how sprawl development—where homes are far from workplaces, shops, restaurants, and other destinations—affects physical activity, weight, and health of community residents.

Sample and Methods

Ewing and colleagues (2003) used the Centers for Disease Control and Prevention Behavioral Risk Factor Surveillance System (BRFSS) surveys (1998–2000) to obtain data on physical activity levels, body mass index and obesity,

hypertension, diabetes, and heart disease of respondents. The BRFSS is a random telephone survey administered to U.S. civilian noninstitutionalized adults. More than 200,000 respondents from 448 counties and more than 175,000 respondents from 83 metropolitan areas were selected from the larger BRFSS samples because they lived in areas for which urban sprawl indexes were available.

Ewing and co-workers (2003) measured the degree of sprawl in 448 counties by using a "sprawl index" based on data available from the U.S. Census Bureau and other federal sources. The bigger the sprawl index, the more compact was the metropolitan area; the smaller the index, the more sprawling was the region.

Findings and Conclusions

Sprawling areas are not conducive to walking or biking, and people who live in sprawling communities have fewer opportunities to walk or bicycle as part of their daily routine. Rather, residents in sprawling areas depend on driving as the most convenient form of transportation.

Therefore it is not surprising that Ewing and colleagues found that people who live in counties with sprawling development are likely to walk less and weigh more than people who live in less-sprawling counties. The people living in the most sprawling areas were likely to weigh 6 pounds more than people in the most compact county.

In addition, after controlling for factors such as age, education, sex, and race and ethnicity, people in more sprawling counties were more likely to suffer from hypertension (high blood pressure). The researchers did not find a statistically significant relationship between level of sprawl and diabetes or cardiovascular disease.

The study by Ewing and colleagues (2003) is the first national study to establish a direct association between the level of sprawl development in a community and the health of the people who live there. The study's findings suggest that we can improve public health by creating more compact communities that offer opportunities and design features (e.g., sidewalks) to include walking in our daily routines.

timated half-million vehicle accidents a year, killing 100 people and injuring thousands more" (Shevis 1999, pp. 2–3).

Suburban sprawl has also brought humans into greater contact with ticks that carry Lyme disease (UNEP 2005). If bitten by an infected tick, humans can develop various symptoms, including skin rash, neurological problems, fatigue, abdominal and joint pain, headache, and heart damage. Other health problems associated with sprawl include lack of physical activity and obesity. As discussed in this chapter's *Social Problems Research Up Close* feature, people who live in high-sprawl areas tend to walk less than people who live in more compact communities.

STRATEGIES FOR ACTION: RESPONDING TO PROBLEMS OF POPULATION GROWTH, POPULATION DECLINE, AND URBANIZATION

Whereas some countries are struggling to slow population growth, others are challenged with maintaining or even increasing their populations. After describing efforts to maintain or increase population in low-fertility countries, we look at strategies to slow population growth, including providing access to family planning services and contraceptive methods, involving men in family planning, implementing a one-child policy as in China, improving the status of women through education and employment, and increasing economic development and improving health status. Because populations increasingly live in urban areas, strategies that address population problems also address, indirectly, problems of urbanization. Additional strategies to alleviate urban problems include those designed to lessen poverty and stimulate economic development in inner cities, implement "smart growth" and "new urbanism" in development, improve transportation and ease traffic congestion, and curb urban growth in developing countries.

Efforts to Maintain or Increase Population in Low-Fertility Countries

In some countries with below-replacement fertility levels, population strategies have focused on *increasing* rather than decreasing the population. For example, Australia's total fertility rate hit a record low of 1.73 in 2001, prompting the government to begin paying a $3,000 bonus in 2004 (which increased to $4,000 in 2005) to families who have babies (Lalasz 2005). The town of Yamatsuri, Japan, offers a $9,200 monetary reward (over a 10-year period) to persuade women who have at least two children to have more (Wiseman 2005). Aside from monetary rewards, many countries encourage childbearing by implementing policies designed to help women combine child rearing with employment. For example, as noted in Chapter 5, many European countries have generous family leave policies and universal child care. Another way to increase population is to increase immigration. Spain, for example, has eased restrictions on immigration as a way to gain population.

What Do You Think? One strategy for encouraging childbearing in low-fertility European countries is to provide work-family supports to make it easier for women to combine childbearing with employment. If the United States offered more generous work-family benefits, such as paid parenting leave and government-supported child care, would the U.S. birth rate increase? Would such policies affect the number of children you would want to have?

Efforts to Curb Population Growth: Reducing Fertility

A number of strategies are associated with efforts to reduce the number of children women have. These include providing access to family planning services, involving men in family planning, implementing a one-child policy as in China, and improving the status of women. To some population experts, stopping population growth is a critical issue. "Zero population growth, which characterized human population for more than 99 percent of its history, must be achieved once again, at least as a long-term average, if the human species is to survive" (McFalls 2007, p. 27).

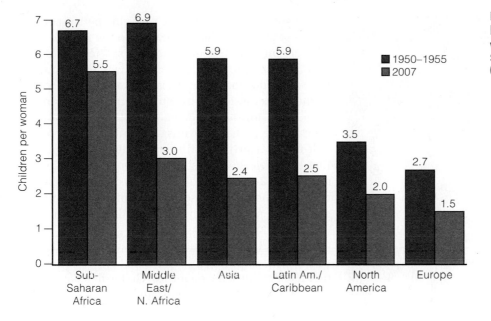

FIGURE 13.5
Fertility rates drop worldwide: 1950s–2007.
Source: Population Reference Bureau
(2007a).

Provide Access to Family Planning Services. Since the 1950s governments and nongovernmental organizations such as the International Planned Parenthood Federation have sought to lower fertility through family planning programs that provide reproductive health services and access to contraceptive information and methods. Such programs, along with developments in contraceptive technology, have achieved the desired result: Globally, the average number of children born to each woman has fallen from 5 in the 1950s to 2.55 in 2007, because more women today want to limit family size and are using modern methods of birth control to control the number and spacing of births (United Nations 2007) (see Figure 13.5). Today, more than half of married women worldwide use some form of modern contraception. Although 68 percent of married women in the United States use modern contraception, in 25 countries the figure is less than 10 percent (Population Reference Bureau 2007b).

Although access to contraceptives has increased worldwide since the 1950s, there is still a significant unmet need for contraception. Worldwide, about 200 million women want to delay or avoid pregnancy but are not using modern contraceptive methods (Cohen 2006). In developing countries, 15 percent of married women and 7 percent of never-married women ages 15–49 want to use contraception, but do not have access to contraceptives. Unmet need for contraceptives is highest in sub-Saharan Africa, where 24 percent of married women and 9 percent of unmarried women ages 15–49 do not have access to contraceptives they desire (Guttmacher Institute 2007).

Family planning programs are effective in increasing use of contraceptives. For example, in rural Pakistan married women living in households served by a "lady health worker," who provides doorstep delivery of contraceptive supplies, are more likely than other married women to use modern contraceptives (Douthwaite & Ward 2005). The provision of doorstep services through female workers is especially valuable in Pakistan, where women's mobility is limited and female modesty is highly valued.

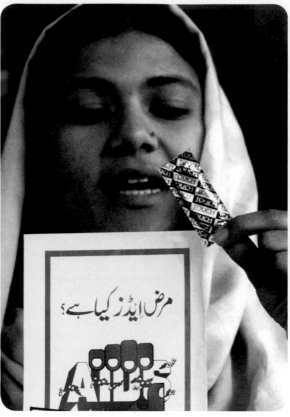

In Pakistan, where women's mobility is limited and female modesty is highly valued, "lady health workers" provide doorstep contraceptive services.

In some countries family planning personnel often refuse or are forbidden by law or policy to make referrals for contraceptive and abortion services for unmarried women. Furthermore, many women throughout the world do not have access to legal, safe abortion. Without access to contraceptives many women who experience unwanted pregnancy resort to abortion—even under illegal and unsafe conditions. More than half of the nearly 80 million unintended pregnancies that occur worldwide every year end in abortion. Research in 12 countries of Central Asia and Eastern Europe has shown that increased contraception use has resulted in significant declines in the rate of abortion (Leahy 2003).

Cuts in U.S. assistance to international family planning programs are largely the result of opposition to abortion practices in some countries. On his first day in office in 2001, George W. Bush reinstated the "global gag rule," which denies U.S. international family planning assistance to organizations that use their own privately raised funds to counsel women on the availability of abortion, advocate change in abortion laws, or provide abortion services.

Thus, U.S. policy—purportedly aimed at reducing abortion—only exacerbates the already daunting challenge of ensuring that all people in the developing world who want to time and space their childbearing without resorting to abortion can actually obtain the contraceptives they need to do so. (Cohen 2006, p. 15)

What Do You Think? In developing countries hundreds of thousands of women die each year as a result of unsafe abortions. The majority of U.S. adults favor U.S. economic aid for family planning programs in developing countries, but only half favor U.S. economic aid to provide voluntary, safe abortions as part of reproductive health care in developing countries that request it (DaVanzo et al. 2000). Do you think that the United States should provide funding to developing countries for medical facilities and equipment and training to health care personnel for the purpose of providing women with access to safe abortion? How would you vote on this issue?

Involvement of Men in Family Planning. Although men play a central role in family planning decisions, they often do not have access to information and services that would empower them to make informed decisions about contraceptive use (Women's Studies Project 2003). Therefore family planning programs need to direct educational programs and health services to men. For example, men need education about the health risks to women when pregnancies are spaced too closely or when pregnancies occur before age 20 and after age 40.

Another important component of family planning and reproductive health programs involves changing traditional male attitudes toward women. According to traditional male gender attitudes, (1) a woman's most important role is being a wife and mother, (2) it is a husband's right to have sex with his wife at his demand, and (3) it is a husband's right to refuse to use condoms and to forbid his wife to use any other form of contraception. A number of programs around the world work with groups of boys and young men to change such traditional male gender attitudes (Schueller 2005).

China's One-Child Policy. In 1979, China initiated a national family planning policy that encourages families to have only one child by imposing a monetary fine on couples that have more than one child. The implementation and enforcement of this policy varies from one province to another, and there are a number of exemptions that allow some couples to have two or even more children. For example, most urban couples are limited to having only one child, but some rural couples are allowed to have a second child if their first child is a girl. Families that need children for labor, ethnic minority populations, and married couples who themselves are only children are also exempt from the one-child rule.

Recognizing that men play a crucial role in family planning decisions, family planning programs are making efforts to include men in family planning education and services.

China has been criticized for using extreme measures to enforce its one-child policy, including steep fines, seizure of property, and forced sterilizations and abortions. In addition, because of a traditional preference for male heirs, many Chinese couples have aborted female fetuses with the hope of having a boy in a subsequent pregnancy. This has led to an imbalanced sex ratio, with many more Chinese males than females. The Chinese government has defended the one-child policy, claiming that it has been effective in preventing at least 300 million births and has contributed to China's recent economic development ("China Is Softening Messages" 2007). China has the highest rate of modern contraceptive use in the world: 86 percent of married women in China use modern contraceptives (Population Reference Bureau 2007b).

Improve the Status of Women: The Importance of Education and Employment. Throughout the developing world the primary status of women is that of wife and mother. Women in developing countries traditionally have not been encouraged to seek education or employment; rather, they are encouraged to marry early and have children.

Improving the status of women by providing educational and occupational opportunities is vital to curbing population growth. Educated women are more likely to marry later, want smaller families, and use contraception. Recent data

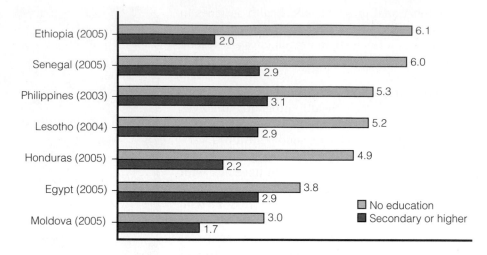

FIGURE 13.6
Lifetime births per woman by highest level of education.
Source: Population Reference Bureau (2007b).

from many countries have shown that women with at least a secondary-level education eventually give birth to one-third to one-half as many children as women with no formal education (Population Reference Bureau 2007b) (see Figure 13.6). The relationship between education and lower fertility is also true among U.S. women. In 2004, U.S. women ages 40–44 with a graduate or professional degree had an average of 1.6 children, compared with 1.9 children for women with only a high school diploma and 2.5 children for women without a high school diploma (McFalls 2007). U.S. women with advanced education are also more likely than women with less education to voluntarily choose to have no children.

Better-educated women tend to delay marriage and exercise more control over their reproductive lives, including decisions about childbearing. In addition, "education can result in smaller family size when the education provides access to a job that offers a promising alternative to early marriage and childbearing" (Population Reference Bureau 2004, p. 18). Providing employment opportunities for women is also important to slow population growth, because high levels of female labor force participation and higher wages for women are associated with smaller family size. Low fertility rates in developed countries are largely due to women postponing having children until they have completed their education and established their careers.

Primary school enrollment of both young girls and boys is also related to declines in fertility rates. In countries where primary school enrollment is widespread or nearly universal, fertility declines more rapidly because (1) schools help spread attitudes about the benefits of family planning and (2) universal education increases the cost of having children, because parents sometimes are required to pay school fees for each child and because they lose potential labor that children could provide (Population Reference Bureau 2004).

In the United States, as in other countries of the world, the cultural norm is for women and couples to, sooner or later, want to have children. However, a small but growing segment of U.S. women and men does not want children and chooses to be childfree. In the United States, 7 percent of women ages 35–44 are voluntarily childless, making voluntary childlessness more common than involuntary childlessness (Hollander 2007). In general, childfree couples are more educated, live in urban areas, are less religious, and do not adhere to traditional gender ideology (Parks 2005). Although one of the motives some childless-by-choice individuals cite is concern for overpopulation and a deep caring for the health of the planet, voluntarily childless individuals are often criticized as being selfish and individualistic, as well as less well-adjusted and less nurturing (Parks 2005). In this chapter's *The Human Side* feature, two women tell their stories about their choice to be childfree.

Increase Economic Development and Improve Health. Although fertility reduction can be achieved without industrialization, economic development may play an important role in slowing population growth. Families in poor countries often rely on having many children to provide enough labor and income to support the family. Economic development decreases the economic value of children and is also associated with more education for women and greater gender equality. As previously noted, women's education and status are related to fertility levels.

Economic development tends to result in improved health status of populations. Reductions in infant and child mortality are important for fertility decline, because couples no longer need to have many pregnancies to ensure that some children survive into adulthood. Finally, the more developed a country is, the more likely women are to be exposed to meanings and values that promote fertility control through their interaction in educational settings and through media and information technologies (Bongaarts & Watkins 1996).

Efforts to Restore Urban Prosperity

Nearly 9 in 10 Americans (86 percent) want their states to fund improvements in existing communities rather than provide incentives for new development (*American Community Survey* 2004). A number of strategies have been proposed and implemented to restore prosperity to U.S. cities and well-being to their residents, businesses, and workers, including strategies to attract new businesses, create jobs, and repopulate cities. The economic development and revitalization of cities also involves improving affordable housing options (see Chapter 6), alleviating urban problems related to HIV/AIDS, addiction, and crime (discussed in Chapters 2–4), and reducing problems of traffic and transportation, which we address later in this chapter.

> **"** Increasing urbanization has the potential for improving human life or increasing human misery. The cities can provide opportunities or frustrate their attainment; promote health or cause disease; empower people to realize their needs and desires or impose on them a simple struggle for basic survival. **"**
>
> United Nations
> 1996 State of the World
> Population Report

Empowerment Zone/Enterprise Community Program. The federal Empowerment Zone/Enterprise Community Initiative, or EZ/EC program, provides tax incentives, grants, and loans to businesses to create jobs for residents living within various designated zones or communities, many of which are in urban areas. Federal money provided to empowerment zones and enterprise communities is also used to train and educate youth and families and to improve child care, health care, and transportation. The EZ/EC program provides grant funding so that communities can design local solutions that empower residents to participate in the revitalization of their neighborhoods.

Infrastructure Improvements. Urban revitalization often involves making improvements in the **infrastructure**—the underlying foundation that enables a city to function. Infrastructure includes things such as water and sewer lines, phone lines, electricity cables, sidewalks, streets, bridges, curbs, lighting, and storm drainage systems. As reflected in this chapter's opening vignette, maintaining adequate infrastructure is a public health and safety issue. Improving infrastruc-

infrastructure The underlying foundation that enables a city to function, including such things as water and sewer lines, phone lines, electricity cables, sidewalks, streets, bridges, curbs, lighting, and storm drainage systems.

My Decision to Be Childfree (by Rebecca Stephens)

© Rebecca Stephens, 2007. Used with permission

As young teens, my girlfriends and I would daydream about the man we'd marry, the type of house we'd live in and what we'd name our children. It was simply assumed that we'd follow the traditional path that our mothers followed. By the time I was a young adult with college behind me, however, I realized that this dream no longer fit the picture of how I wanted to live my life.

Growing up, I played with dolls just like all of my girlfriends. Most of us were raised with siblings in solid families; the majority of our parents remain married to this day. There was no trauma that caused me to rethink living my life with the picket fence, two kids and a dog.

I was raised in a liberal household and taught to think for myself, appreciate different beliefs and cultures and respect the environment. My parents recycled long before most people did; I took my homemade lunch to school in a brown paper bag and was expected to bring the empty bag home every day to reuse. My parents belonged to Zero Population Growth (ZPG), and I often glimpsed the newsletters in our mail. This is the only item from my childhood that even slightly hints of my eventual decision to remain childfree. My parents never sat me down and lectured me on overpopulation. For the most part, I just never had the "urge" to have children. After marrying my husband, who was also fairly ambivalent about becoming a parent, I told myself that I had until the age of 37 to decide if I wanted kids or not, which gave me seven years to change my mind. I never did. We have been happily married and childfree for 11 years now. Being child-free allows my husband and me a great deal of flexibility in our day to day lives and in our careers. We have joined an international organization for childfree individuals and couples called "No Kidding" and have very close friends who have also remained childfree, giving us a very satisfying social life and the support of other people who have embraced the childfree lifestyle.

The decision to not have children was largely based on the kind of lifestyle I wanted. But not having children was also a conscious decision not to further the destruction of our planet and its resources. Even if a couple has just two children, assuming further generations arrive every 25 years with each having two children, within a century two people become 64. Considering the overpopulation in Third World countries, the homeless in our own country, cities filled with smog, urban sprawl and crowded classrooms, my choice seems like an intelligent one to make. Many of the stories in the news today, such as violence and disregard for our planet and its human and animal inhabitants, just depress me, and I feel relieved that I'm not contributing to what I see as a downward spiral. I'm not naive enough to think that *my* not having children will make all the difference to the world, but I can set an example for others that having children is

ture also helps to attract business to an area, increases property values, and renews residents' sense of pride in their neighborhood (Cowherd 2001).

Brownfield Redevelopment. Brownfields are abandoned or undeveloped sites that are located on contaminated land. There are more than 400,000 brownfields throughout the United States and an estimated 5 million acres of abandoned industrial sites in U.S. cities—roughly the same amount of land occupied by 60 of the nation's largest cities (U.S. Department of Housing and Urban Development 2004).

Cleaning up and redeveloping brownfields is not only an important environmental measure but also is a key component of urban revitalization because it provides jobs; increases tax revenues; attracts more businesses, residents, and tourists; and helps to curb urban sprawl. Hundreds of urban brownfield sites across the United States have been successfully redeveloped into residential, business, and recreational property. Recognizing that a major obstacle to brownfield redevelopment is lack of funding, the U.S. Department of Housing and

brownfields Abandoned or undeveloped sites that are located on contaminated land.

a choice, not a mandate, and a choice that should be carefully considered.

Choosing to Be Childfree (by Cynthia Marks Garnant)

I was 28 when I married my first husband. Before we got married we had a "discussion" that lasted about 2 seconds. He said he wanted children someday. I said fine. I wanted to get married and gave this "discussion" less thought than what to have for dinner. After 3 years of marriage, he announced he was ready for children. I said I wasn't. A year later,

he brought it up again. I told him I just wasn't ready yet. For me it was more like a waiting game. Although it seemed like almost every married woman had children at some point, and friends my age were starting to have them, I felt no desire whatsoever to have children. I just assumed that someday a craving would kick in—much like the ones I have for chocolate—to have a squalling infant in my arms.

For the next few years my husband continued to be "ready" and I continued to not be. We got divorced. About a year after the divorce I first heard of the concept "Childless by Choice." By then I was 36 and that "maternal instinct" thing had yet to kick in. Embarrassingly enough, it dawned on me for the first time that it was not that I wasn't "ready" to have children. It was that I didn't want to. Period. I had managed to live in a complete state of denial for years. This is very useful when one wants to feel loved and to fit in. But I should have seen the signs, which were everywhere. I found baby showers inordinately painful. When someone approached with an actual baby, I not only did not want to hold it and coo, I practically ran from the room. When people talked about their children and showed photos, I smiled through gritted teeth.

At last I learned from my mistake. For the next few (painful) years of dating, I announced on every first date that I did not want children. These frequently became last dates. Finally, at age 40, I met the love of my life, who is a conservation biologist. We attended a lecture given by a top environmentalist (Jeremy Rifkin) who explained that having children was one of the worst things you could do to the planet. The biologist and I have been very happily married (and childfree) for 5 years now.

I wish more than anything that I had seriously thought about the childfree choice sooner. I could have saved myself and my first husband a very bad marriage. But I am glad I stuck to my guns: There were times when I thought of giving in to pressure and having a child just to please him. I feel good about my choice to be childfree and proud that I have not added to the overpopulation of the planet. But as bad as having a child is for the environment, I think having one who is not truly wanted, would be even worse. For everyone.

Sources: Garnant, Cynthia Marks. 2007 (September 13) Personal communication. Used by permission.

Stephens, Rebecca. 2007 (August 27). Personal communication. Used by permission.

Urban Development developed a funding program known as the Brownfields Economic Development Initiative, which has provided millions of dollars to communities across the country for the purpose of brownfield redevelopment (U.S. Department of Housing and Urban Development 2004).

Gentrification and Incumbent Upgrading. Gentrification is a type of neighborhood revitalization in which middle- and upper-income individuals buy and rehabilitate older homes in an economically depressed neighborhood. The city provides tax incentives for investing in old housing with the goal of attracting wealthier residents back into these neighborhoods and increasing the tax base. After the house is renovated, the owner may live there, rent, or sell the house. A downside to gentrification is that low-income city residents are often forced into substandard housing because less affordable housing is available. In effect, gentrification often displaces the poor and the elderly (Johnson 1997). However, a survey of residents in two gentrifying neighborhoods in Portland, Oregon, found that most residents—including owners and renters, whites and minorities, newcomers and

gentrification A type of neighborhood revitalization in which middle- and upper-income individuals buy and rehabilitate older homes in an economically depressed neighborhood.

Funding from the U.S. Department of Housing and Urban Development's Brownfields Economic Development Initiative was used to transform an old abandoned factory building in Wheeling, West Virginia, into a new, usable office facility.

long-time residents—liked how their neighborhood had changed and were optimistic that it would continue to improve (Sullivan 2007).

An alternative to gentrification is **incumbent upgrading,** in which aid programs help residents of depressed neighborhoods buy or improve their homes and stay in the community. Both gentrification and incumbent upgrading improve decaying neighborhoods, attracting residents as well as businesses.

Community Development Corporations. In many low- and moderate-income urban neighborhoods, **community development corporations (CDCs)**—nonprofit groups formed by residents, small business owners, congregations, and other local stakeholders—work to create jobs and affordable housing and create or renovate parks and other community facilities, such as child care centers, senior centers, arts/cultural centers, and health care centers. Most CDCs augment their housing and economic development projects with other community building activities, such as budget/credit counseling, immigration services, prisoner re-entry services, education/training, and homeless services. CDCs are funded by a variety of sources, including federal, state, and local governments; banks, foundations, corporations, religious institutions, and individual donors.

The number of CDCs throughout the United States has grown from a handful in the 1970s to 4,600 in 2005 (National Congress for Community Economic Development 2006). An analysis of CDCs found that they have produced dramatic improvements and raised property values in neighborhoods (Galster et al. 2005). By 2005, CDCs had created more than 1 million low-income housing units and created three-quarter of a million jobs (National Congress for Community Economic Development 2006). A major strength of CDCs is that they involve community residents in planning and implementing urban renewal projects, giving residents a sense of empowerment in their communities.

incumbent upgrading Aid programs that help residents of depressed neighborhoods buy or improve their homes and stay in the community.

community development corporations (CDCs) Nonprofit groups formed by residents, small business owners, congregations, and other local stakeholders that work to create jobs and affordable housing and renovate parks and other community facilities.

Improve Transportation and Alleviate Traffic Congestion

An important strategy for reducing traffic congestion involves increasing the use of public transit, such as buses, trains, and subways. A national survey found that the majority of Americans (80 percent) support building more rail systems serving cities, suburbs, and entire regions to give them the option of not driving their cars (U.S. Conference of Mayors 2001).

In Austin, Texas, a community bike program makes bikes available for anyone to use and then leave in a prominent place so someone else can use the bike.

Transportation planners increasingly recognize that building more roads does not necessarily ease traffic problems. The U.S. public agrees: In a poll of randomly selected registered voters the majority (66 percent) said that they do not think traffic congestion will be eased if more roads are built (U.S. Conference of Mayors 2001). More important, concern is growing over the social and environmental problems related to road building and motor vehicle use. According to Jan Lundberg, director of a grassroots group called the Alliance for a Paving Moratorium, nearly half of all urban space is paved; more land is devoted to cars than to housing, and every year nearly 100,000 people are displaced by highway construction (Jensen 2001). The Alliance for a Paving Moratorium advocates a halt to road building. "In the Alliance's view, a paving moratorium would limit the spread of population, redirect investment from suburbs to inner cities, and free up funding for mass transportation and maintenance of existing roads" (Jensen 2001, p. 6).

Another way to ease traffic congestion is to encourage means of transportation other than motor vehicles. In some cities motorists must pay a "congestion charge" for driving in a "congestion charge zone" during weekday high-traffic periods. Congestion charges encourage travelers to use public transport, bicycles, motorcycles, or alternative fuel vehicles, which are exempt from the charge. In 1998 Singapore became the first city to use a congestion charge. Other cities that levy congestion charges include Oslo, Bergen, Trondheim, and London.

Finally, the development of communities that enable residents to walk or ride a bicycle to schools, shops, and other locations can help relieve traffic congestion, as well alleviate air pollution associated with motor vehicles. In Denmark and the Netherlands, cities have created networks of streets for bicycle use where motor vehicles are banned. Bicycle use is increasing in cities that create cycleways. In 2003, in Copenhagen, 36 percent of residents biked to work (27 percent drove, 33 percent used public transit, and 5 percent walked) (Newman & Kenworthy 2007).

> **"** Adding highway capacity to solve traffic congestion is like buying larger pants to deal with your weight problem. **"**
>
> Michael Replogle
> Transportation director
> for Environmental Defense

For each of the following items, select the answer that best represents your attitudes. You can compare your answers with those of a national random sample of U.S. adults who participated in a 2002 telephone survey conducted by Belden, Russonello & Stewart (*Americans' Attitudes Toward Walking* 2003).

1. Which of the following statements describes you more: (A) If it were possible, I would like to walk more throughout the day either to get to specific places or for exercise, or (B) I prefer to drive my car wherever I go.

2. How much of a factor is each of the following in why you do not walk more right now (A, a major reason; B, somewhat of a reason; C, not much of a reason; D, not a reason at all):

2a. Things are too far to get to and it is not convenient to walk.

2b. Not enough time to walk.

2c. Laziness.

2d. It is hard to walk where I live because of traffic and lack of places to walk.

2e. It is hard to walk where I live because there are not enough sidewalks or crosswalks.

2f. Physically I am unable to walk more.

2g. I do not like to walk.

2h. There is too much crime to walk where I live.

3. For each of the following proposals to create more walkable communities, indicate your views using the following key: (A) strongly favor, (B) somewhat favor, (C) somewhat oppose, (D) strongly oppose.

3a. Better enforce traffic laws such as speed limit.

3b. Use part of the transportation budget to design streets with sidewalks, safe crossings, and other devices to reduce speeding in residential areas and make it safer to walk, even if this means driving more slowly.

3c. Use part of the state transportation budget to create more sidewalks and stop signs in communities, to make it safer and easier for children to walk to school, even if this means less money to build new highways.

3d. Increase federal spending on making sure people can safely walk across the street, even if this means less tax dollars go to building roads.

3e. Have your state government use more of its transportation budget for improvements in public transportation, such as trains, buses and light rail, even if this means less money to build new highways.

3f. Design communities so that more stores and other places are within walking distance of homes, even if this means building homes closer together.

These strategies are part of the "smart growth" and "new urbanism" movement discussed in the next section. You can assess your attitudes toward walking and proposals to create more walkable communities in this chapter's *Self and Society* feature.

What Do You Think? What could be done in your community to encourage the use of bicycles as an alternative to motor vehicles? If you lived in a community that had cycleways where motor vehicles were banned, would you be more likely to use a bicycle as a means of transportation?

Responding to Urban Sprawl: Growth Boundaries, Smart Growth, and New Urbanism

In a national survey nearly half of Americans (46 percent) said that slowing development of open space was an "extremely high" or "high" priority (*American Community Survey* 2004). Some cities have tried to manage urban sprawl by establishing growth boundaries. Rather than simply put a limit on urban growth, another

Comparison Data

1. Walk more: 55 percent; drive: 41 percent; don't know: 5 percent

2. Reasons for not walking more:

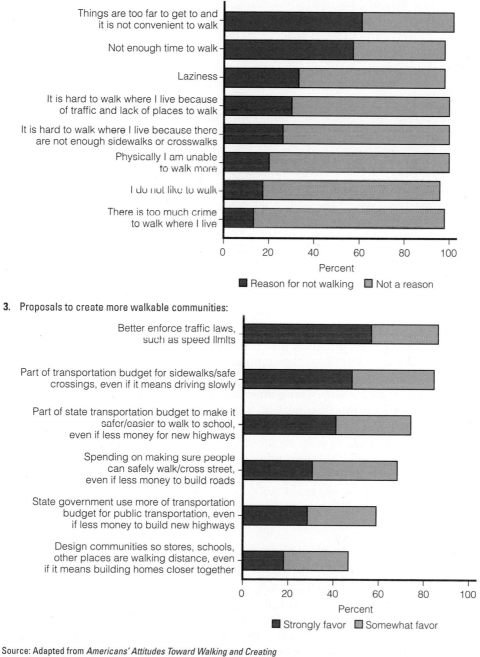

3. Proposals to create more walkable communities:

Source: Adapted from *Americans' Attitudes Toward Walking and Creating Better Walking Communities* (2003).

approach to managing urban sprawl is to develop land according to principles known as **smart growth.** A smart-growth urban development plan entails the following principles (Froehlich 1998; Rees 2003; Smart Growth Network 2007):

- Mixed-use land, which allows homes, jobs, schools, shops, workplaces, and parks to be located within close proximity of each other.
- Ample sidewalks, encouraging residents to walk to jobs and shops.
- Compact building design.
- Housing and transportation choices.
- Distinctive and attractive community design.
- Preservation of open space, farmland, natural beauty, and critical environmental areas.
- Redevelopment of existing communities, rather than letting them decay and building new communities around them.
- Regional planning and collaboration among businesses, private residents, community groups, and policy makers on development and redevelopment issues.

Smart growth is similar to another movement in urban planning called **New Urbanism.** The goals and methods of New Urbanism are similar to those of smart growth, but the impetus for these movements is slightly different. Smart growth approaches the idea of sustainable urban communities with the primary goal of stopping sprawl. The New Urbanism approach is to raise the quality of life for all those in the community by creating compact communities with a sustainable infrastructure. Smart growth and New Urbanism are often impeded by local zoning codes that mandate large housing setbacks, wide streets, and separation of residential and commercial areas (Pelley 1999).

Regionalism

The various social problems that face urban areas may best be addressed through **regionalism**—a form of collaboration among central cities and suburbs that encourages local governments to share common responsibility for common problems. Central cities, declining inner suburbs, and developing suburbs are often in conflict over the distribution of government-funded resources, zoning and land use plans, transportation and transit reform, and development plans. Rather than compete with each other, regional government provides a mechanism for achieving the interests of an entire region. A metropolitan-wide government would handle the inequities and concerns of both suburban and urban areas. As might be expected, suburban officials resist regionalization because they believe it will hurt their neighborhoods economically by draining off money for the cities.

Strategies for Reducing Urban Growth in Developing Countries

In developing countries limiting population growth is essential for alleviating social problems associated with rapidly growing urban populations. Another strategy for minimizing urban growth in less-developed countries involves redistributing the population from urban to rural areas. Such redistribution strategies include the following: (1) promoting agricultural development in rural areas, (2) providing incentives to industries and businesses to relocate from

smart growth A strategy for managing urban sprawl that serves the economic, environmental, and social needs of communities.

New Urbanism A movement in urban planning that approaches the idea of sustainable urban communities with the goal of raising the quality of life for all those in the community by creating compact communities with a sustainable infrastructure.

regionalism A form of collaboration among central cities and suburbs that encourages local governments to share common responsibility for common problems.

urban to rural areas, (3) providing incentives to encourage new businesses and industries to develop in rural areas, and (4) developing the infrastructure of rural areas, including transportation and communication systems, clean water supplies, sanitary waste disposal systems, and social services. Of course, these strategies require economic and material resources, which are in short supply in less-developed countries.

UNDERSTANDING PROBLEMS OF POPULATION GROWTH AND URBANIZATION

What can we conclude from our analysis of population growth and urbanization? First, although fertility rates have declined significantly in recent years and although some countries are experiencing a decline in their population size, world population will continue to grow for several decades. This growth will occur largely in urban areas in developing regions. Given the problems associated with population growth, such as environmental problems and resource depletion, global insecurity, poverty and unemployment, and poor maternal and infant health, most governments recognize the value of controlling population size and support family planning programs. However, efforts to control population must go beyond providing safe, effective, and affordable methods of birth control. Slowing population growth necessitates interventions that change the cultural and structural bases for high fertility rates. Two of these interventions are increasing economic development and improving the status of women, which includes raising their levels of education, their economic position, and their (and their children's) health. Addressing problems associated with population growth also requires the willingness of wealthier countries to commit funds to providing reproductive health care to women, improving the health of populations, and providing universal education for people throughout the world (see Table 13.3).

Attention to urban problems and issues is increasingly important because the United States and the rest of the world are rapidly becoming urbanized. Aside from population concerns, problems affecting urban residents include poverty and unemployment, inferior or unaffordable housing, and traffic and transportation problems.

The social forces affecting urbanization in industrialized countries are different from those in developing countries. Countries such as the United States have experienced urban decline as a result of deindustrialization, deconcentration, and the shift to a service economy in which jobs that pay well and come with full benefits are scarce. At the same time, developing countries have experienced rapid urban growth as a result of industrialization, fueled in part by a global economy in which transnational corporations locate industry in developing countries to gain access to cheap labor, raw materials, and new markets.

Like other social problems, slowing population growth and improving the cities of the world require political will and leadership. When Enrique Penalosa became Mayor of Bogota, Colombia, in 1998, his vision was to make Bogota a city designed for people, not cars. Under Penalosa's leadership, the city created or renovated 1,200 parks, introduced a successful bus transit system, built extensive bicycle paths and pedestrian streets, reduced rush hour traffic by 40 percent, planted 100,000 trees, and involved local citizens directly in improving their neighborhoods (Brown 2006).

TABLE 13.3 Annual Expenditures on Luxury Items Compared with Funding Needed to Meet Selected Basic Needs

PRODUCT	ANNUAL EXPENDITURE
Makeup	$18 billion
Pet food in Europe and the United States	$17 billion
Perfume	$15 billion
Ocean cruises	$14 billion
Ice cream in Europe	$11 billion

SOCIAL OR ECONOMIC GOAL	ADDITIONAL ANNUAL INVESTMENT NEEDED TO ACHIEVE GOAL
Reproductive health care for all women	$12 billion
Elimination of hunger and malnutrition	$19 billion
Universal literacy	$5 billion
Clean drinking water for all	$10 billion
Immunizing every child	$1.3 billion

Source: World Watch Institute Press Release, 2004 (January 1). "State of the World 2004: Consumption by the Numbers." Available at http://www.worldwatch.org

One of the shadows lingering over cities throughout the world is cast by environmental problems, because urban populations consume the largest share of the world's natural resources and contribute most of the pollution and waste that compromise the health of our planet. Global environmental problems, discussed in Chapter 14, are linked to both population growth and urban life. "Key global environmental problems have their roots in cities—from the vehicular exhaust that pollutes and warms the atmosphere, to the urban demand for timber that denudes forests and threatens biodiversity, to the municipal thirst that heightens tensions over water" (Sheehan 2003, p. 131). As we ponder ways to improve cities and as we strive to meet the needs of growing populations, we must include the larger environment in the equation.

CHAPTER REVIEW

- **How long did it take for the world's population to reach 1 billion? How long did it take for it to reach 6 billion?**

  It took thousands of years for the world's population to reach 1 billion, and just another 300 years for the population to grow from 1 billion to 6 billion.

- **Where is most of the world's population growth occurring?**

  Ninety-nine percent of world population growth is in developing countries, mostly in Africa and Asia. Most of this growth is occurring in urban areas.

- **In the history of urbanization, why is 2008 a historic milestone?**

  In the year 2008, for the first time in history more than half of the world's population lived in urban areas.

- **What is "urban sprawl," and why is it a problem?**

  Urban sprawl refers to the ever-increasing outward growth of urban areas. Urban sprawl results in the loss of green, open spaces, the displacement and endangerment of wildlife, traffic congestion and noise, and pollution liabilities. Sprawl is also linked to negative health effects in humans, because people in high-sprawl areas walk less and are more overweight than people who live in low-sprawl areas.

- **What is the demographic transition?**

  The demographic transition is the progression from a population with short lives and large families to one in which people live longer and have smaller families. About one-third of countries have completed the demographic transition.

- **Many countries are experiencing below-replacement fertility (fewer than 2.1 children born to each woman). Why are some countries concerned about a "birth dearth"?**

  In countries with below-replacement fertility, there are or will be fewer workers to support a growing number of elderly retirees and to maintain a productive economy.

- **What kinds of environmental problems are associated with population growth?**

  Population growth places increased demands on natural resources, such as forests, water, cropland, and oil, and results in increased waste and pollution. Although population growth is a contributing factor in environmental problems, patterns of production and consumption are at least as important in influencing the effects of population on the environment.

- **Why is population growth considered a threat to global security?**

  In developing countries rapid population growth results in a "youth bulge"—a high proportion of 15- to 29-year-olds relative to the adult population. The combination of a youth bulge with other characteristics of rapidly growing populations, such as resource scarcity, high unemployment rates, poverty, and rapid urbanization, sets the stage for civil unrest, war, and terrorism, because large groups of unemployed young people resort to violence in an attempt to improve their living conditions.

- **Efforts to curb population growth include what strategies?**

  Efforts to curb population growth include strategies to reduce fertility by providing access to family planning services, involving men in family planning, implementing a one-child policy as in China, and improving the status of women by providing educational and employment opportunities. Achievements in economic development and health are also associated with reductions in fertility.

- **Globally, what was the average number of children born to each woman in the 1950s? In 2007?**

  Globally, the average number of children born to each woman has fallen from 5 in the 1950s to 2.55 in 2007.

- **What are brownfields?**

  Brownfields are abandoned or undeveloped sites that are located on contaminated land. Cleaning up and redeveloping brownfields is a key component of urban revitalization because it provides jobs; increases tax revenues; and potentially attracts more businesses, residents, and tourists.

- **What does the term "mixed-use land" refer to? What are the benefits of mixed-use land?**

  The designation of mixed-use land is a strategy of the smart-growth movement whereby homes, jobs, schools, shops, workplaces, and parks are located within close proximity of each other. Mixed-use land encourages walking as a means of transportation, which minimizes the use of cars.

TEST YOURSELF

1. In 2007, world population was 6.7 billion. In 2050, world population is projected to be _____ billion.
   a. 6.5
   b. 6.8
   c. 7.7
   d. 9.2
2. How many countries have achieved below-replacement fertility rates?
   a. None
   b. One
   c. Five
   d. Fifty
3. What would happen if every country in the world achieved below-replacement fertility rates?
   a. Population growth would stop and world population would remain stable.
   b. World population would immediately begin to decline.
   c. World population would continue to grow for several decades.
   d. World population would decline, but then go up again.

4. By 2015, most megacities—cities with 10 million residents or more—will be located in less developed countries.
   a. True
   b. False
5. Conflict theorists argue that food shortages result primarily from overpopulation of the planet.
   a. True
   b. False
6. Pronatalism is a cultural value that promotes which of the following?
   a. Car ownership
   b. Abstaining from sex until one is married
   c. Having children
   d. Urban living
7. Worldwide, the growth of slums is outpacing the growth of urbanization.
   a. True
   b. False

8. According to the U.S. Fish and Wildlife Service, what is the number one reason that wildlife species are becoming increasingly endangered?
   a. Air pollution from motor vehicles
   b. Habitat loss resulting from urban and suburban sprawl
   c. Hormones in the water supply that result from the widespread use of oral contraceptives
   d. Deaths of wild animals due to being hit by motor vehicles
9. Today, more than half of married women worldwide use some form of modern contraception.
   a. True
   b. False
10. U.S. women with advanced education are more likely than women with less education to voluntarily choose to have no children.
    a. True
    b. False

Answers: 1 d. 2 d. 3 c. 4 a. 5 b. 6 c. 7 a. 8 b. 9 a. 10 a.

KEY TERMS

barrios
brownfields
community development
    corporations (CDCs)
deconcentration
demographic transition theory
doubling time
environmental footprint
gentrification
ghettos
incumbent upgrading

infrastructure
megacities
metropolis
metropolitan area
micropolitan area
New Urbanism
population density
population momentum
pronatalism
regionalism

replacement-level fertility
slums
smart growth
suburbanization
suburbs
total fertility rate
urban area
urbanization
urban sprawl

MEDIA RESOURCES

Understanding Social Problems, Sixth Edition
Companion Website

academic.cengage.com/sociology/mooney
Visit your book companion website, where you will find flash cards, practice quizzes, Internet links, and more to help you study.

Just what you need to know NOW! Spend time on what you need to master rather than on information you already have learned. Take a pre-test for this chapter, and CengageNOW will generate a personalized study plan based on your results. The study plan will identify the topics you need to review and direct you to online resources to help you master those topics. You can then take a post-test to help you determine the concepts you have mastered and what you will need to work on. Try it out! Go to academic.cengage.com/login to sign in with an access code or to purchase access to this product.

"Human activity is putting such a strain on the natural functions of Earth that the ability of the planet's ecosystems to sustain future generations can no longer be taken for granted."

2005 Millennium Ecosystems Assessment Board of Directors

Environmental Problems

The Global Context: Globalization and the Environment | **Sociological Theories of Environmental Problems** | **Environmental Problems: An Overview** | **Social Causes of Environmental Problems** | **Strategies for Action: Responding to Environmental Problems** | **Understanding Environmental Problems** | **Chapter Review**

© AP Photo/Itsuo Inouye

Japan's "Cool Biz" initiative encourages business men and women and government officials to wear lightweight clothing to work so that the use of air conditioning can be lowered. This initiative has reduced Japan's emissions of greenhouse gases.

In 2005, Yuriko Koike, Japan's environment minister, came up with an idea for how to reduce Japan's emissions of greenhouse gases—the main cause of global warming and climate change. The idea was simple: persuade men to stop wearing suits so that office buildings could lower the use of air conditioning. Koike's idea was not easy to implement because it challenged a decades-old Japanese tradition. In Japanese culture, not wearing a jacket and tie was considered rude. But Koike's "Cool Biz" initiative had the support of government officials. In June 2005, the heads of the ministries—Japan's top government officials—went to work without jackets and ties. Air conditioners in all government buildings were set to higher temperatures: 28° C, or 82.4° F. The number of companies and business men and women participating in Japan's "Cool Biz" initiative has increased enormously. In the first 2 years since the initiative began, Japan avoided releasing a half-million tons of carbon dioxide (a greenhouse gas) that would have normally been emitted into the atmosphere (Brooke 2005; Kestenbaum 2007).

In this chapter we focus on environmental problems that threaten the lives and well-being of people, plants, and animals all over the world—today and in future generations. After examining how globalization affects environmental problems, we view environmental issues through the lens of structural functionalism, conflict theory, and symbolic interactionism. We then present an overview of major environmental problems, examining their social causes and exploring strategies that are used in the attempt to reduce or alleviate them.

The Global Context: Globalization and the Environment

In 1992 leaders from across the globe met at the first Earth Summit in Rio de Janeiro, Brazil, to forge agreements to protect the planet's environment and at the same time alleviate world poverty. When world leaders met a decade later in 2002 for the second Earth Summit in Johannesburg, South Africa, the overall state of the environment had deteriorated and poverty had deepened. How is it that the combined efforts of leaders who met at the first Earth Summit were so ineffectual in achieving their goals? A large part of the answer lies in the increasing globalization of the last two decades. Three aspects of globalization that have affected the environment are (1) the permeability of international borders to pollution and environmental problems, (2) cultural and social integration spurred by communication and information technology, and (3) growth of free trade and transnational corporations.

Red fire ants, known for their painful sting, are an example of bioinvasion. They came from Paraguay and Brazil on shiploads of lumber to Mobile, Alabama, in 1957 and have spread throughout the southern states.

© Esan Mahalu/iStockphoto.com

Permeability of International Borders

We recognize now that environmental problems such as global warming and destruction of the ozone layer (discussed later in this chapter) extend far beyond their source to affect the entire planet and its inhabitants. A striking example of the permeability of international borders to pollution is the spread of toxic chemicals (such as polychlorinated biphenyls [PCBs]) from the Southern Hemisphere into the Arctic. In as few as 5 days chemicals from the tropics can evaporate from the soil, ride the winds thousands of miles north, condense in the cold air, and fall on the Arctic in the form of toxic snow or rain (French 2000). This phenomenon was discovered in the mid-1980s, when scientists found high levels of PCBs in the breast milk of Inuit women in the Canadian Arctic region.

Another environmental problem involving permeability of borders is **bioinvasion:** the emergence of organisms in regions where they are not native. Bioinvasion is largely a product of the growth of global trade and tourism (Chafe 2005). Exotic species travel in the ballast water of ships (water taken in to stabilize empty vessels as they cross waterways), in packing material, in shipments of crops and other goods, and in many other ways. Invasive species may compete with native species for food, start an epidemic, or prey on natives, threatening not only their immediate victims but also the entire ecosystem in which the victims live.

Red fire ants are an example of a bioinvasion. They traveled from Paraguay and Brazil on shiploads of lumber to Mobile, Alabama, in 1957 and have since spread throughout the southern states. Fire ants damage gardens, yards, homes, and electrical equipment and invade the food supplies (seeds, young plants, and insects) of animals (Hilgenkamp 2005). And fire ants harm humans with their painful sting.

bioinvasion The emergence of organisms in regions where they are not native, usually as a result of being carried in the ballast water of ships, in packing material, and in shipments of crops and other goods that are traded around the world.

Cultural and Social Integration

As the messages conveyed through mass media infiltrate the world, people across the globe aspire to consume the products and mimic the materially saturated lifestyles portrayed in movies, television, and advertising. As patterns of consumption in China, India, and other developing countries increasingly follow those in wealthier Western nations, so do the problems associated with overconsumption: depletion of natural resources, pollution, and global warming.

On the positive side the Internet and other forms of mass communication have helped to integrate the efforts of diverse environmental groups across the globe. The globalization of the environmental movement has created new opportunities for environmental groups to join forces, share information, and educate the public through mass communication.

The Growth of Transnational Corporations and Free Trade Agreements

As discussed in Chapter 7, the world's economy is dominated by transnational corporations, many of which have established factories and other operations in developing countries where labor and environmental laws are lax. Many transnational corporations have been implicated in environmentally destructive activities—from mining and cutting timber to dumping of toxic waste.

The World Trade Organization (WTO) and free trade agreements such as the North American Free Trade Agreement (NAFTA) and the Free Trade Area of the Americas (FTAA) allow transnational corporations to pursue profits, expand markets, use natural resources, and exploit cheap labor in developing countries while weakening the ability of governments to protect natural resources or to implement environmental legislation. Transnational corporations have influenced the world's most powerful nations to institutionalize an international system of governance that values commercialism, corporate rights, and "free" trade over environment, human rights, worker rights, and human health (Bruno & Karliner 2002).

Under NAFTA's Chapter 11 provisions corporations can challenge local and state environmental policies, federal controlled substances regulations, and court rulings if such regulatory measures and government actions negatively affect the corporation's profits. Any country that decides, for example, to ban the export of raw logs as a means of conserving its forests or, as another example, to ban the use of carcinogenic pesticides can be charged under the WTO by member states on behalf of their corporations for obstructing the free flow of trade and investment. A secret tribunal of trade officials would then decide whether these laws were "trade restrictive" under the WTO rules and should therefore be struck down. Once the secret tribunal issues its edict, no appeal is possible. The convicted country is obligated to change its laws or face the prospect of perpetual trade sanctions (Clarke 2002, p. 44). As of early 2005, 42 cases had been filed by corporate interests and investors, 11 of which had been finalized. Five corporations that won their claims received $35 million paid by taxpayers in Canada and Mexico. For example, in the late 1990s Ethyl, a U.S. chemical company, used NAFTA rules to challenge Canadian environmental regulation of the toxic gasoline additive MMT. Ethyl won the suit, and Canada paid $13 million in damages and legal fees to Ethyl and reversed the ban on MMT (Public Citizen 2005). Although the United States has not, as of this writing, lost a case, it is only a matter of time before a corporation based in Mexico or Canada wins a NAFTA case against the United States (Public Citizen 2005).

SOCIOLOGICAL THEORIES OF ENVIRONMENTAL PROBLEMS

Each of the three main sociological theories—structural functionalism, conflict theory, and symbolic interactionism—provide insights into social causes of and responses to environmental problems.

Structural-Functionalist Perspective

Structural functionalism emphasizes the interdependence between human beings and the natural environment. From this perspective human actions, social patterns, and cultural values affect the environment, and, in turn, the environment affects social life. For example, population growth affects the environment as more people use natural resources and contribute to pollution. However, the environmental impact of population growth varies tremendously according to a society's patterns of economic production and consumption. As discussed in Chapter 13, the impact that each person makes on the environment—each person's **environmental footprint**—is determined by the patterns of production and consumption in that person's culture. The environmental footprint of an average person in a high-income country is about six times bigger than that of someone in a low-income country and many more times larger than that in the least developed nations (United Nations Population Fund 2004).

Structural functionalism focuses on how changes in one aspect of the social system affect other aspects of society. For example, in the 2 years after the terrorist attacks of September 11, 2001, public concern about most environmental problems declined sharply, most likely as a result of increasing concern about the economy and terrorism over the same period (Saad 2002). The effect of oil prices on the economy provides another illustration of how a change in one aspect of the social system affects other aspects of society. When the price of oil skyrocketed after Hurricane Katrina in 2005, businesses suffered and consumers struggled to pay higher prices for food and other goods. Because so much of our economy depends on oil, an oil shortage or price spike affects virtually every aspect of our economy.

The structural-functionalist perspective raises our awareness of latent dysfunctions—negative consequences of social actions that are unintended and not widely recognized. For example, the more than 840,000 dams worldwide provide water to irrigate farmlands and supply some of the world's electricity. Yet dam building has had unintended negative consequences for the environment, including the loss of wetlands and wildlife habitat, the emission of methane (a gas that contributes to global warming) from rotting vegetation trapped in reservoirs, and the alteration of river flows downstream, which kills plant and animal life ("A Prescription for Reducing the Damage Caused by Dams," 2001). Dams have also displaced millions of people from their homes. As philosopher Kathleen Moore pointed out, "Sometimes in maximizing the benefits in one place, you create a greater harm somewhere else. . . . While it might sometimes seem that small acts of cruelty or destruction are justified because

> **"** The uncomfortable truth is that the current scale and character of human activities are decreasing the planet's life-support capacity both in known and in unanticipated ways, and perhaps more rapidly than we suspect. **"**
>
> Carl N. McDaniel
> Author of *Wisdom for a Livable Planet*

environmental footprint The effect that each person makes on the environment.

they create a greater good, we need to be aware of the hidden systematic costs" (quoted by Jensen 2001, p. 11). Being aware of latent dysfunctions means paying attention to the unintended and often hidden environmental consequences of human activities.

Conflict Perspective

The conflict perspective focuses on how wealth, power, and the pursuit of profit underlie many environmental problems. Wealth is related to consumption patterns that cause environmental problems. Wealthy nations have higher per capita consumption of petroleum, wood, metals, cement, and other commodities that deplete the earth's resources, emit pollutants, and generate large volumes of waste. The wealthiest 20 percent of the world's population is responsible for 86 percent of total private consumption (Bright 2003). The United States is responsible for 25 percent of the world's oil consumption, yet the United States produces less than 3 percent of the world's oil supplies (Pope 2005).

The capitalistic pursuit of profit encourages making money from industry regardless of the damage done to the environment. McDaniel (2005) noted that "our culture tolerates environmentalism only so long as it has minimal impact on big business. . . . In an economically centered culture, jobs come first, not the health of people or the environment" (pp. 22–23).

To maximize sales, manufacturers design products intended to become obsolete. As a result of this **planned obsolescence,** consumers continually throw away used products and purchase replacements. Industry profits at the expense of the environment, which must sustain the constant production and absorb ever-increasing amounts of waste.

Industries also use their power and wealth to influence politicians' environmental and energy policies as well as the public's beliefs about environmental issues. ExxonMobil, the world's largest oil company, has spent millions of dollars on lobbying and has funded numerous organizations that have tried to discredit scientific findings that link fossil fuel burning to global climate change (Mooney 2005). Despite the agreement of hundreds of scientists from around the world that global warming is occurring and is largely due to greenhouse gases released by fossil fuel burning, the Bush administration pulled out of the Kyoto Protocol—an international agreement to reduce greenhouse gas emissions— citing "incomplete" science as the reason. The electric utility industry gave President Bush's election campaigns more than $1 million and has spent millions of dollars on lobbying. In return, President Bush pushed for the Clear Skies Initiative, which would save power companies billions of dollars by relaxing pollution emission caps and extending timelines for air pollution reduction, as required by the 1970 Clean Air Act (Clarren 2005). Current government policies that support ethanol (an alternative fuel discussed later in this chapter) have been linked to political financial contributions by major players in the ethanol industry, such as Archer Daniels Midland, the nation's largest ethanol producer (Food & Water Watch and Network for New Energy Choices 2007).

With both President G. W. Bush and Vice President Dick Cheney having personal and family ties to the oil industry, it is not surprising that the Bush administration supported numerous policy decisions that benefit the oil industry at the expense of the environment and public health. The energy bill passed in 2005 included $85 million in subsidies and tax breaks for most forms of energy, but oil, gas, and nuclear energy received the lion's share. The Bush admin-

planned obsolescence The manufacturing of products that are intended to become inoperative or outdated in a fairly short period of time.

istration has appointed industry-friendly officials to environmental posts. For example, Bush appointed former American Petroleum Institute attorney Philip Cooney, who opposed the Kyoto Protocol, as chief of staff of the White House Council on Environmental Quality. Also, Larisa Dobriansky, a former lobbyist who worked on climate change issues for ExxonMobil, was appointed the deputy assistant secretary for national energy policy at the Department of Energy, where she managed the department's Office of Climate Change Policy (Mooney 2005).

Although the Republican Party has been widely criticized for supporting policies that favor industry over environmental protection, the Democratic Party's environmental record is not unblemished. President Bill Clinton supported NAFTA, which, as discussed earlier, can override environmental laws and policies. The Clinton administration also gave out tax breaks to oil companies drilling in the Gulf of Mexico, permitted logging in Alaska's Tongass National Forest and other old-growth areas, failed to raise automobile fuel-efficiency standards, and repealed a ban on the import of tuna caught in nets that also killed dolphins. In the early 1990s, the Clinton Administration failed to stop the approval of a hazardous waste incinerator in East Liverpool, Ohio, that would burn 70,000 tons of hazardous waste a year. This incinerator is located 350 feet from the nearest house and only a few hundred yards from an elementary school. The construction of the incinerator was partially funded by a major financial supporter of the Clinton-Gore campaign ("Clinton's Environmental Policy" 2007).

Symbolic Interactionist Perspective

The symbolic interactionist perspective focuses on how meanings, labels, and definitions learned through interaction and through the media affect environmental problems. Whether an individual recycles, drives a sport utility vehicle (SUV), or joins an environmental activist group is influenced by the meanings and definitions of these behaviors that the individual learns through interaction with others.

Large corporations and industries commonly use marketing and public relations strategies to construct favorable meanings of their corporation or industry. The term **greenwashing** refers to the way in which environmentally and socially damaging companies portray their corporate image and products as being "environmentally friendly" or socially responsible. Greenwashing is commonly used by public relations firms that specialize in damage control for clients whose reputations and profits have been hurt by poor environmental practices. Wal-Mart, for example, has paid millions of dollars in fines and settlements for a number of environmental violations, including excessive storm runoff at construction sites, petroleum storage violations, and Clean Air Act violations (Wal-Mart Watch 2005). When Wal-Mart announced in 2005 that it would donate $35 million (less than 1 percent of its 2004 profits) to the National Fish and Wildlife Foundation over a 10-year period to purchase 1 acre of conservation land for every acre of land developed by Wal-Mart, many environmentalists accused Wal-Mart of greenwashing. Just before Earth Day (April 22) in 2005, Wal-Mart bought full-page ads in at least 20 newspapers touting its new program, Acres for America (Associated Press 2005). Such publicity suggests that Wal-Mart's Acres for America program is designed to improve Wal-Mart's public image. The comments of Eric Olson of the Sierra Club reflect the skepticism that many environmentalists have toward corporate greenwashing: "Wal-Mart thinks it can

greenwashing The way in which environmentally and socially damaging companies portray their corporate image and products as being "environmentally friendly" or socially responsible.

paint over its record with a nice shade of green, but that won't hide its true colors" (quoted by Associated Press 2005).

Ford Motor Company was accused of greenwashing in 2004 when it advertised its Escape Hybrid SUV as an example of Ford's commitment to the environment when, for the fifth consecutive year, Ford's vehicles had the worst overall fuel economy of all major automakers. In fact, Ford's fleetwide fuel economy was worse in 2004 than it was in 1984! And hybrids accounted for less than 1 percent of Ford's annual sales (Johnson 2004).

Greenwashing also occurs in the political arena when political leaders attempt to portray an environmentally damaging policy or program as environmentally friendly. Consider, for example, President Bush's Clear Skies Initiative, which allows five times more mercury emissions, one and a half times more sulfur dioxide emissions, and hundreds of thousands more tons of smog-forming nitrogen oxide emissions than those allowed under the 1970 Clean Air Act (Clarren 2005). Naming this proposal the Clear Skies Initiative is a blatant example of greenwashing.

Although greenwashing involves manipulation of public perception to maximize profits, many corporations make genuine and legitimate efforts to improve their operations, packaging, or overall sense of corporate responsibility toward the environment. For example, in 1990 McDonald's announced that it was phasing out foam packaging and switching to a new, paper-based packaging that is partially degradable. But many environmentalists are not satisfied with what they see as token environmentalism, or as Peter Dykstra of Greenpeace suggested, 5 percent of environmental virtue to mask 95 percent of environmental vice (Hager & Burton 2000).

What Do You Think? In recent years Earth Day events have been sponsored by corporations such as Office Depot, Texas Instruments, Raytheon Missile Systems, and Waste Management (a Houston-based company responsible for numerous hazardous waste sites). Some environmentalists, including Earth Day's founder, former senator Gaylord Nelson, consider the participation of corporations in Earth Day evidence of the celebration's success. But other environmentalists oppose corporate sponsorship of Earth Day events, accusing corporations of using their financial support of Earth Day as a public relations greenwashing strategy. Do you think that organizers of Earth Day events should accept sponsorship from corporations with poor environmental records?

ENVIRONMENTAL PROBLEMS: AN OVERVIEW

ecosystems The complex and dynamic relationships between forms of life and the environments they inhabit.

Over the past 50 years humans have altered **ecosystems**—the complex and dynamic relationships between forms of life and the environments they inhabit—more rapidly and extensively than in any other comparable period of time in history (Millennium Ecosystem Assessment 2005). As a result, humans have created environmental problems, including depletion of natural resources, air, land,

and water pollution, global warming, environmental illness, environmental injustice, threats to biodiversity, and disappearing livelihoods. Because many of these environmental problems are related to the ways that humans produce and consume energy, we begin this section with an overview of global energy use.

Energy Use Worldwide: An Overview

Being mindful of environmental problems means seeing the connections between energy use and our daily lives.

> Everything we consume or use—our homes, their contents, our cars and the roads we travel, the clothes we wear, and the food we eat—requires energy to produce and package, to distribute to shops or front doors, to operate, and then to get rid of. We rarely consider where this energy comes from or how much of it we use—or how much we truly need. (Sawin 2004, p. 25)

Most of the world's energy—86 percent in 2005—comes from fossil fuels, which include petroleum (or oil), coal, and natural gas (Energy Information Administration 2007) (see Figure 14.1). As you continue reading this chapter, notice that the major environmental problems facing the world today—air, land, and water pollution, destruction of habitats, biodiversity loss, global warming, and environmental illness—are linked to the production and use of fossil fuels.

The next most common source of energy is hydroelectric power (6.3 percent), which involves generating electricity from moving water. As water passes through a dam into a river below, energy is produced by a turbine in the dam. Although hydroelectric power is nonpolluting and inexpensive, it is criticized for affecting natural habitats. For example, dams make certain fish unable to swim upstream to reproduce.

Nuclear power, accounting for 6 percent of world energy production in 2005, is associated with a number of problems related to radioactive nuclear waste—problems that are discussed later in this chapter. In the United States, there are 103 commercial nuclear plants operating in 31 states, which produce about 20 percent of the nation's energy (Weeks 2006). Only 1.4 percent of the world's energy comes from clean renewable resources, which include geothermal power (from the heat of the earth), solar power, wind power, and biomass

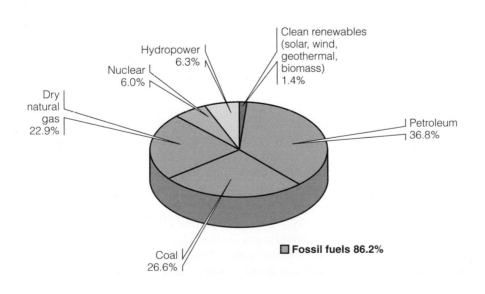

FIGURE 14.1
World energy production by source: 2005.
Source: Energy Information Administration (2007).

(e.g., fuel wood, crops, and animal wastes), which are primarily used by poor populations in developing countries.

Depletion of Natural Resources

Humans have used more of the earth's natural resources since 1950 than in the million years preceding 1950 (Lamm 2006). Population growth, combined with consumption patterns, is depleting natural resources such as forests, water, minerals, and fossil fuels. For example, freshwater resources are being consumed by agriculture, by industry, and for domestic use. More than 1 billion people lack access to clean water (Fornos 2005).

The world's forests are also being depleted. The demand for new land, fuel, and raw materials has resulted in **deforestation**—the conversion of forest land to nonforest land. Global forest cover has been reduced by half of what it was 8,000 years ago (Gardner 2005). Between the 1960s and the 1990s one-fifth of the world's tropical forests were cut or burned (Youth 2003). Between 2000 and 2005, the world had a net loss of about 7.3 million hectares of forest per year, down from 8.9 million in 1990–2000 (1 hectare = 2.47 acres) (Food and Agriculture Organization of the United Nations 2005). The major causes of deforestation are the expansion of agricultural land, human settlements, wood harvesting, and road building.

Deforestation displaces people and wild species from their habitats; soil erosion caused by deforestation can cause severe flooding; and, as we explain later in this chapter, deforestation contributes to global warming. Deforestation also contributes to **desertification**—the degradation of semiarid land, which results in the expansion of desert land that is unusable for agriculture. Overgrazing by cattle and other herd animals also contributes to desertification. The problem of desertification is most severe in Africa (Reese 2001). As more land turns into desert, populations can no longer sustain a livelihood on the land, and so they migrate to urban areas or other countries, contributing to social and political instability.

Air Pollution

Transportation vehicles, fuel combustion, industrial processes (such as the burning of coal and the processing of minerals from mining), and solid waste disposal have contributed to the growing levels of air pollutants, including carbon monoxide, sulfur dioxide, nitrogen dioxide, mercury, and lead. Air pollution levels are highest in areas with both heavy industry and traffic congestion, such as Los Angeles, New Delhi, Jakarta, Bangkok, Tehran, Beijing, and Mexico City.

Air pollution kills about 3 million people a year (Pimentel et al. 2007). Air pollution is linked to heart disease, lung cancer, and respiratory ailments, such as emphysema, chronic bronchitis, and asthma. In Western and Central Europe, air pollution shortens average life expectancy by almost 1 year (European Environment Agency 2007). In all regions of Europe, the estimated annual loss of life due to air pollution is significantly greater than that due to car accidents.

In the United States, nearly half (46 percent) of the population lives in counties that have either short-term or year-round unhealthful levels of either smog or particulate air pollution. Nearly one in five U.S. residents lives in an area with unhealthful levels of particulate pollution (American Lung Association 2007). Indoor air pollution from burning wood and biomass for heating and

deforestation The conversion of forest land to nonforest land.

desertification The degradation of semiarid land, which results in the expansion of desert land that is unusable for agriculture.

cooking, which we discuss next, is a significant cause of respiratory illness, lung cancer, and blindness in developing countries.

Indoor Air Pollution. When we hear the phrase *air pollution,* we typically think of industrial smokestacks and vehicle exhausts pouring gray streams of chemical matter into the air. But indoor air pollution is also a major problem, especially in poor countries.

More than half of the world's population cook food and generate heat by burning dung, wood, crop waste, or coal on open fires or stoves without chimneys (World Health Organization 2005). The resulting indoor smoke contains health-damaging pollutants including small soot or dust particles that are able to penetrate deep into the lungs. Exposure is particularly high among women and children, who spend the most time near the domestic hearth or stove. Every year, indoor air pollution is responsible for the death of 1.6 million people. Exposure to indoor air pollution increases the risk of pneumonia, chronic respiratory disease, asthma, cataracts, tuberculosis, and lung cancer.

Even in affluent countries much air pollution is invisible to the eye and exists where we least expect it—in our homes, schools, workplaces, and public buildings. Sources of indoor air pollution include lead dust (from old lead-based paint); secondhand tobacco smoke; by-products of combustion (e.g., carbon monoxide) from stoves, furnaces, fireplaces, heaters, and dryers; and other common household, personal, and commercial products (American Lung Association 2005). Some of the most common indoor pollutants include carpeting (which emits more than a dozen toxic chemicals); mattresses, sofas, and pillows (which emit formaldehyde and fire retardants); pressed wood found in kitchen cabinets and furniture (which emits formaldehyde); and wool blankets and dry-cleaned clothing (which emit trichloroethylene). Air fresheners, deodorizers, and disinfectants emit the pesticide paradichlorobenzene. Potentially harmful organic solvents are present in numerous office supplies, including glue, correction fluid, printing ink, carbonless paper, and felt-tip markers. Many homes today contain a cocktail of toxic chemicals: "Styrene (from plastics), benzene (from plastics and rubber), toluene and xylene, trichloroethylene, dichloromethane, trimethylbenzene, hexanes, phenols, pentanes and much more outgas from our everyday furnishings, construction materials, and appliances" (Rogers 2002).

Destruction of the Ozone Layer. The ozone layer of the earth's atmosphere protects life on earth from the sun's harmful ultraviolet rays. Yet the ozone layer has been weakened by the use of certain chemicals, particularly chlorofluorocarbons (CFCs), used in refrigerators, air conditioners, spray cans, and other applications. The depletion of the ozone layer allows hazardous levels of ultraviolet rays to reach the earth's surface and is linked to increases in skin cancer and cataracts,

© Ted Spiegel/Corbis

Indoor air pollution is a serious problem in developing countries. As this woman cooks food for her family, she is exposed to harmful air contaminants from the fumes.

weakened immune systems, reduced crop yields, damage to ocean ecosystems and reduced fishing yields, and adverse effects on animals.

Satellite data revealed that the ozone hole in 2007 was 9.7 million square miles—just larger than the size of North America (NASA 2007). The ozone hole was largest in 2006 when it reached a record-breaking area of 11.4 million square miles. Despite measures that have ended production of CFCs, the ozone hole is not expected to shrink significantly for about another decade. This is because CFCs already in the atmosphere remain for 40 to 100 years. Full recovery of the ozone layer is expected in about 2070 (NASA 2007).

Acid Rain. Air pollutants, such as sulfur dioxide and nitrogen oxide, mix with precipitation to form **acid rain.** Polluted rain, snow, and fog contaminate crops, forests, lakes, and rivers. As a result of the effects of acid rain, all the fish have died in a third of the lakes in New York's Adirondack Mountains (Blatt 2005). Because pollutants in the air are carried by winds, industrial pollution in the Midwest falls back to earth as acid rain on southeast Canada and the northeast New England states. Acid rain is not just a problem in North America; it decimates plant and animal species around the globe. In China, most of the electricity comes from burning coal, which creates sulfur dioxide pollution and acid rain that falls on one-third of China, damaging lakes, forests, and crops (Woodward 2007). Acid rain also deteriorates the surfaces of buildings and statues. "The Parthenon, Taj Mahal, and Michelangelo's statues are dissolving under the onslaught of the acid pouring out of the skies" (Blatt 2005, p. 161).

Former Vice President Al Gore's bestselling book *An Inconvenient Truth* and award-winning documentary of the same name increased public awareness of global warming and climate change. Gore shared the 2007 Nobel Peace Prize with the Intergovernmental Panel on Climate Change for their work in the area of global warming and climate change.

Global Warming and Climate Change

Former Vice President Al Gore's bestselling book and award-winning film *An Inconvenient Truth* are widely credited for raising public awareness of global warming and climate change. The seriousness of global warming was underscored in 2007 when the Nobel Peace Prize was awarded to Al Gore and the Intergovernmental Panel on Climate Change for their work to raise awareness of global warming. Before reading further, you may want to assess your attitudes toward global warming in this chapter's *Self and Society* feature.

Global warming refers to the increasing average global air temperature, caused mainly by the accumulation of various gases (greenhouse gases) that collect in the atmosphere. According to the Intergovernmental Panel on Climate Change (IPCC)—a team of more than 1,000 scientists from 113 countries—

acid rain The mixture of precipitation with air pollutants, such as sulfur dioxide and nitrogen oxide.

global warming The increasing average global air temperature, caused mainly by the accumulation of various gases (greenhouse gases) that collect in the atmosphere.

Answer each of the following questions on global warming. Then compare your answers with those of a national representative sample of U.S. adults.

1. How convinced are you that global warming is happening?
   (a) Completely convinced
   (b) Mostly convinced
   (c) Not so convinced
   (d) Not at all convinced

2. Which of the following statements comes closer to your own view?
   (a) Most scientists think global warming is happening.
   (b) Most scientists think global warming is not happening.
   (c) There is a lot of disagreement among scientists about whether or not global warming is happening.
   (d) You do not know enough to say.

3. If global warming is happening, do you think it is
   (a) caused mostly by human activities
   (b) caused mostly by natural changes in the environment

4. How much do you personally worry about global warming?
   (a) A great deal
   (b) A fair amount
   (c) Only a little
   (d) Not at all

5. Do you strongly agree, somewhat agree, somewhat disagree, or strongly disagree with the following statement: "Life on earth will continue without major disruptions only if we take immediate and drastic action to reduce global warming."
   (a) Strongly agree
   (b) Somewhat agree
   (c) Somewhat disagree
   (d) Strongly disagree

6. Do you strongly agree, somewhat agree, somewhat disagree, or strongly disagree with the following statement: "The actions of a single person won't make any difference in reducing global warming."
   (a) Strongly agree
   (b) Somewhat agree
   (c) Somewhat disagree
   (d) Strongly disagree

Comparison Data from a National Sample

A 2007 telephone survey of a nationally representative sample of U.S. adults revealed the following results in response to the six questions on global warming (Yale University/Gallup/ClearVision Institute Poll 2007) (the questions presented here are slightly modified from the actual question asked in the interview).

1. (a) Completely convinced: 38 percent
   (b) Mostly convinced: 34 percent
   (c) Not so convinced: 17 percent
   (d) Not at all convinced: 12 percent

2. (a) Is happening: 48 percent
   (b) Is not happening: 3 percent
   (c) A lot of disagreement: 40 percent
   (d) Don't know enough: 9 percent

3. (a) Human activities: 57 percent
   (b) Natural changes: 30 percent
   (c) Both equally (volunteered): 12 percent
   (d) Don't know: 2 percent

4. (a) A great deal: 15 percent
   (b) A fair amount: 35 percent
   (c) Only a little: 28 percent
   (d) Not at all: 22 percent

5. (a) Strongly agree: 33 percent
   (b) Somewhat agree: 29 percent
   (c) Somewhat disagree: 17 percent
   (d) Strongly disagree: 20 percent
   (e) Don't know: 1 percent

6. (a) Strongly agree: 15 percent
   (b) Somewhat agree: 15 percent
   (c) Somewhat disagree: 20 percent
   (d) Strongly disagree: 49 percent
   (e) Don't know: 1 percent

Source: Based on Yale University/Gallup/ClearVision Institute Poll. 2007. *American Opinions on Global Warming.* Available at http://environment/yale.edu/news/general/

"Warming of the climate system is unequivocal, as is now evident from observations of increases in global average air and ocean temperatures, widespread melting of snow and ice, and rising global average sea level" (2007a, p. 5). Eleven of the 12 years spanning 1995 to 2006 ranked among the 12 warmest years in global surface temperature since 1850. Average global surface temperatures have increased by about 0.74° C over the past century (1906–2005) (Intergovernmental Panel on Climate Change 2007a). If current greenhouse gas emissions trends continue, average global temperatures may rise another 2° C by 2035. Although 2° C may not seem significant, average temperatures during the last Ice Age were only 5° C lower than they are today (Woodward 2007).

Causes of Global Warming. The prevailing scientific view is that **greenhouse gases**—primarily carbon dioxide (CO_2), methane, and nitrous oxide—accumulate in the atmosphere and act like the glass in a greenhouse, holding heat from the sun close to the earth. Most scientists believe that global warming has resulted from the marked increase in global atmospheric concentrations of greenhouse gases since industrialization began. Global increases in carbon dioxide concentration are due primarily to the use of fossil fuels. Deforestation also contributes to increasing levels of carbon dioxide in the atmosphere. Trees and other plant life use carbon dioxide and release oxygen into the air. As forests are cut down or are burned, fewer trees are available to absorb the carbon dioxide. The greenhouse gases methane and nitrous oxide are primarily due to agriculture (Intergovernmental Panel on Climate Change 2007a).

Total world carbon dioxide emissions were 9.9 billion tons in 2006, 35 percent above emissions in 1990 (Global Carbon Project 2007). In 2004, developing and least developed countries, which make up 80 percent of world population, accounted for only 41 percent of total global emissions and only 23 percent of global cumulative emissions since the mid-18th century (Raupach et al. 2007).

But the growth of emissions is strongest in developing countries, particularly China. Nearly three-quarters (73 percent) of global carbon dioxide emissions growth in 2004 took place in developing and least developed countries (Raupach et al. 2007). With less than 5 percent of the world's population, in 2005 the United States produced 21 percent of the world's carbon emissions from fossil fuel burning (Energy Information Administration 2007). In North America, carbon emissions per person far exceed that of other regions of the world (see Figure 14.2).

Effects of Global Warming and Climate Change. Numerous effects of global warming and climate change have been observed and are anticipated in the future (Intergovernmental Panel on Climate Change 2007b; National Assessment Synthesis Team 2000; UNEP 2007) (see this chapter's Photo Essay). Global warming and climate change are projected to affect regions in different ways. As temperature increases, some areas will experience heavier rain, whereas other regions will get drier, although global warming has produced an overall 2.2 percent increase in the air's humidity over a 30-year period (from 1973 to 2002) (Willett et al. 2007). Some regions will experience increased water availability and crop yields, but other regions, particularly tropical and subtropical regions, are expected to experience decreased water availability and a reduction in crop yields. Global warming results in shifts of plant and animal habitats and the increased risk of extinction of some species. Regions that experience increased rainfall as a result of increasing temperatures may face increases in waterborne diseases and diseases transmitted by insects. Over time climate change may produce opposite effects on the same resource. For example, in the short term forest productivity is likely to increase in response to higher levels of carbon

greenhouse gases Gases (primarily carbon dioxide, methane, and nitrous oxide) that accumulate in the atmosphere and act like the glass in a greenhouse, holding heat from the sun close to the earth.

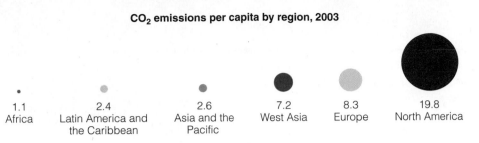

CO$_2$ emissions per capita by region, 2003

1.1	2.4	2.6	7.2	8.3	19.8
Africa	Latin America and the Caribbean	Asia and the Pacific	West Asia	Europe	North America

Units of measurement: Tons per capita

FIGURE 14.2
CO$_2$ emissions per capita by region, 2003.
Source: UNEP (2007)

dioxide in the air, but over the long term forest productivity is likely to decrease because of drought, fire, insects, and disease.

As global warming melts glaciers and permafrost (soil at or below freezing temperature for 2 or more years), the sea level will continue to rise. As sea levels rise, some island countries, as well as some barrier islands off the U.S. coast, are likely to disappear and low-lying coastal areas will become increasingly vulnerable to storm surges and flooding. Global warming thus threatens the lives of the 21.2 billion people who live within 20 miles of the coastline (Bohan 2005).

In urban areas flooding can be a problem where storm drains and waste management systems are inadequate. Increased flooding associated with global warming is expected to result in increases in drownings and in diarrheal and respiratory diseases. Increases in the number of people exposed to insect- and water-related diseases, such as malaria and cholera, are also expected.

Even if greenhouse gases are stabilized, global air temperature and sea level are expected to continue to rise for hundreds of years. That is because global warming that has already occurred contributes to further warming of the planet—a process known as a *positive feedback loop*. For example, the melting of Siberia's frozen peat bog—a result of global warming—could release billions of tons of methane, a potent greenhouse gas, into the atmosphere (Pearce 2005). And the melting of ice and snow—another result of global warming—exposes more land and ocean area, which absorbs more heat than ice and snow, further warming the planet.

Land Pollution

About 30 percent of the world's surface is land, which provides soil to grow the food we eat. Increasingly, humans are polluting the land with nuclear waste, solid waste, and pesticides. In 2007, 1,246 hazardous waste sites were on the National Priority List (also called Superfund sites), with another 66 sites proposed (EPA 2007b).

Nuclear Waste. Nuclear waste, resulting from both nuclear weapons production and nuclear reactors or power plants, contains radioactive plutonium, a substance linked to cancer and genetic defects. Radioactive plutonium has a half-life of 24,000 years, meaning that it takes 24,000 years for the radioactivity to be reduced by half (Mead 1998). Thus nuclear waste in the environment remains potentially harmful to human and other life for thousands of years.

Under the 1982 Nuclear Waste Policy Act, since 1998 the Department of Energy has been responsible for handling the storage of nuclear waste from commercial nuclear power plants. Proposals for disposing of nuclear waste

Global warming and climate change may be the most significant threat facing the world today. In this photo essay, we present just a few examples of the many current and projected effects of global warming and climate change.

Threat of Extinction of Polar Bears and Other Species. As a result of global warming, sea ice in the Arctic is melting earlier and forming later each year. Arctic sea ice during the 2007 melt season was at the lowest level since satellite measures began in 1979. In just 2 years, from 2005 to 2007, sea ice plummeted from 2.14 million square miles to 1.65 million square miles—a 23 percent drop. Scientists say the Arctic Ocean in summer could be ice-free by 2030 (National Snow and Ice Data Center 2007). What does this mean for polar bears? They have less time on the ice to hunt for food and build up their fat stores, and increased time on land where they must survive without food. As their ice habitat shrinks, polar bears struggle to survive. The U.S. Geological Survey (2007) predicts that the entire polar bear population of Alaska may be extinct in the next 43 years. But the threat to polar bears is just the tip of the iceberg (pun intended): Scientists have predicted that in certain areas of the world, global warming will lead to the extinction of up to 43 percent of plant and animal species, representing the potential loss of 56,000 plant species and 3,700 vertebrate species (Malcolm et al. 2006).

Sea-Level Rise. As higher global temperatures melt ice caps, sea level rises. In early 2007, scientists predicted that the global sea-level rise this century would be about 8–24 inches. But more recent observations of Greenland's melting ice caps have led scientists to predict that by the end of the century, sea level will rise 2–6.5 feet (Brown 2007). Scientists flying over Greenland's Ilulissat glacier have observed hundreds of gigantic holes called "moulins," through which swirling masses of meltwater fall to the bottom of the glacier, creating a very deep lake. This water lubricates the glacier, which hastens the glacier's slide into the sea. Rising sea levels pose a threat to coastal populations around the world.

Extreme Weather: Hurricanes, Droughts, and Heat Waves. Rising temperatures are causing drought in some parts of the world, and too much rain in other parts. Warmer tropical ocean temperatures can cause

© Jim Reed/Getty Images

Hurricanes of the intensity of Hurricane Katrina in 2005 that devastated New Orleans and other areas in the south are expected to become more common as ocean temperatures warm.

more intense hurricanes; research has found a recent marked increase in hurricane intensity (Chafe 2006).

Droughts are also expected to increase with rising temperatures. Globally, the proportion of land surface in extreme drought is predicted to increase from 1 to 3 percent (present day) to 30 percent by 2090 (Intergovernmental Panel on Climate Change 2007). Drought threatens agricultural production and food security in areas of the world where hunger is already epidemic (United Nations Development Programme 2007).

With rising temperatures, an increase in the number, intensity, and duration of heat waves is expected, with the accompanying adverse health effects (Intergovernmental Panel on Climate Change 2007). In summer 2003, more than 52,000 Europeans died from heat-related causes (Larsen 2006). In summer 2006, more than 140 Californians died in a triple-digit heat wave that lasted for nearly 2 weeks.

Another effect of global warming is an increase in the number and size of forest fires. In the Western region of the United States, wildfire activity increased suddenly and markedly in the mid-1980s, with higher large-wildfire frequency, longer wildfire du-

© Hans Strand/Corbis

Shrinking Arctic ice threatens the survival of polar bears.

© Torsten Blackwood/AFP/Getty Images

The Polynesian island nation of Tuvalu, located in the Pacific Ocean midway between Hawaii and Australia, is one of many coastal and low-lying areas throughout the world that could be devastated by rises in sea level. Tuvalu's government has asked larger countries such as Australia and New Zealand to consider accepting "climate refugees" in the event that rising sea levels force them to evacuate.

Global warming is expected to produce increases in drought, threatening agricultural production in areas of the world where hunger is already epidemic.

rations, and longer wildfire seasons (Westerling et al. 2006). Warmer temperatures dry out brush and trees, creating ideal conditions for fires to spread. Global warming also means that spring comes earlier, making the fire season longer. In 1988, a massive fire

burned a third of Yellowstone National Park. Since then, fires have broken records in nine states (CBS 2007).

Mosquitoes and the Spread of Diseases. Mosquitoes carry a variety of diseases including encephalitis, dengue fever, yellow fever, West Nile virus, and malaria, which is the leading cause of death in Africa (Knoell 2007). Mosquitoes cannot survive in cool temperatures. With the warming of the planet, mosquitoes are now living in areas in which they previously were not found, placing more people at risk of acquiring one of the diseases carried by the insect. The spread of mosquitoes and the malaria they carry is an example of how climate change is affecting poor populations who contribute the least to global warming.

References

CBS. 2007 (October 21). "Expert: Warming Climate Fuels Mega-Fires." *60 Minutes.* Available at http://www.cbsnews.com

Brown, Paul. 2007. "Melting Ice Caps Triggering Earthquakes." *The Guardian,* September 8. Available at http://www.guardian.co.uk

Chafe, Zoe. 2006. "Weather-Related Disasters Affect Millions." In *Vital Signs,* L. Starke ed. (pp. 44–45). New York: W. W. Norton & Co.

Intergovernmental Panel on Climate Change. 2007. *Climate Change 2007: Impacts, Adaptation and Vulnerability.* Available at http://www.ipcc.ch

Knoell, Carly. 2007 (August 9). "Malaria: Climbing in Elevation as Temperature Rises." *Population Connection.* Available at www.populationconnection.org

Larsen, Janet. 2006 (July 18). "Setting the Record Straight." *Earth Policy Institute.* Available at http://www.earth-policy.org

Malcolm, Jay R., Canran Liu, Ronald P. Neilson, Lara Hansen, and Lee Hannah. 2006. "Global Warming and Extinctions of Endemic Species from Biodiversity Hotspots." *Conservation Biology* 20(2): 538–548.

National Snow and Ice Data Center. 2007 (October 1). *Arctic Sea Ice Shatters All Previous Record Lows.* Available at http://www.nsidc.org

An increase in mega-fires in the U.S. West has been associated with global warming. This fire in San Diego County, California, in October 2007 resulted in more than 1,300 homes being destroyed and residents of hundreds of thousands of homes being ordered to evacuate the area.

United Nations Development Programme. 2007. *Human Development Report 2007/2008: Fighting Climate Change: Human Solidarity in a Divided World.* New York: Palgrave Macmillan.

U.S. Geological Survey. 2007 (September 7). "Future Retreat of Arctic Ice Will Lower Polar Bear Populations and Limit Their Distribution." *USGS Newsroom.* Available at http://www.usgs.gov

Westerling, A. L., H. G. Hidalgo, D. R. Cayan, and T. W. Swetnam. 2006. "Warming and Earlier Spring Increase Western U.S. Forest Wildfire Activity." *Science* 313 (5789): 940–943.

Because of the cooler temperatures in highland villages in Kenya, mosquitoes were rarely found there in the past. But as global temperatures rise, the highland villages are experiencing a malaria epidemic for the first time. This child sleeps under mosquito netting, one of the preventive measures in the fight against malaria and other diseases carried by mosquitoes.

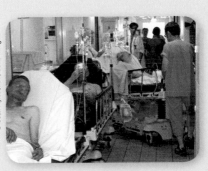

Nearly 15,000 people died in France in a heat wave in August 2003. Hospitals were so overcrowded with individuals who had heat-related illness cases that they were treated in hospital corridors.

have included burying it in rock formations deep below the earth's surface or under Antarctic ice, injecting it into the ocean floor, hurling it into outer space, and storing the waste in huge casks that could be parked in a handful of high-security lots around the country. Each of these options is risky and costly.

In 1978, the U.S. Department of Energy began studying Yucca Mountain in Nevada to determine if it is suitable to house the country's first nuclear waste storage facility, and construction of such a facility is currently under way, with a target opening date of March 2017. The project is widely opposed in Nevada and in 2005 the Western Shoshone National Council filed a lawsuit against the secretaries of the U.S. Departments of the Interior and Energy to try to halt the project. The Western Shoshone National Council claims that the U.S. government has targeted Indian lands for nuclear dumping, which violates an earlier treaty and threatens the health and well-being of the Indian population around Yucca Mountain. Further, according to Western Shoshone legend, if Yucca Mountain (also known as Snake Mountain) is harmed, the snake would rise up as a horrible serpent (Norrell 2005).

Nuclear plants have about 52,000 tons of radioactive spent fuel, with about 10,000 tons of that amount sealed in casks (Vedantam 2005b). Because of inadequate oversight and gaps in safety procedures, radioactive spent fuel is missing or unaccounted for at some U.S. nuclear power plants, which raises serious safety concerns (Vedantam 2005a). Accidents at nuclear power plants, such as the 1986 accident at Chernobyl, and the potential for nuclear reactors to be targeted by terrorists add to the actual and potential dangers of nuclear power plants.

Recognizing the hazards of nuclear power plants and their waste, Germany became the first country to order all of its 19 nuclear power plants shut down by 2020 ("Nukes Rebuked," 2000). Belgium is also phasing out nuclear reactors, and Austria, Denmark, Italy, and Iceland have prohibitions against nuclear energy (International Atomic Energy Agency 2007). Nevertheless, at the end of 2006, there were 435 operating nuclear reactors in 30 countries worldwide, with 29 more under construction (International Atomic Energy Agency 2007). The United States had the most operating reactors (103) followed by France (59), Japan (55), and Russia (31).

Solid Waste. In 1960 each U.S. citizen generated 2.7 pounds of garbage on average every day. This figure increased to 3.7 pounds in 1980 and to 4.54 pounds in 2005 (see Figure 14.3) (EPA 2006). This figure does not include mining, agricultural, and industrial waste; demolition and construction wastes; junked au-

FIGURE 14.3
Pounds of garbage per person: United States.
Source: U.S. Environmental Protection Agency (2006).

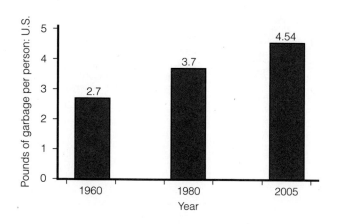

tos; or obsolete equipment wastes. In 2005, 24 percent of solid waste was recycled; more than half was taken to landfills. The availability of landfill space is limited, however. Some states have passed laws that limit the amount of solid waste that can be disposed of; instead, they require that bottles and cans be returned for a deposit or that lawn clippings be used in a community composting program.

Solid waste includes discarded electrical appliances and electronic equipment, known as **e-waste.** Ever think about where your discarded computer ends up when you replace it with a newer model? Most discarded computers end up in landfills, incinerators, or hazardous waste exports; less than 10 percent are recycled (Computer Take Back Campaign 2005). The main concern about dumping e-waste in landfills is that hazardous substances, such as lead, cadmium, barium, mercury, PCBs, and polyvinyl chloride, can leach out of e-waste and contaminate the soil and groundwater. Recycling computers and printers or incinerating them releases toxic emissions into the air (Silicon Valley Toxics Coalition 2001).

What Do You Think? More than 380 billion plastic shopping bags are used in the United States annually, and only a fraction of the plastic bags are recycled (0.6 percent) (Cheeseman 2007). Most plastic bags, which contain toxic chemicals, end up in landfills where it takes 1,000 years for them to degrade. After the Republic of Ireland enacted a 15-cent tax on plastic shopping bags in 2002, the use of plastic bags declined by 90 percent. Bangladesh banned plastic shopping bags in 2002 after it was discovered that the bags blocked drainage systems and were one of the main causes of floods. Taiwan, Singapore, South Africa, and a number of East African countries have also banned plastic shopping bags. In spring 2007, San Francisco became the first U.S. city to ban plastic bags from supermarkets and chain pharmacies. Would you support a ban, or a tax, on plastic bags in your community?

Pesticides. Pesticides are used worldwide for crops and gardens; outdoor mosquito control; the care of lawns, parks, and golf courses; and indoor pest control. Pesticides contaminate food, water, and air and can be absorbed through the skin, swallowed, or inhaled. At least 53 carcinogenic pesticides are applied to major U.S. food crops (Blatt 2005). Many common pesticides are considered potential carcinogens and neurotoxins.

Even when a pesticide is found to be hazardous and is banned in the United States, other countries from which we import food may continue to use it. In an analysis of more than 7,000 domestic and imported food samples, pesticide residues were detected in more than one-third (37 percent) of the domestic samples and in more than one-fourth (28 percent) of the imported samples (Food and Drug Administration 2005). Pesticides also contaminate our groundwater supplies.

e-waste Discarded electrical appliances and electronic equipment.

Water Pollution

Our water is being polluted by a number of harmful substances, including pesticides, vehicle exhaust, acid rain, oil spills, and industrial, military, and agricultural waste. Fertilizer runoff from agricultural lands in the Mississippi River Basin is the main cause of the annual "Dead Zone"—a seasonal phenomenon in which oxygen depletion causes an area of the Gulf of Mexico the size of Massachusetts to become uninhabitable to marine organisms.

Water pollution is most severe in developing countries, where more than 1 billion people lack access to clean water. In developing nations as much as 95 percent of untreated sewage is dumped directly into rivers, lakes, and seas that are also used for drinking and bathing (Pimentel et al. 1998). Mining operations, located primarily in developing countries are notoriously damaging to the environment. Modern gold-mining techniques use cyanide to extract gold from low-grade ore. Cyanide is extremely toxic: One teaspoon of a 2 percent cyanide solution can kill an adult (cyanide was used to kill Jews in Hitler's gas chambers). In 2000 a dam holding cyanide-laced waste at a Romanian gold mine broke, dumping 22 million gallons of cyanide-laced waste into the Tisza River, which flowed into Hungary and Serbia. Some have called this event the worst environmental disaster since the 1986 Chernobyl nuclear explosion. Most gold is used to make jewelry, hardly a necessity warranting the environmental degradation that results from gold mining. To make matters worse, about 300,000 tons of waste are generated for every ton of marketable gold—which translates roughly into 3 tons of waste per gold wedding ring (Sampat 2003).

In the United States one indicator of water pollution is the number of fish advisories issued; these advisories warn against the consumption of certain fish caught in local waters because of contamination with pollutants such as mercury and dioxin. In 2006, there were 3,851 state-issued fish advisories in effect covering more than one-third (38 percent) of total U.S. lake acreage (excluding the Great Lakes) and about one-fourth (26 percent) of river miles (EPA 2007a). The U.S. Environmental Protection Agency (EPA) advises women who may become pregnant, pregnant women, nursing mothers, and young children to avoid eating certain fish altogether (swordfish, shark, king mackerel, and tilefish) because of the high levels of mercury (EPA 2004).

Pollutants also find their way into the water we drink. At Camp Lejeune—a Marine Corps base in Onslow County, North Carolina—as many as 1 million people were exposed to water contaminated with trichloroethylene (TCE), an industrial degreasing solvent, and tetrachloroethylene (PCE), a dry-cleaning agent from 1957 until 1987. The Agency for Toxic Substances and Disease Registry found levels of PCE in Camp Lejeune's drinking water system during the affected years as high as 200 parts per billion, compared with 5 parts per billion, the amount that federal regulators set in 1992 as the maximum allowable level (Sinks 2007). Exposure to TCE may cause nervous system effects; kidney, liver, and lung damage; abnormal heartbeat; coma; and possibly death. Exposure to TCE has been associated with adult cancers (such as kidney and liver cancer), and non-Hodgkin's lymphoma, as well as childhood leukemia and birth defects. Exposure to PCE can cause dizziness, headaches, sleepiness, confusion, nausea, difficulty in speaking and walking, unconsciousness, and death. Exposure to PCE-contaminated drinking water has been linked with non-Hodgkin's lymphoma, leukemia, bladder cancer, and breast cancer (Sinks 2007). In this chapter's *The Human Side* feature, a retired Marine tells the story of his daughter's illness and death that he believes resulted from the contaminated water at Camp Lejeune.

Chemicals, Carcinogens, and Health Problems

When scientists tested journalist Bill Moyers's blood as part of a documentary on the chemical industry, they found traces of 84 of the 150 chemicals they had tested for (PBS 2001). Sixty years ago, when Moyers was 6 years old, only one of these chemicals—lead—would have been present in his blood. During a 2004 World Health Organization convention in Budapest, 44 different hazardous chemicals were found in the bloodstreams of top European Union officials (Schapiro 2004). And in a study of umbilical cord blood of 10 newborns, researchers found an average of 200 industrial chemicals, pesticides, and other pollutants (Environmental Working Group 2005).

In the United States, more than 62,000 chemical substances are in commercial use, 1,500 new chemicals are introduced each year (UNEP 2002), and about 3 million tons of toxic chemicals are released into the environment each year (Pimentel et al. 2007). The average adult uses 9 personal care products a day, with roughly 120 chemicals among them, including those that have known or suspected adverse health effects on users (Zandonella 2007). Complete data on the health and environmental effects are known for only 7 percent of chemicals produced in high volume (UNEP 2002).

Findings of a federal agency responsible for assessing the dangers that chemicals pose to reproductive health may also be influenced by the chemical manufacturing industry. The federal Center for the Evaluation of Risks to Human Reproduction within the National Institutes of Health assesses the dangers of chemicals and helps determine which ones should be regulated. The Center has used Sciences International, an Alexandria, Virginia, consulting firm that

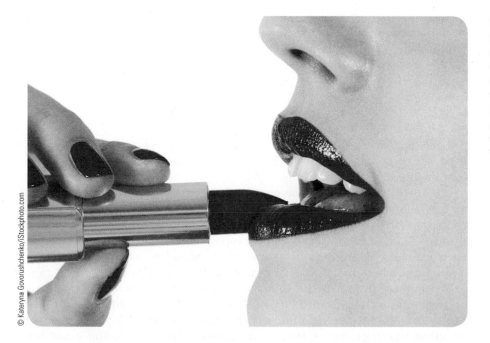

Many of the personal care products we use, including lipstick and other cosmetics, toothpaste, sunscreen, shampoo, soap, and deodorants, contain chemicals with known or suspected adverse health effects. See what chemicals are in the personal care products you use at the Environmental Working Group's website *Skin Deep* at http://www.cosmeticsdatabase.com.

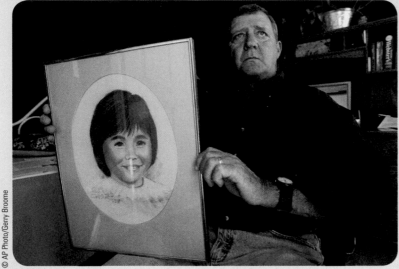

Jerry Ensminger holds a portrait of his daughter Janey, whose death from leukemia is believed to have been caused by contaminated water at Camp Lejeune.

Retired Marine Master Sgt. Jerry Ensminger is one of at least 850 individuals who have filed administrative claims worth $4 billion against Camp Lejeune in Onslow County, North Carolina, for damages that allegedly resulted from drinking and bathing in water contaminated with the dry-cleaning agent PCE and the industrial degreasing solvent TCE (Hefling 2007). In this feature, Jerry Ensminger tells the story of his daughter, Janey, who died in 1985

of leukemia at age 9. His story is excerpted from the book *Poisoned Nation* (2007). Sgt. Ensminger has gathered evidence pointing to the cause of Janey's illness and death: contaminated water at Camp Lejeune that Janey's mother was exposed to during her pregnancy with Janey. As this text goes to print, Jerry Ensminger and other individuals who filed claims against Camp Lejeune are awaiting the completion of a government scientific study

designed to determine whether there is an association between exposure to contaminated water and certain birth defects and cancers among children born between 1968 and 1985 to women who lived at Camp Lejeune during some portion of their pregnancy.

My little girl died in my arms and fifteen years later I found out that the people I had faithfully served for almost 25 years knew she was being poisoned by the water all along. I've spent hundreds of hours on this case. . . . It was 1997 before I finally found out why my daughter died. Even then, it was just by chance. A local TV station had picked up the story. The evening news was turned on in the living room, and I was carrying my dinner in on a plate from the kitchen. All of a sudden I heard the newscaster saying that the water at Camp Lejeune had been highly contaminated from 1968 to 1985 and that the chemicals it contained had been linked to childhood leukemia. When I heard that, I just dropped my plate right on the floor and began shaking. The next day . . . I started reading everything I could find and making contacts with everyone I knew.

There are stages you go through when you lose a child to a catastrophic illness. First you go into shock; then you start won-

has been funded by more than 50 industrial companies, to review the risks of chemicals, prepare reports, and help select members of the scientific review panel and set their agenda (Cone 2007).

The *11th Report on Carcinogens* (U.S. Department of Health and Human Services 2004) lists 246 chemical substances that are "known to be human carcinogens" or "reasonably anticipated to be human carcinogens," meaning that they are linked to cancer. However, the report suggests that these 246 chemical substances may constitute only a fraction of actual human carcinogens. In a review of 152 research studies of environmental pollution and breast cancer, researchers concluded that the evidence supports a link between breast cancer and a number of environmental pollutants (Brody et al. 2007). Many of the chemicals we are exposed to in our daily lives can cause not only cancer but also other health problems, such as infertility, birth defects, and a number of

dering why it happened to your child. So, years ago, I checked my family history and her mother's and found there was nothing on either side. But that nagging question of why Janey got leukemia had stayed with me throughout her illness, her death, and for fourteen and a half years after it. And in that moment, that one moment, when I heard the newscast and dropped the plate of food right out of my hands, it all became clear. I suddenly knew why my little girl died.

I signed up with the Marines to serve my country, but I never signed anything that gave them the right to kill my child, to knowingly poison her. . . . I've said that publicly and the Marine Corps has never refuted anything I've said, including that the contamination went back to the 1950s. I'm sure they know that from the geological studies. . . . The connection between leukemia and contaminated water has been confirmed and I'm leading this fight for everyone. . . .

The day she died . . . Janey was in a lot of pain, so they suggested that she take morphine. She didn't want to, because she had tried it before and it made her so tired. But this time she just couldn't handle the pain. Janey went through hell for nearly two and a half years, and I went through hell with her. . . . I was there with her every step

of the way. Her mother couldn't handle it. Every time Janey went to the hospital, I was the parent who went with her. Sometimes, she was screaming in my ear, "Daddy, don't let them hurt me." Like when she had the bone marrow transplant and the spinal taps. The last time she went into the hospital was the last day of July, just before her ninth birthday. She didn't come out until September 20, and that was in a casket.

About a week before she died, the head of hematology had come in talking about a new form of therapy. He said it would cause severe burns and ulcers, and they didn't recommend it, but Janey looked at them, blinking to control her tears, and said, "This is my life you are talking about, and I'm not giving up. Let's try it."

The ulcers were all over her mouth, her legs, inside her nose and her vagina. The day she died she was in such intense pain she could hardly speak, but finally she managed to whisper, "I want to die peacefully." When Janey said that, I started sobbing. She hugged me and said, "Stop it. Stop crying." I said, "I can't help it, I love you." "I know you do," she replied. "I love you too. But, Daddy, I hurt so bad." "Do you want some morphine?" I asked. She was already being given methadone. "Yes, Daddy" she said, I'm ready."

When the nurse heard that Janey wanted morphine, she knew the time had come. Then, just as they started to give it to her, she said, "Wait. Stop. I want some for my daddy, too."

"This is a very powerful pain medicine. I can't give it to your daddy," the nurse said.

"But my daddy hurts, too," Janey answered. You see, I always took a little of whatever she took to show her I was with her. The morphine killed her. It's a respiratory depressant. . . . As soon as she died a man rushed in to resuscitate her, and I jumped up and put my hands on his shoulders and said, "If you touch her, I'm going to kill you." And then to find out that the organization I served faithfully for twenty-four and a half years knew about this all along, and they never said anything, well, shame on them.

Source: From Hefling, Kimberly. 2007. "Hearing Planned Today in Lejeune Water Case." *Marine Corps Times,* June 12. Available at http://www.marinecorpstimes.com; Sinks, Thomas. 2007 (June 12). Statement by Thomas Sinks, Ph.D., Deputy Director, Agency for Toxic Substances and Disease Registry on ATSDR's Activities at U.S. Marine Corps Base Camp Lejeune before Committee on Energy and Commerce Subcommittee on Oversight and Investigations United States House of Representatives. Available at http://www.hhs.gov; Schwartz-Nobel, Loretta. 2007. *Poisoned Nation.* New York: St. Martin's Press.

childhood developmental and learning problems (Fisher 1999; Kaplan & Morris 2000; McGinn 2000; Schapiro 2007). Some chemicals, such as persistent organic pollutants, accumulate in the food chain and persist in the environment, taking centuries to degrade. Chemicals found in common household, personal, and commercial products can result in a variety of temporary acute symptoms, such as drowsiness, disorientation, headache, dizziness, nausea, fatigue, shortness of breath, cramps, diarrhea, and irritation of the eyes, nose, throat, and lungs. Long-term exposure can affect the nervous system, reproductive system, liver, kidneys, heart, and blood. Fragrances, which are found in many consumer products, may produce sensory irritation, pulmonary irritation, decreases in expiratory airflow velocity, and possible neurotoxic effects (Fisher 1998). Fragrance products can cause skin sensitivity, rashes, headache, sneezing, watery eyes, sinus problems, nausea, wheezing, shortness of breath, inability to concentrate, dizziness, sore throat, cough, hyperactivity, fatigue, and drowsiness (DesJardins 1997).

> **What Do You Think?** Some businesses and local governments are voluntarily limiting fragrances in the workplace or banning them altogether to accommodate employees who experience ill effects from them. For example, Portland, Oregon, has a policy banning all workers in its Bureau of Emergency Communications from wearing fragrances, terming it a "Fragrance Free Policy." The policy states that if an employee is deemed to be wearing too much fragrance, he or she will be sent home to bathe in an effort to remove all evidence of any such odor. What do you think about banning fragrances in the workplace or other public places?

This chapter's *Social Problems Research Up Close* feature describes the Third National Report on Human Exposure to Environmental Chemicals (Centers for Disease Control and Prevention 2005).

Vulnerability of Children. The World Health Organization estimates that 5,500 children die each day from diseases linked to polluted water, air, and food (Mastny 2003). In the United States an estimated 1 in 200 children has developmental or neurological deficits as a result of exposure to toxic substances during gestation or after birth (UNEP 2002). Even low levels of lead found in the blood of children is associated with lower fourth grade reading and math scores; half of U.S. children 1–5 years old nationwide are estimated to have blood lead levels that are linked to adverse effects on cognitive abilities (Miranda et al. 2007). Asthma, the number one childhood illness in the United States, is linked to air pollution. In California and other agricultural states, hundreds, perhaps thousands, of school children in the past decade have been exposed to pesticides and other farm chemicals that are linked to illness, brain damage, birth defects, and death (Associated Press 2007). Exposure to these chemicals occurs from contact with the school grounds and outdoor play equipment, as well as from the chemicals drifting in the air. There are no federal laws prohibiting agricultural spraying near schools. Eight states have policies creating a buffer zone between schools and the spraying of chemicals; however, enforcement and penalties for violations are weak.

Children are more vulnerable than adults to the harmful effects of most pollutants for a number of reasons. For instance, children drink more fluids, eat more food, and inhale more air per unit of body weight than do adults; in addition, crawling and a tendency to put their hands and other things in their mouths provide more opportunities for children to ingest chemical or heavy metal residues.

Multiple Chemical Sensitivity Disorder. Multiple chemical sensitivity (MCS), also known as environmental illness, is a condition whereby individuals experience

multiple chemical sensitivity Also known as "environmental illness," a condition whereby individuals experience adverse reactions when exposed to low levels of chemicals found in everyday substances.

adverse reactions when exposed to low levels of chemicals found in everyday substances (vehicle exhaust, fresh paint, housecleaning products, perfume and other fragrances, synthetic building materials, and numerous other petrochemical-based products). Symptoms of MCS include headache, burning eyes, difficulty breathing, stomach distress or nausea, loss of mental concentration, and dizziness. The onset of MCS is often linked to acute exposure to a high level of chemicals or to chronic long-term exposure. Individuals with MCS often avoid public places and/or wear a protective breathing filter to avoid inhaling the many chemical substances in the environment. Some individuals with MCS build houses made from materials that do not contain the chemicals that are typically found in building materials. Although estimates of the prevalence of chemical sensitivity vary, one recent study reported that in the Atlanta, Georgia, metropolitan area, 12.6 percent of a population sample reported an unusual sensitivity to common chemicals, and 3.1 percent had been diagnosed as having MCS or environmental illness (Caress, Steinemann, & Waddick 2002).

Environmental Injustice

Although environmental pollution and degradation and depletion of natural resources affect us all, some groups are more affected than others. **Environmental Injustice,** also referred to as **environmental racism,** refers to the tendency for socially and politically marginalized groups to bear the brunt of environmental ills.

Environmental Injustice in the United States. In the United States polluting industries, industrial and waste facilities, and transportation arteries (that generate vehicle emissions pollution) are often located in minority communities (Bullard & Johnson 2007; Bullard 2000).

There are 413 commercial hazardous waste sites in the United States. More than half (56 percent) of people living within 3 km (1.8 miles) of one of these hazardous waste facilities are people of color (Bullard et al. 2007). Rates of poverty are also higher among households located near hazardous waste facilities. However, "racial disparities are more prevalent and extensive than socioeconomic disparities, suggesting that race has more to do with the current distribution of the nation's hazardous waste facilities than poverty" (Bullard et al. 2007, p. 60).

An area between New Orleans and Baton Rouge, Louisiana—known as Cancer Alley because of the high rates of cancer in this area—has about 140 chemical plants. The people who live and work in Cancer Alley are disproportionately poor and black. One resident of Norco, Louisiana, who lived next to a Shell Oil refinery and chemical plant, reported that nearly everyone in the community suffered from health problems caused by industry pollution (Bullard 2000). When nearly all the chemical plants in Cancer Alley were damaged by Hurricane Katrina in 2005, residents and their homes were exposed to the toxic poisons that were released from these plants into floodwaters.

A study of air lead concentrations in 3,111 U.S. counties found that counties with the largest proportion of black youth under age 16 have more than 7 percent more lead in the air than counties with no black youth, and counties with the largest proportion of white youth have nearly 10 percent less lead in the air than counties with the smallest proportion of white youth (Stretesky 2003). Consequently, nearly 11 percent of black children suffer from lead poisoning

environmental injustice Also known as **environmental racism,** the tendency for socially and politically marginalized groups to bear the brunt of environmental ills.

In 2001 the Centers for Disease Control and Prevention published a landmark study reporting on measures of human exposure to 27 environmental chemicals, 24 of which had never been measured before in a nationally representative sample of the U.S. population. The *Second National Report on Human Exposure to Environmental Chemicals* and the *Third National Report on Human Exposure to Environmental Chemicals* (Centers for Disease Control and Prevention 2003, 2005) are continuations of the ongoing assessment of the U.S. population's exposure to environmental chemicals. For the *Second Report* human exposure to 116 chemicals was tested. One of the key findings of the *Second Report* was that phthalates—compounds found in products such as soap, shampoo, hair spray, many types of nail polish, and flexible plastics—are absorbed by humans. For the *Third Report* exposure to 148 chemicals was tested.

Methods and Sample

The *Third National Report on Human Exposure to Environmental Chemicals* is based on 2001–2002 data from the National Health and Nutrition Examination Survey (NHANES) conducted by the National Center for Health Statistics. The NHANES is a series of surveys designed to collect data on the health and nutritional status of the U.S. population.

The NHANES is based on a representative sample of the noninstitutionalized, civilian U.S. population, with an oversampling of African Americans, Mexican Americans, adolescents (ages 12–19 years), older Americans (ages 60 years and older), and pregnant women to produce more reliable estimates for these groups. The samples did not specifically target people who were believed to have high or unusual exposures to environmental chemicals.

Assessing research participants' exposure to environmental chemicals involved measuring the chemicals or their metabolites (breakdown products) in the participants' blood or urine—a method known as *biomonitoring*. Blood was obtained by venipuncture for participants ages 1 year and older, and urine specimens were collected for participants ages 6 years and older.

The chemicals measured in the research participants' blood and urine were grouped into the following categories:

1. Metals (including barium, cobalt, lead, mercury, uranium, and others)
2. Cotinine (tracks human exposure to tobacco and tobacco smoke)
3. Organophosphate, organochlorine, and other pesticides
4. Phthalates (chemicals commonly used in such consumer products as soap, shampoo, hair spray, nail polish, and flexible plastics)

5. Polycyclic aromatic hydrocarbons (PAHs)
6. Dioxins, furans, and polychlorinated biphenyls (PCBs)
7. Phytoestrogens
8. Herbicides
9. Pyrethroid, carbamate, and other pesticides

How did the Centers for Disease Control and Prevention (CDC) determine which chemicals to test? In 2002 the CDC solicited nominations from the public for candidate chemicals or categories of chemicals for possible inclusion in future reports and received nominations for hundreds of chemicals. The CDC selected the chemicals to be tested on the basis of several factors, including (1) scientific data that suggested exposure in the U.S. population, (2) the seriousness of health effects known or suspected to result from exposure, (3) the need to assess public health efforts to reduce exposure to a chemical, and (4) the cost of analysis for the chemical.

The researchers conducting the study for the CDC had several objectives, including the following:

- To determine which selected environmental chemicals are getting into the bodies of Americans and at what concentrations.
- For chemicals that have a known toxicity level, to determine the prevalence of people in the U.S. population with levels above those toxicity levels.

(which is associated with deficits in intellectual and academic functioning), compared with 2 percent of white children. In addition, in North Carolina, hog industries—and the associated environmental and health risks associated with hog waste—tend to be located in communities with large black populations, low voter registration, and low incomes (Edwards & Ladd 2000).

Native Americans also experience the effects of environmental injustice. The U.S. federal government approved a plan by a private corporation to store tens of thousands of tons of radioactive nuclear waste on a Native American reservation in Utah (Vedantam 2005b). The reservation leaders agreed to the plan with the hope that the nuclear waste storage facility will provide jobs to Native Americans and bring in needed income.

- To establish reference ranges that can be used by physicians and scientists to determine whether a person or group has an unusually high level of exposure.
- To assess the effectiveness of public health efforts to reduce the exposure of Americans to specific environmental chemicals.
- To track, over time, trends in the levels of exposure of the U.S. population to environmental chemicals.
- To determine whether exposure levels are higher among minorities, children, women of childbearing age, or other potentially vulnerable groups.

Key Findings and Conclusions

Selected findings from the *Third National Report on Human Exposure to Environmental Chemicals* (Centers for Disease Control and Prevention 2005) include the following:

- *New measures for some widely used insecticides*. The *Third Report* presents first-time exposure information for five commonly used pyrethroid insecticides. The findings suggest widespread exposure to pyrethroid insecticides in the U.S. population.
- *Establishment of reference ranges*. The levels of chemicals found in the blood and urine of the research participants provide *reference ranges* (also known as

background exposure levels) for all of the environmental chemicals for which participants were tested. Physicians, researchers, and public health officials use the reference ranges to determine whether individuals or groups are experiencing levels of exposure that are unusually high compared to the level of exposure experienced by the rest of the population. The *Third Report* established reference ranges for 38 of the 148 chemicals tested (reference ranges for the remaining 110 chemicals were established in earlier reports).

- *Reduced exposure of the U.S. population to environmental tobacco smoke*. Cotinine is a metabolite of nicotine that tracks exposure to environmental tobacco smoke (ETS) among nonsmokers; higher cotinine levels reflect more exposure to ETS. Results from the *Third Report* show that from 1988–1991 to 1999–2002, median levels in nonsmokers decreased by 68–69 percent in children and adolescents and by about 75 percent in adults. Levels in blacks were more than twice the levels in Mexican Americans and non-Hispanic whites. Although efforts to reduce ETS exposure show significant progress, ETS exposure remains a major public health concern.
- *Continued progress in reducing blood lead levels among children*. For the period

1999–2002, 1.6 percent of children ages 1–5 had elevated blood lead levels. This percentage has decreased from 4.4 percent in the early 1990s. This significant decrease reflects the success of public health efforts to decrease the exposure of children to lead. Nevertheless, children at high risk for lead exposure (e.g., those living in homes containing lead-based paint or lead-contaminated dust) remain a major public health concern.

Because the presence of an environmental chemical in a person's blood or urine does not necessarily mean that the chemical will cause disease, more studies are needed to determine which levels of these chemicals in people result in disease. Research has already documented adverse health effects of lead and environmental tobacco smoke. However, for many environmental chemicals, few studies are available, and more research is needed to assess health risks associated with different blood or urine levels of these chemicals. Future CDC reports, expected to be released every 2 years, will include more chemicals and will also address the following questions:

1. Are chemical exposure levels increasing or decreasing over time?
2. Are public health efforts to reduce chemical exposure effective?
3. Do certain groups of people have higher levels of chemical exposure than others?

Environmental Injustice around the World. Environmental injustice affects marginalized populations around the world, including minority groups, indigenous peoples, and other vulnerable and impoverished communities, such as peasants and nomadic tribes (Renner 1996). These groups are often powerless to fight against government and corporate powers that sustain environmentally damaging industries. At the Second International Indigenous Youth Conference in Vancouver, a delegate from the Igorot tribal people of the Philippines explained, "Indigenous people are no longer able to plant fruits or vegetables because the . . . mercury poisoning, produced from massive logging and mining operations, inhibits the growth of any plant life" (Sterritt 2005, n.p.). Indigenous groups in Nigeria, such as the Urhobo, Isoko, Kalabare, and Ogoni, are facing environmental threats caused by oil production operations run by transnational corporations. Oil spills, natural gas fires, and leaks from toxic waste pits have polluted the soil, water, and air and have compromised the health of various local tribes. "Formerly lush agricultural land is now covered by oil slicks, and much vegetation and wildlife has been destroyed. Many Ogoni suffer from respiratory diseases and cancer, and birth defects are frequent" (Renner 1996, p. 57). The environmental injustices experienced by the Ogoni and the Igorot reflect only the tip of the iceberg. Renner (1996) warned that "minority populations and indigenous peoples around the globe are facing massive degradation of their environments that threatens to irreversibly alter, indeed destroy, their ways of life and cultures" (p. 59).

Threats to Biodiversity

About 1.6 million species of life have been identified on earth (IUCN 2007). However, there may be 10 times that number and perhaps many times more if we include microbial diversity (Chivian & Bernstein 2004). This enormous diversity of life, known as **biodiversity,** provides food, medicines, fibers, and fuel; purifies air and freshwater; pollinates crops and vegetation; and makes soils fertile.

In recent decades we have witnessed mass extinction rates of diverse life forms. On average, one species of plant or animal life becomes extinct every 20 minutes (Levin & Levin 2002). Human activity is the primary cause of disappearing species today (Hunter 2001). Biologists expect that at least half of the roughly 10 million species alive today will become extinct over the next few centuries as a result of habitat loss caused by deforestation and urban sprawl, overharvesting, and global warming (Cincotta & Engelman 2000). Another threat to biodiversity is overexploitation of species for their meat, hides, horns, or medicinal or entertainment value. A poacher, for example, can earn several hundred dollars for a single rhinoceros horn—equivalent to a year's salary in many African countries. That same horn, ground up and used as a medicinal product, can sell for half a million dollars in Asia (Schmidt 2004). The International Fund for Animal Welfare (2005) checked the Internet for 1 week and found more than 9,000 wild animal products and specimens and live wild animals for sale—predominantly from species protected by law—a figure that represents "only the tip of the iceberg" (p. ii).

As shown in Table 14.1, 16,306 species worldwide are threatened with extinction, 188 more than in 2006. One in four mammals, one in eight birds, one-third of all amphibians, and 70 percent of the world's assessed plants are in jeopardy (IUCN 2007).

Environmental Problems and Disappearing Livelihoods

Environmental problems threaten the ability of the earth to sustain its growing population. Agriculture, forestry, and fishing provide 50 percent of all jobs worldwide and 70 percent of jobs in sub-Saharan Africa, East Asia, and the Pacific (World Resources Institute 2000). As croplands become scarce or degraded, as forests shrink, and as marine life dwindles, millions of people who make their living from these natural resources must find alternative livelihoods. In Canada's Maritime Provinces collapse of the cod fishing industry in the 1990s from overfishing left 30,000 fishers dependent on government welfare payments and decimated the economies of hundreds of communities (World Resources Institute 2000).

biodiversity The diversity of living organisms on earth.

The numbers of vultures have declined in Africa and Asia as a result of them eating the carcasses of livestock given the drug diclofenac. With the disappearing vulture population, humans are exposed to higher health risks posed by decomposing carcasses.

TABLE 14.1 Threatened Species Worldwide: 2007*	
CATEGORY	NUMBER OF THREATENED SPECIES IN 2007
Mammals	1,094
Birds	1,217
Amphibians	1,808
Reptiles	422
Fishes	1,201
Invertebrates (insects, mollusks, etc.)	2,108
Plants	8,447
Other (lichens, mushrooms, brown algae)	9
Total	16,306

*Threatened species include those classified as critically endangered, endangered, and vulnerable.
Source: *IUCN* (2007).

Globally, there are an estimated 30 million **environmental refugees**—individuals who have migrated because they can no longer secure a livelihood as a result of deforestation, desertification, soil erosion, and other environmental problems (Margesson 2005). As individuals lose their source of income, so do nations. In one-fourth of the world's nations, crops, timber, and fish contribute more to the nation's economy than industrial goods do (World Resources Institute 2000).

environmental refugees Individuals who have migrated because they can no longer secure a livelihood as a result of deforestation, desertification, soil erosion, and other environmental problems.

SOCIAL CAUSES OF ENVIRONMENTAL PROBLEMS

Various structural and cultural factors have contributed to environmental problems. These include population growth, industrialization and economic development, and cultural values and attitudes such as individualism, consumerism, and militarism.

Population Growth

As noted in Chapter 13, the world's population doubled from 3 billion in 1960 to 6 billion in 1999, and it is projected to reach more than 9 billion by 2050. Population growth places increased demands on natural resources and results in increased waste. As Hunter (2001) explained:

> Global population size is inherently connected to land, air, and water environments because each and every individual uses environmental resources and contributes to environmental pollution. While the scale of resource use and the level of wastes produced vary across individuals and across cultural contexts, the fact remains that land, water, and air are necessary for human survival. (p. 12)

However, population growth itself is not as critical as the ways in which populations produce, distribute, and consume goods and services. Recall from Chapter 13 the fact that wealthy, industrialized countries have the slowest rates of population growth, yet these countries consume the most natural resources and contribute most to pollution, global warming, and other environmental problems.

Industrialization and Economic Development

Many of the environmental problems confronting the world, including global warming and the depletion of the ozone layer, are associated with industrialization and economic development. Industrialized countries, for example, consume more energy and natural resources and contribute more pollution to the environment than poor countries.

The relationship between level of economic development and environmental pollution is curvilinear rather than linear. For example, industrial emissions are minimal in regions with low levels of economic development and are high in the middle-development range as developing countries move through the early stages of industrialization. However, at more advanced stages of industrialization industrial emissions ease because heavy-polluting manufacturing industries decline and "cleaner" service industries increase and because rising incomes are associated with a greater demand for environmental quality and cleaner technologies. However, a positive linear correlation has been demonstrated between per capita income and national carbon dioxide emissions (Hunter 2001).

In less developed countries environmental problems are largely the result of poverty and the priority of economic survival over environmental concerns. Vajpeyi (1995) explained:

> Policymakers in the Third World are often in conflict with the ever-increasing demands to satisfy basic human needs—clean air, water, adequate food, shelter, education—and to safeguard the environmental quality. Given the scarce economic and technical resources at their disposal, most of these policymakers have ignored long-range environmental concerns and opted for short-range economic and political gains. (p. 24)

❝ Over the past two decades every issue I have been involved in as an ecological activist . . . has revealed that what the industrial economy calls 'growth' is really a form of theft from nature and people. **❞**

Vandana Shiva
Environmental activist

A major environmental concern today is the growing economy of China—the most populated country in the world. The income of China's 1.3 billion people has been growing at a rapid rate, and along with that rising income Chinese consumers are increasingly able to spend money on cars, travel, meat, and other goods and services that use fossil fuels. If China were to burn coal at the U.S. level of 2 tons per person, the country would use 2.8 billion tons per year—more than the current world production of 2.5 billion tons. In addition, if the Chinese were to use oil at the same rate as Americans, by 2031 China would need 99 million barrels of oil a day—more than the current world production of 79 million barrels per day (Aslam 2005). Lester Brown of World Watch Institute said that "China is teaching us that we need a new economic model, one that is based not on fossil fuels but that instead harnesses renewable sources of energy" (quoted by Aslam 2005, n.p.).

In considering the effects of economic development on the environment, note that the ways we measure economic development and the "health" of economies also influence environmental outcomes. Two primary measures of economic development and the health of economies are gross domestic product (GDP) and consumer spending. But these measures overlook the social and environmental costs of the production and consumption of goods and services. Until definitions and measurements of "economic development" and "economic health" reflect these costs, the pursuit of economic development will continue to contribute to environmental problems.

Cultural Values and Attitudes

Cultural values and attitudes that contribute to environmental problems include individualism, consumerism, and militarism.

Individualism. Individualism, which is a characteristic of U.S. culture, puts individual interests over collective welfare. Even though recycling is good for our collective environment, many individuals do not recycle because of the personal inconvenience involved in washing and sorting recyclable items. Similarly, individuals often indulge in countless behaviors that provide enjoyment and convenience for themselves at the expense of the environment: long showers, use of dish-washing machines, recreational boating, frequent meat eating, the use of air conditioning, and driving large, gas-guzzling SUVs, to name just a few.

"In America we have two dominant religions: Christianity and accumulation.**"**

Brian Swimme
California Institute of Integral Studies

Consumerism. Consumerism—the belief that personal happiness depends on the purchasing of material possessions—also encourages individuals to continually purchase new items and throw away old ones. The media bombard us daily with advertisements that tell us life will be better if we purchase a particular product. After the terrorist attacks of September 11, 2001, President Bush advised Americans that it was their patriotic duty to go to the malls and spend money. Consumerism contributes to pollution and environmental degradation by supporting polluting and resource-depleting industries and by contributing to waste.

Militarism. The cultural value of militarism also contributes to environmental degradation (see also Chapter 16). "It is generally agreed that the number one polluter in the United States is the American military. It is responsible each year for the generation of more than one-third of the nation's toxic waste . . . an amount greater than the five largest international chemical companies combined" (Blatt 2005, p. 25). Toxic substances from military vehicles, weapons materials, and munitions pollute the air, land, and groundwater in and around military bases and training areas. The Pentagon has asked Congress to loosen environmental laws for the military, so that training exercises can occur unimpeded (Janofsky 2005). In addition, the EPA is forbidden to investigate or sue the military (Blatt 2005).

STRATEGIES FOR ACTION: RESPONDING TO ENVIRONMENTAL PROBLEMS

One strategy for alleviating environmental problems is to lower fertility rates and slow population growth, as discussed in detail in Chapter 13. Responses to environmental problems also include environmental activism, environmental education, the use of green energy, modifications in consumer products and behavior, and government regulations and legislation. Sustainable economic development, international cooperation and assistance, and institutions of higher education also play important roles in alleviating environmental problems.

Environmental Activism

With more than 10,000 environmental organizations with a combined membership of between 19 million and 41 million, "the U.S. environmental movement is now one of the most powerful social movements in the United States" (Faber 2005, p. 51). The environmental movement has grown in recent years, as environmental groups have reported recent increases in both members and financial contributions (Price 2006). A Gallup survey found that one in five (21 percent) of U.S. adults says that he or she is an active participant in the environmental movement (Gallup Organization 2007) (see Table 14.2).

What Do You Think? Gallup poll results indicate that women are more likely than men to worry about the environment, to be active in or sympathetic toward the environmental movement, and to give precedence to the environment over economic and energy concerns (Gallup Organization 2007). Why do think this gender difference exists?

Environmental organizations exert pressure on government and private industry to initiate or intensify actions related to environmental protection. Environmentalist groups also design and implement their own projects and disseminate information to the public about environmental issues. In a national survey of college freshmen one in five (21.1 percent) said that "becoming involved in programs to clean up the environment" was personally "essential" or "very important" (Pryor et al. 2006).

In the United States environmentalist groups date back to 1892 with the establishment of the Sierra Club, followed by the Audubon Society in 1905. Other environmental groups include the National Wildlife Federation, the World Wildlife Fund, the Environmental Defense, Friends of the Earth, the

TABLE 14.2 Involvement in the Environmental Movement*	
INVOLVEMENT	**PERCENTAGE OF U.S. ADULTS**
Active participant	21
Sympathetic but not active	49
Neutral	23
Unsympathetic	5
No opinion	2

*In a 2007 Gallup survey, a national sample of U.S. adults was asked: "Do you think of yourself as an active participant in the environmental movement, sympathetic towards the movement, but not active, neutral, or unsympathetic towards the movement?" Source: Gallup Organization (2007).

Union of Concerned Scientists, Greenpeace, Environmental Action, the Natural Resources Defense Council, Worldwatch Institute, World Resources Institute, the Rainforest Alliance, and Conservation International.

Online Activism. The Internet and e-mail provide important tools for environmental activism. For example, Action Network, an online environmental activism community sponsored by Environmental Defense, sends e-mail action alerts to 800,000 members informing them when Congress and other decision makers threaten the health of the environment. These members can then send e-mails and faxes to Congress, the president, and business leaders, urging them to support policies that protect the environment. In 2002, with the help of Environmental Defense's Action Network, California passed a bill to mandate stricter greenhouse gas emissions controls for automobiles (Environmental Defense 2004). Other environmental organizations, such as the Sierra Club, also have similar action alert features on their websites.

Religious Environmentalism. From a religious perspective environmental degradation can be viewed as sacrilegious, sinful, and an offense against God (Gottlieb 2003a). "The world's dominant religions—as well as many people who identify with the 'spiritual' rather than with established faiths—have come to see that the environmental crisis involves much more than assaults on human health, leisure, or convenience. Rather, humanity's war on nature is at the same time a deep affront to one of the essentially divine aspects of existence" (Gottlieb 2003b, p. 489). This view has compelled religious groups to take an active role in environmental activism.

For example, the National Association of Evangelicals, an umbrella group of 51 church denominations, adopted a platform called For the Health of the Nation: An Evangelical Call to Civic Responsibility. This platform, which has been signed by nearly 100 evangelical leaders, calls on the government to "protect its citizens from the effects of environmental degradation" (Goodstein 2005). Larry Schweiger, president of the National Wildlife Federation, welcomes evangelicals as allies and explains that conservative lawmakers who might not pay attention to what environmental groups say may be more likely to pay attention to what the faith community is saying. Recently, the faith community has been voicing concerns over global warming and climate change. The magazine *Christianity Today* ran an editorial endorsing a bill (sponsored by Senators McCain and Lieberman) that would curb greenhouse gases (Goodstein 2005).

"The natural world is the larger sacred community to which we belong. To be alienated from this community is to become destitute in all that makes us human."

Thomas Berry
Catholic monk and ecotheologian

Ecoterrorism. An extreme form of environmental activism is **ecoterrorism,** defined as any crime intended to protect wildlife or the environment that is violent, puts human life at risk, or results in damages of $10,000 or more (Denson 2000). The best-known ecoterrorist groups are the Earth Liberation Front (ELF) and the Animal Liberation Front (ALF). ALF and the ELF are international underground movements consisting of autonomous individuals and small groups who engage in "direct action" to (1) inflict economic damage on those profiting from the destruction and exploitation of the natural environment, (2) save animals from places of abuse (e.g., laboratories, factory farms, and fur farms), and (3) reveal information and educate the public on atrocities committed against the earth and all the species that populate it. Direct actions of ALF and ELF have included setting fire to a ski resort in Vail, Colorado, to express opposition to the resort's proposed expansion into the forest habitat of the Canada lynx, setting fire to gas-guzzling Hummers and other SUVs at car dealerships, vandalizing SUVs parked on streets and in driveways, setting fire to construction sites (in opposition to the development), and smashing windows at the homes of fur retailers. The FBI estimates that between 1996 and 2002 ALF and ELF committed more than 600 criminal acts in the United States, resulting in damages in excess of $43 million (Jarboe 2002). Because the direct actions of ELF and ALF are illegal, activists work anonymously in small groups, or individually, and do not have any centralized organization or coordination. Although critics of ecoterrorism cite the damage done by such tactics, members and supporters of ecoterrorist groups claim that the real terrorists are corporations that plunder the earth.

> **What Do You Think?** In Chapters 9 and 11, we discussed hate crimes, noting that hate crime laws impose harsher penalties on the perpetrator of a crime if the motive for that crime was hate or bias. Should motives be considered in imposing penalties on individuals who are convicted of acts of ecoterrorism? For example, should a person who sets fire to a business to protest that business's environmentally destructive activities receive the same penalty as a person who sets fire to a business for some other reason?

The Role of Corporations in the Environmental Movement. Corporations are major contributors to environmental problems and often fight against environmental efforts that threaten their profits. However, some corporations are joining the environmental movement for a variety of reasons, including pressure from consumers and environmental groups, the desire to improve their public image, genuine concern for the environment, and/or concern for maximizing current or future profits.

In 1994, out of concern for public and environmental health, Ray Anderson, CEO of Interface, set a goal of being a sustainable company by 2020—"a company that will grow by cleaning up the world, not by polluting or degrading it" (McDaniel 2005, p. 33). Anderson envisioned recycling all the materials used, not releasing any toxins into the environment, and using solar energy to power

ecoterrorism Any crime intended to protect wildlife or the environment that is violent, puts human life at risk, or results in damages of $10,000 or more.

all production and the company has made significant progress toward these goals.

Car manufacturers can play a role in the environmental movement by developing fuel-efficient and alternative-fuel vehicles. In collaboration with the environmental organization Environmental Defense, FedEx has developed a hybrid electric delivery truck that cuts smog-forming pollution by 65 percent, reduces soot by 96 percent, and goes 57 percent farther on a gallon of fuel (Environmental Defense 2004). However, 4 years after it announced it would deploy 3,000 clean-burning hybrid trucks a year, FedEx had purchased fewer than 100 hybrid trucks, less than one-third of 1 percent of its fleet (Elgin 2007).

Rather than hope that industry voluntarily engages in eco-friendly practices, corporate attorney Robert Hinkley suggested that corporate law be changed to mandate socially responsible behavior. Corporations pursue profit at the expense of the public good, including the environment, because corporate executives are bound by corporate law to try to make a profit for shareholders, said Hinkley (Cooper 2004). Hinkley suggested that corporate law should include a Code for Corporate Citizenship that would say the following: "The duty of directors henceforth shall be to make money for shareholders but not at the expense of the environment, human rights, public health and safety, dignity of employees, and the welfare of the communities in which the company operates" (quoted by Cooper 2004, p. 6).

Environmental Education

One goal of environmental organizations and activists is to educate the public about environmental issues and the seriousness of environmental problems. Some Americans underestimate the seriousness of environmental concerns because they lack accurate information about environmental issues. In a national survey of U.S. adults most (69 percent) rated themselves as having either "a lot" (10 percent) or "a fair amount" (59 percent) of knowledge about environmental issues and problems (National Environmental Education and Training Foundation & Roper Starch Worldwide 1999). Yet on 7 of 10 questions asked about emerging environmental issues, more Americans gave incorrect answers than correct ones. The average was 3.2 correct answers out of 10 questions. Even those who rated themselves as having "a lot" of environmental knowledge, as well as those with a college degree, scored only 4 out of 10 items correctly.

Being informed about environmental issues is important because people who have higher levels of environmental knowledge tend to engage in higher levels of pro-environment behavior. For example, environmentally knowledgeable people are more likely to save energy in the home, recycle, conserve water, purchase environmentally safe products, avoid using chemicals in yard care, and donate funds to conservation (Coyle 2005).

A main source of information about environmental issues for most Americans is the media. However, because the media are owned by corporations and wealthy individuals with corporate ties, unbiased information about environmental impacts of corporate activities may not readily be found in mainstream media channels. Indeed, the public must consider the source in interpreting information about environmental issues. Propaganda by corporations sometimes comes packaged as "environmental education." Hager and Burton (2000) explained: "Production of materials for schools is a growth area for public relations companies around the world. Corporate interests realize the value of getting their spin into the classrooms of future consumers and voters" (p. 107).

"Green" Energy

As noted earlier, many of the environmental problems facing the world today are linked to energy use, particularly the use of fossil fuels. Increasing the use of **green energy**—energy that is renewable and nonpolluting—can help alleviate environmental problems associated with fossil fuels. Also known as clean energy, green energy sources include solar power, wind power, biofuel, and hydrogen. Although clean energy represents just over 1 percent of world energy production, this amount is expected to increase in coming years.

Solar Power. The world's fastest developing clean energy is photovoltaic cells—a form of solar power that converts sunlight to electricity (Sawin 2005b). In Japan, which produces half of all photovoltaic cells, the aim is for photovoltaic cells to generate 10 percent of electricity by 2030. Driven by government incentive programs, Germany and other European countries are also increasing their use of solar cells to provide electricity. In the United States, California leads the nation in photovoltaic cell use.

Another form of solar power is the use of solar thermal collectors, which capture the sun's warmth to heat building space and water. China is the world's leader in solar thermal production and use, followed by Japan, Europe, and Turkey (Sawin 2005b). The United States was once the world leader in solar thermal production but has fallen far behind because of low natural gas prices and the lack of government incentives. Another form of solar power comes from "concentrating solar power plants," which use the sun's heat to make steam to turn electricity-producing turbines.

Wind Power. Wind power is the world's second fastest-growing energy source, after solar power. Wind turbines, which turn wind energy into electricity, are operating in more than 65 countries, although nearly three-fourths of wind power is produced in Europe. In Germany, the leading country in wind power, wind energy meets 6.6 percent of electricity needs. The United States is in third place worldwide (after Spain) in use of wind power (Sawin 2005a).

One disadvantage of wind power is that wind turbines have been known to result in bird mortality. However, this problem has been mitigated in recent years through the use of painted blades, slower rotational speeds, and careful placement of wind turbines.

Biofuel. Biofuels are fuels derived from agricultural crops. Two types of biofuels are ethanol and biodiesel.

Ethanol is an alcohol-based fuel that is produced by fermenting and distilling corn or sugar. Ethanol is blended with gasoline to create E85 (85 percent ethanol and 15 percent gasoline). Vehicles that run on E85, called flexible fuel vehicles, have been used by the government and in private fleets for years and have just recently become available to consumers. However, only a small fraction of passenger vehicles are flexible fuel vehicles, and many owners of these vehicles do not know that their vehicle can operate with E85 and/or do not have access to any of the 850 gas stations that sell E85 (Price 2006).

green energy Also known as clean energy, energy that is nonpolluting and/or renewable, such as solar power, wind power, biofuel, and hydrogen.

Wind energy is harnessed by turbines such as those pictured in this photo of a wind farm in Alamont Pass, California.

Another problem associated with ethanol fuel is that increased demand for corn, which is used to make most ethanol in the United States, has driven up the price of corn, resulting in higher food prices (many processed food items contain corn, and animal feed is largely corn). And as corn prices rise, so too do those of rice and wheat because the crops compete for land. Rising food prices threaten the survival of the world's poorest 2 billion people who depend on grain to survive. The grain it takes to fill a 25-gallon tank with ethanol would feed one person for an entire year (Brown 2007).

Increased corn and/or sugar cane production to meet the demand for ethanol also has adverse environmental effects, including increased use and run-off of fertilizers, pesticides, and herbicides; depletion of water resources; and soil erosion. In addition, tropical forests are being clear-cut to make room for "energy crops" leaving less land for conservation and wildlife (Price 2006), biofuel refineries commonly run on coal and natural gas (which emit greenhouse gases), farm equipment and fertilizer production require fossil fuels, and the use of ethanol involves emissions of several pollutants. Finally, even if 100 percent of the U.S. corn crop were used to produce ethanol, it would only displace less than 15 percent of U.S. gasoline use (Food & Water Watch and Network for New Energy Choices 2007).

Biodiesel fuel is a cleaner burning diesel fuel made from vegetable oils and/or animal fats, including recycled cooking oil. Some individuals who make their own biodiesel fuel obtain used cooking oil from restaurants at no charge. The European Union, which is the leading manufacturer of biodiesel fuel, hopes that biofuels will supply 20 percent of the fuel market in 2020 (Aeck 2005).

Hydrogen Power. Hydrogen, the most plentiful element on earth, is a clean burning fuel that can be used for electricity production, heating, cooling, and transportation. Many see a movement to a hydrogen economy as a long-term solution to the environmental and political problems associated with fossil fuels. Further research is needed, however, to develop nonpolluting and cost-effective ways to extract and transport hydrogen. In the meantime, all the major auto manufacturers are developing hydrogen fuel cell technology. Also, in 2004 Governor Arnold Schwarzenegger issued an executive order calling for the development of a Hydrogen Highway Network Blueprint Plan to make hydrogen a viable alternative transportation fuel in California.

Modifications in Consumer Behavior

In the United States and other industrialized countries consumers are making "green" choices in their behavior and purchases that reflect concern for the environment. In some cases these choices carry a price tag, such as paying more for organically grown food or for clothing made from organic cotton. In many cases, however, consumers are motivated to make green choices in their purchasing behavior because doing so saves money. For example, after gas prices topped $3 a gallon in 2005, sales of gas-guzzling SUVs dropped, and sales of more fuel-efficient cars, such as hybrids, increased. According to Dan Becker, director of the Sierra Club's global warming program, switching to more fuel-efficient vehicles such as hybrids is the single biggest step to reducing dependency on foreign oil (quoted by Gartner 2005).

Consumers often consider their utility bill when they choose energy-efficient appliances and electrical equipment. To assist consumers in choosing energy-efficient products, in 1992 EPA introduced the Energy Star label for products that meet EPA energy-efficiency standards. Globally, 37 countries use an energy-efficient labeling system for appliances and electronic equipment (Makower, Pernick, & Wilder 2005).

Consumers can also choose to purchase "green power"—clean energy from nonpolluting sources (e.g., wind and solar power). Green power is available in every state, but not every utility company offers it. Purchasing green power does not necessarily mean the electricity you actually receive comes directly from a wind turbine or solar panel because the local grid that delivers electricity combines all power—green or not. Instead, the amount of clean power consumers buy is generated on their behalf and added to the larger pool of electricity (Vartan 2005). Buying green power is not widespread because many consumers do not know that it is an option. To find out if it is available in their area, consumers can call their utility company and ask.

Each person in the United States produces 22 tons of carbon dioxide each year. Table 14.3 presents tips for how consumers can reduce the amount of carbon dioxide each of us produces.

What Do You Think? Consumers who want to make environmentally friendly purchasing decisions are sometimes faced with difficult choices. For example, suppose that an individual lives in an area where locally grown organic produce is not available. In this case, is it better to purchase (1) organic produce that has been trucked in from a distant state (no pesticides, but the transportation contributes to fossil fuel emissions) or (2) locally grown produce from a farm that uses pesticides?

TABLE 14.3 Top Ten Things You Can Do to Fight Global Warming

National Geographic's *The Green Guide* (www.thegreenguide.com) lists the following tips for consumers to reduce the amount of carbon dioxide they produce each year. All CO_2 reductions listed below are on an annual basis.

1. Replace five incandescent light bulbs in your home with compact fluorescent bulbs (CFLs): Swapping those 75-watt incandescent bulbs with 19-watt CFLs can cut 275 pounds of CO_2.

2. Instead of short haul flights of 500 miles or so, take the train and bypass 310 pounds of CO_2.

3. Sure it may be hot, but get a fan, set your thermostat to 75° F and blow away 363 pounds of CO_2.

4. Replace refrigerators more than 10 years old with today's more energy-efficient Energy Star models and save more than 500 pounds of CO_2.

5. Shave your 8-minute shower to 5 minutes for a savings of 513 pounds.

6. Caulk, weatherstrip, and insulate your home. If you rely on natural gas heating, you'll stop 639 pounds of CO_2 from entering the atmosphere (472 pounds for electric heating). And this summer, you'll save 226 pounds from air conditioner use.

7. Whenever possible, dry your clothes on a line outside or a rack indoors. If you air dry half your loads, you'll dispense with 723 pounds of CO_2.

8. Trim down on the red meat. Because it takes more fossil fuels to produce red meat than fish, eggs, and poultry, switching to these foods will slim your CO_2 emissions by 950 pounds.

9. Leave the car at home and take public transportation to work. Taking the average U.S. commute of 12 miles by light rail will leave you 1,366 pounds of CO_2 lighter than driving. The standard, diesel-powered city bus can save 804 pounds, and heavy rail subway users save 288.

10. Finally, support the creation of wind, solar, and other renewable energy facilities by choosing green power if offered by your utility. To find a green power program in your state, call your local utility or visit U.S. Department of Energy's Green Power Markets page.

Source: "Top Ten Tips to Fight Global Warming." 2007 (June 5). *The Green Guide*. Available at http://www.thegreenguide.com.

Green Building. The proposed design for New York City's Freedom Tower to be built at the site of the World Trade Towers includes a "wind farm" of turbines 60 floors up that will generate 20 percent of the tower's electricity. The proposed Freedom Tower is an example of a "green building."

The U.S. Green Building Council developed green building standards known as LEED (Leadership in Energy and Environmental Design). These standards consist of 69 criteria to be met by builders in six areas, including energy use and emissions, water use, materials and resource use, and sustainability of the building site. LEED buildings include the Pentagon Athletic Center, the Detroit Lions' football training facility, and the David L. Lawrence Convention Center in Pittsburgh (Makower et al. 2005).

One of the most environmentally friendly green buildings in the world is the Adam Joseph Lewis Center for Environmental Studies at Oberlin College in Oberlin, Ohio. The building is constructed with either recycled or sustainably produced nontoxic materials, bathroom and other liquid wastes are purified on-site and used again to flush toilets or to water outside vegetation, and power and heat are produced by solar and geothermal power (McDaniel 2005).

The Adam Joseph Lewis Center for Environmental Studies, located on the Oberlin College campus, is among the most innovative, environmentally friendly buildings in the world. The building resulted from the vision of David Orr, professor of environmental studies at Oberlin.

Government Policies, Programs, and Regulations

Worldwide, governments spend about $1 trillion (U.S. dollars) per year in subsidies that allow the prices of fuel, timber, metals, and minerals (and products using these materials) to be much lower than they otherwise would be, encouraging greater consumption (Renner 2004). Thus governmental policies can contribute to environmental problems. But government policies, regulations, and funding can also play a role in protecting and restoring the environment. A few examples of environmental government policies, programs, and regulations follow.

Cap and Trade Programs. Cap and trade programs, also known as emissions trading, are a free-market approach used to control pollution by providing economic incentives to power plants and other industries for achieving reductions in the emissions of pollutants. In a cap and trade system, a power plant can exceed its permitted level of emissions by buying credits from a plant in the same region whose emissions are below what is allowed. The European Union Emission Trading Scheme is the largest multinational greenhouse gas emissions trading scheme in the world. In the United States, the Regional Greenhouse Gas Initiative, in which 10 northeastern and mid-Atlantic states are participating (as of 2007), is the first cap and trade program to control carbon dioxide emissions from power plants in the United States.

Critics of the cap and trade approach argue that it fails to achieve the lowest possible emissions because it does not require all plants to use the best available technology to reduce emissions. By allowing some plants to have higher emissions, it also exposes populations living near these high-emissions plants to excessive air pollution.

Commitments to Reduce Greenhouse Gases. In 2006, California legislators passed the Global Warming Solutions Act that commits California to reduce greenhouse gas emissions to 1990 levels by 2020, a cut of about 25 percent. This Act marks the first time that mandatory, comprehensive limits on greenhouse gas emissions have been imposed in the United States. In 2004, California adopted the nation's first regulation to reduce greenhouse gas emissions from cars. Twelve states have pledged to follow California's initiative (with implementation contingent upon an EPA ruling) (Prah 2007).

Policies on Chemical Safety. In 2003 the European Union drafted legislation known as REACH (Registration, Evaluation, and Authorization of Chemicals) that requires chemical companies to conduct safety and environmental tests to prove that the chemicals they are producing are safe. If they cannot prove that a chemical is safe, it will be banned from the market (Rifkin 2004). The European Union has become a world leader in environmental stewardship by placing the "precautionary principle" at the center of EU regulatory policy. The precautionary principle requires industry to prove that their products are safe. In contrast, in the United States chemicals are assumed to be safe unless proven otherwise, and the burden is put on the consumer, the public, or the government to prove that a chemical causes harm. Thus, products that contain chemicals that are banned in Europe are sold in the United States. For example, phthalates (which are added to plastic to make it soft and pliable) are banned in Europe, but are found in numerous products sold in the United States, including toys, food containers, shower curtains, shampoo bottles, raincoats, medical tubing, cars, and perfumes (to help it adhere to the skin) (Schapiro 2007).

What Do You Think? In general, the European Union seems to be more concerned than the United States about health effects of chemicals—as evidenced by their stricter controls and bans on chemicals. If the United States had a national health insurance plan similar to that of European countries, in which the federal government paid for health care, do you think the U.S. government would enact tougher controls on hazardous chemicals and other environmental issues?

Policies and Regulations on Energy Use. In 2004 more than 20 countries committed to specific targets for the renewable share of total energy use (UNEP 2007). A number of states have set goals of producing a minimum percentage of electricity from wind power, solar power, or other renewable sources. In 2007, Minnesota, New Hampshire, and Oregon set the highest goals of producing 25 percent of electricity from renewable sources by 2025 (Prah 2007). In addition, more than 70 mayors and other local leaders from around the world signed the Urban Environmental Accords, pledging to obtain 10 percent of energy from renewable resources by 2012 and to reduce greenhouse gases by 25 percent by 2030 (Stoll 2005).

Taxes. Some environmentalists propose that governments use taxes to discourage environmentally damaging practices and products (Brown & Mitchell 1998). In the 1990s a number of European governments increased taxes on environ-

Small, fuel-efficient cars like this one are common in Europe where, as a result of high gasoline taxes, gas costs up to $8 a gallon.

mentally harmful activities and products (such as gasoline, diesel, and motor vehicles) and decreased taxes on income and labor (Renner 2004). As a result of high gasoline taxes in Europe, gas there costs as much as $8 a gallon, which has increased consumer demand for small, fuel-efficient cars. Raising gasoline taxes in the United States is highly unpopular with voters and consumers.

Fuel Efficiency Standards. The Energy Policy and Conservation Act of 1975 established the Corporate Average Fuel Economy (CAFE) program that required vehicle manufacturers to increase the average fuel economy of passenger cars and light trucks to reduce gas consumption. However, several manufacturers have chosen to pay CAFE penalties rather than comply with the standards. Another problem is that SUVs, minivans, and light trucks have less stringent CAFE standards. For years, environmentalists lobbied the U.S. government to further raise fuel economy standards. An increase in fuel economy of just 1 mile per gallon in all U.S. passenger vehicles would produce significant reductions in oil consumption. Currently, the United States has the lowest fuel efficiency standards and the most permissible standards for tailpipe greenhouse gas emissions compared with most developed countries (Food & Water Watch and Network for New Energy Choices 2007). But a major step forward occurred with the passage of the Energy Independence and Security Act of 2007, which requires a 40 percent increase in fuel economy to 35 miles per gallon by 2020.

Sustainable Economic Development

Achieving global cooperation on environmental issues is difficult, in part, because developed countries (primarily in the Northern Hemisphere) have different economic agendas from those of developing countries (primarily in the Southern Hemisphere). The northern agenda emphasizes preserving wealth and affluent lifestyles, whereas the southern agenda focuses on overcoming mass poverty and achieving a higher quality of life (Koenig 1995). Southern countries are concerned that northern industrialized countries—having already achieved economic wealth—will impose international environmental policies that restrict the economic growth of developing countries just as they are beginning to industrialize. Global strategies to preserve the environment must address both wasteful lifestyles in some nations and the need to overcome overpopulation and widespread poverty in others.

Development involves more than economic growth; it involves sustainability—the long-term environmental, social, and economic health of societies. **Sustainable development** involves meeting the needs of the present world without endangering the ability of future generations to meet their own needs. "The aim here is for those alive today to meet their own needs without making it impossible for future generations to meet theirs. . . . This in turn calls for an economic structure within which we consume only as much as the natural environment can produce, and make only as much waste as it can absorb" (McMichael, Smith, & Corvalan 2000, p. 1067).

Sustainable development requires the use of clean, renewable energy. Renewable energy projects in developing countries have demonstrated that providing affordable access to green energy such as solar power, wind power, and biofuel helps to alleviate poverty by providing energy for creating business and jobs and by providing power for refrigerating medicine, sterilizing medical equipment, and supplying freshwater and sewer services needed to reduce infectious disease and improve health (Flavin & Aeck 2005).

International Cooperation and Assistance

Global environmental concerns such as global warming, destruction of the ozone layer, and loss of biodiversity call for global solutions forged through international cooperation and assistance. For example, the 1987 Montreal Protocol on Substances That Deplete the Ozone Layer forged an agreement made by 70 nations to curb the production of CFCs (which contribute to ozone depletion and global warming). The 1992 Earth Summit in Rio de Janeiro brought together heads of state, delegates from more than 170 nations and nongovernmental organizations, and participants to discuss an international agenda for both economic development and the environment. The 1992 Earth Summit resulted in the Rio Declaration—"a nonbinding statement of broad principles to guide environmental policy, vaguely committing its signatories not to damage the environment of other nations by activities within their borders and to acknowledge environmental protection as an integral part of development" (Koenig 1995, p. 15).

In 1997 delegates from 160 nations met in Kyoto, Japan, and forged the **Kyoto Protocol**—the first international agreement to place legally binding limits on greenhouse gas emissions from developed countries. The United States, the world's largest producer of greenhouse gas emissions, rejected the Kyoto Protocol in 2001. As of October 2007, 175 countries had accepted, approved, or ratified the Kyoto Protocol; the United States and Australia were the only industrialized nations who had refused to join. In November 2007, Kevin Rudd, the newly elected prime minister of Australia, promised he would make signing the Kyoto Protocol one of his first acts in office, leaving the United States as the only Western country not to ratify the pact.

sustainable development Societal development that meets the needs of current generations without threatening the future of subsequent generations.

Kyoto Protocol The first international agreement to place legally binding limits on greenhouse gas emissions from developed countries.

Although the Kyoto Protocol is a step forward, it does not go far enough. Avoiding dangerous climate change will require rich countries to cut carbon emissions by at least 80 percent by the end of the 21st century, with cuts of 30 percent by 2020—significantly more than the cuts required under the Kyoto Protocol (United Nations Development Programme 2007).

Leaders of countries that do not support the Kyoto Protocol often cite cost as an obstacle to reducing greenhouse gas emissions. But a British study concluded that failure to take action to reduce greenhouse gases could cost the global economy as much as 20 percent of its annual income ($7 trillion), whereas achieving significant reductions in greenhouse gas emissions would cost only 1 percent of the world's gross domestic product (about $350 billion) (Woodward 2007).

In another global environmental treaty known as the Stockholm Convention, representatives from 122 countries drafted an agreement to phase out a group of dangerous chemicals known as persistent organic pollutants. More recently, Germany hosted a major intergovernmental conference, Renewables 2004, that resulted in the adoption of the International Action Programme on Renewables; this program supports more than 200 ongoing or planned policies and projects in more than 150 countries (UNEP 2007).

Some countries do not have the technical or economic resources to implement the requirements of environmental treaties. Wealthy industrialized countries can help less-developed countries address environmental concerns through economic aid. Because industrialized countries have more economic and technological resources, they bear the primary responsibility for leading the nations of the world toward environmental cooperation. Jan (1995) emphasizes the importance of international environmental cooperation and the role of developed countries in this endeavor:

> Advanced countries must be willing to sacrifice their own economic well-being to help improve the environment of the poor, developing countries. Failing to do this will lead to irreparable damage to our global environment. Environmental protection is no longer the affair of any one country. It has become an urgent global issue. Environmental pollution recognizes no national boundaries. No country, especially a poor country, can solve this problem alone. (pp. 82–83)

The Role of Institutions of Higher Education

Colleges and universities can play an important role in efforts to protect the environment. In 1990, university leaders from around the world met in Talloires, France, where they established the Talloires Declaration—an action plan for incorporating sustainability and environmental literacy in teaching, research, operations, and outreach at colleges and universities. More than 350 university presidents and chancellors in more than 40 countries (more than 100 in the United States alone) have signed the Talloires Declaration. Colleges and universities that are committed to environmental sustainability engage in a variety of practices, such as encouraging use of bicycles on campus, using hybrid and electric vehicles, establishing recycling programs, using local and renewable building materials for new buildings, involving students in organic gardening to provide food for the campus, using clean energy, and incorporating environmental education into the curricula. However, David Newport, director of the environmental center at the University of Colorado at Boulder, believes that institutions of higher education are not doing enough to promote sustainability. "We're supposed to be on the leading edge, and we're behind the curve. . . . There are, what 4,500 colleges in the United States, and how many of them are really doing something? Less than 100 or 200?" (quoted by Carlson 2006, p. A10). Nevertheless, an assessment of 200 public and private U.S. universities found that more than two of three schools showed improvement in their green practices and policies over the last year (Sustainable Endowments Institute 2008).

UNDERSTANDING ENVIRONMENTAL PROBLEMS

Environmental problems are linked to a number of interrelated features of social life, including corporate globalization, rapid and dramatic population growth, expanding world industrialization, patterns of excessive consumption, and reliance on fossil fuels for energy. Growing evidence of the irreversible effects of global warming and loss of biodiversity, increased concerns about the link between global warming and extreme weather events such as Hurricane Katrina, and adverse health effects of toxic waste and other forms of pollution suggest that we cannot afford to ignore environmental problems. Researchers at Cornell University estimate that 40 percent of deaths worldwide are caused by water, air, and soil pollution (Pimentel et al. 2007).

Many Americans believe in a "technological fix" for the environment—that science and technology will solve environmental problems. Paradoxically, the same environmental problems that have been caused by technological progress may be solved by technological innovations designed to clean up pollution, preserve natural resources and habitats, and provide clean forms of energy. But leaders of government and industry must have the will to finance, develop, and use technologies that do not pollute or deplete the environment. When asked how companies can produce products without polluting the environment, corporate attorney Robert Hinkley suggested that, first, it must become a goal to do so:

> I don't have the technological answers for how it can be done, but neither did President John F. Kennedy when he announced a national goal to land a man on the moon by the end of the 1960s. The point is that, to eliminate pollution, we first have to make it our goal. Once we've done that, we will devote the resources necessary to make it happen. We will develop technologies that we never thought possible. But if we don't make it our goal, then we will never devote the resources, never develop the technology, and never solve the problem. (quoted by Cooper 2004, p. 11)

But the direction of technical innovation is largely in the hands of big corporations that place profits over environmental protection. Unless the global community challenges the power of transnational corporations to pursue profits at the expense of environmental and human health, corporate behavior will continue to take a heavy toll on the health of the planet and its inhabitants. Because oil has been implicated in political and military conflicts involving the Middle East (see Chapter 16), such conflicts are likely to continue so long as oil plays the lead role in providing the world's energy.

Global cooperation is also vital to resolving environmental concerns but is difficult to achieve because rich and poor countries have different economic development agendas: Developing poor countries struggle to survive and provide for the basic needs of their citizens; developed wealthy countries struggle to maintain their wealth and relatively high standard of living. Can both agendas be achieved without further pollution and destruction of the environment? Is

> ❝The world will no longer be divided by the ideologies of 'left' and 'right', but by those who accept ecological limits and those who don't.❞
>
> Wolfgang Sachs
> Wuppertal Institute

The 2004 Nobel Peace Prize was awarded to Wangari Maathai for leading a grassroots environmental campaign called the Green Belt Movement, which is responsible for planting 30 million trees across Kenya. Maathai is the first person to be awarded the Nobel Peace Prize for environmental work.

© Micheline Pelletier/Corbis

sustainable economic development an attainable goal? With mounting concern about climate change, the health impacts of air pollution, rising oil prices, and the need to ensure energy access to all, governments worldwide have strengthened their commitment to sustainable, renewable energy policies and projects (UNEP 2007).

Since September 11, 2001, world leaders, the media, and citizens in countries across the globe have been preoccupied with terrorism. Lester Brown, of the Earth Policy Institute, warns against the environmental dangers of such preoccupation:

> Terrorism is certainly a matter of concern, but if it diverts us from the environmental trends that are undermining our future until it is too late to reverse them, Osama bin Laden and his followers will have achieved their goal of bringing down Western civilization in a way they could not have imagined. (Brown 2003, p. 5)

In 2004 the Nobel Peace Prize was given to Wangari Maathai for leading a grassroots environmental campaign to plant 30 million trees across Kenya. This was the first time ever that the Nobel Peace Prize was awarded to someone for his or her accomplishments in restoring the environment. In her acceptance speech, Maathai explained, "A degraded environment leads to a scramble for scarce resources and may culminate in poverty and even conflict" (quoted by Little 2005, p. 2). With ongoing conflict around the globe, it is time for world leaders to recognize the importance of a healthy environment for world peace and to prioritize environmental protection in their political agendas.

CHAPTER REVIEW

- **What three aspects of globalization have affected the environment?**

  Three aspects of globalization that have affected the environment are (1) the permeability of international borders to pollution and environmental problems, (2) cultural and social integration spurred by communication and information technology, and (3) growth of free trade and transnational corporations.

- **How do free trade agreements pose a threat to environmental protection?**

  Free trade agreements such as NAFTA and the FTAA provide transnational corporations with privileges to pursue profits, expand markets, use natural resources, and exploit cheap labor in developing countries while weakening the ability of governments to protect natural resources or to implement environmental legislation.

- **How did the terrorist attacks of September 11, 2001, affect public concern for environmental issues in the 2 years after the attacks?**

  Gallup poll surveys found that in the 2 years after the terrorist attacks of September 11, 2001, public concern about most environmental problems declined sharply, most likely because of increasing concern about the economy and terrorism over the same period. Perhaps environmental problems look less serious in contrast with the economy and the terrorist threat, or perhaps the environment did not receive as much media attention, because media stories focused on terrorism and war and economic issues.

- **Where does most of the world's energy come from?**

  Most of the world's energy comes from fossil fuels, which include oil, coal, and natural gas. This is significant because many of the serious environmental problems in the world today, including global warming and climate change, biodiversity loss, and pollution, stem from the use of fossil fuels.

- **What are the major causes, and effects, of deforestation?**

  The major causes of deforestation are the expansion of agricultural land, human settlements, wood harvesting, and road building. Deforestation displaces people and wild species from their habitats, contributes to global warming, and contributes to desertification, which results in the expansion of desert land that is unusable for agriculture. Soil erosion caused by deforestation can cause severe flooding.

- **What are the effects of air pollution on human health?**

  Air pollution, which is linked to heart disease, lung cancer, and respiratory ailments, such as emphysema, chronic bronchitis, and asthma kills about 3 million people a year. In Western and Central Europe, air pollution shortens average life expectancy by almost 1 year. In all regions of Europe, the estimated annual loss of life due to air pollution is significantly greater than that due to car accidents.

- **What are some examples of common household, personal, and commercial products that contribute to indoor pollution?**

  Some common indoor air pollutants include carpeting, mattresses, drain cleaners, oven cleaners, spot removers, shoe polish, dry-cleaned clothes, paints, varnishes, furniture polish, potpourri, mothballs, fabric softener, caulking compounds, air fresheners, deodorizers, disinfectants, glue, correction fluid, printing ink, carbonless paper, and felt-tip markers.

- **What is the primary cause of global warming?**

  The prevailing view on what causes global warming is that greenhouse gases—primarily carbon dioxide, methane, and nitrous oxide—accumulate in the atmosphere and act like the glass in a greenhouse, holding heat from the sun close to the earth. The primary greenhouse gas is carbon dioxide, which is released into the atmosphere by burning fossil fuels.

- **How does global warming contribute to further global warming?**

  As global warming melts ice and snow, it exposes more land and ocean area, which absorbs more heat than ice and snow, further warming the planet. The melting of Siberia's frozen peat bog—a result of global warming—could release billions of tons of methane, a potent greenhouse gas, into the atmosphere and cause further global warming. This process, whereby the effects of global warming cause further global warming, is known as a positive feedback loop.

- **What is the relationship between level of economic development and environmental pollution?**

  There is a curvilinear relationship between level of economic development and environmental pollution. In regions with low levels of economic development, industrial emissions are minimal, but emissions rise in countries that are in the middle economic development range as they move through the early stages of industrialization. However, at more advanced stages of industrialization, industrial emissions ease because heavy-polluting manufacturing industries decline and "cleaner" service industries increase and because rising incomes are associated with a greater demand for environmental quality and cleaner technologies.

- **What are some of the concerns about nuclear energy?**

Nuclear waste contains radioactive plutonium, a substance linked to cancer and genetic defects. Nuclear waste in the environment remains potentially harmful to human and other life for thousands of years, and disposing of nuclear waste is problematic. Accidents at nuclear power plants, such as the 1986 accident at Chernobyl, and the potential for nuclear reactors to be targeted by terrorists add to the actual and potential dangers of nuclear power plants.

- **How much garbage did each person in the United States generate in 1960? 1980? 2005?**

In 1960 each U.S. citizen generated 2.7 pounds of garbage on average every day. This figure increased to 3.7 pounds in 1980 and to 4.54 pounds in 2005.

- **Why are women who may become pregnant, pregnant women, nursing mothers, and young children advised against eating certain types of fish?**

The U.S. Environmental Protection Agency advises women who may become pregnant, pregnant women, nursing mothers, and young children to avoid eating certain fish altogether (swordfish, shark, king mackerel, and tilefish) because of the high levels of mercury.

- **Why are children more vulnerable than adults to the harmful effects of most pollutants?**

Children are more vulnerable than adults to the harmful effects of most pollutants for a number of reasons. For instance, children drink more fluids, eat more food, and inhale more air per unit of body weight than do adults; in addition, crawling and a tendency to put their hands and other things in their mouths provide more opportunities to ingest chemical or heavy metal residues.

- **What does the term "environmental injustice" refer to?**

Environmental injustice, also referred to as environmental racism, refers to the tendency for socially and politically marginalized groups to bear the brunt of environmental ills. For example, in the United States polluting industries, industrial and waste facilities, and transportation arteries (which generate vehicle emissions pollution) are often located in minority communities.

- **What percentage of mammals, birds, amphibians, and plants are under threat of extinction?**

One in four mammals, one in eight birds, one-third of all amphibians, and 70 percent of the world's assessed plants are in jeopardy.

- **What social and cultural factors contribute to environmental problems?**

Social and cultural factors that contribute to environmental problems include population growth, industrialization and economic development, and cultural values and attitudes such as individualism, consumerism, and militarism.

- **What are some of the strategies for alleviating environmental problems?**

Strategies for alleviating environmental problems include efforts to lower fertility rates and slow population growth, environmental activism, environmental education, the use of "green" energy, modifications in consumer products and behavior, and government regulations and legislation. Sustainable economic development and international cooperation and assistance also play important roles in alleviating environmental problems.

- **According to 2004 Nobel Peace Prize winner Wangari Maathai, why is environmental protection important for national and international security?**

In her acceptance speech for the 2004 Nobel Peace Prize, Wangari Maathai explained that "a degraded environment leads to a scramble for scarce resources and may culminate in poverty and even conflict."

TEST YOURSELF

1. In June 2005, what did the heads of the ministries—Japan's top government officials—do to help curb greenhouse gas emissions?
   a. Started riding a bicycle to work
   b. Passed a law banning plastic bags in grocery stores
   c. Went to work without a jacket and tie so they could lower the air conditioning
   d. Passed a "one-child" policy to curb population growth

2. Red fire ants are an example of:
   a. a bioinvasion
   b. an extinct species
   c. a threatened species
   d. an alternative fuel
3. Satellite data revealed that the ozone hole in 2007 was just larger than the size of
   a. New York City
   b. Rhode Island
   c. Texas
   d. North America
4. If greenhouse gases were to be stabilized today, global air temperature and sea level would be expected to
   a. remain at their current level
   b. decrease immediately
   c. begin to decrease within 20 years
   d. continue to rise for hundreds of years
5. At the end of 2006, China had more nuclear reactors than any other country.
   a. True
   b. False
6. Most solid waste in the United States is recycled.
   a. True
   b. False
7. In spring 2007, which U.S. city became the first to ban plastic bags from supermarkets and chain pharmacies?
   a. Miami
   b. San Francisco
   c. Atlanta
   d. Portland
8. The average adult uses nine personal care products a day. How many chemicals, roughly, are in those nine personal care products?
   a. 12
   b. 20
   c. 120
   d. 1,200
9. Which of the following is the number one polluter in the United States?
   a. The military
   b. Dow Chemical
   c. Archer Daniels Midland
   d. ExxonMobil
10. The United States has the lowest fuel efficiency standards and the most permissible standards for tailpipe greenhouse gas emissions compared with most developed countries.
    a. True
    b. False

Answers: 1 c, 2 a, 3 d, 4 d, 5 b, 6 b, 7 b, 8 c, 9 a, 10 a.

KEY TERMS

acid rain
biodiversity
bioinvasion
deforestation
desertification
ecosystems
ecoterrorism

environmental footprint
environmental injustice
environmental racism
environmental refugees
e-waste
global warming
green energy

greenhouse gases
greenwashing
Kyoto Protocol
multiple chemical sensitivity
planned obsolescence
sustainable development

MEDIA RESOURCES

Understanding Social Problems, Sixth Edition
Companion Website

academic.cengage.com/sociology/mooney
Visit your book companion website, where you will find flash cards, practice quizzes, Internet links, and more to help you study.

Just what you need to know NOW! Spend time on what you need to master rather than on information you already have learned. Take a pre-test for this chapter, and CengageNOW will generate a personalized study plan based on your results. The study plan will identify the topics you need to review and direct you to online resources to help you master those topics. You can then take a post-test to help you determine the concepts you have mastered and what you will need to work on. Try it out! Go to academic.cengage.com/login to sign in with an access code or to purchase access to this product.

15

Science and Technology

"We have arranged things so that almost no one understands science and technology. This is a prescription for disaster. We might get away with it for a while, but sooner or later this combustible mixture of ignorance and power is going to blow up in our faces."

Carl Sagan, Astronomer and astrobiologist

At 19, Erica Robinson lived in a household of "fighters." Her brothers were both athletes—one a star basketball player, the other a wrestler. Her father's strength was summed up in a plaque hanging from the wall of their home: "Don't quit." When Erica found out she was pregnant, she was determined to do whatever was best for her baby (Kilen 2003).

But something was terribly wrong. At Erica's 6-month checkup the doctor said that the baby wasn't getting enough blood to survive. An emergency caesarean section was performed, and with it E'Maria Robinson-Butler was introduced to the world—10 inches, 11 ounces—one of the tiniest babies ever born. In fact, on any given day in the United States 1,300 premature babies are born. "Preemies" are more likely than full-term infants to have lasting disabilities, such as chronic lung disease, cerebral palsy, mental retardation, and vision and hearing problems. The rate of premature babies has increased 30 percent since 1981 (March of Dimes 2007).

As soon as E'Maria was born, a tube was placed in her windpipe; she was hooked up to a respirator, fed intravenously, and given medicine to help her tiny lungs develop. During her 5-month stay in the neonatal unit, E'Maria had three operations to correct her eyesight, heart, and a hernia. Erica was finally able to bring her little girl home, but not without a heart monitor, multiple medications, and an oxygen tank. There were still problems, but the prognosis is good.

Twenty years ago, E'Maria, born at 25 weeks and weighing little more than two sticks of butter, would not have survived. Ten years ago, over half of babies like E'Maria would not have survived. Today, as a result of medical technology, 85 percent of babies born at 25 weeks' gestation survive. Moreover, the limit of viability is being pushed back 1 week every 5 years (Kilen 2003).

"She always keeps her fist balled up," said Grandpa Robinson. "She's a little fighter."

Many of the medical technologies available today seem futuristic. But such technologies, for example, virtual reality, cloning, and teleportation, are no longer just the stuff of popular science fiction movies. Virtual reality is now used to train workers in occupations as diverse as medicine, engineering, and professional football. The ability to genetically replicate embryos has sparked worldwide debate over the ethics of reproduction, and California Institute of Technology scientists have transported a ray of light from one location to another. Just as the telephone, the automobile, television, and countless other technological innovations have forever altered social life, so will more recent technologies.

Science and technology go hand in hand. **Science** is the process of discovering, explaining, and predicting natural or social phenomena. A scientific approach to understanding acquired immunodeficiency syndrome (AIDS), for example, might include investigating the molecular structure of the virus, the means by which it is transmitted, and public attitudes about AIDS. **Technology,** as a form of human cultural activity that applies the principles of science and mechanics to the solution of problems, is intended to accomplish a specific task—in this case, the development of an AIDS vaccine.

Societies differ in their level of technological sophistication and development. In agricultural societies, which emphasize the production of raw materials, the use of tools to accomplish tasks previously done by hand, or **mechanization,** dominates. As societies move toward industrialization and become more concerned with the mass production of goods, automation prevails. **Automation** involves the use of self-operating machines, as in an automated factory in which

science The process of discovering, explaining, and predicting natural or social phenomena.

technology Activities that apply the principles of science and mechanics to the solutions of a specific problem.

mechanization Dominant in an agricultural society, the use of tools to accomplish tasks previously done by hand.

automation Dominant in an industrial society, the replacement of human labor with machinery and equipment that are self-operating.

autonomous robots assemble automobiles. Finally, as a society moves toward post-industrialization, it emphasizes service and information professions (Bell 1973). At this stage technology shifts toward **cybernation,** whereby machines control machines—making production decisions, programming robots, and monitoring assembly performance.

What are the effects of science and technology on humans and their social world? How do science and technology help to remedy social problems, and how do they contribute to social problems? Is technology, as Postman (1992) suggested, both a friend and a foe to humankind? We address each of these questions in this chapter.

The Global Context: The Technological Revolution

Less than 50 years ago, traveling across state lines was an arduous task, a long-distance phone call was a memorable event, and mail carriers brought belated news of friends and relatives from far away. Today, travelers journey between continents in a matter of hours, and for many, e-mail, faxes, instant messaging, texting, and electronic fund transfers have replaced previously conventional means of communication.

The world is a much smaller place than it used to be, and it will become even smaller as the technological revolution continues. The Internet has 1.1 billion users in more than 200 countries with 208 million users in the United States alone (Internet World Stats 2007a). The most common online language populations are English (28.9 percent), followed by Chinese (14.7 percent), Spanish (8.9 percent), Japanese (7.6 percent), German (5.2 percent), and French (5.0 percent) (Internet World Stats 2007b). Although three-fourths of all Internet users live in industrialized countries, there is some movement toward the Internet's becoming a truly global medium as Africans, Middle Easterners, and Latin Americans increasingly "get online" (Internet World Stats 2007c). Table 15.1 displays the Internet activities of the average global user.

The movement toward globalization of technology is, of course, not limited to the use and expansion of the Internet. The world robot market—and the U.S. share of it—continues to expand; Microsoft's Internet platform and support products are sold all over the world; scientists collect skin and blood samples from remote islanders for genetic research; a global treaty regulating trade of genetically altered products has been signed by more than 100 nations; and Intel computer processing units (CPUs) power an estimated four-fifths of the world's personal computers (PCs).

To achieve such scientific and technological innovations, sometimes called research and development, countries need material and economic resources. *Research* entails the pursuit of knowledge; *development* refers to the production of materials, systems, processes, or devices directed to the solution of practical problems. In 2006 the United States spent $343 billion on research and development. As in most other countries, U.S. funding sources are primarily from private industry (68 percent) followed by the federal government (28 percent) and nonprofit organizations such as research institutes at colleges and universities (4 percent) (National Science Foundation 2007a).

The United States leads the world in science and technology although there is some evidence that we are falling behind (Lemonick 2006a). Federal funding of scientific projects has been shrinking, and there are now six countries that

cybernation Dominant in a postindustrial society; the use of machines to control other machines.

The world was made a smaller place in the late 1800s by the Pony Express. Today, the iPhone, combining a number of technological feats, makes the world even smaller.

TABLE 15.1 Global Internet Use from Home, May 2007	
• Average number of sessions per month	35
• Average number of unique domains visited	71
• Average pages viewed per month	1,509
• Average pages viewed per surfing session	42
• Average time online per month	25 hours, 48 minutes
• Average time spent during surfing session	54 minutes
• Average duration of a page viewed	45 seconds
• Average online population	338,250,261

Source: Nielson/Net Ratings. Available at http://www.nielsennetratings.com

spend a larger proportion of their wealth on science than the United States: Israel, Sweden, Finland, Japan, Iceland, and South Korea. The number of patents granted to U.S. companies as a percentage of all patents has decreased over the last several decades, and publishing in science and engineering journals has lagged behind that of western European scientists as Asian scientists continue an upward surge (Lemonick 2006a).

The decline of U.S. supremacy in science and technology is likely to be the result of three interacting forces (Lemonick 2006a). First, the federal government has been scaling back its investment in research and development in response to fiscal deficits of the last 30 years. Second, corporations, the largest contributors to research and development, have begun to focus on short-term products and higher profits as pressure from stockholders mounts. Finally, there has been a drop in science and math education in U.S. schools both in terms of quality and quantity. For example, although the United States still grants the highest proportion of science and engineering Ph.D. degrees in the world, in recent years the number has declined, whereas other countries' rates have increased.

As Lemonick (2006a, p. 33) noted, "In absolute terms, of course, the U.S. is still the world's leader in scientific research. A half century's worth of momentum is tough to derail." Nevertheless, President Bush has announced a new program called the American Competitiveness Initiative (ACI) as reflected in his 2008 budget (U.S. Department of Education 2007). According to the White House, the ACI will

> keep America the most innovative and competitive economy in the world by encouraging more aggressive investments by businesses through permanent enhanced research and development tax credits . . . , greatly increasing and prioritizing Federal support for vital research . . ., and improving math and science education for American's students. . . . (Department of Education 2007, p. 1)

What Do You Think? Scientific discoveries and technological developments require the support of a country's citizens and political leaders. For example, although abortion has been technically possible for years, millions of the world's citizens live in countries where abortion is either prohibited or permitted only when the life of the mother is in danger. Thus the degree to which science and technology are considered good or bad, desirable or undesirable, is, as social constructionists argue, a function of time and place. Can you name other scientific discoveries or technological developments that are technically possible, but likely to be rejected by large segments of the population?

Postmodernism and the Technological Fix

Many Americans believe that social problems can be resolved through a **technological fix** (Weinberg 1966) rather than through social engineering. For example, a social engineer might approach the problem of water shortages by persuading people to change their lifestyle: Use less water, take shorter showers, and wear clothes more than once before washing. A technologist would avoid the challenge of changing people's habits and motivations and instead concentrate on the development of new technologies that would increase the water supply. Social problems can be tackled through both social engineering and a technological fix. In recent years, for example, social engineering efforts to reduce drunk driving have included imposing stiffer penalties for drunk driving and disseminating public service announcements, such as "Friends don't let friends drive drunk." An example of a technological fix for the same problem is the development of car airbags, which reduce injuries and deaths resulting from car accidents.

technological fix The use of scientific principles and technology to solve social problems.

Not all individuals, however, agree that science and technology are good for society. **Postmodernism,** an emerging worldview, holds that rational thinking and the scientific perspective have fallen short in providing the "truths" they were once presumed to hold. During the industrial era, science, rationality, and technological innovations were thought to hold the promises of a better, safer, and more humane world. Today, postmodernists question the validity of the scientific enterprise, often pointing to the unforeseen and unwanted consequences of resulting technologies. Automobiles, for example, began to be mass-produced in the 1930s in response to consumer demands. But the proliferation of automobiles has also led to increased air pollution and the deterioration of cities as suburbs developed, and, today, traffic fatalities are the number one cause of accident-related deaths.

SOCIOLOGICAL THEORIES OF SCIENCE AND TECHNOLOGY

Each of the three major sociological frameworks helps us to better understand the nature of science and technology in society.

Structural-Functionalist Perspective

Structural functionalists view science and technology as emerging in response to societal needs—that "science was born indicates that society needed it" (Durkheim 1973). As societies become more complex and heterogeneous, finding a common and agreed-on knowledge base becomes more difficult. Science fulfills the need for an assumed objective measure of "truth" and provides a basis for making intelligent and rational decisions. In this regard, science and the resulting technologies are functional for society.

If society changes too rapidly as a result of science and technology, however, problems may emerge. When the material part of culture (i.e., its physical elements) changes at a faster rate than the nonmaterial part (i.e., its beliefs and values), a **cultural lag** may develop (Ogburn 1957). For example, the typewriter, the conveyor belt, and the computer expanded opportunities for women to work outside the home. With the potential for economic independence, women were able to remain single or to leave unsatisfactory relationships and/or establish careers. But although new technologies have created new opportunities for women, beliefs about women's roles, expectations of female behavior, and values concerning equality, marriage, and divorce have lagged behind.

Robert Merton (1973), a structural functionalist and founder of the subdiscipline sociology of science, also argued that scientific discoveries or technological innovations may be dysfunctional for society and may create instability in the social system. For example, the development of time-saving machines increases production, but it also displaces workers and contributes to higher rates of employee alienation. Defective technology can have disastrous effects on society. In 1994 a defective Pentium chip was discovered to exist in 2 million computers in aerospace, medical, scientific, and financial institutions, as well as in schools and government agencies. Replacing the defective chip was a massive undertaking but was necessary to avoid thousands of inaccurate computations and organizational catastrophe.

" I've got gigabytes. I've got megabytes. I'm voice-mailed. I'm e-mailed. I surf the Net. I'm on the Web. I am a Cyber-Man. So how come I feel so out of touch? **"**

Volkswagen television commercial

postmodernism A worldview that questions the validity of rational thinking and the scientific enterprise.

cultural lag A condition in which the material part of culture changes at a faster rate than the nonmaterial part.

Conflict Perspective

Conflict theorists, in general, argue that science and technology benefit a select few. For some conflict theorists technological advances occur primarily as a response to capitalist needs for increased efficiency and productivity and thus are motivated by profit. As McDermott (1993) noted, most decisions to increase technology are made by "the immediate practitioners of technology, their managerial cronies, and for the profits accruing to their corporations" (p. 93). In the United States private industry spends more money on research and development than the federal government does. The Dalkon shield (IUD) and silicone breast implants are examples of technological advances that promised millions of dollars in profits for their developers. However, the rush to market took precedence over thorough testing of the products' safety. Subsequent lawsuits filed by consumers who argued that both products had compromised the physical well-being of women resulted in large damage awards for the plaintiffs.

Science and technology also further the interests of dominant groups to the detriment of others. The need for scientific research on AIDS was evident in the early 1980s, but the required large-scale funding was not made available so long as the virus was thought to be specific to homosexuals and intravenous drug users. Only when the virus became a threat to mainstream Americans were millions of dollars allocated to AIDS research. Hence conflict theorists argue that granting agencies act as gatekeepers to scientific discoveries and technological innovations. These agencies are influenced by powerful interest groups and the marketability of the product rather than by the needs of society.

When the dominant group feels threatened, it may use technology as a means of social control. For example, the use of the Internet is growing dramatically in China, the world's second largest Internet market. In 2005 the government of China announced that every website must be registered "and provide complete information on its organizers . . . or face being declared illegal" (Reuters 2005, p. 1). In 2006, Internet giant Google announced that it would censor search inquiries to conform to Chinese officials' requirements (BBC 2006). China is not alone, however. A study by OpenNet Initiative found that 25 of the 41 countries surveyed engaged in Internet censorship including Vietnam, Saudi Arabia, India, Singapore, and Thailand (Associated Press 2007a). Today, thousands of censors monitor Web pages, blogs, and chat rooms looking for politically incorrect statements.

Finally, conflict theorists as well as feminists argue that technology is an extension of the patriarchal nature of society that promotes the interests of men and ignores the needs and interests of women. As in other aspects of life, women play a subordinate role in reference to technology in terms of both its creation and its use. For example, washing machines, although time-saving devices, disrupted the communal telling of stories and the resulting friendships among women who gathered together to do their chores. Bush (1993) observed that in a "society characterized by a sex-role division of labor, any tool or technique . . . will have dramatically different effects on men than on women" (p. 204).

Symbolic Interactionist Perspective

Knowledge is relative. It changes over time, over circumstances, and between societies. We no longer believe that the world is flat or that the earth is the center of the universe, but such beliefs once determined behavior because individuals responded to what they thought to be true. The scientific process is a social

Motivated by profit, private industry spends more money on research and development than the federal government does.

process in that "truths"—socially constructed truths—result from the interactions between scientists, researchers, and the lay public.

Kuhn (1973) argued that the process of scientific discovery begins with assumptions about a particular phenomenon (e.g., the world is flat). Because unanswered questions always remain about a topic (e.g., why don't the oceans drain?), science works to fill these gaps. When new information suggests that the initial assumptions were incorrect (e.g., the world is not flat), a new set of assumptions or framework emerges to replace the old one (e.g., the world is round). It then becomes the dominant belief or paradigm.

Symbolic interactionists emphasize the importance of this process and the effect that social forces have on it. Conrad (1997), for example, described the media's contribution in framing societal beliefs that alcoholism, homosexuality, and racial inequality are genetically determined. Social forces also affect technological innovations, and their success depends, in part, on the social meaning assigned to any particular product. As social constructionists argue, individuals socially construct reality as they interpret the social world around them, including the meaning assigned to various technologies. If claims makers can successfully define a product as impractical, cumbersome, inefficient, or immoral, the product is unlikely to gain public acceptance. Such is the case with RU486, an oral contraceptive that is widely used in France, Great Britain, and China, but availability of which, although legal in the United States, is opposed by a majority of Americans (Gallup 2000; Gottlieb 2000).

Not only are technological innovations subject to social meaning, but also who becomes involved in what aspects of science and technology is socially defined. Men, for example, outnumber women three to one in earning computer science degrees, and although women make up 46 percent of the general workforce, they make up just 29 percent of computer scientists and 28 percent of computer programmers (Gallaga 2007; U.S. Bureau of the Census 2007). Societal definitions of men as being rational, mathematical, and scientifically minded and as having greater mechanical aptitude than women are, in part, responsible for these differences. This chapter's *Social Problems Research Up Close* feature highlights one of the consequences of the masculinization of technology, as well as the ways in which computer hacker identities and communities are socially constructed.

TECHNOLOGY AND THE TRANSFORMATION OF SOCIETY

A number of modern technologies are considerably more sophisticated than technological innovations of the past. Nevertheless, older technologies have influenced the social world as profoundly as the most mind-boggling modern inventions. Postman (1992) described how the clock—a relatively simple innovation that is taken for granted in today's world—profoundly influenced not only the workplace but also the larger economic institution:

> The clock had its origin in the Benedictine monasteries of the twelfth and thirteenth centuries. The impetus behind the invention was to provide a more or less precise regularity to the routines of the monasteries, which required, among other things, seven periods of devotion during the course of the day. The bells of the monastery were to be rung to signal the canonical hours; the mechanical clock was the technology that could provide precision to these rituals of devotion. . . . What the monks did not foresee was

Cyberstalking, pornography on the Internet, and identity theft are crimes that were unheard of before the computer revolution and the enormous growth of the Internet. One such "high-tech" crime, computer hacking, ranges from childish pranks to deadly viruses that shut down corporations. In this classic study, Jordan and Taylor (1998) enter the world of hackers, analyzing the nature of this illegal activity, hackers' motivations, and the social construction of the "hacking community."

Sample and Methods

Jordan and Taylor (1998) researched computer hackers and the hacking community through 80 semistructured interviews, 200 questionnaires, and an examination of existing data on the topic. As is often the case in crime, illicit drug use, and other similarly difficult research areas, a random sample of hackers was not possible. Snowball sampling is often the preferred method in these cases; that is, one respondent refers the researcher to another respondent, who then refers the researcher to another respondent, and so forth. Through their analysis the investigators provide insight into this increasingly costly social problem and the symbolic interactionist notion of "social construction"—in this case, of an online community.

Findings and Conclusions

Computer hacking, or "unauthorized computer intrusion," is an increasingly serious problem, particularly in a society dominated by information technologies. Unlawful entry into computer networks or databases can be achieved by several means, including (1) guessing someone's password, (2) tricking a computer about the identity of another computer (called "IP spoofing"), or (3) "social engineering," a slang term referring to getting important access information by stealing documents, looking over someone's shoulder, going through their garbage, and so on.

Hacking carries with it certain norms and values, because, according to Jordan and Taylor (1998), the hacking community can be thought of as a culture within a culture. The two researchers identified six elements of this socially constructed community:

- *Technology.* The core of the hacking community is the technology that allows it to occur. As one professor who was interviewed stated, the young today have "lived with computers virtually from the cradle, and therefore have no trace of fear, not even a trace of reverence."
- *Secrecy.* The hacking community must, on the one hand, commit secret acts because their "hacks" are illegal. On the other hand, much of the motivation for hacking requires publicity to achieve the notoriety often sought. In addition, hacking is often a group activity that bonds members together. As one hacker stated, hacking "can give you a real kick some time. But it can give you a lot more satisfaction and recognition if you share your experiences with others."
- *Anonymity.* Whereas secrecy refers to the hacking act, anonymity refers to the importance of the hacker's identity remaining unknown. Thus, for example, hackers and hacking groups take on names such as Legion of Doom, the Inner Circle I, Mercury, and Kaos, Inc.
- *Membership fluidity.* Membership is fluid rather than static, often characterized by high turnover rates, in part, as a response to law enforcement pressures. Unlike more structured organizations, there are no formal rules or regulations.
- *Male dominance.* Hacking is defined as a male activity; consequently, there are few female hackers. Jordan and Taylor (1998) also note, after recounting an incident of sexual harassment, that "the collective identity hackers share and construct . . . is in part misogynist" (p. 768).
- *Motivation.* Contributing to the articulation of the hacking communities' boundaries are the agreed-upon definitions of acceptable hacking motivations, including (1) addiction to computers, (2) curiosity, (3) excitement, (4) power, (5) acceptance and recognition, and (6) community service through the identification of security risks.

Finally, Jordan and Taylor (1998, p. 770) note that hackers also maintain group boundaries by distinguishing between their community and other social groups, including "an antagonistic bond to the computer security industry (CSI)." Ironically, hackers admit a desire to be hired by the CSI, which would not only legitimize their activities but also give them a steady income.

Jordan and Taylor conclude that the general fear of computers and of those who understand them underlies the common, although inaccurate, portrayal of hackers as pathological, obsessed computer "geeks." When journalist Jon Littman asked hacker Kevin Mitnick if he was being demonized because of increased dependence on and fear of information technologies, Mitnick replied, "Yeah. . . . That's why they're instilling fear of the unknown. That's why they're scared of me. Not because of what I've done, but because I have the capability to wreak havoc" (Jordan & Taylor 1998, p. 776).

that the clock is a means not merely of keeping track of the hours but also of synchronizing and controlling the actions of men. And thus, by the middle of the fourteenth century, the clock had moved outside the walls of the monastery, and brought a new and precise regularity to the life of the workman and the merchant. . . . In short, without the clock, capitalism would have been quite impossible. The paradox . . . is that the clock was invented by men who wanted to devote themselves more rigorously to God; it ended as the technology of greatest use to men who wished to devote themselves to the accumulation of money. (pp. 14–15)

Technology has far-reaching effects not only on the economy but also on every aspect of social life. In the following sections we discuss societal transformations resulting from various modern technologies, including workplace technology, computers, the information highway, and science and biotechnology.

Technology in the Workplace

All workplaces, from government offices to factories and from supermarkets to real-estate agencies, have felt the impact of technology. The Office of Technology Assessment of the U.S. Congress estimates that more than 7 million U.S. workers are under some type of computer surveillance, which lessens the need for supervisors and makes control by employers easier.

For example, employees in the Department of Design and Construction in New York City must scan their hands each time they enter or leave the workplace. The use of identifying characteristics such as hands, fingers, and eyes is part of a technology called *biometrics*. Union leaders "called the use of biometrics degrading, intrusive and unnecessary and said that experimenting with the technology could set the stage for a wider use of biometrics to keep tabs on all elements of the workday" (Chan 2007, p. 1).

Technology can also make workers more accountable by gathering information about their performance. Through time-saving devices such as personal digital assistants (PDAs) and battery-powered store-shelf labels, technology can enhance workers' efficiency. Technology can also contribute to worker error, however. In a recent study of a popular hospital computer system, researchers found several ways that the computerized drug-ordering program endangered the health of patients. For example, the software program warned a doctor of a patient's drug allergy only *after* the drug was ordered, and, rather than showing the usual dose of a particular drug, the program showed the dosage available in the hospital pharmacy (DeNoon 2005).

Technology is also changing the location of work. Worldwide, it is estimated that by 2008, 41 million corporate employees will spend one day a week **teleworking,** that is, completing all or part of their work away from the traditional workplace. In the United States, 32 percent of workers report telecommuting at some time, a sharp increase from 1995 when just 9 percent of workers reported telecommuting. Of those who telecommute, a clear majority (81 percent) believe that people who work from home are just as or more productive than people who work in an office (Jones 2006).

Information technologies are also changing the nature of work. Eli Lilly and Company employees communicate by means of their own "intranet," on which all work-related notices are posted. Federal Express not only created a FedEx network for its employees but also allowed customers to enter their package-tracking database, saving the company millions of dollars a year. The importance of such technologies is empirically documented. In a survey of British

teleworking A form of work that allows employees to work part- or full-time at home or at a satellite office.

Automation means that machines can now perform the labor originally provided by human workers, such as the robots that perform tasks on automobile assembly lines.

© Adam Lubroth/Stone/Getty Images

office workers, 25 percent reported that information technologies were more important than office managers (Hayday 2003).

Robotic technology has also revolutionized work. Ninety percent of robots work in factories, and more than half of these are used in heavy industry, such as automobile manufacturing (Tesler 2003). Robots are most commonly used for materials handling and spot welding although recently there has been a trend toward more fully integrating them into the manufacturing process (Pethokoukis 2004). An employer's decision to use robotics depends on direct costs (e.g., initial investment) and indirect costs (e.g., unemployment compensation), the feasibility and availability of robots to perform the desired tasks, and the resulting increased rate of productivity. Use of robotics may also depend on whether labor unions resist the replacement of workers by machines.

What Do You Think? In Japan, "robots are more than mere gadgetry—they're practically family," says journalist Jonathan Skillings (2006, p. 1), and truer words have never been spoken. First there was Paro, a baby harp seal robot covered with fur used in assisted living facilities as a form of animal-assisted therapy. Want to study child development? How about CB2—the toddler robot? Weighing just 73 pounds and standing 4 feet tall, this "humanoid" reacts to touch and sound (Noyes 2007). The Japanese government is now pushing for "a corps of non-threatening robots ready to assist in office tasks, housekeeping, and eldercare" (Skillings 2006, p. 1). It is estimated that by the end of the decade there will be 39 million household robots in Japan. Would you allow your grandmother to be cared for by a robot? Do you think American society would be as accepting of robots as Japanese culture? Why or why not?

The Computer Revolution

Early computers were much larger than the small machines we have today and were thought to have only esoteric uses among members of the scientific and military communities. In 1951 only about a half dozen computers existed (Ceruzzi 1993). The development of the silicon chip and sophisticated microelectronic technology allowed tens of thousands of components to be imprinted on a single chip smaller than a dime. The silicon chip led to the development of laptop computers, cellular phones, digital cameras, the iPod, and portable DVDs. The silicon chip also made computers affordable. Although the first PC was developed only 20 years ago, today more than 62 percent of American homes have a computer (U.S. Bureau of the Census 2007). Ownership varies by a number of factors but is greatest among households with higher incomes (U.S. Bureau of the Census 2007) (see Table 15.2).

Seventy-seven million employees use a computer at work—more than half the labor force (BLS 2005). The percentage of workers who use computers varies by occupation, with 80 percent of managers and professionals using a computer but only 26 percent of, for example, laborers using a computer. Females, Asians, and more educated individuals have higher rates of computer use at work. The most common computer activity is accessing the Internet, sending e-mail, using a search engine, getting the news or e-mail, followed by word processing, working with spreadsheets or databases, and accessing or updating calendars or schedules.

Not surprisingly, computer education has also mushroomed in the last two decades. In 1971, 2,388 U.S. college students earned a bachelor's degree in computer and information sciences; by 2004 that number had increased to 59,488 (U.S. Bureau of the Census 2007). Universities are moving toward requiring their students to have laptop computers, wireless corridors are an increasingly common occurrence, and college and university spending on hardware and software is at an all-time high.

Computers are also big business, and the United States is one of the most successful producers of computer technology in the world, boasting several of the top companies—Dell, Hewlett-Packard, IBM, and Apple. Retail sales of computers exceed $18.2 billion annually with Americans spending more than $5.0 billion on software alone (U.S. Census Bureau 2007). Spending on computers is predicted to grow as consumers increasingly define PCs as a necessity rather than a luxury.

Computer software is, as noted, big business, and in some states it is too big. In 2000 a federal judge found that Microsoft Corporation was in violation of antitrust laws, which prohibit unreasonable restraint of trade. At issue were Microsoft's Windows operating system and the vast array of Windows-based applications (e.g., spreadsheets, word processors, tax software)—applications that *only* work with Windows. The court held that the 70,000 programs written exclusively for Windows made "competing against Microsoft impractical" (Markoff 2000). In an agreement reached between the U.S. Department of Justice and Microsoft Corporation, Microsoft was to be divided into two companies—an operating systems company and an applications company. However, a federal appeals court overturned the lower court's order to split the company in half.

In 2007, Google claimed that Microsoft's latest operating system, Vista, was designed to make using the search engine more difficult and, therefore, was a

> **"** I think there is a world market for maybe five computers. **"**
>
> Thomas Watson
> Chairman of IBM, speaking in 1943

TABLE 15.2 Households with Computers and Internet Access by Selected Characteristic, 2003

CHARACTERISTIC	HOUSEHOLDS WITH COMPUTERS	HOUSEHOLDS WITH INTERNET ACCESS
All households	61.8	54.6
Age of householder		
Under 25 years old	56.5	46.9
25–34 years old	68.6	60.2
35–44 years old	73.2	65.2
45–54 years old	71.9	65.1
55 years old or older	46.6	40.8
Sex		
Male	65.6	58.6
Female	57.4	50.1
Education of householder		
Elementary	20.6	14.0
Some high school	32.7	24.3
High school graduate or GED	51.1	43.0
Some college	70.6	62.4
BA degree or higher	83.3	78.3
Household income		
Under $5,000	35.6	26.8
$5,000–$9,999	26.9	20.0
$10,000–$14,999	31.7	23.7
$15,000–$19,999	38.2	29.4
$20,000–$24,999	46.1	36.7
$25,000–$34,999	55.4	45.6
$35,000–$49,999	71.1	62.8
$50,000–$74,999	81.9	76.0
$75,000–$99,999	88.1	84.1
$100,000–$149,999	92.9	90.4
$150,000 or more	94.7	92.4

Source: U.S. Bureau of the Census (2007).

violation of antitrust laws. In an about face from the Clinton administration, officials appointed by President Bush dismissed the claim as unwarranted and recommended to state attorney generals that they do the same. Microsoft's problems are not over, however. Prosecutors in several states are pursuing Google's claim regardless of the federal government's position (Labaton 2007).

The Internet

Information technology, or IT, refers to any technology that carries information. Most information technologies were developed within a 100-year span: photography and telegraphy (1830s), rotary power printing (1840s), the typewriter (1860s), transatlantic cable (1866), the telephone (1876), motion pictures (1894), wireless telegraphy (1895), magnetic tape recording (1899), radio (1906), and television (1923) (Beniger 1993). The concept of an "information society" dates back to the 1950s, when an economist identified a work sector he called "the production and distribution of knowledge." In 1958, 31 percent of the labor force was employed in this sector—today more than 50 percent is. When this figure is combined with those in service occupations, more than 75 percent of the labor force is involved in the information society.

The development of a national information infrastructure was outlined in the Communications Act of 1994. An information infrastructure performs three functions (Kahin 1993). First, it carries information, just as a transportation system carries passengers. Second, it collects data in digital form that can be understood and used by people. Finally, it permits people to communicate with one another by sharing, monitoring, and exchanging information based on common standards and networks. In short, an information infrastructure facilitates telecommunications, knowledge, and community integration.

The **Internet** is an international information infrastructure—a network of networks—available through universities, research institutes, government agencies, libraries, and businesses. In 2007, 71 percent of all Americans used the Internet from some location. U.S. users were equally likely to be male or female but were more likely to be 18–29 years old, to be college graduates from urban or suburban areas, and to have an annual household income of $75,000 or more (Pew 2007a). The most active Internet users are connected to broadband—that is, services that provide high-speed (DSL and cable) access rather than dial-up service—and are high-income young males (McGann 2005a). About one-third of all Internet users have used wireless connections at home, at the office, or at some other location to log on to the Internet (Pew 2007b).

The Internet or the "web" as it is most commonly known has evolved to what is now called **Web 2.0**—a platform for millions of users to express themselves online in the common areas of cyberspace (Grossman 2006, Pew 2007c). The development of Web 2.0 is a story

> about community and collaboration on a scale never seen before. It's about the cosmic compendium of knowledge Wikipedia and the million-channel people's network YouTube and the online metropolis MySpace. It's about the many wresting power from the few and helping one another for nothing and how that will not only change the world, but also change the way the world changes. (Grossman 2006, p. 1)

The extent to which people avail themselves of Web 2.0 capabilities was measured by the Pew Internet and American Life Project in a survey of 4,000 American adults (Horrigan 2007). Research questions surrounded people's relationships to information and communications technology (ICT)—what they use, what they do, and what they think. The resulting typology of ICT users is displayed in Table 15.3. Note that 31 percent are elite technology users, 20 percent are middle of the road technology users, and 49 percent have few "technology assets," i.e., they have little use for Web 2.0, the Internet, or cell phones.

Internet An international information infrastructure available through universities, research institutes, government agencies, libraries, and businesses.

Web 2.0 A platform for millions of users to express themselves online in the common areas of cyberspace.

TABLE 15.3 Typology of Information and Communication Technology (ICT) Users

GROUP NAME	PERCENTAGE OF ADULT POPULATION	WHAT YOU NEED TO KNOW ABOUT THEM
Elite tech users (31% of American adults)		
Omnivores	8	They have the most information gadgets and services, which they use voraciously to participate in cyberspace and express themselves online and do a range of Web 2.0 activities such as blogging or managing their own Web pages.
Connectors	7	Between feature-packed cell phones and frequent online use, they connect to people and manage digital content using ICTs—all with high levels of satisfaction about how ICTs let them work with community groups and pursue hobbies.
Lackluster veterans	8	They are frequent users of the Internet and less avid about cell phones. They are not thrilled with ICT-enabled connectivity.
Productivity enhancers	8	They have strongly positive views about how technology lets them keep up with others, do their jobs, and learn new things.
Middle-of-the-read tech users (20%)		
Mobile centrics	10	They fully embrace the functionality of their cell phones. They use the Internet, but not often, and like how ICTs connect them to others.
Connected but hassled	10	They have invested in a lot of technology, but they find the connectivity intrusive and information something of a burden.
Few tech assets (49%)		
Inexperienced experimenters	8	They occasionally take advantage of interactivity, but if they had more experience, they might do more with ICTs.
Light but satisfied	15	They have some technology, but it does not play a central role in their daily lives. They are satisfied with what ICTs do for them.
Indifferents	11	They have some technology, but it does not plan a central role in their daily lives. They are satisfied with what ICTs do for them.
Off the network	15	Those with neither cell phones nor Internet connectivity tend to be older adults who are content with old media.

Source: Horrigan (2007).

E-commerce is the buying and selling of goods and services over the Internet. Although initially growing at warp speed, e-commerce has recently slowed down (Richtel & Tedeschi 2007). For example, 2007 online book sales are projected to grow 11 percent compared with 40 percent growth in the preceding year. eBay and Expedia report that online sales have increased just 1 percent in the first quarter of 2007 compared with the first quarter of 2006. Dell, which mastered the art of selling online, is now selling their computers at mall kiosks and discount stores. Not surprisingly, computers are no longer the number one online sale item (O'Donnell 2007). Other sectors that have leveled off include online sales of appliances, sporting goods, tickets, and office supplies.

Nancy Koehn, a professor at the Harvard School of Business, theorizes that the e-commerce slowdown is a result of the consumer's need for the in-store shopping experience and the tendency for online shopping to feel too much like work (Richtel & Tedeschi 2007). One result is what is called "*clicks and*

e-commerce The buying and selling of goods and services over the Internet.

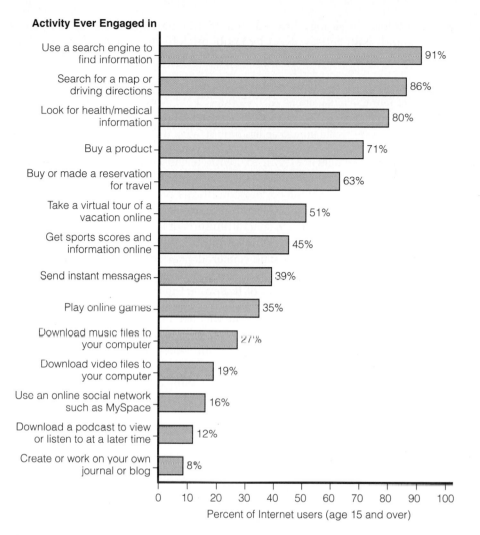

Activity Ever Engaged in

FIGURE 15.1
Common Online Activities by U.S. Adult Internet Users, 2006.
SOURCE: Pew (2007d).

Activity	Percent
Use a search engine to find information	91%
Search for a map or driving directions	86%
Look for health/medical information	80%
Buy a product	71%
Buy or made a reservation for travel	63%
Take a virtual tour of a vacation online	51%
Get sports scores and information online	45%
Send instant messages	39%
Play online games	35%
Download music files to your computer	27%
Download video files to your computer	19%
Use an online social network such as MySpace	16%
Download a podcast to view or listen to at a later time	12%
Create or work on your own journal or blog	8%

Percent of Internet users (age 15 and over)

bricks"—a hybrid model that combines the best of both worlds. Consumers purchase or reserve items online but pick them up at the store. Alternatively, some online retailers are revamping their web pages to make them more user-friendly and others are adding new capabilities to heighten the online shopping experience. Barnes & Noble, for example, has added an online book club, interviews with authors, and reader forums to make "online" seem more like "offline" (Richtel & Tedeschi 2007). Figure 15.1 lists some of the most common activities engaged in by Internet users. Note that 71 percent of Internet users report buying a product online (Pew 2007c).

Science and Biotechnology

Although recent computer innovations and the establishment of the Internet have led to significant cultural and structural changes, science and its resulting biotechnologies have produced not only dramatic changes but also hotly contested issues. In this section we look at some of the issues raised by developments in genetics, food and biotechnology, and reproductive technologies.

Genetics. Molecular biology has led to a greater understanding of the genetic material found in all cells—DNA (deoxyribonucleic acid)—and with it the ability for **genetic screening.**

> If you could uncoil a strip of DNA, it would reach 6 feet in length, a code written in words of four chemical letters: A, T, G, and C. Fold it back up, and it shrinks to trillionths of an inch, small enough to fit in any one of our 100 trillion cells, carrying the recipe for how to create human beings from scratch. (Gibbs 2003, p. 42)

Currently, researchers are trying to complete genetic maps that will link DNA to particular traits. There is some evidence that personality characteristics are inherited; other evidence links certain conditions previously thought of as psychological in nature (e.g., anorexia and autism) as, at least in part, genetically induced (Harmon 2006). Already, specific strands of DNA have been identified as carrying physical traits such as eye color and height as well as such diseases as breast cancer, cystic fibrosis, prostate cancer, depression, and Alzheimer's.

The U.S. Human Genome Project (HGP), a 13-year effort to decode human DNA, is now complete. Conclusion of the project is transforming medicine.

> All diseases have a genetic component whether inherited or resulting from the body's response to environmental stresses like viruses or toxins. The successes of the HGP have . . . enabled researchers to pinpoint errors in genes—the smallest units of heredity—that cause or contribute to disease. The ultimate goal is to use this information to develop new ways to treat, cure, or even prevent the thousands of diseases that afflict humankind. (Human Genome Project 2007, p. 1)

The hope is that if a defective or missing gene can be identified, possibly a healthy duplicate can be acquired and transplanted into the affected cell. This is known as **gene therapy.** Alternatively, viruses have their own genes that can be targeted for removal. Experiments are now under way to accomplish these biotechnological feats.

Food and Biotechnology. Genetic engineering is the ability to manipulate the genes of an organism in such a way that the natural outcome is altered. Genetically modified food, also known as genetically engineered (GE) food, and genetically modified (GM) organisms involve this process of DNA recombination—scientists transferring genes from one plant into the genetic code of another plant.

An estimated 60–70 percent of processed foods in U.S. markets contain some form of GM ingredient, most often corn or soy, followed by canola and cotton (in cottonseed oil). Yet a national survey of U.S. adults found that less than half (41 percent) were aware that foods containing GM ingredients are currently sold in stores, and although most Americans are likely to consume foods with GM ingredients every day, less than one-third (26 percent) of the survey sample said they had consumed food containing GM ingredients (Pew 2006).

Biotechnology companies and other supporters of GM foods commonly cite the alleviation of hunger and malnutrition as a main benefit, claiming that this technology can enable farmers to produce crops with higher yields. Critics of GM foods argue that the world already produces enough food for all people to have a healthy diet. According to the United Nations Development Programme, if all the food produced worldwide were distributed equally, every person would be able to consume 2,760 calories a day (UNDP 2003). Biotechnology,

genetic screening The use of genetic maps to detect predispositions to human traits or disease(s).

gene therapy The transplantation of a healthy gene to replace a defective or missing gene.

genetic engineering The manipulation of an organism's genes in such a way that the natural outcome is altered.

The first genetically engineered crop was introduced for commercial production in 1996. Today, there are more than 200 million acres devoted to these crops with the United States being the largest producer in the world.

critics argue, will not alter the fundamental causes of hunger, which are poverty and lack of access to food and to land on which to grow food.

Biotechnology companies claim that GM foods approved by the Food and Drug Administration are safe for human consumption, and they even cite potential health benefits such as the use of genetic modification to remove allergens that naturally occur in foods such as nuts, making these foods safer to eat for individuals who have allergies to these foods (Bailey 2005). But critics claim that research on the effects of GM crops and foods on human health is inadequate, especially concerning long-term effects. Human health concerns include possible toxicity, carcinogenicity, food intolerance, antibiotic resistance buildup, decreased nutritional value, and food allergens in GM foods.

Biotechnology skeptics are also concerned about the environmental effects of GM crops. Biotechnology companies claim that crops that are genetically designed to repel insects negate the need for chemical (pesticide) control and thus reduce pesticide poisoning of land, water, animals, foods, and farm workers. However, critics are concerned that insect populations can build up resistance to GM plants with insect-repelling traits, which would necessitate increased rather than decreased use of pesticides. Indeed, a recent study showed that although GE crops substantially reduce pesticide use in the first few years of planting, GE crops have increased the overall volume of pesticides applied to corn, soybeans, and cotton over a 9-year period (Benbrook 2004).

Another health and environmental risk is the spread of traits from GM plants to non-GM plants, the effects of which are unknown. In 2003 an analysis of corn grown in nine Mexican states showed that 24 percent of the samples tested positive for contamination by several varieties of GM corn, including Starlink, produced as cattle feed and deemed unfit for human consumption because of the presence of a bacterial protein that is not broken down by the human digestive system and is therefore a potential allergen (ETC Group 2003a; Ruiz-Marrero 2002). Mexican farmers and community members view the contamination of

Mexican corn as an attack on Mexican culture. In the words of one Mexican citizen, "Our seeds, our corn, is [sic] the basis of the food sovereignty of our communities. It's much more than a food, it's part of what we consider sacred, of our history, our present and future" (quoted by ETC Group 2003a, p. 2).

GM seed contamination is of particular concern with regard to seed sterility technology. To maintain control over their products, biotechnology companies have developed "terminator" seeds, which cause the plant to produce sterile seeds. Because of public opposition to terminator seeds, in 1999 Monsanto agreed not to market its terminator technology. However, Monsanto later adopted a positive stance on genetic seed sterilization, suggesting that the commercialization of terminator technology may occur in the future (ETC Group 2003b). Could the seed sterility trait in terminator crops inadvertently contaminate both traditional crops and wild plant life? The possible ramifications of widespread plant sterility could be devastating to life on earth.

Biotechnology critics also raise concerns about insufficient safeguards and regulatory mechanisms. In 2000 Taco Bell taco shells, made by Kraft Foods, were recalled after traces of Starlink corn were found in the taco shells. No one—from farmers to grain dealers to Kraft—could explain how the Starlink corn got mixed into corn meant for taco shells. The traces of unapproved corn were not found by the U.S. Department of Agriculture's Food Safety and Inspection Service or by the Department of Health and Human Service's Food and Drug Administration. Rather, the traces were discovered by Genetically Engineered Food Alert—a coalition of biotechnology skeptics. In addition to taco shells, Starlink contamination caused a recall of more than 300 corn-based foods.

In 2000, worldwide concern about the safety of GM crops resulted in 130 nations signing the landmark Biosafety Protocol, which requires producers of a GM food to demonstrate that it is safe before it is widely used. The Biosafety Protocol also allows countries to ban the import of GM crops based on suspected health, ecological, or social risks. As of 2007, several countries had banned the importation of GM crops, and others had banned the commercial planting of GM crops. For example, a federal judge recently halted the planting of a genetically altered commercial crop that had previously been approved by the U. S. Department of Agriculture, expressing concern that the genetically altered "roundup ready" alfalfa might contaminate organic and conventional alfalfa (Center for Food Safety 2007).

Reproductive Technologies. The evolution of "reproductive science" has been furthered by scientific developments in biology, medicine, and agriculture. At the same time, however, its development has been hindered by the stigma associated with sexuality and reproduction, its link with unpopular social movements (e.g., contraception), and the feeling that such innovations challenge the natural order (Clarke 1990). Nevertheless, new reproductive technologies have been and continue to be developed.

In **in vitro fertilization (IVF)** an egg and a sperm are united in an artificial setting, such as a laboratory dish or test tube. Although the first successful attempt at IVF occurred in 1944, the first test-tube baby, Louise Brown, was not born until 1978. In 2006, there were an estimated 500,000 frozen embryos in U.S. clinics. Sixty percent of infertility patients are willing to donate their surplus frozen embryos to research—three times the number willing to donate their embryos for adoption (Mundy 2006; Kliff 2007). Criticisms of IVF are often based on traditional definitions of the family and the legal complications created

in vitro fertilization (IVF) The union of an egg and a sperm in an artificial setting such as a laboratory dish.

when a child can have as many as five potential parental ties—egg donor, sperm donor, surrogate mother, and the two people who raise the child (depending on the situation, IVF may not involve donors and/or a surrogate). Litigation over who the "real" parents are has already occurred.

Perhaps more than any other biotechnology, abortion epitomizes the potentially explosive consequences of new technologies. **Abortion** is the removal of an embryo or fetus from a woman's uterus before it can survive on its own. Ninety percent of all abortions take place in the first 12 weeks of pregnancy (Sherman 2006).

Since the U.S. Supreme Court's ruling in *Roe v. Wade* in 1973, abortion has been legal in the United States. However, recent Supreme Court decisions have limited the scope of the *Roe v. Wade* decision. For example, in *Planned Parenthood of Southeastern Pennsylvania v. Casey,* the Court ruled that a state may restrict the conditions under which an abortion is granted, such as requiring a 24-hour waiting period or parental consent for minors. In 2005 the Supreme Court agreed to hear a challenge to parental notification laws. Specifically, the Court considered in *Ayotte v. Planned Parenthood* "whether laws requiring parental notification before a minor can get an abortion must make an explicit exception when the minor's health is at stake" (Barbash 2005, p. 1). In 2006, the Court's unanimous decision reaffirmed the need for a medical emergency exception for teenagers seeking an abortion (Greenhouse 2006).

Additional challenges *to Roe v. Wade* occur at the state level where some argue the real fight is being fought (Tumulty 2006). In Missouri there has been a flurry of bills designed to make access to abortion more difficult, including a bill that would allow civil suits against anyone who helps a Missouri teen get an abortion without parental consent. In South Dakota state legislators voted to prohibit all abortions except when the mother's life is at risk—no exception for women who have been victims of rape or incest. Although in South Dakota the issue became a referendum that was eventually turned down by voters, similar bills have been introduced in Georgia, Kentucky, Ohio, Tennessee, and Indiana (Caplan 2006). Although other reasons could be offered—a healthy economy, improved contraceptive methods, and safe sex practices—some would argue that state laws restricting abortion are the reason abortion rates have declined over the last several decades (Tumulty 2006; U.S. Census Bureau 2007).

What Do You Think? If some South Carolina legislators have their way, women seeking abortions will be required to view an ultrasound image of their fetus before receiving the procedure. Presently, seven states have laws governing abortions and ultrasound technology. For example, "women in Oklahoma, Utah, and Wisconsin must be told an ultrasound is available [and] in Arkansas and Michigan, if an ultrasound is performed, women must be given the opportunity to view it" (Adcox 2007, p. 1). Mississippi is considering a proposal that would require women to view an ultrasound image of their fetus or listen to the fetus' heartbeat. What do you think? Should women be required to view an ultrasound and/or listen to the fetal heartbeat before an abortion? What are the arguments for or against such practices?

abortion The intentional termination of a pregnancy.

Many of the state restrictions on abortion are a result of the pro-life movement's success in redefining the abortion issue as one concerning *fetal rights* rather than *women's rights* (Greenhouse 2007). For example, the Unborn Victims of Violence Act of 2004 protects all children in utero, regardless of stage of development. To date, 35 states have criminalized harm to a fetus. Because California defines an 8-week-old fetus as a person, Scott Peterson was convicted in the deaths of his wife, Laci, *and* his unborn son, Conner. Whereas the original intent of the Act was to protect the fetus from violent crimes, observers note that if in law a fetus is defined as a human being with the same rights as women and men, then it may in effect overturn *Roe v. Wade* (Rosenberg 2003).

Most recent debates concern intact dilation and extraction (D&E) abortions, which often take place in the second trimester of pregnancy (Sherman 2006). Opponents refer to such abortions as **partial birth abortions** because the limbs and the torso are typically delivered before the fetus has expired. However, National Organization for Women president Kim Gandy states, "Try as you might, you won't find the term 'partial birth abortion' in any medical dictionary. That's because it doesn't exist in the medical world—it's a fabrication of the anti-choice machine" (U.S. Newswire 2003, p. 1). D&E abortions are performed because the fetus has a serious defect, the woman's health is jeopardized by the pregnancy, or both. In 2003 a federal ban on partial birth abortions was signed into law (White House 2003). Several constitutional challenges to the ban have occurred and, in 2004, a federal judge ruled that the ban was unconstitutional because it imposes an "undue burden on a woman's right to choose an abortion" (Willing 2005). However, in a major victory for the Bush administration, in 2007 the U.S. Supreme Court upheld the Partial-Birth Abortion Ban in a 5–4 decision (Greenhouse 2007). The significance of the case lies in the fact that it is the first time the U.S. Supreme Court has upheld a ban on any type of abortion procedure.

Feminists, including U.S. Supreme Court Justice Ruth Bader Ginsburg, strongly oppose the ban, arguing that it is just one step closer to making all abortions illegal. They are also quick to note that the ban was not supported by "the American Medical Association, the American College of Obstetricians and Gynecologists, the American Medical Women's Association, the American Nurses Association, or the American Public Health Association" (U.S. Newswire 2003, p. 1). Says Eleanor Smeal, president of the Feminist Majority Foundation, "[I]n upholding the Bush Administration's abortion ban . . . the U.S. Supreme Court showed its true colors: that it does not care about the health, well-being, and safety of American woman. . . ." (Smeal 2007). Further, Justice Ginsburg, writing a strongly worded dissent, states, "this way of thinking reflects ancient notions of women's place in the family and under the Constitution—ideas that have long been discredited" (quoted in Greenhouse 2007, p. 1).

Abortion is a complex issue for everyone, but especially for women, whose lives are most affected by pregnancy and childbearing. Women who have abortions are disproportionately poor, unmarried minority women who say that they intend to have children in the future. Abortion is also a complex issue for societies, which must respond to the pressures of conflicting attitudes toward abortion and the reality of high rates of unintended and unwanted pregnancy. Using the most recent available data, Figure 15.2 illustrates the percentage of Americans who support legal abortions under various circumstances.

Attitudes toward abortion tend to be polarized between two opposing groups of abortion activists—pro-choice and pro-life. Advocates of the pro-choice movement hold that freedom of choice is a central human value, that

partial birth abortion Also called an intact dilation and extraction (D & E) abortion, the procedure may entail delivering the limbs and the torso of the fetus before it has expired.

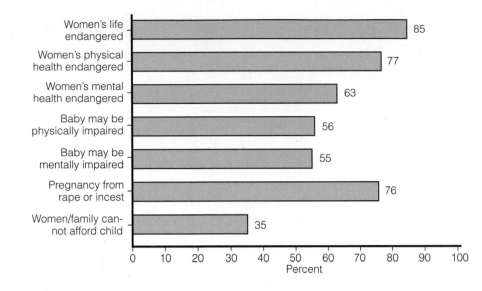

FIGURE 15.2
Support for legal abortions under specific circumstances: 2003.
Source: Gallup (2003).

procreation choices must be free of government interference, and that because the woman must bear the burden of moral choices, she should have the right to make such decisions. Alternatively, pro-lifers hold that the unborn fetus has a right to live and be protected, that abortion is immoral, and that alternative means of resolving an unwanted pregnancy should be found. Assess your attitudes toward abortion in this chapter's *Self and Society* feature.

In July 1996 scientist Ian Wilmut of Scotland successfully cloned an adult sheep named Dolly. To date, cattle, goats, mice, pigs, cats, rabbits, and horses have also been cloned. This technological breakthrough has caused worldwide concern about the possibility of human cloning, leading the United Nations to adopt a declaration that calls for governments to ban all forms of cloning that are at odds with human dignity and the preservation of human life (Lynch 2005). One argument in favor of developing human cloning technology is its medical value; it may potentially allow everyone to have "their own reserve of therapeutic cells that would increase their chance of being cured of various diseases, such as cancer, degenerative disorders and viral or inflammatory diseases" (Kahn 1997, p. 54). Human cloning could also provide an alternative reproductive route for couples who are infertile and for those in which one partner is at risk for transmitting a genetic disease.

Arguments against cloning are largely based on moral and ethical considerations. Critics of human cloning suggest that whether used for medical therapeutic purposes or as a means of reproduction, human cloning is a threat to human dignity (Human Cloning Prohibition Act of 2007). For example, cloned humans would be deprived of their individuality, and as Kahn (1997, p. 119) pointed out, "creating human life for the sole purpose of preparing therapeutic material would clearly not be for the dignity of the life created." **Therapeutic cloning** uses stem cells from human embryos. **Stem cells** can produce any type of cell in the human body and thus can be "modeled into replacement parts for people suffering from spinal cord injuries or degenerative diseases, including Parkinson's and diabetes" (Eilperin and Weiss 2003, p. A6). For example, stem cells have been used for repairing spinal cord injuries in mice, allowing them to walk normally (Weiss 2005). Because the use of stem cells can entail the destruction

therapeutic cloning Use of stem cells from human embryos to produce body cells that can be used to grow needed organs or tissues.

stem cells Undifferentiated cells that can produce any type of cell in the human body.

Self and Society | Abortion and the Law

This is not a test. There are no wrong or right answers to any of the questions, so just answer as honestly as you can. Respond to each question and circle only one response.

1. With respect to the abortion issue, would you consider yourself to be
   a. pro-choice
   b. pro-life

2. Do you think abortion should be
   a. legal under any circumstances
   b. legal only under certain circumstances
   c. illegal in all circumstances

3. Would you like to see the Supreme Court overturn its 1973 *Roe v. Wade* decision concerning abortion or not?
   a. Yes, overturn
   b. No, do not overturn
   c. No opinion

4. "Late term" abortions or "partial birth" abortions are sometimes performed on women during the last few months of pregnancy. Do you think that this procedure should be legal or illegal?
   a. Legal
   b. Illegal
   c. No opinion

5. Thinking about how the abortion issue might affect your vote for major offices, would you
   a. only vote for a candidate who shares your view on abortion
   b. consider a candidate's position on abortion as just one of many important factors
   c. not see abortion as a major issue
   d. no opinion

6. Would you like to see abortion laws in this country
   a. made stricter
   b. made less strict
   c. remain as they are
   d. no opinion

7. Would you favor or oppose a law in your state that would ban all abortions except those necessary to save the life of the mother?
   a. Favor
   b. Oppose
   c. No opinion

Comparison: In 2007, a Gallup poll asked these questions of a national sample of U.S. adults. The distribution of responses is below. Not all responses sum to 100 percent due to rounding error and missing data.

1. With respect to the abortion issue, would you consider yourself to be
   a. pro-choice: 49 percent
   b. pro-life: 45 percent

2. Do you think abortion should be
   a. legal under any circumstances: 26 percent
   b. legal only under certain circumstances: 55 percent
   c. illegal in all circumstances: 18 percent

3. Would you like to see the Supreme Court overturn its 1973 *Roe v. Wade* decision concerning abortion or not?
   a. Yes, overturn: 35 percent
   b. No, do not overturn: 53 percent
   c. No opinion: 12 percent

4. "Late term" abortions or "partial birth" abortions are sometimes performed on women during the last few months of pregnancy. Do you think that this procedure should be legal or illegal?
   a. Legal: 22 percent
   b. Illegal: 72 percent
   c. No opinion: 5 percent

5. Thinking about how the abortion issue might affect your vote for major offices, would you
   a. only vote for a candidate who shares your view on abortion: 16 percent
   b. consider a candidate's position on abortion as just one of many important factors: 59 percent
   c. not see abortion as a major issue: 23 percent
   d. no opinion: 3 percent

6. Would you like to see abortion laws in this country
   a. made stricter: 37 percent
   b. made less strict: 23 percent
   c. remain as they are: 36 percent
   d. no opinion: 3 percent

7. Would you favor or oppose a law in your state that would ban all abortions except those necessary to save the life of the mother?
   a. Favor: 36 percent
   b. Oppose: 60 percent
   c. No opinion: 3 percent

Source: Adapted from Gallup (2007).

of human embryos, many conservatives, including President Bush, are opposed to the practice (see Table 15.4).

The Bush administration is also critical of a process known as SCNT, or somatic cell nuclear transfer, a "complex technique that merges eggs (whose nuclei have been removed) with adult cells to create specialized embryonic stem cell lines" (Kalb 2005, p. 52). U.S. scientists believe that SCNT "will allow them to watch complex diseases in the Petri dish, spot problems, and then test drugs to fix them" (Kalb and Lee 2006, p. 10). The techniques used in SCNT have already been proven to work in mice.

Despite Congress' pledge to relax restrictions, in 2007 President Bush vetoed legislation

© Schwadron. www.cartoonstock.com

TABLE 15.4 Demographics of the Stem Cell Issue (Percentages), 2004			
	WHICH IS MORE IMPORTANT?		
	NEW CURES FROM STEM CELL RESEARCH	**PROTECTING HUMAN EMBRYOS**	**DO NOT KNOW**
Total	**56**	**32**	**12**
Age			
18–29	61	32	7
30–49	58	31	11
50–64	55	34	11
64 and older	50	32	18
Political affiliation			
Republican	45	45	10
Democrat	68	22	10
Independent	58	30	12
Self-described			
Conservative	44	45	11
Moderate	61	27	12
Liberal	77	16	7
Religion			
Protestants	52	38	10
White evangelical	33	58	9
White mainline	69	19	12
Black Protestant	47	36	17
Catholics	63	28	9
Secular	70	16	14

Source: Adapted from Pew Research Center (2005).

© Robert Visser/Corbis Sygma

Christopher Reeve, stage and movie actor best known for his portrayal of Superman, was a long-time advocate of federally funded research on embryonic stem cells. He died on October 10, 2004, at the age of 52.

(for the second time) that would permit federal funding of embryonic stem cell research. At the heart of the controversy is the issue of when life begins. For scientists, stem cell research "is a path to discovering the basic workings of human cells and the causes and therapies for a host of human maladies. But for some religious leaders and social conservatives, who liken the research to abortion, the studies violate the sanctity of human life" (Vestal 2007, p. 2). Although there are other types of stem cells, most research indicates that stem cells from other sources (e.g., adult stem cells) do not have the same regenerative properties as embryonic stem cells (Taylor 2007).

Despite what appears to be a universal race to the future and the indisputable benefits of scientific discoveries such as the workings of DNA and the technology of IVF, some people are concerned about the duality of science and technology. Science and the resulting technological innovations are often life assisting and life giving (see *The Human Side* feature in this chapter); they are also potentially destructive and life-threatening. The same scientific knowledge that led to the discovery of nuclear fission, for example, led to the development of both nuclear power plants and the potential for nuclear destruction. Thus we now turn our attention to the problems associated with science and technology.

SOCIETAL CONSEQUENCES OF SCIENCE AND TECHNOLOGY

Scientific discoveries and technological innovations have implications for all social actors and social groups. As such, they also have consequences for society as a whole. Figure 15.3, for example, displays the percent of Americans, Canadians, and Europeans who believe the listed technologies will improve our way of life over the next 20 years (National Science Foundation 2007b). Technology, however, also has negative consequences as we discuss in the following sections.

Alienation, Deskilling, and Upskilling

As technology continues to play an important role in the workplace, workers may feel that there is no creativity in what they do—they feel alienated (see Chapter 7). For example, a study of California's high-tech "white-collar factories" found that employees were suffering from isolation, job insecurity, and pressure to update skills (BBC 2003). The movement from mechanization to automation to cybernation increasingly removes individuals from the production process, often relegating them to flipping a switch, staring at a computer monitor, or entering data at a keyboard. Many low-paid employees, often women, sit at computer terminals for hours entering data and keeping records for thousands of businesses, corporations, and agencies. The work that takes place in these "electronic sweatshops" is monotonous, solitary, and almost completely devoid of autonomy for the worker.

Not only are these activities routine, boring, and meaningless, but they also promote **deskilling**—that is, "labor requires less thought than before and gives the workers fewer decisions to make" (Perrolle 1990, p. 338). Deskilling stifles

" For a list of all the ways technology has failed to improve the quality of life, please press 3. **"**

Alice Kahn
Humorist

deskilling The tendency for workers in a postindustrial society to make fewer decisions and for labor to require less thought.

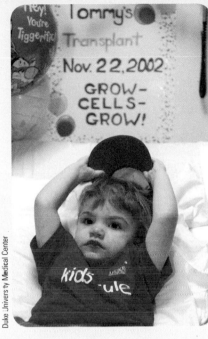

Tommy Bennett's big brown eyes and sweet demeanor make it that much harder to accept his plight. Just 3 years old, he blithely endures the constant barrage of drugs, needles, and tests as though he instinctively knows that they are destined to cure him. The development of new technologies has produced new forms of work and new demands for highly skilled workers in certain segments of the labor market.

Born with a rare, degenerative disease called Sanfilippo syndrome, Tommy lacks a critical enzyme needed for proper organ and brain development. Without enzymes, Tommy will die by adolescence. With the enzymes, Tommy's brain may unlock the potential to allow him to talk, dress, and care for himself. Such skills have eluded his two affected siblings, 4-year-old Hunter and 6-year-old Ciara. Ciara had just been diagnosed with Sanfilippo syndrome when their mom became pregnant with Tommy.

Since that time, they have searched desperately for someone willing to take a chance on helping Ciara, Hunter, and Tommy. The Bennetts found hope at Duke University Medical Center, the only program in the country willing to apply the benefits of stem cells—derived from newborn babies' umbilical cords—to treat this disease.

Proof of a Sanfilippo cure remains elusive, and Tommy is only the sixth Sanfilippo patient ever to have received a stem cell transplant. Yet if the transplant is to help, Tommy is a good candidate. He is young enough that the disease has only just begun to wreak havoc on his brain and organs. His siblings have progressed too far to be helped. Still, the sting of disappointment was palpable when doctors deemed Tommy the only viable candidate.

Thankful as they are for the opportunity, the Bennetts have embarked on a costly gamble—financially, emotionally, and physically. The Bennetts uprooted their kids and moved 900 miles away from family and friends to undergo a series of grueling tests before transplant could begin.

Then came the real test of endurance. Confined to the hospital unit for four straight weeks, Tommy's small body was ravaged by toxic doses of chemotherapy designed to wipe out his immune system and make way for a new one that might provide the crucial enzyme.

Alicia took on hospital duty, caring for Tommy night and day, and catching a few winks of sleep as time permitted on a pull-down cot. John assumed full-time care for Ciara and Hunter at a rented apartment nearby, no easy task given Ciara's penchant for 3 a.m. awakenings. The process is clearly daunting, yet the transplant itself is deceptively simple. It takes just fifteen minutes for a bag of red liquid to drip intravenously into a child's bloodstream. Nurses literally squeeze every last drop from the bag, lest they lose a single stem cell that floats amidst a billion blood and supporting cells.

Every parent knows that stem cells hold the key to their child's survival. If they grow, the child has a fighting chance to live. If they do not, the child has probably exhausted his or her last resort at a cure.

Then comes the waiting, and the familiar refrain: "Grow, Cells, Grow." The words resonate within the halls, grace the walls of every room, and are sprinkled throughout cards of love and hope. Parents recite them like a battle cry designed to incite soldiers in action.

Indeed, stem cells are like tiny soldiers who descend upon bone marrow and rescue it from near-certain demise. So powerful are stem cells that it takes only ten to a hundred of them to restore a child's entire blood-forming and immune system—in Tommy's case providing the missing enzymes. Moreover, they know exactly where to go and what function to perform.

Yet such remarkable power is not without its drawbacks. Stem cells can attack the last remnants of the child's immune system, a complication called graft-versus-host disease. Stem cells take time to grow and mature, leaving the child's developing immune system vulnerable to minor infections that could prove deadly.

Children also suffer mightily from the dangerously high levels of chemotherapy needed to wipe out their immune system. Often, their mucous linings literally slough off from within, causing severe diarrhea and vomiting. Nausea, painful sores, fatigue, and stomach pains also plague the children as the chemo exerts its effects.

Luckily, Tommy endured far less of the usual symptoms of his transplant, but only time will tell if the new cells have become his own. A year must pass before his new immune system will be running at full force. A lot can happen in that time, but hope, prayer, and a will to overcome will be on their side.

Sadly, Tommy Bennett died on November 24, 2003.

Source: Reprinted from *Phi Kappa Phi Forum* 83(1) (Winter 2003). Copyright by Duke University Medical Center. Reprinted by permission of the publisher.

FIGURE 15.3
Percent saying new
technologies will improve
way of life/have positive
impact, by location, 2005.
*No data available for Europe.
Source: National Science Foundation
(2007b).

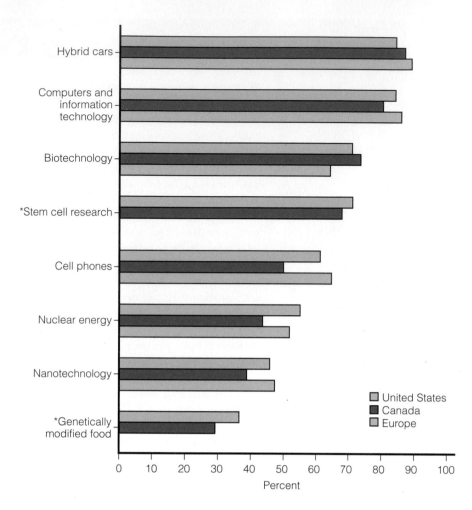

development of alternative skills and limits opportunities for advancement and creativity as old skill sets become obsolete. Conflict theorists believe that deskilling also provides the basis for increased inequality, for "throughout the world, those who control the means of producing information products are also able to determine the social organization of the 'mental labor' which produces them" (Perrolle 1990, p. 337).

Technology in some work environments, however, may lead to **upskilling.** Unlike deskilling, upskilling reduces alienation because employees find their work more rather than less meaningful and have greater decision-making powers as information becomes decentralized. Futurists argue that upskilling in the work-place could lead to a "horizontal" work environment in which "employees do not so much what they are told to do, but what their expansive knowledge of the entire enterprise suggests to them needs doing" (Global Internet Project 1998).

Social Relationships and Social Interaction

Technology affects social relationships and the nature of social interaction. The development of telephones has led to fewer visits with friends and relatives; with the advent of DVRs and cable television, the number of places where social

upskilling The opposite of deskilling; upskilling reduces employee alienation and increases decision-making powers.

life occurs (e.g., movie theaters) has declined. Even the nature of dating has changed as computer networks facilitate instant messaging, cyberdates, and "private" chat rooms. As technology increases, social relationships and human interaction are transformed.

Technology also makes it easier for individuals to live in a cocoon—to be self-sufficient in terms of finances (e.g., Quicken), entertainment (e.g., DVR and YouTube), work (e.g., telecommuting), recreation (e.g., virtual reality), shopping (e.g., eBay), communication (e.g., e-mail and instant messaging), and many other aspects of social life. For example, MySpace, the most popular social networking site on the Internet, brings together 85 million users, most of whom are high school or college students (Rawe 2006). Members cruise the web pages of other users looking for people to meet in a process called "friending." Even "seeing" your doctor may go online. Blue Shield of California is now paying doctors $25 for each e-mail consultation. Contrary to alienating patients, many respond that they feel closer to their doctors, given the often conversational tone of online communication (Freudenheim 2005).

Although technology can bring people together, it can also isolate them from each other (Klotz 2004). For example, children who use a home computer "spend much less time on sports and outdoor activities than non-computer users" (Attewell, Suazo-Garcia, & Battle 2003, p. 277). In addition, a study of more than 1,500 U.S. Internet users between the ages of 18 and 64 found that for every hour a respondent was on the Internet there was a corresponding 23.5-minute reduction in face-to-face interaction with family members (Nie et al. 2004). Some technological innovations replace social roles—an answering machine may replace a secretary, a computer-operated vending machine may replace a waitperson, an automatic teller machine may replace a banker, and closed circuit television may replace a teacher. Parking attendants may even become obsolete as New York City opens its first robotic parking garage (Svensson 2007). These technologies may improve efficiency, but they also reduce necessary human contact.

What Do You Think? Facebook, MySpace, Xanga, and Bebo are just some of the Internet sites available online for cyber-socializing. With their dramatic growth rates (MySpace adds 250,000 accounts a day) there is concern that such sites are potentially dangerous (Rawe 2006). Despite prohibitions about posting last names, addresses, or telephone numbers (see Figure 15.4), in a recent survey of online teens, 63 percent said that they "believe that a motivated person could eventually identify them from the information they publicly provide on their profiles" (Lenhart and Madden 2007, p. 1). Further, teens who use online social networks are more likely to be contacted by a stranger than teens who do not. Do you have a profile online? Is it available to all Internet users or is it restricted access? Have you ever felt threatened by any one who has contacted you? What are some of the benefits and negative aspects of such sites?

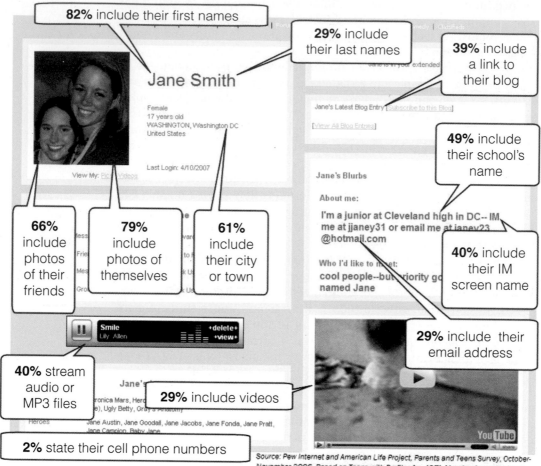

FIGURE 15.4
What teens share and restrict in an online environment, 2006.
Source: Lenhart, Amanda, and Mary Madden. 2007. *Teens, Privacy and Online Social Networks.* Washington, DC: Pew Internet and American Life Project.

Loss of Privacy and Security

Schools, employers, and the government are increasingly using technology to monitor individuals' performance and behavior. A 2005 study reported that 36 percent of companies use "keystroke monitoring so they can read what people type as well as track how much time they spend at the computer" and "55 percent retain and review e-mail messages" (MacMillan 2005, p. 1). Today, the legality of monitoring e-mails is under scrutiny. One in 20 companies has been sued for e-mail–related surveillance, and a federal appeals court recently ruled that e-mails, like letters and phone conversations, are protected under the Fourth Amendment's prohibition against unreasonable search and seizure (Holding 2007). In addition to e-mail monitoring, high-tech machines monitor countless other behaviors, including counting a telephone operator's minutes online, videotaping a citizen walking down a city street, or tracking the whereabouts of a student or faculty member on campus. One group of privacy advocates, using the Freedom of Information Act, has filed for "technical information about a network of video cameras that has been established in their city" (Markoff 2002).

Employers and schools may subject individuals to drug-testing technology (see Chapter 2) and, in 2006, identity theft was the number one complaint filed with the Federal Trade Commission for the seventh year in a row (Bosworth 2007) (see Chapter 4). Through computers individuals can obtain access to someone's phone bills, tax returns, medical reports, credit histories, bank account balances, and driving records. In 2007 alone (Associated Press 2007b; Apuzzo 2007; Jewell 2007; Mearian 2007; McMillan 2007):

- a contractor lost data tapes containing sensitive information about a large number of former IBM employees
- the names and social security numbers of 64,000 Ohio employees were stolen when an intern left the backup storage device in his car
- Transportation Security Administration (a division of the Department of Homeland Security) officials misplaced a computer hard drive that contained the names, social security numbers, bank data, and payroll information of 100,000 employees
- hackers stole data from 45.7 million credit and debit cards used at such discount stores as T J Maxx and Marshalls
- the Georgia Department of Community Health reports that a compact disc containing the names, address, birth dates, and social security numbers of 2.9 million Medicaid recipients was lost while being transported

Although just inconvenient for some, unauthorized disclosure is potentially devastating for others. If a person's medical records indicate that he or she is human immunodeficiency virus (HIV)-positive, for example, that person could be in danger of losing his or her job or health benefits. If DNA testing of hair, blood, or skin samples reveals a condition that could make the person a liability in the insurer's or employer's opinion, the individual could be denied insurance benefits, medical care, or even employment. In response to the possibility of such consequences, Brin (1998), author of *The Transparent Society,* argues that, because it is impossible to prevent such intrusions, "reciprocal transparency," or complete openness, should prevail. If organizations can collect the information, then citizens should have access to it and to its uses.

Our privacy is also disturbed by the intrusion in our e-mail in-boxes of unwanted mail called *spam.* Between 2006 and 2007 the amount of junk mail sent on the Internet is estimated to have increased by 63 percent (Evett 2007). Motivated in part by the "promise of an easy profit, spammers have gone from pests to an invasive species of parasite that threatens to clog the inner workings of the Internet" (Taylor 2003). Recently, spammers have targeted cell phones, sending thousands of unwanted text messages to unsuspecting subscribers. For the average Internet user, about 5 minutes out of every hour online is spent dealing with spam (Nie et al. 2004).

Technology has created threats not only to the privacy of individuals but also to the security of entire nations. Computers can be used (or misused) in terrorism and warfare to cripple the infrastructure of a society and to tamper with military information and communication operations (see Chapter 16). In a 2004 assessment of efforts made to address computer security standards, the Department of Homeland Security and seven other federal agencies received a failing grade for cybersecurity from a congressional oversight committee (Krebs 2005).

Customer service representatives answer queries at a call center in India. India's outsourcing industry employs more than 1.2 million workers—a number that is likely to grow as multinational corporations seek qualified personnel outside the United States at a lower cost.

© Jagadeesh/Reuters/Corbis

Unemployment and Outsourcing

Some technologies replace human workers—robots replace factory workers, word processors displace secretaries and typists, and computer-assisted diagnostics reduce the need for automobile mechanics. Unemployment rates can also increase when companies **outsource** jobs to lower-wage countries.

> The globalization of work tends to start from the bottom up. The first jobs to be moved abroad are typically simple assembly tasks, followed by manufacturing, and later skilled work like computer programming. At the end of this progression is the work done by scientists and engineers in research and development laboratories. (Lohr 2006, p. 1)

For example, officials at Cisco Systems, the largest maker of information and communication equipment, have announced that over the next 5 years 20 percent of its "top talent" will be in India; Dow Chemical is building a new research center in Shanghai that will employ 600 technical workers; pharmaceutical company Eli Lilly is paying Indian scientists $500,000 to $1.5 million a year per scientist to develop new drugs for commercial use; and, since 1992, IBM has reduced its American labor force by 31,000 even as its Indian work force increased from 0 to 52,000 (Giridharadas 2007). As author Thomas Friedman explained in his book *The World Is Flat* (2005), ". . . it is now possible for more people than ever to collaborate and compete in real time with more other people on more different kinds of work from more different corners of the planet on a more equal footing than at any previous time in the history of the world . . ." (p. 8).

What will the impact of outsourcing be? Alan Blinder, former economic advisor to President Clinton and former vice chairman of the Federal Reserve, estimates that outsourcing "poses a risk to the employment of as many as 28 million to 42 million workers in the United States" (Giridharadas 2007, p. 3). Not surprisingly, unemployment rates for information technology workers are fairly high, exceeding the overall jobless rate in 2005 (U.S. Census Bureau 2007). When layoffs occur, older IT workers are more likely to remain unemployed, being less likely than younger workers to take a job outside the field. Downsizing of corpo-

outsource The practice of sending jobs to low-wage countries such as China and India.

rations, business failures, less need for technological support, and an economic recession in addition to outsourcing have contributed to disappearing high-tech jobs (McNair 2003; Schneider 2004; Friedman 2005).

Technology also changes the nature of work and the types of jobs available. For example, fewer semiskilled workers are needed because many of their jobs are now being done by machines. Of those jobs projected to decline the fastest by 2014, a substantial majority are in manufacturing—cut-and-sew apparel manufacturers, footwear manufacturing, tobacco manufacturing, and so forth (U.S. Bureau of the Census 2007). The jobs that remain, often white-collar jobs, require more education and technological skills. Technology thereby contributes to the split labor market as the pay gulf between skilled and unskilled workers continues to grow.

© David Sams/Stock Boston

The development of new technology has produced new forms of work and new demands for highly skilled workers in certain segments of the labor market.

The Digital Divide

One of the most significant social problems associated with science and technology is the increased division between the classes. As Welter (1997) noted,

> It is a fundamental truth that people who ultimately gain access to, and who can manipulate, the prevalent technology are enfranchised and flourish. Those individuals (or cultures) that are denied access to the new technologies, or cannot master and pass them on to the largest number of their offspring, suffer and perish. (p. 2)

The fear that technology will produce a "virtual elite" is not uncommon. Several theorists hypothesize that as technology displaces workers, most notably the unskilled and uneducated, certain classes of people will be irreparably disadvantaged—the poor, minorities, and women. There is even concern that biotechnologies will lead to a "genetic stratification," whereby genetic screening, gene therapy, and other types of genetic enhancements are available only to the rich.

The wealthier the family, for example, the more likely the family is to have a computer. Of American families with an annual income of $150,000 or more, 94.7 percent have at least one computer in the household (see Table 15.2). However, only 26.9 percent of households with annual incomes between $5,000 and $9,999 own a computer (U.S. Bureau of the Census 2007). Furthermore, 74.6 percent of white Americans own a computer, compared with 50.6 percent of African Americans and 48.7 percent of Hispanic Americans. Whites are also twice as likely as blacks or Hispanics to have broadband at home (Fairlie 2005).

Racial and ethnic disparities also exist in Internet access. Although 67.3 percent of whites used the Internet from home, only 40.5 percent of blacks and 38.1 percent of Hispanics have home Internet access (Fairlie 2005). Inner-city neighborhoods are disproportionately populated by racial and ethnic minorities and are simply less likely to be "wired," that is, to have the telecommunications hardware necessary for access to online services. In fact, cable and telephone companies are less likely to lay fiberoptic cables in these areas—a practice called "information apartheid" or "electronic redlining." Urban-rural differences also exist. For example, when survey respondents were asked why they did not have high-speed Internet access, 22.1 percent of the respondents who live in rural areas and 4.7 percent of those who live in urban areas replied that broadband service was not available (ANOL 2004).

The cost of equalizing such differences is enormous, but the cost of not equalizing them may be even greater. Employees who are technologically skilled have higher incomes than those who are not—up to 15 percent higher (Hancock 1995; International Labour Organization 2001). Further, technological disparities exacerbate the structural inequities perpetuated by the split labor force and the existence of primary and secondary labor markets.

Mental and Physical Health

Some new technologies have unknown risks. Biotechnology, for example, has promised and, to some extent, has delivered everything from life-saving drugs to hardier pest-free tomatoes. Limbs are being replaced by bionic devices controlled by the recipient's thoughts (Brown 2006). However, biotechnologies have also created **technology-induced diseases,** such as those experienced by Chellis Glendinning (1990). Glendinning, after using the pill and, later, the Dalkon Shield IUD, became seriously ill.

> Despite my efforts to get help, medical professionals did not seem to know the root of my condition lay in immune dysfunction caused by ingesting artificial hormones and worsened by chronic inflammation. In all, my life was disrupted by illness for twenty years, including six years spent in bed. . . . For most of the years of illness, I lived in isolation with my problem. Doctors and manufacturers of birth control technologies never acknowledged it or its sources. (p. 15)

Other technologies that pose a clear risk to a large number of people include nuclear power plants, the pesticide DDT, automobiles, X-rays, food coloring, and breast implants.

The production of new technologies may also place manufacturing employees in jeopardy. For example, the electronics industry uses thousands of hazardous chemicals:

> The semiconductor industry uses large amounts of toxic chemicals to manufacture the components that make up a computer, including disk drives, circuit boards, video display equipment, and silicon chips themselves, the basic building blocks of computer devices. The toxic materials needed to make the 220 billion silicon chips manufactured annually are staggering in amount and include highly corrosive hydrochloric acid; metals such as arsenic, cadmium, and lead; volatile solvents such as methyl chloroform, toluene, benzene, acetone, and trichloroethylene; and toxic gases such as arsine. Many of these chemicals are known or probable human carcinogens. (Chepesiuk 1999, p. 1)

Technological innovations are, for many, a cause of anguish, stress, and fear, particularly when the technological changes are far-reaching. About one-third of Americans believe that "science is going too far and hurting society" and 31 percent agree with the statement that "technology is making life too complicated for me" (Pew 2007c). Moreover, nearly 60 percent of workers report being "technophobes," that is, fearful of technology, and as many as 10 percent of Internet users are "addicted" to being online (Boles & Sunoo 1998; Papadakis 2000). Researchers at the University of Florida have developed a list of symptoms associated with Internet addiction, as signified by the acronym MOUSE: "**M**ore than intended time spent online; **O**ther responsibilities neglected; **U**nsuccessful attempts to cut down; **S**ignificant relationship discord; and **E**xcessive thoughts or anxieties when not online" (Deutsche Welle 2003, p. 3).

technology-induced diseases
Diseases that result from the use of technological devices, products, and/or chemicals.

What Do You Think? Some evidence suggests that technological gadgetry and the multitasking it engenders is taking a toll "on our ability to think clearly, work effectively, and function as healthy human beings" (Wallis and Steptoe 2006, p. 74). Psychiatrist Edward Hallowell, an expert in attention deficit disorder, has named this syndrome ADT—attention deficit trait. Dr. Hallowell explained that ADT takes "hold when we get so overloaded with incoming messages and competing tasks that we are unable to prioritize" (as quoted in Wallis and Steptoe 2006, p. 74). In the end we feel distracted, impulsive, guilty, and inadequate as we slip farther and farther behind in our work and in our lives. Do you think ADT exists, and if so, do you suffer from it? How much time do you think you lose a day from unwanted technologically based interruptions, and how much time does it takes you to get back on task?

The Challenge to Traditional Values and Beliefs

Technological innovations and scientific discoveries often challenge traditionally held values and beliefs, in part because they enable people to achieve goals that were previously unobtainable. Before recent advances in reproductive technology, for example, women could not conceive and give birth after menopause. Technology that allows postmenopausal women to give birth challenges societal beliefs about childbearing and the role of older women. Macklin (1991) noted that the techniques of egg retrieval, in vitro fertilization, and gamete intrafallopian transfer make it possible for two different women to each make a biological contribution to the creation of a new life. Such technology requires society to reexamine its beliefs about what a family is and what a mother is. Should family be defined by custom, law, or the intentions of the parties involved?

Medical technologies that sustain life lead us to rethink the issue of when life should end. The increasing use of computers throughout society challenges the traditional value of privacy. New weapons systems make questionable the traditional idea of war as something that can be survived and even won. And cloning causes us to wonder about our traditional notions of family, parenthood, and individuality. Toffler (1970) coined the term **future shock** to describe the confusion resulting from rapid scientific and technological changes that unravel our traditional values and beliefs.

STRATEGIES FOR ACTION: CONTROLLING SCIENCE AND TECHNOLOGY

As technology increases, so does the need for social responsibility. Nuclear power, genetic engineering, cloning, and computer surveillance all increase the need for social responsibility: "Technological change has the effect of enhancing the importance of public decision making in society, because technology is continually creating new possibilities for social action as well as new problems that

future shock The state of confusion resulting from rapid scientific and technological changes that unravel our traditional values and beliefs.

have to be dealt with" (Mesthene 1993, p. 85). In the following sections we address various aspects of the public debate, including science, ethics, and the law, the role of corporate America, and government policy.

Science, Ethics, and the Law

Science and its resulting technologies alter the culture of society through the challenging of traditional values. Public debate and ethical controversies, however, have led to structural alterations in society as the legal system responds to calls for action. For example, several states now have what are called genetic exception laws. **Genetic exception laws** require that genetic information be handled separately from other medical information, leading to what is sometimes called patient shadow files (Legay 2001). The logic of such laws rests with the potentially devastating effects of genetic information being revealed to insurance companies, other family members, employers, and the like. Presently, 17 states "require informed consent for a third party either to perform or require a genetic test or to obtain genetic information" and 27 states require informed consent to disclose genetic information (NCSL 2007).

Are such regulations necessary? In a society characterized by rapid technological and thus social change—a society in which custody of frozen embryos is part of the divorce agreement—many would say yes. Cloning, for example, is one of the most hotly debated technologies in recent years. Bioethicists and the public vehemently debate the various costs and benefits of this scientific technique. Despite such controversy, however, the chairman of the National Bioethics Advisory Commission warned nearly 10 years ago that human cloning will be "very difficult to stop" (McFarling 1998). At present, 15 states have laws pertaining to human cloning, with some prohibiting cloning for reproductive purposes, some prohibiting therapeutic cloning, and still others prohibiting both (NCSL 2006).

Should the choices that we make, as a society, be dependent on what we can do or what we should do? Whereas scientists and the agencies and corporations who fund them often determine what we *can* do, who should determine what we *should* do? Although such decisions are likely to have a strong legal component—that is, they must be consistent with the rule of law and the constitutional right of scientific inquiry—legality or the lack thereof often fails to answer the question, What should be done? *Roe v. Wade* (1973) did little to squash the public debate over abortion and, more specifically, the question of when life begins. Thus it is likely that the issues surrounding the most controversial of technologies will continue into the 21st century with no easy answers.

Technology and Corporate America

As philosopher Jean-Francois Lyotard noted, knowledge is increasingly produced to be sold. The development of genetically altered crops, the commodification of women as egg donors, and the harvesting of regenerated organ tissues are all examples of potentially market-driven technologies. Like the corporate pursuit of computer technology, profit-motivated biotechnology creates several concerns.

First is the concern that only the rich will have access to life-saving technologies such as genetic screening and cloned organs. Such fears are justified. Companies with obscure names such as Progenitor International Research,

> **"**Prohibiting scientific and medical activities would also raise troubling enforcement issues. . . . Would the FBI raid research laboratories and universities? Seize and read the private medical records of infertility patients? Burst into operating rooms with their guns drawn? Grill new mothers about how their babies were conceived?**"**
>
> Mark Eibert
> Attorney

genetic exception laws Laws that require that genetic information be handled separately from other medical information.

Millennium Pharmaceuticals, Darwin Molecular, and Myriad Genetics have been patenting human life. Millennium Pharmaceuticals holds the patent on the melanoma gene and the obesity gene, Darwin Molecular controls the premature aging gene, deCODE controls the type 2 diabetes gene, Progenitor controls the gene for schizophrenia, and Myriad Genetics has nine patents on the breast and ovarian cancer genes (Shand 1998; Mayer 2002; Lemonick 2006b).

These patents result in **gene monopolies,** which have led to astronomical patient costs for genetic screening and treatment. At present, more than 20 percent of human genes are privately owned (Crichton 2007). One company's corporate literature candidly states that its patent of the breast cancer gene will limit competition and lead to huge profits (Shand 1998). They were right. Today the cost of a test for the breast cancer gene is $3,000. The biotechnology industry argues that such patents are the only way to recoup research costs that, in turn, lead to further innovations.

The commercialization of technology causes several other concerns, including issues of quality control and the tendency for discoveries to remain closely guarded secrets rather than collaborative efforts (Rabino 1998; Lemonick & Thompson 1999; Mayer 2002; Crichton 2007). In addition, industry involvement has made government control more difficult because researchers depend less and less on federal funding. More than 68 percent of research and development in the United States is supported by private industry using their own company funds (National Science Foundation 2007a).

Finally, although there is little doubt that profit acts as a catalyst for some scientific discoveries, other less commercially profitable but equally important projects may be ignored. As biologist Isaac Rabino states, "Imagine if early chemists had thrown their energies into developing profitable household products before the periodic table was discovered" (Rabino 1998, p. 112).

Runaway Science and Government Policy

Science and technology raise many public policy issues. Policy decisions, for example, address concerns about the safety of nuclear power plants, the privacy of electronic mail, the hazards of chemical warfare, and the ethics of cloning. In creating science and technology, have we created a monster that has begun to control us rather than the reverse? What controls, if any, should be placed on science and technology? And are such controls consistent with existing law? Consider the use of Kazaa to download music files (the question of intellectual property rights and copyright infringement); a Utah law limiting children's access to material on the Internet (free speech issues); and Carnivore (DCS 1000), an FBI surveillance program that can search every message that passes through an Internet service provider (Fourth Amendment privacy issues) (ACLU 2005; Kaplan 2000; Schaefer 2001; White 2000).

The government, often through regulatory agencies and departments, prohibits the use of some technologies (e.g., assisted-suicide devices) and requires others (e.g., seat belts). For example, in 2007 a bill was introduced in the U.S. House of Representatives that would (1) make human cloning illegal, (2) make knowingly shipping or receiving the product of a cloned egg cell illegal, (3) set a criminal penalty of fines, imprisonment of up to 10 years, or both, and (4) set a civil penalty of $1 million or two times the amount of monetary gain made from the cloning, whichever is greater (Human Cloning Prohibition Act of 2007).

66 As president, I will prohibit genetic discrimination, criminalize identity theft, and guarantee the privacy of medical and sensitive financial records. 99

President George W. Bush

gene monopoly Exclusive control over a particular gene as a result of government patents.

The federal government has also instituted several initiatives dealing with technology-related crime. A 2007 bill introduced in the U.S. Senate would amend the federal criminal code to make "phishing" illegal—that is, it would criminalize the practice of fraudulently obtaining personal information [Internet Spyware (I-SPY) Prevention Act of 2007]. As a government report noted, "Many of the attributes of this Internet technology—low cost, ease of use, and anonymous nature, among others make it an attractive medium for fraudulent scams, child sexual exploitation, and . . . cyberstalking" (U.S. Department of Justice 1999, p. 1).

Of late, the issue of online pornography has come to the forefront. A Federal Bureau of Investigation (FBI) report states that "computer telecommunications have become one of the most prevalent techniques used by pedophiles to share illegal photographic images of minors and to lure children into illicit sexual relationships" (FBI 2002, p. 1). For example, it is estimated that there are more than 100,000 websites that offer child pornography, an industry that generates over $4.9 billion a year (Internet Pornography Statistics 2007). A second concern is the ease with which children can access and view online pornography. A University of New Hampshire study found that 42 percent of 10- to 17-year-olds reported being exposed to pornography while online—most often accidentally as they "surf" the web (CBS/Associated Press 2007).

In response to such availability, the federal government enacted the Children's Internet Protection Act, which requires "public schools and libraries to install Internet filters on their computers so children and adults cannot view 'inappropriate' information on the Internet" (MacMillan 2004). Americans support such initiatives. In a 2005 survey of likely American voters, 71 percent responded that Congress needs to pass new laws designed to keep the Internet safe (CNN 2005).

Finally, the government has several science and technology boards, including the National Science and Technology Council, the Office of Science and Technology Policy, and the President's Council of Advisors on Science and Technology. These agencies advise the president on matters of science and technology, including research and development, implementation, national policy, and coordination of different initiatives.

What Do You Think? Craigslist is an online network of communities featuring classified advertisements from more than 190 U.S. and foreign cities (Cole 2006). Many of the ads, however, could never appear in a local paper because of legal restrictions. For example, one Craigslist housing ad in the Chicago area read, "African Americans and Arabians tend to clash with me so that won't work out" (quoted in Cole 2006, p. 8). The question is whether or not Internet sites are *publishing* information or *distributing* information. Whereas the former is subject to the antidiscrimination ban of the Fair Housing Act, the latter may not be. What do you think? Should Internet content be subject to the same federal regulations newspapers are held to?

UNDERSTANDING SCIENCE AND TECHNOLOGY

What are we to understand about science and technology from this chapter? As structural functionalists argue, science and technology evolve as a social process and are a natural part of the evolution of society. As society's needs change, scientific discoveries and technological innovations emerge to meet these needs, thereby serving the functions of the whole. Consistent with conflict theory, however, science and technology also meet the needs of select groups and are characterized by political components. As Winner (1993) noted, the structure of science and technology conveys political messages, including "power is centralized," "there are barriers between social classes," "the world is hierarchically structured," and "the good things are distributed unequally" (Winner 1993, p. 288).

The scientific discoveries and technological innovations that are embraced by society as truth itself are socially determined. Research indicates that science and the resulting technologies have both negative and positive consequences—a **technological dualism.** Technology saves lives and time and money; it also leads to death, unemployment, alienation, and estrangement. Weighing the costs and benefits of technology poses ethical dilemmas, as does science itself. Ethics, however, "is not only concerned with individual choices and acts. It is also and, perhaps, above all concerned with the cultural shifts and trends of which acts are but the symptoms" (McCormick & Richard 1994, p. 16).

Thus society makes a choice by the very direction it follows. These choices should be made on the basis of guiding principles that are both fair and just, such as those listed here (Goodman 1993; Winner 1993; Eibert 1998; Buchanan et al. 2000; Murphie & Potts 2003):

1. Science and technology should be prudent. Adequate testing, safeguards, and impact studies are essential. Impact assessment should include an evaluation of the social, political, environmental, and economic factors.

2. No technology should be developed unless all groups, and particularly those who will be most affected by the technology, have at least some representation "at a very early stage in defining what that technology will be" (Winner 1993, p. 291). Traditionally, the structure of the scientific process and the development of technologies have been centralized (i.e., decisions have been made by a few scientists and engineers); decentralization of the process would increase representation.

3. Means should not exist without ends. Each new innovation should be directed to fulfilling a societal need rather than the more typical pattern in which a technology is developed first (e.g., high-definition television) and then a market is created (e.g., "You'll never watch a regular TV again!"). Indeed, from the space program to research on artificial intelligence the vested interests of scientists and engineers, whose discoveries and innovations build careers, should be tempered by the demands of society.

What the 21st century will hold, as the technological transformation continues, may be beyond the imagination of most of society's members. Technology empowers; it increases efficiency and productivity, extends life, controls the environment, and expands individual capabilities. According to a National Intelligence Council report, "Life in 2015 will be revolutionized by the growing

> **"**One day soon the Gillette company will announce the development of a razor that, thanks to a computer chip, can actually travel ahead in time and shave beard hairs that don't even exist yet.**"**
>
> **Dave Barry**
> Writer and humorist

technological dualism The tendency for technology to have both positive and negative consequences.

effort of multi-disciplinary technology across all dimensions of life: social, economic, political, and personal" (NIC 2003, p. 1).

As we proceed into the first computational millennium, one of the great concerns of civilization will be the attempt to reorder society, culture, and government in a manner that exploits the digital bonanza yet prevents it from running roughshod over the checks and balances so delicately constructed in those simpler precomputer years.

CHAPTER REVIEW

- **What are the three types of technology?**

  The three types of technology, escalating in sophistication, are mechanization, automation, and cybernation. Mechanization is the use of tools to accomplish tasks previously done by hand. Automation involves the use of self-operating machines, and cybernation is the use of machines to control machines.

- **What are some of the reasons the United States may be "losing its edge" in scientific and technological innovations?**

  According to Lemonick (2006a), the decline of U.S. supremacy in science and technology is likely to be the result of three forces. First, the federal government has been scaling back its investment in research and development; second, corporations have begun to focus on short-term products and higher profits; and third, there has been a drop in science and math education in U.S. schools both in terms of quality and quantity.

- **What are some Internet global trends?**

  Globally, English speakers are the largest language group online, but non-English speakers constitute the fastest growing group on the Internet. The clear majority of Internet users live in industrialized countries, although there is some movement toward the Internet's becoming truly global as those in developing countries "get online."

- **According to Kuhn, what is the scientific process?**

  Kuhn describes the process of scientific discovery as occurring in three steps. First are assumptions about a particular phenomenon. Next, because unanswered questions always remain about a topic, science works to start filling in the gaps. Then, when new information suggests that the initial assumptions were incorrect, a new set of assumptions or framework emerges to replace the old one. It then becomes the dominant belief or paradigm until it is questioned and the process repeats.

- **What is meant by the computer revolution?**

  The silicon chip made computers affordable. Today, 62 percent of American homes have a computer. In addition, over half the labor force uses a computer at work. The most common computer activity at work is accessing the Internet or e-mail, followed by word processing, working with spreadsheets or databases, and accessing or updating calendars or schedules.

- **What is the Human Genome Project?**

  The U.S. Human Genome Project is an effort to decode human DNA. The 13-year-old project is now complete, allowing scientists to "transform medicine" through early diagnosis and treatment as well as possibly preventing disease through gene therapy. Gene therapy entails identifying a defective or missing gene and then replacing it with a healthy duplicate that is transplanted to the affected area.

- **How are some of the problems of the Industrial Revolution similar to the problems of the technological revolution?**

  The most obvious example is in unemployment. Just as the Industrial Revolution replaced many jobs with technological innovations, so too has the technological revolution. Furthermore, research indicates that many of the jobs created by the Industrial Revolution, such as working on a factory assembly line, were characterized by high rates of alienation. Rising rates of alienation are also a consequence of increased estrangement as high-tech employees work in "white-collar factories."

- **What is meant by outsourcing and why is it important?**

  Outsourcing refers to sending jobs overseas to low wage countries such as India and China. The problem with outsourcing is that it tends to lead to higher rates of unemployment in the export countries. It is estimated that the United States will lose between 28 and 42 million jobs over the coming years due to outsourcing.

- **What is meant by the acronym MOUSE?**

  MOUSE is a reference to Internet addiction and refers to a list of symptoms, as indicated by each letter: "**M**ore than intended time spent online; **O**ther responsibilities neglected; **U**nsuccessful attempts to cut down; **S**ignificant relationship discord; and **E**xcessive thoughts or anxieties when not online" (Deutsche Welle 2003, p. 3).

- **What is the digital divide?**

  The digital divide is the tendency for technology to be most accessible to the wealthiest and most educated. For example, some fear that there will be "genetic stratification," whereby the benefits of genetic screening, gene therapy, and other genetic enhancements will be available to only the richest segments of society.

- **What is meant by the commercialization of technology?**

  The commercialization of technology refers to profit-motivated technological innovations. Whether it be the isolation of a particular gene, genetically modified organisms, or the regeneration of organ tissues, where there is a possibility for profit, private enterprise will be there.

TEST YOURSELF

1. Which of the following technologies is associated with industrialization?
   a. Mechanization
   b. Cybernation
   c. Hibernation
   d. Automation
2. The most common online language of Internet users is Chinese.
   a. True
   b. False
3. The U.S. government, as part of the technological revolution, has recently increased spending on research and development.
   a True
   b. False
4. Which theory argues that technology is often used as a means of social control?
   a. Structural functionalism
   b. Social disorganization
   c. Conflict theory
   d. Symbolic interactionism
5. The most common computer activity is accessing the Internet.
   a. True
   b. False
6. The ability to manipulate the genes of an organism to alter the natural outcome is called
   a. gene therapy
   b. gene splicing
   c. genetic engineering
   d. genetic screening
7. Genetically modified foods have been documented as harmless to humans by the Food and Drug Administration.
   a. True
   b. False
8. In 2007, the U.S. Supreme Court upheld the Partial Birth Abortion Ban in a 5–4 decision.
   a. True
   b. False
9. The practice of outsourcing entails
   a. creating high-tech jobs in the United States for immigrants
   b. hiring temporary workers to cover for employees who are absent
   c. allowing company workers to work from home
   d. exportation of jobs to low wage countries
10. The Human Cloning Prohibition Act of 2007
    a. makes human cloning illegal in the United States
    b. establishes a criminal penalty for human cloning in the United States
    c. establishes a civil penalty for human cloning in the United States
    d. all of the above

Answers: 1 d. 2 b. 3 b. 4 c. 5 a. 6 c. 7 b. 8 a. 9 d. 10 d.

KEY TERMS

abortion	genetic exception laws	stem cells
automation	genetic screening	technological dualism
cultural lag	in vitro fertilization (IVF)	technological fix
cybernation	Internet	technology
deskilling	mechanization	technology-induced diseases
e-commerce	outsource	teleworking
future shock	partial birth abortion	therapeutic cloning
gene monopoly	postmodernism	upskilling
gene therapy	science	Web 2.0
genetic engineering		

MEDIA RESOURCES

Understanding Social Problems, **Sixth Edition**
Companion Website

academic.cengage.com/sociology/mooney
Visit your book companion website, where you will find flash cards, practice quizzes, Internet links, and more to help you study.

CENGAGENOW™
Just what you need to know NOW! Spend time on what you need to master rather than on information you already have learned. Take a pre-test for this chapter, and CengageNOW will generate a personalized study plan based on your results. The study plan will identify the topics you need to review and direct you to online resources to help you master those topics. You can then take a post-test to help you determine the concepts you have mastered and what you will need to work on. Try it out! Go to academic.cengage .com/login to sign in with an access code or to purchase access to this product.

"Every gun that is made, every warship launched, every rocket fired, signifies in the final sense a theft from those who hunger and are not fed, those who are cold and not clothed."

Dwight D. Eisenhower, former U.S. president and military leader

16

Conflict, War, and Terrorism

Marwan Abu Ubeida was born in Fallujah, Iraq, where he attended high school and enjoyed all the advantages that having a successful father can bring (Ghosh 2005). Although a Sunni Muslim—as was Saddam Hussein—Marwan decided not to join the resistance as so many others had. Surely, after the fall of the Ba'athist regime and the capture of Saddam the Americans would leave his homeland. But they stayed and stayed, and after U.S. soldiers fired and killed 12 demonstrators and injured many more, Marwan, who had been at the demonstration, joined the Iraqi insurgency.

Now, at only age 20, Marwan is a seasoned soldier. He has experienced dozens of assaults against U.S. troops and is considered an expert with the Russian-made PKC machine gun, his weapon of choice. Marwan is a "jihadi foot soldier," a member of Abu Mousab al-Zarqawi's terrorist group, Al-Qaeda (Ghosh 2005, p. 24). But for the past several months Marwan has been training to carry out a suicide mission. "I can't wait," he says, recounting the day that his name was added to the list of volunteers as the happiest day of his life. When he is called to duty, he will spend his final days in seclusion, spiritually and psychologically preparing for his mission. Time permitting, he will call his family to say goodbye.

Marwan has practiced and practiced his final prayer. What is it he is praying for? "First I will ask Allah to bless my mission with a high rate of casualties among the Americans. . . . Then I will ask him to purify my soul so I am fit to see him, and I will ask to see my mujahedin brothers who are already with him. . . . The most important thing is that he should let me kill many Americans" (Ghosh 2005, pp. 23–24).

Religious fanatic or calculated strategy? A recent book by Robert Pape, titled *Dying to Win: The Strategic Logic of Suicide Bombers* (2005), suggests that most suicide bombers are, in fact, not Islamic fundamentalists but soldiers in a "coherent campaign" to rid their homeland of foreign military forces. Although religion may play a part, particularly in recruitment, only 40 percent of the 462 suicide attacks studied were religiously motivated. Pape concluded that suicide attacks, in Iraq and around the world, are on the rise because they are an effective means to accomplish a collective end.

Thus the use of suicide bombers can be seen as a direct result of the continued presence of United States–led troops in Iraq. The pivotal events leading to the war in Iraq and, more generally, the war on terrorism, include the bombing of the World Trade Center and the U.S. Pentagon on September 11, 2001. **War,** the most violent form of conflict, refers to organized armed violence aimed at a social group in pursuit of an objective. Wars have existed throughout human history and continue in the contemporary world.

War is one of the great paradoxes of human history. It both protects and annihilates. It creates and defends nations but also destroys them. Whether war is just or unjust, defensive or offensive, it involves the most horrendous atrocities known to humankind. In this chapter we focus on the causes and consequences of conflict, war, and terrorism. Along with population and environmental problems, conflict, war, and terrorism are among the most serious of all social problems in their threat to societies, the human race, and life on earth.

war Organized armed violence aimed at a social group in pursuit of an objective.

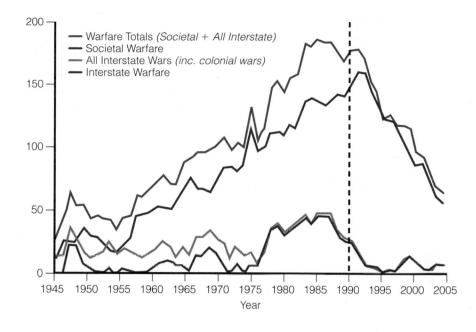

FIGURE 16.1
Global trends in violent conflict, 1946–2004.
Source: Marshall and Gurr (2005, p. 11).

Chart legend:
- Warfare Totals (*Societal + All Interstate*)
- Societal Warfare
- All Interstate Wars (*inc. colonial wars*)
- Interstate Warfare

Y-axis: 0, 50, 100, 150, 200
X-axis (Year): 1945, 1950, 1955, 1960, 1965, 1970, 1975, 1980, 1985, 1990, 1995, 2000, 2005

The Global Context: Conflict in a Changing World

As societies have evolved and changed throughout history, the nature of war has also changed. Before industrialization and the sophisticated technology that resulted, war occurred primarily between neighboring groups on a relatively small scale. In the modern world war can be waged between nations that are separated by thousands of miles as well as between neighboring nations. Increasingly, war is a phenomenon internal to states, involving fighting between the government and rebel groups or among rival contenders for state power. Indeed, Marshall and Gurr (2005) show in Figure 16.1 that today wars between states, that is, interstate wars, make up the smallest percentage of armed conflicts. In the following sections we examine how war has changed our social world and how our changing social world has affected the nature of war in the industrial and postindustrial information age.

> "The significance of wars is not just that they led to major changes during the period of hostilities and immediately after. They produced transformations which have turned out to be of enduring significance."
>
> Anthony Giddens
> Sociologist

War and Social Change

The very act that now threatens modern civilization—war—is largely responsible for creating the advanced civilization in which we live. Before large political states existed, people lived in small groups and villages. War broke the barriers of autonomy between local groups and permitted small villages to be incorporated into larger political units known as chiefdoms. Centuries of warfare between chiefdoms culminated in the development of the state. The **state** is "an apparatus of power, a set of institutions—the central government, the armed forces, the regulatory and police agencies—whose most important functions involve the use of force, the control of territory and the maintenance of internal order" (Porter 1994, pp. 5–6). The creation of the state in turn led to other profound social and cultural changes:

state The organization of the central government and government agencies such as the military, police, and regulatory agencies.

© Bob Mahoney/The Image Works

Industrialization can decrease a society's propensity for war, but it also increases the potential destructiveness of war because, with industrialization, warfare technology becomes more sophisticated and lethal.

And once the state emerged, the gates were flung open to enormous cultural advances, advances undreamed of during—and impossible under—a regimen of small autonomous villages. . . . Only in large political units, far removed in structure from the small autonomous communities from which they sprang, was it possible for great advances to be made in the arts and sciences, in economics and technology, and indeed in every field of culture central to the great industrial civilizations of the world. (Carneiro 1994, pp. 14–15)

Industrialization and technology could not have developed in the small social groups that existed before military action consolidated them into larger states. Thus war contributed indirectly to the industrialization and technological sophistication that characterize the modern world. Industrialization, in turn, has had two major influences on war. Cohen (1986) calculated the number of wars fought per decade in industrial and preindustrial nations and concluded that "as societies become more industrialized, their proneness to warfare decreases" (p. 265). Thus, for example, in 2006 there were 17 major armed conflicts, the majority of which were in less developed countries in Africa and Asia (SIPRI 2007a).

Although industrialization may decrease a society's propensity to war, it also increases the potential destruction of war. With industrialization, military technology became more sophisticated and more lethal. Rifles and cannons replaced the clubs, arrows, and swords used in more primitive warfare and, in turn, were replaced by tanks, bombers, and nuclear warheads. In the postindustrial information age, computer technology has revolutionized the nature of warfare and future warfare capabilities. Today, a "whole range of new technologies are offered for the next generations of weapons and military operations" (Bonn International Center for Conversion 1998, p. 3), including the use of high-performance sensors, information processors, directed energy technologies, precision-guided munitions, and computer worms and viruses (Bonn International Center for Conversion 1998; O'Prey 1995; PBS 2003).

The Economics of Military Spending

The increasing sophistication of military technology has commanded a large share of resources totaling, worldwide, $1.3 trillion in 2006 (Stalenheim, Perdomo, & Skons 2007). Global military spending has been increasing since 1998, with dramatic increases between 2002 and 2008 as a consequence of expenditures for United States–led operations after September 11. (Table 16.1 shows the government's estimates of the costs of the wars in Iraq and Afghanistan, as well as for Operation Noble Eagle, which deploys reservists for homeland defense.) When a representative sample of Americans were asked about national defense and military spending, 20 percent responded "too little" was being spent, 35 percent responded that the amount was "about right," and 43 percent responded that "too much" was being spent (Carroll 2007a).

TABLE 16.1 Annual Costs of Wars in Iraq and Afghanistan, 2001–2008 (Estimated, in Billions)

OPERATION	2001 AND 2002	2003	2004	2005	2006	2007	2008 (REQUESTED)	2001–2008
Iraq	0	53	76	84.5	101	133.6	115.9	564
Afghanistan	20.8	14.7	14.5	20.8	19.7	35.1	29	154.6
Domestic security	13	8	3.7	2.1	.8	0	0	27.6
Unallocated		5.5						
TOTAL	33.8	81.1	94.2	107.5	121.6	168.7	144.9	755.7

Source: Belasco (2007).

The **Cold War,** the state of political tension, economic competition, and military rivalry that existed between the United States and the former Soviet Union for nearly 50 years, provided justification for large expenditures for military preparedness. However, the end of the Cold War, along with the rising national debt, resulted in cutbacks in the U.S. military budget in the 1990s.

Today, military spending has returned to nearly Cold War levels. The United States accounts for almost half of the world's military spending, the largest single percentage of any nation—more than the combined total of the 32 next most powerful nations. The next highest military spenders are England, France, Japan, and China, which, along with the United States, account for 64 percent of the world's military budget (SIPRI 2007b). U.S. national defense outlays, estimated to be $560 billion in 2007, include expenditures for salaries of military personnel, research and development, weapons, veterans' benefits, and other defense-related expenses (Office of Management and Budget 2007).

What Do You Think? According to journalist Mark Thompson, "a volunteer Army reflects the most central and sacred vow that citizens make to one another; soldiers protect and defend the country; in return the country promises to give them the tools they need to complete their mission . . ." (2007, p. 31). Some would argue, however, that the United States has not held up its side of the bargain, particularly when it comes to providing the quality and quantity of military personnel needed to win a war. For example, in an effort to boost numbers recruitment standards have been lowered (meaning that more convicted felons and high school dropouts may be recruited), the maximum enlistment age has been raised from 35 to 42, and the Pentagon's so-called "stop-loss orders" have forced soldiers to remain in uniform and combat past the date of their contractual obligations for service. Those who serve get "abbreviated training" and time between deployments is now measured in months rather than years (Thompson 2007). Given the above, should the military draft be reinstated? Should both men and women have to sign up for the draft? Alternatively, would you advocate universal military service upon high school graduation?

Cold War The state of military tension and political rivalry that existed between the United States and the former Soviet Union from the 1950s through the late 1980s.

The U.S. government not only spends money on its own military and defense but also sells military equipment to other countries, either directly or by helping U.S. companies sell weapons abroad. Although the purchasing countries may use these weapons to defend themselves from hostile attack, foreign military sales may pose a threat to the United States by arming potential antagonists. For example, the United States, the world's leading arms-exporting nation, supplied weapons to Iraq to use against Iran during the 1980–1987 war. These same weapons were then used against Americans in the Gulf and Iraq wars (Silverstein 2007). Similarly, after the Soviet invasion of Afghanistan in 1979, the United States funded Afghan rebel groups. Years after the Soviets left Afghanistan, rebels continued to fight for control of the country. Using weapons supplied by the United States, the Taliban took over much of Afghanistan and sheltered Osama bin Laden—also a former recipient of U.S. support—and Al Qaeda as they planned the attacks on September 11 (Rashid 2000; Bergen 2002).

In 2003, the United States transferred arms to several countries in active conflict, including Chad, Angola, Pakistan, Israel, and Colombia. A 2005 report by the World Policy Institute, titled *U.S. Weapons at War: Promoting Freedom or Fueling Conflict?* concluded that far "from serving as a force for security and stability, U.S. weapons sales frequently serve to empower unstable, undemocratic regimes to the detriment of U.S. and global security" (Berrigan & Hartung 2005).

SOCIOLOGICAL THEORIES OF WAR

Sociological perspectives can help us understand various aspects of war. In this section we describe how structural functionalism, conflict theory, and symbolic interactionism can be applied to the study of war.

Structural-Functionalist Perspective

Structural functionalism focuses on the functions that war serves and suggests that war would not exist unless it had positive outcomes for society. We have already noted that war has served to consolidate small autonomous social groups into larger political states. An estimated 600,000 autonomous political units existed in the world at about 1000 B.C. Today, that number has dwindled to fewer than 200.

Another major function of war is that it produces social cohesion and unity among societal members by giving them a "common cause" and a common enemy. For example, in 2005 *Newsweek* began a new feature about everyday American heroes called "Red, White, and Proud." Unless a war is extremely unpopular, military conflict also promotes economic and political cooperation. Internal domestic conflicts between political parties, minority groups, and special interest groups often dissolve as they unite to fight the common enemy. During World War II, U.S. citizens worked together as a nation to defeat Germany and Japan.

In the short term war may also increase employment and stimulate the economy. The increased production needed to fight World War II helped pull the United States out of the Great Depression. The investments in the manufacturing sector during World War II also had a long-term impact on the U.S.

economy. Hooks and Bloomquist (1992) studied the effect of the war on the U.S. economy between 1947 and 1972 and concluded that the U.S. government "directed, and in large measure, paid for a 65 percent expansion of the total investment in plant and equipment" (p. 304). War can also have the opposite effect, however. In a 2005 restructuring of the military, the Pentagon, seeking a "meaner, leaner fighting machine," recommended shutting down or reconfiguring nearly 180 military installations, "ranging from tiny Army reserve centers to sprawling Air Force bases that have been the economic anchors of their communities for generations" (Schmitt 2005) at a cost of thousands of civilian jobs.

Another function of war is the inspiring of scientific and technological developments that are useful to civilians. For example, innovations in battlefield surgery during World War II and the Korean War resulted in instruments and procedures that later became common practice in civilian hospital emergency wards (Zoroya 2006). Research on laser-based defense systems led to laser surgery, and research in nuclear fission and fusion facilitated the development of nuclear power. The airline industry owes much of its technology to the development of air power by the Department of Defense, and the Internet was created by the Pentagon for military purposes. Other **dual-use technologies,** a term referring to defense-funded innovations that have commercial and civilian applications, include SLICE, a high-speed twin-hull water vessel

© David Pollack/Corbis

originally made for the Office of Naval Research. SLICE has a variety of commercial applications, "including its use as a tour or sport fishing boat, oceanographic research vessel, oil spill response ship, and high-speed ferry" (State of Hawaii 2000).

War also serves to encourage social reform. After a major war members of society have a sense of shared sacrifice and a desire to heal wounds and rebuild normal patterns of life. They put political pressure on the state to care for war victims, improve social and political conditions, and reward those who have sacrificed lives, family members, and property in battle. As Porter (1994) explained, "Since . . . the lower economic strata usually contribute more of their blood in battle than the wealthier classes, war often gives impetus to social welfare reforms" (p. 19).

Finally, the U.S. military has historically provided an alternative for the advancement of poor or disadvantaged groups who otherwise face discrimination or limited opportunities in the formal economy. The military's specialized training, tuition assistance programs for a college education, and preferential hiring practices improve the prospects of veterans to find a decent job or career after their service (Military 2007).

As structural functionalists argue, a major function of war is that it produces unity among societal members. War provides a common cause and a common identity. Societal members feel a sense of cohesion, and they work together to defeat the enemy.

dual-use technologies Defense-funded technological innovations with commercial and civilian use.

Conflict Perspective

Conflict theorists emphasize that the roots of war are often antagonisms that emerge whenever two or more ethnic groups (e.g., Bosnians and Serbs), countries (United States and Vietnam), or regions within countries (the U.S. North and South) struggle for control of resources or have different political, economic, or religious ideologies. In addition, conflict theory suggests that war benefits the corporate, military, and political elites. Corporate elites benefit because war often results in the victor taking control of the raw materials of the losing nations, thereby creating a bigger supply of raw materials for its own industries. Indeed, many corporations profit from defense spending. Under the Pentagon's bid-and-proposal program, for example, corporations can charge the cost of preparing proposals for new weapons as overhead on their Defense Department contracts. Also, Pentagon contracts often guarantee a profit to the developing corporations. Even if the project's cost exceeds initial estimates, called a cost overrun, the corporation still receives the agreed-on profit. In the late 1950s President Dwight D. Eisenhower referred to this close association between the military and the defense industry as the **military-industrial complex.** Conflict theorists would be quick to note that "many former Republican officials and political associates of the Bush administration are associated with the Carlyle Group, an equity investment firm with billions of dollars in military and aerospace assets" (Knickerbocker 2002, p. 2).

The military elite benefits because war and the preparations for it provide prestige and employment for military officials. For example, Military Professional Resources Inc. (MPRI), an organization staffed by former military, defense, law enforcement, and other professionals, serves "the national security needs of the U.S. government, selected foreign governments, international organizations, and the private sector" and lists capabilities such as war gaming, force development and management, and democracy transition assistance (Military Professional Resources Inc. 2007, p. 1). According to the latest information available, in 2004 the total value of MPRI's contracts in Iraq was more than $2.5 million (Center for Public Integrity 2005).

Private contractors such as MPRI and many others contribute more than 180,000 civilians to the occupation of Iraq, about 20,000 more than the U.S. military and government employees (Miller 2007). For example, private security companies—for-profit organizations contracted by the U.S. government to perform security functions formerly provided by the military—account for a significant portion of the U.S. forces. According to estimates from Congressional Research Service, the U.S. government has awarded between 30 and 60 contracts to private security companies that provide between 20,000 and 30,000 private security personnel in Iraq at a cost of more than $4 billion from 2003 to 2007 (CRS 2007).

The North Carolina-based firm Blackwater Worldwide received national attention in 2004 when four of its employees in Iraq were killed by a Sunni mob in Fallujah, where their charred corpses were hung along public streets (in response, President Bush ordered an assault on the city). In September 2007, Blackwater personnel guarding a U.S. diplomatic convoy opened fire at a traffic circle in Baghdad, killing 17 and wounding 24 Iraqi civilians. Company officials initially claimed that their contractors responded proportionately to a nearby attack. In response, and after many years of lodging complaints about alleged indiscriminate firings by private security contractors, the Iraqi government revoked Blackwater's license to operate in Iraq (Tavernise 2007). An FBI investiga-

military-industrial complex
A term used by Dwight D. Eisenhower to connote the close association between the military and defense industries.

tion of the incident found that most of the killings were in violation of the rules for use of deadly force (Johnston and Broder 2007a). The U.S. Department of Justice convened a grand jury investigation into the incident and is considering prosecution against company personnel (Johnston and Broder 2007b).

War also benefits the political elite by giving government officials more power. Porter (1994) observed that "throughout modern history, war has been the level by which . . . governments have imposed increasingly larger tax burdens on increasingly broader segments of society, thus enabling ever-higher levels of spending to be sustained, even in peacetime" (p. 14). Political leaders who lead their country to a military victory also benefit from the prestige and hero status conferred on them.

Finally, feminists argue that war and other conflicts are often justified using the "language of feminism" (Viner 2002). For example, the attack on Afghanistan in 2001 was, in part, to liberate women who had been subjugated by the Taliban regime. Ironically, the position of women in Afghanistan has improved little in the last several years, leading many Muslim women to reject "Western-style feminism" and embrace Muslim feminism.

> Muslim women deplore misogyny just as Western women do, and they know that Islamic societies also oppress them; why wouldn't they? But liberation for them does not encompass destroying their identity, religion, or culture, and many of them want to retain the veil. (Viner 2002, p. 2)

Other differences also exist. Muslim feminism is based on the teachings of Islam, is profamily, and rejects the concept of patriarchy (McElroy 2003).

Symbolic Interactionist Perspective

The symbolic interactionist perspective focuses on how meanings and definitions influence attitudes and behaviors regarding conflict and war (see this chapter's *Social Problems Research Up Close* feature). The development of atti-

The face of patriotism is changing. Although, traditionally, older Americans have been the most patriotic, the spread of patriotism has been extending to all ages. For example, a recent survey of 2005 college graduates found that 83 percent defined themselves as patriotic.

Research indicates that attitudes and values differ between generations. For example, members of Generation X—that is, Americans born in the post–baby boom years—are often thought to be more materialistic, less civically engaged, and less socially trusting compared to previous generations. Rarely studied, however, are the differences within a generation; that is, are there different mind-sets within the same age cohort? Thus, Franke (2001) compares the values and attitudes of cadets at West Point with the values and attitudes of a sample of college students at Syracuse University. The research addresses an important question: To what extent are the attitudes and values of future military officers representative of those of the civilian population?

Sample and Methods

In the fall of 1995 the Future Officer Survey was administered anonymously to a representative sample of West Point cadets (N = 1,233) representing 31 percent of the total student population. The response rate was 48.2 percent, yielding 594 usable questionnaires. A revised version of the same survey, with specific references to military duties and West Point deleted or modified, was administered to 372 students in upper- and lower-division social science classes at Syracuse University (SU).

Using a 5-point Likert scale (from "strongly agree" to "strongly disagree") respondents were asked their level of agreement or disagreement with 36 state- ments, each one an indicator of a dependent variable. The dependent variables include the following: (1) political conservatism (e.g., "The government should provide health insurance for every American"), (2) patriotism (e.g., "All Americans should be willing to fight for their country"), (3) warriorism (e.g., "Sometimes war is necessary to protect the national interest"), and (4) globalism and global institutions (e.g., "World government is the best way to ensure international peace"). A fifth dependent variable was measured by asking respondents whether they defined themselves as conservative or liberal. The independent variables included sample type (i.e., comparing cadets to SU students) and class level in school (e.g., comparing first-year cadets to fourth-year cadets).

Several hypotheses were generated from the literature. First, Franke (2001) hypoth- esized that West Point cadets will have higher levels of political conservatism, patriotism, and warriorism than SU students. Second, Franke hypothesizes that SU students will be more likely to agree with statements that advocate peacekeeping operations and that are supportive of globalism and global institutions. Finally, Franke hypothesized that such differences will be magnified by years in school. Thus, for example, Franke predicted that senior cadets will be more patriotic than first-year cadets and that SU students in their fourth year will be more liberal than those in their first year.

Findings and Conclusions

Looking at statistical differences between and within samples yields some interesting results.

1. *Political conservatism.* Compared with first-year cadets, senior cadets were significantly more likely to agree with only one of the four indicators of political con- servatism ("The United States has gone too far in providing for equal opportunity under the law"). Contrary to predictions, the overall measure of political conser- vatism did not vary significantly between year in school for cadets or SU students. Between-sample differences were highly significant, however, with cadets, as expected, having higher rates of political conservatism than SU students.

2. *Patriotism.* Contrary to predictions, cadets' patriotism scores *decreased* between class cohorts; that is, seniors were significantly less rather than more patriotic than first-year cadets. Between- sample comparisons were also significant. Cadets on each of the six indicators of patriotism scored significantly higher than SU students. For example, 90 percent of cadets agreed with the statement that "an American should always feel that his or her primary allegiance is to his or her country," whereas only 75 percent of SU students felt similarly.

3. *Warriorism.* Warriorism increased across class year for West Point cadets. Alter- natively, in general, SU students' levels of

tudes and behaviors that support war begins in childhood. American children learn to glorify and celebrate the Revolutionary War, which created our nation. Movies romanticize war, children play war games with toy weapons, and vari- ous video and computer games glorify heroes conquering villains.

Symbolic interactionism helps to explain how military recruits and civil- ians develop a mind-set for war by defining war and its consequences as accept- able and necessary. The word *war* has achieved a positive connotation through its use in various phrases—the war on drugs, the war on poverty, and the war on crime. Positive labels and favorable definitions of military personnel facilitate military recruitment and public support of armed forces. In 2005, the Army

warriorism decreased as class increased, although statistical significance was reached for only one of the seven indicators of warriorism. Twice as many SU first-year students (43 percent) as seniors (21 percent) agreed with the statement "The most important role of the U.S. military is preparation for and conduct of war." Not surprisingly, between-sample results were significant, with cadets scoring much higher on this dependent variable than SU students.

4. *Globalism and global institutions.* Support for globalism and global institutions decreased between class cohorts for cadets and increased, although not significantly, between classes for SU students. Between-sample differences were not significant. Overall, neither cadets nor SU students were supportive of the ideals of globalism and global institutions such as a multinational military or the United Nations.

Finally, respondents were asked about their political affiliation. Cadets, in general, and senior cadets in particular, most often categorized themselves as conservative. For example, 58.6 percent of first-year cadets and 67.7 percent of senior cadets self-identified as conservatives. Interestingly, SU students' political affiliation was essentially the opposite. First-year students (49.6 percent) as well as seniors (60.8 percent) defined themselves as liberal.

The results of Franke's study, although mixed, generally support the hypotheses. On average, West Point cadets were more politi-cally conservative, patriotic, and warrioristic than civilian students at a private university. They also were less supportive of global-ism and global institutions. Differences also existed within samples, with, for example, self-identified conservatism increasing between cadets' class level—from first to fourth year.

There are two explanations for the results. The first is called self-selection. The self-selection model argues that people who attend military academies are more politically con-servative, patriotic, and warrioristic *before* they enter the academy; that is, they select a military academy because it is consistent with their atti-tudes and values. The fact that first-year cadets had higher rates of warriorism than first-year SU students would support this perspective. The second model is called the socialization model; it holds that the longer an individual is socialized into an institution, the more that institution influ-ences the individual's attitudes and values. Thus, for example, SU students were more liberal than cadets upon entering college (self-selection), but the extent of their liberal thinking increased with years at SU (socialization).

There are, however, several limitations of the study that need to be addressed. First, Syracuse University is a private university, and the students who attend it may not be representative of all "Gen X" college students. Future research needs to look at students from both private and public colleges and uni-versities. Second, variables other than class position may be responsible for the results.

As Franke noted, differences in maturity may explain the within-sample findings. Seniors not only have undergone more socializa-tion at their respective institutions but also have had more "life experiences," and it is these life experiences that may be respon-sible for the findings. Finally, the research is cross-sectional in nature, which means that it looked at only one point in time. It did not follow respondents through their academic careers but measured different respondents in different class cohorts and then compared the results. Future research should be longitudinal in nature; that is, the same people should be measured at different points in time.

Given these limitations, it is difficult to answer the question posed at the beginning of this article. As Franke (2001) noted:

> While the present analysis cannot provide a definitive answer, the data indicate that a gap between West Point cadets and their civilian generational peers exists already prior to military socialization. The observed correlations between value orientation and the general sociopolitical views could be another manifestation of the growing politicization of the U.S. military. The data also suggest that socialization may widen the gap. (p. 116)

Franke (2001) concluded that his research supports the notion of service academies but, at the same time, questions their role in an increasingly "global security environment."

National Guard launched a \$38 million marketing campaign targeting young men and women with advertisements showing "troops with weapons drawn, helicopters streaking and tanks rolling," all "in an attempt to remind people what the Guard has been about since Colonial Days: fighting wars and protecting the homeland." The new slogan? "The most important weapon in the war on terrorism. You" (Davenport 2005, p. 1).

Many government and military officials convince the masses that the way to ensure world peace is to be prepared for war. Patriotism is a popular sentiment in American society. For example, 62 percent of Americans say they display a U.S. flag at home, at the office, or on their car (Doherty 2007) (see Table 16.2).

TABLE 16.2 Who Flies the Flag?

	DISPLAY FLAG AT HOME, OFFICE, OR ON CAR?*	
	YES	NO
Total	62	38
Men	65	34
Women	59	41
White	67	33
African American	41	59
18–29 years old	51	49
30–49 years old	63	37
50–64 years old	65	35
65 years old and older	71	28
Republican	73	26
Democrat	55	45
Independent	63	37
Northeast	69	31
Midwest	67	33
South	58	41
West	57	43
College graduate	61	39
Some college	65	35
High school graduate	64	36
Less than high school	54	45

*Numbers may not sum to 100 because of refusals or don't know responses.
Source: Doherty (2007).

Governments may use propaganda and appeals to patriotism to generate support for war efforts and to motivate individuals to join armed forces. Salladay (2003), for example, notes that those in favor of the war on Iraq have commandeered the language of patriotism, making it difficult but necessary for peace activists to use the same symbols or phrases.

To legitimize war, the act of killing in war is not regarded as "murder." Deaths that result from war are referred to as casualties. Bombing military and civilian targets appears more acceptable when nuclear missiles are "peacekeepers" that are equipped with multiple "peace heads." Killing the enemy is more acceptable when derogatory and dehumanizing labels such as Gook, Jap, Chink, Kraut, and Haji convey the attitude that the enemy is less than human.

Such labels are socially constructed as images, often through the media, and are presented to the public. Social constructionists, like symbolic interactionists in general, emphasize the social aspects of "knowing." Thus Li and Izard (2003) used content analysis to analyze newspaper and television coverage of the World Trade Center and Pentagon attacks on September 11. The researchers examined the first 8 hours of coverage of the attacks presented on CNN, ABC,

CBS, NBC, and Fox as well as in eight major U.S. newspapers (including the *Los Angeles Times,* the *New York Times,* and the *Washington Post*). Results of the analysis indicated that newspaper articles tended to have a "human interest" emphasis, whereas television coverage was more often "guiding and consoling." Other results suggested that both media relied most heavily on government sources, that newspapers and the networks were equally factual, and that networks were more homogeneous in their presentation than newspapers. One indication of the importance of the media lies in President George W. Bush's creation of the Office of Global Communications—"a huge production company, issuing daily scripts on the Iraq war to U.S. spokesmen around the world, auditioning generals to give media briefings, and booking administration stars on foreign news shows" (Kemper 2003, p. 1).

CAUSES OF WAR

The causes of war are numerous and complex. Most wars involve more than one cause. The immediate cause of a war may be a border dispute, for example, but religious tensions that have existed between the two combatant countries for decades may also contribute to the war. The following section reviews various causes of war.

Conflict Over Land and Other Natural Resources

Nations often go to war in an attempt to acquire or maintain control over natural resources, such as land, water, and oil. Michael Klare, author of *Resource Wars: The New Landscape of Global Conflict* (2001), predicted that wars will increasingly be fought over resources as supplies of the most needed resources diminish. Disputed borders have been common motives for war. Conflicts are most likely to arise when borders are physically easy to cross and are not clearly delineated by natural boundaries, such as major rivers, oceans, or mountain ranges.

Water is another valuable resource that has led to wars. Unlike other resources, water is universally required for survival. At various times the empires of Egypt, Mesopotamia, India, and China all went to war over irrigation rights. In 1998, 5 years after Eritrea gained independence from Ethiopia, forces clashed over control of the port city Assab and with it, access to the Red Sea.

Not only do the oil-rich countries in the Middle East present a tempting target in themselves, but war in the region can also threaten other nations that are dependent on Middle Eastern oil. Thus, when Iraq seized Kuwait and threatened the supply of oil from the Persian Gulf, the United States and many other nations reacted militarily in the Gulf War. In a document prepared for the Center for Strategic and International Studies, Starr and Stoll (1989) warned that soon

> water, not oil, will be the dominant resource issue of the Middle East. According to World Watch Institute, "despite modern technology and feats of engineering, a secure water future for much of the world remains elusive." The prognosis for Egypt, Jordan, Israel, the West Bank, the Gaza Strip, Syria, and Iraq is especially alarming. If present consumption patterns continue, emerging water shortages, combined with a deterioration in water quality, will lead to more competition and conflict. (p. 1)

Despite such predictions, tensions in the Middle East have erupted into fighting repeatedly in recent years—but not over water. In July 2006, Israel and

Lebanon fought a border war that killed a thousand people and displaced a million Lebanese. Civil war erupted in the Palestinian territories between rival parties when Hamas seized control of the Gaza Strip in response to Fatah's refusal to hand over the government after Hamas won legislative elections (BBC 2007a). Further, in the summer of 2007, hundreds were killed when Lebanese security forces attacked Palestinian militants based in Lebanon's refugee camps (BBC 2007b). Negotiations between Israeli Prime Minister Ehud Olmert and Palestinian leader Mahmoud Abbas have stalled, and the prospects for meaningful negotiations on land, settlement, and refugee issues are dim.

Conflict Over Values and Ideologies

Many countries initiate war not over resources but over beliefs. World War II was largely a war over differing political ideologies: democracy versus fascism. The Cold War involved the clash of opposing economic ideologies: capitalism versus communism. Wars over differing religious beliefs have led to some of the worst episodes of bloodshed in history, in part, because some religions are partial to martyrdom—the idea that dying for one's beliefs leads to eternal salvation. The Shiites (one of the two main sects within Islam) in the Middle East represent a classic example of holy warriors who feel divine inspiration to kill the enemy.

Conflicts over values or ideologies are not easily resolved. The conflict between secularism and Islam has lasted for 14 centuries. Conflict over values and ideologies are less likely to end in compromise or negotiation because they are fueled by people's convictions. For example, when a representative sample of American Jews were asked, "Do you agree or disagree with the following statement? 'The goal of Arabs is not the return of occupied territories but rather the destruction of Israel,'" 81 percent agreed, 13 percent disagreed, and 6 percent were unsure (American Jewish Committee 2007).

If ideological differences can contribute to war, do ideological similarities discourage war? The answer seems to be yes; in general, countries with similar ideologies are less likely to engage in war with each other than countries with differing ideological values (Dixon 1994). Democratic nations are particularly disinclined to wage war against one another (Brown, Lynn-Jones, & Miller 1996; Rasler & Thompson 2005).

Racial, Ethnic, and Religious Hostilities

Racial, ethnic, and religious groups vary in their cultural beliefs, values, and traditions. Thus conflicts between racial, ethnic, and religious groups often stem from conflicting values and ideologies. Such hostilities are also fueled by competition over land and other scarce natural and economic resources. Gioseffi (1993) noted that "experts agree that the depleted world economy, wasted on war efforts, is in great measure the reason for renewed ethnic and religious strife. 'Haves' fight with 'have-nots' for the smaller piece of the pie that must go around" (p. xviii). Racial, ethnic, and religious hostilities are also perpetuated by the wealthy minority to divert attention away from their exploitations and to maintain their own position of power. Such **constructivist explanations** of ethnic conflict—those that emphasize the role of leaders of ethnic groups in stirring up intercommunal hostility—differ sharply from **primordial explanations,** or those that emphasize the existence of "ancient hatreds" rooted in deep psychological or cultural differences between ethnic groups.

constructivist explanations Those explanations that emphasize the role of leaders of ethnic groups in stirring up hatred toward others external to one's group.

primordial explanations Those explanations that emphasize the existence of "ancient hatreds" rooted in deep psychological or cultural differences between ethnic groups, often involving a history of grievance and victimization, real or imagined, by the enemy group.

As described by Paul (1998), sociologist Daniel Chirot argued that the recent worldwide increase in ethnic hostilities is a consequence of "retribalization," that is, the tendency for groups, lost in a globalized culture, to seek solace in the "extended family of an ethnic group" (p. 56). Chirot identified five levels of ethnic conflict: (1) multiethnic societies without serious conflict (e.g., Switzerland), (2) multiethnic societies with controlled conflict (e.g., United States and Canada), (3) societies with ethnic conflict that has been resolved (e.g., South Africa), (4) societies with serious ethnic conflict leading to warfare (e.g., Sri Lanka), and (5) societies with genocidal ethnic conflict, including "ethnic cleansing" (e.g., Darfur).

Religious differences as a source of conflict have recently come to the forefront. An Islamic jihad, or holy war, has been blamed for the September 11 attacks on the World Trade Center and Pentagon as well as for bombings in Kashmir, Sudan, the Philippines, Kenya, Tanzania, Saudi Arabia, Spain, and Great Britain. Some claim that Islamic beliefs in and of themselves have led to recent conflicts (Feder 2003). Others contend that religious fanatics, not the religion itself, are responsible for violent confrontations. For example, Islamic leader Osama bin Laden claimed that unjust U.S. Middle East policies are responsible for "dividing the whole world into two sides—the side of believers and the side of infidels" (Williams 2003, p. 18). Despite bin Laden's decree, the terrorist activity in Iraq is increasingly sectarian as Sunni Arabs rebel against the Shiite Arab–dominated government (Tavernise 2005).

> ❝ It is very difficult to get a handle on all of the contours of the current situation in Iraq. This is a civil war on top of an insurgency on top of other conflicts. There is no one simple split between side A and side B. There are numerous subgroups and splinter groups that make it difficult to say any one leader is in charge of those who come under one label. ❞
>
> Paul Pillar
> Former CIA chief intelligence analyst for the Middle East

What Do You Think? The Charter of the United Nations obligates all member countries "to settle their international disputes by peaceful means" except in self-defense. In September 2002, the White House released a document describing the nation's new international security policy in light of the September 11 attacks. In situations of "imminent threat," "[W]e will not hesitate to act alone, if necessary, to exercise our right of self-defense by acting preemptively against such terrorists, to prevent them from doing harm against our people and our country" (National Security Council 2002). The U.S. government later asserted this preemptive right to self-defense in the justification for the war in Iraq. Do you agree with this policy? Should the United States attack countries that have not attacked it first? What if other countries adopt this policy?

Defense Against Hostile Attacks

The threat or fear of being attacked may cause the leaders of a country to declare war on the nation that poses the threat. This is an example of what experts in international relations refer to as the **security dilemma:** "actions to increase one's security may only decrease the security of others and lead them to respond in ways that decrease one's own security" (Levy 2001, p. 7). Such situations may lead to war inadvertently. The threat may come from a foreign country or from a group within the country. After Germany invaded Poland in 1939, Britain and

security dilemma A characteristic of the international state system that gives rise to unstable relations between states; as State A secures its borders and interests, its behavior may decrease the security of other states and cause them to engage in behavior that decreases A's security.

France declared war on Germany out of fear that they would be Germany's next victims. Germany attacked Russia in World War I, in part out of fear that Russia had entered the arms race and would use its weapons against Germany. Japan bombed Pearl Harbor hoping to avoid a later confrontation with the U.S. Pacific fleet, which posed a threat to the Japanese military. In 2001 a United States–led coalition bombed Afghanistan in response to the September 11 terrorist attacks. Moreover, in March 2003 the United States, Great Britain, and a loosely coupled "coalition of the willing" invaded Iraq in response to perceived threats of weapons of mass destruction and the reported failure of Saddam Hussein to cooperate with United Nations weapons inspectors. Yet, in 2005 a presidential commission concluded that the attack on Iraq was based on faulty intelligence and that, in fact, "America's spy agencies were 'dead wrong' in most of their judgments about Iraq's weapons of mass destruction" (Shrader 2005, p. 1). As a result, many Americans (more than 60 percent) favor a partial or complete withdrawal from Iraq (CNN/Opinion Research Corporation Poll 2007).

Revolutions and Civil Wars

Revolutions and civil wars involve citizens warring against their own government and often result in significant political, economic, and social change. The difference between a revolution and a civil war is not always easy to determine. Scholars generally agree that revolutions involve sweeping changes that fundamentally alter the distribution of power in society (Skocpol 1994). The American Revolution resulted from colonists revolting against British control. Eventually, they succeeded and established a republic where none existed before. The Russian Revolution involved a revolt against a corrupt, autocratic, and out-of-touch ruler, Czar Nicholas II. Among other changes, the revolution led to wide-scale seizure of land by peasants who formerly were economically dependent on large landowners.

Civil wars may result in a different government or a new set of leaders but do not necessarily lead to such large-scale social change. Because the distinction between a revolution and a civil war depends upon the outcome of the struggle, it may take many years after the fighting before observers agree on how to classify it. Revolutions and civil wars are more likely to occur when a government is weak or divided, when it is not responsive to the concerns and demands of its citizens, and when strong leaders are willing to mount opposition to the government (Barkan & Snowden 2001; Renner 2000).

In Sri Lanka, a contemporary example of a country undergoing a civil war, the government has fought an insurgency led by separatist militants since 1983. The war has resulted in more than 68,000 deaths (Gardner 2007) and consists primarily of "jungle skirmishes, government air strikes and ambushes by the Liberation of Tamil Tigers (LTTE) which want to carve a separate Tamil state out of Sri Lanka's north and east" (Reuters 2001). Like many civil wars, the war in Sri Lanka is also a struggle between a majority community (in this case, Sinhalese Buddhists) and a relatively poor and disadvantaged minority community (Hindu Tamils). Similarly, Liberia's 14-year on-again off-again civil war pits Liberians United for Reconciliation and Democracy rebels against government officials. In 2003, under heavy pressure from the United States, the president of Liberia resigned. His exile has led to talks between warring factions, the arrival of much-needed U.S. supplies and 15,000 United Nations peacekeepers, and the

❝ Those who make peaceful evolution impossible, make violent revolution inevitable. **❞**

John F. Kennedy
Former U.S. president

beginnings of an interim government (Sengupta 2003; Global Security 2005). Civil wars have also erupted in newly independent republics created by the collapse of communism in Eastern Europe, as well as in Rwanda, Sierra Leone, Chile, Uganda, Liberia, and Sudan.

Nationalism

Some countries engage in war in an effort to maintain or restore their national pride. For example, Scheff (1994) argued that "Hitler's rise to power was laid by the treatment Germany received at the end of World War I at the hands of the victors" (p. 121). Excluded from the League of Nations, punished by the Treaty of Versailles, and ostracized by the world community, Germany turned to nationalism as a reaction to material and symbolic exclusion. Furthermore, some observers have noted that despite the *official* end of the war in Iraq and, with it, Saddam Hussein's rule, the dictator has "become a symbol of nationalist resistance to forces opposing the U.S.-led occupation of Iraq" (Morahan 2003, p. 1).

In the late 1970s Iranian militants seized the U.S. embassy in Tehran and held its occupants hostage for more than 1 year. President Carter's attempt to use military forces to free the hostages was not successful. That failure intensified doubts about America's ability to use military power effectively to achieve its goals. The hostages in Iran were eventually released after President Reagan took office, but doubts about the strength and effectiveness of the U.S. military still called into question America's status as a world power. Subsequently, U.S. military forces invaded the small island of Grenada because the government of Grenada was building an airfield large enough to accommodate major military armaments. U.S. officials feared that this airfield would be used by countries in hostile attacks on the United States. From one point of view, the large-scale "successful" attack on Grenada functioned to restore faith in the power and effectiveness of the U.S. military.

What Do You Think? Dan Frazier, a 42-year-old businessman, had a great idea. To protest what he saw as an unjust war and to make some money, he printed t-shirts that read "Bush Lied" on the front and "They Died" on the back over a list of soldiers who were killed in Iraq (Fischer 2007). Although sales were meager (Frazier donated $1 for each sale to a charity that supports families of soldiers who died in the war), the t-shirts were receiving some media attention, which led to several states passing a law restricting the unauthorized use of soldier's names or images for profit. Eventually, Frazier's home state of Arizona passed a law banning the shirts and authorizing family members to sue for damages. Shortly thereafter, the American Civil Liberties Union petitioned a federal judge to suspend the law on the grounds that it violates First Amendment rights to free speech (Davenport 2007). What would you do if you were the judge in this case? Is this a case of offensive and greedy profiteering from tragedy or an overzealous attack on the right to voice opposition to government policies?

Arizona businessman Dan Frazier wears a version of the t-shirt that is banned in Arizona, Florida, Louisiana, Oklahoma, and Texas. A judge in Arizona will decide whether this is a commercial product legitimately regulated by the state or a form of free speech protected by the Constitution.

© AP Photo/Jake Bacon

TERRORISM

Terrorism is the premeditated use, or threatened use, of violence against civilians by an individual or group to gain a political or social objective (Interpol 1998; Barkan & Snowden 2001; Brauer 2003). Terrorism may be used to publicize a cause, promote an ideology, achieve religious freedom, attain the release of a political prisoner, or rebel against a government. Terrorists use a variety of tactics, including assassinations, skyjackings, suicide bombings, armed attacks, kidnapping and hostage taking, threats, and various forms of bombings. Through such tactics, terrorists struggle to induce fear within a population, create pressure to change policies, or undermine the authority of a government they consider objectionable. Most analysts agree that unlike war—where there is a clear winner and loser—terrorism is unlikely to be completely defeated.

> There can be no final victory in the fight against terrorism, for terrorism (rather than full-scale war) is the contemporary manifestation of conflict, and conflict will not disappear from earth as far as one can look ahead and human nature has not undergone a basic change. But it will be in our power to make life for terrorists and potential terrorists much more difficult. (Laqueur 2006, p. 173)

terrorism The premeditated use or threatened use of violence by an individual or group to gain a political objective.

transnational terrorism Terrorism that occurs when a terrorist act in one country involves victims, targets, institutions, governments, or citizens of another country.

Types of Terrorism

Terrorism can be either transnational or domestic. **Transnational terrorism** occurs when a terrorist act in one country involves victims, targets, institutions, governments, or citizens of another country. The 1988 bombing of Pan Am Flight 103 over Lockerbie, Scotland, exemplifies transnational terrorism. The incident took the lives of 270 people, including 35 Syracuse University undergraduates returning from an overseas studies program in London. After a 10-year investigation, a Libyan intelligence agent (CNN 2001a) was sentenced to life imprisonment for the bombing, although serious questions have been raised recently about the verdict (Cowell 2007). In 2003 the Libyan government agreed to

A jeep is in flames at the main entrance of Glasgow's International Airport, site of a "failed" suicide bombing in June 2007. A suicide note confirmed that the drivers, one a British-born doctor of Iraqi descent, intended to explode the propane fuel bomb in the terminal. Less than 1 month later, police reported over 200 reprisal attacks on Muslims in Scotland.

pay $2.7 billion in compensation to the victims' families (Smith 2004). Other examples of transnational terrorism include attacks on U.S. embassies in Kenya and Tanzania and the bombing of a naval ship, the USS *Cole,* moored in Aden Harbor, Yemen. The 2001 attacks on the World Trade Center, the Pentagon, and Flight 93—the most devastating in U.S. history—are also examples of transnational terrorism. Al-Qaeda, a global alliance of militant Sunni Islamic groups advocating jihad ("holy war") against the West, was also linked to deadly bombings in Bali, Indonesia (2002), Madrid (2004), and London (2005).

Domestic terrorism, sometimes called insurgent terrorism (Barkan & Snowden 2001), is exemplified by the 1995 truck bombing of a nine-story federal office building in Oklahoma City, resulting in 168 deaths and the injury of more than 200 people. Gulf War veteran Timothy McVeigh and Terry Nichols were convicted of the crime. McVeigh is reported to have been a member of a paramilitary group that opposes the U.S. government. In 1997 McVeigh was sentenced to death for his actions, and he was executed in 2001 (Barnes 2004). The 2004 bombing of a Russian school, which killed 324 people, nearly half of them children, is also an act of domestic terrorism, as Chechen rebels continue to fight for an independent state. Further, in December 2006, the Basque separatist group ETA resumed its bombing campaign after an unprecedented declaration of a permanent ceasefire that led to hopes of a peace process to settle the decades-old struggle (Espiau 2006).

Patterns of Global Terrorism

Although estimates vary, a 2007 report by the National Counterterrorism Center described patterns of terrorism around the world (Office of the Coordinator for Counterterrorism 2007). In 2006 (see Figure 16.2):

- There were approximately 14,000 domestic and international terrorist attacks around the world.
- About 20,000 people lost their lives as a result of these incidents.

domestic terrorism Domestic terrorism, sometimes called insurgent terrorism, occurs when the terrorist act involves victims, targets, institutions, governments, or citizens from one country.

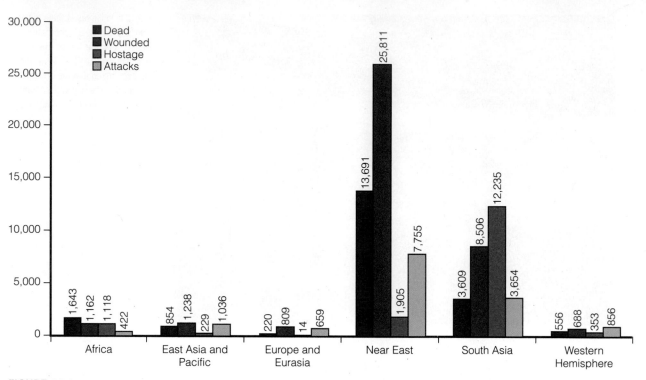

FIGURE 16.2
Comparison of terrorist attacks and victims by region, 2006.
Source: U.S. State Department (2007).

- There was a 25 percent increase in the number of incidents and a 40 percent increase in the number of fatalities compared with 2005.
- An additional 38,000 individuals were wounded by terrorist attacks in 2006.
- More than 50 percent of those killed or wounded were Muslims, and the majority of them lived in Iraq.

Counterterrorism officials report that the number of terrorist threats against the United States, particularly from Al Qaeda, has increased in recent years, leading to a "heightened threat environment" (DeYoung & Pincus 2007, p. 1). When a random sample of U.S. adults was asked, "Who do you think is currently winning the war against terrorism?" 29 percent responded the United States and its allies. However, 50 percent responded "neither side" and 20 percent responded "the terrorists" (Carroll 2007b).

The Roots of Terrorism

In 2003 a panel of terrorist experts came together in Oslo, Norway, to address the causes of terrorism (Bjorgo 2003). Although not an exhaustive list, several causes emerged from the conference:

- A failed or weak state, which is unable to control terrorist operations.
- Rapid modernization, when, for example, a country's sudden wealth leads to rapid social change.
- Extreme ideologies—religious or secular.
- A history of political violence, civil wars, and revolutions.

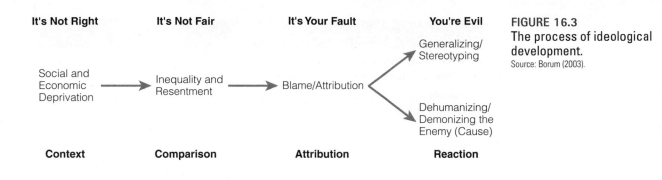

It's Not Right	It's Not Fair	It's Your Fault	You're Evil

FIGURE 16.3
The process of ideological development.
Source: Borum (2003).

Context	Comparison	Attribution	Reaction

- Repression by a foreign occupation (i.e., invaders to the inhabitants).
- Large-scale racial or ethnic discrimination.
- The presence of a charismatic leader.

Note that Iraq has several of the characteristics listed here, including rapid modernization (e.g., oil reserves), extreme ideologies (e.g., Islamic fundamentalism), a history of violence (e.g., invasion of Kuwait), and a weak state that is unable to control terrorist operations (e.g., the newly elected Iraqi government).

The causes of terrorism listed here, however, are macro in nature. What of social-psychological variables? How do individuals choose to join terrorist organizations or use terrorist tactics? Borum (2003) suggested a four-stage micro-level process. The decision to commit a terrorist act begins with an individual's assessment that *something is not right* (e.g., government-imposed restrictions). Next, individuals *define the situation as unfair* in that the "not right" condition does not apply to everyone (e.g., government-imposed restrictions are imposed on some but not on others). Individuals then begin to *blame specific others for the injustice* (e.g., government leaders) and, finally, to *redefine those who are responsible for the injustice as bad or evil* (e.g., a fascist regime). This process of "ideological development," as portrayed in Figure 16.3, often leads to stereotyping and dehumanizing the enemy, which then facilitates violence. Although the process was developed as a heuristic device, Borum noted that "understanding the mind-set" of a terrorist can help in the fight against terrorism.

America's Response to Terrorism

A government can use both defensive and offensive strategies to fight terrorism (see Table 16.3). Defensive strategies include using metal detectors and X-ray machines at airports and strengthening security at potential targets, such as embassies and military command posts. The Department of Homeland Security (DHS) coordinates such defensive tactics for the U.S. government. DHS was created in 2002 from 22 domestic agencies (e.g., the U.S. Coast Guard, the Immigration and Naturalization Service, and the Secret Service) and has more than 180,000 employees. With an estimated budget of $46.4 billion in 2008, the mission of DHS is as follows: "We will prevent and deter terrorist attacks and protect against and respond to threats and hazards to the nation. We will ensure safe and secure borders, welcome lawful immigrants and visitors, and promote the free-flow of commerce" (U.S. Department of Homeland Security 2007).

Offensive strategies include retaliatory raids, such as the U.S. bombing of terrorist facilities in Afghanistan, the invasion of Iraq, group infiltration, and preemptive strikes. New legislation facilitates such offensive tactics. In October

TABLE 16.3 U.S. Policy Responses to Terrorism

- **Diplomacy and Constructive Engagement**—Using diplomacy, including verbal contact and direct negotiations, to create a "global antiterror coalition."

- **Economic Sanctions**—Banning or threatening to ban, for example, trade and investment relations with a nation as a means of control. Other economic sanctions include freezing bank accounts, imposing trade embargoes, and suspending foreign aid.

- **Economic Inducements**—Developing assistance programs to reduce poverty and literacy in countries that are breeding grounds for terrorist activity.

- **Covert Action**—Infiltrating terrorist groups, military operations, and intelligence gathering. Also included in this category are sabotaging weapons facilities and capturing wanted terrorists.

- **Rewards for Information Programs**—Exchanging money for information on or the capture of a wanted terrorist. For example, the reward for capturing Osama bin Laden is $25 million.

- **Extradition/Law Enforcement Cooperation**—Enlisting the worldwide cooperation of law enforcement agencies, including international extradition of terrorists.

- **Military Force**—Using military force in fighting terrorism. For example, military force was used to successfully overthrow the Taliban in Afghanistan.

- **International Conventions**—Entering into agreements with other countries that obligate the signatories to conform to the articles of the convention, including, for example, prosecution and extradition of terrorists.

Source: Perl (2003).

2001 the USA PATRIOT Act (Uniting and Strengthening America by Providing Appropriate Tools Required to Intercept and Obstruct Terrorism) was signed into law. The Act increases police powers both domestically and abroad. Advocates of the Patriot Act argue that during war, some restrictions of civil liberties are necessary. Moreover, the legislation is not "a substantive shift in policy but a mere revitalization of already established precedents" (Smith 2003, p. 25). Critics hold that the Act poses a danger to civil liberties. For example, the original Act provides for the *indefinite* detention of immigrants if the immigrant group is defined as a "danger to national security" (Romero 2003). In 2007, Congress revised the Act to address some of these concerns, particularly to limit the government's authority to conduct wiretaps. Unlike the earlier Patriot Act, which had "sunset" clauses mandating review by Congress before renewal, most of the provisions in the new Act are now permanent features of the law.

Among the most controversial U.S. policies in the war on terrorism is the indefinite detention of "enemy combatants" at a military prison and interrogation camp in Guantanamo, Cuba. Since 2002, as many as 775 detainees have been held at "Gitmo." By mid-2007, 340 had already been released and 110 are scheduled for release contingent upon finding a host government willing to confine or monitor the detainees. Of the remaining 325, according to the Pentagon, about 70 are likely to face trial, 10 of whom have been certified as Al Qaeda members (Dedman 2006). In June 2007, in a reversal of its previous decisions, the U.S. Supreme Court agreed to hear a case that addresses whether constitutional guarantees of a right to a public trial protect the detainees (Barnes 2007). Table 16.4 contains the results of a national survey of 1,464 Americans who were asked whether they thought the reports of mistreatment at Guantanamo Bay were "isolated incidents" or a "wider pattern" of abuse.

AFP/Getty Images

The prisoner abuses at Guantanamo Bay (Cuba) and Abu Ghraib (Iraq) shocked the nation. Several of the photos, transmitted around the world on the Internet, were made into billboards in Arab countries. Army Pfc. Lynndie England, pictured, was convicted on six of seven counts involving prisoner mistreatment and was sentenced to 3 years in prison and given a dishonorable discharge.

TABLE 16.4 Reports of Prisoner Mistreatment at Guantanamo Bay

	ISOLATED (%)	WIDER PATTERN (%)	NEITHER/DON'T KNOW (%)
Total	54	34	12
Men	56	64	10
Women	52	34	14
18–29 years old	43	46	11
30–49 years old	55	36	9
50–64 years old	59	28	13
65 years old and older	56	25	19
White	57	31	12
Black	35	52	13
Hispanic	44	45	11
Republican	76	14	10
Democrat	43	45	12
Independent	45	44	11

Source: Adapted from Pew Research Center (2005).

Combating terrorism is difficult, and recent trends will make it increasingly problematic (Zakaria 2000; Strobel et al. 2001). First, data stored on computers can be easily acquired by hackers who illegally gain access to classified information. For example, interlopers obtained the fueling and docking schedules of the USS *Cole.* Second, the Internet permits groups with similar interests, once separated by geography, to share plans, fund-raising efforts, recruitment strategies, and other coordinated efforts. Worldwide, thousands of terrorists keep in

An Iraqi insurgent takes aim on U.S. positions in Najaf during battle in January 2006. Although the U.S. Army has superior weapons and training, insurgent forces rely on deep knowledge of urban terrain and count on support from the local population.

© Joao Silva/The New York Times

guerrilla warfare Warfare in which organized groups oppose domestic or foreign governments and their military forces; often involves small groups of individuals who use camouflage and underground tunnels to hide until they are ready to execute a surprise attack.

weapons of mass destruction (WMD) Chemical, biological, and nuclear weapons that have the capacity to kill large numbers of people indiscriminately.

touch through Hotmail.com e-mail accounts, and virtually all terrorist groups maintain websites for recruitment, fund-raising, internal communication, and propagandizing (Weimann 2006). Third, globalization contributes to terrorism by providing international markets where the tools of terrorism—explosives, guns, electronic equipment, and the like—can be purchased. Finally, fighting terrorism under guerrilla warfare–like conditions is increasingly a concern. Unlike terrorist activity, which targets civilians and may be committed by lone individuals, **guerrilla warfare** is committed by organized groups opposing a domestic or foreign government and its military forces. In March 2007, an estimated 70,000 Iraqi insurgents were fighting U.S., allied, and Iraqi forces (O'Hanlon & Campbell 2007).

The possibility of terrorists using weapons of mass destruction is the most frightening scenario of all and, as stated earlier, the motivation for the 2003 war with Iraq. **Weapons of mass destruction (WMD)** include chemical, biological, and nuclear weapons. Anthrax, for example, although usually associated with diseases in animals, is a highly deadly disease in humans and, although preventable by vaccine, has a "lethal lag time." In a hypothetical city of 100,000 people, delaying a vaccination program 1 day would result in 5,000 deaths; a delay of 6 days would result in 35,000 deaths. In 2001 trace amounts of anthrax were found in several letters sent to media and political figures, resulting in five deaths and the inspection and closure of several postal facilities (Baliunas 2004).

Other examples of the use of WMD exist. On at least eight occasions Japanese terrorists dispersed aerosols of anthrax and botulism in Tokyo (Inglesby et al. 1999), and in 2000 a religious cult, hoping to disrupt elections in an Oregon county, "contaminated local salad bars with salmonella, infecting hundreds" (Garrett 2001, p. 76). In 2003, the poison ricin was detected on a mail-opening machine in Senate majority leader Bill Frist's Washington, DC, office (Associated Press 2005). Furthermore, in 2006, the hit TV series *24* featured a fictional story about a Russian separatist group that stole chemical weapons from the U.S. military and planted them in a gas distribution facility in downtown Los Angeles, threatening to kill thousands unless special agent Jack Bauer (played by Keifer Sutherland) could catch the terrorists in time. The show builds on real fears among experts, who estimate that Russia possesses about 40,000 metric tons of chemical weapons, the world's largest stockpile by far. The United States, Canada, and the European Union have pledged nearly $2 billion to support Russia's program to secure and destroy all chemical weapons by 2012 (Walker & Tucker 2006).

SOCIAL PROBLEMS ASSOCIATED WITH CONFLICT, WAR, AND TERRORISM

Social problems associated with conflict, war, and terrorism include death and disability; rape, forced prostitution, and displacement of women and children; social-psychological costs; diversion of economic resources; and destruction of the environment.

Death and Disability

Many American lives have been lost in wars, including 53,000 in World War I, 292,000 in World War II, 34,000 in Korea, and 47,000 in Vietnam (U.S. Bureau of the Census 2004). In Iraq, between March 2003 and July 2007, 26,000 U.S. troops were wounded and more than 3,700 were killed (Global Security 2007). Estimates of Iraqi civilian casualties vary dramatically—from 30,000 cited by President Bush (in December of 2005) to 655,000 (in October 2006) as quoted by a group of Iraqi physicians and epidemiologists at Johns Hopkins School of Public Health (Brown 2006; Robinson 2006). Globalization and sophisticated weapons technology combined with increased population density have made it easier to kill a large number of people in a short amount of time. When the atomic bomb was dropped on the Japanese cities of Hiroshima and Nagasaki during World War II, 250,000 civilians were killed.

The impact of war and terrorism extends far beyond those who are killed. Many of those who survive war incur disabling injuries or contract diseases. For example, 1 million people worldwide have been killed or disabled by landmines—a continuing problem in the aftermath of the 2003 war with Iraq (Renner 2005). In 1997 the Mine Ban Treaty, which requires that governments destroy stockpiles within 4 years and clear landmine fields within 10 years, became international law. To date, 153 countries have signed the agreement; 42 countries remain, including China, India, Israel, Russia, and the United States (International Campaign to Ban Landmines 2007).

War-related deaths and disabilities also deplete the labor force, create orphans and single-parent families, and burden taxpayers who must pay for the care of orphans and disabled war veterans (see Chapter 2 for a discussion of military health care). In addition, the killing of unarmed civilians is likely to undermine the credibility of armed forces and make their goals more difficult to defend. In Iraq, for example, the events in Haditha (a city in western Iraq) became international news, outraged Iraqis, and led to intense condemnation of the U.S. mission. In November 2005, after a roadside bomb killed one Marine and wounded two others, "Marines shot five Iraqis standing by a car and went

> " Of all the enemies to public liberty war is, perhaps, the most to be dreaded because it comprises and develops the germ of every other. War is the parent of armies; from these proceed debts and taxes. And armies, and debts, and taxes are the known instruments for bringing the many under the domination of the few. . . . No nation could preserve its freedom in the midst of continual warfare. "
>
> James Madison
> Former U.S. president

house to house looking for insurgents, using grenades and machine guns to clear houses" (Watkins 2007, p. 1). Twenty-four Iraqis were killed, many of them women and children, "shot in the chest and head from close range" (McGirk 2006, p. 3). After *Time* magazine broke the story in March 2006 (McGirk 2006), the military investigated and reversed its claim that the civilians died as a result of the roadside bomb. Three officers have been charged with murder and one with failure to investigate the killings at the time.

Individuals who participate in experiments for military research may also suffer physical harm. U.S. representative Edward Markey of Massachusetts identified 31 experiments dating back to 1945 in which U.S. citizens were subjected to harm from participation in military experiments. Markey charged that many of the experiments used human subjects who were captive audiences or populations considered "expendable," such as elderly individuals, prisoners, and hospital patients. Eda Charlton of New York was injected with plutonium in 1945. She and 17 other patients did not learn of their poisoning until 30 years later. Her son, Fred Shultz, said of his deceased mother:

> I was over there fighting the Germans who were conducting these horrific medical experiments . . . at the same time my own country was conducting them on my own mother. (Miller 1993, p. 17)

Rape, Forced Prostitution, and Displacement of Women and Children

66 The use of rape in conflict reflects the inequalities women face in their everyday lives in peacetime. . . . Women are raped because their bodies are seen as the legitimate spoils of war.99

Amnesty International

Half a century ago the Geneva Convention prohibited rape and forced prostitution in war. Nevertheless, both continue to occur in modern conflicts.

Before and during World War II the Japanese military forced 100,000–200,000 women and teenage girls into prostitution as military "comfort women." These women were forced to have sex with dozens of soldiers every day in "comfort stations." Many of the women died as a result of untreated sexually transmitted diseases, harsh punishment, or indiscriminate acts of torture.

Since 1998, Congolese government forces have fought Uganda and Rwanda rebels. Women have paid a high price for this civil war, in which gang rape is "so violent, so systematic, so common . . . that thousands of women are suffering from vaginal fistula, leaving them unable to control bodily functions and enduring ostracism and the threat of debilitating health problems" (Wax 2003, p. 1). Rapes of Albanian women by Serbian and Yugoslavian paramilitary soldiers were, according to a Human Rights Watch report, "used deliberately as an instrument to terrorize the civilian population, extort money from families, and push people to flee their homes" (quoted in Lorch & Mendenhall 2000, p. 2). Feminist analysis of wartime rape emphasizes that the practice reflects not only a military strategy but also ethnic and gender dominance. In 2001 three Serbian soldiers accused of mass rape and forced prostitution were convicted by a United Nations tribunal and sentenced to a combined total of 60 years in prison (CNN 2001b).

War and terrorism also force women and children to flee to other countries or other regions of their homeland. For example, since 1990, at least 17 million children have been forced to leave their homeland because of armed conflict (Save the Children 2005). Refugee women and female children are particularly vulnerable to sexual abuse and exploitation by locals, members of security forces, border guards, or other refugees. In refugee camps women and children

A child soldier in Liberia points his gun at a cameraman while carting a teddy bear on his back. Although reliable figures are hard to obtain, the UN estimates that there are about 300,000 child soldiers fighting in wars worldwide.

may also be subjected to sexual violation. A 2003 report by Save the Children examined the treatment of women and children in 40 conflict zones. The use of child soldiers was reported in 70 percent of the zones (see this chapter's *The Human Side* feature), and trafficking of women and girls was reported in 85 percent of the zones. Wars are particularly dangerous for the very young—"Nine out of ten countries with the highest under-5 mortality rates are experiencing, or emerging from, armed conflict." These include Sierra Leone, Angola, Afghanistan, Niger, Liberia, Somalia, Mali, Chad, and the Democratic Republic of Congo (Save the Children 2007a, p. 13). Save the Children also estimated that, worldwide, 43 million primary-school-aged children are prevented from attending school because of armed conflict (Save the Children 2007b).

Social–Psychological Costs

Terrorism, war, and living under the threat of war interfere with social-psychological well-being and family functioning. In a study of 269 Israeli adolescents, Klingman and Goldstein (1994) found a significant level of anxiety and fear, particularly among younger females, with regard to the possibility of nuclear and chemical warfare. Similarly, Myers-Brown, Walker, and Myers-Walls (2000) reported that Yugoslavian children suffer from depression, anxiety, and fear as a response to conflicts in that region. Children, as well as adults, had similar emotional responses to the September 11 attacks on the World Trade Center and Pentagon (NASP 2003). Whether it is war or terrorism, "virtually every aspect of a child's development is damaged in such circumstances" (Bellamy 2003).

Guerrilla warfare is particularly costly in terms of its psychological toll on soldiers. In Iraq, "guerrilla insurgents attack with impunity. Friends and foes are indistinguishable, creating the need for constant vigilance. Death is always lurking, and can come from hand grenades thrown by children, earth-rattling bombs in suicide trucks, or snipers hidden in bombed-out buildings" (Waters 2005, p. 1). Soldiers repeatedly engage in intense, high-risk behavior and are exposed to

Smoke from mortars and grenade explosions hung thickly in the midnight air the first time he fought in combat, but he barely noticed. Automatic fire from M60 machine guns provided a symphony of chaos. "I saw a lot of my comrades killed or wounded," recalled Luis, who fired blindly into the ensuing chaos. "After you've been in battle, your blood gets warm," he explained to me, "and you don't feel the cold."

Fear is another matter. Desertion meant certain death at the hands of his guerilla commanders. For courage, he took off the tip of a bullet and ate the lead powder inside. It didn't work.

He and several hundred members of the FARC—Fuerzas Armadas Revolucionarias De Colombia, or the Revolutionary Armed Forces of Colombia, one of the most powerful guerilla groups in Latin America—had been ordered by commanders to lay siege on a rural police station in central Colombia. The guerillas faced a defiant police squad and army reinforce-

ments. Luis began shooting with an AK-47. Aside from 8 months of jungle boot camp, he had never been to school. The young soldier was proficient with an AK-47 but had never even held a pencil.

About 30 insurgents would rotate to the front of the assault line every half hour, collecting discarded guns and wounded before retreating. "I was afraid I'd be killed," he admitted in a quiet, calm voice, eyes focused downward, while sitting at a chipped wooden table with his hands folded in front of him. I was sitting with him in the staff conference room of a rehabilitation center about an hour outside Bogata, the capital. It was 5 years after the incident he described to me. Given the intensity of his story, I hadn't expected him to talk about it as easily as he did. Less than 15 minutes after meeting me, he took me on a journey of his painful memories. We spoke through a translator, and three staff workers at the center sat in the room with us, but we might

as well have been alone sharing stories. His knuckles were tattooed with the initials E, L, N, and F, representing former girlfriends. The frail-looking youth was scrawny and so shy that he couldn't hold eye contact with me for long. In the fleeting moments when I could catch his eyes, the deep sadness behind them was painfully obvious.

"Your training reminds you that you either die or survive," he continued. "That's the law of the jungle. You never know if your shots hit someone because there are thirty of you firing in one direction, and thirty firing back. I thought, "I should find another way of living, of moving forward into the fight. One has to be brave to fight an enemy."

Twenty-four hours after the initial assault, the rebels took the station along with twenty-eight police officers as prisoners. The next time Luis fought, he wasn't scared anymore.

Luis was 12 years old.

Source: Briggs (2005).

a high level of traumatic experiences during counterinsurgency efforts. Not surprisingly, the suicide rate among soldiers is higher for those deployed in combat. For example, the U.S. Army estimates that for every 100,000 soldiers deployed in Iraq, about 16 have taken their own lives. This number compares to about 12 suicides per 100,000 soldiers among those not deployed to Iraq during the same time period (U.S. Army Medical Command 2007). In addition, between 2001 and 2004 the divorce rates of active duty army officers and enlisted personnel have almost doubled (Crary 2005).

Military personnel who engage in combat and civilians who are victimized by war may experience a form of psychological distress known as **posttraumatic stress disorder (PTSD),** a clinical term referring to a set of symptoms that can result from any traumatic experience, including crime victimization, rape, or war. Symptoms of PTSD include sleep disturbances, recurring nightmares, flashbacks, and poor concentration (National Center for Posttraumatic Stress Disorder 2007). For example, Canadian Lt. General Romeo Dallaire, head of the United Nations peacekeeping mission in Rwanda, witnessed horrific acts of genocide. Four years after his return he continued to have images of "being in a valley at sunset, waist deep in bodies, covered in blood" (quoted in Rosenberg 2000, p. 14). PTSD is also associated with other personal problems, such as alcoholism, family violence, divorce, and suicide.

posttraumatic stress disorder (PTSD) A set of symptoms that may result from any traumatic experience, including crime victimization, war, natural disasters or abuses.

One study estimated that about 30 percent of male veterans of the Vietnam War have experienced PTSD and that about 15 percent continue to experience it (Hayman & Scaturo 1993). Another study of 215 Army National Guard and Army Reserve troops who served in the Gulf War and who did not seek mental health services upon return to the states reported that 16–24 percent exhibited symptoms of PTSD (Sutker et al. 1993). Almost 12 percent of a sample of soldiers returning from deployment in Iraq in 2005 reported symptoms of PTSD and 25 percent reported symptoms indicating clinical depression (Vasterling et al. 2006). Between 60 percent and 77 percent of soldiers deployed to Iraq who report symptoms of a mental disorder as a result of the war do not seek mental health treatment. Among those with symptoms, the most common reason cited (65 percent) for not seeking care was fear of being perceived as weak (Hoge et al. 2004). Further, compared with other civilians, refugees have higher rates of PTSD and depression owing to stressors common to refugee camps (De Jong et al. 2000). Finally, research on PTSD in children reveals that females are generally more symptomatic than males and that PTSD may disrupt the normal functioning and psychological development of children (Pfefferbaum 1997; Cauffman et al. 1998; National Center for Posttraumatic Stress Disorder 2007).

Diversion of Economic Resources

As discussed earlier, maintaining the military and engaging in warfare require enormous financial capital and human support. In 2006 worldwide military expenditures totaled more than $1.3 trillion (Stalenheim et al. 2007). This amount exceeds the combined government research expenditures on developing new energy technologies, improving human health, raising agricultural productivity, and controlling pollution. Although military spending by rogue states (Cuba, Iraq, Iran, Libya, North Korea, Sudan, and Syria) was just under $20 billion in 2005, U.S. military spending weighed in at $505 billion in the same year (SIPRI 2007b).

Money that is spent for military purposes could be allocated to social programs. The decision to spend $567 million, equal to the operating cost of the Smithsonian Institution (Center for Defense Information 2003), for one Trident II D-5 missile is a political choice. Similarly, allocating $2.3 billion for a "Virginia" attack submarine while our schools continue to deteriorate is also a political choice. Taxpayers will pay $137.6 billion dollars for the cost of the Iraq War in funding year 2007, the equivalent of providing (1) 39.2 million people with health care, or (2) 2.3 million elementary school teachers, or (3) 18.8 million placements for children in Head Start, or (4) 1.1 million affordable housing units, or (5) 22.7 million scholarships for university students (National Priorities Project 2007). As Figure 16.4 indicates, estimated expenditures for the 2007 fiscal year include more money for national defense than for justice, transportation, veterans' benefits, and natural resources and the environment combined (Office of Management and Budget 2007).

Destruction of the Environment

Traditional definitions of and approaches to national security have assumed that political states or groups constitute the principal threat to national security and welfare. This assumption implies that national defense and security are best served by being prepared for war against other states. The environmental costs of military preparedness are often overlooked or minimized in such discussions.

FIGURE 16.4
Selected federal U.S. outlays by function for 2007
(estimated).
Source: Office of Management and Budget
(2007).

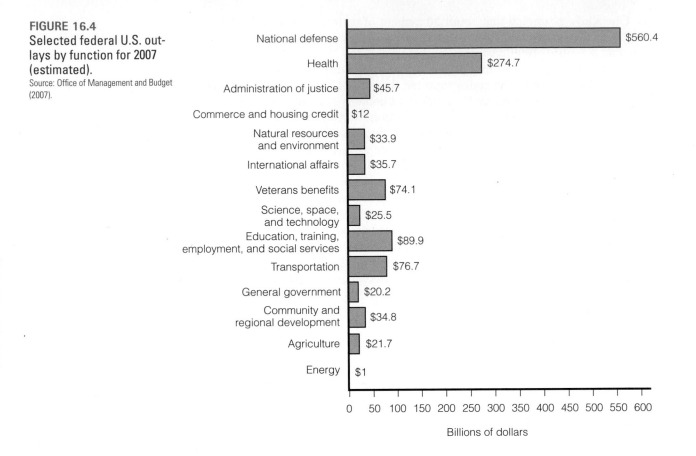

Billions of dollars

Annually, for example, the U.S. Navy Public Works Center in San Diego produces 12,000 tons of paint cans. The depleted cans, enough to fill 12 semitrucks, contain volatile chemicals, making their disposal or any proposed recycling method problematic (Lifsher 1999). Although this is a seemingly trivial matter—paint cans, in 1 year, at one facility—imagine the environmental problems generated by the military worldwide.

Destruction of the Environment During War. The environmental damage that occurs during war continues to devastate human populations long after war ceases. As the casings for landmines erode, poisonous substances—often carcinogenic—leak into the ground (Landmines 2003). In Vietnam 13 million tons of bombs left 25 million bomb craters. Nineteen million gallons of herbicides, including Agent Orange, were spread over the countryside. An estimated 80 percent of Vietnam's forests and swamplands were destroyed by bulldozing or bombing (Funke 1994).

The Gulf War also illustrates how war destroys the environment (Renner 1993; Funke 1994; Environmental Media Services 2002). In 6 weeks 1,000 air missions were flown, dropping 6,000 bombs. In 1991 Iraqi troops set 650 oil wells on fire, releasing oil, which still covers the surface of the Kuwaiti desert and seeps into the ground, threatening to poison underground water supplies. The 6–8 million barrels of oil that spilled into the Gulf waters are threatening marine life. This spill is by far the largest in history—25–30 times the size of the 1989 Exxon Valdez oil spill in Alaska.

The clouds of smoke that hung over the Gulf region for 8 months contained soot, sulfur dioxide, and nitrogen oxides—the major components of acid rain—and a variety of toxic and potentially carcinogenic chemicals and heavy metals. The U.S. Environmental Protection Agency estimates that in March 1991 about 10 times as much air pollution was being emitted in Kuwait as by all U.S. industrial and power-generating plants combined. Acid rain destroys forests and harms crops. It also activates several dangerous metals normally found in soil, including aluminum, cadmium, and mercury. Furthermore, the 900,000 depleted uranium munitions fired at Kuwait and Iraq are thought to have serious environmental and health consequences.

Combatants also often intentionally exploit natural resources to fuel their fighting. The local elephant population was heavily depleted during the civil war in southern Angola. The rebel group UNITA killed the animals to trade ivory for money to buy weapons. In addition, many were killed or fatally crippled by land mines planted by the guerrillas, though scientists reported that, remarkably, the animals seemed to have learned to avoid mined areas (Marshall 2007). Between 1992 and 1997, UNITA also earned $3.7 billion to support its fighting from the sale of "conflict" or "blood" diamonds (GreenCarat 2007). Among all the world's wars in the late 1990s, blood diamonds are thought to have counted for 15 percent of the global market (Partnership Africa Canada 2007). Depending upon the location of diamonds, mining can be highly destructive to riverbed ecosystems or to areas surrounding open-pit mines.

Sometimes environmental degradation is indirect. In Afghanistan two decades of conflict have

> put previous environmental management and conservation strategies on hold, brought about a collapse of local and national governance, destroyed infrastructure, hindered agricultural activity, and driven people into cities already lacking the most basic public amenities. (Haavisto 2005, p. 158)

In urban areas of Afghanistan, for example, safe water may be reaching less than 12 percent of the residents.

The ultimate environmental catastrophe facing the planet is a massive exchange thermonuclear war. Aside from the immediate human casualties, poisoned air, poisoned crops, and radioactive rain, many scientists agree that the dust storms and concentrations of particles would block vital sunlight and lower temperatures in the Northern Hemisphere, creating a **nuclear winter.** In the event of large-scale nuclear war, most living things on earth would die. The fear

© Michel Lipschitz/AP/Wide World Photo

Oil smoke from the 650 burning oil wells left in the wake of the Gulf War contains soot, sulfur dioxide, and nitrogen oxides, the major components of acid rain, along with a variety of toxic and potentially carcinogenic chemicals and heavy metals.

nuclear winter The predicted result of a thermonuclear war whereby thick clouds of radioactive dust and particles would block out vital sunlight, lower temperature in the Northern Hemisphere, and lead to the death of most living things on earth.

of nuclear war has greatly contributed to the military and arms buildup, which, ironically, also causes environmental destruction even in times of peace.

Destruction of the Environment in Peacetime. Even when military forces are not engaged in active warfare, military activities assault the environment. For example, modern military maneuvers require large amounts of land. The use of land for military purposes prevents other uses, such as agriculture, habitat protection, recreation, and housing. More important, military use of land often harms the land. In practicing military maneuvers, the armed forces demolish natural vegetation, disturb wildlife habitats, erode soil, silt up streams, and cause flooding. Military bombing and shooting ranges leave the land pockmarked with craters and contaminate the soil and groundwater with lead and other toxic residues. Furthermore,

> mined areas can restrict access to large areas of agricultural land, forcing populations to use small tracts of land to earn their livelihoods. The limited productive land that is available is over-cultivated, which contributes to long term underproduction, as minerals are depleted from the soil, and the loss of valuable vegetation. (Landmines 2003, p. 1)

Bombs exploded during peacetime leak radiation into the atmosphere and groundwater. From 1945 to 1990, 1,908 bombs were tested—that is, exploded—at more than 35 sites around the world. Although underground testing has reduced radiation, some radioactive material still escapes into the atmosphere and is suspected of seeping into groundwater. Similarly, in Russia decommissioned submarines have been found to emit radioactive pollution ("Foreign Conservationists Under Siege" 2000).

Finally, although arms control and disarmament treaties of the last decade have called for the disposal of huge stockpiles of weapons, no completely safe means of disposing of weapons and ammunition exist. Many activist groups have called for placing weapons in storage until safe disposal methods are found. Unfortunately, the longer the weapons are stored, the more they deteriorate, increasing the likelihood of dangerous leakage. In 2003 a federal judge gave permission, despite objections by environmentalists, to incinerate 2,000 tons of nerve agents and mustard gas left from the Cold War era. Although the army says that it is safe to dispose of the weapons, they have issued protective gear in case of an "accident" to the nearly 20,000 residents who live nearby (CNN 2003).

STRATEGIES FOR ACTION: IN SEARCH OF GLOBAL PEACE

Various strategies and policies are aimed at creating and maintaining global peace. These include the redistribution of economic resources, the creation of a world government, peacekeeping activities of the United Nations, mediation and arbitration, and arms control.

Redistribution of Economic Resources

Inequality in economic resources contributes to conflict and war because the increasing disparity in wealth and resources between rich and poor nations fuels hostilities and resentment. Therefore any measures that result in a more

On January 30, 2005, millions of Iraqis defied the insurgents and voted in their first free election in half a century. Afghanistan's first ever parliamentary election took place on September 18, 2005.

© Majid Saeedi/Getty Images

equal distribution of economic resources are likely to prevent conflict. John J. Shanahan (1995), retired U.S. Navy vice admiral and former director of the Center for Defense Information, suggested that wealthy nations can help reduce the social and economic roots of conflict by providing economic assistance to poorer countries. Nevertheless, U.S. military expenditures for national defense far outweigh U.S. economic assistance to foreign countries. For instance, the U.S. government estimates that its 2008 budget for foreign affairs (including all nonmilitary foreign aid, the U.S. Department of State, the Peace Corps, and scholarly exchange programs) will be $36.5 billion, or about 1 percent of the entire federal budget and 5 percent of the total budget for the Department of Defense, the wars in Afghanistan and Iraq, and the global war on terrorism (Brooks 2007).

As discussed in Chapter 13, strategies that reduce population growth are likely to result in higher levels of economic well-being. Funke (1994) explained that "rapidly increasing populations in poorer countries will lead to environmental overload and resource depletion in the next century, which will most likely result in political upheaval and violence as well as mass starvation" (p. 326). Although achieving worldwide economic well-being is important for minimizing global conflict, it is important that economic development does not occur at the expense of the environment.

Finally, former United Nations Secretary General Kofi Annan in an address to the United Nations observed that it is not poverty per se that leads to conflict but rather the "inequality among domestic social groups" (Deen 2000). Referencing a research report completed by the Tokyo-based United Nations University, Annan argued that "inequality . . . based on ethnicity, religion, national identity or economic class . . . tends to be reflected in unequal access to political power that too often forecloses paths to peaceful change" (Deen 2000).

Global Governance

From the 1950s to the early 1990s world power was primarily divided between the United States and the Soviet Union. With the fall of the communist governments in Eastern Europe in the early 1990s the United States became the world's only remaining superpower, causing observers to debate the proper role of the United States in global politics. Many argued that with the demise of the Soviet Union the world no longer needed a watchdog—a role the United States had historically played. Still others argued that "it is both the moral duty and in the strategic interest of the United States to become involved in regions where there is unrest, and stand as a leading force in international organizations, especially the United Nations" (Social Science Research Council 2003, p. 1).

Although some commentators are pessimistic about the likelihood of a new world order, Lloyd (1998) identified three global trends that, he contends, signify the "ghost of a world government yet to come" (p. 28). First is the increasing tendency for countries to engage in "ecological good behavior," indicating a concern for a global rather than national well-being. Second, despite several economic crises worldwide, major world players such as the United States and Great Britain have supported rather than abandoned nations in need. And, finally, an International Criminal Court has been created with "powers to pursue, arraign, and condemn those found guilty of war and other crimes against humanity" (p. 28). On the basis of these three trends, Lloyd (1998) concluded, "Global justice, a wraith pursued by peace campaigners for over a century, suddenly seems achievable" (p. 28). Despite Lloyd's optimism, the French and Dutch overwhelmingly rejected the proposed constitution of the European Union (EU) in a 2005 referendum. The EU is an organization of 27 member states, all of which must ratify the constitution for it to become law. Presently, 17 countries have ratified the constitution—Austria, Bulgaria, Cyprus, Estonia, Finland, Germany, Greece, Hungary, Italy, Latvia, Lithuania, Luxembourg, Malta, Romania, Slovakia, Slovenia, and Spain (BBC 2007c).

The United Nations

The United Nations (UN), whose charter begins, "We the people of the United Nations—Determined to save succeeding generations from the scourge of war," has engaged in more than 60 peacekeeping operations since 1948 (see Table 16.5) (United Nations 2007). The UN Security Council can use force, when necessary, to restore international peace and security.

> United Nations peacekeepers—military personnel in their distinctive blue helmets or blue berets, civilian police and a range of other civilians—help implement peace agreements, monitor cease fires, create buffer zones, or support complex military and civilian functions essential to maintain peace and begin reconstruction and institution building in societies devastated by war. (United Nations 2003, p. 1)

Recently, the UN has been involved in overseeing multinational peacekeeping forces in Sudan, Burundi, Haiti, Liberia, and Ethiopia (United Nations 2007).

In the last few years, the UN has come under heavy criticism. First, in recent missions, developing nations have supplied more than 75 percent of the troops while developed countries—United States, Japan, and Europe—have contributed 85 percent of the finances. As one UN official commented, "You can't have a situation where some nations contribute blood and others only money" (quoted by Vesely 2001, p. 8). Second, a review of UN peacekeeping

A woman in Darfur scoops up grain that has spilled from bags dropped from a plane by the UN's World Food Program. As in many of today's conflicts, the fighting in Darfur makes it very difficult for humanitarian aid agencies to run regular operations to feed, clothe, and shelter civilians.

© AP Photo/Josphat Kasire

TABLE 16.5 United Nations Peacekeeping Operations: Summary Data, 2007	
Military personnel and civilian police serving in peacekeeping operations	82,873
Countries contributing military personnel and civilian police	118
International civilian personnel	4,782
Local civilian personnel	10,879
UN volunteers	2,049
Total number of fatalities in peacekeeping operations since 1948	2,365
Approved budgets for July 1, 2006, to June 30, 2007	$5.48 billion
Estimated total cost of operations from 1948 to June 2006	$41.74 billion
Source: United Nations (2007).	

operations noted several failed missions, including an intervention in Somalia in which 44 U.S. marines were killed (Lamont 2001). Third, as typified by the debate over the disarming of Iraq, the UN cannot take sides but must wait for a consensus of its members which, if not forthcoming, undermines the strength of the organization (Goure 2003). The consequences of delays can be staggering. For example, under the Genocide Convention of 1948, the UN is obligated to prevent instances of **genocide,** defined as "acts committed with the intent to destroy, in whole or in part, a national, racial, ethnic or religious group" (United Nations 1948). In 2000, the UN Security Council formally acknowledged its failure to prevent the 1994 genocide in Rwanda (BBC 2000). After the death of 10 Belgian soldiers in the days leading up to the genocide, and without a consensus for action among the members, the Security Council ignored warnings from the mission's commander about impending disaster and withdrew its 2,500 peacekeepers.

genocide The deliberate, systematic annihilation of an entire nation, people, or ethnic group.

Finally, the concept of the UN is that its members represent individual nations, not a region or the world. And because nations tend to act in their own best economic and security interests, UN actions performed in the name of world peace may be motivated by nations acting in their own interests.

What Do You Think? The war between the Sudanese Army and pro-government Arab militias on the one side and rebel groups in Darfur on the other shows no sign of coming to an end. The conflict has already claimed at least 200,000 lives (Hagan & Palloni 2006). Another 200,000 have fled across the border to neighboring Chad, where they are living in refugee camps indefinitely and remain vulnerable to cross-border attacks. In July 2007, the UN authorized deployment of up to 26,000 military personnel and civilian police (CNN 2007), but many believe this is not enough to protect civilians in a region the size of France. Although most agree that Sudanese government forces have committed war crimes and atrocities, UN investigators refute claims by the U.S. government and others in the international community that genocide is under way (BBC 2007d). Meanwhile, Sudan claims that it is fighting a legitimate war against internal rebels. Is it appropriate for international forces to intervene on moral and humanitarian grounds? If so, who should bear the responsibility for such a mission? Why intervene in Darfur and not other deadly wars?

As a result of such criticisms, outgoing UN Secretary-General Kofi Annan called on the 191 members of the UN to approve the most far-reaching changes in the 60-year history of the organization (Lederer 2005). One of the most controversial recommendations concerns the composition of the Security Council, the most important decision-making body of the organization. Annan's recommendation that the 15 members of the Security Council—a body dominated by the United States, Great Britain, France, Russia, and China—be changed to include a more representative number of nations could, if approved, shift the global balance of power. Other recommendations include the establishment of a Humans Rights Council and the development of a comprehensive antiterrorism strategy. Ban Ki-moon, former foreign minister of South Korea, campaigned for the Secretary-General position on a platform that included support for the UN reforms. He was elected as the eighth Secretary-General in October 2006 (MacAskill, Pilkington, & Watts 2006).

Mediation and Arbitration

Most conflicts are resolved through nonviolent means (Worldwatch Institute 2003). Mediation and arbitration are just two of the nonviolent strategies used to resolve conflicts and stop or prevent war. In mediation a neutral third party intervenes and facilitates negotiation between representatives or leaders of conflicting groups. Good mediators do not impose solutions but rather help disputing parties generate options for resolving the conflict (Conflict Research Consortium 2003). Ideally, a mediated resolution to a conflict meets at least some of the concerns and interests of each party to the conflict. In other words, mediation attempts to find "win-win" solutions in which each side is satisfied with the solution.

Although mediation is used to resolve conflict between individuals, it is also a valuable tool for resolving international conflicts. For example, in 2000 mediators were used to resolve the escalating violence between Israeli and Palestinian troops, and in 2004 mediation talks between the Sudan Liberation Army and the government led to a temporary cease fire. Using mediation as a means of resolving international conflict is often difficult, given the complexity of the issues. However, research by Bercovitch (2003) shows that there are more mediators available and more instances of mediation in international affairs than ever before. For example, Bercovitch identified more than 250 separate mediation attempts during the Balkans Wars of the early and mid-1990s.

Arbitration also involves a neutral third party who listens to evidence and arguments presented by conflicting groups. Unlike mediation, however, in arbitration the neutral third party arrives at a decision or outcome that the two conflicting parties agree in advance to accept. For instance, the Permanent Court of Arbitration—an intergovernmental organization based in The Hague since 1899—arbitrates disputes about territory, treaty compliance, human rights, commerce, and investment among any of its 107 member states who signed and ratified either of its two founding legal conventions (Permanent Court of Arbitration 2007). Recent cases include a dispute between France, Britain, and Northern Ireland about the Eurotunnel, a boundary dispute between Eritrea and Ethiopia, and a maritime boundary dispute between Guyana and Suriname.

What Do You Think? "We do not negotiate with terrorists" is a common claim of politicians and heads of state. This is a popular notion in societies divided by violent conflict, where many have been victimized by political violence or terrorism. Negotiating with terrorists, it is feared, will confer legitimacy on them and their tactics and encourage other groups to adopt violence to achieve their goals. Yet there seem to be situations in which governments have no other choice but to negotiate with terrorists to end violence and establish a working peace. After all, as the saying goes, "You make peace with your enemies, not your friends." Are there situations when it is appropriate to negotiate with terrorists? Do different situations call for different policies? What do you think?

arbitration Dispute settlement in which a neutral third party listens to evidence and arguments presented by conflicting groups and arrives at a decision or outcome that the parties have agreed to in advance to accept.

mutually assured destruction (MAD) A Cold War doctrine referring to the capacity of two nuclear states to destroy each other, thus reducing the risk that either state will initiate war.

Arms Control and Disarmament

In the 1960s the United States and the Soviet Union led the world in an arms race, with each competing to build a more powerful military arsenal than its adversary. If either superpower were to initiate a full-scale war, the retaliatory powers of the other nation would result in the destruction of both nations. Thus the principle of **mutually assured destruction (MAD)** that developed from nu-

Self and Society | The Nuclear Numbers Game*

Carefully read each of the following questions and select the correct answer. When you are finished, compare your answers to those provided and rank your performance using the scale below.

1. The Manhattan Project was the code name for the secret government project to develop and test the world's first atom bomb. How much did it cost (in 1996 dollars)?
   - (a) $100,000
   - (b) $250,000
   - (c) $5.5 billion
   - (d) $20 billion

2. How many nuclear weapons has the United States built since starting production in 1951?
   - (a) About 67,500
   - (b) About 100,000
   - (c) Nearly 25,000
   - (d) The number is classified

3. In November 1952, the United States tested a 10.4 megaton hydrogen bomb officially named "Mike" for megaton, but dubbed the "Sausage" by its handlers because of its enormous tube-like size. The bomb was detonated in the Marshall Islands about 3,000 miles west of Hawaii. Can it be found on a map today?
   - (a) Yes, it has been repopulated by local inhabitants.
   - (b) No, the island was vaporized as a result of the surface detonation.

4. If you knew where to look, you could find 2,000 nuclear weapons in the state of Georgia.
   - (a) True
   - (b) False

5. How many nuclear weapons did the United States detonate in tests between 1945 and 1992 (the last year it detonated a nuclear weapon)?
   - (a) 13
   - (b) 130
   - (c) 1,030
   - (d) 10,300

6. What is the total volume (in cubic meters) of radioactive waste resulting from weapons activities?
   - (a) 650,000
   - (b) 104,000,000
   - (c) 426,000,000
   - (d) 1.2 billion

7. How many nuclear bombs has the United States lost (and not recovered) in accidents?
   - (a) 0
   - (b) 1
   - (c) 6
   - (d) 11

8. How many nuclear warheads and bombs does the United States currently possess in its stockpile?
   - (a) 3,416
   - (b) 7,982
   - (c) 10,600
   - (d) 32,750

9. With the end of the Cold War and the fall of the Soviet Union, most U.S. nuclear weapons are no longer targeted at specific sites around the world.
   - (a) True: Only about 100 weapons are targeted against other nuclear powers.
   - (b) False: The U.S. arsenal targets 2,500 sites for nuclear destruction around the world.

clear weapons capabilities transformed war from a win-lose proposition to a lose-lose scenario. If both sides would lose in a war, the theory suggested, neither side would initiate war. At its peak year in 1966, the U.S. stockpile of nuclear weapons included more than 32,000 warheads and bombs (see this chapter's *Self and Society* feature).

Because of the end of the Cold War and the growing realization that current levels of weapons literally represented overkill, governments have moved in the direction of arms control, which, at least in the past, involved reducing or limiting defense spending, weapons production, and armed forces. Recent arms control initiatives include SALT (the Strategic Arms Limitation Treaty), START (the Strategic Arms Reduction Treaty), SORT (the Strategic Offensive Reduction Treaty), and NPT (the Nuclear Nonproliferation Treaty).

❝ War doesn't determine who's right, just who's left. **❞**

George Carlin
Comedian

10. If the United States is attacked by nuclear weapons, how many emergency facilities are available for the President to take cover?

(a) 1
(b) 5
(c) 26
(d) 75

11. On U.S. soil, the government restricted its nuclear tests only to the Southwest desert because of the remote terrain and relatively low risks to human populations.

(a) True: There have been no nuclear tests on U.S. soil except those in Nevada and New Mexico.
(b) False: Nuclear weapons have been detonated in Alaska, Colorado, and Mississippi.

12. Which is larger? The total land area occupied by nuclear weapons bases and facilities or the combined square mileage of the District of Columbia, Massachusetts, and New Jersey?

(a) The combined area of the two states and the District of Columbia is much larger than the other.
(b) The nuclear bases require much more space than the states and the District of Columbia.
(c) They are about the same.

13. How many submarines with nuclear weapons does the United States deploy?

(a) Only a few
(b) 71
(c) About 200
(d) More than 1,000

14. Which arm of the U.S. government supervises our nuclear weapons stockpile?

(a) Department of Defense
(b) Department of Energy
(c) National Security Council
(d) Department of State

15. Which was the last year that the United States detonated a nuclear bomb?

(a) 1945
(b) 1962
(c) 1992
(d) 1999

Answers: 1 (d). 2 (a). 3 (b). 4 (a). 5 (c). 6 (b). 7 (d). 8 (c). 9 (b). 10 (d). 11 (b). 12 (c). 13 (b). 14 (b). 15 (c).

Rank Your Performance: Number Correct

14 or 15 Excellent
12 or 13 Good
10 or 11 Average
8 or 9 Fair
7 or less Poor

*Since 1998, the time of this study, the United States has not detonated nuclear weapons for tests and has not added new nuclear weapons to its arsenal (Federation of American Scientists 2007).

Source: Adapted from U.S. Nuclear Weapons Cost Study Project (1998).

Strategic Arms Limitation Treaty. Under the 1972 SALT agreement (SALT I), the United States and the Soviet Union agreed to limit both their defensive weapons and their land-based and submarine-based offensive weapons. Also in 1972, Henry Kissinger drafted the Declaration of Principles, known as **détente,** which means "negotiation rather than confrontation." A further arms-limitation agreement (SALT II) was reached in 1979 but was never ratified by Congress because of the Soviet invasion of Afghanistan. Subsequently, the arms race continued with the development of new technologies and an increase in the number of nuclear warheads.

détente The philosophy of "negotiation rather than confrontation" in reference to relations between the United States and the former Soviet Union; put forth by Secretary of State Henry Kissinger's Declaration of Principles in 1972.

Strategic Arms Reduction and Strategic Offensive Reduction Treaties. Strategic arms talks resumed in 1982, but relatively little progress was made for several years. During this period, President Reagan proposed the Strategic Defense Initiative (SDI), more commonly known as "star wars," which purportedly would be able to block missiles launched by another country against the United States. Although some research was conducted on the system, SDI, which was land based, was never actually built. More recently, a space-based national weapons program has been proposed by the U.S. Air Force. Proponents argue that such a program, although costly, would provide "freedom to attack as well as freedom from attack" from space (Weiner 2005).

In 1991 the international situation changed dramatically. The communist regime in the Soviet Union had fallen, the Berlin Wall had been dismantled, and many Eastern European and Baltic countries were under self-rule. SALT was renamed START and was signed in 1991. A second START agreement, signed in 1993 and ratified by the U.S. Senate, signaled the end of the Cold War. START II called for the reduction of nuclear warheads to 3,500 by the year 2003 but was superseded by SORT, which requires both the United States and Russia to reduce their nuclear arsenals to 1,700–2,200 warheads by December 31, 2012 (Arms Control Association 2003a).

Nuclear Nonproliferation Treaty. The 1970 NPT was renewed in 2000 and adopted by 187 countries. Only India, Israel, and Pakistan have not signed the agreement (Dickey 2005). The treaty holds that countries without nuclear weapons will not try to get them; in exchange, the countries with nuclear weapons (the United States, United Kingdom, France, China, and Russia) agree that they will not provide nuclear weapons to countries that do not have them. In January 2003 North Korea, under suspicion of secretly producing nuclear weapons, announced that it was withdrawing from the treaty, effective immediately (Arms Control Association 2003b). In July 2005, after "more than a year of stalemate, North Korea agreed . . . to return to disarmament talks . . . and pledged to discuss eliminating its nuclear-weapons program" (Brinkley & Sanger 2005, p. 1). However, in October 2006, North Korea detonated a nuclear device underground and, in 2007 successfully tested a ballistic missile, heightening fears that North Korea would launch a nuclear attack against South Korea, China, or Japan (Burns 2007).

Even if military superpowers honor agreements to limit arms, the availability of black-market nuclear weapons and materials presents a threat to global security. Kyl and Halperin (1997) noted that U.S. security is threatened more by nuclear weapons falling into the hands of a terrorist group than by a nuclear attack from an established government. One of the most successful nuclear weapons brokers was Pakistan's Abdul Qadeer Khan. Khan, the "father" of Pakistan's nuclear weapons program, sold the technology and equipment required to make nuclear weapons to rogue states such as Iran and North Korea. He is also suspected of selling nuclear weapons to Al-Qaeda. At present, his activities are under investigation by the United States and the International Atomic Energy Agency (Powell & McGirk 2005).

In 2003 the United States accused Iran of operating a covert nuclear weapons program. Although Iranian officials responded that their nuclear program was solely for the purpose of generating electricity, concerns continue. Hostile public statements about Israel by Iran's president Mahmoud Ahmadinejad deepen already grave concerns about Iran's possible acquisition of nuclear weapons. Analysts fear the development of nuclear weapons by Iran would spur a nuclear competition with Israel, the only other state in the Middle East to possess nuclear weapons, and provoke other states in the region to acquire them (Salem 2007). In 2005 the United States announced that it would freeze the assets of any company doing business with Iran's Atomic Energy Organization or any other organizations thought to be involved in nuclear proliferation (Dickey 2005). Iran announced its uranium enrichment program in 2006 and invited the International Atomic Energy Agency (IAEA) to inspect its facilities. IAEA officials could not conclusively confirm Iran's claim that the program was for peaceful purposes. The UN Security Council passed a resolution demanding an immediate halt to Iran's enrichment program. Iran ignored the resolution, and in December of 2006, the UN Security Council imposed sanctions against Iran's trade in nuclear technology and materials. In March 2007, the Security Council intensified sanctions by banning sales of major military systems to Iran, export of arms from Iran, and international financial transactions with the Iran state bank and many of its top officials and firms, and restricting foreign loans to Iran except for humanitarian purposes (BBC News 2007e). Iran continues to assert its authority to pursue a nuclear program while unofficial talks continue (Global Policy Forum 2007).

Many observers consider South Asia the world's most dangerous nuclear rivalry. India and Pakistan are the only two nuclear powers that share a border and have repeatedly fired upon each other's armies while in possession of these weapons (Stimson Center 2007). Both countries have detonated nuclear weapons. India's first detonation was in 1974. Pakistan detonated six weapons in 1998, a few weeks after India's second round of nuclear tests. Precise figures are hard to determine, but experts believe that India possesses about 50 nuclear bombs and that Pakistan has between 40 and 70 nuclear bombs (Center for Defense Information 2007). Many fear that a conventional military confrontation between these two countries may some day escalate to an exchange of nuclear weapons.

As states that want to obtain nuclear weapons are quick to point out, nuclear states that advocate for nonproliferation possess well over 25,000 weapons, a huge reduction in the world's arsenal from Cold War days but still a massive potential threat to the earth. "Do as I say not as I do" is a weak bargaining position. Recognizing this situation and with concern about the possibility of new arms races, many high-level experts have begun to advocate a more comprehensive approach to banning nuclear weapons—with the United States leading the nuclear powers toward complete nuclear disarmament. In January 2007, three former U.S. Secretaries of State (George Schultz, William Perry, and Henry Kissinger) and former chairman of the Senate Armed Services Committee Sam Nunn published an editorial in the *Wall Street Journal* advocating that the United States "take the world to the next stage" of disarmament to "a world free of nuclear weapons" (Schultz et al. 2007). As a start, their proposal advocated the elimination of all short-range nuclear missiles, a complete halt in the production of weapons-grade uranium, and continued reduction of nuclear forces.

The Problem of Small Arms

Although the devastation caused by even one nuclear war could affect millions, the availability of conventional weapons fuels many active wars around the world. Small arms and light weapons include handguns, submachine guns

A stockpile of assault rifles and light machine guns is ready to be inspected and prepared for destruction in 2006 through a U.S. government–funded program in Angola. Collected from ex-combatants and the Angolan army, these weapons are left over from Angola's 27-year-old civil war that formally ended in 2002.

© AP Photo/Arnulfo Franco

and automatic weapons, grenades, mortars, landmines, and light missiles. The Small Arms Survey estimated that in 2006 there were 639 million firearms in circulation, with about 59 percent of them being owned legally (Geneva Graduate Institute for International Studies 2006). Civilians purchased 80 percent of these weapons and were the victims of 90 percent of the casualties caused by them. Half a million people die each year as a result of small arms use, 200,000 in homicides and suicides and the rest during wars and other armed conflicts.

Unlike control of weapons of mass destruction such as chemical and biological weapons, controlling the flow of small arms is not easy because they have many legitimate uses by the military, by law enforcement officials, and for recreational or sporting activities (Schroeder 2007). Small arms are easy to afford, to use, to conceal, and to transport illegally. The top five conventional arms suppliers between 1998 and 2005 were the United States, Russia, France, Germany, and the United Kingdom (Shah 2007). During this time the United States made $97.1 billion in arms sales, about 36 percent of total global sales (Shah 2007). Despite these enormous figures, the U.S. State Department's Office of Weapons Removal and Abatement administers a program that has supported the destruction of 800,000 weapons and 81 million rounds of ammunition in 23 countries since its founding in 2001 (Schroeder 2007). The availability of these weapons fuels terrorist groups and undermines efforts to promote peace after wars have formally concluded. "If not expeditiously destroyed or secured, stocks of arms and ammunition left over after the cessation of hostilities frequently re-circulate into neighboring regions, exacerbating conflict and crime" (U.S. Department of State 2007, p.1). Running on a budget of $8.6 million in 2007, the Department proposes to expand the program to $440.7 million in 2008.

UNDERSTANDING CONFLICT, WAR, AND TERRORISM

As we come to the close of this chapter, how might we have an informed understanding of conflict, war, and terrorism? Each of the three theoretical positions discussed in this chapter reflects the realities of global conflict. As structural functionalists argue, war offers societal benefits—social cohesion, economic prosperity, scientific and technological developments, and social change. Furthermore, as conflict theorists contend, wars often occur for economic reasons because corporate elites and political leaders benefit from the spoils of war—land and water resources and raw materials. The symbolic interactionist perspective emphasizes the role that meanings, labels, and definitions play in creating conflict and contributing to acts of war.

The September 11 attacks on the World Trade Center and the Pentagon and the aftermath—the battle against terrorism, the bombing of Afghanistan, and the war with Iraq—forever changed the world we live in. For some theorists these events were inevitable. Political scientist Samuel P. Huntington argued that such conflict represents a **clash of civilizations.** In *The Clash of Civilizations and the Remaking of World Order* (1996), Huntington argued that in the new world order

> the most pervasive, important and dangerous conflicts will not be between social classes, rich and poor, or economically defined groups, but between people belonging to different cultural entities . . . the most dangerous cultural conflicts are those along the fault lines between civilizations . . . the line separating peoples of Western Christianity, on the one hand, from Muslim and Orthodox peoples on the other. (p. 28)

Although not without critics, the clash-of-civilizations hypothesis has some support. In an interview of almost 10,000 people from nine Muslim states representing half of all Muslims worldwide, only 22 percent had favorable opinions toward the United States (CNN 2003). Even more significantly, 67 percent saw the September 11 attacks as "morally justified," and the majority of respondents found the United States to be overly materialistic and secular and having a corrupting influence on other nations. Conversely, 39 percent of Americans report being prejudiced against Muslims, 46 percent believe that they are too extreme in their religious beliefs, and 39 percent agree that Muslims in the United States should carry a special ID (Saad 2006).

Ultimately, we are all members of one community—Earth—and have a vested interest in staying alive and protecting the resources of our environment for our own and future generations. But, as we have seen, conflict between groups is a feature of social life and human existence that is not likely to disappear. What is at stake—human lives and the ability of our planet to sustain life—merits serious attention. World leaders have traditionally followed the advice of philosopher Karl von Clausewitz: "If you want peace, prepare for war." Thus nations have sought to protect themselves by maintaining large military forces and massive weapons systems. These strategies are associated with serious costs, particularly in hard economic times. In diverting resources away from other social concerns, defense spending undermines a society's ability to improve the overall security and well-being of its citizens. Conversely, defense-spending cutbacks, although unlikely in the present climate, could potentially free up resources for other social agendas, including lowering taxes, reducing the national debt, addressing environmental concerns, eradicating hunger and

❝I hate war as only a soldier who has lived it can, only as one who has seen its brutality, its futility, its stupidity.**❞**

Dwight D. Eisenhower
Former U.S. president, military leader, hero

clash of civilizations A hypothesis that the primary source of conflict in the 21st century has shifted away from social class and economic issues and toward conflict between religious and cultural groups, especially those between large-scale civilizations such as the peoples of Western Christianity and Muslim and Orthodox peoples.

poverty, improving health care, upgrading educational services, and improving housing and transportation. Therein lies the promise of a "peace dividend." The hope is that future dialogue on the problems of war and terrorism will redefine national and international security to encompass social, economic, and environmental well-being.

CHAPTER REVIEW

- **What is the relationship between war and industrialization?**

  War indirectly affects industrialization and technological sophistication because military research and development advances civilian-used technologies. Industrialization, in turn, has had two major influences on war: The more industrialized a country is, the lower the rate of conflict, and if conflict occurs, the higher the rate of destruction.

- **What are the latest trends in armed conflicts?**

  Since World War II, wars between two or more states make up the smallest percentage of armed conflicts. In the contemporary era, the majority of armed conflicts have occurred between groups in a single state, who compete for the power to control the resources of the state or to break away and form their own state.

- **In general, how do feminists view war?**

  Feminists are quick to note that wars are part of the patriarchy of society. Although women and children may be used to justify a conflict (e.g., improving women's lives by removing the repressive Taliban in Afghanistan), the basic principles of male dominance and control are realized through war.

- **What are some of the causes of war?**

  The causes of war are numerous and complex. Most wars involve more than one cause. Some of the causes of war are conflict over land and natural resources; values or ideologies; racial, ethnic, and religious hostilities; defense against hostile attacks; revolution; and nationalism.

- **What is terrorism, and what are the different types of terrorism?**

  Terrorism is the premeditated use, or threatened use, of violence by an individual or group to gain a political or social objective. Terrorism can be either transnational or domestic. Transnational ter-

rorism occurs when a terrorist act in one country involves victims, targets, institutions, governments, or citizens of another country. Domestic terrorism involves only one nation, such as the 1995 truck bombing of a nine-story federal office building in Oklahoma City.

- **What are some of the macrolevel "roots" of terrorism?**

  Although not an exhaustive list, some of the macrolevel "roots" of terrorism include (1) a failed or weak state, (2) rapid modernization, (3) extreme ideologies, (4) a history of violence, (5) repression by a foreign occupation, (6) large-scale racial or ethnic discrimination, and (7) the presence of a charismatic leader.

- **What problems do small arms pose?**

  Even after a conflict ends, these weapons circulate in society, making crime worse or falling into the hands of terrorists. Trade in small arms is legal because they have many legitimate uses—for example, by the military, police, and hunters. Because they are small and simple to handle, these weapons are easily concealed and transported, making it difficult to control them.

- **How has the United States responded to the threat of terrorism?**

  The United States has used both defensive and offensive strategies to fight terrorism. Defensive strategies include using metal detectors and X-ray machines at airports and strengthening security at potential targets, such as embassies and military command posts. The Department of Homeland Security coordinates such defensive tactics. Offensive strategies include retaliatory raids such as the U.S. bombing of terrorist facilities in Afghanistan, group infiltration, and preemptive strikes.

- **What is meant by "diversion of economic resources"?**

  Worldwide, the billions of dollars used on defense could be channeled into social programs dealing with, for example, education, health, and poverty. Thus defense monies are economic resources diverted from other needy projects.

- **What are some of the criticisms of the United Nations?**

First, in recent missions, developing nations have supplied more than 75 percent of the troops. Second, several recent UN peacekeeping operations have failed. Third, the UN cannot take sides but must wait for a consensus of its members, which, if not forthcoming, undermines the strength of the organization. Finally, the concept of the UN is that its members represent individual nations, not a region or the world. Because nations tend to act in their own best economic and security interests, UN actions performed in the name of world peace may be motivated by nations acting in their own interests.

TEST YOURSELF

1. The majority of suicide bombers are Islamic fundamentalist.
   a. True
   b. False
2. War between states is still the most common form of warfare.
   a. True
   b. False
3. The rise of the modern state is most directly a result of
   a. industrialization and the creation of national markets
   b. innovations in communications technology
   c. the development of armies to control territory
   d. the development of police to control a population
4. Structural-functionalist explanations about war emphasize
   a. that war is a biological necessity
   b. that despite its destructive power, war persists because it fulfills social needs

   c. that war is an anachronism that will eventually disappear
   d. that war is necessary because it benefits political and military elites
5. On the whole, conflicts over values and ideologies are more difficult to resolve than those over material resources.
   a. True
   b. False
6. All wars are a result of unequal distribution of wealth.
   a. True
   b. False
7. Next to defense spending, transfers to foreign governments are the most expensive item in the U.S. government budget.
   a. True
   b. False
8. Which of the following factors is a likely cause of revolutions or civil wars?
   a. A weak or failed state
   b. An authoritarian government that ignores major demands from citizens
   c. The availability of strong opposition leaders
   d. All of the above
9. Primordial explanations of ethnic conflict suggest that
   a. ethnic leaders instigate hostilities to serve their own interests
   b. people become hostile when they blame their frustration with economic hardship on competing ethnic groups
   c. ancient hatreds compel ethnic groups to continue fighting
   d. None of the above
10. The consequences of war and the military for the environment
   a. are prevalent only during wartime
   b. persist in peacetime or for many years after a war is over
   c. do not significantly impact the environment
   d. have been mostly reduced by technological innovations

Answers: 1 b. 2 b. 3 c. 4 b. 5 a. 6 b. 7 b. 8 d. 9 c. 10 b.

KEY TERMS

arbitration
clash of civilizations
Cold War
constructivist explanations
détente
domestic terrorism
dual-use technologies
genocide

guerrilla warfare
military-industrial complex
mutually assured destruction (MAD)
nuclear winter
posttraumatic stress disorder (PTSD)
primordial explanations

security dilemma
state
terrorism
transnational terrorism
war
weapons of mass destruction (WMD)

MEDIA RESOURCES

Understanding Social Problems, **Sixth Edition**
Companion Website

academic.cengage.com/sociology/mooney
Visit your book companion website, where you will find flash cards, practice quizzes, Internet links, and more to help you study.

 Just what you need to know NOW! Spend time on what you need to master rather than on information you already have learned. Take a pre-test for this chapter, and CengageNOW will generate a personalized study plan based on your results. The study plan will identify the topics you need to review and direct you to online resources to help you master those topics. You can then take a post-test to help you determine the concepts you have mastered and what you will need to work on. Try it out! Go to academic.cengage .com/login to sign in with an access code or to purchase access to this product.

Epilogue

Today, there is a crisis—a crisis of faith: faith in the ideals of equality and freedom, faith in political leadership, faith in the American dream, and, ultimately, faith in the inherent goodness of humankind and the power of one individual to make a difference. A 2007 survey found that more than half of U.S. adults agreed with the following statements (Harris Interactive 2007):

- "The rich get richer and the poor get poorer" (73 percent)
- "The people running the country don't really care what happens to you" (59 percent)
- "Most people with power try to take advantage of people like you" (57 percent)
- "What you think doesn't count very much any more" (55 percent)

To some extent, faith is shaken by texts such as this one. The global gap between the rich and the poor is widening, war and conflict around the world continues to take lives and use resources that otherwise could be spent on improving health and education, profit-seeking multinational corporations are increasingly influencing world affairs, political corruption is everywhere, and the environment is killing us—if we don't kill it first. Social problems are everywhere, and what is worse, many solutions seem only to create more problems.

The transformation of American society in recent years has been dramatic. With the exception of the Industrial Revolution, no other period in human history has seen such rapid social change. The structure of U.S. society, forever altered by such macrosociological processes as corporate multinationalization, deindustrialization, and globalization, continues to be characterized by social inequities—in our schools, in our homes, in our cities, and in our salaries.

The culture of society has also undergone rapid change, leading many politicians and laypersons alike to call for a return to traditional values and beliefs and to emphasize the need for moral education. The implications are that somehow things were better in the "good old days" and that if we could somehow return to those times, things would be better again.

Fifty years ago, there were fewer divorces and less crime. AIDS and crack cocaine were unheard of, and violence in schools was almost nonexistent. At the same time, however, in 1950, the infant mortality rate was more than three times what it is today; racial and ethnic discrimination flourished in an atmosphere of bigotry and hate and millions of Americans were routinely denied the right to vote because of the color of their skin; more than half of all Americans smoked cigarettes; and people older than age 25 had completed a median of 6.8 years of school.

The social problems of today are the cumulative result of structural and cultural forces over time. Today's problems are not necessarily more or less serious than those of generations ago—they are different and, perhaps, more diverse as a result of the increased complexity of social life. But, as surely as we brought the infant mortality rate down; passed laws to prohibit racial discrimination in education, housing, and employment; increased educational levels; and reduced

> "Never doubt that a small group of thoughtful, committed citizens can change the world. Indeed, it's the only thing that ever has."
>
> Margaret Mead, Anthropologist

> "A pessimist sees the difficulty in every opportunity; an optimist sees the opportunity in every difficulty."
>
> Sir Winston Churchill
> British statesman

the number of smokers, we can continue to meet the challenges of today's social problems. But how does positive social change occur? How does one alter something as amorphous as society? The answer is really quite simple. All social change takes place because of the acts of individuals. Every law, every regulation and policy, every social movement and media exposé, and every court decision began with one person.

Rosa Parks is an example of how one person can make a difference. Rosa Parks was a seamstress in Montgomery, Alabama, in the 1950s. As for almost everything else in the South in the 1950s, public transportation was racially segregated. On December 1, 1955, Rosa Parks was on her way home from work when the "white section" of the bus she was riding became full. The bus driver told black passengers in the first row of the black section to relinquish their seats to the standing white passengers. Rosa Parks refused. She was arrested and put in jail, but her treatment so outraged the black community that a boycott of the bus system was organized by a new minister in town—Martin Luther King, Jr. The Montgomery bus boycott was a success. Just 11 months later, in November 1956, the U.S. Supreme Court ruled that racial segregation of public facilities was unconstitutional. Rosa Parks had begun a process that in time would echo her actions—the civil rights movement, the March on Washington, the 1963 Equal Pay Act, the 1964 Civil Rights Act, the 1965 Voting Rights Act, regulations against discrimination in housing, and affirmative action.

Was social change accomplished? In 1957 just 18 percent of blacks 25 years old and older had completed high school, compared with 43 percent of whites. By 2006, 81 percent of blacks 25 years old and older had completed high school, compared with 86 percent of whites (U.S. Census Bureau 2007). Whereas many would point out that such changes and thousands like them have created other problems that need to be addressed, who among us would want to return to the "good old days" of the 1950s in Montgomery, Alabama?

Throughout this text we have mentioned just a few of the millions of individuals who make a difference in our world, including the following:

- Muhammad Yunus, who won the 2006 Nobel Peace Prize for his pioneering work in starting the Grameen Bank, which has more than 2,000 branches in more than 65,000 villages. It has provided loans to more than 6 million poor people (see Chapter 6).
- The many college students across the country who have participated in boycotts against Coca-Cola in protest of the violence against union leaders at Colombian Coca-Cola plants (see Chapter 7).
- The thousands of protesters who marched in Jena, Louisiana, in September 2007 to protest the unfair and harsh charges against six black students dubbed the "Jena Six" (see Chapter 9).
- Myka Held, who was involved in a high school campaign to promote gay tolerance by having teachers and students wear t-shirts that say, "gay? fine by me." The "gay? fine by me" t-shirt campaign was started by students at Duke University (see Chapter 11).
- Cynthia Marks Garnant and Rebecca Stephens, whose awareness of population growth and the environmental problems associated with overpopulation contributed to their decision to not have children (see Chapter 13's *The Human Side* feature).

Although only a fraction of the readers of this text will occupy social roles that directly influence social policy, one need not be a politician or member of a social reform group to make a difference. We, the authors of this text, challenge you, the reader, to make individual decisions and take individual actions to make the world a more humane, just, and peaceful place for all. If you feel that your own actions cannot alleviate problems that are bigger than you are, consider the perspective of former California state senator Tom Hayden:

> The action we take, successful or not, reminds people that progress is possible. If we don't take action, we give the appearance that it's impossible to change things. Any action, while not guaranteeing change, creates possibilities that weren't there before. (quoted by McKee 2006, p. 6)

So, where should you begin? Where Rosa Parks and others like her began—with a simple individual act of courage, commitment, and faith.

Appendix

Methods of Data Analysis

DESCRIPTION, CORRELATION, CAUSATION, RELIABILITY AND VALIDITY, AND ETHICAL GUIDELINES IN SOCIAL PROBLEMS RESEARCH

There are three levels of data analysis: description, correlation, and causation. Data analysis also involves assessing reliability and validity.

Description

Qualitative research involves verbal descriptions of social phenomena. Having a homeless and single pregnant teenager describe her situation is an example of qualitative research.

Quantitative research often involves numerical descriptions of social phenomena. Quantitative descriptive analysis may involve computing the following: (1) means (averages), (2) frequencies, (3) mode (the most frequently occurring observation in the data), (4) median (the middle point in the data; half of the data points are above the median and half are below), and (5) range (the highest and lowest values in a set of data).

Correlation

Researchers are often interested in the relationship between variables. *Correlation* refers to a relationship between or among two or more variables. The following are examples of correlational research questions: What is the relationship between poverty and educational achievement? What is the relationship between race and crime victimization? What is the relationship between religious affiliation and divorce?

If there is a correlation or relationship between two variables, then a change in one variable is associated with a change in the other variable. When both variables change in the same direction, the correlation is positive. For example, in general, the more sexual partners a person has, the greater the risk of contracting a sexually transmissible disease (STD). As variable A (number of sexual partners) increases, variable B (chance of contracting an STD) also increases. Similarly, as the number of sexual partners decreases, the chance of contracting an STD decreases. Notice that in both cases, the variables change in the same direction, suggesting a positive correlation (see Figure A.1).

When two variables change in opposite directions, the correlation is negative. For example, there is a negative correlation between condom use and contracting STDs. In other words, as condom use increases, the chance of contracting an STD decreases (see Figure A.2).

The relationship between two variables may also be curvilinear, which means that it varies in both the same and opposite directions. For example, suppose a researcher finds that after drinking one alcoholic beverage, research participants are more prone to violent behavior. After two drinks, violent behavior

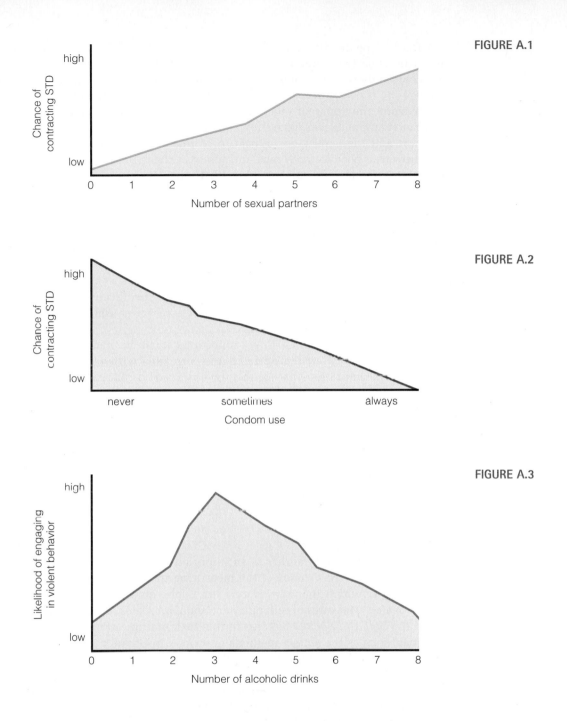

is even more likely, and this trend continues for three and four drinks. So far, the correlation between alcohol consumption and violent behavior is positive. After the research participants have five alcoholic drinks, however, they become less prone to violent behavior. After six and seven drinks, the likelihood of engaging in violent behavior decreases further. Now the correlation between alcohol consumption and violent behavior is negative. Because the correlation changed from positive to negative, we say that the correlation is curvilinear (the correlation may also change from negative to positive) (see Figure A.3).

A fourth type of correlation is called a spurious correlation. Such a correlation exists when two variables appear to be related, but the apparent relationship occurs only because each variable is related to a third variable. When the third variable is controlled through a statistical method in which the variable is held constant, the apparent relationship between the first two variables disappears. For example, blacks have a lower average life expectancy than whites do. Thus, race and life expectancy appear to be related. This apparent correlation exists, however, because both race and life expectancy are related to socioeconomic status. Because blacks are more likely than whites to be impoverished, they are less likely to have adequate nutrition and medical care.

Causation

If the data analysis reveals that two variables are correlated, we know only that a change in one variable is associated with a change in another variable. We cannot assume, however, that a change in one variable causes a change in the other variable unless our data collection and analysis are specifically designed to assess causation. The research method that best assesses causality is the experimental method (discussed in Chapter 1).

To demonstrate causality, three conditions must be met. First, the data analysis must demonstrate that variable A is correlated with variable B. Second, the data analysis must demonstrate that the observed correlation is not spurious. Third, the analysis must demonstrate that the presumed cause (variable A) occurs or changes before the presumed effect (variable B). In other words, the cause must precede the effect.

It is extremely difficult to establish causality in social science research. Therefore, much social research is descriptive or correlative, rather than causative. Nevertheless, many people make the mistake of interpreting a correlation as a statement of causation. As you read correlative research findings, remember the following adage: "Correlation does not equal causation."

Reliability and Validity

Assessing reliability and validity is an important aspect of data analysis. *Reliability* refers to the consistency of the measuring instrument or technique; that is, the degree to which the way information is obtained produces the same results if repeated. Measures of reliability are made on scales and indexes (such as those in the *Self and Society* features in this text) and on information-gathering techniques, such as the survey methods described in Chapter 1.

Various statistical methods are used to determine reliability. A frequently used method is called the "test-retest method." The researcher gathers data on the same sample of people twice (usually 1 or 2 weeks apart) using a particular instrument or method and then correlates the results. To the degree that the results of the two tests are the same (or highly correlated), the instrument or method is considered reliable.

Measures that are perfectly reliable may be absolutely useless unless they also have a high validity. *Validity* refers to the extent to which an instrument or device measures what it intends to measure. For example, police officers administer "Breathalyzer" tests to determine the level of alcohol in a person's system. The Breathalyzer is a valid test for alcohol consumption.

Validity measures are important in research that uses scales or indexes as measuring instruments. Validity measures are also important in assessing the accuracy of self-report data that are obtained in survey research. For example, survey research on high-risk sexual behaviors associated with the spread of HIV relies heavily on self-report data on topics such as number of sexual partners, types of sexual activities, and condom use. Yet how valid are these data? Do survey respondents underreport the number of their sexual partners? Do people who say they use a condom every time they engage in intercourse really use a condom every time? Because of the difficulties in validating self-reports of number of sexual partners and condom use, we may not be able to answer these questions.

Ethical Guidelines in Social Problems Research

Social scientists are responsible for following ethical standards designed to protect the dignity and welfare of people who participate in research. These ethical guidelines include the following:

1. *Freedom from coercion to participate.* Research participants have the right to decline to participate in a research study or to discontinue participation at any time during the study. For example, professors who are conducting research using college students should not require their students to participate in their research.

2. *Informed consent.* Researchers are required to inform potential participants of any aspect of the research that might influence a subject's willingness to participate. After informing potential participants about the nature of the research, researchers typically ask participants to sign a consent form indicating that the participants are informed about the research and agree to participate in it.

3. *Deception and debriefing.* Sometimes the researcher must disguise the purpose of the research to obtain valid data. Researchers may deceive participants as to the purpose or nature of a study only if there is no other way to study the problem. When deceit is used, participants should be informed of this deception (debriefed) as soon as possible. Participants should be given a complete and honest description of the study and why deception was necessary.

4. *Protection from harm.* Researchers must protect participants from any physical and psychological harm that might result from participating in a research study. This is both a moral and a legal obligation. It would not be ethical, for example, for a researcher studying drinking and driving behavior to observe an intoxicated individual leaving a bar, getting into the driver's seat of a car, and driving away.

   Researchers are also obligated to respect the privacy rights of research participants. If anonymity is promised, it should be kept. Anonymity is maintained in mail surveys by identifying questionnaires with a number coding system rather than with the participants' names. When such anonymity is not possible, as is the case with face-to-face interviews, researchers should tell participants that the information they provide will be treated as confidential. Although interviews may be summarized and excerpts quoted in published material, the identity of the individual participants is not revealed. If a research participant experiences either physical or psychological harm as a result of participation in a research study, the researcher is ethically obligated to provide remediation for the harm.

5. *Reporting of research.* Ethical guidelines also govern the reporting of research results. Researchers must make research reports freely available to the public. In these reports, a researcher should fully describe all evidence obtained in the study, regardless of whether the evidence supports the researcher's hypothesis. The raw data collected by the researcher should be made available to other researchers who might request it for purposes of analysis. Finally, published research reports should include a description of the sponsorship of the research study, its purpose, and all sources of financial support.

Glossary

abortion The intentional termination of a pregnancy.

absolute poverty The lack of resources that leads to hunger and physical deprivation.

acculturation The process of adopting the culture of a group different from the one in which a person was originally raised.

achieved status A status assigned on the basis of some characteristic or behavior over which the individual has some control.

acid rain The mixture of precipitation with air pollutants, such as sulfur dioxide and nitrogen oxide.

acquaintance rape Rape committed by someone known to the victim.

activity theory A theory that the elderly disengage, in part, because they are structurally segregated and isolated with few opportunities to engage in active roles.

adaptive discrimination Discrimination that is based on the prejudice of others.

affirmative action A broad range of policies and practices to promote equal opportunity as well as diversity in the workplace and on campuses.

age grading The assignment of social roles and opportunities given an individual's chronological age.

ageism The belief that age is associated with certain psychological, behavioral, and/or intellectual traits.

alienation A sense of powerlessness and meaninglessness in people's lives. The condition that results when workers perform repetitive, monotonous work tasks, and they become estranged from their work, the product they create, other people, and themselves.

alternative certification programs A program whereby college graduates with degrees in fields other than education can become certified if they have "life experience" in industry, the military, or other relevant jobs.

androgyny Having both feminine and masculine characteristics.

anomie A state of normlessness in which norms and values are weak or unclear.

antimiscegenation laws Laws banning interracial marriage until 1967, when the Supreme Court (in *Loving v. Virginia*) declared these laws unconstitutional.

arbitration Dispute settlement in which a neutral third party listens to evidence and arguments presented by conflicting groups and arrives at a decision or outcome that the parties have agreed in advance to accept.

ascribed status A status that society assigns to an individual on the basis of factors over which the individual has no control.

assimilation The process by which minority groups gradually adopt the cultural patterns of the dominant majority group.

automation Dominant in an industrial society, the replacement of human labor with machinery and equipment that is self-operating.

aversive racism A subtle form of prejudice that involves feelings of discomfort, uneasiness, disgust, fear, and pro-white attitudes.

barrios In the United States, slums that are occupied primarily by Latinos.

behavior-based safety programs A strategy used by business management that attributes health and safety problems in the workplace to workers' behavior, rather than to work processes and conditions.

beliefs Definitions and explanations about what is assumed to be true.

bigamy The criminal offense of marrying one person while still legally married to another.

bilingual education In the United States, teaching children in both English and their non-English native language.

binge drinking As defined by the U.S. Department of Health and Human Services, drinking five or more drinks on the same occasion on at least 1 day in the past 30 days prior to the National Survey on Drug Use and Health.

biodiversity The diversity of living organisms on earth.

bioinvasion The emergence of organisms in regions where they are not native, usually as a result of being carried in the ballast water of ships, in packing material, and in shipments of crops and other goods that are traded around the world.

biomedicalization The view that medicine can not only control particular conditions but also transform bodies and lives.

biphobia Negative attitudes and emotions toward bisexuality and people who identify as bisexual.

bisexuality Emotional and sexual attraction to members of both sexes.

boy code A set of societal expectations that discourage males from expressing emotion, weakness, or vulnerability, or asking for help.

brownfields Abandoned or undeveloped sites that are located on contaminated land.

bullying Inherent in a relationship between individuals, groups, or individuals and groups, bullying entails an imbalance of power that exists over a long period of time in which the more powerful intimidate or belittle others.

capital punishment The state (the federal government or a state) takes the life of a person as punishment for a crime.

capitalism An economic system characterized by private ownership of the means of production and distribution of goods and services for profit in a competitive market.

character education Education that emphasizes the moral and interpersonal aspects of an individual.

charter schools Schools that originate in contracts, or charters, which articulate a plan of instruction that must be approved by local or state authorities.

chattel slavery An old form of slavery in which slaves are considered property that could be bought and sold.

chemical dependency A condition in which drug use is compulsive and users are unable to stop because of physical and/or psychological dependency.

child abuse The physical or mental injury, sexual abuse, negligent treatment, or maltreatment of a child younger than age 18 by a person who is responsible for the child's welfare.

child labor Involves children performing work that is hazardous, that interferes with a child's education, or that harms a child's health or physical, mental, social, or moral development.

civil union A legal status that entitles same-sex couples who apply for and receive a civil union certificate to nearly all of the benefits available to married couples.

clash of civilizations A hypothesis that the primary source of conflict in the 21st century has shifted away from social class and economic issues and toward conflict between religious and cultural groups, especially those between large-scale civilizations such as the peoples of Western Christianity and Muslim and Orthodox peoples.

classic rape Rape committed by a stranger, with the use of a weapon, resulting in serious bodily injury to the victim.

clearance rate The percentage of crimes in which an arrest and official charge have been made and the case has been turned over to the courts.

club drugs A general term for illicit, often synthetic, drugs commonly used at nightclubs or all-night dances called "raves."

Cold War The state of military tension and political rivalry that existed between the United States and the former Soviet Union from the 1950s through the late 1980s.

colonialism Occurs when a racial or ethnic group from one society takes over and dominates the racial or ethnic group(s) of another society.

community development corporations (CDCs) Nonprofit groups formed by residents, small business owners, congregations, and other local stakeholders that work to create jobs and affordable housing and renovate parks and other community facilities.

comparable worth The belief that individuals in occupations, even in different occupations, should be paid equally if the job requires "comparable" levels of education, training, and responsibility.

comprehensive primary health care An approach to health care that focuses on the broader social determinants of health, such as poverty and economic inequality, gender inequality, environment, and community development.

comprehensive sexuality education Sex education that includes topics such as contraception, sexually transmitted diseases, HIV/AIDS, and disease-prevention methods as well as the benefits of abstinence.

compressed workweek A work arrangement that allows employees to condense their work into fewer days (e.g., four 10-hour days each week).

computer crime Any violation of the law in which a computer is the target or means of criminal activity.

constructivist explanations Those explanations that emphasize the role of leaders of ethnic groups in stirring up hatred toward others external to one's group.

corporal punishment The intentional infliction of pain for a perceived misbehavior.

corporate violence The production of unsafe products and the failure of corporations to provide a safe working environment for their employees.

corporate welfare Laws and policies that benefit corporations.

corporatocracy A system of government that serves the interests of corporations and that involves ties between government and business.

couch-homeless Individuals who do not have a home of their own and who stay at the home of family or friends.

covenant marriage A type of marriage (offered in a few states) that requires premarital counseling and that permits divorce only under condition of fault or after a marital separation of more than 2 years.

crack A crystallized illegal drug product produced by boiling a mixture of baking soda, water, and cocaine.

crime An act, or the omission of an act, that is a violation of a federal, state, or local criminal law for which the state can apply sanctions.

crime rate The number of crimes committed per 100,000 population.

cultural imperialism The indoctrination into the dominant culture of a society.

cultural lag A condition in which the material part of culture changes at a faster rate than the nonmaterial part.

cultural sexism The ways in which the culture of society perpetuate the subordination of an individual or group based on the sex classification of that individual or group.

culture The meanings and ways of life that characterize a society including beliefs, values, norms, sanctions, and symbols.

culture of poverty The set of norms, values, and beliefs and self-concepts that contribute to the persistence of poverty among the underclass.

cumulative trauma disorders Also known as repetitive motion disorders, muscle, tendon, vascular, and nerve injuries that result from repeated or sustained actions or exertions of different body parts.

cyberbullying The use of electronic devices (e.g., websites, e-mail, instant messaging, text messaging) to send or post negative or hurtful messages or images about an individual or a group.

cybernation Dominant in a postindustrial society; the use of machines to control other machines.

cycle of abuse A pattern of abuse in which a violent or abusive episode is followed by a makeup period when the abuser expresses sorrow and asks for forgiveness and "one more chance," before another instance of abuse occurs.

date-rape drugs Drugs that are used to render victims incapable of resisting sexual assaults.

deconcentration The redistribution of the population from cities to suburbs and surrounding areas.

decriminalization The removal of criminal penalties for a behavior, as in the decriminalization of drug use.

Defense of Marriage Act Federal legislation that states that marriage is a legal union between one man and one woman and denies federal recognition of same-sex marriage.

deforestation The conversion of forest land to nonforest land.

deinstitutionalization The removal of individuals with psychiatric disorders from mental hospitals and large residential institutions to outpatient community mental health centers.

demand reduction One of two strategies in the U.S. war on drugs (the other is supply reduction), demand reduction focuses on reducing the demand for drugs through treatment, prevention, and research.

demographic transition theory A theory that attributes population growth patterns to changes in birth rates and death rates associated with the process of industrialization.

dependency ratio The number of societal members who are under 18 or are 65 and older compared to the number of people who are between 18 and 64.

dependent variable The variable that the researcher wants to explain; the variable of interest.

deregulation The reduction of government control over, for example, certain drugs.

desertification The degradation of semiarid land, which results in the expansion of desert land that is unusable for agriculture.

deskilling The tendency for workers in a postindustrial society to make fewer decisions and for labor to require less thought.

détente The philosophy of "negotiation rather than confrontation" in reference to relations between the United States and the former Soviet Union; put forth by Secretary of State Henry Kissinger's Declaration of Principles in 1972.

deterrence The use of harm or the threat of harm to prevent unwanted behaviors.

devaluation hypothesis The hypothesis that women are paid less because the work they perform is socially defined as less valuable than the work performed by men.

developed countries Countries that have relatively high gross national income per capita and have diverse economies made up of many different industries.

developing countries Countries that have relatively low gross national income per capita, with simpler economies that often rely on a few agricultural products.

discrimination Actions or practices that result in differential treatment of categories of individuals.

disengagement theory A theory that the elderly disengage from productive social roles in order to relinquish these roles to younger members of society.

distance learning Learning in which, by time or place, the learner is separated from the teacher.

divorce mediation A process in which divorcing couples meet with a neutral third party (mediator) who assists the individuals in resolving issues such as property division, child custody, child support, and spousal support in a way that minimizes conflict and encourages cooperation.

divorce rate The number of divorces per 1,000 population.

domestic partners A status granted to unmarried couples, including gay and lesbian couples, by some states, counties, cities, and workplaces that conveys various rights and responsibilities.

domestic partnership A status granted to unmarried couples, including gay and lesbian couples, by some states, counties, cities, and workplaces that conveys various rights and responsibilities.

domestic terrorism Domestic terrorism, sometimes called insurgent terrorism, occurs when the terrorist act involves victims, targets, institutions, governments, or citizens from one country.

double or triple (multiple) jeopardy The disadvantages associated with being a member of two or more minority groups.

doubling time The time required for a population to double in size from a given base year if the current rate of growth continues.

downsizing The practice of laying off large numbers of employees.

drug Any substance other than food that alters the structure or functioning of a living organism when it enters the bloodstream.

drug abuse The violation of social standards of acceptable drug use, resulting in adverse physiological, psychological, and/or social consequences.

dual-use technologies Defense-funded technological innovations with commercial and civilian use.

earned income tax credit (EITC) A refundable tax credit based on a working family's income and number of children.

e-commerce The buying and selling of goods and services over the Internet.

economic institution The structure and means by which a society produces, distributes, and consumes goods and services.

ecosystems The complex and dynamic relationships between forms of life and the environments they inhabit.

ecoterrorism Any crime intended to protect wildlife or the environment that is violent, puts human life at risk, or results in damages of $10,000 or more.

egalitarian relationship A relationship in which partners share decision making and assign family roles based on choice rather than on traditional beliefs about gender.

elder abuse The physical or psychological abuse, financial exploitation, or medical abuse or neglect of the elderly.

emotion work Work that involves caring for, negotiating, and empathizing with people.

Employment Nondiscrimination Act (ENDA) A bill that would make it illegal to discriminate based on sexual orientation; as of August 2007 this bill had failed to pass the Senate.

environmental footprint The effect that each person makes on the environment.

environmental injustice Also known as **environmental racism**, the tendency for socially and politically marginalized groups to bear the brunt of environmental ills.

environmental racism See *environmental injustice*.

environmental refugees Individuals who have migrated because they can no longer secure a livelihood as a result of deforestation, desertification, soil erosion, and other environmental problems.

epidemiological transition A societal shift from low life expectancy and predominance of parasitic and infectious diseases to high life expectancy and predominance of chronic and degenerative diseases.

Equal Employment Opportunity Commission (EEOC) A U.S. federal agency charged with ending employment discrimination in the United States that is responsible for enforcing laws against discrimination, including Title VII of the 1964 Civil Rights Act that prohibits employment discrimination on the basis of race, color, religion, sex, or national origin.

ergonomics The designing or redesigning of the workplace to prevent and reduce cumulative trauma disorders.

ethnicity A shared cultural heritage or nationality.

e-waste Discarded electrical appliances and electronic equipment.

experiments A research method that involves manipulating the independent variable to determine how it affects the dependent variable.

expressive roles Roles into which women are traditionally socialized, i.e., nurturing and emotionally supportive roles.

expulsion Occurs when a dominant group forces a subordinate group to leave the country or to live only in designated areas of the country.

Faith-Based and Community Initiative A program in which faith-based (i.e., religious) organizations compete for federal funding for programs that serve the needy, such as homeless services and food aid programs.

familism The view that the family unit is more important than individual interests.

family A kinship system of all relatives living together or recognized as a social unit, including adopted persons.

Family and Medical Leave Act (FMLA) A federal law that requires companies with 50 or more employees to provide eligible workers (who work at least 25 hours a week and have been working for at least 1 year) with up to 12 weeks of job-protected, unpaid leave so that they can care for an ill child, spouse, or parent; stay home to care for their newborn, newly adopted, or newly placed child; or take time off when they are seriously ill.

family household As defined by the U.S. Census Bureau, two or more individuals related by birth, marriage, or adoption who reside together.

family preservation programs In-home interventions for families who are at risk of having a child removed from the home because of abuse or neglect.

feminism The belief that men and women should have equal rights and responsibilities.

feminization of poverty The disproportionate distribution of poverty among women.

fetal alcohol syndrome A syndrome characterized by serious physical and mental handicaps as a result of maternal drinking during pregnancy.

field research Research that involves observing and studying social behavior in settings in which it occurs naturally.

flextime A work arrangement that allows employees to begin and end the workday at different times so long as 40 hours per week are maintained.

forced labor Also known as slavery, any work that is performed under the threat of punishment and is undertaken involuntarily.

free trade agreements Pacts between two countries or among a group of countries that make it easier to trade goods across national boundaries by reducing or eliminating restrictions on exports and tariffs (or taxes) on imported goods and protecting intellectual property rights.

future shock The state of confusion resulting from rapid scientific and technological changes that unravel our traditional values and beliefs.

gateway drug A drug (e.g., marijuana) that is believed to lead to the use of other drugs (e.g., cocaine).

gay-straight alliances (GSAs) School-sponsored clubs for gay teens and their straight peers.

gender The social definitions and expectations associated with being female or male.

gender tourism The recent tendency for definitions of masculinity and femininity to become less clear, resulting in individual exploration of the gender continuum.

gene monopolies Exclusive control over a particular gene as a result of government patents.

gene therapy The transplantation of a healthy gene to replace a defective or missing gene.

genetic engineering The manipulation of an organism's genes in such a way that the natural outcome is altered.

genetic exception laws Laws that require that genetic information be handled separately from other medical information.

genetic screening The use of genetic maps to detect predispositions to human traits or disease(s).

genocide The deliberate, systematic annihilation of an entire nation, people, or ethnic group.

gentrification A type of neighborhood revitalization in which middle- and upper-income individuals buy and rehabilitate older homes in an economically depressed neighborhood.

gerontophobia Fear or dread of the elderly.

ghettos In the United States, slums that are occupied primarily by African Americans.

glass ceiling An invisible barrier that prevents women and other minorities from moving into top corporate positions.

glass elevator effect The tendency for men seeking or working in traditionally female occupations to benefit from their minority status.

GLBT See *LGBT*.

global economy An interconnected network of economic activity that transcends national borders.

global warming The increasing average global air temperature, caused mainly by the accumulation of various gases (greenhouse gases) that collect in the atmosphere.

globalization The growing economic, political, and social interconnectedness among societies throughout the world.

green energy Also known as clean energy, energy that is nonpolluting and/or renewable, such as solar power, wind power, biofuel, and hydrogen.

greenhouse gases Gases (primarily carbon dioxide, methane, and nitrous oxide) that accumulate in the atmosphere and act like the glass in a greenhouse, holding heat from the sun close to the earth.

greenwashing The way in which environmentally and socially damaging companies portray their corporate image and products as being "environmentally friendly" or socially responsible.

guerrilla warfare Warfare in which organized groups oppose domestic or foreign governments and their military forces; often involves small groups of individuals who use camouflage and underground tunnels to hide until they are ready to execute a surprise attack.

harm reduction A recent public health position that advocates reducing the harmful consequences of drug use for the user as well as for society as a whole.

hate crime An unlawful act of violence motivated by prejudice or bias.

Head Start Begun in 1965 to help preschool children from the most disadvantaged homes, Head Start provides an integrated program of health care, parental involvement, education, and social services for qualifying children.

health According to the World Health Organization, "a state of complete physical, mental, and social well-being."

heterosexism The institutional and societal reinforcement of heterosexuality as the privileged and powerful norm, based on the belief that heterosexuality is superior to homosexuality.

heterosexuality The predominance of emotional and sexual attraction to individuals of the other sex.

home-schooling The education of children at home instead of in a public or private school.

homophobia Negative attitudes and emotions toward homosexuality and those who engage in same-sex sexual behavior.

homosexuality The predominance of emotional and sexual attraction to individuals of the same sex.

human capital The skills, knowledge, and capabilities of the individual.

human capital hypothesis The hypothesis that female-male pay differences are a function of differences in women's and men's levels of education, skills, training, and work experience.

human poverty index (HPI) A measure of poverty based on measures of deprivation of a long, healthy life; deprivation of knowledge; and deprivation in decent living standards.

hypothesis A prediction or educated guess about how one variable is related to another variable.

identity theft The use of someone else's identification (e.g., social security number, birth date) to obtain credit or other economic rewards.

in vitro fertilization (IVF) The union of an egg and a sperm in an artificial setting such as a laboratory dish.

incapacitation A criminal justice philosophy that argues that recidivism can be reduced by placing the offender in prison so that he or she is unable to commit further crimes against the general public.

incumbent upgrading Aid programs that help residents of depressed neighborhoods buy or improve their homes and stay in the community.

independent variable The variable that is expected to explain change in the dependent variable.

index offenses Crimes identified by the FBI as the most serious including personal or violent crimes (homicide, assault, rape, and robbery) and property crimes (larceny, motor vehicle theft, burglary, and arson).

individual discrimination The unfair or unequal treatment of individuals because of their group membership.

individualism The tendency to focus on one's individual self-interests and personal happiness rather than on the interests of one's family and community.

industrialization The replacement of hand tools, human labor, and animal labor with machines run by steam, gasoline, and electric power.

infant mortality rate The number of deaths of live-born infants under 1 year of age per 1,000 live births (in any given year).

infantilizing elders The portrayal of the elderly in the media as childlike in terms of clothes, facial expressions, temperament, and activities.

infrastructure The underlying foundation that enables a city to function, including such things as water and sewer lines, phone lines, electricity cables, sidewalks, streets, bridges, curbs, lighting, and storm drainage systems.

institution An established and enduring pattern of social relationships.

institutional discrimination Discrimination in which the normal operations and procedures of social institutions result in unequal treatment of minorities.

instrumental roles Roles into which men are traditionally socialized, i.e., task-oriented roles.

integration hypothesis A theory that the only way to achieve quality education for all racial and ethnic groups is to desegregate the schools.

intergenerational poverty Poverty that is transmitted from one generation to the next.

internalized homophobia A sense of personal failure and self-hatred among some lesbians and gay men due to social rejection and stigmatization.

Internet An international information infrastructure available through universities, research institutes, government agencies, libraries, and businesses.

intimate partner violence (IPV) Actual or threatened violent crimes committed against individuals by their current or former spouses, cohabiting partners, boyfriends, or girlfriends.

Jim Crow laws Between 1890 and 1910, a series of U.S. laws enacted to separate blacks from whites by prohibiting blacks from using "white" buses, hotels, restaurants, and drinking fountains.

job burnout Prolonged job stress that can cause or contribute to high blood pressure, ulcers, headaches, anxiety, depression, and other health problems.

job exportation The relocation of jobs to other countries where products can be produced more cheaply.

Kyoto Protocol The first international agreement to place legally binding limits on greenhouse gas emissions from developed countries.

labor unions Worker advocacy organizations originally developed to protect workers and represent them at negotiations between management and labor.

latent functions Consequences that are unintended and often hidden.

least developed countries The poorest countries of the world.

legalization Making prohibited behaviors legal; for example, legalizing drug use or prostitution.

lesbigay population A collective term sometimes used to refer to lesbians, gays, and bisexuals.

LGBT or GLBT Terms often used to refer collectively to lesbians, gays, bisexuals, and transgendered individuals.

life expectancy The average number of years that individuals born in a given year can expect to live.

living apart together (LAT) relationships An emerging family form in which couples—married or unmarried—live apart in separate residences.

living wage laws Laws that require state or municipal contractors, recipients of public subsidies or tax breaks, or, in some cases, all businesses to pay employees wages that are significantly above the federal minimum, enabling families to live above the poverty line.

long-term unemployment rate The share of the unemployed who have been out of work for 27 weeks or more.

MADD (Mothers Against Drunk Driving) A social action group committed to reducing drunk driving.

mailbox economy The tendency for a substantial portion of local economies to be dependent on pension and Social Security checks received in the mail by older residents.

managed care Any medical insurance plan that controls costs through monitoring and controlling the decisions of health care providers.

manifest functions Consequences that are intended and commonly recognized.

marital decline perspective A pessimistic view of the current state of marriage that includes the beliefs that (1) personal happiness has become more important than marital commitment and family obligations and (2) the decline in lifelong marriage and the increase in single-parent families have contributed to a variety of social problems.

marital resiliency perspective A view of the current state of marriage that includes the beliefs that (1) poverty, unemployment, poorly funded schools, discrimination, and the lack of basic services (such as health insurance and child care) represent more serious threats to the well-being of children and adults than does the decline in married two-parent families and (2) divorce provides a second chance for happiness for adults and an escape from dysfunctional and aversive home environments for many children.

master status The status that is considered the most significant in a person's social identity.

maternal mortality rate A measure of deaths that result from complications associated with pregnancy, childbirth, and unsafe abortion.

McDonaldization The process by which principles of the fast food industry (efficiency, calculability, predictability, and control through technology) are being applied to more and more sectors of society, particularly the workplace.

means-tested programs Assistance programs that have eligibility requirements based on income.

mechanization Dominant in an agricultural society, the use of tools to accomplish tasks previously done by hand.

Medicaid A public health insurance program, jointly funded by the federal and state governments, that provides health insurance coverage for the poor who meet eligibility requirements.

medicalization Defining or labeling behaviors and conditions as medical problems.

Medicare A federally funded public insurance program that provides health insurance benefits to the elderly, disabled workers, and their dependents, and persons with advanced kidney disease.

medigap The difference between Medicare benefits and the actual cost of medical care.

megacities Urban areas with 10 million residents or more.

mental health The successful performance of mental function, resulting in productive activities, fulfilling relationships with other people, and the ability to adapt to change and to cope with adversity.

mental illness All mental disorders, which are health conditions that are characterized by alterations in thinking, mood, and/or behavior associated with distress and/or impaired functioning and that meet specific criteria (such as level of intensity and duration) specified in *The Diagnostic and Statistical Manual of Mental Disorders*.

metropolis From the Greek meaning "mother city." See also *metropolitan area*.

metropolitan area Also known as a metropolis, a densely populated core area together with adjacent communities.

microcredit programs The provision of loans to people who are generally excluded from traditional credit services because of their low socioeconomic status.

micropolitan area A small city (between 10,000 and 50,000 people) located beyond congested metropolitan areas.

military-industrial complex A term used by Dwight D. Eisenhower to connote the close association between the military and defense industries.

Millennium Development Goals Eight goals that comprise an international agenda for reducing poverty and improving lives.

minority group A category of people who have unequal access to positions of power, prestige, and wealth in a society and who tend to be targets of prejudice and discrimination.

modern racism A subtle form of racism that involves the belief that serious discrimination in America no longer exists, that any continuing racial inequality is the fault of minority group members, and that the demands for affirmative action for minorities are unfair and unjustified.

modernization theory A theory claiming that as society becomes more technologically advanced, the position of the elderly declines.

monogamy Marriage between two partners; the only legal form of marriage in the United States.

morbidity Illnesses, symptoms, and the impairments they produce.

mortality Death.

multicultural education Education that includes all racial and ethnic groups in the school curriculum thereby promoting awareness and appreciation for cultural diversity.

multiple chemical sensitivity Also known as "environmental illness," a condition whereby individuals experience adverse reactions when exposed to low levels of chemicals found in everyday substances.

mutually assured destruction (MAD) A Cold War doctrine referring to the capacity of two nuclear states to destroy each other, thus reducing the risk that either state will initiate war.

naturalized citizens Immigrants who apply for and meet the requirements for U.S. citizenship.

needle exchange programs Programs designed to reduce transmission of HIV by providing intravenous drug users with new, sterile syringes in exchange for used, contaminated syringes.

neglect A form of abuse involving the failure to provide adequate attention, supervision, nutrition, hygiene, health care, and a safe and clean living environment for a minor child or a dependent elderly individual.

New Urbanism A movement in urban planning that approaches the idea of sustainable urban communities with the goal of raising the quality of life for all those in the community by creating compact communities with a sustainable infrastructure.

no-fault divorce A divorce that is granted based on the claim that there are irreconcilable differences within a marriage (as opposed to one spouse being legally at fault for the marital breakup).

nonfamily household A household that consists of one person who lives alone, two or more people as roommates, or cohabiting heterosexual or homosexual couples involved in a committed relationship.

norms Socially defined rules of behavior including folkways, mores, and laws.

nuclear winter The predicted result of a thermonuclear war whereby thick clouds of radioactive dust and particles would block out vital sunlight, lower temperature in the Northern Hemisphere, and lead to the death of most living things on earth.

nursing homes Public or private residential facilities that provide full-time nursing care for residents.

objective element of a social problem Awareness of social conditions through one's own life experiences and through reports in the media.

occupational sex segregation The concentration of women in certain occupations and men in other occupations.

organized crime Criminal activity conducted by members of a hierarchically arranged structure devoted primarily to making money through illegal means.

outsource The practice of sending jobs to low-wage countries such as China and India.

overt discrimination Discrimination that occurs because of an individual's own prejudicial attitudes.

parental alienation syndrome (PAS) An emotional and psychological disturbance in which children engage in exaggerated and unjustified denigration and criticism of a parent.

parity In health care, a concept requiring equality between mental health care insurance coverage and other health care coverage.

partial birth abortions Also called an intact dilation and extraction (D & E) abortion, the procedure may entail delivering the limbs and the torso of the fetus before it has expired.

patriarchy A male-dominated family system that is reflected in the tradition of wives taking their husband's last name and children taking their father's name.

phased retirement Allows workers to ease into retirement by reducing hours worked per day, days worked per week, or months worked per year.

pink collar jobs Jobs that offer few benefits, often have low prestige, and are disproportionately held by women.

planned obsolescence The manufacturing of products that are intended to become inoperative or outdated in a fairly short period of time.

pluralism A state in which racial and ethnic groups maintain their distinctness but respect each other and have equal access to social resources.

polyandry The concurrent marriage of one woman to two or more men.

polygamy A form of marriage in which one person may have two or more spouses.

polygyny A form of marriage in which one husband has more than one wife.

population density The number of people per unit of land area.

population momentum Continued population growth as a result of past high fertility rates that have resulted in a large number of young women who are currently entering their childbearing years.

post-industrialization The shift from an industrial economy dominated by manufacturing jobs to an economy dominated by service-oriented, information-intensive occupations.

postmodernism A worldview that questions the validity of rational thinking and the scientific enterprise.

posttraumatic stress disorder (PTSD) A set of symptoms that may result from any traumatic experience, including crime victimization, war, natural disasters, or abuses.

prejudice Negative attitudes and feelings toward or about an entire category of people.

primary aging Biological changes associated with aging due to physiological variables such as cellular and molecular variation.

primary deviance Deviant behavior committed before a person is caught and labeled an offender.

primary group Usually small numbers of individuals characterized by intimate and informal interaction.

primary prevention strategies Family violence prevention strategies that target the general population.

primordial explanations Those explanations that emphasize the existence of "ancient hatreds" rooted in deep psychological or cultural differences between ethnic groups, often involving a history of grievance and victimization, real or imagined, by the enemy group.

privatization A practice in which states hire businesses to provide services or operate local schools.

probation The conditional release of an offender who, for a specific time period and subject to certain conditions, remains under court supervision in the community.

pronatalism A cultural value that promotes having children.

public housing Federally subsidized housing that is owned and operated by local public housing authorities (PHAs).

race A category of people who are believed to share distinct physical characteristics that are deemed socially significant.

racial profiling The law enforcement practice of targeting suspects on the basis of race.

racism The belief that race accounts for differences in human character and ability and that a particular race is superior to others.

regionalism A form of collaboration among central cities and suburbs that encourages local governments to share common responsibility for common problems.

registered partnerships Federally recognized relationships that convey most but not all the rights of marriage.

rehabilitation A criminal justice philosophy that argues that recidivism can be reduced by changing the criminal through such programs as substance abuse counseling, job training, education, etc.

relative poverty A deficiency in material and economic resources compared with some other population.

reparative therapy or conversion therapy Various therapies that are aimed at changing homosexuals' sexual orientation.

replacement-level fertility 2.1 births per woman (slightly more than 2 because not all female children will live long enough to reach their reproductive years).

restorative justice A philosophy primarily concerned with reconciling conflict between the victim, the offender, and the community.

role The set of rights, obligations, and expectations associated with a status.

sample A portion of the population, selected to be representative so that the information from the sample can be generalized to a larger population.

sanctions Social consequences for conforming to or violating norms.

sandwich generation The generation that has the responsibility of simultaneously caring for their children and their aging parents.

school vouchers Tax credits that are transferred to the public or private school that parents select for their child.

science The process of discovering, explaining, and predicting natural or social phenomena.

second shift The household work and child care that employed parents (usually women) do when they return home from their jobs.

secondary aging Biological changes associated with aging that can be attributed to poor diet, lack of exercise, increased stress, and the like.

secondary deviance Deviance that results from being caught and labeled as an offender.

secondary group Involving small or large numbers of individuals, groups that are task oriented and are characterized by impersonal and formal interaction.

secondary prevention strategies Family violence prevention strategies that target groups thought to be at high risk for family violence.

Section 8 housing A housing assistance program in which federal rent subsidies are provided either to tenants (in the form of certificates and vouchers) or to private landlords.

security dilemma A characteristic of the international state system that gives rise to unstable relations between states; as State A secures its borders and interests, its behavior may decrease the security of other states and cause them to engage in behavior that decreases A's security.

segregation The physical separation of two groups in residence, workplace, and social functions.

selective primary health care An approach to health care that focuses on using specific interventions to target specific health problems.

self-fulfilling prophecy A concept referring to the tendency for people to act in a manner consistent with the expectations of others.

senescence The biology of aging.

serial monogamy A succession of marriages in which a person has more than one spouse over a lifetime but is legally married to only one person at a time.

sex A person's biological classification as male or female.

sexism The belief that innate psychological, behavioral, and/or intellectual differences exist between women and men and that these differences connote the superiority of one group and the inferiority of the other.

sexual harassment In reference to workplace harassment, when an employer requires sexual favors in exchange for a promotion, salary increase, or any other employee benefit and/or the existence of a hostile environment that unreasonably interferes with job performance.

sexual orientation The classification of individuals as heterosexual, bisexual, or homosexual, based on their emotional and sexual attractions, relationships, self-identity, and lifestyle.

shaken baby syndrome A form of child abuse whereby the caretaker shakes a baby to the point of causing the child to experience brain or retinal hemorrhage.

slums Concentrated areas of poverty and poor housing in urban areas.

smart growth A strategy for managing urban sprawl that serves the economic, environmental, and social needs of communities.

social group Two or more people who have a common identity, interact, and form a social relationship.

social movement An organized group of individuals with a common purpose to either promote or resist social change through collective action.

social problem A social condition that a segment of society views as harmful to members of society and in need of remedy.

social promotion The passing of students from grade to grade even if they are failing.

socialism An economic system characterized by state ownership of the means of production and distribution of goods and services.

socialized medicine Also known as **universal health care,** a state-supported system of health care delivery in which health care is purchased by the government and sold to the consumer at little or no additional cost.

sociological imagination The ability to see the connections between our personal lives and the social world in which we live.

sodomy Oral and anal sexual acts.

state The organization of the central government and government agencies such as the military, police, and regulatory agencies.

State Children's Health Insurance Program (SCHIP) A public health insurance program, jointly funded by the federal and state governments, that provides health insurance coverage for children whose families meet income eligibility standards.

status A position that a person occupies within a social group.

stem cells Undifferentiated cells that can produce any type of cell in the human body.

stereotypes Exaggerations or generalizations about the characteristics and behavior of a particular group.

stigma A discrediting label that affects an individual's self-concept and disqualifies that person from full social acceptance.

structural sexism The ways in which the organization of society, and specifically its institutions, subordinate individuals and groups based on their sex classification.

structure The way society is organized including institutions, social groups, statuses, and roles.

structured choice Choices that are limited by the structure of society.

subjective element of a social problem The belief that a particular social condition is harmful to society, or to a segment of society, and that it should and can be changed.

suburbanization The development of urban areas outside central cities, and the movement of populations into these areas.

suburbs Urban areas surrounding central cities.

supply reduction One of two strategies in the U.S. war on drugs (the other is demand reduction), supply reduction concentrates on reducing the supply of drugs available on the streets through international efforts, interdiction, and domestic law enforcement.

survey research A research method that involves eliciting information from respondents through questions.

sustainable development Societal development that meets the needs of current generations without threatening the future of subsequent generations.

sweatshops Work environments that are characterized by less-than-minimum-wage pay, excessively long hours of work (often without overtime pay), unsafe or inhumane working conditions, abusive treatment of workers by employers, and/or the lack of worker organizations aimed to negotiate better working conditions.

symbol Something that represents something else.

technological dualism The tendency for technology to have both positive and negative consequences.

technological fix The use of scientific principles and technology to solve social problems.

technology Activities that apply the principles of science and mechanics to the solutions of a specific problem.

technology-induced diseases Diseases that result from the use of technological devices, products, and/or chemicals.

telework A work arrangement involving the use of information technology that allows employees to work part- or full-time at home or at a satellite office.

Temporary Assistance to Needy Families (TANF) A federal cash welfare program that involves work requirements and a 5-year lifetime limit.

terrorism The premeditated use or threatened use of violence by an individual or group to gain a political objective.

tertiary prevention strategies Family violence prevention strategies that target families who have experienced family violence.

therapeutic cloning Use of stem cells from human embryos to produce body cells that can be used to grow needed organs or tissues.

therapeutic communities Organizations in which approximately 35–500 individuals reside for up to 15 months to abstain from drugs, develop marketable skills, and receive counseling.

total fertility rates The average lifetime number of births per woman in a population.

total immersion An educational program, in which students, particularly elementary age students, receive literacy and communication instruction entirely in a foreign language, usually Spanish.

transgendered individuals A range of people whose gender identities do not conform to traditional notions of masculinity and femininity, including homosexuals, bisexuals, cross-dressers, transvestites, and transsexuals.

transnational corporations Also known as multinational corporations, corporations that have their home base in one country and branches, or affiliates, in other countries.

transnational crime Criminal activity that occurs across one or more national borders.

transnational terrorism Terrorism that occurs when a terrorist act in one country involves victims, targets, institutions, governments, or citizens of another country.

under-5 mortality rate The rate of deaths of children under age 5.

underclass A persistently poor and socially disadvantaged group.

underemployment Unemployed workers as well as (1) those working part-time but who wish to work full-time, (2) those who want to work but have been discouraged from searching by their lack of success, and (3) others who are neither working nor seeking work but who want and are available to work and have looked for employment in the last year.

unemployment To be currently without employment, actively seeking employment, and available for employment, according to U.S. measures of unemployment.

union density The percentage of workers who belong to unions.

universal health care See *socialized medicine.*

upskilling The opposite of deskilling; upskilling reduces employee alienation and increases decision-making powers.

urban area A spatial concentration of people whose lives are centered around nonagricultural activities.

urban sprawl The ever increasing outward growth of urban areas that results in the loss of green open spaces, the displacement and endangerment of wildlife, traffic congestion and noise, and pollution.

urbanization The transformation of a society from a rural to an urban one.

values Social agreements about what is considered good and bad, right and wrong, desirable and undesirable.

variable Any measurable event, characteristic, or property that varies or is subject to change.

victimless crimes Illegal activities that have no complaining participant(s) and are often thought of as crimes against morality such as prostitution.

war Organized armed violence aimed at a social group in pursuit of an objective.

wealth The total assets of an individual or household minus liabilities.

wealthfare Laws and policies that benefit the rich.

weapons of mass destruction (WMD) Chemical, biological, and nuclear weapons that have the capacity to kill large numbers of people indiscriminately.

Web 2.0 A platform for millions of users to express themselves online in the common areas of cyberspace.

white-collar crime Includes both *occupational crime,* in which individuals commit crimes in the course of their employment, and *corporate crime,* in which corporations violate the law in the interest of maximizing profit.

Workers' compensation Also known as workers' comp, an insurance program that provides medical and living expenses for people with work-related injuries or illnesses.

working poor Individuals who spend at least 27 weeks per year in the labor force (working or looking for work) but whose income falls below the official poverty level.

References

Chapter 1

Alvarado, Diana. 2000. "Student Activism Today." *Diversity Digest.* Available at http://www.diversityweb.org/Digest/Sm99/activism.html

Associated Press. 2006, September 7. "Florida Appeals Court Upholds Ban of Veil in Driver's License Photo." Religious News. Pew Forum on Religion and Public Life. Available at http://pewforum.org/news

Batsell, Jake. 2002. "USA: Students Campaign for Coffee in Good Conscience." *The Seattle Times,* March 17.

Beeby, Dean. 2005. "Census Includes Same Sex Marriage Question." *Globe and Mail,* April 18. Available at http://globeandmail.com

Berkowitz, Elana. 2005. "They Say Tomato, Students Say Justice." *Campus Progress,* March 18.

Blumer, Herbert. 1971. "Social Problems as Collective Behavior." *Social Problems* 8(3):298–306.

Boston College. 2003. *The Anti-Sweatshop Movement and Boston College.* Available at http://www.bc.edu/bc_org/cas/soc/justice

Caldas, Stephen, and Carl L. Bankston III. 1999. "Black and White TV: Race, Television Viewing, and Academic Achievement." *Sociological Spectrum* 19:39–61.

Canedy, Dana. 2003. "Lifting Veil for Photo ID Goes Too Far, Driver Says." *New York Times,* June 27. Available at http://www.nytimes.com/2003/06/27/national

Carroll, Joseph. 2006. "Americans' Mood Sours Some this Month." *Gallup Poll News Service,* August 15. Available at http://www.galluppoll.com/content

Centers for Disease Control and Prevention. 2006 (August 11). *Trends in HIV-Related Risk Behaviors Among High School Students—United States, 1991–2005.* Morbidity and Mortality Weekly Report 55, no. 31:851–854.

Dordick, Gwendolyn A. 1997. *Something Left to Lose: Personal Relations and Survival Among New York's Homeless.* Philadelphia: Temple University Press.

Eitzen, Stanley, and Maxine Baca Zinn. 2000. *Social Problems.* Boston, MA: Allyn and Bacon.

Engel, Robin Shepard, and Robert E. Worden. 2003. "Police Officers' Attitudes, Behavior, and Supervisory Influences: An Analysis of Problem Solving." *Criminology* 41:131–166.

Ferrara, Leigh, Ann Friedman, April Rabkin, Amaya Rivera, Cameron Scott, Marisa Taylor, and Marcus Wohlsen. 2006. "Extra Credit: Campus Activism 2006." Mother Jones News, September/October. Available at http://motherjones.com

Fleming, Zachary. 2003. "The Thrill of It All." In *In Their Own Words,* Paul Cromwell, ed. (pp. 99–107). Los Angeles, CA: Roxbury.

Gee, Gilbert, Barbara Curbow, Margaret Ensminger, Joan Griffin, et al. 2006. "Are You Positive? The Relationship of Minority Composition to Workplace Drug and Alcohol Testing." *Journal of Drug Issues* 35(4):755–779.

Hewlett, Sylvia Ann. 1992. When the Bough Breaks: The Cost of Neglecting Our Children. New York: Harper Perennial.

Human Development Report. 2005. *2006 Human Development Report: Overview.* Available at http://hdr.undp.org/reports/global/2005/pdf.HDR05_overview

Jacobs, Bruce A. 2003. "Researching Crack Dealers." In *In Their Own Words,* Paul Cromwell, ed. (pp. 1–11). Los Angeles, CA: Roxbury.

James, Susan Donaldson. 2007. "Students Use Civil Rights Tactics to Combat Global Warming." *ABC News,* January 19. Available at http://abcnews.go.com/technology

Jekielek, Susan M. 1998. "Parental Conflict, Marital Disruption, and Children's Emotional Well-Being." *Social Forces* 76:905–935.

Killer Coke. 2005. Available at http://killercoke.org/grincamp.htm

Kmec, Julie A. 2003. "Minority Job Concentration and Wages." *Social Problems* 50:38–59.

Merton, Robert K. 1968. *Social Theory and Social Structure.* New York: Free Press.

Mills, C. Wright. 1959. *The Sociological Imagination.* London: Oxford University Press.

Newport, Frank. 2006. "Jobs Loom Large When Americans Ponder the Economy." Gallup Poll, October 2. Available at http://www.galluppoll.com

Palacios, Wilson R., and Melissa E. Fenwick. 2003. "'E' Is for Ecstasy." In *In Their Own Words,* Paul Cromwell, ed. (pp. 277–283). Los Angeles, CA: Roxbury.

Pryor, John, and Sylvia Hurtado, Victor Saenz, Jessica Korn, Jose Luis Santos, and William S. Korn. 2006. *The American Freshman: National Norms for Fall 2006.* Los Angeles, CA: Higher Education Research Institute.

Reiman, Jeffrey. 2007. *The Rich Get Richer and the Poor Get Prison,* 8th ed. Boston, MA: Allyn and Bacon.

Romer, D., R. Hornik, B. Stanton, M. Black, X. Li, I. Ricardo, and S. Feigelman. 1997. "Talking Computers: A Reliable and Private Method to Conduct Interviews on Sensitive Topics with Children." *Journal of Sex Research* 34:3–9.

Saad, Lydia. 2007 (January 5). "Local TV is No. 1 Source of News for Americans." Gallup Poll. The Gallup Organization. Available at http://www.galluppoll.com

Schwalbe, Michael. 1998. *The Sociologically Examined Life: Pieces of the Conversation.* Mountain View, CA: Mayfield.

Thomas, W. I. 1966 [1931]. "The Relation of Research to the Social Process." In *W. I. Thomas on Social Organization and Social Personality,* Morris Janowitz, ed. (pp. 289–305). Chicago: University of Chicago Press.

Troyer, Ronald J., and Gerald E. Markle. 1984. "Coffee Drinking: An Emerging Social Problem." *Social Problems* 31:403–416.

Ukers, William H. 1935. *All About Tea,* Vol. 1. New York: Tea and Coffee Trade Journal Co.

Weir, Sara, and Constance Faulkner. 2004. *Voices of a New Generation: A Feminist Anthology.* Boston, MA: Pearson Education Inc.

Wilson, John. 1983. *Social Theory.* Englewood Cliffs, NJ: Prentice Hall.

Chapter 2

"Access to Free Health Care Means Crime Does Pay for Some Sick Inmates." 2001. *The Independent,* April 20. Available at http://theIndependent.com

*The authors and Wadsworth acknowledge that some of the Internet sources may have become unstable; that is, they are no longer hot links to the intended reference. In that case, the reader may want to access the article through the search engine or archives of the homepage cited (e.g., fbi.gov, cbsnews.com).

Alan Guttmacher Institute. 2004. *New Evidence from Africa, Asia, and Latin America Underscores Impact of Violence Against Women*. News release, December 10. Available at http://www.guttmacher.org

Allen, P. L. 2000. *The Wages of Sin: Sex and Disease, Past and Present*. Chicago: University of Chicago Press.

Allen, Terry J. 2007 (March 7). "Merck's Murky Dealings: HPV Vaccine Lobby Backfires." *CorpWatch*. Available at www.corpwatch.org

Altman, Dennis. 2003. "Understanding HIV/AIDS as a Global Security Issue." In *Health Impacts of Globalization*, Kelley Lee, ed. (pp. 33–46). New York: Palgrave MacMillan.

American College Health Association. 2007. "American College Health Association National College Health Assessment Spring 2006 Reference Group Data Report" (Abridged). *Journal of American College Health* 55(4):195–226.

American Psychiatric Association. 2000. *Diagnostic and Statistical Manual of Mental Disorders*, 4th ed, Text Revision. Washington, DC: American Psychiatric Association.

American Psychological Association. 2007 (February). *The Psychological Needs of U.S. Military Service Members and Their Families: A Preliminary Report*. American Psychological Association. Available at http://www.apa.org

Associated Press. 2007. "Mom: Social Workers Threaten to Take Away 254-Pound Son." *The Daily Reflector*, March 23, p. B2.

Bailey, Eric. 2006. *Food Choice and Obesity in Black America: Creating a New Cultural Diet*. Westport, CT: Praeger Publishers.

Barker, Kristin. 2002. "Self-Help Literature and the Making of an Illness Identity: The Case of Fibromyalgia Syndrome (FMS)." *Social Problems* 49(3):279–300.

Bartley, Mel, Jane Ferrie, and Scott M. Montgomery. 2001. "Living in a High-Unemployment Economy: Understanding the Health Consequences." In *Social Determinants of Health*, M. Marmot and R. G. Wilkinson, eds. (pp. 81–104). New York: Oxford University Press.

Capriccioso, Rob. 2006 (March 13). "Counseling Crisis." *Inside Higher Ed*. Available at http://insidehighered.com/news

Carroll, Joseph. 2006 (November 16). "Healthcare Costs Top Americans' Financial Concerns." *Gallup Brain*. The Gallup Organization. Available at http://www.gallup.com

Centers for Disease Control and Prevention. 2004. *Defining Overweight and Obesity*. Available at http://www.cdc.gov

Centers for Disease Control and Prevention. 2005a. *HIV/AIDS Among African Americans*. Fact sheet. Available at http://www.cdc.gov

Centers for Disease Control and Prevention. 2005b. "HIV Prevalence, Unrecognized Infection, and HIV Testing Among Men Who Have Sex with Men: Five U.S. Cities, June 2004–April 2005." *Morbidity and Mortality Weekly Report* 54(24):597–601.

Centers for Disease Control and Prevention. 2005c (July 15). "Update: Syringe Exchange Programs—United States, 2002." *Morbidity and Mortality Weekly Report* 54(27):673–676.

Centers for Disease Control and Prevention. 2006. *HIV/AIDS Surveillance Report, 2005* (Vol. 17). Available at http://www.cdc.gov

Centers for Disease Control and Prevention. 2007a (January). *A Glance at the HIV/AIDS Epidemic*. Available at http://www.cdc.gov

Centers for Disease Control and Prevention. 2007b. "West Nile Virus Activity." *Morbidity and Mortality Weekly Report* 56(22):556–559.

Centers for Medicare & Medicaid Services. 2007 (January 18). *National Health Expenditure Data. U.S. Department of Health & Human Services*. Available at http://www.cms.hhs.gov

Chandra, Anita, and Cynthia S. Minkovitz. 2006. "Stigma Starts Early: Gender Differences in Teen Willingness to Use Mental Health Services." *Journal of Adolescent Health* 38(6):754.e1–754.e8.

Children's Defense Fund. 2000. *The State of America's Children Yearbook 2000*. Available at http://www.childrensdefense.org

Children's Defense Fund. 2006. *Improving Children's Health: Understanding Children's Health Disparities and Promising Approaches to Address Them*. Washington, D.C.: Children's Health Fund. Available at http://www.childrensdefense.org

Clarke, Adele E., Laura Mamo, Jennifer R. Fishman, Janet K. Shim, and Jennifer Ruth Fosket. 2003. "Biomedicalization: Technoscientific Transformations of Health, Illness, and U.S. Biomedicine." *American Sociological Review* 68:161–194.

Claxton, G., I. Gil, B. Finder, B. DiJulio, S. Hawkins, J. Pickreign, H. Whitmore, and J. Gabel. 2006. *Employer Health Benefits 2006 Annual Survey*. Kaiser Family Foundation and Health Research and Education Trust. Available at http://www.kff.org

Cockerham, William C. 2004. *Medical Sociology*, 9th ed. Upper Saddle River, NJ: Prentice Hall.

Cockerham, William C. 2005. *Sociology of Mental Disorder*, 7th ed. Upper Saddle River, NJ: Prentice Hall.

Combes, Marie-Laure. 2007 (March 2). "French Government Wants Food Warning." Associated Press. Available at http://www.washingtonpost.com

Conyers, John. 2003. "A Fresh Approach to Health Care in the United States: Improved and Expanded Medicare for All." *American Journal of Public Health* 93(2):193.

Cotton, Ann, Kenneth R. Stanton, Zoltan J. Acs, and Mary Lovegrove. 2007. *2006 UB Obesity Report Card*. Baltimore, MD: University of Baltimore. Available at http://www.ubalt.edu

Cowley, Geoffrey. 2006 (May 15). "The Life of a Virus Hunter." *Newsweek*, pp. 54–60, 63–65.

Davis, Karen, Cathy Schoen, Stephen C. Schoenbaum, Michelle M. Doty, Alyssa L. Holmgren, Jennifer L. Kriss, and Katherine K. Shea. 2007 (May). *Mirror, Mirror on the Wall: An International Update on the Comparative Performance of American Health Care*. Commonwealth Fund. Available at http://commonwealthfund.org

Delva, J., P. M. O'Malley, and L. D. Johnston. 2007. "Racial/Ethnic and Socioeconomic Status Differences in Overweight and Health-Related Behaviors among American Students: National Trends 1986–2003." *Journal of Adolescent Health* 39(4):536–545.

DeNavas-Walt, Carmen, Bernadette D. Proctor, and Jessica Smith. 2007 (August). *Income, Poverty, and Health Insurance in the United States: 2006*. Current Population Reports P60-233. U.S. Census Bureau. Available at http://www.census.gov

Diamond, Catherine, and Susan Buskin. 2000. "Continued Risky Behavior in HIV-Infected Youth." *American Journal of Public Health* 90(1):115–118.

Durkheim, Emile. 1951 [1897]. *Suicide: A Study in Sociology*, J. A. Spaulding and G. Simpson, trans.; G. Simpson, ed. New York: Free Press.

Egwu, Igbo N. 2002. "Health Promotion Challenges in Nigeria: Globalization, Tobacco Marketing, and Policies." In *Health Communication in Africa: Contexts, Constraints, and Lessons*, A. O. Alali and B. A. Jinadu, eds. (pp. 40–55). Lanham, MD: University Press of America.

Ehrenreich, Barbara. 2005. "A Society That Throws the Sick Away." *Los Angeles Times*, April 28. Available at http://www.latimes.com

Ezzell, Carol. 2003. "Why? The Neuroscience of Suicide." *Scientific American* 288(2):45–51.

Families USA. 2005 (February). "Have Health Insurance? Think You're Well Protected?" *Think Again! Fact sheet*. Available at http://www.familiesusa.org

Families USA. 2007 (January 9). No Bargain: Medicare Drug Plans Deliver Higher Prices. Available at http://www.familiesusa.org

Feachum, Richard G. A. 2000. "Poverty and Inequality: A Proper Focus for the New Century." *International Journal of Public Health* 78:1–2.

Feldman, Debra S., Dennis H. Novack, and Edward Gracely. 1998. "Effects of Managed

Care on Physician Patient Relationships, Quality of Care, and the Ethical Practice of Medicine: A Physician Survey." *Archives of Internal Medicine* 158:1626–1632.

Fischer, Edward, and Amerigo Farina. 1995. "Attitudes Toward Seeking Professional Psychological Help: A Shortened Form and Considerations for Research." *Journal of College Student Development* 36(4):368–373.

Gallagher, Robert P. 2006. *National Survey of Counseling Center Directors*. The International Association of Counseling Services. Available at http://www.iacsinc.org

Garfinkel, P. E., and D. S. Goldbloom. 2000. "Mental Health: Getting Beyond Stigma and Categories." *Bulletin of the World Health Organization* 78(4):503–505.

Garson, Arthur "Tim," Jr. 2007. "Heart of the Uninsured." *Health Affairs* 26(1):227–231.

Gilmore, Todd, and Richard Kronick. 2005. "It's the Premiums, Stupid: Projections of the Uninsured Through 2013." *Health Affairs*, April 5. Available at http://www.healthaffairs.org

Goering, Laurie. 2005. "Beauty Tames Beast of HIV Stigma." *Chicago Tribune*, February 28. Available at http://www.chicagotribune.com

Goldberg, Lisa. 2005. "Educating Seniors on Safe Sex." *Baltimore Sun*, March 1. Available at http://www.baltimoresun.com

Goldstein, Michael S. 1999. "The Origins of the Health Movement." In *Health, Illness, and Healing: Society, Social Context, and Self*, Kathy Charmaz and Debora A. Paterniti, eds. (pp. 31–41). Los Angeles, CA: Roxbury.

Gruttadaro, Darcy. 2005. "Federal Leaders Call on Schools to Help." *NAMI Advocate*, Winter, pp. 7–9.

Hadley, Jack. 2007. "Insurance Coverage, Medical Care Use, and Short-Term Health Changes Following an Unintentional Injury or the Onset of a Chronic Condition." *JAMA* 297(10):1073–1084.

Halperin, D. T., M. J. Steiner, M. M. Cassell, E. C. Green, N. Hearst, D. Kirby, H. D. Gayle, and W. Cates. 2004. "The Time Has Come for Common Ground on Preventing Sexual Transmission of HIV." *The Lancet* 364: 1913–1915.

Hawkes, Corinna. 2006. "Uneven Dietary Development: Linking the Policies and Processes of Globalization with the Nutrition Transition, Obesity, and Diet-Related Chronic Diseases." *Globalization and Health* 2:4.

Harrison, Mary M. 2002. "Silence That Promotes Stigma." *Teaching Tolerance* 21:52–53.

Himmelstein, D. U., E. Warren, D. Thorne, and S. Woolhandler. 2005. "MarketWatch: Illness and Injury as Contributors to Bankruptcy." *Health Affairs*, February 2. Available at http://www.healthaffairs.org

Hippert, Christine. 2002. "Multinational Corporations, the Politics of the World Economy, and Their Effects on Women's Health in the Developing World: A Review." *Health Care for Women International* 23:861–869.

HIV & AIDS Discrimination and Stigma. 2004 (November 30). Available at http://www.avert.org

Hollander, D. 2003. "Having Two Parents Helps." *Perspectives on Sexual and Reproductive Health* 35(2):60.

Hollander, D. 2005. "Unprotected Anal Sex Is Not Uncommon Among Men with HIV Infection." *Perspectives on Sexual and Reproductive Health* 37(1). Available at http://www.guttmacher.org

Honberg, Ron. 2005. "Decriminalizing Mental Illness." *NAMI Advocate*, Winter, pp. 4–5, 7.

Hudson, C. G. 2005. "Socioeconomic Status and Mental Illness: Tests of the Social Causation and Selection Hypothesis." *American Journal of Orthopsychiatry* 75(1):3–18.

Hughes, Mary Elizabeth, and Linda J. Waite. 2002. "Health in Household Context: Living Arrangements and Health in Late Middle Age." *Journal of Health and Social Behavior* 43:1–21.

Hull, Anne, and Dana Priest. 2007. "'It Is Just Not Walter Reed': Soldiers Share Troubling Stories of Military Health Care across U.S." *Washington Post*, March 5. Available at http://washingtonpost.com

Human Rights Watch. 2003. *Ill-Equipped: U.S. Prisons and Offenders with Mental Illness*. Available at http://www.hrw.org

Human Rights Watch. 2005 (March 3). *U.S. Gag on Needle Exchange Harms U.N. AIDS Efforts*. Available at http://www.hrw.org

Institute of Medicine. 2004. *Insuring America's Health: Principles and Recommendations*. Washington, DC: National Academies Press.

James, Cara, Megan Thomas, Marsha Lillie-Blanton, and Rachel Garfield. 2007. *Key Facts: Race, Ethnicity, & Medical Care*. Kaiser Family Foundation. Available at http://www.kff.org

Jha, A. K., E. S. Fisher, Z. Li, E. J. Orav, and A. M. Epstein. 2005. "Racial Trends in the Use of Major Procedures among the Elderly." *New England Journal of Medicine* 353(7):683–691.

Johnson, Cary Alan. 2007. *Off the Map: How HIV/AIDS Programming is Failing Same-Sex Practicing People in Africa*. New York: International Gay and Lesbian Human Rights Commission.

Johnson, Tracy L., and Elizabeth Fee. 1997. "Women's Health Research: An Introduction." In *Women's Health Research: A Medical and Policy Primer*, Florence P. Haseltine and Beverly Greenberg Jacobson,

eds. (pp. 3–26). Washington, DC: Health Press International.

Joint United Nations Programme on HIV/AIDS. 2006. *Global Summary of the AIDS Epidemic: December 2006*. Available at http://www.unaids.org

Kaiser Commission on Medicaid and the Uninsured. 2004 (November). *The Uninsured: A Primer Key Findings about Americans without Health Insurance*. Washington, DC: Kaiser Family Foundation.

Kaiser Commission on Medicaid and the Uninsured. 2007 (March). *Medicaid: A Primer*. Kaiser Family Foundation. Available at http://www.kff.org

Kaiser Health Poll Report. 2004 (July-August). *Views of Managed Care and Customer Service*. Available at http://www.kff.org

Kaiser Family Foundation. 2004a (June). *The HIV/AIDS Epidemic in the U.S.* HIV/AIDS Policy Fact Sheet. Available at http://www.kff.org

Kaiser Family Foundation. 2004b (August). *Survey of Americans on HIV/AIDS. Part 3. Experiences and Opinions by Race/Ethnicity and Age*. Available at http://www.kff.org

Kaiser Family Foundation. 2004c. *Survey of Americans on HIV/AIDS. Part 2. HIV Testing*. Available at http://www.kff.org

Kaiser Family Foundation. 2005a (January). *November/December 2004 Health Poll Report Survey*. Available at http://www.kff.org

Kaiser Family Foundation. 2005b. *Trends and Indicators in the Changing Health Care Marketplace*. Available at http://www.kff.org

Kaiser Family Foundation. 2007a (January). "Health Care Spending in the United States and OECD Countries." Available at http://www.kff.org

Kaiser Family Foundation. 2007b. *How Changes in Medical Technology Affect Health Care Costs*. Available at http://www.kff.org

Kolata, Gina. 2007 (January 3). "A Surprising Secret to a Long Life: Stay in School." *New York Times*. p. A1, A16.

Komiya, Noboru, Glenn E. Good, and Nancy B. Sherrod. 2000. "Emotional Openness as a Predictor of College Students' Attitudes Toward Seeking Psychological Help." *Journal of Counseling Psychology* 47(1):138–143.

Krugman, Paul. 2005. "Passing the Buck." *New York Times Online*, April 22. Available at http://www.nytimes.com

Kumar, Ruma. 2007 (May 18). "Ban on Smoking Becomes Md. Law." *Baltimore Sun*. Available at http://www.baltimoresun.com

Lalasz, Robert. 2004 (November). *World AIDS Day 2004: The Vulnerability of Women and Girls*. Population Reference Bureau. Available at http://www.prb.org

Lantz, Paula M., James S. House, James M. Lepkowski, David R. Williams, Richard P. Mero, and Jieming Chen. 1998. "Socioeconomic Factors, Health Behaviors, and Mortality: Results from a Nationally Representative Prospective Study of U.S. Adults." *Journal of the American Medical Association* 279:1703–1708.

Lee, Christopher. 2007 (June 21). "Study Finds 1.8 Million Veterans Are Uninsured." *Washington Post*: A09.

Lee, Kelley. 2003. "Introduction." In *Health Impacts of Globalization*, Kelley Lee, ed. (pp. 1–10). New York: Palgrave MacMillan.

Leech, Marie. 2007 (June 15). "High-Calorie Drinks Banned from Schools." *The Birmingham News*. Available at http://www.al.com/news/birminhamnews/

Lerer, Leonard B., Alan D. Lopez, Tord Kjellstrom, and Derek Yach. 1998. "Health for All: Analyzing Health Status and Determinants." *World Health Statistics Quarterly* 51:7–20.

Lightfoot, M., D. Swendeman, M. J. Rotham-Borus, W. S. Comulada, and R. Weiss. 2005. "Risk Behaviors of Youth Living with HIV: Pre- and Post HAART." *American Journal of Health Behavior* 29(2):162–171.

Link, Bruce G., and Jo Phelan. 2001. "Social Conditions as Fundamental Causes of Disease." In *Readings in Medical Sociology*, 2nd ed., William C. Cockerham, Michael Glasser, and Linda S. Heuser, eds. (pp. 3–17). Upper Saddle River, NJ: Prentice Hall.

Mahapatra, Rajesh. 2005. "Groups Slam Indian Decision on Patent Law." *Business Week Online*, March 23. Available at http://www.businessweek.com

Mahar, Maggie. 2006. *Money-Driven Medicine*. New York: Harper-Collins.

Mastny, L. and R. P. Cincotta. 2005. "Examining the Connection between Population and Security." In *State of the World 2005: Redefining Global Security*, M. Renner, H. French, and E. Assadourian, eds. (pp. 22–39). New York: W. W. Norton.

McGinn, Anne P. 2003. "Combating Malaria." In *State of the World 2003*, L. Starke, ed. (pp. 62–83). New York: W. W. Norton.

Mercy, James A., Etienne G. Krug, Linda L. Dahlberg, and Anthony B. Zwi. 2003. "Violence and Health: The United States in a Global Perspective. *American Journal of Public Health* 93(2):256–261.

Millennium Ecosystem Assessment. 2005. *Millennium Ecosystem Assessment Synthesis Report*. Pre-Publication Final Draft. Available at http://www.milleniumassessment.org

Miller, Joel E. 2005. "Assessing the Future Needs of Veterans with PTSD." *NAMI Advocate*, Winter, pp. 14–15.

Mirowsky, John, and Catherine E. Ross. 2003. *Social Causes of Psychological Distress*, 2nd ed. New York: Walter de Gruyter.

Mui, Y. Q. 2005. "Md. Senate Backs School Snacks Limit." *Washington Post*, March 19, p. B10.

Murphy, Elaine M. 2003. "Being Born Female Is Dangerous for Your Health." *American Psychologist* 58(3):205–210.

Murray, C., and A. Lopez, eds. 1996. *The Global Burden of Disease*. Boston: Harvard University Press.

National Center for Chronic Disease Prevention and Health Promotion. 2004. *Physical Activity and Good Nutrition: Essential Elements to Prevent Chronic Diseases and Obesity*. Available at http://www.cdc.gov/nccdphp

National Center for Health Statistics. 2005 (March 23). *Early Release of Selected Estimates Based on Data from the January–September 2004 National Health Interview Survey*. Available at http://www.cdc.gov/nchs

National Center for Health Statistics. 2006. *Health, United States, 2006 with Chartbook on Trends in the Health of Americans*. Hyattsville, MD: U.S. Government Printing Office.

National Council on Disability. 2002. *The Well-Being of Our Nation: An Inter-Generational Vision of Effective Mental Health Services and Supports*. Washington DC: National Council on Disability.

National Institute of Mental Health. 2006. *The Numbers Count: Mental Disorders in America*. Available at http://www.nimh.nih.gov

Newport, Frank. 2004 (December 2). *Assessing American's Mental Health*. Gallup Organization. Available at http://www.gallup.com

Niemeier, H. M., H. A. Raynor, El E. Lloyd-Richardson, M. L. Rogers, and R. R. Wing. 2006. "Fast Food Consumption and Breakfast Skipping: Predictors of Weight-Gain from Adolescence to Adulthood in a Nationally Representative Sample." *Journal of Adolescent Health* 39(6):842–849.

Ninan, Ann. 2003 (March). "*Without My Consent": Women and HIV-Related Stigma in India*. Population Reference Bureau. Available at http://prb.org

"Obese Boy Stays with Mother." 2007 (February 27). *Guardian Unlimited*. Available at http://www.guardian.co.uk

Olshansky, S. J., D. J. Pasaro, R. C. Hershow, J. Layden, B. A. Carnes, J. Brody, L. Hayflick, R. N. Butler, D. B. Allison, and D. S. Ludwig. 2005. "A Potential Decline in Life Expectancy in the United States in the 21st Century." *New England Journal of Medicine* 352(11):1138–1145.

Oxfam GB. 2004 (March 19). *The Cost of Childbirth*. Press release. Available at http://www.oxfam.org.uk/

Parsons, Talcott. 1951. *The Social System*. New York: Free Press.

Peeno, Linda. 2000. "Taking On the System." *Hope* (22):18–21.

Pereira, M. A., A. I. Kartashov, C. B. Ebbeling, L. Van Horn, M. I. Slattery, D. R. Jacobs, and D. S. Ludwig. 2005. "Fast-Food Habits, Weight Gain, and Insulin Resistance." *The Lancet* 365(9453):36–42.

PNHP (Physicians for a National Health Plan). 2005 (February 2). *Bankruptcy Study Highlights Need For National Health Insurance*. Available at http://www.pnph.org/news

Population Reference Bureau. 2006. *2006 World Population Data Sheet*. Available at http://www.prb.org

Quadagno, Jill. 2004. "Why the United States Has No National Health Insurance: Stakeholder Mobilization Against the Welfare State 1945–1996." *Journal of Health and Social Behavior* 45:25–44.

Robbins, John. 2006. *Healthy at 100*. New York: Random House.

Rosenberg, Tina. 2005. "Think Again: AIDS." *Foreign Policy*, March-April. Available at http://www.foreignpolicy.com

Sachs, Jeffrey D. 2005. "A Practical Plan to End Poverty." *Washington Post*, January 17, p. A17.

Sager, Alan, and Deborah Socolar. 2004 (October 28). *2003 U.S. Prescription Drug Prices 81 Percent Higher Than in Other Wealthy Nations*. Data Brief No. 7. Boston University School of Public Health.

Sampson, Zinie Chen. 2007 (February 23). "Bill Would Bar Punishing Students Solely for Attempting Suicide." Associated Press. Available at http://www.timesdispatch.com

Sanders, David, and Mickey Chopra. 2003. "Globalization and the Challenge of Health for All: A View from Sub-Saharan Africa." In *Health Impacts of Globalization*, Kelley Lee, ed. (pp. 105–119). New York: Palgrave Macmillan.

Sanders, Jim. 2005. "Bill Would OK Condoms in Prisons." *The Sacramento Bee*, February 26. Available at http://www.sacbee.com

Save the Children. 2002. *State of the World's Mothers, 2002*. Available at http://www.savethechildren.org

Scal, Peter, and Robert Town. 2007. "Losing Insurance and Using the Emergency Department: Critical Effect of Transition to Adulthood for Youth with Chronic Conditions." *Journal of Adolescent Health* 40 (2 Suppl 1): S4.

Schoenborn, Charlotte A. 2004 (December 15). "Marital Status and Health: United States, 1999–2002." Advance data from *Vital and*

Health Statistics, no. 351. Hyattsville, MD: National Center for Health Statistics.

Sered, Susan Starr, and Rushika Fernandopulle. 2005. *Uninsured in America: Life and Death in the Land of Opportunity*. Berkeley and Los Angeles, CA: University of California Press.

Shaffer, E. R., H. Waizkin, J. Brenner, and R. Jasso-Aguilar. 2005. "Global Trade and Public Health." *American Journal of Public Health* 95(1):23–34.

Sidel, Victor W., and Barry S. Levy. 2002. "The Health and Social Consequences of Diversion of Economic Resources to War and Preparation for War." In *War or Health: A Reader*, Ilkka Taipale, P. Helena Makela, Kati Juva, and Vappu Taipale, eds. (pp. 208–221). New York: Palgrave MacMillan.

SIECUS. 2004. *Public Support for Comprehensive Sexuality Education*. SIECUS Fact Sheet. Available at http://www.siecus.org

SmokeFree Educational Services. 2003 (June 5). *5 Million Deaths a Year Worldwide from Smoking*. Corp-Watch. Available at http://www.corpwatch.org

Spector, Mike, and John D. Stoll. 2007 (February 26). "GM, Ford to see escalating health care costs in 2007." *MarketWatch*. Available at http://www.marketwatch.com

Springen, Karen. 2006. "Health Hazards: How Mounting Medical Costs Are Plunging More Families into Debilitating Debt and Why Insurance Doesn't Always Keep Them out of Bankruptcy." *Newsweek* (August 25). Available at www.msnbc.com/newsweek

Stanton, Mark W. 2006 (June). The High Concentration of U.S. Health Care Expenditures. *Research in Action* Issue #19. Rockville, MD: U.S. Department of Health and Human Services, Public Health Service Agency for Healthcare Research and Quality, pp. 1–9.

"Stark Introduces Constitutional Amendment to Establish Right to Health Care." 2005 (March 3). Press release. Available at http://www.house.gov

Stein, R., and C. Connolly. 2004. "Medicare Changes Policy on Obesity, Some Treatments May Be Covered." *Washington Post*, July 16. Available at http://www.washingtonpostonline

StigmaBusters Alert. 2004 (April 27). Available at http://www.nami.org

Stockard, Jean, and Robert M. O'Brien. 2002. "Cohort Effects on Suicide Rates: International Variations." *American Sociological Review* 67(6):854–872.

Sturm, Roland. 2005 (April). "Childhood Obesity: What We Can Learn from Existing Data on Social Trends, Part 2." *Preventing Chronic Disease* 2(2). Available at http://www.cdc.gov/pcd

Substance Abuse and Mental Health Services Administration. 2003 (July 21). *Treatment of Adults with Serious Mental Illness*. NHSDA

(National Household Survey on Drug Abuse) Report. Available at http://www.samhsa.gov

Suris, J. C., Ralph Thomas, Jean-Jacques Cheseaux, Pierre-Andre Michaud, and Swiss Mother-Child HIV Cohort Study. 2007. "I don't want to be disappointed: HIV(+) adolescents' reasons for not disclosing their disease: A qualitative study." *Journal of Adolescent Health* 40 (2 Suppl 1):S41–S42.

Szasz, Thomas. 1970 [1961]. *The Myth of Mental Illness: Foundations of a Theory of Personal Conduct*. New York: Harper & Row.

Thomas, Caroline. 2003. "Trade Policy, the Politics of Access to Drugs, and Global Governance for Health." In *Health Impacts of Globalization*, Kelley Lee, ed. (pp. 177–191). New York: Palgrave MacMillan.

Thomson, George, and Nick Wilson. 2005. "Policy Lessons from Comparing Mortality from Two Global Forces: International Terrorism and Tobacco." *Globalization and Health* 1:18.

Timberg, C. 2005. "In S. Africa, Stigma Magnifies Pain of AIDS." *Washington Post*, January 14, p. A14.

Toner, Robin, and Janet Elder. 2007 (March 2). "Most support U.S. guarantee of health care." *New York Times*. Available at http://www.nytimes.com

Trust for America's Health. 2007. *F as in Fat: How Obesity Policies are Failing in America*. Available at http://healthyamericans.org

UNICEF. 2006. *The State of the World's Children 2007*. Available at http://www.unicef.org

United Nations Population Fund. 2002. *The State of World Population 2002: People, Poverty, and Possibilities*. Available at http://www.unfpa.org

United Nations Population Fund. 2004. *The State of World Population 2004*. Available at http://www.unfpa.org

UNOG (United Nations of Geneva). 2005. "Commission Adopts Five Resolutions and Three Decisions on Economic, Social, and Cultural Rights." *News and Media*, April 15. Available at http://www.unog.ch

Urban Institute. 2007 (February 9). "1.8 Million Children Are Eligible for the State Children's Health Insurance Program." The Urban Institute. Available at http://www.urban.org

U.S. Department of Health and Human Services. 1999. *Mental Health: A Report of the Surgeon General—Executive Summary*. Rockville, MD: U.S. Government Printing Office.

U.S. Department of Health and Human Services. 2001. *Mental Health: Culture, Race, and Ethnicity—A Supplement to Mental Health: A Report of the Surgeon General*. Rockville, MD: U.S. Government Printing Office.

Vaccarino, V., S. S. Rathore, N. K. Wenger, P. D. Frederick, J. L. Abramson, H. V. Barron, A. Manhapra, S. Mallik, and H. M. Krumholz. 2005 (August 18). "Sex and Racial Differences in the Management of Acute Myocardial Infarction, 1994 through 2002." *New England Journal of Medicine* 353(7):671–682.

Veugelers, Paul J., and Angela L. Fitzgerald. 2005. "Effectiveness of School Programs in Preventing Childhood Obesity: A Multilevel Comparison." *American Journal of Public Health* 95(3):432–435.

Vock, Daniel C. 2007. "Health Care: States Think Big on Health Reform." *State of the States: A Report on State Policy Developments and Trends* (pp. 18–21). Washington, DC. Available at http://www.Stateline.org

Waxman, Henry A. 2004 (December). *The Context of Federally Funded Abstinence-Only Education Programs*. U.S. House of Representatives Committee on Government Reform. Minority Staff, Special Investigation Division. Available at http://www.democrats.reform.house.gov

Weitz, Rose. 2006. *The Sociology of Health, Illness, and Health Care: A Critical Approach*, 4th ed. Belmont, CA: Wadsworth.

White, Frank. 2003. "Can International Public Health Law Help to Prevent War?" *Bulletin of the World Health Organization* 81(3):228.

WHO International Consortium in Psychiatric Epidemiology. 2000. "Cross-National Comparisons of the Prevalences and Correlates of Mental Disorders." *Bulletin of the World Health Organization* 78(4):413–426.

Williams, David R. 2003. "The Health of Men: Structured Inequalities and Opportunities." *American Journal of Public Health* 93(5):724–731.

World Health Organization. 1946. *Constitution of the World Health Organization*. New York: World Health Organization Interim Commission.

World Health Organization. 2000. *The World Health Report 2000*. Available at http://www.who.int

World Health Organization. 2002. *The World Health Report 2002*. Available at http://www.who.int

World Health Organization. 2004. *The World Health Report 2004: Changing History*. Available at http://www.who.int

World Health Organization. 2005. *The World Health Report 2005: Make Every Mother and Child Count*. Available at http://www.who.int

World Health Organization. 2006. *World Health Statistics 2006*. Available at http://www.who.int

World Health Organization. 2007. *World Health Report 2007: A Safer Future: Global Public Health Security in the 21st Century*. Available at http://www.who.int

Zeldin, Cindy, and Mark Rukavina. 2007. *Borrowing to Stay Healthy: How Credit Card Debt is Related to Medical Expenses*. New York: Demos.

Chapter 3

Abadinsky, Howard. 2004. *Drugs: An Introduction*. Belmont, CA: Wadsworth.

Alcoholics Anonymous. 2007. *AA at a Glance*. Available at http://www.alcoholics-anonymous.org

American Cancer Society. 2006. *Questions about Smoking, Tobacco, and Health. Prevention and Early Detection*. Available at http://www.cancer.org

American Heart Association (AHA). 2003 (July 30). *How Does the Tobacco Industry Target Youth?* Available at http://www.americanheart.org

American Heart Association. 2007 (January 20). *Cigarette Smoking and Cardiovascular Diseases*. Available at http://www.americanheart.org

Apuzzo, Matt. 2006. "Court Blocks Ruling on Tobacco Industry." Associated Press. October 31. Available at http://www.breitbart.com/news/2006/10/31

Armstrong, Elizabeth, and Christina McCarroll. 2004. "Girls Lead in Teen Alcohol Use." *Seattle Times*, August 14. Available at http://seattletimes.nwsource.com

Associated Press. 1999 (December 30). *Alcoholism Touches Millions*. Available at http://www.abcnewsgo.com

Boseley, Sarah. 2007. "Vouchers Seen as Aid to Kicking Heroin Habit." *Guardian Unlimited,* January 27. Available at http://www.guardian.co.uk

BBC (British Broadcasting Company). 2004. *Teens "Fake Drug Use to Fit In."* Available at http://newsvote.bbc.co.uk/mpapps/pagetoolsc

Becker, H. S. 1966. *Outsiders: Studies in the Sociology of Deviance*. New York: Free Press.

Bogart, L. M., R. L. Collins, P. L. Ellickson, and D. J. Klein. 2006. "Are Adolescent Substance Users Less Satisfied with Life and, If So, Why?" *Social Indicators Research*, September 19.

Bureau of Justice Statistics. 2006. *Criminal Victimization in the United States 2004 Statistical Tables*, Table 32. NCJ 213257. Washington D.C.: U.S. Department of Justice.

Burke, Jason. 2006. "Europeans Turn to Cocaine and Alcohol as Cannabis Loses Favour." *Guardian Unlimited,* October 15. Available at http://observer.guardian.co.uk

Campaign for Tobacco-Free Kids. 2005 (September 19). *Special Report: Big Tobacco Still Targeting Kids*. Available at http://tobaccofreekids.org

Carroll, Joseph. 2005. *Crystal Meth, Child Molestation Top Crime Concerns*. Gallup Poll, May 3. Available at http://www.galluppoll.com

Center for Substance Abuse Treatment. 2006. *Children and Addiction*. Available at http://csat.samhsa.gov/publications/youcanhelp.aspx

Centers for Disease Control and Prevention. 2004. *Alcohol Use and Pregnancy*. Available at http://www.cdc.gov/ncbddd/fas/factsheets.htm

Charon, Joel. 2002. *Social Problems: Readings with Four Questions*. Belmont, CA: Wadsworth.

Cloud, John. 2000. "The Lure of Ecstasy." *Time*, June 5, pp. 63–72.

CORE. 2006. CORE Alcohol and Drug Survey. Available at: http://www.siu.edu/departments/coreinst/public

Crittenden, Jules. 2002. "MIT Fraternity Settles Lawsuit in Freshman's Drinking Death." *Boston Herald*. Available at http://www.groups.yahoo.com/group/fraternalnews

DEA (Drug Enforcement Administration). 2000. *An Overview of Club Drugs*. Drug Intelligence Brief, February, pp. 1–10. Washington, DC: U.S. Department of Justice.

Defao, Janine. 2007. "Proposed Car-smoking Ban Angers Foes of 'nanny laws'." *San Francisco Chronicle*. March 26, A1.

Drug Policy Alliance. 2003. *Drug Policy around the World: The Netherlands*. Available at http://www.drugpolicy.org/global

Drug Policy Alliance. 2005 (January 6). *European Parliament on Record Admitting Drug War a Failure*. Available at http://www.dpf.org.news

Drug Policy Alliance. 2006 (November 8). *Drug Policy News*. Available at http://www.drugpolicy.org/news/

Drug Policy Alliance. 2007. *State by State*. Available at http://www.drugpolicy.org/statebystate/

Duke, Steven, and Albert C. Gross. 1994. *America's Longest War: Rethinking Our Tragic Crusade against Drugs*. New York: G. P. Putnam.

Easley, Margaret, and Norman Epstein. 1991. "Coping with Stress in a Family with an Alcoholic Parent." *Family Relations* 40:218–224.

Eisner, Robin. 2005. "Marijuana Abuse: Age of Initiation, Pleasure of Response Foreshadow Young Adult Outcomes." *NIDA Notes* 9(5):1, 6–7.

Feagin, Joe R., and C. B. Feagin. 1994. *Social Problems*. Englewood Cliffs, NJ: Prentice Hall.

Firshein, Janet. 2003. "The Role of Biology and Genetics." *Moyers on Addiction*. Public Broadcasting System (PBS). Available at http://www.pbs.org/wnet/closetohome/

Fox, Maggie. 2006. "Manufacturers' Anti-Smoking Ads Ineffective: Study." *Reuters,* October 31. Available at http://today.reuters.com

Freeman, Dan, Merrie Brucks, and Melanie Wallendorf. 2006. "Young Children's Understanding of Cigarette Smoking." *Addiction* 100(10):1537–1545.

Gentry, Cynthia. 1995. "Crime Control through Drug Control." In *Criminology*, 2nd ed, Joseph F. Sheley, ed. (pp. 477–493). Belmont, CA: Wadsworth.

Goldstein, Adam, Rachel Sobel, and Glen Newman. 1999. "Tobacco and Alcohol Use in G-Rated Children's Animated Films." *Journal of the American Medical Association* 281:1121–1136.

Grant, Tavia. 2005. "Cannabis Spray Gets Go-Ahead." *The Globe and Mail*, April 19.

Gusfield, Joseph. 1963. *Symbolic Crusade: Status Politics and the American Temperance Movement*. Urbana, IL: University of Illinois Press.

Hanson, Glen R. 2002. "New Vistas in Drug Abuse Prevention." *NIDA Notes* 16:3–4.

Hays, Tom. 2006. "Judge Allows Class Action Tobacco Suit." Associated Press, September 25. Available at http://www.breitbart/com/news/2006/09/25

Healy, Patrick. 2005. "Sobriety Tests Are Becoming Part of the School Day." *New York Times*, March 3. Available at http://www.nytimes.com

How Does Alcohol Affect the World of a Child. 2003. Available at http://www.alcoholfreechildren.org

Human Rights Watch. 2007. "Reforming the Rockefeller Drug Laws." Available at http://www.hrw.org/campaigns/drugs

Jamieson, Bob. 2007. "Maine Initiative Keeps Kids out of Smoke-filled Autos." January 19. Available at http://abcnews.go.com

Jarvik, M. 1990. "The Drug Dilemma: Manipulating the Demand." *Science* 250:387–392.

Jernigan, David. 2005. "National Experiences, the U.S.A.: Alcohol and Young People Today." *Addiction* 100(3):271–273.

Johnson, Lynn M. 2003. *Alcohol, Drugs and Violence: Detrimental Effects on Children*. Available at http://www.babyparenting.about.com

Jones, Jeffrey. 2006. "U.S. Drinkers Consuming Alcohol More Regularly." Gallup Poll, July 31. Available at http://www.galluppoll.com

Katel, P. 2006 (June 2). "War on Drugs." June 2. *CQ Researcher* 16 (21). Available at http://library.cqpress.com/cqresearcher

Kelly, J. F., J. W. Finney, and R. Moos. 2005. "Substance Use Disorder Patients Who Are Mandated to Treatment: Characteristics, Treatment Process, and 1 and 5 Year Outcomes." *Journal of Substance Abuse Treatment* 28(3):213–223.

Kelly, John, Robert Stout, William Zywiak, and Robert Schneider. 2006. "A 3-Year Study of Addiction Mutual-Help Group Participation Following Intensive Outpatient Treatment." *Alcoholism: Clinical and Experimental Research* 30:1381–1392.

Klutt, Edward C. 2000. *Pathology of Drug Abuse.* Available at http://www.library.med.utah.edu/webpath

Kohnle, Diana. 2006. "Alcohol a Bigger Threat to U.S. Youth than Drugs." *HealthDay* (June):1–2.

Lee, Yoon, and M. Abdel-Ghany. 2004. "American Youth Consumption of Licit and Illicit Substances." *International Journal of Consumer Studies* 28(5):454–465.

Leonard, K. E., and H. T. Blane. 1992. "Alcohol and Marital Aggression in a National Sample of Young Men." *Journal of Interpersonal Violence* 7:19–30.

Leshner, Alan. 2003. "Using Science to Counter the Spread of Ecstasy Abuse." *NIDA Notes* 16:3–4.

Lindholm, Eric. 2006. "Percentage of Total Health Costs attributable to Smoking." Available at http://www.tobaccofreekids.org

Lyons, Linda. 2005. *Drug Use Still Among Americans' Top Worries.* Gallup Poll, April 5. Available at http://www.galluppoll.com

MacCoun, Robert J., and Peter Reuter. 2001. "Does Europe Do It Better? Lessons from Holland, Britain and Switzerland." In *Solutions to Social Problems*, D. Stanley Eitzen and Craig S. Leedham, eds. (pp. 260–264). Boston: Allyn and Bacon.

MADD (Mothers Against Drunk Driving). 2003. *Alcohol Advertising.* Available at http://www.madd.org/stats

MADD. 2006. "MADD Programs." Available at http://www.madd.org

McCaffrey, Barry. 1998. *Remarks by Barry McCaffrey, Director, Office of National Drug Control Policy, to the United Nations General Assembly: Special Session on Drugs.* Office of National Drug Control Policy. Available at http://www.whitehousedrugpolicy.gov/news/speeches

Mickleburgh, Rod, and Gloria Galloway. 2007. "Storm Brews over Drug Strategy." *The Globe and Mail.* January 15. Available at http://www.theglobeandmail.com

Monitoring the Future. 2006. *National Results on Adolescent Drug Use.* The University of Michigan Institute for Social Research. Available at http://monitoringthefuture.org

Moore, Martha T. 1997. "Binge Drinking Stalks Campuses." *USA Today*, October 1, p. A3.

Moreau, Ron, and Sami Yousafzai. 2006. "A Harvest of Treachery." *Newsweek.* January 9:32–35.

Morgan, Patricia A. 1978. "The Legislation of Drug Law: Economic Crisis and Social Control." *Journal of Drug Issues* 8:53–62.

National Drug Threat Assessment 2007. National Drug Intelligence Center. U.S. Department of Justice. Available at http:www.usdoj.gov/ndic/pubs21/21137

National Highway Traffic Safety Administration. 2006. *Alcohol-Related Crashes and Fatalities.* National Center for Statistics and Analysis. Available at http://www.nhtsa.gov

National Institute on Alcohol Abuse and Alcoholism. 2000 (April). *Mechanisms of Addiction.* Alcohol Alert. National Institute on Alcohol Abuse and Alcoholism no. 47. Available at http://www.niaaa.nih.gov/publications/aa47.htm

Neergaard, Lauren. 2006. "Surgeon General Warns of Secondhand Smoke." Associated Press, June 27. Available at http://breitbart.com/news/2006/06/27

NIDA (National Institute on Drug Abuse). 2000. *Heroin Abuse and Addiction.* Research Report Series, NIH Publication 00-4165. Available at http://www.nida.nih.gov/ResearchReports/Heroin/Heroin.html

NIDA. 2006a. *Principles of Drug Addiction Treatment: A Research Based Guide.* Available at http://drugabuse.gov/PODAT

NIDA. 2006b. Commonly Abused Drugs. Available at http://www.nida.nih.gov/DrugPages/DrugsofAbuse.html

NSDUH (National Survey on Drug Use and Health). 2004 (February 13). *Alcohol Dependence or Abuse among Parents with Children Living at Home.* Office of Applied Studies, Substance Abuse and Mental Health Services Administration. Available at http://www.DrugAbuseStatistics.samhsa.gov

NSDUH. 2006. "Illicit Drug Use among Persons Arrested for Serious Crimes." *The NSDUH Report,* December 16.

ODCCP (Office of Drug Control and Crime Prevention). 2003. *Global Illicit Drug Trends 2002.* New York: United Nations.

OJJDP (Office of Juvenile Justice and Delinquency Prevention). 2005. "Youthful Drinking Rates and Problems: A Comparison of European Countries and the U.S." Pacific Institute for Research and Evaluation. Available at http://www.pire.org/

ONDCP (Office of National Drug Control Policy). 2003a. *Drug Facts: Club Drugs.* Available at http://www.whitehousedrugpolicy.gov/club

ONDCP. 2003b. *Rohypnol.* Fact sheet. Available at http://www.whitehousedrugpolicy.gov/publications/factsht/Rohypnol

ONDCP. 2003c. *Drug Data Summary.* Available at http://www.whitehousedrugpolicy.gov/publications/factsht/summary

ONDCP. 2004. *The Economic Costs of Drug Abuse in the U.S.: 1992–2002.* Section IV. Executive Office of the President. Available at http://www.whitehousedrugpolicy.gov

ONDCP. 2005a. *Inhalants.* Available at http://www.whitehousedrugpolicy.gov/inhalants/index.html

ONDCP. 2005b. *Marijuana.* Available at http://www.whitehousedrugpolicy.gov/drugfact/marijuana/index.html

ONDCP. 2005c. *Media Campaign: What's New.* Available at http://www.mediacampaign.org

ONDCP. 2005d. (May 3). *White House Drug Czar, Research and Mental Health Communities Warn Parents That Marijuana Use Can Lead to Depression, Suicidal Thoughts, and Schizophrenia.* Available at http://www.whitehousedrugpolicy.gov/news/press05

ONDCP. 2006a. *Methamphetamine.* Available at http://www.whitehousedrugpolicy.gov/drugfact/methamphetamine

ONDCP. 2006b. (April 5). "White House Drug Czar Praises Dutch Disruption of Global Ecstasy Market. . . ." Available at http://www.whitehousedrugpolicy.gov/NEWS/press06

ONDCP. 2006c. *Club Drugs.* Available at http://www.whitehousedrugpolicy.gov/drugfact/club/index.html

ONDCP. 2006d. *Heroin.* Available at http://www.whitehousedrugpolicy.gov/drugfact/heroin/index.html

ONDCP. 2006e. *Inhalants.* Available at http://www.whitehousedrugpolicy.gov/drugfact/inhalants/index.html

ONDCP. 2006f. *Source Countries and Drug Transit Zones: Colombia.* Available at http://www.whitehousedrugpolicy.gov/international/colombia

ONDCP. 2006g. *White House Anti-Drug Office Begins Posting Videos on YouTube.* Available at http://www.whitehousedrugpolicy.gov/news/press06

ONDCP. 2007. *National Priorities.* Available at http://www.whitehousedrugpolicy.gov

Park, Michael. 2006. "College Alcohol Abuse Sparks Drinking Prevention Debate." Fox News, August 28. Available at http://www.foxnews.com

Pew Research Center. 2002. *Illegal Drugs.* Pew Research Center for the People and Press Survey. Available at http://www.pollingreport.com/drugs

Pacific Institute for Research and Evaluation. 2007. *New Study Shows a "Tidal Wave" of Underage Drinking Costs.* Press Release. Available at http://www.pire.org

Phoenix House. 2007. Available at http://www.phoenixhouse.org

Quinlan, K. P., R. D. Brewer, P. Siegel, D. A. Sleet, A. H. Mokdad, R. A. Shults, N. Flowers. 2005. Alcohol-impaired driving among U.S. adults, 1993–2002. *American Journal of Preventive Medicine* 28(4): 345–350.

Reiman, Jeffrey. 2004. *The Rich Get Richer and the Poor Get Prison*. Boston: Allyn and Bacon.

Ritter, John. 2007. "Adults Penalized for Teen Drinking." *USA Today*. January 4. Available at http://usatoday.com

Rorabaugh, W. J. 1979. *The Alcoholic Republic: An American Tradition*. New York: Oxford University Press.

Saad, Lydia. 2006. *Families of Drug and Alcohol Abusers Pay an Emotional Toll*. Gallup Poll, August 25. Available at http://www.galluppoll.com

SAMHSA (Substance Abuse and Mental Health Services Administration). 2006. *Youth Drug Use Continues Downward Slide. . . .* Press Release, September 7. Available at http://www.samhsa.gov/news/newsreleases

SAMHSA. 2007. *The fact is. . . .* Available at http://ncadi.samhsa.gov/govpubs/ph318

Sargant, James, Thomas Wills, Mike Stoolmiller, Jennifer Gibson, and Frederick Gibbons. 2006. "Alcohol Use in Motion Pictures and Its Relation with Early Onset Teen Drinking." *Journal of Studies on Alcohol* 77(1):54–66.

Schemo, Diana Jean. 2002. "Study Calculates the Effects of College Drinking in the U.S." *New York Times,* April 10. Available at http://www.nytimes.com

Sheldon, Tony. 2000. "Cannabis Use among Dutch Youth." *British Medical Journal* 321:655.

Shenfeld, Hilary. 2006. "The Best High They've Ever Had." *Newsweek,* June 12, p. 8.

Siegel, Larry. 2006. *Criminology*. Belmont, CA: Wadsworth.

Singley, Devin. 2004. "Collegiate Alcohol Awareness Week Speaker Focuses on the Dangers of Hazing." *The Carolinian Online*, October 25. Available at http://carolinianonline.com/news/2004/10/25

Smith, Stephen. 2007. "Nicotine Boost Was Deliberate, Study Says." *Boston Globe*. January 18. Available at http://www.boston.com/news/local/articles/2007/01/18

Taifa, Nkechi. 2006 (May). "The 'Crack/Powder Disparity': Can the International Race Convention Provide a Basis for Relief?" American Constitution Society for Law and Policy White Paper. Available at http://acslaw.org

Thio, Alex. 2007. *Deviant Behavior*. Boston: Allyn and Bacon.

Time/CNN. 2003. "The New Politics of Pot." *Time*, November 4, pp. 56–66.

UNODC (United Nations Office on Drugs and Crime). 2004. Drugs and Crime Trends in Europe and Beyond. Available at http://unodc.org/pdf/factsheets

UNODC (United Nations Office on Drugs and Crime). 2006a. "UN Drugs Chief Warns European Mayors about Risk of Overdoses from Bumper Afghan Opium Crop." Available at http://unodc.org/unodc/en/press_release_2006_11_22

UNODC (United Nations Office on Drugs and Crime). 2006b. "Who is Using Drugs?" Available at http://unodc.org/unodc/en/drug_demand_who

U.S. Department of Health and Human Services. 2006. *2005 National Survey on Drug Use and Health*. Substance Abuse and Mental Health Service Administration. Washington, DC: U.S. Government Printing Office.

U.S. Department of Justice. 2007. *National Drug Threat Survey, 2006*. National Drug Threat Assessment. Available at http://www.usdoj.gov

Van Dyck, C., and R. Byck. 1982. "Cocaine." *Scientific American* 246:128–141.

Van Kammen, Welmoet B., and Rolf Loeber. 1994. "Are Fluctuations in Delinquent Activities Related to the Onset and Offset in Juvenile Illegal Drug Use and Drug Dealing?" *Journal of Drug Issues* 24:9–24.

Weitzman, Elissa, Henry Wechsler, and Toben F. Nelson. 2003. *Environment, Not Education, a Stronger Predictor of Binge Drinking Behavior among College Freshman*. Press Release, Harvard School of Public Health, January 21.

Whitten. Lori. 2006. "Studies Identify Factors Surrounding a Rise in Abuse of Prescription Drugs by College Students." *NIDA Notes* 20(4):1, 6.

Willing, Richard. 2002. "Study Shows Alcohol Is Main Problem for Addicts." *USA Today*, October 3, p. B4.

Witters, Weldon, Peter Venturelli, and Glen Hanson. 1992. *Drugs and Society*, 3rd ed. Boston: Jones & Bartlett.

World Drug Report. 2006. *Executive Summary*. Office of Drug Control and Crime Prevention. United Nations. Available at http://www.unodc.org/unodc/world_drug_report.html

World Health Organization. 2006. *Regulation Urgently Needed to Control Growing List of Deadly Tobacco Products*. United Nations: World Health Organization. Available at http://www.who.int/mediacentre/news/releases/2006

Wysong, Earl, Richard Aniskiewicz, and David Wright. 1994. "Truth and Dare: Tracking Drug Education to Graduation and as Symbolic Politics." *Social Problems* 41:448–468.

Zailckas, Koren. 2005. *Smashed: Story of a Drunken Girlhood*. New York: Viking.

Zickler, Patrick. 2003. "Study Demonstrates That Marijuana Smokers Experience Significant Withdrawal." *NIDA Notes* 17:7, 10.

Chapter 4

Amnesty International. 2007. *Death Sentences and Executions*. Available at http://web.amnesty.org

Anderson, David. 1999. "The Aggregate Burden of Crime." *Journal of Law and Economics* 42:611–642.

Associated Press. 2005a. "Fla. Gov. Jeb Bush Signs Lunsford Act." Fox News, May 2. Available at http://www.foxnews.com

Associated Press. 2005b, April 18. *UN Conference Takes Aim at Crime*. Available at http://theglobeandmail.com

Associated Press. 2007a. "Austria Uncovers Vast Pornography Ring." *International Herald Tribune*, February 7. Available at http://www.iht.com/articles/2007/02/07

Associated Press. 2007b. "Man Pleads Guilty to Smuggling Women into U.S. for Prostitution." Fox News, February 10. Available at http://www.foxnews.com

Associated Press. 2007c. "Special License Plates for Sex Predators?" MSNBC, March 1. Available at http://www.msnbc.msn.com

Barkan, Steven. 2006. *Criminology*. Upper Saddle River, NJ: Prentice Hall.

Barkan, Steven, and Steven Cohn, 2005. "Why Whites Favor Spending More Money to Fight Crime: The Role of Racial Prejudice." *Social Problems* 52(2):300–314.

Becker, Howard S. 1963. *Outsiders: Studies in the Sociology of Deviance*. New York: Free Press.

Buechner, Maryanne Murray. 2006. "Shock Absorbers." *Time*, March, p. A8.

Bureau of Justice Statistics. 2006a. *Victim Characteristics*. Available at http://ojp.usdoj.gov/bjs/cvict_v.htm

Bureau of Justice Statistics. 2006b. *Homicide Trends in the U.S.: Trends by Gender*. Available at http://ojp.usdoj.gov/bjs/homicide/gender.htm

Bureau of Justice Statistics. 2006c. *Prison and Jail Inmates at Midyear 2005*. Available at http://ojp.usdoj.gov/bjs/

Bureau of Justice Statistics. 2006d. *Corrections Statistics*. Available at http://www.ojp.usdoj.gov/bjs/correct.htm

Bureau of Justice Statistics. 2006e. *One in Every 32 Adults Was in Prison, Jail, on Probation, or on Parole at the End of 2005*. Available at http://www.ojp.usdoj.gov/bjs/

Butterfield, Fox. 2002. "Tight Budgets Force States to Reconsider Crime and Penalties." *New York Times*, January 21. Available at http://www.nytimes.com

Campo-Flores, Arian. 2005. "The Most Dangerous Gang in America." *Newsweek*, March 28, pp. 23–25.

Carlson, Darren K. 2005. "Americans Deal with Crime by Steering Clear." Gallup Poll, November 22. Available at http://www.galluppoll.com

Carroll, Joseph. 2006. "Americans' Crime Worries." Gallup Poll, October 19. Available at http://www.galluppoll.com

Centers for Disease Control and Prevention. 2006. "Youth Risk Behavior Surveillance." *Morbidity and Mortality Weekly Report* June 9, pp. 1–108.

Center for Health and Gender Equity. 2006. *Prostitution Loyalty Oath: A U.S. Global AIDS Policy Restriction in Violation of First Amendment.* Available at http://www.genderhealth.org/loyalty oath.php

Chesney-Lind, Meda, and Randall G. Shelden. 2004. *Girls, Delinquency, and Juvenile Justice.* Belmont, CA: Wadsworth.

Chicago Tribune. 2003. "Ford Fined, Ordered to Hand over Van Data." *Tribune News Services.* Available at http://www.lieffcabraser.com

CNN. 2007a. "After Two Years, Sex Offender to Stand on Trial for Girl's Death." *CNN,* February 12. Available at http://www.cnn.com/2007/law/02/12/jury

CNN. 2007b. "Boy Tries to Reconnect with Old Life after 4 Years." *CNN,* January 15. Available at http://www.cnn.com/2007/US/01/15/missouri.boys

Coates, Sam. 2005. "Rader Gets 175 Years for BTK Slayings." *Washington Post,* August 19, p. A3. Available at http://www.washingtonpost.com

Cohen, Jon. 2005. *Poll: Identity Theft Concerns Rise. ABC News,* May 18. Available at http://abcnews.go.com/Business

Conklin, John E. 1998. *Criminology,* 6th ed. Boston: Allyn and Bacon.

Conklin, John E. 2007. *Criminology,* 9th ed. Boston: Allyn and Bacon

COPS. 2007. *What Is Community Policing?* Available at http://www.cops.usdoj.gov

D'Alessio, David, and Lisa Stolzenberg. 2002. "A Multilevel Analysis of the Relationship Between Labor Surplus and Pretrial Incarceration." *Social Problems* 49:178–193.

D'Alessio, David, and Lisa Stolzenberg. 2003. "Race and the Probability of Arrest." *Social Forces* 81:1381–1397.

"Death Behind Bars." 2005. Editorial. *New York Times,* March 10. Available at http://www.nytimes.com

Death Penalty Information Center. 2007. *Costs of Death Penalty.* Available at http://www.deathpenaltyinfo.org

Dickerson, Debra. 2000. "Racial Profiling: Are We All Really Equal in the Eyes of the Law?" *Los Angeles Times,* July 16. Available at http://www.latimes.com

Dixon, Travis L., and Daniel Linz. 2000. "Race and the Misrepresentation of Victimization on Local Television News." *Communication Research* 27:547–574.

Eckholm, Erik. 2006. "States Are Growing More Lenient in Allowing Felons to Vote." *The New York Times,* October 12. Available at http://www.nytimes.com/2006/10/12/us

Eitle, David, Stewart D'Alessio, and Lisa Stolzenberg. 2002. "Racial Threat and Social Control: A Test of the Political, Economic, and Threat of Black Crime Hypothesis." *Social Forces* 81:557–576.

Erikson, Kai T. 1966. *Wayward Puritans.* New York: Wiley.

Europol. 2006. "Frequently asked Questions." Available at http://www.europol.europa.eu

FBI (Federal Bureau of Investigation). 2006. *Crime in the United States, 2005. Uniform Crime Reports.* Washington, DC: U.S. Government Printing Office.

Fellner, Jamie. 2006. "U.S. Addiction to Incarceration Puts 2.3 Million in Prison." *Human Rights Watch.* Available at http://hrw.org/english/docs/2006

Felson, Marcus. 2002. *Crime and Everyday Life,* 3rd ed. Thousand Oaks, CA: Sage.

Fight Crime. 2007. *Fight Crime: Invest in Kids.* Available at http://www.fightcrime.org

Fletcher, Michael A. 2000. "War on Drugs Sends More Blacks to Prison than Whites." *Washington Post,* June 8, p. A10.

Florida Criminal Code. 2006. Available at http://www.leg.state.fl.us/statutes

Friedrichs, David O. 2004. *Trusted Criminals.* Belmont, CA: Wadsworth.

Gallup Poll. 2007a. *Gallup's Pulse of Democracy: Crime.* Available at http://www.galluppoll.com

Gallup Poll. 2007b. *Gallup's Pulse of Democracy: Gun Laws.* Available at http://institution.gallup.com

Gest, Ted, and Dorian Friedman. 1994. "The New Crime Wave." *U.S. News and World Report,* August 29, pp. 26–28.

Harrow, Robert O. 2005. "ID Data Conned from Firm." *Washington Post,* February 16. Available at http://www.washingtonpost.com

Harvard School of Public Health. 2002. "American Females at Highest Risk of Murder." Press release. Harvard School of Public Health, April 17.

Heimer, Karen, and Stacy DeCoster. 1999. "The Gendering of Violent Delinquency." *Criminology,* May, pp. 377–389.

Herbert, Bob. 2002. "The Fatal Flaws." *New York Times,* February 11. Available at http://www.nytimes.com

Hirschi, Travis. 1969. *Causes of Delinquency.* Berkeley, CA: University of California Press.

ICCC (Internet Crime Complaint Center). 2006. *IC3 2005 Internet Crime Report.* Washington, DC: National White Collar Crime Center and the Federal Bureau of Investigation. Available at http://www.ic3.gov/media/annualreports.aspx

ICPC (International Centre for the Prevention of Crime). 2007. *About ICPC.* Available at http://www.crime-prevention-intl.org/

Interpol. 2007a. *Interpol's Five Priority Crime Areas.* Available at http://www.interpol.int/

Interpol. 2007b. *About Interpol.* Available at http://www.interpol.int/

Interpol. 2007c. *Interpol's Three Core Functions.* Available at http://www.interpol.int/

Irwin, John. 2005. *The Warehouse Prison.* Los Angeles, CA: Roxbury.

Jacobs, David, Zhenchao Qian, Jason Carmichael, and Stephanie Kent. 2007. "Who Survives on Death Row? An Individual and Contextual Analysis." *American Sociological Review* 72: 610–632.

Jessica Lunsford Foundation. 2006. *About Jessica Marie Lunsford.* Available at http://www.jmlfoundation.com

Jordan, Lara Jakes. 2007. "Big City Murders Way up Since 04." *MSNBC.* Available at http://www.msnbc.msn.com

Kaplan, Esther. 2005. "Just Say Não." *The Nation.* May 30. Available at http://www.thenation.com

Kerley, Kent, Michael Benson, Matthew Lee, and Francis Cullen. 2004. "Race, Criminal Justice Contact, and Adult Position in the Social Stratification System." *Social Problems* 51(4):549–568.

Kingsbury, Kathleen. 2006. "The Next Crime Wave." *Time,* December 11, pp. 71–77.

Klapper, Bradley. 2005. "UN Told Governments Must Combat Internet Child Pornography." Associated Press, April 14. Available at http://informationweek.com

Kong, Deborah, and Jon Swartz. 2000 "Experts See Rash of Hack Attacks Coming." *USA Today,* September 27, p. 1B.

Kruse, Michael. 2007. "Jessie's Story: An Ordinary Kid." *St. Petersburg Times,* February 10. Available at http://www.sptimes.com

Kubrin, Charis E. 2005. "Gangsters, Thugs, and Hustlas: Identity and the Code of the Street in Rap Music." *Social Problems* 52(3): 360–378.

Kubrin, Charis, and Ronald Weitzer. 2003. "Retaliatory Homicide: Concentrated Disadvantage and Neighborhood Culture." *Social Problems* 50:157–180.

Lane, Charles. 2004. "Less Support for Death Sentence Cited for Decline in Executions." *Washington Post,* December 15. Available at http://www.washingtonpost.com

Laub, John, Daniel S. Nagan, and Robert Sampson. 1998. "Trajectories of Change in Criminal Offending: Good Marriages and the Desistance Process." *American Sociological Review* 63:225–238.

Lee, Jennifer. 2003. "Identity Theft Complaints Double in 2002." *New York Times,* January 22. Available at http://www.nytimes.com

Levy, Steven. 2006. "An Identity Heist the Size of Texas." *Newsweek,* June 12, p. 18.

Lichtblau, Eric. 2003. "Panel Clears Harsher Terms in Corporate Crime Cases." *New York Times,* January 9. Available at http://www.nytimes.com

Lott, John R., Jr. 2003. "Guns Are an Effective Means of Self-Defense." In *Gun Control,* Helen Cothran, ed. (pp. 86–93). Farmington Hills, MI: Greenhaven Press.

Maclachlan, Malcolm. 2006. "Romero, Cooley Heading for Another Try on Three Strikes." *Capital Weekly.* July 27. Available at http://www.capitolweekly.net/news/article

MAD DADS. 2007. *About Us.* Available at http://www.maddads.com

"Major Rulings of the 2002–2003 Term." 2003. *USA Today,* June 27, p. 4A

Marks, Alexandria. 2006. "Prosecutions Drop for US White Collar Crime." *The Christian Science Monitor,* August 31. Available at http://www.csmonitor.com

Maxwell, Lesli A. 2006. "The Long Arm of the Law." *Education Weekly,* October 18, pp. 25–28.

McGuire, David. 2004. "Bush Signs Identity Theft Bill." *The Washington Post,* July 15. Available at http://www.washingtonpost.com

Merton, Robert. 1957. *Social Theory and Social Structure.* Glencoe, IL: Free Press.

Moore, David. 2005. "Young Adults Most Likely to Be Crime Victims." Gallup Organization, March 21. Available at http://galluppoll.com

Morse, Jane. 2006. *Journalist Urges More Enforcement of Laws against Trafficking.* Available at http://usinfo.state.gov.utils

MSNBC. 2007. "Jury Selection in Child Killer Case." MSNBC, February 12. Available at http://www.msnbc.msn.com

Murray, Mary E., Nancy Guerra, and Kirk Williams. 1997. "Violence Prevention for the Twenty-First Century." In *Enhancing Children's Awareness,* Roger P. Weissberg, Thomas Gullota, Robert L. Hampton,

Bruce Ryan, and Gerald Adams, eds. (pp. 105–128). Thousand Oaks, CA: Sage.

Napolitano, Jo. 2004. *Top Illinois Court Upholds Total Amnesty of Death Row.* Available at http://www.nytimes.com

NGA (National Governor's Association). 2007. *Prisoner Reentry.* Available at http://www.nga.org

National Night Out. 2007. *What is National Night Out?* Available at http://www.nationalnightout.org/nno/about.html

National Research Council. 1994. *Violence in Urban America: Mobilizing a Response.* Washington, DC: National Academy Press.

NCMEC (National Center on Missing and Exploited Children). 2007. *Online Enticement Laws Vary between States.* Available at http://www.missingkids.com

NCPC (National Crime Prevention Council). 2007. Available at http://www.ncpc.org

NWCCC (National White Collar Crime Center). 2006. *2005 NW3C National Survey.* Available at http://www.nw3c/pressroom/current_releases.cfm

News Highlights. 2007. *World Prison Population Still Growing.* Available at http://www.klc.ac.uk/phpnews

Ohlemacher, Stephen. 2007. *Study: Prison Population on the Rise.* Associated Press, February 14. Available at http://www.commondreams.org

Orator. 2007. *News and Information.* Available at http://www.theorator.com/bills/110/text

Pasha, Shaheen. 2006. *Enron Founder Ken Lay Dies.* CNN Money, July 5. Available at http://money.cnn.com

Pertossi, Mayra. 2000 (September 27). *Analysis: Argentine Crime Rate Soars.* Available at http://news.excite.com

Peterson, Kavan. 2007 (January 17). *Death Penalty: Lethal Injection on Trial.* Available at http://www.stateline.org

Petrillo, Lisa. 2006. "Parolees Beat Odds at Second Chance." *The San Diego Union Tribune,* September 16. Available at http://signonsandiego.com

Philips, Julie. 2002. "White, Black, and Latino Homicide Rates: Why the Difference?" *Social Problems* 49:349–374.

Pickler, Nedra. 2000 (September 6). *Documents Point to Tire Problem.* Available at http://news.excite.com

Quindlen, Anna. 2006. "The Failed Experiment." *Time,* June 26, p. 64.

Reid, Sue Titus. 2003. *Crime and Criminology,* 10th ed. Boston: McGraw-Hill.

Reiman, Jeffrey. 2007. *The Rich Get Richer and the Poor Get Prison.* Boston: Allyn and Bacon.

Reisig, Michael, and Roger Parks. 2004. "Community Policing and Quality of Life." In

Community Policing, Wesley Skogan, ed. (pp. 207–227). Belmont, CA: Wadsworth.

Richtel, Matt. 2002. "Credit Card Theft Thrives Online as Global Market." *New York Times.* Available at http://www.nytimes.com

Richtel, Matt. 2003. "Mayhem, and Far from the Nicest Kind." *New York Times,* February 10. Available at http://www.nytimes.com

Ripley, Amanda. 2003. "The Night Detective." *Time,* January 6, pp. 45–50.

Romano, Lois. 2005. "More Complete Portrait of BTK Suspect Is Emerging." *Washington Post,* March 5. Available at http://www.washingtonpost.com.

Rosoff, Stephen, Henry Pontell, and Robert Tillman. 2002. "White Collar Crime." In *Social Problems: Readings with Four Questions,* Joel M. Charon, ed. (pp. 339–350). Belmont, CA: Wadsworth.

Rubin, Paul H. 2002. "The Death Penalty and Deterrence." *Forum,* Winter, pp. 10–12.

Saad, Lydia. 2006. "Worry about Crime Remains at Last Year's Elevated Levels." Gallup Poll, October 19. Available at http://www.galluppoll.com

Sasseen, Jane. 2006. "White-Collar Crime: Who Does Time?" *Business Week Online,* February 6. Available at http://www.businessweek.com

Schelzig, Erik. 2007. "Court Ruling Halts Tennessee Executions." September 19. Available at http://www.wral.com/news/national_world/national/story/1836857/

Schiesel, Seth. 2005. "Growth of Wireless Internet Opens New Paths for Thieves." *New York Times,* March 19. Available at http://www.nytimes.com

Shelden, Randall, Sharon Tracy, and William Brown. 2004. *Youth Gangs in American Society.* Belmont, CA: Wadsworth.

Shelley, Louise. 2007. "Terrorism, Transnational Crime and Corruption Center." American University. Available at http://www.american.edu/traccc/

Sherman, Lawrence. 2003. "Reasons for Emotion." *Criminology* 42:1–37.

Siegel, Diane. 2005. "Recent Trends in Women Trafficking and Voluntary Prostitution: Russian-Speaking Sex Workers in the Netherlands." In *Transnational Crime,* Jay Albanese, ed. (pp. 4–23). Whitby, Ontario, Canada: de Sitter Publications.

Siegel, Larry. 2006. *Criminology.* Belmont, CA: Wadsworth.

Situation in the Netherlands. 2003. *Prostitution in Holland.* Available at http://www.ex.ac.uk/politics/por_data

Skogan, Wesley, and Jeffrey Roth. 2004. "Introduction." In *Community Policing: Can It Work?,* Wesley Skogan, ed. (pp. xvii–xxiv). Belmont, CA: Wadsworth.

Steffensmeier, Darryl, and Stephen Demuth. 2000. "Ethnicity and Sentencing Outcomes in U.S. Federal Courts: Who Is Punished More Harshly?" *American Sociological Review* 65:705–729.

Surgeon General. 2002. "Cost-Effectiveness." In *Youth Violence: A Report of the Surgeon General.* Available at http://www.mentalhealth.org/youthviolence/surgeongeneral

Sutherland, Edwin H. 1939. *Criminology.* Philadelphia: Lippincott.

Thio, Alex. 2004. *Deviant Behavior,* 7th ed. Boston: Allyn and Bacon.

Thio, Alex. 2007. *Deviant Behavior,* 9th ed. Boston: Allyn and Bacon.

Thomas, Landon. 2005. "On Wall Street, a Rise in Dismissals over Ethics." *New York Times,* March 29. Available at http://www.nytimes.com/2005/03/29/business

United Nations. 2003. *Creating Guidelines for Restorative Justice Programmes.* Available at http://www.restorativejustice.org/rj3/feature/2003

UNODC (United Nations Office on Drugs and Crime). 2007. *Crime Programme.* Available at http://www.unodc.org/unodc/crime_cicp.html

U.S. Bureau of the Census. 2007. *Statistical Abstract of the United States,* 127th edition. Washington, DC: U.S. Government Printing Office.

U.S. Department of Justice. 2003. "Global Crime Issues." In *International Center Global Crimes Issues,* National Institute of Justice. Washington, DC: U.S. Government Printing Office.

U.S. State Department. 2006. *Victim Stories.* Available at http://www.state.gov/g/tip/c16482.htm

Van Wilsem, Johan, Karin Wittebrood, and Nan Dirk De Graaf. 2006. "Socioeconomic Dynamics of Neighborhoods and the Risk of Crime Victimization: A Multilevel Study of Improving, Declining, and Stable Areas in the Netherlands." *Social Problems* 53(2):226–247.

Victim-Offender Reconciliation Program. 2006. *About Victim-Offender Mediation and Reconciliation.* Available at http://www.vorp.com

Vu, Pauline. 2007. "Executions Halted as Doctors Balk." *Stateline,* March 21. Available at http://www.stateline.org

Wagley, John R. 2006. *Transnational Organized Crime: Principal Threats and U.S. Responses.* Congressional Research Service: The Library of Congress.

Wallis, Claudia. 2005. "Too Young to Die." *Time,* March 14, p. 40.

Warner, Barbara, and Pamela Wilcox Rountree. 1997. "Local Social Ties in a Community and Crime Model." *Social Problems* 4(4):520–536.

Warren, Jenifer. 2007. "Prisoners Resist Moving Out of State." *Los Angeles Times.* January 28. Available at http://www.latimes.com/news/local

Weed and Seed. 2007. *Weed and Seed.* Available at http://www.ojp.usdoj.gov/ccdo/ws/welcome.html

Williams, Linda. 1984. "The Classic Rape: When Do Victims Report?" *Social Problems* 31:459–467.

Williams, Pete. 2006. "MySpace, Facebook Attract Online Predators." MSNBC, February 3. Available at http://www.msnbc.msn.com/id/11165576

Williams, Pete. 2007. "Court overturns D.C. Handgun Ban." MSNBC, March 9. Available at http://www.msnbc.msn.com/id/17538139

Wilgoren, Jodi. 2003. "Governor Empties Illinois Death Row." *New York Times,* January 12. Available at http://www.nytimes.com

Winslow, Robert W., and Sheldon X. Zhang. 2008. *Criminology: A Global Perspective.* Upper Saddle River, NJ: Prentice Hall.

Wright, Darlene and Kevin Fitzpatrick. 2006. "Violence and Minority Youth: The Effects of Risk and Asset Factors among African American Children and Adolescents." *Adolescence* 41(162):251–263.

Zhang, Sheldon X., Robert Roberts, and Valeria J. Callanan. 2006. "Preventing Parolees from Returning to Prison through Community-Based Reintegration." *Crime and Delinquency* 52(4):551–571.

Chapter 5

Adams, Bert N. 2004. "Families and Family Study in International Perspective." *Journal of Marriage and Family* 66(5):1076–1088.

Administration for Children and Families. 2003. *Prevention Pays: The Costs of Not Preventing Child Abuse and Neglect.* U.S. Dept. of Health and Human Services. Available at http://www.acf.hhs.gov

Ahrons, C. 2004. *We're Still Family: What Grown Children Have to Say about Their Parents' Divorce.* New York: HarperCollins.

Amato, Paul. 1999. "The Postdivorce Society: How Divorce Is Shaping the Family and Other Forms of Social Organization." In *The Postdivorce Family: Children, Parenting, and Society,* R. A. Thompson and P. R. Amato, eds. (pp. 161–190). Thousand Oaks, CA: Sage.

Amato, Paul. 2003. "The Consequences of Divorce for Adults and Children." In *Family in Transition,* 12th ed., Arlene S. Skolnick and Jerome H. Skolnick, eds. (pp. 190–213). Boston: Allyn and Bacon.

Amato, Paul. 2004. "Tension between Institutional and Individual Views of Marriage." *Journal of Marriage and Family* 66:959–965.

Amato, P. R., and J. Cheadle. 2005 "The Long Reach of Divorce: Divorce and Child Well-Being across Three Generations." *Journal of Marriage and the Family* 67:191–206.

Amato, P. R., A. Booth, D. R. Johnson, and S. J. Rogers. 2007. *Alone Together: How Marriage in America Is Changing.* Cambridge MA: Harvard University Press.

American Council on Education and University of California. 2006. *The American Freshman: National Norms for Fall, 2006.* Los Angeles: Los Angeles Higher Education Research Institute.

Anderson, Kristin L. 1997. "Gender, Status, and Domestic Violence: An Integration of Feminist and Family Violence Approaches." *Journal of Marriage and the Family* 59:655–669.

Applewhite, Ashton. 2003. "Covenant Marriage Would Not Benefit the Family." In *The Family: Opposing Viewpoints,* Auriana Ojeda, ed. (pp. 189–195). Farmington Hill, MI: Greenhaven Press.

As-Sanie, S., A. Gantt, and M. S. Rosenthal. 2004. "Pregnancy Prevention in Adolescents." *American Family Physician* 70:1517–1519.

Babcock, J. C., S. A. Miller, and C. Siard. 2003. "Toward a Typology of Abusive Women: Differences between Partner-Only and Generally Violent Women in the Use of Violence." *Psychology of Women Quarterly* 27.153–161.

Bernstein, Nina. 2007. "Polygamy, practiced in secrecy, follows Africans to New York." *New York Times,* March 23, p. A1.

Block, Nadine. 2003. "Disciplinary Spanking Should Be Banned." In *Child Abuse: Opposing Viewpoints,* L. I. Gerdes, ed. (pp. 182–190). Farmington Hills, MI: Greenhaven Press.

Browning, Christopher R., and Edward O. Laumann. 1997. "Sexual Contact between Children and Adults: A Life Course Perspective." *American Sociological Review* 62:540–560.

Browning, Don S. 2003. *Marriage and Modernization: How Globalization Threatens Marriage and What to Do About It.* Grand Rapids, MI: William B. Eerdmans.

Bureau of Labor Statistics. 2006. *Employment Characteristics of Families in 2005.* Available at http://www.bls.gov

Bureau of Justice Statistics. 2007. *Crime Characteristics.* Available at http://www.ojp.usdoj.gov/bjs

Carrington, Victoria. 2002. *New Times: New Families.* Dordrecht, the Netherlands: Kluwer Academic.

Catalano, Shannan. 2006. *Intimate Partner Violence in the United States.* Bureau of Justice Statistics. Available at http://www.ojp.usdoj.gov/bjs/

Centers for Disease Control and Prevention. 2006. "Surveillance Summaries: Youth Risk Behavior Surveillance—United States, 2005". *Morbidity and Mortality Weekly Report* 55(SS-5):1–112.

Cherlin, Andrew J., Linda M. Burtin, Tera R. Hurt, and Diane M. Purvin. 2005. "The Influence of Physical and Sexual Abuse on Marriage and Cohabitation." *American Sociological Review* 69:768–789.

Coontz, Stephanie. 1992. *The Way We Never Were: American Families and the Nostalgia Trap.* New York: Basic.

Coontz, Stephanie. 1997. *The Way We Really Are.* New York: Perseus.

Coontz, Stephanie. 2000. "Marriage: Then and Now." *Phi Kappa Phi Journal* 80:10–15.

Coontz, Stephanie. 2004. "The World Historical Transformation of Marriage." *Journal of Marriage and Family* 66(4):974–979.

Coontz, Stephanie. 2005a. *Marriage, a History.* New York: Penguin Books.

Coontz, Stephanie. 2005b. "For Better, for Worse." *Washington Post*, May 1. Available at http://www.washingtonpost.com

Daniel, Elycia. 2005. "Sexual Abuse of Males." In *Sexual Assault: The Victims, the Perpetrators, and the Criminal Justice System*, Frances P. Reddington and Betsy Wright Kreisel, eds. (pp. 133–140). Durham, NC: Carolina Academic Press.

Davis, Shannon N., and Theodore N. Greenstein. 2004. "Cross-National Variations in the Division of Household Labor." *Journal of Marriage and Family* 66:1260–1271.

de Anda, Diane. 2006. "Baby Think It Over: Evaluation of an Infant Simulation Intervention for Adolescent Pregnancy Prevention." *Health and Social Work* 31(1):26–35.

Decuzzi, A., D. Knox, and M. Zusman. 2004. "The Effect of Parental Divorce on Relationships with Parents and Romantic Partners of College Students." Roundtable Discussion, Southern Sociological Society, Atlanta, April 17.

Demian. 2004 (December 12). *Civil Solidarity Pact.* Partners Task Force for Gay and Lesbian Couples. Available at http://www.buddybuddy.com

Demo, David H., Mark A. Fine, and Lawrence H. Ganong. 2000. "Divorce as a Family Stressor." In *Families and Change: Coping with Stressful Events and Transitions*, 2nd ed., P. C. McKenry and S. J. Price, eds. (pp. 279–302). Thousand Oaks, CA: Sage.

DiLillo, D., G. C. Tremblay, and L. Peterson. 2000. "Linking Childhood Sexual Abuse and Abusive Parenting: The Mediating Role of Maternal Anger." *Child Abuse and Neglect* 24:767–769.

Edin, Kathryn. 2000. "What Do Low-Income Single Mothers Say About Marriage?" *Social Problems* 47(1):112–133.

Edin, Kathryn, and Laura Lein. 1997. *Making Ends Meet: How Single Mothers Survive Welfare and Low-Wage Work.* New York: Russell Sage Foundation.

Edleson, J. L., L. F. Mbilinyi, S. K. Beeman, and A. K. Hagemeister. 2003. "How Children Are Involved in Adult Domestic Violence." *Journal of Interpersonal Violence* 18:18–32.

Emery, Robert E. 1999. "Postdivorce Family Life for Children: An Overview of Research and Some Implications for Policy." In *The Postdivorce Family: Children, Parenting, and Society*, R. A. Thompson and P. R. Amato, eds. (pp. 3–27). Thousand Oaks, CA: Sage.

Emery, Robert E., David Sbarra, and Tara Grover. 2005. "Divorce Mediation: Research and Reflections." *Family Court Review* 43(1):22–37.

Family Court Reform Council of America. 2000. *Parental Alienation Syndrome.* Rancho Santa Margarita, CA: Family Court Reform Council of America.

Family Violence Prevention Fund. 2005 (February 24). *Testimony of the Family Violence Prevention Fund on Welfare Reform and Marriage Promotion Initiatives: Submitted to the House Ways and Means Committee.* Available at http://endabuse.org

Few, April L., and Karen H. Rosen. 2005. "Victims of Chronic Dating Violence: How Women's Vulnerabilities Link to Their Decision to Stay." *Family Relations* 54(2):265–279.

Fields, Jason. 2004. *America's Families and Living Arrangements: 2003.* Current Population Reports. Washington, DC: U.S. Census Bureau.

Finkelhor, D., R. K. Ormrod, and H. A. Turner. 2007. "Poly-victimization: A neglected component in child victimization." *Child Abuse & Neglect* 31:7–26.

Fisher, Bonnie S., Francis T. Cullen, and Michael G. Turner. 2000. *The Sexual Victimization of College Women.* National Institute of Justice and Bureau of Justice Statistics. Washington, DC: U.S. Department of Justice.

Fogle, Jean M. 2003. "Domestic Violence Hurts Dogs, Too." *Dog Fancy*, April, p. 12.

Forum on Child and Family Statistics. 2000. *America's Children: Key National Indicators of Well-Being, 2000.* Available at http://www.childstats.gov

Fox, James A., and Marianne W. Zawitz. 2006. *Homicide Trends in the United States: 2006.* Bureau of Justice Statistics. Available at http://www.ojp.usdoj.gov/bjs

Gelles, Richard J. 1993. "Family Violence." In *Family Violence: Prevention and Treatment*, Robert L. Hampton, Thomas P. Gullotta, Gerald R. Adams, Earl H. Potter III, and Roger P. Weissberg, eds. (pp. 1–24). Newbury Park, CA: Sage.

Gelles, Richard J. 2000. "Violence, Abuse, and Neglect in Families." In *Families and Change: Coping with Stressful Events and Transitions*, 2nd ed., P. C. McKenry and S. J. Price, eds. (pp. 183–207).Thousand Oaks, CA: Sage.

Gilbert, Neil. 2003. "Working Families: Hearth to Market." In *All Our Families*, 2nd ed., M. A. Mason, A. Skolnick, and S. D. Sugarman, eds. (pp. 220–243). New York: Oxford University Press.

Global Initiative to End All Corporal Punishment of Children. 2007. *States with Full Abolition.* Available at http://www.endcorporalpunishment.org

Goldstein, Joshua R., and Catherine T. Kenney. 2001. "Marriage Delayed or Marriage Forgone? New Cohort Forecasts of First Marriage for U.S. Women." *American Sociological Review* 66:506–519.

Gore, Al, and Tipper Gore. 2002. *Joined at the Heart.* New York: Henry Holt.

Grogan-Kaylor, Andrew, and Melanie Otis. 2007. "The Predictors of Parental Use of Corporal Punishment." *Family Relations* 56:80–91.

Grych, John H. 2005. "Interparental Conflict as a Risk Factor for Child Maladjustment: Implications for the Development of Prevention Programs." *Family Court Review* 43(1):97–108.

Guttmacher Institute. 2006. *Facts on Sex Education in the United States.* Available at http://www.guttmacher.org

Guttmacher Institute. 2007a (April). *State Policies in Brief: Sex and STD/HIV Education.* Available at http://www.guttmacher.org

Guttmacher Institute. 2007b (April). *State Policies in Brief: Minors' Access to Contraceptive Services.* Available at http://www.guttmacher.org

Guttmacher Institute. 2007c (April). *State Policies in Brief: Emergency Contraception.* Available at http://www.guttmacher.org

Hacker, Andrew. 2003. *Mismatch: The Growing Gulf between Women and Men.* New York: Scribner.

Hackstaff, Karla B. 2003. "Divorce Culture: A Quest for Relational Equality in Marriage." In *Family in Transition*, 12th ed., Arlene S. Skolnick and Jerome H. Skolnick, eds. (pp. 178–190). Boston: Allyn and Bacon.

Hamilton, Brady E., Joyce A. Martin, and Paul D. Sutton. 2004. "Births: Preliminary Data for 2003." *National Vital Statistics Reports* 53(9). Available at http://www.cdc.gov/nchs

Hamilton, Brady E., Joyce A. Martin, and Stephanie J. Ventura. 2006. "Births: Preliminary Data for 2005." *National Vital Statistics Reports* 55(11). Available at http://www.cdc.gov/nchs

Harrington, Donna, and Howard Dubowitz. 1993. "What Can Be Done to Prevent Child Maltreatment?" In *Family Violence: Prevention and Treatment*, Robert L. Hampton, Thomas P. Gullotta, Gerald R. Adams, Earl H. Potter III, and Roger P. Weissberg, eds. (pp. 258–280). Newbury Park, CA: Sage.

Hawkins, Alan J., Jason S. Carroll, William J. Doherty, and Brian Willoughby. 2004. "A Comprehensive Framework for Marriage Education." *Family Relations* 53(5):547–558.

Hendy, H. M., D. Eggen, C. Gustitus, K. C. McLeod, and P. Ng. 2003. "Decision to Leave Scale: Perceived Reasons to Stay in or Leave Violent Relationships." *Psychology of Women Quarterly* 27:162–173.

Hewlett, Sylvia Ann, and Cornel West. 1998. *The War against Parents: What We Can Do for Beleaguered Moms and Dads*. Boston: Houghton Mifflin.

Heyman, R. E., and A. M. S. Slep. 2002. "Do Child Abuse and Interpersonal Violence Lead to Adult Family Violence?" *Journal of Marriage and Family* 64:864–870.

Hochschild, Arlie Russell. 1989. *The Second Shift: Working Parents and the Revolution at Home*. New York: Viking.

Hochschild, Arlie Russell. 1997. *The Time Bind: When Work Becomes Home and Home Becomes Work*. New York: Henry Holt.

Huffstutter, P. J. 2007 (April 9). "States Reject Abstinence Classes, Funds." *The News & Observer*. Available at http://www.newsobserver.com

Jackson, Shelly, Lynette Feder, David R. Forde, Robert C. Davis, Christopher D. Maxwell, and Bruce G. Taylor. 2003 (June). *Batterer Intervention Programs: Where Do We Go from Here?* U.S. Department of Justice. Available at http://www.usdoj.gov

Jalovaara, M. 2003. "The Joint Effects of Marriage Partners' Socioeconomic Positions on the Risk of Divorce." *Demography* 40:67–81.

Jasinski, J. L., L. M. Williams, and J. Siegel. 2000. "Childhood Physical and Sexual Abuse as Risk Factors for Heavy Drinking Among African-American Women: A Prospective Study." *Child Abuse and Neglect* 24:1061–1071.

Jekielek, Susan M. 1998. "Parental Conflict, Marital Disruption, and Children's Emotional Well-Being." *Social Forces* 76:905–935.

Johnson, Michael P. 2001. "Patriarchal Terrorism and Common Couple Violence: Two Forms of Violence Against Women." In *Men and Masculinity: A Text Reader*, T. F. Cohen, ed. (pp. 248–260). Belmont, CA: Wadsworth.

Johnson, Michael P., and Kathleen Ferraro. 2003. "Research on Domestic Violence in the 1990s: Making Distinctions." In *Family in Transition*, 12th ed., A. S. Skolnick and J. H. Skolnick, eds. (pp. 493–514). Boston: Allyn and Bacon.

Jones, Rachel K., Alison Purcell, Sushella Singh, and Lawrence B. Finer. 2005. "Adolescents' Reports of Parental Knowledge of Adolescents' Use of Sexual Health Services and Their Reactions to Mandated Parental Notification for Prescription Contraception." *JAMA* 293(3):340–348.

Jorgensen, Stephen R. 2000. "Adolescent Pregnancy Prevention: Prospects for 2000 and Beyond." Presidential address at the National Council on Family Relations, 62nd Annual Conference, Minneapolis, November 11.

Kaiser Family Foundation. 2005 (January). "U.S. Teen Sexual Activity." Available at http://www.kff.org

Kalmijn, Matthijs, and Christiaan W. S. Monden. 2006. "Are the Negative Effects of Divorce on Well-Being Dependent on Marital Quality?" *Journal of Marriage and the Family* 68:1197–1213.

Kaufman, Joan, and Edward Zigler. 1992. "The Prevention of Child Maltreatment: Programming, Research, and Policy." In *Prevention of Child Maltreatment: Developmental and Ecological Perspectives*, Diane J. Willis, E. Wayne Holden, and Mindy Rosenberg, eds. (pp. 269–295). New York: John Wiley.

Kimmel, Michael S. 2004. *The Gendered Society*, 2nd ed. New York: Oxford University Press.

Kitzmann, K. M., N. K. Gaylord, A. R. Holt, and E. D. Kenny. 2003. "Child Witnesses to Domestic Violence: A Meta-Analytic Review." *Journal of Clinical and Consulting Psychology* 71:339–352.

Knox, David (with Kermit Leggett). 1998. *The Divorced Dad's Survival Book: How to Stay Connected with Your Kids*. New York: Insight Books.

LaFraniere, Sharon. 2005. "Entrenched Epidemic: Wife-Beatings in Africa." *New York Times*, August 11, pp. A1 and A8.

Lara, Adair, 2005 (June 29). "One for the Price of Two: Some Couples Find Their Marriages Thrive When They Share Separate Quarters." *San Francisco Chronicle*. Available at http://www.sfgate.com

Laungani, P. 2005. "Changing Patterns of Family Life in India." In *Families in Global Perspective*, J. L. Roopnarine and U. P. Gielen, eds. (pp. 85–103). Boston: Pearson, Allyn and Bacon.

Levin, Irene. 2004. "Living Apart Together: A New Family Form." *Current Sociology*; 52(2):223–240.

Levine, Judith A., Clifton R. Emery, and Harold Pollack. 2007. "The Well-Being of Children Born to Teen Mothers." *Journal of Marriage and Family* 69:105–122.

Lewin, Tamar. 2000. "Fears for Children's Well-Being Complicates a Debate over Marriage." *New York Times*, November 4. Available at http://www.nytimes.com

Lindsey, Linda L. 2005. *Gender Roles: A Sociological Perspective*, 4th ed. Upper Saddle River, NJ: Pearson Prentice Hall.

Lloyd, Sally A. 2000. "Intimate Violence: Paradoxes of Romance, Conflict, and Control." *National Forum* 80(4):19–22.

Lloyd, Sally A., and Beth C. Emery. 2000. *The Dark Side of Courtship: Physical and Sexual Aggression*. Thousand Oaks, CA: Sage.

Loy, E., L. Machen, M. Beaulieu, and G. L. Greif. 2005. "Common Themes in Clinical Work with Women Who Are Domestically Violent." *American Journal of Family Therapy* 33:33–42.

Mason, Mary Ann. 2003. "The Modern American Step-Family: Problems and Possibilities." In *All Our Families*, 2nd ed., Mary Ann Mason, Arlene Skolnick, and Stephen D. Sugarman, eds. (pp. 96–116). New York: Oxford University Press.

Mason, Mary Ann, Arlene Skolnick, and Stephen D. Sugarman. 2003. "Introduction." In *All Our Families*, 2nd ed., Mary Ann Mason, Arlene Skolnick, and Stephen D. Sugarman, eds. (pp. 1–13). New York: Oxford University Press.

Mauldon, Jane. 2003. "Families Started by Teenagers." In *All Our Families*, 2nd ed., Mary Ann Mason, Arlene Skolnick, and Stephen D. Sugarman, eds. (pp. 40–65). New York: Oxford University Press.

McKenry, P. C., J. M. Scrovich, T. L. Mason, and K. E. Mosack. 2004 (November). "Perpetration of Gay and Lesbian Violence: A Disempowerment Perspective." Paper presented at the Annual Conference of the National Council on Family Relations, Orlando, Florida.

Mental Health America. 2003. *Effective Discipline Techniques for Parents: Alternatives to Spanking*. Strengthening Families Fact Sheet. Available at http://www.nmha.org

Mindel, Charles H., Robert W. Habenstein, and Roosevelt Wright, Jr. 1998. *Ethnic Families in America: Patterns and Variations*. Upper Saddle River, NJ: Prentice Hall.

Mollborn, Stephanie. 2007. "Making the Best of a Bad Situation: Material Resources and Teenage Parenthood." *Journal of Marriage and Family* 69:92–104.

Moore, David W. 2003. "Family, Health Most Important Aspects of Life." Gallup News Service, January 3. Available at http://www.galluppoll.org

Munson, M. L., and P. D. Sutton. 2006. "Births, Marriages, Divorces, and Deaths: Provisional Data for 2005." *National Vital Statistics Reports* 54(20). Hyattsville, MD: National Center for Health Statistics.

National Campaign to Prevent Teen Pregnancy. 2005. *What If: How Declines in Teen Births Have Reduced Poverty and Increased Child Well-Being.* Available at http://www.teenpregnancy.org

National Center for Injury Prevention and Control. 2006. *Intimate Partner Violence: Overview.* Available at http://www.cdc.gov/ncipc

National Center for Injury Prevention and Control. 2007. *Child Maltreatment: Fact Sheet.* Available at http://www.cdc.gov/ncipc

Nelson, B. S., and K. S. Wampler. 2000. "Systemic Effects of Trauma in Clinic Couples: An Exploratory Study of Secondary Trauma Resulting from Childhood Abuse." *Journal of Marriage and Family Counseling* 26:171–184.

Newton, James P. 2005. "Abuse in the Elderly: A Perennial Problem." *Gerontology* 22(1):1.

Nock, Steven L. 1995. "Commitment and Dependency in Marriage." *Journal of Marriage and the Family* 57:503–514.

O'Reilly, S., D. Knox, M. Zusman, and H. R. Thompson. 2005 (March 18). "What Women Want." Poster, Annual Meeting of the Eastern Sociological Society, Washington, DC.

Parker, Marcie R., Edward Bergmark, Mark Attridge, and Jude Miller-Burke. 2000. "Domestic Violence and Its Effect on Children." *National Council on Family Relations Report* 45(4):F6–F7.

Pasley, K. 2000. "Stepfamilies Doing Well Despite Challenges." National Council on Family *Relations Report* 45:6–7.

Pasley, Kay, and Carmelle Minton. 2001. "Generative Fathering After Divorce and Remarriage: Beyond the 'Disappearing Dad.'" In *Men and Masculinity: A Text Reader*, T. F. Cohen, ed. (pp. 239–248). Belmont CA: Wadsworth.

Pew Research Center. 2007 (March 22). *Trends in Political Values and Core Attitudes: 1987–2007.* Washington DC: Pew Research Center for People and the Press.

Rand, M. R. 2003. "The Nature and Extent of Recurring Intimate Partner Violence Against Women in the United States." *Journal of Comparative Family Studies* 34:137–146.

Ricci, L., A. Giantris, P. Merriam, S. Hodge, and T. Doyle. 2003. "Abusive Head Trauma in Maine Infants: Medical, Child Protective, and Law Enforcement Analysis." *Child Abuse and Neglect* 27:271–283.

Rubin, D. M., C. W. Christian, L. T. Bilaniuk, K. A. Zaxyczny, and D. R. Durbin. 2003. "Occult Head Injury in High-Risk Abused Children." *Pediatrics* 111:1382–1386.

Russell, D. E. 1990. *Rape in Marriage.* Bloomington: Indiana University Press.

Saad, Lydia. 2006 (May 30). "Americans Have Complex Relationship with Marriage." *Gallup Poll Briefing.* Gallup Organization. Available at http://www.gallup.com

Schacht, Thomas E. 2000. "Protection Strategies to Protect Professionals and Families Involved in High-Conflict Divorce." *UALR Law Review* 22(3):565–592.

Scott, K. L., and D. A. Wolfe. 2000. "Change among Batterers: Examining Men's Success Stories." *Journal of Interpersonal Violence* 15:827–842.

Shepard, Melanie F., and James A. Campbell. 1992. "The Abusive Behavior Inventory: A Measure of Psychological and Physical Abuse." *Journal of Interpersonal Violence* 7(3):291–305.

SIECUS (Sex Information and Education Council of the United States). 2004. *Public Support for Comprehensive Sexuality Education.* Fact sheet. Available at http://www.siecus.org

Sigle-Rushton, W., and S. McLanahan. 2002. "The Living Arrangements of New Unmarried Mothers." *Demography* 39:415–433.

Simmons, T., and M. O'Connell. 2003. *Married-Couple and Unmarried Partner Households: 2000.* Census 2000 Special Reports. Available at http://www.census.gov

Simonelli, C. J., T. Mullis, A. N. Elliott, and T. W. Pierce. 2002. "Abuse by Siblings and Subsequent Experiences of Violence within the Dating Relationship." *Journal of Interpersonal Violence* 17:103–121.

Singh, Susheela, and Jacqueline E. Darroch. 2000. "Adolescent Pregnancy and Childbearing: Levels and Trends in Developed Countries." *Family Planning Perspectives* 32(1):14–23.

Smith, J. 2003. "Shaken Baby Syndrome." *Orthopaedic Nursing.* 22:196–205.

Sorensen, Elaine and Helen Oliver. 2002 (April). *Policy Reforms Are Needed to Increase Child Support from Poor Families.* Washington, DC: Urban Institute.

Spiegel, D. 2000. "Suffer the Children: Long-Term Effects of Sexual Abuse." *Society* 37:18–20.

Steimle, Brynn M., and Stephen F. Duncan. 2004. "Formative Evaluation of a Family Life Education Web Site." *Family Relations* 53(4): 367–376.

Stock, J. L., M. A. Bell, D. K. Boyer, and F. A. Connell. 1997. "Adolescent Pregnancy and Sexual Risk-Taking among Sexually Abused Girls." *Family Planning Perspectives* 29:200–203.

Stone, R. D. 2004. *No Secrets, No Lies: How Black Families Can Heal from Sexual Abuse.* New York: Broadway Books.

Straus, Murray. 2000. "Corporal Punishment and Primary Prevention of Physical Abuse." *Child Abuse and Neglect* 24:1109–1114.

Teaster, Pamela B., Tyler A. Dugar, Marta S. Mendiondo, Erin L. Abner, Kara A. Cecil, and Joanne M. Otto. 2006 (February). *The 2004 Survey of State Adult Protective Services: Abuse of Adults 60 Years of Age and Older.* National Center on Elder Abuse. Available at http://www.elderabusecenter.org

Terry-Humen, Elizabeth, Jennifer Manlove, and Kristin A. Moore. 2005. *Playing Catch-Up: How Children Born to Teen Mothers Fare.* Washington, DC: National Campaign to Prevent Teen Pregnancy.

Thakkar, R. R., P. M. Gutierrez, C. L. Kuczen, and T. R. McCanne. 2000. "History of Physical and/or Sexual Abuse, and Current Suicidality in College Women." *Child Abuse and Neglect* 24:1345–1354.

Thornton, A., and L. Young-DeMarco. 2001. "Four Decades of Trends in Attitudes toward Family Issues in the United States: The 1960s through the 1990s." *Journal of Marriage and the Family* 63:1009–1037.

Trenholm, Christopher, Barbara Devaney, Ken Fortson, Lisa Quay, Justin Wheeler, and Melissa Clark. 2007 (April). *Impacts of Four Title V, Section 510 Abstinence Education Programs.* Princeton, NJ: Mathematica Policy Research, Inc.

Ulman, A. 2003. "Violence by Children against Mothers in Relation to Violence between Parents and Corporal Punishment by Parents." *Journal of Comparative Family Studies* 34:41–56.

Umberson, D., K. L. Anderson, K. Williams, and M. D. Chen. 2003. "Relationship Dynamics, Emotion State, and Domestic Violence: A Stress and Masculine Perspective." *Journal of Marriage and the Family* 65:233–247.

United Nations Development Programme. 2000. *Human Development Report 2000.* New York: Oxford University Press.

U.S. Census Bureau. 2000. "America's Families and Living Arrangements." *Current Population Reports*, Series P20-537. Washington, DC: U.S. Government Printing Office.

U.S. Census Bureau 2006a (May 26). "Americans Marrying Older, Living Alone More, See Households Shrinking." Press Release. Available at http://www.census.gov

U.S. Census Bureau. 2006b. *America's Families and Living Arrangements: 2006.* Table MS-2 Estimated Median Age at First Marriage, by Sex, 1890 to the Present. Available at http://www.census.gov

U.S. Census Bureau. 2007. *Statistical Abstract of the United States: 2007.* Washington, DC: U.S. Government Printing Office.

U.S. Department of Health and Human Services. 2000 (June 17). HHS Fatherhood Initiative. Available at http://www.hhs.gov/news/press/2000pres/20000617.html

U.S. Department of Health and Human Services, Administration on Children, Youth, and Families. 2007. *Child Maltreatment 2005*.Washington, DC: U.S. Government Printing Office.

Ventura, Stephanie. 1995 (June). "Births to Unmarried Mothers: United States, 1980–92." *Vital and Health Statistics,* Series 21(no.53). Available at http://www.cdc.gov/nchs

Ventura, S. J., and Christine A. Bachrach. 2000 (October 18). "Nonmarital Childbearing in the United States, 1940–99." *National Vital Statistics Report* 48(16).

Ventura, Stephanie J., Sally C. Curtin, and T. J. Mathews. 2000 (April 24). "Variations in Teenage Birth Rates, 1991–1998: National and State Trends." *National Vital Statistics Report* 48(6).

Viano, C. Emilio. 1992. "Violence among Intimates: Major Issues and Approaches." In *Intimate Violence: Interdisciplinary Perspectives,* C. E. Viano, ed. (pp. 3–12). Washington, DC: Hemisphere.

Walker, Alexis J. 2001. "Refracted Knowledge: Viewing Families through the Prism of Social Science." In *Understanding Families into the New Millennium: A Decade in Review,* Robert M. Milardo, ed. (pp. 52–65). Minneapolis, MN: National Council on Family Relations.

Wallerstein, Judith S. 2003. "Children of Divorce: A Society in Search of Policy." In *All Our Families,* 2nd ed., Mary Ann Mason, Arlene Skolnick, and Stephen D. Sugarman, eds. (pp. 66–95). New York: Oxford University Press.

Waxman, Henry A. 2004 (December). The Context of Federally Funded Abstinence-Only Education Programs. U.S. House of Representatives Committee on Government Reform, Minority Staff, Special Investigation Division. Available at http://www.democrats.reform.house.gov

Whiffen, V. E., J. M. Thompson, and J. A. Aube. 2000. "Mediators of the Link between Childhood Sexual Abuse and Adult Depressive Symptoms." *Journal of Interpersonal Violence* 15:1100–1120.

Whitehead, Barbara Dafoe, and David Popenoe. 2005. The State of Our Unions: The Social Health of Marriage in America. National Marriage Project, Rutgers University. Available at http://marriage.rutgers.edu

Williams, K. and A. Dunne-Bryant. 2006. "Divorce and Adult Psychological Well-Being: Clarifying the Role of Gender and Child Age." *Journal of Marriage and the Family* 68:1178–1196.

Chapter 6

Administration for Children and Families. 2002. *Early Head Start Benefits Children and Families.* U.S. Department of Health and Human Services. Available at http://www.acf.hhs.gov

Albelda, Randy, and Chris Tilly. 1997. *Glass Ceilings and Bottomless Pits: Women's Work, Women's Poverty.* Boston: South End Press.

Alex-Assensoh, Yvette. 1995. "Myths About Race and the Underclass." *Urban Affairs Review* 31:3–19.

Anderson, Sarah, John Cavanagh, Chris Hartman, Scott Klinger, and Stacey Chan. 2004. *Executive Excess: 11th Annual CEO Compensation Survey.* Boston: Institute for Policy Studies and United for a Fair Economy.

Anderson, Sarah, John Cavanagh, Chuck Collins, and Eric Benjamin. 2006. *Executive Excess: 13th Annual CEO Compensation Survey.* Boston: Institute for Policy Studies and United for a Fair Economy.

Associated Press. 2005. "Judge Halts Grant Over Religion." *New York Times,* January 16. Available at http://www.nytimes.com

Barrett, A. 2006. *Characteristics of Food Stamp Households: Fiscal Year 2005.* U.S. Department of Agriculture, Food & Nutrition Service. Available at http://www.fns.usda.gov/oane

Bickel, G., M. Nord, C. Price, W. Hamilton, and J. Cook. 2000. *United States Department of Agriculture Guide to Measuring Household Food Security.* Alexandria, VA: U.S. Department of Agriculture, Food and Nutrition Service.

Boston, Rob. 2005 (February 15). "Faith-Based 'Flim-Flam' Initiative Didn't Have a Prayer Says Former White House Aid." Americans United for a Separation of Church and State. Available at http://blog.au.org

Briggs, Vernon M., Jr. 1998. "American-Style Capitalism and Income Disparity: The Challenge of Social Anarchy." *Journal of Economic Issues* 32(2):473–481.

Bureau of Labor Statistics. 2007. *May 2006 National Occupational Employment and Wage Estimates: United States.* Available at http://www.bls.gov

Cauthen, Nancy K., and Sarah Fass. 2007 (April). *Measuring Income and Poverty in the United States.* National Center for Children in Poverty. Available at http://nccp.org/

Center on Budget and Policy Priorities. 2005 (May 31). *New Study Finds Poor Medicaid Beneficiaries Face Growing Out-of-Pocket Medical Costs.* Available at http://www.cbpp.org

Children's Defense Fund. 2003. Children in the United States. Available at http://www.childrensdefense.org

Citizens for Tax Justice. 2005 (April 21). *International Tax Comparisons, 1965–2003 (Federal, State, and Local).* Available at http://www.ctj.org

Citizens for Tax Justice. 2007. *The Bush Tax Cuts: The Latest CTJ Data.* Available at http://www.ctj.org

Corcoran, Mary, and Terry Adams. 1997. "Race, Sex, and the Intergenerational Transmission of Poverty." In *Consequences of Growing Up Poor,* Greg J. Duncan and Jeanne Brooks-Gunn, eds. (pp. 461–517). New York: Russell Sage Foundation.

Davies, James B., Susanna Sandstrom, Anthony Shorrocks, and Edward N. Wolff. 2006 (December 5). *The World Distribution of Household Wealth.* United Nations University-World Institute for Development Economics Research. Available at http://www.wider.unu.edu

Davis, Kingsley, and Wilbert Moore. 1945. "Some Principles of Stratification." *American Sociological Review* 10:242–249.

DeNavas-Walt, Carmen, Bernadette D. Proctor, and Cheryl Hill Lee. 2006. *Income, Poverty, and Health Insurance in the United States: 2005.*U.S. Census Bureau, Current Population Reports P60-231. Washington, DC: U.S. Government Printing Office.

Deng, Francis M. 1998. "The Cow and the Thing Called 'What': Dinka Cultural Perspectives on Wealth and Poverty." *Journal of International Affairs* 52(1):101–115.

Dordick, Gwendolyn. 1997. *Something Left to Lose: Personal Relations and Survival Among New York's Homeless.* Philadelphia: Temple University Press.

Dowd, Maureen. 2005. "United States of Shame." *New York Times,* September 3. Available at http://www.nytimes.com

Duncan, Greg J., and Jeanne Brooks-Gunn. 1997. "Income Effects Across the Life Span: Integration and Interpretation." In *Consequences of Growing Up Poor,* Greg J. Duncan and Jeanne Brooks-Gunn, eds. (pp. 596–610). New York: Russell Sage Foundation.

Duncan, Greg J., and P. Lindsay Chase-Lansdale. 2001. "Welfare Reform and Child Well-Being." Paper presented at the Blank/Haskins conference, *The New World of Welfare Reform.* Washington, DC, February 1–2.

Ebaugh, Helen Rose, Janet Saltzman Chafetz, and Paula F. Pipes. 2005. "Faith-Based Social Service Organizations and Government Funding: Data From a National Survey." *Social Science Quarterly* 86(2):273–292.

Edin, Kathryn, and Laura Lein. 1977. *Making Ends Meet.* New York: Russell Sage Foundation.

Epstein, William M. 2004. "Cleavage in American Attitudes Toward Social Welfare." *Journal of Sociology and Social Welfare* 31(4):177–201.

Federal Interagency Forum on Child and Family Statistics. 2006. *America's Children in Brief: Key National Indicators of Well-Being.* Available at http://www.childstats.gov

Fremstad, Sean. 2004 (January 30). *Recent Welfare Reform Research Findings: Implications for TANF Reauthorization and State TANF Policies.* Center on Budget and Policy Priorities. Available at http://www.cbpp.org

Goesling, Brian. 2001. "Changing Income Inequalities Within and Between Nations: New Evidence." *American Sociological Review* 66:745–761.

Gonzalez, David. 2005. "From Margins of Society to Center of Tragedy." *New York Times,* September 2. Available at http://www.nytimes.com

Grunwald, Michael. 2006 (August 27). "The Housing Crisis Goes Suburban." *Washington Post,* August 27. Available at http://www.washingtonpost.com

Halweil, Brian. 2005. "Grain Harvest and Hunger Both Grow." In *Vital Signs 2005*, Linda Starke, ed. (pp. 22–23). New York: W. W. Norton & Co.

Hill, Lewis E. 1998. "The Institutional Economics of Poverty: An Inquiry into the Causes and Effects of Poverty." *Journal of Economic Issues* 32(2):279–286.

Hoback, Alan, and Scott Anderson. 2007. "Proposed Method for Estimating Local Population of Precariously Housed." National Coalition for the Homeless. Available at http://www.nationalhomeless.org

hooks, bell. 2000. *Where We Stand: Class Matters.* New York: Routledge.

Jargowsky, Paul A. 1997. *Poverty and Place: Ghettos, Barrios, and the American City.* New York: Russell Sage Foundation.

Johnston, David Cay. 2007 (March 29). "Income Gap Is Widening, Data Shows." *New York Times.* Available at http:www.nytimes.com

Jones, Jeffrey M. 2007 (February 9). "Public: Family of 4 Needs to Earn Average of $52,000 to Get By." Gallup News Service. Available at http://www.galluppoll.com

Kennedy, Bruce P., Ichiro Kawachi, Roberta Glass, and Deborah Prothrow-Stith. 1998. "Income Distribution, Socioeconomic Status, and Self-Rated Health in the U.S.: Multilevel Analysis." *British Medical Journal* 317(7163):917–921.

Kingsley, G. Thomas, and Kathryn L. S. Pettit. 2003 (May). *Concentrated Poverty: A Change in Course. Urban Institute.* Available at http://www.urban.org/nnip

Kraut, Karen, Scott Klinger, and Chuck Collins. 2000. *Choosing the High Road: Businesses That Pay a Living Wage and Prosper.* Boston: United for a Fair Economy.

Leventhal, Tama, and Jeanne Brooks-Gunn. 2003. "Moving to Opportunity: An Experimental Study of Neighborhood Effects on Mental Health." *American Journal of Public Health* 93(9):1576–1585.

Lewan, Todd. 2007 (April 8). "Unprovoked Beatings of Homeless Soaring." Associated Press. Available at http://www.breitbart.com

Lewis, Oscar. 1966. "The Culture of Poverty." *Scientific American* 2(5):19–25.

Lewis, Oscar. 1998. "The Culture of Poverty: Resolving Common Social Problems." *Society* 35(2):7–10.

Llobrera, Joseph, and Bob Zahradnik. 2004. *A HAND UP: How State Earned Income Tax Credits Helped Working Families Escape Poverty in 2004.* Center on Budget and Policy Priorities. Available at http://www.cbpp.org

Luker, Kristin. 1996. *Dubious Conceptions: The Politics of Teenage Pregnancy.* Cambridge, MA: Harvard University Press.

Malatu, Mesfin Samuel, and Carmi Schooler. 2002. "Causal Connections Between Socioeconomic Status and Health: Reciprocal Effects and Mediating Mechanisms." *Journal of Health and Social Behavior* 43:22–41.

Mann, Judy. 2000 (May 15). "Demonstrators at the Barricades Aren't Very Subtle, But They Sometimes Win." *Washington Spectator* 26(10):1–3.

Massey, D. S. 1991. "American Apartheid: Segregation and the Making of the American Underclass." *American Journal of Sociology* 96:329–357.

Mayer, Susan E. 1997. *What Money Can't Buy: Family Income and Children's Life Chances.* Cambridge, MA: Harvard University Press.

McIntyre, Robert. 2005a (February 2). *Corporate Tax Avoidance in the States Even Worse Than Federal. Citizens for Tax Justice.* Available at http://www.ctj.org

McIntyre, Robert. 2005b (April 12). Tax Cheats and Their Enablers. Citizens for Tax Justice. Available at http://www.ctj.org

Mishel, Lawrence. 2005. "The Economy Is a Values Issue." *EPI Journal,* Winter, pp. 1 and 6.

Mishel, Lawrence, Jared Bernstein, and Sylvia Allegretto. 2007. *The State of Working America* 2006–2007. New York: Cornell University Press.

Narayan, Deepa. 2000. *Voices of the Poor: Can Anyone Hear Us?* New York: Oxford University Press.

National Association of Child Care Resource and Referral Agencies. 2006. *Breaking the Piggy Bank: Parents and the High Price of Child Care.* Available at http://www.naccrra.org

National Coalition for the Homeless. 2007. *Hate, Violence, and Death on Main Street USA: A Report on Hate Crimes and Violence against People Experiencing Homelessness.* National Coalition for the Homeless. Available at http://www.nationalhomeless.org

Newman, Katherine S. 1999. *No Shame in My Game: The Working Poor in the Inner City.* New York: Alfred A. Knopf and The Russell Sage Foundation.

Nord, Mark, Margaret Andrews, and Steven Carlson. 2006. *Household Food Security in the United States, 2005.* Economic Research Report No. 29. USDA Economic Research Service. Available at http://www.ers.usda.gov

Office of Family Assistance. 2007. Characteristics and Financial Circumstances of TANF Recipients: Fiscal Year 2005. Available at http://www.acf.hhs.gov

Oxfam. 2005. *Paying the Price: Why Rich Countries Must Invest Now in a War on Poverty.* Available at http://www.oxfam.org

Oxfam. 2006 (November). *Our Generation's Choice.* Oxfam Briefing Paper. Available at http://www.oxfam.org

Parisi, Domenico, Steven Michael Grice, and Michael Taquino. 2003. "Poverty and Inequality in the Context of Welfare Reform." *Social Problems Forum: The SSSP Newsletter,* 34(2):19–21.

Pew Research Center. 2007 (March 22). *Trends in Political Values and Core Attitudes: 1987–2007.* Washington, DC: Pew Research Center for People & the Press.

Pugh, Tony. 2007 (February 22). "U.S. Economy Leaving Record Numbers in Severe Poverty." *McClatchy Newspapers.* Available at http://www.realcities.com/mld/krwashington

Roberts, John. 2005 (January 17). *Thai Government Puts Tourism Ahead of the Poor in Tsunami Relief Effort.* World Socialist Web Site. Available at http://www.wsws.org

Roseland, Mark, and Lena Soots. 2007. "Strengthening Local Economies." In *2007 State of the World*, Linda Starke (ed.), pp. 152–169. New York: W. W. Norton & Co.

Rothstein, Richard. 2004. *Class and Schools.* Washington, DC: Economic Policy Institute.

Sanderson, Stephen K., and Arthur S. Alderson. 2005. *World Societies: The Evolution of Human Social Life.* Boston: Pearson Education.

Schifferes, Steve. 2004. "Can Globalization Be Tamed?" *BBC News Online,* February 24. Available at http://www.bbc.co.uk

Schiller, Bradley R. 2004. *The Economics of Poverty and Discrimination*, 9th ed. Upper Saddle River, NJ: Pearson Prentice Hall.

Seccombe, Karen. 2001. "Families in Poverty in the 1990s: Trends, Causes, Consequences, and Lessons Learned." In *Understanding Families into the New Millennium: A Decade in Review*, Robert M. Milardo, ed. (pp. 313–332). Minneapolis, MN: National Council on Family Relations.

Sobolewski, Juliana M., and Paul R. Amato. 2005. "Economic Hardship in the Family of Origin and Children's Psychological Well-Being in Adulthood." *Journal of Marriage and Family* 67(1):141–156.

Stocking, Barbara. 2005 (January 5). *The Tsunami and the Bigger Picture.* Oxfam. Available at http://www.oxfam.org

Streeten, Paul. 1998. "Beyond the Six Veils: Conceptualizing and Measuring Poverty." *Journal of International Affairs* 52(1):1–8.

Susskind, Yifat. 2005 (May). *Ending Poverty, Promoting Development: MADRE Criticizes the United Nations Millennium Development Goals.* Available at http://www.madre.org

Turner, Margery Austin, Susan J. Popkin, G. Thomas Kingsley, and Deborah Kaye. 2005 (April). *Distressed Public Housing: What it Costs to Do Nothing.* The Urban Institute. Available at http://www.urban.org

UNDP (United Nations Development Programme). 1997. *Human Development Report 1997.* New York: Oxford University Press.

UNDP. 2000. *Human Development Report 2000.* New York: Oxford University Press.

UNDP. 2003. *Human Development Report 2003.* New York: Oxford University Press.

UNDP. 2006. *Human Development Report 2006.* New York: Palgrave Macmillan.

United Nations. 2005. *Report on the World Social Situation 2005.* New York: United Nations.

United Nations Population Fund. 2002. *State of World Population 2002: People, Poverty, and Possibilities.* New York: United Nations.

U.S. Census Bureau. 2006a. Annual Social and Economic Supplement. *Current Population Survey.* Available at http://pubdb3.census.gov/macro/032006/pov/toc.htm

U.S. Census Bureau. 2006b. *Historical Poverty Tables: People.* Available at http://www.census.gov/hhes/poverty/histpov/perindex.htm

U.S. Census Bureau. 2007. *Poverty Thresholds 2006.* Available at http://www.census.gov

U.S. Conference of Mayors–Sodexho, Inc. 2006 (December). *Hunger and Homelessness Survey: A Status Report on Hunger and Homelessness in America's Cities.* Washington, DC: U.S. Conference of Mayors.

U.S. Department of Agriculture. 2006 (June). *Food Stamp Program Participation Rates: 2004.* Available at http://www.fns.usda.gov

U.S. Department of Health and Human Services. 2006. "Welfare Rolls Continue to Fall." *HHS News,* February 9. Available at http://www.hhs.gov

Van Kempen, Eva T. 1997. "Poverty Pockets and Life Chances: On the Role of Place in Shaping Social Inequality." *American Behavioral Scientist* 41(3):430–450.

Wagstaff, Adam. 2003. "Child Health on a Dollar a Day: Some Tentative Cross-Country Comparisons." *Social Science and Medicine* 57:1529–1538.

Washburn, Jennifer. 2004. "The Tuition Crunch." *Atlantic Monthly* 293(1):140.

Whitehead, Barbara Dafoe, and David Popenoe. 2004. The State of Our Unions: The Social Health of Marriage in America. The National Marriage Project, Rutgers University. Available at http://marriage.rutgers.edu

Wilson, William J. 1987. *The Truly Disadvantaged: The Inner City, the Underclass, and Public Policy.* Chicago: University of Chicago Press.

Wilson, William J. 1996. *When Work Disappears: The World of the New Urban Poor.* New York: Knopf.

World Bank. 2001. *World Development Report: Attacking Poverty, 2000/2001.* Herndon, VA: World Bank and Oxford University Press.

World Bank. 2005. Global Monitoring Report 2005. Available at http://www.worldbank.org

World Bank. 2007. *Global Monitoring Report 2007.* Washington D.C.: The World Bank

World Health Organization. 2002. *The World Health Report 2002.* Available at http://www.who.int/pub/en

World Population News Service. 2003. "Reducing Poverty Is Key to Global Stability." *Popline,* May–June, p. 4.

Zedlewski, Sheila R. 2003. Work and Barriers to Work among Welfare Recipients in 2002. Urban Institute. Available at http://www.urban.org

Zedlewski, Sheila R., Sandi Nelson, Kathryn Edin, Heather L. Koball, and Kate Roberts. 2003. *Families Coping Without Earnings or Government Cash Assistance.* Urban Institute. Available at http://www.urbaninstitute.org

Zedlewski, Sheila R., and Kelly Rader. 2005 (March 31). *Feeding America's Low-Income Children.* New Federalism: National Survey of America's Families, No. B-65. Urban Institute. Available at http://www.urban.org

Chapter 7

AFL-CIO. 2004. *Ask a Working Woman Survey Report.* Available at http://www.aflcio.org

AFL-CIO. 2005. *Death on the Job: The Toll of Neglect,* 14th ed. Available at http://www.aflcio.org

Aron-Dine, Aviva, and Isaac Shapiro. 2007 (March 29). "Share of National Income Going to Wages and Salaries at Record Low in 2006." Center for Budget and Policy Priorities. Available at http://www.cbpp.org

Austin, Colin. 2002. "The Struggle for Health in Times of Plenty." In *The Human Cost of Food: Farmworkers' Lives, Labor, and Advocacy,* C. D. Thompson Jr. and M. F. Wiggins, eds. (pp. 198–217). Austin: University of Texas Press.

Bales, Kevin. 1999. *Disposable People: New Slavery in the Global Economy.* Berkeley: University of California Press.

Barling, J. E., K. Kelloway, and M. R. Fron. 2005. *Handbook of Work Stress.* Thousand Oaks, CA: Sage.

Barstow, David, and Lowell Bergman. 2003. "Deaths on the Job, Slaps on the Wrist." *New York Times Online,* January 10. Available at http://www.nytimes.com

Bassi, Laurie J., and Jens Ludwig. 2000. "School-to-Work Programs in the United States: A Multi-Firm Case Study of Training, Benefits, and Costs." *Industrial and Labor Relations Review* 53(2):219–239.

Bello, Walden. 2001. "Lilliputians Rising: 2000: The Year of Global Protest Against Corporate Globalization." *Multinational Monitor* 22(1–2). Available at http://www.essential.org/monitor

Benjamin, Medea. 1998. *What's Fair About Fair Labor Association (FLA)?* Sweatshop Watch. Available at http://www.sweatshopwatch.org

Bernstein, Jared. 2006. *All Together Now: Common Sense for a Fair Economy.* San Francisco: Berrett-Koehler Publishers, Inc.

"Big Business for Reform." 2000. *Multinational Monitor* 21(11). Available at http://www.essential.org/monitor

Bond, James T., Ellen Galinsky, Stacy S. Kim, and Erin Brownfield. 2005. *2005 National Study of Employers.* Families and Work Institute. Available at http://www.familiesandwork.org

Bond, James T., Cindy Thompson, Ellen Galinsky, and David Prottas. 2002. *Highlights of the National Study of the Changing Workforce.* Families and Work Institute. Available at http://www.familiesandwork.org

Bonior, David. 2006. "Undermining Democracy: Worker Repression in the United States." *Multinational Monitor* 27(4). Available at http://www.essential.org/monitor

Brenner, Mark. 2007 (June). "Give Your Union Dues a Checkup." *Labornotes,* No. 339. Available at http://labornotes.org

Bronfenbrenner, Kate. 2000. "Raw Power: Plant-Closing Threats and the Threat to Union Organizing." *Multinational Monitor* 21(12). Available at http://www.essential.org/monitor

Buncombe, Andrew. 2007 (March 14). *Independent: Halliburton: From Bush's Favorite to a National Disgrace.* Corpwatch. Available at http://www.corpwatch.org

Bureau of Labor Statistics. 2004. *Workplace Injuries and Illnesses in 2003.* Available at http://www.bls.gov

Bureau of Labor Statistics. 2005a. (March 30). *Lost-Worktime Injuries and Illnesses: Characteristics and Resulting Days Away From Work, 2003.* Available at http://www.bls.gov

Bureau of Labor Statistics. 2005b. *Women in the Labor Force: A Databook.* Available at http://www.bls.gov

Bureau of Labor Statistics. 2006. *National Census of Fatal Occupational Injuries in 2005.* Available at http://www.bls.gov

Bureau of Labor Statistics. 2007a. *Manufacturing.* Available at http://www.bls.gov

Bureau of Labor Statistics. 2007b. *Employment Characteristics of Families in 2006.* Available at http://www.bls.gov

Bureau of Labor Statistics. 2007c. *Union Members Summary.* Available at http://www.bls.gov

Cantor, David, Jane Waldfogel, Jeffrey Kerwin, Mareena McKinley Wright, Kerry Levin, John Rauch, Tracey Hagerty, and Martha Stapelton Kudela. 2001. *Balancing the Needs of Families and Employers: The Family and Medical Leave Surveys, 2000 Update.* U.S. Department of Labor. Available at http://www.dol.gov

Caston, Richard J. 1998. *Life in a Business-Oriented Society: A Sociological Perspective.* Boston: Allyn and Bacon.

Center for Responsive Politics. 2005a. *The Bush Administration Corporate Connections.* Available at http://www.opensecrets.org

Center for Responsive Politics. 2005b. *Tracking the Payback.* Available at http://www.opensecrets.org

Cernasky, Rachel. 2002 (December). *Slavery: Alive and Thriving in the World Today—The Satya Interview with Kevin Bales.* Available at http://www.satyamag.com

Cockburn, Andrew. 2003. "21st Century Slaves." *National Geographic,* September, pp. 2–11, 18–24.

Davis, Amy E., and Arne L. Kalleberg. 2006. "Family-Friendly Organizations? Work and Family Programs in the 1990s." *Work and Occupations* 33(2):191–223.

Durkheim, Emile. 1966 [1893]. *On the Division of Labor in Society,* G. Simpson, trans. New York: Free Press.

"Editorial: What Is Society Willing to Spend on Human Beings?" 2000. *Multinational Monitor* 21(11). Available at http://www.essential.org

FLA Watch. 2007a. *FLA Watch: Monitoring the Fair Labor Association.* Available at http://www.flawatch.org

FLA Watch 2007b. *About FLA Watch.* Available at http://www.flawatch.org

Fram, Alan. 2007 (June 3). Poll: *A Fifth Vacation with Laptop.* Associated Press. Yahoo News. Available at http://news.yahoo.com

Frederick, James, and Nancy Lessin. 2000. "Blame the Worker: The Rise of Behavior-Based Safety Programs." *Multinational Monitor* 21(11). Available at http://www.essential.org/monitor

Galinsky, E., J. T. Bond, and E. J. Hill. 2004. *When Work Works: A Status Report on Workplace Flexibility.* New York: Families and Work Institute.

Galinsky, Ellen, James T. Bond, Stacy S. Kim, Lois Backon, Erin Brownfield, and Kelly Sakai. 2005. *Overwork in America: When the Way We Work Becomes Too Much.* New York: Families and Work Institute.

Gallup Organization. 2005. *Race Relations.* The Gallup Brain. Available at http://www.gallup.com

Gallup Organization. 2006. *Work and Workplace.* Available at http://www.galluppoll.com

"The Garment Industry." 2001. Sweatshop Watch. Available at http://www.sweatshopwatch.org

George, Kathy. 2003 (December 1). *Myanmar: Unocal Faces Landmark Trial over Slavery.* CorpWatch. Available at http://www.corpwatch.org

Gordon, David M. 1996. *Fat and Mean: The Corporate Squeeze of Working Americans and the Myth of Managerial "Downsizing."* New York: Free Press.

Greenhouse, Steven. 2000. "Anti-Sweatshop Movement Is Achieving Gains Overseas." *New York Times,* January 26. Available at http://www.nytimes.com

Heymann, Jody. 2006. *Forgotten Families: Ending the Growing Crisis Confronting Children and Working Parents in the Global Economy.* New York: Oxford University Press.

Human Rights Watch. 2000. *Unfair Advantage: Workers' Freedom of Association in the United States Under International Human Rights Standards.* Available at http://www.hrw.org

Human Rights Watch. 2007 (May). *Discounting Rights: Wal-Mart's Violation of US Workers' Right to Freedom of Association.* Volume 19, No. 2 (G). Available at http://hrw.org.

Inglehart, Ronald. 2000. "Globalization and Postmodern Values." *Washington Quarterly,* Winter, pp. 215–228.

International Confederation of Free Trade Unions. 2006. *2006 Annual Survey of Violations of Trade Union Rights.* Available at http://www.icftu.org

International Labour Organization. 2003. *Global Employment Trends: 2003.* Geneva: International Labour Organization.

International Labour Organization. 2005a. *A Global Alliance Against Forced Labour.* Geneva: International Labour Office.

International Labour Organization. 2007. *Global Employment Trends 2006.* Geneva: International Labour Office.

Jensen, Derrick. 2002. "The Disenchanted Kingdom: George Ritzer on the Disappearance of Authentic Culture." *The Sun,* June, pp. 38–53.

"Jobless Robber Gets What He Wanted—Prison." 2006. *USA Today,* October 13, p. 3A.

Kenworthy, Lane. 1995. *In Search of National Economic Success.* Thousand Oaks, CA: Sage.

"Labor's 'Female Friendly' Agenda." 1998. *Labor Relations Bulletin* no. 690, p. 2.

Lee, S., D. McCann, and J. C. Messenger. 2007. *Working Time Around the World: Trends in Working Hours, Laws, and Policies in a Global Comparative Perspective.* New York: Routledge and International Labour Organization.

Lenski, Gerard, and J. Lenski. 1987. *Human Societies: An Introduction to Macrosociology,* 5th ed. New York: McGraw-Hill.

Leonard, Bill. 1996 (July). "From School to Work: Partnerships Smooth the Transition." *HR Magazine* (Society for Human Resource Management). Available at http://www.shrm.org

Lovelace, Glenn. 2000. "The Nuts and Bolts of Telework." Paper presented at the symposium "Telework and the New Workplace of the 21st Century," Xavier University, New Orleans, October 16. U.S. Department of Labor. Available at http://www.dol.gov/dol/asp/public/telework/main.htm

Lovell, Vicky. 2005 (April). *Valuing Good Health: An Estimate of Costs and Savings for the Healthy Families Act.* Institute for Women's Policy Research. Available at http://www.iwpr.org

MacEnulty, Pat. 2005 (September). "An Offer They Can't Refuse: John Perkins on His Former Life as an Economic Hit Man." *The Sun* 357:4–13.

Martinson, Karin, and Pamela Holcomb. 2007. *Innovative Employment Approaches and Programs for Low-Income Families.* Washington, DC: The Urban Institute.

Mehta, Chirag, and Nik Theodore. 2005 (December). *Undermining the Right to Organize: Employer Behavior during Union Representation Campaigns.* American Rights at Work. Available at http://www.americanrightsatwork.org

Miers, Suzanne. 2003. *Slavery in the Twentieth Century: The Evolution of a Global Problem.* Walnut Creek, CA: AltaMira Press.

Mishel, Lawrence, Jared Bernstein, and Sylvia Allegretto. 2006. *The State of Working America, 2006/2007.* Ithaca, NY: ILR Press.

Moen, Phyllis. 2003. "Epilogue: Toward a Policy Agenda." In *It's About Time: Couples and Careers,* P. Moen, ed. (pp. 333–337). Ithaca, NY: Cornell University Press.

Moloney, Anastasia. 2005. "Terror as Anti-Union Strategy: The Violent Suppression of Labor Rights in Colombia." *Multinational Monitor* 26(3–4).

National Labor Committee. 2001 (January 16). *Nightmare at J. C. Penney Contractor.* Available at http://www.nlcnet.org

National Labor Committee. 2007. "Senate Minority Leader Senator Harry Reid, Congressman Bernie Sanders, AFL-CIO and others endorse anti-sweatshop bill." Available at http://www.nlcnet.org

National League of Cities. 2004. *The American Dream in 2004: A Survey of the American People.* Washington, DC: National League of Cities.

"New OSHA Policy Relieves Employees." 1998. *Labor Relations Bulletin* no. 687, p. 8.

OECD. 2004 (October). *Clocking in and Clocking Out: Recent Trends in Working Hours.* OECD Policy Brief. Available at http://www.oecd.org

O'Rourke, Dara. 2003. "Outsourcing Regulation: Analyzing Nongovernmental Systems of Labor Standards and Monitoring," *Policy Studies Journal* 31(1):1–29.

Parenti, Michael. 2007 (February 16). *Mystery: How Wealth Creates Poverty in the World.* Common Dream News Center. Available at http://www.commondream.org

Passaro, Jamie. 2003 (January). "Fingers to the Bone: Barbara Ehrenreich on the Plight of the Working Poor." *The Sun,* pp. 4–10.

Perkins, John. 2004. *Confessions of an Economic Hit Man.* San Francisco: Berrett-Koehler Publishers, Inc.

Pew Research Center. 2007 (March 22). *Trends in Political Values and Core Attitudes: 1987-2007.* Washington DC: The Pew Research Center for People & the Press.

Ritzer, George. 1995. *The McDonaldization of Society: An Investigation into the Changing Character of Contemporary Social Life.* Thousand Oaks, CA: Pine Forge Press.

Roach, Brian. 2007. *Corporate Power in a Global Economy.* Medford, MA: Global Development and Environment Institute, Tufts University.

Schaeffer, Robert K. 2003. *Understanding Globalization: The Social Consequences of Political, Economic, and Environmental Change,* 2nd ed. Lanham, MD: Rowman & Littlefield.

Scott, Robert E., and David Ratner. 2005 (July 20). *NAFTA'S Cautionary Tale.* Economic Policy Institute Briefing Paper 214. Available at http://www.epi.org

"Sex Trade Enslaves Millions of Women, Youth." 2003. *Popline* 25:6.

Shipler, David K. 2005. *The Working Poor.* New York: Vintage Books.

Stettner, Andrew, and Sylvia A. Allegretto. 2005 (May 26). *The Rising Stakes of Job Loss: Stubborn Long-Term Joblessness Amid Falling Unemployment Rates.* Briefing Paper 162. Washington, DC: Economic Policy Institute.

Still, Mary C., and David Strang. 2003. "Institutionalizing Family-Friendly Policies." In *It's About Time: Couples and Careers,* Phyllis Moen, ed. (pp. 288–309). Ithaca, NY: Cornell University Press.

Students and Scholars Against Corporate Misbehavior. 2005 (August 12). *Looking for Mickey Mouse's Conscience: A Survey of the Working Conditions of Disney Factories in China.* Available at http://www.nlcnet.org

"Sweat-Free Update." 2007. *Sweatshop Watch* 13(1):6.

Tate, Deborah. 2007 (February 14). *US Lawmakers Seek to Crack Down on Foreign Sweatshops.* The National Labor Committee. Available at http://www.nlcnet.org

Thompson, Charles D., Jr. 2002. "Introduction." In *The Human Cost of Food: Farm workers' Lives, Labor, and Advocacy,* C. D. Thompson, Jr., and M. F. Wiggins, eds. (pp. 2–19). Austin: University of Texas Press.

Uchitelle, Louis. 2006. *The Disposable American: Layoffs and Their Consequences.* New York: Knopf.

"Unions Forge Global Network." 2002 (April 22). CorpWatch. Available at http://www.corpwatch.org

United Nations. 2005. *The Millennium Development Goals Report.* New York: United Nations.

U.S. Department of Labor. 2001 (January 16). *OSHA Statement: Statement of Assistant Secretary Charles N. Jeffress on Effective Date of OSHA Ergonomic Standard.* Available at http://www.osha.gov

Vandivere, Sharon, Kathryn Tout, Martha Zaslow, Julia Calkins, and Jeffrey Cappizzano. 2003. *Unsupervised Time: Family and Child Factors Associated with Self-Care. Assessing the New Federalism.* Occasional Paper No. 71. Urban Institute. Available at http://www.urbaninstitute.org

Watkins, Marilyn. 2002. *Building Winnable Strategies for Paid Family Leave in the United States.* Economic Opportunity Institute. Available at http://www.econop.org

Went, Robert. 2000. Globalization: Neoliberal Challenge, Radical Responses. Sterling, VA: Pluto Press.

"Workers at Risk." 2003. *Multinational Monitor* 24(6). Available at http://www.essential.org/monitor

WorldatWork. 2007 (February). *Telework Trendlines for 2006.* Available at http://www.workingfromanywhere.org

Chapter 8

Abdul-Alim, Jamaal. 2006. "Internet Cheating Clicks with Students." *Milwaukee Journal Sentinel,* December 13. Available at http://www.jsonline.com

Action for Children. 2007 (March). "What Stands between North Carolina Students and a Sound Basic Education." Raleigh, NC: UNC Center for Civil Right Education. *Issue Brief.*

Addington, Lynne, Sally Ruddy, Amanda Miller, Jill Devoe, and Kathryn Chandler. 2004. "Are America's Schools Safe? Students Speak Out." In *Schools and Society,* Jeanne Ballantine and Joan Spade, eds. (pp. 161–164). Belmont, CA: Thomson Wadsworth.

AFT (American Federation of Teachers). 2006. *Building Minds, Minding Buildings.* Available at http://www.aft.org

Alliance for Excellent Education. 2005 (August). "Teacher Attrition: A Costly Loss to the Nation and to the States." *Issue Brief.* Washington, DC.

Allen, I. Elaine, and Jeff Seaman. 2006. *Making the Grade: Online Education in the United States.* Needham, MA: Sloan Consortium.

Andersen, Erin. 2006. "How to Stop Bullying." *Lincoln Journal Star,* October 6, p. D1.

Apuzzo, Matt. 2007 (April 19). "Va. Tech Shooter was picked on at School." Available at http://www.breitbart.com

Arbelo, Enid, and Meaghan McDermott. 2005. "Curing the Pain of Bullying." *Democrat and Chronicle,* February 13. Available at http://www.democratandchronicle.com

Archer, Jeff. 2005a. "Connecticut Files Court Challenge to NCLB." *Education Week,* August 31. Available at http://www.edweek.org

Archer, Jeff. 2005b. "Connecticut Pledges First State Legal Challenge to NCLB Law." *Education Week,* April 13. Available at http://www.edweek.org

ASCE (American Society of Civil Engineers). 2005. "Schools." In *America's Crumbling Infrastructure Eroding the Quality of Life.* Available at http://www.asce.org/reportcard/2005

Attorney General. 2007. "Attorney General's Legal Action in the No Child Left Behind (NCLB) Act." Attorney General of Connecticut Richard Blumenthall. Available at http://www.ct.gov

Baker, David P., and Deborah P. Jones. 1993. "Creating Gender Equality: Cross-National Gender Stratification and Mathematical Performance." *Sociology of Education* 66:91–103.

Bankston, Carl, and Stephen Caldas. 1997. "The American School Dilemma: Race and Scholastic Performance." *Sociological Quarterly* 8(3):423–429.

Barton, Paul E. 2004. "Why Does the Gap Persist?" *Educational Leadership* 62(3):9–13.

Barton, Paul E. 2005. *One Third of a Nation: Rising Drop out Rates and Declining Opportunities.* Princeton, NJ: Educational Testing Service.

Battistich, Vitor. 2005. *Featured Report: Character Education, Prevention, and Positive Youth Development.* Character Education Partnership. Available at http://www.character.org

Bausell, Carole Viongrad, and Elizabeth Klemick. 2007. "Tracking U.S. Trends." *Education Week* 26(30):42–44.

Berger, Joseph. 2007. "Fighting Over When Public Pay Private Tuition for Disabled." *New York Times,* March 21. Available at http://www.nytimes.com

BEST (Building Educational Success Together). 2006. *Growth and Disparity: A Decade of U.S. Public School Construction.* Available at http://citiesandschools.berkeley.edu

Boaz, David, and Morris Barrett. 2004. "What Would a School Voucher Buy? The Real Cost of Private Schools." Cato Institute Briefing Paper No. 25. Available at http://www.cato.org

Bridgeland, John M., John DiIulio, Jr., and Karen Burke Morison. 2006. *The Silent Epidemic.* A Report by Civic Enterprises in association with Peter D. Hart Research Association. The Bill & Melinda Gates Foundation.

Burros, Marian, and Melanie Warner. 2006. "Bottlers Agree to a School Ban on Sweet Drinks." *The New York Times,* May 4. Available at http://www.nytimes.com

Bushweller, Kevin. 1995. "Turning Our Backs on Boys." *Education Digest,* January, pp. 9–12.

Callan, Patrick M. 2007. *Measuring Up: The National Report Card on Higher Education, Commentary, Current Year.* Available at http://measuringup.highereducation.org

CAPE (Council for American Private Education). 2007. *Facts and Studies.* Available at http://www.capenet.org/facts.html

Carvin, Andy. 2007 (June 25). *Supreme Court Rules against Student in "Bong hits 4 Jesus" Case.* Available at http://www.pbs.org/teachers/learning.now

Cavanaugh, Cathy, Kathy Jo Gillan, Jeff Kromrey, Melinda Hess, and Robert Blomeyer. 2004. *The Effects of Distance Education on K-12 Outcomes: A Meta-Analysis.* Naperville, IL: Learning Point Associates.

Cavanagh, Sean. 2007. "Survey Finds Interest in Blend of Traditional and Online Courses." *Education Week* 26(26):11.

Chaker, Anne Marie. 2007. "Schools Move to Stop Spread of Cyberbullying." *The Wall Street Journal,* January 24. Available at http://online.wsj

CDF (Children's Defense Fund). 2003a. *Administration Proposal to Test Head Start Children.* Available at http://www.childrensdefense.org

CDF. 2003b. *New TV Ad Tells Congress: "Head Start's Not Broken, Don't Break It."* Available at http://www.childrensdefense.org

CDF. 2004. *Educational Resource Disparities for Minority and Low Income Children.* Available at http://www.childrensdefense.org/education/resources-disparities

CDF. 2006. *Cradle to Prison Pipeline: Education.* Available at http://www.childrensdefense.org

CER (Center for Education Reform). 2006. *National Charter School Data.* Available at http://www.edreform.com

Chaney, B., and L. Lewis. 2007. *Public School Principals Report on Their School Facilities: Fall 2005.* U.S. Department of Education. Washington, DC: National Center for Educational Statistics.

Coleman, James S., J. E. Campbell, L. Hobson, J. McPartland, A. Mood, F. Weinfield, and R. York. 1966. *Equality of Educational Opportunity.* Washington, DC: U.S. Government Printing Office.

Cronin, John, Gage Kingbury, Martha McCall, and Branin Bowe. 2005 (April). *The Impact of No Child Left Behind Act on Student Achievement and Growth.* Northwest Evaluation Association. Research Brief. Available at http://www.nwea.org

DeBell, Matthew, and Chris Chapman. 2006. *Computer and Internet Use by Students in 2003.* Washington DC: U.S. Department of Education, NCES 2006-065.

DeNeui, Daniel, and Tiffany Dodge. 2006. "Asynchronous Learning Networks and Student Outcomes." *Journal of Instructional Psychology* 33(4):256–259.

Dobbs, Michael. 2005. "Youngest Students Most Likely to Be Expelled." *Washington Post,* May 16. Available at http://www.washingtonpost.com

DPI (Department of Public Instruction). 2007. *No Child Left Behind: Overview.* Available at http://www.ncpublicschools.org/nclb

EFA Global Monitoring Report. 2007. *Education for All: The Quality Imperative.* UNESCO Publishing. Available at http://unesdoc.unesco.org

Elam, Stanley M., Lowell C. Rose, and Alec M. Gallup. 1994. "The 26th Annual Phi Delta Kappa/Gallup Poll of the Public's Attitudes Toward the Public Schools." *Phi Delta Kappa,* September, pp. 41–56.

Entwisle, Doris, Carl Alexander, and Linda Olson. 2004. "Temporary as Compared to Permanent High School Drop-Out." *Social Forces* 82(3):1181–1205.

ETS (Educational Testing Service). 2007. *The Praxis Series.* Available at http://www.ets.org

Evans, Lorraine, and Kimberly Davies. 2000. "No Sissy Boys Here." *Sex Roles,* February, pp. 255–271.

Fact Sheet on the Major Provisions of the Conference Report to H.R. 1, the No Child Left Behind Act.2003. Available at http://www.ed.gov

Fagan, Amy. 2007. "Big Man on Campus Likely a Woman." *The Washington Times,* March 27. Available at http://www.washtimes.com

Fletcher, Robert S. 1943. *History of Oberlin College to the Civil War.* Oberlin, OH: Oberlin College Press.

Flexner, Eleanor. 1972. *Century of Struggle: The Women's Rights Movement in the United States.* New York: Atheneum.

Foulkes, Imogen. 2005. "Girls Missing out on Schooling." BBC, April 18. Available at http://news.bbc.co.uk

Fry, Rick. 2006. *The Changing Landscape of American Public Education: New Students, New Schools.* Pew Hispanic Center Report. Available at http://pewhispanic.org

Gallup Poll. 2007. *The People's Priorities.* Available at http://www.galluppoll.com

Goldberg, Carey. 1999. "After Girls Get Attention, Focus Is on Boys' Woes." In *Themes of the Times: New York Times* (p. 6). Upper Saddle River, NJ: Prentice-Hall.

Greenhouse, Linda. 2007. "Supreme Court Votes to Limit the Use of Race in Integration Plans." *New York Times,* June 29. Available at http://www.nytimes.com

Hall, Daria, and Shana Kennedy. 2006. *Primary Progress, Secondary Challenge: A State-by-State Look at Student Achievement Patterns.* Washington, DC: The Education Trust.

Hanushek, Eric A., Steven G. Rivkin, and John J. Kain. 2005. "Teachers, Schools and Academic Achievement." *Econometrics* 73(2):417–458.

Head Start. *Facts.* 2007. Available at http://www.acf.hhs.gov

Honawar, Vaishali. 2007. "Teacher Colleges Urged to Pay Heed to Child Development." *Education Week,* May 1. Available at http://www.edweek.org

Hurst, Marianne. 2005. "When It Comes to Bullying, There Are No Boundaries." *Education Week,* February 8. Available at http://www.edweek.org

IDEAData. 2006. Available at http://www.ideadata.org

Independent Sector. 2002. *Engaging Youth in Lifelong Service.* Newsroom. Available at http://www.independentsector.org

Jacobs, Joanne. 1999. "Gov. Davis to Make Volunteering Mandatory for Students." *Jose Mercury News*, April 23.

Jencks, Christopher, and Meredith Phillips. 1998. "America's Next Achievement Test: Closing the Black-White Test Score Gap." *The American Prospect*, September–October, pp. 44–53.

Jenkins, Robert F. 2006. "The Pendulum of Change." Unpublished essay.

Kahlenberg, Richard D. 2006. "A New Way on School Integration." *Issue Brief*. The Century Foundation. Available at http://www.equaleducation.org

Kanter, Rosabeth Moss. 1972. "The Organization Child: Experience Management in a Nursery School." *Sociology of Education* 45:186–211.

Keller, Bess. 2006. "NEA: Earn a Diploma or Stay in School until Age 21." *Education Week* 26(7):5, 14.

Kharfen, Michael. 2006 (August 17). 1 of 3 Teens and 1 of 6 Pre-Teens Are Victims of Cyberbullying. Available at http://www.fightcrime.org

Kluger, Jeffery. 2006. "How Bill Put the Fizz in the Fight Against Fat." *Time*, May 15, pp. 21–25.

Kozol, Jonathan. 1991. *Savage Inequalities: Children in America's Schools*. New York: Crown.

Kozol, Jonathan. 2005. *The Shame of the Nation*. New York: Crown.

Lareau, Annette. 1989. *Home Advantage: Social Class and Parental Intervention in Elementary Education*. Philadelphia: Falmer Press.

Lareau, Annette, and Erin Horvat. 2004. "Moments of Social Inclusion and Exclusion." In *Schools and Society*, Jeanne Ballantine and Joan Spade, eds. (pp. 276–286). Belmont, CA: Thomson Wadsworth.

Leandro v. State, 346 N.C. 336, 488 S.E.2d 249 (1997); aff'd. *Hoke Cty Bd. of Educ. v. State*, 530 PA02, Filed July 30, 2004.

Leonhardt, David. 2005. "The College Drop Out Boom." *New York Times*, May 24. Available at http://www.nytimes.com

Levin, Henry, Clive Belfield, Peter Muennig, and Cecilia Rouse. 2006. *The Costs and Benefits of an Excellent Education for All of America's Children*. Available at http://www.nd.edu/~econplcy/

Manzo, Kathleen K. 2005. "College-Based High Schools Fill Growing Need." *Education Week*, May 25. Available at http://www.edweek.org

Mathews, Jay. 2007. "As Push for Longer Hours Forms, Intriguing Models Arise in D.C." *Washington Post*, February 5. Available at http://www.washingtonpost.com

Mead, Sara. 2006. *The Truth about Boys and Girls*. Education Sector, June. Available at http://www.educationsector.org

Mendez, Teresa. 2005. "Public Schools: Do They Outperform Private Ones?" *Christian Science Monitor*. May 10, p. 11.

Merton, Robert K. 1968. *Social Theory and Social Structure*. New York: Free Press.

MetLife. 2005. *The MetLife Survey of the American Teacher*. Available at http://www.metlife.com

Morse, Jodie. 2002. "Learning While Black." *Time*, May 27, pp. 50–52.

MPR (Mathematica Policy Research, Inc.). 2007. "Early Head Start Research and Evaluation." Available at http://mathematica-mpr.com

Muller, Chandra, and Katherine Schiller. 2000. "Leveling the Playing Field?" *Sociology of Education* 73:196–218.

NAAL (National Assessment of Adult Literacy). 2007 (April 4). *Literacy in Everyday Life: Results from the 2003 National Assessment of Adult Literacy*. Available at http://nces.ed.gov/naal

NABE (National Association of Bilingual Education). 2005. *Head Start Tests Neither "Valid" or "Reliable," GAO Study Finds*. Available at http://www.nabe.org

NationMaster. 2006. *Average Years of Schooling of Adults by Country*. Educational Statistics. Available at http://www.nationmaster.com

NBPTS (National Board for Professional Teaching Standards). 2007. *Impact of National Board Certification*. Available at http://www.nbpts.org/resources/research

NCCAI (North Carolina Child Advocacy Institute). 2005. *What the Courts Decided in the Leandro Case: A Brief History*. Available at http://www.ncchild.org

NCD (National Council on Disability). 2000. *Executive Summary*. Available at http://www.ncd.gov

NCES (National Center for Educational Statistics). 2003a. *Academic Background of College Graduates Who Enter and Leave Teaching. Contexts of Elementary and Secondary Education*. Available at http://nces.ed.gov

NCES. 2003b. *Young Children's Access to Computers in the Home and at School in 1999 and 2000: Executive Summary*. Available at http://nces.ed.gov

NCES. 2003c. *Special Analysis 2002: Private Schools—A Brief Portrait*. Available at http://nces.ed.gov

NCES. 2005. *Indicators of School Safety and Crime: Executive Summary*. U.S. Department of Education. Available at http//nces.ed.gov

NCES. 2006a. *Digest of Education Statistics, 2005*. U.S. Department of Education. Available at http://nces.ed.gov

NCES. 2006b. *The Condition of Education, 2005*. U.S. Department of Education. Available at http://www.nces.ed.gov

NCES. 2006c. *Home Schooling in the United States: 2003*. U.S. Department of Education. Available at http://www.nces.ed.gov

NCES. 2007a. *Revenues and Expenditures for Public Elementary and Secondary Education: School Year 2004–05 (2005 Fiscal Year)*. Available at http://nces.ed.gov/pubs2007/expenditures

NCES. 2007b. *Indicators of School Crime and Safety: 2006*. Washington DC: U.S. Department of Education. NCES 2007-003.

NCES. 2007c. *Effectiveness of Reading and Mathematics Software Products*. Report to Congress. Washington, DC: U.S. Department of Education. NCES 2007-4005.

NEA (National Education Association). 2006a. *Attracting and Keeping Quality Teachers*. Available at http://www.nea.org

NEA. 2006b (July 18). *National School Voucher Legislation Announced by Congressional Leaders and Education Secretary*. Available at http://www.nea.org/newsreleases

NEA. 2007a. *Stand Up for Children: Pontiac v. Spellings*. Available at http://www.nea.org/lawsuit

NEA. 2007b. *Urge Congress: Cut the Interest Rate on Student Loans*. Available at http://www.nea.org

Nelson, Anne. 2006. "Overcoming the Income Gap." *ASCD InfoBrief* 47(Fall):1–8. The Association for Supervision and Curricular Development.

NORC (National Opinion Research Center). 2007. *Americans Want to Spend More on Education, Health*. General Social Survey: University of Chicago. Available at http://www-news.uchicago.edu/releases

NPR (National Public Radio). 2003. *NPR/Kaiser/Kennedy School Education Survey*. Available at http://www.npr.org/programs

NSBA (National School Board Association). 2007. *Cleveland Voucher Program*. Available at http://www.nsba.org

OECD (Organization for Economic Cooperation and Development). 2006. *Education at a Glance 2006*. Available at http://www.oecd.org

Official Press Release. 2005 (August 22). *State Sues Federal Government Over Illegal Unfunded Mandates Under No Child Left Behind Act*. Connecticut Attorney General's Office. Available at http://www.cslib.org

Olson, Lynn. 2007a. "Paying Attention Early On." *Education Week* 26(17):29–31.

Olson, Lynn. 2007b. "Moving Beyond Grade 12." *Education Week* 26(17):54–58.

Olson, Lynn, and David Hoff. 2006. "Framing the Debate." *Education Week* 26(15):22.

Orfield, Gary, and Chungmei Lee. 2006. *Racial Transformation and the Changing Nature of Segregation.* Cambridge, MA: Harvard University: The Civil Rights Project.

Peskin, Melissa Fleschler, Susan R. Tortolero, and Christine M. Markham. 2006. "Bullying and Victimization among Black and Hispanic Adolescents." *Adolescence* 41(163):467–484.

Pinson, Nicola. 2006. "Soda Contract: Who Really Benefits? *Rethinking School Online* 20(4):1–3. Available at http://www .rethinkingschools.org

Poliakoff, Anne Rogers. 2006. "Closing the Gap: An Overview." *ASCD InfoBrief* 44 (January):1–10. The Association for Supervision and Curricular Development.

Price, Hugh B. 2007. "Diversity Goals Help Kids in School—and Later in Life." *The Press-Enterprise,* February 16. The Brookings Institution. Available at http://www .brookings.edu

A Quality Teacher in Every Classroom. 2002. White House Fact Sheet. Available at http:// www.ed.gov

Ramierz-Valles, Jesus, and Amanda Brown. 2003. "Latinos' Community Involvement in HIV/AIDS: Organizational and Individual Perspectives on Volunteering." *AIDS Education and Prevention* 15(Suppl 1):90–104.

Rand. 2003. Rand Study Finds California Charter Schools Produce Achievement Gains Similar to Conventional Public Schools. Available at http://www.rand.org

Riehl, Carolyn. 2004. "Bridges to the Future: Contributions of Qualitative Research to the Sociology of Education." In *Schools and Society,* Jeanne Ballantine and Joan Spade, eds. (pp. 56–72). Belmont, CA: Thomson Wadsworth.

Rodriguez, Richard. 1990. "Searching for Roots in a Changing World." In *Social Problems Today,* James M. Henslin, ed. (pp. 202–213). Englewood Cliffs, NJ: Prentice-Hall.

Rolnick, Arthur J., and Rob Grunewald. 2007. "Early Intervention on a Large Scale." *Education Week* 26(17):32–36.

Roscigno, Vincent. 1998. "Race and the Reproduction of Educational Disadvantage." *Social Forces* 76(3):1033–1060.

Rosenbaum, James. 2002. "Beyond College for All: Career Paths for the Forgotten Half." In *Schools and Society,* Jeanne Ballantine and Joan Spade, eds. (pp. 485–490). Belmont, CA: Thomson Wadsworth.

Rosenthal, Robert, and Lenore Jacobson. 1968. *Pygmalion in the Classroom: Teacher Expectations and Pupils' Intellectual Development.* New York: Holt, Rinehart & Winston.

Rothstein, Richard. 2004. *Class and Schools.* Washington DC: Economic Policy Institute.

Sadovnik, Alan. 2004. "Theories in the Sociology of Education." In *Schools and Society,* Jeanne Ballantine and Joan Spade, eds. (pp. 7–26). Belmont, CA: Thomson Wadsworth.

Safe School Initiative. 2004. *The Final Report and Findings of the Safe School Initiative.* Washington, DC: U.S. Secret Service and U.S. Department of Education.

Samuels, Christina. 2007. "Unilateral Placements Face Review." *Education Week,* March 7. Available at http://www.edweek .org

Samuels, Christina. 2006. "Schools Respond to Federal 'Wellness' Requirement." *Education Week,* June 14. Available at: http:// www.edweek.org

Saporito, Salvatore. 2003. "Private Choices, Public Consequences: Magnet School Choice and Segregation by Race and Poverty." *Social Problems* 50:181–203.

Schemo, Diana. 2006. "Federal Rules Back Single-Sex Public Education." *New York Times,* October 25. Available at http://www .nytimes.com

Schleicher, Andreas. 2006. "Elevating Performance in a 'Flat World.'" *Education Week* 26(17):79–82.

Shin, Annys. 2007. "Removing Schools' Soda is Sticky Point." *The Washington Post,* March 22, p. D3.

Slobogin, Kathy. 2002. "Survey: Many Students Say Cheating's OK." CNN News. Available at http://www.cnn.com

Sommers, Christina. 2000. *The War Against Boys.* New York: Simon and Schuster.

Spade, Joan. 2004. "Gender and Education in the United States." In *Schools and Society,* Jeanne Ballantine and Joan Spade, eds. (pp. 287–295). Belmont, CA: Thomson Wadsworth.

Spence, David. S. 2007. "A Road Map to College and Career Readiness." *Education Week* (26)17:59–61.

Takach, Joe. 2006. "Cheating is Common among College Students." *The Newark Targum,* December 6, p. A1.

Thomas, Pierre, and Jack Date. 2006. "Students Dropping out of High School Reaches Epidemic Levels." ABC News, November 20. Available at http://abcnews.go.com

Thornburgh, Nathan. 2006. "Dropout Nation." *Time,* April 17, pp. 30–39.

Toppo, Greg, 2004. "Low-Income College Students Are Increasingly Left Behind." *USA Today,* January 14. Available at http://www .usatoday.com

Toppo, Greg. 2006 "Student 'Superstar' on Brink of Dropping Out." *USA Today,* December 4. Available at http://usatoday.com

Toppo, Greg. 2007. "Many Teachers See Failure in Student's Future." *USA Today,* March 26. Available at http://usatoday.com

UCLA. 2005. *Williams v. California.* Brochure. Institute for Democracy, Education and Access. Los Angeles: University of California.

UNESCO (United Nations Educational, Scientific, and Cultural Organization). 2005. *United Nations Literacy Decade: 2003–2012.* Available at http://portal.unesco.org/ education/en

UNESCO. 2006. *Literacy Statistics.* Available at http://www.uis.unesco.org

U.S. Census Bureau. 2007. *Statistical Abstracts of the United States.* Washington, DC: U.S. Government Printing Office.

U.S. Department of Education. 2000. *The Baby Boom Echo: No End in Sight, Figure 2.* Available at http://www.ed.gov

U.S. Department of Education. 2006. *Highly Qualified Teachers for Every Child.* Available at http://www.ed.gov/nclb

U.S. Department of Education. 2007. *10 Facts about K-12 Education Funding.* Available at http://www.ed.gov

Viadero, Debra. 2005. "Study Sees Positive Effects of Teacher Certification." *Education Week,* April 27. Available at http://www .edweek.org

Viadero, Debra. 2006. "Rags to Riches in U.S. Largely a Myth, Scholars Write." *Education Week,* October 25. Available at http://www .edweek.org

Webb, Julie. 1989. "The Outcomes of Home-Based Education: Employment and Other Issues." *Educational Review* 41:121–133.

WEP (Wake Education Partnership). 2005. *The Issues.* Raleigh, NC: Wake County Schools.

What Is Character Education? 2003. Boston College: Center for the Advancement of Ethics and Character.

White House Fact Sheet. 2007. *Fact Sheet: The No Child Left Behind Act: Five Years of Results for America's Children.* January 8. Available at http://www.whitehouse .gov/news

Wiener, Ross, and Eli Pristoop. 2006. "How States Shortchange The Districts that Need the Most Help." In *Funding Gaps 2006* (pp. 5–6). Education Trust. Available at http://www2.edtrust.org

Wilgoren, Jodi. 2001. "Calls for Change in the Scheduling of the School Day." *New York Times,* January 10. Available at http://www .nytimes.com

Winter, Greg. 2005. "Study Finds Poor Performance by Nations' Education Schools." *New York Times,* March 15. Available at http://www.nytimes.com

Winters, Rebecca. 2001. "From Home to Harvard." *Time*, September 11. Available at http://www.time.com

Zigler, Edward, Sally Styfco, and Elizabeth Gilman. 2004. "The National Head Start Program for Disadvantaged Preschoolers." In Schools and Society, Jeanne Ballantine and Joan Spade, eds. (pp. 341–346). Belmont, CA: Thomson Wadsworth.

Zimmerman, Frederick, Gwen Glew, Dimitri Christakis, and Wayne Katon. 2005. "Early Cognitive Stimulation, Emotional Support, and Television Watching as Predictors of Subsequent Bullying Among Grade School Children." *Archives of Pediatric and Adolescent Medicine* 159(4):384–388.

Chapter 9

Allport, G. W. 1954. *The Nature of Prejudice*. Cambridge, MA: Addison-Wesley.

Amato, P. R., A. Booth, D. R. Johnson, and S. J. Rogers. 2007. *Alone Together: How Marriage in America Is Changing*. Cambridge MA: Harvard University Press.

American Council on Education and American Association of University Professors. 2000. Does Diversity Make a Difference? Three Research Studies on Diversity in College Classrooms. Washington, DC: American Council on Education and American Association of University Professors.

Bazar, Emily. 2007 (July 11). "Local Laws Target Immigrant Ills." *USA Today*. Available at http://www.usatoday.com

Beeman, Mark, Geeta Chowdhry, and Karmen Todd. 2000. "Educating Students About Affirmative Action: An Analysis of University Sociology Texts." *Teaching Sociology* 28(2):98–115.

Beirich, Heidi. 2007 (Summer). "Getting Immigration Facts Straight." *Intelligence Report*. Available at http://www.splcenter.org

Brace, C. Loring. 2005. *"Race" Is a Four-Letter Word*. New York: Oxford University Press.

Brooks, Roy. 2004. *Atonement and Forgiveness: A New Model for Black Reparations*. Berkeley: University of California Press.

Brown University Steering Committee on Slavery and Justice. 2007. *Slavery and Justice*. Providence, RI: Brown University.

Buchanan, Susy. 2007 (Winter). "Indian Blood." *Intelligence Report*. Available at http://www.splcenter.org

Capps, Randolph, Michael E. Fix, Jason Ost, Jane Reardon-Anderson, and Jeffrey S. Passel. 2005 (February 8). *The Health and Well-Being of Young Children of Immigrants*. Urban Institute. Available at http://www.urban.org

Carroll, Joseph. 2005 (July 12). Who Are the People in Your Neighborhood? Gallup Organization. Available at http://www.gallup.com

Clemmitt, M. 2007 (June 1). "Shock Jocks." *CQ Researcher* 17(21):481–504.

Cohen, Mark Nathan. 1998. "Culture, Not Race, Explains Human Diversity." *Chronicle of Higher Education* 44(32):B4–B5.

Conley, Dalton. 1999. *Being Black, Living in the Red: Race, Wealth, and Social Policy in America*. Berkeley: University of California Press.

Conley, Dalton. 2002. "The Importance of Being White." *Newsday*, October 13. Available at http://www.newsday.com

Day, Jennifer Cheeseman, and Eric Newburger. 2002. *The Big Payoff: Educational Attainment and Synthetic Estimates of Work-Life Earnings*. Washington, DC: U.S. Census Bureau.

Dees, Morris. 2000 (December 28). Personal correspondence. Morris Dees, co-founder of the Southern Poverty Law Center, 400 Washington Avenue, Montgomery, AL 36104.

EEOC (Equal Employment Opportunity Commission). 2003 (July 17). "Muslim Pilot Fired Due to Religion and Appearance, EEOC Says in Post-9/11 Backlash Discrimination Suit." Press release. U.S. Equal Employment Opportunity Commission. Available at http://www.eeoc.gov

EEOC. 2004. *Retail Distribution Centers: How New Business Processes Impact Minority Labor Markets*. U.S. Equal Employment Opportunity Commission. Available at http://www.eeoc.gov

EEOC. 2007a. *Why Do We Need E-RACE?* Available at http://www.eeoc.gov

EEOC. 2007b. "EEOC and Walgreens Resolve Lawsuit." Press release. Available at http://www.eeoc.gov

Etzioni, Amitai. 1997. "New Issues: Re-Thinking Race." *Public Perspective*, June–July, pp. 39–40. Available at http://www.ropercenter.uconn.edu

FBI (Federal Bureau of Investigation). 2006. *Hate Crime Statistics 2005*. Available at http://www.fbi.gov

Fields, Jason. 2004 (November). *America's Families and Living Arrangements: 2003*. Current Population Reports P20-553. U.S. Census Bureau. Available at http://www.census.gov

Fitrakis, Bob, and Harvey Wasserman. 2004. "Hearings on Ohio Voting Put 2004 Election in Doubt." *The Free Press*, November 18. Available at http://www.thefreepress.org

Fix, Michael E., and Randolph Capps. 2002. *The Dispersal of Immigrants in the 1990s*. Washington DC: Urban Institute.

Foust, Dean, Brian Grow, and Aixa M. Pascual. 2002. "The Changing Heartland." *Business Week*, September 9, pp. 80–84.

Gaertner, Samuel L., and John F. Dovidio. 2000. *Reducing Intergroup Bias: The Common Ingroup Identity Model*. Philadelphia: Taylor & Francis.

Gallup Organization. 2007a. *Race Relations*. Available at http://www.gallup.com

Gallup Organization. 2007b. Immigration. Available at http://www.gallup.com

Glaser, Jack, Jay Dixit, and Donald P. Green. 2002. "Studying Hate Crime with the Internet: What Makes Racists Advocate Racial Violence?" *Journal of Social Issues* 58(1):177–193.

Glaubke, Christina Roman, and Katharine Heintz-Krowles. 2004. *Fall Colors: Prime Time Diversity Report 2003–04*. Oakland, CA: Children Now & the Media Program.

Goldstein, Joseph. 1999. "Sunbeams." *The Sun* 277, January, p. 48.

Goodnough, Abby. 2001. "New York City Is Short-Changed in School Aid, State Judge Rules." *New York Times*, January 11. Available at http://www.nytimes.com

Greenberg, Daniel S. 2003. "Supreme Court Sets Showdown on Affirmative Action." *The Lancet* 361(March 1):762. Available at http://www.thelancet.com

Greenhouse, Steven. 2003. "Suit Claims Discrimination Against Hispanics on Job." *New York Times*, February 9. Available at http://www.nytimes.com

Greenhouse, Steven. 2005. "Wal-Mart to Pay U.S. $11 Million in Lawsuit on Illegal Workers." *New York Times*, March 19. Available at http://www.nytimes.com

Gurin, Patricia. 1999. "New Research on the Benefits of Diversity in College and Beyond: An Empirical Analysis." *Diversity Digest*, Spring, pp. 5–15.

"Hate on Campus." 2000 (Spring). *Intelligence Report* 98:6–15.

"Hate Websites." 2007 (Spring). *Intelligence Report* 125:59–65.

Healey, Joseph F. 1997. *Race, Ethnicity, and Gender in the United States: Inequality, Group Conflict, and Power*. Thousand Oaks, CA: Pine Forge Press.

"Hear and Now." 2000 (Fall). *Teaching Tolerance*, p. 5.

Hodgkinson, Harold L. 1995. "What Should We Call People? Race, Class, and the Census for 2000." *Phi Delta Kappa*, October, pp. 173–179.

Holthouse, David. 2006 (July 7). "A Few Bad Men." *Intelligence Report*. Available at http://www.spl.org

Holzer, Harry. 2005 (May 14). *New Jobs in Recession and Recovery: Who Are Getting Them and Who Are Not?* Urban Institute. Available at http://www.urban.org

Holzer, Harry, and David Neumark. 2000. "Assessing Affirmative Action." *Journal of Economic Literature* 38(3):483–568.

hooks, bell. 2000. *Where We Stand: Class Matters.* New York: Routledge.

Humphreys, Debra. 1999. "Diversity and the College Curriculum: How Colleges and Universities Are Preparing Students for a Changing World." *Diversity-Web.* Available at http://www.inform.umd.edu

Humphreys, Debra. 2000 (Fall). "National Survey Finds Diversity Requirements Common Around the Country." *Diversity Digest.* Available at http://www.diversityweb.org

Jackson, Camille. 2005 (September 19). "Katrina: Decoding the Language of Race and Class." Tolerance in the News. *Teaching Tolerance.* Available at http://www.tolerance.org

Jay, Gregory. 2005 (March 17). "Introduction to Whiteness Studies." Available at http://www.uwm.edu

Jensen, Derrick. 2001. "Saving the Indigenous Soul: An Interview with Martin Prechtel." *The Sun,* 304(April):4–15.

Kaplan, David E., and Lucian Kim. 2000. "Nazism's New Global Threat." *U.S. News Online,* September 25. Available at http://www.usnews.com

Keita, S. O. Y., and Rick A. Kittles. 1997. "The Persistence of Racial Thinking and the Myth of Racial Divergence." *American Anthropologist* 99(3):534–544.

King, Joyce E. 2000 (Fall). "A Moral Choice." *Teaching Tolerance* 18:14–15.

Kozol, Jonathan. 1991. *Savage Inequalities: Children in America's Schools.* New York: Crown.

Larsen, Luke J. 2004 (August). *The Foreign-Born Population in the United States: 2003.* Current Population Reports P20-551. U.S. Census Bureau. Available at http://www.census.gov

Lawrence, Sandra M. 1997. "Beyond Race Awareness: White Racial Identity and Multicultural Teaching." *Journal of Teacher Education* 48(2):108–117.

Lee, Christopher. 2006. "EEOC is Hobbled, Groups Contend." *Washington Post,* June 14, p. A21.

Levin, Jack, and Jack McDevitt. 1995. "Landmark Study Reveals Hate Crimes Vary Significantly by Offender Motivation." *Klanwatch Intelligence Report,* August, pp. 7–9.

Lollock, Lisa. 2001 (March). *The Foreign-Born Population in the United States: March 2000.* Current Population Reports P20-534. Washington DC: U.S. Bureau of the Census.

Maril, Robert Lee. 2004. *Patrolling Chaos: The U.S. Border Patrol in Deep South Texas.* Lubbock, TX: Texas Tech University Press.

Massey, Douglas, and Nancy Denton. 1993. *American Apartheid: Segregation and the Making of an American Under-Class.* Cambridge, MA: Harvard University Press.

Massey, Douglas S., and Garvey Lundy. 2001. "Use of Black English and Racial Discrimination in Urban Housing Markets: New Methods and Findings." *Urban Affairs Review* 36(4):452–469.

McIntosh, Peggy. 1990 (Winter). "White Privilege: Unpacking the Invisible Knapsack." *Independent School* 49(2):31–35.

Mock, Brentin. 2007 (Spring). "Sharing the Hate." *Intelligence Report* 125:15–16.

Morrison, Pat. 2002. "September 11: A Year Later—American Muslims Are Determined Not to Let Hostility Win." *National Catholic Reporter,* 38(38):9–10.

Moser, Bob. 2004. "The Battle of 'Georgiafornia.'" *Intelligence Report* 116:40–50.

Mukhopadhyay, Carol C., Rosemary Henze, and Yolanda T. Moses. 2007. *How Real is Race?* Lanham, MD: Rowman & Littlefield Education.

NAACP. 2000 (November 11). *NAACP Voting Irregularities Public Hearing.* Press release. National Association for the Advancement of Colored People. Available at http://www.naacp.org

NAACP. 2001 (January 10). *NAACP National Civil Rights Groups File Florida Voting Rights Lawsuit to Eliminate Unfair Voting Practices.* Press release. National Association for the Advancement of Colored People. Available at http://www.naacp.org

"National Briefing." 2005. *New York Times,* August 11, p. A19.

National Immigration Law Center. 2006. Basic Facts About In-State Tuition for Undocumented Immigrant Students. Available at http://www.nilc.org

National Immigration Law Center. 2007. *Facts about Immigrant Workers.* Available at http://www.nilc.org

Navarro, Mireya. 2003. "For New York's Black Latinos, a Growing Racial Awareness." *New York Times,* April 28. Available at http://www.nytimes.com

Nemes, Irene. 2002. "Regulating Hate Speech in Cyberspace: Issues of Desirability and Efficacy." *Information and Communication Technology Law* 11(3):193–220.

"Neo-Nazi Label Woos Teens With Hate-Music Sampler." 2004 (Winter). *Intelligence Report* 116:5.

NPR, Kaiser Family Foundation, and Kennedy School of Government. 2005. *Immigration: Summary of Findings.* NPR/Kaiser/Kennedy School Poll. Available at http://www.npr.org/news/

Olson, James S. 2003. *Equality Deferred: Race, Ethnicity, and Immigration in America Since 1945.* Belmont, CA: Wadsworth/Thomson.

Orfield, Gary. 2001 (July). *Schools More Separate: Consequences of a Decade of Resegregation.* Cambridge, MA: Harvard University, Civil Rights Project.

Ossorio, Pilar, and Troy Duster. 2005. "Race and Genetics." *American Psychologist* 60(1):115–128.

Pager, Devah. 2003. "The Mark of a Criminal Record." *American Journal of Sociology* 108(5):937–975.

Passel, Jeffrey S., Randolph Capps, and Michael Fix. 2004 (January 12). "Undocumented Immigrants: Facts and Figures." Urban Institute. Available at http://www.urban.org

Paul, Pamela. 2003. "Attitudes Toward Affirmative Action." *American Demographics* 25(4):18–19.

Pew Research Center. 2005 (September 8). *Huge Racial Divide over Katrina and Its Consequences.* Available at http://www.people-press.org

Pew Research Center. 2007 (March 22). *Trends in Political Values and Core Attitudes: 1987–2007.* Washington, DC: Pew Research Center for People & the Press.

Picca, Leslie, and Joe R. Feagin. 2007. *Two-Faced Racism.* New York: Routledge.

Plous, S. 2003. "Ten Myths About Affirmative Action." In *Understanding Prejudice and Discrimination,* S. Plous, ed. (pp. 206–212). New York: McGraw-Hill.

"Poll: Most Americans See Lingering Racism." 2006 (December 12). CNN.com. Available at http:/www.cnn.com

Posner, Eric A., and Adrian Vermeule. 2003. "Reparations for Slavery and Other Historical Injustices." *Columbia Law Review* 103(689):689–748.

Potok, Mark. 2005 (September 20). "In Katrina's Wake, White Supremacists Spew Hatred." Southern Poverty Law Center. Available at http://www.splcenter.org/news/new.jsp

Pryor, John H., Sylvia Hurtado, Victor B. Saenz, Jessica S. Korn, Jose Luis Santos, and William S. Korn. 2006. *The American Freshman: National Norms for Fall 2006.* Los Angeles: Higher Education Research Institute, UCLA.

Ramirez, Roberto R., and G. Patricia de la Cruz. 2003 (June). *The Hispanic Population in the United States: March 2002.* Current Population Reports P20-545. Washington, DC: U.S. Census Bureau.

Rosenfeld, Michael J., and Byung-Soo Kim. 2005 (August). "The Independence of Young Adults and the Rise of Interracial and Same-Sex Unions." *American Sociological Review* 70:541–562.

Rothenberg, Paula S. 2002. *White Privilege.* New York: Worth.

Saad, Lydia. 2007 (April 25). "Most Americans Favor Giving Illegal Immigrants a Chance." Gallup Poll News Service. Available at http://www.galluppoll.com

Schaefer, Richard T. 1998. *Racial and Ethnic Groups,* 7th ed. New York: HarperCollins.

Schiller, Bradley R. 2004. *The Economics of Poverty and Discrimination,* 9th ed. Upper Saddle River, NJ: Pearson Education.

Schmidt, Peter. 2004 (January 30). "New Pressure Put on Colleges to End Legacies in Admissions." *Chronicle of Higher Education* 50(21):A1.

Schmitt, Eric. 2001a. "Analysis of Census Finds Segregation Along with Diversity." *New York Times,* April 4. Available at http://www.nytimes.com

Schmitt, Eric. 2001b. "Blacks Split on Disclosing Multiracial Roots." *New York Times,* March 31. Available at http://www.nytimes.com

Schuman, Howard, and Maria Krysan. 1999. "A Historical Note on Whites' Beliefs About Racial Inequality." *American Sociological Review* 64:847–855.

Shipler, David K. 1998. "Subtle vs. Overt Racism." *Washington Spectator* 24(6):1–3.

SPLC (Southern Poverty Law Center). 2005a (June). "Center Wins $1.35 Million Judgment Against Violent Border Vigilantes." *SPLC Report* 35(2):3.

SPLC. 2005b (June). "New Lawsuits Seek Reform of Abusive Labor Practice." *SPLC Report* 35(2):1.

SPLC. 2007a. *Close to Slavery: Guestworker Programs in the United States.* Montgomery, AL: Southern Poverty Law Center.

SPLC. 2007b. "Hate Group Numbers Continue Increase." *SPLC Report* 37(1):3.

Tanneeru, Manav. 2007 (May 11). "Asian-Americans' Diverse Voices Share Similar Stories." CNN.com. Available at http://www.cnn.com

"The Year in Hate." 2007 (Summer). *Intelligence Report.* Available at http://www.splcenter.org

Tolbert, Caroline J., and John A. Grummel. 2003. "Revisiting the Racial Threat Hypothesis: White Voter Support for California's Proposition 209." *State Politics and Policy Quarterly* 3(2):183–202, 215–216.

Turner, Margery Austin, Stephen L. Ross, George Galster, and John Yinger. 2002. *Discrimination in Metropolitan Housing Markets.* Washington, DC: Urban Institute.

Turner, Margery Austin, and Felicity Skidmore. 1999. *Mortgage Lending Discrimination: A Review of Existing Evidence.* Washington, DC: Urban Institute.

U.S. Census Bureau. 2005. *2005 American Community Survey.* Available at http://www.census.gov

U.S. Citizenship and Immigration Services. 2003. *General Naturalization Requirements.* Available at http://www.uscis.gov

U.S. Customs and Border Security. 2007. *2006: A Year of Accomplishment.* Available at http://www.cbp.gov

U.S. Department of Labor. 2002. Facts on Executive Order 11246 Affirmative Action. Available at http://www.dol.gov

Van Ausdale, Debra, and Joe R. Feagin. 2001. *The First R: How Children Learn Race and Racism.* Lanham, MD: Rowman & Littlefield.

"White Power Bands." 2002 (January). *Hate in the News.* Southern Poverty Law Center. Available at http://www.tolerance.org

Williams, Eddie N., and Milton D. Morris. 1993. "Racism and Our Future." In *Race in America: The Struggle for Equality,* Herbert Hill and James E. Jones Jr., eds. (pp. 417–424). Madison: University of Wisconsin Press.

Williams, Richard, Reynold Nesiba, and Eileen Diaz McConnell. 2005. "The Changing Face of Inequality in Home Mortgage Lending." *Social Problems* 52(2):181–208.

Willoughby, Brian. 2003. "Hate on Campus." *Tolerance in the News,* June 13. Available at http://www.tolerance.org

Wilson, William J. 1987. *The Truly Disadvantaged: The Inner City, the Underclass and Public Policy.* Chicago: University of Chicago Press.

Winter, Greg. 2003a. "Schools Resegregate, Study Finds." *New York Times,* January 21. Available at http://www.nytimes.com.

Winter, Greg. 2003b. "U. of Michigan Alters Admissions Use of Race." *New York Times,* August 29. Available at http://www.nytimes.com

Zack, Naomi. 1998. *Thinking About Race.* Belmont, CA: Wadsworth.

Zinn, Howard. 1993. "Columbus and the Doctrine of Discovery." In *Systemic Crisis: Problems in Society, Politics, and World Order,* William D. Perdue, ed. (pp. 351–357). Fort Worth, TX: Harcourt Brace Jovanovich.

Chapter 10

AAUW (American Association of University Women). 2006 (January 24). *Drawing the Line: Sexual Harassment on Campus.* Available at http://www.aauw.org

AAUW. 2007 (April 23). *Pay Gap Exists as Early as One Year out of College, Research Says.* Available at http://www.aauw.org

Abernathy, Michael. 2003. *Male Bashing on TV. Tolerance in the News.* Available at http://www.tolerance.org/news

ACE (American Council on Education). 2006 (July 11). *College Enrollment Gender Gap Widens for White and Hispanic Students, but Race and Income Disparities Still Most Significant New ACE Report Finds.* Available at http://www.acenet.edu

Adams, Jimi. 2007. "Stained Glass Makes the Ceiling Visible." *Gender and Society* 21(1):80–105.

Alley, Thomas R., and Catherine M. Hicks. 2005. "Peer Attitudes Towards Adolescent Participants in Male- and Female-Oriented Sports." *Adolescence* 40(158):273–280.

American Sociological Association. 2004 (July). *The Best Time to Have a Baby: Institutional Resources and Family Strategies among Early Career Sociologists.* ASA Research Brief.

Amnesty International. 2005 (April 26). Afghanistan: Stoned to Death—Human Rights Scandal. Press release. Available at http://www.amnesty.org

Amnesty International. 2007. *About This Campaign.* Available at http://www.amnestyusa.org

Anderson, David A., and Mykol Hamilton. 2005. "Gender Role Stereotyping of Parents in Children's Picture Books: The Invisible Father." *Sex Roles* 52:145–151.

Anderson, Margaret L. 1997. *Thinking About Women,* 4th ed. New York: Macmillan.

Athreya, Bama. 2003. "Trade Is a Women's Issue." *ATTAC,* February 20. Available at http://www.globalpolicy.org/socecon

Baker, Robin, Gary Kriger, and Pamela Riley. 1996. "Time, Dirt, and Money: The Effects of Gender, Gender Ideology, and Type of Earner Marriage on Time, Household Task, and Economic Satisfaction among Couples with Children." *Journal of Social Behavior and Personality* 11:161–177.

Banister, Judith. 2003. *Shortage of Girls in China: Causes, Consequences, International Comparisons, and Solutions.* Population Reference Bureau online. Available at http://www.prb.org

Bannon, Lisa. 2000. "Why Girls and Boys Get Different Toys." *Wall Street Journal,* February 14, p. B1.

Barak, Azy. 2005. "Sexual Harassment on the Internet." *Social Science Computer Review* 23(1):77–92.

Barko, Naomi. 2003. "Equal Pay for Equal Work." In *Women's Rights,* Shasta Gaughen, ed. (pp. 43–48). Farmington, MA: Greenhaven Press.

Basow, Susan A. 1992. *Gender: Stereotypes and Roles,* 3rd ed. Pacific Grove, CA: Brooks/Cole.

Beeman, Mark, Geeta Chowdhry, and Karmen Todd. 2000. "Educating Students about Affirmative Action: An Analysis of University Sociology Texts." *Teaching Sociology* 28(2):98–115.

Begley, Sharon. 2000. "The Stereotype Trap." *Newsweek,* November 6, pp. 66–68.

Bianchi, Suzanne M., Melissa A. Milkie, Liana C. Sayer, and John Robinson. 2000. "Is Anyone Doing the Housework? Trends in the Gender Division of Household Labor." *Social Forces* 79:191–228.

Bittman, Michael, and Judy Wajcman. 2000. "The Rush Hour: The Character of Leisure Time and Gender Equity." *Social Forces* 79:165–189.

Black, Linda James. 2006. "Textbooks, Gender and World History." *World History Connected.* 3(2). Available at http://www.historycooperative.org/whcindex.htm

Bly, Robert. 1990. *Iron John: A Book about Men.* Boston: Addison-Wesley.

Bowker, Anne, Shannon Gadbois, and Becki Cornock. 2003. "Sports Participation and Self-Esteem: Variations as a Function of Gender and Gender Role Orientation." *Sex Roles* 49(1/2):47–58.

Budig, Michelle. 2003. "Male Advantage and the Gender Composition of Jobs: Who Rides the Glass Escalator?" *Social Problems* 49:258–277.

Burger, Jerry M., and Cecilia H. Solano. 1994. "Changes in Desire for Control over Time: Gender Differences in a Ten-Year Longitudinal Study." *Sex Roles* 31:465–472.

CAWP (Center for American Women and Politics). 2006. *Women's Votes Pivotal in Shifting Control of U.S. Senate to Democrats.* Available at http://www.cawp.rutgers.edu

CAWP. 2007. *Women in Elected Offices, 2005.* Available at http://www.cawp.rutgers.edu

CBS News. 2006 (February 5). *Ready for a Woman President?* Available at http://www.cbsnews.com/

Cejka, Mary Ann, and Alice Eagly. 1999. "Gender Stereotypic Images of Occupations Correspond to the Sex Segregation of Employment." *Personality and Social Psychology Bulletin* 25:413–423.

Chavez, Linda. 2000. *The Color Bind.* Berkeley: University of California Press.

Cherney, Isabelle, and Kamala London. 2006. "Gender Linked Differences in the Toys, Television Shows, Computer Games, and Outdoor Activities of 5- to 13-year old Children." *Sex Roles* 54: 717–726.

Clark, Hannah. 2006. "Are Women Happy Under the Glass Ceiling?" *Forbes,* March 8. Available at http://www.forbes.com

Cohen, Philip, and Matt Huffman. 2003. "Individuals, Jobs, and Labor Markets: The Devaluation of Women's Work." *American Sociological Review* 68:443–463.

Cohen, Theodore. 2001. *Men and Masculinity.* Belmont, CA: Wadsworth.

Common Dreams. 2005 (March 24). *Media Advisory: Women's Opinions Also Missing on Television.* Available at http://www.commondreams.org/news2005/0324-07.htm

Connolly, Ceci. 2005. "Access to Abortion Pared at State Level." *Washington Post,* August 29, p. A1.

Cullen, Lisa Takeuchi. 2003. "I Want Your Job, Lady!" *Time,* May 12, pp. 52–56.

Cunningham, Mick. 2001. "Parental Influences on the Gendered Division of Housework." *American Sociological Review* 66:184–203.

Davis, James A., Tom W. Smith, and Peter V. Marsden. 2002. *General Social Surveys, 1972–2002: 2nd ICPSR Version.* Chicago: National Opinion Research Center.

Davis, Michelle R. 2007. "Gender Gaps in GPA Seen as Linked to Self-Discipline." *Education Week* 26(23):8.

Dee, Thomas S. 2006 (Fall). "How a Teacher's Gender Affects Boys and Girls." *Education Next.* Available at http://www.educationnext.org

Dittrich, Liz. 2002. *About-Face Facts on the Media.* Available at http://www.aboutface.org

Dominus, Susan. 2006. "A Girly-Girl Joins the 'Sesame' Boys." *New York Times,* August 6. Available at http://nytimes.com

Dumais, Susan A. 2002. "Cultural Capital, Gender, and School Success: The Role of Habitus." *Sociology of Education* 75: 44–68.

EEOC (Equal Employment Opportunity Commission). 2007a. *Sex-Based Discrimination.* Available at http://www.eeoc.gov

EEOC. 2007b. *Sexual Harassment.* Available at http://www.eeoc.gov

EEOC. 2003 (July 31). *Women of Color Make Gains in Employment and Job Status.* Available at http://www.eeoc.gov

England, Paul, and Su Li. 2006. "Desegregation Stalled: The Changing Gender Composition of College Majors, 1971–2002." *Gender and Society* 20(5):657–677.

EOC (Equal Opportunities Commission). 2006. *Investigation: Free to Choose: Tackling Gender Barriers to Better Jobs.* Available at http://www.eoc.org.uk

EOC. 2007. *The Gender Equality Duty.* Available at http://www.eoc.org.uk

Epstein, Cynthia Fuchs. 2007. "Great Divides: The Cultural, Cognitive, and Social Bases of the Global Subordination of Women." *American Sociological Review* 72:1–22.

ERA (Equal Rights Amendment). 2007. *The Equal Rights Amendment.* Available at http://www.eracampaign.net

Erickson, Jan. 2007. "Conservatives Push to End Affirmative Action." National Organization for Women. *National NOW Times,* Winter, p. 1.

Evans, Lorraine, and Kimberly Davies. 2000. "No Sissy Boys Here." *Sex Roles,* February, pp. 255–271.

Evertsson, Marie. 2006. "The Reproduction of Gender: Housework and Attitudes Toward Gender Equality in the Home among Swedish Boys and Girls." *The British Journal of Sociology* 57(3):415–436.

Faiola, Anthony. 2007. "Japanese Working Women Still Serve the Tea." *The Washington Post,* March 2, p. A09.

Faludi, Susan. 1991. *Backlash: The Undeclared War Against American Women.* New York: Crown.

Fitzpatrick, Catherine. 2000. "Modern Image of Masculinity Changes with Rise of New Celebrities." *Detroit News,* June 24. Available at http://www.detnews.com/

Fridkin, Kim L., and Patrick J. Kenney. 2007. "Examining the Gender Gap in Children's Attitudes toward Politics." *Sex Roles* 56:133–140.

Gallagher, Margaret. 2006. *Who Makes the News?: Global Media Monitoring Project 2005.* Available at http://www.whomakesthenews.org

Gallagher, Sally K. 2004. "The Marginalization of Evangelical Feminism." *Sociology of Religion* 65:215–237.

Goffman, Erving. 1963. *Stigma.* Englewood Cliffs, NJ: Prentice Hall.

Goldman, Lea, and Kiri Blakeley (eds.). 2006. "The Celebrity 100." *Forbes,* June 12. Available at http://www.forbes.com

Guiller, J., and A. Durndell. 2006. "'I totally agree with you': Gender Interactions in Educational Online Discussion Groups." *Journal of Computer Assisted Learning* 22(5):368–381.

Gupta, Sanjay. 2003. "Why Men Die Young." *Time,* May 12, p. 84.

Guy, Mary Ellen, and Meredith A. Newman. 2004. "Women's Jobs, Men's Jobs: Sex Segregation and Emotional Labor." *Public Administration Review* 64:289–299.

Hack, Damon. 2006. "Golf: Augusta's Payne Has No Plans to Meet Burk." *New York Times,* May 9, p. D3.

Haines, Errin. 2006 (November 1). *Father Convicted in Genital Mutilation.* Available at http://www.breitbart.com/news

Hausmann, Ricardo, Laura Tyson, and Saadia Zahidi. 2007. *The Global Gender Gap Report 2006.* Geneva: World Economic Forum.

Hawkesworth, Mary, Kathleen J. Casey, Krista Jenkins, and Kathleen E. Kleeman. 2001. *Legislating by and for Women: A Comparison of the 103rd and 104th Congress.* Center for American Women and Politics, Rutgers University.

Hewlett, Sylvia Ann. 2002. "Executive Women and the Myth of Having It All." *Harvard Business Review,* April, pp. 66–73.

Heyzer, Noeleen. 2003. "Enlisting African Women to Fight AIDS." *Washington Post,* July 8. Available at http://www.globalpolicy.org/socecon/inequal

Hochschild, Arlie. 1989. *The Second Shift.* London: Penguin.

Hochschild, Arlie. 2001. *The Time Bind: When Work Becomes Home and Home Becomes Work.* New York: Owl Books.

Hoffnung, Michele. 2004. "Wanting It All: Career, Marriage, and Motherhood During College-Educated Women's 20s." *Sex Roles* 50:711–723.

Hook, Jennifer L. 2006. "Care in Context: Men's Unpaid Work in 20 Countries, 1965–2003," *American Sociological Review* 71:639–660.

HRC (Human Rights Campaign). 2005. *Transgender Americans: A Handbook for Understanding.* Available at http://www.hrc.org

IDEA (International Institute for Democracy and Electoral Assistance). 2006. "Quotas for Women." Available at http://www.idea.int

ILO (International Labor Office). 2007a (March). *Global Employment Trends for Women Brief.* Available at http://www.ilo.org

ILO. 2007b. *Equality at Work: Tackling the Challenges.* International Labor Conference, 96th Session. Geneva.

IWPR (Institute for Women's Policy Research). 2006a (December). *Running Faster to Stay in Place.* Available at http://www.iwpr.org

IWPR. 2006b (December). *Best and Worst Economies for Women.* Briefing Paper R334. Available at http://www.iwpr.org

IWPR. 2007 (April). *The Gender Wage Ratio: Women and Men's Earnings.* Fact Sheet #C350. Available at http://www.iwpr.org

International Women's Rights Project. 2000. *The First CEDAW Impact Study.* Available at http://www.yorku.ca/iwrp/cedawReport

Ison, Elizabeth R. 2005 (August). *California Supreme Court Says Workers Who Are Not Involved in Office Romance Can Sue Over the Affair.* HR California. Available at http://www.hrcalifornia.com

Jackson, Robert, and Meredith A. Newman. 2004. "Sexual Harassment in the Federal Workplace Revisited: Influences on Sexual Harassment." *Public Administration Review* 64(6):705–717.

Jones, Jeffrey M. 2005 (March 10). *Public Believes Men, Women Have Equal Abilities in Math, Science.* Gallup News Services, Gallup Organization.

Kalev, Alexandra, Erin Kelly, and Frank Dobbin. 2006. "Best Practices or Best Guesses? Assessing the Efficacy of Corporate Affirmative Action and Diversity Policies." *American Sociological Review* 71:589–617.

Kilbourne, Barbara S., Georg Farkas, Kurt Beron, Dorothea Weir, and Paula England. 1994. "Returns to Skill, Compensating Differentials, and Gender Bias: Effects of Occupational Characteristics on the Wages of White Women and Men." *American Journal of Sociology* 100:689–719.

Kilbourne, Jean. 2000. *Killing Us Softly 3: Advertising's Image of Women.* Northampton, MA: Media Education Foundation.

Kimmel, Michael. 2006. "A War Against Boys?" *Dissent* 53(4):65–70.

Kravets, David. 2007 (February 6). *Court Says Wal-Mart Must Face Bias Trial.* Available at http://www.breitbart.com/news

Lawlor, Andrew. 1999. "Tenured Women Battle to Make It Less Lonely at the Top." *Science* 286:1272–1279.

Lewis, Tyler. 2007. *Michigan Releases New Guidance for Compliance with Proposal 2.* Available at http://www.civilrights.org

Long, J. Scott, Paul D. Allison, and Robert McGinnis. 1993. "Rank Advancement in Academic Careers: Sex Differences and the Effects of Productivity." *American Sociological Review* 58:703–722.

Lopez-Claros, Augusto, and Saadia Zahidi. 2005. *Women's Empowerment: Measuring the Global Gender Gap.* World Economic Forum.

Lorber, Judith. 1998. "Night to His Day." In *Reading between the Lines,* Amanda Konradi and Martha Schmidt, eds. (pp. 213–220). Mountain View, CA: Mayfield.

"Major Rulings of the 2002–2003 Term." 2003. *Daily Reflector,* June 27, p. 4A.

Mannino, Clelia Anna, and Francine M. Deutsch. 2007. "Changing the Division of Household Labor: A Negotiated Process between Partners." *Sex Roles* 56:309–324.

Martin, Patricia Yancey. 1992. "Gender, Interaction, and Inequality in Organizations." In *Gender, Interaction, and Inequality,* Cecilia Ridgeway, ed. (pp. 208–231). New York: Springer-Verlag.

Mattingly, Marybeth J., and Suzanne M. Bianchi. 2003. "Gender Differences in the Quantity and Quality of Free Time: The U.S. Experience." *Social Forces* 81(3):999–1030.

McBrier, Debra Branch. 2003. "Gender and Career Dynamics within a Segmented Professional Labor Market: The Case of Law Academic." *Social Forces* 81:1201–1266.

McGregor, Liz. 2003. Women Bear Brunt of AIDS Toll. Available at http://www.globalpolicy.org/socecon/develop

Mead, Sara. 2006. *The Truth about Boys and Girls.* Available at http://www.educationsector.org

Mensch, Barbara, and Cynthia Lloyd. 1997. *Gender Differences in the Schooling Experiences of Adolescents in Low-Income Countries: The Case of Kenya.* Policy Research Working Paper 95. New York: Population Council.

Messner, Michael A., and Jeffrey Montez de Oca. 2005. "The Male Consumer as Loser: Beer and Liquor Ads in Mega Sports Media Events." *Signs* 30:1879–1909.

Miller, Carol D. 2004. "Participating but Not Leading: Women's Underrepresentation in Student Government Leadership Positions." *College Student Journal,* September, pp. 423–428.

Moen, Phyllis, and Yan Yu. 2000. "Effective Work/Life Strategies: Working Couples, Working Conditions, Gender, and Life Quality." *Social Problems* 47:301–326.

Morin, Richard, and Megan Rosenfeld. 2000. "The Politics of Fatigue." In *Annual Editions: Social Problems,* Kurt Finsterbusch, ed. (pp. 152–154). Guilford, CT: Dushkin/McGraw-Hill.

Nanda, Serena. 2000. *Gender Diversity: Cross-Cultural Variations.* Long Grove, IL: Wavelane Press.

National Council of Women's Organizations. 2005. *National Council of Women's Organizations Homepage.* Available at http://www.womensorganizations.org

National Science Foundation. 2007. *Doctorates Awarded to Women: 1996–2005.* NSF 7-305. National Science Foundation, Division of Science Resource Studies. Available at http://www.nsf.gov

National Women's Law Center. 2004. "Commission Finds University of Colorado Knew of Alleged Sexual Harassment in Football Recruiting Program and Failed to Take Appropriate Action." *NWLC News Room.* Available at http://www.nwlc.org/details.cfm?id=1884&section=newsroom

NEDA (National Eating Disorders Association). 2006. *Statistics: Eating Disorders and Their Precursors.* Available at http://www.edap.org

Nelson, Robert, and William Bridges. 1999. *Legalizing Gender Inequality.* New York: Cambridge University Press.

Nichols-Casebolt, Ann, and Judy Krysik. 1997. "The Economic Well-Being of Never and Ever-Married Mother Families." *Journal of Social Service Research* 23(1):19–40.

NOMAS (The National Organization for Men Against Sexism). 2007. "Statement of Principles." Available at http://www.nomas.org

NOW (National Organization for Women). 2007a (April 24). "Women Lose Millions Due to Wage Gap, NOW Calls for Passage of Pay Equity Legislation." Available at http://now.org

NOW. 2007b. *EEOC Clarifies Prohibition on Maternal Profiling.* Available at http://now.org/issues/mothers

NOW. 2007c. Strong Civil Rights Bill Passes House. Available at http://now.org

Orecklin, Michelle. 2003. "Now She's Got Game." *Time,* March 3, pp. 57–59.

Orenstein, Peggy. 2001. *Flux: Women on Sex, Work, Love, Kids, and Life in a Half-Changed World.* New York: Anchor.

Padavic, Irene, and Barbara Reskin. 2002. *Men and Women at Work,* 2nd ed. Thousand Oaks, CA: Pine Forge Press.

Palmberg, Elizabeth. 2005. "Teach a Women to Fish and Everyone Eats: Why Women Are the Key to Global Poverty." *Sojourner's Magazine,* June, pp. 1–4. Available at http://www.sojo.net

Parker, Kathleen. 2000. "It's Time for Women to Get Angry but Not at Men." *Greensboro News Record,* March 14, p. F4.

Partlow, Joshua, and Lonnae O'Neal Parker. 2006. "West Point Mourns a Font of Energy, Laid to Rest by War." *The Washington Post,* September 27. A01.

Plan International. 2007. *Because I am a Girl: The State of the World's Girls 2007.* Surrey, England. Available at http://www.plan-international.org

Pollack, William. 2000a. Real Boys' Voices. New York: Random House.

Pollack, William. 2000b. "The Columbine Syndrome." *National Forum* 80:39–42.

Portraits of Sacrifice. 2006. "U.S. Causalities in Iraq." *San Francisco Chronicle,* n.d. Available at http://www.sfgate.com

Purcell, Piper, and Lara Stewart. 1990. "Dick and Jane in 1989." *Sex Roles* 22:177–185.

Quist-Areton, Ofeibea. 2003. "Fighting Prejudice and Sexual Harassment of Girls in Schools." *All Africa,* June 12. Available at http://www.globalpolicy.org/socecon

Rao, Kavitha. 2006 (March 16). "Missing Daughters on an Indian Mother's Mind." March 16. *Women'se-News.* Available at http://www.womensenews.org

Reid, Pamela T., and Lillian Comas-Diaz. 1990. "Gender and Ethnicity: Perspectives on Dual Status." *Sex Roles* 22:397–408.

Renzetti, Claire, and Daniel Curran. 2003. *Women, Men and Society.* Boston: Allyn and Bacon.

Reskin, Barbara, and Debra McBrier. 2000. "Why Not Ascription? Organizations' Employment of Male and Female Managers." *American Sociological Review* 65:210–233.

Reuters. 2007 (May 11). *Self-Esteem Tied to Body Image for Most Teens.* Available at http://www.msnbc.msn.com

Robinson, John P., and Suzanne Bianchi. 1997. "The Children's Hours." *American Demographics,* December, pp. 1–6.

Ross, Catherine E., and Marieke Van Willigen. 1996. "Gender, Parenthood, and Anger." *Journal of Marriage and the Family* 58(3):572–584.

Rossiter, Molly. 2007. "For Female Clergy, Expectations Come with Territory." *Chicago Tribune,* April 18. Available at http://www.chicagotribune.com

Sadker, Myra, and David Sadker. 1990. "Confronting Sexism in the College Classroom." In *Gender in the Classroom: Power and Pedagogy,* S. L. Gabriel and I. Smithson, eds. (pp. 176–187). Chicago: University of Illinois Press.

Sanchez, Diana T., and Jennifer Crocker. 2005. "How Investment in Gender Ideals Affects Well-Being: The Role of External Contingencies of Self-Worth." *Psychology of Women Quarterly* 29:63–77.

Sapiro, Virginia. 1994. *Women in American Society.* Mountain View, CA: Mayfield.

Sax, Linda, Sylvia Hurtado, Jennifer Lindholm, Alexander Astin, William Korn, and Kathryn Mahoney. 2004. *The American Freshman: National Norms for 2004.* Los Angeles: Higher Education Research Institute.

Schwalbe, Michael. 1996. *Unlocking the Iron Cage: The Men's Movement, Gender Politics, and American Culture.* New York: Oxford University Press.

Sheehan, Molly. 2000. "Women Slowly Gain Ground in Politics." In *Vital Signs: The Environmental Trends That Are Shaping Our Future,* Linda Starke, ed. (pp. 152–153). New York: W. W. Norton.

Simpson, Ruth. 2005. "Men in Non-Traditional Occupations: Career Entry, Career Orientation, and Experience of Role Strain." *Gender Work and Organization* 12(4):363–380.

Skuratowicz, Eva, and Larry W. Hunter. 2004. "Where Do Women's Jobs Come From? Job Resegregation in an American Bank." *Work and Occupations* 31:73–110.

Smith, Melanie. 2006. "Is Church Too Feminine for Men?" *The Decatur Daily,* July1. Available at http://www.decaturdaily.com

Smolken, Rachael. 2000. "Girls' SAT Scores Still Lag Boys'." *Post Gazette.* Available at http://www.post-gazette.com/headlines/20000830sat2.asp

Snell, William E., Jr. 1997. *The Beliefs About Women Scale.* Department of Psychology, College of Liberal Arts, Southeast Missouri State University.

Sommers, Christina. 2000. *The War Against Boys.* New York: Simon and Schuster.

Sondhaus, Elizabeth L., Richard M. Kurtz, and Michael J. Strube. 2001. "Body Attitude, Gender, and Self-Concept: A 30-Year Perspective." *Journal of Psychology* 135:413–429.

Thornburgh, Nathan. 2006. "A Death in the Class of 9/11." *Time,* September 28. Available at http://time.com

Tinklin, Teresa, Linda Croxford, Alan Ducklin, and Barbara Frame. 2005. "Gender and Attitudes to Work and Family Roles: The Views of Young People at the Millennium." *Gender and Education* 17: 129–143.

Uggen, C., and A. Blackstone. 2004. "Sexual Harassment as a Gendered Expression of Power." *American Sociological Review* 69:64–92.

UNICEF. 2005. *Female Genital Mutilation/Cutting.* United Nations: United Nations Children's Fund. Available at http://www.unicef.org/publications

UNICEF. 2007. *The State of the World's Children: 2007.* United Nations: United Nations Children's Fund. Available at http://www.unicef.org/sowc07

United Nations. 2005. *Final Report and Appraisal of the Beijing Declaration and Platform for Action and the Outcome Document of the 23rd Special Session of the General Assembly.* Available at http://un.org/womenwatch/daw/Review

United Nations. 2000. *The World's Women 2000: Trends and Statistics.* New York: United Nations Statistics Division.

United Nations Population Fund. 2003. "Women and Gender Inequality." In *State of the World Population, 2002.* Available at http://www.unfpa.org/swp

UM (University of Michigan) News Service. 2005. *Working Moms Need to Negotiate Better Terms on Childcare Burden.* Available at http://www.umich.edu

Urban Institute. 2003 (September 2). *Gender Gap in Higher Education Focus of New Urban Institute Research.* Available at http://www.urban.org

U.S. Bureau of Labor Statistics. 2000. *Report on the Youth Labor Force.* Washington, DC: U.S. Department of Labor.

U.S. Bureau of Labor Statistics. 2005. *Kids' Careers.* Available at http://www.bls.gov

U.S. Bureau of Labor Statistics. 2007 (March 1). "Characteristics of Minimum Wage Workers." Available at http://www.bls.gov/

U.S. Bureau of the Census. 2005. *Statistical Abstract of the United States: 2004,* 124th ed. Washington, DC: U.S. Government Printing Office.

U.S. Bureau of the Census. 2007. *Statistical Abstract of the United States: 2006,* 126th ed. Washington, DC: U.S. Government Printing Office.

U.S. Department of Labor. 2006. *Women in the Labor Force: A Databook. Bureau of Labor Statistics.* Report 996. Washington, DC: U.S. Government Printing Service.

Vincent, Norah. 2006. *Self-Made Man.* New York: Viking Penguin.

White, Gayle. 2000. "Carter Cuts Ties to 'Rigid' Southern Baptists." *Atlanta Constitution-Journal,* October 20. Available at http://www.ajc.com

WHO (World Health Organization). 2006. *Gender and Women's Mental Health.* Available at http://www.who.int/mental_health

WHO. 2007. *Women and HIV/AIDS.* Available at http://www.who.int/gender

Williams, Christine L. 2007. "The Glass Escalator: Hidden Advantages for Men in the "Female" Occupations." In *Men's Lives,* 7th ed., Michael S. Kimmel and Michael Messner, eds. (pp. 242–255). Boston: Allyn and Bacon.

Williams, Christine L. 1995. *Still a Man's World: Men Who Do Women's Work.* Berkeley: University of California Press.

Williams, Joan. 2000. *Unbending Gender: Why Family and Work Conflict and What to Do About It.* Oxford: Oxford University Press.

Williams, John E., and Deborah L. Best. 1990. *Sex and Psyche: Gender and Self Viewed Cross-Culturally.* London: Sage.

Williams, Kristi, and Debra Umberson. 2004. "Marital Status, Marital Transitions, and Health: A Gendered Life Course Perspective." *Journal of Health and Social Behavior* 45:81–98.

Winfield, Nicole. 2000. "Activists Give U.S. Mixed Rating for Efforts on Gender." *Boston Globe,* June 8, p. A21.

Winters, Rebecca. 2005. "Harvard's Crimson Face." *Time,* January 31, p. 52.

Witt, S. D. 1996. "Traditional or Androgynous: An Analysis to Determine Gender Role Orientation of Basal Readers." *Child Study Journal* 26:303–318.

Women's International Network. 2000. "Reports from Around the World." *WIN News,* Autumn, pp. 50–58.

World Bank. 2003 (April). *Gender Equality and the Millennium Development Goals.* Gender and Development Group, World Bank.

World Global Issues. 2003. *Women.* Available at http://www.osearth.com

Yoder, Janice D., and Patricia Aniakudo. 1997 "Outsiders Within the Firehouse: Subordination and Difference in the Social Interactions of African American Women Firefighters." *Gender and Society* 11(3):324–341.

Yoder, P. Stanley, N. Abderrahim, and A. Zhuzhuni. 2004. *Female Genital Cutting in the Demographic and Health Surveys: A Critical and Comparative Analysis.* DHS Comparative Reports 7. Calverton, MD: ORC Macro.

Yumiko, Ehara. 2000. "Feminism's Growing Pains." *Japan Quarterly* 47:41–48.

Zakrzewski, Paul. 2005. "Daddy, What Did You Do in the Men's Movement?" Boston Globe, June 19. Available at http://www.bostonglobe.com/news/globe

Zimmerman, Marc A., Laurel Copeland, Jean Shope, and T. E. Dielman. 1997. "A Longitudinal Study of Self-Esteem: Implications for Adolescent Development." *Journal of Youth and Adolescence* 26(2):117–141.

Chapter 11

ACLU. 1999 (January 14). *The Rights of Lesbian, Gay, Bisexual, and Transgendered People.* Available at http://www.aclu.org

The Advocate. 2002 (August 20). *Advocate Sex Poll,* pp. 28–43.

Allport, G. W. 1954. *The Nature of Prejudice.* Cambridge, MA: Addison-Wesley.

Amato, Paul R. 2004. "Tension Between Institutional and Individual Views of Marriage." *Journal of Marriage and Family* 66:959–965.

American Anthropological Association. n.d. *Statement on Marriage and Family from the American Anthropological Association.* Available at http://www.aaanet.org

American College Health Association. 2007. "American College Health Association National College Health Assessment Spring 2006 Reference Group Data Report (Abridged)." *Journal of American College Health* 55(4):195–226.

Amnesty International. 2005. *Stonewalled: Police Abuse and Misconduct Against Lesbian, Gay, Bisexual, and Transgender People in the United States.* Available at http://www.amnesty.org

Bayer, Ronald. 1987. *Homosexuality and American Psychiatry: The Politics of Diagnosis,* 2nd ed. Princeton, NJ: Princeton University Press.

Besen, Wayne. 2000. "Introduction." In *Feeling Free: Personal Stories—How Love and Self-Acceptance Saved Us from "Ex-Gay" Ministries,* Human Rights Campaign, ed. (p. 7). Washington, DC: Human Rights Campaign Foundation.

Besen, Wayne. 2007. "The Politics of the Ex-Gay Movement." *Gay & Lesbian Review,* July–August, pp. 19–21.

Bobbe, Judith. 2002. "Treatment with Lesbian Alcoholics: Healing Shame and Internalized Homophobia for Ongoing Sobriety." *Health and Social Work* 27(3):218–223.

Bontempo, Daniel E., and Anthony R. D'Augelli. 2002. "Effects of At-School Victimization and Sexual Orientation on Lesbian, Gay, or Bisexual Youths' Health Risk Behavior." *Journal of Adolescent Health* 30:364–374.

Bradford, Judith, Kirsten Barrett, and Julie A. Honnold. 2002. *The 2000 Census and Same-Sex Households: A User's Guide.* New York: National Gay and Lesbian Task Force Policy Institute, Survey and Evaluation Research Laboratory, and Fenway Institute. Available at http://www.thetaskforce.org

Buhl, Larry. 2005 (June 16). "Youth's Blog Stirs Uproar Over 'Ex-Gay' Camp." *Planet Out.* Available at http://www.planetout.com

Button, James W., Barbara A. Rienzo, and Kenneth D. Wald. 1997. *Private Lives, Public Conflicts: Battles over Gay Rights in American Communities.* Washington, DC: CQ Press.

Cahill, Sean. 2005. *The Glass Nearly Half Full: 47% of U.S. Population Lives in Jurisdiction with Sexual Orientation Nondiscrimination Law.* National Gay and Lesbian Task Force. Available at http://www.thetaskforce.org

Cahill, Sean. 2007. "The Coming GLBT Senior Boom." *The Gay & Lesbian Review,* January–February, pp. 19–21.

Cahill, Sean, Ellen Mitra, and Sarah Tobias. 2002. *Family Policy: Issues Affecting Gay, Lesbian, Bisexual, and Transgender Families.* National Gay and Lesbian Task Force Policy Institute. Available at http://www.thetaskforce.org

Cahill, Sean, and Kenneth T. Jones. 2003. *Child Sexual Abuse and Homosexuality: The Long History of the 'Gays as Pedophiles' Fallacy.* National Gay and Lesbian Task Force. Available at http://www.thetaskforce.org

Cahill, Sean, and Samuel Slater. 2004. *Marriage: Legal Protections for Families and Children.* Policy brief. Washington, DC: National Gay and Lesbian Task Force Policy Institute.

Chase, Bob. 2000. *NEA President Bob Chase's Historic Speech from 2000 GLSEN Conference.* Available at http://www.glsen.org

Cianciotto, Jason, and Sean Cahill. 2003. *Educational Policy: Issues Affecting Lesbian, Gay, Bisexual, and Transgender Youth.* Washington, DC: National Gay and Lesbian Task Force Policy Institute.

Cianciotto, Jason, and Sean Cahill. 2007. "Anatomy of a Pseudo-Science." *The Gay & Lesbian Review,* July–August, pp. 22–24.

Colvin, Roddrick. 2004. "The Extent of Sexual Orientation Discrimination in Topeka, Kansas." National Gay and Lesbian Task Force Policy Institute. Available at http://www.thetaskforce.org

Curtis, Christopher. 2003 (October 7). "Poll: U.S. Public is 50–50 on Gay Marriage." *PlanetOut.* Available at http://www.planetout.com

Curtis, C. 2004. "Poll: 1 in 20 High School Students Is Gay." *PlanetOut.* Available at http://www.planetout.com

Curtis, Christopher. 2005 (August 31). "Group Links Katrina to Annual Gay Party." *Planet-Out*. Available at http://www.planetout.com

"Custody and Visitation." 2000. Human Rights Campaign FamilyNet. Available at http://familynet.hrc.org

D'Emilio, John. 1990. "The Campus Environment for Gay and Lesbian Life." *Academe* 76(1):16–19.

Diamond, Lisa M. 2003. "What Does Sexual Orientation Orient? A Biobehavioral Model Distinguishing Romantic Love and Sexual Desire." *Psychological Review* 110(1):173–192.

Dozetos, Barbara. 2001 (March 7). "School Shooter Taunted as 'Gay.'" *PlanetOut*. Available at http://www.planetout.com

Durkheim, Emile. 1993. "The Normal and the Pathological." In *Social Deviance*, Henry N. Pontell, ed. (pp. 33–63). Englewood Cliffs, NJ: Prentice-Hall. (Originally published in *The Rules of Sociological Method*, 1938.)

Esterberg, K. 1997. *Lesbian and Bisexual Identities: Constructing Communities, Constructing Selves*. Philadelphia: Temple University Press.

FBI (Federal Bureau of Investigation). 2006. *Hate Crime Statistics 2005*. Available at http://www.fbi.gov

Firestein, B. A. 1996. "Bisexuality as Paradigm Shift: Transforming Our Disciplines." In *Bisexuality: The Psychology and Politics of an Invisible Minority*, B. A. Firestein, ed. (pp. 263–291). Thousand Oaks, CA: Sage.

Fone, Byrne. 2000. *Homophobia: A History*. New York: Henry Holt.

Foreman, Matt. 2007. "Big Gains in GLBT Rights from Coast to Coast." *Gay & Lesbian Review Worldwide*, July–August, p. 5.

Frank, Barney. 1997. "Foreword." In *Private Lives, Public Conflicts: Battles over Gay Rights in American Communities*, by J. W. Button, B. A. Rienzo, and K. D. Wald (pp. i–xi). Washington, DC: CQ Press.

Frank, David John, and Elizabeth H. McEneaney. 1999. "The Individualization of Society and the Liberalization of State Policies on Same-Sex Relations, 1984–1995." *Social Forces* 77(3):911–944.

Franklin, Karen. 2000. "Antigay Behaviors Among Young Adults." *Journal of Interpersonal Violence* 15(4):339–362.

Garnets, L., G. M. Herek, and B. Levy. 1990. "Violence and Victimization of Lesbians and Gay Men: Mental Health Consequences." *Journal of Interpersonal Violence* 5:366–383.

Gates, Gary J., M. V. Lee Badgett, Jennifer Ehrle Macomber, and Kate Chambers. 2007 (March). *Adoption and Foster Care by Gay and Lesbian Parents in the United States*. Washington, DC: The Urban Institute.

Gilman, Stephen E., Susan D. Cochran, Vickie M. Mays, Michael Hughes, David Ostrow, and Ronald C. Kessler. 2001. "Risk of Psychiatric Disorders Among Individuals Reporting Same-Sex Sexual Partners in the National Comorbidity Survey." *American Journal of Public Health* 91(6):933–939.

GLSEN (Gay, Lesbian, and Straight Education Network). 2000. *Homophobia 101: Teaching Respect for All. Gay, Lesbian, and Straight Education Network*. Available at http://www.glsen.org

GLSEN. 2007 (July 2). *Number of Gay-Straight Alliance Registrations Passes 3,500*. Available at http://www.glsen.org

Goode, Erica E., and Betsy Wagner. 1993. "Intimate Friendships." *U.S. News and World Report,* July 5, pp. 49–52.

Green, John C. 2004. *The American Religious Landscape and Political Attitudes: A Baseline for 2004*. Pew Forum on Religion and Public Life. Available at http://pewforum.org

Harris Interactive and GLSEN. 2005. *From Teasing to Torment: School Climate in America, A Survey of Students and Teachers*. New York: GLSEN.

Held, Myka. 2005 (March 16). *Mix It Up: T-Shirts and Activism*. Available at http://www.tolerance.org/teens

Holthouse, David. 2005 (Spring). "Curious Cures." *Intelligence Report* 117:14.

Human Rights Campaign. 2000. *Feeling Free: Personal Stories—How Love and Self-Acceptance Saved Us from "Ex-Gay" Ministries*. Washington, DC: Human Rights Campaign Foundation.

Human Rights Campaign. 2005a (March 22). "Recruiting Age Bump Highlights Consequences of "Don't Ask, Don't Tell." HRC press release. Available at http://www.hrc.org

Human Rights Campaign. 2005b. *Statewide Laws or Policies Affecting Schools & Educational Institutions*. Available at http://www.hrc.org

Human Rights Campaign. 2007a. *The State of the Workplace for Lesbian, Gay, Bisexual, and Transgendered Americans, 2006–2007*. Washington, DC: Human Rights Campaign. Available at http://www.hrc.org

Human Rights Campaign. 2007b. *State Prohibitions on Marriage for Same-Sex Couples*. Available at http://www.hrc.org

Human Rights Campaign. 2007c. *Parenting Laws: Joint Adoption*. Available at http://www.hrc.org

Human Rights Campaign. 2007d. *Parenting Laws: Second Parent Adoption*. Available at http://www.hrc.org

Human Rights Watch. 2001 (March). *World Report 2001*. Available at http://www.hrw.org

Intelligence Briefs. 2006. "Phelps' Latest Tactic Shocks Americans." *SPLC Report*, p. 3.

International Gay and Lesbian Human Rights Commission. 1999. *Antidiscrimination Legislation*. Available at http://www.iglhrc.org

International Gay and Lesbian Human Rights Commission. 2004 (January 30). *IGLHRC Calls for Global Mobilization to Help Pass the United Nations Resolution on Sexual Orientation and Human Rights*. Available at http://www.iglhrc.org

Israel, Tania, and Jonathan J. Mohr. 2004. "Attitudes Toward Bisexual Women and Men: Current Research, Future Directions." In *Current Research on Bisexuality*, Ronald C. Fox, ed. (pp. 117–134). New York: Harrington Park Press.

Johnston, Eric. 2003 (October 14). "Clinton Blasts Bush's Gay Marriage Attack." *Planet-Out*. Available at http://www.planetoutcom

Johnston, Eric. 2005 (May 5). "Mass. Releases Data on Same-Sex Marriages." *PlanetOut*. Available at http://www.planetout.com

Kilman, Carrie. 2007 (Spring). "This Is Why We Need a GSA." *Teaching Tolerance*, pp. 30–37.

Kinsey, A. C., W. B. Pomeroy, and C. E. Martin. 1948. *Sexual Behavior in the Human Male*. Philadelphia: W. B. Saunders.

Kinsey, A. C., W. B. Pomeroy, C. E. Martin, and P. H. Gebhard. 1953. *Sexual Behavior in the Human Female*. Philadelphia: W. B. Saunders.

Kirkpatrick, R. C. 2000. "The Evolution of Human Sexual Behavior." *Current Anthropology* 41(3):385–414.

Kosciw, Joseph G., and Elizabeth M. Diaz. *The 2005 National School Climate Survey*. New York: GLSEN.

Landis, Dan. 1999 (February 17). "Mississippi Supreme Court Made a Tragic Mistake in Denying Custody to Gay Father, Experts Say." American Civil Liberties Union News. Available at http://www.aclu.org

LaSala, M. 2004. "Extradyadic Sex and Gay Male Couples: Comparing Monogamous and Nonmonogamous Relationships." *Families in Society: The Journal of Contemporary Human Services* 85:405–415.

LaSala, M. 2005. "Monogamy of the Heart: A Qualitative Study of Extradyadic Sex among Gay Male Couples." *Journal of Gay and Lesbian Social Services* 17:1–24.

LAWbriefs. 2000 (Fall). "Recent Developments in Sexual Orientation and Gender Identity Law." *LAWbriefs* 3(3).

LAWbriefs. 2003 (Spring). "Recent Developments in Sexual Orientation and Gender Identity Law." *LAWbriefs* 6(1).

LAWbriefs. 2005a (April). "Recent Developments in Sexual Orientation and Gender Identity Law." *LAWbriefs* 7(1).

LAWbriefs. 2005b (July). "Recent Developments in Sexual Orientation and Gender Identity Law." *LAWbriefs* 7(2).

Lever, Janet. 1994. "The 1994 Advocate Survey of Sexuality and Relationships: The Men." *The Advocate*, August 23, pp. 16–24.

Loftus, Jeni. 2001. "America's Liberalization in Attitudes Toward Homosexuality, 1973 to 1998." *American Sociological Review* 66:762–782.

Louderback, L. A., and B. E. Whitley. 1997. "Perceived Erotic Value of Homosexuality and Sex-Role Attitudes as Mediators of Sex Differences in Heterosexual College Students' Attitudes Toward Lesbians and Gay Men." *Journal of Sex Research* 34:175–182.

Lyall, Sarah. 2005. "New Course by Royal Navy: A Campaign to Recruit Gays." *New York Times*, February 22. Available at http://www.nytimes.com

McCammon, Susan, David Knox, and Caroline Schacht. 2004. *Choices in Sexuality*, 2nd ed. Cincinnati, OH: Atomic Dog.

McKinney, Jack. 2004 (February 8). "The Christian Case for Gay Marriage." Pullen Memorial Baptist Church. Available at http://www.pullen.org

Michael, Robert T., John H. Gagnon, Edward O. Laumann, and Gina Kolata. 1994. *Sex in America: A Definitive Survey*. Boston: Little, Brown.

Miller, Patti, McCrae A. Parker, Eileen Espejo, and Sarah Grossman-Swenson. 2002. *Fall Colors: Prime Time Diversity Report 2001–02*. Oakland CA: Children Now and The Media Program.

Mohipp, C., and M. M. Morry. 2004. "Relationship of Symbolic Beliefs and Prior Contact to Heterosexuals' Attitudes Toward Gay Men and Lesbian Women." *Canadian Journal of Behavioral Science* 36(1):36–44.

Moltz, Kathleen. 2005 (April 13). *Testimony of Kathleen Moltz*. Testimony given before the United States Senate Judiciary Committee, Subcommittee on the Constitution, Civil Rights and Property Rights. Human Rights Campaign. Available at http://www.hrc.org

Moser, Bob. 2005 (Spring). "Holy War." *Intelligence Report* 117:8–21.

National Coalition of Anti-Violence Programs. 2007. *Anti-Lesbian, Gay, Bisexual and Transgender Violence in 2006*. New York: National Coalition of Anti-Violence Programs.

National Gay and Lesbian Task Force. 2007 (May). *Hate Crime Laws in the U.S. National Gay and Lesbian Task Force*. Available at http://www.thetaskforce.org

"NCLR Wins Equal Tax Benefits for Nonbiological Lesbian Mother." 2000 (Fall). *NCLR Newsletter* (National Center for Lesbian Rights), pp. 1 and 10. Available at http://www.nclrights.org

Neidorf, Shawn and Rich Morin. 2007 (May 23). "Four-in-ten Americans Have Close Friends or Relatives Who Are Gay." Pew Research Center for People & the Press. Available at http://pewresearch.org

Newport, Frank. 2002 (September). "In-Depth Analysis: Homosexuality." Gallup Organization. Available at http://www.gallup.com

Newport, Frank. 2007 (May 17). "Public Favors Expansion of Hate Crime Law to Include Sexual Orientation." Gallup News Service. Available at http://www.gallup.com

NPR/Kaiser/Kennedy School Poll. 2004. *Sex Education in America*. Available at http://www.kff.org

Ottosson, Daniel. 2006. *LGBT World Legal Wrap Up Survey*. International Lesbian and Gay Association. Available at http://www.ilga.org

Ottosson, Daniel. 2007. *A World Survey of Laws Prohibiting Same-Sex Activity between Consenting Adults*. International Lesbian and Gay Association. Available at http://www.ilga.org

Page, Susan. 2003. "Gay Rights Tough to Sharpen into Political 'Wedge Issue.'" *USA Today*, July 28, p. 10A.

Parker, Laura. 2007. "Case Involves a Collision of Rights." *USA Today*, August 3, p. 3A.

Patel, S. 1989. *Homophobia: Personality, Emotional, and Behavioral Correlates*. Master's thesis, East Carolina University, Greenville, NC.

Patel, S., T. E. Long, S. L. McCammon, and K. L. Wuensch. 1995. "Personality and Emotional Correlates of Self-Reported Antigay Behaviors." *Journal of Interpersonal Violence* 10:354–366.

Pathela, Preeti, Anjum Hajat, Julia Schillinger, Susan Blank, Randall Sell, and Farzad Mostashari. 2006. "Discordance between Sexual Behavior and Self-Reported Sexual Identity: A Population-Based Survey of New York City Men." *Annals of Internal Medicine* 145(6):416–425.

Patterson, Charlotte J. 2001. "Family Relationships of Lesbians and Gay Men." In *Understanding Families into the New Millennium: A Decade in Review*, Robert M. Milardo, ed. (pp. 271–288). Minneapolis, MN: National Council on Family Relations.

Paul, J. P. 1996. "Bisexuality: Exploring/Exploding the Boundaries." In *The Lives of Lesbians, Gays, and Bisexuals: Children to Adults*, R. Savin-Williams and K. M. Cohen, eds. (pp. 436–461). Fort Worth, TX: Harcourt Brace.

Pew Research Center. 2006 (March 22). "Less Opposition to Gay Marriage, Adoption, and Military Service." Available at http://people-press.org

Pew Research Center. 2007 (March 22). *Trends in Political Values and Core Attitudes: 1987–2007*. Washington, DC: Pew Research Center for People and the Press.

Platt, Leah. 2001. "Not Your Father's High School Club." *American Prospect* 12(1): A37–A39.

Plugge-Foust, C., and George Strickland. 2000. "Homophobia, Irrationality, and Christian Ideology: Does a Relationship Exist?" *Journal of Sex Education and Therapy* 25:240–244.

Potok, Mark. 2005 (Spring). "Vilification and Violence." *Intelligence Report* 117:1.

Price, Jammie, and Michael G. Dalecki. 1998. "The Social Basis of Homophobia: An Empirical Illustration." *Sociological Spectrum* 18:143–159.

Pryor, John H., Sylvia Hurtado, Victor B. Saenz, Jessica S. Korn, Jose Luis Santos, and William S. Korn. 2006. *The American Freshman: National Norms for Fall 2006*. Los Angeles: Higher Education Research Institute, UCLA.

Puccinelli, Mike. 2005 (April 19). "Students Support, Decry Gays with T-Shirts." CBS 2 Chicago. Available at http://cbs2chicago.com

Rankin, Susan R. 2003. *Campus Climate for Gay, Lesbian, Bisexual, and Transgendered People: A National Perspective*. New York: National Gay and Lesbian Task Force Policy Institute. Available at http://www.thetaskforce.org

Roderick, T. 1994. *Homonegativity: An Analysis of the SBS-R*. Master's thesis, East Carolina University, Greenville, NC.

Roderick, T., S. L. McCammon, T. E. Long, and L. J. Allred. 1998. "Behavioral Aspects of Homonegativity." *Journal of Homosexuality* 36:79–88.

Rosario, M., E. W. Schrimshaw, J. Hunter, and L. Braun. 2006. "Sexual Identity Development among Lesbian, Gay, and Bisexual Youth: Consistency and Change over Time." *Journal of Sex Research* 43:46–58.

Rosin, Hanna, and Richard Morin. 1999 (January 11). "In One Area, Americans Still Draw a Line on Acceptability." *Washington Post National Weekly Edition* 16(11):8.

Saad, Lydia. 2005 (May 20). "Gay Rights Attitudes a Mixed Bag." Gallup Organization. Available at http://www.gallup.com

Saad, Lydia. 2007 (May 29). "Tolerance for Gay Rights at High Water Mark." Gallup News Service. Available at http://www.gallup.com

Safe Zone Resources. 2003. *National Consortium of Directors of LGBT Resources in Higher Education.* Available at http://www.lgbtcampus.org

Sanday, Peggy R. 1995. "Pulling Train." In *Race, Class, and Gender in the United States*, 3rd ed., P. S. Rothenberg, ed. (pp. 396–402). New York: St. Martin's Press.

Sanders, Douglas. 2007 (May 25). "What Role Has International Law Played in the LGBT Movement?" International Lesbian and Gay Association. Available at http://www.ilga.org

Savin-Williams, R. C. 2006. "Who's Gay? Does It Matter?" *Current Directions in Psychological Science* 15:40–44.

Schiappa, E., P. B. Gregg, and D. E. Hewes. 2005. "The Parasocial Contact Hypothesis." *Communication Monographs* 72(1):92–115.

Servicemembers Legal Defense Network. 2007. *Fact Sheet: The Military Readiness Enhancement Act (HR 1246).* Available at http://www.sldn.org

Shernoff, Michael. 2006. "Negotiated Nonmonogamy and Male Couples." *Family Process* 45(4):407–418.

SIECUS (Sexuality Information and Education Council of the United States). 2000. *Fact Sheets: Sexual Orientation and Identity.* New York: SIECUS.

SIECUS. 2006. *State Profiles (2006): A Portrait of Sexuality Education and Abstinence-Only-Until-Marriage Programs in the States.* Available at http://www.siecus.org

Simmons, Tavia, and Martin O'Connell. 2003 (February). *Married-Couple and Unmarried-Partner Households: 2000.* Washington, DC: U.S. Census Bureau.

Singh, Daniel P., and Heather D. Wathington. 2003. "Valuing Equity: Recognizing the Rights of the LGBT Community." *Diversity Digest* 7(1–2):8–9.

Smith, D. M., and G. J. Gates. 2001. *Gay and Lesbian Families in the United States: Same-Sex Unmarried Partner Households—Preliminary Analysis of 2000 U.S. Census Data.* Washington, DC: Human Rights Campaign. Available at http://www.hrc.org

Stotzer, Rebecca. 2007 (June). "Comparison Of Hate Crime Rates across Protected and Unprotected Groups." The Williams Institute at UCLA Law School. Available at https://www.law.ucla.edu/williamsinstitute

Sullivan, A. 1997. "The Conservative Case." In *Same-Sex Marriage: Pro and Con*, A. Sullivan, ed. (pp. 146–154). New York: Vintage Books.

Sullivan, A. 2003. "Legalizing Same-Sex Marriage Would Strengthen Marriage." In *The Family: Opposing Viewpoints*, A. Ojeda, ed. (pp. 196–200). Farmington Hill, MI: Greenhaven Press.

Sylivant, S. 1992. *The Cognitive, Affective, and Behavioral Components of Adolescent Homonegativity.* Master's thesis, East Carolina University, Greenville, N.C.

Thompson, Cooper. 1995. "A New Vision of Masculinity." In *Race, Class, and Gender in the United States*, 3rd ed., P. S. Rothenberg, ed. (pp. 475–481). New York: St. Martin's Press.

Tobias, Sarah, and Sean Cahill. 2003. *School Lunches, the Wright Brothers, and Gay Families.* National Gay and Lesbian Task Force. Available at http://www.thetaskforce.org

The United Methodist Church and Homosexuality. 1999. Available at www.religioustolerance.org/hom_umc.htm

White, Mel. 2005 (Spring). "A Thorn in Their Side." *Intelligence Report* 117:27–30.

Wilcox, Clyde, and Robin Wolpert. 2000. "Gay Rights in the Public Sphere: Public Opinion on Gay and Lesbian Equality." In *The Politics of Gay Rights*, Craig A. Rimmerman, Kenneth D. Wald, and Clyde Wilcox, eds. (pp. 409–432). Chicago: University of Chicago Press.

Yang, Alan. 1999. *From Wrongs to Rights 1973 to 1999: Public Opinion on Gay and Lesbian Americans Moves Toward Equality.* New York: Policy Institute of the National Gay and Lesbian Task Force.

Yvonne. 2004 (December 1). *Devout Christian Finds a Reason to Stand Up for Equality.* Human Rights Campaign. Available at http://www.hrc.org

Chapter 12

AAHSA (American Association of Homes and Service for the Aging). 2007. *Aging Services: The Facts.* Available at http://www.aahsa.org

AARP. 2003. "Fighting Ageism." *AARP Magazine*, July-August. Available at http://www.aarpmagazine.org

AARP. 2006 (October 26). "AARP Announces 37 Millionth Member." News Release. Available at http://www.aarp.org

AARP. 2007. *Aging in Asia and Oceania: AARP Multinational Survey of Opinion Leaders 2006, United States Country Report.* Washington, DC: Princeton Survey Research Associates.

Abrahms, Sally. 2006. "Not Home Alone." *Time*, Oct. 23, pp. F3–F5.

ACF (Administration for Children and Families). 2006. *The AFCARS Report.* Available at http://www.acf.hhs.gov

ACF. 2007. *Summary: Child Maltreatment 2005.* Available at http://www.acf.hhs.gov

AHA (American Heart Association). 2005. *Childhood Obesity: Early Prevention Offers Best Solution.* Available at http://www.americanheart.org

"AIDS Creating Global 'Orphans Crisis.'" 2003. CBS News, July 10. Available at http://www.cbsnews.com

Allen, Claude A. 2002. *Speech to Trafficking Convention.* Speech presented in Honolulu, Hawaii, November 8. Available at http://www.iabolish.com

Anderson, Craig, and Karen Dill. 2000. "Video Games and Aggressive Thoughts, Feelings and Behavior in the Laboratory and in Life." *Journal of Personality and Social Psychology* 78(4):772–790.

Anderson, Mark H. 2007 (June 4). "U.S. Supreme Court to Review FedEx Discrimination Lawsuit." Market Watch. Available at http://www.marketwatch.com

Angel, Jacqueline, Maren Jimenez, and Ronald Angel. 2007. "The Economic Consequences of Widowhood for Older Minority Women." *The Gerontologist* 47(2):224–235.

Annual Report. 2007. *Summary of Oregon's Death with Dignity Act—2006.* Available at http://egov.oregon.gov/DHS

AOA (Administration on Aging). 2001. *The Older Americans Act.* Available at http://www.aoa.gov

AOA. 2003. *Older Americans Act. Fact Sheets.* Available at http://www.aoa.gov

AOA. 2004. *A Profile of Older Americans: 2004.* Washington, DC: U.S. Department of Health and Human Services. Available at http://www.aoa.dhhs.gov

AOA. 2007. *A Profile of Older Americans: 2006.* Washington, DC: U.S. Department of Health and Human Services. Available at http://www.aoa.gov

APA (American Psychological Association). 2007. *Elder Abuse and Neglect: In Search of Solutions.* Available at http://www.apa.org

AP. 2006. "Study: Drugs Better for Elderly Depression." *USA Today*, March 16. Available at http://www.usatoday.com

Atchley, Robert C. 2000. *Social Forces and Aging.* Belmont, CA: Wadsworth.

Avert. 2007. *AIDS Orphans.* Available at http://www.avert.org

Bauder, David. 2006 (November 19). *Study: TV's Youth Obsession Backfiring.* Available at http://www.breitbart.com

Benoit, Bertrand. 2007. "Germany to Crack Down on Violent Video Games." *Financial Times*, December 6. Available at http://www.ft.com

Breton, Tracy. 2007. "A Trust Betrayed: How Two Women in Their 90s Were Fleeced by a Helper." *Providence Journal*, January 9. Available at http://www.projo.com

Binstock, Robert H. 1986. "Public Policy and the Elderly." *Journal of Geriatric Psychiatry* 19:115–143.

Birnbaum, Jeffrey. 2005. "AARP Leads with Wallet in Fight over Social Security." *Washington Post,* March 29. Available at http://www.washingtonpost.com

BLR (Business and Legal Reports). 2006 (April 26). "Survey Finds Generational Gap Among Workers." Available at http://hr.blr .com

Bradley, Ed. 2005. "Can a Video Game Lead to Murder?" CBS News, March 6. Available at http://www.cbsnews.com

Brazzini, D. G., W. D. McIntosh, S. M. Smith, S. Cook, and C. Harris. 1997. "The Aging Woman in Popular Film: Underrepresented, Unattractive, Unfriendly, and Unintelligent." *Sex Roles* 36:531–543.

Brooks-Gunn, Jeanne, and Greg Duncan. 1997. "The Effects of Poverty on Children." *Future of Children* 7 (2):55–70.

Butler, Robert. 2007. "Who Will Care for You?" *AARP Bulletin,* January, p. 41.

Carasso, Adam, C. Eugene Steuerle, and Gillian Reynolds. 2007. "Kids' Share: Executive Summary." Urban Institute. Available at http://www.urban.org

Carroll, Joseph. 2005 (May 17). *Retirement Age Expectations and Realities.* Gallup Organization. Available at http://gallup.com

Casa Alianza. 2003. *Living in the Streets.* Available at http://www.casa-alianza.org

CDF (Children's Defense Fund). 2000. *Child Care Now!* Available at http://www .childrensdefense.org

CDF. 2003a. *Among Industrialized Countries, the United States Ranks . . . .* Available at http://www.childrensdefense.org

CDF. 2003b. *Media Violence. Fact Sheet.* Available at http://www.childrensdefense .org

CDF. 2004 (October). *2003 Facts on Child Poverty in America.* Available at http://www .childrensdefense.org

CDF. 2006. *Poverty.* Available at http://www .childrensdefensefund.org/

CDF. 2007a. *Children's Defense Fund Releases Annual Gun Violence Report.* Available at http://www.childrensdefense.org

CDF. 2007b. *Each Day in America.* Available at http://www.childrensdefense.org

CDF. 2007c. *The All Health Children Act (H.R.1688): An Achievable, Smart and Right Goal.* Healthy Child Campaign. Available at http://childrensdefense.org

Child Trends. 2006. *Food Stamp Receipt.* Child Trends Databank. Available at http:// www.childtrendsdatabank.org

Children's Rights. 2007. *Children's Rights Today: Federal Court Approves Settlement for Mississippi's Foster Children, Agreement Calls for System Overhaul.* Available at http://www.childrensrights.org

Clemmitt, Marcia. 2006. "Caring for the Elderly." *Congressional Researcher* 16(36):1–12.

CWLA (Children's Welfare League of America). 2005. *Quick Facts About Foster Care.* Available at http://www.cwla.org

Death with Dignity Act. 2006. *FAQs about the Death with Dignity Act.* Available at http:// www.oregon.gov/DHS/ph/pas

Dewan, Shaila. 2005. "Abused Children Are Found Entitled to Legal Aid." *New York Times,* February 9. Available at http://www .nytimes.com

Donlan, Maggie M., Ori Ashman, and Becca R. Levy. 2005. "Re-Vision of Older Television Characters: A Stereotype-Awareness Intervention." *Journal of Social Issues* 61(2):307–319.

Dorman, Peter. 2001. *Child Labour in the Developed Economies.* Geneva: International Labour Office.

Dove. 2007. *Campaign for Real Beauty Uncovers the Beauty of Women over 50.* Available at http://www.campaignforrealbeauty.com

Duncan, Greg, W. Jean Yeung, Jeanne Brooks-Gunn, and Judith Smith. 1998. "How Much Does Childhood Poverty Affect the Life Chance of Children?" *American Sociological Review* 63:402–423.

Elder Justice Act of 2007. *Summary.* Available at http://www.elderjusticeact.org

"Elderly Become 'Muti' Targets." 2003. *African Eye News,* April 15.

EEOC (Equal Employment Opportunity Commission). 2007a. *Age Discrimination.* Available at http://www.eeoc.gov/types

EEOC (Equal Employment Opportunity Commission). 2007b. *Age Discrimination in Employment Charges.* Available at http://www .eeoc.gov/stats

EurekAlert. 2003 (April 22). *Children's Stereotypes of Aging Start Early.* Available at http://www.globalaging.org

EWC (Elder Wisdom Circle). 2007. *The Elder Wisdom Circle Guide for a Meaningful Life.* Available at http://www.elderwisdomcircle .org

FBI (Federal Bureau of Investigation). 2006. *Crime in the United States, 2005. Uniform Crime Reports.* Washington, DC: U.S. Government Printing Office.

Federal Elections Commission. 2005. *Voter Registration and Turnout by Age, Gender, and Race.* Available at http://www.fec.gov

Federal Interagency Forum on Aging-Related Statistics. 2007. *Older American Update 2006: Key Indicators of Well-Being.* Washington DC: U.S. Government Printing Office.

FHEO (Federal Housing and Economic Opportunity Office). 2006. "About FHEO." U.S. Department of Housing and Urban Development. Available at http://www.hud.gov

Flaherty, Kristina. 2004. "Abused Seniors Turn to Alameda Court for Help." *California Bar Journal,* October. Available at http:// www.calbar.ca.gov

Forum on Child and Family Statistics. 2006. *America's Children in Brief: Key National Indicators of Well-Being, 2006.* Washington, DC: Federal Interagency Forum on Child and Family Statistics.

Foundation for Child Development. 2007. *The Foundation for Child Development and Youth Well-Being Index (CWI), 1975–2005, with Projections for 2006.* Kenneth Land, Project Coordinator. Durham, NC.

Gallup. 2004. *Children and Violence.* Gallup Poll. Available at http://www.galluppoll .com

Gallup. 2005. *Social Security.* Gallup Poll. Available at http://brain.galluppoll.com

Generations United. 2007a. *About Us.* Available at http://www.gu.org

Generations United. 2007b. *Congress Reauthorizes Older American Act.* Available at http://www.gu.org

Goldberg, Beverly. 2000. *Age Works.* New York: Free Press.

Green, Shane. 2003. "Hidden Abuse of Elderly Emerging Problem for Japan." *Sydney Morning Herald,* June 21. Available at http:// www.globalaging.org/elderrights

Greenhouse, Linda. 2005. "Supreme Court Removes Hurdle to Age Bias Suits." *New York Times,* March 31. Available at http://www .nytimes.com

Harder, Ben. 2006. "Here's to Good Buddies, Good Health." *Los Angeles Times,* October 16. Available at http://www.latimes.com

Harris, Diana K. 1990. *Sociology of Aging.* New York: Harper and Row.

Harris, Kathleen, and Jeremy Marmer. 1996. "Poverty, Paternal Involvement, and Adolescent Well-Being." *Journal of Family Issues* 17(5):614–640.

Hillier, Susan, and Georgia Barrow. 2007. *Aging, the Individual, and Society.* Belmont, CA: Wadsworth.

Hooyman, Nancy R., and H. Asuman Kiyak. 1999. *Social Gerontology: A Multidisciplinary Perspective,* 2nd ed. Boston: Allyn and Bacon.

Human Rights Watch. 2001 (January). *Underage and Unprotected: Child Labor in Egypt's Cotton Fields. Human Rights Watch* 13 (1). Available at http://www.hrw.org/ reports/2001/Egypt

Human Rights Watch. 2003. *Child Farmworkers.* Available at http://www.hrw.org/ campaigns

Human Rights Watch. 2005. *Orphans and Abandoned Children.* Available at http:// www.hrw.org

Human Rights Watch. 2007 (April 17). *California Takes Step to End Life without Parole for Children*. Available at http://www.hrw.org/

ILO (International Labour Office). 2006. "The End of Child Labour: Within Reach." International Labour Conference, 95th Session. Report I (B). Geneva, Switzerland.

Jackson, Maggie. 2003. "More Sons Are Juggling Jobs and Care for Parents." *New York Times,* June 15. Available at http://www.globalaging.org/elderrights

Jones, Jeffrey M. 2005 (May 2). "Retirement Expectations, Reality Not Entirely Consistent." Gallup Organization. Available at http://galluppoll.com

Kaiser Family Foundation. 2005 (March 9). *Media Multi-Tasking Changing the Amount and Nature of Young People's Media Use.* Available at http://www.kff.org/entmedia

Kaiser Family Foundation. 2007. "Total Number of Medicare Beneficiaries." State Health Facts. Available at http://www.statehealthfacts.org

Liedtke, Michael. 2007. "New Search Engine for Aging Boomers." Associated Press, January 10. Available at http://www.breitbart.com

Lyst, Catherine. 2007. "Hard Life for African Street Children." British Broadcasting Corporation. May 31. Available at http://newsvote.bbc.co.uk

Mack, Brandy, and Kathi Jones. 2003 (August 5). "Elder Abuse: Identification and Prevention." *ASHA Leader* 8:10A–12A.

Matras, Judah. 1990. *Dependency, Obligations, and Entitlements: A New Sociology of Aging, the Life Course, and the Elderly.* Englewood Cliffs, NJ: Prentice-Hall.

McPherson, Miller, Lynn Smith-Lovin, and Matthew E. Brashears. 2006. "Social Isolation in America: Changes in Core Discussion Networks over Two Decades." *American Sociological Review* 71(3):353–375.

Miner, Sonia, John Logan, and Glenna Spitze. 1993. "Predicting Frequency of Senior Center Attendance." *The Gerontologist* 33:650–657.

Mittal, Vicki, Jules Rosen, Rahul Govind, Howard Degenholtz, Sunil Shingala, Shelley Hulland, Young Joo Rhee, Kari B. Kastango, Benoit H. Mulsant, Nick Castle, Fred Rubin, and David Nace. 2007. "Perception Gap in Quality of Life Ratings: An Empirical Investigation of Nursing Home Residents and Caregivers." *The Gerontologist* 47(2):159–169.

Morin, Richard, and Jim VandeHei. 2005. "Social Security Plan's Support Dwindling." *Washington Post,* June 9. Available at http://www.washingtonpost.com

NBC News/*Wall Street Journal.* 2007. *Problems and Priorities.* Available at http://www.pollingreport.com

NCAL (National Center for Assisted Living). 2006. *Assisted Living Resident Profile.* Available at http://www.ncal.org/about

NCAL (National Center for Assisted Living). 2007. *What Is Assisted Living?* Available at http://www.ncal.org/consumer

NCL (National Consumer's League). 2007. *NCL's 2007 Five Worse Teen Jobs.* Available at http://nclnet.org/labor

NIMH (National Institute of Mental Health). 2006. *Suicide in the U.S.: Statistics and Prevention.* Available at http://www.nimh.nih.gov

NMHA (National Mental Health Association). 2003. *Children's Mental Health Statistics.* Available at http://www.nimh.org

NYS Division of Parole. 2007. *Murtaugh, Patricia Decision.* Albany, NY: New York State Division of Parole.

ORC (Ombudsmen Resource Center). 2007. "About the Ombudsman Resource Center." National Long Term Care Ombudsmman Resource Center. Available at http://www.ltcombudsman.org

Palmore, Erdman B. 1984. "The Retired." In *Handbook on the Aged in the United States,* Erdman B. Palmore, ed. (pp. 63–75). Westport, CT: Greenwood Press.

Palmore, Erdman B. 1999. *Ageism: Negative and Positive.* New York: Springer.

PBS. 2005. *The Last Hope.* Available at http://www.pbs.org

PRB (Population Reference Bureau) 2007. "Prescription Drugs and Medicare." *Today's Research on Aging* 2:1–4.

Pear, Robert. 2002. "9 out of 10 Nursing Homes Lack Adequate Staff, Study Finds." *New York Times,* February 18. Available at http://www.globalaging.org/elderrights

Pear, Robert. 2007a. "Oversight of Nursing Homes Is Criticized." *New York Times,* April 22. Available at http://www.nytimes.com

Pear, Robert. 2007b. "AARP Says It Will Become Major Medicare Insurer While Remaining a Consumer Lobby." *New York Times,* April 17. Available at http://www.nytimes.com

Peterson, Karen. 2003. "Kids Are Better Off Than Adults Believe." *USA Today,* July 2. Available at http://www.usatoday.com

Pew. 2004. "Foster Care: Voices from the Inside." Commissioned by the Pew Commission on Children in Foster Care. Pew Charitable Trust, St. Louis, MO.

Pew. 2007. *Time for Reform: Aging Out and on Their Own.* Pew Charitable Trust and Jim Casey Youth Opportunities Initiative, St. Louis, MO.

Porter, Eduardo, and Mary Walsh. 2005. "Retirement Turns into a Rest Stop as Benefits Dwindle." *New York Times,* February 9. Available at http://www.nytimes.com

Quadagno, Jill. 2005. *Aging and the Life Course.* Boston: McGraw-Hill.

"Queens: Woman Sentenced for Stealing from the Elderly." 2006. *New York Times,* September 2. Available at http://www.nytimes.com

Ramirez, Eddy. 2002. "Ageism in the Media Is Seen as Harmful to Health of Elderly." *Los Angeles Times,* September 5. Available at http://www.globalaging.org

Reio, Thomas G., and Joanne Sanders-Reto. 1999. "Combating Workplace Ageism." *Adult Learning* 11:10–13.

Riley, Matilda White. 1987 (February). "On the Significance of Age in Sociology." *American Sociological Review* 52:1–14.

Riley, Matilda W., and John W. Riley. 1992. "The Lives of Older People and Changing Social Roles." In *Issues in Society,* Hugh Lena, William Helmreich, and William McCord, eds. (pp. 220–231). New York: McGraw-Hill.

Robertson, Anne. 2006. "Monday's with Mother: An Alzheimer's Story." Available at http://www.annerobertson.com/blog3.html

Runyan, Carol W., Michael Schulman, Janet Dal Santo, Michael Bowling, Robert Agans, and Ta Myduc. 2007. "Work Related Hazards and Workplace Safety of U.S. Adolescents Employed in the Retail and Service Sectors. Pediatrics 119(3):526–534.

Saad, Lydia. 2006 (November 30). "Growing Old Doesn't Necessarily Mean Growing Infirm." Gallup Poll. Available at http://www.galluppoll.com

Saad, Lydia. 2007 (June 4). "Americans Rate the Morality of 16 Social Issues." Gallup Poll. Available at http://www.galluppoll.com

Safire, William. 2007. "Donut Hole." *New York Times,* February 25. Available at http://www.nytimes.com

Saul, Stephanie. 2007. "Taking on Alzheimer's." *New York Times,* June 10. Available at http://www.nytimes.com

Save the Children. 2007. *State of the World's Mothers.* Available at http://www.savethechildren.org

Seeman, Teresa E., and Nancy Adler. 1998 (Spring). "Older Americans: Who Will They Be?" *National Forum,* pp. 22–25.

Senior Nutrition Act of 2007. *Summary.* Available at http://thomas.loc.gov

Shanks, Jonathan. 2007. *Letter for Parole Hearing for Patricia Murtaugh.* Available at http://www.projo.com

Sloan, Allan, 2005. "Social Security: A Daring Leap." *Newsweek,* February 14, pp. 41–45.

Sloan, Allan, 2006. "Bush's Social Security Sleight of Hand." *Washington Post,* February 8, p. D2.

"Snapshots of the Elderly." 2000. *Washington Post.* Available at http://www.washingtonpost.com

Social Security. 2007. "2006 Social Security/SSI Information." Social Security Online. Available at http://www.ssa.gov

SOWC (State of the World's Children). 2007. *United Nations Children's Fund.* Available at http://www.unicef.org

Statistical Handbook of Japan. 2006. "Population." Ministry of Internal Affairs and Communications. Statistics Bureau and Statistical Research and Training Institute, Japan.

Stich, Sally. 2006. "Online Wisdom." *Time,* November 30, F7.

Taylor, Curtis. 2006. "Routine Path to Long Life." *Newsday,* July 12. Available at http://www.newsday.com

Thurow, Lester C. 1996. "The Birth of a Revolutionary Class." *New York Times Magazine,* May 19, pp. 46–47.

Toner, Robin. 1999. "A Majority over 45 Say Sex Lives Are Just Fine." *New York Times,* August 4, p. A10.

Torres-Gil, Fernando. 1990. "Seniors React to Medicare Catastrophic Bill: Equity or Selfishness?" *Journal of Aging and Social Policy* 2(1):1–8.

Totenberg, Nina. 2005. "Supreme Court Ends Death Penalty for Juveniles." National Public Radio, March 22. Available at http://www.npr.org

UNICEF. 2007a. "An Overview of Child Well-Being in Rich Countries." The United Nations Children's Fund. Available at http://www.unicef.org

UNICEF. 2007b (January 12). *Child Rights.* Available at http://www.unicefusa.org

United Nations. 1998. *The First Nearly Universally Ratified Human Rights Treaty in History.* Status. Washington, DC: UNICEF. Available at http://www.unicef.org/crc/status

United Nations. 2003a. *Facts on Children: Child Protection.* Available at http://www.unicef.org/media

United Nations. 2003b. *Introduction to Convention on the Rights of the Child.* Available at http://www.unicef.org

U.S. Bureau of the Census. 1980. *1980 Census of Population, General Population Characteristics, United States Summary.* Publication PC80-1-B1. Washington, DC: U.S. Government Printing Office.

U.S. Bureau of the Census. 2002. *International Data Base.* Washington, DC: U.S. Department of Commerce.

U.S. Bureau of the Census. 2007a (May 17). *Minority Population Tops 100 Million.* Press Release. Available at http://www.census.gov

U.S. Bureau of the Census. 2007b. *Statistical Abstract of the United States, 2006,* 126th ed. Washington, DC: U.S. Government Printing Office.

U.S. Department of Health and Human Services. 1999. "Children's Health Insurance Program National Back-to-School Kick Off." *HHS News,* September 22. Available at http://www.hcfa.gov

U.S. State Department. 2007. *Why Population Aging Matters.* In conjunction with the National Institute on Aging, the National Institute of Health, U.S. Department of Health and Human Services. Washington DC: U.S. Government Printing Office.

Wade, Nicholas. 2007. "Study Detects a Gene Linked to Alzheimer's." *New York Times,* January 15. Available at http://www.nytimes.com

White House. 2006. *Fact Sheet: The Adam Walsh Child Protection and Safety Act of 2006.* Available at http://www.whitehouse.gov

World Youth Report. 2005. *General Assembly Economic and Social Council: Report of the Secretary-General.* New York: United Nations.

YRBSS (Youth Risk Behavior Surveillance System). 2006 (June 9). *Morbidity and Mortality Weekly Report* 55(SS-52).

Zhang, Ho, and J. Woo. 2003. "Association between Income and Health Status in the Elderly Chinese in Hong Kong." *Academy Health Meeting* 20(Abstract 604).

Chapter 13

American Community Survey. 2004. Washington, DC: Belden Russonello & Stewart.

Americans' Attitudes Toward Walking and Creating Better Walking Communities. 2003 (April). Washington, DC: Belden Russonello & Stewart Research and Communications.

Baochang, Gu, Wang Feng, Guo Zhigang, and Zhang Erl. 2007. "China's Local and National Fertility Policies at the End of the Twentieth Century." *Population and Development Review* 33(1):129–148.

Blatt, Harvey. 2005. *America's Environmental Report Card: Are We Making the Grade?* Cambridge, MA: MIT Press.

Bongaarts, John, and Susan Cotts Watkins. 1996. "Social Interactions and Contemporary Fertility Transitions." *Population and Development Review* 22(4):639–682.

"The Bridge to the 21st Century Leads to Gridlock In and Around Decaying Cities." 1997. *Washington Spectator* 23(12).

Brown, Lester. 2000. "Overview: The Acceleration of Change." In *Vital Signs: The Environmental Trends That Are Shaping Our Future,* Linda Starke, ed. (pp. 17–29). New York: W. W. Norton.

Brown, Lester. 2006. *Plan B 2.0: Rescuing a Planet Under Stress and a Civilization in Trouble.* New York: W. W. Norton.

Butz, Bill. 2005. "The World's Next Population Problem." Population Reference Bureau. Available at http://www.prb.org

Catley-Carlson, Margaret, and Judith A. M. Outlaw. 1998. "Poverty and Population Issues: Clarifying the Connections." *Journal of International Affairs* 52(1):233–243.

"China Is Softening Messages." 2007. *Popline,* September-October, pp. 1, 3.

Cincotta, Richard, Robert Engelman, and Daniele Anastasion. 2003. *The Security Demographic: Population and Civil Conflict After the Cold War.* Washington, DC: Population Action International.

Clark, David. 1998. "Interdependent Urbanization in an Urban World: An Historical Overview." *Geographical Journal* 164(1):85–96.

Cohen, Susan A. 2006. "The Global Contraceptive Shortfall: U.S. Contributors and U.S. Hindrances." *Guttmacher Policy Review* 9(2):15–18.

Cowherd, Phil. 2001. "What Is the Business Case for Investing in Inner-City Neighborhoods?" *Public Management* 83(1):12–14.

DaVanzo, Julie, David M. Adamson, Nancy Belden, and Sally Patterson. 2000. How Americans View World Population Issues: A Survey of Public Opinion. Santa Monica, CA: Rand Corporation.

Davis, Mike. 2006. *Planet of Slums.* New York: Verso.

DeNavas-Walt, Carmen, Bernadette D. Proctor, and Robert J. Mills. 2004. *Income, Poverty, and Health Insurance Coverage in the United States: 2003.* U.S. Census Bureau, Current Population Reports P60-226. Washington, DC: U.S. Government Printing Office.

Douthwaite, M., and P. Ward. 2005. "Increasing Contraceptive Use in Rural Pakistan: An Evaluation of the Lady Health Worker Programme." *Health Policy and Planning* 20(2):117–123.

Durning, Alan. 1996. *The City and the Car.* Northwest Environment Watch. Seattle, WA: Sasquatch Books.

Egan, Timothy. 2005. "Vibrant Cities Find One Thing Missing: Children." *New York Times,* March 24. Available at http://www.nytimes.com

Ewing, Reid, Tom Schmid, Richard Killingsworth, Amy Zlot, and Stephen Raudenbush. 2003. "Relationship between Urban Sprawl and Physical Activity, Obesity, and Morbidity." *American Journal of Health Promotion* 18(1):47–57.

Fisher, Claude. 1982. *To Dwell Among Friends: Personal Networks in Town and City*. Chicago: University of Chicago Press.

Froehlich, Maryann. 1998. "Smart Growth: Why Local Governments Are Taking a New Approach to Managing Growth in Their Communities." *Public Management* 80(5):5–9.

Galster, George, Diane Levy, Noah Sawyer, Kenneth Temkin, and Christopher Walker. 2005 (June 30). *The Impact of Community Development Corporations on Urban Neighborhoods.Urban Institute*. Available at http://www.urban.org

Gans, Herbert. 1984. *The Urban Villagers*, 2nd ed. New York: Free Press.

Gardner, Gary. 1999. "Cities Turning to Bicycles to Cut Costs, Pollution, and Crime." *Public Management* 81(1):23.

Gardner, Gary. 2003. "Bicycle Production Seesaws." In *Vital Signs*, M. Renner and M. O. Sheehan, eds. (pp. 58–59). New York: W. W. Norton.

Gardner, Gary. 2005. "Bicycle Production Recovers." In *Vital Signs*, L. Starke, ed. (pp. 58–59). New York: W. W. Norton.

Guttmacher Institute. 2007 (May). *Facts About the Unmet Need for Contraception in Developing Countries*. Available at http://www.guttmacher.org

Hollander, Dore. 2007. "Women Who Are Fecund but Do Not Wish to Have Children Outnumber the Involuntarily Childless." *Perspectives on Sexual and Reproductive Health* 39(2):120.

Jensen, Derrick. 2001. "Road to Ruin: An Interview with Jan Lundberg." *The Sun* 302:4–13.

Johnson, William C. 1997. *Urban Planning and Politics*. Chicago: American Planning Association and Planners Press.

Kimuna, Sitawa R., and Donald J. Adamchak. 2001. "Gender Relations: Husband-Wife Fertility and Family Planning Decisions in Kenya." *Journal of Biosocial Science* 33:13–23.

Lalasz, Robert. 2005 (July). *Baby Bonus Credited with Boosting Australia's Fertility Rate. Population Reference Bureau*. Available at http://www.prb.org

Leahy, Elizabeth. 2003. "As Contraceptive Use Rises, Abortions Decline." *Popline*, November-December, pp. 3, 8.

Lindstrom, Matthew J., and Hugh Bartling. 2003. "Introduction." In *Suburban Sprawl: Culture, Theory, and Politics*, M. J. Lindstrom and H. Bartling, eds. (pp. xi–xxvii). Lanham, MD: Rowman & Littlefield.

Livernash, Robert, and Eric Rodenburg. 1998. "Population Change, Resources, and the Environment." *Population Bulletin* 53(1):1–36.

Lohn, Martiga. 2007 (August 13). "Divers Locate Remains at Bridge Collapse Site." *The Boston Globe*. Available at http://www.boston.com

McFalls, Joseph A. 2007 (March). "Population: A Lively Introduction, 5th ed." *Population Bulletin* 62(1).

McLaughlin, Eliot C. 2007 (August 3). "Experts: Leadership, Money Keys to Building Bridges." CNN. Available at http://www.CNN.com

Millennium Ecosystem Assessment. 2005. *Ecosystems and Human Well-Being: Synthesis*. Washington, DC: Island Press.

National Congress for Community Economic Development. 2006. *Reaching New Heights: Trends and Achievements for Community-Based Development Organization*. Washington, DC: National Congress for Community Economic Development.

Nelson, Mara. 2004. "Silence and Myths: A Response to the 'Birth Dearth.'" *The Reporter*, Winter, pp. 17–21, 24. Population Connection. Available at http://www.popconnect.org

Newman, Peter, and Jeff Kenworthy. 2007. "Greening Urban Transportation." In L. Starke, ed. *2007 State of the World: Our Urban Future*. (pp. 66–85). New York: W.W. Norton.

Orfield, Myron. 1997. *Metropolitics: A Regional Agenda for Community and Stability*. Washington, DC: Brookings Institution Press, and Cambridge, MA: Lincoln Institute of Land Policy.

Parks, Kristin. 2005. "Choosing Childlessness: Weber's Typology of Action and Motives of the Voluntarily Childless." *Sociological Inquiry* 75(3):372–402.

Pelley, Janet. 1999. "Building Smart-Growth Communities." *Environmental Science and Technology News* 33(1):28A–32A.

Population Reference Bureau. 2004. "Transitions in World Population." *Population Bulletin* 59(1).

Population Reference Bureau. 2007a. *World Population Highlights. Population Bulletin* 62(3).

Population Reference Bureau. 2007b. *World Population Data Sheet*. Washington, DC: Population Reference Bureau. Available at http://www.prb.org

Rees, Amanda. 2003. "New Urbanism: Visionary Landscapes in the Twenty-First Century." In *Suburban Sprawl: Culture, Theory, and Politics*, M. J. Lindstrom and H. Bartling, eds. (pp. 93–114). Lanham, MD: Rowman & Littlefield.

Rodriquez, Luis. 2000. "Urban Renewal: The Resurrection of an Ex-Gang Member" (in an interview with Derrick Jensen). *The Sun*, April, pp. 4–13.

Schrank, David, and Tim Lomax. 2007. *The 2007 Urban Mobility Report*. TexasTransportation Institute and Texas A&M University System. Available at http://mobility.tamu.edu

Schueller, Jane. 2005 (August). "Boys and Changing Gender Roles." YouthNet. *YouthLens* 16. Available at http://www.fhi.org

Sheehan, M. O. 2001. "Making Better Transportation Choices." In *State of the World 2001*, L. Starke, ed. (pp. 103–122). New York: W. W. Norton.

Sheehan, M. O. 2003. "Uniting Divided Cities." In *State of the World 2003*, L. Starke, ed. (pp. 130–151). New York: W. W. Norton.

Shevis, Jim. 1999. "More Affluent Than Their Inner-City Neighbors, Suburbanites Still Have Growth Problems." *Washington Spectator* 25(19):1–3.

Smart Growth Network. 2007. *About Smart Growth*. Available at http://www.smartgrowth.org

Sullivan, Daniel Monroe. 2007. "Reassessing Gentrification: Measuring Residents' Opinions Using Survey Data." *Urban Affairs Review* 42(4):583–592.

Tittle, Charles. 1989. "Influences on Urbanism: A Test of Predictions from Three Perspectives." *Social Problems* 36(3):270–288.

UNEP (United Nations Environment Programme). 2005. *GEO Yearbook: An Overview of Our Changing Environment 2004/5*. Available at http://www.unep.org

UNICEF. 2006. *State of the World's Children 2007: Women and Children: The Double Dividend of Gender Equality*. New York: UNICEF.

Union of Concerned Scientists. 2003. *Cars and Trucks and Air Pollution*. Available at http://www.ucsusa.org

United Nations. 2006. *World Urbanization Prospects: The 2005 Revision*. New York: United Nations.

United Nations. 2007. *World Population Prospects: The 2006 Revision*. New York: United Nations.

United Nations Population Fund. 2004. *The State of World Population 2004*. New York: United Nations

United Nations Population Fund. 2007. *State of World Population 2007: Unleashing the Potential of Urban Growth*. New York: United Nations.

U.S. Conference of Mayors. 2001. *Traffic Congestion and Rail Investment*. Washington, DC: Global Strategy Group.

U.S. Department of Housing and Urban Development. 2004. "Revitalizing Neighborhoods by Redeveloping Brownfields: The HUD Brownfields Economic Development Initiative." *Research Works* 1(3):2.

Warren, Roxanne. 1998. *The Urban Oasis: Guideways and Greenways in the Human Environment.* New York: McGraw Hill.

Weeks, John R. 2005. *Population: An Introduction to Concepts and Issues,* 9th ed. Belmont, CA: Wadsworth/Thomson.

Weiland, Katherine. 2005. *Breeding Insecurity: Global Security Implications of Rapid Population Growth.* Washington, DC: Population Institute.

Wirth, Louis. 1938. "Urbanism as a Way of Life." *American Journal of Sociology* 44:8–20.

Wiseman, Paul. 2005. "Towns Hope Cash-for-Babies Incentives Boost Populations." *USA Today,* July 28. Available at http://www .usatoday.com

Wolff, Kurt H. 1978. *The Sociology of George Simmel.* Toronto, ON, Canada: Free Press.

Women's Studies Project. 2003. *Women's Voices, Women's Lives: The Impact of Family Planning.* Family Health International. Available at http://www.fhi.org

World Health Organization. 2003. *World Health Report 2003: Shaping the Future.* Geneva: World Health Organization.

World Watch Institute. 2004 (January 7). "State of the World 2004: Consumption by the Numbers." Press Release. Available at http://www.worldwatch.org

Zabin, L. S., and K. Kiragu. 1998 (June 29). "The Health Consequences of Adolescent Sexual and Fertility Behavior in Sub-Saharan Africa." *Studies in Family Planning* 2:210–232.

Chapter 14

Aeck, Molly. 2005. "Biofuel Use Growing Rapidly." In *Vital Signs 2005,* Linda Starke, ed. (pp. 38–39). New York: W. W. Norton.

American Lung Association. 2005. Lung Disease Data in Culturally Diverse Communities: 2005. Available at http://www .lungusa.org

American Lung Association. 2007. *State of the Air: 2007.* Available at http://lungaction.org

Aslam, Abid. 2005. "Planet Faces Nightmare Forecasts as Chinese Consumption Grows and Grows." *Common Dreams News Center,* October 9. Available at http://www .commondreams.org

Associated Press. 2005 (April 13). "Wal-Mart Pledges $35 Million for Wildlife." MSNBC. com. Available at http://www.msnbc.msn .com

Associated Press. 2007 (May 13). "Report: Students Exposed to Pesticides Used Near Schools." CNN.com. Available at http:// www.cnn.com

Blatt, Harvey. 2005. *America's Environmental Report Card: Are We Making the Grade?* Cambridge, MA: MIT Press.

Bohan, Caren. 2005 (August 23). "The Illulissat Glacier, a Wonder of the World Melting Away." *Common Dreams News Center,* August 23. Available at http://www .common dreams.org

Bright, Chris. 2003. "A History of Our Future." In *State of the World 2003,* Linda Starke, ed. (pp. 3–13). New York: W. W. Norton.

Brody, Julia Gree, Kirsten B. Moysich, Olivier Humblet, Kathleen R. Attfield, Gregory P. Beehler, and Ruthann A. Rudel. 2007. "Environmental Pollutants and Breast Cancer." *Cancer* 109(S12):2667–2711.

Brooke, James. 2005. "Japan's Salarymen Shed Their Suits?" *International Herald Tribune,* May 21. Available at http://www.iht.com

Brown, Lester R. 2003. *Plan B: Rescuing a Planet Under Stress and a Civilization in Trouble.* New York: W. W. Norton.

Brown, Lester R. 2007. "Distillery Demand for Grain to Fuel Cars Vastly Understated: World May Be Facing Highest Grain Prices in History." *Earth Policy News,* January 4. Available at http://www.earthpolicy.org

Brown, Lester R., and Jennifer Mitchell. 1998. "Building a New Economy." In *State of the World 1998,* Lester R. Brown, Christopher Flavin, and Hilary French, eds. (pp. 168–187). New York: W. W. Norton.

Bruno, Kenny, and Joshua Karliner. 2002. *Earthsummit.biz: The Corporate Takeover of Sustainable Development.* CorpWatch and Food First Books. Available at http://www .corpwatch.org

Bullard, Robert D. 2000. *Dumping in Dixie: Race, Class, and Environmental Quality,* 3rd ed. Boulder, CO: Westview Press.

Bullard, Robert D., and Glenn S. Johnson. 1997. "Just Transportation." In *Just Transportation: Dismantling Race and Class Barriers to Mobility,* Robert D. Bullard and Glenn S. Johnson, eds. (pp. 1–21). Stony Creek, CT: New Society.

Bullard, Robert D., Paul Mohai, Robin Saha, and Beverly Wright. 2007 (March). *Toxic Wastes and Race at Twenty 1987–2007.* Cleveland Ohio: United Church of Christ.

Caress, Stanley, Anne C. Steinemann, and Caitlin Waddick. 2002. "Symptomatology and Etiology of Multiple Chemical Sensitivities in the Southeastern United States." *Archives of Environmental Health* 57(5):429–436.

Carlson, Scott. 2006. "In Search of the Sustainable Campus." *The Chronicle of Higher Education* LIII (9):A10–A12, A14.

Centers for Disease Control and Prevention, National Center for Environmental Health. 2003. *Second National Report on Human Exposure to Environmental Chemicals.* Atlanta, GA: Centers for Disease Control and Prevention.

Centers for Disease Control and Prevention, National Center for Environmental Health. 2005. *Third National Report on Human Exposure to Environmental Chemicals.* Atlanta, GA: Centers for Disease Control and Prevention.

Chafe, Zoe. 2005. "Bioinvasions." In *State of the World 2005,* L. Starke, ed. (pp. 60–61). New York: W. W. Norton.

Cheeseman, Gina-Marie. 2007. "Plastic Shopping Bags Being Banned." *The Online Journal,* June 27. Available at http://www .onlinejournal.com

Chivian, Eric, and Aaron S. Bernstein. 2004. "Embedded in Nature: Human Health and Biodiversity." *Environmental Health Perspectives* 112(1). Available at http://ehp .niehs.nih.gov

Cincotta, Richard P., and Robert Engelman. 2000. *Human Population and the Future of Biological Diversity.* Washington, DC: Population Action International.

Clarke, Tony. 2002. "Twilight of the Corporation." In *Social Problems, Annual Editions 02/03,* 30th ed., Kurt Finsterbusch, ed. (pp. 41–45). Guilford, CT: McGraw-Hill/Dushkin.

Clarren, Rebecca. 2005. "Dirty Politics, Foul Air." *The Nation,* March 14, pp. 6, 8.

"Clinton's Environmental Policy." 2007 (June 16). *Media Mouse, Grand Rapids Independent Media.* Available at http://www .mediamouse.org

Computer Take Back Campaign. 2005. *Fifth Annual Computer Report Card.* Available at http://www.computertakeback.com

Cone, Marla. 2007. "Public Health Agency Linked to Chemical Industry." *Los Angeles Times,* March 4. Available at http://www .latimes.com

Cooper, Arnie. 2004. "Twenty-Eight Words That Could Change the World: Robert Hinkley's Plan to Tame Corporate Power." *The Sun* 345(September):4–11.

Coyle, Kevin. 2005. *Environmental Literacy in America.* Washington, DC: The National Environmental Education and Training Foundation.

Denson, Bryan. 2000. "Shadowy Saboteurs." *IRE Journal* 23(May-June):12–14.

DesJardins, Andrea. 1997. Sweet Poison: What Your Nose Can't Tell You About the Dangers of Perfume. Available at http://members.aol .com/enviroknow/perfume/sweet-poison .htm

Edwards, Bob, and Anthony Ladd. 2000. "Environmental Justice, Swine Production, and Farm Loss in North Carolina." *Sociological Spectrum* 20(3):263–290.

Elgin, Ben. 2007 (October 29). "Little Green Lies." *Business Week.* Available at http:// businessweek.com

Energy Information Administration. 2007. *International Energy Annual 2005*. Washington DC: Energy Information Administration.

Environmental Defense. 2004. *2004 Annual Report*. Available at http://www.environmentaldefense.org

Environmental Working Group. 2005 (July 14). *Body Burden: The Pollution in Newborns*. Available at http://www.ewg.org

EPA (U.S. Environmental Protection Agency). 2004. *What You Need to Know About Mercury in Fish and Shellfish*. Available at http://www.epa.gov

EPA. 2006. *Municipal Solid Waste in the United States: 2005 Facts and Figures*. Available at http://www.epa.gov

EPA. 2007a (July). *National Listing of Fish Advisories*. Available at http://www.epa.gov

EPA. 2007b. *NPL Site Totals by Status and Milestone*. Available at http://www.epa.gov/superfund/sites

European Environment Agency. 2007. *Europe's Environment: The Fourth Assessment*. Copenhagen, Denmark: European Environment Agency.

Faber, Daniel. 2005. "Building a Transnational Environmental Justice Movement: Obstacles and Opportunities in the Age of Globalization." In *Coalitions Across Borders*, Joe Bandy and Jackie Smith, eds. (pp. 43–68). Lanham, MA: Rowman & Littlefield.

Fisher, Brandy E. 1998. "Scents and Sensitivity." *Environmental Health Perspectives* 106(12). Available at http://ehpnet1.niehs.nih.gov/docs/1998/106-12/focus-abs.html

Fisher, Brandy E. 1999. "Focus: Most Unwanted." *Environmental Health Perspectives* 107(1). Available at http://ehpnet1.niehs.nih.gov/docs/1999/107-1/focus-abs.html

Flavin, Chris, and Molly Hull Aeck. 2005 (September 15). "Cleaner, Greener, and Richer." Tom.Paine.com. Available at http://www.tompaine.com/articles/2005/09/15/cleaner_greener_and_richer.php

Food and Agriculture Organization of the United Nations. 2005. *Global Forest Resources Assessment: 2005*. Available at http://www.fao.org

Food and Drug Administration. 2005 (May). *Pesticide Program Residue Monitoring 2003*. Available at http://www.cfsan.fda.gov

Food & Water Watch and Network for New Energy Choices. 2007. *The Rush to Ethanol: Not All Biofuels Are Created Equal*. Available at http://www.newenergychoices.org

Fornos, Werner. 2005. "Homo Sapiens: An Endangered Species?" *Popline* 27:4.

French, Hilary. 2000. *Vanishing Borders: Protecting the Planet in the Age of Globalization*. New York: W. W. Norton.

Gallup Organization. 2007. *Gallup's Pulse of Democracy*. Environment. Available at http://www.gallup.com/poll

Gardner, Gary. 2005. "Forest Loss Continues." In *Vital Signs 2005*, Linda Starke, ed. (pp. 92–93). New York: W. W. Norton.

Gartner, John. 2005 (September 13). "Sierra Club Gets Behind the Wheel." *Wired News*. Available at http://www.wired.com

Global Carbon Project. 2007 (Oct. 20). *Recent Carbon Trends and the Global Carbon Budget*. Available at http://www.globalcarbonproject.org

Goodstein, Laurie. 2005. "Evangelical Leaders Swing Influence Behind Effort to Combat Global Warming." *New York Times*, March 10. Available at http://www.nytimes.com

Gottlieb, Roger S. 2003a. "Saving the World: Religion and Politics in the Environmental Movement." In *Liberating Faith*, Roger S. Gottlieb, ed. (pp. 491–512). Lanham, MD: Rowman & Littlefield.

Gottlieb, Roger S. 2003b. "This Sacred Earth: Religion and Environmentalism." In *Liberating Faith*, Roger S. Gottlieb, ed. (pp. 489–490). Lanham, MD: Rowman & Littlefield.

Hager, Nicky, and Bob Burton. 2000. *Secrets and Lies: The Anatomy of an Anti-Environmental PR Campaign*. Monroe, ME: Common Courage Press.

Hilgenkamp, Kathryn. 2005. *Environmental Health: Ecological Perspectives*. Sudbury, MA: Jones and Bartlett Publishers.

Hunter, Lori M. 2001. *The Environmental Implications of Population Dynamics*. Santa Monica, CA: Rand Corporation.

Intergovernmental Panel on Climate Change. 2007a. *Climate Change 2007: The Physical Science Basis*. United Nations Environmental Programme and the World Meteorological Organization. Available at http://www.ipcc.org

Intergovernmental Panel on Climate Change. 2007b. *Climate Change 2007: Impacts, Adaptation and Vulnerability*. United Nations Environmental Programme and the World Meteorological Organization. Available at http://www.ipcc.org

International Atomic Energy Agency. 2007. *Energy, Electricity and Nuclear Power for the Period up to 2030*. Available at http://www.iaea.org

International Fund for Animal Welfare. 2005. *Caught in the Web: Wildlife Trade on the Internet*. Available at http://www.ifaw.org

IUCN 2007. *2007 IUCN Red List of Threatened Species*. Available at http://www.iucnredlist.org

Jan, George P. 1995. "Environmental Protection in China." In *Environmental Policies in the Third World: A Comparative Analysis*, O. P. Dwivedi and Dhirendra K. Vajpeyi, eds. (pp. 71–84). Westport, CT: Greenwood Press.

Janofsky, Michael. 2005. "Pentagon Is Asking Congress to Loosen Environmental Laws." *New York Times*, May 11. Available at http://www.nytimes.com

Jarboe, James F. 2002 (February 12). *The Threat of Eco-Terrorism. Congressional Statement. Federal Bureau of Investigation*. Available at http://www.fbi.gov

Jensen, Derrick. 2001. "A Weakened World Cannot Forgive Us: An Interview with Kathleen Dean Moore." *The Sun* 303 (March):4–13.

Johnson, Geoffrey. 2004 (April). "Don't Be Fooled: America's Ten Worst Greenwashers." *The Green Life*. Available at http://www.thegreenlife.org

Kaplan, Sheila, and Jim Morris. 2000. "Kids at Risk." *U.S. News and World Report*, June 19, pp. 47–53.

Kestenbaum, David. 2007 (Oct. 2). "Morning Edition." *National Public Radio*. Available at http://www.npr.org

Koenig, Dieter. 1995. "Sustainable Development: Linking Global Environmental Change to Technology Cooperation." In *Environmental Policies in the Third World: A Comparative Analysis*, O. P. Dwivedi, and Dhirendra K. Vajpeyi, eds. (pp. 1–21). Westport, CT: Greenwood Press.

Lamm, Richard. 2006. "The Culture of Growth and the Culture of Limits." *Conservation Biology* 20(2): 269–271.

Levin, Phillip S., and Donald A. Levin. 2002. "The Real Biodiversity Crisis." *American Scientist* 90(1):6.

Little, Amanda Griscom. 2005. "Maathai on the Prize: An Interview with Nobel Peace Prize Winner Wangari Maathai." *Grist Magazine*, February 15. Available at http://www.grist.org

Makower, Joel, Ron Pernick, and Clint Wilder. 2005. *Clean Energy Trends 2005. Clean Edge*. Available at http://www.cleanedge.com

Margesson, Rhoda. 2005. "Environmental Refugees." In *State of the World 2005*, Linda Starke, ed. (pp. 40–41). New York: W. W. Norton.

Mastny, Lisa. 2003. "State of the World: A Year in Review." In *State of the World 2003*, Linda Starke, ed. (pp. xix–xxiii). New York: W. W. Norton.

McDaniel, Carl N. 2005. *Wisdom for a Livable Planet*. San Antonio, TX: Trinity University Press.

McGinn, Anne Platt. 2000. "Endocrine Disrupters Raise Concern." In *Vital Signs 2000*, Lester R. Brown, Michael Renner, and Brian Halweil, eds. (pp. 130–131). New York: W. W. Norton.

McMichael, Anthony J., Kirk R. Smith, and Carlos F. Corvalan. 2000. "The Sustainability Transition: A New Challenge." *Bulletin of the World Health Organization* 78(9):1067.

Mead, Leila. 1998. "Radioactive Wastelands." *The Green Guide* 53(April 14):1–3.

Millennium Ecosystem Assessment. 2005. *Ecosystems and Human Well-Being: Synthesis*. Washington, DC: Island Press.

Miranda, Marie Lynn, Dohyeong Kim, Alicia Overstreet Galeano, Christopher J. Paul, Andrew P. Hull, and S. Philip Morgan. 2007. "The Relationship between Early Childhood Blood Lead Levels and Performance on End-of-Grade Tests." *Environmental Health Perspectives* 115(8):1242–1247.

Mooney, Chris. 2005. "Some Like It Hot." *Mother Jones*, May-June. Available at http://www.MotherJones.com

NASA. 2007 (October 18). *NASA Data Reveals 'Average' Ozone Hole in 2007*. Available at http://www.nasa.gov

National Assessment Synthesis Team. 2000. *Climate Change Impacts on the United States: The Potential Consequences of Climate Variability and Change*. Washington, DC: U.S. Global Change Research Program.

National Environmental Education and Training Foundation, and Roper Starch Worldwide. 1999. *1999 NEETF/Roper Report Card*. Washington, DC: National Environmental Education and Training Foundation.

Norrell, Brenda. 2005 (March 11). "Yucca Mountain Lawsuit Filed." *Indian Country Today*. Available at http://www.indiancountry.com

"Nukes Rebuked." 2000. *Washington Spectator* 26(13):4.

PBS. 2001. *Trade Secrets: A Moyers Report*. Available at http://www.pbs.org/tradesecrets/program/program.html

Pearce, Fred. 2005. "Climate Warming as Siberia Melts." *New Scientist*, August 11. Available at http://www.NewScientist.com

Pimentel, David, Maria Tort, Linda D'Anna, Anne Krawic, Joshua Berger, Jessica Rossman, Fridah Mugo, Nancy Don, Michael Shriberg, Erica Howard, Susan Lee, and Jonathan Talbot. 1998. "Ecology of Increasing Diseases: Population Growth and Environmental Degradation." *BioScience* 48(October):817–827.

Pimentel, D., S. Cooperstein, H. Randell, D. Filiberto, S. Sorrentino, B. Kaye, C. Nicklin, J. Yagi, J. Brian, J. O'Hern, A. Habas and C. Weinstein. 2007. "Ecology of Increasing Diseases: Population Growth and Environmental Degradation." *Human Ecology* 35(6):653–668.

Pope, Carl. 2005 (July 7). "America Needs Real Energy Solutions." *Sierra Club press release*. Available at http://www.sierraclub.org

Prah, Pamela M. 2007 (July 30). "States Forge Ahead on Immigration, Global Warming." *Stateline*. Available at http://www.stateline.org

"A Prescription for Reducing the Damage Caused by Dams." 2001. *Environmental Defense* 32(2).

Price, Tom. 2006. "The New Environmentalism." *CQ Researcher* 16(42):987–1007.

Pryor, John H., S. Hurtado, V. B. Saenz, J. S. Korn, J. L. Santos, and W. S. Korn. 2006. *The American Freshman: National Norms for Fall 2006*. Los Angeles: Higher Education Research Institute, UCLA.

Public Citizen. 2005 (February). *NAFTA Chapter 11 Investor-State Cases: Lessons for the Central America Free Trade Agreement. Public Citizens Global Trade Watch Publication E9014*. Available at http://www.citizen.org

Raupach, Michael, Gregg Marland, Philippe Ciais, Corinne Le Quere, Josep G. Canadell, Gernot Klepper, and Christopher B. Field. 2007. "Global and Regional Drivers of Accelerating CO_2 Emissions." *Proceedings of the National Academy of Sciences of the United States of America*. 104:10288–10293.

Reese, April. 2001 (February). *Africa's Struggle with Desertification. Population Reference Bureau*. Available at http://www.prb.org

Renner, Michael. 1996. *Fighting for Survival: Environmental Decline, Social Conflict, and the New Age of Insecurity*. New York: W. W. Norton.

Renner, Michael. 2004. "Moving Toward a Less Consumptive Economy." In *State of the World 2004*, Linda Starke, ed. (pp. 96–119). New York: W. W. Norton.

Rifkin, Jeremy. 2004. *The European Dream: How Europe's Vision of the Future Is Quietly Eclipsing the American Dream*. New York: Tarcher/Penguin.

Rogers, Sherry A. 2002. *Detoxify or Die*. Sarasota, FL: Sand Key.

Saad, Lydia. 2002. "Americans Sharply Divided on Seriousness of Global Warming." *Gallup News Service*. Available at http://www.gallup.org

Sampat, Payal. 2003. "Scrapping Mining Dependence." In *State of the World 2003*, Linda Starke, ed. (pp. 110–129). New York: W. W. Norton.

Sawin, Janet. 2004. "Making Better Energy Choices." In *State of the World 2004*, Linda Starke, ed. (pp. 24–43). New York: W. W. Norton.

Sawin, Janet. L. 2005a. "Global Wind Growth Continues." In *Vital Signs 2005*, Linda Starke, ed. (pp. 34–35). New York: W. W. Norton.

Sawin, Janet L. 2005b. "Solar Energy Markets Booming." In *Vital Signs 2005*, Linda Starke, ed. (pp. 36–37). New York: W. W. Norton.

Schapiro, Mark. 2004. "New Power for 'Old Europe.'" *The Nation*, December 27, pp. 11–16.

Schapiro, Mark. 2007. *Exposed: The Toxic Chemistry of Everyday Products and What's at Stake for American Power*. White River Junction, Vermont: Chelsea Green Publishing.

Schmidt, Charles W. 2004. "Environmental Crimes: Profiting at the Earth's Expense." *Environmental Health Perspectives* 112(2): A96–A103.

Silicon Valley Toxics Coalition. 2001. *Just Say No to E-Waste: Background Document on Hazards and Waste from Computers*. Available at http://www.svtc.org

Sinks, Thomas. 2007 (June 12). *Statement by Thomas Sinks, Ph.D., Deputy Director, Agency for Toxic Substances and Disease Registry on ATSDR's Activities at U.S. Marine Corps Base Camp Lejeune before Committee on Energy and Commerce Subcommittee on Oversight and Investigations United States House of Representatives*. Available at http://www.hhs.gov

Sterritt, Angela. 2005. "Indigenous Youth Challenge Corporate Mining." *Cultural Survival, Weekly Indigenous News*, July 15. Available at http://209.200.101.189/home.cfm

Stoll, Michael. 2005 (September-October). "A Green Agenda for Cities." *E Magazine* 16(5). Available at http://www.emagazine.com

Stretesky, Paul. 2003. "The Distribution of Air Lead Levels Across U.S. Counties: Implications for the Production of Racial Inequality." *Sociological Spectrum* 23:91–118.

Sustainable Endowments Institute. 2008. *College Sustainability Report Card 2008*. Available at http://www.endowmentinstitute.org

United Nations Development Programme. 2007. *Human Development Report 2007/2008: Fighting Climate Change: Human Solidarity in a Divided World*. New York: Palgrave Macmillan.

UNEP (United Nations Environment Programme). 2002. *North America's Environment: A Thirty-Year State of the Environment and Policy Retrospectives*. Available at http://www.na.unep.net

UNEP. 2007. *GEO Yearbook 2007: An Overview of Our Changing Environment*. Available at http://www.unep.org

United Nations Population Fund. 2004. *The State of World Population 2004*. New York: United Nations.

U.S. Department of Health and Human Services. 2004. *11th Report on Carcinogens*. Washington, DC: Public Health Service.

Vajpeyi, Dhirendra K. 1995. "External Factors Influencing Environmental Policymaking: Role of Multilateral Development Aid Agencies." In *Environmental Policies in the Third World: A Comparative Analysis*, O. P. Dwivedi and Dhirendra K. Vajpeyi, eds. (pp. 24–45). Westport, CT: Greenwood Press.

Vartan, Starre. 2005. "Money Matters: Buying Cleaner Energy." *E Magazine* 16(5). Available at http://www.emagazine.com

Vedantam, Shankar. 2005a. "Nuclear Plants Not Keeping Track of Waste." *Washington Post*, April 19. Available at http://www.washingtonpost.com

Vedantam, Shankar. 2005b. "Storage Plan Approved for Nuclear Waste." *Washington Post*, September 10. Available at http://www.washingtonpost.com

Wal-Mart Watch. 2005 (April 6). *Wal-Mart Watch*. Press release. Available at http://walmartwatch.com

Weeks, Jennifer. 2006. "Nuclear Energy." *CQ Researcher* 16(10):217–240.

Willett, Katharine M., Nathan P. Gillett, Philip D. Jones and Peter W. Thorne. 2007. "Attribution of Observed Surface Humidity Changes to Human Influence." *Nature* 449(October 11):710–712.

Woodward, Colin. 2007. Curbing Climate Change. *CQ Global Researcher* 1(2):27–50. Available at http://www.globalresearcher.com

World Health Organization. 2005. *Indoor Air Pollution and Health*. Fact Sheet. Available at http://www.who.org

World Resources Institute. 2000. *World Resources 2000–2001: People and Ecosystems —The Fraying Web of Life*. Washington, DC: World Resources Institute.

Youth, Howard. 2003. "Watching Birds Disappear." In *State of the World 2003*, Linda Starke, ed. (pp. 85–109). New York: W. W. Norton.

Zandonella, Catherine. 2007. "The Dirty Dozen Chemicals in Cosmetics." *The Greene Guide* No. #122 (October/November). Available at http://www.thegreenguide.com

Chapter 15

ACLU (American Civil Liberties Union). 2005. *Utah Businesses, Free Speech Groups, and Individuals Challenge Restrictions on Internet Speech. Privacy and Technology*. Available at http://www.aclu.org/Privacy

Adcox, Seanna. 2007. "S.C. Lawmakers Weigh Abortion Bill." Associated Press, March 16. Available at http://www.breitbart.com

ANOL (A Nation Online). 2004. *A Nation Online: Entering the Broadband Age*. Washington, DC: National Telecommunications and Information Administration.

Apuzzo, Matt. 2007. "TSA Loses Hard drive with Personal Info." *Washington Post*, May 4. Available at http://www.washingtonpost.com

Associated Press. 2007a (June 24). *Google Fights Global Internet Censorship*. Available at http://www.breitbart.com

Associated Press. 2007b. *Ohio Changes Policies After Data Theft*. Available at http://www.msnbc.msn.com

Attewell, Paul, Belkis Suazo-Garcia, and Juan Battle. 2003. "Computers and Young Children: Social Benefit or Social Problem?" *Social Forces* 82:277–296.

Bailey, Ronald. 2005. "Scientific Arguments Against Biotechnology Are Fallacious." In *Genetically Engineered Foods*, Nancy Harris, ed. (pp. 80-93). Farmington Hills, MI: Greenhaven Press.

Barbash, Fred. 2005. "High Court Takes Up Abortion Consent Case." *Washington Post*, May 25. Available at http://www.washingtonppost.com

BBC. 2003. "Hi-Tech Workplaces No Better Than Factories." *BBC News*, November 27. Available at http://www.bbc.co.uk

BBC. 2006. "Google Censors Itself for China." *BBC News*, January 25. Available at http://newsvote.bbc.co.uk

Bell, Daniel. 1973. *The Coming of Post-Industrial Society: A Venture in Social Forecasting*. New York: Basic Books.

Benbrook, Charles M. 2004. *Impacts of Genetically Engineered Crops on Pesticide Use in the United States: The First Nine Years*. Ag BioTech InfoNet, Technical Paper 7. Available at http://www.biotech-info.net

Beniger, James R. 1993. "The Control Revolution." In *Technology and the Future*, Albert H. Teich, ed. (pp. 40– 65). New York: St. Martin's Press.

BLS (Bureau of Labor Statistics). 2005. "Computer and Internet Use at Work Summary." Press Release, August 2. Available at http://www.bls.gov

Boles, Margaret, and Brenda Sunoo. 1998. "Do Your Employees Suffer from Techno-Phobia?" *Workforce* 77(1):21.

Bosworth, Martin H. 2007 (February 9). *FTC Findings Undercut Industry Claims That Identity Theft Is Declining*. Available at http://www.consumeraffairs.com

Brin, David. 1998. *The Transparent Society: Will Technology Force Us to Choose between Privacy and Freedom?* Reading, MA: Addison Wesley.

Brown, David. 2006. "For 1st Woman with Bionic Arm, a New Life Is Within Reach." *Washington Post*, September 14, p. A01.

Buchanan, Allen, Dan Brock, Norman Daniels, and Daniel Wikler. 2000. *From Chance to Choice: Genetics and Justice*. New York: Cambridge University Press.

Bush, Corlann G. 1993. "Women and the Assessment of Technology." In *Technology and the Future*, Albert H. Teich, ed. (pp. 192–214). New York: St. Martin's Press.

Caplan, Jeremy. 2006. "The New Front Line in the Abortion Wars." *Time*, March 6, p. 8.

CBS/Associated Press. 2007. "Study: Children Bombarded with On-Line Porn." CBS News. Available at http://www.cbsnews.com

Center for Food Safety. 2007 (May 3). *Federal Judge Orders First Ever Halt to Planting of a Commercially Genetically-Altered Crop*. Available at http://www.centerforfoodsafety.org

Ceruzzi, Paul. 1993. "An Unforeseen Revolution." In *Technology and the Future*, Albert H. Teich, ed. (pp. 160–174). New York: St. Martin's Press.

Chan, Sewell. 2007. "New Scanners for Tracking City Workers." *New York Times*, January 23. Available at http://www.nytimes.com

Chepesiuk, Ron. 1999. "Where the Chips Fall: Environmental Health in the Semiconductor Industry." *Environmental Health Perspectives* 107(9). Available at http://www.junkscience.com/aug99/chips.htm

Clarke, Adele E. 1990. "Controversy and the Development of Reproductive Sciences." *Social Problems* 37(1):18–37.

CNN. 2005 (June 15) *Poll: Most Want Congress to Make Sure Internet Safe*. Available at http://cnn.com/technology

Cole, Wendy. 2006. "Who's Liable for Illegal Posts Online?" *Time*, February 27, p. 8

Conrad, Peter. 1997. "Public Eyes and Private Genes: Historical Frames, New Constructions, and Social Problems." *Social Problems* 44:139–154.

Crichton, Michael. 2007. "Patenting Life." *New York Times*, February 13. Available at http://www.nytimes.com

DeNoon, Daniel. 2005 (March 8). *Study: Computer Design Flaws May Create Dangerous Hospital Errors*. Available at http://my.webmd.com

Deutsche Welle. 2003. *German Kids Go to Camp for Internet Addiction*. Available at http://www.dw-world.de/dw/article

Durkheim, Emile. 1973 [1925]. *Moral Education*. New York: Free Press.

Eibert, Mark D. 1998. "Clone Wars." *Reason* 30(2):52–54.

Eilperin, Juliet, and Rick Weiss. 2003. "House Votes to Prohibit All Human Cloning."

Washington Post, February 28. Available at http://www.washingtonpost.com

ETC Group. 2003a. *Contamination by Genetically Modified Maize in Mexico Much Worse Than Feared.* Available at http://www.etcgroup.org

ETC Group. 2003b. *Broken Promise? Monsanto Promotes Terminator Seed Technology.* News release, April 23. Available at http://www.etcgroup.org

Evett, Don. 2007. Spam Statistics 2006. Available at http://spam-filter-review.toptenreviews.com

Fairlie, Robert W. 2005. "Are We Really a Nation On-Line?" Report for the Leadership Conference on Civil Rights Education Fund. Santa Cruz, CA: University of California.

FBI (Federal Bureau of Investigation). 2002. *Online Child Pornography: Innocent Images National Initiative.* Available at http://www.fbi.gov

Freudenheim, Milt. 2005. "Digital Rx: Take Two Aspirins and E-mail Me in the Morning." *New York Times,* March 2. Available at http://www.nytimes.com

Friedman, Thomas L. 2005. *The World Is Flat.* New York: Farrar, Strauss, and Giroux.

Gallaga, Omar L. 2007. "The IT Girls." *American Statesman,* June 23. Available at http://www.statesman.com

Gallup. 2000. *Abortion Issues.* Available at http://www.gallup.com/poll/indicators

Gallup. 2003. *Abortion Issues.* Available at http://www.brain.gallup.com

Gallup. 2007. *Gallup's Pulse of Democracy: Abortion.* Available at http://www.galluppoll.com

Gibbs, Nancy. 2003. "The Secret of Life." *Time,* February 17, pp. 42–45.

Giridharadas, Anand. 2007. "India's Edge Goes Beyond Outsourcing." *New York Times,* April 4. Available at http://www.nytimes.com

Glendinning, Chellis. 1990. *When Technology Wounds: The Human Consequences of Progress.* New York: William Morrow.

Global Internet Project. 1998. *The Workplace.* Available at http://www.gip.org/gip2g.html

Goodman, Paul. 1993. "Can Technology Be Humane?" In *Technology and the Future,* Albert H. Teich, ed. (pp. 239–255). New York: St. Martin's Press.

Gottlieb, Scott. 2000. "Abortion Pill Is Approved for Sale in United States." *BMJ* 321:851.

Greenhouse, Linda. 2006. "Justices Reaffirm Emergency Access to Abortion." *New York Times,* January 19. Available at http://www.nytimes.com

Greenhouse, Linda. 2007. "Justices Back Ban on Method of Abortion." *New York Times,*

April 19. Available at http://www.nytimes.com

Grossman, Lev. 2006. "Time's Person of the Year: You." *Time,* December 13. Available at http://www.time.com

Hancock, LynNell. 1995. "The Haves and the Have-Nots." *Newsweek,* February 27, pp. 50–53.

Harmon, Amy. 2006. "That Wild Streak? Maybe It Runs in the Family." *New York Times,* June 15. Available at http://www.nytimes.com

Hayday, Graham. 2003. *Technology in the Workplace: More Important Than Managers?* Available at http://news.zdnet.co.uk

Holding, Reynolds. 2007. "E-Mail Privacy Gets a Win in Court." *Time,* June 21. Available at http://www.time.com

Horrigan, John. 2007 (May 7). "A Typology of Information and Communication Technology Users." Washington, DC: Pew Internet and American Life Project.

Human Cloning Prohibition Act of 2007. U.S. House of Representatives, Washington, DC.

Human Genome Project. 2007. *Medicine and the New Genetics.* Available at http://www.ornl.gov

International Labour Organization. 2001. *Digital Divide Is Wide and Getting Wider.* World Employment Report. Geneva: International Labour Organization. Available at http://www.ilo.org

Internet Pornography Statistics. 2007. Available at http://www.internet-filter-review.toptenreviews.com

Internet Spyware (I-SPY) Prevention Act of 2007. U.S. House of Representatives, Washington, DC.

Internet World Stats. 2007a. *Internet Usage Statistics for the Americas.* Available at http://www.internetworldstats.com

Internet World Stats. 2007b. *Internet World Users by Language.* Available at http://www.internetworldstats.com

Internet World Stats. 2007c. *Internet Usage Statistics: The Big Picture.* Available at http://www.internetworldstats.com

Jewell, Mark. 2007. "T.J. Maxx Theft Believed Largest Hack Ever." Associated Press, March 30. Available at http://www.msnbc.msn.com

Jones, Jeffrey M. 2006. "One in Three U.S. Workers Have "Telecommuted" to Work." Gallup Poll. August 16. Available at http://poll.gallup.com

Jordan, Tim, and Paul Taylor. 1998. "A Sociology of Hackers." *Sociological Review,* 46(4):757–778.

Kahin, Brian. 1993. "Information Technology and Information Infrastructure." In *Empowering Technology: Implementing a*

U.S. Strategy, Lewis M. Branscomb, ed. (pp. 135–166). Cambridge, MA: MIT Press.

Kahn, A. 1997. "Clone Mammals . . . Clone Man?" *Nature,* 386:119.

Kalb, Claudia. 2005. "Ethics, Eggs, and Embryos." *Newsweek,* June 20, pp. 52–53.

Kalb, Claudia, and B.J. Lee. 2006. "Sizing Up SCNT." *Newsweek,* January 16, p. 10.

Kaplan, Carl S. 2000 (December 22). *The Year in Technology Law.* Available at http://www.nytimes.com

Kilen, Mike. 2003. "She's a Little Fighter." *Des Moines Register,* February 23. Available at http://www.desmoinesregister.com/news

Kliff, Sarah. 2007. "Stem Cells: Many Willing Embryo Donors." *Newsweek,* June 20. Available at http://www.msnbc.msn.com

Klotz, Joseph. 2004. *The Politics of Internet Communication.* Lanham, MD: Rowman and Littlefield.

Krebs, Brian. 2005. "Uncle Sam Gets 'D-Plus' on Cyber-Security." *Washington Post,* February 16. Available at http://www.washingtonpost.com

Kuhn, Thomas. 1973. *The Structure of Scientific Revolutions.* Chicago: University of Chicago Press.

Labaton, Stephen. 2007. "Microsoft Finds Legal Defender in Justice Depart." *New York Times,* June 10. Available at http://nytimes.com

Legay, F. 2001 (January). "Genethics: Should Genetic Information Be Treated Separately?" *Virtual Mentor,* p. E5. Available at http://virtualmentor.ama-assn.org/site/archives.html

Lemonick, Michael. 2006a. "Are We Losing our Edge?" *Time,* February 13, pp. 22–33.

Lemonick, Michael. 2006b. "The Iceland Experiment." *Time,* February 20, pp. 49–52.

Lemonick, Michael, and Dick Thompson. 1999. "Racing to Map Our DNA." *Time Daily* 153:1–6. Available at http://www.time.com

Lenhart, Amanda, and Mary Madden. 2007. *Teens, Privacy and Online Social Networks.* Washington, DC: Pew Internet and American Life Project.

Lohr, Steve. 2006. "Outsourcing is Climbing Skills Ladder." *New York Times,* February 16. Available at http://www.nytimes.com

Lynch, Colum. 2005. "U.N. Backs Human Cloning Ban." *Washington Post,* March 8. Available at http://www.washingtonpost.com

Macklin, Ruth. 1991. "Artificial Means of Reproduction and Our Understanding of the Family." *Hastings Center Report,* January–February, pp. 5–11.

MacMillan, Robert. 2004. "Primer: Children, the Internet and Pornography." *Washington Post*, June 29. Available at http://www.washingtonpost.com

MacMillan, Robert. 2005. "My Cubicle, My Cell." *Washington Post*, May 19. Available at http://www.washingtonpost.com

March of Dimes. 2007 (May 2). *Nearly 28,000 U.S. Infants Died in 2004. . . .* Available at http://www.marchofdimes.com

Markoff, John. 2000. "Report Questions a Number in Microsoft Trial." *New York Times*, August 28. Available at http://www.nytimes.com

Markoff, John. 2002. "Protesting the Big Brother Lens, Little Brother Turns an Eye Blind." *New York Times*, October 7. Available at http://www.nytimes.com

Mayer, Sue. 2002. "Are Gene Patents in the Public Interest?" *BIO-IT World*, November 12. Available at http://www.bioitworld.com

McCormick, S. J., and A. Richard. 1994. "Blastomere Separation." *Hastings Center Report*, March-April, pp. 14–16.

McDermott, John. 1993. "Technology: The Opiate of the Intellectuals." In *Technology and the Future*, Albert H. Teich, ed. (pp. 89–107). New York: St. Martin's Press.

McFarling, Usha L. 1998. "Bioethicists Warn Human Cloning Will Be Difficult to Stop." *Raleigh News and Observer*, November 18, p. A5.

McGann, Rob. 2005 (January 6). "Most Active Web Users Are Young, Affluent." *Jupiter Research*. Available at http://www.clickz.com

McMillan, Robert. 2007. "IBM Contractor Loses Employee Data." *Infoworld*, May 15. Available at http://www.infoworld.com

McNair, James. 2003. "Rising Unemployment Affects Even Tech Sector." *Cincinnati Enquirer*, March 26, p. 1.

Mearian, Lucas. 2007. "Georgia Agency Loses Private Data of 2.9 M Medicaid Recipients." *Computerworld*, April 10. Available at http://www.computerworld.com

Merton, Robert K. 1973. "The Normative Structure of Science." In *The Sociology of Science*, Robert K. Merton, ed. Chicago: University of Chicago Press.

Mesthene, Emmanuel G. 1993. "The Role of Technology in Society." In *Technology and the Future*, Albert H. Teich, ed. (pp. 73–88). New York: St. Martin's Press.

Mundy, Liza. 2006. "Souls on Ice: American's Embryo Glut and the Wasted Promise of Stem Cell Research." *Mother Jones*, July/August. Available at http://www.motherjones.com

Murphie, Andrew, and John Potts. 2003. *Culture and Technology*. New York: Palgrave Macmillan.

National Science Foundation. 2007a. "U.S. R&D Increased 6.0% in 2006 According to NSF." Available at http://www.nsf.gov

National Science Foundation. 2007b. "Science and Engineering Indicators, 2006." Available at http://www.nsf.gov

NCSL (National Conference of State Legislatures). 2006. "State Human Cloning Laws." Available at http://ncsl.org

NCSL. 2007. "State Genetic Privacy Laws." Available at http://ncsl.org

NIC (National Intelligence Council). 2003. "The Global Technology Revolution, Preface and Summary." Rand Corporation. Available at http://www.rand.org

Nie, Norman, Alberto Simpser, Irena Stepanikova, and Lu Zheng. 2004 (December). *Ten Years After the Birth of the Internet: How Do Americans Use the Internet in Their Daily Lives?* Stanford Institute for the Quantitative Study of Society.

Noyes, Katherine. 2007. "Researchers Create 'Creepy' Child Robot." *TechNewsWorld*, June 8. Available at http://www.technewsworld.com

O'Donnell, Jayne. 2007. "Computers Bumped from Top of Online Sales." *USA Today*, May 13. Available at http://www.usatoday.com

Ogburn, William F. 1957. "Cultural Lag as Theory." *Sociology and Social Research* 41:167–174.

Papadakis, Maria. 2000 (March 31). *Complex Picture of Computer Use in Home Emerges*. Washington, DC: National Science Foundation, NSF000-314.

Perrolle, Judith A. 1990. "Computers and Capitalism." In *Social Problems Today*, James M. Henslin, ed. (pp. 336–342). Englewood Cliffs, NJ: Prentice-Hall.

Pethokoukis, James M. 2004. "Meet Your New Co-Worker." *U.S. News and World Report*, March 7. http://www.usnews.com

Pew. 2006. *Awareness of Genetically Modified Food Has Declined Over the Last Five Years*. Available at http://pewagbiotech.org/research

Pew. 2007a. *Demographics of Internet Users*. Available at http://www.pewinternet.org

Pew. 2007b. *34% of Internet Users Have Logged on with a Wireless Internet Connection. . . .* Available at http://www.pewinternet.org

Pew. 2007c. *Trends In Political Values and Core Attitudes: 1987–2007*. Washington, DC: The Pew Research Center.

Pew. 2007d. *Internet Activities*. Available at http://www.pewinternet.org

Pew Research Center. 2005 (May 23). *More See Benefits of Stem Cell Research*. Available at http://people-press.org/commentary

Postman, Neil. 1992. *Technopoly: The Surrender of Culture to Technology*. New York: Alfred A. Knopf.

Rabino, Isaac. 1998. The Biotech Future." *American Scientist* 86(2):110–112.

Rawe, Julie. 2006. "How Safe is MySpace?" *Time*, July 3, pp. 34–36.

Reuters. 2005. *Report: China's New Bid to Gag Web*. Available at http://www.cnn.com/technology

Richtel, Matt, and Bob Tedeschi. 2007. "Online Sales Lose Steam." *New York Times*, June 17. Available at http://www.nytimes.com

Rosenberg, Debra. 2003. "The War over Fetal Rights." *Newsweek*, June 9, pp. 40–44.

Ruiz-Marrero, Carmelo. 2002. *Genetic Pollution: Starlink Corn Invades Mexico. CorpWatch*. Available at http://www.corpwatch.org

Schaefer, Naomi. 2001. "The Coming Internet Privacy Scrum." *American Enterprise* 12:50–51.

Schneider, Greg. 2004. "Slowdown Forces Many to Wander for Work." *Washington Post*, November 9, p. A01.

Shand, Hope. 1998. "An Owner's Guide." *Mother Jones*, May-June, p. 46.

Sherman, Mark. 2006. "Supreme Court to Consider Abortion Issue." Associated Press. November 8. Available at http://www.breitbart.com

Skillings, Jonathan. 2006. "In Japan, Robots Are People, Too." C/Net News.Com, October 11. Available at http://news.com.com

Smeal, Eleanor. 2007. "Supreme Court Ruling in Abortion Ban Case Disastrous for Women's Health and Safety." Press Release, April 18. Available at http://www.feminist.org

Svensson, Peter. 2007 (January 30). *Robot Parking Garage to Open in New York*. Available at http://www.breitbart.com

Taylor, Andrew. 2007. "House Debates Stem Cell Research Bill." Associated Press, January 11. Available at http://www.breitbart.com

Taylor, Chris. 2003. "Spam's Big Bang." *Time*, June 16, pp. 50–53.

Tesler, Pearl. 2003. "Universal Robots: The History and Workings of Robots." Tech Museum of Innovation. Available at http: www.thetech.org/robotics

Toffler, Alvin. 1970. *Future Shock*. New York: Random House.

Tumulty, Karen. 2006. "Where the Real Action Is. . . ." *Time*, January 30, pp. 50–53.

UNDP (United Nations Development Programme) 2003. *Human Development Report 2003*. New York: Oxford University Press.

U.S. Bureau of the Census. 2007. *Statistical Abstract of the United States, 2006,* 126th ed. Washington, DC: U.S. Government Printing Office.

U.S. Department of Education. 2007. *The American Competitiveness Initiative: A Continued Commitment to Leading the World in Innovation.* Available at http://www.ed.gov

U.S. Department of Justice. 1999. *1999 Report on Cyberstalking: A New Challenge for Law Enforcement and Industry.* Washington, DC: Department of Justice.

U.S. Newswire. 2003 (October 2). *Feminists Condemn House Passage of Deceptive Abortion Ban, Urge Activists to March on Washington.* Available at http://www.usnewswire.com

Vestal, Christine. 2007 (June 21). *Embryonic Stem Cell Research Divides States.* Available at http://www.stateline.org

Wallis, Claudia, and Sonja Steptoe. 2006. "E-Mail and Cellphones Help Us Multi-Task. . . ." *Time,* January 16, pp.73–76.

Weinberg, Alvin. 1966. "Can Technology Replace Social Engineering?" *University of Chicago Magazine* 59(October):6–10.

Weiss, Rick. 2005. "Stem Cell Injections Repair Spinal Cord Injuries in Mice." *Washington Post,* September 20. Available at http://www.washingtonpost.com

Welter, Cole H. 1997. "Technological Segregation: A Peek Through the Looking Glass at the Rich and Poor in an Information Age." *Arts Education Policy Review* 99(2):1–6.

White House. 2003 (November 5). *President Bush Signs Partial Birth Abortion Ban Act of 2003.* Available at http://www.whitehouse.gov

White, Lawrence. 2000. "Colleges Must Protect Privacy in the Digital Age." *Chronicle of Higher Education,* June 30, pp. B5, 6.

Willing, Richard. 2005. "Federal Judge Blocks 'Partial Birth Abortion' Ban." *USA Today,* June 1. Available at http://www.usatoday.com

Winner, Langdon. 1993. "Artifact/Ideas as Political Culture." In *Technology and the Future,* Albert H. Teich, ed. (pp. 283–294). New York: St. Martin's Press.

Chapter 16

American Jewish Committee. 2007. *2006 Annual Survey of American Jewish Opinion.* Available at http://www.ajc.org

Arms Control Association. 2003a (January). *Start II and Its Extension at a Glance.* Available at http://www.armscontrol.org/factsheets

Arms Control Association. 2003b (May). *The Nuclear Proliferation Treaty at a Glance.*

Available at http://www.armscontrol.org/factsheets

Associated Press. 2005 (March 15). *Pentagon Tests Negative for Anthrax.* MSNBC. Available at http://www.msnbc.msn.com

Baliunas, Sallie. 2004. "Anthrax Is a Serious Threat." In *Biological Warfare,* William Dudley, ed. (pp. 53–58). Farmington Hills, MA: Greenhaven Press.

Barkan, Steven, and Lynne Snowden. 2001. *Collective Violence.* Boston: Allyn and Bacon.

Barnes, Robert. 2007. "Justices Weigh Detainee Rights." *Washington Post,* June 30. Available at http://www.washingtonpost.com

Barnes, Steve. 2004. "No Cameras in Bombing Trial." *New York Times,* January 29, p. 24.

BBC. 2000. "UN Admits Rwanda Genocide Failure." *BBC News,* April 15. Available at http://news.bbc.co.uk

BBC. 2007a. "Hamas Takes Full Control of Gaza." *BBC News,* June 15. Available at http://news.bbc.co.uk

BBC. 2007b. "Fresh Clashes Engulf Lebanon Camp." *BBC News,* June 1. Available at http://news.bbc.co.uk

BBC. 2007c. *EU Constitution: Where Member States Stand.* Available at http://news.bbc.co.uk

BBC. 2007d. "Q & A: Sudan's Darfur Conflict." *BBC News,* May 29. Available at http://news.bbc.co.uk

BBC. 2007e. "Q & A: Iran and the Nuclear Issue." *BBC News,* May 24. Available at http://bbc.news.co.uk

Belasco, Amy. 2007. *The Cost of Iraq, Afghanistan, and Other Global War on Terror Operations Since 9/11.* Congressional Research Service. Available at http://www.opencrs.com

Bellamy, Carol. 2003. *Children Are War's Greatest Victims.* Organization for Economic Cooperation and Development. Available at http://www.oecdobserver.org

Bercovitch, Jacob, ed. 2003. *Studies in International Mediation: Advances in Foreign Policy Analysis.* New York: Palgrave Macmillan.

Bergen, Peter. 2002. *Holy War, Inc.: Inside the Secret World of Osama bin Laden.* New York: Free Press.

Berrigan, Frida, and William Hartung. 2005. *U.S. Weapons at War: Promoting Freedom or Fueling Conflict?* World Policy Institute Report. Available at http://www.worldpolicy.org/projects

Bjorgo, Tore. 2003. *Root Causes of Terrorism.* Paper presented at the International Expert Meeting, June 9–11. Oslo, Norway: Norwegian Institute of International Affairs.

Bonn International Center for Conversion. 1998. *Conversion Survey, 1998,* ch. 6. Bonn, Germany: Bonn International Center for Conversion.

Borum, Randy. 2003. "Understanding the Terrorist Mind-Set." *FBI Law Enforcement Bulletin,* July, pp. 7–10.

Brauer, Jurgen. 2003. "On the Economics of Terrorism." *Phi Kappa Phi Forum,* Spring, pp. 38–41.

Briggs, Jimmie. 2005. *Innocents Lost: When Child Soldiers Go To War.* New York: Basic Books.

Brinkley, Joel, and David E. Sanger. 2005. "North Koreans Agree to Resume Nuclear Talks." *New York Times,* July 10. Available at http://www.nytimes.com

Brooks, Rosa. 2007. "To the Rest of the World, We're Cheapskates." *The Los Angeles Times,* April 13. Available at http://www.latimes.com

Brown, David. 2006. "Study Claims Iraq's 'Excess' Death Toll Has Reached 655,000." *Washington Post,* October 11. Available at http://www.washingtonpost.com

Brown, Michael E., Sean M. Lynn-Jones, and Steven E. Miller, eds. 1996. *Debating the Democratic Peace.* Cambridge, MA: MIT Press.

Burns, Robert. 2007. "U.S. Official: North Korea Has New Missile." *Washington Post,* July 6. Available at http://www.washingtonpost.com

Carneiro, Robert L. 1994. "War and Peace: Alternating Realities in Human History." In *Studying War: Anthropological Perspectives,* S. P. Reyna and R. E. Downs, eds. (pp. 3–27). Langhorne, PA: Gordon & Breach.

Carroll, Joseph. 2007a (June 22). *Only 29% of Americans Say U.S. Is Winning the War on Terrorism.* Available at http://www.galluppoll.com

Carroll, Joseph. 2007b (March 2). *Perceptions of 'Too Much' Military Spending at 15-Year High.* Available at http://www.galluppoll.com

Cauffman, Elizabeth, Shirley Feldman, Jaime Waterman, and Hans Steiner. 1998. "Post-Traumatic Stress Disorder Among Female Juvenile Offenders." *Journal of the American Academy of Child and Adolescent Psychiatry* 37:1209–1217.

Center for Defense Information. 2003. *Military Almanac.* Available at http://www.cdi.org

Center for Defense Information. 2007 (April 30). *The World's Nuclear Arsenals.* Available at http://ww.cdi.org

Center for Public Integrity. 2005. *Military Professional Resources, Inc. Background.* Available at http://www.publicintegrity.org

CNN. 2001a (January 31). *Libyan Bomber Sentenced to Life.* Available at http://www.europe.cnn.com

CNN. 2001b (February 23). *Rape War Crime Verdict Welcomed.* Available at http://www.cnn.com

CNN. 2003 (February 26). *Poll: Muslims Call U.S. Ruthless, Arrogant.* Available at http://www.cnn.com

CNN. 2007 (July 31). "UN Troops Will Go To War-Ravaged Darfur." Available at http://www.cnn.com/2007/WORLD/africa

CNN/Opinion Research Corporation Poll. 2007 (June 22–24). *Iraq.* Available at http://www.pollingreport.com/iraq3.htm

Cohen, Ronald. 1986. "War and Peace Proneness in Pre- and Post-industrial States." In *Peace and War: Cross-Cultural Perspectives*, M. L. Foster and R. A. Rubinstein, eds. (pp. 253–267). New Brunswick, NJ: Transaction Books.

Conflict Research Consortium. 2003. *Mediation.* Available at http://www.colorado.edu/conflict/peace

Cowell, Alan. 2007. "Scottish Panel Challenges Lockerbie Conviction." *New York Times*, June 29. Available at http://www.nytimes.com

Crary, David. 2005. "Army Marriages Dying in the War on Terror." *Salt Lake Tribune*, June 30. Available at http://www.sltrib.com

CRS (Congressional Research Service). 2007 (June 21.) *Private Security Contractors in Iraq: Background, Legal Status, and Other Issues.* Available at http://www.opencrs.com

Davenport, Christian. 2005. "Guard's New Pitch: Fighting Words." *Washington Post*, April 28. Available at http://www.washingtonpost.com

Davenport, Paul. 2007. "States Crack Down on T-Shirts With Fallen Soldiers' Names." *Associated Press*, May 18. Available at http://content.hamptonroads.com

De Jong, J., W. Scholte, M. Koeter, and A. Hart. 2000. "The Prevalence of Mental Health Problems in Rwandan and Burundese Refugee Camps." *Acta Psychiatrica Scandinavica* 102:171–177.

Dedman, Bill. 2006 (October 24). *In Limbo: Cases Are Few Against Gitmo Detainees.* Available at http://www.msnbc.msn.com

Deen, Thalif. 2000 (September 9). *Inequality Primary Cause of Wars, Says Annan.* Available at http://www.hartford-hwp.com/archives

DeYoung, Karen, and Walter Pincus. 2007. "Al-Qaeda's Gains Keep U.S. at Risk, Report Says." *Washington Post*, July 18, A01.

Dickey, Christopher. 2005. "Iran's Nuclear Lies." *Newsweek*, July 11, pp. 36–40.

Dixon, William J. 1994. "Democracy and the Peaceful Settlement of International Conflict." *American Political Science Review* 88(1):14–32.

Doherty, Carroll. 2007 (June 27). "Who Flies the Flag? Not Always Who You Might Think." *Pew Research Center for the People and the Press.* Available at http://www.pewresearch.org

Environmental Media Services. 2002 (October 7). *Environmental Impacts of War.* Available at http://www.ems.org

Espiau, Gorka. 2006. *The Basque Conflict: New Ideas and Prospects for Peace. Special Report*, No. 161, United States Institute of Peace. Available at http://www.usip.org

Feder, Don. 2003. "Islamic Beliefs Led to the Attack on America." In *The Terrorist Attack on America*, Mary E. Williams, ed. (pp. 20–23). Farmington Hills, MA: Greenhaven Press.

Federation of American Scientists. 2007. *Estimate of the U.S. Nuclear Weapons Stockpile, 2007 and 2012.* Available at http://www.fas.org

Fischer, Howard. 2007. "ACLU Seeks To End Ban on Sales of Shirt Listing Iraq War Dead." *The Arizona Daily Star*, June 29. Available at http://www.azstarnet.com

"Foreign Conservationists Under Siege." 2000. Editorial. *New York Times*, April 1, p. A14.

Franke, Volker C. 2001. "Generation X and the Military: A Comparison of Attitudes and Values between West Point Cadets and College Students." *Journal of Political and Military Sociology* 29:92–119.

Funke, Odelia. 1994. "National Security and the Environment." In *Environmental Policy in the 1990s: Toward a New Agenda*, 2nd ed., Norman J. Vig and Michael E. Kraft, eds. (pp. 323–345). Washington, DC: Congressional Quarterly.

Gardner, Simon. 2007 (February 22). *Sri Lanka Says Sinks Rebel Boats on Truce Anniversary.* Available at http://www.reuters.com

Garrett, Laurie. 2001. "The Nightmare of Bioterrorism." *Foreign Affairs* 80:76.

Geneva Graduate Institute for International Studies. 2006. *Small Arms Survey, 2006.* New York: Oxford University Press.

Ghosh, Aparisim. 2005. "Inside the Mind of an Iraqi Suicide Bomber." *Time*, July 4, pp. 22–29.

Gioseffi, Daniela. 1993. "Introduction." In *On Prejudice: A Global Perspective*, Daniela Gioseffi, ed. (pp. xi-l). New York: Anchor Books, Doubleday.

Global Policy Forum. 2007. *UN Sanctions Against Iran?* Available at http://www.globalpolicy.org

Global Security. 2005. *Liberian Conflict.* Available at http://www.globalsecurity.org

Global Security. 2007. *U.S. Casualties in Iraq.* Available at http://www.globalsecurity.org

Goure, Don. 2003. *First Casualties? NATO, the U.N.* MSNBC News, March 20. Available at http://www.msnbc.com/news

GreenKarat. 2007. *Mining for Gems.* Available at http://www.greenkarat.com

Haavisto, Pekka. 2005. "Environmental Impacts of War." In *State of the World: 2005*, Linda Starke, ed. (pp. 158–159). New York: W. W. Norton.

Hagan, John, and Alberto Palloni. 2006. "Social Science: Death in Darfur." *Science* 313:1578–1579. Available at http://www.sciencemag.org

Hayman, Peter, and Douglas Scaturo. 1993. "Psychological Debriefing of Returning Military Personnel: A Protocol for Post-Combat Intervention." *Journal of Social Behavior and Personality* 8(5):117–130.

Hoge, Charles, Carl A. Castro, Stephen C. Messer, Dennis McGurk, Dave I. Cotting, and Robert L. Koffman. 2004. "Combat Duty in Iraq and Afghanistan, Mental Health Problems, and Barriers to Care." *New England Journal of Medicine* 351:13–22. Available at http://content.nejm.org

Hooks, Gregory, and Leonard E. Bloomquist. 1992. "The Legacy of World War II for Regional Growth and Decline: The Effects of Wartime Investments on U.S. Manufacturing, 1947–72." *Social Forces* 71(2):303–337.

Huntington, Samuel. 1996. *The Clash of Civilizations and the Remaking of World Order.* New York: Simon and Schuster.

Inglesby, Thomas, Donald Henderson, John Bartlett, Michael Archer, et al. 1999. "Anthrax as a Biological Weapon: Medical and Public Health Management." *Journal of the American Medical Association* 281:1735–1745.

International Campaign to Ban Landmines. 2007. *What is the Mine Ban Treaty?* Available at http://www.icbl.org

Interpol. 1998. *Frequently Asked Questions About Terrorism.* Available at http://www.kenpubs.co.uk/interpol.com/English/faq

Johnston, David, and John Broder. 2007a. "F.B.I. Says Guards Killed 14 Iraqis Without Cause." *The New York Times*, November 14. Available at http://www.nytimes.com

Johnston, David, and John Broder. 2007b. "U.S. Prosecutors Subpoena Blackwater Employees." *The New York Times*, November 14. Available at http://www.nytimes.com

Kemper, Bob. 2003. "Agency Wages Media Battle." *Chicago Tribune*, April 7. Available at http://www.chicagotribune.com

Klare, Michael. 2001. *Resource Wars: The New Landscape of Global Conflict.* New York: Metropolitan Books

Klingman, Avigdor, and Zehara Goldstein. 1994. "Adolescents' Response to Unconventional War Threat Prior to the Gulf War." *Death Studies* 18:75–82.

Knickerbocker, Brad. 2002. "Return of the Military-Industrial Complex?" *Christian Science Monitor*, February 13. Available at http://www.csmonitor.com

Kyl, Jon, and Morton Halperin. 1997. "Q: Is the White House's Nuclear-Arms Policy on the Wrong Track?" *Insight on the News* 42:24–28.

Lamont, Beth. 2001. "The New Mandate for UN Peacekeeping." *The Humanist* 61:39–41.

Landmines. 2003. *The Problem: Impact of Landmines.* Available at http://www.landmines.org

Laqueur, Walter. 2006. "The Terrorism to Come." In *Annual Editions 05–06,* Kurt Finsterbusch, ed. (pp. 169–176), Dubuque, IA: McGraw-Hill/Dushkin.

Lederer, Edith. 2005 (May 20). "Annan Lays out Sweeping Changes to U.N." Associated Press. Available at http://www.apnews

Levy, Jack S. 2001. "Theories of Interstate and Intrastate War: A Levels of Analysis Approach." In *Turbulent Peace: The Challenges of Managing International Conflict,* Chester A. Crocker, Fen Osler Hampson, and Pamela Aall, eds. (pp. 3–27), Washington, DC: U.S. Institute of Peace.

Li, Xigon, and Ralph Izard. 2003. "Media in a Crisis Situation Involving National Interest: A Content Analysis of Major U.S. Newspapers' and TV Networks' Coverage of the 9/11 Tragedy." *Newspaper Research Journal* 24:1–16.

Lifsher, Marc. 1999. "Critics Assail Plans to Dispose of Paint Cans." *Wall Street Journal,* June 16, p. CA1.

Lloyd, John. 1998. "The Dream of Global Justice." *New Statesman* 127:28–30.

Lorch, Donatella, and Preston Mendenhall. 2000. "A War's Hidden Tragedy." *Newsweek,* August 8. Available at http://www.msnbc.com/news

MacAskill, Ewen, Ed Pilkington, and Jon Watts. 2006. "Despair at UN over Selection of 'Faceless' Ban Ki-moon as General Secretary." *The Guardian,* October 7. Available at http://www.guardian.co.uk

Marshall, Leon. 2007 (July 16). *Elephants 'Learn' to Avoid Land Mines in War-Torn Angola.* National Geographic News. Available at http://www.nationalgeographic.com

Marshall, Monty G., and Ted Robert Gurr. 2005. *Peace and Conflict: A Global Survey of Armed Conflicts, Self-Determination Movements, and Democracy.* College Park, MD: Center for International Development and Conflict Management.

McElroy, Wendy. 2003 (March 18). *Iraq War May Kill Feminism as We Know It.* Fox News Channel. Available at http://www.foxnews.com

McGirk, Tim. 2006. *Collateral Damage or Civilian Massacre in Haditha? Time,* March 19. Available at http://www.time.com

Military. 2007. *Tuition Assistance (TA) Program Overview.* Available at http://education.military.com

Military Professional Resources Inc. 2007. *About MPRI.* Available at http://www.mpri.com

Miller, Susan. 1993. "A Human Horror Story." *Newsweek,* December 27, p. 17.

Miller, T. Christian. 2007. "Private Contractors Outnumber U.S. Troops in Iraq." *Los Angeles Times,* July 4. Available at http://www.latimes.com

Morahan, Lawrence. 2003. "Hussein Threatens to Become Symbol of Nationalism in Iraq." *Townhall,* July 22. Available at http://www.townhall.com/news

Myers-Brown, Karen, Kathleen Walker, and Judith A. Myers-Walls. 2000. "Children's Reactions to International Conflict: A Cross-Cultural Analysis." Paper presented at the National Council of Family Relations, Minneapolis, November 20.

NASP (National Association of School Psychologists). 2003. *Children and Fear of War and Terrorism.* Available at http://www.nasponline.org

National Center for Posttraumatic Stress Disorder. 2007. *What Is Post Traumatic Stress Disorder?* Available at http://www.ncptsd.va.gov

National Priorities Project. 2007. *Federal Budget Trade-Offs.* Available at http://database.nationalpriorities.org

National Security Council. 2002. *The National Security Strategy of the United States of America.* Available at http://www.whitehouse.gov/nsc

Office of Management and Budget. 2007. *Table S-10: Outlays by Function.* Available at http://www.whitehouse.gov

Office of the Coordinator for Counterterrorism. 2007. *Country Reports on Terrorism, 2006.* Washington, DC: U.S. Department of State.

O'Hanlon, Michael E., and Jason H. Campbell. 2007 (July 16). *Iraq Index: Tracking Variables of Reconstruction and Security in Post-Saddam Iraq.* Available at http://www.brookings.edu

O'Prey, Kevin P. 1995. *The Arms Export Challenge: Cooperative Approaches to Export Management and Defense Conversion.* Washington, DC: Brookings Institution.

Pape, Robert A. 2005. *Dying to Win: The Strategic Logic of Suicide Bombers.* New York: Random House.

Partnership Africa Canada. 2007. *The Kimberly Controls: How Effective?* Available at http://blooddiamond.pacweb.org/kimberlyprocess

Paul, Annie Murphy. 1998. "Psychology's Own Peace Corps." *Psychology Today* 31:56–60.

PBS (Public Broadcasting System). 2003. *FAQ: Cybersecurity.* Available at http://www.pbs.org

Perl, Raphael. 2003. *Terrorism, the Future, and U.S. Policy.* Congressional Research Service. Washington, DC: Library of Congress.

Permanent Court of Arbitration. 2007. *About Us* and *Cases.* Available at http://www.pca-cpa.org

Pew Research Center. 2005 (July 13). *Guantanamo Prisoner Mistreatment Seen as an Isolated Incident.* Available at http://www.pewtrust.org

Pfefferbaum, Betty. 1997. "Post-Traumatic Stress Disorder in Children: A Review of the Last Ten Years." *Journal of the American Academy of Child and Adolescent Psychiatry* 36:1503–1512.

Porter, Bruce D. 1994. *War and the Rise of the State: The Military Foundations of Modern Politics.* New York: Free Press.

Powell, Bill, and Tim McGirk. 2005. "The Man Who Sold the Bomb." *Time,* February 14, pp. 22–31.

Rashid, Ahmed. 2000. *Taliban: Militant Islam, Oil, and Fundamentalism in Central Asia.* New Haven, CT: Yale University Press.

Rasler, Karen, and William R. Thompson. 2005. *Puzzles of the Democratic Peace: Theory, Geopolitics, and the Transformation of World Politics.* New York: Palgrave Macmillan.

Renner, Michael. 1993. "Environmental Dimensions of Disarmament and Conversion." In *Real Security: Converting the Defense Economy and Building Peace,* Karl Cassady and Gregory A. Bischak, eds. (pp. 88–132). Albany: State University of New York Press.

Renner, Michael. 2000. "Number of Wars on Upswing." In *Vital Signs: The Environmental Trends That Are Shaping Our Future,* Linda Starke, ed. (pp. 110–111). New York: W. W. Norton.

Renner, Michael. 2005. "Disarming Postwar Societies." In *State of the World: 2005,* Linda Starke, ed. (pp. 122–123). New York: W. W. Norton.

Reuters. 2001 (November 26). *Death Toll Climbs Despite Lull in Sri Lanka Conflict.* Available at http://www.in.news.yahoo.com

Robinson, Eugene. 2006. "Counting the Iraqi Dead." *Washington Post,* October 13. Available at http://www.washingtonpost.com

Romero, Anthony. 2003. "Civil Liberties Should Not Be Restricted During Wartime." In *The Terrorist Attack on America*, Mary Williams, ed. (pp. 27–34). Farmington Hills, MA: Greenhaven Press.

Rosenberg, Tina. 2000. "The Unbearable Memories of a U.N. Peacekeeper." *New York Times*, October 8, pp. 4, 14.

Saad, Lydia. 2006. "Anti-Muslim Sentiments Fairly Commonplace." August 10. Available at http://poll.gallup.com

Salem, Paul. 2007. "Dealing with Iran's Rapid Rise in Regional Influence." *The Japan Times*, February 22. Available at http://www.carnegieendowment.org

Salladay, Robert. 2003 (April 7). *Anti-War Patriots Find They Need to Reclaim Words, Symbols, Even U.S. Flag from Conservatives*. Available at http://www.commondreams.org

Save the Children. 2005. *One World, One Wish: The Campaign to Help Children and Women Affected by War*. Available at http://www.savethechildren.org

Save the Children. 2007a. *State of the World's Mothers: Saving the Lives of Children Under Five*. Available at http://www.savethechildren.org

Save the Children. 2007b. *Rewrite the Future: Education for Children in Conflict-Affected Countries*. Available at http://www.savethechildren.org

Scheff, Thomas. 1994. *Bloody Revenge*. Boulder, CO: Westview Press.

Schmitt, Eric. 2005. "Pentagon Seeks to Shut Down Bases Across Nation." *New York Times*, May 14. Available at http://www.nytimes.com

Schroeder, Matt. 2007. "The Illicit Arms Trade." Washington, DC: Federation of American Scientists. Available at http://www.fas.org

Schultz, George P., William J. Perry, Henry A. Kissinger, and Sam Nunn. 2007. "A World Free of Nuclear Weapons." *Wall Street Journal*, January 4 Available at http://www.wsj.com

Sengupta, Somini. 2003. "Taylor Steps Down as President of Liberia." *New York Times*, August 11. Available at http://www.nytimes.com

Shah, Anup. 2007. *The Arms Trade Is Big Business*. Available at http://www.globalissues.org

Shanahan, John J. 1995. "Director's Letter." *Defense Monitor* 24(6):8.

Shrader, Katherine. 2005. "WMD Commission Releases Scathing Report." *Washington Post*, March 31. Available at http://www.washingtonpost.com

Silverstein, Ken. 2007. "Six Questions for Joost Hiltermann on Blowback from the Iraq-Iran War." *Harper's Magazine*, July 5. Available at http://www.harpers.org

Skocpol, Theda. 1994. *Social Revolutions in the Modern World*. Cambridge, UK: Cambridge University Press.

Smith, Craig. 2004. "Libya to Pay More to French in '89 Bombing." *New York Times*, January 9, p. 6.

Smith, Lamar. 2003. "Restricting Civil Liberties During Wartime Is Justifiable." In *The Terrorist Attack on America*, Mary Williams, ed. (pp. 23–26). Farmington Hills, MA: Greenhaven Press.

Social Science Research Council. 2003. *Introduction to New World Order*. Available at http://www.ssrc.org/sept11/essays

Stalenheim, Petter. Catalina Perdomo, and Elis Skons. 2007. *Chapter 8: Military Expenditure*. Available at http://yearbook2007.sipri.org/chap8

Starr, J. R., and D. C. Stoll. 1989. *U.S. Foreign Policy on Water Resources in the Middle East*. Washington, DC: Center for Strategic and International Studies.

State of Hawaii. 2000. *Dual-Use Technologies*. Available at http://www.state.hi.us

Stimson Center. 2007. *Reducing Nuclear Dangers in South Asia*. Available at http://www.stimson.org

SIPRI 2007a. *SIPRI Yearbook 2007: Armaments, Disarmament, and International Security*. Oxford, UK: Oxford University Press.

SIPRI. 2007b. *Recent Trends in Military Expenditures*. Available at http://www.sipri.org

Strobel, Warren, David Kaplan, Richard Newman, Kevin Whitelaw, and Thomas Grose. 2001. "A War in the Shadows." *U.S. News and World Report* 130:22.

Sutker, Patricia B., Madeline Uddo, Karen Brailey, and Albert N. Allain, Jr. 1993. "War Zone Trauma and Stress-Related Symptoms in Operation Desert Storm (ODS) Returnees." *Journal of Social Issues* 40(4):33–50.

Tavernise, Sabrina. 2005. "Many Iraqis See Sectarian Roots in New Killing." *New York Times*, May 27. Available at http://www.nytimes.com

Tavernise, Sabrina. 2007. "U.S. Contractor Banned by Iraq Over Shootings." *New York Times*, September 18. Available at http://www.nytimes.com

Thompson, Mark. 2007. "Broken Down" *Time*, April 16, pp. 28–33.

United Nations. 1948. *Convention on the Prevention and Punishment of the Crime of Genocide*. Available at http://www.un.org

United Nations. 2003. *Some Questions and Answers*. Available at http://www.unicef.org

United Nations. 2007. *United Nations Peacekeeping Operations*. Available at http://www.un.org/Depts/dpko/dpko/bnote.htm

U.S. Army Medical Command. 2007 (May). *Mental Health Advisory Team IV Findings*. Available at http://www.armymedicine.army.mil

U.S. Bureau of the Census. 2004. *Statistical Abstract of the United States: 2004–2005*, 124th ed. Washington, DC: U.S. Government Printing Office.

U.S. Department of Homeland Security. 2007. *Securing Our Homeland: Strategic Plan*. Available at http://www.dhs.gov

U.S. Department of State. 2007. *Small Arms/Light Weapons Destruction*. Available at http://www.state.gov

U.S. Nuclear Weapons Cost Study Project. 1998. *50 Facts About U.S. Nuclear Weapons*. Available at http://www.brookings.edu

Vasterling, Jennifer J., Susan P. Proctor, Paul Amoroso, Robert Kane, Timothy Heeren, and Roberta White. 2006. "Neuropsychological Outcomes of Army Personnel Following Deployment to the Iraq War." *Journal of the American Medical Association* 296(5):219–529. Available at http://jama.ama-assn.org

Vesely, Milan. 2001. "UN Peacekeepers: Warriors or Victims?" *African Business* 261:8–10.

Viner, Katharine. 2002. "Feminism as Imperialism." *The Guardian*, September 21. Available at http://www.guardian.co.uk

Walker, Paul F., and Jonathan B. Tucker. 2006. "The Real Chemical Threat." *The Los Angeles Times*, April 1. Available at http://cns.miis.edu

Waters, Rob. 2005. "The Psychic Costs of War." *Psychotherapy Networker*, March–April, pp. 1–3.

Watkins, Thomas. 2007. *Haditha Hearings Enter Fourth Day*. Time, May 11. Available at http://www.time.com

Wax, Emily. 2003. "War Horror: Rape Ruining Women's Health." *Miami Herald*, November 3. Available at http://www.miami.com

Weimann, Gabriel. 2006. "Terror on the Internet: The New Arena, the New Challenges." Washington, DC: U.S. Institute of Peace Press.

Weiner, Tim. 2005. "Air Force Seeks Bush's Approval for Space Weapons Program." *New York Times*, May 18. Available at http://www.nytimes.com

Williams, Mary E., ed. 2003. *The Terrorist Attack on America*. Farmington Hills, MA: Greenhaven Press.

Worldwatch Institute. 2003. *Vital Signs: The Trends That Are Shaping Our Future*. New York: W. W. Norton.

Zakaria, Fareed. 2000. "The New Twilight Struggle." *Newsweek,* October 12. Available at http://www.msnbc.com/news

Zoroya, Gregg. 2006. "Lifesaving Knowledge, Innovation Emerge In War Clinic." *USA Today,* March 27. Available at http://www.usatoday.com

Epilogue

Harris Interactive. 2007 (November 8). *The Harris Poll's 'Alienation Index' Rises Slightly to Highest Level in Presidency of George W. Bush.* The Harris Poll #110. Available at http://www.harrisinteractive.com

McKee, Tim. 2006 (January). "A More Perfect Union: Tom Hayden on Democracy and Redemption." *The Sun* 361:4–11.

U.S. Census Bureau. 2007. "Historical Tables. Table A-2. Percent of People 25 Years and Over Who Have Completed High School or College, by Race, Hispanic Origin and Sex: Selected Years 1940 to 2006." *Current Population Survey.* Available at http://www.census.gov

Credits

Chapter 1 1: © Scott Picunko. Cover reprinted by permission of Oxford University Press; **2:** © Eduardo Verdugo/AP Photo; **4:** © AP/Wide World Photos; **20:** from the film *Obedience*, © 1968 by Stanley Milgram. Copyright renewed © 1993 by Alexandra Milgram and distributed by Penn State; **26:** © Paul Fusco/Magnum Photos.

Chapter 2 29: © Gideon Mendel/Corbis; **30:** © Suzy Allman/Getty Images; **38:** © AP Photo/Kelley McCall; **38:** © Photowood Inc./Corbis; **38:** © AP Photo/Gary Kazanjian; **39:** © Nigel Dickinson/Peter Arnold, Inc.; **39:** © AP Photo/E.B. McGovern; **39:** © AP Photo/Yomiuri Shimbun; **39:** © AP Photo/Douglas C. Pizac; **41:** © AP Photo/Khalil Senosi; **43:** © AP Photo/Scott Heppell; **59:** © AP Photo/Cheryl Senter; **62:** © Posh Journeys and Medaccess India; **66:** © Steven Lunetta Photography, 2007; **69:** © Per Anderson-Petersson/Getty Images; **74:** NIMH.

Chapter 3 80: © Kwame Zikomo/SuperStock; **82:** © AP/Wide World Photos; **85:** Partnership for a Drug-Free America; **92:** Campaign for Tobacco-Free Kids; **105:** Multnomah County Sheriff's Office; **110:** NHTLA.

Chapter 4 119: © Rich H. Legg/iStockphoto; **120:** © AP Photo/Steve Cannon; **128:** © Michael Newman/PhotoEdit; **135:** © Dave Einsel/Getty Images; **139:** © A. Ramey/PhotoEdit; **149:** © AP Photo/Lee Celano; **151:** © AP/Wide World Photos.

Chapter 5 160: © Roy Morsch/Corbis; **162:** © HBO/The Kobal Collection; **166:** © AP Photo/Laurent Emmanuel; **168:** © Getty Images; **183:** Office of Attorney General Jim Petro and the Ohio Department of Aging; **201:** The East Los Angeles Men's Health Center, a project of Bienvenidos; **203:** © AP Photo/Independent Record, Mark Goldstein.

Chapter 6 209: © Irving Olson, 2007; **210:** © Bettmann/Corbis; **215:** © Reme de la Mauvineire/AP Photos; **215:** U.S. Army, Ken Rich, photographer; **224:** © Desmond Boylan/Reuters/Corbis; **219:** Photo courtesy of the author; **223:** Texas A&M, Damian Medina; **224:** © Sebastian D'Souza/AFP/Getty Images; **224:** © Rob Elliott/AFP/Getty Images; **224:** © James Pomerantz/Corbis; **225:** © Neil Cooper/Alamy; **225:** © Caroline Penn/Corbis; **225:** © AP Photo/Luis Romero; **225:** © AP Photo/Ajit Solanki; **227:** © Reuters/Landov; **228:** © AP/Wide World Photos; **231:** © Elena Rooraid/PhotoEdit; **235:** USDA Food Stamp Program; **236:** ©Lillis Werder/iStockphoto.com; **244:** © AP Photo/Fabian Bimmer.

Chapter 7 249: © 2007, Sulabh Sanitation and Social Reform Movement. Dr. Bindeshwar Pathak; **249:** © AP Photo/The Hearald, Michael V. Martina; **250:** Ranklin County, OH; **255:** © AP Photo/Pat Sullivan; **260:** © Mark Peterson/Corbis; **265:** Caroline Schacht and David Knox; **262:** © Guenter Stand/VISUM/The Image Works; **266:** © Kimberly Butler/Time Life Pictures/Getty Images; **277:** from *United Students Against Sweatshops International Intern Report, 2005*. Used with permission; **285:** Kirschner/United Students Against Sweatshops; **286:** © AP Photo/Yonhap, Seo Myung-kon.

Chapter 8 291: © AP Photo/CTK, Petra Masova, **294:** © AP Photo/Denis Farrell; **297:** © AP Photo/Michael Conroy; **302:** © James D. Wilson/Woodfin Camp & Associates; **302:** © Dan Habib; **305:** © Michael Newman/PhotoEdit; **314:** © AP Photo/Carolyn Kaster; **318:** © Mark Richards/PhotoEdit; **321:** © AP Photo/Scott Audette; **327:** © AP Photo/Microsoft, Tim Shaffer.

Chapter 9 334: © AP Photo/Kiichiro Sato; **335:** © AP Photo/Richard Drew, File; **338:** © David Tumley/Corbis; **338:** © AP Photo/Lawrence Jackson; **338:** © AP Photo/Joe Marquette/File; **341:** © AP Photo/Alan Diaz; **342:** © AP/Wide World Photos; **344:** © AP Photo/The Plain Delaer, Lisa DeJong; **345:** © AP Photo/Bebeto Matthews; **353:** © AP Photo/Julio Cortez; **362:** © AP Photo/The Minnesota Public Radio, Jeff Horwich; **370:** © AP/Wide World Photos; **377:** © Southern Poverty Law Center.

Chapter 10 385: © Meredith Mullins/iStockphoto; **389:** © David Turnley/Corbis; **393:** © Richard Levine/Alamy; **405:** Corbis Premium RF/Alamy; **409:** © AP Photo/Mike Derer; **411:** Roman Catholic Womenpriests; **413:** © AP Photo/David Longstreath; **416:** © Bob Sacha; **416:** © From Self-Made Man by Norah Vincent. Copyright © 2006 by Norah Vincent. Used by permission of Viking Penguin, a division of Penguin Group (USA) Inc.; **421:** © Phil Schermeister/Corbis; **428:** © AP/Wide World Photos.

Chapter 11 432: © AP Photo/Canadian Press, Aaron Harris; **433:** Fine by Me, Michael Yaksich; **435:** © AP/Wide World Photos; **436:** © Vincent Yu/AP Photos; **438:** Palms of Manasota; **442:** © AP Photo/Douglas C. Pizac; **447:** © AP Photo/Herald-Leader, Pablo Alcala; **448:** © AP/Wide World Photos; **452:** © Kevin Winter/Getty Images for The Billies; **453:** © AP Photo/Jennifer Graylock; **461:** © Brennan Linsley/AP Photo; **475:** © AP/Wide World Photos.

Chapter 12 480: © jozef sedmak/iStockphoto.com; **481:** © The Pew Charitable Trusts, 2007. Kidsarewaiting.org; **487:** © Reuters/NewMedia Inc./Corbis; **491:** © Tomas Bravo/Reuters/Corbis; **494:** © AP Photo/The Plain Dealer, Lynn Ischay/The Plain Dealer; **496:** NDOT Photograph; **500:** Advertising Archives, Inc.; **509:** National Center on Elder Abuse; **513:** © Frank Fournier/Contact Press.

Chapter 13 521: © AP Photo/Bikas Das; **522:** © AP Photo/KMSP-TV; **526:** © Kim Steele/The Image Bank/Getty Images; **534:** © David Turnley/Corbis; **535:** © Peter Johnson/Corbis; **536:** © EGDigital/iStockphoto; **540:** © AP Photo/K.M. Chaudary; **541:** © Dave G. Houser/Corbis; **544:** © Rebecca Stephens, 2007. Used with permission; **545:** © Cynthia Marks Garnant, 2007. Used with permission; **546:** Courtesszy of Wheeling Naitonal Heritage Area; **547:** © AP Photo/George Bridges.

Chapter 14 555: © Derek Dammann/iStockphoto.com; **556:** © AP Photo/Itsuo Inouye; **557:** © Eran Mahalu/iStockphoto.com; **565:** © Ted Spiegel/Corbis; **566:** © AP Photo/Francois Mori; **570:** © Hans Strand/Corbis; **570:** © Torsten Blackwood/AFP/Getty Images; **570:** © Jim Reed/Getty Images; **571:** © AP Photo/Eric Miller; **571:** © JEAN AYISSI/AFP/Getty Images; **571:** Eric Thayer/Getty Images; **571:** © Andy Crump, TDR, WHO/Photo Researchers, Inc.; **575:** © Kateryna Govorushchenko/ iStockphoto.com; **576:** © AP Photo/Gerry Broome; **583:** © Tom Uhlman/Alamy; **591:** © Morton Beebe/Corbis; **594:** Adam Joseph Lewis Center for Environmental Studies, Ed Hancock, courtesy of DOE/NREL; **596:** © AP Photo/fr/stf/Fritz Reiss; **600:** © Micheline Pelletier/Corbis.

Chapter 15 604: © Emrah Turudu/iStockphoto.com; **607:** © Bettmann/Corbis; **607:** © AP Photo/Jason DeCrow, file; **610:** © Corbis; **614:** © Adam Lubroth/Stone/Getty Images; **621:** © AP Photo/The Southern Illinoisan, Ceasar Maragni; **628:** © Robert Visser/Corbis Sygma; **629:** Duke University Medical Center; **634:** © Jagadeesh/Reuters/Corbis; **635:** © David Sams/Stock Boston.

Chapter 16 645: © Ap Photo/Jaim MacMillan; **648:** © Bob Mahoney/The Image Works; **651:** © David Pollack/Corbis; **653:** I am a Human Becoming, Inc.; **662:** © AP Photo/Jake Bacon; **663:** © AP Photo/Alistair Robertson/PA; **667:** AFP/Getty Images; **668:** © Joao Silva/The New York Times; **671:** © GEORGES GOBET/AFP/Getty Images; **675:** © Michael Lipschitz/AP/Wide World Photo; **677:** © Majid Saeedi/Getty Images; **679:** © AP Photo/Josphat Kasire; **686:** Deborah Netland, Office of Weapons Removal and Abatement, U.S. Dept. of State.

Name Index

Page numbers in italics refer to information in tables, figures, illustrations and captions.

Subject Index

Page numbers in italics refer to information in tables, figures, illustrations and captions.

Education (continued)
class and family background of, 299–303; socialization function of, 295; sociological theories of, 295–98; structural-functionalist perspective of, 295–96; and structural sexism, 394–95; symbolic interactionist perspective of, 298; understanding problems in, 329–31; working-class parents value, 300
Educational attainment: inequality of, 298–310; by OECD, 294
Educational discrimination, and segregation, 367–68
Educational strategies, 376
Educational technology, use of, 326
Educational Testing Service, 312
EEOC. See Equal Employment Opportunity Commission
EFA Global Monitoring Report, 307–8
Egalitarian relationship, 162; definition of, 162
Eighteenth Amendment to the U.S. Constitution, 89
EITC. See Earned income tax credit
Elder abuse, 508; common form of, 183; definition of, 183, 508; in Japan, 509; parent abuse, and sibling abuse, 183–84; risk factors associated with, 183; on U.S. adults, 183
Elder Wisdom Circle (EWC), 498
Elderhood, 483–84
Elderly: defined as, 486; Japanese on, 482; in Malawi, 482; myths and facts about, 486; problems of, 501–12, 519; Scandinavian countries on, 482; stereotypes of, 487
Electronic benefits transfer (EBT), 235
Electronic fund transfer, 606
"electronic sweatshops", 628
Elementary and secondary education, total U.S. expenditures for, 301
Eli Lilly, 215
Emergency Medical Treatment and Active Labor Act, 60
Emotion work, 400; definition of, 400
Employed parents, 266
Employee Free Choice Act, 285
Employer-Provided Work-Family Policy, 281
Employment: service-focused, 283; subsidized, 283
Employment discrimination, 364–65; contribution of, 364
Employment Nondiscrimination Act (ENDA), 466; definition of, 466
Empowerment zone, 543
Energy Information Administration, 563, 568
Energy Policy and Conservation Act, 596
Energy use worldwide, overview, 563–64
England, cannabis in, 82
English major, 296

Enhancing Education through Technology program, 326
Enron, 134–35
Enterprise Community Initiative, 543
Enterprise Community Program, 543
Entry: forcible, 132; unlawful, 132
Environment, destruction of, 673–76
Environmental Action, 587
Environmental activism, tools for, 587
Environmental Defense, 586, 589
Environmental footprint, 532, 559; definition of, 532, 559
Environmental injustice, 579–82; around world, 581; definition of, 579
Environmental movement: involvement in, 587; role of Corporations in, 588
Environmental organizations, goal of, 589
Environmental problems, 555–603; and disappearing livelihoods, 582–83; overview of, 562–83; and resource scarcity, 532–33; responding to, 586–99; social causes of, 584–86; sociological theories of, 559–62; understanding, 599–600
Environmental Protection Agency (EPA), 92, 574
Environmental racism, 579
Environmental refugees, 583; definition of, 583
Environmentalism, religious, 587
Environmentalists versus industrialists, value conflicts between, 12
Epidemiological transition, 31; definition of, 31
Episcopal Church, of United States, 411
Equal Credit Opportunity Act, 365
Equal Employment Opportunity Commission (EEOC), 363, 372–73, 382, 402, 514
Equal Pay Act, 402
Equal rights amendment (ERA), 419; supporters of, 420
Equality of Educational Opportunity, 305
ERA. See Equal rights amendment
Eradicating poverty, cost of, 246
Ergonomics, 279, 289; definition of, 279
Essence magazine, 409
Ethanol, 590; and biodiesel, 590
Ethnic cleansing, 337
Ethnic diversity, in United States, 342–43
Ethnicity, 342; definition of, 342; race and, 303–7; race, immigration and, 334–84
Ethyl, U.S. chemical company, 558
Euphoria, intoxication and health effects, 97
European Environment Agency, 564

European immigrants, 527
European Renaissance, 523
European Union Emission Trading Scheme, 594
European Union (EU), 250
Europol, 156
Evangelical Call to Civic Responsibility, 587
Event dropout rates, 312; by race and ethnicity, 312
EWC. See Elder Wisdom Circle
Executioner, 101
Exodus international, 441
Exorcists, 101
Experimental method, major strength of, 19
Experiments, definition of, 17
Explanations: constructivist, 658; primordial, 658
Exposure, among women and children, 565
Expressive roles, 393; definition of, 393
Expulsion, 338; definition of, 338
Extracurricular activities, 407
Extreme poverty, 242
ExxonMobil, 560–61
EZ/EC program, 543

Facebook, 138
Factories, development of, 251
Factory farms, 38
Fair Labor Association (FLA), 276–77
Fair Labor Standards Act, 495
Faith-Based and Community Initiative, 240; definition of, 240
Faith-based groups, 240
False fraternization, 252
Families USA, 63
Familism, 191; definition of, 191
Family, 161; definition of, 161; gender relations within, 174; measured by degree, 170; role of women in, 162; U.S. census defines, 161; violence and abuse in, 184
Family and Medical Leave Act, 174, 422; definition of, 280
Family Court Reform Council of America, 194–95
Family forms, diversification of, 169
Family-friendly policy, 280
Family household, 164; definition of, 164
Family Leave Insurance bill, 281
Family leave policy, 281
Family planning, 33; component of, 541; involvement of men in, 540; reduces maternal mortality, 67
Family planning programs, 539
Family preservation programs, 188; definition of, 188
Family problems, 160–208; cause of, 229; conflict and feminist perspectives on, 171–74; impact of, 204; sociological theory of, 171–75; strategies for action, 186–88; structural-functionalist explanations of, 171; structural-functionalist perspective on, 171; symbolic interactionist perspective on, 174–75; understanding, 204–6

Family relations, and cultural sexism, 404–6
Family structure: and poverty, 220; U.S. poverty rates by, 221
Family Support Act, for child care services, 237
Father-Child relationships, effects on, 193
Fathers' rights group, 421
Fathers4Justice, in Great Britain and Canada, 421
Faxes, 606
FBI, 63, 130. See Federal Bureau of Investigation
FDA. See Food and Drug Administration
Fear, and grief, 465
Federal and State Family and Medical Leave Initiatives, 281
Federal and State Health Care Reform, U.S., 72–73
Federal Bureau of Investigation (FBI), 105, 123, 640
Federal Case Registry, 238
Federal criminal prosecutions, number of, 136
Federal Death Penalty Abolition Act, 155
Federal drug control, function, 112
Federal Empowerment zone, 543
Federal Express, 613
Federal Fair Housing Act, 365
Federal funding, 313
Federal health care reform, 72
Federal Healthy Marriage Initiative, 196
Federal Housing Assistance Programs, 235
Federal Interagency Forum on Aging-Related Statistics, 503
Federal Interagency Forum on Child and Family Statistics, 220
Federal Marriage Amendment, 455
Federal minimum wage, 240
Federal Poverty Income Guidelines, 516
Federal SSI, 234
Federal tax, 216
Federal Trade Commission, 137, 633
FedEx, 589
Female circumcision, 389
Female genital cutting (FGC), 389; forms of, 389; prevalence of, 390
Females: in blended course, 327; sexual assault and rape, 124
Feminine Mystique, 419
Feminism: definition of, 419; new wave of, 420; rebirth of, 419; and women's movement, 419
Feminist criminology, 141; on public policy, 141
Feminist theory, 11, 393
Feminization of poverty, 220, 412–13; definition of, 220
Fertility rates, 525; dropping, 539; highest, in Africa, 526; by region, 525
Fertility, reducing, 538–43
Fetal alcohol syndrome, 107; definition of, 107